THEORY OF DISLOCATIONS

THEORY OF DISLOCATIONS

Second Edition

John Price Hirth

Department of Metallurgical Engineering,
The Ohio State University

Jens Lothe

Institute of Physics, Oslo University

A Wiley-Interscience Publication

JOHN WILEY & SONS

New York Chichester Brisbane Toronto Singapore

Library of Congress Cataloging in Publication Data:

Hirth, John Price, 1930-
  Theory of dislocations.

  "A Wiley-Interscience publication."
  Includes index.
  1. Dislocations in crystals. I. Lothe, Jens.
II. Title.
QD921.H56 1982 548′.842 81-15939
ISBN 0-471-09125-1 AACR2

Printed in the United States of America

10 9 8 7 6 5 4 3 2 1

Preface

This book is based on the lecture notes developed by the authors for courses on the theory of dislocations at Carnegie Institute of Technology, The Ohio State University, and Oslo University. The book is intended to be primarily a comprehensive text in the field of dislocations. For this reason, an exhaustive literature survey has not been attempted, although throughout the book key references are cited.

In recent years, studies of dislocations and their effects on material properties have expanded greatly, resulting in the apparent need for a detailed approach to the various problems in the field. We have attempted both to provide sufficient detail that much of the book could effectively be used as an undergraduate text and to extend the treatment of specific problems sufficiently to stimulate the advanced graduate student. As a result, some sections of the book are intended only for advanced students.

The book is comprised of two general groupings. Parts 1 and 2 essentially consider only fundamentals of dislocation theory. These fundamentals are solidly grounded and can be presented without problem. Parts 3 and 4 treat both fundamentals and applications of the fundamentals to the understanding of physical phenomena. Those aspects of the theory that are discussed in detail are well founded. However, the treatment of some subjects, such as work hardening, in the applications section is still moot. In such cases, we have briefly outlined current theories, pointed out their shortcomings, and suggested approaches to a general solution to the problem in question.

Throughout the book, we assume a background in mathematics through differential equations. More advanced mathematical topics are treated in the text. However, in such cases, we have attempted to outline the derivations in sufficient detail for the student to derive them either directly or with the aid of a reference book such as I. S. Sokolnikoff and R. M. Redheffer, "Mathematics of Physics and Modern Engineering," McGraw-Hill, New York, 1966. Because of the variety of topics treated, we have encountered a problem in notation. Rather than use unfamiliar symbols, we have used the same symbol for different quantities in some cases. For example, F denotes both Helmholtz free energy and force, and v denotes both velocity and volume. The context makes the definitions clear.

We are indebted to many authors and publishers for kindly supplying illustrations and permission to publish them; acknowledgements are presented in the figure captions. A number of coworkers and students contributed valuable discussions of the subject matter, as cited in the first edition of this book. For the present edition, we are grateful for the helpful comments of V. I. Alshits, R. W. Balluffi, W. A. T. Clark, A. K. Head, U. F. Kocks, I. G. Ritchie, L. M. Slifkin, and R. W. Whitworth and for the careful reading of the manuscript by F. A. Nichols. We are pleased to acknowledge the support of the research contributing to this edition by the U.S. National Science Foundation, the U.S. Office of Naval Research, and the Norges Teknisk Naturvitenskapelige Forskningsrad.

JOHN PRICE HIRTH
JENS LOTHE

Columbus, Ohio
Oslo, Norway
January 1982

Contents

Part 1
Dislocations In Isotropic Continua

THEORY OF DISLOCATIONS

PART 1

DISLOCATIONS IN ISOTROPIC CONTINUA

1

Introductory Material

1-1. INTRODUCTION

This chapter deals largely with the historical development of the concept of a dislocation. Physical phenomena that led to the discovery of dislocations are discussed, together with early mathematical work that eventually contributed to dislocation theory. Today, of course, there is a large variety of observations that indicate directly the presence of dislocations in crystals; selected examples of these observations are presented. In the final portion of the chapter dislocations are defined formally in terms of their geometric properties. Some simple axioms follow directly from this definition.

1-2. PHYSICAL BASIS FOR DISLOCATIONS

Early Work

Probably the first suggestion of dislocations was provided by observations[1,2] in the nineteenth century that the plastic deformation of metals proceeded by the formation of slip bands or slip packets, wherein one portion of a specimen sheared with respect to another. Initially the interpretation of this phenomenon was obscure, but with the discovery that metals were crystalline, it was appreciated that such slip must represent the shearing of one portion of a crystal with respect to another upon a rational crystal plane.

Volterra[3] and others, notably Love,[4] in treating the elastic behavior of homogeneous, isotropic media, considered the elastic properties of a cut cylinder (Fig. 1-1*a*) deformed as shown (Figs. 1-1*b* to 1-1*g*). Some of the deformation operations clearly correspond to slip, and some of the resulting configurations correspond to dislocations. However, the relation of the work of

[1] O. Mügge, *Neues Jahrb. Min.*, 13 (1883).

[2] A. Ewing and W. Rosenhain, *Phil. Trans. Roy. Soc.*, **A193:** 353 (1899).

[3] V. Volterra, *Ann. Ecole Norm. Super.*, **24**, 400 (1907).

[4] A. E. H. Love, "The Mathematical Theory of Elasticity," Cambridge University Press, Cambridge, 1927.

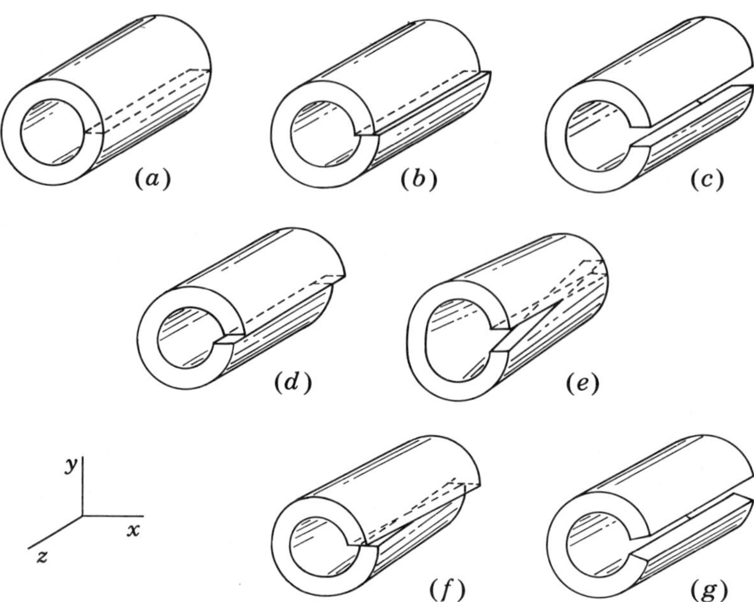

FIGURE 1-1. A cylinder (*a*) as originally cut, and (*b*) to (*g*), as deformed to produce the six types of dislocations as proposed by Volterra.

the elasticians to crystalline slip remained unnoticed until the late 1930s, after dislocations had been postulated as crystalline defects. Configurations (*b*) and (*c*) in Fig. 1-1 correspond to edge dislocations, and (*d*) corresponds to a screw dislocation. Configurations (*e*), (*f*), and (*g*), which correspond to disclinations,[5] are not discussed further here. They do not appear as isolated defects in metal crystals because the displacements produced by these configurations are proportional to the outer cylinder radius and hence do not vanish as the radius tends to infinity.

Following the discovery of x-rays and of x-ray diffraction, establishing crystallinity, Darwin[6] and Ewald[7] found that the intensity of x-ray beams reflected from crystals was about 20 times greater than that expected for a beam reflected from a perfect crystal. In a perfect crystal the intensity would be low because of the long absorption path provided by multiple internal reflections. In addition, the width of the reflected beam was about 1 to 30 minutes of arc, whereas that expected for the perfect crystal was a few seconds. To account for these results, the theory evolved that real crystals consisted of small, roughly equiaxed crystallites, 10^{-4} to 10^{-5} cm in diameter, slightly

[5] W. F. Harris, "Surfaces and Defect Properties of Solids," vol. 3, The Chemical Society, London, 1974, p. 57.

[6] C. G. Darwin, *Phil. Mag.*, **27**: 315, 675 (1914).

[7] P. P. Ewald, *Ann. Phys.*, **54**: 519 (1917).

misoriented with respect to one another, with the boundaries between them consisting of amorphous material.

In this "mosaic-block" theory, the crystallite size limits the absorption path, accounting for the intensity effect, while the misorientation accounts for the beam width. The crystallite boundaries actually consist of arrays of dislocation lines, but this was not appreciated until recent times.

Crystal growth is another area of study implying the presence of dislocations. Volmer's[8] work on nucleation, following the ideas of Gibbs,[9] indicated that the layer growth of perfect crystals should not be appreciable until supersaturations of about 1.5, sufficient for nucleation of new layers, were attained. Experimentally, on the other hand, crystals were observed to grow under nearly equilibrium conditions; see, for example, the work of Volmer and Schultze[10] on iodine. This discrepancy between theory and experiment remained a puzzle until Frank[11] resolved it by postulating that growth could proceed at lower supersaturations by the propagation of ledges associated with the point of emergence of a dislocation at a surface.

A number of other cases could be cited. For example, the rapid equilibration of point defects in a crystal subjected to a change in temperature suggests the presence of internal sources and sinks for point defects in crystals. It is now established that dislocations and arrays thereof can provide such sources and sinks.[12] However, these other examples in general either were developed at a later time or were less striking than those cited above, and therefore are not discussed further here.

The final case, involving the consideration of the strength of a perfect crystal, provided the major impetus for the development of dislocation theory, and serves to terminate the early work on dislocations. Because of its importance in stimulating work on dislocations, and because it involves a phenomenological approach that is applicable in many other dislocation problems, this topic is treated in detail in the following section.

Theoretical Shear Strength of a Perfect Crystal

Once it was realized that metals were crystalline, interest developed in the computation of the strength of perfect crystals. The classical work in this area

[8] M. Volmer, "Kinetik der Phasenbildung," Steinkopff, Dresden and Leipsig, 1939.

[9] J. W. Gibbs, "Collected Works," vol. 1, "Thermodynamics," Yale University Press, New Haven, Conn., 1948.

[10] M. Volmer and W. Schultze, *Z. phys. Chem.*, **156**: 1 (1931).

[11] F. C. Frank, *Disc. Faraday Soc.*, **5**: 48, 67 (1949).

[12] D. N. Seidman and R. W. Balluffi [*Phys. Rev.*, **139**: A1824 (1965)] have shown that dislocations act as vacancy sources in up-quenched gold. On the other hand, R. S. Barnes [*Phil. Mag.*, **5**: 635 (1960)] showed that in α-bombarded copper, helium bubbles did *not* nucleate near single dislocations (vacancies are required for such nucleation), but that vacancies were produced at grain boundaries.

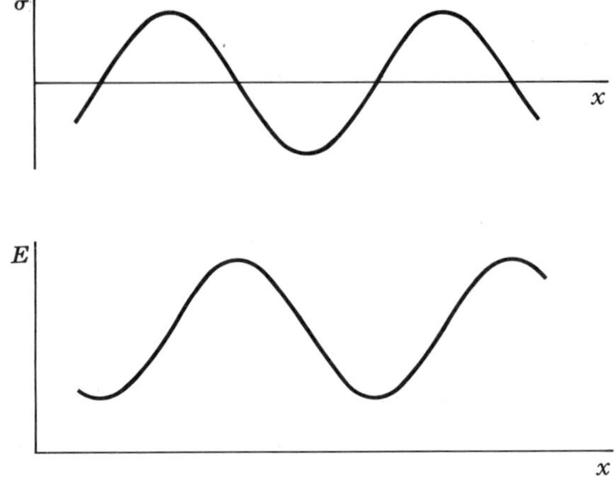

FIGURE 1-2. The periodic lattice potential, and the equivalent value of the shear stress, accompanying the shear of a perfect lattice.

was that of Frenkel,[13] whose model for the shear-stress–shear-displacement relation is shown in Fig. 1-2. He supposed that a crystal plastically shearing on a rational plane passed through equivalent configurations with equal energies and with a period equal to b, the magnitude of a simple lattice-translation vector. He thus neglected the small end effects associated with the formation of surface steps by the shear. The applied shear stress required to accomplish a shear translation x is proportional to dW/dx, where W is the energy of translation per unit area of the plane. As a first approximation, Frenkel took the periodicity of the energy to be sinusoidal, so that

$$\sigma = \sigma_{\text{theor}} \sin \frac{2\pi x}{b} \tag{1-1}$$

In the limit of small shear strain x/d, where d is the interplanar spacing, Hooke's law applies in the form

$$\sigma = \mu \frac{x}{d} \tag{1-2}$$

where μ is the shear modulus. Equating (1-1) and (1-2) in the small-strain limit, where $\sin(2\pi x/b) \cong 2\pi x/b$, one obtains for σ_{theor}, the theoretical shear stress, the value

$$\sigma_{\text{theor}} = \frac{\mu b}{2\pi d} \cong \frac{\mu}{5} \tag{1-3}$$

[13] J. Frenkel, *Z. Phys.*, **37**: 572 (1926).

In marked disagreement with the prediction of Eq. (1-3), the experimental values for the maximum resolved shear stress required to initiate plastic flow in metals were in the range of 10^{-3} to 10^{-4} μ at about the time of Frenkel's work.

To digress for a moment, later investigators noted that Eq. (1-3) was probably an overestimate of σ_{theor}, because the various semiempirical interatomic-force laws indicated that the attractive forces decreased much more rapidly with distance than did the sinusoidal force in Fig. 1-2, and because of the possibility of additional minima in the *W-x* plot corresponding to twin or other special orientations. Mackenzie,[14] using central forces for the case of close-packed lattices, found that σ_{theor} could be reduced to a value of $\mu/30$. This value is likely to be an underestimate, because of the neglect of the small directional forces that are also present in such lattices. Also, the contributions of thermal stresses, treated later in detail, reduce σ_{theor} below $\mu/30$ only near the melting point. Thus at room temperature σ_{theor} should be in the range $\mu/5 > \sigma_{\text{theor}} > \mu/30$, say $\sim \mu/15$. All theoretical estimates tend to fall within this range.[15] In excellent agreement with this estimate, the maximum values of the resolved shear stress for the initiation of plastic flow in (presumably perfect) whiskers of various metals[16] is $\sim \mu/15$.

Experimental work on bulk copper[17] and zinc[18] on the other hand, indicates that plastic deformation begins at stresses of the order of 10^{-9} μ. Thus, except for whiskers, the discrepancy between σ_{theor} and experimental values is even larger than was first supposed.

Exercise 1-1. Carry out a calculation, analogous to that of Frenkel, using a Morse function for the interatomic potential. Such functions, given by

$$W = W_0\left(e^{-2a(r-r_0)} - 2e^{-a(r-r_0)}\right) \tag{1-4}$$

where W_0 and a are parameters and r_0 is the equilibrium separation of atoms, give good fits to *P-V* data, compressibility, and elastic constants in fcc crystals.[19] Taking $r_0 \sim b$, the typical value $a = 5/r_0$, and $x \sim r - r_0$, show that the use of Eq. (1-4) in place of Eq. (1-1) gives the result $\sigma_{\text{theor}} = \mu b/20d$, and that the maximum stress occurs at $x = 0.138b$ versus $x = 0.25b$ for the case shown in Fig. 1-2. The close correspondence of this result to that of Frenkel demonstrates the reasonableness of the order of magnitude of his estimate of σ_{theor}.

[14] J. K. Mackenzie, unpublished doctoral dissertation, University of Bristol, Bristol, 1949.

[15] A. Kelly, "Strong Solids," Oxford University Press, Fair Lawn, N.J., 1973.

[16] S. S. Brenner, in R. H. Doremus *et al.* (eds.), "Growth and Perfection of Crystals," Wiley, New York, 1958, p. 3.

[17] R. F. Tinder and J. Washburn, *Acta Met.*, **12**: 129 (1964).

[18] R. F. Tinder, *J. Metals*, **16**: 94 (1964).

[19] L. A. Girifalco and V. G. Weizer, *Phys. Rev.*, **114**: 687 (1959); **120**: 837 (1960).

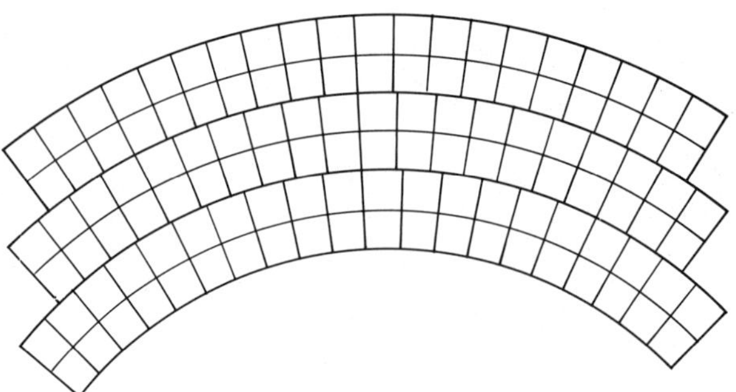

FIGURE 1-3. Imperfections in a crystal deformed by bending, according to Masing and Polanyi.

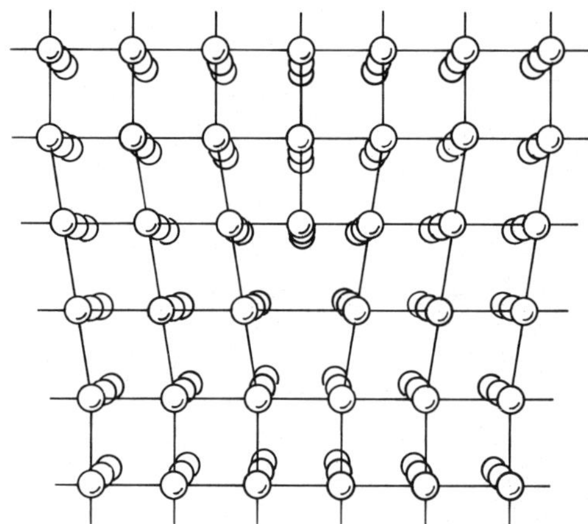

FIGURE 1-4. An edge dislocation in a simple cubic crystal.

After Frenkel's work, Masing and Polanyi,[20] Prandtl,[21] and Dehlinger[22] proposed various defects which were precursors of the dislocation. For example, the defects proposed by Masing and Polanyi, shown in Fig. 1-3, resemble a polygonized structure of dislocations in a crystal composed of hard atoms bound by weak directional bonds. Finally, in 1934, the edge dislocation, shown in Fig. 1-4, was proposed by Orowan,[23] Polanyi,[24] and Taylor[25] to explain the discrepancy between σ_{theor} and experiment, as discussed above. In 1939,

[20] G. Masing and M. Polanyi, *Ergeb. exact. Naturwiss.*, **2**: 177 (1923).

[21] L. Prandtl, *Z. ang. Math. Phys.*, **8**: 85 (1928).

[22] U. Dehlinger, *Ann. Phys.*, **2**: 749 (1929).

[23] E. Orowan, *Z. Phys.*, **89**: 605, 634 (1934).

[24] M. Polanyi, *Z. Phys.*, **89**: 660 (1934).

[25] G. I. Taylor, *Proc. Roy. Soc.*, **A145**: 362 (1934).

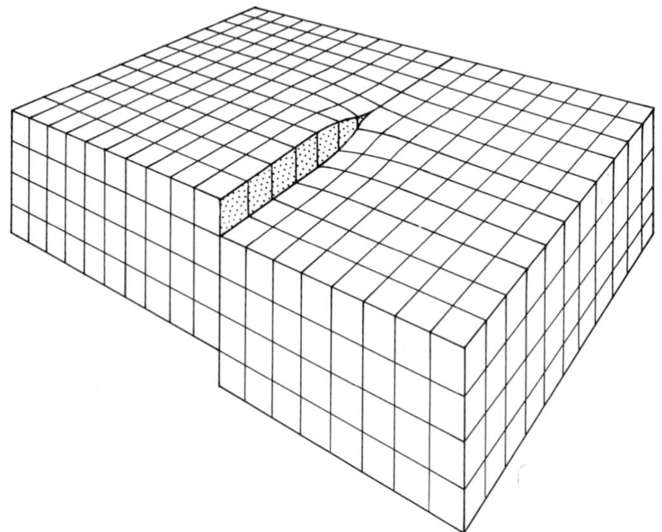

FIGURE 1-5. A screw dislocation in a simple cubic crystal.

Burgers[26] advanced the description of the screw dislocation, depicted in Fig. 1-5.

Observations of Dislocations

In the past two decades an overwhelming number of observations have been made which in summary provide unequivocal evidence of the existence of dislocations in crystals. There is such an abundance of these observations that we can cite only a few examples here. The reader is referred to the reviews of such observations listed at the end of this chapter for an extensive survey.

Figure 1-6, due to Bragg and Nye, shows an edge dislocation in the two-dimensional lattice of a bubble raft. Figure 1-7 depicts a growth spiral associated with a dislocation emergent at its center in *n*-nonatriacontane; the dislocation slipped out of the crystal after growth was completed, leaving behind a slip trace. A similar spiral on a $\{100\}$ face of silver is shown in Fig. 1-8; in this picture the spiral steps are one atom layer in height and are revealed by decoration and phase contrast microscopy. Both spirals provide confirmation of Frank's postulate,[27] discussed earlier.

The strain energy of a dislocation often leads to a faster rate of chemical etching at the point of emergence of a dislocation at a free surface, resulting in the formation of etch pits at such points. Similarly, because of the strain

[26] J. M. Burgers, *Proc. Kon. Ned. Akad. Wetenschap.*, **42**: 293, 378 (1939).

[27] F. C. Frank, *Disc. Faraday Soc.*, **5**: 48, 67 (1949).

FIGURE 1-6. A dislocation in a two-dimensional bubble raft [W. L. Bragg and J. F. Nye, *Proc. Roy. Soc.*, **A190**:474 (1947)].

energy, dislocations act as catalytic sites for precipitation from solid solution. Dash[28] formed etch pits at dislocations in silicon, decorated the dislocations with copper, and viewed the resulting structure by infrared transmission. Both techniques revealed the same dislocation structure as shown in Fig. 1-9. Figure 1-10 shows dislocations decorated by the precipitation of silver in silver halide; the prismatic loops are generated by the stresses arising from differential thermal contraction at glass spheres embedded in the material. Figure 1-11 is an example from Gilman and Johnston's extensive work on etch pits at dislocation sites on {100} LiF surfaces.

Radiation can be selectively diffracted near a dislocation because of its strain field, so that dislocations can be revealed by various optical techniques. Figure 1-12 is a direct lattice image of a dislocation taken in electron transmission microscopy. Figure 1-13 is Newkirk's reflection x-ray micrograph revealing dislocations in lithium fluoride, while Fig. 1-14 is Nost's x-ray transmission micrograph showing dislocations in aluminum. Figure 1-15 is one of the earliest electron-transmission micrographs from the extensive work of

[28] W. C. Dash, in J. C. Fisher *et al.* (eds.), "Dislocations and Mechanical Properties of Crystals," Wiley, New York, 1957, p. 57.

FIGURE 1-7. Growth spiral on an *n*-nonatriacontane crystal. The dislocation at the spiral center has slipped out of the crystal following growth, leaving behind a slip trace, 10,000× [N. G. Anderson and I. M. Dawson, *Proc. Roy. Soc.*, **A218**:255 (1953).]

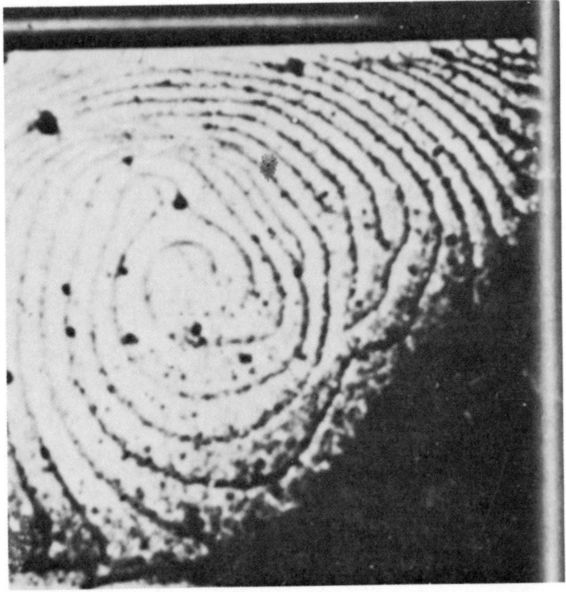

FIGURE 1-8. Decorated spiral-growth ledges on a silver crystal, indicating the presence of a dislocation at the spiral center during crystal growth, 1200× [F. C. Frank and A. J. Forty, *Proc. Roy. Soc.*, **A217**:262 (1953)].

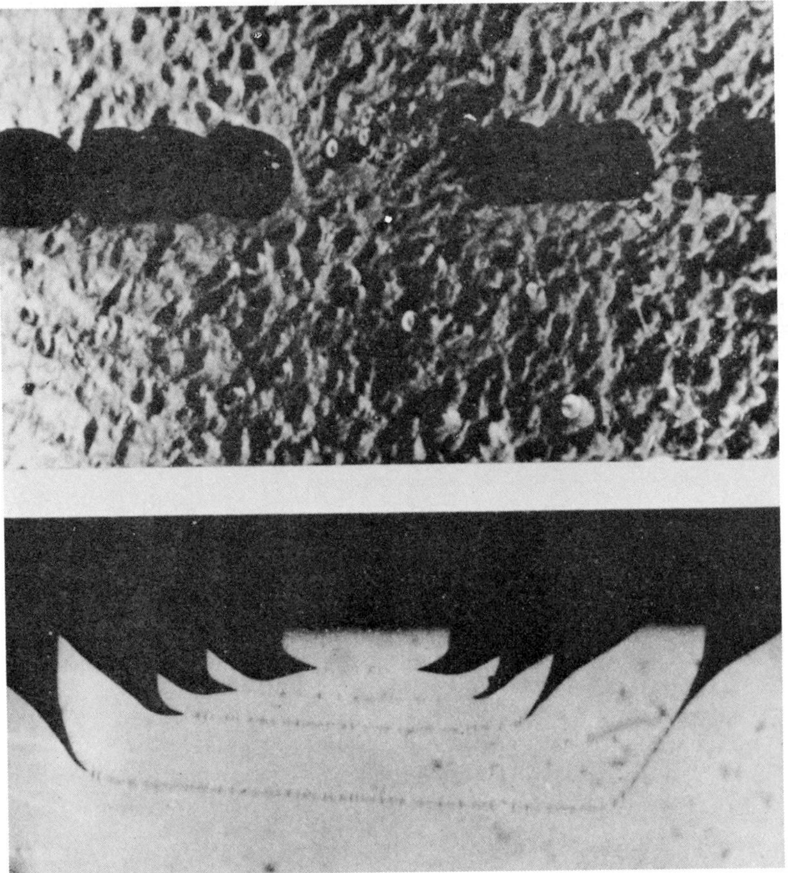

FIGURE 1-9. Chemical etch pits at copper-decorated dislocations observed by infrared transmission microscopy, ~80×, (W. C. Dash, in "Dislocations and Mechanical Properties of Crystals," Wiley, New York, 1957, p. 57).

Hirsch et al.,[29] showing polygonization in aluminum. Figure 1-16 is a transmission micrograph showing dislocations in a Cu-7% Al alloy. Figure 1-17, from Pashley, is a moiré pattern, in electron transmission, showing dislocations in overlying Pd and Au {111} layers.[30] Finally, Fig. 1-18, due to Müller, is an ion-emission pattern indicating the presence of dislocations in a platinum ion-emission tip.[31] Several other examples of dislocation observations are given later in the text.

[29]P. B. Hirsch, R. W. Horne, and M. J. Whelan, *Phil. Mag.*, **1**: 677 (1956); in J. C. Fisher *et al.* (eds.), "Dislocations and Mechanical Properties of Crystals," Wiley, New York, 1957, p. 92; P. B. Hirsch, A. Howie, and M. J. Whelan, *Phil. Trans. Roy. Soc.*, **A252**: 499 (1960).

[30]D. W. Pashley, private communication. See D. W. Pashley, J. W. Menter, and G. A. Bassett, *Nature*, **179**: 752 (1957); *Proc. Roy. Soc.*, **A246**: 345 (1958).

[31]E. Müller, private communication. See E. Müller, in J. B. Newkirk and J. H. Wernick (eds.), "Direct Observation of Imperfections in Crystals," Interscience, New York, 1962, p. 77.

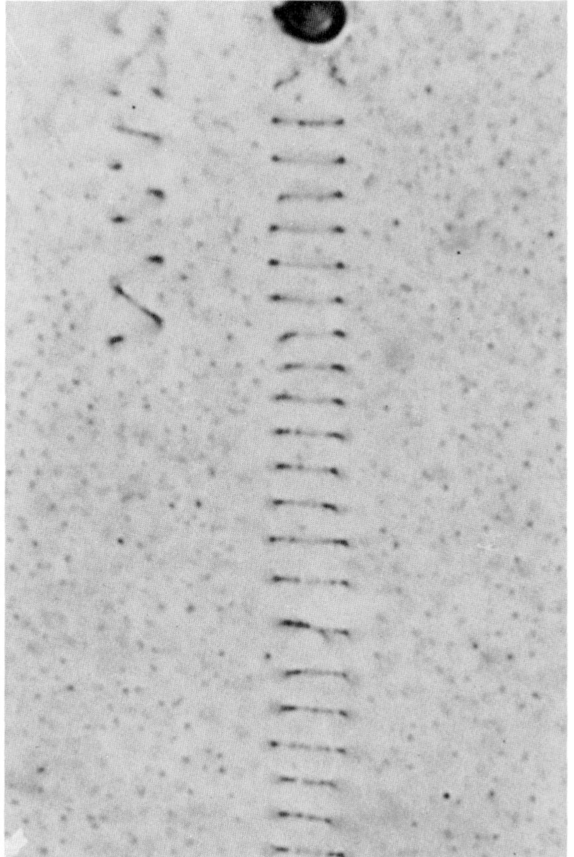

FIGURE 1-10. Dislocation loops in silver halide decorated by a precipitate of silver, 3000× [D. A. Jones and J. W. Mitchell, *Phil. Mag.*, **3**:1 (1958)].

1-3. SOME ELEMENTARY GEOMETRIC PROPERTIES OF DISLOCATIONS

Displacement associated with a dislocation

Consider a perfect crystal cube, acted on by a shear stress as shown in Fig. 1-19*a*. The edge dislocation in Fig. 1-19*b* is produced by slicing the cube from left to right normal to z and displacing the cube to comply with the stress. Evidently, the dislocation is a line representing the boundary of the slipped region, and its strength is characterized by the displacement b. Similarly, the cut and displacement shown in Fig. 1-19*b* produce a right-handed screw dislocation, which is again the boundary of an area over which slip has occurred, and is again characterized by b. The opposite displacement of the cut

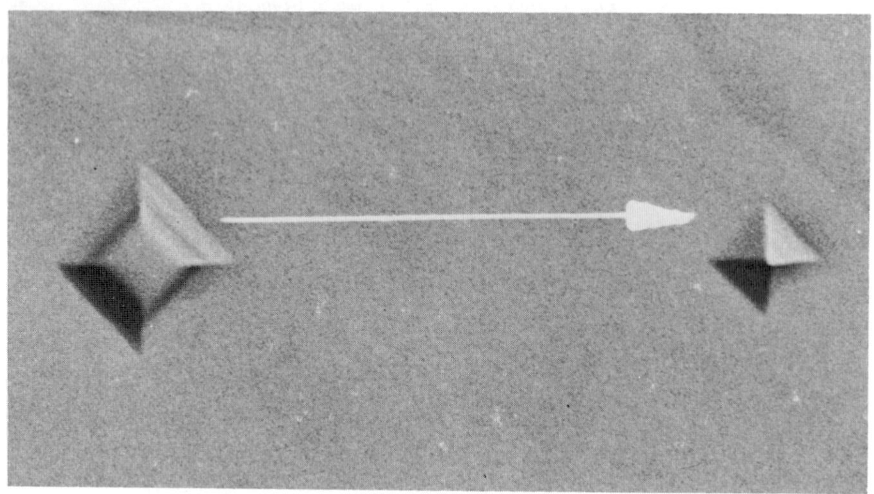

FIGURE 1-11. Dislocation motion as revealed by etch pits on a {100} surface of lithium fluoride. The original dislocation was etched, forming a large pit. The crystal was then deformed and re-etched, showing the translated position of the dislocation, 4000× [J. J. Gilman and W. G. Johnston, in J. C. Fisher et al. (eds.), "Dislocations and Mechanical Properties of Crystals," Wiley, New York, 1957, p. 116.]

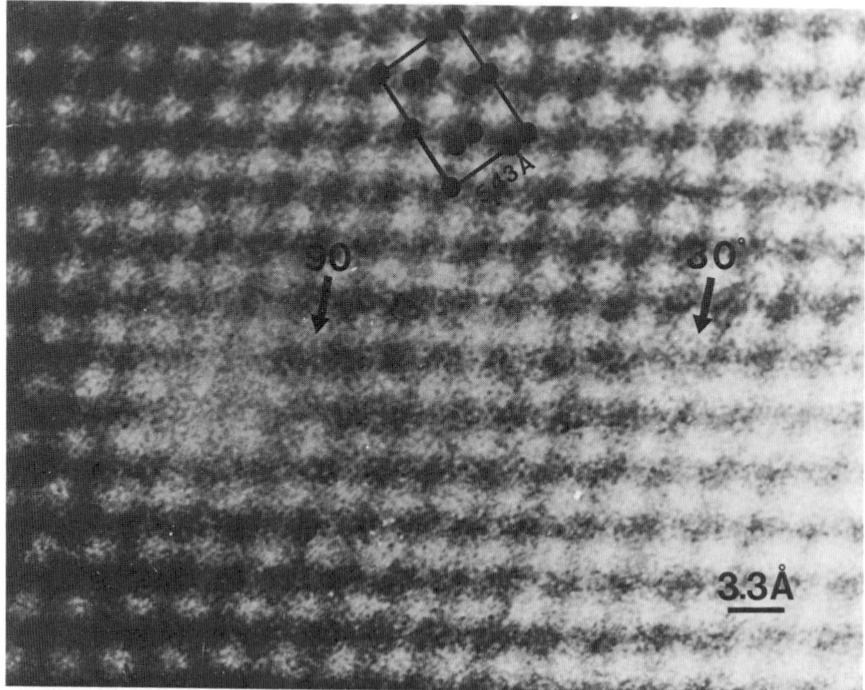

FIGURE 1-12. Direct lattice image of a dissociated dislocation in Si (view figure at a glancing angle). ξ points out of the page. Each dot represents a row of atoms [A. Olsen and J. C. H. Spence, *Phil. Mag.*, **43**:945 (1981)].

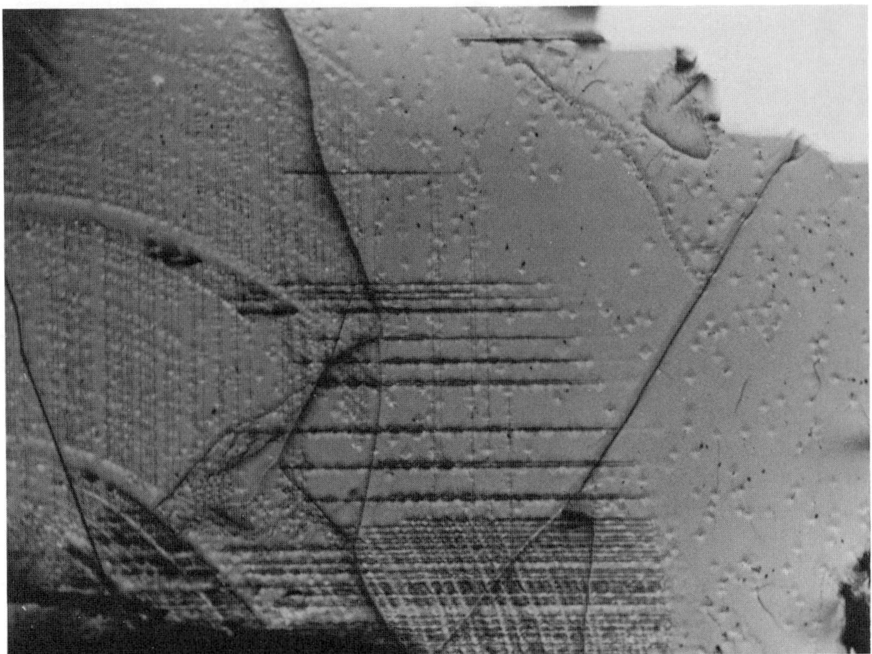

FIGURE 1-13. (220) Berg-Barrett x-ray-diffraction micrograph (Cr Kα) of an etched lithium fluoride crystal. Note (1) etch pits at dislocations, (2) subboundaries, and (3) dislocation lines, 80× [J. B. Newkirk, *J. Metals*, **14**:661 (1962).]

would produce a left-handed screw dislocation. Continuation of the motion of either dislocation produces the identical shear of the cube, as noted in Fig. 1-19*c*. Such shear displacement is conventionally called *glide* when it is produced by a single dislocation and *slip* when it is produced by a number of dislocations. As can be seen in the figure, glide or slip is produced by the conservative motion of dislocations. Here *conservative motion* means that the total number of atoms and lattice sites is conserved.

If a cut is made and opened up under a normal stress, as shown in Fig. 1-19*e*, and then extra material is inserted to fill the cut, the edge dislocation in Fig. 1-19*f*, identical to that of Fig. 1-19*b*, is produced. For a single dislocation in a lattice, the added material corresponds to an extra *plane* of atoms. This nonconservative process of addition of matter is associated with the process of *climb* of the dislocation, wherein the edge dislocation moves up or down by gaining or losing vacancies or interstitials.

The symbol adopted to represent a general dislocation is $\perp$. For an edge dislocation the "leg" of the inverted T points in the direction of the added material, as depicted in Fig. 1-19*f*.

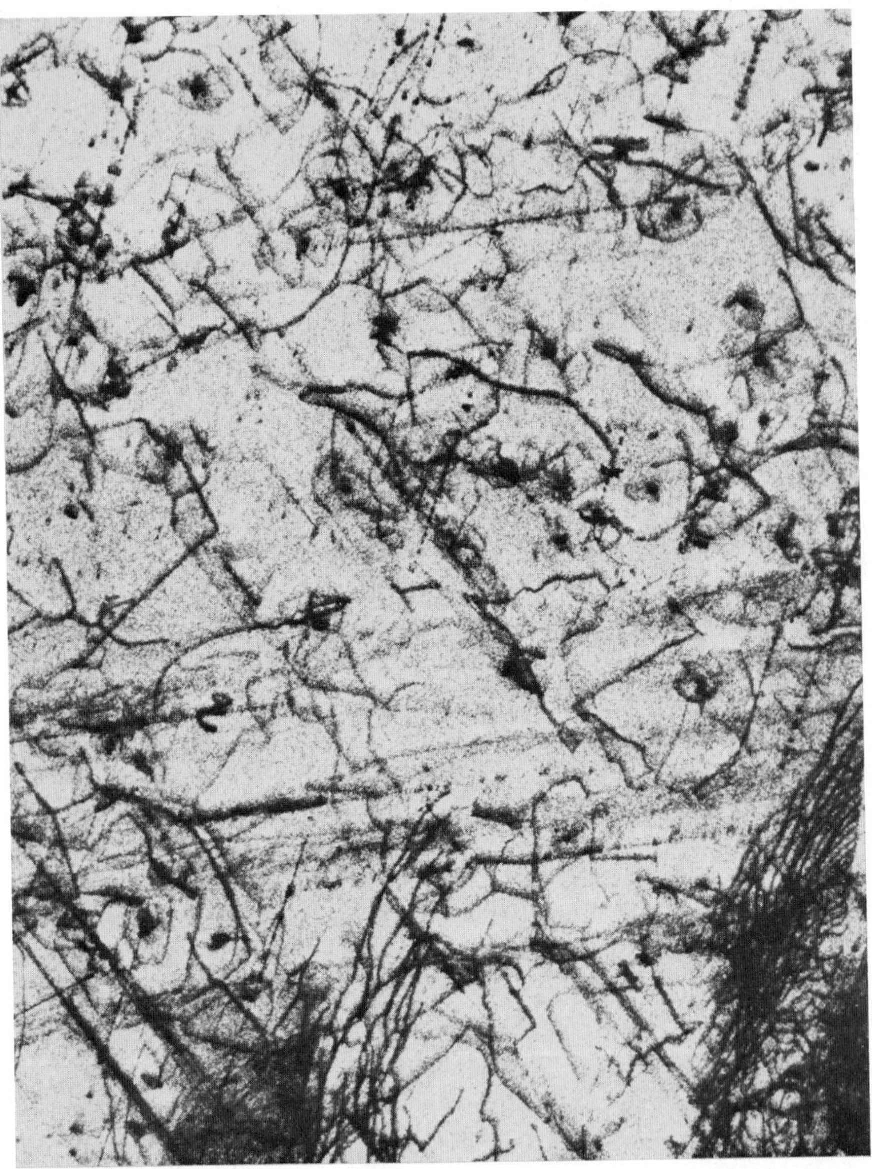

FIGURE 1-14. (220) Lang x-ray-transmission micrograph (Ag Kα) of a 1-mm-thick aluminum single crystal showing a very low density, $\sim 10^2/\text{cm}^2$, of dislocation lines, 35× [B. Nøst, *Phil. Mag.*, **11**:183 (1965)].

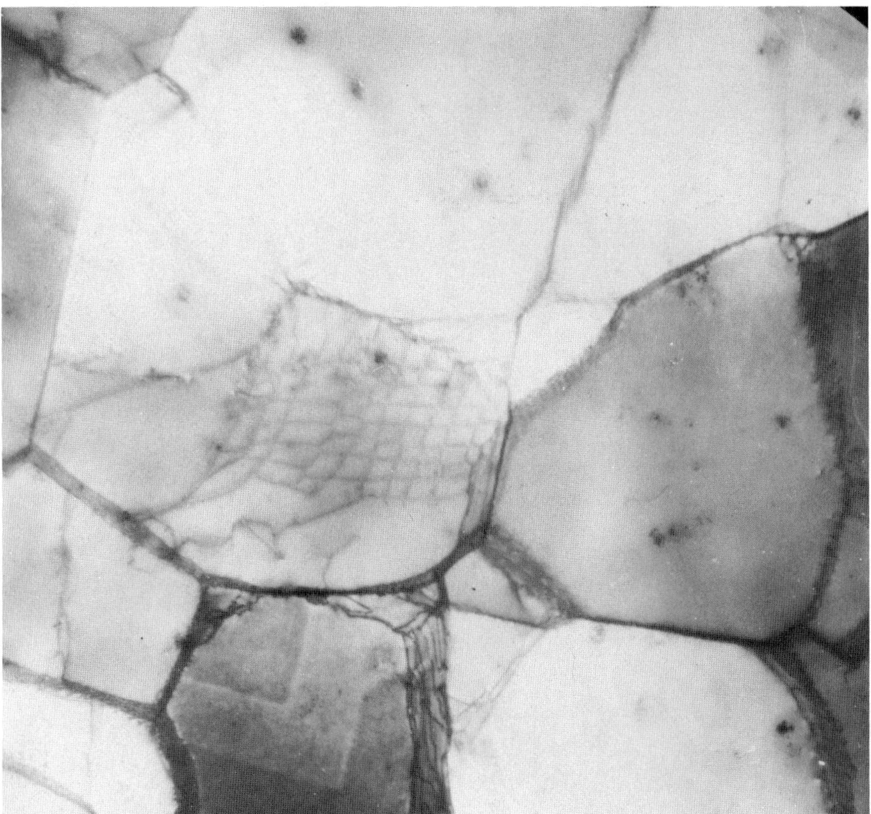

FIGURE 1-15. Electron-transmission micrograph showing polygonized subboundaries in aluminum, 65,000× (courtesy of P. B. Hirsch).

The Burgers Vector

We define the dislocation-displacement vector, the *Burgers vector* **b**, by the procedure suggested by Frank.[32] This definition is developed in considerable detail because of its importance in later chapters, and because of occasional confusion about it in the literature. This formal definition of the dislocation strength in a region holds for any elastic medium containing any distribution of dislocations, but it is usually applied to single dislocations or small numbers of dislocations.

Let us consider a dislocation in a simple cubic lattice. Figure 1-20*a* shows a section normal to a cube plane intersected by an edge dislocation. Figure 1-20*b* depicts a perfect reference lattice, where the lattice points are related by the translation vector $m\mathbf{t}_1 + n\mathbf{t}_2$, with $m = 1, 2, 3, \ldots$ and $n = 1, 2, 3, \ldots$. In Fig.

[32] F. C. Frank, *Phil. Mag.*, **42**: 809 (1951).

FIGURE 1-16. Electron-transmission micrograph of a Cu-7% Al alloy showing dislocations, partial dislocations, and stacking faults. Extinction contours indicate changes in thickness in the wedge-shaped crystal. Dashed and dotted oscillations are diffraction contrast effects, 50,000× [A. Howie and M. J. Whelan, *Proc. Roy. Soc.*, **A267**:206 (1962).]

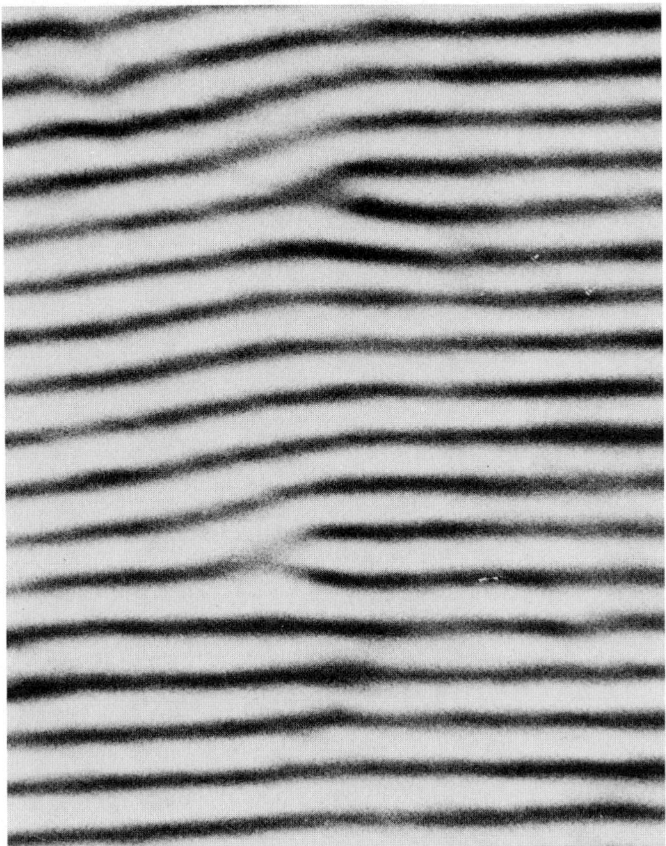

FIGURE 1-17. Electron-transmission moiré pattern from overlying palladium and gold {111} layers, indicating dislocations in one of the layers, 1,800,000× (courtesy of D. W. Pashley).

1-20*a* the crystal is divided into a "good" region, where the displacements from the ideal reference packing are produced only by small elastic strains and by thermal vibrations, and a "bad" region near the dislocation line, where the displacements are large.

The definition of **b** depends on the *sense* of a dislocation line. The sense is arbitrary. When a mirror plane exists normal to the line, there is no rule that can be devised to avoid this arbitrariness. We define the sense of the dislocation line by assigning a *unit vector* $\boldsymbol{\xi}$ tangent to the dislocation line and taking the positive sense in the positive direction of $\boldsymbol{\xi}$. **b** now can be defined by the procedure illustrated in Fig. 1-20, where the positive sense of $\boldsymbol{\xi}$ is taken to be into the paper. First we form in the real crystal a closed, clockwise *Burgers circuit* *S*-1-2-3-*F*, which lies entirely in good material and encloses the dislocation (Fig. 1-20*a*). Then we draw the same circuit in the perfect reference lattice, as shown in Fig. 1-20*b*. The vector required to close the latter circuit,

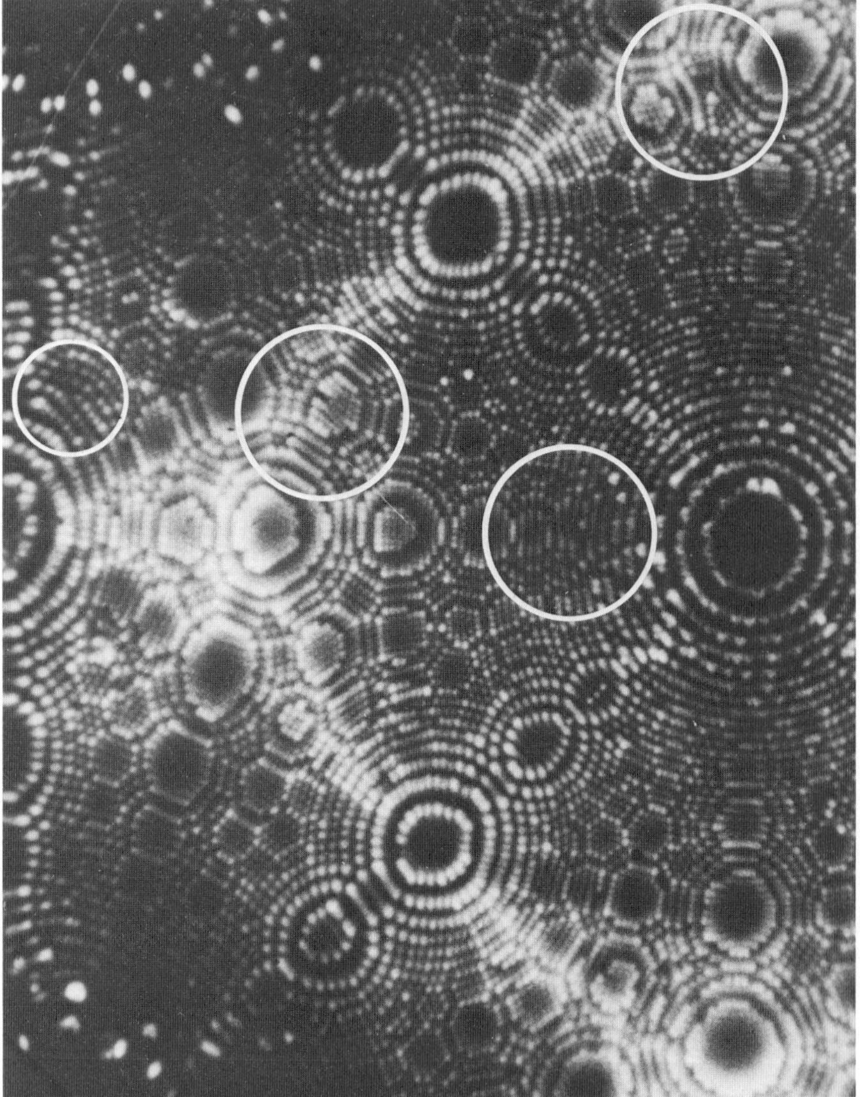

FIGURE 1-18. Ion-emission pattern of a platinum tip, showing dislocations intersecting the surface, 10,000,000× (courtesy of E. Müller).

that drawn from finish F to start S in Fig. 1-20*b*, is defined as the *true Burgers vector* **b**. Since the sense of the circuit is that of a right-handed screw RH, this convention for **b** is called the **FS**/RH convention.[33]

A simpler method, which is useful in dealing with single dislocations, is illustrated in Fig. 1-21. We draw in the real crystal a clockwise circuit which

[33] B. A. Bilby, R. Bullough, and E. Smith, *Proc. Roy. Soc.*, **A231**: 263 (1955).

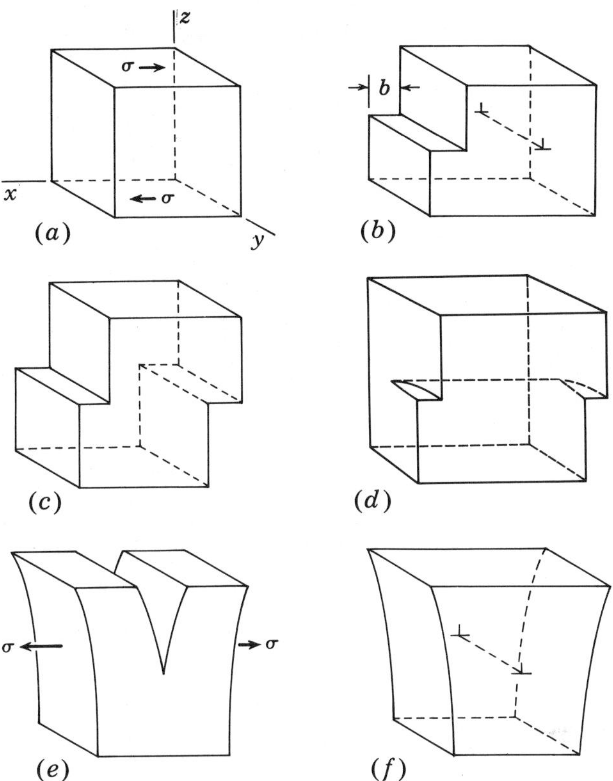

FIGURE 1-19. Operations on a perfect crystal cube (*a*), forming dislocations and slip steps (*b*) to (*d*) by shear, and (*e*) and (*f*), by dilatation.

would be closed in the perfect reference crystal; the closure vector **SF** is the *local Burgers vector* **b**. The local Burgers vector[33] in general differs from the true Burgers vector, because the former is affected by elastic strains and thermal vibrations. However, the two approach one another as the Burgers circuit is made very large and lies in nearly perfect crystal. Note that for the two definitions to be self-consistent, the local Burgers vector must be defined by an **SF**/*RH* convention. Equivalently, as shown in Fig. 1-22, the local Burgers vector is given by the line integral, taken in a right-handed sense relative to $\boldsymbol{\xi}$, of the elastic displacement **u** around the dislocation

$$\mathbf{b}=\oint_C \frac{\partial \mathbf{u}}{\partial l}\,dl \tag{1-5}$$

There is no general agreement in the literature on the convention for **b**. Since one of the aims of this text is to provide a synthesis of continuum

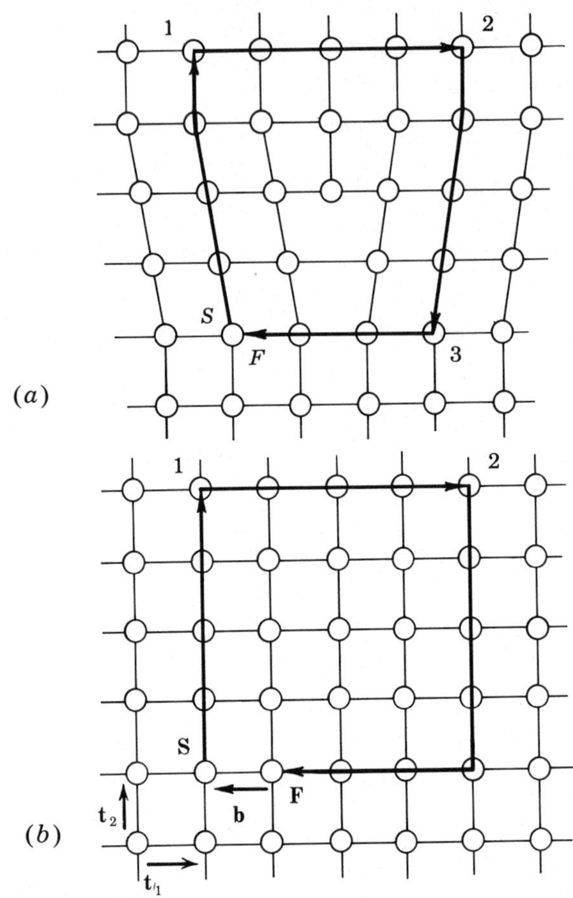

FIGURE 1-20. **FS**/*RH* Burgers circuits in (*a*) a real crystal and (*b*) a perfect reference crystal. ξ points into the paper.

dislocation theory and discrete lattice theory, we have adopted the above convention, which is used in the continuum theory.[34]

On the basis of the above definitions, an axiom follows directly.

[34] This convention is specifically employed in M. O. Peach and J. S. Koehler, *Phys. Rev.*, **80**: 436 (1950); B. A. Bilby, *Research*, **4**: 387 (1951); B. A. Bilby, R. Bullough, and E. Smith, *Proc. Roy. Soc.*, **A231**: 263 (1955); F. C. Frank, *Phil. Mag.*, **42**: 809 (1951); J. D. Eshelby, *Solid State Phys.*, **3**: 79 (1956); J. Friedel, "Les Dislocations," Gauthiers-Villars, Paris, 1956; E. Kröner, *Ergeb. angew. Math.*, **5** (1958); N. Thompson, *Proc. Phys. Soc.*, **B66**; 481 (1953); R. deWit, *Solid State Phys.*, **10**: 249 (1960). Thompson's simplified vector notation, which is used extensively for fcc crystals, also employs this convention. However, the *opposite* convention has been used in J. M. Burgers, *Proc. Kon. Ned. Akad. Wetenschap.*, **42**: 293 (1939); F. R. N. Nabarro, *Adv. in Phys.*, **1**: 269 (1952); W. T. Read, Jr., "Dislocations in Crystals," McGraw-Hill, New York, 1953; A. Seeger, *Handbuch der Physik*, **7**: 383 (1955); J. Weertman and J. R. Weertman, "Elementary Dislocation Theory," Macmillan, New York, 1964. Thus one must be careful about the defining convention for **b** in using results from the literature.

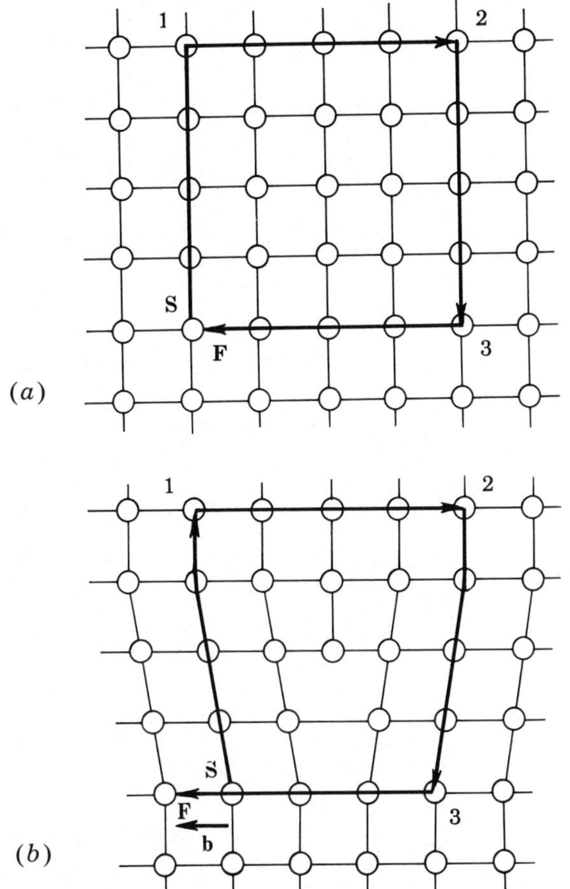

FIGURE 1-21. **FS**/*RH* Burgers circuits in (*a*) a perfect reference crystal and (*b*) a real crystal. $\boldsymbol{\xi}$ points into the paper.

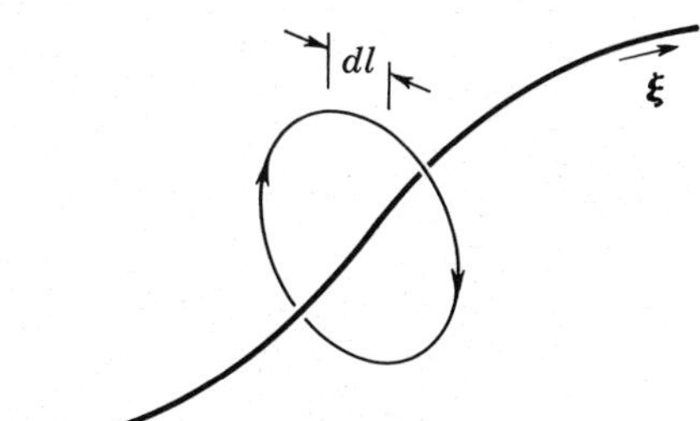

FIGURE 1-22. A displacement circuit.

Axiom 1-1: Reversing the sense of the dislocation line by reversing the direction of $\boldsymbol{\xi}$ causes $\mathbf{b}$ to reverse its direction.

This axiom is readily verified by taking $\boldsymbol{\xi}$ out of the paper in Fig. 1-20 and repeating the Burgers circuit operation.

Referring to the previous section, we can now define a dislocation more formally as being characterized by a line direction $\boldsymbol{\xi}$ and a Burgers vector $\mathbf{b}$. For an edge dislocation $\mathbf{b}\cdot\boldsymbol{\xi}=0$; for a right-handed screw dislocation $\mathbf{b}\cdot\boldsymbol{\xi}=b$; for a left-handed screw dislocation $\mathbf{b}\cdot\boldsymbol{\xi}=-b$.

Continuity of a Dislocation

Because, as noted above, a dislocation line is the boundary of an area over which slip displacement has occurred, a dislocation line evidently cannot end within an otherwise perfect region of crystal, but must terminate at a free surface, another dislocation line, a grain boundary, or some other defect. This has been proved as a *reductio ad absurdum* by Nabarro,[35] with formal elasticity theory, but it is also clear from the above definition of $\mathbf{b}$. If one attempts to end the dislocation shown in Fig. 1-20a by superposing a perfect-lattice plane over the plane shown there, one finds that this is possible only with the introduction of a tube of "bad" material which runs out at the top or bottom of the page. In fact, this tube represents an edge dislocation with its line perpendicular to the initial dislocation.

Consider the dislocation in the extension of Fig. 1-19 depicted in Fig. 1-23. The dislocation is continuous within the crystal and bounds a region of slip. The definition of $\mathbf{b}$ indicates that $\mathbf{b}$ is invariant along the dislocation line, although the dislocation changes continuously from screw character at A to edge character at C. The dislocation line at B is between pure edge and pure screw character and is called *mixed*. As shown in Fig. 1-23, its Burgers vector can be resolved into a screw component

$$\mathbf{b}_s=(\mathbf{b}\cdot\boldsymbol{\xi})\boldsymbol{\xi} \tag{1-6}$$

and an edge component

$$\mathbf{b}_e=\boldsymbol{\xi}\times(\mathbf{b}\times\boldsymbol{\xi}) \tag{1-7}$$

Equivalent Burgers Circuits

Two Burgers circuits are defined as being equivalent if one can be translated or deformed to coincide with the other in space without cutting through any "bad" material during translation. It follows from the preceding section that the total strength of dislocation, i.e., the resultant $\mathbf{b}$, within equivalent Burgers circuits is the same. Consider Fig. 1-24, which shows three dislocations meeting

[35] F. R. N. Nabarro, *Adv. Phys.*, **1**: 284 (1952).

The matrix $\{c_{ijkl}\}$ is 9×9, relating the nine elements σ_{ij} to the nine elements ϵ_{kl}. According to Eq. (2-13), the matrix $\{c_{ijkl}\}$ is symmetric about the diagonal. The elastic coefficients are often written in a contracted matrix notation as c_{mn}, where m and n are each indices corresponding to a pair of indices ij or kl, according to the following reduction:

$$\begin{array}{ccccccccccccc} ij & \text{or} & kl & 11 & 22 & 33 & 23 & 31 & 12 & 32 & 13 & 21 \\ m & \text{or} & n & 1 & 2 & 3 & 4 & 5 & 6 & 7 & 8 & 9 \end{array} \qquad (2\text{-}16)$$

Thus, by definition,

$$\begin{aligned} c_{11} &= c_{1111} \quad & c_{12} &= c_{1122} \\ c_{44} &= c_{2323} \quad & c_{46} &= c_{2312} \\ &\cdots\cdots & &\cdots\cdots \end{aligned} \qquad (2\text{-}17)$$

Equations (2-6) and (2-13) indicate that because of symmetry there are only 21 independent elastic constants among the 81 in c_{mn}. Thus Eq. (2-15) reduces to

$$\begin{bmatrix} \sigma_{11} \\ \sigma_{22} \\ \sigma_{33} \\ \sigma_{23} \\ \sigma_{31} \\ \sigma_{12} \\ \sigma_{32} \\ \sigma_{13} \\ \sigma_{21} \end{bmatrix} = \begin{bmatrix} c_{11} & c_{12} & c_{13} & c_{14} & c_{15} & c_{16} & c_{14} & c_{15} & c_{16} \\ c_{12} & c_{22} & c_{23} & c_{24} & c_{25} & c_{26} & c_{24} & c_{25} & c_{26} \\ c_{13} & c_{23} & c_{33} & c_{34} & c_{35} & c_{36} & c_{34} & c_{35} & c_{36} \\ c_{14} & c_{24} & c_{34} & c_{44} & c_{45} & c_{46} & c_{44} & c_{45} & c_{46} \\ c_{15} & c_{25} & c_{35} & c_{45} & c_{55} & c_{56} & c_{45} & c_{55} & c_{56} \\ c_{16} & c_{26} & c_{36} & c_{46} & c_{56} & c_{66} & c_{46} & c_{56} & c_{66} \\ c_{14} & c_{24} & c_{34} & c_{44} & c_{45} & c_{46} & c_{44} & c_{45} & c_{46} \\ c_{15} & c_{25} & c_{35} & c_{45} & c_{55} & c_{56} & c_{45} & c_{55} & c_{56} \\ c_{16} & c_{26} & c_{36} & c_{46} & c_{56} & c_{66} & c_{46} & c_{56} & c_{66} \end{bmatrix} \begin{bmatrix} \epsilon_{11} \\ \epsilon_{22} \\ \epsilon_{33} \\ \epsilon_{23} \\ \epsilon_{31} \\ \epsilon_{12} \\ \epsilon_{32} \\ \epsilon_{13} \\ \epsilon_{21} \end{bmatrix} \qquad (2\text{-}18)$$

Because of the symmetry in Eq. (2-18), it usually is reduced to the following 6×6 representation:

$$\begin{bmatrix} \sigma_{11} \\ \sigma_{22} \\ \sigma_{33} \\ \sigma_{23} \\ \sigma_{31} \\ \sigma_{12} \end{bmatrix} = \begin{bmatrix} c_{11} & c_{12} & c_{13} & c_{14} & c_{15} & c_{16} \\ c_{12} & c_{22} & c_{23} & c_{24} & c_{25} & c_{26} \\ c_{13} & c_{23} & c_{33} & c_{34} & c_{35} & c_{36} \\ c_{14} & c_{24} & c_{34} & c_{44} & c_{45} & c_{46} \\ c_{15} & c_{25} & c_{35} & c_{45} & c_{55} & c_{56} \\ c_{16} & c_{26} & c_{36} & c_{46} & c_{56} & c_{66} \end{bmatrix} \begin{bmatrix} \epsilon_{11} \\ \epsilon_{22} \\ \epsilon_{33} \\ \gamma_{23} \\ \gamma_{31} \\ \gamma_{12} \end{bmatrix} \qquad (2\text{-}19)$$

The c_{mn} in the 6×6 matrix are the same as the c_{mn} occurring in the 9×9 matrix. The 6×6 matrix is also symmetric about the diagonal. *Notice that in the reduced scheme the γ_{ij} [Eq. (2-4)] are used instead of the ϵ_{ij} for shear strains.* This is necessary to preserve diagonal symmetry in the matrix relating stress and strain.

When the axes of reference are rotated, stresses, strains, and elastic constants must be transformed accordingly. *Transformations are performed most conveniently in the complete* 9×9 *scheme*. As is evident from a comparison of Eqs. (2-18) and (2-19), the transposition of terms from a given 6×6 matrix to a complete 9×9 matrix is obvious. Specific examples of such transformations are given in Chap. 13.

For most crystals the number of independent elastic constants is reduced further from 21 because of crystal symmetry. As an example, only three constants are independent for cubic crystals, as shown in the following development. Let the coordinate axes coincide with the cubic axes. Since cubic crystals are invariant in symmetry to 90° rotations about the cube axes, then obviously $c_{iiii} = c_{11}$, $c_{iijj} = c_{12}$ for $i \neq j$, and $c_{ijij} = c_{44}$ for $i \neq j$. All other constants are zero, because the sign of ϵ_{ij} for $i \neq j$ cannot affect the value of σ_{ii}; thus $c_{iiij} = 0$, etc. Written out fully for cubic crystals, the 6×6 elastic-constant matrix is

$$c_{mn} = \begin{bmatrix} c_{11} & c_{12} & c_{12} & 0 & 0 & 0 \\ c_{12} & c_{11} & c_{12} & 0 & 0 & 0 \\ c_{12} & c_{12} & c_{11} & 0 & 0 & 0 \\ 0 & 0 & 0 & c_{44} & 0 & 0 \\ 0 & 0 & 0 & 0 & c_{44} & 0 \\ 0 & 0 & 0 & 0 & 0 & c_{44} \end{bmatrix} \tag{2-20}$$

and the relation between σ and ϵ is, by Eqs. (2-19) and (2-20),

$$\begin{aligned} \sigma_{11} &= c_{11}\epsilon_{11} + c_{12}\epsilon_{22} + c_{12}\epsilon_{33} & \sigma_{23} &= 2c_{44}\epsilon_{23} \\ \sigma_{22} &= c_{12}\epsilon_{11} + c_{11}\epsilon_{22} + c_{12}\epsilon_{33} & \sigma_{31} &= 2c_{44}\epsilon_{31} \\ \sigma_{33} &= c_{12}\epsilon_{11} + c_{12}\epsilon_{22} + c_{11}\epsilon_{33} & \sigma_{12} &= 2c_{44}\epsilon_{12} \end{aligned} \tag{2-21}$$

2-3. TRANSFORMATION OF STRESS, STRAIN, AND ELASTIC CONSTANTS

Transformations of coordinate axes are often useful, particularly in the theory of dislocation interactions. Consider two orthogonal coordinate systems $x_1 x_2 x_3$ and $x_1' x_2' x_3'$, with a common origin. The rotation of the system x_i' relative to system x_i is given in terms of the three eulerian angles θ, ϕ, and κ, as indicated in Fig. 2-3. These eulerian coordinates[8] are not commonly used but are presented here as an example because they simplify the problem of resolution

[8]See H. Goldstein, "Classical Mechanics," Addison-Wesley, Cambridge, Mass., 1950, for a detailed discussion of these coordinates.

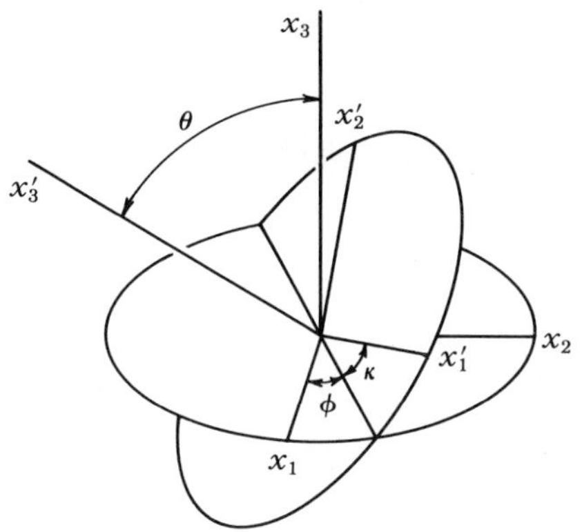

FIGURE 2-3. General rotation of coordinate axes in Euler coordinates.

of stresses in Chap. 9. Some straightforward but tedious geometry reveals that the full transformation reads

$$\begin{aligned} x'_1 &= (\cos\kappa\cos\phi - \cos\theta\sin\phi\sin\kappa)x_1 \\ &\quad + (\cos\kappa\sin\phi + \cos\theta\sin\kappa\cos\phi)x_2 + \sin\theta\sin\kappa x_3 \\ x'_2 &= (-\sin\kappa\cos\phi - \cos\theta\cos\kappa\sin\phi)x_1 \\ &\quad + (-\sin\kappa\sin\phi + \cos\theta\cos\kappa\cos\phi)x_2 + \sin\theta\cos\kappa x_3 \\ x'_3 &= \sin\theta\sin\phi x_1 - \sin\theta\cos\phi x_2 + \cos\theta x_3 \end{aligned} \tag{2-22}$$

In brief, the transformation is written

$$x'_i = T_{ij}x_j \tag{2-23}$$

where $\{T\}$ is a unitary orthogonal matrix; the components satisfy the relations

$$T_{il}T_{jl} = \delta_{ij} \tag{2-24}$$

where δ_{ij} is the Kronecker delta with the properties

$$\delta_{ij} = \begin{cases} 0 & i \neq j \\ 1 & i = j \end{cases} \tag{2-25}$$

Note that T_{ij} is the direction cosine between the x'_i and x_j axes. A commonly used alternative to Eq. (2-22) is to write the transformation directly in terms of the direction cosines.

Exercise 2-1. Derive Eq. (2-22).

Exercise 2-2. Verify Eq. (2-24) directly from the explicit expressions for T_{ij} [Eq. (2-22)].

The inverse of a unitary orthogonal transformation is obtained by transposition,[9] so that

$$T_{ji} = \{\tilde{T}\}_{ij} = \{T^{-1}\}_{ij} \tag{2-26}$$

that is, the ij component in the matrix for $\{\tilde{T}\}$ or $\{T^{-1}\}$ is the ji component in the matrix for $\{T\}$. Thus the inversion of Eq. (2-23) is

$$x_i = T_{ji} x'_j \tag{2-27}$$

With interchange of dummy indices, the differentiation of Eq. (2-27) yields

$$\frac{\partial}{\partial x'_i} = \frac{\partial x_j}{\partial x'_i}\frac{\partial}{\partial x_j} = T_{ij}\frac{\partial}{\partial x_j} \tag{2-28}$$

and similarly,

$$\frac{\partial}{\partial x_i} = T_{ji}\frac{\partial}{\partial x'_j} \tag{2-29}$$

By definition [Eq. (2-3)],

$$\epsilon'_{ij} = \frac{1}{2}\left(\frac{\partial u'_i}{\partial x'_j} + \frac{\partial u'_j}{\partial x'_i}\right) \tag{2-30}$$

The displacements u'_i are expressed by the u_j as

$$u'_i = T_{ij} u_j \tag{2-31}$$

Thus, by the insertion of Eq. (2-31) into Eq. (2-30),

$$\epsilon'_{ij} = \frac{1}{2} T_{jm}\frac{\partial}{\partial x_m} T_{il} u_l + \frac{1}{2} T_{il}\frac{\partial}{\partial x_l} T_{jm} u_m$$

which can be contracted to

$$\epsilon'_{ij} = T_{il} T_{jm} \epsilon_{lm} \tag{2-32}$$

summed over l and m. Quantities which transform according to a law of the

[9] I. S. Sokolnikoff and R. M. Redheffer, "Mathematics of Physics and Modern Engineering," McGraw-Hill, New York, 1966, p. 297.

form of Eq. (2-32) are defined as *second-rank tensors*; ϵ_{ij} is such a tensor. The product $T_{il}T_{jm}$ represents the 81 elements in the 9×9 matrix for the transformation of the nine elements ϵ_{lm}. This 9×9 matrix is also a unitary orthogonal matrix, for the product of the rows *ij* and *gh* is

$$T_{il}T_{jm}T_{gl}T_{hm}=T_{il}T_{gl}T_{jm}T_{hm}=\delta_{ig}\delta_{jh} \tag{2-33}$$

with the use of Eq. (2-24) in the last step. Thus, by the properties of unitary orthogonal matrices, the inverse matrix is found simply by transposition. As an example, the inverse of Eq. (2-32) must be

$$\epsilon_{ij}=T_{li}T_{mj}\epsilon'_{lm} \tag{2-34}$$

Now consider the force components

$$F_i=\sigma_{ij}A_j \tag{2-35}$$

where the A_j are the components of a surface element **A**, given by

$$\mathbf{A}=A\mathbf{n}$$

where A is the area and **n** is the outward unit normal vector. Since F_i transforms like a vector, σ_{ij}, which relates the vectors **F** and **A**, must transform like a second-rank tensor [*cf*. Eq. (2-32)],

$$\sigma'_{ij}=T_{il}T_{jm}\sigma_{lm} \tag{2-36}$$

In a contracted notation Eq. (2-35) would become[10]

$$\mathbf{F}=\boldsymbol{\sigma}\cdot\mathbf{A}=\mathbf{A}\cdot\boldsymbol{\sigma} \tag{2-37}$$

[10]The contracted notation of Eq. (2-37) requires the use of Gibbs' *dyadic* notation for $\boldsymbol{\sigma}$ (see P. Chadwick, "Continuum Mechanics," Wiley, New York, 1976, p. 25). In this notation

$$\boldsymbol{\sigma}=\begin{bmatrix}\sigma_{11}\mathbf{i}\otimes\mathbf{i} & \sigma_{12}\mathbf{i}\otimes\mathbf{j} & \sigma_{13}\mathbf{i}\otimes\mathbf{k}\\ \sigma_{21}\mathbf{j}\otimes\mathbf{i} & \sigma_{22}\mathbf{j}\otimes\mathbf{j} & \sigma_{23}\mathbf{j}\otimes\mathbf{k}\\ \sigma_{31}\mathbf{k}\otimes\mathbf{i} & \sigma_{32}\mathbf{k}\otimes\mathbf{j} & \sigma_{33}\mathbf{k}\otimes\mathbf{k}\end{bmatrix}$$

The products $\mathbf{i}\otimes\mathbf{i}$, etc., are *dyads*. To illustrate multiplication, consider the dyadic product of a two-dimensional tensor **B** with a vector **c**,

$$\mathbf{B}\cdot\mathbf{c}=\begin{bmatrix}B_{11}\mathbf{i}\otimes\mathbf{i} & B_{12}\mathbf{i}\otimes\mathbf{j}\\ B_{21}\mathbf{j}\otimes\mathbf{i} & B_{22}\mathbf{j}\otimes\mathbf{j}\end{bmatrix}\begin{bmatrix}c_1\mathbf{i}\\ c_2\mathbf{j}\end{bmatrix}$$

$$=(B_{11}c_1+B_{12}c_2)\mathbf{i}+(B_{21}c_1+B_{22}c_2)\mathbf{j}$$

When multiplied on the right side, as in the above expression, the vector must be written as a

Finally, consider the transformation of the elastic constants. Starting with the relationships

$$\sigma_{gh} = c_{ghmn}\epsilon_{mn} \tag{2-38}$$

and inserting

$$\epsilon_{mn} = T_{km}T_{ln}\epsilon'_{kl}$$

and (2-39)

$$\sigma_{gh} = T_{ig}T_{jh}\sigma'_{ij}$$

one obtains

$$T_{ig}T_{jh}\sigma'_{ij} = c_{ghmn}T_{km}T_{ln}\epsilon'_{kl} \tag{2-40}$$

which, by virtue of the orthogonality of the transformation, can be written as

$$\sigma'_{ij} = T_{ig}T_{jh}c_{ghmn}T_{km}T_{ln}\epsilon'_{kl} \tag{2-41}$$

A comparison with the equation

$$\sigma'_{ij} = c'_{ijkl}\epsilon'_{kl} \tag{2-42}$$

reveals that

$$c'_{ijkl} = T_{ig}T_{jh}c_{ghmn}T_{km}T_{ln} \tag{2-43}$$

with summation over g, h, m, and n understood. The elastic constants are tensors of rank 4; that is, they transform as fourth-rank tensors, according to Eq. (2-43).

---

column matrix and its dot product taken term by term with the right-hand member of the dyad. When multiplied on the left-hand side, the vector must be written as a row matrix and its dot product taken with the left-hand member of the dyad,

$$\mathbf{c}\cdot\mathbf{B} = \begin{bmatrix} c_1\mathbf{i} & c_2\mathbf{j} \end{bmatrix} \begin{bmatrix} B_{11}\mathbf{i}\otimes\mathbf{i} & B_{12}\mathbf{i}\otimes\mathbf{j} \\ B_{21}\mathbf{j}\otimes\mathbf{i} & B_{22}\mathbf{j}\otimes\mathbf{j} \end{bmatrix}$$

$$= (B_{11}c_1 + B_{21}c_2)\mathbf{i} + (B_{12}c_1 + B_{22}c_2)\mathbf{j} \neq \mathbf{B}\cdot\mathbf{c}$$

As an exercise, verify that $\mathbf{A}\cdot\boldsymbol{\sigma} = \boldsymbol{\sigma}\cdot\mathbf{A}$.

The use of matrix, tensor, and dyadic notation interchangeably might be somewhat confusing, but as each notation is most useful in a particular section of the book, all are presented here.

2-4. ELASTICITY THEORY FOR ISOTROPIC SOLIDS

With cubic symmetry, three independent elastic constants are needed. If the medium is elastically isotropic, i.e., if the elastic properties are independent of direction, only two independent elastic constants are required. As an illustration of this property, consider the reduction from cubic symmetry to isotropic behavior. Let only $\epsilon_{11} \neq 0$ among ϵ_{ij}, so that Eq. (2-21) gives the results

$$\sigma_{11} = c_{11}\epsilon_{11}$$

$$\sigma_{22} = \sigma_{33} = c_{12}\epsilon_{11} = \frac{c_{12}}{c_{11}}\sigma_{11}$$

Now rotate the coordinate system 45° about the z axis (Fig. 2-4). The transformation matrix is

$$T_{ij} = \begin{bmatrix} \sqrt{2}/2 & \sqrt{2}/2 & 0 \\ -(\sqrt{2}/2) & \sqrt{2}/2 & 0 \\ 0 & 0 & 1 \end{bmatrix}$$

Thus Eq. (2-32) yields

$$\epsilon'_{12} = T_{11}T_{21}\epsilon_{11} = -\tfrac{1}{2}\epsilon_{11}$$

and Eq. (2-36) yields

$$\sigma'_{12} = -\tfrac{1}{2}\sigma_{11} + \tfrac{1}{2}\sigma_{22} = -\tfrac{1}{2}c_{11}\epsilon_{11} + \tfrac{1}{2}c_{12}\epsilon_{11} = (c_{11} - c_{12})\epsilon'_{12}$$

But by Eq. (2-21), $\sigma'_{12} = 2c_{44}\epsilon'_{12}$ so that

$$2c_{44} = c_{11} - c_{12} \tag{2-44}$$

and only two of the three elastic constants are independent. For anisotropic

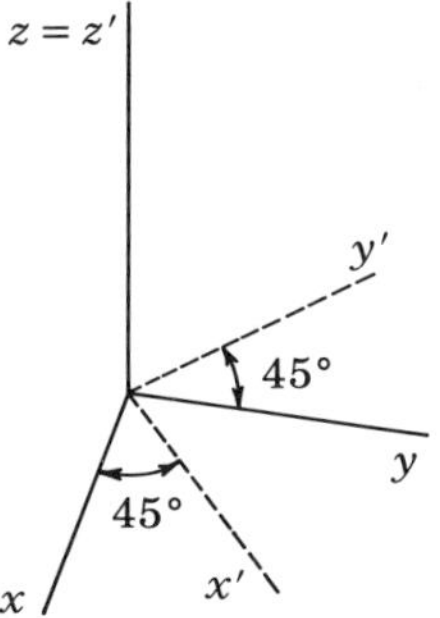

FIGURE 2-4. Rotation of the coordinate axes about the z axis.

cubic crystals, the factor

$$A=\frac{2c_{44}}{c_{11}-c_{12}} \tag{2-45}$$

is called the *anisotropy ratio*. The degree of anisotropy is measured by the deviation of A from the value $A=1$ characteristic of an isotropic medium. Some values of A for various materials are listed in Appendix 1.

When the relation (2-45) is satisfied, it is usual to express Eq. (2-21) in terms of the two elastic constants λ (the Lamé constant) and μ (the shear modulus) as follows:

$$\begin{aligned}
\sigma_{11}&=(\lambda+2\mu)\epsilon_{11}+\lambda\epsilon_{22}+\lambda\epsilon_{33} & \sigma_{23}&=2\mu\epsilon_{23}\\
\sigma_{22}&=\lambda\epsilon_{11}+(\lambda+2\mu)\epsilon_{22}+\lambda\epsilon_{33} & \sigma_{31}&=2\mu\epsilon_{31}\\
\sigma_{33}&=\lambda\epsilon_{11}+\lambda\epsilon_{22}+(\lambda+2\mu)\epsilon_{33} & \sigma_{12}&=2\mu\epsilon_{12}
\end{aligned} \tag{2-46}$$

where

$$\mu=c_{44}=\tfrac{1}{2}(c_{11}-c_{12}) \qquad \lambda=c_{12}$$

$$\lambda+2\mu=c_{11}$$

Formulas expressed in a general way in terms of c_{ijkl} are conveniently reduced to the isotropic case by the substitution

$$c_{ijkl}=\mu(\delta_{ik}\delta_{jl}+\delta_{il}\delta_{jk})+\lambda\delta_{ij}\delta_{kl} \tag{2-47}$$

with the Kronecker delta given by Eq. (2-25). In the isotropic case, the strain-energy density w [Eq. (2-14)] then has the form

$$\begin{aligned}
w&=\tfrac{1}{2}(\lambda+2\mu)(\epsilon_{11}+\epsilon_{22}+\epsilon_{33})^2\\
&\quad+2\mu\left(\epsilon_{23}{}^2+\epsilon_{31}{}^2+\epsilon_{12}{}^2-\epsilon_{11}\epsilon_{22}-\epsilon_{11}\epsilon_{33}-\epsilon_{22}\epsilon_{33}\right)\\
&=\frac{1}{2E}(\sigma_{11}+\sigma_{22}+\sigma_{33})^2\\
&\quad+\frac{1+\nu}{E}\left(\sigma_{23}{}^2+\sigma_{31}{}^2+\sigma_{12}{}^2-\sigma_{11}\sigma_{22}-\sigma_{11}\sigma_{33}-\sigma_{22}\sigma_{33}\right)
\end{aligned} \tag{2-48}$$

Some other relations between elastic constants follow directly from the above equations. Equation (2-46) indicates that the compressibility K, the ratio of the negative of the dilatation $e=\epsilon_{11}+\epsilon_{22}+\epsilon_{33}$ to the pressure $p=-\frac{1}{3}(\sigma_{11}+\sigma_{22}+\sigma_{33})$, must be

$$K=\frac{3}{3\lambda+2\mu}=-\frac{e}{p}=\frac{1}{B} \tag{2-49}$$

where B is the bulk modulus. Two other commonly used isotropic elastic constants are Young's modulus E, the ratio of simple tensile stress to strain, and Poisson's ratio ν, the ratio of transverse contraction to elongation in simple tension. Some useful relationships among these constants are

$$E=\frac{\mu(3\lambda+2\mu)}{\mu+\lambda}=\frac{9\mu B}{3B+\mu}=2\mu(1+\nu)$$

$$\nu=\frac{3B-2\mu}{2(3B+\mu)}=\frac{\lambda}{2(\mu+\lambda)}=\frac{E-2\mu}{2\mu}$$

$$\mu=\frac{E}{2(1+\nu)}$$

$$\lambda=\frac{\nu E}{(1+\nu)(1-2\nu)}=\frac{2\nu\mu}{1-2\nu} \qquad (2\text{-}50)$$

σ can be expressed in terms of ϵ by any combination of these constants, but only two of them will be independent. In terms of these constants, the inverse of Eq. (2-46) is given by

$$\begin{aligned}
\epsilon_{11}&=\frac{1}{E}\left[\sigma_{11}-\nu(\sigma_{22}+\sigma_{33})\right] & \epsilon_{23}&=\frac{1}{2\mu}\sigma_{23}\\
\epsilon_{22}&=\frac{1}{E}\left[\sigma_{22}-\nu(\sigma_{11}+\sigma_{33})\right] & \epsilon_{31}&=\frac{1}{2\mu}\sigma_{31} \qquad (2\text{-}51)\\
\epsilon_{33}&=\frac{1}{E}\left[\sigma_{33}-\nu(\sigma_{11}+\sigma_{22})\right] & \epsilon_{12}&=\frac{1}{2\mu}\sigma_{12}
\end{aligned}$$

2-5. PLANE STRAIN

Even in isotropic elasticity theory, and in particular for finite bodies where image effects must be considered, the mathematical complexity is considerable. For plane stress or strain the treatment simplifies a great deal. Of these simple cases, plane strain is discussed as an example because of its importance in the theory of straight dislocations. Plane strain is produced in a body under a state of stress such that there is no dependence of the displacements on one coordinate, say x_3, and such that the displacement u_3 is zero. Thus

$$u_1=u_1(x_1,x_2) \qquad u_2=u_2(x_1,x_2)$$

$$u_3=0 \quad \text{and} \quad \frac{\partial}{\partial x_3}=0 \qquad (2\text{-}52)$$

With the conditions of Eqs. (2-52), the essential description of plane strain derives from the equilibrium equations (2-2). In applications we are interested primarily in the state of *internal strain* in an infinitesimal volume element lying within the isotropic continuum and not acted on by body forces. Equations (2-2) and (2-52) then combine to the form (in cartesian or polar coordinates)

$$\frac{\partial \sigma_{11}}{\partial x_1}+\frac{\partial \sigma_{12}}{\partial x_2}=0 \quad \text{or} \quad \frac{\partial \sigma_{rr}}{\partial r}+\frac{1}{r}\frac{\partial \sigma_{r\theta}}{\partial \theta}+\frac{\sigma_{rr}-\sigma_{\theta\theta}}{r}=0$$
$$\frac{\partial \sigma_{12}}{\partial x_1}+\frac{\partial \sigma_{22}}{\partial x_2}=0 \quad \text{or} \quad \frac{\partial \sigma_{r\theta}}{\partial r}+\frac{1}{r}\frac{\partial \sigma_{\theta\theta}}{\partial \theta}+\frac{2\sigma_{r\theta}}{r}=0 \tag{2-53}$$

Equations (2-53) are fulfilled automatically if σ_{11}, σ_{22}, and σ_{12} are expressed in terms of a stress function ψ (the Airy stress function) as

$$\sigma_{11}=\frac{\partial^2\psi}{\partial x_2^{\,2}} \quad \text{or} \quad \sigma_{rr}=\frac{1}{r}\frac{\partial \psi}{\partial r}+\frac{1}{r^2}\frac{\partial^2\psi}{\partial \theta^2}$$
$$\sigma_{22}=\frac{\partial^2\psi}{\partial x_1^{\,2}} \quad \text{or} \quad \sigma_{\theta\theta}=\frac{\partial^2\psi}{\partial r^2} \tag{2-54}$$
$$\sigma_{12}=-\frac{\partial^2\psi}{\partial x_1\partial x_2} \quad \text{or} \quad \sigma_{r\theta}=-\frac{\partial}{\partial r}\frac{\partial \psi}{r\,\partial \theta}$$

The differentiation of Eqs. (2-3) produces for the case of plane strain the compatibility equation

$$\frac{\partial^2\epsilon_{11}}{\partial x_2^{\,2}}+\frac{\partial^2\epsilon_{22}}{\partial x_1^{\,2}}=2\frac{\partial^2\epsilon_{12}}{\partial x_1\partial x_2} \tag{2-55}$$

This equation expresses the fact that ϵ_{11}, ϵ_{22}, and ϵ_{12} cannot be considered independent, but must be derivable from just two functions u_1 and u_2. Inserting Eq. (2-51) into Eq. (2-55) and making use of relations (2-50) and (2-54), one finds that

$$\frac{\partial^4\psi}{\partial x_1^{\,4}}+2\frac{\partial^4\psi}{\partial x_1^{\,2}\partial x_2^{\,2}}+\frac{\partial^4\psi}{\partial x_2^{\,4}}=0$$

or

$$\nabla^4\psi=\nabla^2(\nabla^2\psi)=\left(\frac{\partial^2}{\partial x_1^{\,2}}+\frac{\partial^2}{\partial x_2^{\,2}}\right)^2\psi=0 \tag{2-56}$$

or, in polar coordinates,

$$\left(\frac{1}{r}\frac{\partial}{\partial r}r\frac{\partial}{\partial r}+\frac{1}{r^2}\frac{\partial^2}{\partial\theta^2}\right)^2\psi=0$$

If Eq. (2-56) is solved for ψ, the stresses σ_{11}, σ_{22}, and σ_{12} are then given by Eq. (2-54), and the strains from Eq. (2-51). Also, since $\epsilon_{33}=0$, Eq. (2-51) gives the result $\sigma_{33}=\nu(\sigma_{11}+\sigma_{22})$. Specific examples of such solutions of Eq. (2-56) are presented in Chap. 3, in the discussion of edge dislocations.

An even simpler situation arises from the conditions

$$u_1=u_2=0 \qquad u_3=u_3(x_1,x_2) \tag{2-57}$$

In this case, when no body forces are present, Eqs. (2-2) and (2-3) combine directly to the form

$$\nabla^2 u_3=0 \tag{2-58}$$

The displacement field around a pure screw dislocation can be treated by Eq. (2-58).

2-6. RESPONSE OF AN ISOTROPIC MEDIUM TO A POINT FORCE

A powerful method for the solution of problems in the continuum theory of elasticity, in particular for the continuum theory of dislocations in Chaps. 4 to 6, derives from a consideration of point forces. If the response of a body to a point force is known, the deformation caused by *any distribution of forces* can be obtained by integration. The problem is similar to that of finding the electrostatic potential $\mathcal{V}$ caused by a distribution of charge ρ. Solutions are derived by the Green's function method of electrostatics.[11] Because of the general analogy between electromagnetic theory and dislocation theory, we present the electrostatic analogy here; it is familiar to engineers and physicists.

Electrostatic Analog

The gradient of the electrostatic potential $\mathcal{V}(\mathbf{r})$ is proportional to the field, and the divergence of the field is proportional to the charge density $\rho(\mathbf{r})$, leading to Poisson's equation

$$\nabla^2\mathcal{V}(\mathbf{r})=-4\pi\rho(\mathbf{r}) \tag{2-59}$$

[11]See J. A. Stratton, "Electromagnetic Theory," McGraw-Hill, New York, 1941, p. 167.

The general solution of Eq. (2-59) is[11]

$$\mathcal{V}(\mathbf{r})=\int\frac{\rho(\mathbf{r}')}{|\mathbf{r}-\mathbf{r}'|}dV' \tag{2-60}$$

Physically, Eq. (2-60) expresses the fact that the potential at $\mathbf{r}$ is the sum of individual potential contributions

$$d\mathcal{V}=\frac{\rho(\mathbf{r}')\,dV'}{|\mathbf{r}-\mathbf{r}'|}$$

produced by charges $\rho(\mathbf{r}')\,dV'$. For a point charge e at $\mathbf{r}_0$,

$$\rho(\mathbf{r}')=e\,\delta(\mathbf{r}'-\mathbf{r}_0) \tag{2-61}$$

where $\delta(\mathbf{r}'-\mathbf{r}_0)$ is the Dirac delta function, with the property

$$\int f(\mathbf{r}')\,\delta(\mathbf{r}'-\mathbf{r}_0)\,dV'=f(\mathbf{r}_0) \tag{2-62}$$

with $f(\mathbf{r}')$ any function. Inserting Eq. (2-61) into (2-60) gives the Green's function for the potential arising from a point charge,

$$\mathcal{V}(\mathbf{r})=\frac{e}{|\mathbf{r}-\mathbf{r}_0|} \tag{2-63}$$

and from Eq. (2-59),

$$\nabla^2\frac{1}{|\mathbf{r}-\mathbf{r}_0|}=-4\pi\,\delta(\mathbf{r}-\mathbf{r}_0) \tag{2-64}$$

Thus in the Green's function method the potential at any point can be determined by the integration of Eq. (2-60) over a continuous distribution of charge, each charge element $\rho(\mathbf{r}')\,dV'$ contributing to the potential like a point charge situated at $\mathbf{r}'$.

Application to Elasticity

In the elasticity case the elastic displacement $\mathbf{u}$ is analogous to the electrostatic potential $\mathcal{V}$, and the body forces $\mathbf{f}$ to the charge density ρ. Our aim is to develop an expression in the form of Poisson's equation (2-59), so that the displacements can be determined from an expression like Eq. (2-60). The procedure is somewhat more complicated for elasticity than in the preceding cases, because vector quantities are involved instead of scalars.

For the case of isotropy, the substitution of Eq. (2-47) into (2-8) yields the result

$$(\lambda+\mu)\frac{\partial}{\partial x_i}\frac{\partial u_j}{\partial x_j}+\mu\frac{\partial^2 u_i}{\partial x_j^2}+f_i=0$$

or, in vector notation,

$$(\lambda+\mu)\nabla(\nabla\cdot\mathbf{u})+\mu\nabla^2\mathbf{u}+\mathbf{f}=0 \tag{2-65}$$

Let us specialize to only one component of the body force and then generalize the result. Suppose that a point force $f_1\delta(\mathbf{r})$ is acting at the origin. Equation (2-65) then becomes

$$(\lambda+\mu)\frac{\partial}{\partial x_1}(\nabla\cdot\mathbf{u})+\mu\nabla^2 u_1+f_1\delta(\mathbf{r})=0$$

$$(\lambda+u)\frac{\partial}{\partial x_2}(\nabla\cdot\mathbf{u})+\mu\nabla^2 u_2=0 \tag{2-66}$$

$$(\lambda+\mu)\frac{\partial}{\partial x_3}(\nabla\cdot\mathbf{u})+\mu\nabla^2 u_3=0$$

The results of potential theory[12] show that a vector $\mathbf{u}$ can be represented in terms of a scalar potential ϕ and a vector potential $\mathbf{A}$,

$$\mathbf{u}=\nabla\phi+\operatorname{curl}\mathbf{A} \tag{2-67}$$

In this representation, Eq. (2-66) becomes, after the sequence of some derivatives is interchanged,

$$(\lambda+2\mu)\nabla^2\frac{\partial\phi}{\partial x_1}+\mu\nabla^2\left(\frac{\partial A_3}{\partial x_2}-\frac{\partial A_2}{\partial x_3}\right)+f_1\delta(\mathbf{r})=0$$

$$(\lambda+2\mu)\nabla^2\frac{\partial\phi}{\partial x_2}+\mu\nabla^2\left(\frac{\partial A_1}{\partial x_3}-\frac{\partial A_3}{\partial x_1}\right)=0 \tag{2-68}$$

$$(\lambda+2\mu)\nabla^2\frac{\partial\phi}{\partial x_3}+\mu\nabla^2\left(\frac{\partial A_2}{\partial x_1}-\frac{\partial A_1}{\partial x_2}\right)=0$$

[12]See I. S. Sokolnikoff and R. M. Redheffer, "Mathematics of Physics and Modern Engineering," McGraw-Hill, New York, 1966, p. 408.

It is easily verified that $\nabla^2|\mathbf{r}|=2/|\mathbf{r}|$, so that Eq. (2-64) is equivalent to

$$\nabla^2\nabla^2|\mathbf{r}|=-8\pi\delta(\mathbf{r}) \tag{2-69}$$

Now, if Eqs. (2-68) and (2-69) can be shown to be equivalent, then the Green's function analog will be established. A solution which gives this equivalency, as can be verified by direct substitution, is

$$\phi=\frac{f_1}{8\pi(\lambda+2\mu)}\frac{\partial r}{\partial x_1} \qquad A_2=-\frac{f_1}{8\pi\mu}\frac{\partial r}{\partial x_3}$$

$$A_1=0 \qquad A_3=\frac{f_1}{8\pi\mu}\frac{\partial r}{\partial x_2}$$

Substituting these definitions into (2-67) and generalizing the result to a point force $f_j\delta(\mathbf{r})$, one finds that the ith component of the displacement $u_{ij}(\mathbf{r})$ caused by a unit point force $f_j=1$ applied in the jth direction at the origin is

$$u_{ij}(\mathbf{r})=\frac{1}{8\pi\mu}\left(\delta_{ij}\nabla^2 r-\frac{\lambda+\mu}{\lambda+2\mu}\frac{\partial^2 r}{\partial x_i\,\partial x_j}\right) \tag{2-70}$$

Also, $u_{ij}(\mathbf{r})=u_{ji}(\mathbf{r})$, by symmetry. $u_{ij}(\mathbf{r})$ is called the *tensor Green's function for the elastic displacements*. A continuous distribution of forces $f_j(\mathbf{r})$ in an elastic medium causes displacements

$$u_i(\mathbf{r})=\int u_{ij}(\mathbf{r}-\mathbf{r}')f_j(\mathbf{r}')\,dV' \tag{2-71}$$

Equation (2-71) is analogous to Eq. (2-60), and the analogy with the electrostatic case is complete.

Equation (2-70) gives the response of an infinite body to a point force. In a finite body, boundary conditions at the surface must be satisfied. For example, no forces can act on a free surface,

$$\sigma_{ij}n_j=0 \tag{2-72}$$

where the n_j are the components of $\mathbf{n}$, the local surface normal. The displacements in a finite body subjected to a point force can be described as a superposition of the displacements (2-70) and displacements caused by "image" stresses applied on the external surface of the body in order to satisfy boundary conditions. The image displacements are continuous throughout the entire body. At a point sufficiently close to that at which the point force is applied, Eq. (2-70) gives the dominant part of the nonuniform displacement giving rise to stress.

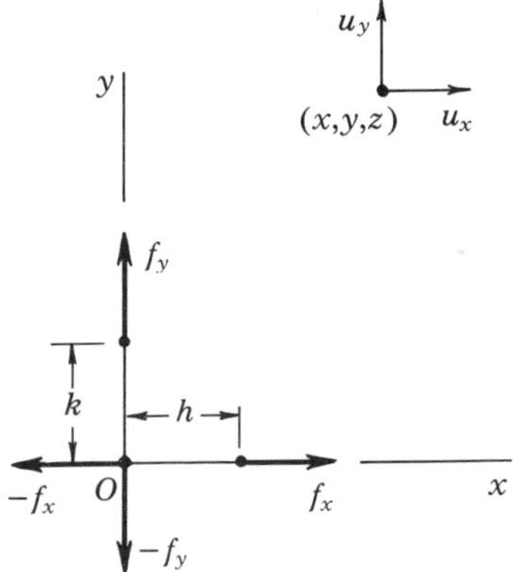

FIGURE 2-5. Displacements u_x and u_y associated with point forces f_x and f_y.

Point Source of Expansion

As an example of the application of Eq. (2-70), consider the displacement field of a point source of expansion. Three perpendicular double forces produce a stress field typical of such a source. In Fig. 2-5, let h, k, and l tend to zero, while $f_x h = f_y k = f_z l = M$ is kept constant. The force $-f_x$ at the origin produces a displacement $-f_x u_{11}(\mathbf{r})$, given by Eq. (2-70). The opposite force f_x at h produces a displacement $f_x u_{11}(\mathbf{r}) - f_x(\partial u_{11}/\partial x)h$. Thus the force pair produces a displacement $-(\partial u_{11}/\partial x) f_x h$. Proceeding similarly for the other force couples, one finds that the total displacement in the x direction is, from Eq. (2-70),

$$u_x = -M\frac{\partial}{\partial x}u_{11} - M\frac{\partial}{\partial y}u_{12} - M\frac{\partial}{\partial z}u_{13} = -\frac{M}{8\pi(\lambda+2\mu)}\frac{\partial}{\partial x}\nabla^2 r$$

$$= -\frac{M}{4\pi(\lambda+2\mu)}\frac{\partial}{\partial x}\frac{1}{r} = \frac{M}{4\pi(\lambda+2\mu)}\frac{x}{r^3} \tag{2-73}$$

u_y and u_z follow from symmetry. Inspection of Eq. (2-73) reveals that u_x, u_y, and u_z can be derived from a purely radial displacement

$$u_r = \frac{M}{4\pi(\lambda+2\mu)}\frac{1}{r^2} \tag{2-74}$$

This displacement is representative of a point of expansion with a strength

$$\delta v = 4\pi r^2 u_r = \frac{M}{\lambda+2\mu} \tag{2-75}$$

Suppose that the source of expansion is at the origin of a free sphere of radius R. In order to obtain a displacement field u_r consistent with the free surface condition [Eq. (2-72)], a term αr must be added to make σ_{rr} vanish at $r=R$:

$$u_r = \frac{\delta v}{4\pi r^2} + \alpha r \tag{2-76}$$

In spherical coordinates, the appropriate form of Eq. (2-7) is[13]

$$\sigma_{rr} = (\lambda+2\mu)\frac{\partial u_r}{\partial r} + \frac{2\lambda u_r}{r}, \qquad \sigma_{\theta\theta} = \sigma_{\phi\phi} = 2(\lambda+\mu)\frac{u_r}{r} + \lambda\frac{\partial u_r}{\partial r} \tag{2-77}$$

Combining Eqs. (2-76) and (2-77) and setting $\sigma_{rr}=0$ at $r=R$, one finds

$$\alpha = \frac{4\mu}{2\mu+3\lambda}\,\frac{\delta v}{4\pi R^3} \tag{2-78}$$

The displacement at the surface is then

$$u_r(R) = \frac{3\lambda+6\mu}{3\lambda+2\mu}\,\frac{\delta v}{4\pi R^2} = \frac{3B+4\mu}{3B}\,\frac{\delta v}{4\pi R^2}$$

This displacement produces an expansion of the total volume $V=(4\pi/3)R^3$, given by[14]

$$\delta V = \frac{3B+4\mu}{3B}\,\delta v \tag{2-80}$$

[13] L. D. Landau and E. M. Lifshitz, "Theory of Elasticity," Pergamon, New York, 1959, p. 3.

[14] The relation between point force distributions and external volume changes is general, as shown by J. D. Eshelby, *Solid State Phys.*, **3**: 79 (1956). We briefly outline the derivation. The total external volume change δV is related to the local dilatation ϵ_{ii} by

$$\delta V = \int \epsilon_{ii}\, dV = s_{iijk}\int \sigma_{jk}\, dV \tag{a}$$

where Hooke's law is used in the second step. An identity is

$$\frac{\partial \sigma_{jl} x_k}{\partial x_l} = x_k \frac{\partial \sigma_{jl}}{\partial x_l} + \sigma_{jl}\delta_{kl} = x_k\frac{\partial \sigma_{jl}}{\partial x_l} + \sigma_{jk} \tag{b}$$

Substituting (b) in (a) and using Eq. (2-2) gives

$$\delta V = s_{iijk}\left[\int \frac{\partial \sigma_{jl} x_k}{\partial x_l}\, dV + \int f_j x_k\, dV\right] \tag{c}$$

Recognizing $s_{iijk}\,\sigma_{jl}x_k$ as a vector v_l, one sees that the first integral is the volume integral of the divergence of **v**, so Stoke's theorem can be used to transform the integral to an integral over a closed surface A:

$$\delta V = s_{iijk}\left[\int \sigma_{jl} n_l x_k\, dA + \int f_j x_k\, dV\right] \tag{d}$$

Near the singularity, for $r \ll R$, the first term in Eq. (2-76) is dominant. In this region the dominant part of the internal stress is found from Eq. (2-77) to be

$$-\sigma_{rr} = \frac{\mu\,\delta v}{\pi r^3} \qquad \sigma_{\theta\theta} = \sigma_{\phi\phi} = \frac{\mu\,\delta v}{2\pi r^3} \tag{2-81}$$

The results of this section are useful in the treatment of point defects and are also applied in the next section. Other specific examples of the use of Eqs. (2-70) and (2-71) are given in Chap. 4. In closing this section we note that the analogy between electromagnetics and elasticity is more general than the special case invoked above. The general analogy is sometimes used in the continuum theory of dislocations.[14–17]

2-7. INTERACTION BETWEEN INTERNAL STRESS AND EXTERNAL STRESS

Approximations in Linear Elasticity

The main purpose of this section is to develop a very important theorem, valid in linear elasticity, about the elastic energy of a body containing internal stresses and subjected to external forces. The desired theorem follows directly from a discussion of the limitations of linear elasticity, without the complicated mathematical considerations of a formal development from the equations of linear elasticity.

In linear elasticity the superposition principle is assumed to hold true. The stresses and displacements caused by a set of forces acting on a body are assumed to be the sum of those caused by the individual forces. The linear-elasticity assumption clearly is invalid for large strains, where force-distance

---

where **n** is a unit vector normal to A. If the surface is a free surface $\sigma_{jl}n_l$, which represents the tractions on the free surface, must be zero and

$$\delta V = s_{iijk}\int f_j x_k \, dV \tag{e}$$

For an isotropic elastic solid, $s_{iijk} = 0$ unless $j = k$ and $s_{iijj} = \frac{1}{3}B$, so

$$\delta V = \frac{1}{3B}\int f_j x_j \, dV \tag{2-79}$$

Thus a set of point forces **f** at positions **x** produce a volume change as in the special case of Eq. (2-80).

[15] R. deWit, *Solid State Phys.*, **10**: 249 (1960).

[16] J. D. Eshelby, *Phys. Rev.*, **90**: 248 (1953).

[17] F. R. N. Nabarro, *Phil. Mag.*, **42**: 1224 (1951).

relations are no longer linear; in addition, it lacks precision in description of a geometrical nature. The stress, strain, and displacement at a point (x, y, z) in a body were discussed quite indiscriminately in the preceding sections. In a precise description one must specify whether the point (x, y, z) in a deformed body means that point which has the coordinates xyz relative to an external fixed coordinate system or that point in the body that has the coordinates xyz when the body is unstressed. In the first description, positions are referred to an external coordinate system; in the second description positions are referred to an embedded coordinate system, which deforms with the body. *Linear elasticity theory does not discriminate between the two descriptions; only first-order terms common to both descriptions are retained*. In the limit of small stresses and strains, the deformation of an embedded coordinate system would be negligible, and the two descriptions would be identical. *Problems in linear elasticity theory are treated as if all displacements and stresses were vanishingly small.*

When stresses, strains, and displacements are large, as is the case near a dislocation core, linear elasticity theory must be used with caution. In linear elasticity, the position to which one ascribes a calculated stress and strain is unspecified within an interval of the order of the displacement. Similarly, the stress is imprecise; no distinction is made between force per unit area as measured externally and force per unit area with the area measured in an embedded coordinate system.

The following description of the operations illustrated in Fig. 2-6 exemplify the imprecision in linear elasticity. For generality, the medium can be supposed

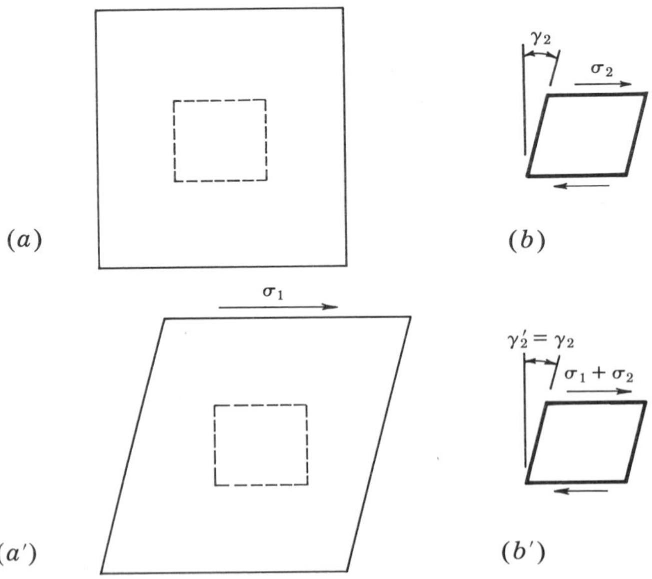

FIGURE 2-6. (a) Shear γ_2 of a volume element from an undeformed block. (b) Shear γ_2' of a volume element from a previously elastically deformed block.

to be elastically anisotropic. Figure 2-6*a* is an unstressed block of material, and 2-6*a*′ is an identical piece under stress σ_1. A piece, 2-6*b*, is cut out of 2-6*a* and a stress σ_2 is put on it. A piece of the same geometry and orientation is then cut out of the "as-stressed" material 2-6*a*′ and subjected to additional stress σ_2. The strain produced by σ_2 is the same in both cases, $\gamma_2 = \gamma_2'$. In linear elasticity theory, matter occupying some part of space can be considered as a medium filling that space with completely linear properties, like the ether in classical electricity theory.

Consider next two free bodies of the same material and with the same external shape and orientation. The body of Fig. 2-7*a* is stress free and the body of Fig. 2-7*b* contains internal stresses originating from point, line, and surface sources of internal stress, S_1, S_2, and S_3. These sources are considered to be *locked in* if they do not move relative to the medium in response to an external force. In the approximation of linear elasticity, the surfaces of Fig. 2-7 both bound media of identical response to externally applied stress. For the same externally applied stress σ, the shaded portions in parts (*a*) and (*b*) are identical. *The stresses and strains caused by external and internal sources of stress can be superposed*, provided that the internal stresses are locked in.

The elastic energy of a body in a given state of stress can be calculated as the work done by applied forces which produce the given stresses. The externally applied forces do the same work on both parts of Fig. 2-7, leading to the following important theorem:

Theorem 2-1: In the expression for the total elastic energy of a stressed body, there is no cross term between internal stress (locked-in stresses present when no external forces act) and stress caused by externally applied forces.

That is, the elastic energy in the body of Fig. 2-7*b* is the sum of the elastic energy it had before the application of external stress, and the elastic energy in Fig. 2-7*a*.

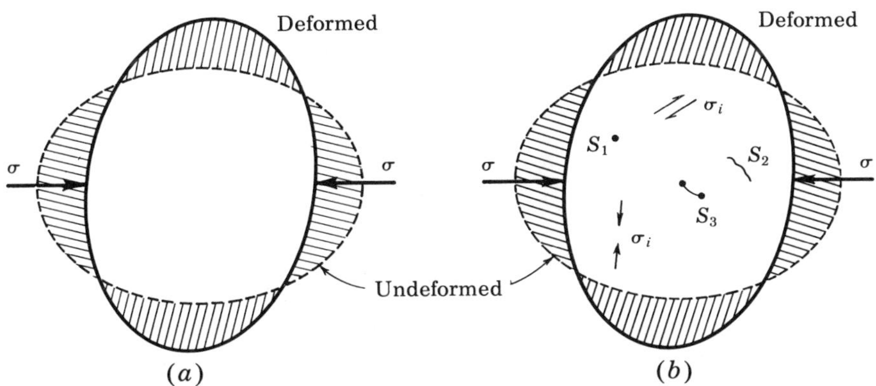

FIGURE 2-7. Deformation of free bodies, (*a*) initially stress free, and (*b*) containing internal stress sources S_1, S_2, and S_3 representing point, line, and surface sources, respectively.

Although Theorem 2-1 seems obviously true in the approximation of linear elasticity, from the simple way in which it was derived, it must be interpreted with caution. For some types of sources of internal stress, a finite amount of interaction energy between external and internal stress is concentrated in the source of internal stress itself. This interaction energy is equal and opposite to the total amount of interaction energy distributed continuously throughout the bulk.

Point Source of Dilatation

As an illustration of Theorem 2-1, let us consider a point source of dilatation, which is the limiting case of a sphere put into a hole that is too small for it. As in Fig. 2-8, cut out a spherical hole of radius r_0, apply a pressure p_0 to the surface of the hole that opens it up by a volume δv, fill the hole with material of the same substance under pressure p_0, and glue the surfaces together. Because the system is in equilibrium, upon removing the constraints there is no relaxation which would be hindered by a nonslipping interface. When $r_0 \ll R$, the pressure in the inserted material is

$$p_0 = \frac{\mu\,\delta v}{\pi r_0^3} \tag{2-82}$$

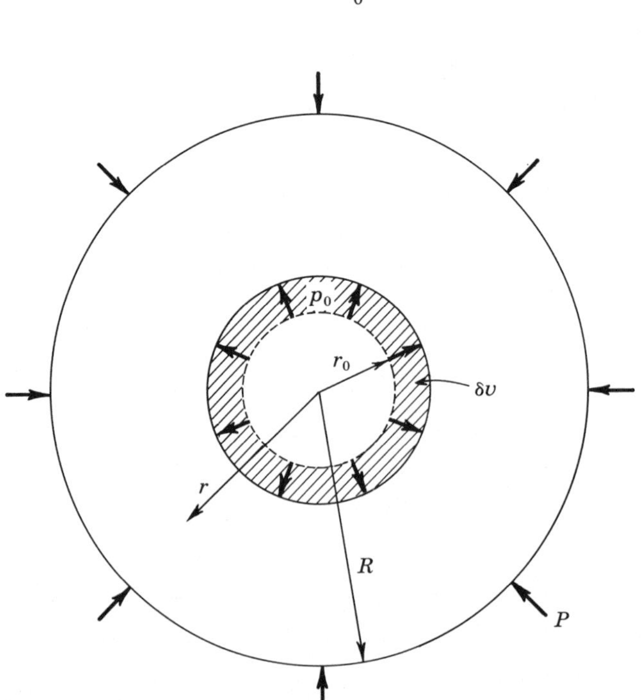

FIGURE 2-8. Sphere of radius R with a concentric spherical hole originally of radius r_0 in the undeformed state.

by Eq. (2-81). Next the entire system is subjected to external pressure P. By Theorem 2-1, there is no cross term between the internal stresses and the external pressure P in the elastic energy of the entire system. The cross term for the compressed central sphere is equal and opposite in sign to the cross term for the surrounding matter. The elastic energy in the central sphere is

$$\frac{(p_0+P)^2}{2B}\frac{4\pi r_0^3}{3}=\frac{p_0^2+P^2}{2B}\frac{4\pi r_0^3}{3}+\frac{p_0 P}{B}\frac{4\pi r_0^3}{3} \tag{2-83}$$

By Eq. (2-82), the cross term, or interaction energy (the last term on the right side in the above equation), can be written as

$$\frac{p_0 P}{B}\frac{4\pi r_0^3}{3}=\frac{4\mu}{3B}P\,\delta v \tag{2-84}$$

Thus, if $r_0\to 0$, while δv is kept constant to maintain constant strength of the singularity, then, with a pressure P on the system, a finite amount of interaction energy proportional to $P\,\delta v$ will concentrate in a point.

If one starts directly with the abstraction of a pure point singularity of expansion, this effect manifests itself as a flow of interaction energy into the point singularity; the singularity acts as a *sink* for interaction energy. When image forces are neglected, which is valid for $r \ll R$, the stress σ_{rr} at a distance r from the singularity is given by Eq. (2-81). An external pressure P causes radial displacements,

$$u_r=-\frac{Pr}{3B} \tag{2-85}$$

as can be verified with Eqs. (2-49) and (2-77). Thus, as P is applied, energy flows toward the singularity through a spherical surface with the singularity as origin, in an amount

$$4\pi r^2\sigma_{rr}u_r=\frac{4\mu}{3B}P\,\delta v \tag{2-86}$$

Of course, this flow of energy out of the region bounding the singularity exactly goes to provide the cross-term energy [Eq. (2-84)] residing in the singularity.

Dislocation Source of Stress

As in the previous case, a line source of expansion, the abstraction for a rod squeezed into a cylindrical hole of too small a bore, constitutes a sink of interaction energy. On the other hand, dislocations are line singularities which do not concentrate interaction energy. Consider a straight dislocation perpendicular to the paper (Fig. 2-9). As shown later, the internal stresses arising

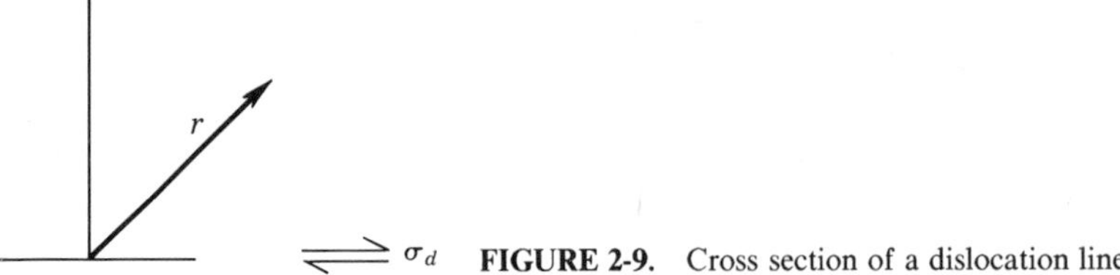

FIGURE 2-9. Cross section of a dislocation line.

from a locked-in dislocation have the general form

$$\sigma_d \propto r^{-1} \tag{2-87}$$

An external stress σ causes displacements

$$u_r \propto \sigma r \tag{2-88}$$

relative to the core at $r=0$. Thus the flow of energy per unit length toward the dislocation line as external stress is applied could be at most

$$W = \sigma_d u_r 2\pi r \propto r \tag{2-89}$$

so that $W \to 0$ as $r \to 0$. *For the case of internal stress caused by dislocations, Theorem 2-1 can be used without ambiguity.*

Exercise 2-3. Derive the expansion δV of the system of a sphere with a singularity of expansion δv at the origin from interaction energy considerations.

Solution. If the singularity is created while an external pressure P acts, an additional energy increment $P\delta v$ is expended. The potential energy of the system maintaining the pressure P is then raised by an amount $P\delta V$. The difference

$$P(\delta v - \delta V)$$

appears in the bulk sphere as interaction energy, which by the preceding theory must be the negative of the interaction energy in the singularity. Thus, by Eq. (2-84),

$$P(\delta v - \delta V) = -\frac{4\mu}{3B} P\delta v$$

or

$$\delta V = \frac{3B + 4\mu}{3B} \delta v \tag{2-90}$$

in agreement with Eq. (2-80).

Exercise 2-4. Consider a point source of dilatation produced in the following way. A hole of radius r_0 is cut out at the origin of a sphere of radius R. A ball of the same material and with a volume $4\pi r_0^3/3+\delta v'$ in the uncompressed state is squeezed into the hole and then released. Show that when $R \gg r_0$, then $\delta V=\delta v'$. Note that the strength of the source of dilatation, δv, is smaller than $\delta v'$ because the squeezed-in ball is compressed.

Exercise 2-5. Repeat Exercise 2-4 for the case when the inserted ball has the elastic constants μ', B', etc. Show that

$$\delta V=\frac{1+(4\mu/3B)}{1+(4\mu/3B')}\delta v' \tag{2-91}$$

2-8. SOME QUALITATIVE PRINCIPLES

St. Venant's Principle

A most useful principle was first stated by St. Venant[18] as the "principle of the elastic equivalence of statically equipollent systems of load." In effect, this principle asserts that if a set of surface forces acting on a region of the surface of an elastic body is replaced by another statically equivalent set of forces acting on the same surface region, the internal stress distribution in the body is changed appreciably only within a distance from the surface region in question equal to the linear dimension of the surface region. For example, in a long cylinder with a torque applied at the end, the stress distribution at distances from the end greater than the cylinder diameter are essentially unaffected by the distribution of the surface forces on the cylinder end which produce the torque. For a specific quantitative evaluation of how rapidly the stress distribution changes with distance from a surface region, and what is an "appreciable" change, see Timoshenko and Goodier.[19]

This principle is useful in the treatment of image-force problems and in the estimation of the interaction energies of complex dislocation arrays.

Principle of Superposition

This principle is often enunciated in elasticity theory and involves the superposition of stresses, strains, and displacements. Here we wish simply to emphasize that it follows directly from the assumptions of the linear theory of elasticity.

[18] B. St. Venant, *Mém. Savants étrangers*, **14**, (1855).

[19] S. Timoshenko and J. N. Goodier, "Theory of Elasticity," McGraw-Hill, New York, 1951, p. 52.

PROBLEMS

2-1. Consider a state of plane strain, with strains ϵ_{xx}, ϵ_{yy}, and ϵ_{xy}. What would be the strains referred to a coordinate system $x'y'$ rotated an angle ϕ about a common z axis?

2-2. Consider a state of strain with ϵ_{xz} and ϵ_{yz} the only nonvanishing components. What would be the strains referred to a coordinate system $x'y'$ rotated an angle ϕ about a common z axis?

2-3. For a rotation ϕ about the z axis, can one simply add the transformations in Probs. 2-1 and 2-2 to obtain the transformation for the general case of strain?

2-4. Write out the general matrix of elastic coefficients for the case where there is a reflection plane normal to the z axis.

2-5. Show that a compressible isotropic medium, for which Poisson's ratio $\nu = \frac{1}{2}$, is a compressible *liquid*. *Hint*: Show that the shear modulus is zero, $\mu = 0$.

2-6. Show that the matrix $\partial u_i / \partial x_j$ consists of a symmetric component, the strain matrix ϵ_{ij}, and an antisymmetric component, the rotation matrix ω_{ij}.

2-7. For an isotropic substance, derive the stress-strain, strain-displacement, and stress-displacement relations in cylindrical coordinates.

2-8. Deduce the number of independent elastic constants for each of the six crystal systems from symmetry considerations.

2-9. With a coordinate system such that the z axis is normal to the surface, determine by Eq. (2-35) the force acting on the surface arising from a general stress σ_{ij}.

2-10. Derive expressions for the stresses and displacements produced by a row of point-force pairs distributed along the axis of a cylinder and acting normal to the axis. Use these results to determine the stress field of a cylindrical rod forced into a smaller cylindrical hole coaxially positioned in a larger cylinder.

BIBLIOGRAPHY

1. Eshelby, J. D., *Solid State Phys.*, **3**: 79 (1956).
2. Landau, L. D., and E. M. Lifshitz, "Theory of Elasticity," Pergamon, New York, 1959.
3. Love, A. E. H., "The Mathematical Theory of Elasticity," Cambridge University Press, Cambridge, 1927.
4. Nye, J. F., "Physical Properties of Crystals," Oxford University Press, Fair Lawn, N.J., 1957.
5. Sokolnikoff, I. S., "Mathematical Theory of Elasticity," McGraw-Hill, New York, 1956.
6. Sokolnikoff, I. S., and R. M. Redheffer, "Mathematics of Physics and Modern Engineering," McGraw-Hill, New York, 1966.
7. Timoshenko, S., and J. N. Goodier, "Theory of Elasticity," McGraw-Hill, New York, 1951.

3

The Theory of Straight Dislocations

3-1. INTRODUCTION

In this chapter elastic properties of straight dislocations are developed from the elasticity theory of Chap. 2. For straight dislocations in infinite, continuous, isotropic media, the treatment is straightforward. In a finite body the displacements and stresses around a dislocation depend on the external surface. Often the boundary conditions can be satisfied by placing an "image dislocation" outside the crystal in such a manner that its stress field cancels that of the real dislocation at the surface.[1] This image method is demonstrated for some specific geometries where the results are simple. For other geometries, or for complex dislocation arrays, the exact solution to the elastic problem is quite difficult. However, rough approximate solutions in these cases can be deduced from the simpler results. Furthermore, by considering the exact solution for the interactions of straight dislocations, one can form a qualitative appreciation of more complex interactions.

Hence in this chapter we discuss the displacements, stresses, and energies of straight, screw, edge, and mixed dislocations. The concept of a force on a dislocation is introduced and applied to some simple cases. Image forces are developed from qualitative arguments based on St. Venant's principle and from formal elasticity theory.

3-2. SCREW DISLOCATIONS

Stresses and Displacements

As the dimensions of a body increase, image stresses caused by surface boundary conditions decrease at a given distance from a dislocation. In the limit, the stress field is characteristic of the dislocation in an infinite medium.

[1]This concept has a direct analog in the "image charges" in electrostatics; see J. A. Stratton, "Electromagnetic Theory," McGraw-Hill, New York, 1941, p. 193.

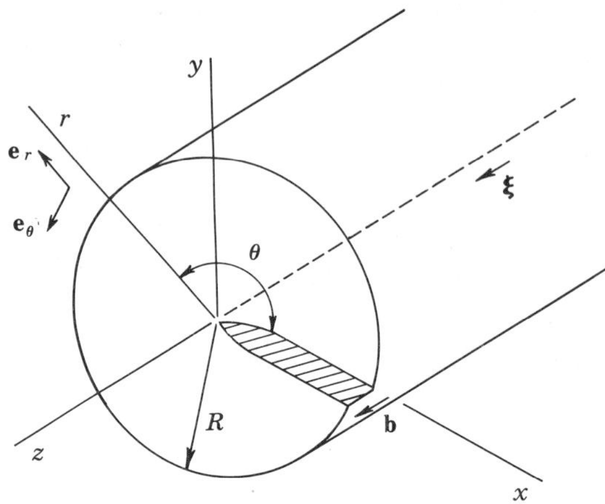

FIGURE 3-1. A right-handed screw dislocation along the axis of a cylinder of radius R and length L.

This stress field is termed the *self-stress* of the dislocation. Let us derive the self-stress and displacements of a screw dislocation. Consider a right-handed screw dislocation along the axis of a cylinder of radius R and length L (Fig. 3-1). This dislocation corresponds to the Volterra dislocation (Fig. 1-1) and can be produced from a perfect cylinder by shear displacements in the z direction across the xz glide plane. A Burgers circuit operation shows that this dislocation has a positive Burgers vector **b**. Alternatively, the Burgers vector can be defined in terms of Eq. (1-5) to emphasize its relation to the displacements. As is evident in Fig. 3-1, $u_x = u_y = 0$, and the displacement u_z is discontinuous at the cut surface defined by $y=0$, $x>0$:

$$\lim_{\substack{\eta \to 0 \\ x>0}} u_z(x,-\eta) - u_z(x,\eta) = b_z \qquad \eta \text{ positive} \tag{3-1}$$

x, y, and z replace x_1, x_2, and x_3 as coordinates to give the notation common for simple problems. It is reasonable to assume that in an isotropic medium the displacement u_z increases uniformly with the angle θ to give this discontinuity,

$$u_z(r,\theta) = b\frac{\theta}{2\pi} = \frac{b}{2\pi}\tan^{-1}\frac{y}{x} \tag{3-2}$$

In fact, Eq. (3-2) really satisfies the equations of elasticity, Eq. (2-58), in the bulk of the rod, as can be checked readily.

The stresses associated with this displacement are determined from Eqs. (2-3) and (2-46) to be

$$\sigma_{xz} = \frac{-\mu b}{2\pi}\frac{y}{x^2+y^2} \qquad \sigma_{yz} = \frac{\mu b}{2\pi}\frac{x}{x^2+y^2}$$

$$\sigma_{xy} = \sigma_{xx} = \sigma_{yy} = \sigma_{zz} = 0 \tag{3-3}$$

or, written out in polar coordinates,

$$\sigma_{\theta z} = \frac{\mu b}{2\pi r}$$

$$\sigma_{rz} = \sigma_{r\theta} = \sigma_{rr} = \sigma_{\theta\theta} = \sigma_{zz} = 0 \tag{3-4}$$

where, for example, $\sigma_{\theta z}$ means the stress in the z direction acting over a radial plane, i.e., a plane normal to $\mathbf{e}_\theta$ (Fig. 3-1). From Sec. 2-7, one sees that the screw dislocation is a source of purely internal stress; no back force is exerted on the source, nor is the source a sink for interaction energy.

A Cylinder Containing a Coaxial Screw Dislocation

In some cases the above solution is inadequate, and additional stresses must be superposed. As an example, consider a screw dislocation in a cylinder. In a free rod no stresses can act over the external surface, according to Eq. (2-72). The solution (3-4) fulfills the boundary condition that $\sigma_{rr} = \sigma_{r\theta} = \sigma_{rz} = 0$ on the cylindrical surface $r = R$. However, on the end surfaces the solution (3-4) yields shear stresses $\sigma_{z\theta} = \sigma_{\theta z}$ which produce a torque

$$M_z = \int_0^R \int_0^{2\pi} r(\sigma_{z\theta} r\,dr\,d\theta) = \frac{\mu b R^2}{2} \tag{3-5}$$

Thus Eq. (3-4) is not the correct solution for a finite rod. However, it can be modified to give the correct solution by superposing stresses to cancel the fictitious stresses $\sigma_{\theta z}$ at the ends, provided that this procedure does not introduce stresses which violate the boundary conditions on the surface $r = R$. The correct solution for the stress-free rod is obtained if stresses

$$\sigma'_{z\theta} = \frac{-\mu b}{2\pi r} \tag{3-6}$$

which are equal and opposite to those of Eq. (3-4), are applied to the cylinder ends (see Fig. 3-2). For a free rod with a screw along the axis, one must then superpose on the displacements (3-2) the twist that would result from the torque M'_z associated with $\sigma'_{z\theta}$, where

$$M'_z = -M_z = \frac{-\mu b R^2}{2} \tag{3-7}$$

According to St. Venant's principle (Sec. 2-8), the twist at distances greater than $2R$ from the end produced by the stress distribution $\sigma'_{z\theta}$ is the same as that produced in a uniformly twisted rod where $\sigma_{z\theta} \propto r$. Thus, except at the ends, one can use the elementary result that the twist per unit length, ϕ/L,

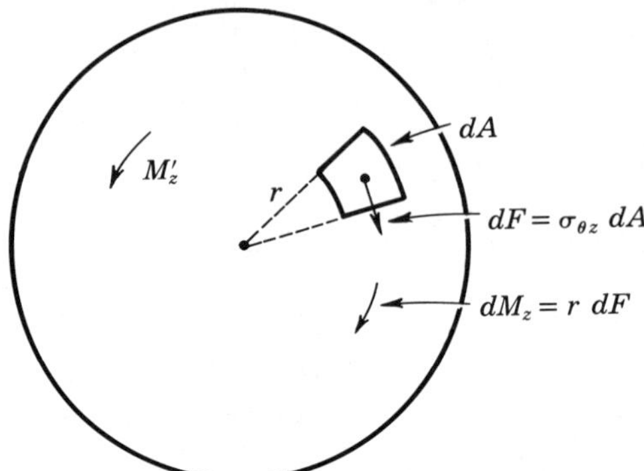

FIGURE 3-2. End view of the cylinder of Fig. 3-1, showing the fictitious stresses $\sigma_{\theta z}$ that lead to a net torque M_z, and the superposed torque M_z' that twists the cylinder in an anticlockwise sense.

produced by M_z' is given by the equation[2]

$$\frac{\phi}{L} = \frac{M_z'}{\mu I} = \frac{-b}{\pi R^2} \tag{3-8}$$

where I is the polar moment of inertia

$$I = \int_A r^2\, dA = \frac{\pi R^4}{2}$$

The twist ϕ is the so-called Eshelby twist,[3] observable in thin, long whiskers containing a screw dislocation.[4]

The twist (3-8) produces displacements and stresses given by

$$u_\theta = \frac{-brz}{\pi R^2} \qquad \sigma_{z\theta} = \frac{-\mu br}{\pi R^2} \tag{3-9}$$

Thus the total displacements and stresses for the axial screw in a free rod are given by

$$u_\theta(r, z) = \frac{-brz}{\pi R^2} \qquad u_z(r, \theta) = \frac{b\theta}{2\pi} \tag{3-10}$$

$$\sigma_{\theta z} = \frac{\mu b}{2\pi r} - \frac{\mu br}{\pi R^2} \tag{3-11}$$

[2] S. Timoshenko and J. N. Goodier, "Theory of Elasticity," McGraw-Hill, New York, 1951, p. 249.

[3] J. D. Eshelby, *J. Appl. Phys.*, **24**: 176 (1953).

[4] W. W. Webb, *J. Appl. Phys.*, **33**: 1961 (1962); R. D. Dragsdorf and W. W. Webb, *J. Appl. Phys.*, **29**: 817 (1958); C. M. Drum, *J. Appl. Phys.*, **36**: 816, 824 (1965).

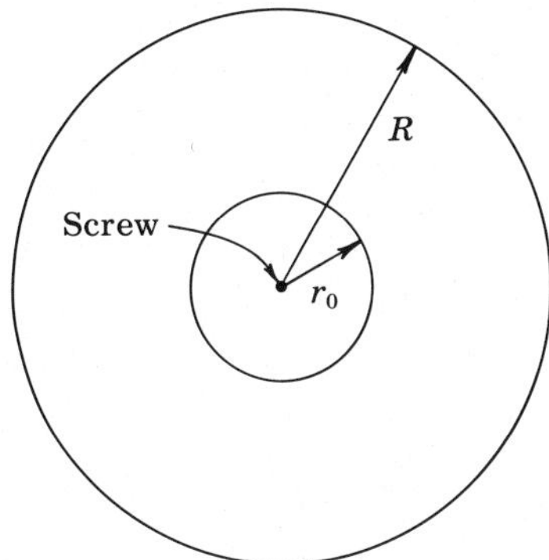

FIGURE 3-3. A screw in an infinite medium. Two coaxial cylindrical surfaces of radii r_0 and R are also shown.

As the radius R of the rod increases, the twist effect decreases in magnitude. The above results then converge to those of Eqs. (3-2) and (3-4), which give the displacements and stresses characteristic of a screw in an infinite medium. Near the $z=0$ surface, terms that were neglected as end effects in the foregoing derivation become important. For the case $R=\infty$, the field that satisfies the free-surface boundary condition at $z=0$ is[5]

$$u_\theta = \frac{br}{2\pi(\rho+z)} \qquad \sigma_{r\theta} = \frac{-\mu br^2}{2\pi\rho(\rho+z)^2} \qquad \sigma_{\theta z} = \frac{-\mu br}{2\pi\rho(\rho+z)} \tag{3-12}$$

where $\rho=(x^2+y^2+z^2)^{1/2}$. For intermediate values of R, image forces associated with $\sigma_{r\theta}$ must be considered, leading to a solution in the form of an infinite series of Bessel functions.[5] The latter solution is representative of more complex solutions associated with the presence of free surfaces.

Strain Energy

Consider the energy stored per unit length in the elastic stress field of the infinite screw dislocation, in a region bounded by cylinders of radius r_0 and R (Fig. 3-3). By Eq. (3-4)

$$\frac{W}{L} = \int_{r_0}^{R} \frac{\sigma_{\theta z}{}^2}{2\mu} 2\pi r\, dr = \frac{\mu b^2}{4\pi} \ln \frac{R}{r_0} \tag{3-13}$$

The energy W/L diverges as $R\to\infty$ or as $r_0\to 0$. The divergence with R shows that one cannot ascribe to the dislocation a definite characteristic energy; the energy depends on the size of the crystal. For the case of one dislocation in a crystal, an approximate choice for R is $\sim l$, where l is the shortest distance

[5] J. D. Eshelby and A. N. Stroh, *Phil. Mag.*, **42**: 1401 (1951). Some minor errors in this treatment are corrected above.

from the dislocation to the free surface. At larger distances from the dislocation the image stresses largely cancel the dislocation stresses. Similarly, in a crystal containing many dislocations of both signs a reasonable approximation would be to take roughly half the average distance between dislocations for R. Because of the logarithmic dependence, the energy is insensitive to the precision of choice of a value of R.

The divergence with r_0 arises from the inadequacy of linear elasticity theory to deal with the severe lattice distortion near the core of the dislocation. Thus the treatment leading to Eq. (3-13) is restricted to the matter outside a cylinder of radius $r_0 \sim b$, where linear elasticity theory should apply. The total energy is obtained by adding an energy of misfit for the matter inside the core cylinder. Atomistic calculations for NaCl crystals indicate that the core energy in these cases does not exceed $\sim 0.2\mu b^2$ per unit length,[6,7] and for the case of glide dislocations in close-packed structures the core energy is expected to be ~ 0.1 to $0.05\mu b^2$ per unit length. Formally, we include all contributions in Eq. (3-13) by choosing r_0 small enough to make the following formula give the expected total energy:

$$\frac{W}{L} = \frac{\mu b^2}{4\pi} \ln \frac{\alpha R}{b} \tag{3-14}$$

where $r_0 = b/\alpha$ and $\alpha \sim 1$. This choice for α is discussed in Chap. 8, where discrete atomic forces are considered.

Free-surface terms also influence the strain energy. With the twist effect included for a screw dislocation in a finite cylinder of outer radius R, Eqs. (2-48) and (3-11) give

$$\frac{W}{L} = \frac{\mu b^2}{4\pi} \ln \frac{\alpha R}{b} - \frac{\mu b^2}{4\pi} \tag{3-15}$$

The last term includes a term $\mu b^2/4\pi$, which would be the self-energy associated with the last term in Eq. (3-11) arising solely from the elastic twist, and a cross term $-\mu b^2/2\pi$ arising when the right-hand side of Eq. (3-11) is squared.

Force on the Screw Dislocation

In mechanics a general force F_η is defined relative to a general configurational coordinate η by the requirement that

$$\delta W_t = -F_\eta \delta\eta \tag{3-16}$$

be the change in total energy W_t when η undergoes the change $\delta\eta$. An even

[6] H. B. Huntington, J. E. Dickey, and R. Thomson, *Phys. Rev.*, **100**: 1117 (1955).

[7] A. Maradudin, *J. Phys. Chem. Solids*, **9**: 1 (1959).

more general definition is given in thermodynamics, with the free energy taking the place of mechanical energy, and with parameters such as composition put on equal footing with configurational coordinates. The coordinates of a system tend to change spontaneously to lower its total free energy. The definition of the force on a dislocation follows this scheme: *forces on dislocations are virtual forces representing the change in free energy of the system with displacement of the dislocation*. Physically, the elastic forces are, of course, distributed throughout the elastically strained continuum rather than acting on the dislocation line.

For the present, let us restrict the discussion to purely elastic and mechanical effects. The total energy to be considered is then the elastic energy in the crystal and the potential energy of the mechanisms exerting external forces on the crystal,

$$W_t = W_e + W_p \tag{3-17}$$

Since W_t depends on the position η of the dislocation, the total force on the dislocation is defined by Eq. (3-16). For example, the energy of a screw dislocation parallel to a free surface depends on the distance l from the surface as

$$\frac{W}{L} = \frac{\mu b^2}{4\pi} \ln \frac{\alpha l}{b}$$

according to the discussion following Eq. (3-13). Consequently, the force per unit length tending to move the screw toward the surface is

$$\frac{F}{L} = -\frac{\partial(W/L)}{\partial(-l)} = \frac{\mu b^2}{4\pi l} \tag{3-18}$$

This is an image force; it arises solely from the change in elastic energy in the crystal.

As another example, consider a right-handed screw dislocation in a crystal slab much longer than it is thick (Fig. 3-4). When the dislocation is far removed from the end faces, so that end image forces are negligible, the energy of the dislocation is independent of its position along the x coordinate. If it moves from x_1 to x_2, then the upper and lower faces are displaced by $\mathbf{b}$ relative to each other over an area $L(x_2 - x_1)$, and if a uniform shear stress σ_{yz} is applied externally, the shear stress does work $\sigma_{yz} L(x_2 - x_1)b$ as the dislocation moves. This work is done at the expense of the potential energy of the external mechanism providing the shear stress,

$$\Delta W_{\text{pot}} = -\sigma_{yz} L(x_2 - x_1)b \tag{3-19}$$

Since Eq. (3-19) represents the change in total energy of the system, Eq. (3-16)

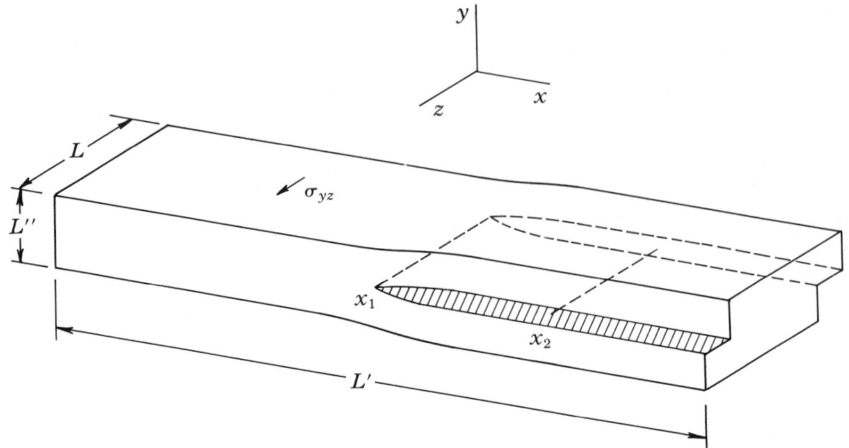

FIGURE 3-4. A right-handed screw dislocation in a long crystal slab, $L' \gg L \cong L''$.

gives the force per unit length on the screw dislocation as

$$\frac{F_x}{L} = \sigma_{yz} b \tag{3-20}$$

In the above example, where the elastic energy in the crystal did not depend on the position of the dislocation, $-\Delta W_{\text{pot}}$ [Eq. (3-19)] represents the energy dissipated in (or lost to) the slip plane as the dislocation moves from x_1 to x_2. In general, as a dislocation moves and changes the slipped area, work is done on the slip plane by the stresses in the crystal, to be dissipated as heat. The work is done at the expense of the total energy of the system and is thus directly related to the total force on the dislocation. The mechanism of energy dissipation via the scattering of phonons by a moving dislocation is discussed in Chap. 7.

Absence of Self-Stress Forces

Let us divide the total stress in the region of the dislocation core σ_T into two contributions.

$$\sigma_T = \sigma_d + \sigma \tag{3-21}$$

where σ_d would be the self-stress of the dislocation in an infinite medium. σ is caused by image effects, external loads, and internal stress sources such as other crystal defects. In a small region near the dislocation core, σ can be considered to be constant. Figure 3-5 shows stresses and relative displacements over a slip plane for the pure screw. σ_d is given by Eqs. (3-3). The singularity at the origin is a feature of simple elasticity theory; in real crystals the core has a

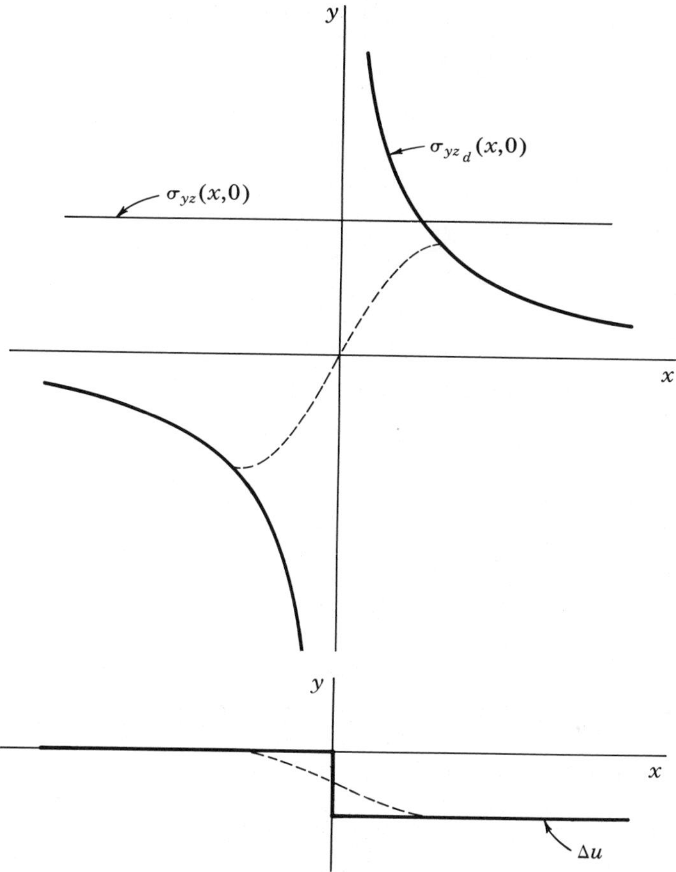

FIGURE 3-5. Stresses σ_{yz} and relative displacements Δu along the slip plane of a screw dislocation.

finite width. The dotted curves show how displacements and stresses vary continuously and symmetrically through the core in real crystals with asymmetry caused by crystal structure being neglected.[8]

If the dislocation moves a distance δx to the right, the self-stress of the dislocation does no work on the slip plane, to the first order in δx,

$$\frac{\delta W}{L} = \delta x \int_{-\infty}^{\infty} \sigma_{yz_d} \frac{\partial}{\partial x} \Delta u \, dx = 0$$

[8]In a real crystal the stresses and displacements in the core are not distributed completely symmetrically for all possible positions of the dislocation. Asymmetries are present that tend to keep the dislocation in valleys of minimum core energy, occurring periodically along the slip plane. These effects give rise to a lattice resistance to dislocation motion, which is neglected here but considered in Chap. 8. The smoothed-out curves of Fig. 3-5 can be taken to represent the average overall parallel positions of the dislocation.

This is so because σ_{yz_d} is antisymmetric about $x=0$, while $\partial(\Delta u)/\partial x$ is symmetric, so that the integral from 0 to ∞ is equal and opposite to that from $-\infty$ to 0. However, the *external* stress field σ_{yz} does do work

$$\frac{\delta W}{L}=\delta x\,\sigma_{yz}\int_{-\infty}^{\infty}\frac{\partial}{\partial x}\Delta u\,dx=\delta x\,\sigma_{yz}b$$

at the expense of the total energy of the system

$$\frac{\delta W_T}{L}=-\,\delta x\,\sigma_{yz}b \tag{3-22}$$

Thus, in the presence of both external stresses and self-stress, the force per unit length of the screw dislocation is

$$\frac{F_x}{L}=\sigma_{yz}b \tag{3-23}$$

and similarly, for motion in the y direction,

$$\frac{F_y}{L}=-\,\sigma_{xz}b \tag{3-24}$$

Hence the external stress does work on a moving dislocation, but the self-stress does not.

Exercise 3-1. Equations (3-23) and (3-24) refer to dislocations of the same sign and with the same orientation with respect to the coordinate system as the dislocation in Fig. 3-1. Show how a general definition of the force on a pure screw must involve both **b** and $\boldsymbol{\xi}$, and how it can be made independent of the orientation of the coordinate system when expressed in terms of **b** and $\boldsymbol{\xi}$.

Solution. Equation (3-90).

All the above results apply for a right-handed screw dislocation. For a left-handed screw, the energy would be the same, but the Burgers vector, the displacements, and the forces would all change sign.

3-3. IMAGE FORCES ON SCREW DISLOCATIONS

In Eq. (3-18) we obtained an expression for the image force on a screw dislocation by an indirect argument based on the self-energy of a dislocation. We shall now derive the same result in a more straightforward manner and illustrate the extension of the reasoning to more complicated cases. Consider a right-handed screw dislocation parallel to a planar free surface, as shown in Fig. 3-6. The boundary condition at the free surface [Eq. (2-72)] leads to the

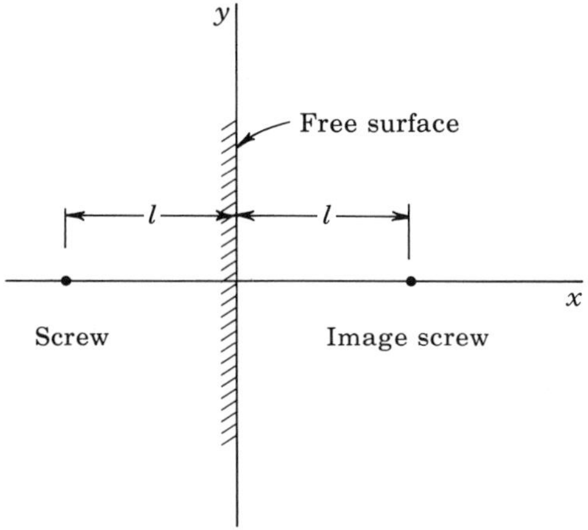

FIGURE 3-6. A screw dislocation parallel to a free surface, and an image dislocation.

requirement that $\sigma_{xz}=0$ at the free surface. This boundary condition is met if one superposes the self-stress of the screw and that of an imaginary screw of the same strength and opposite sign at the mirror position outside the solid. According to Eq. (3-3), the image stress at the core of the dislocation then is

$$\sigma_{yz}=\frac{\mu b}{4\pi l}$$

and consequently, by Eq. (3-23), the screw is drawn toward the surface by a force

$$\frac{F_x}{L}=\frac{\mu b^2}{4\pi l}$$

per unit length. This result is identical to that of Eq. (3-18).

A more complicated image-force problem[9] is illustrated in Fig. 3-7, which shows a cylindrical crystal with a right-handed screw dislocation parallel to the cylinder axis but displaced along the x axis by a distance λ from the cylinder axis. The free-surface boundary condition [Eq. (2-72)] requires that the shear stress component acting on the cylindrical surface $\sigma_{\rho z}$, in the coordinates fixed on the cylinder axis in Fig. 3-7, be zero. Yet the dislocation produces a stress $\sigma_{\theta z}$, given by Eq. (3-4), which has a component $\sigma_{\rho z}=\sigma_{\theta z}\sin(\phi-\theta)$ acting on the surface at $\rho=R$. The boundary condition in this case is satisfied if the self-stress of an imaginary screw of equal strength and opposite sign at a position R^2/λ along the x axis is superposed on the self-stress of the real screw dislocation. The displacement associated with the two dislocations is

$$u_z=\frac{b}{2\pi}(\theta-\theta')=\frac{b}{2\pi}\left(\tan^{-1}\frac{y}{x-\lambda}-\tan^{-1}\frac{y}{x-R^2/\lambda}\right) \tag{3-25}$$

[9] J. D. Eshelby, *J. Appl. Phys.*, **24**: 176 (1953).

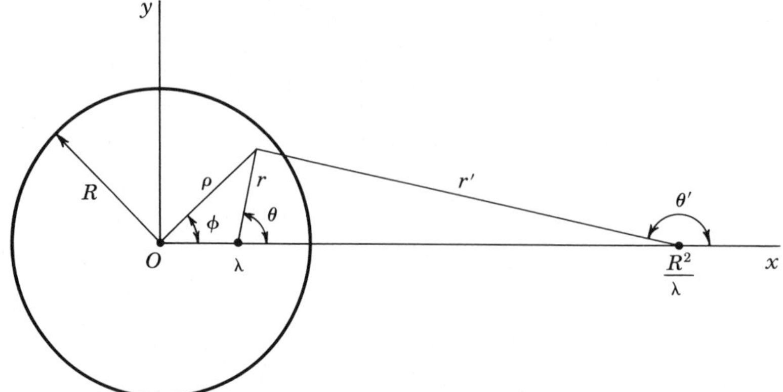

FIGURE 3-7. A screw dislocation displaced by a distance λ along the x axis from the axis of a cylinder and an image dislocation at R^2/λ. (ρ,ϕ) coordinates refer to the cylinder axis, (r,θ) coordinates to the screw dislocation, and (r',θ') coordinates to the image dislocation.

In this example the determination of the volume strain energy in the cylinder by direct integration is tedious, and a simpler approach is used, based upon Theorem 2-1; the method is generally useful in the calculation of dislocation energies. If the dislocation is produced by making the cut shown in Fig. 3-8 and then displacing the bottom surface of the cut by a distance b in the z direction, surface forces are applied only on the cut surface, with the cylinder surface remaining free. Thus only stresses $\sigma_{yz}(x,0)$ need be applied to produce the displacement. The total work done by the surface forces must equal the total volume strain energy introduced by the displacement. From Eqs. (2-3), (2-46), and (3-25) one finds

$$\sigma_{yz}(x,0)=\frac{\mu b}{2\pi}\left(\frac{1}{x-\lambda}-\frac{1}{x-R^2/\lambda}\right) \tag{3-26}$$

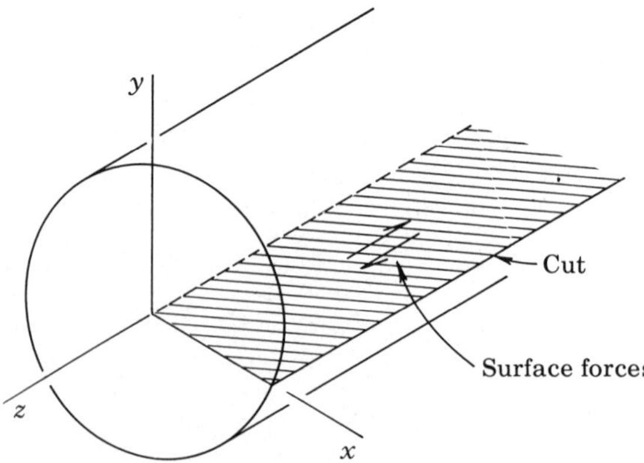

FIGURE 3-8. A cylinder cut radially to enable a right-handed screw dislocation to be formed by the displacement of the cut surface in the z direction by surface forces.

First consider the work done on an infinitesimal element of cut surface in producing the displacement $\Delta u_z = b$ by a sum of infinitesimal displacements $\delta b' = \delta u_z$. From Eq. (3-26), this work is given by

$$\delta W = \int_0^b \sigma_{yz}(x,0)\,\delta A\,db'$$

$$= \frac{\mu}{2\pi}\delta A\left(\frac{1}{x-\lambda} - \frac{1}{x-R^2/\lambda}\right)\int_0^b b'\,db'$$

$$= \frac{\mu b^2}{4\pi}\delta A\left(\frac{1}{x-\lambda} - \frac{1}{x-R^2/\lambda}\right) \qquad (3\text{-}27)$$

Integrating over the area of the cut surface gives the total energy per unit length as

$$\frac{W}{L} = \frac{\mu b^2}{4\pi}\int_{\lambda+r_0}^{R}\left(\frac{1}{x-\lambda} - \frac{1}{x-R^2/\lambda}\right)dx$$

$$\cong \frac{\mu b^2}{4\pi}\ln\frac{R^2-\lambda^2}{r_0 R} \qquad (3\text{-}28)$$

The moment corresponding to Eq. (3-5) is given in the present case by

$$M_z = \int_A \sigma_{\phi z}\rho\,dA = \mu\int_A\left(x\frac{\partial u_z}{\partial y} - y\frac{\partial u_z}{\partial x}\right)dA$$

$$= \mu\oint_C (u_z x\,dx + u_z y\,dy) \qquad (3\text{-}29)$$

Here Green's theorem has been applied in the last step. Again, the only contributions to the line integral in (3-29) are those on the cut surface,[10] so that

$$M_z = \mu b\int_{\lambda+r_0}^{R} x\,dx \cong \frac{\mu b}{2}(R^2-\lambda^2) \qquad (3\text{-}30)$$

Proceeding as in the development of Eq. (3-5), one finds that the added work term on the cut surface arising from stresses associated with the twist of the

[10] The path of integration must be specified as shown in Fig. 3-9, so that it encloses an area where u_z is single-valued. The outer and inner circular contours do not contribute. The integrand can be written as

$$u_z x\,dx + u_z y\,dy = \tfrac{1}{2}u_z\,d(x^2+y^2) = \tfrac{1}{2}u_z\,d\rho^2$$

However, since ρ is constant on the contours $\rho = R$ and $\rho = r_0$, then $d\rho^2 = 0$.

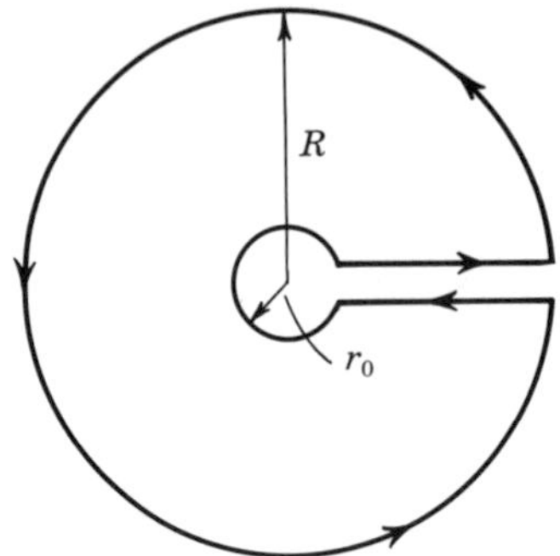

FIGURE 3-9. The path of integration for Eq. (3-29).

cylinder is given by[11]

$$\frac{W}{L}=\frac{M_z^2}{\pi R^4}=\frac{\mu b^2}{4\pi R^4}(R^2-\lambda^2)^2 \tag{3-31}$$

This is a work of relaxation. Equation (3-28) overestimates the total energy, because it includes the virtual torque at the free ends which is removed by the superposed torque. Thus the true total energy is given by the difference between Eqs. (3-28) and (3-31),

$$\frac{W}{L}=\frac{\mu b^2}{4\pi}\left[\ln\frac{R^2-\lambda^2}{r_0 R}-\frac{(R^2-\lambda^2)^2}{R^4}\right] \tag{3-32}$$

The energy in Eq. (3-32) can be regarded as either the work done on the cut surface in creating the dislocation or, the equivalent by Stokes theorem, the stored elastic strain energy in the cylinder containing the dislocation. In the latter case, as for Eq. (3-15), the last term in Eq. (3-32) includes both the strain energy associated with the twist alone and the cross term between the twist stresses and those of Eq. (3-26).

Exercise 3-2. Show from Eqs. (3-28) and (3-31) that the screw dislocation is in a metastable-equilibrium configuration when $\lambda=0$, that is, when it lies along the cylinder axis. Compute the energy that must be supplied to move the dislocation out of the cylinder.

Exercise 3-3. Demonstrate that the image displacements [Eq. (3-25)] indeed yield stresses that have no components acting on the free surfaces of the cylinder.

[11] S. Timoshenko and J. N. Goodier, "Theory of Elasticity," McGraw-Hill, New York, 1951, p. 164.

The results of these exercises indicate that a screw dislocation which is constrained to remain straight is stable with respect to small displacements from the cylinder axis.[12] These results are important in the theory of whisker growth by the screw-dislocation mechanism.[13–15] In the physical problem of determining the equilibrium configuration of a screw dislocation in a cylindrical whisker one must minimize the *total* free energy of the system, in this case the whisker. Thus, in addition to the elastic strain energy of the dislocation, one must consider[15] the surface energy associated with the surface step produced by the screw dislocation (Fig. 3-1). This step surface energy causes the step to have an effective line tension pulling on the dislocation. Formally, one should also consider entropy terms, but as is shown in Chap. 14, these are generally negligible except for configurational line shape terms.

3-4. EDGE DISLOCATIONS

Much of what has been said about the screw dislocation can be applied directly in the discussion of the edge dislocation. Where the analogy is obvious, detailed explanations are not presented.

Stresses and Displacements in an Infinite Medium

The straight-edge dislocation produces planar strain defined by $u_z=0$ and $\partial u_i/\partial z=0$ (Fig. 3-10). In polar coordinates, the differential equation for the Airy stress function, Eq. (2-56), becomes

$$\left(\frac{\partial^2}{\partial r^2}+\frac{1}{r}\frac{\partial}{\partial r}+\frac{1}{r^2}\frac{\partial^2}{\partial\theta^2}\right)^2\psi=0 \tag{3-33}$$

Let us briefly review the derivation[16] of a general solution to Eq. (3-33) and then specifically apply it to the edge dislocation in an infinite solid. In terms of the function

$$\phi=(\sigma_{xx}+\sigma_{yy})=\nabla^2\psi$$

Eq. (3-33) becomes the Laplace equation,

$$\nabla^2\phi=\left(\frac{\partial^2}{\partial r^2}+\frac{1}{r}\frac{\partial}{\partial r}+\frac{1}{r^2}\frac{\partial^2}{\partial\theta^2}\right)\phi=0 \tag{3-34}$$

[12] J. D. Eshelby, *J. Appl. Phys.*, **24**: 176 (1953).

[13] J. D. Eshelby, *Phys. Rev.*, **91**: 755 (1953).

[14] F. C. Frank, *Phil. Mag.*, **44**: 854 (1953).

[15] J. P. Hirth and F. C. Frank, *Phil. Mag.*, **3**: 1110 (1958).

[16] J. H. Michell, *Proc. London Math. Soc.*, **31**: 100 (1899).

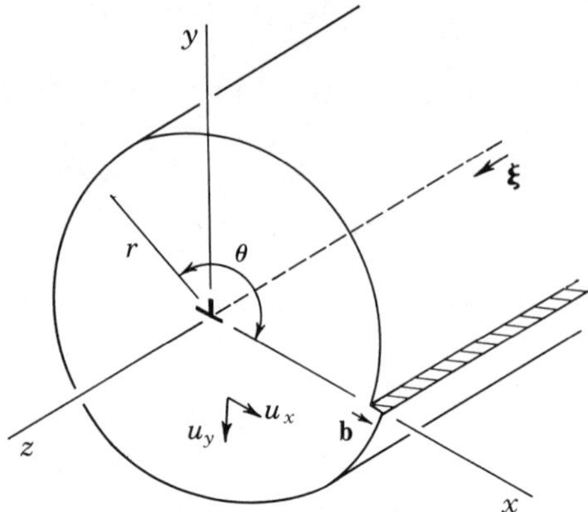

FIGURE 3-10. An edge dislocation. The plane-strain displacements are indicated.

Since the harmonic equation (3-34) is separable, the solution is of the form

$$\phi = \sum_n R_n(r)\phi_n(\theta) \tag{3-35}$$

which is separable term by term. With the requirement that the solution be single-valued in the plane, Eq. (3-35) becomes

$$\phi = (\alpha_0 + \beta_0 \ln r) + \sum_{n=1}^{\infty} (\alpha_n r^n + \beta_n r^{-n}) \sin n\theta$$
$$+ \sum_{n=1}^{\infty} (\gamma_n r^n + \delta_n r^{-n}) \cos n\theta \tag{3-36}$$

A qualitative inspection of Fig. 3-10 indicates that $\phi = \sigma_{xx} + \sigma_{yy}$ should have a maximum for $\theta = -\pi/2$ and a minimum for $\theta = \pi/2$, and that ϕ should decrease with increasing r. The simplest term in Eq. (3-36) that typifies this behavior is

$$\phi_e = \beta_1 r^{-1} \sin\theta \tag{3-37}$$

The following reasoning shows that this is the only term characteristic of the long-range stress field of the dislocation.

The terms r^n correspond to various conditions of externally applied surface forces. The logarithmic term involves a constant shear traction over half the external surface of a half-infinite solid,[17] and cannot be relevant to the problem of the dislocation. Terms r^{-n}, $n \geqslant 2$, correspond to various line

[17]A. E. H. Love, "The Mathematical Theory of Elasticity," Cambridge University Press, Cambridge, 1927, p. 214, par. 152, example C.

singularities other than the dislocation. For example, terms r^{-2} can be combined to represent a line of expansion. The matter along the dislocation core could tend to expand because of anharmonicities and the particular structure assumed in the core in a particular case, so that terms of the type r^{-n}, $n \geqslant 2$, could be present. These terms depend on details in or near the core and are not related to the general characteristics of a dislocation. Also, such contributions diminish rapidly away from the dislocation; at large distances from the core Eq. (3-37) must dominate.

Thus the stress-field characteristic of the edge dislocation must be determined from

$$\left(\frac{\partial^2}{\partial r^2}+\frac{1}{r}\frac{\partial}{\partial r}+\frac{1}{r^2}\frac{\partial^2}{\partial\theta^2}\right)\psi=\beta_1 r^{-1}\sin\theta \tag{3-38}$$

A particular solution is

$$\psi_e=\frac{\beta_1}{2}r\sin\theta\ln r \tag{3-39}$$

Solutions of the homogeneous equation[18] can be ignored, as is the case in reducing Eq. (3-36) to (3-37), thus excluding core terms for the moment.

With b the magnitude of the Burgers vector, one can envision the dislocation as being produced by a half-infinite slab of inserted material of thickness b, as in Fig. 1-19. Thus b must equal the difference in the integral elastic strain above and below the slip plane in opening up the cut, as in Fig. 1-19e,

$$b=-\int_{-\infty}^{\infty}\left[\epsilon_{xx}(x,\eta)-\epsilon_{xx}(x,-\eta)\right]dx \qquad \eta\rightarrow 0, \eta \text{ positive} \tag{3-40}$$

Deriving the stresses from ψ_e by Eq. (2-54) and then inserting them in (2-51) gives the ϵ_{xx} to be used in Eq. (3-40). Equation (3-40) then reduces to

$$b=-\frac{2\pi(1-\nu)}{\mu}\frac{\beta_1}{2} \qquad \beta_1=\frac{-\mu b}{\pi(1-\nu)} \tag{3-41}$$

independent of η, as it should be, according to the Burgers circuit criterion.

Exercise 3-4. Derive Eq. (3-41).

With Eq. (3-41), the stress function (3-39) becomes

$$\psi_e=-\frac{\mu b y}{4\pi(1-\nu)}\ln(x^2+y^2) \tag{3-42}$$

[18] The most general solution is given in S. Timoshenko and J. N. Goodier, "Theory of Elasticity," McGraw-Hill, New York, 1951, p. 116.

The stresses which derive from ψ_e are given by Eq. (2-54) as

$$\sigma_{xx} = -\frac{\mu b}{2\pi(1-\nu)} \frac{y(3x^2+y^2)}{(x^2+y^2)^2}$$

$$\sigma_{yy} = \frac{\mu b}{2\pi(1-\nu)} \frac{y(x^2-y^2)}{(x^2+y^2)^2}$$

$$\sigma_{xy} = \frac{\mu b}{2\pi(1-\nu)} \frac{x(x^2-y^2)}{(x^2+y^2)^2} \tag{3-43}$$

$$\sigma_{zz} = \nu(\sigma_{xx}+\sigma_{yy}) = -\frac{\mu b \nu}{\pi(1-\nu)} \frac{y}{x^2+y^2}$$

$$\sigma_{xz} = \sigma_{yz} = 0$$

Expressed in polar coordinates, these stresses become

$$\sigma_{rr} = \sigma_{\theta\theta} = -\frac{\mu b \sin\theta}{2\pi(1-\nu)r}$$

$$\sigma_{r\theta} = \frac{\mu b \cos\theta}{2\pi(1-\nu)r}$$

$$\sigma_{zz} = \nu(\sigma_{rr}+\sigma_{\theta\theta}) = -\frac{\mu b \nu \sin\theta}{\pi(1-\nu)r}$$

$$\sigma_{rz} = \sigma_{\theta z} = 0 \tag{3-44}$$

The σ_{zz} term is determined from Eq. (2-51) for the present plane-strain case, where $\epsilon_{zz}=0$. Simple diagrams of the stress distribution around an edge dislocation are presented in Figs. 3-11 and 3-12. In analogy to the screw-dislocation case, the above stresses do not exert a back force on their source; the dislocation is a source of purely internal stress. Also, the dislocation core is not a sink for interaction energy.

Given the above stresses, the displacements can be derived from Eq. (2-3) by integration,[19] which gives

$$u_x = \frac{b}{2\pi}\left[-\tan^{-1}\frac{x}{y} + \frac{xy}{2(1-\nu)(x^2+y^2)}\right] + C$$

[19]See J. N. Timoshenko and S. Goodier, "Theory of Elasticity," McGraw-Hill, New York, 1951, p. 36.

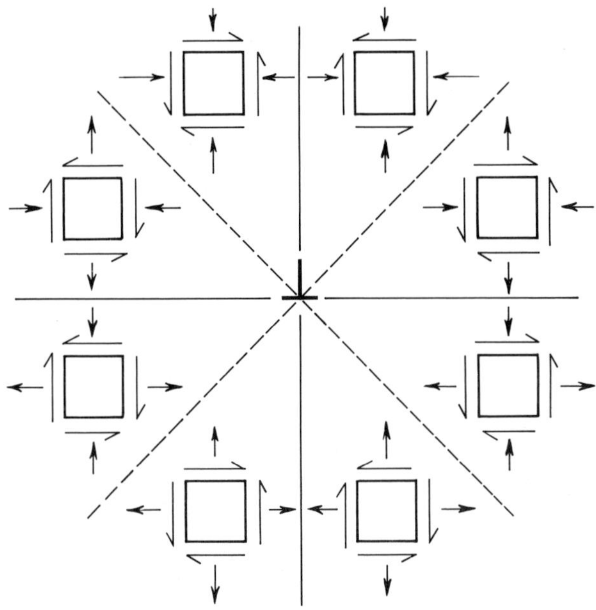

FIGURE 3-11. Schematic representation of the stress field about an edge dislocation.

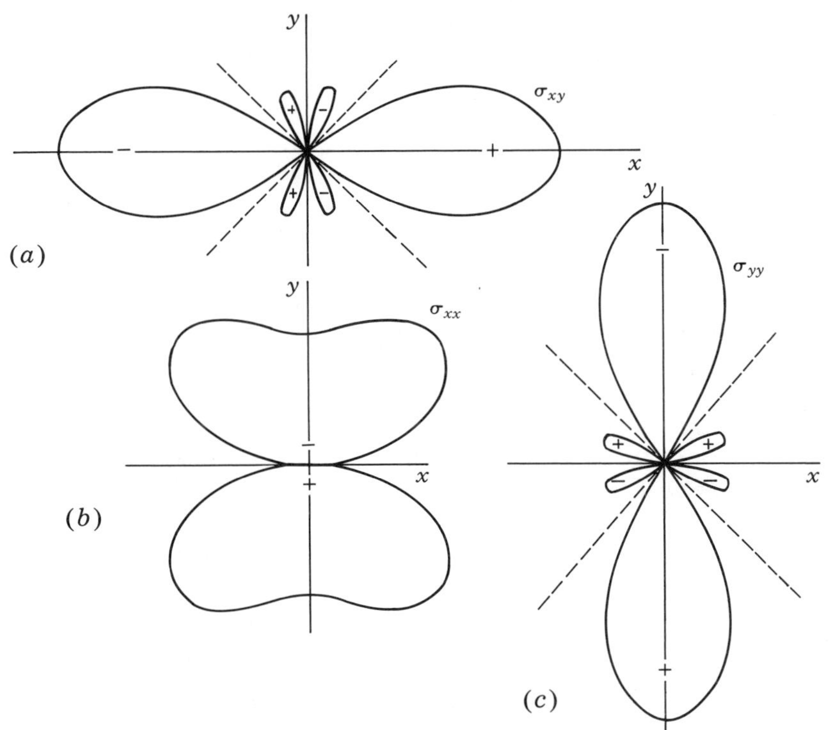

FIGURE 3-12. Graphs showing constant stress contours about an edge dislocation: (*a*) σ_{xy}, (*b*) σ_{xx}, (*c*) σ_{yy}.

The constant of integration is determined by the condition that $u_x = 0$ at $y = 0$, which yields $C = b/4$, so that

$$u_x = \frac{b}{2\pi}\left[\tan^{-1}\frac{y}{x} + \frac{xy}{2(1-\nu)(x^2+y^2)}\right] \tag{3-45}$$

Similarly,

$$u_y = -\frac{b}{2\pi}\left[\frac{1-2\nu}{4(1-\nu)}\ln(x^2+y^2) + \frac{x^2-y^2}{4(1-\nu)(x^2+y^2)}\right] \tag{3-46}$$

u_y includes a constant of integration $C = b/4\pi(1-\nu)$, introduced to yield a symmetric expression[20] in x and y. Notice that u_y contains a logarithmic term divergent in r. This divergence can be rationalized physically (see Fig. 1-4) as being associated with the bending of lattice planes about the z axis caused by the introduction of the extra plane forming the dislocation.

In polar coordinates, the displacements are

$$u_r = \frac{b}{2\pi}\left[-\frac{(1-2\nu)}{2(1-\nu)}\sin\theta\ln r + \frac{\sin\theta}{4(1-\nu)} + \theta\cos\theta\right]$$

$$u_\theta = \frac{b}{2\pi}\left[-\frac{(1-2\nu)}{2(1-\nu)}\cos\theta\ln r - \frac{\cos\theta}{4(1-\nu)} - \theta\sin\theta\right] \tag{3-47}$$

Edge Dislocation in a Cylinder

Now consider a hollow dislocation along the axis of a right circular cylinder with inner and outer radii r_0 and R, respectively. This model is sometimes used instead of introducing the parameter α, as in Eq. (3-14), to account for the nonlinear elastic effects near the core; the core is considered hollow, and a compensatory core energy is added to the dislocation energy.[21, 22] Furthermore, this model corresponds to the unusual but real physical case of dislocations, with very large Burgers vectors, which are hollow because the large strain energy near the core is sufficient to supply the surface energy of the hollow tube.[23, 24] For such a dislocation the first-order solution, Eq. (3-42), gives

[20] This is identical to the result of W. T. Read, Jr. ("Dislocations in Crystals," McGraw-Hill, New York, 1953, p. 116). F. R. N. Nabarro [*Advan. Phys.*, **1**: 278 (1952)] gives a different choice for the constant of integration.

[21] A. H. Cottrell, "Dislocations and Plastic Flow in Crystals," Oxford University Press, Fair Lawn, N.J., 1953, p. 39.

[22] A. E. H. Love, "The Mathematical Theory of Elasticity," Cambridge University Press, Cambridge, 1927, p. 227.

[23] F. C. Frank, *Acta Cryst.*, **4**: 497 (1951).

[24] A. R. Verma, "Crystal Growth and Dislocations," Butterworths, London, 1953, pp. 166–169.

stresses σ_{rr} and $\sigma_{r\theta}$, which do not vanish at either the external free surface $r=R$ or the inner surface $r=r_0$. Note, however, that these nonvanishing fictitious stresses exert no *net* force on the cylinder. The stress function which removes the fictitious stresses at the external cylinder surface and thus fulfills the boundary condition [Eq. (2-72)] is

$$\psi_R=\frac{\mu b r^3}{4\pi(1-\nu)R^2}\sin\theta \tag{3-48}$$

That which leads to zero forces normal to the inner cutoff surface at $r=r_0$ is

$$\psi_{r_0}=-\frac{\mu b r_0^2}{4\pi(1-\nu)r}\sin\theta \tag{3-49}$$

ψ_R corresponds to the surface traction term $\alpha_1 r\sin\theta$ in Eq. (3-36). ψ_{r_0} is a core term arising from the solution to the homogeneous equation in Eq. (3-38).

Exercise 3-5. Determine the stress components corresponding to ψ_R and ψ_{r_0}. Show that with these stresses superposed on those of Eq. (3-44) the boundary conditions at the free surface and at the core cutoff surface are satisfied to first order in r_0/R. To second order in r_0/R, a small residual arising from Eq. (3-48) is present at $r=r_0$, and one from Eq. (3-49) is present at $r=R$. Show that the stresses and displacements resulting from ψ_R and ψ_{r_0} are

$$\sigma_{rr}=\frac{\mu b}{2\pi(1-\nu)}\sin\theta\left(\frac{r}{R^2}+\frac{r_0^2}{r^3}\right),\qquad \sigma_{\theta\theta}=\frac{\mu b}{2\pi(1-\nu)}\sin\theta\left(\frac{3r}{R^2}-\frac{r_0^2}{r^3}\right)$$

$$\sigma_{r\theta}=-\frac{\mu b}{2\pi(1-\nu)}\cos\theta\left(\frac{r}{R^2}+\frac{r_0^2}{r^3}\right),\qquad \sigma_{zz}=\nu(\sigma_{rr}+\sigma_{\theta\theta}) \tag{3-50}$$

$$u_r=\frac{b}{8\pi(1-\nu)}\sin\theta\left[(1-4\nu)\frac{r^2}{R^2}-\frac{r_0^2}{r^2}\right]$$

$$u_\theta=\frac{b}{8\pi(1-\nu)}\cos\theta\left[-(5-4\nu)\frac{r^2}{R^2}+\frac{r_0^2}{r^2}\right] \tag{3-51}$$

Strain Energy of an Edge Dislocation

The strain energy per unit length of dislocation between two cylindrical surfaces of radius r_0 and R about an edge dislocation in an infinite medium is

$$\frac{W}{L}=\int_{r_0}^{R} r\,dr\int_0^{2\pi}d\theta\left[\frac{1}{2\mu}\sigma_{xy}^2+\frac{1}{2E}\left(\sigma_{xx}^2+\sigma_{yy}^2-2\nu\sigma_{xx}\sigma_{yy}-\sigma_{zz}^2\right)\right]$$

The above integral yields

$$\frac{W}{L}=\frac{\mu b^2}{4\pi(1-\nu)}\ln\frac{R}{r_0} \tag{3-52}$$

which is identical in form with the result for the screw dislocation [Eq. (3-13)] except for the factor $(1-\nu)$.

In addition to the generation of strain energy in the cylindrical section bounding an edge dislocation, the creation of the dislocation is accompanied by radial energy flow through the cylinder. The stresses of Eq. (3-44) produce forces on the cylindrical surfaces at $r=R$ and $r=r_0$ as they are displaced. These forces respectively do work given by

$$\frac{W}{L}=\frac{1}{2}\int_0^{2\pi}[\sigma_{rr}u_r]_{r=R}R\,d\theta+\frac{1}{2}\int_0^{2\pi}[\sigma_{r\theta}u_\theta]_{r=R}R\,d\theta \tag{3-53}$$

and an equivalent expression with R replaced by r_0. Substitution of the results of Eqs. (3-44) and (3-47) into (3-53) reveals that work

$$\frac{W}{L}=\frac{\mu b^2(1-2\nu)}{16\pi(1-\nu)^2} \tag{3-54}$$

is done on the cylinder at $R=r$ with an equal amount of work done by the cylinder at $r=r_0$.

With a real core inside the smaller cylinder, the work done by the inner cylinder would be retained in the core for a reversible process, i.e., the inward radial energy would accumulate as part of the core energy. Formally, the core energy is incorporated into the result of Eq. (3-52) by replacing r_0 by b/α and adjusting α so that the total strain energy within the cylinder bounded by $r=R$ is given by

$$\frac{W}{L}=\frac{\mu b^2}{4\pi(1-\nu)}\ln\frac{R\alpha}{b} \tag{3-55}$$

This treatment is consonant with that for the screw dislocation leading to Eq. (3-14).

The finite cylinder result can be derived in alternate ways. The strain energy in the final state represents the free-energy change of the crystal in creating the dislocation. Since the free-energy change is a state function, it is independent of the path of the change. Suppose that the cylinder is cut on $r=R$ with stresses equal and opposite to those in Eq. (3-44) acting to keep the cylinder fixed. These stresses are then relaxed, permitting the displacements of Eq. (3-51) that make the cylinder traction free. The substitution of Eq. (3-44) and (3-51) into Eq. (3-53) shows that the work done on the crystal in this relaxation is

$$\frac{W}{L} = \frac{-\mu b^2(3-4\nu)}{16\pi(1-\nu)^2} \tag{3-56}$$

which gives a total energy for the finite cylinder of

$$\frac{W}{L} = \frac{\mu b^2}{4\pi(1-\nu)} \ln \frac{R\alpha}{b} - \frac{\mu b^2(3-4\nu)}{16\pi(1-\nu)^2} \tag{3-57}$$

Another model of the dislocation is that with a hollow core. Let us modify Eq. (3-52) directly for the situation of a hollow core and a free outer surface. First, the dislocation is created in the state to which Eq. (3-52) refers. Then the outer surface is relaxed, giving the energy reduction in Eq. (3-56). Then the hollow core surface is relaxed giving a similar energy reduction, $W/L = -\mu b^2/16\pi(1-\nu)^2$. Altogether, there is an energy change $-\mu b^2/4\pi(1-\nu)$, and the total energy of the dislocation with a hollow core and a free outer surface is

$$\frac{W}{L} = \frac{\mu b^2}{4\pi(1-\nu)} \left[\ln \frac{R}{r_0} - 1 \right] \tag{3-58}$$

In general, in a finite crystal image forces are important, but a rough estimate of the energy is obtained by setting $R \sim l$, the distance from the dislocation to the surface. Energy contributions caused by surface ledges associated with the dislocation should also be included in an exact treatment. The term arising from Eq. (3-49) is included in α.

Exercise 3-6. Consider an unstrained cylinder with free surfaces at r_0 and R. Create a dislocation by making a radial cut on the θz plane at $\theta = 0$ and displacing the surfaces in the radial direction by a relative shear b. Calculate the necessary work performed on the cut surfaces using Eq. (3-50) and show that it gives the result Eq. (3-58).

Force Considerations

For the edge dislocation in Fig. 3-13, let us split up the total stress field near the core σ_T into two terms, σ_d and σ, where σ_d is the self-stress as given by Eqs. (3-43). As for the screw dislocation, only the stress σ does work on the slip plane as the dislocation moves. The force per unit length is produced by the stress component σ_{xy}:

$$\frac{F_x}{L} = \sigma_{xy} b \tag{3-59}$$

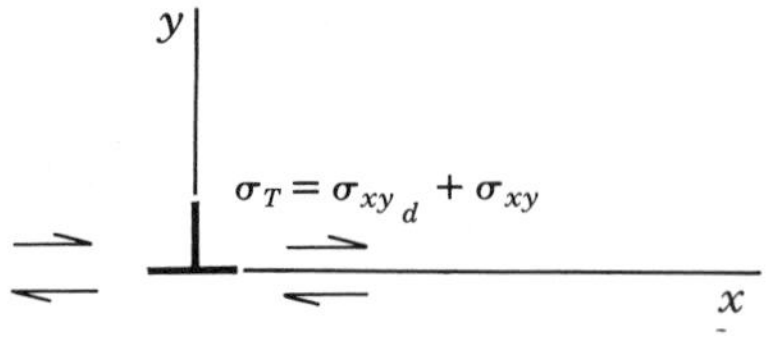

FIGURE 3-13. An edge dislocation.

If matter is removed from the edge of the inserted plane to make the edge dislocation climb, both the elastic strain energy and the potential energy of the external sources of stress for the remaining system change. In complete analogy with the definitions employed for the force for glide, the force for climb is defined from the derivative of the sum of strain energy and potential energy with respect to the position y of the dislocation. The total energy change involved is, by definition, that of removing matter from the inserted plane to deposit it on a stress-free perfect crystal (Fig. 3-14). If less energy is needed to remove an atom from the inserted plane than is gained on depositing it on a stress-free crystal, the decrease in strain energy and potential energy of external mechanisms comprises the difference. At the expense of strain energy and potential energy, energy flows to the dislocation core and is dissipated there as the dislocation climbs.

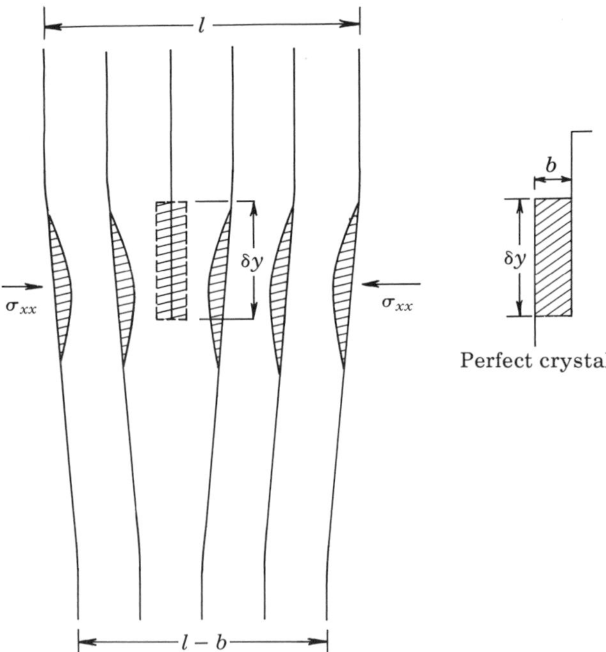

FIGURE 3-14. The removal of material from an edge dislocation and the replacement of the material at a free surface, the net process producing climb. Hatched regions represent displacements that accompany the climb.

We now develop the expression for the climb force in greater detail, presenting general, detailed considerations, which also corroborate previous results. These considerations also are valid for elastically anisotropic solids. The only result from isotropic elasticity theory to be used is the fact that the self-stress around a dislocation diminishes as $1/r$ away from the dislocation, which must also hold true for anisotropic media.

Let us first establish rigorously that the self-stress of an edge dislocation does not exert a climb force on the dislocation. Consider the edge dislocation in Fig. 3-15. The lines AA, BB, etc., are fixed in space. The total distortion energy per unit length of dislocation line between AA and BB, for example, must be independent of y (with a finite core energy this distortion energy is finite). If a climb force F/L acts on unit length of the dislocation, an energy $(F/L)\,\delta y$ is absorbed at the dislocation as it climbs a distance δy. It follows that this energy has to be transported through the planes AA and BB to the dislocation, and also through the planes $A'A'$ and $B'B'$, since the energy between say AA and $A'A'$ remains constant.

The climb δy makes lines initially coincident with AA and BB, but embedded in the crystal, bend to enclose a volume decreased by $\delta v/L$ per unit depth. The work done on the moving walls is of the order

$$\frac{\delta W}{L} \sim \sigma_d(h)\frac{\delta v}{L} \propto \frac{\delta v}{Lh}$$

Adding to this the elastic energy in $\delta v/L$, of the order

$$\frac{\delta W}{L} \propto \frac{\delta v}{Lh^2}$$

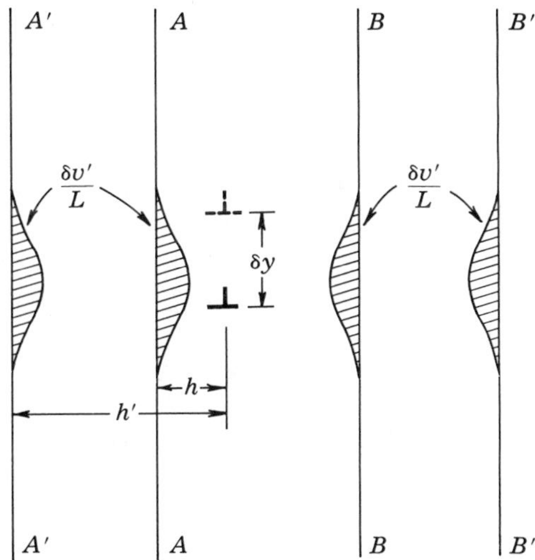

FIGURE 3-15. Scribe lines about an edge dislocation, and their displacement following dislocation climb.

one obtains the total energy flux through AA and BB,

$$\frac{\delta W}{L} = \frac{F_y}{L}\delta y = \alpha\frac{\delta v}{Lh} + \beta\frac{\delta v}{Lh^2} \tag{3-60}$$

where α and β are constants. Since the strain goes to zero as $y \to \pm\infty$, it must be true that

$$b = \int_{-\infty}^{\infty}\left[\frac{\partial u_x(y,h)}{\partial y} - \frac{\partial u_x(y,-h)}{\partial y}\right] dy \tag{3-61}$$

Thus

$$\frac{\delta v}{L} = \delta y \int_{-\infty}^{\infty}\left[\frac{\partial u_x(y,h)}{\partial y} - \frac{\partial u_x(y,-h)}{\partial y}\right] dy = b\,\delta y \tag{3-62}$$

Combining Eqs. (3-60) and (3-62) gives

$$\frac{F_y}{L}\delta y = \left(\frac{\alpha b}{h} + \frac{\beta b}{h^2}\right)\delta y \to 0 \qquad \text{as} \quad h \to \infty \tag{3-63}$$

Because $(F_y/L)\,\delta y$ is independent of h, the only possibility is that

$$\frac{F_y}{L} = 0 \tag{3-64}$$

i.e., *the self-stress of the dislocation exerts no climb force on the dislocation*. The generality of the argument is realized; it is not necessary to discuss the stress distribution and the displacement distribution near the core, as is the case in the discussion of Eq. (3-21).

If, in addition to σ_d, a compressive stress $-\sigma_{xx}$ acts over the region of the dislocation core, the dislocation experiences a climb force. The preliminary discussion leading to Eq. (3-23) did serve to make clear that with a net glide force F_x/L on the screw, an energy $(F_x/L)\,\delta x$ disappears locally at the screw as it glides by δx. Physically, as the dislocation moves, the local state of stress over the dislocation core determines the force on the dislocation, and thus the energy disappearing at the dislocation, at the expense of the total energy of the system. A local stress $\sigma_d + \sigma_{xx}$ in the core region thus exerts the same force on the dislocation as a homogeneous stress σ_{xx} superimposed on σ_d. In the presence of a homogeneous stress σ_{xx}, the symmetry of the situation indicates that the energy absorbed as the dislocation climbs by δy must pass through AA and BB, and through $A'A'$ and $B'B'$. For any given σ_{xx}, the value of h can be set large enough so that σ_d is negligible compared to σ_{xx}. In such a case, by analogy with Eq. (3-60),

$$\frac{\delta W}{L} = \frac{F_y}{L}\delta y = -\sigma_{xx} b\,\delta y + \frac{\sigma_{xx}^2}{2E} b\,\delta y \tag{3-65}$$

The last term in Eq. (3-65) should be ignored, since it is of the same order of magnitude as the imprecision in the linear-elasticity description (see Exercise 3-7). Thus, generally, in the approximation of linear elasticity, the climb force is

$$\frac{F_y}{L} = -\sigma_{xx} b \tag{3-66}$$

The methods employed above evidently could be used without modification to derive the glide forces developed previously by more expedient methods. The force expressions derived in this chapter are generally valid for straight dislocations, in either isotropic or anisotropic media. They are related directly to the work done on the dislocation, at the expense of the total energy of the system, as the dislocation moves.

Exercise 3-7. Discuss the validity of Eqs. (3-64), (3-65), and (3-66). Show that the second-order term in Eq. (3-65) is only one of several second-order terms that strictly should be included. Show that in the precision of linear elasticity, Eq. (3-66) is adequate.

Outline of discussion. Equation (3-64) is rigorous; the derivation depends only on symmetry and on the long-range behavior of the self-stresses. The finite second-order term reflects the inadequacy of linear elasticity theory to give a precise description.

One second-order term neglected in Eq. (3-65) is evident. With a homogeneous compressive stress σ_{xx}, the inserted plane is compressed to the thickness $(1+\sigma_{xx}/E)b$, which is the "effective" Burgers vector to be substituted for b in the first-order term, giving rise to an additional second-order term $-(\sigma_{xx}^2/E)b\,\delta y$.

The neglect of these second-order terms is consistent with the approximation of linear elasticity; in linear elasticity no distinction is made between σ_{xx} referring to force per unit original area or referring to force per unit area of the as-stressed matter (Sec. 2-7). This indistinction leads to an imprecision in σ_{xx} of $\sim 2\nu\sigma_{xx}^2/E$, which leads to an imprecision in Eq. (3-65) of the order of magnitude of the second-order terms discussed above. Even for a stress $\sigma_{xx} \sim 1/10E$ (roughly the theoretical strength), the second-order terms are only of the order $1/20$ of the first-order term. Externally applied stresses rarely exceed $\sim 10^{-3}E$ in actual crystals.

3-5. IMAGE FORCES ON EDGE DISLOCATIONS

As an example of an image problem for the edge dislocation, consider a dislocation line parallel to a free surface and with its slip plane normal to the surface. Even for this relatively simple problem, the mathematics is somewhat lengthy and tedious. However, the treatment is worthwhile in indicating the method of approach for more complex problems. Also, working out the mathematics in this case provides an appreciation of the justification for an approximate treatment in more complicated cases.

For the edge dislocation of Fig. 3-16, let us first attempt a simple image construction. If the stresses from the dislocation and its image are added together, all stresses acting on the surface vanish except σ_{xy}. Hence the simple image construction is insufficient to give the correct answer. With the simple image, the shear stresses σ_{xy} at the free surface add up to

$$\sigma_{xy}(0, y)=\frac{\mu b}{\pi(1-\nu)}\frac{l(l^2-y^2)}{r^4} \tag{3-67}$$

where $r=(l^2+y^2)^{1/2}$. Thus, to obtain the correct solution, the fictitious stress of Eq. (3-67) must be canceled by the addition of a stress function ψ, which gives

$$\sigma_{xy}(0, y)=-\frac{\mu b}{\pi(1-\nu)}\frac{l(l^2-y^2)}{r^4} \tag{3-68}$$

and $\sigma_{xx}=\sigma_{xz}=0$ at the boundary and vanishing stress at $x=-\infty$. A separable solution to the Airy stress function [Eq. (2-56)] is

$$\psi=X(x)Y(y) \tag{3-69}$$

As can be verified by substitution into Eq. (2-56), X and Y must satisfy the

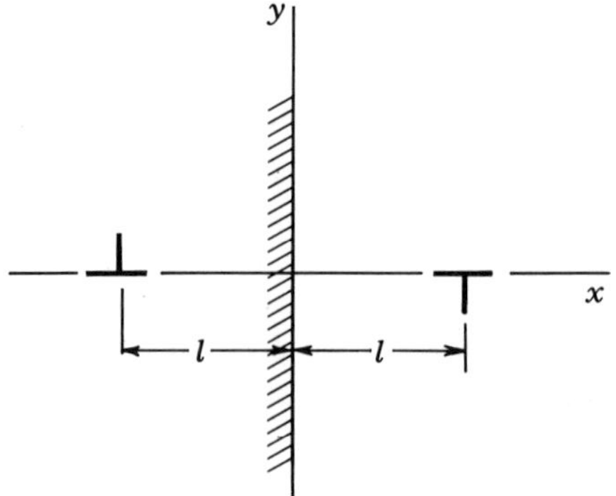

FIGURE 3-16. Edge dislocation parallel to a free surface, and an image dislocation.

equation

$$\frac{\partial^4 X}{\partial x^4}+\frac{2}{Y}\frac{\partial^2 X}{\partial x^2}\frac{\partial^2 Y}{\partial y^2}+\frac{X}{Y}\frac{\partial^4 Y}{\partial y^4}=0 \tag{3-70}$$

Requiring that $(\partial^2 Y/\partial y^2)/Y$ and $(\partial^4 Y/\partial y^4)/Y$ be independent of y, so that the variables are separable, Y must be of the form

$$Y=a\sin ky+b\cos ky \tag{3-71}$$

Hence the differential equation for X becomes

$$\frac{\partial^4 X}{\partial x^4}-2k^2\frac{\partial^2 X}{\partial x^2}+k^4X=0 \tag{3-72}$$

which has the solution

$$X=(a_0+a_1x)e^{kx}+(c_0+c_1x)e^{-kx} \tag{3-73}$$

The condition that $X\to 0$ as $x\to -\infty$ is fulfilled if $c_0=c_1=0$. The condition $\sigma_{xx}(0, y)=0$ is satisfied if $X(0, y)=0$, so that $a_0=0$. By superposition of Eq. (3-71) and (3-73), the desired solution is

$$\psi=\int_0^\infty a(k)xe^{kx}\sin ky\,dk \tag{3-74}$$

The $\cos ky$ term has been dropped because $\sigma_{xy}(0, y)$ must be even in y [Eq. (3-68)].

The boundary condition for σ_{xy} now yields

$$-\sigma_{xy}(0, y)=\left(\frac{\partial^2\psi}{\partial x\,\partial y}\right)_{x=0}=\int_0^\infty a(k)k\cos ky\,dk$$

$$=\frac{\mu b}{\pi(1-\nu)}\frac{l(l^2-y^2)}{(l^2+y^2)^2} \tag{3-75}$$

By Fourier inversion,

$$ka(k)=\frac{\mu b}{\pi^2(1-\nu)}\int_{-\infty}^{\infty}\frac{l(l^2-y^2)}{(l^2+y^2)^2}\cos ky\,dy \tag{3-76}$$

After one integration by parts, Eq. (3-76) becomes

$$ka(k)=\frac{\mu b}{\pi^2(1-\nu)}\frac{kl}{i}\int_{-\infty}^{\infty}\frac{ye^{iky}}{l^2+y^2}dy \tag{3-77}$$

Now, taking the residue[25] at $y=il$, one finds

$$a(k)=\frac{\mu bl}{\pi(1-\nu)}e^{-kl} \tag{3-78}$$

so that

$$\begin{aligned}\psi&=\frac{\mu bl}{\pi(1-\nu)}\int_0^\infty xe^{k(x-l)}\sin ky\,dk\\&=\frac{\mu blxy}{\pi(1-\nu)\left[(x-l)^2+y^2\right]}\end{aligned} \tag{3-79}$$

At $(-l,0)$ this stress function yields a shear stress

$$\sigma_{xy}=-\frac{\partial^2\psi}{\partial x\,\partial y}(-l,0)=-\frac{\mu bl}{\pi(1-\nu)}\int_0^\infty k(1-kl)e^{-2kl}\,dk=0 \tag{3-80}$$

Since the stresses from Eq. (3-79) exert no force component F_x, the attraction toward the surface is simply obtained from the stress of the image dislocation alone,[26–28]

$$\frac{F_x}{L}=\frac{\mu b^2}{4\pi(1-\nu)l} \tag{3-81}$$

The simplicity of the result is not fortuitous[28]; see Sec. 5-9.

The complete stress distribution for the edge dislocation in Fig. 3-16 is given by the superposition of the stress fields of the edge dislocation, the image dislocation, and the stresses derived from Eq. (3-79). The stresses given by Eq. (3-79) are

$$\sigma_{xx}=-\frac{2\mu blxy}{\pi(1-\nu)r^6}\left[3(l-x)^2-y^2\right]$$

$$\sigma_{yy}=\frac{\mu bl}{\pi(1-\nu)r^6}\left[4(l-x)^3y+6(l-x)^2xy+4(l-x)y^3-2xy^3\right]$$

$$\sigma_{xy}=\frac{-\mu bl}{\pi(1-\nu)r^6}\left[(l-x)^4+2x(l-x)^3-6xy^2(l-x)-y^4\right]$$

$$\sigma_{zz}=\frac{4\mu bl\nu}{\pi(1-\nu)r^6}\left[(l-x)^3y+(l-x)y^3\right] \tag{3-82}$$

[25] For a discussion of the method of residues, see I. S. Sokolnikoff and R. M. Redheffer, "Mathematics of Physics and Modern Engineering," McGraw-Hill, New York, 1966, pp. 564–594. Equation (3-76) can be evaluated directly by the method of residues, but it involves a second-order singular pole, whereas after integration Eq. (3-77) contains a simple pole and can be evaluated more straightforwardly.

[26] J. S. Koehler, *Phys. Rev.*, **60**: 397 (1941).

[27] A. K. Head, *Proc. Phys. Soc.*, **66B**: 793 (1953).

[28] J. Lothe, *Physica Norvegica*, **2**: 153 (1967).

These stress fields are important in image-force problems such as those involving dislocations in thin films, as observed in electron-transmission microscopy.

Exercise 3-8. The stresses associated with the core cutoff stress function ψ_{r_0} [Eq. (3-49)] do not vanish at the free surface for the dislocation in Fig. 3-16. A simple image superposition, analogous to the above case, leaves a residual stress σ_{xy} at the free surface

$$\sigma_{xy}(0, y) = -\frac{\mu b r_0^2}{\pi(1-\nu)} \frac{x^3 - 3xy^2}{r^6}$$

Proceed as above to derive the stress function that would cause this surface stress to vanish at the free surface. Show that in this case the resulting stress function gives rise to a shear stress at $(-l,0)$ given by

$$\sigma_{xy}(-l,0) = \frac{\mu b r_0^2}{16\pi l^3(1-\nu)}$$

Thus in this case the attractive force is *not* given by the simple image construction.

For an edge dislocation normal to a free surface, added stresses arise analogous to those for the screw dislocation [Eq. (3-12)]. Consider the half-crystal $z>0$ with a free surface at $z=0$ and in the limit $R \to \infty$. The problem is a three-dimensional one that requires more background than others treated in this text so only the results are given.[29] The problem is a special case of a general stress function Φ treated by Green and Zerna[30]:

$$\Phi = y \log(\rho + z) + \frac{yz}{\rho + z} \tag{3-83}$$

where $\rho = (x^2 + y^2 + z^2)^{1/2}$. The solution[29] yields two stresses of particular importance for later discussion

$$\sigma_{zz} = \frac{\nu\mu b}{\pi(1-\nu)} \left[\frac{yz}{\rho^3} + \frac{y}{\rho(\rho+z)} \right]$$

$$\sigma_{yz} = \frac{\nu\mu b}{\pi(1-\nu)} \left[\frac{-z}{\rho(\rho+z)} + \frac{y^2 z}{\rho^3(\rho+z)} + \frac{y^2 z}{\rho^2(\rho+z)^2} \right] \tag{3-84}$$

[29]E. H. Yoffe, *Phil. Mag.*, **6**: 1147 (1961). She gives a different sign for z. However, the present usage agrees with that used by Eshelby and Stroh in deriving Eq. (3-12) and gives the proper limiting behavior of vanishing stress when $\rho = z = \infty$.

[30]A. E. Green and W. Zerna, "Theoretical Elasticity," Clarendon Press, Oxford, 1968, p. 168.

When $z=0$, the stresses σ_{zz} cancel the stresses of Eq. (3-43) to satisfy the free-surface boundary condition. For a finite value of R, added stresses from the stress function ψ_R also produce stresses σ_{zz} and lead to complicated end effects for the problem in Fig. 3-10. Far from the end, $|z| \gg R$, the end effect stresses from ψ_R cancel those of Eq. (3-84) and restore the results of Eqs. (3-43) and (3-50).

3-6. THE GENERAL STRAIGHT DISLOCATION

Consider a straight dislocation, ascribe a sense $\boldsymbol{\xi}$ to it, and construct the Burgers vector $\mathbf{b}$ of the dislocation (Fig. 3-17). In general, the Burgers vector $\mathbf{b}$ has one component $\mathbf{b}_s$, the screw component, parallel to the line, and one component $\mathbf{b}_e$, the edge component, normal to the line. When the dislocation is neither pure screw nor pure edge, it is defined to be of *mixed character*. The glide plane of the mixed dislocation is determined by the glide plane of its edge component; the screw component has no definite glide plane. *The glide plane is that plane which contains both the dislocation line and the Burgers vector.*

If the dislocation climbs a distance δh normal to the glide plane, an amount of matter per unit length

$$\frac{\delta v}{L} = b_e \, \delta h \tag{3-85}$$

is removed. In vector notation,

$$\frac{\delta v}{L} = (\boldsymbol{\xi} \times \mathbf{b}) \cdot \delta \mathbf{r} \tag{3-86}$$

represents the matter removed when the dislocation is displaced normal to itself by a distance $\delta \mathbf{r}$. Since $\boldsymbol{\xi} \times \mathbf{b}$ is normal to the slip plane, only the

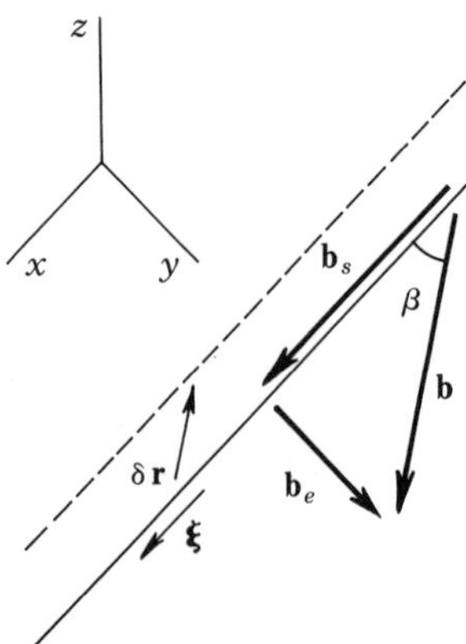

FIGURE 3-17. A mixed dislocation with its Burgers vector inclined at an angle β to the dislocation line.

component of $\delta\mathbf{r}$ normal to the slip plane is important in Eq. (3-86). Also, arbitrarily changing the sign of $\boldsymbol{\xi}$ changes the sign of $\mathbf{b}$, leaving $\boldsymbol{\xi}\times\mathbf{b}$ unchanged. Thus for a given dislocation Eq. (3-86) unambiguously gives the correct result.

By the superposition principle, the self-stress of the mixed dislocation is the sum of the self-stress of a screw dislocation of Burgers vector with magnitude $b_s = b\cos\beta$ and an edge dislocation with $b_e = b\sin\beta$. In an isotropic medium the displacements for the components are orthogonal and, with the same core cutoff parameter for both components, the energy per unit length within a cylinder of radius R is, from Eqs. (3-14) and (3-55).

$$\frac{W}{L} = \frac{\mu b^2}{4\pi}\left(\cos^2\beta + \frac{\sin^2\beta}{1-\nu}\right)\ln\frac{\alpha R}{b} = E(\beta)\ln\frac{\alpha R}{b} \tag{3-87}$$

where $E(\beta)$ is called the *prelogarithmic energy factor*.

The dislocation moves by $\delta\mathbf{r}$ if a cut is made of area $\boldsymbol{\xi}\times\delta\mathbf{r}$ per unit length, material is removed uniformly over the cut by the amount Eq. (3-86), and the continuity of the material is then restored by displacing the opposite surfaces of the cut relative to one another by $\mathbf{b}$. If a uniform stress exists over the core, it does work during the operation in the amount

$$\frac{\delta W}{L} = \frac{\mathbf{F}}{L}\cdot\delta\mathbf{r} = [\boldsymbol{\sigma}\cdot(\boldsymbol{\xi}\times\delta\mathbf{r})]\cdot\mathbf{b} \tag{3-88}$$

$\boldsymbol{\sigma}\cdot(\boldsymbol{\xi}\times\delta\mathbf{r})$ is the force caused by the stress tensor $\boldsymbol{\sigma}$ on the surface $\boldsymbol{\xi}\times\delta\mathbf{r}$. Equation (3-88) can be rewritten as

$$\frac{\mathbf{F}}{L}\cdot\delta\mathbf{r} = (\mathbf{b}\cdot\boldsymbol{\sigma})\cdot(\boldsymbol{\xi}\times\delta\mathbf{r}) = [(\mathbf{b}\cdot\boldsymbol{\sigma})\times\boldsymbol{\xi}]\cdot\delta\mathbf{r} \tag{3-89}$$

Hence the force per unit length is[31]

$$\frac{\mathbf{F}}{L} = (\mathbf{b}\cdot\boldsymbol{\sigma})\times\boldsymbol{\xi} \tag{3-90}$$

with a glide component

$$\frac{F_{\mathrm{gl}}}{L} = \frac{[(\mathbf{b}\cdot\boldsymbol{\sigma})\times\boldsymbol{\xi}]\cdot[\boldsymbol{\xi}\times(\mathbf{b}\times\boldsymbol{\xi})]}{|\mathbf{b}\times\boldsymbol{\xi}|} \tag{3-91}$$

and a climb component

$$\frac{F_{\mathrm{cl}}}{L} = \frac{[(\mathbf{b}\cdot\boldsymbol{\sigma})\times\boldsymbol{\xi}]\cdot(\mathbf{b}\times\boldsymbol{\xi})}{|\mathbf{b}\times\boldsymbol{\xi}|} \tag{3-92}$$

[31]As has been noted by R. de Wit [*Acta Met.*, **13**: 1210 (1965)], a number of sign errors have appeared in the literature relative to Eq. (3-90). Therefore one should be careful to check the sign by physical reasoning when using results in the literature.

These are the general force expressions. Although Eq. (3-92) gives the force in the climb direction for the edge component of a dislocation, it also contains a glide contribution from the screw component.

Alternatively, Eq. (3-90) can be written in terms of indices as

$$F_k = -\epsilon_{ijk}\xi_i\sigma_{jl}b_l \tag{3-93}$$

Here ϵ_{ijk} is the permutation operator, with components

$$\begin{aligned} &\epsilon_{123}=\epsilon_{231}=\epsilon_{312}=1 \\ &\epsilon_{132}=\epsilon_{321}=\epsilon_{213}=-1 \\ &\epsilon_{ijk}=0, \qquad \text{for} \quad i=j, \quad j=k, \quad \text{or} \quad i=k \end{aligned} \tag{3-94}$$

3-7. STRAIGHT DISLOCATIONS IN ANISOTROPIC MEDIA

The force expressions of the preceding section are generally valid for anisotropic crystals within the assumptions of linear elasticity. The calculation of the self-stress of dislocations is quite complicated for anisotropic crystals. For most purposes, a calculation assuming isotropy is adequate in view of the inaccuracy in observations with which theory is to be compared and of sources of inaccuracy in the theories other than the neglect of anisotropy. However, there is one class of problems for which it is both feasible and important to include anisotropy: reactions and interactions between straight dislocations in low-index crystallographic directions. In some cases considerations of anisotropy can be essential in deciding whether or not a particular type of dislocation reaction takes place.

The elasticity theory for dislocations in anisotropic media is outlined in Chap. 13. The more important formulas for stresses and displacements, which would be used in the above force formulas, are given there.

PROBLEMS

3-1. In Fig. 3-6, insert a screw dislocation of the same sign midway between the original screw and the surface. Compute the force exerted on this second screw. Will this force tend to move the inserted dislocation away from or toward the surface?

3-2. Discuss the interaction between two edge dislocations on parallel glide planes (Fig. 3-18). If dislocation A is fixed, which are the possible equilibrium positions of dislocation B in glide?

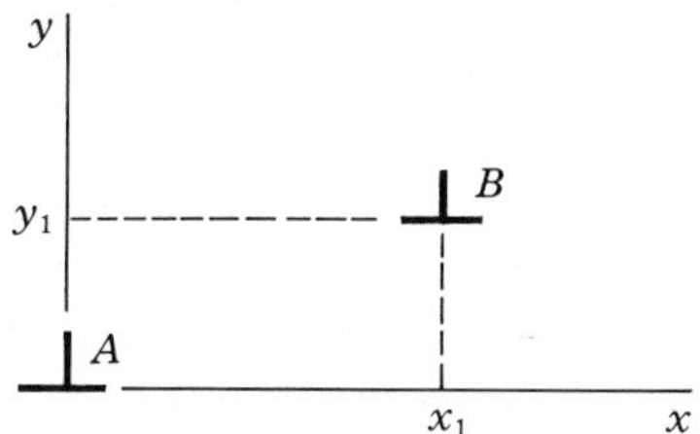

FIGURE 3-18. Two parallel straight edge dislocations.

3-3. Consider an edge dislocation with ξ along the z axis and **b** inclined at 45° to the x and y axes. Let an external stress with components σ_{xx}, σ_{xy}, and σ_{xz} be present. Use Eq. (3-90) to determine the total force per unit length on the dislocation. Discuss the physical significance of the various terms.

3-4. Consider two straight edge dislocations with mutually orthogonal Burgers vectors of the same magnitude (Fig. 3-19). Calculate the energy required to move dislocation A in its slip plane from $x=\infty$ to $x=a$. (This problem serves to show that dislocations with nonparallel, and even orthogonal, Burgers vectors interact in general.)

3-5. Consider a straight screw dislocation parallel to a *rigid* surface, constrained not to deform. Show that the screw is repelled from the surface by a force equivalent to an image dislocation of the *same* sign and magnitude. (This example is relevant to the situation of a dislocation near a surface with a hard oxide layer upon it.)

3-6. Suppose that the dislocations in Prob. 3-2 are in a copper crystal. Compute the glide and climb forces on dislocation B if $x_1=7$ nm, $y_1=3$ nm. Compare these forces with those produced by homogeneous external stresses $\sigma_{xy}=10^{-3}\mu$, $\sigma_{xx}=10^{-3}E$.

3-7. Prove that the stresses of Eqs. (3-43) and (3-82) satisfy the equilibrium equations (2-2).

3-8. Application of the results for a screw dislocation in an infinite medium to the case of a screw in a finite cylinder gave virtual stresses on the ends [Eq. (3-5)] which were removed by superposition. Show that no such long-range

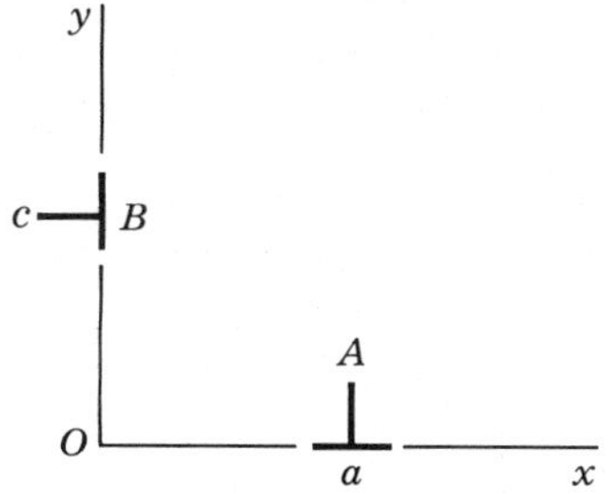

FIGURE 3-19. Two parallel straight edge dislocations.

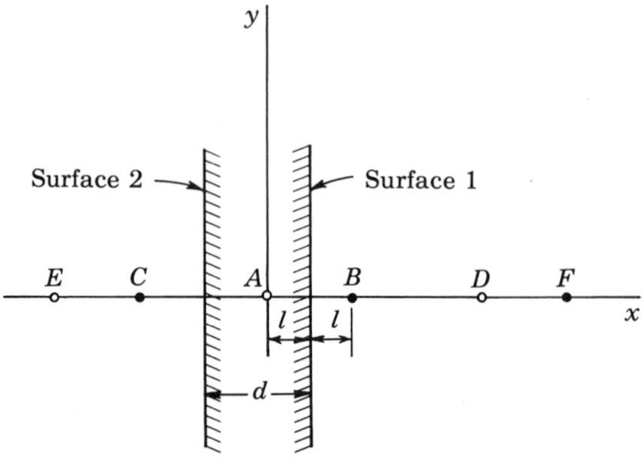

FIGURE 3-20. A right-handed screw dislocation A parallel to the surfaces of a thin plate. Image dislocations B, C, D, etc., are also shown. The symbol ○ indicates a right-handed image, ● a left-handed image.

stresses are present for an edge dislocation coaxially positioned in a *finite* right circular cylinder. Show that there are *local* forces on the cylinder end, however, and estimate the length over which the perturbation caused by these forces extends. *Hint:* Inclusion of the stresses from ψ_R is essential.

3-9. With the use of the results of Exercise 3-5, predict the equilibrium radius r_0 of a hollow dislocation. Assume that the surface energy γ of the hollow tube is the same as that of a bulk surface. For copper, with $\gamma = 1.7\ \text{J/m}^2$, how large a Burgers vector would be required to give a hollow dislocation of radius $r_0 = 1$ nm?[32]

3-10. Consider a right-handed screw dislocation parallel to the surfaces of an infinite plate (Fig. 3-20). The image dislocation B satisfies the boundary condition on surface 1 but leaves a residual stress acting on surface 2, so that an additional image E is required, which in turn requires an image F, etc. A similar set is associated with image C. The result is an infinite set of image dislocations. Show that the sum of all the image stresses acting at the origin on dislocation A is

$$\sigma_{yz} = \frac{\mu b}{-4\pi d} \sum_{n=-\infty}^{\infty} \frac{1}{n-(l/d)} = \frac{\mu b}{4\pi d} \sum_{n=-\infty}^{\infty} \frac{1}{n+(l/d)}$$

The solution to this sum is[33]

$$\sigma_{yz} = \frac{\mu b}{4d} \cot \pi l/d$$

[32] F. C. Frank, *Acta. Cryst.*, **4**: 497 (1951).

[33] P. M. Morse and H. Feshbach, "Methods of Theoretical Physics," vol. 1, McGraw-Hill, New York, 1953, p. 383.

Note that the set of right-handed image dislocations is symmetric about the origin, so that σ_{xy} at the origin is produced completely by the left-handed set.

3-11. Show that the external volume change for an edge dislocation located on the axis of a right-circular cylinder is zero. Since one can always imagine creating the dislocation by applying equal and opposite tractions to opposite sides of a glide plane cut, this result is true for a dislocation in a finite body of any shape, as can be demonstrated from Eq. (2-79).

BIBLIOGRAPHY

1. Burgers, J. M., *Proc. Kon. Ned. Akad. Wetenschap.*, **42**: 293, 378 (1939).
2. Love, A. E. H., "The Mathematical Theory of Elasticity," Cambridge University Press, 1927.
3. Nabarro, F. R. N., *Advan. Phys.*, **1**: 269 (1952).

4

The Theory of Curved Dislocations

4-1. INTRODUCTION

This chapter is concerned with an extension to generally curved dislocations of the ideas developed for straight dislocations. The development mainly involves the mathematical formalism required to derive four key equations in dislocation theory: the Burgers formula for the displacements produced by an infinitesimal element of dislocation line; the two Peach-Koehler formulas for the stress produced by such an element and for the force on it produced by an external stress; and the Blin formula for the interaction energy between two such elements. These formulas are all very important in providing tools to handle interactions between complex arrays of dislocations as discussed in the subsequent two chapters. Because of their importance, we present their derivation in sufficient detail that it can be followed through step by step.

4-2. CONSERVATIVE AND NONCONSERVATIVE MOTION

Let us extend the concepts of Chap. 3 to generally curved dislocations. Consider a closed loop C, bounding some surface A (Fig. 4-1). Ascribe a sense $\boldsymbol{\xi}$ to C. The positive normal $\mathbf{n}$ to an element dA is defined by the requirement that if C were made to shrink continuously in A until it just bounded dA, it would encircle $\mathbf{n}$ in the positive sense, by the right-hand rule. Also $d\mathbf{A}=\mathbf{n}\,dA$.

C becomes a dislocation line of Burgers vector $\mathbf{b}$ if, over the surface A, one removes (or inserts) material

$$\delta V=\mathbf{b}\cdot d\mathbf{A} \tag{4-1}$$

displaces the surface on the negative side of the cut by $\mathbf{b}$ relative to the positive side, and then pastes the surfaces together. The surface A is perfect again, and a pure line defect, the dislocation line C, results. Any surface A bounded by C could be used for the operation. For example, the identical dislocation could

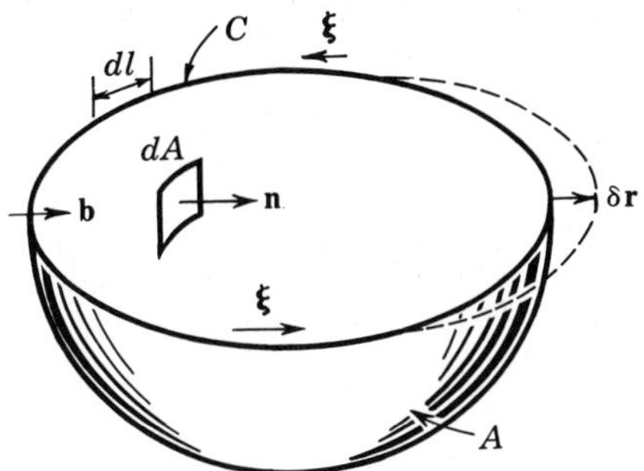

FIGURE 4-1. Closed dislocation loop C bounding surface A.

be produced by cutting and displacing any of the three surfaces shown in Fig. 4-2.

If the dislocation moves by $\delta\mathbf{r}$, matter

$$\delta V = \oint_C \mathbf{b}\cdot(\delta\mathbf{r}\times d\mathbf{l}) \qquad d\mathbf{l} = \boldsymbol{\xi}\, dl \tag{4-2}$$

must be removed, according to Eq. (4-1). $\mathbf{r}$ is variable along C. Surface elements $\delta\mathbf{A} = \delta\mathbf{r}\times d\mathbf{l}$ are added to A by the motion.

If everywhere along C the quantity $\mathbf{b}\cdot(\delta\mathbf{r}\times d\mathbf{l})$ is zero, the motion is one of pure slip. This occurs if $\delta\mathbf{r}$ is perpendicular to $\mathbf{b}\times d\mathbf{l}$, since $\mathbf{b}\cdot(\delta\mathbf{r}\times d\mathbf{l}) = -(\mathbf{b}\times d\mathbf{l})\cdot\delta\mathbf{r}$. *The dislocation can move conservatively, by pure slip, on the cylindrical surface containing C and* **b**.

Equation (4-2) can be interpreted geometrically as follows:
Project the dislocation onto a screen normal to **b** *(Fig. 4-3). The total mass transport to the dislocation during dislocation motion is given as the magnitude of the Burgers vector* **b** *times the change in projected enclosed area, counted negative if the projection of C encircles* **b** *in the positive sense.*

Exercise 4-1. The above condition for conservative motion, i.e., that $\delta\mathbf{r}$ be perpendicular to $\mathbf{b}\times d\mathbf{l}$, is sufficient but not necessary. A segment for which

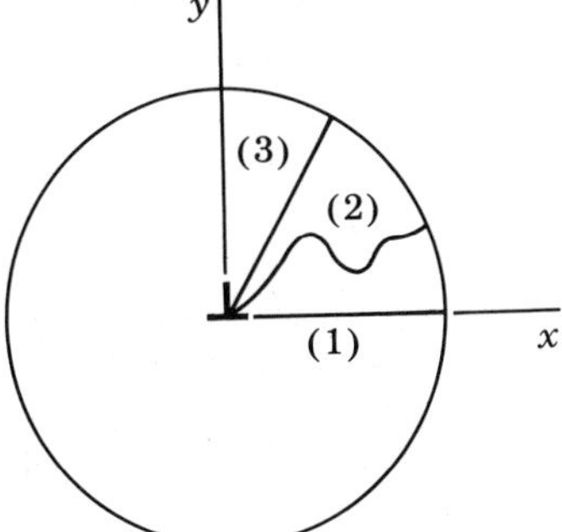

FIGURE 4-2. Three possible surface cuts, 1, 2, and 3, for producing a pure edge dislocation in a cylinder. Projection is along the z axis.

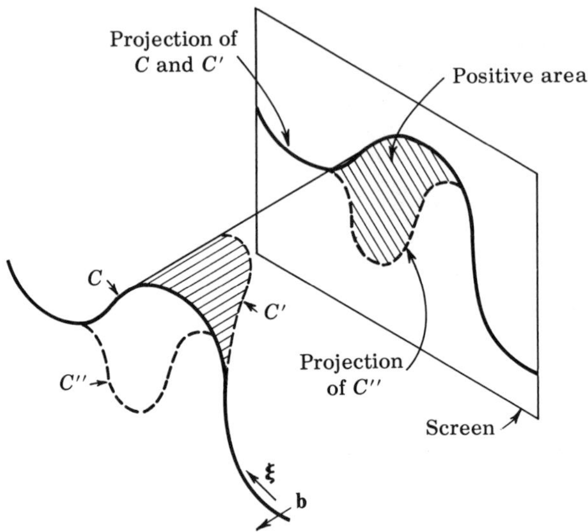

FIGURE 4-3. Projection on a screen normal to **b** of a dislocation line C which has undergone glide to position C' and climb to position C''.

$\mathbf{b}\times d\mathbf{l}=0$ ($d\mathbf{l}$ being a pure screw segment) is not restricted to glide on the cylindrical surface. Discuss how the statements of the preceding section should be qualified to take this special case into account. How will such conservative motion out of the cylindrical surface appear on the screen?

4-3. DISPLACEMENTS CAUSED BY CURVED DISLOCATIONS

Application of Green's Function Method

The derivation of the displacements associated with a dislocation loop of arbitrary shape involves the application of the theory, and Theorem 2-1, presented in Sec. 2-7. Consider a material of infinite extent, and suppose that a closed dislocation loop C of Burgers vector **b** is created (Fig. 4-4). The creation of the dislocation produces some displacement **u(r)** at **r**. Imagine for the moment that a point force acts at **r**. If a point force **F** acts at **r** while the dislocation is created, it does work

$$W=\mathbf{F}\cdot\mathbf{u}(\mathbf{r})=F_m u_m(\mathbf{r}) \tag{4-3}$$

where u_m and F_m are the components of **u** and **F**, respectively. If the displacements relieve the point force, they decrease the energy of the mechanism producing the point force by an amount W, the interaction energy. Therefore, a positive W represents a decrease in the system energy. *Since, by Theorem* 2-1, *there is no cross term in the elastic energy between the stress field of the*

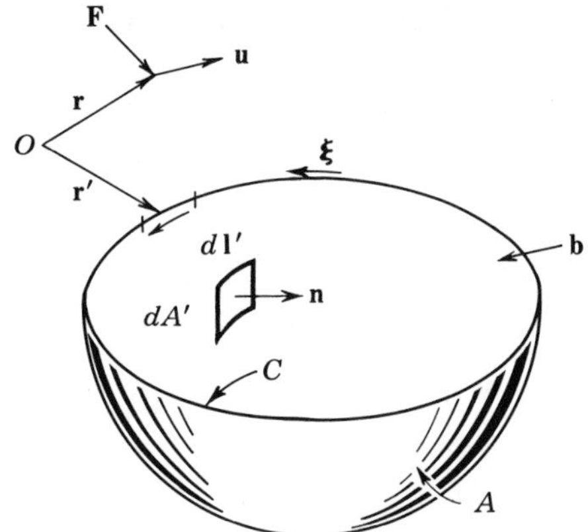

FIGURE 4-4. A point force **F** acting within an elastic continuum containing a closed dislocation loop.

dislocation and the stress field produced by the force **F**, *and since the dislocation line is not a sink of interaction energy, the entire energy W is spent as work done on the surface A*,[1]

$$W = -\int_A dA_j b_i F_m \sigma_{ij_m}(\mathbf{r}' - \mathbf{r}) \tag{4-4}$$

where $F_m \sigma_{ij_m}(\mathbf{r}'-\mathbf{r})$ is the stress σ_{ij} at $\mathbf{r}'$ caused by the components F_m of a point force **F** at **r**. The b_i are the components of **b**, dA_j are the components of $d\mathbf{A}$, etc.

In terms of the displacement functions $u_{mk}(\mathbf{r}'-\mathbf{r})$ introduced in Eq. (2-70), the so-called Green's functions of elasticity, Eq. (4-4), can be written as

$$W = -\int_A dA_j b_i c_{ijkl} \frac{\partial}{\partial x_l'} F_m u_{mk}(\mathbf{r}' - \mathbf{r}) \tag{4-5}$$

with summation over i, j, k, and m understood. Equating Eqs. (4-3) and (4-5) with only one nonvanishing component of **F**, $F_{m'} = 0$ when $m' \neq m$, one obtains the displacement field u_m, caused by the dislocation, by canceling F_m,

$$u_m(\mathbf{r}) = -\int_A dA_j b_i c_{ijkl} \frac{\partial}{\partial x_l'} u_{mk}(\mathbf{r}' - \mathbf{r}) \tag{4-6}$$

In other words, we use the point force **F** as a *test probe* to determine, via Eq. (2-71), the displacement field of the dislocation. The surface work is written in terms of the surface displacement b_i and the stress in A, produced by the point

[1] That is, less energy is introduced at the cut surface, and since there is no cross term, less by exactly the amount W if the point force has such a sign that it contributes to the energy of formation of the loop. The total energy contributed by the external mechanisms producing the surface displacements and the point force equals the elastic energy of the self-stress field of the loop.

force, at the point $\mathbf{r}'$. By the elasticity equations (2-3) and (2-15), this stress can in turn be written in terms of the displacements at $\mathbf{r}'$ produced by the point force. Thus, component by component in the point force, one can equate the work done by the point force at $\mathbf{r}$ and the surface work, cancel F_m, and deduce the displacement field of the dislocation.

Mura[2] has demonstrated that the Burgers integral expression for $u_m(\mathbf{r})$ can be transformed to a line integral for the gradient of $u_m(\mathbf{r})$. Because they are relative coordinates (Fig. 4-4) $dx'_s = -dx_s$, so the derivative of Eq. (4-6) can be written, with $\mathbf{R}=\mathbf{r}'-\mathbf{r}$, as

$$\frac{\partial u_m(\mathbf{r})}{\partial x_s} = b_i c_{ijkl} \int_A dA_j \frac{\partial^2 u_{mk}(\mathbf{R})}{\partial x'_s \partial x'_l} \tag{4-7}$$

This quantity enters the strains [Eq. (2-3)], which must be continuous across the cut A.

In accord with Eq. (2-8), we obtain

$$c_{ijkl} \frac{\partial^2 u_{mk}(\mathbf{R})}{\partial x'_j \partial x'_l} = 0 \tag{4-8}$$

when $R \neq 0$. This condition is similar to the requirement for mechanical equilibrium in the strain field of a point force. Since A can always be chosen so that $R \neq 0$ on the cut surface, Eqs. (4-7) and (4-8) can be combined to give

$$\frac{\partial u_m(\mathbf{r})}{\partial x_s} = b_i c_{ijkl} \int_A \left[dA_j \frac{\partial^2 u_{mk}}{\partial x'_s \partial x'_l} - dA_s \frac{\partial^2 u_{mk}}{\partial x'_j \partial x'_l} \right] \tag{4-9}$$

Stokes' theorem[3] has the form

$$\int_A \left(\frac{\partial \phi}{\partial x_j} dA_i - \frac{\partial \phi}{\partial x_i} dA_j \right) = \epsilon_{ijk} \oint_C \phi \, dx_k \tag{4-10}$$

where, in terms of the orthogonal unit vectors, the Einstein permutation operator [Eq. (3-94)] is given by

$$\epsilon_{ijk} = \mathbf{e}_i \cdot (\mathbf{e}_j \times \mathbf{e}_k) \tag{4-11}$$

[2] T. Mura, *Phil. Mag.*, **8**; 843 (1963).

[3] Stokes' theorem is widely known in the form

$$\int_A \operatorname{curl} \mathbf{M} \cdot d\mathbf{A} = \oint_C \mathbf{M} \cdot d\mathbf{l}$$

See I. S. Sokolnikoff and E. S. Sokolnikoff, "Higher Mathematics for Engineers and Physicists," McGraw-Hill, New York, 1941, p. 421. With $\mathbf{M} = \phi \mathbf{e}_k$ and $d\mathbf{l} = dx_i \mathbf{e}_i$, Eq. (4-10) follows directly.

Application of Stokes' theorem to Eq. (4-9) yields the line integral

$$\frac{\partial u_m(\mathbf{r})}{\partial x_s} = \epsilon_{jsn} b_i c_{ijkl} \oint_C \frac{\partial}{\partial x_l'} u_{mk}(\mathbf{R})\, dx_n' \tag{4-12}$$

which is Mura's formula.

Equations (4-6) and (4-12) are generally valid for an anisotropic material. If the Green's function $u_{mk}(\mathbf{r}'-\mathbf{r})$ is known, the displacement and strain fields caused by a dislocation of any shape can in principle be calculated. In the further development we shall assume elastic isotropy. Much of what follows is based on a review article by R. de Wit.[4]

The Burgers Displacement Equation

With the use of Eq. (2-47), Eq. (4-6) becomes, in a more extended form,

$$u_m(\mathbf{r}) = -\lambda \int_A dA_j b_j \frac{\partial u_{mk}}{\partial x_k'} - \mu \int_A dA_j b_i \frac{\partial u_{mi}}{\partial x_j'} - \mu \int_A dA_j b_i \frac{\partial u_{mj}}{\partial x_i'} \tag{4-13}$$

In order to derive the first term on the right-hand side of this equation, a change of dummy indices $dA_j b_j = dA_i b_i$ is required. Next, introducing Eq. (2-70) for the Green's functions into Eq. (4-13), one obtains

$$\begin{aligned} u_m(\mathbf{r}) = -\frac{1}{8\pi\mu} \int_A \Bigg[& \left(\lambda b_j \frac{\partial}{\partial x_m'} \nabla'^2 R\, dA_j + \mu b_m \frac{\partial}{\partial x_j'} \nabla'^2 R\, dA_j + \mu b_i \frac{\partial}{\partial x_i'} \nabla'^2 R\, dA_m \right) \\ & - \frac{\lambda+\mu}{\lambda+2\mu} \left(\lambda b_j \frac{\partial^3 R}{\partial x_m' \partial^2 x_k'} dA_j \right. \\ & \left. + 2\mu b_i \frac{\partial^3 R}{\partial x_m' \partial x_i' \partial x_j'} dA_j \right) \Bigg] \end{aligned} \tag{4-14}$$

Here

$$R = |\mathbf{r}' - \mathbf{r}| \tag{4-15}$$

and

$$\mathbf{R} = \mathbf{r}' - \mathbf{r} \tag{4-16}$$

[4] R. de Wit, *Solid State Phys.*, **10**: 249 (1960).

Finally, since $\partial^2/\partial x_i'^2 = \nabla'^2$, etc., one can rewrite Eq. (4-14) in a more symmetrical form suitable for the introduction of vector notation,

$$u_m(\mathbf{r}) = -\frac{1}{8\pi}\int_A b_m \frac{\partial}{\partial x_j'} \nabla'^2 R\, dA_j - \frac{1}{8\pi}\int_A \left(b_i \frac{\partial}{\partial x_i'} \nabla'^2 R\, dA_m - b_i \frac{\partial}{\partial x_m'} \nabla'^2 R\, dA_i \right)$$

$$+\frac{1}{4\pi}\frac{\lambda+\mu}{\lambda+2\mu}\int_A \left(b_i \frac{\partial}{\partial x_i'} \frac{\partial^2 R}{\partial x_m' dx_j'} dA_j - b_i \frac{\partial}{\partial x_j'} \frac{\partial^2 R}{\partial x_m' \partial x_j'} dA_i \right) \qquad (4\text{-}17)$$

Again some changes in dummy indices are required. Equations (4-10) and (4-17) combine to give

$$u_m(\mathbf{r}) = \frac{-1}{8\pi}\int_A b_m \frac{\partial}{\partial x_j'} \nabla'^2 R\, dA_j - \frac{1}{8\pi}\oint_C b_i \epsilon_{mik} \nabla'^2 R\, dx_k'$$

$$-\frac{1}{8\pi(1-\nu)}\oint_C b_i \epsilon_{ijk} \frac{\partial^2 R}{\partial x_m' \partial x_j'} dx_k' \qquad (4\text{-}18)$$

where use is made of Eq. (2-50) in the form $(\lambda+\mu)/(\lambda+2\mu) = 1/2(1-\nu)$. Since

$$\nabla'^2 R = \frac{2}{R}$$

and

$$\text{grad}' \frac{1}{R} = -\frac{\mathbf{R}}{R^3} \qquad (4\text{-}19)$$

the vector form of Eq. (4-18) is

$$\mathbf{u}(\mathbf{r}) = -\frac{\mathbf{b}}{4\pi}\Omega - \frac{1}{4\pi}\oint_C \frac{\mathbf{b}\times d\mathbf{l}'}{R} + \frac{1}{8\pi(1-\nu)} \text{grad} \oint_C \frac{(\mathbf{b}\times\mathbf{R})\cdot d\mathbf{l}'}{R} \qquad (4\text{-}20)$$

Here

$$\Omega = -\int_A \frac{\mathbf{R}\cdot d\mathbf{A}}{R^3} \qquad (4\text{-}21)$$

is the solid angle through which the positive side of A is seen from $\mathbf{r}$. The change in sign in the last term from Eq. (4-18) to Eq. (4-20) comes about because the variable with respect to which differentiation is performed has been changed; grad $= -$grad$'$. Equation (4-20) was first derived by Burgers[5] (with a difference in sign because of his different definition of $\mathbf{b}$).

[5] J. M. Burgers, *Proc. Kon. Ned. Akad. Wetenschap.*, **42**: 293, 378 (1939).

The first term in Eq. (4-20) gives a discontinuity $\Delta\mathbf{u}=\mathbf{b}$ over the surface A, consistent with the operation of producing the dislocation by cutting and displacing A. The other two terms are continuous except at the dislocation line.

With the use of Eq. (4-20), the displacement produced at a point $\mathbf{r}$ by an arbitrarily curved dislocation, or array of dislocations, can be determined by integration over the dislocation line.

4-4. SELF-STRESS OF A CURVED DISLOCATION

The stresses for the isotropic case could be derived starting with Eq. (4-12). Instead, the stresses are derived following deWit's derivation,[6] developed before Mura's formula[7] was known. The stresses are obtained by the differentiation of Eq. (4-18) and insertion of the result into Eq. (2-46). In the formula for the displacements, there is a discontinuity over the surface A. In the formula for the stresses, this discontinuity must disappear for continuity and equilibrium to be maintained. Hence, unlike the case for displacements, the stresses can be expressed in terms of line integrals alone; the dislocation is defined by an operation that leaves the material perfect and continuous except at the dislocation core.

The only term that is not expressed as a line integral in Eq. (4-20) is the one involving the solid angle Ω. Consider $\partial\Omega/\partial x_j$. Evidently, $\delta x_j(\partial\Omega/\partial x_j)$ can be interpreted as the change in solid angle as seen from $\mathbf{r}$ if C is displaced by $\delta\mathbf{r}'=-\delta x_j\mathbf{e}_j$ (Fig. 4-5). Thus Eq. (4-21) indicates that

$$\frac{\partial\Omega}{\partial x_j}=\oint_C\frac{\mathbf{R}\cdot(\mathbf{e}_j\times d\mathbf{l})}{R^3} \tag{4-22}$$

or, written out in the notation of Eq. (4-18),

$$\frac{\partial\Omega}{\partial x_j}=-\frac{1}{2}\oint_C\epsilon_{ijk}\frac{\partial}{\partial x_i'}\nabla'^2R\,dx_k' \tag{4-23}$$

The stresses are obtained from Eqs. (2-7) and (2-47) in the form

$$\sigma_{\alpha\beta}=\left[\lambda\delta_{\alpha\beta}\delta_{ml}+\mu\left(\delta_{\alpha l}\delta_{\beta m}+\delta_{\alpha m}\delta_{\beta l}\right)\right]\frac{\partial u_m}{\partial x_l} \tag{4-24}$$

A simple but tedious rearrangement of Eqs. (4-18), (4-20), and (4-24), involving

[6] R. deWit, *Solid State Phys.*, **10**: 249 (1960).

[7] T. Mura, *Phil. Mag.*, **8**: 843 (1963).

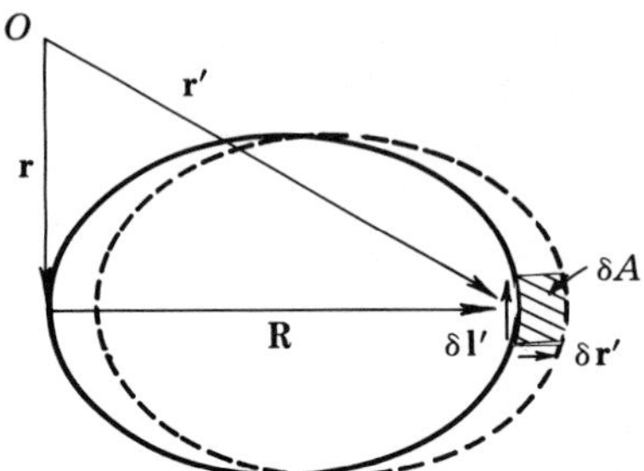

FIGURE 4-5. A dislocation loop.

changes of dummy indices, leads to the expression

$$\begin{aligned}\sigma_{\alpha\beta} = {} & \frac{\mu}{8\pi}\oint_C \big(\overset{(1)}{\delta_{\alpha l}\delta_{\beta m}\epsilon_{ilk}} + \overset{(2)}{\delta_{\alpha i}\delta_{\beta l}\epsilon_{lmk}} + \overset{(3)}{\delta_{\alpha m}\delta_{\beta l}\epsilon_{ilk}} + \overset{(4)}{\delta_{\alpha l}\delta_{\beta i}\epsilon_{lmk}}\big) b_m \frac{\partial}{\partial x'_i}\nabla'^2 R\, dx'_k \\ & + \frac{\mu(\lambda+\mu)}{2\pi(\lambda+2\mu)}\oint_C b_i \epsilon_{ijk}\frac{\partial^3 R}{\partial x'_\alpha \partial x'_\beta \partial x'_j} dx'_k \\ & + \frac{\mu\lambda\delta_{\alpha\beta}}{4\pi(\lambda+2\mu)}\oint_C b_m \epsilon_{imk}\frac{\partial}{\partial x'_i}\nabla'^2 R\, dx'_k \end{aligned} \tag{4-25}$$

This relation can be simplified by the following procedure of de Wit.[8] By use of the relation[9]

$$\epsilon_{ijk}\epsilon_{klm} = \delta_{il}\delta_{jm} - \delta_{im}\delta_{jl} \tag{4-26}$$

one can transform the first term in Eq. (4-25) to leave a more symmetrical form

[8] R. de Wit, *Solid State Phys.*, **10**: 249 (1960).

[9] Relation (4-26) is proved as follows: Equation (4-11) can be rewritten as

$$\epsilon_{ijk} = \mathbf{e}_i \cdot (\mathbf{e}_j \times \mathbf{e}_k) = \mathbf{e}_k \cdot (\mathbf{e}_i \times \mathbf{e}_j)$$

Taking the dot product of $\mathbf{e}_k$ and the above equation, one finds

$$\mathbf{e}_i \times \mathbf{e}_j = \epsilon_{ijk}\mathbf{e}_k \qquad \text{so that} \qquad \mathbf{e}_l \times \mathbf{e}_m = \epsilon_{klm}\mathbf{e}_k$$

Multiplying the two last equations, one obtains

$$\begin{aligned}\epsilon_{ijk}\epsilon_{klm} &= (\mathbf{e}_i \times \mathbf{e}_j)\cdot(\mathbf{e}_l \times \mathbf{e}_m) = \mathbf{e}_i \cdot [\mathbf{e}_j \times (\mathbf{e}_l \times \mathbf{e}_m)] \\ &= \mathbf{e}_i \cdot [(\mathbf{e}_j \cdot \mathbf{e}_m)\mathbf{e}_l - (\mathbf{e}_j \cdot \mathbf{e}_l)\mathbf{e}_m] \\ &= (\mathbf{e}_i \cdot \mathbf{e}_l)(\mathbf{e}_j \cdot \mathbf{e}_m) - (\mathbf{e}_i \cdot \mathbf{e}_m)(\mathbf{e}_j \cdot \mathbf{e}_l)\end{aligned}$$

which proves Eq. (4-26).

for $\sigma_{\alpha\beta}$. With the aid of Eq. (4-26), the four subterms (1), (2), (3), and (4) in the first term of Eq. (4-25) can be expanded as follows:

$$\overset{(1)}{\delta_{\alpha l}\delta_{\beta m}\epsilon_{ilk}} = \overset{(1)'}{\epsilon_{\beta lp}\epsilon_{pm\alpha}\epsilon_{ilk}} + \delta_{\alpha\beta}\delta_{lm}\epsilon_{ilk}$$

$$\overset{(2)}{\delta_{\alpha i}\delta_{\beta l}\epsilon_{lmk}} = \overset{(2)'}{\epsilon_{\beta ip}\epsilon_{pl\alpha}\epsilon_{lmk}} + \delta_{\alpha\beta}\delta_{il}\epsilon_{lmk}$$

$$\overset{(3)}{\delta_{\alpha m}\delta_{\beta l}\epsilon_{ilk}} = \overset{(3)'}{\epsilon_{\beta mp}\epsilon_{pl\alpha}\epsilon_{ilk}} + \delta_{\alpha\beta}\delta_{lm}\epsilon_{ilk}$$

$$\overset{(4)}{\delta_{\alpha l}\delta_{\beta i}\epsilon_{lmk}} = \overset{(4)'}{\epsilon_{\beta lp}\epsilon_{pi\alpha}\epsilon_{lmk}} + \delta_{\alpha\beta}\delta_{il}\epsilon_{lmk} \qquad \text{(4-27)}$$

A second application of Eq. (4-26) to the factor $\epsilon_{\beta lp}\epsilon_{ilk}$ in (1)′ gives

$$\epsilon_{\beta lp}\epsilon_{ilk} = \delta_{pk}\delta_{\beta i} - \delta_{pi}\delta_{\beta k} \qquad \text{(4-28)}$$

and a term

$$\delta_{\beta i}\delta_{pk}\epsilon_{pm\alpha} = \delta_{\beta i}\epsilon_{km\alpha}$$

is obtained, which is evidently the negative of term (4),

$$\delta_{\alpha l}\delta_{\beta i}\epsilon_{lmk} = \delta_{\beta i}\epsilon_{\alpha mk} = -\delta_{\beta i}\epsilon_{km\alpha}$$

In this fashion, (1)′ is decomposed to yield a term $-(4)$, (2)′ yields a term $-(3)$, (3)′ yields a term $-(2)$, and (4)′ yields a term $-(1)$. The sum of terms $(1)+(2)+(3)+(4)$ is

$$(1)+(2)+(3)+(4) = -\epsilon_{im\alpha}\delta_{\beta k} - \epsilon_{im\beta}\delta_{\alpha k} + 2\delta_{\alpha\beta}\epsilon_{imk} \qquad \text{(4-29)}$$

After the substitution of Eq. (4-29), and of ν for λ, Eq. (4-25) takes the neater form

$$\sigma_{\alpha\beta} = -\frac{\mu}{8\pi}\oint_C b_m\epsilon_{im\alpha}\frac{\partial}{\partial x_i'}\nabla'^2 R\,dx_\beta' - \frac{\mu}{8\pi}\oint_C b_m\epsilon_{im\beta}\frac{\partial}{\partial x_i'}\nabla'^2 R\,dx_\alpha'$$

$$-\frac{\mu}{4\pi(1-\nu)}\oint_C b_m\epsilon_{imk}\left(\frac{\partial^3 R}{\partial x_i'\partial x_\alpha'\partial x_\beta'} - \delta_{\alpha\beta}\frac{\partial}{\partial x_i'}\nabla'^2 R\right)dx_k' \qquad \text{(4-30)}$$

This equation was first derived by Peach and Koehler[10] (with a different sign because of their different definition of **b**). Equation (4-30) enables one to determine the stress field of an arbitrarily curved dislocation by line integration. Specific examples of such a procedure are given in Chap. 5. A more contracted form of Eq. (4-30) is given in dyadic notation

$$\sigma = \frac{\mu}{4\pi}\oint_C (\mathbf{b}\times\nabla')\frac{1}{R}\otimes d\mathbf{l}' + \frac{\mu}{4\pi}\oint_C d\mathbf{l}'\otimes(\mathbf{b}\times\nabla')\frac{1}{R} - \frac{\mu}{4\pi(1-\nu)}\oint_C \nabla'\cdot(\mathbf{b}\times d\mathbf{l}')(\nabla\otimes\nabla - \mathbf{I}\nabla^2)R \tag{4-31}$$

where **I** is the unit dyadic or idemfactor, given by $\mathbf{e}_1\otimes\mathbf{e}_1+\mathbf{e}_2\otimes\mathbf{e}_2+\mathbf{e}_3\otimes\mathbf{e}_3$.

Exercise 4-2. Derive the stress σ_{xx} of a straight pure edge dislocation in an infinite medium, $\mathbf{b}=(b_x,0,0)$, from Eq. (4-30). The result is given in Eq. (3-43). Notice that the derivatives in (4-30) are expressed in the $\mathbf{r}'$ coordinates of Fig. 4-4, and that $\mathbf{R}=\mathbf{r}'-\mathbf{r}$, $R=[(x'-x)^2+(y-y)^2+(z'-z)^2]^{1/2}$. *Hint*: Even though $x'=y'=0$ in this problem, in terms involving $\partial/\partial x'$ and $\partial/\partial y'$, the derivatives must be taken before setting $x'=y'=0$. Equation (4-30) formally applies to a closed loop of dislocation, yet here line integrals are taken only over the z' axis. Imagine that the dislocation is a portion of a closed loop with the portions other than that along z' infinitely far removed, and rationalize the fact that these portions do not contribute to the integral.

4-5. ENERGY OF INTERACTION BETWEEN TWO DISLOCATION LOOPS

If loop 1 is created while loop 2 is present, the stresses originating from loop 2 do work $-W_{12}$, where W_{12} is the interaction energy between the two loops (Fig. 4-6). Since, by Eq. (4-3), the work done on the surface of loop 1 is $W=-W_{12}$, and since, by Theorem 2-1, no energy flows in or out of loop 2, the work done on loop 1 represents a decrease in the strain energy of the total system. Therefore, if W_{12} is negative, the energy of the system decreases if loop 1 is created in the presence of loop 2, and an attractive force exists between the loops.

The interaction energy is, from Eq. (4-4),

$$W_{12} = \int_{A_1} dA_{1_\beta} b_{1_\alpha} \sigma_{2_{\alpha\beta}} \tag{4-32}$$

[10] M. O. Peach and J. S. Koehler, *Phys. Rev.*, **80**: 436 (1950).

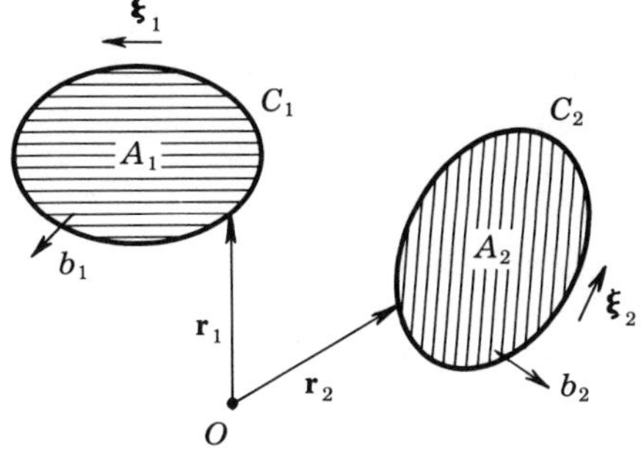

FIGURE 4-6. Two dislocation loops within the same elastic continuum.

If one inserts Eq. (4-30) into this formula, recalling that $\partial/\partial x_{1_i} = -(\partial/\partial x_{2_i})$, one finds

$$W_{12} = \frac{\mu}{8\pi}\int_{A_1}\oint_{C_2} dA_{1_\beta}\, dx_{2_\beta}\, b_{1_\alpha} b_{2_m} \epsilon_{im\alpha} \frac{\partial}{\partial x_{1_i}} \nabla^2 R \tag{4-33a}$$

$$+ \frac{\mu}{8\pi}\int_{A_1}\oint_{C_2} dA_{1_\beta}\, dx_{2_\alpha}\, b_{1_\alpha} b_{2_m} \epsilon_{im\beta} \frac{\partial}{\partial x_{1_i}} \nabla^2 R \tag{4-33b}$$

$$+ \frac{\mu}{4\pi(1-\nu)}\int_{A_1}\oint_{C_2} dA_{1_\beta}\, dx_{2_k}\, b_{1_\alpha} b_{2_m} \epsilon_{imk} \frac{\partial^3 R}{\partial x_{1_i}\,\partial x_{1_\alpha}\,\partial x_{1_\beta}} \tag{4-33c}$$

$$- \frac{\mu}{4\pi(1-\nu)}\int_{A_1}\oint_{C_2} dA_{1_\beta}\, dx_{2_k}\, b_{1_\alpha} b_{2_m} \epsilon_{imk} \delta_{\alpha\beta} \frac{\partial}{\partial x_{1_i}} \nabla^2 R \tag{4-33d}$$

By Stokes' theorem, Eq. (4-10), term (4-33a) becomes

$$(4\text{-}33a) = \frac{\mu}{8\pi}\int_{A_1}\oint_{C_2} dA_{1_i}\, dx_{2_\beta}\, b_{1_\alpha} b_{2_m} \epsilon_{im\alpha} \frac{\partial}{\partial x_{1_\beta}} \nabla^2 R$$

$$+ \frac{\mu}{8\pi}\oint_{C_1}\oint_{C_2} \epsilon_{\beta il}\, dx_{1_l}\, dx_{2_\beta}\, b_{1_\alpha} b_{2_m} \epsilon_{im\alpha} \nabla^2 R \tag{4-34}$$

The first term in (a) vanishes because

$$\oint_{C_2} dx_{2_\beta} \frac{\partial}{\partial x_{1_\beta}} \nabla^2 R = -\oint_{C_2} dx_{2_\beta} \frac{\partial}{\partial x_{2_\beta}} \nabla^2 R = -\oint_{C_2} d(\nabla^2 R) = 0$$

Thus (a) is given by the second term in Eq. (4-34), which is

$$(4\text{-}33a) = -\frac{\mu}{4\pi}\oint_{C_1}\oint_{C_2} \frac{(\mathbf{b}_1 \times \mathbf{b}_2)\cdot(d\mathbf{l}_1 \times d\mathbf{l}_2)}{R} \tag{4-35}$$

More simply for (b),

$$(4\text{-}33b)=\frac{\mu}{8\pi}\oint_{C_2}\mathbf{b}_1\cdot d\mathbf{l}_2\int_{A_1}(\text{grad}_1\nabla^2R\times\mathbf{b}_2)\cdot d\mathbf{A}_1$$

$$=\frac{\mu}{8\pi}\oint_{C_2}\mathbf{b}_1\cdot d\mathbf{l}_2\int_{A_1}\text{curl}_1\nabla^2R\,\mathbf{b}_2\cdot d\mathbf{A}_1$$

$$=\frac{\mu}{4\pi}\oint_{C_1}\oint_{C_2}\frac{(\mathbf{b}_1\cdot d\mathbf{l}_2)(\mathbf{b}_2\cdot d\mathbf{l}_1)}{R} \qquad (4\text{-}36)$$

The subscript 1 on grad and curl is inserted as a reminder that the derivatives are to be taken in terms of the $\mathbf{r}_1$ coordinates in Fig. (4-6).

For term (c), Stokes' theorem, in the form of Eq. (4-10), yields[11]

$$(4\text{-}33c)=\frac{\mu}{4\pi(1-\nu)}\int_{A_1}\oint_{C_2}dA_{1_\alpha}dx_{2_k}b_{1_\alpha}b_{2_m}\epsilon_{imk}\frac{\partial^3R}{\partial x_{1_i}\partial^2x_{1_\beta}}$$

$$+\frac{\mu}{4\pi(1-\nu)}\oint_{C_1}\oint_{C_2}dx_{1_l}dx_{2_k}b_{1_\alpha}b_{2_m}\epsilon_{imk}\epsilon_{\beta\alpha l}\frac{\partial^2R}{\partial x_i\partial x_\beta} \qquad (4\text{-}37)$$

The first term in (c) exactly cancels term (d). Thus

$$(4\text{-}33c)+(4\text{-}33d)=\frac{\mu}{4\pi(1-\nu)}\oint_{C_1}\oint_{C_2}(\mathbf{b}_1\times d\mathbf{l}_1)\cdot\mathbf{T}\cdot(\mathbf{b}_2\times d\mathbf{l}_2) \qquad (4\text{-}38)$$

where $\mathbf{T}$ is a tensor with components

$$T_{ij}=\frac{\partial^2R}{\partial x_i\partial x_j} \qquad (4\text{-}39)$$

Collection of terms and use of the relation

$$(\mathbf{b}\times\mathbf{b}_2)\cdot(d\mathbf{l}_1\times d\mathbf{l}_2)=(\mathbf{b}_1\cdot d\mathbf{l}_1)(\mathbf{b}_2\cdot d\mathbf{l}_2)-(\mathbf{b}_2\cdot d\mathbf{l}_1)(\mathbf{b}_1\cdot d\mathbf{l}_2)$$

to eliminate (4-36) now yields

$$W_{12}=-\frac{\mu}{2\pi}\oint_{C_1}\oint_{C_2}\frac{(\mathbf{b}_1\times\mathbf{b}_2)\cdot(d\mathbf{l}_1\times d\mathbf{l}_2)}{R}+\frac{\mu}{4\pi}\oint_{C_1}\oint_{C_2}\frac{(\mathbf{b}_1\cdot d\mathbf{l}_1)(\mathbf{b}_2\cdot d\mathbf{l}_2)}{R}$$

$$+\frac{\mu}{4\pi(1-\nu)}\oint_{C_1}\oint_{C_2}(\mathbf{b}_1\times d\mathbf{l}_1)\cdot\mathbf{T}\cdot(\mathbf{b}_2\times d\mathbf{l}_2) \qquad (4\text{-}40)$$

[11]We simply write $\partial^2R/\partial x_i\partial x_\beta$, since $\partial^2R/\partial x_{1_i}\partial x_{1_\beta}=\partial^2R/\partial x_{2_i}\partial x_{2_\beta}$. In symmetrical terms, where the subscript does not matter, it is left out.

The above formula was first obtained by Blin.[12] An alternative form[13] for the last term in Eq. (4-40) is

$$\frac{\mu}{4\pi(1-\nu)}\oint_{C_1}\oint_{C_2}\left\{-(\mathbf{b}_1\cdot\mathbf{T}\cdot\mathbf{b}_2)(d\mathbf{l}_1\cdot d\mathbf{l}_2)+\frac{2}{R}(\mathbf{b}_1\times d\mathbf{l}_1)\cdot(\mathbf{b}_2\times d\mathbf{l}_2)\right\}$$

The integrands are not identical but differ by terms which give no net contribution after integration over *complete* loops C_1 and C_2. They would differ for integrals over segments; Chap. 6. The form derived by Blin is always used in this book. Equation (4-40) has a much more extensive application than simply to two loops of dislocation; in Chap. 6, Eq. (4-40) is extended to yield the interaction energy between two arbitrarily positioned *segments* of dislocation line.

4-6. FORCE PRODUCED BY AN EXTERNAL STRESS ACTING ON A DISLOCATION LOOP

Let $\boldsymbol{\sigma}$ denote the stress tensor in the medium, excluding the self-stress of the dislocation loop under consideration. As the loop is created, the stress does work

$$W=\int_A -\mathbf{b}\cdot(\boldsymbol{\sigma}\cdot d\mathbf{A}) \tag{4-41}$$

If every line element $d\mathbf{l}$ of the loop is displaced by some distance, the area A changes by increments $\delta\mathbf{r}\times d\mathbf{l}$ and the stress $\boldsymbol{\sigma}$ does addition work

$$\delta W=\oint_C d\mathbf{F}\cdot\delta\mathbf{r}=-\oint_C \mathbf{b}\cdot[\boldsymbol{\sigma}\cdot(\delta\mathbf{r}\times d\mathbf{l})]$$

$$=-\oint_C(\mathbf{b}\cdot\boldsymbol{\sigma})\cdot(\delta\mathbf{r}\times d\mathbf{l})=\oint_C[(\mathbf{b}\cdot\boldsymbol{\sigma})\times d\mathbf{l}]\cdot\delta\mathbf{r} \tag{4-42}$$

Thus, since $\delta\mathbf{r}$ is arbitrarily variable along C,

$$d\mathbf{F}=(\mathbf{b}\cdot\boldsymbol{\sigma})\times d\mathbf{l} \tag{4-43}$$

Each element $d\mathbf{l}$ is acted upon by a force $d\mathbf{F}$, as given in Eq. (4-43), in agreement with Eq. (3-90). Equation (4-43) also was derived first by Peach and Koehler.[14] Together with Eq. (4-30), Eq. (4-43) can be used to determine the interaction force between dislocation segments, as demonstrated in Chap. 5.

[12] J. Blin, *Acta. Met.*, **3**: 199 (1955).

[13] R. Fuentes-Samaniego, research in progress.

[14] M. O. Peach and J. S. Koehler, *Phys. Rev.*, **80**: 436 (1950).

4-7. SELF-ENERGY OF A DISLOCATION LOOP

In the preceding section we considered the work done by stresses other than the self-stresses when a loop is created. In forming the dislocation, work must also be done against the self-stress of the loop. Each element $d\mathbf{l}$ of the loop is acted upon by a force caused by the stress originating from all other parts of the loop, and the work done against all these forces is the work done to supply the self-energy. Only when the total force on each element $d\mathbf{l}$ of the loop is zero is the loop in equilibrium, and only then does one find an extremum in the total energy of the system.

Formally, the self-energy W_s is obtained if in Eq. (4-40) one inserts $C_1 = C_2 = C$ and $\mathbf{b}_1 = \mathbf{b}_2 = \mathbf{b}$ and then divides by 2:

$$W_s = \frac{\mu}{8\pi} \oint_{C_1 = C} \oint_{C_2 = C} \frac{(\mathbf{b} \cdot d\mathbf{l}_1)(\mathbf{b} \cdot d\mathbf{l}_2)}{R}$$

$$+ \frac{\mu}{8\pi(1-\nu)} \oint_{C_1 = C} \oint_{C_2 = C} (\mathbf{b} \times d\mathbf{l}_1) \cdot \mathbf{T} \cdot (\mathbf{b} \times d\mathbf{l}_2) \qquad (4\text{-}44)$$

$$T_{ij} = \frac{\partial^2 R}{\partial x_i \partial x_j}$$

The factor of 2 can be justified by the following reasoning, which is generally useful in the consideration of dislocation interactions. Imagine that the loop is created in infinitesimal increments of the Burgers vector **b**; for each increment the self-energy increases by the interaction energy of a loop of fractional Burgers vector $\mathbf{b}f$ and a loop of vector $\mathbf{b}\,df$. The sum of all interactions leads to an average factor

$$\int_0^1 f\,df = \tfrac{1}{2}$$

with which Eq. (4-40) must be multiplied.

Alternatively, one can regard the self-energy as the interaction energy between all segments of the loop. Since the integrations in Eq. (4-44) count the interactions between two given elements twice, one again concludes that the result must be divided by 2.

As is expected from the discussion of the core region of straight dislocations, Eq. (4-44) diverges as the separation between elements, R, approaches zero. Cutoff procedures must be introduced to avoid the divergence, as discussed in detail in Chap. 6. This problem is characteristic of all types of interaction between two adjacent segments of the same dislocation line.

PROBLEMS

4-1. Consider the originally pure screw dislocation lying along ABC in Fig. 4-7. If AB is moved conservatively through positions $A'B'$, $A''B''$, etc., while BC is held fixed, a screw with a loop results. Is the loop a vacancy or an interstitial loop?

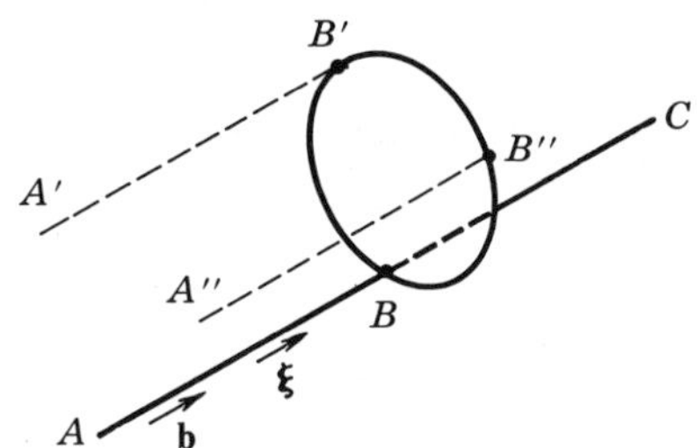

FIGURE 4-7. A screw dislocation along ABC. AB is moved conservatively in a clockwise manner viewed along $\boldsymbol{\xi}$ to form a closed loop $BB'B''B$.

4-2. The formation of the loop in Prob. 4-1 requires material transport. Where is the source or sink for this matter? *Hint*: Consider that the screw emerges normal to a free surface at A and study the configuration at A as the loop is formed.

4-3. Derive the stress field for a pure screw dislocation in an infinite medium from Eq. (4-30).

4-4. Demonstrate that the displacements $\mathbf{u}(\mathbf{r})$, given by Eq. (4-20), change discontinuously by $\Delta\mathbf{u} = \pm\mathbf{b}$ if the point $\mathbf{r}$ is intersected by the surface.

4-5. Show that the displacements given by Eq. (4-20) are dependent only on the configuration of the dislocation *line* forming the loop, and not on the shape of the surface A, provided the surface A is not intersected in the manner of Prob. 4-4.

BIBLIOGRAPHY

1. Eshelby, J. D., *Solid State Phys.*, **3**: 79 (1956).
2. Kröner, E., *Ergeb. angew. Math.*, **5** (1958).
3. de Wit, R., *Solid State Phys.*, **10**: 249 (1960).
4. Teodosiu, C., "Modele Elastice ale Defectelor Cristaline," Editura Acad. Rep. Soc. Romania, Bucarest, 1977.

5

Applications to Dislocation Interactions

5-1. INTRODUCTION

The preceding chapter deals in general terms with the interaction between closed dislocation loops. When the loops are composed of straight line segments, the total interaction splits into interactions between the individual segments, so that the integrations can be performed explicitly. Although an isolated dislocation segment has no physical meaning, formal expressions for segment-segment interactions enable us to determine by summation the total interaction with any piecewise straight dislocation array.

An extension of the force development leads to expressions for the stress at a point in a medium caused by a segment of dislocation. This result is related to several configurations widely studied in the literature, for example, the angular dislocation and the infinitesimal dislocation loop. The in-plane elastic fields of dislocation loops are expressed in terms of the prelogarithmic energy factor and its derivatives. The displacement field of a straight dislocation segment is presented. Finally, the results are used in a more qualitative discussion of general image forces associated with complex dislocation arrays.

5-2. INTERACTION ENERGY AND FORCE BETWEEN TWO PARALLEL STRAIGHT DISLOCATIONS

End Effects

Consider two dislocations parallel with the z axis, and with Burgers vectors $\mathbf{b}_1$ and $\mathbf{b}_2$. The positive z direction is chosen as the sense $\boldsymbol{\xi}$ of both dislocations. The coordinates of the first dislocation are (x_1, y_1) and those of the second one are (x_2, y_2). Equation (4-40) applies strictly to the interaction between two closed loops. Therefore the two dislocations are considered to be segments of closed loops, so that only the two segments are close enough to contribute appreciably to Eq. (4-40). In order that the other interaction terms, called *end*

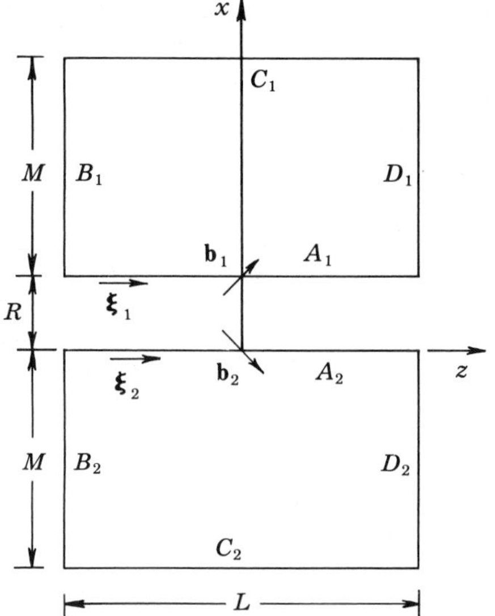

FIGURE 5-1. Two coplanar dislocation loops.

effects, have a negligible effect on the interaction energy as a function of R, the length of the segments L must be much larger than the separation R between the dislocations,

$$L \gg R \tag{5-1}$$

To demonstrate explicitly that the end effects produce at most small constant terms in the interaction energy, consider two simple cases of the interaction of two glide loops constrained to remain rectangular and coplanar, as shown in Fig. 5-1. In the first case let $\mathbf{b}_1 = \mathbf{b}_2 = (0, 0, b_z)$; that is, suppose that the two segments of principal interest are pure screw. For the interactions between A_1 and A_2, A_1 and C_2, A_1 and C_1, A_2 and C_1, and A_2 and C_2, only the second term in Eq. (4-40) is nonzero. For interactions of the type D-B, B-B, and D-D, only the third term is nonzero. All other interactions have zero interaction energies. The interaction energy per unit length between A_1 and A_2 is

$$\frac{W(A_1\text{-}A_2)}{L} = \frac{\mu b_z^2}{4\pi L}\left\{2\left[R-(L^2+R^2)^{1/2}\right] + L\ln\frac{L+(L^2+R^2)^{1/2}}{-L+(L^2+R^2)^{1/2}}\right\} \tag{5-2}$$

In the limit of $L \gg R$, the first term in Eq. (5-2), representing the end effect, is

$$\left[\frac{W(A_1\text{-}A_2)}{L}\right]_{\text{end}} \cong \frac{-\mu b_z^2}{2\pi}\left(1-\frac{R}{L}\right) \cong \frac{-\mu b_z^2}{2\pi} \tag{5-3}$$

The interaction energy per unit length between A_2 and C_1 is

$$\frac{W(A_2\text{-}C_1)}{L}=\frac{-\mu b_z^2}{4\pi L}\left(2\left\{M+R-\left[(M+R)^2+L^2\right]^{1/2}\right\}\right.$$

$$\left.+L\ln\frac{L+\left[L^2+(M+R)^2\right]^{1/2}}{-L+\left[L^2+(M+R)^2\right]^{1/2}}\right) \tag{5-4}$$

In the limit of $M\gg L\gg R$ this becomes

$$\frac{W(A_2\text{-}C_1)}{L}\cong\frac{-\mu b_z^2}{4\pi}\left(-\frac{L}{M}+\ln\frac{1+L/M}{1-L/M}\right)\cong 0$$

while in the limit of $L\gg M\gg R$ it becomes

$$\frac{W(A_2\text{-}C_1)}{L}\cong\frac{\mu b_z^2}{2\pi}\left[\left(1-\frac{M}{L}\right)-\ln\frac{2L}{M}\right]$$

$$\cong\frac{\mu b_z^2}{2\pi}\left(1-\ln\frac{2L}{M}\right) \tag{5-5}$$

The other type of end-effect term arises when $\mathbf{b}_1=(0,0,b_{1_z})$, $\mathbf{b}_2=(b_{2_x},0,0)$. The nonzero interaction terms now act between A and D, A and B, C and B, and C and D. Consider A_2-D_1, for which both the first and second term in Eq. (4-40) contribute to W. The interaction energy per unit length of A_2 is

$$\frac{W(A_2\text{-}D_1)}{L}=-\frac{\mu b_{1_z}b_{2_x}}{2\pi L}\left((M+R)\ln\left\{\frac{1+\left[1+4(M+R)^2/L^2\right]^{1/2}}{2(M+R)/L}\right\}\right.$$

$$\left.-R\ln\left\{\frac{1+\left[1+(4R^2/L^2)\right]^{1/2}}{2R/L}\right\}\right) \tag{5-6}$$

In the limit of $L\gg M\gg R$ this becomes

$$\frac{W(A_2\text{-}D_1)}{L}\cong-\frac{\mu b_{1_z}b_{2_x}}{2\pi}\left(\frac{M}{L}\ln\frac{L}{M}-\frac{R}{L}\ln\frac{M}{R}\right)\cong 0 \tag{5-7}$$

upon the application of l'Hospital's rule. In the limit of $M\gg L\gg R$, Eq. (5-6) becomes

$$\frac{W(A_2\text{-}D_1)}{L} \cong -\frac{\mu b_{1_z} b_{2_x}}{2\pi}\left[\frac{M}{L}\ln\left(1+\frac{L}{2M}\right)-\frac{R}{L}\ln\frac{L}{R}\right]$$

$$\cong -\frac{\mu b_{1_z} b_{2_x}}{4\pi} \tag{5-8}$$

The other terms similarly give values either about equal to Eq. (5-8) or equal to zero, and some of the nonzero terms cancel when summed to give the total energy. Thus the end effects contribute constant terms of the order of magnitude of the core energy to the total interaction energy.

Now consider the *variation* of the separation R. The only term in the interaction energy that varies with R, in the limit of $L \gg R$, is the second term in Eq. (5-2), which becomes

$$\frac{W(A_1\text{-}A_2)}{L} = \frac{\mu b_z^2}{2\pi}\ln\frac{2L}{R} \tag{5-9}$$

One can now express the interaction energy between dislocations 1 and 2, as a function of the variable R, as the increase in the energy of the system when the dislocations are brought together from some value R_a to a separation R,

$$\frac{W_{12}}{L} = W_{(A_1\text{-}A_2)}(R) - W_{(A_1\text{-}A_2)}(R_a) = -\frac{\mu b_z^2}{2\pi}\ln\frac{R}{R_a} \tag{5-10}$$

It is readily verified that the force obtained by the negative of the variation of W_{12}/L with respect to R agrees with the force per unit length exerted on one dislocation by the other as obtained from Eqs. (3-4) and (3-90).

Thus both the interaction energy and the interaction force between A_1 and A_2 are insensitive to the end effects to the first order in a power-series expansion in terms of R/L, provided Eq. (5-1) holds. We have treated this case in some detail because confusion has frequently arisen about the uncertainties introduced by the neglect of end effects. The above principle can be extended to any general dislocation in that the change in interaction energy produced by a local variation in dislocation configuration does not contain contributions from segments far removed from the local configuration.

General Parallel Dislocations

Because the end effects are negligible, we can proceed to derive an expression for the interaction energy between parallel dislocations with arbitrary Burgers vectors, dropping end-effect terms throughout. Consider two dislocations with

$\boldsymbol{\xi}$ parallel to the z axis. From Eq. (4-40), it follows that

$$W_{12}=\frac{\mu}{4\pi}(\mathbf{b}_1\cdot\boldsymbol{\xi})(\mathbf{b}_2\cdot\boldsymbol{\xi})\int_{-L/2}^{L/2}dz_2\int_{-L/2}^{L/2}\left[R^2+(z_1-z_2)^2\right]^{-1/2}dz_1$$

$$+\frac{\mu}{4\pi(1-\nu)}(\mathbf{b}_1\times\boldsymbol{\xi})\cdot\mathbf{T}\cdot(\mathbf{b}_2\times\boldsymbol{\xi}) \qquad (5\text{-}11)$$

where $\mathbf{T}$ is the tensor with components

$$T_{ij}=\frac{\partial^2}{\partial X_i\partial X_j}\int_{-L/2}^{L/2}dz_2\int_{-L/2}^{L/2}\left[R^2+(z_1-z_2)^2\right]^{1/2}dz_1$$

$$X_1=x_2-x_1 \qquad (5\text{-}12)$$

$$X_2=y_2-y_1$$

$\mathbf{b}_1\cdot\boldsymbol{\xi}$ and $\mathbf{b}_2\cdot\boldsymbol{\xi}$ are the screw components. In the last term of Eq. (5-11), obviously only the edge components contribute. Integration of Eq. (5-11) yields

$$\int_{-L/2}^{L/2}\left[R^2+(z_1-z_2)^2\right]^{-1/2}dz_1=\ln\left\{\frac{L/2-z_2+\left[R^2+(L/2-z_2)^2\right]^{1/2}}{-L/2-z_2+\left[R^2+(L/2+z_2)^2\right]^{1/2}}\right\}$$

$$(5\text{-}13)$$

In the numerator in the logarithm, the dependence on R is an end effect; over the greater range of z_2 one can put

$$\frac{L}{2}-z_2+\left[R^2+\left(\frac{L}{2}-z_2\right)^2\right]^{1/2}\cong L-2z_2$$

Similarly, in the denominator, the two terms nearly cancel,

$$-\frac{L}{2}-z_2+\left[R^2+\left(\frac{L}{2}+z_2\right)^2\right]^{1/2}\cong\frac{R^2}{L+2z_2}$$

In this approximation, the only term in Eq. (5-13) that depends on R when end effects are ignored is $-\ln R^2=-2\ln R$. Then $-2\ln(R/R_a)$ is the only remaining part in Eq. (5-13); here R_a [see Eq. (5-10)] has been inserted to make the argument dimensionless without affecting the derivatives of the expression.

Similarly, in the integral

$$\int_{-L/2}^{L/2}\left[R^2+(z_1-z_2)^2\right]^{1/2}dz_1$$

the only term other than end effects can be written

$$-R^2 \ln \frac{R}{R_a}$$

Terms simply proportional with R^2 have been neglected, since, by Eq. (5-12), they only give a constant difference in the interaction energy. Since $R^2 = X_1^2 + X_2^2$, it follows that

$$\frac{\partial^2 R^2}{\partial X_i \partial X_j} = 2\delta_{ij} \tag{5-14}$$

Hence the tensor components T_{ij} are

$$T_{ij} = L \frac{\partial^2}{\partial X_i \partial X_j}\left(-R^2 \ln \frac{R}{R_a}\right) = -2L \ln \frac{R}{R_a}\delta_{ij} - 2L\frac{X_i X_j}{R^2} - L\delta_{ij} \tag{5-15}$$

and again the last term on the right-hand side can be ignored.

Insertion of all important terms into Eq. (5-11) yields for the energy of interaction per unit length

$$\frac{W_{12}}{L} = -\frac{\mu(\mathbf{b}_1 \cdot \boldsymbol{\xi})(\mathbf{b}_2 \cdot \boldsymbol{\xi})}{2\pi} \ln \frac{R}{R_a} - \frac{\mu}{2\pi(1-\nu)}[(\mathbf{b}_1 \times \boldsymbol{\xi}) \cdot (\mathbf{b}_2 \times \boldsymbol{\xi})] \ln \frac{R}{R_a}$$
$$- \frac{\mu}{2\pi(1-\nu)R^2}[(\mathbf{b}_1 \times \boldsymbol{\xi}) \cdot \mathbf{R}][(\mathbf{b}_2 \times \boldsymbol{\xi}) \cdot \mathbf{R}] \tag{5-16}$$

This equation was first developed by Nabarro[1] and is very useful in the discussion of reactions between parallel dislocations. Note that if polar coordinates are used with the first dislocation as the origin, and the second dislocation at (R, θ), then the first two terms in Eq. (5-16) represent the variation of W_{12}/L with R, while the last term, independent of R, represents the variation of W_{12}/L with θ.

The radial component of the interaction force per unit length between two parallel dislocations is obtained by differentiation to be

$$\frac{F_R}{L} = -\frac{\partial(W_{12}/L)}{\partial R}$$
$$= \frac{\mu}{2\pi R}(\mathbf{b}_1 \cdot \boldsymbol{\xi})(\mathbf{b}_2 \cdot \boldsymbol{\xi}) + \frac{\mu}{2\pi(1-\nu)R}[(\mathbf{b}_1 \times \boldsymbol{\xi}) \cdot (\mathbf{b}_2 \times \boldsymbol{\xi})] \tag{5-17}$$

[1] F. R. N. Nabarro, *Adv. Phys.*, **1**: 269 (1952).

The θ component of the interaction force is

$$\frac{F_\theta}{L} = -\frac{1}{R}\frac{\partial(W_{12}/L)}{\partial\theta} = \frac{\mu}{2\pi(1-\nu)R^3}\{(\mathbf{b}_1\cdot\mathbf{R})[(\mathbf{b}_2\times\mathbf{R})\cdot\boldsymbol{\xi}] + (\mathbf{b}_2\cdot\mathbf{R})[(\mathbf{b}_1\times\mathbf{R})\cdot\boldsymbol{\xi}]\} \tag{5-18}$$

Exercise 5-1. Consider two parallel edge dislocations $\mathbf{b}_1 = \mathbf{b}_2 = \mathbf{b}$ and $\boldsymbol{\xi}_1 = \boldsymbol{\xi}_2 = \boldsymbol{\xi}$. Derive the force per unit length exerted on one dislocation by the other one as a function of their relative position, by variation of W_{12}/L [Eq. (5-16)]. Show that the result agrees with that obtained from Eqs. (3-44) and (3-90).

Exercise 5-2. Integrate explicitly and completely the integrals in Eqs. (5-11) and (5-12), and demonstrate, by series development, that as R/L approaches zero, Eq. (5-16) results within an insignificant constant.

5-3. ENERGY OF INTERACTION BETWEEN TWO COAXIAL CIRCULAR LOOPS

Consider two coaxial loops of the same radius a, with the same Burgers vector $\mathbf{b}$, parallel with the axis, and with the same sense $\boldsymbol{\xi} = d\mathbf{l}_1/dl_1 = d\mathbf{l}_2/dl_2$, as shown in Fig. 5-2. Only the last term in Eq. (4-40) contributes to the interaction energy. Derivatives of R involving z will not contribute, since the vectors $\mathbf{b}_1\times d\mathbf{l}_1$ and $\mathbf{b}_2\times d\mathbf{l}_2$ are perpendicular to the z axis. Because of symmetry, if one first integrates over C_2, all elements $d\mathbf{l}_1$ give the same contribution in the second integration. Thus only the situation where $d\mathbf{l}_1$ is parallel to the y axis, as depicted in Fig. 5-2, need be considered, in which case only the tensor components

$$\frac{\partial^2 R}{\partial x^2} = \frac{1}{R} - \frac{(a-x)^2}{R^3} = \frac{1}{R} - \frac{a^2(1-\cos\phi)^2}{R^3}$$

$$\frac{\partial^2 R}{\partial x\,\partial y} = \cdots = \frac{a^2(1-\cos\phi)\sin\phi}{R^3} \tag{5-19}$$

contribute to Eq. (4-40). The interaction energy is found by integration to be

$$W_{12} = \frac{\mu b^2}{4\pi(1-\nu)}2\pi a^2\int_0^{2\pi}\left[\frac{1}{R}\cos\phi - \frac{a^2(1-\cos\phi)^2}{R^3}\cos\phi + \frac{a^2(1-\cos\phi)\sin\phi}{R^3}\sin\phi\right]d\phi \tag{5-20}$$

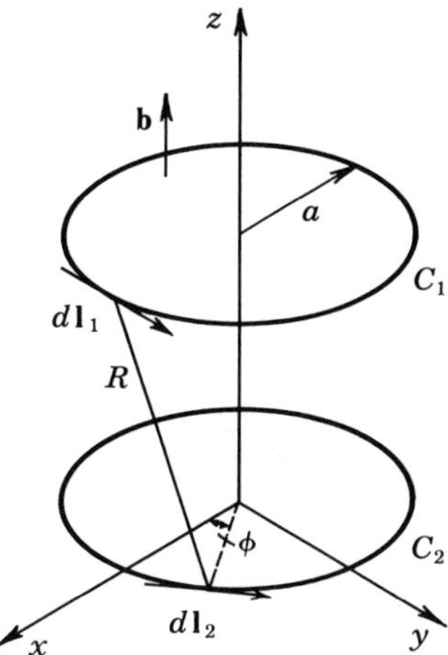

FIGURE 5-2. Two coaxial, circular dislocation loops.

R is expressed in terms of z and ϕ as

$$R=\left(z^2+4a^2\sin^2\frac{\phi}{2}\right)^{1/2}$$

With the symbols

$$k^2=\frac{4a^2}{z^2+4a^2} \qquad k'^2=\frac{z^2}{z^2+4a^2} \qquad k^2+k'^2=1 \qquad (z^2+4a^2)^{1/2}=\frac{2a}{k} \tag{5-21}$$

and with the substitution

$$\eta=\frac{\pi}{2}-\frac{\phi}{2} \tag{5-22}$$

Eq. (5-20) can be transformed to

$$W_{12}=\frac{\mu b^2 ak}{1-\nu}\int_0^{\pi/2}\left[\frac{k^2-1}{(1-k^2\sin^2\eta)^{1/2}}+\frac{\sin^2\eta}{(1-k^2\sin^2\eta)^{1/2}}\right.$$
$$\left.+\frac{(1-2k^2+k^4)\sin^2\eta}{(1-k^2\sin^2\eta)^{3/2}}\right]d\eta \tag{5-23}$$

The integrals occurring in Eq. (5-23) are elliptical.

Some of the definitions and properties of these functions, as listed in the text by Whittaker and Watson,[2] are

$$E=\int_0^{\pi/2}(1-k^2\sin^2\eta)^{1/2}\,d\eta$$

$$K=\int_0^{\pi/2}(1-k^2\sin^2\eta)^{-1/2}\,d\eta$$

$$\frac{dE}{dk}=-\int_0^{\pi/2}k\sin^2\eta(1-k^2\sin^2\eta)^{-1/2}\,d\eta=\frac{E-K}{k}$$

$$\frac{dK}{dk}=\int_0^{\pi/2}k\sin^2\eta(1-k^2\sin^2\eta)^{-3/2}\,d\eta=\frac{E}{kk'^2}-\frac{K}{k} \tag{5-24}$$

$$\frac{K-E}{k^2}\to\tfrac{1}{4}\pi \qquad \text{when } k\to 0$$

$$K\to E\to\frac{\pi}{2} \qquad \text{when } k\to 0$$

$$K\to\tfrac{1}{2}\ln\frac{16}{1-k^2} \qquad \text{when } k\to 1$$

$$E\to 1 \qquad \text{when } k\to 1$$

K and E are the complete elliptic integrals of the first and second kind, respectively. In terms of these functions, Eq. (5-23) becomes[3,4]

$$W_{12}=\frac{\mu b^2 ak}{1-\nu}(K-E) \tag{5-25}$$

When $z\ll a$ (that is, $k^2\sim 1$), the asymptotic formula

$$W_{12}\simeq\frac{\mu b^2 a}{1-\nu}\left(\ln\frac{8a}{z}-1\right) \tag{5-26}$$

can be used, and when $z\gg a$ (that is, $k\ll 1$), then

$$W_{12}\simeq\frac{2\pi\mu b^2 a}{1-\nu}\frac{a^3}{z^3} \tag{5-27}$$

[2]E. T. Whittaker and G. N. Watson, "A Course of Modern Analysis," Cambridge University Press, New York, 1952, p. 521.

[3]E. Kröner, *Ergeb. angew. Math.*, **5**: 163 (1958).

[4]R. de Wit, *Solid State Phys.*, **10**: 249 (1960).

The self-energy of a circular dislocation loop is obtained by setting $z = z_0 \ll a$ and dividing Eq. (5-26) by 2 (since, as discussed in Sec. 4-7, each dl has been counted twice),

$$W_s = \frac{\mu b^2 a}{2(1-\nu)}\left(\ln\frac{8a}{z_0} - 1\right) \tag{5-28}$$

where W_s becomes negative in the limit that a tends to $z_0/8$. However, this is outside the range of validity $z \ll a$ for the basis formula (5-26). In addition, continuum theory is not meaningful for configurations that extend over dimensions of the order of the core size.

Applications of the above results occur in treating prismatic dislocation loops punched out by precipitate particles. The theory of this section is also important in the theory of spiral dislocations, which occur in quenched materials.

5-4. INTERACTION ENERGY AND FORCE BETWEEN TWO NONPARALLEL STRAIGHT DISLOCATIONS

Consider the two dislocations depicted in Fig. 5-3. The origin O of the coordinate system is a fixed point on dislocation (1). Dislocation (1) is contained in the xz plane. The y axis is perpendicular out of the paper. The position of dislocation (2) relative to dislocation (1) is given by the coordinates y and θ. A point on dislocation (2) has the relative cartesian coordinates $(x_2 - x_1, y, z_2 - z_1)$ with respect to a point on dislocation (1). The treatment is restricted to the consideration of only parallel displacements of the two dislocations, so that only changes in y can cause energy changes. Formally, the energy of interaction is infinite, but the derivative with respect to y is finite. The following development is best understood if one pictures two very long straight dislocation segments which are parts of two different loops, with only the straight segments close enough to interact strongly.

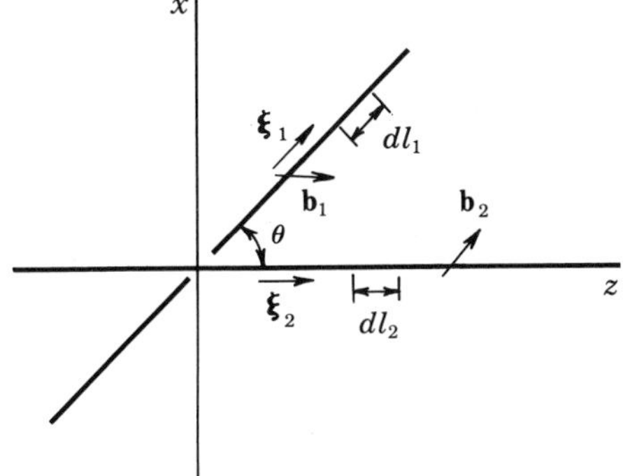

FIGURE 5-3. Two nonparallel, noncoplanar, straight dislocations. The unit vector **j** points out of the plane of the page.

The total force F_{12_y} exerted by dislocation (1) on dislocation (2) is given by

$$F_{12_y} = -\frac{\partial W_{12}}{\partial y} \tag{5-29}$$

From Eq. (4-40) it follows that

$$\begin{aligned}F_{12_y} = & \frac{\mu}{2\pi}(\mathbf{b}_1\times\mathbf{b}_2)\cdot(\xi_1\times\xi_2)\int_{-\infty}^{\infty}dl_1\int_{-\infty}^{\infty}\frac{\partial}{\partial y}\frac{1}{R}dl_2\\ & -\frac{\mu}{4\pi}(\mathbf{b}_1\cdot\xi_1)(\mathbf{b}_2\cdot\xi_2)\int_{-\infty}^{\infty}dl_1\int_{-\infty}^{\infty}\frac{\partial}{\partial y}\frac{1}{R}dl_2\\ & -\frac{\mu}{4\pi(1-\nu)}[(\mathbf{b}_1\times\xi_1)\cdot\mathbf{j}][(\mathbf{b}_2\times\xi_2)\cdot\mathbf{j}]\int_{-\infty}^{\infty}dl_1\int_{-\infty}^{\infty}\frac{\partial^3}{\partial y^3}R\,dl_2\end{aligned} \tag{5-30}$$

In the last term of the above equation, all terms in the tensor T_{ij} of Eq. (4-40) involving $\partial/\partial x_1$ and $\partial/\partial z_1$ have been dropped, since the function R depends only on the relative coordinates x_2-x_1, etc., so that

$$\begin{aligned}\frac{\partial}{\partial x_1} &= \frac{1}{\sin\theta}\frac{\partial}{\partial l_1}+\frac{1}{\tan\theta}\frac{\partial}{\partial l_2}\\ \frac{\partial}{\partial z_1} &= -\frac{\partial}{\partial l_2}\end{aligned} \tag{5-31}$$

and integrations taken over l_1 and l_2 to give zero can immediately be effected.[5]

[5]Equation (5-30) can be verified directly by performing partial differentiations of R and substituting into Eq. (4-40). As an example of the integrations, consider the term

$$\int_{-\infty}^{\infty}dl_1\int_{-\infty}^{\infty}\frac{\partial}{\partial y}\frac{\partial^2}{\partial x_1{}^2}R\,dl_2 = y\int_{-\infty}^{\infty}dl_1\int_{-\infty}^{\infty}\frac{\partial^2}{\partial x_1{}^2}\frac{1}{R}dl_2$$

By Eq. (5-31), the latter integral becomes

$$\begin{aligned}\int_{-\infty}^{\infty}dl_1\int_{-\infty}^{\infty}\frac{\partial^2}{\partial x_1{}^2}\frac{1}{R}dl_2 &= \int_{-\infty}^{\infty}dl_1\int_{-\infty}^{\infty}\left(\frac{1}{\sin\theta}\frac{\partial}{\partial l_1}+\frac{1}{\tan\theta}\frac{\partial}{\partial l_2}\right)\frac{\partial}{\partial x_1}\frac{1}{R}dl_2\\ &= \int_{-\infty}^{\infty}dl_2\int_{-\infty}^{\infty}\frac{\partial}{\partial l_1}\left(\frac{1}{\sin\theta}\frac{\partial}{\partial x_1}\frac{1}{R}\right)dl_1\\ &\quad+\int_{-\infty}^{\infty}dl_1\int_{-\infty}^{\infty}\frac{\partial}{\partial l_2}\left(\frac{1}{\tan\theta}\frac{\partial}{\partial x_1}\frac{1}{R}\right)dl_2\end{aligned}$$

since the order of integration is immaterial. Now

$$\int_{-\infty}^{\infty}\frac{\partial}{\partial l_1}\left(\frac{1}{\sin\theta}\frac{\partial}{\partial x_1}\frac{1}{R}\right)dl_1 = \left[\frac{1}{\sin\theta}\frac{\partial}{\partial x_1}\frac{1}{R}\right]_{-\infty}^{\infty} = 0$$

and similarly, the other term integrates to zero.

The distance R is determined from

$$R^2 = l_1^2 + l_2^2 - 2l_1 l_2 \cos\theta + y^2 \tag{5-32}$$

Computation of the two integrals yields

$$\int_{-\infty}^{\infty} dl_1 \int_{-\infty}^{\infty} \frac{\partial}{\partial y}\frac{1}{R}\, dl_2 = -\frac{2\pi}{|\sin\theta|}$$

$$\int_{-\infty}^{\infty} dl_1 \int_{-\infty}^{\infty} \frac{\partial^3}{\partial y^3} R\, dl_2 = -\frac{4\pi}{|\sin\theta|} \tag{5-33}$$

Since ξ_1 and ξ_2 are unit vectors, it follows that

$$\frac{1}{|\sin\theta|} = \frac{1}{|\xi_1 \times \xi_2|} \tag{5-34}$$

With the use of Eqs. (5-33) and (5-34), the total force from dislocation (1) on dislocation (2), $\mathbf{F}_{12}$, can be expressed in the general form

$$\mathbf{F}_{12} = \frac{\mu}{|\xi_1 \times \xi_2|}\frac{\mathbf{R}_{12}}{R_{12}}\left\{\tfrac{1}{2}(\mathbf{b}_1 \cdot \xi_1)(\mathbf{b}_2 \cdot \xi_2) - (\mathbf{b}_1 \times \mathbf{b}_2)\cdot(\xi_1 \times \xi_2)\right.$$
$$\left. + \frac{1}{1-\nu}\left[(\mathbf{b}_1 \times \xi_1)\cdot\frac{\mathbf{R}_{12}}{R_{12}}\right]\left[(\mathbf{b}_2 \times \xi_2)\cdot\frac{\mathbf{R}_{12}}{R_{12}}\right]\right\} \tag{5-35}$$

where $\mathbf{R}_{12}$ is the shortest vector from dislocation (1) to dislocation (2).[6]

Note that for constant relative orientation, and with the constraint that the dislocations remain straight, $\mathbf{F}_{12}$ is constant; it does not depend on the separation between the dislocations. As the dislocations approach one another, the stresses increase in the region of close approach, but they act over smaller distances, leaving the magnitude of the total force unchanged. This point is discussed further in the following section.

The theory of the present section is important in the discussion of dislocation intersections occurring in large-scale plastic deformation. Some applications are presented in Chap. 22.

Exercise 5-3. Show that two pure screw dislocations making an angle of 45° with each other do not repel or attract each other.

[6] This result was first obtained by F. Kroupa, *Czech. J. Phys.*, **B11**: 847 (1961) and was extended to the anisotropic elastic case in a simple and elegant way by S. S. Orlov and V. L. Indenbom, *Kristallografiya*, **14**: 780 (1969).

Exercise 5-4. Discuss in more detail the assumptions underlying the development in this section. It was assumed sufficient to consider the two segments of closest approach, but the final result shows that the total force between two infinite dislocations is independent of their separation, indicating that the initial assumption may be invalid. Resolve this apparent discrepancy.

5-5. INTERACTION FORCE BETWEEN A STRAIGHT SEGMENT OF DISLOCATION AND A DIFFERENTIAL ELEMENT OF ANOTHER DISLOCATION

There are three possible approaches to consideration of the interaction force between dislocation segments. One involves calculation of the interaction energy and its differentiation with respect to the coordinate in question. A second involves the determination of the stress field of one dislocation, transformation of it to coordinates pertinent to the other dislocation, and the use of Eq. (3-90) or (4-43). Indeed, for many simple configurations, such as an infinite straight dislocation interacting with a straight segment of another dislocation, either of these methods can be applied straightforwardly, and in general they are simpler than the third method. However, for a complicated configuration, it is easier to derive the interaction force directly by integration, chiefly because all stress components are not required in general.

Suppose the dislocation arrangement indicated in Fig. 5-4, where the oblique coordinate axes x and y lie along dislocations (1) and (2) and z lies along the line of closest approach of the two dislocations. These axes are defined by the three unit vectors $\boldsymbol{\xi}_1$, $\boldsymbol{\xi}_2$, and

$$\mathbf{e}_3 = (1/\sin\theta)(\boldsymbol{\xi}_1 \times \boldsymbol{\xi}_2).$$

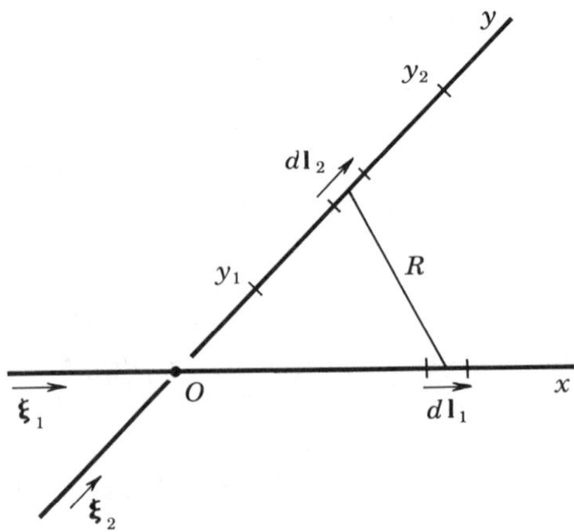

FIGURE 5-4. Two straight dislocation segments.

We wish to find the force on an element $d\mathbf{l}_1$ caused by the segment y_1y_2. The force has two components, one in the $\mathbf{e}_3$ direction and one in the direction $\mathbf{e}_3 \times \boldsymbol{\xi}_1$.

In a real crystal, of course, a dislocation line cannot end within a perfect crystal region, so that the force caused by a single segment has little meaning. For generally curved dislocation lines, however, one can obtain an excellent approximation of the interaction force with a dislocation element by approximating the line by a sequence of straight segments and then determining the interaction force between each segment and the differential element in question. This procedure is possible only because the stress field [Eq. (4-30)] of a segmented dislocation line transforms like a tensor *for each segment independently*. Thus the contribution to the force from each individual segment is independent of the orientation of the coordinate system used and hence is unique. Upon integrating explicitly[7] along the segment y_1y_2, the force on $d\mathbf{l}_1$ is given as

$$\delta\mathbf{F} = \delta\mathbf{F}(y_2, x_1) - \delta\mathbf{F}(y_1, x_1) \tag{5-36}$$

Following is a calculation of $\delta\mathbf{F}(y,x)$. Integration constants, i.e., terms not depending on y, are not important in $\delta\mathbf{F}$.

Combining the Peach-Koehler equation (4-30) with Eq. (3-90) in the form

$$\delta\mathbf{F} = b_{1_\alpha}\sigma_{\alpha\beta}(\mathbf{e}_\beta \times d\mathbf{l}_1) \tag{5-37}$$

one obtains the formula

$$\delta\mathbf{F} = -\frac{\mu}{8\pi}\int[(\mathbf{b}_2\times\mathbf{b}_1)\cdot\nabla(\nabla^2 R)](d\mathbf{l}_1\times d\mathbf{l}_2) \tag{5-38a}$$

$$-\frac{\mu}{8\pi}\int\{[\mathbf{b}_2\times\nabla(\nabla^2 R)]\times d\mathbf{l}_1\}(\mathbf{b}_1\cdot d\mathbf{l}_2) \tag{5-38b}$$

$$-\frac{\mu}{4\pi(1-\nu)}\int[(\mathbf{b}_2\times d\mathbf{l}_2)\cdot\nabla](d\mathbf{l}_1\times\mathbf{b}_1\mathbf{T}) \tag{5-38c}$$

$$+\frac{\mu}{4\pi(1-\nu)}\int[(\mathbf{b}_2\times d\mathbf{l}_2)\cdot\nabla(\nabla^2 R)](d\mathbf{l}_1\times\mathbf{b}_1) \tag{5-38d}$$

where $\mathbf{T}$ is the tensor[8]

$$\mathbf{T} = \frac{\partial^2 R}{\partial x_\alpha \partial x_\beta}\mathbf{e}_\alpha\otimes\mathbf{e}_\beta \tag{5-39}$$

[7]The possibility of such integrations for straight-segment dislocation configurations is not limited to isotropic theory. The theory developed by Lothe, Brown, Indenbom and Orlov, developed in Sec. 5-7, can be applied in a similar fashion to polygons in anisotropic media.

[8]The reader is reminded that the coordinates x_α and x_β are those of dislocation (1) with respect to dislocation (2); see Fig. 4-4 and the discussion in Exercise 4.2.

and in terms of the dislocation coordinates the gradient operator is given by

$$\nabla = \boldsymbol{\xi}_1\left(\frac{1}{\sin^2\theta}\frac{\partial}{\partial x} + \frac{\cos\theta}{\sin^2\theta}\frac{\partial}{\partial y}\right) + \boldsymbol{\xi}_2\left(-\frac{\cos\theta}{\sin^2\theta}\frac{\partial}{\partial x} - \frac{1}{\sin^2\theta}\frac{\partial}{\partial y}\right) + \mathbf{e}_3\frac{\partial}{\partial z} \tag{5-40}$$

For the tensor $\mathbf{T}$ we need the components[9]

$$\begin{aligned}\mathbf{T} = \Bigg[&\boldsymbol{\xi}_2\otimes\boldsymbol{\xi}_2\left(\frac{\cos^2\theta}{\sin^4\theta}\frac{\partial^2}{\partial x^2} + 2\frac{\cos\theta}{\sin^4\theta}\frac{\partial^2}{\partial x\,\partial y} + \frac{1}{\sin^4\theta}\frac{\partial^2}{\partial y^2}\right)\\ &+\boldsymbol{\xi}_1\otimes\boldsymbol{\xi}_2\left(-\frac{\cos\theta}{\sin^4\theta}\frac{\partial^2}{\partial x^2} - \frac{1+\cos^2\theta}{\sin^4\theta}\frac{\partial^2}{\partial x\,\partial y} - \frac{\cos\theta}{\sin^4\theta}\frac{\partial^2}{\partial y^2}\right)\\ &+(\mathbf{e}_3\otimes\boldsymbol{\xi}_2 + \boldsymbol{\xi}_2\otimes\mathbf{e}_2)\left(-\frac{\cos\theta}{\sin^2\theta}\frac{\partial^2}{\partial x\,\partial z} - \frac{1}{\sin^2\theta}\frac{\partial^2}{\partial y\,\partial z}\right)\\ &+\boldsymbol{\xi}_1\otimes\mathbf{e}_3\left(\frac{1}{\sin^2\theta}\frac{\partial^2}{\partial x\,\partial z} + \frac{\cos\theta}{\sin^2\theta}\frac{\partial^2}{\partial y\,\partial z}\right) + \mathbf{e}_3\otimes\mathbf{e}_3\frac{\partial^2}{\partial z^2}\Bigg]R\end{aligned} \tag{5-41}$$

The Burgers vectors are decomposed by the following scheme:

$$\begin{aligned} &\mathbf{b}_1\cdot\boldsymbol{\xi}_1 = b_{1s} &\qquad &\mathbf{b}_2\cdot\boldsymbol{\xi}_2 = b_{2s}\\ &\mathbf{b}_1\cdot(\mathbf{e}_3\times\boldsymbol{\xi}_1) = b_{1e} &\qquad &\mathbf{b}_2\cdot(\mathbf{e}_3\times\boldsymbol{\xi}_2) = b_{2e}\\ &\mathbf{b}_1\cdot\mathbf{e}_3 = b_{1n} &\qquad &\mathbf{b}_2\cdot\mathbf{e}_3 = b_{2n}\end{aligned} \tag{5-42}$$

Note that the Burgers vectors are not decomposed in the same coordinate system, but rather into screw and edge components, with b_e and b_n referring to the edge components in the plane of Fig. 5-4 and normal to it, respectively.

All the line integrals in the resulting form of Eq. (5-38) are elementary,[10] and all the terms of Eq. (5-38) can be systematically developed as follows:

[9]Here the terms $\boldsymbol{\xi}_1\otimes\mathbf{e}_3$, etc., are dyadic products. That is, the term $\boldsymbol{\xi}_1\otimes\mathbf{e}_3$ must be dotted into on the left by the vector term containing $\boldsymbol{\xi}_1$ and on the right by the term containing $\mathbf{e}_3$ (see Sec. 2-3).

[10]For example, the term g_{333} defined below is given by

$$\begin{aligned} g_{333} &= \int\frac{\partial^3 R}{\partial z^3}dy = \frac{\partial^2}{\partial z^2}\int\frac{\partial R}{\partial z}dy\\ &= \frac{\partial^2}{\partial z^2}z\ln(y - x\cos\theta + R)\\ &= \frac{z}{(y - x\cos\theta + R)R}\left[3 - \frac{z^2}{R^2} - \frac{z^2}{(y - x\cos\theta + R)R}\right]\end{aligned}$$

Integration constants, such as a $\ln 2$ term in the indefinite integral in g_{333}, which do not give a net contribution to Eq. (5-36), are dropped.

Term (a) is given by

$$\frac{\mu}{4\pi}\Big[(b_{2s}b_{1s}+b_{2e}b_{1e})h_3\sin^2\theta+(b_{2e}b_{1s}-b_{2s}b_{1e})h_3\sin\theta\cos\theta$$

$$+b_{2s}b_{1n}\left(-\frac{\cos\theta}{R}-h_1\right)+b_{2n}b_{1s}\left(h_1\cos\theta+\frac{1}{R}\right)$$

$$+b_{2e}b_{1n}\frac{\sin\theta}{R}+b_{2n}b_{1e}h_1\sin\theta\Big]\mathbf{e}_3\,dl_1$$

Term (b) is given by

$$\frac{\mu}{4\pi}\Bigg\{\Bigg[b_{2s}b_{1s}\left(\frac{h_1\cos\theta}{\sin\theta}+\frac{\cos^2\theta}{R\sin\theta}\right)+b_{2s}b_{1e}\left(h_1+\frac{\cos\theta}{R}\right)-b_{2e}b_{1s}\frac{\cos\theta}{R}$$

$$-b_{2e}b_{1e}\frac{\sin\theta}{R}\Bigg](\mathbf{e}_3\times\boldsymbol{\xi}_1)+\Big[(b_{2e}b_{1s}-b_{2s}b_{1e})h_3\sin\theta\cos\theta+b_{2s}b_{1s}$$

$$\times(-h_3\cos^2\theta)+b_{2e}b_{1e}h_3\sin^2\theta+b_{2n}b_{1s}h_1\cos\theta+b_{2n}b_{1e}h_1\sin\theta\Big]\mathbf{e}_3\Bigg\}dl_1$$

Term (c) is given by

$$\frac{\mu}{4\pi(1-\nu)}\Bigg\{\Bigg[-b_{2e}b_{1s}g_{133}+b_{2e}b_{1e}\left(\frac{\cos\theta}{\sin\theta}g_{133}+\frac{1}{\sin\theta}\frac{\partial^2R}{\partial z^2}\right)$$

$$-b_{2e}b_{1n}g_{333}+b_{2n}b_{1s}\left(-\frac{1}{\sin\theta}g_{113}-\frac{\cos\theta}{\sin\theta}\frac{\partial^2R}{\partial x\,\partial z}\right)$$

$$+b_{2n}b_{1e}\left(\frac{\cos\theta}{\sin^2\theta}g_{113}+\frac{1+\cos^2\theta}{\sin^2\theta}\frac{\partial^2R}{\partial x\,\partial z}+\frac{\cos\theta}{\sin^2\theta}\frac{\partial^2R}{\partial y\,\partial z}\right)$$

$$+b_{2n}b_{1n}\left(-\frac{1}{\sin\theta}g_{133}-\frac{\cos\theta}{\sin\theta}\frac{\partial^2R}{\partial z^2}\right)\Bigg](\mathbf{e}_3\times\boldsymbol{\xi}_1)$$

$$+\Bigg[b_{2e}b_{1s}\left(-\frac{\cos\theta}{\sin\theta}g_{113}-\frac{1}{\sin\theta}\frac{\partial^2R}{\partial x\,\partial z}\right)$$

$$+b_{2e}b_{1e}\left(\frac{\cos^2\theta}{\sin^2\theta}g_{113}+2\frac{\cos\theta}{\sin^2\theta}\frac{\partial^2R}{\partial x\,\partial z}+\frac{1}{\sin^2\theta}\frac{\partial^2R}{\partial y\,\partial z}\right)$$

$$+b_{2e}b_{1n}\left(-\frac{\cos\theta}{\sin\theta}g_{133}-\frac{1}{\sin\theta}\frac{\partial^2R}{\partial z^2}\right)$$

$$+b_{2n}b_{1s}\left(-\frac{\cos\theta}{\sin^2\theta}g_{111}-\frac{1+\cos^2\theta}{\sin^2\theta}\frac{\partial^2 R}{\partial x^2}-\frac{\cos\theta}{\sin^2\theta}\frac{\partial^2 R}{\partial x\,\partial y}\right)$$

$$+b_{2n}b_{1e}\left(\frac{\cos^2\theta}{\sin^3\theta}g_{111}+\cos\theta\frac{2+\cos^2\theta}{\sin^3\theta}\frac{\partial^2 R}{\partial x^2}\right.$$

$$\left.+\frac{1+2\cos^2\theta}{\sin^3\theta}\frac{\partial^2 R}{\partial x\,\partial y}+\frac{\cos\theta}{\sin^3\theta}\frac{\partial^2 R}{\partial y^2}\right)$$

$$+b_{2n}b_{1n}\left(-\frac{\cos\theta}{\sin^2\theta}g_{113}-\frac{1+\cos^2\theta}{\sin^2\theta}\frac{\partial^2 R}{\partial x\,\partial z}\right.$$

$$\left.\left.\left.-\frac{\cos\theta}{\sin^2\theta}\frac{\partial^2 R}{\partial y\,\partial z}\right)\right]\mathbf{e}_3\right\}dl_1$$

Term (d) is given by

$$\frac{\mu}{4\pi(1-\nu)}\left\{\left[b_{2e}b_{1n}2h_3+b_{2n}b_{1n}\left(\frac{2h_1}{\sin\theta}+\frac{2\cos\theta}{R\sin\theta}\right)\right](\mathbf{e}_3\times\boldsymbol{\xi}_1)\right.$$

$$\left.+\left[-b_{2e}b_{1e}2h_3+b_{2n}b_{1e}\left(-\frac{2h_1}{\sin\theta}-\frac{2\cos\theta}{R\sin\theta}\right)\right]\mathbf{e}_3\right\}dl_1$$

The symbols are the following:

$$g_{111}=\int\frac{\partial^3 R}{\partial x^3}dy=\frac{\rho'}{R\rho}\left(2\sin^2\theta+\frac{y^2\sin^2\theta+z^2}{R^2}-\frac{x\rho'\sin^2\theta}{\rho R}\right)$$

$$g_{113}=\int\frac{\partial^3 R}{\partial x^2\,\partial z}dy=\frac{z}{R\rho}\left[\sin^2\theta-\frac{\rho'(x-y\cos\theta)}{R^2}-\frac{x\rho'\sin^2\theta}{R\rho}\right]$$

$$g_{133}=\int\frac{\partial^3 R}{\partial x\,\partial z^2}dy=\frac{\rho'}{R\rho}\left(1-\frac{z^2}{R^2}\right)-\frac{xz^2\sin^2\theta}{R^2\rho^2}$$

$$g_{333}=\int\frac{\partial^3 R}{\partial z^3}dy=\frac{z}{R\rho}\left(3-\frac{z^2}{R^2}-\frac{z^2}{R\rho}\right)$$

$$h_1=\int\frac{\partial}{\partial x}\frac{1}{R}dy=-\frac{\cos\theta}{R}+\frac{x\sin^2\theta}{R\rho}$$

$$h_3=\int\frac{\partial}{\partial z}\frac{1}{R}dy=\frac{z}{R\rho}$$

$$R = (x^2 + y^2 - 2xy\cos\theta + z^2)^{1/2}$$

$$\rho' = x - y\cos\theta - R\cos\theta$$

$$\rho = y - x\cos\theta + R$$

Terms (a), (b), (c), and (d) must be added to give the result in Eq. (5-38). Some contraction of the result is possible in general, but we have presented the terms separately as an aid in deriving the final result. These results appear quite formidable, but as mentioned above, it is generally easier to apply Eq. (5-38) directly for a complex interaction than either to differentiate an energy expression or to transform an expression for the stress tensor.

For parallel dislocations the coordinate system shown in Fig. 5-4 breaks down, and instead the system shown in Fig. 5-5 is adopted. By limiting procedures, $\theta \to 0$, it is found that term (a) in Eq. (5-38) vanishes, term (b) becomes

$$\frac{\mu}{4\pi}\left\{\left[b_{2s}b_{1s}\left(-\frac{d\lambda}{h^2R}\right) + b_{2e}b_{1s}\left(-\frac{1}{R}\right)\right]\mathbf{e}_2 + \left\{b_{2s}b_{1s}\frac{z\lambda}{h^2R} + b_{2n}b_{1s}\left(-\frac{1}{R}\right)\right]\mathbf{e}_3\right\}dl_1$$

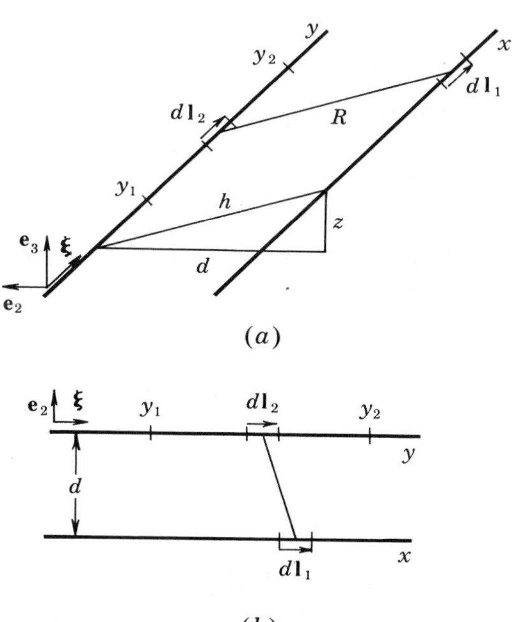

FIGURE 5-5. Coordinates for parallel dislocation interaction: (a) perspective view; (b) viewed along the $\mathbf{e}_3$ axis.

term (c) becomes

$$\frac{\mu}{4\pi(1-\nu)}\left(\left\{b_{2e}b_{1s}\left(\frac{1}{R}-\frac{z^2}{R^3}\right)+b_{2e}b_{1e}\frac{d}{R(R+\lambda)}\left[1-\frac{z^2}{R^2}-\frac{z^2}{R(R+\lambda)}\right]\right.\right.$$

$$-b_{2e}b_{1n}\frac{z}{R(R+\lambda)}\left[3-\frac{z^2}{R^2}-\frac{z^2}{R(R+\lambda)}\right]-b_{2n}b_{1s}\frac{zd}{R^3}$$

$$+b_{2n}b_{1e}\frac{z}{R(R+\lambda)}\left[1-\frac{d^2}{R^2}-\frac{d^2}{R(R+\lambda)}\right]$$

$$\left.-b_{2n}b_{1n}\frac{d}{R(R+\lambda)}\left[1-\frac{z^2}{R^2}-\frac{z^2}{R(R+\lambda)}\right]\right\}\mathbf{e}_2$$

$$+\left\{-b_{2e}b_{1s}\frac{zd}{R^3}+b_{2e}b_{1e}\frac{z}{R(R+\lambda)}\left[1-\frac{d^2}{R^2}-\frac{d^2}{R(R+\lambda)}\right]\right.$$

$$-b_{2e}b_{1n}\frac{d}{R(R+\lambda)}\left[1-\frac{z^2}{R^2}-\frac{z^2}{R(R+\lambda)}\right]+b_{2n}b_{1s}\left(\frac{1}{R}-\frac{d^2}{R^3}\right)$$

$$+b_{2n}b_{1e}\frac{d}{R(R+\lambda)}\left[3-\frac{d^2}{R^2}-\frac{d^2}{R(R+\lambda)}\right]$$

$$\left.\left.-b_{2n}b_{1n}\frac{z}{R(R+\lambda)}\left[1-\frac{d^2}{R^2}-\frac{d^2}{R(R+\lambda)}\right]\right\}\mathbf{e}_3\right)dl_1$$

and term (d) becomes

$$\frac{\mu}{4\pi(1-\nu)}\left[\left(-b_{2e}b_{1n}\frac{2z\lambda}{h^2R}-b_{2n}b_{1n}\frac{2d\lambda}{h^2R}\right)\mathbf{e}_2\right.$$

$$\left.+\left(b_{2e}b_{1e}\frac{2z\lambda}{h^2R}+b_{2n}b_{1e}\frac{2d\lambda}{h^2R}\right)\mathbf{e}_3\right]dl_1$$

Here

$$R^2=d^2+z^2+(y-x)^2$$

$$\lambda=y-x$$

$$h^2=d^2+z^2=R^2-\lambda^2$$

The simpler coplanar case can be obtained by setting $z=0$.

This completes the formulas. Note that Eq. (5-38) can be used to calculate the resolved shear stress at any point in any direction. The element $d\mathbf{l}_1$ is effectively a "test piece," which enables one to measure the stress by Eq. (5-37). As mentioned at the outset, one of the principal uses of the theory of this section lies in the determination of the interaction force between a complex dislocation configuration and a differential element (or integral element) of another dislocation by approximating the configuration by a sequence of straight segments and summing the interaction forces of each segment and the element in question. Of course, if the configuration of a generally curved dislocation can be given analytically, its interaction force with another dislocation element can be determined exactly by numerical integration of Eq. (5-38).

Self-Force on a Segment

Equation (5-38) also can be used to determine the force on an element in a dislocation caused by other segments of the *same* dislocation line. However, an additional term must be added to the force. Consider the problem of determining the total force on an element $d\mathbf{l}$ in the piecewise straight configuration of Fig. 5-6. The interaction force between $d\mathbf{l}$ and the segments AE, ED, DC, and BC can be obtained from the preceding formulas, but a separate formula is needed for the force on $d\mathbf{l}$ from the segment AB to which $d\mathbf{l}$ belongs. The required formula can be obtained by the following procedure.

First calculate by Eq. (5-38) the force $d\mathbf{F}'$ on an element $d\mathbf{l}'$, parallel to AB, with the same Burgers vector $\mathbf{b}$, and separated from AB by a small distance h (Fig. 5-7). Then subtract from $d\mathbf{F}'$ the force that an infinite straight dislocation containing the segment AB would exert on $d\mathbf{l}'$. In the limit $h \to 0$, the difference is the force $d\mathbf{F}$ on $d\mathbf{l}$ caused by the segment to which $d\mathbf{l}$ belongs. The difference depends exactly upon the amount by which the stress field along AB deviates from that of an infinite straight dislocation with the same Burgers vector. These stress differences are the stresses which exert forces on AB.

By such a procedure, one finds that the self-force $d\mathbf{F}$ on an element $d\mathbf{l}$ at y in a segment $y_1 y_2$ is given by

$$d\mathbf{F} = \frac{\mu\nu b_s}{4\pi(1-\nu)}\left(\frac{1}{\lambda_1} + \frac{1}{\lambda_2}\right)(b_e\mathbf{e}_2 + b_n\mathbf{e}_3)\, dl \tag{5-43}$$

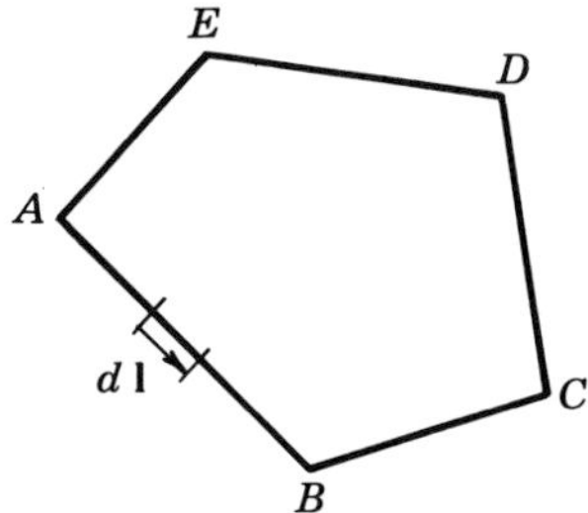

FIGURE 5-6. Piecewise straight-dislocation configuration.

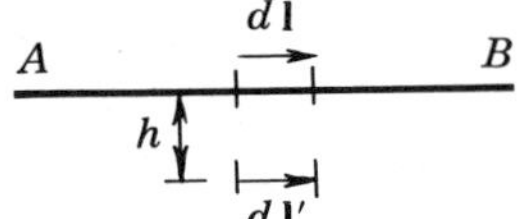

FIGURE 5-7. Dislocation segment AB with an element dl' parallel to it.

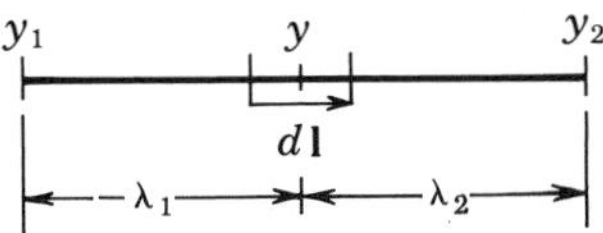

FIGURE 5-8. Dislocation segment $y_1 y_2$.

Here, as shown in Fig. 5-8, $\lambda_1 = y_1 - y$ and $\lambda_2 = y_2 - y$. Thus the total force on a segment such as AB in Fig. 5-6 can be determined from Eqs. (5-38) and (5-43). In computing the self-force for a finite segment singular behavior arises. For the example of Fig. 5-8 the self force contains factors $\ln(y_1 - y)$ and $\ln(y_2 - y)$ to be evaluated between limits y_1 and y_2. To avoid singularities cutoffs $y_1 - \rho$ and $y_2 - \rho$, are introduced. As discussed in Chap. 8, the cutoff parameter ρ, of the order of core dimensions, is closely related to r_0.

Exercise 5-5. Use Eq. (5-38) to compute the forces on an element $d\mathbf{l}_1$ from a perpendicular dislocation. In case (a) let both dislocations be pure screw dislocations and in case (b) let both dislocations be pure edge dislocations with their Burgers vectors parallel to $\mathbf{e}_3$. Let $y_2 \to \infty$ and $y_1 \to -\infty$, and compare the results with those computed directly for such a case.[11] Integrate the force $\delta \mathbf{F}$ from $x = -\infty$ to $+\infty$ and verify that the result is identical to that of Eq. (5-35).

5-6. THE STRESS FIELD ABOUT A STRAIGHT SEGMENT OF DISLOCATION

For problems involving the stress field of a complex dislocation configuration, a useful approximation that greatly reduces the labor of calculation and leads to tractable analytic solutions is approximation of the configuration by a sequence of straight line segments and summation of the stress fields of the individual segments, as for the preceding force development. This method is possible because the line integrals in $\mathbf{u}$ [Eq. (4-20)] transform like vectors for *each* segment $d\mathbf{l}'$, as does the $\mathbf{b}\Omega$ term. By extension, $\sigma_{\alpha\beta}$ [Eq. (4-30)] transforms like a tensor for each dislocation line element. Thus the stresses of the individual line segments can be transformed as tensors to a common coordinate system and summed to give the total physical stress. As discussed in the preceding section, one can use a test element $d\mathbf{l}_1$ and determine the stresses

[11] C. S. Hartley and J. P. Hirth, *Acta Met.*, **13**: 79 (1965); F. Kroupa, *Czech. J. Phys.*, **11B**: 847 (1961); R. Bullough and J. V. Sharp, *Phil. Mag.*, **11**: 605 (1965).

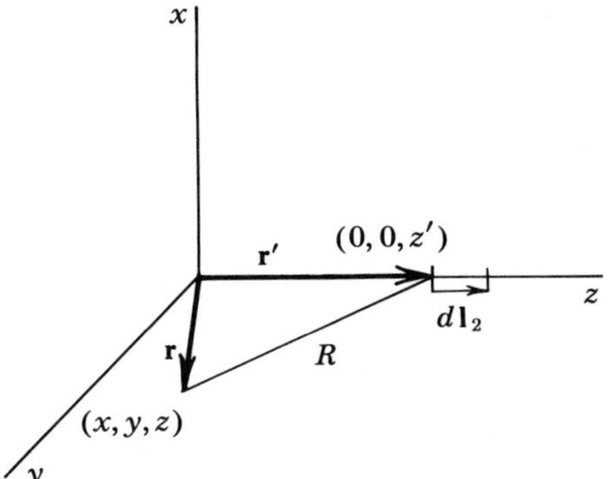

FIGURE 5-9. Coordinates for stress calculations.

from Eqs. (5-37) and (5-38). However, just as it was less tedious to determine the forces directly by integration, it is generally easier to obtain the stresses directly when they are of interest.

In order to obtain stresses consistent with the notation of Chap. 3, the coordinate system in Fig. 5-9 is adopted. The **r**′ coordinates are fixed on the dislocation line, $x'=y'=0$ and $z'\neq 0$. We wish to determine the stresses at **r**, with coordinates x, y, and z. Here

$$R^2=x^2+y^2+(z-z')^2$$

The stress at **r** from a straight segment $z'_A z'_B$ will be given by

$$\sigma_{ij}=\sigma_{ij}(z'_B)-\sigma_{ij}(z'_A) \tag{5-44}$$

The calculation of $\sigma_{ij}(z')$ proceeds as follows. Recalling that $\partial/\partial x_i=-\partial/\partial x'_i$, one can write the term σ_{xx} from Eq. (4-30) in the form

$$\sigma_{xx}=\frac{\mu}{4\pi(1-\nu)}\int b_m\epsilon_{imz}\left(\frac{\partial^3 R}{\partial x_i\,\partial x^2}-\frac{\partial}{\partial x_i}\nabla^2 R\right)dz'$$

$$=\frac{\mu}{4\pi(1-\nu)}\int\left[b_x\left(-\frac{\partial^3 R}{\partial x^2\,\partial y}+\frac{\partial}{\partial y}\nabla^2 R\right)+b_y\left(\frac{\partial^3 R}{\partial x^3}-\frac{\partial}{\partial x}\nabla^2 R\right)\right]dz'$$

Once again, the line integrals are elementary.[12]

[12] For example, dropping terms independent of z', one finds

$$\int\frac{\partial}{\partial y}\nabla^2 R\,dz'=\int\frac{\partial}{\partial y}\frac{2}{R}dz'=-\int\frac{2y}{R^3}dz'$$

$$=\frac{-2y(z'-z)}{(x^2+y^2)R}=\frac{-2y(z'-z)}{R\left[R^2-(z'-z)^2\right]}$$

Proceeding similarly for the other stress components, one obtains

$$\frac{\sigma_{xx}}{\sigma_0} = b_x \frac{y}{R(R+\lambda)}\left[1+\frac{x^2}{R^2}+\frac{x^2}{R(R+\lambda)}\right]$$

$$+ b_y \frac{x}{R(R+\lambda)}\left[1-\frac{x^2}{R^2}-\frac{x^2}{R(R+\lambda)}\right]$$

$$\frac{\sigma_{yy}}{\sigma_0} = - b_x \frac{y}{R(R+\lambda)}\left[1-\frac{y^2}{R^2}-\frac{y^2}{R(R+\lambda)}\right]$$

$$- b_y \frac{x}{R(R+\lambda)}\left[1+\frac{y^2}{R^2}+\frac{y^2}{R(R+\lambda)}\right]$$

$$\frac{\sigma_{zz}}{\sigma_0} = b_x\left[\frac{2\nu y}{R(R+\lambda)}+\frac{y\lambda}{R^3}\right]+b_y\left[-\frac{2\nu x}{R(R+\lambda)}-\frac{x\lambda}{R^3}\right] \tag{5-45}$$

$$\frac{\sigma_{xy}}{\sigma_0} = - b_x \frac{x}{R(R+\lambda)}\left[1-\frac{y^2}{R^2}-\frac{y^2}{R(R+\lambda)}\right]$$

$$+ b_y \frac{y}{R(R+\lambda)}\left[1-\frac{x^2}{R^2}-\frac{x^2}{R(R+\lambda)}\right]$$

$$\frac{\sigma_{xz}}{\sigma_0} = - b_x \frac{xy}{R^3} + b_y\left(-\frac{\nu}{R}+\frac{x^2}{R^3}\right)+b_z\frac{y(1-\nu)}{R(R+\lambda)}$$

$$\frac{\sigma_{yz}}{\sigma_0} = b_x\left(\frac{\nu}{R}-\frac{y^2}{R^3}\right)+b_y\frac{xy}{R^3}-b_z\frac{x(1-\nu)}{R(R+\lambda)}$$

Here $\sigma_0 = \mu/4\pi(1-\nu)$ and $\lambda = z'-z$. Two equivalent forms[13] for $\sigma_{ij}(z')$ are

$$\frac{\sigma_{xx}}{\sigma_0} = - b_x \frac{y}{R(R-\lambda)}\left[1+\frac{x^2}{R^2}+\frac{x^2}{R(R-\lambda)}\right]\cdots \tag{5-46}$$

and

$$\frac{\sigma_{xx}}{\sigma_0} = - b_x \frac{y\lambda}{\rho^2 R}\left(1+\frac{2x^2}{\rho^2}+\frac{x^2}{R^2}\right)\cdots \tag{5-47}$$

Furthermore, one can add the term $2y/(x^2+y^2)$, which is independent of z' and write the result in the form

$$\int \frac{\partial}{\partial y}\nabla^2 R\, dz' = \frac{2y}{R[R+(z'-z)]}$$

[13] The other forms are not equal term by term to Eq. (5-45); they differ by integration constants but give the same result when substituted into Eq. (5-44).

Here $\rho^2 = x^2 + y^2$. Notice that only one change of sign occurs in terms containing the factor $R+\lambda$; no change of sign occurs in the terms that do not include $R+\lambda$. Equation (5-47) is the most convenient form to use in limiting cases where $z' \to \infty$ or $z' \to 0$. Equation (5-45) is the form that follows directly from the force formula, Eq. (5-38).

Exercise 5-6. Set $z'_B = \infty$ and $z'_A = -\infty$ and show that Eqs. (5-44) and (5-45) or (5-47) give results identical to Eqs. (3-3) and (3-43). Set $z'_B = \infty$ and $z'_A = 0$ (a nonphysical case, of course) and show that when $z \gg (x^2 + y^2)^{1/2}$, then the dominant term in all stresses is the same as the stress from an infinite dislocation. This illustrates the principle that much closer to a dislocation segment than the length of the segment or the distance to other segments, the dominant part of the stress is the same as the stress from an infinite straight dislocation containing the segment.

Exercise 5-7. Prove that Eqs. (5-38) and (5-45) give identical results for a segment $z'_B = Z$ and $z'_A = 0$ for the terms σ_{xz} and σ_{yz}.

Dyadic Expressions for the Stress Field

When Eqs. (5-45) are applied to a complex dislocation configuration, the contributions from the various segments must finally be transformed to a common coordinate system. A more convenient scheme in many cases is to use general dyadic expressions not relying on a particular choice of coordinates. Such a representation also is consistent with the energy expressions in Chap. 6. With $\mathbf{r}' = (0,0,z')$ fixed on the dislocation line as in Fig. 5-9, we can select an orthogonal set of unit vectors $\mathbf{e}_1, \mathbf{e}_2, \mathbf{e}_3 = \boldsymbol{\xi}$, with $\boldsymbol{\xi}\, dl = d\mathbf{l}$. Without loss of generality the vectors can be chosen so that $\mathbf{R} = (R_1, 0, R_3)$. The result can be expressed in any other coordinate set with unit vectors $\mathbf{i}, \mathbf{j}, \mathbf{k}$ by expressing $\mathbf{e}_1, \mathbf{e}_2, \boldsymbol{\xi}$ in the new set, e.g., $\boldsymbol{\xi} = \xi_1 \mathbf{i} + \xi_2 \mathbf{j} + \xi_3 \mathbf{k}$.

In the $\mathbf{e}_\alpha$ coordinates the integral in Eq. (4-31) can be evaluated directly[14] for a segment extending from z'_A to z'_B. The stress at $\mathbf{r}$ is again given by Eq.

[14] The use of Eulers' theorem for homogeneous functions simplifies the integration over $d\mathbf{l}' = \boldsymbol{\xi}\, dz'$ in Eq. (4-31). For example, since

$$R_1 \frac{\partial}{\partial x'} \frac{1}{R} + R_2 \frac{\partial}{\partial y'} \frac{1}{R} + R_3 \frac{\partial}{\partial z'} \frac{1}{R} = \frac{1}{R}$$

$R_2 = 0$, R_1 is constant, and $\partial/\partial z' = -\partial/\partial z$, we can write

$$\frac{\partial}{\partial x'} \frac{1}{R} = \frac{1}{R_1}\left(\frac{1}{R} + R_3 \frac{\partial}{\partial z} \frac{1}{R}\right) = \frac{1}{R_1} \frac{\partial}{\partial z}\left(\frac{R_3}{R}\right) = -\frac{1}{R_1} \frac{\partial}{\partial z'}\left(\frac{R_3}{R}\right)$$

Thus the first two integrals in Eq. (4-31) are of the form $\int \frac{\partial}{\partial z'} f(z')\, dz' = f(z')$ and can be evaluated directly.

(5-44) with $\boldsymbol{\sigma}(z')$ given by

$$\boldsymbol{\sigma}(z')/[\mu/4\pi(1-\nu)] = -[(\mathbf{b}\times\boldsymbol{\xi})\otimes\boldsymbol{\xi}+\boldsymbol{\xi}\otimes(\mathbf{b}\times\boldsymbol{\xi})]\frac{\nu}{R}$$

$$-[(\mathbf{b}\times\mathbf{e}_1)\otimes\boldsymbol{\xi}+\boldsymbol{\xi}\otimes(\mathbf{b}\times\mathbf{e}_1)]\frac{R_3(1-\nu)}{R_1R}$$

$$+[\mathbf{e}_1\otimes\boldsymbol{\xi}+\boldsymbol{\xi}\otimes\mathbf{e}_1]\frac{b_2R_1^{\,2}}{R^3}-[\mathbf{e}_1\otimes\mathbf{e}_2+\mathbf{e}_2\otimes\mathbf{e}_1]\frac{b_1R_3}{R_1R} \tag{5-48}$$

$$-\mathbf{I}\frac{b_2R_3}{R_1R}-[\mathbf{e}_1\otimes\mathbf{e}_1]\frac{b_2R_1R_3}{R^3}-[\boldsymbol{\xi}\otimes\boldsymbol{\xi}]\frac{b_2R_3^{\,3}}{R_1R^3}$$

Here $\mathbf{I}$ is the idemfactor; see Eq. (4-31).

Angular Dislocation

The angular dislocation shown in Fig. 5-10 is treated as a specific example of the utility of Eq. (5-45). This problem has been solved directly from Eq. (4-20) by Yoffe.[15] The coordinate system relative to dislocation (2) is xyz, and that relative to dislocation (1) is myp. The stresses $\sigma_{ij}(2)$ of dislocation (2) can be written out directly from Eq. (5-45) in xyz coordinates. The stresses $\sigma_{ij}(1)$ can be written out directly in myp coordinates and then transformed to xyz coordinates by Eq. (2-36) The appropriate transformation matrix is

$$T_{ij} = \begin{array}{c|ccc} & m & y & p \\ \hline x & \cos\theta & 0 & \sin\theta \\ y & 0 & 1 & 0 \\ z & -\sin\theta & 0 & \cos\theta \end{array} \tag{5-49}$$

The Burgers vectors transform by the scheme

$$b_i' = T_{ij}b_j \tag{5-50}$$

The sense vectors of the two dislocations are oppositely directed. Thus, to obtain the total stress for the total angular dislocation, one consistent sense must be used. Reversing $\boldsymbol{\xi}_1$ reverses $\mathbf{b}_1$ and thus $\sigma_{ij}(1)$ also. Thus, when $\mathbf{b}_1$ and $\mathbf{b}_2$ are chosen to be compatible with $\boldsymbol{\xi}_2$, the total stress is given by

$$\sigma_{ij} = \sigma_{ij}(2) - \sigma_{ij}(1) \tag{5-51}$$

[15] E. H. Yoffe, *Phil. Mag.*, **5**: 161 (1960).

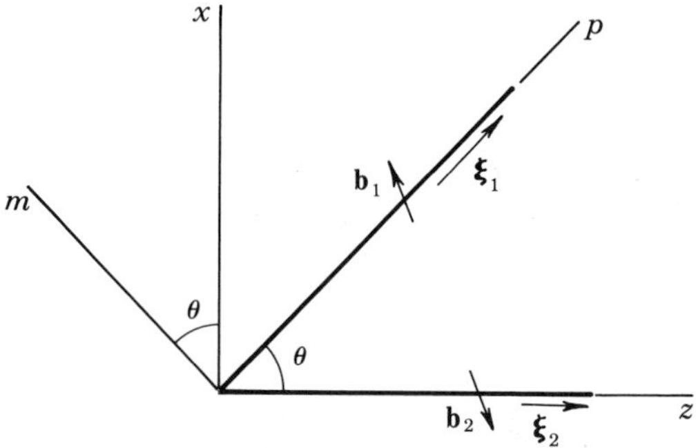

FIGURE 5-10. Angular dislocation. The y axis points out of the page.

Consider the term σ_{yz}, with $b = b_y$. From Eq. (5-45), setting $z'_B = \infty$ and $z'_A = 0$, one obtains

$$\sigma_{yz}(2) = -\sigma_0 b_y \frac{xy}{r^3} \tag{5-52}$$

where $r^2 = x^2 + y^2 + z^2$. Equation (2-36) indicates that

$$\sigma_{yz}(1) = -\sigma_{ym} \sin\theta + \sigma_{yp} \cos\theta \tag{5-53}$$

Equation (5-45) yields

$$\sigma_{ym} = -\sigma_0 b_y \frac{y}{r(r-p)} \left[1 - \frac{m^2}{r^2} - \frac{m^2}{r(r-p)} \right] \tag{5-54}$$

$$\sigma_{yp} = -\sigma_0 b_y \frac{my}{r^3} \tag{5-55}$$

Combining Eqs. (5-50) to (5-54), one obtains

$$\sigma_{yz} = \sigma_0 b_y \left[\frac{-xy}{r^3} + \frac{my\cos\theta}{r^3} - \frac{y\sin\theta}{r(r-p)} + \frac{m^2 y \sin\theta}{r^2(r-p)^2} + \frac{m^2 y \sin\theta}{r^3(r-p)} \right] \tag{5-56}$$

The other stress components follow in an analogous fashion. This result is identical to that of Yoffe.[16] She used the angular-dislocation result in superposing various angular segments to form dislocation configurations bounded by straight line segments. The same result for such configurations can be determined somewhat more simply by the direct application of Eq. (5-45).

[16] E. H. Yoffe, *Phil. Mag.*, **5**: 161 (1960), as corrected in E. H. Yoffe, *Phil. Mag.*, **6**: 1147 (1961). Yoffe used different coordinates. The transformation from Eq. (5-56) to her system is $x \to y$, $y \to -x$, $z \to z$, $p \to \zeta$, $m \to \eta$, $\theta \to \alpha$.

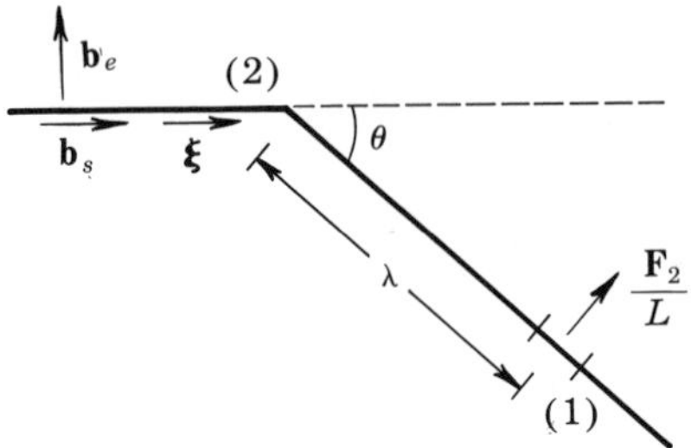

FIGURE 5-11. A dislocation bend. The component b_n points out of the page.

Exercise 5-8. Consider a dislocation bend of angle θ as in Fig. 5-11. Use Eqs. (5-38) and (5-43) to show that the force per unit length on dislocation (1) at a distance λ from the bend is given by the two components[17]

$$\frac{F_2}{L}=\frac{\mu}{4\pi\lambda}\left[b_s^2\left(\frac{\cos\theta-1}{\sin\theta}-\frac{\nu\sin\theta\cos\theta}{1-\nu}\right)+b_sb_e\frac{2\nu\sin^2\theta}{1-\nu}\right.$$
$$\left.+b_e^2\frac{\cos\theta-1+\nu\sin^2\theta\cos\theta}{(1-\nu)\sin\theta}+b_n^2\frac{\cos\theta-1}{(1-\nu)\sin\theta}\right]$$

and

$$\frac{F_n}{L}=\frac{\mu}{4\pi(1-\nu)\lambda}\left\{b_nb_s(1+\nu)(1-\cos\theta)+b_nb_e\left[\frac{-(1-\cos\theta)^2}{\sin\theta}+\nu\sin\theta\right]\right\}$$

Here F_n and b_n refer to components normal to the page of Fig. 5-11. Discuss how Eqs. (5-45) or the Yoffe angular-dislocation formulas could be applied to give the same result.

Stress Field of an Infinitesimal Loop

Like the force formulas, Eq. (5-45) can be applied to an infinitesimal element and the result numerically integrated over a generally curved dislocation line to give an exact result. Another useful application involves the determination of the stress field of an infinitesimal loop. Such stresses can then be integrated over the area of a finite closed loop to give its stress field; Kroupa[18] has performed extensive applications of this kind. As another example of the use of Eq. (5-45), let us derive his result.

Consider the infinitesimal loop in Fig. 5-12, with Burgers vector $\mathbf{b}=(0,0,b_z)$. Let us compute the stress σ_{yz} at the point $(x,0,z)$ in the slip plane $y=0$.

[17] J. Lothe, *Phil. Mag.*, **15**: 353 (1967).

[18] See, for example, F. Kroupa, *Czech. J. Phys.*, **10B**: 284 (1960); *Czech. J. Phys.*, **12B**: 191 (1962); *Phil. Mag.*, **7**: 783 (1962); *Phys. Stat. Solidi*, **9**: 27 (1965).

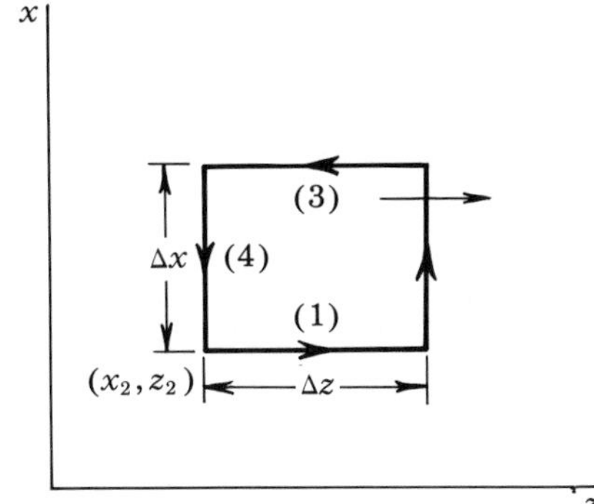

FIGURE 5-12. An infinitesimal dislocation loop. Arrows on the loop indicate the sense of ξ.

Equation (5-45) gives the result for σ_{yz} as a function of $x' = x_2$ and $z' = z_2$,

$$\sigma_{yz} = -\frac{\sigma_0 b_z (1-\nu)(z-z_2)}{(x-x_2)\left[(x-x_2)^2+(z-z_2)^2\right]^{1/2}} \tag{5-57}$$

The sum of the stresses from segments (1) and (3) in Fig. (5-12) is

$$\begin{aligned}
\sigma_{yz}(1,3) &= -\sigma_{yz}(z_2+\Delta z, x_2+\Delta x) + \sigma_{yz}(z_2, x_2+\Delta x) \\
&\quad + \sigma_{yz}(z_2+\Delta z, x_2) - \sigma_{yz}(z_2, x_2) \\
&= -\frac{\partial^2 \sigma_{yz}}{\partial x_2 \partial z_2} dx_2\, dz_2 \qquad (5\text{-}58)\\
&= \frac{\sigma_0 b_z (1-\nu)}{\left[(x-x_2)^2+(z-z_2)^2\right]^{3/2}} \left[1 - \frac{3(x-x_2)^2}{(x-x_2)^2+(z-z_2)^2}\right] dx_2\, dz_2
\end{aligned}$$

The coordinates of Fig. 5-12 must be transformed to a set rotated by $\pi/2$ in order to use Eq. (5-45) for segments (2) and (4). Transforming the coordinates, proceeding as above, and transforming the result back to the coordinates of Fig. 5-12, one finds that the stresses from segments (2) and (4) are

$$\sigma_{yz}(2,4) = \frac{\sigma_0 b_z \left[(x-x_2)^2 - 2(z-z_2)^2\right]}{\left[(x-x_2)^2+(z-z_2)^2\right]^{5/2}} dx_2\, dz_2 \tag{5-59}$$

Thus the total stress at the point (x, z) is

$$\sigma_{yz} = \frac{\sigma_0 b_z \left[(-1+2\nu)(x-x_2)^2 - (1+\nu)(z-z_2)^2\right]}{\left[(x-x_2)^2+(z-z_2)^2\right]^{5/2}} dx_2\, dz_2 \tag{5-60}$$

which is identical to Kroupa's result.[19] The other stress components follow analogously.

For finite loops with shapes similar to circles of radius R, the long-range stress fields at distances from the loop greater than $2R$ converge rapidly to the stress field of the infinitesimal loop, in accord with expectation from St. Venant's principle. Thus the above results can be used to determine the long-range interactions of dislocation loops with other dislocation arrays. Note that the stresses in Eq. (5-59) decrease as $1/R^3$ in comparison to $1/R$ for the stress field around an infinite straight dislocation.

5-7. BROWN'S FORMULA

An alternative method for the determination of the stresses and strains in the plane of a planar configuration was developed by Brown,[20] who showed that these stresses and strains could be expressed by simple line integrals where the integrand is determined by the theory of straight dislocations. This possibility was suggested by earlier work on line tension by Lothe,[21] and Brown presented his work as a proof of Lothe's theorem. Indenbom and Orlov[22] independently presented a more general theory, for curved dislocations in terms of straight dislocations, which is not limited to the planar case. The development of the latter theory is beyond the scope of this book, so we limit consideration to the simpler, special case treated by Brown. The results of both theories apply generally to the anisotropic elastic case and are useful in determining the equilibrium configurations of dislocations in the presence or absence of forces from external stresses.[23] When applied to configurations made up of sequences of piecewise straight segments, both methods apply to two- or three-dimensional arrays. From Eq. (4-6), the elastic distortions leading to the strains are given by

$$\frac{\partial u_q}{\partial x_s} = -\frac{\partial u_q}{\partial x'_s} = \int_A b_i c_{ijkl} \frac{\partial^2}{\partial x'_l \, \partial x'_s} u_{qk}(\mathbf{r}' - \mathbf{r}) \, dA_j \tag{5-61}$$

For a given direction of $\mathbf{R} = \mathbf{r}' - \mathbf{r}$, u_{qk} varies as R^{-1} with $R = |\mathbf{r}' - \mathbf{r}|$; see Eq. (2-70). Thus, u_{qk} can be written as

$$u_{qk} = R^{-1} G_{qk}\left(\frac{X}{R}, \frac{Y}{R}, \frac{Z}{R}\right) \tag{5-62}$$

[19] F. Kroupa, *Czech. J. Phys.*, **12B**: 191 (1962). Kroupa's coordinates differ from those used here.

[20] L. M. Brown, *Phil. Mag.*, **15**: 363 (1967).

[21] J. Lothe, *Phil. Mag.*, **15**: 353 (1967).

[22] V. L. Indenbom and S. S. Orlov, *Sov. Phys. Crystallogr.*, **12**: 849 (1968).

[23] D. J. Bacon, D. M. Barnett, and R. O. Scattergood, *Progr. Mat. Sci.*, **23**: 51 (1978).

where

$$X = -X' = x - x', Y = -Y' = y - y', Z = -Z' = z - z' \tag{5-63}$$

The **R** dependence of u_{qk} shows that it satisfied the symmetries

$$u_{qk}(\mathbf{R}) = u_{qk}(-\mathbf{R}) = u_{kq}(\mathbf{R}) \tag{5-64}$$

The form of Eq. (5-62) indicates that u_{qk} is a homogeneous function of degree -1.

We now apply Eq. (5-61) to the distortion in the plane of a planar loop, for simplicity lying in the $x-y$ plane so that $z = z' = Z = Z' = 0$. Performing the differentiation in Eq. (5-61) and then setting $Z = 0$ shows that the distortions are of the form

$$\frac{\partial u_q}{\partial x_s} = \int_A B_{qs}(X', Y')\, dX'\, dY' \tag{5-65}$$

where B_{qs} is a homogeneous function of degree -3,

$$B_{qs}(X, Y) = R^{-3} a_{qs}\left(\frac{X}{R}, \frac{Y}{R}\right) \tag{5-66}$$

In a power series development, a homogeneous function of degree p in the variables X and Y can be written[24]

$$H = \sum_{m,n} d_{mn} R^{p-m-n} X^m Y^n \tag{5-67}$$

where the d_{mn} are constants. The form in Eq. (5-67) satisfies the so-called Euler identity[24]

$$X\frac{\partial H}{\partial X} + Y\frac{\partial H}{\partial Y} = pH \tag{5-68}$$

The function $B_{qs} = H$ satisfies this identity with $p = -3$; so, rearranged, it becomes

$$\frac{\partial}{\partial X} X B_{qs} + \frac{\partial}{\partial Y} Y B_{qs} = -B_{qs} \tag{5-69}$$

Hence Eq. (5-65) can be rewritten in the form

$$\frac{\partial u_q}{\partial x_s} = -\int_A \left[\frac{\partial}{\partial X'} X' B_{qs} + \frac{\partial}{\partial Y'} Y' B_{qs}\right] dX'\, dY' \tag{5-70}$$

[24] I. S. Sokolnikoff and R. M. Redheffer, "Mathematics of Physics and Modern Engineering," McGraw-Hill, New York, 1966, p. 325.

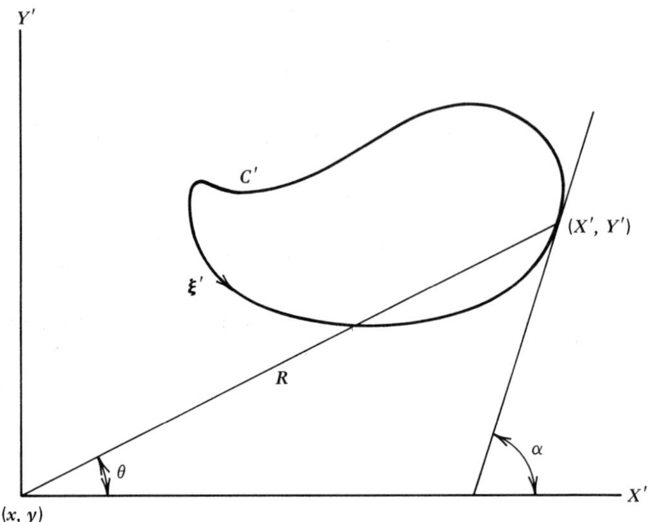

FIGURE 5-13. Coordinates in plane of dislocation loop C'.

The integral can now be partly integrated by Green's theorem to give the line integral

$$\frac{\partial u_q}{\partial x_s} = -\oint_{C'} (X'\,dY' - Y'\,dX')B_{qs} \tag{5-71}$$

The form of the line integral in the plane is shown in Fig. 5-13.

From Fig. 5-13, one sees that $X' = R\cos\theta$, $Y' = R\sin\theta$, $dX' = \cos\alpha\,dl$, $dY' = \sin\alpha\,dl$. Thus after insertion of Eq. (5-66), (5-71) takes the form

$$\frac{\partial u_q}{\partial x_s} = -\oint_{C'} \frac{1}{R^2}\sin(\theta - \alpha)a_{qs}(\theta)\,dl \tag{5-72}$$

This is a line integral of the form desired for simplicity in the further procedure. Starting with Mura's Eq. (4-12) instead, one would be led to a line integral of the preceding form, but with an added term $R^{-2}\cos(\theta - \alpha)g_{qs}(\theta)$ in the integrand, which would present some problems in the further development.

Let us apply Eq. (5-72) to a straight infinite dislocation line with the same Burgers vector, in the same plane and with orientation α as depicted in Fig. 5-14. The distortion is inversely proportional to distance h from the dislocation line:

$$\frac{\partial u_q}{\partial x_s} = \frac{1}{h}S_{qs}(\alpha) \tag{5-73}$$

However, according to Eq. (5-72), since $R\,d\theta = \sin(\theta - \alpha)dl$ and $h = R\sin(\theta - \alpha)$,

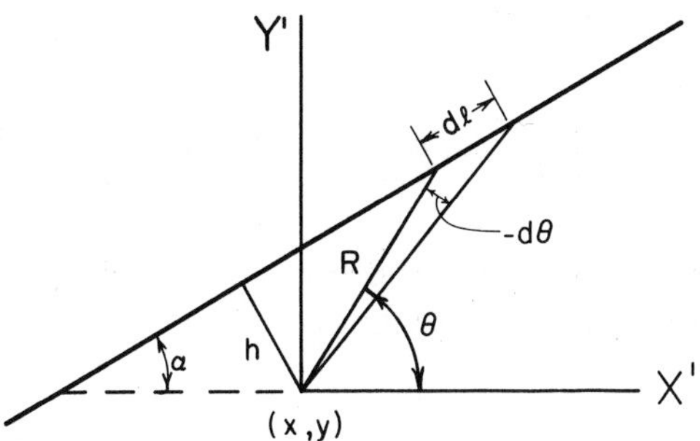

FIGURE 5-14. Coordinates in plane of dislocation line. Unit vector **n** points out of the page.

we obtain

$$\frac{\partial u_q}{\partial x_s} = \frac{1}{h} \int_{\alpha}^{\alpha+\pi} \sin(\theta-\alpha) a_{qs}(\theta) d\theta \tag{5-74}$$

Thus, by comparison with Eq. (5-73), we have

$$S_{qs}(\alpha) = \int_{\alpha}^{\alpha+\pi} \sin(\theta-\alpha) a_{qs}(\theta)\, d\theta \tag{5-75}$$

Now, the trick of the Brown theory is to invert Eq. (5-75) so that a_{qs} is expressed in terms of the straight dislocation function S_{qs}. Operating on Eq. (5-75) with $\partial/\partial\alpha$, we obtain

$$\frac{\partial S_{qs}}{\partial \alpha} = -\int_{\alpha}^{\alpha+\pi} \cos(\theta-\alpha) a_{qs}(\theta)\, d\theta$$

Operating with $\partial/\partial\alpha$ once more, and using Leibniz' formula because the limits of integration are functions of α, we find

$$\partial^2 S_{qs}/\partial\alpha^2 = -\left[\cos(\theta-\alpha) a_{qs}(\theta)\right]_{\theta=\alpha+\pi} + \left[\cos(\theta-\alpha) a_{qs}(\theta)\right]_{\theta=\alpha} - \int_{\alpha}^{\alpha+\pi} \sin(\theta-\alpha) a_{qs}(\theta)\, d\theta$$

or, using Eq. (5-52) in the form $a_{qs}(\alpha+\pi) = a_{qs}(\alpha)$,

$$\frac{\partial^2 S_{qs}}{\partial \alpha^2} = 2a_{qs}(\alpha) - S_{qs}(\alpha) \tag{5-76}$$

which solves the inversion problem as

$$a_{qs}(\alpha) = \frac{1}{2}\left[S_{qs}(\alpha) + \frac{\partial^2 S_{qs}(\alpha)}{\partial \alpha^2}\right] \tag{5-77}$$

Now Eq. (5-72) can be written

$$\frac{\partial u_q}{\partial x_s} = \frac{1}{2} \oint \frac{1}{R^2} \sin(\theta - \alpha) \left[S_{qs}(\theta) + \frac{d^2 S_{qs}(\theta)}{d\theta^2} \right] dl \tag{5-78}$$

which is Brown's formula for the distortions. Because the stresses depend linearly on the strains, Eq. (5-78) implies that the stresses in the plane of the loop are

$$\sigma_{ij} = \frac{1}{2} \oint \frac{1}{R^2} \sin(\theta - \alpha) \left[\Sigma_{ij}(\theta) + \frac{d^2 \Sigma_{ij}(\theta)}{d\theta^2} \right] dl \tag{5-79}$$

where the Σ_{ij} are the straight dislocation stress factors analogous to the distortion factors S_{qs} in Eq. (5-73). For the straight dislocation in Fig. 5-14, we obtain

$$\sigma_{ij} = \frac{1}{h} \Sigma_{ij}(\alpha) \tag{5-80}$$

Analogous to Σ_{ij}, there are prelogarithmic factors $E(\alpha)$ in the energy, for example, in Eq. (3-87). The results in Chapter 3 show that the in-plane stress factor is related to $E(\alpha)$ by[25]

$$\tfrac{1}{2} b_i \Sigma_{ij}(\alpha) n_j = E(\alpha) \tag{5-81}$$

where $\mathbf{n}$ is a unit vector normal to the plane $X'Y'$ defined in Fig. 5-14. The in-plane stress field is thus given by

$$b_i \sigma_{ij} n_j = \oint \frac{1}{R^2} \sin(\theta - \alpha) \left[E(\theta) + \frac{d^2 E}{d\theta^2} \right] dl \tag{5-82}$$

The quantity $\sigma_{ij} n_j$ can be regarded as the in-plane stress resolved in the direction of $\mathbf{b}$. Equation (5-82) is the best-known form of Brown's formula.

All of the Eqs. (5-78), (5-79) and (5-82) have the remarkable feature of describing the field of the real dislocation defined by dl in Fig. 5-14 by the energy factors of a fictitious dislocation with another orientation. Specifically, the fictitious dislocation would lie along R in Fig. 5-14 but would have the same $\mathbf{b}$ as the real dislocation.

[25]For example, consider the simple case where $\mathbf{b}$ lies in the plane of Fig. 5-14. A convenient choice for α is then along the projection of $\mathbf{b}$, so θ becomes equal to the character angle β, $\mathbf{n}=(0,0,1)$, and $\mathbf{b}=(b,0,0)$. From Eqs. (3-3) and (3-43), $\Sigma_{13}=(\mu/2\pi)\cos\beta+[\mu/2\pi(1-\nu]\sin\beta$. With these quantities, Eq. (5-81) then gives an expression for $E(\alpha)$ that agrees with the directly derived result in Eq. (3-87) for the specific choice $\alpha=\beta$.

Exercise 5-9. Derive the forces at a bend in Exercise 5-8 from Eq. (5-82).

Mura's formula [Eq. (4-12)] can be transformed to the form

$$\frac{\partial u_j}{\partial x_s} = b_k c_{kpim} \epsilon_{qpn} \oint (x_n - x'_n) \frac{\partial^2}{\partial x_m \partial x_s} u_{ij}(\mathbf{r}-\mathbf{r}')\, dx'_q \tag{5-83}$$

This form, first obtained by Indenbom and Orlov,[26] and the starting point of their theory, leads to an equation similar to Eq. (5-72) and would be an alternative starting point for the Brown theory as well. However, the main application of the formulas is to piecewise straight segments of the type considered in the previous section. For such a model, the Brown formulas can be applied to a planar configuration or to a three-dimensional configuration, with the plane for each segment then defined relative to the field point as in Fig. 5-14. Converting the line integral by the substitutions $\sin(\theta-\alpha)\, dl = R\, d\theta$ and $h = R\sin(\theta-\alpha)$ integrating by parts for a straight segment extending from θ_1 to θ_2 and in the configuration in Fig. 5-14, we find that the segment form of Eq. (5-79) is

$$\sigma_{ij} = \frac{1}{2h}\left[-\Sigma_{ij}(\theta)\cos(\theta-\alpha) + \frac{d\Sigma_{ij}}{d\theta}\sin(\theta-\alpha)\right]_{\theta_1}^{\theta_2} \tag{5-84}$$

whereas that for Eq. (5-82) is

$$b_i \sigma_{ij} n_j = \frac{1}{h}\left[-E(\theta)\cos(\theta-\alpha) + \frac{dE}{d\theta}\sin(\theta-\alpha)\right]_{\theta_1}^{\theta_2} \tag{5-85}$$

as can be verified directly by differentiation. Thus the segment stresses are defined by the properties of infinite fictitious straight dislocations connecting the field point and the real segment ends.

The results in Eqs. (5-84) and (5-85) can be used in conjunction with Eq. (3-90) to determine the force on a straight dislocation segment from other segments. When used to determine the force balance on a segment, the self-force must be included. One can manipulate Eq. (5-85), for example, in the same manner as that illustrated in Fig. 5-7. Expanding the trigonometric functions in the limit $(\theta-\alpha)\to 0$, consistent with $h\to 0$, we find the self-force expression for a segment of length dl

$$dF_k = -\frac{1}{2}\epsilon_{ijk}\xi_i b_m \left(\frac{d\Sigma_{jm}}{d\theta}\right)_{\theta=\alpha}\left(\frac{1}{\lambda_1}+\frac{1}{\lambda_2}\right) dl \tag{5-86}$$

As discussed for Eq. (5-43), a cutoff parameter ρ is needed when the self-force is integrated for a finite segment.

[26] V. L. Indenbom and S. S. Orlov, *Sov. Phys. Crystallogr.*, **12**: 849 (1968).

A more detailed discussion of Brown's formula and of self-forces is given in the review by Bacon et al.[27]

5-8. THE DISPLACEMENT FIELD OF A STRAIGHT DISLOCATION SEGMENT

The displacement field of a segment can be used to determine interaction energies between dislocations and solute atoms represented as point forces.[28] Of greater utility, the displacement fields can be used to generate the total field of loops or other configurations represented by a sequence of straight segments. Such total fields, in turn, can be used in programs to give computer-simulated images of thin film transmission electron micrographs of importance in identifying dislocation arrays. Salden and Whelan,[29] for example, have simulated the images of extended dislocation nodes using the displacement fields for angular dislocations.[30] As discussed in the Sec. 5-6, angular dislocation fields can be derived from the single-segment fields, so only the latter are presented here.

Coordinates are fixed on the dislocation line in the same manner as for the derivation of Eq. (5-48). Equation (4-20) can then be integrated directly for a segment extending from z'_A to z'_B using the same simplifications as in the derivation of Eq. (5-48). The displacement field of the segment is given by

$$\mathbf{u}(\mathbf{r}) = -\frac{\mathbf{b}\Omega_{AB}}{4\pi} + \mathbf{u}(z'_B) - \mathbf{u}(z'_A) \tag{5-87}$$

The function $\mathbf{u}(z')$ is given by

$$\mathbf{u}(z') = -\frac{(1-2\nu)}{8\pi(1-\nu)}(\mathbf{b}\times\boldsymbol{\xi})\ln(R+R_3) + \frac{1}{8\pi(1-\nu)}(\mathbf{R}\times\mathbf{e}_2)\frac{b_2}{R} \tag{5-88}$$

The quantity Ω contains a discontinuity over the cut surface. Complete specification of Ω requires the cut surface to be defined. For a planar configuration, the cut surface is conveniently chosen to coincide with the plane of the loop. Let P be the normal projection of the field point onto this surface (Fig. 5-15). Thus P is a function of the field point. By Ω_{AB} [Eq. (5-87)], we mean the solid angle through which ABP is seen from the field point. A planar triangular configuration ABC will have a common P for a given field point, so that the triangle is seen through the solid angle $\Omega = \Omega_{AB} + \Omega_{BC} + \Omega_{CA}$.

[27] D. J. Bacon, D. M. Barnett, and R. O. Scattergood, *Progr. Mat. Sci.*, **23**: 51 (1978).

[28] R. A. Masamura and G. Sines, *J. Appl. Phys.*, **41**: 3930 (1970).

[29] D. K. Salden and M. J. Whelan, *Phil. Trans. Roy. Soc.* (*Lond.*), **292**: 513 (1979); D. K. Salden, A. Y. Stathopoulos, and M. J. Whelan, ibid., p. 523.

[30] E. H. Yoffe, *Phil. Mag.*, **5**: 161 (1960).

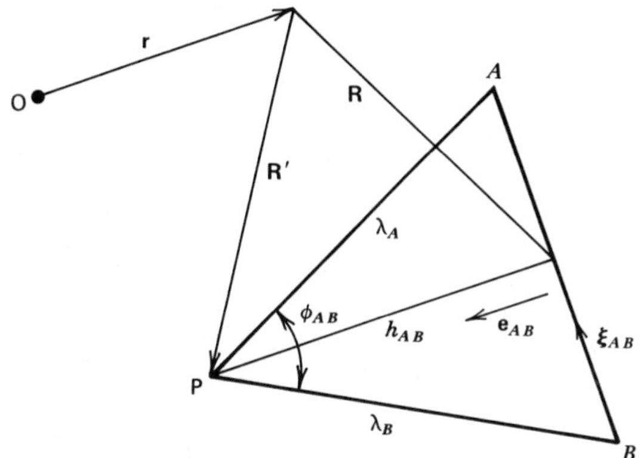

FIGURE 5-15. Perspective view of triangular dislocation loop ABP.

For a nonplanar, piecewise straight configuration $ABCDE, \ldots,$ a more complex cut surface must be selected. An additional (arbitrary) fixed point Q is chosen. Then a pyramidal cut surface consisting of triangles ABQ, BCQ, $CDQ, \ldots$ is defined. The solid angle through which each triangle is seen in projection from a field point can be calculated as in the planar case.

The coordinates $\mathbf{r}$, $\mathbf{R}$ and a normal $\mathbf{n}$, specified relative to the circulation of ξ, are defined as in Fig. 4-4, with $\mathbf{R}$ the vector from the general field point $\mathbf{r}$ to a point on the projected segment AB. In-plane vectors $\mathbf{e}_{AB}$, $\mathbf{e}_{BC}$, and so forth, are defined by $\mathbf{e}_{AB}=\mathbf{n}\times\xi_{AB}$. Because $\mathbf{e}_{AB}$ is the same if the signs of $\mathbf{n}$ and ξ are both changed, it follows that $\mathbf{e}_{AB}=\mathbf{e}_{BA}$. In terms of these vectors, the geometry of Fig. 5-15 shows that h_{AB}, the shortest distance from P to AB, the λ values, and Z, the shortest distance from the plane containing P to the general field point $\mathbf{r}$ are given by

$$\begin{aligned}
h_{AB} &= (\mathbf{R}'-\mathbf{R})\cdot\mathbf{e}_{AB}=(\mathbf{R}'-\mathbf{R})\cdot\mathbf{e}_{BA}=h_{BA}\\
\lambda_A &= |\mathbf{R}_A-\mathbf{R}'|,\ \lambda_B=|\mathbf{R}_B-\mathbf{R}'| \qquad (5\text{-}89)\\
Z &= -\mathbf{R}\cdot\mathbf{n}=-\mathbf{R}'\cdot\mathbf{n}
\end{aligned}$$

and ϕ_{AB} is the angle subtended by λ_A and λ_B, defined as positive in the sense of circulation of ξ_{AB}.

With this geometry, the integral in Eq. (4-21) over the area defined by ABP is elementary, giving

$$\begin{aligned}
\Omega_{AB} &= (\phi_{AB}-\pi)\,\mathrm{sgn}\,Z+\tan^{-1}\frac{h_{AB}}{Z}\frac{(\lambda_A^2+Z^2)^{1/2}}{(\lambda_A^2-h_{AB}^2)^{1/2}}\\
&\quad+\tan^{-1}\frac{h_{BA}}{Z}\frac{(\lambda_B^2+Z^2)^{1/2}}{(\lambda_B^2-h_{BA}^2)^{1/2}} \qquad (5\text{-}90)
\end{aligned}$$

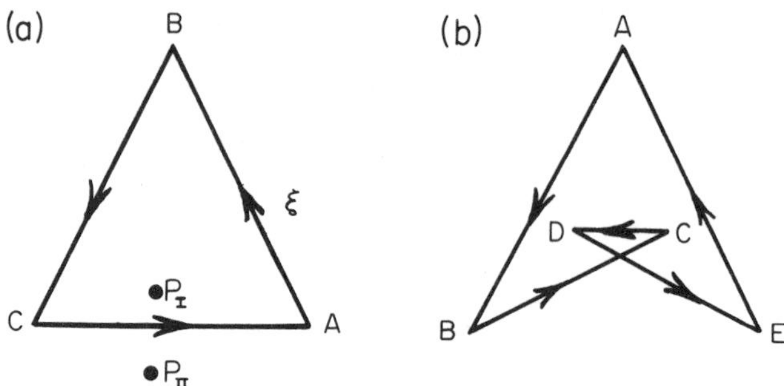

FIGURE 5-16. Dislocation loops.

where sgn indicates "sign of," the $\tan^{-1}$ values are defined in the interval $-(\pi/2) \leqslant \phi = \tan^{-1}(\ ) \leqslant \pi/2$, and the square roots are the positive roots.

The total value of Ω for a configuration is determined by summing the segment terms, as illustrated by the examples in Fig. 5-16. With Z positive and P_{I}; ϕ_{AB}, ϕ_{BC} and ϕ_{CA} are all acute and sum to 2π, which together with the three $-\pi$ contributions give a total contribution to Ω of $-\pi$ for Fig. 5-16*a*. The remainder of Ω is given by six $\tan^{-1}$ contributions, all positive. With P_{II}, ϕ_{AB}, and ϕ_{BC} remain acute but ϕ_{CA} becomes obtuse, still summing to 2π as before. The answer is thus the same except that $h_{CA} = h_{AC}$ becomes negative as do the two corresponding $\tan^{-1}$ contributions. Note that if the sign of $\boldsymbol{\xi}_{AB}$ is changed, so is that of $\mathbf{n}$ so that the sign of h_{AB} is unchanged but that of Z is changed. Thus Ω_{AB} changes sign, but so does $\mathbf{b}$, so the displacement given by Eq. (4-21) is invariant to the arbitrary selection of $\boldsymbol{\xi}$, as it must be. Thus the specified procedure gives the correct answer for a configuration that includes a spiral loop as in Fig. 5-16*b*, provided that $\mathbf{b}$ and $\mathbf{n}$ are consistently defined by $\boldsymbol{\xi}$.

With the geometry of Fig. 4-4, the area enclosed by ABC in Fig. 5-16*a* is the cut plane. Thus Ω should change discontinuously by $\pm 4\pi$ on passing the field point $\mathbf{r}$ through the cut plane. In Eq. (5-90) this corresponds to $\mathbf{r}$ passing through P_{I} from positive Z to negative Z in the limit $Z \to 0$. For Z positive, the sum of the first terms from Eq. (5-90) is $-\pi$ as before, and there are six terms $\tan^{-1}\infty = \pi/2$, altogether giving $\Omega = 2\pi$. For Z negative, all terms change sign, so $\Omega = -2\pi$ and $\Delta\Omega = -4\pi$ as required. Note that in the application of Eq. (5-90) to a triangle ABC, $\phi_{AB} + \phi_{BC} + \phi_{CA} = 2\pi$; therefore, the individual values such as ϕ_{AB} need not be evaluated.

This completes a scheme for calculating the displacement field of a piecewise straight dislocation configuration, equivalent to the work of Yoffe[31] on angular arrays. For configurations that cannot be approximated by piecewise straight segments, there is no alternative to direct numerical integration of Eq. (4-20), except for simple geometries such as circles. In such a case one can choose a cylindrical cut surface (analogous to the cylindrical slip surface for

[31]E. H. Yoffe, *Phil. Mag.*, **5**: 161 (1960); ibid., **6**: 1147 (1961).

prismatic loops; see Fig. 4-3.) Then the Ω integral can be transformed to a line integral,[32-34] making two-dimensional numerical integration unnecessary. However, this method does not provide added simplicity in the piecewise straight case and hence is not discussed further.

5-9. GENERAL IMAGE FORCES

In Chap. 3, we considered some simple dislocation geometries and found that in one instance, Exercise 3-8, the analysis of the image stresses is quite complicated. In general, the treatment of less simple geometries is still more difficult; examples include straight dislocations inclined to a free surface, for which analytical solutions are very complicated, and curved dislocations, for which analytic solutions are intractable. The procedure in these cases is the same as that outlined above: first to superpose the simple image, and then to devise a stress function which will cancel the remaining forces acting at free surfaces and thus which will satisfy the boundary conditions. Groves and Bacon[35] have given the explicit solution for an infinitesimal dislocation loop in an infinite half-space below a planar free surface. The general case of a curved dislocation below a planar free surface can be solved in principle by integration of this result, but the integration can be achieved analytically only for the simplest geometries.

Yoffe[36] has developed the general image stresses and displacements for a straight dislocation inclined to a free surface at an arbitrary angle. The image system consists of the image dislocation and a few additional elastic singularities. The stress field of the inclined dislocation and its image may be determined directly, of course, from Eq. (5-45), or from Yoffe's angular-dislocation formulas.

More recently, Maurissen and Capella[37] have integrated the Groves-Bacon formula to obtain the solution for a piecewise straight configuration consisting of segments respectively parallel with and normal to a planar surface. This

[32] M. O. Peach and J. S. Koehler, *Phys. Rev.*, **80**: 436 (1950).

[33] R. deWit, in M. F. Ashby, R. Bullough, C. S. Hartley, and J. P. Hirth (eds.), "Dislocation Models of Physical Phenomena," Pergamon, Oxford, 1981.

[34] The form is

$$\Omega = \int \frac{-(\mathbf{R}\times\mathbf{e})\cdot\boldsymbol{\xi}}{R}\,\frac{dl'}{(R+\mathbf{R}\cdot\mathbf{e})}$$

where $\mathbf{e}$ is an arbitrary, constant vector. The discontinuities in Ω occur when $R+\mathbf{R}\cdot\mathbf{e}=0$, which defines the cylindrical projection discussed above.

[35] P. P. Groves and D. J. Bacon, *Phil. Mag.*, **22**: 83 (1970).

[36] E. H. Yoffe, *Phil. Mag.*, **6**: 1147 (1961); see also J. L. Hokanson, *J. Appl. Phys.*, **34**: 2337 (1963), for a complete listing of the stress components of the doubly angular dislocation treated by Yoffe.

[37] Y. Maurissen and L. Capella, *Phil. Mag.*, **29**: 1227 (1974); **30**: 679 (1974).

result is an important step toward extension of the methods available for piecewise straight configurations in infinite media. However, this method cannot be applied to give an exact solution for a straight dislocation with skew orientation relative to a planar surface. The method can only represent a skew dislocation by a zig-zag approximation. Obviously, extension of the method to skew segments would greatly increase its applicability.

Validity and Limitations of the Simple Image Construction

As just discussed, the exact solution for dislocations near free surfaces is usually considerably more complicated than a simple image construction would yield. However, in many cases the image dislocation gives the dominant part of the solution and, in some cases, gives exact results.

To give a simple example of this concept of image forces, let us reconsider the screw dislocation of Fig. 3-7. The image force at $(x=\lambda, y=0)$ is given by Eq. (5-17) to be

$$\frac{F}{L}=\frac{\mu b^2}{2\pi\left[(R^2/\lambda)-\lambda\right]}=\frac{\mu b^2\lambda}{2\pi(R^2-\lambda^2)} \tag{5-91}$$

This is precisely the result found in Chap. 3, as can be verified by differentiating Eq. (3-28) with respect to λ. Integration shows the energy of the configuration to be

$$\frac{W}{L}=-\int_{R-\rho'}^{\lambda}\frac{F}{L}\,d\lambda=\frac{\mu b^2}{4\pi}\ln\frac{R^2-\lambda^2}{2\rho' R} \tag{5-92}$$

A cutoff parameter ρ' has been introduced to avoid a logarithmic infinity in W/L. Comparing Eq. (5-92) with Eq. (3-28), one sees that the surface cutoff parameter $2\rho'$ is equivalent to the core cutoff parameter r_0.

As another example, recall the case of an edge dislocation with its line parallel to a free surface (Sec. 3-5). In that case, or for a mixed dislocation with its line parallel to the surface, the image dislocation is not sufficient to yield the exact solution. However, the image force [Eqs. (3-18) and (3-81)] is given exactly by the image construction.[38]

Contrariwise, there are other simple situations where the image construction fails completely, both in satisfying free-surface boundary conditions and in estimating the force attracting the dislocation to the surface. Consider a mixed dislocation lying perpendicular to a planar free surface as in Fig. 5-17 with $\theta=\pi/2$. In a simple image construction there would be no force on the dislocation. However, rotating the dislocation into skew orientation such that β decreases and the dislocation assumes more screw character gives a decrease in

[38] This result is also valid in the anisotropic elastic case as shown by J. Lothe, *Phys. Norvegica* **2**: 153 (1967).

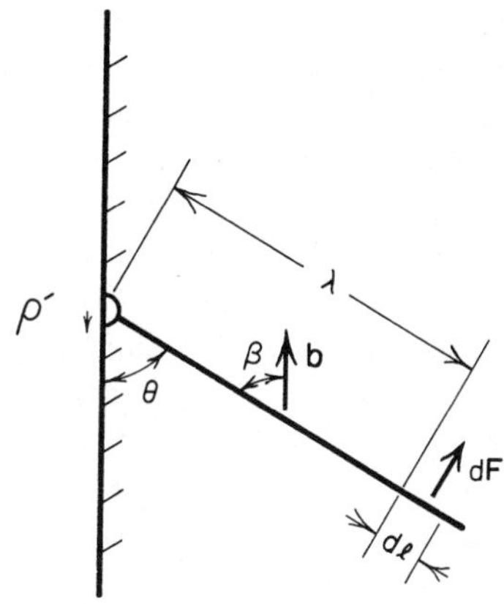

FIGURE 5-17. Dislocation line emerging at an angle θ with respect to a planar free surface.

the energy per unit length. Thus in the configuration of Fig. 5-17, surface-related forces should exist, consistent with such an effect. These forces can be found from Yoffe's solution[39] by procedures similar to those leading to Eq. (5-43). The calculation shows that indeed there is a force distribution along the dislocation

$$dF = \frac{\nu\mu b^2 \sin 2\beta}{4\pi(1-\nu)\lambda}\, dl \tag{5-93}$$

This force tends to rotate the dislocation into screw orientation. As it rotates, image forces develop that oppose the rotation. More generally for the case where $\theta \neq \pi/2$, the total force is given by[40]

$$dF = \frac{1}{\lambda}\left[-E(\beta)\cot\theta + \frac{\partial E(\beta)}{\partial \theta} \right] dl \tag{5-94}$$

This result is valid for the isotropic or anisotropic elastic case. As a specific example, consider the isotropic case with the prefactor $E(\beta)$ given by Eq. (3-87) and the dislocation skewing in its glide plane so that $d\theta = d\beta$ and $\beta = \theta + \beta_0 - \pi/2$, where β_0 is the value of β when $\theta = \pi/2$. The first term on the right side of Eq. (5-94) is the simple image term, which vanishes when $\theta = \pi/2$. The second term is seen to be precisely the force increment in Eq. (5-93), as it should be. In general, when $dF = 0$, Eq. (5-94) is a transcendental equation that can be solved to give the equilibrium value of β. The above effects are important in determining slip systems activated by surface dislocation sources.[41, 42]

[39] E. H. Yoffe, *Phil. Mag.*, **5**: 161 (1960).

[40] J. Lothe, in J. A. Simmons, R. deWit, and R. Bullough (eds), "Fundamental Aspects of Dislocation Theory," National Bureau of Standards Special Publication 317, 1970, p. 11.

[41] R. Gevers, S. Amelinckx, and P. Delavignette, *Phil. Mag.*, **6**: 1515 (1961).

[42] O. Lohne, in R. M. Latanision and J. T. Fourie (eds.), "Surface Effects in Crystal Plasticity," Noordhoff, Leyden, 1977, p. 487.

These examples show that one should not indiscriminately use the simple image construction as a first approximation. Rather, one should try to split the configuration for a given problem into components that are close to those in the various simple special cases that have been studied.

The integrated force of Eq. (5-94) diverges. The divergence at $\lambda \to \infty$ disappears when finite crystals are considered instead of semi-infinite crystals. The divergence as $\lambda \to 0$ is a singularity at the point of emergence of a dislocation at the surface. Thus, as illustrated in Fig. 5-17, one again must cut out material of radius ρ' around the point of emergence in calculations of force or energy.

Image Stresses at Internal Surfaces

In all the foregoing treatment, images at free surfaces only have been considered. Utilizing the analog between electrostatic image treatments[43, 44] and those of elasticity, one can also develop image forces for internal surfaces, as suggested by Head.[45]

Consider the arrangement in Fig. 5-18. A represents a right-handed screw dislocation parallel to the z axis and to the interface $x=0$ between materials 1 and 2 of shear moduli μ_1 and μ_2. The image dislocation of A is at C. To account for the finite strain in material 2, an additional image is placed at B. Both images are virtual images.[46] The dislocations A, B, and C have Burgers vectors of magnitude b, βb, and γb, respectively.

In view of the required boundary conditions that the displacements and stresses must be continuous across the boundary,[47] the three dislocations must lie in the plane $y=0$, and in fact, the results verify that such an arrangement fulfills the boundary conditions. Also, in order that the displacements at the boundary vary in the same way with varying y, it is necessary that $r=r_1=r_2$. Equation (3-2) then gives the displacements

$$u_{z_1}=\frac{b}{2\pi}\left(\tan^{-1}\frac{y}{x-a}+\gamma\tan^{-1}\frac{y}{x+a}\right)$$

$$u_{z_2}=\frac{b\beta}{2\pi}\tan^{-1}\frac{y}{x-a} \qquad (5\text{-}95)$$

[43] J. A. Stratton, "Electromagnetic Theory," McGraw-Hill, New York, 1941.

[44] B. I. Bleany and B. Bleany, "Electricity and Magnetism," Oxford University Press, Fair Lawn, N.J., 1957, p. 46.

[45] A. K. Head, *Phil. Mag.*, **44**: 92 (1953).

[46] In other words, the strain field in material 1 is the same as that which would be produced by dislocation A and the image dislocation C, while the strain field in material 2 is the same as that produced by dislocation B.

[47] Note that these conditions exclude nonelastic deformation of the boundary as well as of the bulk materials.

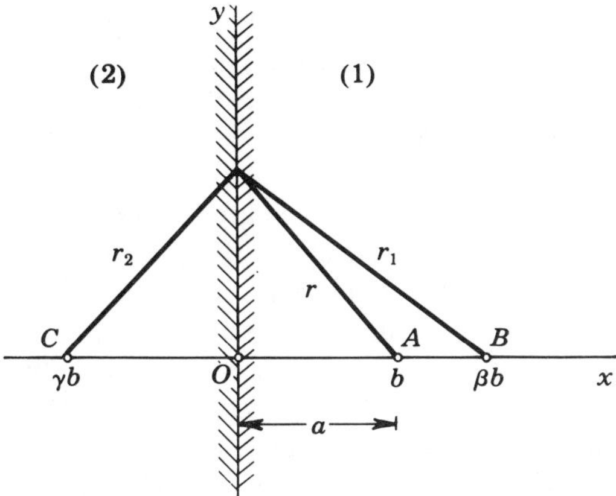

FIGURE 5-18. Image dislocations at an internal surface.

At $x=0$, setting $u_{z_1}=u_{z_2}$, one finds

$$\beta=1-\gamma \tag{5-96}$$

The stresses corresponding to Eq. (5-95) are

$$\sigma_{xz}(1)=-\frac{\mu_1 yb}{2\pi[(x-a)^2+y^2]}-\frac{\gamma\mu_1 yb}{2\pi[(x+a)^2+y^2]}$$

$$\sigma_{xz}(2)=-\frac{\beta\mu_2 by}{2\pi[(x-a)^2+y^2]} \tag{5-97}$$

The condition $\sigma_{xz}(1)=\sigma_{xz}(2)$ at $x=0$ yields the relation

$$\mu_1(1+\gamma)=\mu_2\beta \tag{5-98}$$

Simultaneous solution of Eqs. (5-96) and (5-98) gives the result

$$\gamma=\frac{\mu_2-\mu_1}{\mu_1+\mu_2} \qquad \beta=\frac{2\mu_1}{\mu_1+\mu_2} \tag{5-99}$$

This result is identical to that found by Head[48] by analogy to electrostatic images. In the limit $\mu_2\to 0$, the result approaches the solution for the image stress at a free surface Eq. (3-82), as it should.

Head also treated the edge-dislocation interaction at an internal surface. Although this case is considerably more complicated, he was able to show that the interaction with the boundary was given within 15 percent by an image

[48]A. K. Head, *Phil. Mag.*, **44**: 92 (1953).

construction equivalent to that in Fig. 5-18. Dundurs and Sendeckyj[49] have also performed detailed calculations of this type.

Equations (5-97) and (5-99) indicate that the screw dislocation is attracted to the boundary if $\mu_1 > \mu_2$ but repelled if $\mu_2 > \mu_1$. It follows that a glissile screw dislocation of this orientation is never attracted toward the interface from both sides and that, if in the interface, it is repelled to the side with smaller μ. In a similar treatment for an edge dislocation, Dundurs and Sendeckyj[49] showed that (1) attraction from one side and repulsion from the other and (2) repulsion from *both* sides were possible, whereas (3) attraction from both sides was not. For all cases of repulsion, dislocations can be trapped at the interface if their motion would create stacking faults with larger forces or if it would require climb and the repulsive force were insufficient to overcome the climb viscosity. Also, if, in addition to image forces, forces are included that are produced by the elastic incompatibilities at the interface causing redistribution of applied stresses, attraction from both sides is possible.[50] For mixed dislocations, image forces alone can produce attraction from both sides in rare cases.[51]

In addition, Head considered the interaction of a screw dislocation with a surface covered with a uniform film of thickness d and modulus μ_2. With $\mu_1 > \mu_2$, the screw is always attracted to the surface. If $\mu_2 > \mu_1$, qualitatively the screw should be attracted to the surface for $a \gg d$, when most of the image strain field is in vacuum, but repelled for $a \lesssim d$, when most of the image stress field resides in the harder surface film. Head's numerical results are in rough agreement with this expectation.

Alternative Image Constructions

For problems of the type illustrated in Figs. 3-16 and 5-18, there are several alternative methods of solution. In Sec. 3-5, for an edge dislocation parallel to a planar free surface, the elastic field additional to a simple dislocation image was determined directly by solution of the biharmonic equation. Dundurs and Mura[52] have solved Eq. (3-33) for a class of half-space problems and have listed stress functions which can be superposed to give the edge dislocation solution. Hirth et al.[53] have shown that the total solution[52] can be represented by a set of dislocation images together with images of line force sets at positions A and C in Fig. 5-18.

For more general two dimensional problems such as a straight dislocation parallel to a surface with one non-zero curvature, the problem can be solved in terms of Cauchy integrals, given the tractions produced on the surface by the

[49] J. Dundurs and G. P. Sendeckyj, *J. Appl. Phys.*, **36**: 3353 (1965).

[50] T. W. Chou and J. P. Hirth, *J. Comp. Mat.*, **4**: 102 (1970).

[51] D. M. Barnett and J. Lothe, unpublished research.

[52] J. Dundurs and T. Mura, *J. Mech. Phys. Solids*, **12**: 177 (1964).

[53] J. P. Hirth, D. M. Barnett, and J. Lothe, *Phil. Mag.*, **40A**: 39 (1979).

defect which is the source of stress.[54] Rice[55] has shown how this type of problem can be solved by superposing a continuous set of infinitesimal dislocations on the surface, facilitating the integration, and has given explicit solutions using such a method for a planar crack. Since the fields of line forces in the presence of half-spaces or cracks are known,[56, 57] a continuous set of line forces on the surface can also be used to solve such problems. For closed dislocation loops, solutions can be obtained by distributing infinitesimal point sources over a surface bounded by the loop and integrating their fields, which are also known for half-spaces.[58]

Marcinkowski[59] has used the continuous distribution of dislocations method extensively. We emphasize that all the above methods are correct and that choosing one is a matter of taste or convenience. Certainly, for the screw dislocation problem, the simple image method in Fig. 5-18 is the simplest. For the edge problem in Fig. 3-16, the Cauchy integral methods have some advantages. For an arbitrarily curved surface one can write down the proper integral, but it cannot be solved analytically in the general case, nor can the proper image construction be deduced. Such problems can be solved in a nonanalytical manner by constructing the electrostatic or magnetostatic analog and measuring the relevant field directly.

PROBLEMS

5-1. Consider two parallel edge dislocations A and B, with their Burgers vectors inclined at an angle α. Let dislocation A be constrained to remain at the origin of the coordinate system, and let both dislocations be constrained to remain parallel.

- **a.** Explicitly express the interaction force on B as a function of R, θ, and α.
- **b.** Is there an equilibrium position for dislocation B? If so, where?
- **c.** With R constant, show that F_θ vanishes when $\theta = \frac{1}{2}\alpha$.

5-2. Graphically compare the interaction force per unit length between the two coaxial dislocation loops of Sec. 5-3 as a function of z for the case $a = 200b$ with the interaction force per unit length between two parallel edge dislocations in the same glide plane as a function of their separation in the glide plane. In which case does the interaction force decrease more rapidly with increasing separation? Why?

[54] N. I. Muskhelishvili, "Mathematical Theory of Elasticity," Nordhoff, Leyden, 1975, Chap. 14.

[55] J. R. Rice, in H. Liebowitz (ed.), "Fracture," vol. II, Academic, New York, 1968, p. 191.

[56] J. Dundurs and M. Hetenyi, *J. Appl. Mech.*, **28**: 103 (1961); ibid., **29**: 362 (1962).

[57] J. P. Hirth and R. H. Wagoner, *Int. J. Solids Struct.*, **12**: 117 (1976).

[58] P. P. Groves and D. J. Bacon, *Phil. Mag.*, **22**: 83 (1970).

[59] M. J. Marcinkowski, "Unified Theory of the Mechanical Behavior of Matter," Wiley, New York, 1979.

5-3. Determine the local interaction force as a function of position on the dislocation line for two perpendicular screw dislocations. If one screw is right-handed and the other left-handed and they are forced together by uniform external stresses, qualitatively deduce the configuration that the dislocations must assume when the interaction forces balance those caused by the applied stresses.

5-4. Solve for the local interaction forces between a screw dislocation and a perpendicular edge dislocation whose Burgers vector is parallel to the line of closest approach between the two. Show that the interaction forces on one dislocation are perpendicular to the interaction forces on the other, so that there is no equal and opposite action and reaction. Discuss this seeming violation of Newton's law.

5-5. Compute the maximum local interaction force between the loop of Fig. 5-12 and a straight screw dislocation lying in the same glide plane and with its Burgers vector parallel to the z axis.

5-6. Compute the five stress components other than Eq. (5-56) for the angular dislocation.

5-7. Compute the six general stress components around an infinitesimal glide dislocation loop like that in Fig. 5-12.

5-8. Consider four parallel edge dislocations lying on the same glide plane, under no external stress, but blocked at both ends of an interval L by barriers which exert very short range repulsive forces, extending over atomic dimensions only. Determine the equilibrium configuration of the array if the dislocations are constrained not to climb.

5-9. Compute the long-range stress field around a dislocation dipole consisting of two parallel edge dislocations with equal and opposite Burgers vectors, in glide equilibrium, and whose glide planes are separated by a distance $2b$. *Hint*: Use the interaction force on an element as a "test probe" for the stress.

5-10. Discuss the role of grain boundaries in real crystals as sources of image stresses.

BIBLIOGRAPHY

1. Eshelby, J. D., in F. R. N. Nabarro (ed.), "Dislocations in Solids," vol. 1, North-Holland, Amsterdam, 1979, p. 167.
2. Eshelby, J. D., *Phil. Trans. Roy. Soc.*, **244A**: 87 (1951).
3. Head, A. K., *Phil. Mag.*, **44**: 92 (1953).
4. Kröner, E., *Ergeb. angew. Math.*, **5** (1958).
5. Yoffe, E. H., *Phil. Mag.*, **6**: 1147 (1961).

6

Applications to Self-Energies

6-1. INTRODUCTION

Energy calculations in dislocation theory are often made in the approximation of a dislocation line tension. The dislocation is considered to be a smooth flexible string with a line tension. If core irregularities are neglected, the dislocations do behave like smooth, taut strings in many cases. However, the analogy is not exact. The self-energy of a segment of dislocation line depends on its interaction with the other portions of the line, and hence depends on the dislocation configuration, so that the effective line tension is difficult to define with much accuracy.

A more accurate procedure would be to use Eq. (4-40) in energy calculations. In practice, unless the dislocation configuration is simple, the integrals in Eq. (4-40) are quite complex. For a fixed configuration, Eq. (4-40) can be solved numerically. However, one is often interested in determining an equilibrium configuration by the variation of its energy with configuration, and in such cases the use of Eq. (4-40) becomes enormously more complicated.

As with the previous treatments of interaction force and of stress, Eq. (4-40) is easier to use if the true dislocation configuration is approximated by a piecewise straight configuration and then the energy of this approximate configuration is calculated accurately. The principle of minimization of energy indicates that an approximate configuration, if not too wrong, has an energy quite close to that of the true configuration. By the introduction of variational parameters into an approximate piecewise straight configuration, it should be possible to obtain quite good values for the minimum-energy configuration if the energies of the approximate configurations can be calculated accurately. There is always the uncertainty of the core energy; in most cases, however, the elastic energy comprises the more important part of the total energy.

The purpose of this chapter is to provide explicit expression for the elastic energy of any configuration made up of straight segments, in terms of the corner coordinates of the configuration. We hope that this will enable researchers to make quantitative estimates which at first might appear formidable. Line tensions appropriate for the various cases are developed from the more general theory. As is the case for the force formulas of Chap. 5, the results of this chapter can be applied to infinitesimal segments and the total energy computed numerically if the equation of an arbitrarily curved dislocation line is known.

6-2. ENERGY OF A PIECEWISE STRAIGHT-DISLOCATION CONFIGURATION

Consider any piecewise straight-dislocation configuration—for example, the one in Fig. 6-1. The Burgers vector is conserved at the nodes; that is,

$$\mathbf{b}_1 = \mathbf{b}_2 + \mathbf{b}_4 \tag{6-1}$$

$$\mathbf{b}_3 = \mathbf{b}_2 + \mathbf{b}_5 \tag{6-2}$$

etc. (Axiom 1-2). Thus the above configuration can be considered to be made up of simple loops, say one loop $C_1C_2C_3$ with Burgers vector $\mathbf{b}_1$, one loop $C_4C_2C_5$ with Burgers vector $\mathbf{b}_2 - \mathbf{b}_1$, etc. With formal application of Eqs. (4-40) and (4-44) to the self-energies of the simple loops and their mutual interaction, one can write the total energy of a piecewise straight configuration made up of segments C_i with Burgers vector $\mathbf{b}_i$ as

$$W = \sum_i W_{s_i} + \sum_{i<j} W_{ij} \tag{6-3}$$

where the notation $i<j$ in the last sum indicates that each interaction energy should occur only once.

The interaction energies W_{ij} [Eq. (4-40)] converge. The self-energies diverge logarithmically, as noted in Sec. 4-7. To remove this divergence, let us tentatively postulate that two elements $d\mathbf{l}_1$ and $d\mathbf{l}_2$ *do not interact* when they are closer than some distance ρ. On the basis of this convention, the self-energy of a screw segment of length L is, from Eq. (4-39),

$$W_s = \frac{\mu b^2}{8\pi} \int_0^L dl_1 \left(\int_0^{l_1-\rho} \frac{dl_2}{l_1 - l_2} + \int_{l_1+\rho}^L \frac{dl_2}{l_2 - l_1} \right) \tag{6-4}$$

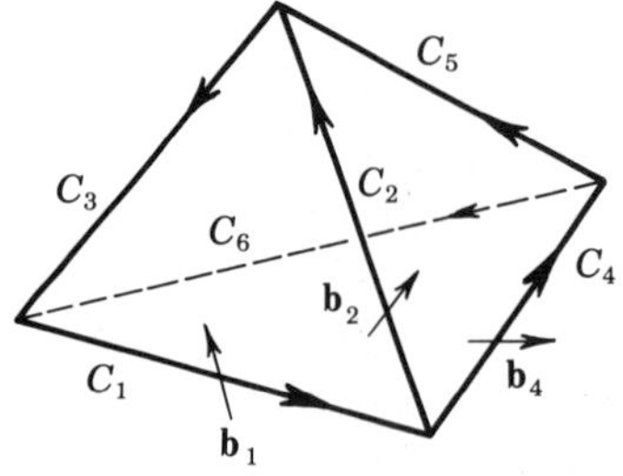

FIGURE 6-1. A piecewise straight-dislocation array. Arrows on lines indicate the sense of ξ.

The integrated form of Eq. (6-4) is

$$W_s = \frac{\mu b^2 L}{4\pi} \ln \frac{L}{e\rho} \tag{6-5}$$

For this cutoff procedure to be sensible, the energy of a straight segment of length $L = L_1 + L_2$ must be

$$W_s = W_{s_1} + W_{s_2} + W_{12} \tag{6-6}$$

where

$$W_{12} = \frac{\mu b^2}{4\pi} \int_0^{L_1} dl_1 \int_0^{L_2} \frac{dl_2}{l_1 + l_2} \tag{6-7}$$

Direct substitution provides a check that Eqs. (6-5) and (6-7) really fulfill Eq. (6-6). Thus our choice of subdivision into straight segments does not lead to ambiguity. This is true for a straight dislocation of any character. The energy per unit length of a screw dislocation along the axis of an infinite-length cylinder of radius R is, from Eq. (3-14),

$$\frac{W}{L} = \frac{\mu b^2}{4\pi} \ln \frac{\alpha R}{b} \tag{6-8}$$

The axis of the cylinder is an equilibrium position for the screw; thus the energy per unit length would not change to the *first order* in displacements from the center position. Now consider the configuration of Fig. 6-2: two screws of opposite Burgers vector, separated by a small distance $r \ll R$, with one at the center. As dislocation (2) is pushed in from the surface to r, it does work against its image stresses, including that caused by the core and its creation. This work is given by Eq. (5-92), which to the first order in r/R is equal to the result of Eq. (6-8). The dislocation already at the center does work on the second dislocation as it moves in, calculable from Eq. (5-17). The energy

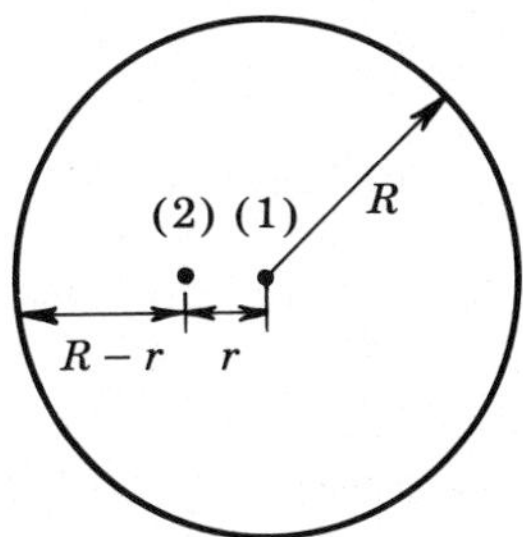

FIGURE 6-2. Two opposite-sense screw dislocations parallel to the cylinder axis of a right-circular cylinder viewed along the axis. Dislocation (1) lies along the axis.

per unit length of the final configuration thus must be

$$\frac{W}{L}=2\frac{\mu b^2}{4\pi}\ln\frac{\alpha R}{b}+\int_R^r\frac{\mu b^2}{2\pi x}dx=\frac{\mu b^2}{2\pi}\ln\frac{\alpha r}{b} \tag{6-9}$$

The external radius R has disappeared from the result, as it should.

When end effects are ignored, the energy of the same configuration, calculated from Eqs. (6-3) and (6-5), is

$$\frac{W}{L}=\frac{\mu b^2}{2\pi}\ln\frac{L}{e\rho}+\frac{W_{12}}{L} \tag{6-10}$$

where

$$W_{12}=-\frac{\mu b^2}{4\pi}\int_{-L/2}^{L/2}dz_2\int_{-L/2}^{L/2}\frac{dz_1}{\left[r^2+(z_1-z_2)^2\right]^{1/2}} \tag{6-11}$$

In the limit of large L, Eqs. (6-10) and (6-11) yield

$$\frac{W}{L}=\frac{\mu b^2}{2\pi}\ln\frac{r}{2\rho} \tag{6-12}$$

Equation (6-12), derived without consideration of end effects, should equal Eq. (6-9), derived for a pair of infinite-length screws. A comparison of Eqs. (6-9) and (6-12) reveals that one must choose

$$\rho=\frac{b}{2\alpha} \tag{6-13}$$

to achieve consistency.

An analogous development can be performed for the edge dislocation. The self-energy of a dislocation at the center of the cylinder in Fig. 6-2 is given by Eq. (3-57) as

$$\frac{W_s}{L}=\frac{\mu b^2}{4\pi(1-\nu)}\ln\frac{R\alpha}{b}-\frac{\mu b^2(3-4\nu)}{16\pi(1-\nu)^2} \tag{6-14}$$

Pushing in a second edge dislocation with the same Burgers vector on the common glide plane, a second self-energy term is added. Also, work is done by the force on the second dislocation, calculable from Eq. (5-17), arising from the stress field of the first dislocation. The force produced by the self-stresses [Eq. (3-44)] give a contribution $-[\mu b^2/2\pi(1-\nu)]\ln(R/r)$. Those from the surface relaxation stresses at $r=R$ [Eq. (3-50)] give a contribution $[\mu b^2/4\pi(1-\nu)]$.

Altogether, the energy of the pair of dislocations is

$$\frac{W}{L} = \frac{\mu b^2}{2\pi(1-\nu)}\left[\ln\frac{\alpha r}{b} - \gamma\right] \tag{6-15}$$

where $\gamma = (1-2\nu)/4(1-\nu)$. The energy of the same configuration, calculated from Eqs. (6-3) and (6-5), is

$$\frac{W}{L} = \frac{\mu b^2}{2\pi(1-\nu)}\ln\frac{r}{2\rho} \tag{6-16}$$

A comparison of Eqs. (6-15) and (6-16) shows that for consistency in the edge dislocation case,

$$\rho = \frac{b}{2\alpha}\exp\gamma \tag{6-17}$$

For typical values of ν from 0.2 to 0.33, $\exp\gamma$ varies from 1.13 to 1.21, close to unity. Thus, the result for the edge is numerically almost the same as that [Eq. (6-13)] for the screw. Since the ρ values appear in the logarithmic term, any uncertainties introduced in energies are completely negligible in comparison with those of the approximations of isotropic elasticity, St. Venant's principle, and the replacement of curved dislocations by straight line segments. The convention is adopted for energy calculations that ρ is constant, independent of dislocation character. Conversely, according to Eqs. (6-13) and (6-17), this corresponds to a small change in α with dislocation character.

With this convention, the self-energy of a straight-dislocation segment of any character is found to be

$$W_s = \frac{\mu}{4\pi}\left[(\mathbf{b}\cdot\boldsymbol{\xi})^2 + \frac{|(\mathbf{b}\times\boldsymbol{\xi})|^2}{1-\nu}\right]L\ln\frac{L}{e\rho} \tag{6-18}$$

Exercise 6-1. Show that the connection between ρ and α in Eq. (6-17) is independent of the plane on which the second dislocation is brought toward the first in Fig. 6-2. Suppose that it is pushed in on a plane inclined at an angle θ to the glide plane. Demonstrate that the work against the surface relaxation stresses of Eq. (3-50) then is greater than that when $\theta = 0$ by an amount $\mu b^2 \sin^2\theta/2\pi(1-\nu)$. All other terms are the same, so the result equivalent to Eq. (6-15) is also greater by this amount. Calculate W_{12} from Eq. (4-40) and show that it is also larger than the result of Eq. (6-16) by the same amount. Thus the two changes compensate, and Eq. (6-17) holds for any value of θ.

This formally completes a scheme for the calculation of self-energies. A precaution must be taken, however, when very acute angles are involved; if proper corrections are not made, quite wrong corner energies may result. One

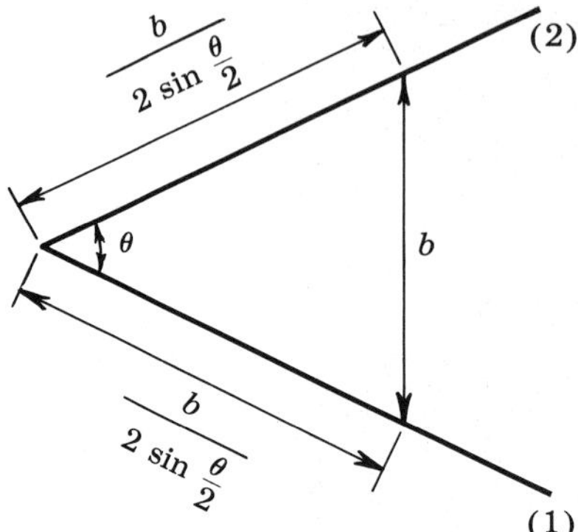

FIGURE 6-3. Corner configuration for intersecting dislocations (1) and (2).

should check in a specific calculation that the errors make up only a small fraction of the total energy to be calculated. Consider the typical corner in a dislocation configuration shown in Fig. 6-3. Near the corner, where the separation between the dislocations is less than $\sim b$, the interaction between the two dislocations is not purely elastic; rather, the length $l = b/2\sin(\theta/2)$ has an energy per unit length of the order of a typical core energy. When these effects are ignored and a purely elastic calculation is carried out, one obtains a spurious corner contribution

$$\Delta W_{\text{cor}} \cong \frac{\mu b^2}{4\pi} \int_0^l \int \frac{dx\,dy}{\left(x^2 + y^2 - 2xy\cos\theta\right)^{1/2}} \tag{6-19}$$

For very acute angles $\theta \ll 1$, this equation becomes

$$\Delta W_{\text{cor}} \cong \frac{\mu b^3}{4\pi} \frac{\ln(2/\theta)}{\theta} \tag{6-20}$$

which gives a catastrophic behavior as $\theta \to 0$. For typical angles, $\theta \sim \pi/2$, the energy uncertainty per corner is of the order

$$\Delta W_{\text{cor}} \sim \frac{\mu b^3}{4\pi} \tag{6-21}$$

In the region $\theta \sim \pi$ these effects can safely be ignored.

6-3. EXPLICIT EXPRESSIONS FOR W_{ij}

Consider two dislocation segments, x_1x_2 and y_1y_2, oriented relative to each other as shown in Fig. 6-4. A nonorthogonal, oblique coordinate system xyz is

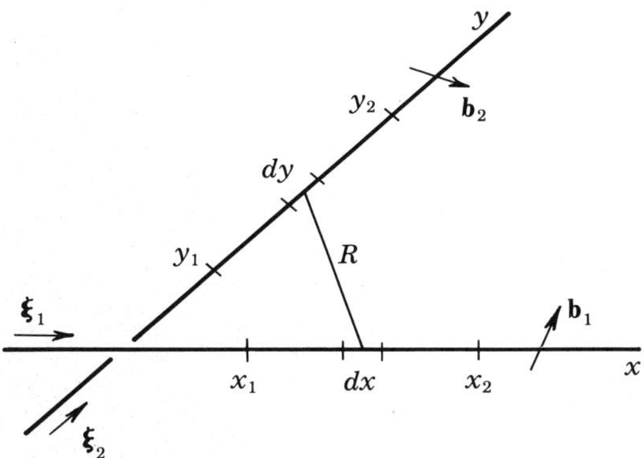

FIGURE 6-4. Coordinates for interaction calculations for dislocations (1) and (2).

constructed which consists of the two senses ξ_1 and ξ_2 and a third perpendicular unit vector

$$\mathbf{e}_3 = \frac{\xi_1 \times \xi_2}{|\xi_1 \times \xi_2|} \tag{6-22}$$

The origin lies on dislocation (2), dislocation (1) is at the position z, and θ is the angle between ξ_1 and ξ_2.

In this coordinate system, the interaction energy W_{12} between the two segments [Eq. (4-40)] takes the form

$$W_{12} = \left[-\frac{\mu}{2\pi}(\mathbf{b}_1 \times \mathbf{b}_2)\cdot(\xi_1 \times \xi_2) + \frac{\mu}{4\pi}(\mathbf{b}_1 \cdot \xi_1)(\mathbf{b}_2 \cdot \xi_2) \right] \int_{x_1}^{x_2} dx \int_{y_1}^{y_2} \frac{dy}{R} + \frac{\mu}{4\pi(1-\nu)}(\mathbf{b}_1 \times \xi_1)\cdot \mathbf{T} \cdot (\mathbf{b}_2 \times \xi_2) \tag{6-23}$$

where $\mathbf{T}$ is the tensor

$$\begin{aligned}\mathbf{T} = \int_{x_1}^{x_2} dx \int_{y_1}^{y_2} dy \Bigg[& \xi_2 \otimes \xi_1 \left(-\frac{\cos\theta}{\sin^4\theta}\frac{\partial^2}{\partial x^2} - \frac{1+\cos^2\theta}{\sin^4\theta}\frac{\partial^2}{\partial x\,\partial y} - \frac{\cos\theta}{\sin^4\theta}\frac{\partial^2}{\partial y^2} \right) \\ & + \mathbf{e}_3 \otimes \xi_1 \left(\frac{1}{\sin^2\theta}\frac{\partial^2}{\partial x\,\partial z} + \frac{\cos\theta}{\sin^2\theta}\frac{\partial^2}{\partial y\,\partial z} \right) \\ & + \xi_2 \otimes \mathbf{e}_3 \left(-\frac{\cos\theta}{\sin^2\theta}\frac{\partial^2}{\partial x\,\partial z} - \frac{1}{\sin^2\theta}\frac{\partial^2}{\partial y\,\partial z} \right) + \mathbf{e}_3 \otimes \mathbf{e}_3 \frac{\partial^2}{\partial z^2} \Bigg] R(x, y)\end{aligned} \tag{6-24}$$

Here

$$R^2(x,y)=x^2+y^2-2xy\cos\theta+z^2 \tag{6-25}$$

All the integrals occurring in Eqs. (6-23) and (6-24) can be calculated explicitly. For example,

$$\iint \frac{dx\,dy}{R}=I(x,y) \tag{6-26}$$

which is to be used in Eqs. (6-23) and (6-24) in the form

$$\int_{x_1}^{x_2}\int_{y_1}^{y_2}\frac{dx\,dy}{R}=I(x_2,y_2)-I(x_1,y_2)-I(x_2,y_1)+I(x_1,y_1)$$

$$=I(x_\alpha,y_\beta)\qquad \alpha,\beta=1,2 \tag{6-27}$$

Similarly, the functions J, K, and L are defined to satisfy the relations

$$\int_{x_1}^{x_2}\int_{y_1}^{y_2}\frac{dx\,dy}{R^3}=J(x_\alpha,y_\beta) \tag{6-28}$$

$$\int_{x_1}^{x_2}\int_{y_1}^{y_2}\frac{\partial}{\partial y}\frac{1}{R}\,dx\,dy=K(x_\alpha,y_\beta) \tag{6-29}$$

and

$$\int_{x_1}^{x_2}\int_{y_1}^{y_2}\frac{\partial^2 R}{\partial y^2}\,dx\,dy=L(x_\alpha,y_\beta) \tag{6-30}$$

Some manipulation reveals that

$$L(x,y)=-\cos\theta\,R(x,y)+y\sin^2\theta\,K(x,y) \tag{6-31}$$

In terms of the above functions, Eqs. (6-23) and (6-24) become

$$W_{12}=\left[-\frac{\mu}{2\pi}(\mathbf{b}_1\times\mathbf{b}_2)\cdot(\boldsymbol{\xi}_1\times\boldsymbol{\xi}_2)+\frac{\mu}{4\pi}(\mathbf{b}_1\cdot\boldsymbol{\xi}_1)(\mathbf{b}_2\cdot\boldsymbol{\xi}_2)\right]I(x_\alpha,y_\beta)$$

$$+\frac{\mu}{4\pi(1-\nu)}(\mathbf{b}_1\times\boldsymbol{\xi}_1)\cdot\mathbf{T}\cdot(\mathbf{b}_2\times\boldsymbol{\xi}_2) \tag{6-32}$$

where

$$\mathbf{T}=-\boldsymbol{\xi}_2\otimes\boldsymbol{\xi}_1\frac{\cos\theta}{\sin^4\theta}\left[L(x_\alpha,y_\beta)+L(y_\alpha,x_\beta)\right]$$

$$-\boldsymbol{\xi}_2\otimes\boldsymbol{\xi}_1\frac{1+\cos^2\theta}{\sin^4\theta}R(x_\alpha,y_\beta)+(\mathbf{e}_3\otimes\boldsymbol{\xi}_1-\boldsymbol{\xi}_2\otimes\mathbf{e}_3\cos\theta)\frac{z}{\sin^2\theta}K(y_\alpha,x_\beta)$$

$$+(\mathbf{e}_3\otimes\boldsymbol{\xi}_1\cos\theta-\boldsymbol{\xi}_2\otimes\mathbf{e}_3)\frac{z}{\sin^2\theta}K(x_\alpha,y_\beta)+\mathbf{e}_3\otimes\mathbf{e}_3\left[I(x_\alpha,y_\beta)-z^2J(x_\alpha,y_\beta)\right] \tag{6-33}$$

The calculation of the integrals and the reduction of the results are laborious and are not repeated here. For details the reader is referred to the work of Jøssang et al.[1] The following abbreviations are introduced:

$$\begin{aligned} s&=y\cos\theta-x+R \qquad & v&=y-x+R \\ t&=x\cos\theta-y+R \qquad & w&=x+y+R \\ u&=x-y+R \end{aligned} \tag{6-34}$$

In terms of these symbols, the integrals are

$$I(x,y)=\frac{x}{2}\ln\frac{z^2+v^2\operatorname{cotan}^2(\theta/2)}{st}+\frac{y}{2}\ln\frac{z^2+u^2\operatorname{cotan}^2(\theta/2)}{st}-z^2J(x,y) \tag{6-35}$$

$$J(x,y)=\frac{\tan(\theta/2)}{2z}\left[\tan^{-1}\left(\frac{u}{z}\operatorname{cotan}\frac{\theta}{2}\right)+\tan^{-1}\left(\frac{v}{z}\operatorname{cotan}\frac{\theta}{2}\right)\right]$$

$$-\frac{\operatorname{cotan}(\theta/2)}{z}\tan^{-1}\left(\frac{w}{z}\tan\frac{\theta}{2}\right) \tag{6-36}$$

$$K(x,y)=-\ln s \qquad K(y,x)=-\ln t \tag{6-37}$$

$$L(x,y)=-R(x,y)\cos\theta-y\sin^2\theta\ln s \tag{6-38}$$

and

$$L(y,x)=-R(x,y)\cos\theta-x\sin^2\theta\ln t$$

[1]T. Jøssang, J. Lothe, and K. Skylstad, *Acta. Met.*, **13**: 271 (1965); Eqs. (30) and (31) of this paper contain some misprints and omissions that are corrected here and by T. Jøssang, *Phys. Stat. Solidi*, **27**: 579 (1968). Jøssang also discusses later treatments of the interaction of straight dislocation segments by P. Humble, R. L. Segall, and A. K. Head, *Phil. Mag.*, **15**: 281 (1967) and R. deWit, *Phys. Stat. Solidi*, **20**: 575 (1967) and shows that they are both equivalent to that of Jøssang and colleagues when a misprint in deWit's paper is corrected.

The Coplanar Case $z = 0$

The method of using the preceding general formulas is not obvious in limiting cases where singularities appear. The use of the theory in such cases is briefly stated as follows. In the coplanar case, $J(x, y)$ is singular. But since it is multiplied by z^2, it vanishes from Eq. (6-33), which reduces to

$$\mathbf{T} = -\boldsymbol{\xi}_2 \otimes \boldsymbol{\xi}_1 \frac{\cos\theta}{\sin^4\theta}\left[L(y_\beta, x_\alpha) + L(x_\alpha, y_\beta)\right]$$

$$-\boldsymbol{\xi}_2 \otimes \boldsymbol{\xi}_1 \frac{1+\cos^2\theta}{\sin^4\theta} R(x_\alpha, y_\beta) + \mathbf{e}_3 \otimes \mathbf{e}_3 I(x_\alpha, y_\beta) \tag{6-39}$$

The interaction energy, Eq. (6-32), in this case becomes, after the insertion of Eqs. (6-38) and (6-39),

$$W_{12} = \frac{\mu}{4\pi}\left\{(\mathbf{b}_1 \cdot \boldsymbol{\xi}_1)(\mathbf{b}_2 \cdot \boldsymbol{\xi}_2) - 2\left[(\mathbf{b}_1 \times \mathbf{b}_2) \cdot (\boldsymbol{\xi}_1 \times \boldsymbol{\xi}_2)\right]\right.$$

$$\left. + \frac{1}{1-\nu}\left[\mathbf{b}_1 \cdot (\boldsymbol{\xi}_1 \times \mathbf{e}_3)\right]\left[\mathbf{b}_2 \cdot (\boldsymbol{\xi}_2 \times \mathbf{e}_3)\right]\right\}$$

$$\times I(x_\alpha, y_\beta) + \frac{\mu}{4\pi(1-\nu)}\left[(\mathbf{b}_1 \cdot \mathbf{e}_3)(\mathbf{b}_2 \cdot \mathbf{e}_3)\right]$$

$$\times\left\{R(x_\alpha, y_\beta) - \cos\theta\left[x_\alpha \ln t(x_\alpha, y_\beta) + y_\beta \ln s(x_\alpha, y_\beta)\right]\right\} \tag{6-40}$$

Here

$$I(x, y) = x \ln\frac{R + y - x\cos\theta}{x} + y \ln\frac{R + x - y\cos\theta}{y} \tag{6-41}$$

and $t(x, y)$ and $s(x, y)$ remain as defined in Eq. (6-34). Furthermore, with $x = 0$ or $y = 0$ and with only positive x and y (the coordinate system can always be selected so that x and y are positive), Eq. (6-35) reduces quite straightforwardly to

$$I(0, y) = y \ln\left(2\sin^2\frac{\theta}{2}\right) \qquad I(x, 0) = x \ln\left(2\sin^2\frac{\theta}{2}\right)$$

$$I(0,0) = 0 \tag{6-42}$$

Also,

$$L(x,0) = -R\cos\theta \qquad L(y,0) = -R\cos\theta$$

$$L(0,0) = 0 \tag{6-43}$$

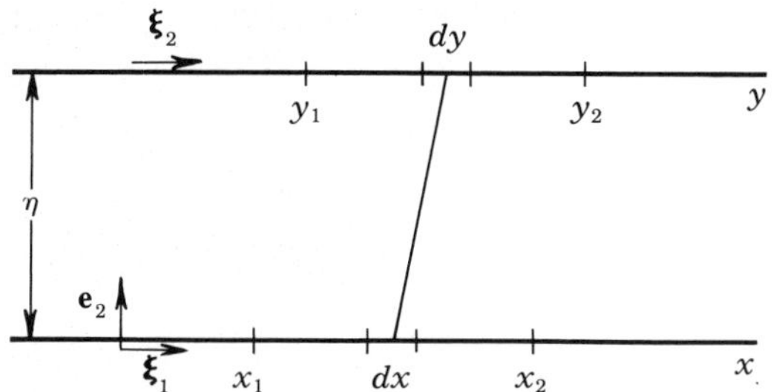

FIGURE 6-5. Coordinates for the interaction between parallel dislocations (1) and (2).

For $y>0$, there is no singular behavior in $L(x, y)$ as $x \to 0$, and similarly, none for $L(y, x)$ as $y \to 0$ when $x>0$.

Parallel Dislocations

The coordinate system of Fig. 6-4 breaks down, in the case of parallel dislocations, since $\boldsymbol{\xi}_1$ and $\boldsymbol{\xi}_2$ coincide. Instead, the coordinate system of Fig. 6-5 is adopted: $\mathbf{e}_3$ is perpendicular to the paper, $\mathbf{e}_2$ points from dislocation (1) to (2), and the separation between the two segments is denoted by η.

This case can be treated as a limiting case of the coplanar arrangement discussed above. The result is

$$W_{12} = \left(\frac{\mu}{4\pi}(\mathbf{b}_1 \cdot \boldsymbol{\xi}_1)(\mathbf{b}_2 \cdot \boldsymbol{\xi}_2) + \frac{\mu}{4\pi(1-\nu)} \right.$$

$$\left. \{(\mathbf{b}_1 \cdot \mathbf{e}_3)(\mathbf{b}_2 \cdot \mathbf{e}_3) + [(\mathbf{b}_1 \times \boldsymbol{\xi}_1) \cdot \mathbf{e}_3][\mathbf{e}_3 \cdot (\mathbf{b}_2 \times \boldsymbol{\xi}_2)]\} \right) I(x_\alpha, y_\beta)$$

$$+ \frac{\mu}{4\pi(1-\nu)} (\mathbf{b}_1 \cdot \mathbf{e}_3)(\mathbf{b}_2 \cdot \mathbf{e}_3) R(x_\alpha, y_\beta) \tag{6-44}$$

where now

$$I(x, y) = R - \tfrac{1}{2}(y-x)\ln s - \tfrac{1}{2}(x-y)\ln t \tag{6-45}$$

and

$$R(x, y) = [(x-y)^2 + \eta^2]^{1/2}$$

$$t = R + x - y \qquad s = R + y - x \tag{6-46}$$

Equation (6-45) can also be written in the simpler nonsymmetrical forms,[2]

$$I(x, y) = R - (y-x)\ln s \qquad \text{or} \qquad I(x, y) = R - (x-y)\ln t \tag{6-47}$$

[2]As η approaches zero, one must choose $\lim \eta \to 2\rho$ so that $W_{12} = -2LW_s$ for the annihilation and $+2LW_s$ for the combination cases.

Exercise 6-2. Derive Eqs. (6-44) and (6-45) directly from Eq. (4-40). This agreement with direct calculation provides some verification of the preceding theory.

6-4. ENERGIES OF DISLOCATION LOOPS

The Circular Loop

As an illustration of the preceding theory, let us calculate the energies of several specific configurations. Consider first the circular slip loop of radius R, as depicted in Fig. 6-6. Without lack of generality, the Burgers vector **b** can be taken to be parallel to the x axis. From Eq. (4-44), and with a cutoff procedure consistent with the one introduced in Sec. 6-2, the energy of the configuration must be

$$W=\frac{\mu b^2}{8\pi}\oint_C \sin\phi_0\, dl_0\int_{l_0+\rho}^{l_0+2\pi R-\rho}\frac{\sin\phi}{R'}\,dl$$

$$+\frac{\mu b^2}{8\pi(1-\nu)}\oint_C \cos\phi_0\, dl_0\int_{l_0+\rho}^{l_0+2\pi R-\rho}\frac{\cos\phi}{R'}\,dl \qquad (6\text{-}48)$$

It is easily verified that only a term $\mathbf{e}_3\otimes\mathbf{e}_3/R$ in $\mathbf{T}$ contributes to the last term in this formula.

Introducing

$$R'=2R\sin\frac{\phi-\phi_0}{2} \qquad dl=R\,d\phi \qquad dl_0=R\,d\phi_0 \qquad (6\text{-}49)$$

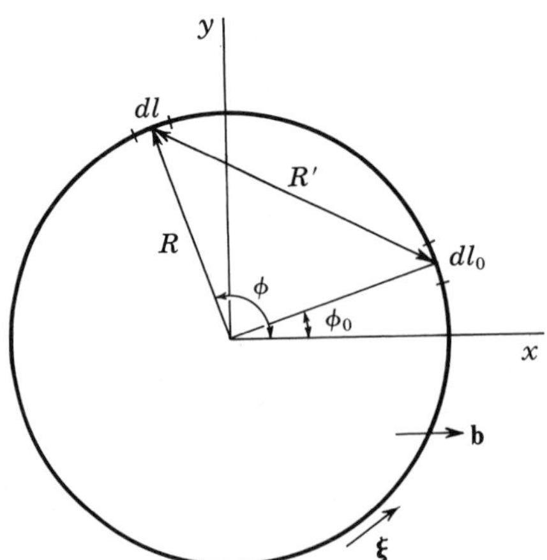

FIGURE 6-6. A circular dislocation loop.

one can integrate Eq. (6-48) explicitly to yield

$$W = 2\pi R \frac{2-\nu}{2(1-\nu)} \frac{\mu b^2}{4\pi} \left[-\ln\left(\tan \frac{\rho}{4R} \right) - 2\cos \frac{\rho}{2R} \right] \tag{6-50}$$

In most applications $R \gg \rho$, so that $\tan(\rho/4R) \sim (\rho/4R)$ and $\cos(\rho/2R) \sim 1$, and Eq. (6-50) can be simplified to the form

$$W = 2\pi R \frac{2-\nu}{2(1-\nu)} \frac{\mu b^2}{4\pi} \left(\ln \frac{4R}{\rho} - 2 \right) \tag{6-51}$$

The fact that W becomes negative when $R \leqslant \rho e^2/4$, that is, when the loop approaches atomic dimensions, simply indicates that continuum theory breaks down in describing the dislocation core.

Exercise 6-3. Consider a circular dislocation loop with the Burgers vector normal to the plane of the loop. Proceed as above and show that for this case the energy of the loop is given by

$$W = 2\pi R \frac{\mu b^2}{4\pi(1-\nu)} \left(\ln \frac{4R}{\rho} - 1 \right) \tag{6-52}$$

The Hexagonal Loop

In the treatments of force and energy in Chaps. 5 and 6, curved configurations are approximated by a set of straight line segments. A possible circular-loop equivalent of this type is the hexagonal arrangement shown in Fig. 6-7. The energy of this configuration is given by the sum of the self-energies of the six segments plus the interaction energies. The total self-energy is given by Eq. (6-18) to be

$$W_s = 6L \left[\frac{2-\nu}{2(1-\nu)} \right] \frac{\mu b^2}{4\pi} \ln \frac{L}{e\rho} \tag{6-53}$$

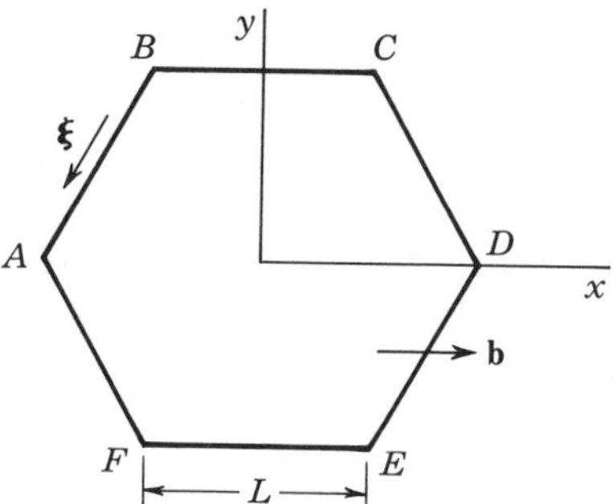

FIGURE 6-7. A hexagonal dislocation loop.

By symmetry, $W_{int}(AB,EF)=W_{int}(CD,EF)$, etc., so that the total interaction energy can be expressed as

$$W_{int}=W_{int}(BC,EF)+2W_{int}(AB,DE)+4W_{int}(AF,EF)$$
$$+2W_{int}(AB,AF)+4W_{int}(AB,EF)+2W_{int}(AF,DE)$$

Evaluating the various interaction terms by Eqs. (6-40) and (6-44), one finds that they sum to

$$W_{int}=6L\frac{2-\nu}{2(1-\nu)}\frac{\mu b^2}{4\pi}\left[2-\sqrt{3}-\tfrac{7}{2}\ln 3+2\ln\left(2\sqrt{3}+3\right)\right] \qquad (6\text{-}54)$$

Thus the total energy of the hexagonal loop is

$$W=6L\frac{2-\nu}{2(1-\nu)}\frac{\mu b^2}{4\pi}\left[\ln\frac{L}{\rho}+1-\sqrt{3}-\frac{7}{2}\ln 3+2\ln\left(2\sqrt{3}+3\right)\right]$$
$$=6L\frac{2-\nu}{2(1-\nu)}\frac{\mu b^2}{4\pi}\left(\ln\frac{L}{\rho}-0.84\right) \qquad (6\text{-}55)$$

It is left as an exercise to show that the energy of a similar hexagonal loop with its Burgers vector normal to the loop is

$$W=6L\frac{\mu b^2}{4\pi(1-\nu)}\left(\ln\frac{L}{\rho}+0.16\right) \qquad (6\text{-}56)$$

Yoffe,[3] who used a force method similar to that of Chap. 5, obtained nearly the same result for this case, but found a constant term of -0.14 instead of 0.16 [in terms of Eq. (6-54), the difference is $1-\ln 2$]. The difference may be associated with the different choice of a cutoff parameter ρ by Yoffe. With self-consistent choices for the cutoff parameter, the force method of Chap. 5 and the present method give identical results.

To compare the results for the energy per unit length for the circle and the hexagon, the loop areas are set equal, so that $R=(3\sqrt{3}/2\pi)^{1/2}L$. With this substitution, Eqs. (6-51) and (6-52) reduce to the same form as Eqs. (6-55) and (6-56), multiplied by $(\pi\sqrt{3}/6)^{1/2}=0.95$, except that the numerical factors are -0.80 and 0.20, respectively, instead of -0.84 and 0.16. Equivalently, if the circumferences are set equal, the numerical factors are -0.66 and 0.34. In an actual energy-minimization problem, the self-energy is equated to the work done by an applied stress, so that the equal-areas method is more meaningful. Thus, as suggested by the discussion in Sec. 5-1, the results for the circle agree with the results for the circle as approximated by a hexagon to within a small constant term that is a fraction of a core energy in magnitude.

[3] E. H. Yoffe, *Phil. Mag.*, **5**: 161 (1960).

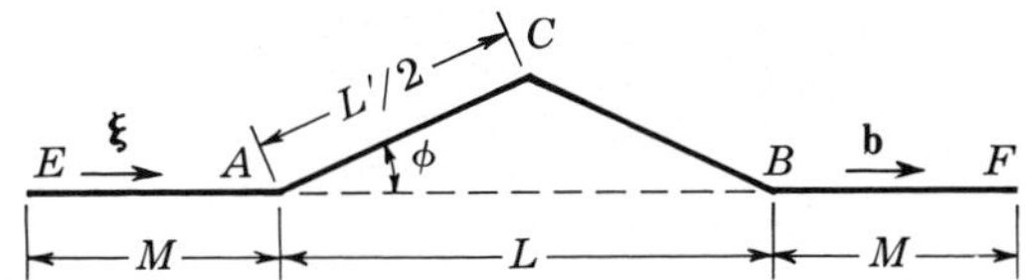

FIGURE 6-8. A small bow-out configuration on a screw dislocation.

Small Bow-Out

As an example of another type of application of the energy formula, consider a segment AB in an infinite screw dislocation bowed out to ACB, as shown in Fig. 6-8. The energy increase caused by the bow-out must be[4]

$$\Delta W = \lim_{M \to \infty} \Delta W_{(M)} \tag{6-57}$$

where $\Delta W_{(M)}$ is given in terms of the self-energies ΔW_s and interaction energies ΔW_{int} among the segments as

$$\begin{aligned}\Delta W_{(M)} = {} & W_s(AC) + W_s(CB) - W_s(AB) + W_{\text{int}}(AC, EA) \\ & + W_{\text{int}}(AC, CB) + W_{\text{int}}(AC, BF) + W_{\text{int}}(CB, EA) \\ & + W_{\text{int}}(CB, BF) - W_{\text{int}}(EA, AB) - W_{\text{int}}(BF, AB)\end{aligned} \tag{6-58}$$

Because of symmetry, Eq. (6-58) can be simplified to

$$\begin{aligned}\Delta W_{(M)} = {} & 2W_s(AC) - W_s(AB) + 2W_{\text{int}}(AC, EA) \\ & + 2W_{\text{int}}(AC, BF) + W_{\text{int}}(AC, CB) - 2W_{\text{int}}(EA, AB)\end{aligned} \tag{6-59}$$

The self-energies are calculable from Eq. (6-18) and the interaction energies are calculable from Eq. (6-40) in this coplanar case.

Let us first consider the contribution

$$\Delta W_1 = 2W_s(AC) + W_{\text{int}}(AC, CB) - W_s(AB) \tag{6-60}$$

Here

$$W_s(AC) = \frac{\mu b^2}{4\pi}\left(\cos^2\phi + \frac{\sin^2\phi}{1-\nu}\right)\frac{L'}{2}\ln\left(\frac{L'}{2e\rho}\right) \tag{6-61}$$

[4]Actually, the line must close back on itself or end at a free surface. However, by the reasoning of Sec. 5-1, the other segments do not appreciably contribute to ΔW, provided they are removed from the bow-out by distances large compared to L.

where

$$L' = \frac{L}{\cos\phi} \tag{6-62}$$

and

$$W_s(AB) = \frac{\mu b^2 L}{4\pi} \ln \frac{L}{e\rho} \tag{6-63}$$

The interaction energy is

$$W_{\text{int}}(AC,CB) = \frac{\mu b^2}{4\pi}\left(\cos^2\phi - \frac{\sin^2\phi}{1-\nu}\right) L \frac{1}{\cos\phi} \ln \frac{1+\cos\phi}{\cos\phi} \tag{6-64}$$

Expanding Eq. (6-64) in a Taylor's series about $\phi = 0$ and taking the limit $\phi \to 0$, one obtains

$$\Delta W_1 = \frac{\mu b^2 L \phi^2}{8\pi}\left(\frac{1+\nu}{1-\nu} \ln \frac{L}{4e\rho} + \frac{3}{2} - 2\ln 2\right) \tag{6-65}$$

In the limit $M \to \infty$, the remaining terms

$$\Delta W_2 = 2W_{\text{int}}(AC, EA) + 2W_{\text{int}}(AC, BF) - 2W_{\text{int}}(EA, AB) \tag{6-66}$$

are similarly calculated to yield

$$\Delta W_2 = \frac{\mu b^2 L \phi^2}{8\pi}(-2 + 2\ln 2) \tag{6-67}$$

The total energy increase given by Eq. (6-57), then, is

$$\Delta W = \frac{\mu b^2 L \phi^2}{8\pi}\left(\frac{1+\nu}{1-\nu} \ln \frac{L}{4e\rho} - 0.5\right)$$

$$= \frac{\mu b^2 L \phi^2}{8\pi(1-\nu)}\left[(1+\nu)\ln \frac{L}{\rho} - 1.89\nu - 2.89\right] \tag{6-68}$$

Exercise 6-4. Show that for an initially straight edge dislocation with its Burgers vector in the plane of the page in Fig. 6-8, and for a bow-out identical with that considered for the screw, the energy increase caused by the bow-out is

$$\Delta W = \frac{\mu b^2 L \phi^2}{8\pi(1-\nu)}\left[(1-2\nu)\ln \frac{L}{4e\rho} - 0.5\right]$$

$$= \frac{\mu b^2 L \phi^2}{8\pi(1-\nu)}\left[(1-2\nu)\ln \frac{L}{\rho} + 4.78\nu - 2.89\right] \tag{6-69}$$

Note that the leading term in the above formula is smaller than the corresponding logarithmic term in Eq. (6-68). This indicates that the energy per unit length of a screw dislocation is smaller than the energy per unit length of an edge dislocation under comparable conditions. With regard to the energy required for bow-out, discuss the significance of the fact that an initially straight edge dislocation develops a screw component upon bowing out.

The energy of a mixed dislocation bowing out in glide is given by the result of Eq. (6-68) times $\cos^2\beta$ plus the result of Eq. (6-63) times $\sin^2\beta$, where β is the angle between the Burgers vector and the original straight line.

For an edge dislocation with its Burgers vector *normal* to the page in the configuration of Fig. 6-8, the energy increase is

$$\Delta W = \frac{\mu b^2 L \phi^2}{8\pi(1-\nu)} \ln\frac{L}{4e\rho} = \frac{\mu b^2 L \phi^2}{8\pi(1-\nu)}\left(\ln\frac{L}{\rho} - 2.39\right) \tag{6-70}$$

Large Bow-Out

A final result, which is utilized later, is that for the energy increase in forming the large bow-out shown in Fig. 6-9. Since the procedures are the same as those discussed above, only the results are presented. For an initially screw dislocation,

$$\Delta W = \frac{\mu b^2 L}{4\pi(1-\nu)}\left[\left(1+\frac{\nu}{2}\right)\ln\frac{L}{\rho} - 0.04\nu - 2.05\right] \tag{6-71}$$

For an edge dislocation with **b** in the plane of the page in Fig. 6-9,

$$\Delta W = \frac{\mu b^2 L}{4\pi(1-\nu)}\left[\left(1-\frac{3\nu}{2}\right)\ln\frac{L}{\rho} + 2.00\nu - 2.05\right] \tag{6-72}$$

The energy of a mixed dislocation which has bowed out by glide is given by the result of Eq. (6-71) times $\cos^2\beta$ plus the result of Eq. (6-72) times $\sin^2\beta$. For an edge dislocation with **b** normal to the page,

$$\Delta W = \frac{\mu b^2 L}{4\pi(1-\nu)}\left(\ln\frac{L}{\rho} - 1.59\right) \tag{6-73}$$

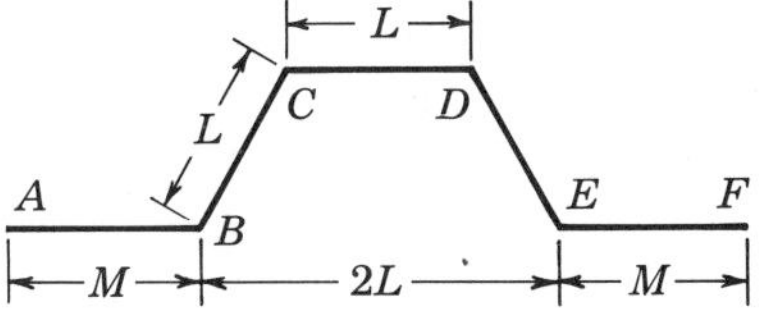

FIGURE 6-9. A large bow-out configuration, semi-hexagon.

6-5. THE CONCEPT OF LINE TENSION

Formally, the line tension in a taut string is defined by the incremental increase in stored energy with an infinitesimal increase in length,

$$\mathcal{S} = \frac{\delta W}{\delta \mathfrak{L}} \tag{6-74}$$

This concept, and its description by Eq. (6-74), is often extended by analogy to dislocation lines, but as mentioned at the outset of this chapter, the analogy is not exact, and the meaning of a line tension for a dislocation is somewhat nebulous.[5] To illustrate this uncertainty, let us consider the dislocation configurations described by Eqs. (6-56), (6-60), and (6-73). These dislocations are edge in character at all times during bow-out, so that complications caused by change in dislocation character are absent. The bow-out process in such cases would require *climb* of the dislocations

For the hexagonal loop,

$$\delta \mathfrak{L} = 6\,\delta L \tag{6-75}$$

Together with Eqs. (6-56) and (6-74), this yields the line tension for this case

$$\mathcal{S} = \frac{\mu b^2}{4\pi(1-\nu)}\left(\ln\frac{L}{\rho} + 1.16\right) \tag{6-76}$$

For the semihexagon of Eq. (6-73), $\delta \mathfrak{L} = \delta L$ and

$$\mathcal{S} = \frac{\mu b^2}{4\pi(1-\nu)}\left(\ln\frac{L}{\rho} - 0.59\right) \tag{6-77}$$

For the small bow-out represented by Eq. (6-70), $\delta \mathfrak{L}$ can be expanded in a Taylor's series in ϕ. The first nonzero term is the second-order term giving

$$\delta \mathfrak{L} = \frac{1}{2}\,\frac{\partial^2 L'}{\partial \phi^2}\,\delta\phi^2 = \tfrac{1}{2} L \phi^2$$

which in turn yields

$$\mathcal{S} = \frac{\mu b^2}{4\pi(1-\nu)}\left(\ln\frac{L}{\rho} - 2.39\right) \tag{6-78}$$

Evidently, the line tension is greatest for the closed hexagon, where the segments which interact with a given segment are near to one another, while the line tension becomes less as the interacting segments are removed from the

[5] J. P. Hirth, T. Jøssang, and J. Lothe, *J. Appl. Phys.*, **37**: 110 (1966).

vicinity of the bow-out, as in the semihexagon and small bulge. Only in the limit $L \gg \rho$ do the line tensions of the various configurations converge to a common value. Even for bow-outs where L is as large as $10^5\rho$, the result of Eq. (6-76) is 40 percent larger than that of Eq. (6-78).

Since the straight-line results are quite good approximations of those for curved loops, as shown in the previous section, Eqs. (6-76) and (6-77) should also apply to a circular loop and a semicircular loop, respectively.[6] Because both of these have the same radius of curvature, the above results also indicate that, unlike the case of a stretched string, the effective line tension does not depend on curvature alone. Considering the curved equivalent of the small bow-out, described by Eq. (6-78), one sees that the line tension in this case is a logarithmic function of the *length* of the bow-out, rather than of the curvature; this was first pointed out by Mott and Nabarro.[7]

Orientation-Dependent Line Tension

A further complication arises in cases where the screw-edge character of the dislocation changes as it bows out, as it would in the case of *glide bow-out*. As indicated by Eq. (6-18), the self-energy of a dislocation segment varies with its orientation in such a case, so that in the line-tension analogy, one must deal with an orientation-dependent line tension. Such an orientation dependence has been dealt with by Herring,[8] Frank,[9] Mullins,[10] and others in their treatments of the anisotropic orientation-dependent surface energy of solids. They show that for a two-dimensional surface, the surface tension, which corresponds to a line tension, is given by[11]

$$\mathcal{S} = \gamma + \frac{\partial^2 \gamma}{\partial \theta^2} \tag{6-80}$$

where γ is the surface energy, which is dependent on the orientation θ.

[6] Indeed, for the circular loop described by Eq. (6-52), $\delta\mathcal{L} = 2\pi\delta R$, so that with the equal-areas condition $R = (3\sqrt{3}/2\pi)^{1/2}L$,

$$\mathcal{S} = \frac{\mu b^2}{4\pi(1-\nu)} \ln\frac{4R}{\rho} = \frac{\mu b^2}{4\pi(1-\nu)}\left(\ln\frac{L}{\rho} + 1.29\right) \tag{6-79}$$

The agreement with Eq. (6-76) is within the same numerical factor as the agreement between Eqs. (6-52) and (6-56).

[7] N. F. Mott and F. R. N. Nabarro, in "Report on Strength of Solids," Physical Society, London, 1948, p. 1.

[8] C. Herring, in W. E. Kingston (ed.), "The Physics of Powder Metallurgy," McGraw-Hill, New York, 1951, p. 143.

[9] F. C. Frank, in N. A. Gjostein and W. D. Robertson (eds.), "Metal Surfaces," American Society of Metals, Cleveland, Ohio, 1963, p. 1.

[10] W. W. Mullins, ibid., p. 17.

[11] This result may be succinctly derived as follows. Suppose the bow-out of a two-dimensional surface as shown in Fig. 6–10. Expanded about the point $\theta = 0$, the orientation-dependent surface

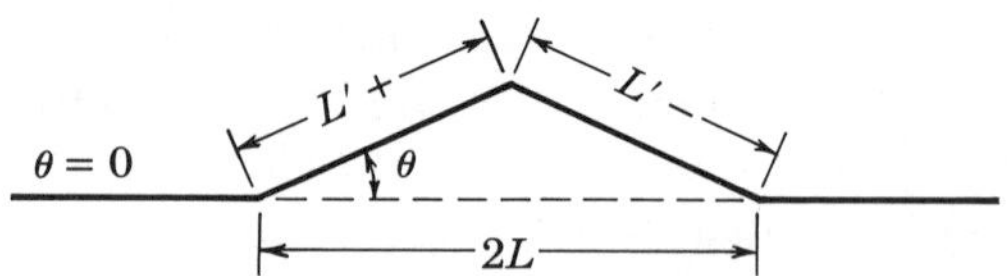

FIGURE 6-10. A bulged surface. The original flat surface has orientation $\theta = 0$.

The self-energy per unit length of a dislocation line W_s/L depends on orientation through the angle β between the line and the Burgers vector. Thus, as first noted by de Wit and Koehler,[12] the equivalent of Eq. (6-80) in the case of a dislocation is

$$\mathbb{S} = \frac{W_s}{L} + \frac{\partial^2 (W_s/L)}{\partial \beta^2} \tag{6-81}$$

With Eq. (6-18) inserted into Eq. (6-81), the simple line-tension analogy yields[13]

$$\mathbb{S} = \frac{\mu b^2}{4\pi(1-\nu)} \left[(1+\nu)\cos^2\beta + (1-2\nu)\sin^2\beta\right] \ln \frac{L}{e\rho} \tag{6-82}$$

---

energy is

$$\gamma(\theta) = \gamma(0) + \frac{\partial \gamma}{\partial \theta}\delta\theta + \frac{\partial^2 \gamma}{\partial \theta^2}\frac{(\delta\theta)^2}{2} + \cdots$$

For this bow-out, $\delta\theta = \theta$ for L'_+, $\delta\theta = -\theta$ for L'_-, and

$$L' \cong L\left[1 + \frac{(\delta\theta)^2}{2}\right] \cong L\left(1 + \frac{\theta^2}{2}\right)$$

so that to second order in θ, the change in surface energy is

$$\delta W = \left(\gamma + \frac{\partial \gamma}{\partial \theta}\theta + \frac{\partial^2 \gamma}{\partial \theta^2}\frac{\theta^2}{2}\right)L'_+ + \left[\gamma + \frac{\partial \gamma}{\partial \theta}(-\theta) + \frac{\partial^2 \gamma}{\partial \theta^2}\frac{(-\theta)^2}{2}\right]L'_- - 2\gamma L$$

$$= 2\left(\gamma + \frac{\partial^2 \gamma}{\partial \theta^2}\frac{\theta^2}{2}\right)L\left(1 + \frac{\theta^2}{2}\right) - 2\gamma L = \theta^2 L\left(\gamma + \frac{\partial^2 \gamma}{\partial \theta^2}\right)$$

Since $\delta\mathfrak{L} = L\theta^2$,

$$\mathbb{S} = \frac{\delta W}{\delta \mathfrak{L}} = \gamma + \frac{\partial^2 \gamma}{\partial \theta^2}$$

[12] G. de Wit and J. S. Koehler, *Phys. Rev.*, **116**: 1113 (1959).

[13] In the literature the logarithmic term is often written $\ln(R/r_0)$, where R is the radius of curvature and r_0 is the core radius.

Now consider the small bow-out described by Eqs. (6-68) and (6-69). Here $\delta\mathcal{L}=\frac{1}{2}L\phi^2$, so that for the initially screw dislocation,

$$\mathcal{S}=\frac{\mu b^2}{4\pi(1-\nu)}\left[(1+\nu)\ln\frac{L}{\rho}-1.89\nu-2.89\right] \tag{6-83}$$

and for the initially edge dislocation,

$$\mathcal{S}=\frac{\mu b^2}{4\pi(1-\nu)}\left[(1-2\nu)\ln\frac{L}{\rho}+4.78\nu-2.89\right] \tag{6-84}$$

Comparison of these results with Eq. (6-82) shows that once again the results of the line-tension analogy converge to the actual results only in the limit $L\gg\rho$. Inspection of Eqs. (6-55), (6-71), and (6-72) shows that the effective line tensions are increasingly larger for the semihexagon and hexagon, as was the case for the edge dislocation treated earlier. For a mixed dislocation the disagreement between the results of Eq. (6-82) and those of (6-83) and (6-84) depend on the value of β, but then can be as much as a factor of 2 for typical bow-out lengths $L\sim10^3b$ with $\nu=\frac{1}{3}$.[14] Thus, although it predicts the correct expression for the logarithmic term in the effective line tension, Eq. (6-82) is only an approximation which becomes relatively more accurate the larger L is.

In anisotropic solids, instead of Eq. (6-18), one has more generally

$$\frac{W_s}{L}=E(\beta)\ln\left(\frac{L}{e\rho}\right) \tag{6-85}$$

where $E(\beta)$ is the anisotropic elastic prelogarithmic energy factor. Methods for calculating $E(\beta)$ are presented in Chap. 13. Equation (6-81) applies also for the anisotropic case and the generalization of Eq. (6-82) becomes

$$\mathcal{S}=\left[E(\beta)+\frac{\partial^2E(\beta)}{\partial\beta^2}\right]\ln\frac{L}{e\rho} \tag{6-86}$$

As for Eqs. (6-80) and (6-81), the absence of $(\partial E/\partial\beta)$ terms arises from cancellation of opposite sign terms for the two segments in the bulge (Fig. 6-10). A dislocation segment ending at a free surface, however, can be rotated without a compensating opposite rotation of another segment. In such a case $(\partial E/\partial\beta)$ terms appear in the restoring force (and hence in the effective line tension) as discussed in connection with Eq. (5-94).

Energy Flow along Dislocations

Consider now that a straight dislocation ending at a free surface is moved forward by a parallel displacement δx (Fig. 6-11). The work supplied to move

[14] The variation of $\mathcal{S}$ with orientation is discussed in J. P. Hirth, T. Jøssang, and J. Lothe, *J. Appl. Phys.*, **37**: 110 (1966), and L. M. Brown, *Phil. Mag.*, **10**: 441 (1964).

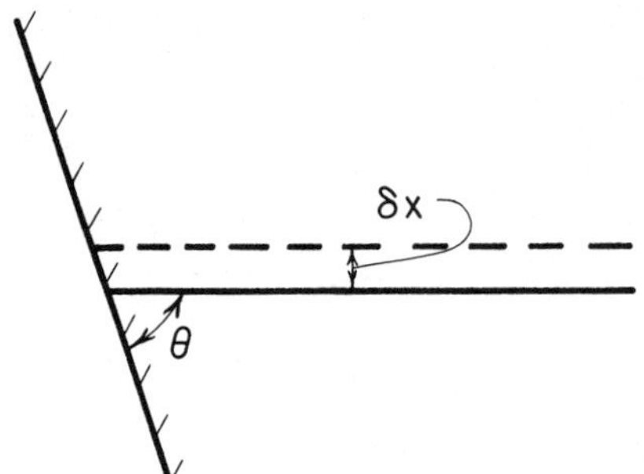

FIGURE 6-11. View parallel to free surface showing dislocation emerging at an angle θ.

the dislocation against the force given by the first term on the right side of Eq. (5-94) is that needed to increase the dislocation line length. Work is also needed to overcome the second [*viz.*, $(\partial E/\partial\theta)$] term, although the dislocation does not change orientation. This work does not appear as configurational energy at the surface since the local configuration is the same before and after the displacement. The only reasonable interpretation is that the second work increment is transported away along the dislocation as it moves forward.

Thus we have arrived at the concept that a parallel displacement of a dislocation leads to transport not only of energy in the direction of the displacement, but also along the line direction of the dislocation. Since the energy transport involves work against forces acting to prevent the displacement, the concept is connected with that of line tension. Let us now formally describe the transport effect.

A dislocation in an infinite medium with its line $\boldsymbol{\xi}//\mathbf{k}$, a unit vector lying along the z axis, is supposed to be displaced by δx. The displacement field produced by the dislocation at a given position then changes by $\delta\mathbf{u}=-(\partial\mathbf{u}/\partial x)\,\delta x$. During this process an energy increment is transported along the z axis, through an area A perpendicular to $\mathbf{k}$, in the amount

$$P\,\delta x=-\int_A \mathbf{k}\cdot\boldsymbol{\sigma}\cdot\delta\mathbf{u}\,dA=\int_A \mathbf{k}\cdot\boldsymbol{\sigma}\cdot\frac{\partial\mathbf{u}}{\partial x}\,\delta x\,dA \tag{6-87}$$

where $\boldsymbol{\sigma}$ is the stress tensor of the dislocation. Since $\boldsymbol{\sigma}$ and $(\partial\mathbf{u}/\partial x)$ are both $\propto r$ while $dA=r\,dr\,d\phi$ in cylindrical coordinates fixed on the dislocation, with the cancellation of δx, application of Eq. (6-87) to the area between two circles of radii r_1 and r_2 gives an expression for P of the form

$$P=p\ln\frac{r_2}{r_1} \tag{6-88}$$

where the prelogarithmic factor p is

$$p=\int_0^{2\pi} r^2\,\mathbf{k}\cdot\boldsymbol{\sigma}\cdot\frac{\partial\mathbf{u}}{\partial x}\,d\phi \tag{6-89}$$

with the integrand independent of r.

Let $E(\theta)$ now be the prelogarithmic energy factor as a function of the angle θ between the z axis and the dislocation line, rotated in the $x-z$ plane. The precise statement of the energy flow theorem[15] is

$$p=\left(\frac{\partial E}{\partial \theta}\right)_{\theta=0} \tag{6-90}$$

Theorem (6-90) can be checked readily in the isotropic elastic case. The Burgers vector of the dislocation has components $b_x=b_e$ and $b_z=b_s$ when $\theta=0$. The extended form of Eq. (6-89) is

$$p=\int_0^{2\pi} r^2\left[\sigma_{xz}^s\frac{\partial u_x^e}{\partial x}+\sigma_{yz}^s\frac{\partial u_y^e}{\partial x}+\sigma_{zz}^e\frac{\partial u_z^s}{\partial x}\right]d\phi \tag{6-91}$$

As denoted by the superscripts s and e, indicating screw and edge, respectively, only cross terms between the screw and edge components contribute to p in the isotropic elastic case. Inserting into Eq. (6-91) from Eqs. (3-2), (3-3), (3-43), (3-45) and (3-46), we obtain

$$p=\frac{\mu\nu}{2\pi(1-\nu)}b_e b_s \tag{6-92}$$

For an arbitrary θ, the screw component is $(b_s\cos\theta-b_e\sin\theta)$ and the edge component is $(b_s\sin\theta+b_e\cos\theta)$ in terms of the *original* screw and edge components b_s and b_e for $\theta=0$. Thus Eq. (3-87) gives

$$E(\beta)=E(\theta)=\frac{\mu}{4\pi(1-\nu)}\left[(1-\nu)(b_s\cos\theta-b_e\sin\theta)^2+(b_s\sin\theta+b_e\cos\theta)^2\right] \tag{6-93}$$

Hence

$$\left(\frac{dE}{d\theta}\right)_{\theta=0}=\frac{\mu\nu}{2\pi(1-\nu)}b_e b_s \tag{6-94}$$

A comparison of Eqs. (6-92) and (6-94) proves that theorem (6-90) is valid for the isotropic elastic case.

Dislocations in Complex Configurations

All the above disagreements with the simple concept of line tension reflect the fact that a dislocation segment interacts with *all other segments* of a given configuration, so that the line-tension concept, where only the local configuration is considered, is only approximate. For a specific simple configuration it is

[15]A complete proof is given in the original work by J. Lothe, *Phil. Mag.*, **15**: 9 (1967).

therefore preferable to use the exact energy formulas. However, often one is interested in the energy of a dislocation segment in a complex array or tangle of dislocations, where an exact calculation of the energy is too difficult and tedious to be undertaken. Qualitatively, the stress field of a given dislocation is largely canceled out by the stress fields of the remainder of the complex array of dislocations outside a region, approximately cylindrical with radius R, bounding the dislocation. R is taken as roughly the interdislocation spacing in this approximation. In such a case a working approximation for the self-energy per unit length of dislocation or line tension is given by Eq. (6-82) as

$$\mathcal{S} \sim \frac{\mu b^2}{4\pi(1-\nu)}\left[(1+\nu)\cos^2\beta+(1-2\nu)\sin^2\beta\right]\ln\frac{R}{\rho}$$
$$\sim \mu b^2 \tag{6-95}$$

The uncertainties in the use of Eq. (6-82) here are no worse than those entailed in fixing a value for R.

Line-tension reasoning sometimes gives exact results. Such a case arises when one is interested in minimizing the energy of a configuration where local interactions are of minor importance. An example is a threefold dislocation node. Lothe[16] and others[17] have shown that simple line-tension models give the same equilibrium result as do the complex force calculations of Chap. 5, except for a region of size $\sim 50b$ adjacent to the node.

PROBLEMS

6-1. Compute the energy of a semihexagonal dislocation loop lying normal to and terminating at a free surface. The Burgers vector is parallel to the free surface. Use a simple image construction. Compare the results with those of Eqs. (6-71) and (6-72).

6-2. Suppose that each loop in Prob. 6-1 assumes an unstable equilibrium configuration under the same force σb. In which case will the equilibrium value of L be larger?

6-3. For equal-area loops in each case, determine the energy of regular polygons with 3, 4, 8, 12, and n sides. Compare the results with those given for the circle and the hexagon. Take $\mathbf{b}$ normal to the plane of the loop in each case.

6-4. Calculate the energy of the square loop of Fig. 6-12 as a function of the angle ϕ.

6-5. Compute the interaction energy between two perpendicular screw dislocations. Differentiate this energy with respect to the distance of separation

[16]J. Lothe, *Phil. Mag.*, **15**: 353 (1967).

[17]V. L. Indenbom and G. N. Dubnova, *Sov. Phys. Solid State*, **9**: 915 (1967).

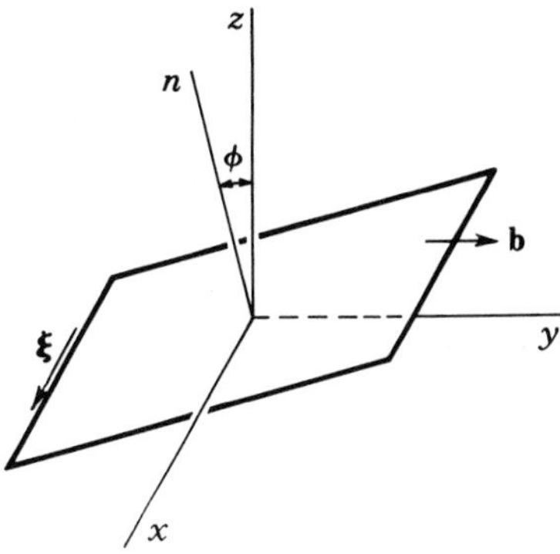

FIGURE 6-12. A square dislocation loop with the origin at the center of the square. **b** is parallel to the y axis. The normal to the square is rotated about the x axis by an angle ϕ from the z axis.

between the two and show that the resulting force agrees with that of Exercise 5-5.

6-6. Under what conditions could a straight dislocation line spontaneously break up into a zigzag dislocation line? *Hint*: The problem is analogous to that of breaking up a flat surface into a hill-and-valley structure.[18] In the isotropic approximation, can any type of dislocation break up in this manner?

BIBLIOGRAPHY

1. Jøssang, T., J. Lothe, and K. Skylstad, *Acta Met.*, **13**: 271 (1965).
2. Kröner, E., *Erg. ang. Math.*, **5** (1958).

[18] C. Herring, *Phys. Rev.*, **82**: 87 (1949).

7

Dislocation Dynamics

7-1. INTRODUCTION

The preceding chapters have dealt with the theory for the displacement fields and stresses generated by static dislocation configurations. Under most dynamic conditions, dislocations move so slowly that the dynamic stresses and displacements are approximated quite accurately by these static solutions. Hence this material provides sufficient background for most dynamic applications. However, there are some special fields of experimental study in which dynamic effects are thought to be important. In the theory for megacycle internal friction, the dislocations are considered as damped vibrating strings.[1] The inertia of the oscillating string is determined by the effective mass per unit length of a moving dislocation, a dynamic property. The damping constant is related to the interaction between dislocations and lattice vibrations, another topic in dislocation dynamics.[2, 3] Also very high speed dislocations are of interest in connection with shock loading and crack formation.[4]

Many of the theories developed for the explanation of specific experiments are still quite tentative. Therefore, the theory of dislocation dynamics is presented here in quite a general approach. The bulk of the chapter is devoted to moving straight dislocations and inertial forces on them. The present status of the theory for dislocation mobility is briefly discussed in the final section.

7-2. THE MOVING SCREW DISLOCATION

Displacements and Stresses

Consider a screw dislocation parallel to the z axis, with sense and Burgers vector as shown in Fig. 7-1. For a straight screw dislocation in an infinite

[1]A. Granato and K. Lücke, *J. Appl. Phys.*, **27**: 789 (1956).

[2]G. Leibfried, *Z. Phys.*, **127**: 344 (1950).

[3]F. R. N. Nabarro, *Proc. Roy. Soc.*, **209A**: 278 (1951).

[4]J. Weertman, in P. G. Shewmon, and V. F. Zackay (eds.), "Response of Metals to High Velocity Deformation," Interscience, New York, 1961, p. 205.

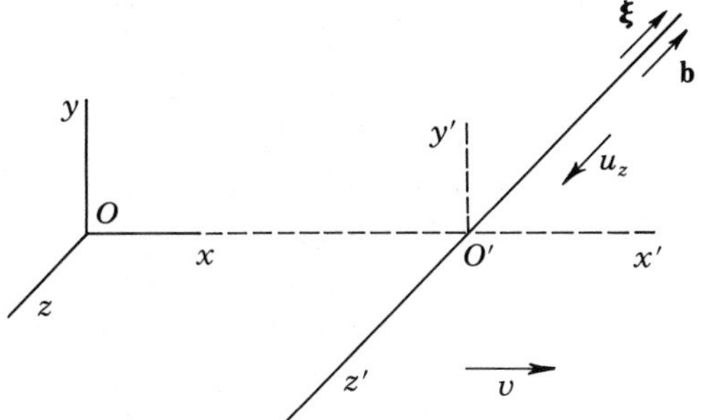

FIGURE 7-1. Moving coordinates for a screw dislocation.

medium, the only displacements are those parallel to the dislocation, whether the screw is stationary or moving.

In the case of a time-dependent displacement field, the body forces f_i in the equilibrium equations (2-2) are replaced by inertial terms caused by acceleration of the volume element; they are given by $-\rho_0(\partial^2 u_i/\partial t^2)$, where ρ_0 is the density. Combining this statement with Eq. (2-3) and Hooke's law, Eq. (2-46), one finds that the equilibrium equations are represented in terms of the displacements by the expression

$$\rho_0 \frac{\partial^2 u_i}{\partial t^2} = \mu \frac{\partial^2 u_i}{\partial x_j^2} + (\mu+\lambda) \frac{\partial}{\partial x_i} \frac{\partial u_j}{\partial x_j} \tag{7-1}$$

or, in vector notation,

$$\rho_0 \frac{\partial^2 \mathbf{u}}{\partial t^2} = \mu \nabla^2 \mathbf{u} + (\mu+\lambda) \nabla(\nabla \cdot \mathbf{u})$$

For a screw dislocation, the displacement components are $u_i = (0, 0, u_z)$. Thus Eq. (7-1) reduces to the following time-dependent generalization of Eq. (2-58), written in the common *xyz*-coordinate notation:

$$\left(\frac{\partial^2}{\partial x^2} + \frac{\partial^2}{\partial y^2} \right) u_z = \frac{\rho_0}{\mu} \frac{\partial^2 u_z}{\partial t^2} \tag{7-2}$$

This equation is recognized as the familiar wave equation[5] for transverse shear waves. Hence

$$\left(\frac{\mu}{\rho_0} \right)^{1/2} = C_t \tag{7-3}$$

[5]See, for example, C. Kittel, "Introduction to Solid State Physics," Wiley, New York, 1956, p. 93.

where C_t is the transverse-sound-wave velocity, i.e., the velocity of a transverse shear wave.

Let the coordinate system $x'y'z'$ in Fig. 7-1 move with the dislocation, and impose the condition that the two origins O and O' coincide at $t=0$. For a uniformly moving screw $v_x = v$, a convenient transform is the "relativistic" one,

$$x' = \frac{x - vt}{\left(1 - v^2/C_t^2\right)^{1/2}}$$

$$y' = y$$

$$z' = z$$

$$t' = \frac{t - vx/C_t^2}{\left(1 - v^2/C_t^2\right)^{1/2}} \tag{7-4}$$

which by substitution into Eq. (7-2) yields a wave equation of the same form as before,

$$\left(\frac{\partial^2}{\partial x'^2} + \frac{\partial^2}{\partial y'^2}\right) u_z = \frac{1}{C_t^2} \frac{\partial^2 u_z}{\partial t'^2} \tag{7-5}$$

No deep significance should be attached to the analogy with relativity. For example, the "time" t' is, of course, not on equal footing with t. Equation (7-4) is simply a transform that preserves the form of the wave equation, as is well known from the theory of relativity—hence the term "relativistic" transform.

By definition, in the case of uniform motion, where $v_x = v =$constant, the displacement field surrounding the dislocation must appear to be constant with time in the moving coordinate system, so that Eq. (7-5) simplifies to

$$\left(\frac{\partial^2}{\partial x'^2} + \frac{\partial^2}{\partial y'^2}\right) u_z = 0 \tag{7-6}$$

This equation has precisely the same form as Eq. (2-58), which was solved for the stationary screw, and the boundary condition is the same:

$$\lim_{\substack{\epsilon \to 0 \\ x' > 0}} u_z(x', \epsilon) - u_z(x', -\epsilon) = -b \qquad \epsilon \text{ positive} \tag{7-7}$$

Therefore the solution is the same as Eq. (3-2),

$$u_z(x, y, t) = \frac{b}{2\pi} \tan^{-1} \frac{y'}{x'} \tag{7-8}$$

or, written out in the xyz coordinates,

$$u_z(x, y, t) = \frac{b}{2\pi}\tan^{-1}\frac{\gamma y}{x - vt} \tag{7-9}$$

which contains the abbreviation

$$\gamma = \left(1 - \frac{v^2}{C_t^2}\right)^{1/2} \tag{7-10}$$

The stresses are

$$\sigma_{xz} = \mu\frac{\partial u_z}{\partial x} = -\frac{\mu b}{2\pi}\frac{\gamma y}{(x - vt)^2 + \gamma^2 y^2}$$

$$\sigma_{yz} = \mu\frac{\partial u_z}{\partial y} = \frac{\mu b}{2\pi}\frac{\gamma(x - vt)}{(x - vt)^2 + \gamma^2 y^2} \tag{7-11}$$

which reduce to the static result, Eq. (3-3), as they must at $v = 0$.

Consider the case of high speeds approaching the velocity of sound, $v \to C_t$, that is, where $\gamma \to 0$. When $|x - vt| \gg |\gamma y|$, the stresses are smaller by a factor γ when compared with those around a stationary screw. In contrast, when $|x - vt| \ll |\gamma y|$, the stresses are larger by $\sim 1/\gamma$. Thus the stress field is contracted toward the y' axis (where $x = vt$), and intensified in that region, as depicted in Fig. 7-2. In the limit $v = C_t$, the stresses vanish everywhere except on the y' axis, where the stresses are infinite. As discussed in the next section, the energy factor for the screw dislocation also diverges at $v = C_t$. Of course, the validity of linear theory breaks down for velocities smaller than C_t, and the question of whether velocities reaching or exceeding C_t can be achieved, cannot

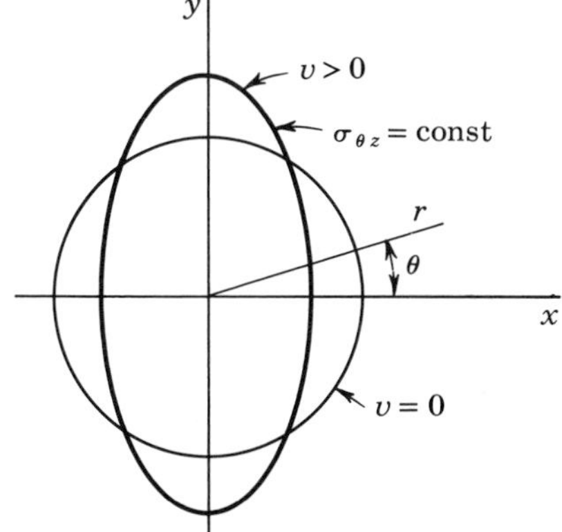

FIGURE 7-2. Shear stress field $\sigma_{\theta z}$ for a moving screw dislocation ($v > 0$) and a static screw ($v = 0$). One constant stress contour is shown for each case.

be decided conclusively on the basis of simple linear theory alone. Nevertheless, we follow the traditional usage and call C_t the limiting velocity.

The linear elasticity solution is valid at small velocities and is useful as an indication of the behavior for large velocities approaching C_t.

An important application of these results involves the interaction force between moving screw dislocations. The above treatment indicates that the interaction force between two screws moving at the same uniform velocity on the same glide plane decreases to zero as $v \to C_t$, while that between closely spaced screws on parallel glide planes increases. The former behavior could lead to collapse of screw partials to a perfect screw (e.g., in bcc crystals), while the latter would lead to cross slip of one of the screws.

Energy

With the inclusion of the kinetic energy density $\frac{1}{2}\rho_0(\partial u_z/\partial t)^2$, Eq. (2-14) for the energy density becomes

$$w = \frac{\mu}{2}\left[\left(\frac{\partial u_z}{\partial x}\right)^2 + \left(\frac{\partial u_z}{\partial y}\right)^2 + \frac{1}{C_t^2}\left(\frac{\partial u_z}{\partial t}\right)^2\right] \tag{7-12}$$

or, in the moving coordinate system,

$$w = \frac{\mu}{2\gamma^2}\left\{\left(\frac{\partial u_z}{\partial x'}\right)^2 + \left(\frac{\partial u_z}{\partial y'}\right)^2 + \frac{v^2}{C_t^2}\left[\left(\frac{\partial u_z}{\partial x'}\right)^2 - \left(\frac{\partial u_z}{\partial y'}\right)^2\right]\right\} \tag{7-13}$$

Consider the energy per unit depth (in the z direction) in an infinite strip parallel to the x axis,

$$\frac{dW}{L} = dy\int_{-\infty}^{\infty} w\,dx = \gamma\,dy\int_{-\infty}^{\infty} w\,dx' \tag{7-14}$$

With the solution of Eq. (7-9), only the first two terms in Eq. (7-13) contribute to the above integral, so that

$$\begin{aligned}\frac{dW}{L} &= \frac{dy}{\gamma}\int_{-\infty}^{\infty}\frac{\mu}{2}\left[\left(\frac{\partial u_z}{\partial x'}\right)^2 + \left(\frac{\partial u_z}{\partial y'}\right)^2\right]dx' \\ &= \frac{dW_0}{L\gamma}\end{aligned} \tag{7-15}$$

where W_0 is the energy of the stress field of a stationary screw. Equation (7-15) shows that as $v \to C_t$, $\gamma \to 0$, and the ratio of the energy in identical strips for a moving screw and a stationary screw, respectively, becomes infinite,

$$\frac{dW}{dW_0} = \frac{1}{\left(1 - v^2/C_t^2\right)^{1/2}} \to \infty \qquad \text{as} \quad v \to C_t \tag{7-16}$$

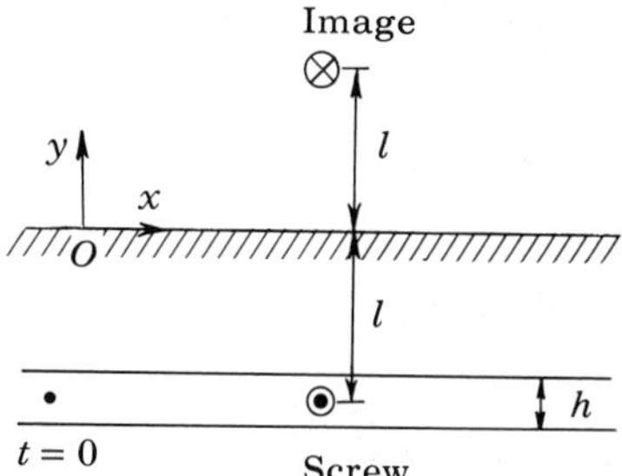

FIGURE 7-3. A screw in semi-infinite solid moving within a slab of thickness h. The original position of the screw at $t=0$ is shown.

This relation is the same as that between energy and rest energy in relativistic mechanics. Such simplicity does not generally occur in dislocation theory; the above form is valid only for simple regions of integration in which the last term in Eq. (7-13) does not contribute. The latter condition is *not* fulfilled, for example, in the case of some image distributions associated with the presence of free surfaces.

Effect of a Free Surface

The preceding results for a screw in an infinite medium also give a rough estimate of the stress field for a moving screw in a finite crystal. In particular, divergence at the velocity C_t also occurs for a dislocation in a finite crystal.

Consider the half-infinite solid shown in Fig. 7-3, with a screw moving parallel to the surface with a velocity v, and at a distance l below the surface. The problem is readily solved by the image method. Superposing solutions of the type of Eq. (7-11) for the screw and its image, one obtains

$$\sigma_{xz} = -\frac{\mu b}{2\pi}\left[\frac{\gamma(l+y)}{(x-vt)^2+\gamma^2(l+y)^2}+\frac{\gamma(l-y)}{(x-vt)^2+\gamma^2(l-y)^2}\right]$$

$$\sigma_{yz} = \frac{\mu b}{2\pi}\left[\frac{\gamma(x-vt)}{(x-vt)^2+\gamma^2(l+y)^2}-\frac{\gamma(x-vt)}{(x-vt)^2+\gamma^2(l-y)^2}\right] \tag{7-17}$$

When $y=0$, then $\sigma_{yz}=0$, as it should at the free surface.

Next, the energy is determined from Eq. (7-13) by integration over the half-infinite solid, except for a strip of width h in the glide plane. Those terms which do not lead to divergence could also be integrated over the strip with no significant change in the results. For this simple image problem, once again the last term in Eq. (7-13) does not contribute to the result, so that

$$\frac{W}{L} = \frac{W_0}{L\gamma} \tag{7-18}$$

Here W_0 is the energy of the stationary system, with the energy in the cutoff strip excluded. A direct calculation gives

$$\frac{W_0}{L} = \frac{\mu b^2}{4\pi} \ln \frac{2l}{h} \tag{7-19}$$

With an appropriate choice of h, of the order $\sim b$, Eq. (7-19) can be made to represent the total rest energy of the system, and Eq. (7-18) would then appear to give the total energy for uniform motion. This result is not without ambiguity; if a different cutoff procedure had been employed, such as the cutting out of a small cylinder around the screw, the energy for the moving screw would not be a simple relativistic expression resembling Eq. (7-18) (this is true because the cylinder has an ellipsoidal cross section in the moving coordinate system). However, when $l \gg b$, the exact value of h does not matter much, and with $h \sim b$, Eq. (7-18) gives the dominant part of the variation of total energy with v. Leibfried and Dietze[6] have shown that in the Peierls model of the screw, which provides a model for the dislocation core structure, the total energy, including the misfit energy in the core, behaves relativistically in the simple fashion of Eq. (7-18).

If $1/\gamma$ is expanded to the first power in v^2/C_t^2, Eq. (7-18) becomes

$$\frac{W}{L} = \frac{W_0}{L} + \frac{1}{2C_t^2}\frac{W_0}{L}v^2 \tag{7-20}$$

By analogy with the expression for kinetic energy in mechanics, it is natural to define an "effective mass" per unit length of dislocation line in terms of the factor multiplying v^2 in the above expression,

$$m^* = \frac{W_0/L}{C_t^2} \tag{7-21}$$

To further illustrate this analogy, let us suppose that a uniform stress σ_{yz} is applied externally to the half-infinite solid in Fig. 7-3, and that the acceleration is so small that the dislocation motion may be regarded as quasi-uniform. The stress does work on the dislocation, per unit length, at a rate $\sigma_{yz}bv$, and this work must appear as an increase in the total energy of the dislocation

$$\sigma_{yz}bv = \frac{d(W/L)}{dt} \tag{7-22}$$

Substituting Eqs. (7-18) and (7-21) into Eq. (7-22) and noting that the force per unit length is $F_x/L = \sigma_{yz}b$, one finds that

$$\frac{F_x}{L} = m^*\frac{dv}{dt} \tag{7-23}$$

in complete analogy with Newton's law in mechanics.

[6] G. Leibfried and H. D. Dietze, *Z. Phys.*, **126**: 790 (1949).

All the essential points in this theory for the moving screw were first discussed by Frank[7]. In particular, he suggested the treatment of the moving screw dislocation by an analog of relativistic mechanics. Limitations of the linear model for high velocities and the possibility of supersonic dislocations were discussed later.[8–11]

Exercise 7-1. Demonstrate that the last term in Eq. (7-12) gives no net contribution to W/L for the screw in the half-infinite solid.

7-3. THE MOVING EDGE DISLOCATION

The displacement field of a moving edge dislocation contains longitudinal as well as transverse components. For an edge dislocation parallel to the z axis and moving uniformly with velocity v in the slip plane xz, (Fig. 7-4), the only displacement components are $u_i=(u_x, u_y, 0)$, and the solution of Eq. (7-1) can then be found quite straightforwardly. We do not give the derivation,[12] but only quote the results for the stresses:

$$\sigma_{xy}=\frac{\mu b C_t^2}{2\pi v^2}\left[\frac{(1+\gamma_t^2)^2 x_t}{\gamma_t^2 r_t^2}-\frac{4x_l}{r_l^2}\right]$$

$$=\frac{\mu b C_t^2}{2\pi v^2}\left\{\frac{(1+\gamma_t^2)^2(x-vt)}{\gamma_t\left[(x-vt)^2+\gamma_t^2 y^2\right]}-\frac{4\gamma_l(x-vt)}{(x-vt)^2+\gamma_l^2 y^2}\right\} \tag{7-24}$$

$$\sigma_{xx}=\frac{byC_t^2}{\pi v^2}\left[\frac{(\lambda+2\mu)-\gamma_l^2\lambda}{\gamma_l r_l^2}-\frac{\mu(1+\gamma_t^2)}{\gamma_t r_t^2}\right] \tag{7-25}$$

$$\sigma_{yy}=\frac{byC_t^2}{\pi v^2}\left[\frac{\lambda-\gamma_l^2(\lambda+2\mu)}{\gamma_l r_l^2}+\frac{\mu(1+\gamma_t^2)}{\gamma_t r_t^2}\right] \tag{7-26}$$

$$\sigma_{zz}=\nu(\sigma_{xx}+\sigma_{yy}) \tag{7-27}$$

In the above expressions,

$$C_t=\left(\frac{\mu}{\rho_0}\right)^{1/2}\qquad C_l=\left(\frac{\lambda+2\mu}{\rho_0}\right)^{1/2} \tag{7-28}$$

[7] F. C. Frank, *Proc. Phys. Soc.*, **62A**: 131 (1949).

[8] J. D. Eshelby, *Proc. Phys. Soc.*, **B69**: 1013 (1956).

[9] A. N. Stroh, *J. Math. Phys.*, **41**: 77 (1962).

[10] R. Thomson, in P. G. Shewmon and V. F. Zackay (eds.), "Response of Metals to High Velocity Deformation," Interscience, New York, 1961, p. 246.

[11] J. Weertman, *J. Mech. Phys. Solids*, **11**: 197 (1963).

[12] J. Hirth and J. Lothe, "Theory of Dislocations," 1st ed.

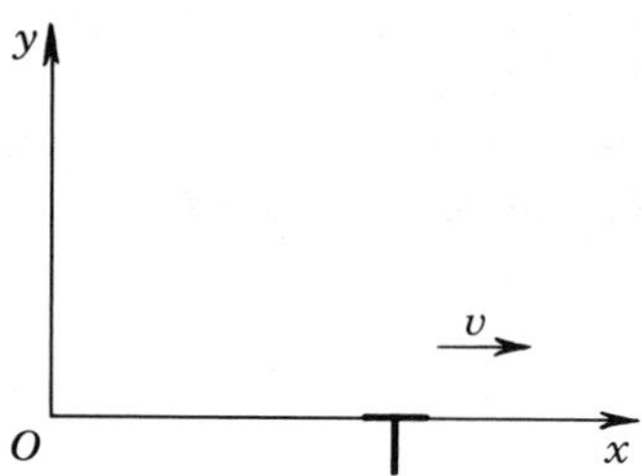

FIGURE 7-4. A moving edge dislocation.

are the transverse and longitudinal sound velocity, respectively. Further, $x_t = (x - vt)/\gamma_t$, $x_l = (x - vt)/\gamma_l$,

$$\gamma_t = \left(1 - \frac{v^2}{C_t^2}\right)^{1/2} \qquad \gamma_l = \left(1 - \frac{v^2}{C_l^2}\right)^{1/2} \tag{7-29}$$

and

$$r_t^2 = \frac{(x - vt)^2}{\gamma_t^2} + y^2 \qquad r_l^2 = \frac{(x - vt)^2}{\gamma_l^2} + y^2 \tag{7-30}$$

The displacements can be obtained from the stresses by integration.[13] Limiting procedures show that the above results reduce to those for the static edge [Eq. (3-43)], as they must.

A graph of σ_{xy} is presented in Fig. 7-5. As indicated by the graph and by Eq. (7-24), the region of positive shear stress contracts towards the x axis as v increases; eventually the stress in the slip plane changes sign. In the slip plane $y=0$, Eq. (7-24) reduces to

$$(\sigma_{xy})_{y=0} = \frac{\mu b}{2\pi\gamma_t(x - vt)} \frac{C_t^2}{v^2}\left[(1+\gamma_t^2)^2 - 4\gamma_t\gamma_l\right] \tag{7-31}$$

Indeed, $(\sigma_{xy})_{y=0}$ changes sign when $(1+\gamma_t^2)^2 = 4\gamma_l\gamma_t$, that is, when

$$\left(2 - \frac{v^2}{C_t^2}\right)^4 = 16\left(1 - \frac{v^2}{C_t^2}\right)\left(1 - \frac{v^2}{C_l^2}\right) \tag{7-32}$$

Equation (7-32) is well known in the theory of surface waves, or so-called Rayleigh waves.[14] Rayleigh waves are characterized as waves which propagate parallel to the surface of a semi-infinite solid and whose amplitude is greatest

[13] See J. D. Eshelby, *Proc. Phys. Soc.*, **62A**: 307 (1949).

[14] J. D. Eshelby (ibid.) has developed a theory for the moving edge dislocation which shows clearly how the properties of Rayleigh waves enter into the problem.

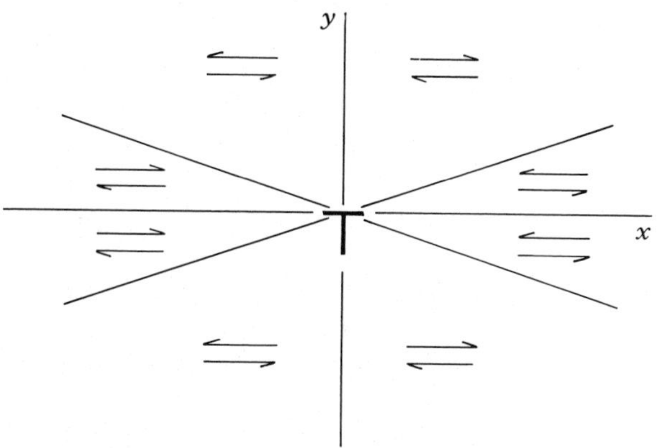

FIGURE 7-5. Distribution of shear stress σ_{xy} in moving coordinates fixed on a moving edge dislocation.

at the surface and decreases exponentially with distance from the surface, as indicated in Fig. 7-6. A Rayleigh wave is neither purely longitudinal nor purely transverse, and its velocity v is given by Eq. (7-32). The only possible real solution of Eq. (7-32) is the Rayleigh sound velocity.

$$v = aC_t \tag{7-33}$$

with the parameter a ranging from 0.874 to 0.955 for Poisson's ratio ν in the range[15] 0 to 0.5.

In the limit $v = C_t$, divergences appear in the edge dislocation solution, i.e., C_t is a limiting velocity also for the edge dislocation. The energy factor diverges, even more strongly than for the screw, as $(1 - v^2/C_t^2)^{-3/2}$. The complete expression for the energy factor is derived by Weertman.[16]

The behavior at the Rayleigh velocity, however, does not involve divergences. The Rayleigh velocity aC_t is not a limiting velocity, as has been emphasized by Weertman.[16]

At small velocities, $v \ll C_t$, the moving edge dislocation can be assigned an effective mass in the same way as the screw dislocation [Eq. (7-21)]. The effective mass for the edge dislocation[17] is the effective mass for a screw with the same magnitude of Burgers vector, multiplied by a factor $(1 + C_t^4/C_l^4)$.

The solution for an edge dislocation climbing with a velocity v normal to its slip plane has also been found.[18] Because of the greater damping forces for climb, achievement of large velocities in climb is less likely than for glide

[15] L. D. Landau and E. M. Lifshitz, "Theory of Elasticity," Pergamon, New York, 1959, p. 105.

[16] J. Weertman, in P. G. Shewmon and V. F. Zackay (eds.), "Response of Metals to High Velocity Deformation," Interscience, New York, 1961, p. 205.

[17] J. Weertman, ibid., p. 212.

[18] J. Weertman, *J. Appl. Phys.*, **38**, 2612 (1967).

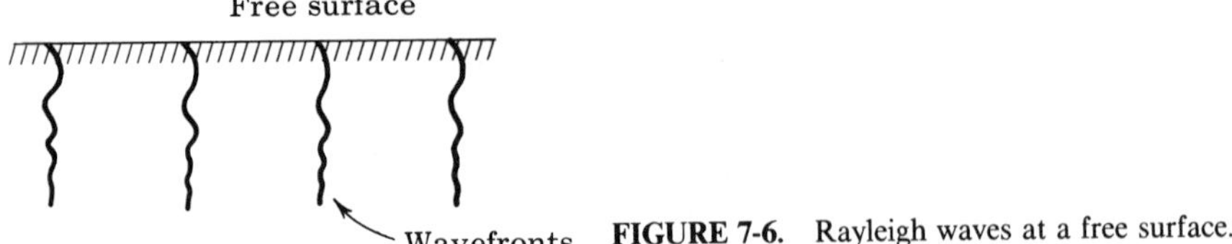

FIGURE 7-6. Rayleigh waves at a free surface.

except in shock fronts where nonlinear effects become important, so we do not consider the climb case further.

Surface Effects at the Rayleigh Velocity

In a practical case the edge dislocation will not be in an infinite medium. Consider, as a more realistic case, an edge dislocation moving parallel to the face of a semi-infinite solid (Fig. 7-7). The solution of the problem must then be modified to fulfill boundary conditions at the free surface; this involves superposing terms of a type that one may call "forced Rayleigh waves." When the dislocation velocity v equals aC_t, resonance with the surface waves occurs, resulting in Rayleigh waves of infinite amplitude. Thus, with a uniform external shear stress σ_{xy} causing accelerations sufficiently slowly that the motion can be considered as quasi-uniform, $v = aC_t$ *would* appear as a limiting velocity, because the energy in the solid would approach infinity as v approached the value aC_t. But it should be possible to give the dislocation a rapid acceleration, so that the core attains a velocity $v > aC_t$ before the dislocation motion is felt at the surface and the braking action caused by the excitation of surface waves commences.

Thus, for impulsive loads, which should accelerate dislocations rapidly, the behavior might resemble that of a dislocation in an infinite medium. One of the most important consequences of the peculiar behavior of edges at the Rayleigh velocity involves the shear stress on the slip plane. The effect is such that the edge dislocation appears to change sign at $v = aC_t$; for velocities $v > aC_t$ the stress field of the dislocation in its slip plane is equivalent to that of a stationary dislocation of opposite sign. This curious behavior is caused in part by a dynamic contraction of the material above the slip plane and a corresponding expansion of the material below the slip plane; at $v = aC_t$ these effects compensate for the effect of the inserted half-infinite plane in determining σ_{xy} in the slip plane. Nonetheless, Eq. (3-1) is always fulfilled, so by an

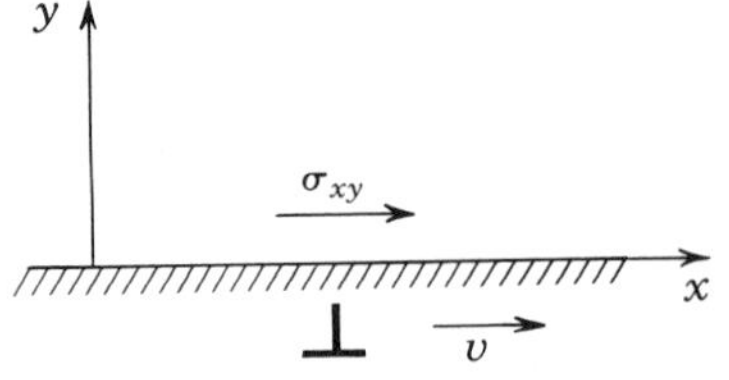

FIGURE 7-7. An edge dislocation in a semi-infinite solid.

argument similar to that preceding Eq. (3-23), the force per unit length associated with an external shear stress σ_{xy} is always

$$\frac{F_x}{L} = \sigma_{xy} b \tag{7-34}$$

The dynamic effects influence the stress field of a dislocation, and hence its interaction force with other dislocations, but do *not* influence the force on the dislocation produced by an applied stress.

Accepting (7-34) as valid for the interaction force between edge dislocations, a remarkable consequence is that a pair of edge dislocations moving with the same velocity in the same slip plane mutually *attract* if the velocity is greater than aC_t.[19] That is, for like-sign, fast moving edge dislocations, there is a tendency for coalescence that could lead to crack formation.

It might appear that such a process violates energy conservation. However, it must be understood that during a spontaneous coalescence process, the *overall* velocity of the total configuration decreases. Weertman[20] and Hirth and Lothe[21] have discussed the energy balance of such processes and the relation with the force expression [Eq. (7-34)].

Throughout the preceding discussion, stable dislocation core structures have been tacitly assumed to exist throughout the entire velocity range up to C_t. This assumption can be questioned by arguments similar to those used for the coalescence problem.[21] Consider a single dislocation b moving at a velocity $aC_t < v < C_t$. A possible core fluctuation could lead to the incipient reaction

$$b \rightarrow -b + 2b \tag{7-35}$$

But $-b$ and $2b$ *repel* one another in this velocity range, so the reactions retard the overall configuration. Instabilities of the type in Eq. (7-35) can cause aC_t to be effectively the largest attainable velocity for edge dislocations, in contrast to a prediction on the basis of inertial effects. Instabilities strikingly similar to these have been observed by Weiner and Pear in atomic simulations of dislocation motion.[22]

7-4. ACCELERATION AND RADIATION

Up to this point essentially only uniform motion of straight dislocations was discussed. Acceleration only was considered in a quasistationary approach

[19] J. Weertman, in P. G. Shewmon and V. F. Zackay (eds.), "Response of Metals to High Velocity Deformation," Interscience, New York, 1961, p. 205.

[20] J. Weertman, *J. Appl. Phys.*, **37**: 4925 (1966).

[21] J. P. Hirth and J. Lothe, in "Dislocations Dynamics," A. R. Rosenfield, G. T. Hahn, A. L. Bement, and R. I. Jaffee (eds.), McGraw-Hill, New York, 1967, p. 231.

[22] J. H. Weiner and M. Pear, *Phil. Mag.*, **31**: 679 (1975).

(Sec. 7-2), with neglect of all radiation. A formal general dynamic theory for arbitrary nonuniform motion of an arbitrary loop was developed by Nabarro.[23] Eshelby[24] derived a more explicit integral expression for an arbitrarily moving screw dislocation, a theme that recently has been studied in great detail by Markenscoff,[25] who has also studied arbitrary nonuniform motion of edge dislocations.[26] All these theories refer to subsonic motion at velocities below C_t.

We do not attempt a detailed account of these theories here. An extensive review of the original work by Nabarro and Eshelby was presented earlier.[27] For our present purpose of studying some special cases and situations, simple expedient methods suffice. Although expedient, the methods give valid results, as can be checked by comparison with rigorous theory.

Let us turn to the problem of the accelerating screw dislocation. In general, an accelerated motion can be represented by a series of impulsive changes in velocity. If the response to the impulsive change in velocity is known, the response to an arbitrary acceleration can be found by integration.

Consider a screw dislocation moving at uniform velocity $v \ll C_t$ up to $t=0$ and that then *abruptly* starts to move with a higher constant velocity $v+\Delta v$. The back stress over the core depends on the strain field set up in the region of the core. However, rather than determining the strain field, we determine σ_{yz} indirectly as follows. If at $t=0$ the velocity changes by Δv, then at a later time t the displacements and the motions within a radius $r=C_t t$ of the dislocation are essentially those of a dislocation moving with the higher velocity $v+\Delta v$, while in the region $r>C_t t$ the change in velocity is not yet felt. Thus at time t there has been a change in energy per unit depth

$$\frac{\Delta W}{L} \simeq \frac{\mu b^2 v \Delta v}{4\pi C_t^2} \ln \frac{C_t t}{b} \tag{7-36}$$

Since, by definition, the core of a free dislocation can sustain no shear stress, the back shear stress acting must be balanced by an external uniform shear stress if the dislocation is not to slow down. This external shear stress does work per unit length at a rate $\Delta\sigma_{yz} bv$ that supplies the energy for the change given by Eq. (7-36),

$$\Delta\sigma_{yz} bv = \frac{\partial}{\partial t}\frac{\Delta W}{L} \tag{7-37}$$

[23] F. R. N. Nabarro, *Phil. Mag.*, **42**: 1224 (1957).

[24] J. J. Eshelby, *Phil. Trans. Roy. Soc.*, **A244**: 87 (1951).

[25] X. Markenscoff, *J. Elast.*, **10**: 193 (1980).

[26] X. Markenscoff, in "Dislocation Modeling of Physical Systems," Pergamon, Oxford, 1981.

[27] J. P. Hirth and J. Lothe, "Theory of Dislocations," 1st ed., McGraw-Hill, New York, 1968.

or

$$\Delta\sigma_{yz} = \frac{\mu b \Delta v}{4\pi C_t^2 t} \tag{7-38}$$

This equation is really an asymptotic formula valid for large t: at a distance $\sim b$ from the core, it is valid for $C_t t \gg b$. For a more detailed theory, knowledge of the detailed core structure would be required, and such knowledge is unavailable.

A cutoff is therefore introduced, so that the integral of $\Delta\sigma b v\, dt$ from $t=0$ to t yields the same result as Eq. (7-36),

$$\Delta\sigma_{yz} = 0, \qquad t < \frac{b}{C_t}$$

$$\Delta\sigma_{yz} = \frac{\mu b \Delta v}{4\pi C_t^2 t}, \qquad t > \frac{b}{C_t} \tag{7-39}$$

Eshelby[28] suggests the alternative approximation of an approximately constant value of $\Delta\sigma_{yz}$ in the interval $0 \leqslant t \leqslant b/C_t$, joined by the asymptotic form for $t > b/C_t$. However, for slow motions and small accelerations, the exact behavior for small t does not matter. The important point is that the integral

$$\int_0^t \Delta\sigma\, dt$$

in the case of $t \sim b/C_t$ can be adjusted to be consistent with the expected energy change. Both forms of $\Delta\sigma$ fulfill this requirement and yield essentially the same result for the effective mass of the dislocation.

For an accelerating screw, then, provided $v \ll C_t$,

$$\sigma_{yz}(t) = \frac{\mu b}{4\pi C_t^2} \int_{-\infty}^{t-b/C_t} \frac{dv_x(\tau)}{d\tau} \frac{d\tau}{t-\tau} \tag{7-40}$$

where $dv_x(\tau)/d\tau$ is the acceleration. It should be possible to develop formulas similar to Eq. (7-40) for edge dislocations and for the more general case of curved dislocations with different velocities along the dislocation.[29] Kiusalaas and Mura[30] have considered the case of the accelerating edge dislocation, also treated by Beltz et al.[31] for the anisotropic case.

Let us now apply (7-40) to a vibrating screw dislocation. Let the amplitude of vibration in the $x-z$ plane be $x = X \sin \Omega t$, and suppose that $|X\Omega| \ll C_t$ and

[28] J. D. Eshelby, *Phys. Rev.*, **90**: 248 (1953).

[29] T. Mura, *Phil. Mag.*, **8**: 843 (1963).

[30] J. Kiusalaas and T. Mura, *Phil. Mag.*, **9**: 1 (1964).

[31] R. J. Beltz, T. L. Davis, and K. Malén, *Phys. Stat. Solidi*, **26**: 621 (1968).

that $\Omega b \ll C_t$, so that (7-40) applies. For this case, Eq. (7-40) takes the form

$$\sigma_{yz} = \frac{\mu b X \Omega^2}{4\pi C_t^2} \int_{-\infty}^{t-b/C_t} \frac{\sin \Omega\tau}{t-\tau} d\tau \tag{7-41}$$

One can put

$$\sin \Omega\tau = \sin \Omega t \cos(\Omega\tau - \Omega t) + \cos \Omega t \sin(\Omega\tau - \Omega t) \tag{7-42}$$

so that

$$\sigma_{yz} = \frac{\mu b X \Omega^2}{4\pi C_t^2} \left(\sin \Omega t \int_{\Omega b/C_t}^{\infty} \frac{\cos z}{z} dz - \cos \Omega t \int_{\Omega b/C_t}^{\infty} \frac{\sin z}{z} dz \right) \tag{7-43}$$

Since $\Omega b/C_t \ll 1$, one can make the approximation

$$\int_{\Omega b/C_t}^{\infty} \frac{\sin z}{z} dz \cong \int_0^{\infty} \frac{\sin z}{z} dz = \frac{\pi}{2} \tag{7-44}$$

and further,[32]

$$\int_{\Omega b/C_t}^{\infty} \frac{\cos z}{z} dz \cong \ln \frac{C_t}{\gamma \Omega b} \qquad \gamma = 1.78 \tag{7-45}$$

Thus

$$\sigma_{yz} = \frac{\mu b X \Omega^2}{4\pi C_t^2} \ln \frac{C_t}{\gamma \Omega b} \sin \Omega t - \frac{\mu b X \Omega^2}{8 C_t^2} \cos \Omega t \tag{7-46}$$

The first term in Eq. (7-46) is equivalent to a force

$$\frac{F_x}{L} = -\sigma_{yz} b = \frac{\mu b^2}{4\pi C_t^2} \ln \frac{C_t}{\gamma \Omega b} (-\Omega^2 X \sin \Omega t) \tag{7-47}$$

which, since it is proportional to the acceleration $-\Omega^2 X \sin \Omega t$, must be interpreted as the force needed to overcome the inertia. Thus the inertia, or the effective mass per unit length of dislocation, is given by

$$m^* = \frac{\mu b^2}{4\pi C_t^2} \ln \frac{C_t}{\gamma \Omega b} \tag{7-48}$$

This result agrees with the earlier one of Eq. (7-21) if one assumes that the matter within a radius of $\sim C_t/\gamma\Omega$ oscillates with the dislocation.

[32] See E. Jahnke and F. Emde, "Tables of Functions," Dover, New York, 1945, p. 3.

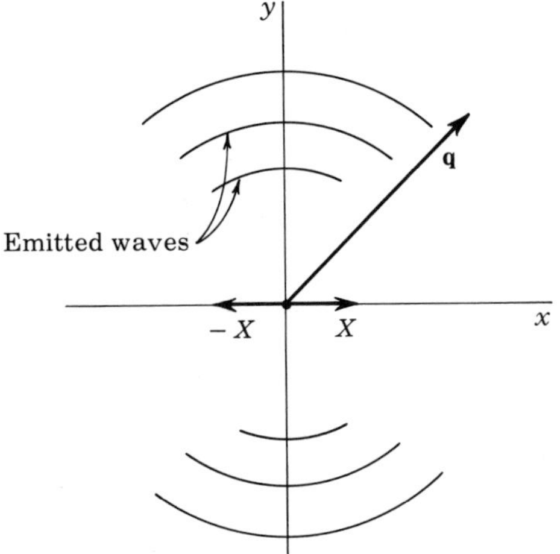

FIGURE 7-8. An oscillating screw dislocation lying nominally along the z axis.

The second term in Eq. (7-46) is proportional to the velocity $\Omega X \cos \Omega t$. Thus the damping constant is $\mu b^2 \Omega / 8C_t^2$, and the damping term gives rise to an energy input

$$\frac{\mu b^2 \Omega}{8C_t^2}(\Omega X \cos \Omega t)^2$$

which on the average yields

$$\frac{\dot{W}}{L} = \frac{\mu b^2 X^2 \Omega^3}{16 C_t^2} \tag{7-49}$$

This is the energy which is radiated out per unit time per unit length as shear waves, as depicted in Fig. 7-8. This result in Eq. (7-49) was first obtained by Eshelby.[33]

The vibrating element is sometimes a kink in a dislocation rather than the entire dislocation. Consider a screw dislocation with a kink of height h, and let the kink vibrate with amplitude X (amplitude of displacement of the kink along the screw) and frequency Ω. When the frequency is sufficiently low that $2\pi C_t / \Omega \gg h$, and when one ignores the radiation of longitudinal waves, the energy radiated per unit time,[33, 34] is

$$\dot{W} = \frac{\mu b^2 h^2 \Omega^4 X^2}{20 \pi C_t^3} \tag{7-50}$$

[33] J. D. Eshelby, *Proc. Roy. Soc.*, **266A**: 222 (1962).

[34] J. Lothe, *J. Appl. Phys.*, **33**: 2116 (1962).

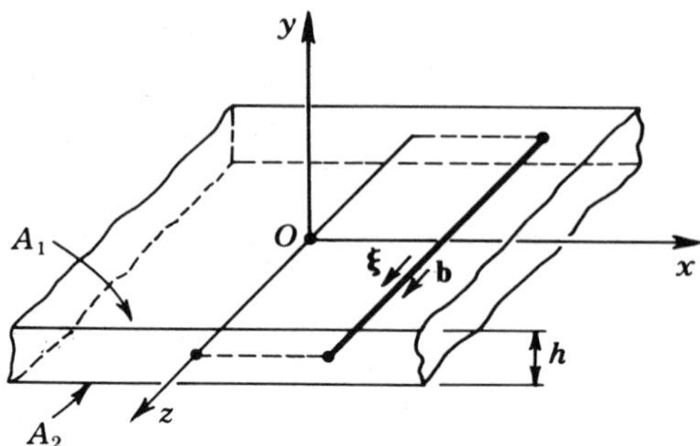

FIGURE 7-9. A screw dislocation moving within a slab of thickness h in an infinite medium.

This is a low-frequency expression, and the detailed kink geometry (whether it is wide or abrupt) is not important.

In the preceding discussion of dislocation dynamics it was often necessary to make approximations to obtain reasonably manageable expressions. As a consequence, some details of a general nature were lost. These details are treated in this final section dealing with dislocation momentum and the interaction between dislocations and elastic waves. The results of this more exact theory can be evaluated explicitly by means of the previously developed approximate theory (see Exercise 7-2).

7-5. DISLOCATION MOMENTUM AND RADIATION FORCES

Consider a straight screw allowed to move freely in one slip plane only, the xz plane (Fig. 7-9). Let A_1 and A_2 be two surfaces on either side of and parallel with the slip plane, A_1 bounding the upper half-infinite solid and A_2 bounding the lower half-infinite solid. According to Sec. 3-2, the force on the screw is

$$F_x = \lim_{h\to 0}\left(\int_{A_1}\sigma_{yz}\frac{\partial u_z}{\partial x}dA_y - \int_{A_2}\sigma_{yz}\frac{\partial u_z}{\partial x}dA_y\right) \tag{7-51}$$

The integrals in Eq. (7-51) are of the type

$$\int(\boldsymbol{\sigma}\cdot d\mathbf{A})\cdot\frac{\partial \mathbf{u}}{\partial x}$$

Now consider the case $\mathbf{u}=(0,0,u_z)$, which can be treated in two dimensions, and that is sufficient for the formulation of the dynamic theory for a screw. Let A be a curve representing a surface which encloses perfect material (Fig. 7-10).

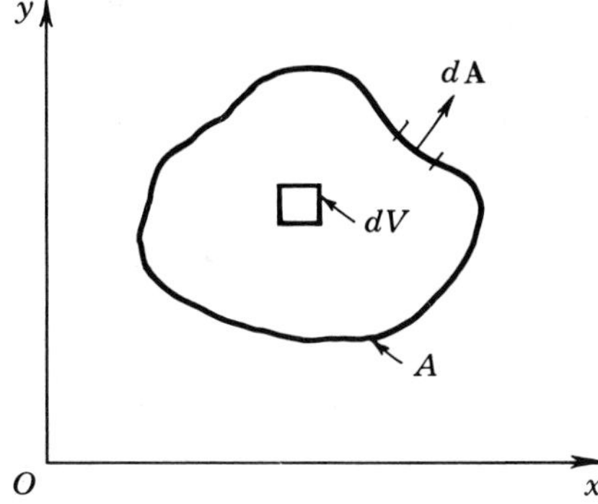

FIGURE 7-10. Surface A enclosing a perfect region of volume V.

One can write for this case

$$\int_A (\boldsymbol{\sigma}\cdot d\mathbf{A})\cdot\frac{\partial \mathbf{u}}{\partial x}=\int_A\left(\frac{\partial \mathbf{u}}{\partial x}\cdot\boldsymbol{\sigma}\right)\cdot d\mathbf{A}=\int_V \nabla\cdot\left(\frac{\partial \mathbf{u}}{\partial x}\cdot\boldsymbol{\sigma}\right)dV$$

$$=\int\left(\frac{\partial\sigma_{xz}}{\partial x}+\frac{\partial\sigma_{yz}}{\partial y}\right)\frac{\partial u_z}{\partial x}dV+\int\left(\sigma_{xz}\frac{\partial^2 u_z}{\partial x^2}+\sigma_{yz}\frac{\partial^2 u_z}{\partial x\,\partial y}\right)dV \tag{7-52}$$

Since $\partial u_z/\partial x=\sigma_{xz}/\mu$, and so on, the last term in Eq. (7-52) can be transformed to

$$\int\frac{\partial}{\partial x}\frac{1}{2\mu}\left(\sigma_{xz}{}^2+\sigma_{yz}{}^2\right)dV=\int\frac{\partial w_{\text{el}}}{\partial x}dV \tag{7-53}$$

where w_{el} is the elastic-energy density. Also, the equation of motion

$$\frac{\partial\sigma_{xz}}{\partial x}+\frac{\partial\sigma_{yz}}{\partial y}=\rho_0\ddot{u}_z \tag{7-54}$$

must be fulfilled, so that the first term on the right-hand side of Eq. (7-52) can be written as

$$\int\rho_0\ddot{u}_z\frac{\partial u_z}{\partial x}dV=\frac{d}{dt}\int\rho_0\dot{u}_z\frac{\partial u_z}{\partial x}dV-\int\frac{\partial}{\partial x}\frac{\rho_0\dot{u}_z{}^2}{2}dV \tag{7-55}$$

After one partial integration, it follows from the substitution of Eqs. (7-53) and (7-55) into (7-52) that

$$\int_A\left(\boldsymbol{\sigma}\cdot\frac{\partial\mathbf{u}}{\partial x}\right)\cdot d\mathbf{A}-\int_A\left(w_{\text{el}}-\tfrac{1}{2}\rho_0\dot{u}_z{}^2\right)dA_x=\frac{d}{dt}\int\rho_0\dot{u}_z\frac{\partial u_z}{\partial x}dV \tag{7-56}$$

Equation (7-56) is a special case of the more general conservation equation[35]

$$\int_A \mathbf{T}\cdot d\mathbf{A}=\frac{d}{dt}\int_V \mathbf{g}\,dV=\frac{d\mathbf{G}}{dt} \tag{7-57}$$

where **T** has the components

$$T_{ij}=\left(w_{\text{el}}-\tfrac{1}{2}\rho_0\dot{u}_m^{\;2}\right)\delta_{ij}-\sigma_{jm}\frac{\partial u_m}{\partial x_i} \tag{7-58}$$

and

$$\mathbf{g}=-\rho_0\dot{u}_m\nabla u_m \tag{7-59}$$

T is the analog in elasticity to the Maxwell tensor in electrodynamics,[36] and **g** is the analog to the electromagnetic-field-momentum density. As will soon be apparent, an appropriate dislocation momentum can be defined in terms of **g**.

Consider a screw dislocation in an infinite slab of material, as shown in Fig. 7-11; this is similar to the situation in Fig. 7-9, except that the slab is bounded by two planes normal to the y axis. A thin sheet of thickness h about the slip plane is separated out. Except for a strip of width d, which contains the dislocation along its axis, the stress, strain, and energy density are continuous from A_1 to A_2. Across the strip d, $\partial u_z/\partial y$, w, and σ_{yz} are continuous from A_1 to A_2, while $\partial u_z/\partial x$ changes sign from A_1 to A_2 but has the same magnitude on the two sides. $\partial u_z/\partial x$ is symmetric about the dislocation line, while σ_{yz} is antisymmetric (see Fig. 3-5).

With an externally applied shear stress present, the screw is postulated to move with an acceleration such that the above conditions are satisfied. This assumption should be representative for the general screw. A screw free to move in one slip plane is characterized as being unable to sustain a net shear stress in the slip plane over the core, and this is so for the above model. The inertia in the thin strip, of thickness h, is neglected. Again, a more detailed theory would have to include the detail of the actual core structure.

Integrating Eq. (7-56) over the two volumes V_1 and V_2, one finds

$$\int_{A_e}\left(\boldsymbol{\sigma}\cdot\frac{\partial\mathbf{u}}{\partial x}\right)\cdot d\mathbf{A}+\int_{A_1}\left(\boldsymbol{\sigma}\cdot\frac{\partial\mathbf{u}}{\partial x}\right)\cdot d\mathbf{A}+\int_{A_2}\left(\boldsymbol{\sigma}\cdot\frac{\partial\mathbf{u}}{\partial x}\right)\cdot d\mathbf{A}$$

$$=\frac{d}{dt}\int_V \rho_0\dot{u}_z\frac{\partial u_z}{\partial x}\,dV \tag{7-60}$$

Here A_e is the external surface of the slab and $V=V_1+V_2$. The integrals over

[35]See J. D. Eshelby, *Solid State Phys.*, **3**: 79 (1956).

[36]See L. D. Landau and E. M. Lifshitz, "Classical Theory of Fields," Pergamon, New York, 1951.

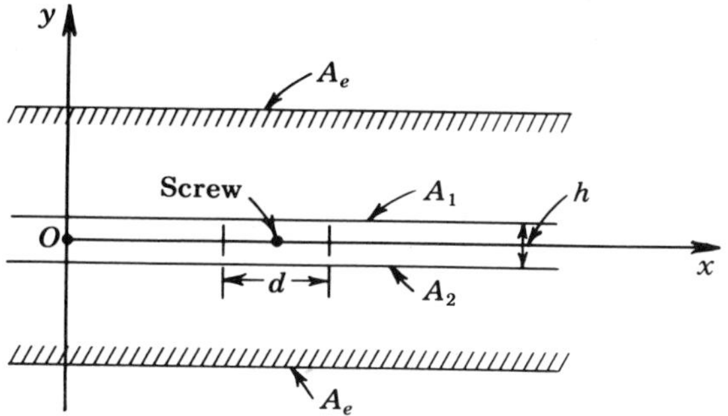

FIGURE 7-11. A screw dislocation lying parallel to the z axis in an infinite slab.

A_1 and A_2, cancel when the imposed conditions are fulfilled, so that

$$\int_{A_e}\left(\boldsymbol{\sigma}\cdot\frac{\partial\mathbf{u}}{\partial x}\right)\cdot d\mathbf{A}=\frac{d}{dt}\int_V \rho_0\dot{u}_z\frac{\partial u_z}{\partial x}\,dV=-\frac{dM_x}{dt} \tag{7-61}$$

Here, according to the electrodynamic analog of Eq. (7-57), M_x is interpreted as the dislocation momentum.[37] For that matter, this interpretation is also consistent with the classical mechanics analog since the left-hand side of Eq. (7-61) is $-F_x$, by Eq. (7-51). In verifying this finding, one must note that the surface normals of the two external surfaces A_e are opposite to the respective normals of surfaces A_1 and A_2 [Eq. (7-60)].

In the present case $dA_x=0$, and only σ_{yz} contributes to the surface integral. With no external applied stress present, $\sigma_{yz}=0$ at the free surface, and Eq. (7-61) becomes $dM_x/dt=0$. Thus when no external force F_x acts on the screw, its momentum M_x is conserved.

If a uniform shear stress acts on the slab, the surface integral in Eq. (7-61) becomes

$$\sigma_{yz}bL=-F_x \tag{7-62}$$

At this juncture, the above results can usefully be compared with the previous approximate results of this chapter. For quasi-uniform motion,

$$\dot{u}_z=-v\frac{\partial u_z}{\partial x}$$

[37]Note that M_x as defined above exists even though the *momentum of the displacement field*, defined by

$$\rho_0\int_V \dot{u}_z\,dV$$

is equal to zero because the material above the slip plane is moving in the opposite direction to that below it. See the discussion in J. Weertman, in P. G. Shewmon and V. F. Zackay (eds.), "Response of Metals to High Velocity Deformation," Interscience, New York, 1961, p. 205.

so that

$$M_x = \frac{2}{v}\int_V \tfrac{1}{2}\rho_0 \dot{u}_z^{\,2}\, dV \tag{7-63}$$

For small v the increase in energy with an increase in velocity is all in the form of kinetic energy; so, by Eq. (7-20),

$$\int_V \tfrac{1}{2}\rho_0 \dot{u}_z^{\,2}\, dV = W - W_0 = \frac{1}{2}\frac{W_0}{C_t^2}v^2$$

Thus

$$M_x = v\frac{W_0}{C_t^2} = m^* vL$$

and the effective mass per unit length is

$$m^* = \frac{W_0}{LC_t^2} \tag{7-64}$$

This result agrees with Eq. (7-21) and Eq. (7-61) agrees with Eq. (7-23), indicating that the simpler approximate theory is in agreement with the general theory.

Phonon Radiation and Momentum Transfer

Electromagnetic radiation can transfer momentum to charges; this well-known fact is evidenced, for example, in Compton scattering between photons and electrons. Similarly, elastic waves, i.e., phonons, impinging on a dislocation and scattered by it, transfer "momentum" to the dislocation. This process provides a mechanism for dislocation damping at finite temperatures resulting from the scattering of thermal elastic waves.

Consider a plane shear wave

$$u_z = Z\cos(\mathbf{k}\cdot\mathbf{r} - \omega t) \tag{7-65}$$

where $\omega = C_t k$ and $\mathbf{k}$ is the wave vector. The quasi-momentum density in this wave is

$$\mathbf{g} = \tfrac{1}{2}\omega Z^2 \mathbf{k} \tag{7-66}$$

or, expressed in terms of the average energy density in the wave, $w = \overline{w}_{\text{el}} + \frac{1}{2}\rho_0 \overline{\dot{u}}^2$, $\mathbf{g}$ becomes

$$\mathbf{g} = \frac{w}{C_t}\frac{\mathbf{k}}{k} \tag{7-67}$$

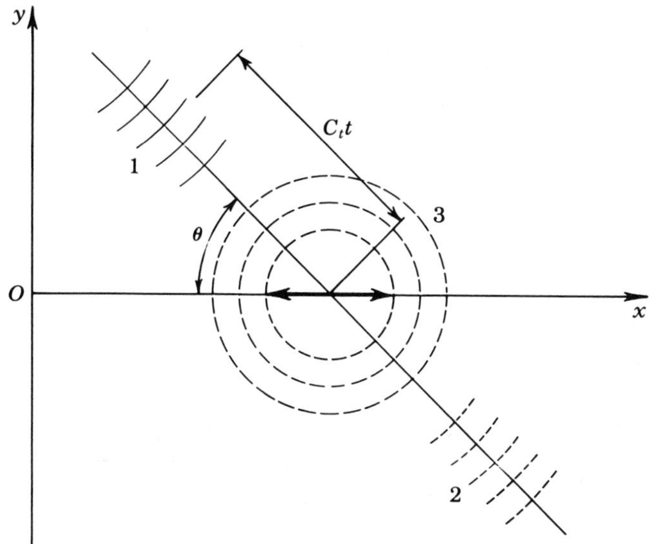

FIGURE 7-12. A phonon wave train impinging on a screw dislocation.

Now consider again the screw defined in Fig. 7-9, and suppose that a wave train (1) of length $C_t t$ impinges on the dislocation as illustrated in Fig. 7-12. Let the energy density in the incident wave be w. As the wave passes the dislocation it is driven into forced vibration, and secondary waves (3) are emitted [Eq. (7-49)]. If the scattering cross section per unit length of dislocation is D, the transmitted wave train (2) contains less energy than it did at (1) by an amount $DC_t tw$. Thus the difference in momentum, ΔM_x, between waves before and after scattering is

$$\frac{\Delta M_x}{L} = Dtw\cos\theta \tag{7-68}$$

per unit length L. This momentum must have been imparted to the dislocation since, as discussed above, the total momentum M_x is conserved. During the scattering process, the force on the dislocation per unit length is then

$$\frac{F_x}{L} = \frac{\Delta M_x}{Lt} = Dw\cos\theta \tag{7-69}$$

The scattering by the induced dislocation oscillations as in Fig. 7-12 is the so-called fluttering mechanism. Near the dislocation core, linear elasticity theory is not strictly valid, and the strain field of the dislocation also scatters sound by second-order elasticity effects (anharmonicity effects); that is, the dislocation scatters sound even in the absence of induced flutter. As explained by Eshelby,[38] the conservation theorem [Eq. (7-57)], which is the basis for the

[38] J. D. Eshelby, *Solid State Phys.*, **3**: 134 (1956).

derivation of Eq. (7-69), is also valid for finite deformations, with proper definitions of all quantities σ, etc. For screw dislocations, for which the displacement u_z is perpendicular to the coordinate net (x, y) so that the imbedded coordinate system need not be distinguished from one that is fixed externally, it follows directly that Eq. (7-69) is valid also for anharmonicity scattering. Although the edge dislocation is more difficult to treat rigorously, we assume that Eq. (7-69) holds also in this case for both types of scattering. Leibfried's[39] original considerations indicate convincingly that Eq. (7-69) is applicable for edge dislocation scattering by the flutter mechanism.

Exercise 7-2. Derive the scattering cross section D for the process depicted in Fig. 7-12. Use Eqs. (7-48) and (7-49).

7-6. SUPERSONIC DISLOCATIONS

The preceding sections refer to the subsonic velocity range. The solutions show divergent behavior as $v \to C_t$. Possible solutions for $v > C_t$, for the *supersonic* range, must be constructed in a different way; they must belong to a different regime of solutions.

Solutions involving shock fronts, or more specifically stated, traveling displacement discontinuity surfaces, are indeed possible. The shock fronts transport energy away from the plane on which the dislocation moves, and steady-state motion requires that this energy be supplied in some manner. This is possible only if the passage of the dislocation *transforms* the slip plane from a state of higher energy to a state of lower energy. An example would be a single partial dislocation removing a stacking fault by its passage.

Steady-state supersonic motion of a complete dislocation on a "perfect" slip plane would be impossible because of the severe energy radiation by shock fronts.

We now describe the supersonic shock front solution in more detail for the simple case of a screw dislocation.

A solution of the equation of motion (7-2) is

$$u_z = f(x\cos\alpha + y\sin\alpha - C_t t) \tag{7-70}$$

with f any continuous differentiable function. In particular, let $f(\phi)$ be the step function

$$f(\phi) = A\Delta(\phi) \tag{7-71}$$

where the delta function has the properties

$$\begin{aligned} \Delta(\phi) &= 1, \qquad \phi<0 \\ \Delta(\phi) &= 0, \qquad \phi>0 \end{aligned} \tag{7-72}$$

[39]G. Leibfried, *Z. Phys.*, **127**: 344 (1950).

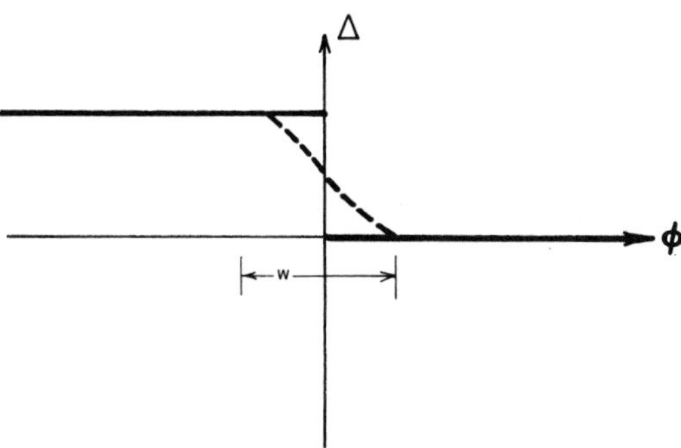

FIGURE 7-13. Relative displacement function $\Delta(\phi)$.

The term A is the amplitude of the discontinuity; $\Delta(\phi)$ is not differentiable, but we can regard it as the limiting case of a continuous smooth step, as indicated in Fig. 7-13.

With $f(\phi)$ as defined by (7-71), Eq. (7-70) describes a shock front with velocity C_t normal to the shock front. The direction of propagation is inclined an angle α to the $x-z$ plane (Fig. 7-14). Referred to the x direction, the configuration is traveling with a supersonic velocity

$$v=\frac{C_t}{\cos\alpha}>C_t$$

Both in front of and behind the shock front the material is unstrained and at rest. All of the strain energy and kinetic energy resides in the discontinuity

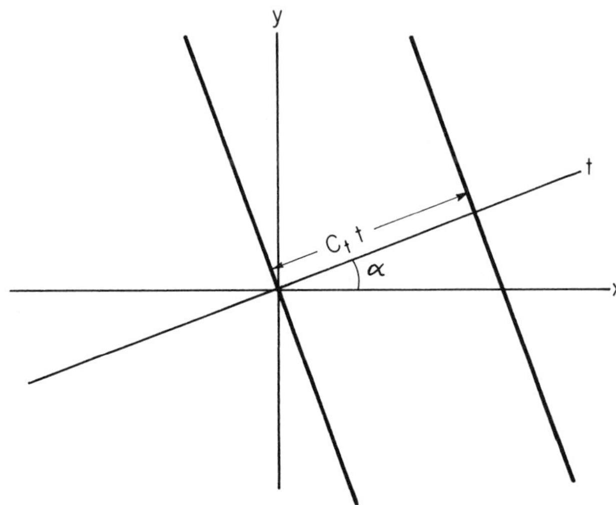

FIGURE 7-14. Shock front positions at time $t=0$ and $t=t$.

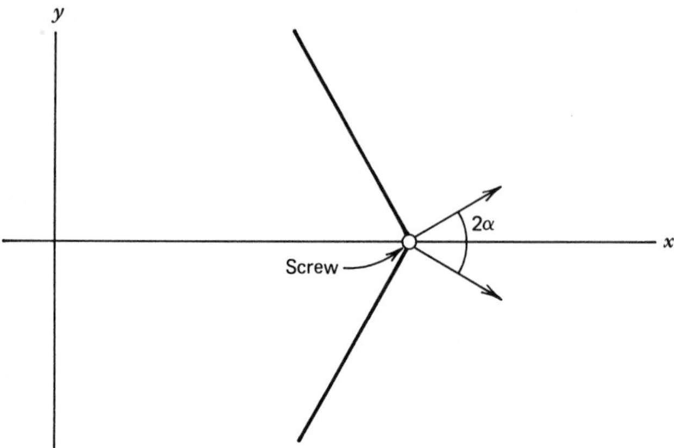

FIGURE 7-15. Two shock fronts emanating from a screw dislocation.

surface, or within the step, with a continuous smooth step as depicted in Fig. 7-13. The supersonic screw dislocation now can be constructed readily. The solution

$$u_z = \tfrac{1}{2} b \Delta(x \cos\alpha + y \sin\alpha - C_t t), \qquad y > 0$$
$$u_z = -\tfrac{1}{2} b \Delta(x \cos\alpha - y \sin\alpha - C_t t), \qquad y < 0 \qquad (7\text{-}73)$$

satisfies the Burgers circuit criterion. Also there is no net external force at the core. The material is unstrained and at rest everywhere except at the two shock fronts. Equation (7-73) describes a supersonic screw dislocation moving at velocity $v = C_t/\cos\alpha$ as two half-infinite shock waves diverge by an angle 2α and thus carry energy away from the slip plane (Fig. 7-15).

The above description refers to the idealization of a line singularity joining two sharp displacement fronts. A more realistic model would be a spread out core[40] and corresponding continuous smooth displacement steps instead of sharp discontinuity fronts (see Fig. 7-13). The energy radiation is inversely proportional to the step width D. Weertman and Weertman[41] have investigated what sort of force laws between the two half-crystals joined in the slip plane are consistent with steady-state motion and show explicitly that the force law must correspond to transformation of the slip plane to lower energy by passage of the dislocation.

The solution for the supersonic gliding edge dislocation is similar, although more complicated.[42–44] In the velocity range $C_t < v < C_l$ it is supersonic with

[40] J. D. Eshelby, *Proc. Phys. Soc.*, **69B**: 1013 (1956).

[41] J. Weertman and J. R. Weertman, "Dislocations in Solids," vol. 3, F. R. N. Nabarro (ed.), North Holland, Amsterdam, 1981, p. 1.

[42] J. Weertman, *J. Appl. Phys.*, **38**: 2612 (1967).

[43] J. Weertman, In "Mathematical Theory of Dislocations," T. Mura (ed.), American Society of Mechanical Engineering, New York, 1969, p. 178.

[44] J. Weertman, *J. Geophys. Res.*, **76**: 1171 (1971).

respect to shear waves and subsonic with respect to longitudinal waves, and the solution involves both shock fronts and a subsonic component. An interesting peculiarity is that at the particular velocity $\sqrt{2}\,C_t$, the character of the solution is purely subsonic and the dislocation can move without radiation of energy.[45]

Callias and Markenscoff[46] have generalized the moving dislocation problem to supersonic *nonuniform* motion in isotropic media. Stroh[47] has considered the uniformly moving supersonic dislocation in anisotropic media.

7-7. DISLOCATION MOBILITY

For linear elastic media, subsonic dislocations can move uniformly in glide without energy radiation and without friction. However, the considerations in Sec. 7-5 show that at finite temperatures, the moving dislocations interact with thermal sound waves (phonons) and exchange quasimomentum with the phonons by the flutter mechanism and by anharmonic strain field scattering, giving rise to a viscous drag on the moving dislocations.[48–51] Similarly, dislocations also scatter conduction electrons in metals, producing a further contribution to the drag.[52]

For these mechanisms to control mobility, other mechanisms not yet considered must be insignificant by comparison. These other processes include retardation of supersonic dislocations by shock-wave radiation demonstrated in Sec. 7-6. Another effect that must be considered is dispersion, the dependence of sound velocity on wavelength. The group velocity of short-wave length sound waves is smaller than the phase velocity, so that a moving dislocation becomes "supersonic" with respect to part of the phonon spectrum for velocities lower than C_t. According to Eshelby,[53] the dislocation can move without energy radiation up to $\sim C_t/2$, whereas in the range $C_t/2 < v < C_t$ "supersonic" radiation is important.

Eshelby included the discreteness of the crystal lattice, but only through its effect on the phonon spectrum. A full account of discreteness must also include the effect of the Peierls barrier and of phonon radiation from the core. We consider such effects in close-packed metals, which are the most studied metals, but the trends for these metals should apply more generally.

The Peierls barrier is small in close-packed metals. Also, the kinetic energy acquired by a dislocation dropping from the top of the potential barrier can be

[45] J. D. Eshelby, *Proc. Phys. Soc.*, **62A**: 307 (1949).

[46] C. Callias and X. Markenscoff, *Quart. Appl. Math*, **38**: 323 (1980).

[47] A. N. Stroh, *J. Math. Phys.*, **41**: 77 (1962).

[48] G. Leibfried, *Z. Phys.*, **127**: 344 (1950).

[49] F. R. N. Nabarro, *Proc. Roy. Soc.*, **A209**: 278 (1951).

[50] J. Lothe, *J. Appl. Phys.*, **33**: 2116 (1962).

[51] W. P. Mason, *J. Acoust. Soc. Am.*, **32**: 458 (1960).

[52] B. Tittman and H. Bommel, *Phys. Rev. Lett.*, **14**: 296 (1965).

[53] J. D. Eshelby, *Proc. Phys. Soc.*, **69B**: 1013 (1956).

used to overcome the next barrier. For low stresses and low velocities, the dislocations move by kink motion, and in fcc metals the barrier for such motion appears to be negligible. For large velocities, the *radiation* connected with the motion rather than the barrier height itself *must be* the important factor. However, the radiation problem is more difficult. Rogula[54] and Celli and Flytzanis[55] calculated the radiation into a discrete lattice from a moving perturbation representing the core. Taking umklapp processes into account, they concluded that there are strong radiation effects with infinite resonances even at low velocities.[56] This was a most disconcerting result. Experiments[57] have failed to reveal such resonance damping. Nor do computer experiments[58] with dislocations moving in discrete lattices show such resonance damping. Instead, they support Eshelby's original analysis of dislocations moving quite freely up to $C_t/2$. Alshits and Indenbom[59] argue that the Celli-Flytzanis result is artificial, resulting because a given motion of the perturbation is assumed, whereas in a complete treatment, the modulated motion of the perturbation must be included self-consistently in the solution.

In conclusion, the evidence is against resonance damping at low velocities. However, a completely satisfactory explanation of why the Celli-Flytzanis result does not apply is lacking. Celli and Flytzanis have posed a most interesting problem.

Accepting the overall evidence, we can assume that the dislocations are quite free up to velocities $\sim C_t/2$, except for viscous phonon and electron damping. Indeed, this point of view rationalizes a large amount of data.

The dislocation viscous drag coefficient is conventionally called B and is defined by the equation

$$\frac{F}{L} = Bv \tag{7-74}$$

Here F/L is the force per unit length that must be applied to the dislocation to keep it in uniform motion v. A variant of (7-74) that perhaps is more instructive is that the resolved shear stress for uniform motion is given by

$$\sigma = \eta \frac{v}{C_t} \tag{7-75}$$

where

$$\eta = \frac{BC_t}{b} \tag{7-76}$$

[54] D. Rogula, *Proc. Vibr. Problems*, **8**: 215 (1967).

[55] V. Celli and N. Flytzanis, *J. Appl. Phys.*, **41**: 4443 (1970).

[56] Similar results have been obtained in a one-dimensional model; see W. Atkinson and N. Cabrera, *Phys. Rev.*, **138**: 763 (1965).

[57] V. R. Parameswaran and J. Weertman, *Met. Trans.*, **2**: 1233 (1971).

[58] S. Ishioka, *J. Phys. Soc. Jap.*, **34**: 462 (1973).

[59] V. I. Alshits and V. L. Indenbom, *Sov. Phys. Usp.*, **18**: 1, (1975).

Since the original work by Leibfried[60] in 1950 and the papers of the following decade that essentially confirmed Leibfried's estimate, over the last 15 years a number of new detailed theoretical investigations have appeared that have significantly revised the earlier estimates of η. All these contributions cannot be reviewed and referenced here. The subsequent exposition is based on the discussion by Alshits and Indenbom[59] for the phonon contributions to η and on the paper by Kaganov, Kravchenko and Natsik[61] for the electron contribution to η. Comprehensive reference lists can be found in these papers.

At moderate temperatures, the phonon scattering effect dominates over the electron scattering effect. The two most important phonon mechanisms are the anharmonic strain field scattering and the flutter effect, the former contributing more to η by a factor ~ 10. The anharmonic scattering contribution has a T^5 temperature dependence at low temperatures, and the flutter mechanism has a T^3 dependence, so at sufficiently low temperatures the flutter mechanism is more important than strain field scattering. At high temperatures $T \gg \theta$, where θ is the Debye temperature, both effects have a linear T dependence.[62] For copper, the two effects add to give

$$\sigma \cong 2 \cdot 10^{-2} \mu \frac{v}{C_t}, \qquad \text{at} \quad T = \theta \tag{7-77}$$

This estimate is also in rough agreement with experiment[63] but is larger by a factor of 10 than the original Leibfried[64] prediction.

The temperature dependence of η is somewhat complicated because the different contributions have different temperature dependences. Let us, for copper, write (7-75) in terms of a temperature-dependent function $g(T/\theta)$ as

$$\sigma = 2 \cdot 10^{-2} \mu g\left(\frac{T}{\theta}\right) \frac{V}{C_t} \tag{7-78}$$

where $g(1) = 1$. The $g(T/\theta)$ curve is drawn in Fig. 7-16 as the full curve.

The electron scattering contribution of η is predicted to be temperature-independent.[65–67]

[60] G. Leibfried, *Z. Phys.*, **127**: 344 (1950).

[61] M. I. Kaganov, V. Ya. Kravchenko, and V. D. Natsik, *Sov. Phys. Usp.*, **16**: 878 (1974).

[62] For a correct estimate of the high-temperature behavior, slow phonons, i.e., phonons with small group velocity, must be included in the dispersive phonon spectrum. The temperature dependence of η is of the type $aT + b$, where the constant b is the slow phonon effect. See V. I. Alshits and V. L. Indenbom, *Sov. Phys. Usp.*, **18**: 1 (1975) and J. W. Martin and R. Paetsch, *Phys. Stat. Solidi*, **74b**: 761 (1976).

[63] K. M. Jassby and T. Vreeland, *Phil. Mag.*, **21**: 1147 (1970).

[64] G. Leibfried, *Z. Phys.*, **127**: 344 (1950).

[65] V. Ya. Kravchenko, *Sov. Phys. Sol. State*, **8**: 740 (1966).

[66] T. Holstein, *Phys. Rev.*, **151**: 187 (1966).

[67] Anomalous deviations from temperature independence are sometimes observed. See W. P. Mason, in "Dislocation Dynamics," McGraw-Hill, New York, 1968, p. 487, or V. R. Parameswaran, N. Urabe, and J. Weertman, *J. Appl. Phys.*, **43**: 2982 (1972). V. I. Alshits, *Sov. Phys. JETP*, **40**: 1099 (1975), has offered an explanation in terms of Fermi surface topography.

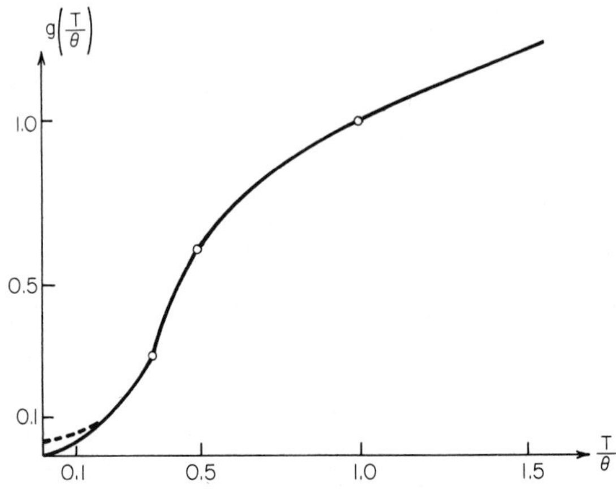

FIGURE 7-16. The parameter g as a function of T/θ.

With the values for copper inserted into the theoretical formula, we derive

$$\sigma \sim 5 \cdot 10^{-4} \mu \frac{v}{C_t} \tag{7-79}$$

for the electronic contribution for the drag stress. Adding this to the phonon part, the dashed curve in Fig. 7-16 results. Figure 7-17 presents an enlarged view of the low-temperature part of $g(T/\theta)$. For temperatures below $\sim 0.1\theta$, the electronic contribution is seen to be dominant. The curves in Fig. 7-17 should be considered neither as universal nor as exact, but as typical. Clearly the relative significance of the electron-scattering must depend on valency effects.

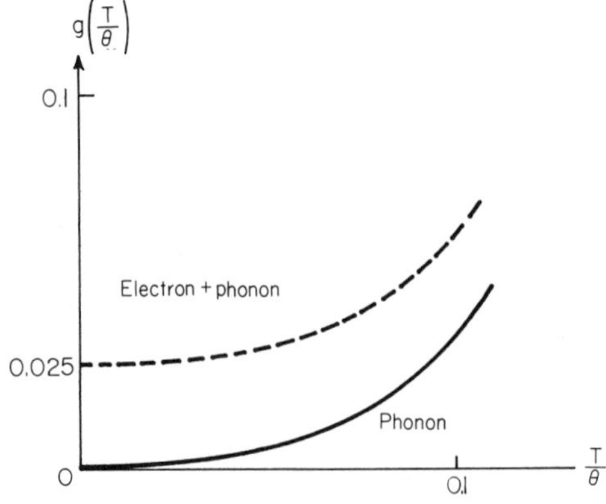

FIGURE 7-17. Low-temperature portion of Fig. 7-16.

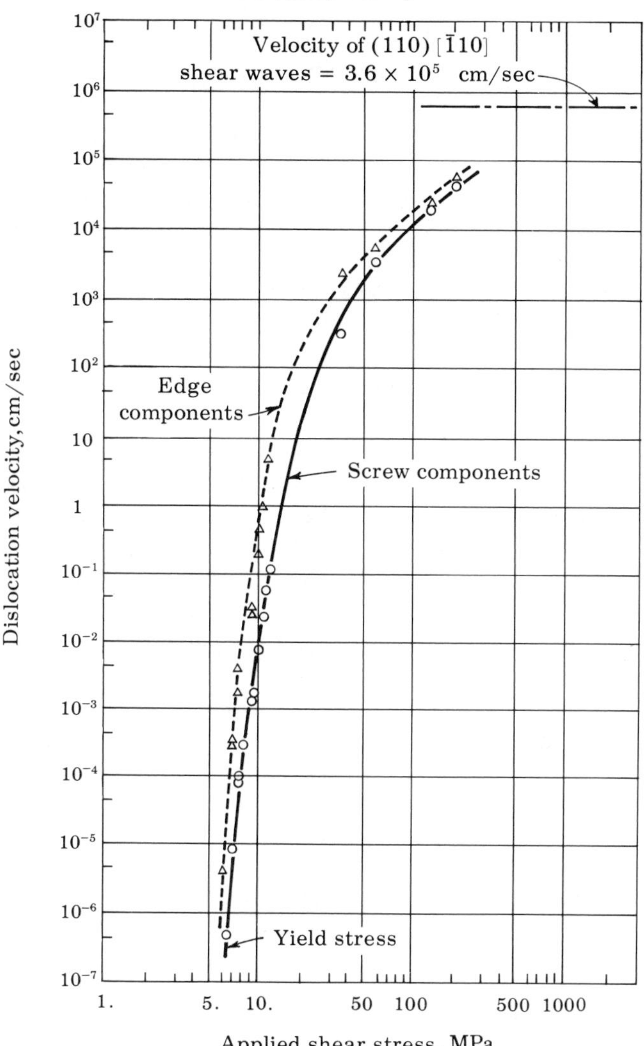

FIGURE 7-18. Dislocation velocity versus resolved shear stress for lithium fluoride [after W. G. Johnston and J. J. Gilman, *J. Appl. Phys.*, **30**:129 (1959)].

Figure 7-18 depicts the result of Johnston and Gilman[68] for dislocation mobility in LiF. At stresses approaching $\sigma \sim 5 \times 10^{-3}\mu$ the slope of the curve decreases rapidly at $v \sim C_t/10$, indicating an increase in friction forces. A reasonable interpretation is that viscous damping becomes important above $v \sim 10^{-3}$ or 10^{-2} C_t. The required stresses are in rough agreement with Eq. (7-77). At velocities lower than about 10^{-3} C_t, where the slope is large,

[68]W. G. Johnston and J. J. Gilman, *J. Appl. Phys.*, **30**, 129 (1959).

dislocation-dislocation and dislocation-defect interaction processes are rate-controlling, as discussed in Chaps. 15 and 16. Consistent with the above interpretation, the steep portion of the curve in Fig. 7-18 is very purity dependent and softens with increasing temperature, indicating a thermally activated process. Contrariwise, the high-velocity region has an impurity-independent slope that *decreases* with increasing temperature, consistent with the phonon viscosity mechanism. Many more recent experiments, on metals and on ionic crystals, have given similar results. Internal friction experiments interpreted in terms of the Granato-Lücke[69, 70] theory are also consistent with the above theory for dislocation mobility,[71] indicating that the theory applies also for the intrinsic dislocation mobility at low velocities.

Interesting softening effects are observed on cooling metals through the normal-superconducting transition temperature.[72, 73] This is the expected behavior when the electronic drag freezes out. The interpretation is not straightforward, however; the connection between yield stress and intrinsic mobility is not obvious. The explanation may be that even in dislocation motion involving breakaway and the overcoming of obstacles, the intrinsic mobility enters when the dislocations go through saddle point configurations.[74, 75] Further, and perhaps more importantly, the intrinsic mobility can enter through its control of inertial effects such as overcoming obstacles dynamically. The inertia would be associated with kinetic energy imparted to the dislocation by external stresses or gained by accelerating through a potential drop from a previous obstacle.[76] Recent experiments[77] demonstrating a jump in amplitude independent internal friction in lead at the superconducting transition demonstrate more directly the freezing out of electron drag.

Under certain conditions, to be discussed in Chap. 15, dislocation motion must be described in terms of the motion of kinks. Thus, knowledge of the kink mobility also becomes important. Eshelby[78] and Lothe[79] considered the flutter mechanism for kinks. Later, Seeger and Engelke[80] and Seeger and

[69]A. Granato and K. Lücke, *J. Appl. Phys.*, **27**: 583, 789 (1956).

[70]K. Lücke and A. Granato, in J. C. Fisher *et al.* (eds.), "Dislocations and Mechanical Properties of Crystals," Wiley, New York, 1957, p. 452.

[71]V. I. Alshits and V. L. Indenbom, *Sov. Phys. Usp.*, **18**: 1 (1975).

[72]G. A. Alers, O. Buck, and B. R. Tittman, *Phys. Rev. Lett.*, **23**: 290 (1969).

[73]V. Soldatov, V. Startsev, and T. Varinblat, *J. Low Temp. Phys.*, **2**: 641 (1970).

[74]V. I. Alshits and V. L. Indenbom, *Sov. Phys. Usp.*, **18**: 1 (1975).

[75]M. I. Kaganov, V. Ya. Kravchenko, and V. D. Natsik, *Sov. Phys. Usp.*, **16**: 878 (1974).

[76]M. Suenega and J. M. Galligan, *Scripta Met.*, **4**: 697 (1970). Later work is discussed by R. D. Isaac, R. B. Schwarz, and A. V. Granato, *Phys. Rev. B*, **18**: 4143 (1978) and V. L. Indenbom and V. M. Chernov, *Sov. Phys. Solid State*, **21**: 759 (1979).

[77]N. P. Kobelev and J. M. Soifer, *Phys. Stat. Solidi*, **50a**: K185 (1979).

[78]J. D. Eshelby, *Proc. Roy. Soc.*, **A266**: 222 (1962).

[79]J. Lothe, *J. Appl. Phys.*, **33**: 2116 (1962).

[80]A. Seeger and H. Engelke, "Dislocation Dynamics," McGraw-Hill, New York, 1968, p. 623.

Schiller[81] showed that, just as for straight dislocations, strain field scattering should be the more important factor at moderate temperatures, whereas the flutter mechanism should predominate at low temperatures.

PROBLEMS

7-1. Consider a screw dislocation moving at a velocity $v = C_t/10$, parallel to a free surface and a distance l below it. Is the dislocation more likely to cross slip out of the crystal because of its image interaction than a static screw in the same position, or less likely to do so?

7-2. Consider a screw dislocation moving parallel to the surface of a semi-infinite slab and a distance l below the surface. With an applied stress $\sigma = 10^{-3}\mu$, how long will it take for the dislocation to reach a velocity $0.9C_t$, starting at test? How far has it then moved? Assume quasi-uniform motion and ignore all friction.

7-3. Plot the distribution of shear stress σ_{xy} around an edge moving at a velocity $v = C_t/10$ and compare the result with the static case. Use the physical parameters for copper, with the elastic constants given in Appendix 1. How will two like-sign edge dislocations interact if they are moving uniformly on the same glide plane? On parallel glide planes?

7-4. Consider a screw segment of length L, pinned at the ends. If the segment is visualized as a string with a line tension $\mathcal{S}$ and effective mass per unit length m^*, what is its basic frequency of vibration? Determine the appropriate effective mass.

7-5. Assuming that the radiation per unit length from the vibrating string in Prob. 7-4 is given approximately by Eq. (7-49), how rapidly will the fundamental vibration damp out? Ignore all other dissipative mechanisms.

7-6. Assume that the effective mass of an edge dislocation is given by Eq. (7-64), with the outer cutoff radius $R \sim 10^4 b$. Determine the kinetic energy of an edge dislocation in copper moving at a velocity $v = C_t/10$. Suppose that the dislocation approaches a grain boundary. Determine whether the kinetic energy is sufficient to supply the surface energy of the step formed on the grain boundary if the edge intersects it. Assume the surface energy of the grain-boundary step is $\gamma = 0.6\ \mathrm{J/m^2}$. Neglect image forces.

7-7. Prove that the elastic displacements for a uniformly moving screw satisfy Eqs. (2-2) if one includes an inertial term

$$f_i = \rho_0 \frac{\partial^2 u_i}{\partial t^2}$$

[81] A. Seeger and P. Schiller, in W. P. Mason (ed.), "Physical Acoustics", vol. 3A, Academic, New York, 1966, p. 361.

BIBLIOGRAPHY

1. Eshelby, J. D., *Solid State Phys.*, **3**: 79 (1956); *Proc. Phys. Soc.*, **62A**: 307 (1949).
2. Leibfried, G. *Z. Phys.*, **127**: 344 (1950).
3. Lothe, J., *J. Appl. Phys.*, **33**: 2116 (1962).
4. Nabarro, F. R. N., "Dislocations," Oxford University Press, Fair Lawn, N.J., 1967.
5. Ninomiya, T., in H. Herman (ed.), "Treatise on Materials Science," vol. 8, Academic, New York, 1975, p. 1.
6. Seeger, A., and P. Schiller, in W. P. Mason (ed.), "Physical Acoustics", vol. 3A, Academic, New York, 1966, p. 361.
7. Weertman, J., in P. G. Shewmon and V. F. Zackay (eds.), "Response of Metals to High Velocity Deformation," Interscience, New York, 1961, p. 205.
8. Weertman, J., and J. R. Weertman, in F. R. N. Nabarro (ed.), "Dislocations in Solids," vol. 3, North Holland, Amsterdam, 1981, p. 1.

PART 2

EFFECTS OF CRYSTAL STRUCTURE ON DISLOCATIONS

8

The Influence of Lattice Periodicity

8-1. INTRODUCTION

Chapters 2 to 7 have dealt with the properties of dislocations in continuous, homogeneous, isotropic media. In Chaps. 8 to 13 the influence of the lattice periodicity of real crystals on dislocation properties is considered. Initially the approximation of isotropic elasticity is retained; modifications associated with anisotropic elasticity are introduced in Chap. 13. This chapter deals mainly with the influence of crystal structure on the atomic configurations of dislocations, and with the stresses and energies associated with them.

As was discussed in Chap. 1, the Frenkel treatment of the shear strength of a perfect crystal involves a potential energy of displacement that is a periodic function, with a period related to the atomic spacing in the crystal. Similarly, one expects a moving dislocation to experience a potential energy, or more formally a free energy, of displacement that reflects the lattice periodicity. Dehlinger and Kochendörfer[1] pointed out that the work of Frenkel and Kontorova[2] on a one-dimensional array of spring-connected balls lying on a periodic substrate could be applied as a simplified model of a dislocation. Subsequent developments of this phenomenological Frenkel-Kontorova model have proved useful in that the model is easy to analyze and provides a qualitative description of the slip process.

A more formal solution for the displacement potential was first presented, rather tersely, by Peierls[3] and was later elucidated and extended by Nabarro.[4] As we shall show, this model is useful in determining the *width* of a dislocation and in estimating the core energy of a dislocation. However, the extension of

[1]U. Dehlinger and A. Kochendörfer, *Z. Phys.*, **116**: 576 (1940).

[2]J. Frenkel and T. Kontorova, *Phys. Z. Sowj.*, **13**: 1 (1938).

[3]R. E. Peierls, *Proc. Phys. Soc.*, **52**: 23 (1940). The problem was suggested to Peierls by Orowan. Interesting accounts are given by E. Orowan in C. S. Smith Ed., "The Sorby Centennial Symposium on the History of Metallurgy," Gordon and Breach, New York, 1965, p. 255 and by R. E. Peierls, in A. R. Rosenfield, G. T. Hahn, A. L. Bement, and R. I. Jaffee (eds.), "Dislocation Dynamics," McGraw-Hill, New York, 1968, p. xiii.

[4]F. R. N. Nabarro, *Proc. Phys. Soc.*, **59**: 256 (1947).

the model to predict the lattice displacement potential resisting dislocation motion rests on quite tenuous assumptions. In view of the uncertainties in the assumed model, and the possibility that thermal vibrations can suffice to eliminate the free-energy-displacement barrier at moderate temperatures, one probably does as well to assume a phenomenological periodic lattice displacement potential. Nonetheless, the formal developments are treated briefly because they are useful in anticipating more refined predictions of the displacement potential.

The Peierls model does have the great merit of providing an *analytical* nonlinear elastic model of a dislocation core. Eventually, it probably will be supplanted by atomic calculations. A brief discussion is given of the present status of such calculations, which, although quite useful in modeling classes of behavior, suffer from nonanalyticity, large computation time, and approximate interatomic potentials.

The concept of a Peierls energy, or a variable lattice displacement potential, leads naturally to the concept of *kinks* and *jogs* in dislocation lines. These defects are in essence details of the dislocation core structure and are important in low-temperature dislocation glide and in dislocation climb processes. The treatment of these defects concludes this chapter.

8-2. THE PEIERLS-NABARRO DISLOCATION MODEL

Edge Dislocation

Consider two semi-infinite simple cubic crystals, as shown in Fig. 8-1*a*, with their cube axes parallel, but with an initial disregistry in the x direction across the plane $y=0$. Initially the disregistry of the bottom half-crystal with respect to the top one is

$$\phi_x{}^0 = \begin{cases} \dfrac{b}{2} & x>0 \\ -\dfrac{b}{2} & x<0 \end{cases}$$

The displacements $u(x)$, antisymmetric about the plane $y=0$,[5] are now imposed on the two half-crystals, which are then joined to form the edge dislocation shown in Fig. 8-1*b*. The disregistry in the latter case is

$$\phi_x(x) = \begin{cases} 2u_x(x)+\dfrac{b}{2} & x>0 \\ 2u_x(x)-\dfrac{b}{2} & x<0 \end{cases} \tag{8-1}$$

[5]An analogous treatment with the disregistry across $x=0$, i.e., normal to the glide plane, is given by J. H. van der Merwe, *Proc. Phys. Soc.* **63A**: 616 (1950).

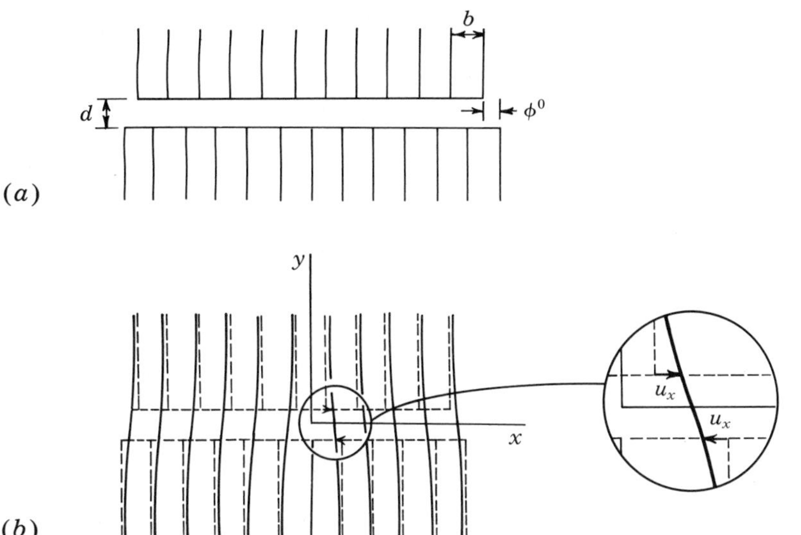

FIGURE 8-1. (*a*) Two semi-infinite simple cubic crystals, with disregistry $b/2$. (*b*) The same crystals displaced to form an edge dislocation.

with the boundary conditions $u_x(\infty) = -u_x(-\infty) = -b/4$. Here the displacements are referred to the bottom half-crystal.[6] By our convention for the Burgers vector, $\boldsymbol{\xi} = -\mathbf{k}$ and $\mathbf{b} = b\mathbf{i}$. Here $\mathbf{k}$ and $\mathbf{i}$ are unit vectors in the z and x directions, respectively. The displacements associated with the dislocation must have the general form shown in Fig. 8-2.

Associated with the displacements are restoring forces, connected with the distorted bonds across the plane $y=0$. As a first approximation, these restoring forces are assumed to lead to a local value of σ_{xy} at the plane $y=0$ which is a sinusoidal function of the bond disregistry $\phi(x)$ (*cf.* the Frenkel treatment in Chap. 1). This assumption requires that the displacements in the y direction be small compared to those in the x direction.

Thus, for the bottom half-crystal, the stress associated with the restoring forces is

$$\sigma_{xy}(x,0) = \text{const}\sin\frac{2\pi\phi_x}{b}$$

$$= -\text{const}\sin\frac{4\pi u_x}{b} \tag{8-2}$$

The constant is evaluated by requiring that Hooke's law be satisfied for small strain ϵ [Eq. (2-46)]. Hence

$$\sigma_{xy}(x,0) = 2\mu\epsilon_{xy} = \frac{\mu\phi_x}{d} \tag{8-3}$$

[6] Throughout this treatment the displacements and stresses are referred to the bottom half-crystal. Those relating to the top half-crystal are related symmetrically to these.

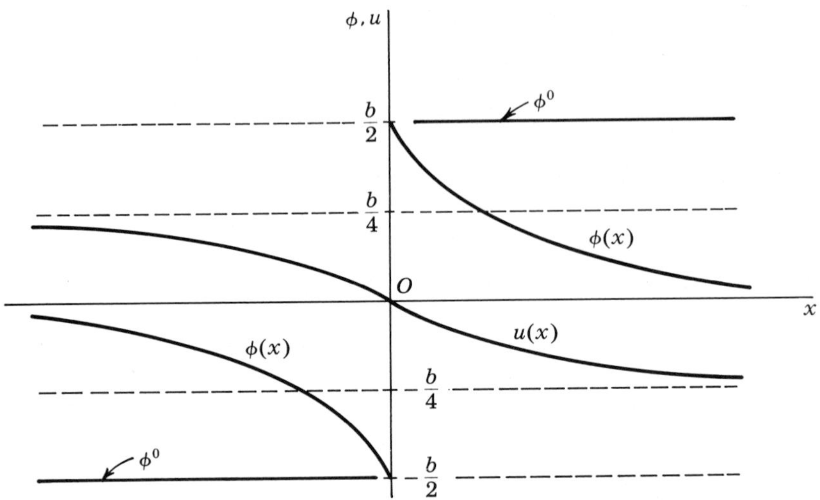

FIGURE 8-2. The displacement function for the edge dislocation.

where d is the interplanar spacing. Combining Eqs. (8-2) and (8-3), one obtains

$$\sigma_{xy}(x,0) = -\frac{\mu b}{2\pi d} \sin \frac{4\pi u_x}{b} \tag{8-4}$$

As originally suggested by Eshelby,[7] a continuous distribution of infinitesimal edge dislocation along the x axis satisfies these requirements, producing displacements u_x, as in Fig. 8-2, and leading to small displacements u_y. Let $\mathbf{b}'(x')\,dx'$ be the Burgers vector of the infinitesimal dislocation lying between x' and $x'+dx'$. The corresponding displacement is given by $-2(du_x/dx)\,dx'$, so that

$$b = \int_{-\infty}^{\infty} b'(x')\,dx' = -2\int_{-\infty}^{\infty} \left(\frac{du_x}{dx}\right)_{x=x'} dx' \tag{8-5}$$

According to Eq. (3-43), this distribution of dislocation produces a shear stress at $(x,0)$ given by

$$\sigma_{xy}(x,0) = -\frac{\mu}{2\pi(1-\nu)} \int_{-\infty}^{\infty} \frac{b'\,dx'}{x-x'} = \frac{\mu}{\pi(1-\nu)} \int_{-\infty}^{\infty} \frac{(du_x/dx)_{x=x'}\,dx'}{x-x'} \tag{8-6}$$

[7]J. D. Eshelby, *Phil. Mag.*, **40**: 903 (1949).

where the integral is defined by its principal value (see below). At equilibrium the net stress at the point $(x,0)$ must vanish; so σ_{xy} according to Eq. (8-6) must be equal and opposite to that given by Eq. (8-4). The combination of these two equations leads to the following integral equation for the displacements u_x:

$$\int_{-\infty}^{\infty} \frac{(du_x/dx)_{x=x'}\,dx'}{x-x'} = \frac{b(1-\nu)}{2d}\sin\frac{4\pi u_x}{b} \tag{8-7}$$

The solution of Eq. (8-7) is[8]

$$u_x = -\frac{b}{2\pi}\tan^{-1}\frac{x}{\zeta} \tag{8-8}$$

where $\zeta = d/2(1-\nu)$. Equation (8-8) has the expected form of Fig. 8-2, and it satisfies the boundary conditions $u_x(\infty) = -u_x(-\infty) = -b/4$. Also, $u_x(\zeta) = -b/8 = \frac{1}{2}u_x(\infty)$. Thus the *width* of the dislocation, defined by $2\zeta = d/(1-\nu)$, includes the region $-\zeta < x < \zeta$, wherein the disregistry is greater than one-half its maximum value at $x=0$. The width of the dislocation thus gives a rough measure of the extent of the core region that cannot be described by linear elasticity.

Substituting Eq. (8-8) into (8-6) gives the result

$$\sigma_{xy}(x,0) = -\frac{\mu b}{2\pi(1-\nu)}\frac{x}{x^2+\zeta^2} \tag{8-9}$$

[8] J. D. Eshelby, *Phil. Mag.*, **40**: 903 (1949). To verify that Eq. (8-8) is a solution of (8-7), take its derivative,

$$\left(\frac{du_x}{dx}\right)_{x=x'} = \frac{-b(1-\nu)d}{\pi\left[d^2+4(1-\nu)^2x'^2\right]} = \frac{p}{d^2+m^2x'^2}$$

The integral in Eq. (8-7) is defined by its principal value, given by

$$\int_{-\infty}^{\infty}\frac{p\,dx'}{(x-x')(d^2+m^2x'^2)} = \lim_{\Delta\to 0}\left(\int_{-\infty}^{x-\Delta}+\int_{x+\Delta}^{\infty}\right)\frac{p\,dx'}{(x-x')(d^2+m^2x'^2)}$$

$$= 2\pi i(\text{residue at } id/m) + \pi i(\text{residue at } x)$$

$$= \frac{\pi x m p}{d(d^2+m^2x^2)} = \frac{-2b(1-\nu)^2x}{d^2+4(1-\nu)^2x^2}$$

Substituting Eq. (8-8) into the right side of Eq. (8-7) gives the identical result. See I. S. Sokolnikoff and R. M. Redheffer, "Mathematics of Physics and Modern Engineering," McGraw-Hill, New York, 1966, pp. 564–569 and 590–594, for a discussion of the method of residues.

Also, from Eqs. (8-8) and (8-5),

$$b' = \frac{b}{\pi} \frac{\zeta}{x'^2 + \zeta^2} \tag{8-10}$$

More generally, the continuous distribution of dislocation [Eq. (8-5)] is expected to correspond to a stress function which, by analogy with Eq. (3-42), is given by

$$\begin{aligned} \psi &= \frac{\mu}{2\pi(1-\nu)} \int_{-\infty}^{\infty} b' y \ln\left[(x-x')^2 + y^2\right]^{1/2} dx' \\ &= \frac{\mu b \zeta y}{4\pi^2(1-\nu)} \int_{-\infty}^{\infty} \frac{\ln\left[(x-x')^2 + y^2\right]}{x'^2 + \zeta^2} dx' \end{aligned} \tag{8-11}$$

The solution to Eq. (8-11) is[9]

$$\psi = \frac{\mu b}{4\pi(1-\nu)} y \ln\left[x^2 + (y \pm \zeta)^2\right] \tag{8-12}$$

where the plus sign in $y \pm \zeta$ is taken in the half-plane $y > 0$ and the minus sign is taken for $y < 0$, so that *there is no divergency in the displacements and strains.*

[9]To solve the integral in Eq. (8-11), rewrite the integrand as

$$\left[\ln(x - x' + iy) + \ln(x - x' - iy)\right]\left(-\frac{1}{2i\zeta}\right)\left(\frac{1}{x' + i\zeta} - \frac{1}{x' - i\zeta}\right)$$

Thus the integral can be split up into parts such as

$$\int_{-\infty}^{\infty} \frac{\ln(x - x' + iy)}{x' + i\zeta} dx'$$

The logarithm has a singularity in the upper plane at $x' = x + iy$, so that for this term the residue is taken in the bottom half-plane. The residue of the pole at $x' = -i\zeta$ is

$$\ln[x + i(y + \zeta)]$$

Two of the other parts of the integral are zero, while the third has a residue in the upper half-plane of

$$\ln[x - i(y + \zeta)]$$

Adding the two contributions and multiplying by the factor $2\pi i(-1/2i\zeta)$, one obtains Eq. (8-12).

According to Eq. (2-54), the complete stress distribution of the Peierls dislocation is found by differentiation to be[10]

$$\sigma_{xy}=\frac{-\mu b}{2\pi(1-\nu)}\left\{\frac{x}{x^2+(y+\zeta)^2}-\frac{2xy(y+\zeta)}{[x^2+(y+\zeta)^2]^2}\right\}$$

$$\sigma_{xx}=\frac{\mu b}{2\pi(1-\nu)}\left\{\frac{(3y+2\zeta)}{x^2+(y+\zeta)^2}-\frac{2y(y+\zeta)^2}{[x^2+(y+\zeta)^2]^2}\right\} \tag{8-13}$$

$$\sigma_{yy}=\frac{\mu b}{2\pi(1-\nu)}\left\{\frac{y}{x^2+(y+\zeta)^2}-\frac{2x^2y}{[x^2+(y+\zeta)^2]^2}\right\}$$

$$\sigma_{zz}=\nu(\sigma_{xx}+\sigma_{yy})=\frac{\mu b\nu}{\pi(1-\nu)}\frac{y+\zeta}{x^2+(y+\zeta)^2}$$

Exercise 8-1. Verify the result for σ_{yy} in Eq. (8-13) by integrating the expression $\sigma_{yy}=\partial^2\psi/\partial x^2$ as generated by Eq. (8-11).

A comparison of Eqs. (8-13) and (3-43) reveals that the Peierls dislocation stress field reduces to that of the Volterra dislocation for $r=(x^2+y^2)^{1/2}\gg\zeta$. The parameter ζ has the interesting effect of removing the singularity at the origin $r=0$ that is present for the Volterra dislocation. Thus the Peierls model, which takes account of the discreteness of the crystal lattice, *removes the artificial divergency at the core* that is associated with the idealized continuum dislocation of Volterra. Nonetheless, for the Peierls dislocation the strains in the core region are so large that it is questionable whether Hooke's law applies. From Eq. (8-13), ϵ_{xy} has a maximum value of $b/4\pi d\sim 0.1$ at $y=0$ and $x=\zeta$; ϵ_{yy} has a maximum value of ~ 0.2; and ϵ_{xx} has a maximum value of $2b/\pi d\sim 0.7$ at $x=y=0$. Corrections of the elastic energy of the Peierls dislocation to account for the likely deviation from Hooke's law in the core region are probably not justifiable, however, in view of the approximations in the Peierls model.

The higher-order terms corresponding to the image corrections of Chap. 3 can be developed directly by analogy with the treatment there, but again it is uncertain that such corrections are justifiable.

Exercise 8-2. For the Peierls dislocation, develop the image stress function corresponding to Eq. (3-79). Discuss the applicability of Eq. (3-79) to the Peierls dislocation.

[10] For ease of depiction, the plus-or-minus notation for ζ is omitted. In the region $y<0$ the factor ζ becomes $-\zeta$.

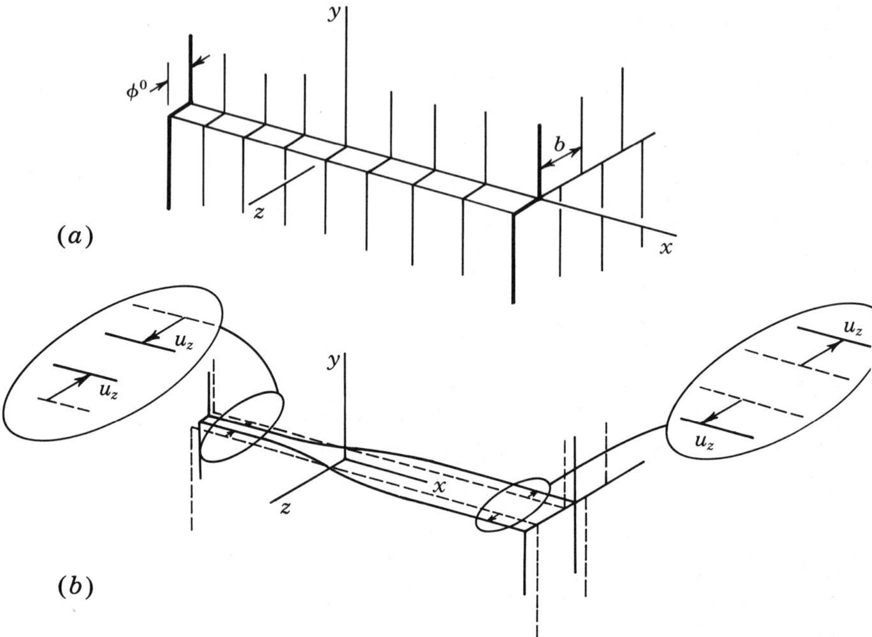

FIGURE 8-3. Two semi-infinite crystals (a) with disregistry $b/2$, and (b) displaced to form a screw dislocation.

Screw Dislocation

As indicated in Fig. 8-3a, the screw dislocation can be treated similarly to the edge dislocation. Initially the two semi-infinite half-crystals have a disregistry

$$\phi_z^0 = \begin{cases} \dfrac{b}{2} & x > 0 \\ -\dfrac{b}{2} & x < 0 \end{cases}$$

The right-handed screw dislocation shown in Fig. 8-3b is then formed by the displacement u_z of the crystals. Figure 8-3 shows clearly that the form of the functional dependence of u_z on x is the same as that presented in Fig. 8-2 for the dependence of u_x on x.

The analysis[11] proceeds exactly as for the edge dislocation. The displacements are assumed to be produced by a continuous distribution of dislocation $b' = -2(\partial u_z/\partial x)_{x=x'}$, where u_z is referred to the bottom half-crystal. The

[11] The results for the screw dislocation were first obtained by J. D. Eshelby [*Phil. Mag.*, **40**: 903 (1949)].

shear stress acting on the plane is, from Eq. (3-3),

$$\sigma_{yz}(x,0)=\frac{\mu}{2\pi}\int_{-\infty}^{\infty}\frac{b'\,dx'}{x-x'}=\frac{-\mu}{\pi}\int_{-\infty}^{\infty}\frac{(\partial u_z/\partial x)_{x=x'}}{x-x'}dx' \qquad (8\text{-}14)$$

where again the integral is defined by its principal value. The shear stress caused by the misfit is

$$\sigma_{yz}(x,0)=\frac{-\mu b}{2\pi d}\sin\frac{4\pi u_z}{b} \qquad (8\text{-}15)$$

Exercise 8-3. Proceed as outlined above for the edge dislocation and show that the results for the screw dislocation are the following:

$$u_z(x,0)=\frac{-b}{2\pi}\tan^{-1}\frac{x}{\eta} \qquad (8\text{-}16)$$

where $\eta=\frac{d}{2}$

$$\sigma_{yz}(x,0)=\frac{\mu b}{2\pi}\frac{x}{x^2+\eta^2} \qquad (8\text{-}17)$$

$$b'=\frac{b}{\pi}\frac{\eta}{x'^2+\eta^2} \qquad (8\text{-}18)$$

The only notable difference for the screw is that η replaces ζ, where $\eta=(1-\nu)\zeta$. Thus the width of the screw dislocation, 2η, is *less* than the width of the edge dislocation.

Generally, the extended form of σ_{yz} for the continuous distribution of infinitesimal screw dislocations is

$$\sigma_{yz}=\frac{\mu}{2\pi}\int_{-\infty}^{\infty}\frac{b'(x-x')}{(x-x')^2+y^2}dx' \qquad (8\text{-}19)$$

Inserting Eq. (8-18) into Eq. (8-19) and taking the residues, one finds[12]

$$\sigma_{yz}=\frac{\mu b}{2\pi}\frac{x}{x^2+(y\pm\eta)^2} \qquad (8\text{-}20)$$

With the use of Hooke's law in the form $\sigma_{yz}=\mu(\partial u_z/\partial y)$, u_z is determined by

[12]Again the term $\pm\eta$ is to be taken as plus for $y>0$ and minus for $y<0$.

integration to be

$$u_z = -\frac{b}{2\pi}\tan^{-1}\frac{x}{y\pm\eta} + f(x) \tag{8-21}$$

Inspection of the boundary conditions reveals that $f(x)=0$. σ_{xz} can now be determined by differentiation

$$\sigma_{xz} = \mu\frac{\partial u_z}{\partial x} = -\frac{\mu b}{2\pi}\frac{y\pm\eta}{x^2+(y\pm\eta)^2} \tag{8-22}$$

After transformation of σ_{xz} and σ_{yz} to polar coordinates and expansion of the resulting stresses in powers of d/r, the stress field of the screw dislocation is found to be

$$\sigma_{\theta z} = \frac{\mu b}{2\pi r}\left(1 \mp \frac{d\sin\theta}{2r}\right)$$

$$\sigma_{rz} = \frac{\mp\mu b}{4\pi r}\frac{d\cos\theta}{r} \tag{8-23}$$

Again the stresses reduce to those for the equivalent Volterra screw dislocation when $\eta = d/2 \ll r$. Also, the parameter η has the effect of removing the singularity in the stresses at $r=0$. Notice that, unlike the Volterra dislocation, the Peierls screw dislocation develops a stress component σ_{rz} near the core. This component arises because the screw dislocation is spread out on the plane $y=0$, as indicated by Eq. (8-18). This spreading phenomenon anticipates the dissociation of a dislocation on its glide plane; it tends to restrict the screw dislocation to glide on the plane $y=0$.

Elastic Energy of the Peierls Dislocation

The energy associated with the Peierls dislocation is divided into two portions: the elastic strain energy stored in the two half-crystals, corresponding to the strain energy of the Volterra dislocation (Chap. 3), and the misfit energy associated with the distorted bonds across the plane $y=0$, corresponding to the core energy of the Volterra dislocation. In view of the continuous distribution expected for the elastic displacements around a dislocation, this division is rather artificial, but it is consistent with the assumptions of the Peierls model. Recall that the equilibrium configuration of the Peierls dislocation was obtained by setting the *stress* arising from the misfit of the bonds in the glide plane equal to the *stress* which acts on the surface of a half-crystal and which is required to produce the displacements in the half-crystal. The surface stress on the half-crystal has a linear elastic response to the displacements (Hooke's law), while the stress in the misfit bonds has a sinusoidal response to displacement [Eq. 8-4)].

The elastic strain energy stored in the upper half-crystal equals the work done by surface forces in generating the displacements u_x. On an element of area δx times unit length in the z direction, this work is, from Eq. (8-4),

$$\delta W = \int_0^{u_x} \sigma_{xy}(x,0)\,\delta x\, du = \tfrac{1}{2}\sigma_{xy}(x,0)u_x\,\delta x \tag{8-24}$$

Integrating over both top and bottom surfaces from $x=-r$ to $x=r$ to obtain a result equivalent to Eq. (3-52), after substituting for σ_{xy} and u_x from Eqs. (8-8) and (8-9), one obtains[13]

$$W = \frac{\mu b^2}{4\pi(1-\nu)} \ln \frac{r}{2\zeta} \tag{8-25}$$

which is quite similar to Eq. (3-52). However, to obtain the total elastic energy with a cylinder of radius r about the dislocation, one must add the work done by the stresses acting on the surface of the cylinder r, that is, the integral work done by the forces at the surface r arising from the stresses [Eq. (3-44)] as the surface is displaced by the amount given by Eqs. (3-45) and (3-46).[14]

$$W = -\frac{1}{2}\int_0^{2\pi} (\sigma_{rr}u_r + \sigma_{r\theta}u_\theta)r\,d\theta = \frac{\mu b^2 \gamma}{4\pi(1-\nu)} \tag{8-26}$$

where $\gamma = (1-2\nu)/4(1-\nu)$ [see Eq. (6-15)].

Similarly, for the screw dislocation the elastic energy is

$$W = \frac{\mu b^2}{4\pi} \ln \frac{r}{2\eta} \tag{8-27}$$

In this case there are no stresses which do work on the cylinder surface.

[13] The integral of Eq. (8-24) over both surfaces is

$$W = \int_{-r}^{r} \sigma_{xy} u_x\, dx = \frac{\mu b^2}{4\pi^2(1-\nu)} \int_{-r}^{r} \frac{x \tan^{-1}(x/\zeta)\, dx}{x^2+\zeta^2}$$

Letting $z = x/\zeta$ and integrating by parts, the integral becomes

$$\int_{-r/\zeta}^{r/\zeta} \frac{z \tan^{-1} z\, dz}{1+z^2} = \frac{1}{2}\int_{-r/\zeta}^{r/\zeta} \tan^{-1} z\, d\ln(1+z^2)$$
$$= \left[\frac{1}{2}\ln(1+z^2)\tan^{-1} z\right]_{-r/\zeta}^{r/\zeta} - \frac{1}{2}\int_{-r/\zeta}^{r/\zeta} \frac{\ln(1+z^2)\, dz}{1+z^2}$$

In the limit $r \gg \zeta$ the first term in this result reduces to $\pi \ln(r/\zeta)$. The second term can be evaluated by the procedure leading to Eq. (8-12), yielding the result $-\pi \ln 2$. Equation (8-25) then follows.

[14] Equation (8-26) applies only for the cut surface along $\theta = 0$. If the cut is instead at $\theta = \beta$, a term $-\mu b^2 \sin^2\beta / 4\pi(1-\nu)$ must be added to the result in Eq. (8-26).

Misfit Energy

The local shear strain on the slip surface of the edge dislocation (Fig. 8-1) is

$$\epsilon_{xy}(x,0)=-\frac{\phi_x}{2d}=-\frac{2u_x+(b/2)}{2d}$$

The local strain energy contains contributions $\sigma_{xy}d\epsilon_{xy}+\sigma_{xy}d\epsilon_{yx}=2\sigma_{xy}d\epsilon_{xy}$. Thus the misfit energy stored in a volume element of height d, width δx, and unit depth at the glide surface $y=0$ is given by

$$\delta W(x)=-2\int d\,\delta x\sigma_{xy}d\epsilon_{xy}=2\int_{b/4}^{u_x}\delta x\,\sigma_{xy}du_x$$

$$=-\frac{\mu b\,\delta x}{\pi d}\int_{b/4}^{u_x}\sin\frac{4\pi u_x}{b}du_x=\frac{\mu b^2\,\delta x}{4\pi^2 d}\left(1+\cos\frac{4\pi u_x}{b}\right) \quad (8\text{-}28)$$

with Eq. (8-4) used for σ_{xy}. The total misfit energy in the glide plane is

$$W_m=\frac{\mu b^2}{4\pi^2 d}\int_{-\infty}^{\infty}\left(1+\cos\frac{4\pi u_x}{b}\right)dx=\frac{\mu b^2}{4\pi(1-\nu)} \quad (8\text{-}29)$$

Similarly, for the Peierls screw dislocation the misfit energy is given by

$$W_m=\frac{\mu b^2}{4\pi} \quad (8\text{-}30)$$

8-3. ATOMIC CALCULATIONS

Methods

Although it has the advantage of yielding analytical results, the treatment of the nonlinear region as a thin strip in the glide plane in the Peierls-Nabarro model is evidently physically unrealistic in view of the cylindrical symmetry of dislocation fields. In recent years there has been an increasing number of atomic calculations that employ more relevant cylindrical boundary conditions. However, they do not provide analytic solutions for the core region. The calculations all involve the use of pair potentials to describe atomic or ionic interaction and are thus limited by the validity of the potentials.[15–17]

[15] M. P. Puls, in "Dislocation Modeling of Physical Systems," Pergamon, New York, 1981, p. 249.

[16] R. Bullough and V. K. Tewary, "Dislocations in Solids," vol. II, North-Holland, Amsterdam, 1979, p. 1.

[17] P. C. Gehlen, J. R. Beeler, Jr., and R. I. Jaffee, "Interatomic Potentials and Simulation of Lattice Defects," Plenum, New York, 1972.

At present there are problems with first-principle, perfect-crystal calculations, in extending them to highly strained dislocation cores; with quantum-mechanical cluster calculations, in defining the boundary conditions; and with chemical-bond methods such as the Miedema[18] and Engel-Brewer[19] models, in their semi-empirical nature; and little agreement among them. Consequently, for metals, use is generally made of empirical, central-force potentials, represented by spline-fit polynomials[20] and truncated after a few interatomic distances, or of pseudopotentials.[21] The longer-range potentials require much more computation time, so the simpler empirical potentials are often used. These have the problem that, no matter how many properties they are fitted with, they are not very successful in predicting nonfitted properties. Thus, they must be regarded as approximate in representing real materials; yet they are useful in testing classes of behavior at the atomic level.

Other potentials that have been used are the semi-empirical Morse, Born-Mayer and Lennard-Jones potentials, the latter of which accurately represents the inert gases.[22] For ionic crystals, where the major part of the energy arises from the Coulomb interaction of ions, the situation is also improved compared to metals. The short-ranged core interaction is given by simple Born-Mayer or van der Waals potentials, the long-range interaction by the Coulomb interaction and polarization effects can be included in the so-called breathing shell model.[23]

The computational procedures involve the minimization of energy of an atom array containing a dislocation with respect to configuration. Most calculations have been performed at 0°K and have used both lattice dynamic and lattice static approaches with the latter treatment describing the atoms in either Fourier inverse space or real space.[24, 25] According to comparative studies by Puls and coworkers,[24] the most efficient method is the lattice statics approach using the conjugate-gradient method in real space.

The optimum method, from a computational time basis is to divide the crystal into an atomic region I and a surrounding elastic region II, which contains atoms or ions out to the range of the interatomic potential (see Fig. 8-4). Rigid boundary calculations have been used in which the Volterra elastic displacements are imposed and region I is then relaxed. This method has the disadvantage of either suppressing nonlinear effects if the size of region I is

[18]A. R. Miedema, E. R. de Boer, and P. F. de Chatel, *J. Phys. F*, **3**: 1558 (1973).

[19]L. Brewer, *Acta Met.*, **15**: 553 (1967).

[20]For example, see the iron potential of R. A. Johnson, [*J. Phys. F*, **3**: 295 (1973)].

[21]W. A. Harrison, "Pseudopotentials in the Theory of Metals," Benjamin, New York, 1966.

[22]A. Paskin, A. Gohar, and G. J. Dienes, *Phys. Rev. Lett.*, **44**: 940 (1980).

[23]V. Nusslein and V. Schroder, *Phys. Stat. Solidi*: **21**: 309 (1967).

[24]M. P. Puls, in M. F. Ashby, R. Bullough, C. S. Hartley and J. P. Hirth Eds. "Dislocation Modeling of Physical Systems," Pergamon, New York, 1981, p. 249.

[25]R. Bullough and V. K. Tewary, "Dislocations in Solids," vol. II, North-Holland, Amsterdam, 1979, p. 1.

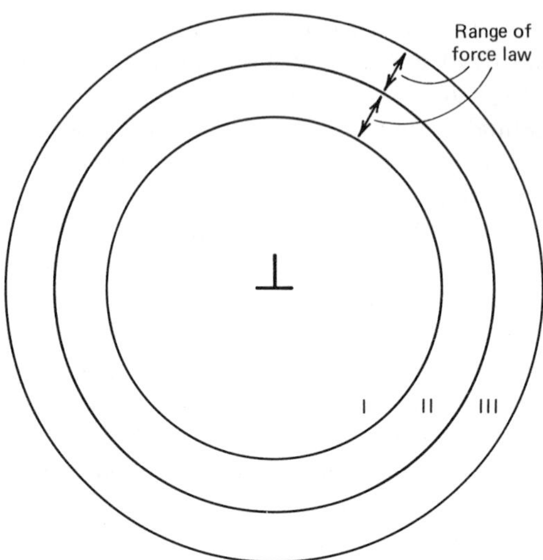

FIGURE 8-4. Regions concentric about a dislocation for atomic calculations. They need not be cylindrical.

small, or of requiring excessive computational time if the size is large. Alternatively, flexible boundaries can be used to ensure compatible forces and displacements at the region I–II boundary. According to Woo and Puls[26] and Sinclair et al.,[27] the most efficient scheme is that called Flex II by the latter authors, developed by them. Rigid boundary conditions are imposed, region I is relaxed, incompatibility forces are determined at the boundary, displacements (tensor Green functions) to accommodate these forces are imposed in regions I and II, and the procedure is iterated. Green's function is available in either the isotropic elastic approximation, Eq. (2-70), or the anisotropic one.[28]

Core Configuration

Numerous calculations of core configurations have been performed, as reviewed by Puls.[24] For undissociated dislocations, or for cases where the region I size was large in the rigid case, both rigid and flexible methods give about the same core configuration. The results have confirmed that dislocations are dissociated in model fcc materials, whereas they are not in ionic crystals, and that nascent dissociations occur in bcc crystals. These results have important implications for glide in these crystals as discussed in Chaps. 10 to 12.

Another important finding is that edge dislocations, in particular, have associated displacement fields of nonlinear elastic origin, reflected by the displacements of the region I–II boundaries in the flexible methods. A major

[26] C. H. Woo and M. P. Puls, *Phil. Mag.*, **35**: 727 (1977).

[27] J. E. Sinclair, P. C. Gehlen, R. G. Hoagland, and J. P. Hirth, *J. Appl. Phys.*, **49**, 3890 (1978).

[28] J. P. Hirth and J. Lothe, *J. Appl. Phys.*, **44**: 1029 (1973).

feature of these displacement fields is a volume expansion per unit length of dislocation (Problem 2-10), which is not present for the linear elastic Volterra field (Problem 3-11). Analogous to the ball in hole case of Exercise 2-5, the field is characterized by an infinite medium volume change per unit length $\delta v/L$ and an image relaxation term associated with free surfaces of a finite body, summing to a total volume change per unit length $\delta V/L$. Simulations of the $\frac{1}{2}[110]\{110\}$ edge dislocation in KCl,[29] MgO,[30] NaCl,[31,32] and of the $\frac{1}{2}[111]$ screw dislocation[33] and the $\frac{1}{2}[111]\ \{110\}$ edge dislocation[34] in α iron all give $\delta v/L$ values of about $0.50b^2$, which would lead to $\delta V/L$ values of about $1.0b^2$ when surface relaxation is included.[27]

The volume change, as well as the other components of the nonlinear field, has several implications with respect to dislocation behavior. For small spacings, the interaction force and interaction energy will be modified. Also, the dislocation will interact with an external stress through a $P\,\delta V$ interaction that will produce forces additional to the Peach-Koehler force.

Core Energy

In contrast to core configuration determinations, there have been only a few cases where the energy was computed. For ionic crystals, the total energy in region I as a function of R has been calculated for several cases. According to the prescription of Eqs. (3-14) and (3-55), these results can be used to incorporate the true total energy of a dislocation in an infinite medium into the linear elastic relations by adjusting $r_0 = b/\alpha$ so that the above equations and the atomic calculations give the same result. The core radii so calculated are listed in Table 8-1. The results show that the rigid boundaries lock in the energy associated with the nonlinear dilatational field and give a gross overestimate of the total energy (an underestimate of r_0). The results are in fair agreement for the different crystals and suggest that $\alpha \approx 3$ for the ionic crystals. Comparative studies have been made, using flexible boundaries, for screw and edge dislocations in NaCl.[35] Although r_0 values cannot be determined from the results presented, they do indicate that r_0 is smaller for the screw dislocation, suggesting $\alpha \approx 5$.

A rigid boundary calculation for a diamond cubic crystal[36] indicates $\alpha \sim 4$ for a screw dislocation. In this structure with covalent bonding one might expect less nonlinear dilatation, so the indication is that α is similar to that for

[29] R. G. Hoagland, J. P. Hirth, and P. C. Gehlen, *Phil. Mag.*, **34**: 413 (1976).

[30] C. H. Woo and M. P. Puls, *Phil. Mag.*, **35**: 727 (1977).

[31] F. Granzer, V. Belzner, M. Bucher, P. Petrasch, and C. Teodosiu, *J. Phys.*, **34**: C9-359 (1973).

[32] M. P. Puls and C. B. So, *Phil. Mag.*, **98**: 87 (1980).

[33] C. Wüthrich, *Phil. Mag.*, **35**: 325 (1977).

[34] B. L. Adams, J. P. Hirth, P. C. Gehlen, and R. G. Hoagland, *J. Phys. F*, **7**: 2021 (1977).

[35] W. Raupach, *Phys. Stat. Solidi*, **56a**: 535 (1979).

[36] A. Maradudin, *J. Phys. Chem. Solids*, **9**: 1 (1959).

TABLE 8-1 Values of the Core Parameter r_0 Required to Make the Dislocation Energy Computed in Atomic Calculations Equal that of Linear Anisotropic Elasticity Theory*

Crystal	Boundary	r_0	$\sigma_p/\mu(\times 10^3)$	Reference
KCl	Flexible	0.37b	1.0–2.5	1
NaCl	Flexible	0.28–0.38b	0.7–3.5	2
MgO	Flexible	0.28–0.37b	0.9–1.5	3
MgO	Rigid	0.10–0.12b	1.6–3.8	3,4

*All results for $\frac{1}{2}[110]\ \{110\}$ edge dislocation. Also listed are the Peierls stresses σ_p.
[1]R. G. Hoagland, J. P. Hirth, and P. C. Gehlen, *Phil. Mag.*, **34**: 413 (1976).
[2]M. P. Puls and C. B. So, *Phys. Stat. Solidi*, **98a**: 87 (1980).
[3]C. H. Woo and M. P. Puls, *Phil. Mag.*, **35**: 1641 (1977).
[4]M. P. Puls and M. J. Norgett, *J. Appl. Phys.*, **47**: 466 (1976).

ionic crystals. In metals, no calculations are available. However, comparisons with experiments in Table 10-4 suggest an α in the range 0.5 to 2 for partial dislocations and a direct estimate of the line energy of a bowing-out loop in zinc suggests a value of α in the same range for a perfect dislocation.[37]

In the absence of more definitive data for metals, a comparison of the analytical results for the Peierls dislocation with the continuum results for the core energy of the Volterra dislocation is of interest. As pointed out in Sec. 8-2, the stresses of the Peierls dislocation reduce to the simpler Volterra values except near the core, so that one is justified in most applications in using the simpler continuum results. Correspondingly, the energies of the two are quite similar, as evidenced by a comparison of Eqs. (3-14), (3-55), (8-25), and (8-27), provided that a consistent value of the core radius is selected. For complete equivalency, the core parameter α must be evaluated. If Eq. (3-55) is equated to the sum of Eqs. (8-25), (8-26), and (8-29), one finds that the energy expressions are equivalent if one chooses[38]

$$r_0 = \frac{b}{\alpha_e} = \frac{2\zeta}{e^{1+\gamma}} = \frac{d}{e^{1+\gamma}(1-\nu)} \tag{8-31}$$

where e is the naperian base of logarithms. Similarly for a screw dislocation,

$$r_0 = \frac{b}{\alpha_s} = \frac{2\eta}{e} = \frac{d}{e} \tag{8-32}$$

[37]N. Nagata, K. M. Jassby, and T. Vreeland, Jr., *Phil. Mag.*, **41A**: 829 (1980).

[38]E. Kröner [*Ergeb. angew. Math.*, **5** (1958), p. 84], considers the interaction of parallel dislocations separated by r_0 and finds $r_0 = 2\zeta/e^{3/2}$, where his ζ differs slightly from that used above, but is essentially the same. F. R. N. Nabarro [*Advan. Phys.*, **1**: 269 (1952)] gives a result that differs from Kröner's by a factor of 2. This difference appears to arise from an error in the expansion of his integral solution.

for the core cutoff parameter. For a mixed dislocation where the Burgers vector is inclined to the line at an angle β, by superposition one finds

$$r_0 = \frac{d}{e}\left[\frac{\sin^2\beta}{e^{\gamma}(1-\nu)} + \cos^2\beta\right] \tag{8-33}$$

As an example, for a $\frac{1}{2}[110]$ dislocation on a $\{111\}$ plane in an fcc crystal, $\alpha = 3.3$ for a screw dislocation and $\alpha = 4.5$ for an edge (with $\nu = \frac{1}{3}$). The trend in these α values is consistent with the trend determined in atomic calculations. However, the results indicate that the sinusoidal potential in the Peierls-Nabarro treatment is too hard a potential, consonant with physical expectation as discussed in Sec. 1-2.

We established the connection between ρ and α in Eqs. (6-13) and (6.17). In terms of the dislocation widths, insertion into Eqs. (8-31) and (8-32) shows that

$$\rho_e = \frac{\zeta}{e}, \qquad \rho_s = \frac{\eta}{e} \tag{8-34}$$

Thus the convention of a constant ρ independent of dislocation character is equivalent to assuming a constant width of dissociation in the glide plane in the Peierls-Nabarro model. This assumption is not justified in the latter model, which instead predicts $(1-\nu)\eta = \zeta$, but the model itself is uncertain, as just discussed. We thus adopt the convention of Chap. 6 of a constant ρ, being aware that it refers to a *standard* model for core energy and core dissociation. When more information about the core becomes available, the difference between the standard core and the real core can be reconciled.

With the definition of ρ in Eqs. (6-13) and (6-17), the methods of Chap. 6 give *unique* results for the energies of dislocation configurations. An alternative description, which also gives unique results for dislocation energies, has been presented by Bullough and Foreman.[39] They consider the elastic energy outside a radius r_0, across which material and force continuity are maintained and that encloses the dislocation. Therefore, as a dislocation is formed by surface forces acting on a cut, as for the Volterra dislocation, work is also done against core tractions acting over the surface r_0 in their model [similar to the work done on the outer cylinder r in Eq. (8-26)]. In comparing the results for the energies of regular polyhedra by using the above two methods,[40, 41] one finds that the results are equal within a small constant in the logarithmic factor in the energy expressions. A comparison of the original results of Bullough and Foreman[39] for a rhombus-shaped dislocation loop with those obtained[42] by the method of Chap. 6 also indicates that the methods are equivalent.

[39] R. Bullough and A. J. E. Foreman, *Phil. Mag.*, **9**: 315 (1964).

[40] D. J. Bacon and A. G. Crocker, *Phil. Mag.*, **12**: 195 (1965).

[41] J. P. Hirth, T. Jøssang, and J. Lothe, *J. Appl. Phys.*, **37**: 110 (1966).

[42] T. Jøssang, private communication.

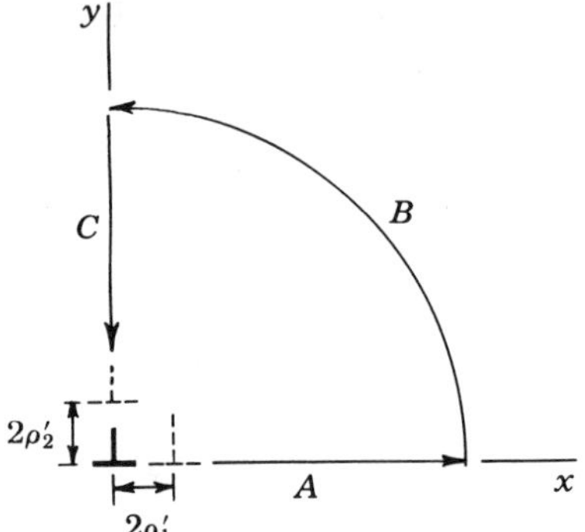

FIGURE 8-5. Reversible work cycle ABC for two edge dislocations, one at the origin and one originally displaced by $2\rho_1'$ from the origin.

Core parameters determined from interaction forces, on the other hand, are not unique. With a constant core parameter, incorrect results obtain for the variation of interaction energy with orientation; alternatively, for correct results, the parameter must vary with orientation. The assumption of a constant-interaction core parameter is equivalent to the incorrect neglect of core tractions in the variation of energy with orientation of the cut, in disagreement with the specification of the Bullough-Foreman model. To illustrate the interaction-parameter problem, consider the work cycle in Fig. 8-5, where one edge dislocation is moved in the stress field of another. If $\rho_1'=\rho_2'$, the work terms in steps A and C are equal and opposite. However, net work $\mu b^2/2\pi(1-\nu)$ is done in step B [Eq. (5-18)]. Because the total net work must be zero for a reversible cycle, cycle ABC cannot, therefore, be reversible; the energy of the configuration with core separation ρ_1' differs from that with core separation ρ_2'. Thus, in order that the interaction energy be unique, the interaction core parameter ρ' must vary with orientation. Interaction models for energy calculations such as that of Yoffe,[43] discussed in Chap. 5, therefore contain some arbitrariness associated with the core parameter.[44]

In conclusion, let us summarize the various core parameters used here, with the understanding that all interaction parameters strictly vary with orientation, as discussed above. In Chap. 3, the core radius is related to b by $r_0=b/\alpha$ [Eq. (3-14)]. The development of the self-energy of a screw dislocation in a rod requires the introduction of a cutoff separation $2\rho'$ between the dislocation and its image at a free surface. Furthermore, a comparison of the energy of Eq. (5-92) with that of (3-28) reveals that one must choose $2\rho'=r_0$ for self-consistency. Extending this interaction cutoff of $2\rho'$ to the interaction of two concentric loops [Eq. (5-28)], one must consistently choose $z_0=2\rho'=r_0$. A comparison of Eqs. (5-28) and (6-52), both of which give the self-energy of a

[43]E. Yoffe, *Phil. Mag.*, **5**: 161 (1960).

[44]An interesting discussion of problems with core descriptions is given in the panel discussion, pp. 720–723 in J. A. Simmons, R. deWit, and R. Bullough (eds.) "Fundamental Aspects of Dislocation Theory," NBS Special Publication 317, 1970.

circular loop, shows that $r_0 = z_0 = 2\rho' = 2\rho$; the linear cutoff parameter ρ is identical to the interaction cutoff parameter ρ'. Consistent with this, a direct evaluation [Eq. (6-13)] shows that $2\rho = r_0$. Thus the cutoff parameters are all self-consistent, and for the screw dislocation,

$$r_0 = z_0 = 2\rho = 2\rho' = \frac{b}{\alpha} \tag{8-35}$$

A similar expression would hold for the edge dislocation but with the terms 2ρ and $2\rho'$ multiplied by $\exp(-\gamma)$.

8-4. THE PEIERLS ENERGY

Review of Original Treatment

As a dislocation glides, the potential energy of the crystal changes periodically because of the *variation* of the misfit energy of the dislocation. Figure 8-6 indicates the sequence of dislocation configurations as the edge dislocation translates by a distance $b/2$ from symmetric configuration (a) to symmetric configuration (c). The net force acting on the dislocation in (a) and (c) is zero, by symmetry, because a mirror plane normal to the x axis and passing through the center of the dislocation relates all the crystal atoms that exert forces on the dislocation. The asymmetric configuration (b) is expected to correspond to a higher potential energy of the crystal than configuration (a), so that the shear stress σ_{xy} at the plane $y=0$ should vary periodically. Similarly, a screw

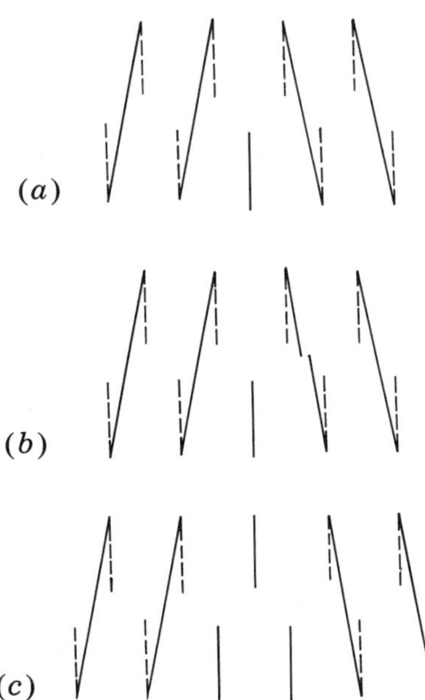

FIGURE 8-6. Schematic representation of the core region of a moving edge dislocation; (a) and (c) are symmetric and (b) is asymmetric.

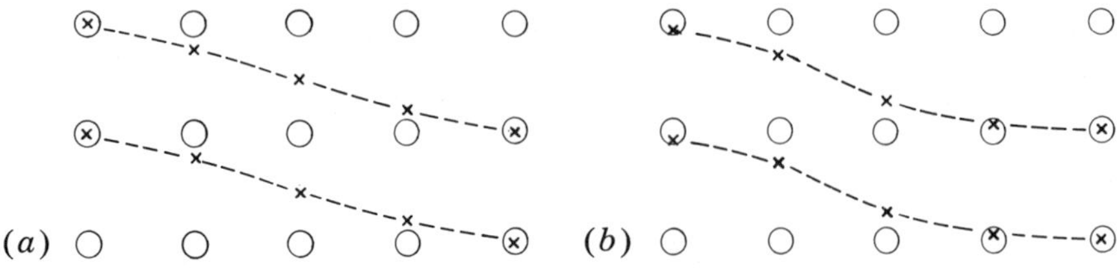

FIGURE 8-7. Symmetric configurations of a moving screw dislocation. The symbols ○ and × represent atoms below and above the glide plane, respectively.

dislocation has the two symmetrical configurations shown in Fig. 8-7. These configurations are separated by a dislocation translation distance $b/2$ and are expected to represent energy minima.

Peierls[45] and Nabarro[46] have extended the above results for the Peierls edge dislocation and have predicted the periodic displacement potential experienced by a dislocation. They assume that the elastic strain energy in the two half-crystals in Fig. 8-6 does not vary as the dislocation translates, so that the entire variation in the lattice potential as the dislocation translates is associated with changes in the shear misfit energy. Actually, one expects changes in local displacements normal to the glide plane as the dislocation moves over atomic "hills." Because such displacements lead to an additional potential variation caused by a dilatational misfit energy, the Peierls-Nabarro assumption is expected to result in an underestimate of the displacement potential. The assumption is consistent, however, with the earlier-mentioned assumption that the u_y displacements are small for the Peierls dislocation. The integral expression for the misfit energy, Eq. (8-29), did not indicate the expected periodic variation because the effects of lattice discreteness were lost in smoothing out the displacements in a continuous function. Instead, the misfit energy must be summed over the atom rows in the glide surface.

Peierls and Nabarro chose to sum the misfit energy per row of atoms on one side of the glide surface over all atom rows on both surfaces. The energy per row is thus one half the value of Eq. (8-28) with $\delta x = b$,

$$W(x) = \frac{\mu b^3}{8\pi^2 d}\left(1 + \cos\frac{4\pi u_x}{b}\right) = \frac{\mu b^3}{4\pi^2 d}\,\frac{\zeta^2}{\zeta^2 + x^2} \tag{8-36}$$

On the bottom and top half-crystals, the sums are, respectively, of the type

$$\sum_{m=-\infty}^{\infty} \frac{1}{\zeta^2 + [(b/2)2m]^2} \qquad \text{and} \qquad \sum_{m=-\infty}^{\infty} \frac{1}{\zeta^2 + [(b/2)(2m+1)]^2}$$

Peierls and Nabarro then supposed that a translation of the dislocation by a

[45] R. E. Peierls, *Proc. Phys. Soc.*, **52**: 23 (1940).

[46] F. R. N. Nabarro, *Proc. Phys. Soc.*, **59**: 256 (1947).

distance αb resulted in the shift of the point x from $(b/2)2m$ to $(b/2)(2m+2\alpha)$ on the bottom half-crystal, and from $(b/2)(2m+1)$ to $(b/2)(2m-2\alpha+1)$ on the top half-crystal. The misfit energies were then summed independently over the top and bottom half-crystals.

With the addition of the translation αb, they combined the sums into a single sum for the misfit energy.[47]

$$W=\frac{\mu b^3}{4\pi^2 d}\frac{4\zeta^2}{b^2}\sum_{-\infty}^{\infty}\frac{1}{(2\zeta/b)^2+(2\alpha+n)^2} \tag{8-37}$$

Cottrell and Nabarro[48] have solved for the sum in Eq. (8-37) directly, giving the result obtained earlier by Nabarro and in essence by Peierls,

$$\begin{aligned} W(\alpha)&=\frac{\mu b^2}{4\pi(1-\nu)}+\frac{\mu b^2}{2\pi(1-\nu)}\exp-\frac{4\pi\zeta}{b}\cos 4\pi\alpha \\ &=\frac{\mu b^2}{4\pi(1-\nu)}+\frac{W_p}{2}\cos 4\pi\alpha \end{aligned} \tag{8-38}$$

where W_p is generally called the Peierls energy and represents the desired periodic displacement potential energy. The stress required to surmount this potential barrier is called the Peierls stress, σ_p, given by

$$\sigma_p=\frac{1}{b^2}\left[\frac{\partial W(\alpha)}{\partial\alpha}\right]_{\max}=\frac{2\pi W_p}{b^2}=\frac{2\mu}{1-\nu}\exp-\frac{4\pi\zeta}{b} \tag{8-39}$$

The result in Eq. (8-38) is not at all the result anticipated from the discussion of Fig. 8-6. Equation (8-38) predicts that the symmetrical configurations (a) and (c) in Fig. 8-6 are *maximum* energy configurations, and that the configuration of greatest asymmetry, $\alpha=\frac{1}{4}$, corresponds to the minimum energy. This point has not been appreciated generally, and has led to some confusion in the literature.

Modifications of the Peierls Model

The unexpected form of the Peierls potential energy is attributable in part to the method of summing Eq. (8-36) *independently* over the top and bottom half-crystals. The method lacks basic physical justification and involves too much smoothing out of the lattice discreteness near the core. As a consequence,

[47] The displacements on the top and bottom half-crystals are, of course, of opposite sign. However, since the sums are from $-\infty$ to $+\infty$, one of the sums can be transformed by $m\rightarrow -m'$, so that the two can be readily combined into Eq. (8-37).

[48] A. H. Cottrell and F. R. N. Nabarro, in A. H. Cottrell, "Dislocations and Plastic Flow in Crystals," Oxford University Press, Fair Lawn, N.J., 1953, p. 98.

a loss of resolution of positive contributions to the misfit energy for narrow dislocations might result. It appears more consistent with the original model embodied in Eq. (8-2) to sum the misfit energy as a function of the disregistry between pairs of atom rows, because the change in bond spacing of pairs is the cause of the energy variation. Elastic displacements and the dislocation translation are both included in the variable disregistry. The quantitative development of the bond-pair approach[49] indeed shows that the symmetry position is one of minimum energy. The model also retains the feature of the original treatment that the lattice forces involved in dislocation motion decrease with increasing dislocation width. However, the bond-pair result does not give an analytical result except in a limiting case, thus losing the greatest advantage of the original model, and is not repeated here. Instead, atomic calculations with a more physically realistic core region are becoming available to supplant the glide-plane-cut, nonlinear region of the original model and its modifications, which we now mention briefly.

Other criticisms of the Peierls-Nabarro treatment, and efforts to refine it, have largely involved modifications of the various assumptions noted above. These assumptions include the use of a continuum elasticity approach in conjunction with a substrate potential associated with discrete lattice points, the use of a sinusoidal potential which must overestimate the distance to the inflection point in the energy-displacement curve in view of the markedly short range of actual interatomic repulsive forces, the assumption of linear elasticity in the core region, and the neglect of temperature effects. That is, the model strictly applies only to the potential energy of a dislocated crystal at 0°K and with no zero-point vibrational energy, with variation in the displacements normal to the glide plane neglected.

Foreman et al.[50] have modified the Peierls-Nabarro treatment by introducing phenomenological force laws to approximate more realistic lattice potentials, i.e., those with shorter-range inflection points and a lower maximum than the sinusoidal function.[51] As expected, their softer potentials give greater dislocation widths and lower Peierls energies than the hard sinusoidal law. Huntington[51] considered the role of displacements normal to the glide plane, which increase the Peierls energy, particularly in close-packed crystals, and introduced nonlinear elastic corrections for the near-core region. Next there is the role of temperature, which has been considered to some extent by Dietze[52] and by Kuhlmann-Wilsdorf.[53] First, a dislocation lying either in a Peierls valley or in an activated position on a Peierls hill has local vibrational modes associated with it. A corresponding number of normal lattice modes vanish, because degrees of freedom are conserved. The differential heat capacity

[49] J. P. Hirth and J. Lothe, "Theory of Dislocations," 1st ed., pp. 217–220.

[50] A. J. E. Foreman, M. A. Jaswon, and J. K. Wood, *Proc. Phys. Soc.*, **64A**: 156 (1951).

[51] H. B. Huntington, *Proc. Phys. Soc.*, **68B**: 1043 (1955).

[52] H. D. Dietze, *Z. Phys.*, **132**: 107 (1952).

[53] D. Kuhlmann-Wilsdorf, *Phys. Rev.*, **120**: 773 (1960).

associated with these local modes lowers the free energy of a dislocation in the Peierls valley and changes the magnitude of the Peierls free-energy barrier. Second, thermal activation can occur, but as is shown later, this is important mainly in the problem of kink formation. Finally, coupled thermal vibrations of the atoms can serve to lower the misfit energy, which is summed over all atoms from $x = -\infty$ to $x = \infty$ at 0°K, and which involves vanishingly small displacements in the limit. This effect emerges as a lowering of the effective shear modulus in the expression for the Peierls free energy, perhaps more rapidly than the lowering of the bulk shear modulus. All of these effects are interesting qualitative guides to expected behavior and anticipated results inherent in atomic calculations.

The above modifications qualitatively retain the basic assumption of the Peierls-Nabarro treatment: that only conservative displacements occur, i.e., only glide rearrangements of the atoms are considered. For the edge dislocation, this leads to the treatment of the problem as one of plane strain involving only u_x and u_y displacements. A final model, which has been suggested by Haasen,[54] involves nonconservative rearrangements and drastically departs from the above approaches. In his model the misfit energy, or core energy, of an edge dislocation is minimized by the periodic removal of an atom along the edge of the extra plane which terminates at the dislocation line. The other atoms in the core relax around these core vacancies, and the relaxations involve local displacements u_z in the z direction, in addition to u_x and u_y displacements. For such a model the bulk of the Peierls-Nabarro approach is no longer applicable.

A number of interesting possibilities are presented by Haasen's approach: (1) the initial (static) Peierls barrier might be large, requiring the filling of the empty sites and the attendant formation of lattice vacancies, but the dynamic Peierls barrier might be considerably lower if the relaxation time to form the minimum energy configuration is long compared to the stay time in the Peierls valley; (2) the empty sites might move along with the dislocation by short-range diffusion, the activation energy for the diffusion contributing to the Peierls energy; (3) motion of the dislocation might require the filling of the empty sites with subsequent relaxation back to the minimum energy configuration in the next Peierls valley.

A phenomenological one-dimensional model of a dislocation was introduced by Frenkel and Kontorova[55] and has been studied extensively by others. This model has a number of direct applications to one-dimensional problems and is important in the prediction of the structure of epitaxial interfaces.[56] The

[54] P. Haasen, *Acta Met.*, **5**: 598 (1957).

[55] J. Frenkel and T. Kontorova, *Phys. Z. Sowj.*, **13**: 1 (1938).

[56] This model has been applied extensively to the theory of interface dislocations in a series of papers by J. H. Van der Merwe and F. C. Frank. For a recent survey of these papers see J. H. Van der Merwe, in M. H. Francombe and H. Sato (eds.), "Single Crystal Films," Pergamon, New York, 1964, p. 139.

model has features that resemble those expected for line dislocations in crystals and it is amenable to numerical analysis, so that we briefly review it here.

In the Frenkel-Kontorova analog of an edge dislocation, the atoms above the slip plane are replaced by a series of mass points connected by identical springs, and the atoms on the bottom of the slip plane are replaced by a sinusoidal potential substrate. One can think of this configuration as a series of chain-connected balls relaxing onto a corrugated trough under the force of gravity. With one more ball than there are potential minima, a one-dimensional dislocation is formed, having stable and unstable equilibrium configurations just as for the Peierls energy. The Frenkel-Kontorova model also has the feature that the Peierls stress drops markedly with increasing dislocation width.[57] In addition, it is discrete from the outset, restores the expected coincidence of symmetry positions with minima in energy, and contains relaxation features that resemble the expected dilatational relaxation that was missing in the Peierls-Nabarro model. However, it is equally unrealistic from a physical viewpoint compared to atomic calculation results.

Thus there are large quantitative uncertainties in all these approaches. They all agree, however, in predicting that the Peierls energy should decrease with increasing d, that is, for low-index planes, and with increasing dislocation width. Also, edge dislocations should have a lower Peierls energy than screw dislocations.

Atomic Calculations and Empirical Representation

As discussed further in Chap. 15, the interpretation of experiments suggest that a Peierls free-energy barrier does exist and that σ_p is likely to vary from $\sim 10^{-4}$ to $\sim 10^{-2}\mu$, the higher values associated with covalent crystals and the lower values with close-packed metals. Moreover, as reviewed by Vitek[58] and Puls,[59] atomic calculations, mainly at 0°K, verify the presence of a Peierls barrier with σ_p values in this range. All the metal calculations have used rigid boundaries, so they are likely to have overestimated σ_p. The bcc calculations[58] have revealed core dissociation of screw dislocations (see Chap. 11) unlikely to be as much affected by the boundary conditions as discussed previously, which has a dramatic effect in increasing σ_p.

The studies using flexible boundaries have mainly been limited to ionic crystals. Some results for σ_p are listed in Table 8-1. The values are in good agreement with experimental estimates of σ_p from internal friction and dislocation velocity measurements.[59] The range of values arises from the use of different ionic potentials and not from scatter in the results. Of significance is the lower value (by a factor of 2 for a given potential) of σ_p for MgO determined by using flexible boundaries compared to that found with rigid

[57] R. H. Hobart, *J. App. Phys.*, **36**: 1944, 1948, (1965).

[58] V. Vitek, *Cryst. Lattice Defects*, **5**: 1 (1974); *Proc. Roy. Soc.* (*Lond.*), **A352**: 109 (1976).

[59] M. P. Puls, in "Dislocation Modeling of Physical Systems," Pergamon, New York, 1980, p. 249.

boundaries. As with the volume changes discussed previously, this difference reflects the constraint imposed on the dislocation by rigid boundary conditions and suggest that calculations of σ_p using them are overestimates, more so the smaller is region I in Fig. 8-4. As another indication of this constraint, early estimates[60, 61] using rigid boundaries suggested that both of the two possible symmetry positions for the $\frac{1}{2}\langle 110\rangle\{110\}$ edge dislocation corresponded to energy minima, whereas the studies summarized in Table 8-1 all showed that one position corresponded to a minimum, the other to a maximum in the Peierls energy.

The detailed shape of the Peierls energy as a function of displacement of this edge dislocation has also been determined for MgO by Woo and Puls.[62] They found that the barrier had a periodicity of b and roughly a sinusoidal shape. More generally for lower symmetries, one might expect asymmetric shapes and subsidiary minima in the Peierls energy.

Guided by the above sparse atomic calculations and in view of the uncertainties in the simpler models, one does as well in an analytical description to assume that the Peierls energy W_p is given by a phenomenological equation for the displacement free energy as a function of the displacement $\alpha = x/b$,

$$W(\alpha) = \frac{W_p}{2}\left(1 - \cos\frac{2\pi\alpha b}{a}\right) = W_p \sin^2\frac{\pi\alpha b}{a} \tag{8-40}$$

where $a = b/2$ or b, depending on the energetic equivalency of the symmetry positions (a) and (c) in Fig. 8-6, and represents the period of the variation in energy. Associated with $W(\alpha)$ there is a periodic lattice resistance stress

$$\sigma(\alpha) = \frac{1}{b^2}\frac{\partial W(\alpha)}{\partial\alpha} = \frac{\pi W_p}{ab}\sin\frac{2\pi\alpha b}{a} \tag{8-41}$$

The functions $W(\alpha)$ and $\sigma(\alpha)$ are depicted in Fig. 8-8*a*. The maximum value of this stress is the Peierls stress

$$\sigma_p = \frac{\pi W_p}{ab} \tag{8-42}$$

σ_p is to be regarded as a phenomenological parameter. As is discussed further in Chap. 15, the interpretation of experiments suggests that a Peierls free-energy barrier does exist and that σ_p is likely to vary from $\sim 10^{-4}$ to $\sim 10^{-2}\mu$, the higher values associated with covalent crystals and the lower values with close-packed metals. It is specified that σ_p should decrease with increasing temperature and with increasing dislocation width within a given crystal class, and that σ_p should be lower for edge dislocations than for screw dislocations.

[60] H. B. Huntington, J. E. Dickey, and R. Thomson, *Phys. Rev.*, **100**: 1117 (1955).

[61] F. Granzer, G. Wagner, and J. Eisenblatter, *Phys. Stat. Solidi*, **30**: 587 (1968).

[62] C. H. Woo and M. P. Puls, *Phil. Mag.*, **35**: 727 (1977).

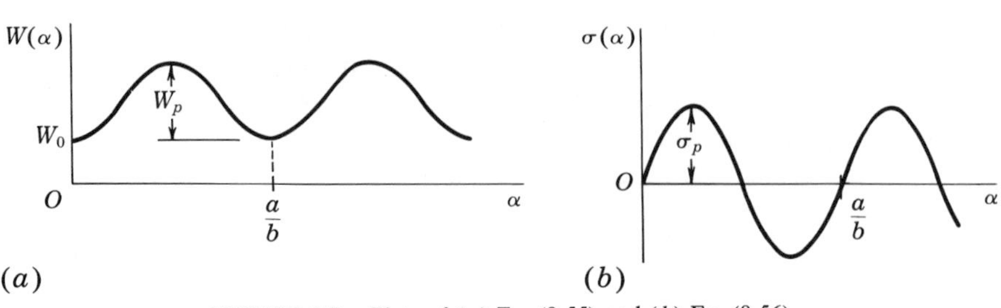

FIGURE 8-8. Plots of (*a*) Eq. (8-55) and (*b*) Eq. (8-56).

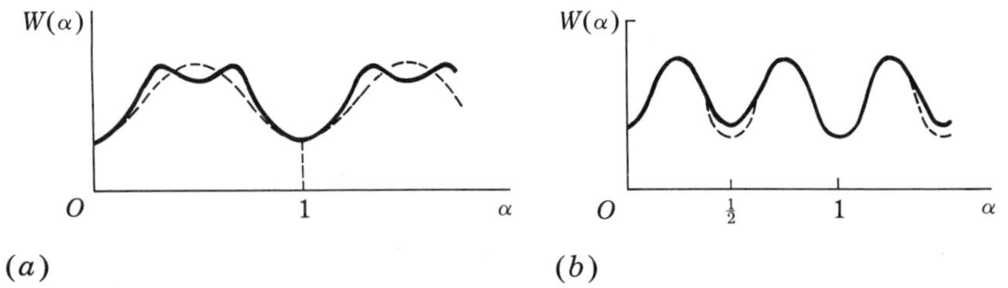

FIGURE 8-9. Possible actual variable lattice potentials (solid lines) and their representation by Eq. (8-55) (dotted lines).

Two possible actual lattice potentials, indicating the likely nonequivalency of symmetry positions, are indicated in Fig. 8-9, along with the smoothed-out approximations of the curve, represented by Eq. (8-40). These are presented to illustrate the phenomenological nature of Eqs. (8-40) and (8-41). Other minima can be introduced into actual potential curves as the translating half-crystals pass through twin orientations, coincidence lattice orientations,[63] or other special orientations.

8-5. ELASTIC ENERGIES OF KINKS

Consider Fig. 8-10, which depicts a segment of dislocation lying in a glide plane between the points *A* and *E*, which can be either pinning points or sites where the dislocation line leaves the glide plane in question. By geometrical necessity, there are configurations such as that at *B* where the dislocation line lies across a Peierls energy hill. These configurations are defined as *kinks*. In addition, there are *kink pairs*, such as the pair at *C* and *D* in the figure, which are formed by thermal fluctuations in the crystal. It is important in the theory of low-temperature deformation to know the energies and equilibrium configuration of such kinks. In this section we use the continuum approach to determine the elastic energies of kink pairs and of both sharp and oblique single kinks. The results are applied in Chap. 15.

[63] M. L. Kronberg and F. H. Wilson, *Trans. AIME*, **185**: 501 (1949).

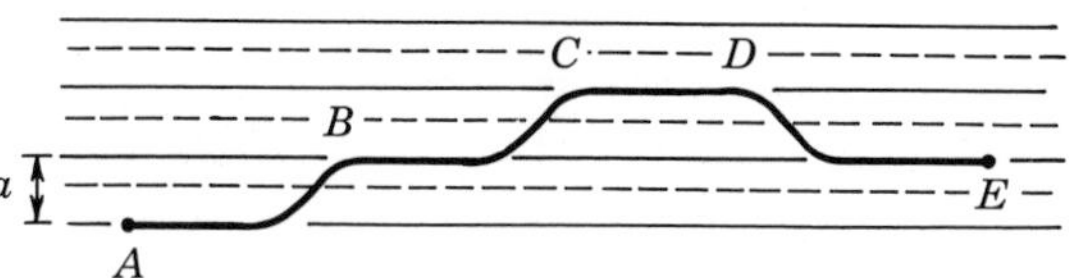

FIGURE 8-10. Kink configurations. Light dashed lines represent Peierls hills; solid lines, Peierls valleys.

The elastic energy of the kink pair in a screw dislocation is treated first. It is clear that there are great uncertainties in the self-energy of dislocation segments as small as a kink. However, the energy of interaction between two kinks as a function of their separation L can be determined quite accurately when they are well separated. The method of Chap. 6 is followed, in that the kink pair is regarded as being composed of straight line segments, as shown in Fig. 8-11.

The energy of the two kinks is

$$W = \lim_{M \to \infty} \Delta W \tag{8-43}$$

where

$$\begin{aligned} \Delta W = {} & W_{\text{int}}(C, A) - W_{\text{int}}(C', A) \\ & + W_{\text{int}}(C, B) - W_{\text{int}}(C', B) + W_{\text{int}}(C, D) + W_{\text{int}}(C, E) \\ & + W_{\text{int}}(D, A) + W_{\text{int}}(D, B) + W_{\text{int}}(E, B) + W_{\text{int}}(E, A) \\ & + W_{\text{int}}(D, E) + W_s(D) + W_s(E) \end{aligned} \tag{8-44}$$

Taking advantage of the symmetry of the problem and noticing that the interaction between orthogonal segments vanishes, one can reduce the above equation to

$$\Delta W = 2W_{\text{int}}(C, A) - 2W_{\text{int}}(C', A) + W_{\text{int}}(D, E) + 2W_s(D)$$

Using Eqs. (6-18) and (6-40), taking the limit $M \to \infty$, and substituting into Eq. (8-43), one finds

$$\begin{aligned} W = {} & \frac{\mu b^2}{2\pi}\left[(L^2 + a^2)^{1/2} - L - a + L \ln \frac{2L}{L + (L^2 + a^2)^{1/2}}\right] \\ & - \frac{\mu b^2}{4\pi(1-\nu)}\left[2L - 2(L^2 + a^2)^{1/2} + 2a \ln \frac{a + (L^2 + a^2)^{1/2}}{L}\right] \\ & + \frac{\mu b^2 a}{2\pi(1-\nu)} \ln \frac{a}{e\rho} \end{aligned} \tag{8-45}$$

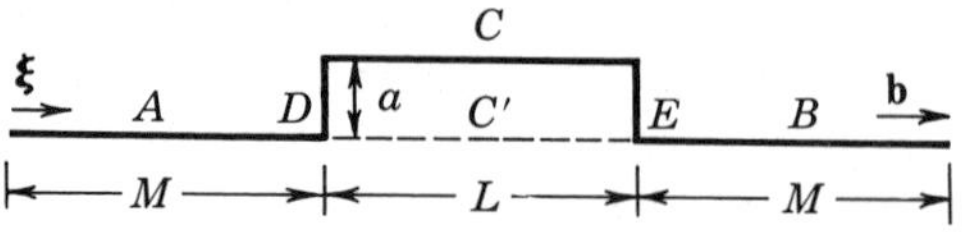

FIGURE 8-11. Segments of a kinked-screw-dislocation line.

Developed to the first power in a/L, this equation simplifies to

$$W = 2W_f + W_{\text{int}} \tag{8-46}$$

where

$$W_f = \frac{\mu b^2 a}{4\pi(1-\nu)}\left[\ln\frac{a}{e\rho} - (1-\nu)\right] \tag{8-47}$$

is the average self-energy of a single kink. As stated at the outset, there is great uncertainty in the self-energy of the kinks because of their small length. Equation (8-47) is accurate for multiple kinks, where $a \gg \rho$. For single kinks, $a \sim \rho$, and the self-energy is of the same order of magnitude as the uncertainties associated with corner effects (Chap. 6). Thus, for a single kink, one is justified only in taking the self-energy to be equal to the energy of a segment of dislocation core of length a.

This effect is not present in the kink-kink interaction energy. The last term is Eq. (8-46) is a negative kink-kink interaction energy,

$$W_{\text{int}} = -\frac{\mu b^2 a^2}{8\pi L}\frac{1+\nu}{1-\nu} \tag{8-48}$$

and the attractive force between kinks is

$$F_{\text{int}} = \frac{dW_{\text{int}}}{dL} = \frac{\mu b^2 a^2}{8\pi L^2}\frac{1+\nu}{1-\nu} \tag{8-49}$$

This attractive force is of the same form as the attractive force between two electrical charges of opposite sign. Note that although the kinks resemble two short segments of opposite-sign dislocations, their interaction force falls off much more rapidly, $\propto 1/L^2$, than the interaction force between parallel dislocations, $\propto 1/L$.

In a similar way, two kinks of the same sign separated by a distance L repel, with a force again given by Eq. (8-49). The equivalent of Eq. (8-45) differs in this case, because the segments A and B in Fig. 8-11 are no longer coaxial. However, this is a second-order effect and drops out in the power-series expansion in terms of a/L.

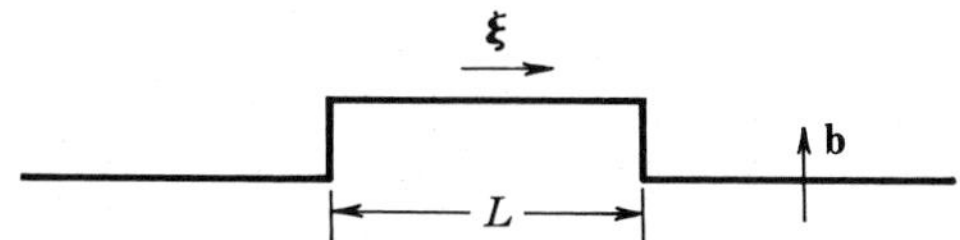

FIGURE 8-12. A kinked edge dislocation.

Exercise 8-4. Consider a pair of kinks in a pure edge dislocation (Fig. 8-12). Show that the energy of the kink pair is given by Eq. (8-46), with

$$W_f = \frac{\mu b^2 a}{4\pi}\left(\ln\frac{a}{e\rho} - \frac{1}{1-\nu}\right)$$

and

$$W_{\text{int}} = -\frac{\mu b^2 a^2}{8\pi L}\frac{1-2\nu}{1-\nu} \tag{8-50}$$

Exercise 8-5. Superpose the double kink of Fig. 8-11 with $|\mathbf{b}_s| = b\cos\beta$, and that of Fig. 8-12 with $|\mathbf{b}_e| = b\sin\beta$. Compute the energy of the resulting mixed dislocations. Notice that the screw component of segment D will interact with the screw component of segment A. Show that W_f and W_{int} are given by the superposition of the values for the screw and edge components. Why do the extra interaction terms vanish?

Exercise 8-6. Show by means of Brown's theorem, Eq. (5-82), that for large separations L the interaction force between kinks is

$$F_{\text{int}} = \frac{a^2}{2L^2}\left(E + \frac{\partial^2 E}{\partial\beta^2}\right) \tag{8-51}$$

Show that this result, which is valid in the general anisotropic elastic case, agrees with the isotropic elastic theory results of Eqs. (8-48) and (8-50).

Finally, consider the double-kink configuration in Fig. 8-13. As one can verify by calculation, the kink-kink interaction energy is again given by the above equations, provided that $L > w$, the *kink width*. Qualitatively, one would predict that this is so, based on St. Venant's principle (Sec. 2-8).

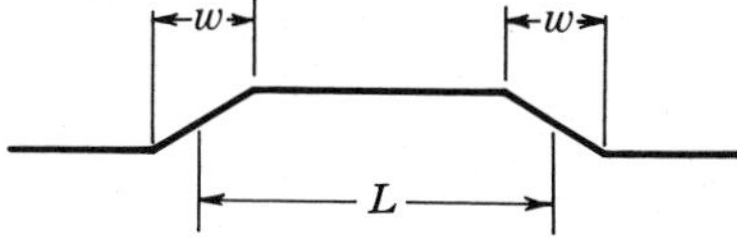

FIGURE 8-13. A dislocation with double oblique kinks.

The theory of kink-kink interactions was originally developed by Eshelby[64] and by Seeger and Schiller.[65] The formation and motion of double kinks are thought to be important in low-temperature deformation processes, including uniform straining, creep, and high-frequency internal-friction processes.

Elastic Energy of a Single Kink

In the preceding section there were no complications caused by surface effects; the kink pairs were considered to be formed on a dislocation that was initially lying completely within a single Peierls valley, and that was infinite in length. One must consider a single kink to form at a free surface in determining its formation energy. This is the only formation process in which the single kink is the sole defect introduced into the crystal. Thus image effects are present, and these are more complicated than those for the straight dislocations treated in Chap. 3. In fact, there are often long-range and intermediate-range components in the image forces that make a definition of single-kink energies difficult in the surface region.

Simple line-tension arguments lead to some preliminary remarks. Suppose a dislocation running from A to B along a Peierls trough in a wedge-shaped crystal, as in Fig. 8-14, so that it emerges *obliquely* at the surface. It is obvious that, other factors being equal, the kink moving A to A' has higher energy than the one moving it to A'', because, not counting the kink itself, in the first case the total dislocation length increases and in the second case it decreases. With a surface ledge present, another asymmetrical factor is apparent.

Even with the dislocation lying normal to the surface, asymmetries are present for mixed dislocations, and the image effects have long-range components as mentioned above. In order to demonstrate this effect, consider a single kink in a mixed dislocation lying along the z axis in an infinite crystal, as shown in Fig. 8-16. There is no net force on the kink in the infinite dislocation line. If the crystal is cut on the plane $z=0$ and external surface forces are maintained, as in Fig. 8-16a, so that the part that is cut out does not relax, the kink is still under no net force. After the imposition of a set of superposed forces[66] that are equal and opposite to the first set, as in Fig. 8-16b, the surface becomes stress free. Only the stresses created by the superposed forces give rise to a force on the kink at P, because only these forces create stresses which are different from those present in the infinite crystal.

The stresses at P caused by superposed forces relating to the straight unkinked dislocation line at O' in Fig. 8-16b differ from those caused by the kinked line at O by a small term proportional to a, and since the kink is of

[64] J. D. Eshelby, *Proc. Roy. Soc.*, **266A**: 222 (1962).

[65] A. Seeger and P. Schiller, *Acta Met.*, **10**: 348 (1962).

[66] These correspond to the stresses of Eq. (3-12) for the screw component and those of Eq. (3-84) for the edge component.

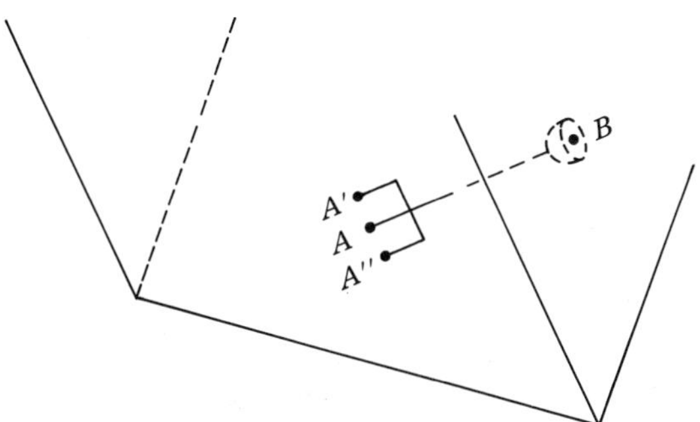

FIGURE 8-14. Dislocation in a wedge-shaped crystal.

length a, this would only result in a second-order term in the force proportional to a^2. Therefore, to first order in a, one need not distinguish between external forces associated with the kinked or unkinked dislocation. The shear stresses on the kink can now be determined as follows. Since there is no cross term between external and internal stresses (Sec. 2-7), the work $b_z\sigma_{yz}\,\delta A_y$ done in creating a little loop δA_y of vector b_z at P, as in Fig. 8-16c, must be equal to the work done by the superposed forces at the surface A,

$$b_z\sigma_{yz}\,\delta A_y = \int_A F_z\,\delta u_z\,dA$$

where δu_z are the z displacements at the surface caused by the little loop. The analogous situation for a loop of vector b_x is shown in Fig. 8-16d.

It is now evident that the image forces produced by the edge component b_x of the dislocation interacts with the edge component b_z of a square kink. The interaction of the screw segment b_z with the screw component b_x of the square kink is less easy to analyze qualitatively, but a formal analysis shows it to be negligible compared to the edge interaction as discussed below.

The screw dislocation produces stresses σ_{xy}, which interact with the screw component of the kink, and σ_{yz}, which interact with the edge component of the kink, producing forces $\propto l^{-2}$, where l is the distance of the kink from the surface. These stresses are given by Eq. (3-12) in the limit $y=0$, $x=a$, $\rho=z=l$. Similarly, the stress field from Eq. (3-83) produces stresses $\sigma_{zz}\propto l^{-2}$, $\sigma_{xy}\propto l^{-2}$, and $\sigma_{xz}\propto l^{-3}$, which interact with the kink. These forces are all localized to the near surface region. However, Eq. (3-84) yields a limiting form of the stress

$$\sigma_{yz} = \frac{\mu b_x \nu}{2\pi(1-\nu)l} \tag{8-52}$$

This produces a force on the kink

$$F_z = \frac{\mu\nu a b_x b_z}{2\pi(1-\nu)l} \tag{8-53}$$

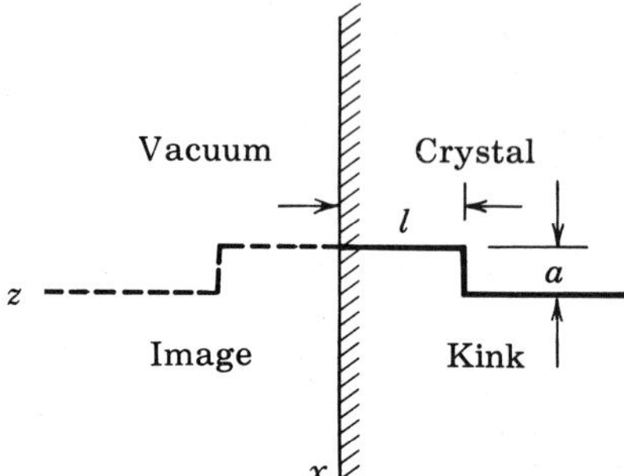

FIGURE 8-15. Kink image construction.

This force is attractive relative to the surfaces when the kink is such as to move the average dislocation away from screw orientation as in Fig. 8-16. For the opposite sign kink, the force would be repulsive. The force is also, in principle, long range, of the same range as interaction forces among dislocations.

However, for a finite body (rather than half-infinite) other image forces are present that cancel the forces of Eq. (8-53) for large l, as discussed in connection with Eq. (3-83). Hence in this case the total image force is of intermediate range, of range $\sim R$.

Creating a kink in a half-infinite body as in Fig. 8-15 yields a kink energy

$$W_f^I = W_f \pm \frac{\mu\nu a b_x b_z}{2\pi(1-\nu)} \ln \frac{l}{\rho} \tag{8-54}$$

where ρ is the appropriate cutoff parameter. If the kink moves into the interior of a finite body, the stresses arising from other image effects, such as those connected with ψ_R, Eq. (3-83), give another term $\sim \ln(R/l)$. This term largely cancels the term $\sim \ln(l/\rho)$ in Eq. (8-54), leaving a difference $\sim \ln(R/\rho)$. Thus, for a kink in the interior of a finite body, we obtain

$$W_f^I = W_f \pm \frac{\mu\nu a b_x b_z}{2\pi(1-\nu)} \ln \frac{R}{\rho} \tag{8-55}$$

instead of Eq. (8-54).

The logarithmic term in Eq. (8-54) is related to the concept of energy flow along a dislocation, discussed in Sect. 6-5. Consider the formation process of the kink in Fig. 8-15, with the kink at constant depth and formed by gradual displacements of the segment l in the $-x$ direction instead of movements of the kink in from the surface in the $-z$ direction. The work W_k is performed locally in the region of the kink. However, energy is also transmitted from the surface to the kink region by the parallel energy flow of Eq. (6-90), and this appears as the logarithmic terms in Eq. (8-54) or (8-55). This energy is supplied by work done by the image forces. The energy in the logarithmic term in Eq. (8-55) is spread over a region of extent $\sim R$ centered about the kink. Thus there is a cutoff problem for single-kink energies similar to that for straight dislocations with R corresponding to specimen dimensions or to typical interdislocation spacings.

Despite these difficulties with an external cutoff, the above discussion indicates a method for a meaningful definition of a *local* single kink energy, which we show corresponds to W_f in Eq. (8-54) or its equivalent for oblique kinks W_f'. Consider that the oblique kink in Fig. 8-17 is created from an infinite straight dislocation in an infinite medium by displacing one semiinfinite arm EF by a. The segments BE and EF are acted on by forces similar to those for dislocation bends as discussed in Chap. 5 and the work done locally against these forces while the kink is formed is called the *local elastic kink energy* W_f'. It is a finite, well-defined quantity except for the problem of an inner cutoff ρ; there is no external cutoff problem.

This definition gives a precise meaning to the single kink energy and circumvents the image problems discussed above. However, additional energy is transported to or from the kink region by the energy transport along the dislocation line. Thus the actual energy increase in the region of the kink within a sphere of radius R is W_f^I as given by Eq. (8-55).

In a double kink configuration in an infinite crystal, R should be chosen equal to $L/2$ where L is the separation of the opposite sign kinks. The total kink pair energy is the sum of the *local* energies plus the kink-kink interaction energy as in Eq. (8-55); the $\pm$ logarithmic terms cancel. The local elastic kink energy W_f' approaches zero as the kink width w increases to infinity. This fact can be deduced from the bend formulas in Chap. 5. This behavior facilitates the identification of local kink energies in the calculations that follow. Thus the cutoff parameter ρ and the energy datum for the logarithmic terms are selected so that the local kink energy separates out as in Eqs. (8-54) and (8-55) and $W_f' \to 0$ as $w \to \infty$.

Thus, in a mixed dislocation, the surface image forces favor the formation of kinks which rotate the line into screw orientation. This leads to a lower formation energy for kinks which rotate the line into screw orientation than for the opposite-sign kinks. From St. Venant's principle, one also expects this asymmetry in a finite crystal, provided that the kink is much closer to one crystal surface than to the others. In the crystal interior the image forces from the other surfaces tend to offset the forces discussed above.[67]

Elastic Energy of an Oblique Kink

As discussed in the last section, the energies of single kinks are not well defined unless their mode of formation is specified, whereas the formation energy of a kink pair is unique. For any given kink, however, the *change* in local energy as a function of ψ, the inclination of the kink to the Peierls valley, can be determined accurately. Such calculations provide a basis for a convenient definition of single-kink energies.

[67]See J. Lothe, *Physica Norvegica*, **2**: 153 (1967).

The energy of difference between an oblique kink of width w and a sharp kink with $\psi=\pi/2$ (Fig. 8-17) is

$$\Delta W=[W_s(BE)-W_s(CD)]+[2W_s(AB)-2W_s(AC)]$$
$$+[2W_{\text{int}}(BE,AB)-2W_{\text{int}}(AC,CD)]$$
$$+[W_{\text{int}}(AB,EF)-W_{\text{int}}(AC,DF)] \tag{8-56}$$

For the screw dislocation in Fig. 8-16, application of Eqs. (6-8) and (6-40), in the limit $M\to\infty$, yields for the total elastic energy of the oblique kink the expression

$$W_s=W_f+\Delta W$$
$$=\frac{\mu b^2a}{4\pi}\left[(\cos\psi-1)\cot\psi\ln\frac{a}{e\rho\sin\psi}-2\cot\psi\ln\left(\cos\frac{\psi}{2}\right)+(\cot\psi-\csc\psi)\right]$$
$$+\frac{\mu b^2a}{4\pi(1-\nu)}\sin\psi\ln\frac{a}{e\rho\sin\psi} \tag{8-57}$$

When $\psi\to\pi/2$, then $W\to W_f$ [Eq. (8-47)], and when $\psi\to 0$, then $W\to 0$, as it should. In view of the symmetry of Eqs. (8-47) and (8-50), it is evident that the elastic energy of an oblique kink in an edge dislocation is given by Eq. (8-57) if one exchanges the factors 1 and $1/(1-\nu)$. Thus one can write directly for the edge

$$W_e=W_f+\Delta W=\frac{\mu b^2a}{4\pi(1-\nu)}$$
$$\times\left[(\cos\psi-1)\cot\psi\ln\frac{a}{e\rho\sin\psi}-2\cot\psi\ln\left(\cos\frac{\psi}{2}\right)+(\cot\psi-\csc\psi)\right]$$
$$+\frac{\mu b^2a}{4\pi}\sin\psi\ln\frac{a}{e\rho\sin\psi} \tag{8-58}$$

For a mixed dislocation the result can be obtained by superposing the results for the screw and edge components (*cf.* Exercise 8-5). In addition, one must add the energy terms arising from the additional interactions between the screw and edge portions of segments BE and CD (Fig. 8-17) and the edge and screw portions, respectively, of AB and AC. Also, cross terms arise from the self-energy of the oblique segment. The additional terms yield the contribution to the energy

$$\Delta W_{\text{mix}}=\frac{\nu\mu b^2a\cos\beta\sin\beta}{2\pi(1-\nu)}\left[(1-\cos\psi)\ln\frac{a}{e\rho\sin\psi}-\ln\frac{a}{2e\rho\cos^2(\psi/2)}\right] \tag{8-59}$$

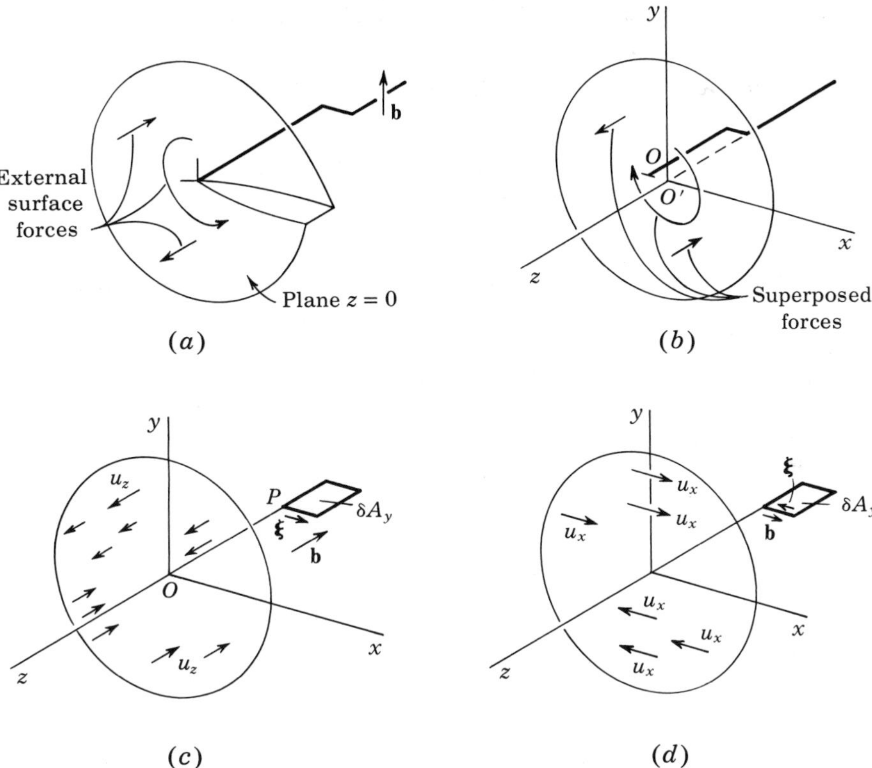

FIGURE 8-16. A kink in a mixed dislocation emerging normal to the free surface (xy): (a) surface forces given by first-order elastic solution, (b) superposed equal and opposite forces, (c) and (d) surface displacements δu_z and δu_x associated with the formation of a loop of area δA_y by the motion of the kink in the z direction.

ΔW_{mix} vanishes for $\psi = \pi/2$, meaning that half the value of the kink-pair energy (Exercise 8-5) has been assigned *arbitrarily* to the single kink when $\psi = \pi/2$. With this datum of energy, ΔW_{mix} approaches *not zero*, but a finite limit as $\psi \to 0$, namely.

$$\Delta W_{\text{mix}} = \frac{\nu \mu b^2 a \cos\beta \sin\beta}{2\pi(1-\nu)} \ln \frac{a}{2e\rho}$$

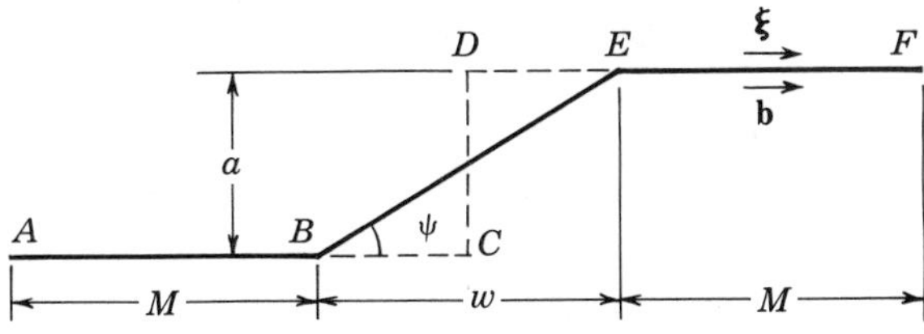

FIGURE 8-17. An oblique kink.

with the sign depending on the sign of the kink. However, according to the specifications of the preceding section, the *local* kink energy W_f' is defined so that it approaches zero as $\psi \to 0$. Changing the datum of energy accordingly, we obtain

$$W_f' = W \pm \frac{\nu\mu b^2 a|\cos\beta\sin\beta|}{2\pi(1-\nu)} \ln\frac{a}{2e\rho} \tag{8-60}$$

where W is the kink energy as defined above. The plus and minus signs apply to kinks which bring the dislocation closer to screw and edge orientation, respectively.[68] This is an explicit demonstration that positive and negative kinks in a mixed dislocation are not equal in local energy, either.

It is instructive to consider Eq. (8-60) in the limit of small ψ. Written out in detail, it becomes

$$W_f' = W_s\cos^2\beta + W_e\sin^2\beta + \Delta W_{\text{mix}} \pm \frac{\nu\mu b^2 a|\cos\beta\sin\beta|}{2\pi(1-\nu)} \ln\frac{a}{2e\rho} \tag{8-61}$$

where W_s, W_e, and ΔW_{mix} are given by Eqs. (8-57), (8-58), and (8-59), respectively. To the lowest order, only the first two terms contribute, giving

$$W_f' = \frac{\mu b^2 a}{8\pi(1-\nu)}\left[(1+\nu)\cos^2\beta + (1-2\nu)\sin^2\beta\right]\psi\ln\frac{a}{e\rho\psi} \tag{8-62}$$

Thus to first order, for very wide kinks there is no asymmetry between positive and negative kinks in the local energy, and the result agrees with line tension considerations as in Eq. (6-82).

The asymmetric term

$$W_{\text{as}} = \Delta W_{\text{mix}} \pm \frac{\nu\mu b^2 a|\cos\beta\sin\beta|}{2\pi(1-\nu)} \ln\frac{a}{2e\rho} \tag{8-63}$$

is of the order $a\psi^2\ln(a/e\rho\psi)$ for wide kinks and becomes more important the more abrupt the kinks, that is, the larger is ψ.

8-6. KINK ENERGY AND LOCAL CONFIGURATION

Constant Line Tension

The simplest derivation of the local kink energy involves the assumption of a constant line tension. With the assumption of a constant line energy W_0 per

[68] Thus the rule for the sign is *opposite* to the rule that applies for the long-range, logarithmic term in Eq. (8-54).

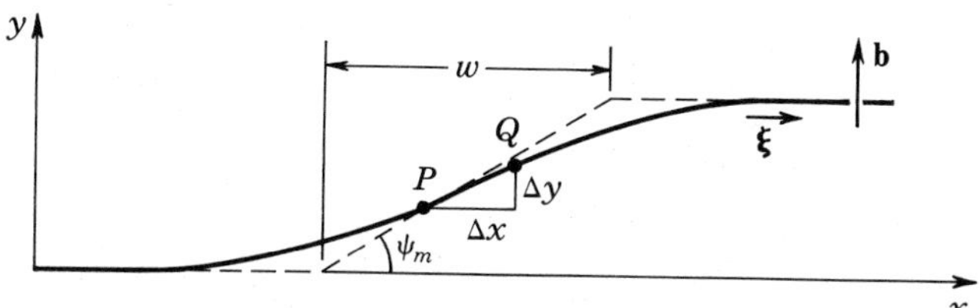

FIGURE 8-18. A kinked edge dislocation.

unit length of dislocation, corresponding to the line tension in a taut string or to the line tension $\mathcal{S}$ of Chap. 6, Read[69] has shown that the equilibrium equation for the line has the same form as the familiar solution for a vibrating string. Consider the element Δl in an edge dislocation, as shown in Fig. 8-18. At P the line tension results in a y component of force

$$W \sin\psi = W_0\left(\frac{\partial y}{\partial l}\right)_l \cong W_0\left(\frac{\partial y}{\partial x}\right)_x$$

while at Q the component is $W_0(\partial y/\partial x)_{x+\Delta x}$, the approximations holding in the limit of small ψ, which corresponds to the actual case for kinks. The resultant force is balanced by forces per unit length $\partial W(y)/\partial y$, associated with the Peierls barrier [Eq. (8-41)]; σb, because of the applied stress σ; and $m^*(\partial^2/y/\partial t^2)$, arising from acceleration, where m^* is the effective mass of the dislocation [Eq. (7-21)]. Thus the equilibrium equation is

$$W_0\frac{\partial^2 y}{\partial x^2} = \frac{\partial W(y)}{\partial y} - \sigma b + m^*\frac{\partial^2 y}{\partial t^2} \tag{8-64}$$

For a dislocation at rest, and with the assumption that the Peierls stress greatly exceeds the applied stress, Eq. (8-64) reduces to

$$\frac{dW(y)}{dy} = W_0\frac{d^2 y}{dx^2} \tag{8-65}$$

With the reasonable assumption $W_0 \cong W(y)$, the variables are separable to

$$d\ln W(y) = \frac{dy}{dx}\, d\,\frac{dy}{dx} \cong \psi\, d\psi \tag{8-66}$$

which integrates to

$$\ln\left[\frac{W(y)}{W_m}\right]^2 = \psi^2 - \psi_m{}^2 \tag{8-67}$$

[69] W. T. Read, Jr., "Dislocation in Crystals," McGraw-Hill, New York, 1953, p. 53.

where W_m is the maximum energy W_0+W_p (Fig. 8-18) and ψ_m is the maximum slope of the kink.

In the limit of small ψ, $W(y)\cong W_0$, and Eq. (8-67) yields for ψ_m the expression

$$\left(\frac{a}{w}\right)^2=\psi_m{}^2=2\ln\frac{W_m}{W_0}=2\ln\frac{W_0+W_p}{W_0}\cong\frac{2W_p}{W_0}$$

or

$$w\cong a\left(\frac{W_0}{2W_p}\right)^{1/2} \tag{8-68}$$

With a typical value of $W_p\sim10^{-3}W_0$, this expression predicts a kink width $w\sim22a$. If a dislocation is constrained to contain a number of like-sign kinks, such as B and C in Fig. 8-10, they can be considered to be independent kinks if the separation between kinks is greater than the kink width w. When the separation between kinks is less than w, the line energy is the dominant term determining the equilibrium configuration, which now is essentially a straight line with only small "ripples" corresponding to the Peierls valleys.

The total kink energy can be obtained from the line integral of the elastic energy of the kink plus the Peierls contribution, or

$$W_k=\int_{-\infty}^{\infty}W_0(dl-dx)+\int_{-\infty}^{\infty}\left[W(y)-W_0\right]dx \tag{8-69}$$

Noting that $dl-dx=\frac{1}{2}\psi(dy/dx)\,dx$, and changing variables, one finds

$$W_k=\int_0^a\frac{W_0}{2}\psi\,dy+\int_0^a\frac{W(y)-W_0}{\psi}dy \tag{8-70}$$

The integral of Eq. (8-66) from $\psi_0=0$ to ψ yields

$$\psi^2=2\ln\frac{W(y)}{W_0}$$

$$\cong\frac{2W_p}{W_0}\sin^2\frac{\pi y}{a} \tag{8-71}$$

with the use of Eq. (8-40). Finally, substituting Eq. (8-71) in (8-70) and integrating, one obtains

$$W_k=\frac{2a}{\pi}\left(2W_pW_0\right)^{1/2} \tag{8-72}$$

An expression for the configuration of overlapping kinks could be obtained by removing the approximations for w in these results. However, it is problematical whether such amendments would improve the results in view of the approximation of constant line tension. This approximation is removed in the treatment in the next section.

Variable Line Energy

Let us now compute the kink energy, using the exact expression for the variable line energy. Once again the model of Fig. 8-17 is invoked. According to Eq. (8-40), the contribution of the Peierls energy to this configuration is

$$\int W_p \sin^2\frac{\pi y}{a}\,dl = \int_0^a W_p \sin^2\frac{\pi y}{a}\csc\psi\,dy$$

$$= \frac{W_p a}{2}\csc\psi \tag{8-73}$$

The elastic energy is given by Eq. (8-57). Pertinent to the concluding remarks in the last section, Eq. (8-57) would lead to a line energy per unit length that is proportional to $\ln\psi$ in the limit of $\psi \to 0$. Thus the neglect of the variation of line energy with ψ is a poor approximation even in the limiting case.

Combining these two energy contributions, taking the limit of ψ small, and setting $\nu = \frac{1}{3}$, one obtains for the energy of a kink in a screw dislocation

$$W_k(\psi) = \frac{W_p a}{2\psi} + \frac{\mu b^2 a\psi}{4\pi}\ln\frac{a}{e^{1.2}\rho\psi} \tag{8-74}$$

Notice that for small ψ the uncertainties associated with the corner effects that were present in Eq. (8-47) are removed, so that Eq. (8-74) gives an accurate expression for the oblique-kink formation energy.

Minimizing Eq. (8-74) to determine the equilibrium configuration, one finds for the kink width and kink energy

$$w \cong \frac{a}{\psi} = a\left(\frac{\mu b^2}{2\pi W_p}\ln\frac{a}{e^{2.2}\rho\psi}\right)^{1/2} \tag{8-75}$$

and

$$W_k = \frac{aW_p}{2}\left(\frac{\mu b^2}{2\pi W_p}\ln\frac{a}{e^{2.2}\rho\psi}\right)^{1/2}\left[1+\frac{\ln(a/e^{1.2}\rho\psi)}{\ln(a/e^{2.2}\rho\psi)}\right] \tag{8-76}$$

Equation (8-75) is an implicit equation for ψ. In practice, one puts a rough estimate for ψ in the right-hand side and solves to obtain a better estimate. When a self-consistent ψ is found, W_k is estimated from Eq. (8-76).

For a kink in an edge dislocation, the equivalent expressions are

$$w = a\left(\frac{\mu b^2}{8\pi W_p}\ln\frac{a}{e^{3.5}\rho\psi}\right)^{1/2} \tag{8-77}$$

and

$$W_k = \frac{W_p a}{2}\left(\frac{\mu b^2}{8\pi W_p}\ln\frac{a}{e^{3.5}\rho\psi}\right)^{1/2}\left[1+\frac{\ln(a/e^{2.5}\rho\psi)}{\ln(a/e^{3.5}\rho\psi)}\right] \tag{8-78}$$

We emphasize that these expressions depend upon ψ being small; if ψ is not small enough that the logarithmic factor exceeds 2.5 in Eq. (8-78), the more complicated Eq. (8-58) must be used instead of its limiting form in computing W_k. Also, corner-effect uncertainties are introduced when $\psi \to \pi/2$.

Comparison of these results with Eqs. (8-68) and (8-72) shows that the assumption of a constant line tension $\sim \mu b^2$ leads to fairly similar results. Only the variable-line-energy approach, however, reveals the large differences between screw and edge dislocations, associated with the lower relative elastic energy of a dislocation segment in screw orientation.

Since W_p is larger for a screw dislocation than for an edge, W_k is larger for the screw by a factor of $\sim$4 to 8, depending on the magnitude of the logarithmic term. For wide kinks, the logarithmic terms predominate in determining the relative kink widths. If $4W_p(\text{edge}) < W_p(\text{screw})$, the kinks in the edge dislocation will be wider than those in the screw dislocation, and vice versa.

In the case of a mixed dislocation, Eq. (8-59) contributes a term ΔW_{mix}, to be added to the sum of the energies ascribed to the superposed screw and edge components. Rather than present the lengthy expressions for W_k and w in this case, we simply note that the results depend on the sign of β. The kink that rotates the dislocation line toward screw orientation is wider, as shown in Fig. 8-19, and has a higher kink energy than the one that rotates the line away from screw orientation. This again illustrates the asymmetry in single-kink energies in a double kink.

The results of this section agree with those of the preceding section in indicating that a line containing like-sign kinks is composed of discrete kinks if the kink separation exceeds the kink width but becomes a straight line, and hence loses its kink structure, if the kink width is greater than the kink separation.

The problem of kink configuration has been dealt with at length not only because of the applicability to low-temperature-deformation problems, but also because the above developments illustrate the differences between the line-tension method and the method of Chap. 6 in treating a configurational problem by energy minimization. The removal of the straight-line-segment approximation to further improve the accuracy of the results presents a problem of forbidding complexity.

8-19. Two positions of a kink pair in a mixed dislocation.

8-7. PERIODICITY OF THE KINK ENERGY

In analogy to the Peierls barrier associated with lattice periodicity, the energy and configuration of a kink might be expected to fluctuate periodically as the kink is translated along the Peierls hill. In view of the uncertainties and complications involved in the Peierls energy itself, the application of such an analysis to the kink is not undertaken here. Qualitatively, the facts that the kink lies in the region of highest misfit associated with the presence of the dislocation and that the elastic energy associated with the kink falls off inversely as the square of the distance from the kink lead to the expectation that the periodic-displacement-energy barrier for the kink must be several orders of magnitude smaller than the Peierls energy.[70] Particularly for wide kinks, the kink translation involves small displacements superposed on the large displacements already present in the "soft" misfit region.

A helpful analog is illustrated in Fig. 8-20. Imagine the dislocation line to be a heavy spring relaxing under the force of gravity onto a substrate made up of hard, close-packed balls. For a wide kink (relatively low Peierls energy), the line tension dominates, and only small ripples in the kink are produced by the periodic substrate variation in the x direction. On the other hand, for a narrow kink (Fig. 8-20*b*, very large Peierls energy), the kink lies in a "valley." In this case the kink experiences a periodic substrate variation in the x direction, so that a Peierls type of barrier must be overcome in its lateral translation.

In most crystals, where $w \gg a$, thermal fluctuations should suffice to overcome the kink Peierls barrier except at temperatures approaching 0°K. Indeed, the zero-point vibrational energy possible could be large enough to destroy such a barrier.

These qualitative arguments break down only when the misfit energy completely dominates the elastic-energy term in determining the kink configuration. This case, corresponding to a very large Peierls barrier, seems feasible only in crystals bound by strong covalent bonds. Some of the properties of narrow kinks in such a case have been considered by Brailsford[71] and Ninomiya et al.[72] Dislocation velocities in various covalently bonded crystals have been interpreted in terms of narrow kink behavior as reviewed by Steinhardt and Haasen.[73] However, as they discuss, electron microscopic studies show the dislocations to be dissociated so that the resistance to kink motion is related instead to extended defects on the dislocation.[73, 74]

[70]Atomic calculations support this expectation; see W. T. Sanders, *J. Appl. Phys.*, **36**: 2822 (1965).

[71]A. D. Brailsford, *Phys. Rev.*, **122**: 778 (1961).

[72]T. Ninomiya, R. Thomson, and F. Garcia-Moliner, *J. Appl. Phys.*, **35**: 3607 (1964).

[73]H. Steinhardt and P. Haasen, *Phys. Stat. Solidi.*, **49a**: 93 (1978).

[74]P. Haasen, *J. Phys.*, **40**: C6-111 (1979).

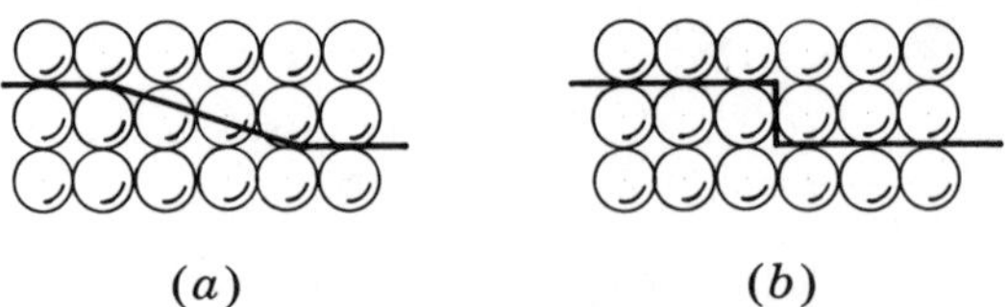

FIGURE 8-20. A heavy spring on a substrate of hard, close-packed balls: (*a*) high spring tension, (*b*) low spring tension.

8-8. LINE TENSION IN THE KINK MODEL

When the kink model of a line is appropriate, i.e., when the spacing between kinks is greater than the kink width w, the kink-kink interaction does not depend on the details of the kink configuration, as shown by Eqs. (8-48) to (8-50). Consider, therefore, the kinked dislocation line in Fig. 8-21, in which the detailed structure of the kinks is unimportant. Let us impose upon the line a bow-out $y = X \sin kx$, corresponding to kink displacements

$$u = \frac{lX}{a} \sin \frac{\pi x}{L} \tag{8-79}$$

At equilibrium the net force on a kink at x_0 in the undisplaced line is zero, or, by Eq. (8-49),

$$0 = \sum_{-\infty}^{\infty} \frac{\mu b^2 a^2}{8\pi} \frac{1+\nu}{1-\nu} \frac{\pm 1}{(x_n - x_0)^2} = \sum_{-\infty}^{\infty} \frac{\pm B}{2(x_n - x_0)^2} \tag{8-80}$$

where the minus and plus signs are to be used when $x_n > x_0$ and $x_n < x_0$, respectively. For the displaced line, the force on the kink at x_0 arising from its interaction with the nth kink changes by

$$\Delta F = \pm B \frac{u(x_n) - u(x_0)}{(x_n - x_0)^3} \tag{8-81}$$

The net force on the kink at x_0 is given by the sum of Eq. (8-81) over all x_n,

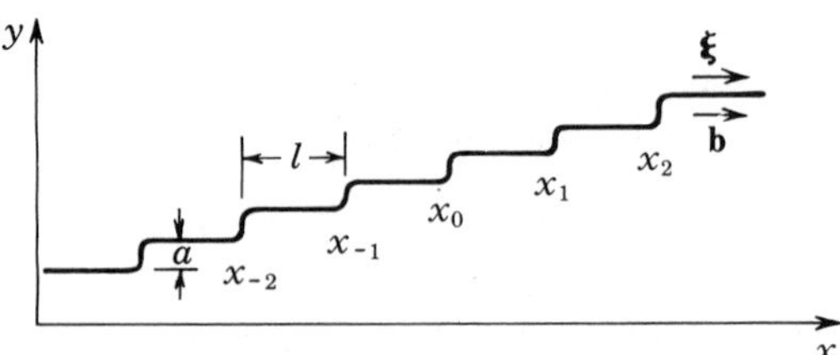

FIGURE 8-21. A kinked line of length L.

which can be written as the integral

$$F=-\frac{BX}{a}\left(\int_{-\infty}^{x_0-\epsilon}-\int_{x_0+\epsilon}^{\infty}\right)\frac{\sin(\pi x/L)-\sin(\pi x_0/L)}{(x-x_0)^3}dx \tag{8-82}$$

The cutoff parameter ϵ is determined by the condition

$$l^2\int_\epsilon^\infty \frac{dx}{x^3}=\sum_{n=1}^{\infty}\frac{1}{n^3}=1.202$$

so that the integral is a good representation of the discrete sums, giving

$$\epsilon=0.644l \tag{8-83}$$

In the limit of small k, integration[75] of Eq. (8-82) yields

$$F=\frac{-XB\pi^2}{aL^2}\sin\frac{\pi x_0}{L}\ln\frac{L}{l}$$

$$=\frac{-B\pi^2 u_0}{lL^2}\ln\frac{L}{l} \tag{8-84}$$

[75] With $k=\pi/L$, $\sin kx-\sin kx_0=\cos kx_0 \sin k(x-x_0)+[\cos k(x-x_0)-1]\sin kx_0$ so that Eq. (8-82) involves integrals of the type

$$\int_\epsilon^\infty \frac{\sin kx'}{x'^3}dx' \qquad \text{and} \qquad \int_\epsilon^\infty \frac{(\cos kx'-1)\,dx'}{x'^3}$$

even and odd, respectively, in x.

Addition of the integrals from $-\infty$ to $-\epsilon$ to those from ϵ to ∞ leaves a net contribution only from the odd integrals. Integrating twice by parts, one finds that the odd integral is

$$\int_\epsilon^\infty \frac{\cos kx'-1}{x'^3}dx'=\frac{\cos kx'-1}{2x'^2}\Big]_\epsilon^\infty+\frac{k\sin kx'}{2x'}\Big]_\epsilon^\infty-\frac{k^2}{2}\int_\epsilon^\infty\frac{\cos kx'}{x'}dx'$$

The last term is now a standard integral, and in the limit of small k, the solution reduces to

$$-\frac{3k^2}{4}-\frac{k^2}{2}\ln\frac{1}{\gamma k\epsilon}$$

where $\gamma=1.78$ (Euler's number). With $\epsilon=0.644l$ and $k=\pi/L$, the solution reduces to

$$-\frac{\pi^2}{2L^2}\left(\ln\frac{L}{l}+0.2\right)\cong-\frac{\pi^2}{2L^2}\ln\frac{L}{l}$$

which gives the above result.

The energy stored in the nth kink in the displaced loop is given by the integral of Eq. (8-84) for the nth kink,

$$W(n) = -\int_0^{u_n} F(u_n)\,du_n = \frac{B\pi^2}{lL^2}\ln\frac{L}{l}\int_0^{u_n} du_n$$

$$= \frac{B\pi^2}{2lL^2}u_n^2\ln\frac{L}{l} = \frac{X^2Bl\pi^2}{2a^2L^2}\sin^2\frac{\pi x_n}{L}\ln\frac{L}{l} \tag{8-85}$$

where Eq. (8-79) has been used in the last step.

The total bow-out energy in a half-loop of length $L=\pi/k$ is given by the sum of Eq. (8-85) over the L/l kinks in the half-loop. Replacing the sum by an integral and noting that $x_n = ln$, one finds

$$W = \frac{X^2Bl\pi^2}{2a^2L^2}\ln\frac{L}{l}\int_0^{L/l}\sin^2\frac{\pi ln}{L}\,dn$$

$$= \frac{B\pi^2X^2}{4a^2L}\ln\frac{L}{l} \tag{8-86}$$

In the approximation that the dislocation line is smooth, the bow-out energy of a half-loop L is given by Eq. (6-74) as the line tension $\mathcal{S}$ times the increase in arc length,

$$W = \frac{\mathcal{S}\pi^2X^2}{4L}$$

Comparison of this with Eq. (8-86) shows the effective line tension of a kinked dislocation line for small bow-outs to be

$$\mathcal{S} = \frac{\mu b^2}{4\pi}\frac{1+\nu}{1-\nu}\ln\frac{L}{l} \tag{8-87}$$

It follows from an inspection of Eq. (8-50) that the equivalent formulas for a kinked edge dislocation are given by replacing the factor $1+\nu$ by the factor $1-2\nu$ in the above formula,

$$\mathcal{S} = \frac{\mu b^2}{4\pi}\frac{1-2\nu}{1-\nu}\ln\frac{L}{l} \tag{8-88}$$

Comparing Eq. (8-88) with Eq. (6-84), one sees that the kink model gives a result resembling that of the continuum model, but that the kink spacing l takes the role of the cutoff parameter ρ. This result is important in the theory of internal friction caused by the oscillation of a dislocation loop.

8-9. JOGS

Screw-Jog Configuration

The preceding discussion has dealt with the configuration of dislocation line segments lying in a specific glide plane. Segments of dislocation line that have a component of their sense vector normal to the glide plane are termed *jogs*. When the normal component extends over only a single interplanar spacing d, as for A and B in Fig. 8-22, the jogs are *unit jogs*, or for brevity just jogs; if the normal component extends over more than one interplanar spacing, as at C in the figure, the jog is called a *superjog*. As is the case for kinks, jogs can be present singly, because of geometrical constraints, or they can be present as jog pairs (D in the figure) formed by thermal fluctuations.

Consider the screw dislocation lying in the glide plane $y=0$, and suppose that it contains a jog lying in the plane $z=0$, as illustrated in Fig. 8-23. Both planes $y=0$ and $z=0$ are possible glide planes for the screw dislocation, because their line of intersection contains both $\mathbf{b}$ and $\boldsymbol{\xi}$. Thus, at equilibrium, the jog relative to the glide plane $y=0$ is simply a kink with respect to the glide plane $z=0$, and the width and energy of the jog are given by the preceding results for kinks in screw dislocations.

Notice that the jog itself has an edge component in its glide plane $z=0$. Thus if the dislocation line advances in the z direction, the jog is sessile; i.e., it cannot glide in the z direction normal to its glide plane. The jog can move in the z direction only by climb involving the generation of lattice vacancies, a process that is considered in detail in Chap. 16. The climb process might require that the dislocation assume the dotted configuration in Fig. 8-23 during the vacancy generation process. If this is the case, the work done in forming the dotted configuration from the relaxed configuration, given by Eq. (8-58), contributes to the activation energy of the climb process.

Edge-Jog Configuration

A jog in an edge dislocation, depicted for a simple cubic lattice in Fig. 8-24, is expected to have a quite different configuration from the screw jog. The jog

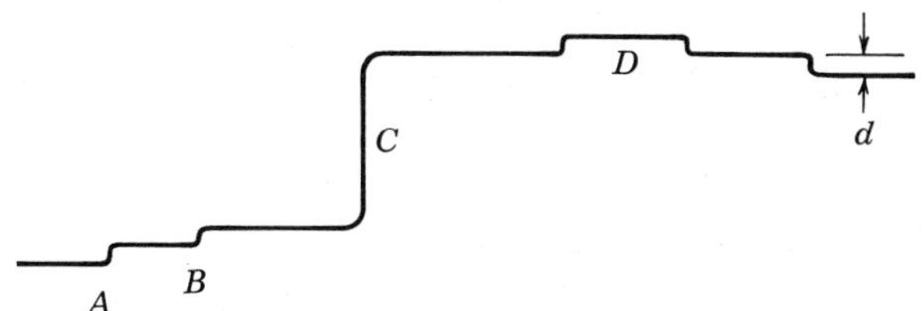

FIGURE 8-22. Jog configurations: $\mathbf{b}$ is normal to the plane of the paper.

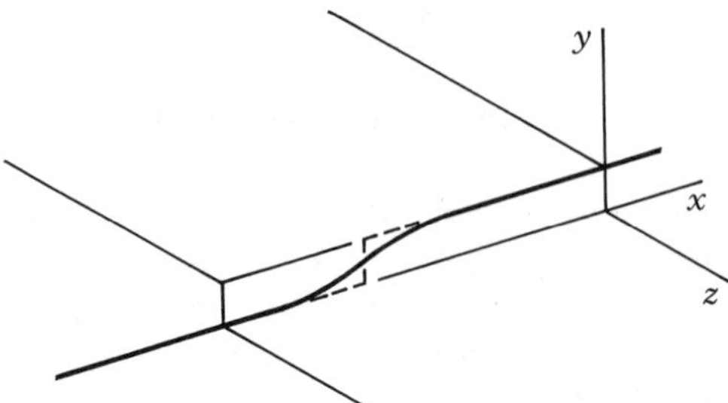

FIGURE 8-23. Relaxed jog in a screw dislocation (solid line) and a contracted position of the jog (dashed line). The traces of the glide surface (xz) are also shown.

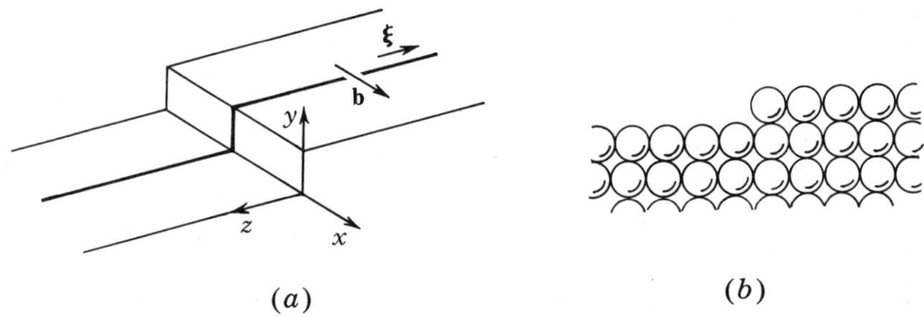

FIGURE 8-24. (a) Jog in an edge dislocation. (b) The corresponding extra half-plane.

now represents a step in the extra plane of atoms which terminates at the dislocation line, as shown in Fig. 8-24b. In the model of the Peierls dislocation, the dislocation is spread out in the xz plane, giving displacements u_x and u_y; the displacements u_z in the z direction are zero. Any smoothing out of the jog to an oblique configuration like the relaxed kink would involve large u_z displacements, and in addition would disrupt the glide extension given by the Peierls dislocation. Thus in the Peierls model one expects the misfit energy associated with the rotation of an edge jog into an oblique configuration to be much higher than is the case for the kink, so that edge jogs should be much narrower than kinks.

On the other hand, if the relaxed configuration of the dislocation line involves the presence of vacant core sites, as suggested by Haasen,[76] u_z displacements are present in the core, and one can envision relaxation of the jog into an extended configuration. The misfit associated with the jog is centered in the core region of the dislocation, and as a result, is less amenable to calculation than the misfit energy in the glide plane, which is quite uncertain itself. The question of the jog misfit energy will probably be resolved only by calculations involving lattice sums of atomic-interaction potentials. In the absence of such calculations, let us adopt the qualitative picture suggested by

[76] P. Haasen, *Acta Met.*, **5**: 598 (1957).

the Peierls model and suppose a very high misfit energy for an oblique jog, so that the jogs are quite narrow and sharp. Qualitatively, this approximation is expected to be most valid in the case of strong covalently or ionically bonded crystals. For such crystals, one can envision that the strong covalent bonds constrain the jog to the configuration in Fig. 8-24*b*. Also, as is discussed in Chap. 10, sharp jogs are to be expected in crystals with low stacking-fault energies.

Unlike the screw jog, the edge jog is glissile in a direction normal to the edge-dislocation line but is sessile with respect to translation along the line. As illustrated in Fig. 8-24*a*, glide motion of the entire configuration in the x direction can be accomplished by motion of the edge-dislocation line on its glide plane $y=0$ and motion of the jog on its glide plane $z=0$.

Energy of a Jog Pair

The energy of a pair of jogs in a screw dislocation is similar to the result for a kink pair. For a jog pair in an edge dislocation, the energy of the pair is given by Eqs. (8-43) and (8-44), where now, of course, the Burgers vector of the dislocation in Fig. 8-10 is normal to the plane of the paper. All the terms in Eq. (8-44) contribute to the energy in this case. Equations (6-8) and (6-40), in the limit $M \to \infty$, give for this case

$$2W_j = \Delta W = 2W_f + W_{\text{int}} \tag{8-89}$$

where the average self-energy of the jog is given by

$$W_f = \frac{\mu b^2 a}{4\pi(1-\nu)} \ln \frac{a}{e\rho} \tag{8-90}$$

This result is subject to the same uncertainties as Eq. (8-47). Since in the present case the jog does not relax to a wider configuration, which would reduce the uncertainty in the elastic portion of the energy attributable to corner effects, a reasonable approximation is to set W_f equal to the energy of a core segment of length a,

$$W_f \cong \frac{\mu b^2 a}{4\pi(1-\nu)} \tag{8-91}$$

The jog-jog interaction energy is given by

$$W_{\text{int}} = -\frac{\mu b^2 a^2}{8\pi L(1-\nu)} \tag{8-92}$$

so that the attractive force between opposite-sign jogs (or the repulsive force between like-sign jogs) is

$$F_{\text{int}} = \frac{\mu b^2 a^2}{8\pi L^2(1-\nu)} \tag{8-93}$$

Single-Jog Energy

The asymmetries of Fig. 8-15 do not arise in the formation of a single jog in an edge dislocation in an isotropic medium. Thus, one-half the energy of a well-separated jog pair is a meaningful estimate of the energy of a single jog. This is also the total energy required to form a jog at a surface perpendicular to the dislocation and to move it into the interior of the crystal against image forces. In the approximation of a simple image construction, since there are no long-range image stresses in this case, the work to move the jog from a distance $\sim a$ below the surface to infinity is, by Eq. (8-92),

$$W_{\text{im}} = \frac{\mu b^2 a^2}{8\pi 2a(1-\nu)} = \frac{\mu b^2 a}{16\pi(1-\nu)} \tag{8-94}$$

or typically about 25 percent of the total self-energy. The other $\sim$75 percent of W_f is the energy of forming the "incipient" jog just below the surface; since long-range stresses are not involved, W_f clearly can be estimated as typically the energy of a core segment of length a. W_f, estimated from a well-separated pair, is the *total* energy required to bring a single jog from the surface into the crystal interior. The use of the simple image construction in the above development is somewhat tenuous in view of the results of Eq. (3-84). In an anisotropic elastic treatment, long-range image effects similar to those discussed for kinks in mixed dislocations are also present for jogs.

Line Tension

The calculation of the line tension of a jogged dislocation line proceeds analogously to the development of Eq. (8-88).

Exercise 8-7. Show that the effective line tension for a bowed-out dislocation line containing single jogs with an interjog spacing of l is given by

$$\mathcal{S} = \frac{\mu b^2}{4\pi(1-\nu)} \ln \frac{L}{l} \tag{8-95}$$

Comparing the magnitudes of $\mathcal{S}$ for the typical case $\nu = \frac{1}{3}$, one finds that they vary in the ratio 4 : 3 : 1 for a bowed line containing screw kinks, jogs, and edge kinks, respectively.

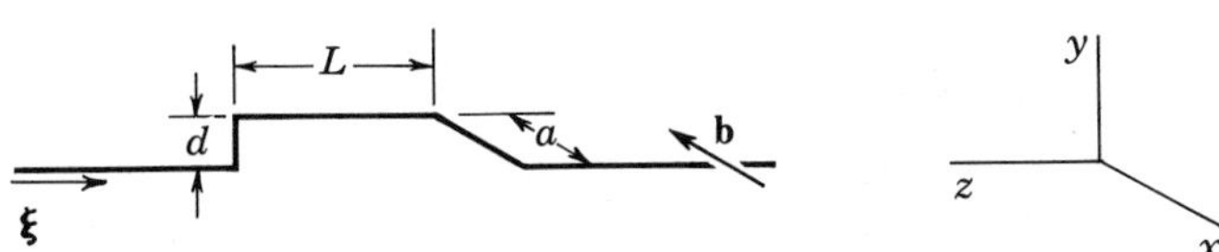

FIGURE 8-25. A jog-kink-interaction configuration.

A consideration of either Eqs. (8-93) or (8-95) reveals that strong repulsive forces exist between jogs. These forces oppose the formation of superjogs. Thus, in the isotropic elasticity approximation, *superjogs are not stable configurations on a dislocation line*. This result has important consequences in both glide and climb deformation mechanisms.

Jog-Kink Interaction

The jog segments discussed above are supposed to be in a plane normal to the glide plane of the remainder of the dislocation. It is possible that jog segments can have a component inclined to such a normal plane. In the formalism of our definitions, such an inclined segment can be regarded as a jog and a kink superposed.[77] To evaluate the likelihood of such an arrangement, consider the interaction of a jog and a kink as shown in Fig. 8-25. As one can verify by the methods of Chap. 6, there is no interaction-energy term to the first order in either a/L or d/L. There are short-range interaction forces that tend to rotate the kink into a jog configuration, but these can be considered to arise from core relaxations. To the first order, one expects that the kink distribution along a dislocation line is determined by kink-kink interactions only, and that jog-jog interactions alone control the jog distribution.

PROBLEMS

8-1. Graphically compare the results of Eqs. (3-43) and (8-13) for the stress σ_{xy} about an edge dislocation. For the example of copper, at what distance from the core do the results differ by 20 percent? What is the value of the elastic strain at this position?

8-2. Compare the tendency to cross slip for a dislocation in the Peierls model with that for one in the Volterra model.

8-3. Atomic binding is often described in terms of central forces acting between atom centers and resisting changes in bond length, and directional forces resisting changes in bond angle. The former are predominant in close-packed metals. Discuss how each of these types of forces would affect the width ζ of a Peierls dislocation and the width w of a kink. Classify, in order of

[77] E. Lear might evoke a jink or a kog here.

increasing Peierls barrier, fcc metals, bcc metals, ionic crystals, covalent crystals, and van der Waals crystals.

8-4. Give a qualitative discussion of the differences in the magnitudes of the kink width and kink energy between kinks as described by the approximate energy relations in Fig. 8-9 and those given by the exact relations.

8-5. Calculate the equilibrium separation between the kinks in a kink pair in a pure screw dislocation for an applied resolved shear stress $\sigma = 10^{-5}\mu$.

8-6. Consider three like-sign kinks in an edge dislocation, piled up against a pinning point that locks the first kink. The applied stress is $\sigma = 10^{-4}\mu$. What are the kink separations?

8-7. Estimate the width of kinks in screw dislocations in aluminum. Assume that $\sigma_p = 10^{-3}\mu$.

8-8. With the assumptions of elastic isotropy and of the Peierls-Nabarro analysis, would the core of a $\langle 110 \rangle$ screw dislocation dissociate on a (001) plane or on a $(1\bar{1}0)$ plane in a NaCl structure? Which slip system would have the lower Peierls barrier? Which is observed experimentally (see Chap. 12)?

8-9. Discuss whether the difference in energy between positive and negative kinks in mixed dislocations will be important for phenomena other than those involving interaction with the external surface.

8-10. By considering the symmetry on an atomic scale, show that the Peierls barrier for $\langle 111 \rangle$ screw dislocations in $\{110\}$ and $\{112\}$ planes is asymmetric.[78]

8-11. Discuss the physical justification for defining Eq. (8-6) by its principal value.

8-12. Compute the elastic interaction energy between an oblique kink and a right-angle jog.

8-13. Would it be physically meaningful to invoke a Peierls-type model for dislocation climb?

BIBLIOGRAPHY

1. Christian, J. W., and V. Vitek, *Rep. Progr. Phys.*, **33**: 307 (1970).
2. Frank, F. C., and J. H. Van der Merwe, *Proc. Roy. Soc.*, **198A**: 205, 216 (1949); *Proc. Roy. Soc.*, **200A**: 261 (1950).
3. Gehlen, P. C., J. R. Beeler, Jr., and R. I. Jaffee (eds.), "Interatomic Potentials and Simulation of Lattice Defects," Plenum, New York, 1972.
4. Hartley, C. S., M. F. Ashby, R. Bullough, and J. P. Hirth (eds.), "Dislocation Modeling of Physical Systems," Pergamon, New York, 1981.
5. Nabarro, F. R. N., *Proc. Phys. Soc.*, **59B**: 256 (1947); *Adv. Phys.*, **1**: 269 (1952).
6. Nabarro, F. R. N. (ed.), "Dislocation in Solids," vol. II, North-Holland, Amsterdam, 1979.
7. Van der Merwe, J. H., in M. H. Francombe and H. Sato (eds.), "Single Crystal Films," Pergamon, New York, 1964, p. 139.

[78] See J. P. Hirth and J. Lothe, *Phys. Stat. Solidi*, **15**: 487 (1966).

9

Slip Systems of Perfect Dislocations

9-1. INTRODUCTION

In the continuum theory of dislocations, treated in Part 1, the Burgers vector was defined formally by means of the Burgers circuit and related to the displacements associated with a dislocation. In Chap. 8 the Burgers vector was related to a lattice-translation vector in a simple cubic crystal lattice.

The first part of this chapter indicates that in general the Burgers vectors of dislocations are associated with lattice vectors, and that they should be given by those of the possible lattice vectors which are smallest in magnitude. Dislocations with such Burgers vectors are called *perfect dislocations*.

The slip systems of these dislocations are discussed, together with the resolution of shear stress on slip planes and the requirement for independent slip systems in crystals. All these topics are important in theories of flow stress, of work hardening in single crystals and polycrystals, and somewhat less directly in theories of such processes as twinning and brittle fracture.

9-2. PERFECT DISLOCATIONS

Crystallographic Notation

For an extensive treatment of crystallography the reader is referred to the work by Barrett.[1] To review the notation briefly, every crystal lattice can be generated by the translation of a lattice point by multiples of three non-coplanar translation vectors $\mathbf{a}_1, \mathbf{a}_2$, and $\mathbf{a}_3$, which together define a unit cell in the lattice. The lattice points coincide with *atoms* in simple crystals such as fcc (face-centered cubic), with *groups of atoms* in crystals with a basis, such as hcp (hexagonal close-packed). The direction notation $[n_1 n_2 n_3]$ represents the vector $n_1\mathbf{a}_1 + n_2\mathbf{a}_2 + n_3\mathbf{a}_3$, where the n_i are integers. A plane with Miller indices $(m_1 m_2 m_3)$ is parallel to a plane cutting the axes at $1/m_1$, $1/m_2$, $1/m_3$, where

[1]C. S. Barrett, "Structure of Metals," McGraw-Hill, New York, 1952.

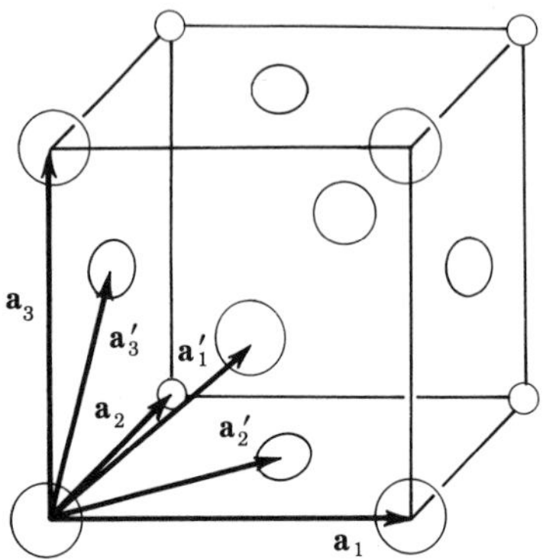

FIGURE 9-1. An fcc crystal with cubic unit-cell vectors $\mathbf{a}_i$ and primitive unit-cell vectors $\mathbf{a}'_i$.

the m_i are integers. In the special case of the cubic system, a plane $(m_1m_2m_3)$ has the direction vector $[m_1m_2m_3]$ as a normal. The notation $\langle n_1n_2n_3\rangle$ represents the set of all directions of the type $[n_1n_2n_3]$ which are equivalent by symmetry, and similarly $\{m_1m_2m_3\}$ represents the set of equivalent planes of the type $(m_1m_2m_3)$.

In the case of primitive unit cells with an atom at each lattice point, the translational vectors $\mathbf{a}_i$ represent the vector translations from one atom center to its nearest-neighbor atom centers. For example, the primitive vector $\mathbf{a}'_1$ for the fcc lattice shown in Fig. 9-1 connects nearest-neighbor atoms. Often it is convenient to use a multiple unit cell for better depiction of the lattice symmetry. For example, the fcc lattice is usually represented by the orthogonal vectors $\mathbf{a}_1$, $\mathbf{a}_2$, and $\mathbf{a}_3$, which generate a unit cell containing four atoms. These vectors, shown in Fig. 9-1, have equal magnitudes, $|\mathbf{a}_1|=|\mathbf{a}_2|=|\mathbf{a}_3|=a_0$. The primitive vector $\mathbf{a}'_1$ connecting nearest-neighbor atoms becomes, in the cubic notation, the vector $\frac{1}{2}\mathbf{a}_1+\frac{1}{2}\mathbf{a}_3$. If the magnitudes of the vectors $\mathbf{a}_i$ are defined to have unit length,[2] this direction vector can be written simply as $\frac{1}{2}[101]$.

The Burgers Vector

Suppose that a dislocation is produced by making a cut on a plane AB and displacing the two sides as shown in Fig. 9-2 for a simple cubic lattice. If the displacement is a primitive lattice vector, or a vector that is the sum of primitive lattice vectors, i.e., if the vector is one that would connect equivalent atom sites in the perfect crystal lattice, then one obtains a configuration such

[2]Sometimes in the cubic lattice the vector is written instead as $\frac{1}{2}a_0[110]$ to indicate that the magnitude of the components of the vector is expressed in terms of the lattice parameter a_0. However, the more compact notation adopted here is more useful in that it can be directly extended to other crystal systems. For example, in the tetragonal system, the vector with components $\frac{1}{2}\mathbf{a}_1$, 0, and $\mathbf{a}_3$ is written $\frac{1}{2}[102]$ even though $|\mathbf{a}_1|\neq|\mathbf{a}_3|$. In the hcp lattice, it is common to use the Miller-Bravais coordinates $\mathbf{a}_1$, $\mathbf{a}_2$, $\mathbf{a}_3$, and $\mathbf{c}$, where the $\mathbf{a}_1$, $\mathbf{a}_2$, and $\mathbf{a}_3$ vectors lie along crystallographically equivalent directions in the basal plane. In this lattice the abbreviated notation for the vector with components $\mathbf{a}_1$, $-2\mathbf{a}_2$, $\mathbf{a}_3$, and $\mathbf{c}$ is $[1\bar{2}11]$.

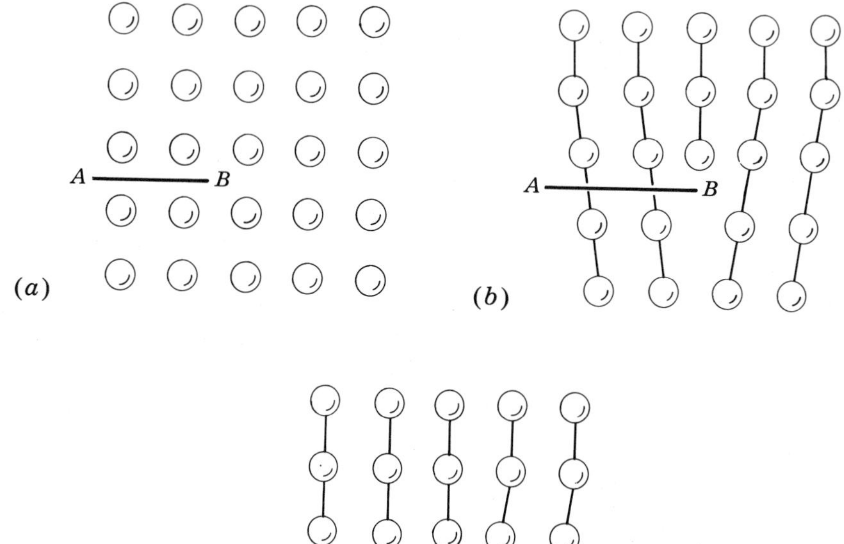

FIGURE 9-2. (*a*) Cut crystal, which is displaced along *AB* to form either (*b*) a perfect dislocation or (*c*) an imperfect dislocation.

as Fig. 9-2*b*. This configuration is again the Peierls dislocation, which has a misfit energy mainly localized near the core and an elastic strain energy. If the displacement cannot be decomposed to a sum of primitive lattice vectors, a different type of configuration results. For example, consider Fig. 9-2*c*, where the Burgers vector is one half of a primitive lattice vector. Apparently, in addition to the misfit energy and elastic strain energy associated with the Peierls dislocation, this configuration has a long-range, divergent misfit-energy contribution caused by the mismatch of atoms across the displaced surface *AB*. Because of this misfit energy, all dislocations in crystals are expected to be perfect dislocations, with the exception of special cases to be treated in the next chapter; their Burgers vectors are perfect-lattice vectors.

The Frank Energy Criterion

Of the possible perfect dislocations defined by the above criterion, only those with the one or two shortest possible Burgers vectors are stable. Equation (3-87) shows that the energy of a dislocation is proportional to b^2. Frank[3] proposed that the variation of the dislocation energy with β be neglected, so

[3]F. C. Frank, *Physica*, **15**: 131 (1949).

that the value of b^2 can be used to assess the stability of a dislocation. His criterion is that if a perfect dislocation with Burgers vector $\mathbf{b}_1$ can dissociate into perfect dislocations $\mathbf{b}_2 + \mathbf{b}_3$, dissociation will occur if

$$b_1^2 > b_2^2 + b_3^2 \tag{9-1}$$

For example, in a reaction like that in Fig. 1-24, if the dislocation $\mathbf{b}_1 = 2\mathbf{b}_2 = 2\mathbf{b}_3$, it dissociates into $\mathbf{b}_2 + \mathbf{b}_3$.

An equivalent procedure involves the application of the interaction-force equation, (5-17), to the product dislocations. If the interaction force between $\mathbf{b}_2$ and $\mathbf{b}_3$ is attractive, $\mathbf{b}_1$ will be stable, while if the interaction force is repulsive, $\mathbf{b}_1$ will dissociate.

A more exact procedure would be to include the screw-edge character and express the above conditions in terms of the energy factors $E(\beta)$ given by Eq. (3-87) for the isotropic case or by Eq. (13-188) for the anisotropic case:

$$E_1(\beta) > E_2(\beta) + E_3(\beta) \tag{9-2}$$

These methods show immediately that any dislocation that has a Burgers vector which is a multiple of a perfect-dislocation Burgers vector is unstable. For most crystal systems only the primitive-lattice vectors, and in some cases their pairwise combinations, correspond to dislocations that are stable against dissociation. In marginal cases one can improve on criterion (9-1) in the isotropic elasticity approximation by including the β variation of Eq. (3-87) in Eq. (9-2). Of course, anisotropic elasticity effects could lead to the stability of dislocations that would be predicted to be unstable by isotropic criteria, or vice versa. Also in anisotropic elasticity the edge dislocation has different values of $E(\beta)$ depending on line direction and hence, on glide plane.

The stable dislocations for several prominent crystal structures are listed in Table 9-1.

TABLE 9-1 Burgers Vector of Stable Dislocations in Several Crystal Structures

Crystal structure	Stable **b**	Marginal Stability **b**	Relative $E(\beta)$ for Glide Plane, Anisotropic Edge Dislocation
fcc	$\frac{1}{2}\langle 110\rangle$	$\langle 100\rangle$	$\{110\}<\{111\}<\{100\}$[1]
bcc	$\frac{1}{2}\langle 111\rangle$, $\langle 100\rangle$		$\{110\}<\{112\}<\{123\}$
hcp	$\frac{1}{3}\langle 11\bar{2}0\rangle$, $\langle 0001\rangle$	$\frac{1}{3}\langle 11\bar{2}3\rangle$	(0002) or $\{1\bar{1}00\}$[2]
Diamond cubic	$\frac{1}{2}\langle 110\rangle$	$\langle 100\rangle$	$\{110\}<\{111\}<\{100\}$
NaCl	$\langle 110\rangle$	$\langle 200\rangle$	$\{110\}$ or $\{100\}$[3]

[1] $\{111\}<\{110\}$ for aluminum and diamond.
[2] (0002) for cadmium and zinc.
[3] J. W. Steeds, "Anisotropic Elastic Theory of Dislocations," Clarendon, Oxford, 1973.

9-3. SLIP SYSTEMS IN CRYSTALS

As discussed in Chap. 1, single crystals deform by slip in close-packed directions on planes that are close-packed planes in general. With the advent of dislocation theory, the slip direction and plane were associated with the Burgers vector and glide plane, respectively, of gliding dislocations, and macroscopic slip lines were assumed to be produced by sets of dislocations with the same Burgers vector. The results of electron-transmission microscopy studies in recent years have verified this supposition.

Dislocation theory predicts that only a few low-index glide planes and directions should be observed for a given crystal system. Frank's criterion, Eq. (9-1), predicts that only the perfect dislocation Burgers vectors listed in Table 9-1 are elastically stable in crystals, and hence that only these could contribute to slip. Furthermore, according to Eq. (8-39), the Peierls stress is least for the dislocations with the smallest-magnitude Burgers vectors, so that these should slip more easily than the larger ones among the stable Burgers vectors.

Similarly, Eq. (8-39) indicates that for a given Burgers vector the glide plane with the largest d spacing has the lowest Peierls stress.[4] The planes with the largest d spacing are the closest-packed planes in the crystal structure in question, and in turn, these planes have the lowest $\{hkl\}$ indices. For example, in cubic crystals,[5]

$$d_{hkl}=\frac{a_0}{\sqrt{h^2+k^2+l^2}} \tag{9-3}$$

In Eq. (9-3), h, k, and l are the reciprocals of the intercepts of the plane of the type $\{hkl\}$ that is closest to the origin fixed on a corner of the unit cell. Thus in fcc structures the closest-packed plane is $\{111\}$, with $d=a_0/\sqrt{3}$; the second-closest-packed plane is of the type $\{001\}$, with specifically $\{002\}$ nearest the origin, giving $d=a_0/2$.

Motohashi and Ohtake[6] summarize anisotropic calculations and present data that enable one to give the $E(\beta)$ values for various glide planes as indicated in Table 9-1, except for NaCl. The range of values for $E(\beta)$ differs by 10 to 20 percent for the particular crystal structures. These data can be used with criterion (9-2) to predict the active glide planes.

In some crystal structures perfect dislocations can dissociate into imperfect partial dislocations bounding a stacking fault, as discussed in detail in the next chapter. Such dissociated dislocations are constrained in general to glide on the

[4] When directional bonding is important, it is possible that the Peierls stress might be lower for a plane other than the closest-packed plane. This possibility is masked in the original Peierls model because of the assumptions in the model.

[5] Equivalent relations between d_{hkl} and the indices of planes are listed for other crystal systems in C. S. Barrett, "Structure of Metals," McGraw-Hill, New York, 1952, p. 633.

[6] Y. Motohashi and S. Ohtake, *Phys. Stat. Solidi*, **50a**: 449 (1978).

fault plane. Thus the existence of a low-energy-fault plane in a crystal provides an indication that the fault plane should be a slip plane. Low-energy-fault planes in general are also low-index planes.

In summary, dislocation theory predicts that the shortest of the possible perfect-dislocation Burgers vectors should correspond to the slip direction, based on Frank's energy criterion and the Peierls stress criterion, while the glide plane should be either the plane with the lowest Peierls stress or a stacking-fault plane, a lesser factor being the $E(\beta)$ variation. Experimentally observed slip systems for several prominent crystal structures are briefly reviewed in the following discussion and compared with the above predictions. Prior to the development of electron-transmission microscopy, slip systems were determined by optical microscopy, replica electron-microscopy techniques, x-ray studies of lattice rotations, and the analysis of Laue spot asterism. For reviews of slip systems as determined by these methods see Schmid and Boas,[7], Barrett,[8] and Maddin and Chen.[9] Following the discovery of the electron-transmission technique for dislocation observation[10, 11] the technique was used to study slip systems in various materials: investigations of fcc, hcp, and bcc metals are reviewed, respectively, by Swann,[12] Price,[13] and Keh and Weissman,[14] and Motohashi and Ohtake.[6] The other technique for slip-system studies is that of dislocation etch pitting. This method has been reviewed by Johnston[15] and Amelinckx.[16]

FCC Slip Systems

The results of all the methods of study consistently indicate that $\{111\}\langle 110\rangle$ is the major operative slip system in fcc structures. An example of $\{111\}$ slip in aluminum is shown in Fig. 1-15. The $\{111\}$ planes are the most densely packed planes in fcc metals, and also the planes upon which stacking faults form. Moreover, $\frac{1}{2}\langle 110\rangle$ is the smallest possible perfect-dislocation Burgers vector. Thus the observed slip systems in fcc structures are consistent with the rough predictions of dislocation theory.

Two unusual slip systems are sometimes encountered. The first is $\{100\}$ $[110]$ slip at elevated temperatures which has been observed by optical studies.[17]

[7] E. Schmid and W. Boas, "Plasticity of Crystals," Hughes, London, 1950.

[8] C. S. Barrett, "Structure of Metals," McGraw-Hill, New York, 1952.

[9] R. Maddin and N. K. Chen, *Prog. Met. Phys.*, **5**: 53 (1954).

[10] W. Bollman, *Phys. Rev.*, **103**: 1588 (1956).

[11] P. B. Hirsch, R. W. Horne, and M. J. Whelan, *Phil. Mag.*, **1**: 677 (1956).

[12] P. R. Swann, in G. Thomas and J. Washburn (eds.), "Electron Microscopy and Strength of Crystals," Interscience, New York, 1963, p. 131.

[13] P. B. Price, in *ibid.*, p. 41; see also P. G. Partridge, *Met. Rev.*, **12**: 169 (1967).

[14] A. S. Keh and S. Weissman, in *ibid.*, p. 231.

[15] W. G. Johnston, *Prog. Ceramic Sci.*, **2**: 1 (1962).

[16] S. Amelinckx, *Solid State Phys., Suppl.* 6 (1964).

[17] P. Lacombe and L. Beaujard, *J. Inst. Met.*, **74**: 1 (1947).

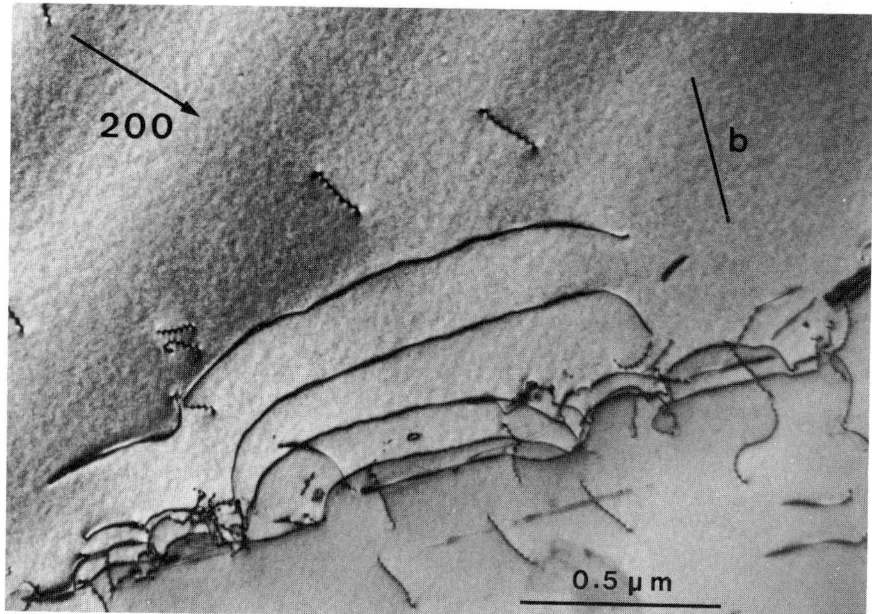

FIGURE 9-3. (001) foil of disordered Ni_3Fe (a high stacking-fault energy fcc alloy), plastically deformed at room temperature. Total dislocations ($b = \pm a/2$ [110]) with a character near edge orientation are bowing out on (001) planes. The diffraction vector **g** = [200] and **b** are indicated. (H. P. Karnthaler, unpublished research.)

A likely source for such slip has been revealed in electron microscopic observations by Korner and Karnthaler,[18] who found that Lomer or Lomer-Cottrell dislocations, which form by the reactions of Eqs. (22-1) and (22-2) as 1/2[110] dislocations on {100}, are mobile. The mobility is greater in high-stacking-fault alloys, as illustrated for Ni_3Fe in Fig. 9-3. The rarity of this type of slip is consistent with the expected greater Peierls stress and $E(\beta)$ factor for {100} slip. The Lomer dislocations form in an orientation where it is difficult for them to cross slip onto {111}, so once formed they persist despite the unfavorable factors.

Second, {110} [110] slip has been observed optically at high temperatures,[19] after impact loading,[20] and in electron microscopy.[21] Some of these observations may reflect composite slip on {111} planes.[22] However, some[21] definitely indicate pure {110} slip. Such slip could be consistent with the lower $E(\beta)$ factor in Table 9-1, despite the high expected Peierls stress. Alternatively, it could be associated with preferential surface nucleation as influenced by image

[18] A. Korner and H. P. Karnthaler, *Phil. Mag.*, **42**: 753 (1980).

[19] R. LeHazif and J. P. Poirier, *Acta Met.*, **23**: 865 (1975).

[20] A. L. Stevens and L. E. Pope, *Scripta Met.*, **5**: 981 (1971).

[21] B. Pichaud and F. Minari, *Scripta Met.*, **14**: 1171 (1980).

[22] R. T. Bhatt, P. A. Thrower, and W. R. Bitler, *Scripta Met.*, **10**: 19 (1976).

interactions,[21] considered in Sec. 9-6. Finally, we note one report of {212} slip in fcc crystals.[22] Dislocations of the type ⟨100⟩ have been observed in electron-transmission microscopy,[23] in agreement with the prediction of Table 9-1, but have never been observed to slip. This finding is consistent with the prediction that the large Burgers vector ⟨100⟩ dislocations should have a higher Peierls stress than the $\frac{1}{2}$⟨110⟩ type.

BCC Slip Systems

As indicated in Table 9-2, the slip direction in bcc structures is almost always ⟨111⟩, which corresponds to the smallest possible perfect-dislocation Burgers vector $\frac{1}{2}$⟨111⟩. The one exception to this is in the work of Reid et al.,[24] who report ⟨100⟩ slip in niobium single crystals with ⟨111⟩ tensile axes. This unusual result may be a result of elastic anisotropy. Niobium and molybdenum differ from other cubic metals in that their anisotropy factors, $A = 2c_{44}/(c_{11} - c_{12})$, are less than unity. Modifications of the Peierls dislocation model to account for anisotropy by Eshelby[25] and others[26–28] indicate that ⟨100⟩ dislocations are wider than $\frac{1}{2}$⟨111⟩ dislocations for niobium and molybdenum, where $A < 1$. Thus Eq. (8-39) would predict a lower Peierls stress for ⟨100⟩ dislocations in these metals.[24] However, ⟨100⟩ slip is certainly not predominant in niobium and molybdenum, so that, pending electron-transmission verification, the suggestion of ⟨100⟩ slip should be regarded as tentative; the modified Peierls model prediction is uncertain because of the approximations in the model.

As in the fcc case, larger Burgers vector dislocations have been observed in electron transmission microscopy,[29–34] in this case with Burgers vectors ⟨100⟩. However, once again, the larger Burgers vector ⟨100⟩ dislocations have never been observed to slip, with the possible exception of the work on niobium, discussed above.

The most densely packed planes are {110}, while the main stacking-fault plane is {112}. Thus either could be favored, depending on the magnitude of

[23] P. B. Hirsch, *J. Inst. Met.*, **87**: 406 (1958).

[24] C. N. Reid, A. Gilbert, and G. T. Hahn, *Acta Met.*, **14**: 975 (1966).

[25] J. D. Eshelby, *Phil. Mag.*, **40**: 903 (1949).

[26] A. J. E. Foreman, *Acta Met.*, **3**: 322 (1955).

[27] A. J. E. Foreman and W. M. Lomer, *Phil. Mag.*, **46**: 73 (1955).

[28] C. N. Reid, *Acta Met.*, **14**: 13 (1966).

[29] A. Berghezan and A. Fourdeux, "Coll. Met. Centre Etudes Nucl. Saclay, 1960," Presse Univ. de France, Paris, 1961, p. 127.

[30] B. R. Banerjee, J. M. Capenos, J. J. Hauser, and J. P. Hirth, *J. Appl. Phys.*, **32**: 556 (1961).

[31] W. Carrington, K. F. Hale, and D. McLean, *Proc. Roy. Soc.*, **259A**: 203 (1960).

[32] A. S. Keh and S. Weissman, in G. Thomas and J. Washburn (eds.), "Electron Microscopy and Strength of Crystals," Interscience, New York, 1963, p. 231.

[33] D. Hull, in *ibid.*, p. 291.

[34] S. M. Ohr and D. N. Beshers, *Phil. Mag.*, **8**: 1343 (1963).

TABLE 9-2 Slip Planes in BCC Metals*

Metal	Predominant Slip Planes at Room Temperature	Slip Planes at Low Temperature	Slip Planes at High Temperatures or Low Strain Rates
(1) Fe	{112}[1](*T*)[2]	{110}[1](*T*)[3,4]	{123}[1]
	{110}[1](*T*)[2]	{112}[1](*T*)[3,4]	{*hkl*}[5,6]
(2) Fe–3+%Si	{112}[1,7,8](*T*)[2]	{110}[1,8,10](*T*)[4]	{123}[1]
	{110}[1,8](*T*)[9]	{112}(*T*)[4]	{*hkl*}[10–13](*T*)[11]
(3) Na	{123}[14]		
(4) K	{123}[14]		
(5) Nb	{110}[15,16](*T*)[17]	{110}[18]	
(6) Ta	{110}[19,20]	{112}[18]	
	{112}[20]		
(7) V	{110}(*T*)[21]	{110}(*T*)[21]	
	{112}(*T*)[21]	{112}(*T*)[21]	
(8) W	{110}[22]	{110}[23]	{112}[24]
	{112}[23]		
(9) Mo	{110}[15,25]		
Mo-Re	{112}[25]		
(10) Cr	{123}[26]	{112}[26]	

*In all cases the slip direction is ⟨111⟩. (*T*) indicates electron-transmission microscopy. {*hkl*} indicates any plane for which ⟨111⟩ is the zone axis.

1. C. S. Barrett, G. Ansel, and R. F. Mehl, *Trans. ASM*, **25:** 702 (1937).
2. A. S. Keh, *Phil. Mag.*, **12:** 9 (1965).
3. R. Priestner and W. C. Leslie, *Phil. Mag.*, **11:** 895 (1965).
4. R. Priestner, in R. E. Reed-Hill, J. P. Hirth, and H. C. Rogers (eds.), "Deformation Twinning," Gordon and Breach, New York, 1964, p. 321.
5. F. L. Vogel, Jr., and R. M. Brick, *Trans. AIME*, **197:** 700 (1953).
6. G. I. Taylor and C. F. Elam, *Proc. Roy. Soc.*, **112A:** 337 (1926).
7. D. F. Stein and P. D. Gorsuch, *Acta. Met.*, **9:** 904 (1961).
8. J. S. Erikson, *J. Appl. Phys.*, **33:** 2499 (1962).
9. J. R. Low and A. M. Turkalo, *Acta. Met.*, **10:** 215 (1962).
10. F. W. Noble and D. Hull, *Phil. Mag.*, **12:** 777 (1965).
11. B. Sěsták and S. Libovický, *Czech. J. Phys.*, **12B:** 131 (1962); *Acta. Met.*, **11:** 1190 (1963).
12. R. P. Steijn and R. M. Brick, *Trans. ASM*, **46:** 1406 (1954).
13. J. J. Cox, G. T. Horne, and R. F. Mehl, *Trans. ASM*, **49:** 118 (1957).
14. E. N. da C. Andrade and L. C. Tsien, *Proc. Roy. Soc.*, **163A:** 912 (1937).
15. R. Maddin and N. K. Chen, *Prog. Metal Phys.*, **5:** 53 (1954).
16. C. J. McHargue, *Trans. Met. Soc. AIME*, **224:** 334 (1962).
17. L. I. Van Torne and G. Thomas, *Acta. Met.*, **11:** 881 (1963).
18. C. N. Reid, A. Gilbert, and G. T. Hahn, *Acta Met.*, **14:** 975 (1966).
19. P. P. Ferris, R. M. Rose, and J. Wulff, *Trans. Met. Soc. AIME*, **224:** 975 (1962).
20. B. L. Mordike, *Z. Met.*, **52:** 587 (1961).
21. J. W. Edington, private communication.
22. R. G. Garlick and H. B. Probst, *Trans. Met. Soc. AIME*, **230:** 1120 (1964).
23. H. W. Shadler, *Trans. AIME*, **218:** 649 (1960); *Acta. Met.*, **12:** 861 (1964).
24. F. S. Goucher, *Phil. Mag.*, **48:** 229 (1924).
25. A. Lawley and R. Maddin, *Trans. Met. Soc. AIME*, **224:** 573 (1962).
26. A. Gilbert, private communication.

the stacking-fault energy. As can be seen in Table 9-2, these two slip planes are preponderant in the observed bcc glide systems, but {123} and noncrystallographic {*hkl*} slip planes are also found. It has been suggested[35] that {123}, {*hkl*}, and perhaps {112} are actually composed of short "composite" slip steps on {110} planes. On the basis of work to date, the question of whether slip is composite {110} slip or actually slip on other planes remains open. In a series of papers involving both the etch-pit technique and electron-transmission microscopy, Šesták and Libovický[36] found that noncrystallographic slip occurred (1) at high temperatures and (2) at low strain rates at either high or low temperatures. At any rate, if the slip were composite, the {110} steps were too small to be resolved in the electron microscope. As noted in Table 9-2, numerous optical slip-line studies also indicate noncrystallographic slip. Finally, as also indicated in the table, {112} slip in addition to {110} slip has been confirmed by electron-transmission microscopy, so that apparently at least {112} slip does not occur by composite {110} slip.

On the other hand, Edington[37] has observed slip in chromium in the electron microscope which is near {110} but which indicates that small composite cross-slip steps can occur, leading to a slip trace that would appear to be of high index at a lower magnification. Figure 9-4 is an example of his results and shows an example of such cross slip. In summary, crystallographic slip has been verified on {110} and {112}. Whether slip occurs on other planes, or whether such slip is composite {110}–{112} slip is in most cases an open question which will require high-resolution techniques for a definitive answer.

Andrade and Chow[38] suggested that with an increasing ratio of T/T_m, where T_m is the melting point, the operative slip planes should be {112}, {110}, and {123}, in that order. However, as shown in Table 9-2, evidence seems to indicate that the order should be {110}, {112}, {123}, and {*hkl*}. Indeed, Taoka et al.[39] found that the temperature dependence of the critical resolved shear stress for glide on {112} exceeds that for {110} in Fe–3%Si, so that at low temperature {110} predominates and at high temperature {112} predominates, confirming the suggestion of Smoluchowski and Opinsky.[40] Erikson[41] had indicated earlier that the stress to move a dislocation at a given velocity increased more rapidly with decreasing temperature for {112} slip than for {110} slip in silicon iron. Also, the observed sequence in Table 9-2 is consistent with the $E(\beta)$ sequence in Table 9-1.

[35] R. Maddin and N. K. Chen, *Prog. Metal Phys.*, **5**: 53 (1954).

[36] B. Šesták and S. Libovický, *Czech. J. Phys.*, **12B**: 131 (1962); *Acta Met.*, **11**: 1190 (1963).

[37] J. W. Edington, private communication.

[38] E. N. da C. Andrade and Y. S. Chow, *Proc. Roy. Soc.*, **175A**: 290 (1940).

[39] T. Taoka, S. Takeuchi, and E. Furubayashi, *J. Phys. Soc. (Japan)*, **19**: 701 (1964).

[40] R. Smoluchowski and A. Opinsky, *J. Appl. Phys.*, **22**: 1488 (1951).

[41] J. S. Erikson, *J. Appl. Phys.*, **33**: 2499 (1962).

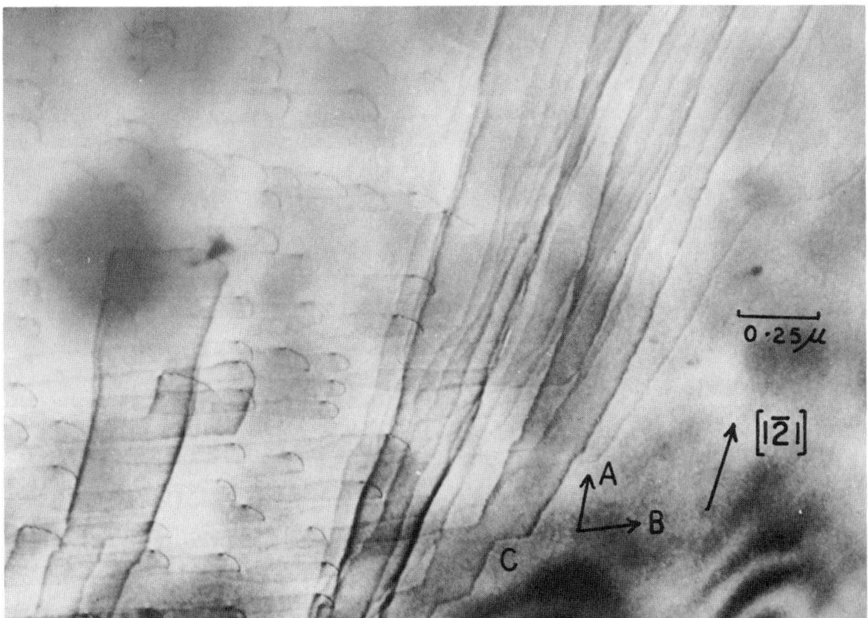

FIGURE 9-4. Cross slip in chromium deformed at 50°C in the tensile stage in the electron microscope. The grain is oriented for slip on two systems, and one set of dislocations has originated at the edge of the foil just out of the bottom of the picture. Slip has occurred on plane *A* and cross slip on plane *B* at point *C*. The fine-scale cross slip in other areas also occurs on these systems. The foil plane is close to (113) and slip in the direction *A* is occurring on the (101) plane at 31° to the foil plane, with cross slip on the *B* (0$\bar{1}$1) plane at 64° to the foil plane. (Courtesy of J. W. Edington.)

Thus the picture that emerges is that crystallographic slip on {110} and {112} occurs at low temperatures, but that {*hkl*} slip, undoubtedly representing cross slip of screw dislocations, occurs at higher temperatures or lower strain rates. The simple Peierls energy and stacking-fault criteria are inadequate to explain this behavior completely. One can speculate that the change in glide plane is associated with the temperature dependence of the Peierls stress or with a temperature-dependent reassociation of extended partial dislocations, but further work is required to resolve this problem. Some discussion of core structure and possible extensions of dislocations in bcc lattices, particularly relevant to screw dislocation motion, is presented in Chap. 11.

HCP Slip Systems

A variety of slip systems have been observed in hcp crystals as noted in Table 9-3. The observation of ⟨11$\bar{2}$0⟩ and ⟨11$\bar{2}$3⟩ as slip directions is consistent with the prediction in Table 9-1. The infrequency of observation of ⟨0001⟩, which appears only for beryllium at high temperatures, is puzzling at first inspection but, as is discussed presently, is probably related to the core structure of such a

TABLE 9-3 Slip Systems for HCP Metals*

Metal	Predominant Slip System at Room Temperature	Less Prominent Systems: High Temperatures	Less Prominent Systems: Favorable Resolved Shear Stress
(1) Cd $\left(\frac{c}{a}=1.89\right)$	$\langle 11\bar{2}0\rangle\{0001\}$[1–4]	$\langle 11\bar{2}0\rangle\{10\bar{1}0\}$[5]	$\langle ?\rangle\{11\bar{2}2\}$[6]
			$\langle 11\bar{2}0\rangle\{10\bar{1}0\}$[7]
			$\langle 11\bar{2}0\rangle\{10\bar{1}1\}(T)$[8]
			$\langle 11\bar{2}3\rangle\{11\bar{2}2\}(T)$[8]
			$\langle 11\bar{2}3\rangle\{hkil\}(T)$[8, 9]
(2) Zn $\left(\frac{c}{a}=1.86\right)$	$\langle 11\bar{2}0\rangle\{0001\}$[1–4]	$\langle 11\bar{2}0\rangle\{10\bar{1}0\}$[10]	$\langle 11\bar{2}3\rangle\{11\bar{2}2\}$[11]
			$\langle 11\bar{2}0\rangle\{10\bar{1}0\}$[12]
			$\langle 11\bar{2}3\rangle\{11\bar{2}2\}$[8]
			$\langle 11\bar{2}3\rangle\{hkil\}$[8, 9]
(3) Mg $\left(\frac{c}{a}=1.62\right)$	$\langle 11\bar{2}0\rangle\{0001\}$[1–4]$(T)$[16]	$\langle 11\bar{2}0\rangle\{10\bar{1}1\}$[1–4]	$\langle 11\bar{2}0\rangle\{10\bar{1}1\}$[13, 14]$(T)$[16]
			$\langle 10\bar{1}0\rangle\{11\bar{2}2\}$[14]
			$\langle 11\bar{2}0\rangle\{10\bar{1}0\}$[14, 15]
			$\langle 11\bar{2}0\rangle\{hkil\}$[15]
(4) Co $\left(\frac{c}{a}=1.62\right)$	$\langle 11\bar{2}0\rangle\{0001\}$[17]$(T)$[18]		$\langle ?\rangle\{11\bar{2}2\}$[17]
(5) Re $\left(\frac{c}{a}=1.62\right)$	$\langle 11\bar{2}0\rangle\{0001\}$[19]		
	$\langle 11\bar{2}0\rangle\{10\bar{1}0\}$[19]		

*Studies using electron-transmission microscopy are indicated by (T). The other studies mainly involve surface slip trace or resolved shear stress criteria.

1. E. Schmid and W. Boas, "Plasticity of Crystals," Hughes, London, 1950.
2. C. S. Barrett, "Structure of Metals," McGraw-Hill, New York, 1952.
3. R. Maddin and N. K. Chen, *Prog. Metal. Phys.*, **5:** 53 (1954).
4. L. M. Clarebrough and M. E. Hargreaves, *Prog. Metal. Phys.*, **8:** 1 (1959).
5. J. J. Gilman, *Trans. Met. Soc. AIME*, **221:** 456 (1961).
6. N. S. Stoloff and M. Gensamer, *Trans. Met. Soc. AIME*, **224:** 732 (1962).
7. A. F. Brown, *Advan. Phys.*, **1:** 427 (1952).
8. P. B. Price, in G. Thomas and J. Washburn (eds.), "Electron Microscopy and Strength of Crystals," Interscience, New York, 1963, p. 41.
9. The notation $\{hkil\}$ indicates that cross slip occurs on any plane for which the slip direction is a zone axis; such slip is also called banal.
10. J. J. Gilman, *Trans. AIME*, **206:** 1326 (1956).
11. R. L. Bell and R. W. Cahn, *Proc. Roy. Soc.*, **239A:** 494 (1957).
12. A. Seeger and H. Trauble, *Z. Met.*, **51:** 435 (1960).
13. E. C. Burke and W. R. Hibbard, Jr., *Trans. AIME*, **194:** 295 (1952).
14. R. E. Reed-Hill and W. D. Robertson, *Trans. AIME*, **212:** 256 (1958).
15. A. R. Chadhuri, J. R. Patel, and N. J. Grant, *Trans. AIME*, **203:** 682 (1955).
16. G. Thomas, R. B. Benson, and J. Nadeau, *Proc. European Regional Conf. Electron Microscopy* (Delft), 1961, p. 447.
17. A. Seeger, H. Kronmüller, O. Boser, and M. Rapp, *Phys. Stat. Solidi*, **3:** 1107 (1963).
18. E. Votava, *J. Inst. Met.*, **90:** 129 (1961).
19. A. T. Churchman, *Trans. Met. Soc., AIME*, **218:** 262 (1960).

TABLE 9-3 (*Continued*)

Metal	Predominant Slip System at Room Temperature	Less Prominent Systems	
		High Temperatures	Favorable Resolved Shear Stress
(6) Ti $\left(\frac{c}{a}=1.59\right)$	$\langle 11\bar{2}0\rangle\{10\bar{1}0\}$[20, 21]		$\langle 11\bar{2}0\rangle\{10\bar{1}1\}$[20, 21]
			$\langle 11\bar{2}0\rangle\{0001\}$[20, 21]
(7) Zr $\left(\frac{c}{a}=1.59\right)$	$\langle 11\bar{2}0\rangle\{10\bar{1}0\}$[22]$(T)$[23]		$\langle 11\bar{2}0\rangle\{0001\}$[21]$(T)$[25]
			$\langle ?\rangle\{10\bar{1}3\}$[25]
			$\langle ?\rangle\{11\bar{2}1\}$[25]
(8) Be $\left(\frac{c}{a}=1.57\right)$	$\langle 11\bar{2}0\rangle\{0001\}$[26, 27]	$\langle 11\bar{2}0\rangle\{1\bar{1}04\}$[27]	$\langle 0001\rangle\{10\bar{1}0\}$[28]
		$\langle 10\bar{1}0\rangle\{?\}$[29]	$\langle 11\bar{2}0\rangle\{10\bar{1}0\}$[26, 27]
		$\langle 11\bar{2}3\rangle\{11\bar{2}2\}$[29]	$\langle 11\bar{2}0\rangle\{10\bar{1}1\}$[27]
(9) Yt $\left(\frac{c}{a}=1.57\right)$	$\langle 11\bar{2}0\rangle\{10\bar{1}0\}$[30]		$\langle 11\bar{2}0\rangle\{0001\}$[30]

20. F. D. Rosi, F. C. Perkins, and L. L. Seigle, *Trans. AIME*, **206:** 115 (1956).
21. A. T. Churchman, *Proc. Roy. Soc.*, **226A:** 216 (1954).
22. E. J. Rapperport and C. S. Hartley, *Trans. AIME*, **218:** 869 (1960).
23. L. M. Howe, J. L. Whitton, and J. F. McGurn, *Acta. Met.*, **10:** 773 (1962).
24. J. L. Martin and R. E. Reed-Hill, *Trans. Met. Soc. AIME*, **230:** 780 (1964).
25. J. E. Bailey, *Acta. Met.*, **11:** 267 (1963).
26. G. L. Tuer and A. R. Kaufmann, in D. W. White, Jr., and J. E. Burke (eds.), "The Metal Beryllium," American Society of Metals, Cleveland, Ohio, 1955, p. 372.
27. E. D. Levine, D. F. Kaufman, and L. R. Aronin, *Trans. Met. Soc. AIME*, **230:** 260 (1964).
28. R. I. Garber, I. A. Gindin, and Y. V. Shubin, *Sov. Phys. Solid State*, **3:** 832 (1961).
29. P. Pointu, P. Azou, and P. Bastien, *Comp. Rend. Acad. Sci.*, **252:** 1984 (1961).
30. E. J. Rapperport and C. S. Hartley, *Trans. AIME*, **215:** 1071 (1959).

dislocation. Consistent with the elastic-energy calculations in Table 9-1, $\langle 0001\rangle$ dislocations have been observed[42] in loop reactions and as products of dissociation of $\frac{1}{3}\langle 11\bar{2}3\rangle$ dislocations, but they did not glide. The $\langle 10\bar{1}0\rangle$ slip direction undoubtedly represents duplex slip of different $\frac{1}{3}\langle 11\bar{2}0\rangle$ dislocations.

Extended dislocations on the (0001) plane have been observed in cobalt,[43] indicating a low stacking-fault energy in cobalt. The splitting of dislocations under stress to form stacking faults on (0001) has been reported[42] for zinc and cadmium but not for the remaining metals, suggesting a lower fault energy for the former pair. In addition, Seeger[44] predicts theoretically that rhenium should have a low fault energy on the basal plane. Higher-energy stacking

[42] P. B. Price, in G. Thomas and J. Washburn (eds.), "Electron Microscopy and Strength of Crystals," Interscience, New York, 1963, p. 41.

[43] E. Votava, *J. Inst. Met.*, **90**: 129 (1961).

[44] A. Seeger, in "Defects in Crystalline Solids," Physical Society, London, 1955, p. 328.

TABLE 9-4 d_{hkil} Values for Various Planes

Plane	(0001)	$\{10\bar{1}0\}$	$\{10\bar{1}1\}$	$\{11\bar{2}2\}$
d_{hkil}	$0.816a$	$0.866a$	$0.765a$	$0.428a$
"Normal" spacing	$0.816a$	$0.816a$	$0.706a$	$0.428a$

faults are predicted for $\{11\bar{2}2\}$ and $\{10\bar{1}0\}$ planes in hcp metals based on hard-ball atom models for slip and twinning.[45]

The d spacing of the planes is somewhat complicated in hcp crystals and requires some further discussion. In hcp structures the interplanar spacing is related to the Miller-Bravais indices by the expression

$$\frac{1}{d^2} = \frac{4}{3}\frac{h^2 + hk + k^2}{a^2} + \frac{l^2}{c^2} \tag{9-4}$$

For an ideal c/a ratio of 1.633, corresponding to the close packing of spheres, Eq. (9-4) yields the d spacings listed in Table 9-4. The d spacings alone, however, are misleading. As can be seen in Fig. 9-5, the $\{10\bar{1}0\}$ planes are actually zigzag planes, with a normal spacing between the planes of $0.816a$. Similarly, the pyramidal planes $\{10\bar{1}1\}$ are zigzag and have a normal spacing of $0.706a$. In actual crystals, the spacing of the basal planes, (0001), is lowered with respect to the others for $c/a < 1.633$ and raised for $c/a > 1.633$.

The above discussion provides a basis for the rationalization for the observed slip systems. In zinc and cadmium, (0001) is the favored slip plane because it is a plane of relatively low stacking-fault energy, the plane of largest d spacing, and the plane of lowest $E(\beta)$. In cobalt, and probably in rhenium,[46] the low stacking-fault energy on (0001) favors this plane as the slip plane. In magnesium the c/a ratio is slightly less than ideal, so that the spacing of $\{10\bar{1}0\}$ is slightly greater than (0001). However, either because of the zigzag nature of $\{10\bar{1}0\}$, which could affect the Peierls stress, or because of some tendency for dislocation dissociation and the attendant fault formation on (0001), the latter is still the predominant slip plane. In titanium, zirconium, and yttrium, with their lower c/a ratios, the ratio of the d spacing of $\{10\bar{1}0\}$ to (0001) is greater, so that $\{10\bar{1}0\}$ becomes the favored slip plane. The only exception to this generalization is beryllium, which would be expected to show $\{10\bar{1}0\}$ as the favored slip system but instead exhibits basal slip. In all cases the slip direction for the most prominent slip system is $\langle 11\bar{2}0\rangle$, which corresponds to the shortest possible perfect-dislocation Burgers vector, $\frac{1}{3}\langle 11\bar{2}0\rangle$.

In magnesium, titanium, zirconium, yttrium, and to some extent beryllium, which apparently have relatively high fault energies, $\langle 11\bar{2}0\rangle$ remains the slip

[45] H. S. Rosenbaum, in R. E. Reed-Hill, J. P. Hirth and H. C. Rogers (eds.), "Deformation Twinning," Gordon and Breach, New York, 1964, p. 43.

[46] A. T. Churchman, *Trans. Met. Soc. AIME*, **218**: 262 (1960).

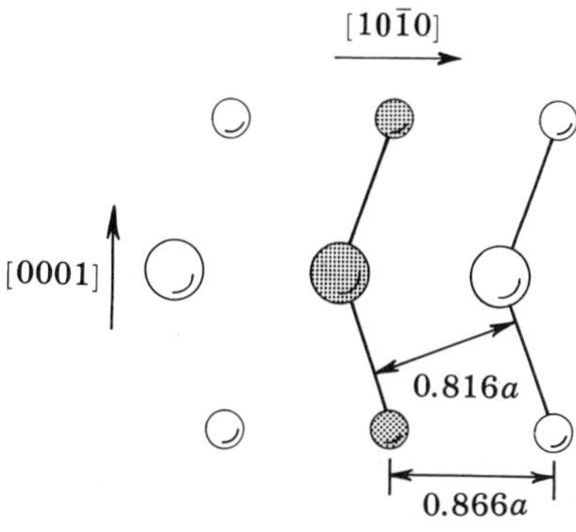

FIGURE 9-5. View of the hcp lattice along a [1$\bar{2}$10] direction, showing the zigzag nature of {1010} planes.

direction in the secondary slip systems that become operative at high temperatures or when the resolved shear stress is low on the basal plane. Cross slip of $\frac{1}{3}\langle 11\bar{2}0\rangle$ dislocations onto planes with higher Peierls stresses can account for all these observed secondary slip systems.

Zinc and cadmium, on the other hand, have relatively low stacking-fault energies, so that it is energetically unfavorable for $\frac{1}{3}\langle 11\bar{2}0\rangle$ dislocations that are dissociated on the basal plane to cross slip, since to do so they must first contract, as discussed in Chap. 11. Thus for these metals the favored secondary slip systems at ambient temperatures involve Burgers vectors not lying in the basal plane.

In Price's extensive and elegant electron-transmission studies[42] of the deformation of cadmium platelets oriented so that the resolved shear stress on (0001) was zero, he observed that the most favored secondary slip system corresponded to the motion of $\frac{1}{3}\langle 11\bar{2}3\rangle$ dislocations on $\{11\bar{2}2\}$ or planes near them in the $\langle 11\bar{2}3\rangle$ zone. This slip system is illustrated in Fig. 9-6. The simple Peierls model would predict rather that $\langle 0001\rangle\{10\bar{1}0\}$ should be favored as the

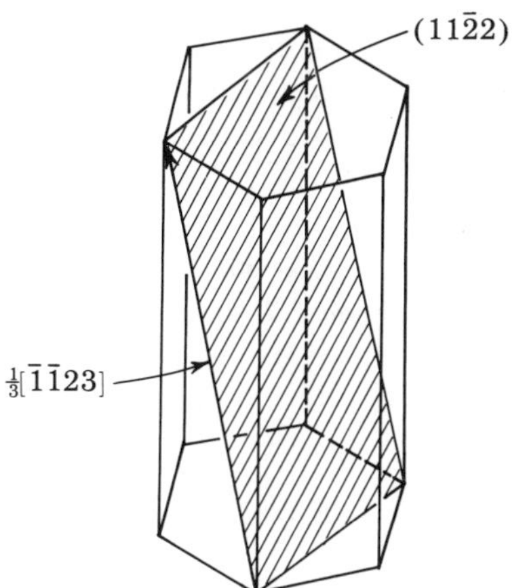

FIGURE 9-6. $\langle 11\bar{2}3\rangle\{1122\}$ slip system in the hcp lattice.

secondary slip system. However, the simple Peierls model is probably inadequate in this case because of the details of the core structure for $\frac{1}{3}\langle 11\bar{2}3\rangle$ versus $\langle 0001\rangle$ dislocations. As can be seen in Figs. 9-5 and 9-6, while neither Burgers vector is a close-packed direction, both can be represented by sums of two vectors connecting nearest neighbors, forming a rumpled, or zigzag, close-packed path. The figures show clearly that the degree of rumpling is less for the $\frac{1}{3}\langle 11\bar{2}3\rangle$ direction. Rosenbaum,[45] basing his work on the zonal-dislocation concept introduced by Kronberg,[47] has considered slip in a hard-ball model of the hcp lattice. As is discussed in greater detail in Chap. 11 in connection with partial dislocations, he finds that $\frac{1}{3}\langle 11\bar{2}3\rangle$ slip indeed should be favored over $\langle 0001\rangle$ slip, based on the large core atom repulsions encountered in the hard-ball model.

Diamond Cubic Slip Systems

Dislocation dissociation on $\{111\}$ has been observed in silicon[48] and germanium,[49] showing that it is a low-energy stacking-fault plane. Also, the d spacing is largest for $\{111\}$ planes. In agreement with the expectation based on these findings, $\{111\}$ was found to be the slip plane in silicon,[50] germanium,[50] and diamond.[51] Electron-transmission studies for diamond[52] and germanium[53] support these findings.

The slip direction is invariably $\langle 110\rangle$, corresponding to the shortest perfect-dislocation Burgers vector $\frac{1}{2}\langle 110\rangle$.

NaCl-type Crystal Slip Systems

The slip direction in NaCl-type crystals is almost always $\langle 110\rangle$, corresponding to the shortest possible perfect-dislocation Burgers vector $\langle 110\rangle$. It is quite likely that the observed $\langle 001\rangle$ slip direction in lead sulfide and lead telluride arises from two sets of $\langle 110\rangle$ dislocations whose shear sums to $\langle 002\rangle$. $\langle 001\rangle$ Burgers vector dislocations have been observed in dislocation networks,[54–56] in agreement with the predictions of Table 9-1, but have never been seen to glide.

The slip planes are mainly $\{110\}$ and $\{100\}$, with the latter becoming more prevalent the higher the ionic polarizability. Pencil glide has been observed for silver halides; a striking example of the pencil glide of a prismatic loop is

[47] M. L. Kronberg, *J. Nucl. Mat.*, **1**: 85 (1959); *Acta Met.*, **9**: 970 (1961).

[48] E. Aerts, P. Delavignette, R. Siems, and S. Amelinckx, *J. Appl. Phys.*, **33**: 3078 (1962).

[49] A. Art, E. Aerts, P. Delavignette, and S. Amelinckx, *Appl. Phys. Letters*, **2**: 40 (1963).

[50] C. J. Gallagher, *Phys. Rev.*, **88**: 721 (1952).

[51] S. Tolansky and M. Omar, *Phil. Mag.*, **44**: 514 (1953).

[52] T. Evans and C. Phaal, *Phil. Mag.*, **7**: 843 (1962).

[53] H. Alexander, *Z. Met.*, **52**: 344 (1961).

[54] S. Amelinckx and W. Dekeyser, *Solid State Phys.*, **8**: 325 (1959).

[55] J. T. Bartlett and J. W. Mitchell, *Phil. Mag.*, **5**: 799 (1960).

[56] J. W. Matthews and K. Isebeck, *Phil. Mag.*, **8**: 469 (1963).

shown in Fig. 1-10, taken from the work of Jones and Mitchell.[57] Slip-plane observations on lithium fluoride were among the first observations made with two other experimental techniques. Figure 1-13, due to Newkirk,[58] reveals $\{110\}$ slip by x-ray reflection, and Fig. 9-7, taken from the extensive work of Gilman and Johnston,[59] illustrates $\{110\}$ slip as revealed by a double-etch-pitting technique.

The rationalization of slip-plane observations was provided by Gilman[60] in a treatment that is the basis for the following discussion. The simple Peierls model, Eq. (8-39), always would predict $\{100\}$ as the glide plane, because of its larger d spacing, and hence can be dismissed immediately as inapplicable. The $E(\beta)$ values predict either $\{100\}$ or $\{110\}$, but in a way inconsistent with the trends in Table 9-5 [$E(\beta)$ is least for $\{100\}$ for LiF]. Also, stacking faults are unlikely in ionic crystals and should not be a factor in determining glide planes. The Peierls model, as refined for anisotropic elasticity,[61,62] does predict the correct primary glide plane for lead sulfide [$A=2c_{44}/(c_{11}-c_{12})=0.78$], and lithium fluoride ($A=1.82$), but gives an incorrect prediction for sodium chloride[63] ($A=0.70$), and other crystals. The numerical computations of Huntington et al.[63] for sodium chloride, using quasi-empirical atomic-force laws, indicate that the "core energy" of a $\{100\}\langle 110\rangle$ edge dislocation exceeds that of a $\{110\}\langle 110\rangle$ edge dislocation by a factor of 2.5. Further, they suggest that ion core repulsions should lead to a larger Peierls stress for the $\{100\}$ dislocation. These atomic calculations give the most consistent rationalization of the slip-plane behavior.

Fontaine[64] suggested that dislocations are dissociated into partials on $\{110\}$ planes, bounding stacking faults with energies of 195 and 161 mJ/m^2, respectively, for NaCl and KCl, for example. Such a dissociation would favor $\{110\}$ glide analogous to effects in other crystal systems. However, atomic calculations[65] for dislocations in KCl, NaCl, and MgO using flexible boundary conditions reveal no dissociations (see Sec. 8-3). Similarly, planar fault calculations[66] do not support the concept of dissociation.

These results can be qualitatively understood with reference to Fig. 9-8. In the expected maximum-energy configuration, or half-glided position, the $\{110\}$

[57] D. A. Jones and J. W. Mitchell, *Phil. Mag.*, **3**: 1 (1958).

[58] J. B. Newkirk, *J. Metals*, **14**: 661 (1962).

[59] J. J. Gilman and W. G. Johnston, in J. C. Fisher et al. (eds.), "Dislocations and Mechanical Properties of Crystals," Wiley, New York, 1957, p. 116.

[60] J. J. Gilman, *Acta Met.*, **7**: 608 (1959). See also C. M. van der Walt and M. J. Sole, *Acta Met.*, **15**: 459 (1967).

[61] J. D. Eshelby, *Phil. Mag.*, **40**: 903 (1949).

[62] A. J. E. Foreman, *Acta Met.*, **3**: 322 (1955).

[63] H. B. Huntington, J. E. Dickey, and R. Thomson, *Phys. Rev.*, **100**: 1117 (1955).

[64] G. Fontaine, *J. Phys. Chem. Solids*, **29**: 209 (1968).

[65] Reviewed by M. P. Puls, in "Dislocation Modeling of Physical Systems," Pergamon, Oxford, 1981, p. 249. See also T. Matsuo and H. Suzuki, *J. Phys. Soc. Jap.*, **46**: 594 (1979).

[66] P. W. Tasker and T. J. Bullough, *Phil. Mag.*, **43A**: 313 (1981).

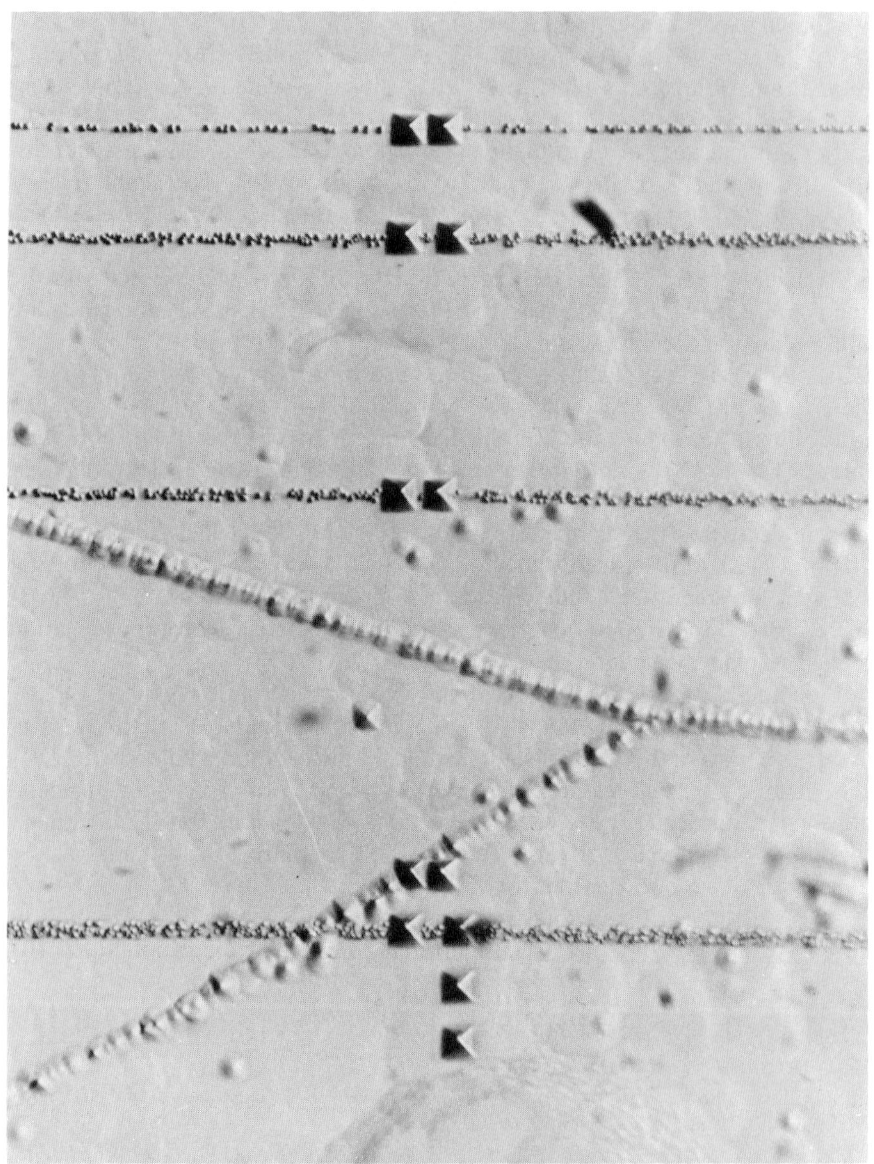

FIGURE 9-7. Double-etch-pitted {001} surface of a lithium fluoride crystal at ~500×. Slip bands are seen to have developed from single half-loops, represented by the large etch pits formed prior to the deformation that produced the small etch pits at dislocation sites. (J. J. Gilman and W. G. Johnston, in J. C. Fisher et al. (eds.), "Dislocations and Mechanical Properties of Crystals," Wiley, New York, 1957, p. 116.)

TABLE 9-5 Slip Systems in Rock-Salt-Type Crystals*

Crystal	Primary Slip Plane[1–3]	Less Prominent Slip Planes Observed under Favorable Shear-Stress Distributions or at High Temperatures
LiF	{110}	{100}[2]
NaF	{110}	
MgO	{110}[4]	
NaCl	{110}	{100},[1,5] {111},[1,6] {*hkl*}[14]
KCl	{110}	{100}[2]
NaBr	{110}	{100},[1] {111}[1]
RbCl	{110}[7]	
AgCl	{110}	{100},[8] {*hkl*}[9]
KBr	{110}	{100}[2]
AgBr	{110}	{100},[8] {*hkl*}[10](*T*)[11]
NaI	{110}	{100},[1] {111}[1]
KI	{110}	{100}[2]
PbS	{100}	{100}⟨001⟩,[1] {110}[12]
PbTe	{100}	{100}⟨001⟩[2, 13]

*Slip direction is ⟨110⟩ unless otherwise noted; {*hkl*} indicates slip on any plane in the ⟨110⟩ zone. The crystals are listed in order of increasing polarizability.[69] For unit cell see Fig. 12-1.

1. M. J. Buerger, *Amer. Min.*, **15:** 21, 35 (1930).
2. J. J. Gilman, *Acta. Met.*, **7:** 608 (1959).
3. J. J. Gilman, *Prog. Ceramic Sci.*, **1:** 146 (1961).
4. O. Mügge, *Neues Jahr.*, 29 (1920).
5. S. Dommerich, *Z. Phys.*, **90:** 189 (1934).
6. A. V. Stepanov and V. P. Bobrikov, *Zh. Tekh. Fiz.*, **26:** 795 (1956).
7. A. Johnsen, *Neues Jahr.*, 147 (1902).
8. A. V. Stepanov, *Phys. Z. Sowj.*, **6:** 312 (1934).
9. J. F. Nye, *Proc. Roy. Soc.*, **198A:** 190 (1949).
10. R. W. Christy, *Acta. Met.*, **2:** 284 (1954).
11. Glide of a circular prismatic loop on {*hkl*} planes was directly observed in optical transmission microscopy by D. A. Jones and J. W. Mitchell [*Phil. Marg.*, **3:** 1 (1958)].
12. J. W. Matthews and K. Isebeck, *Phil. Mag.*, **8:** 469 (1963).
13. W. A. Rachinger, *Acta. Met.*, **4:** 647 (1956).
14. H. Strunk, *Phys. Stat. Solidi*, **28:** 119 (1975).

dislocation core still has appreciable net electrostatic attraction between nearest-neighbor ions on opposite sides of the glide plane. Conversely, in the half-glided position (Fig. 9-8*d*), the {100} dislocation core has zero net nearest-neighbor electrostatic attraction between the opposite sides of the glide planes.[67, 68] For example, the repulsion between ions *A* and *B* is equal and

[67]M. J. Buerger, *Amer. Min.*, **15**: 21, 35 (1930).

[68]J. J. Gilman, *Prog. Ceramic Sci.*, **1**: 146 (1961).

[69]A more extensive listing, including ionic crystals with different structures, is given by M. T. Sprackling (The Plastic Deformation of Simple Ionic Crystals", Academic, New York, 1976).

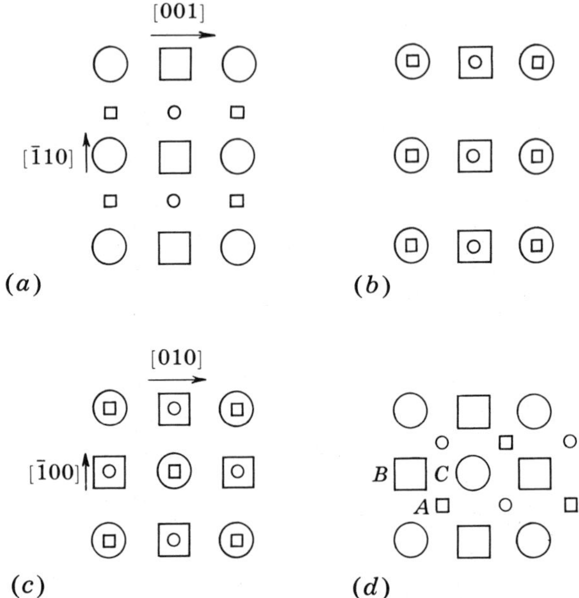

FIGURE 9-8. A $\{110\}\langle 110\rangle$ slip system in a rock-salt crystal, (*a*) unslipped and (*b*) in the half-glided position. (*c*) and (*d*) show the same positions for the $\{100\}\langle 110\rangle$ slip system. The symbols ○ and □ represent cations and anions, respectively, in the plane of the paper, while ○ and □ represent cations and anions, respectively, in the adjacent plane below the plane of the paper.

opposite to the attraction between *A* and *C*. Thus more work must be done against the electrostatic attraction for $\{100\}$ glide than for $\{110\}$ glide.

Therefore $\{110\}$ should be the glide plane for crystals with low ionic polarizability, while the tendency for $\{100\}$ glide should increase with increasing polarizability (which weakens the electrostatic binding), as indicated in Table 9-5. So-called pencil glide, glide on $\{hkl\}$ planes, should be more likely in crystals with high polarizability.

General Slip Systems

The crystal systems discussed above are the ones on which the most work has been performed. Other crystal systems are not discussed here. In general though, from the above discussion one would expect glide planes to correspond to low-energy fault planes, planes of weak atomic bonding in layer structures such as mica, and one among the largest two or three interplanar spacing planes. The slip direction in general should correspond to one of the smallest two or three perfect-dislocation Burgers vectors in the lattice.

In summary, the predictions of dislocation theory are in substantial agreement with observations of slip systems. The only discrepancies relate to the relative Peierls stresses of possible slip planes, and these serve to illustrate the inadequacies of the Peierls model arising from the approximations in its development. Nonetheless, the Peierls model provides a useful rough prediction of the slip systems in many crystals.

9-4. RESOLVED SHEAR STRESS

Tensile Tests

Frequently, particularly when testing single crystals, the shear stress on the various possible slip systems must be resolved. As indicated by Eq. (3-90), only the resolved shear stress on the glide plane and in the slip direction produces a glide force on a dislocation. Thus, among slip systems of a given $\langle h'k'l' \rangle\{hkl\}$ type, the one with the greatest resolved shear stress acting upon it will predominate in the slip process.

Consider the single crystal under simple tension, as shown in Fig. 9-9. With the x_1 axis parallel to the tensile axis, the stress tensor is simply

$$\sigma_{ij} = \begin{vmatrix} \sigma_{11} & 0 & 0 \\ 0 & 0 & 0 \\ 0 & 0 & 0 \end{vmatrix} \tag{9-5}$$

The resolved shear stress on a given slip system can be determined by transforming σ_{ij} to a coordinate system in which x_1' corresponds to the glide direction and x_2' to the glide plane normal. In these coordinates, Eq. (2-36)

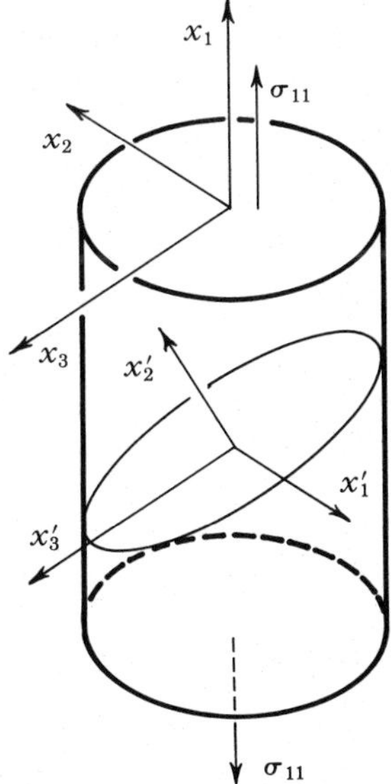

FIGURE 9-9. Slip-system coordinates for a single crystal under simple tension.

yields for σ'_{12}, the shear stress on the glide system, the result

$$\sigma'_{12} = T_{11}T_{21}\sigma_{11} = \cos\alpha\cos\beta\,\sigma_{11} = m\sigma_{11} \tag{9-6}$$

where α is the angle between the tensile axis x_1 and the glide direction x'_1 and β is the angle between x_1 and the glide plane normal x'_2. Equation (9-6) was first presented by Schmid,[70] and the factor m is often called the *Schmid factor*.

For a given tensile axis, one can compute the resolved shear stresses directly from Eq. (9-6). It is often easier, particularly for noncubic crystals, to read the stresses from a stereographic plot.[71] The projected point corresponding to the tensile axis can be determined by manipulation of the results of a Laue pattern, and the resolved stress can be read off directly without bothering to index the tensile axis.

Figure 9-10 presents a stereographic projection normal to the general glide plane $\{hkl\}$ and with the general glide direction $\langle h'k'l'\rangle$ at the north pole.[72] For a tensile axis at any point on the projection, the resolved shear stress on $\langle h'k'l'\rangle(hkl)$ can be obtained by interpolation between the isostress contours. Suppose one is interested in the Schmid factor for $[1\bar{1}0](111)$ in an fcc crystal with a tensile axis [511]. One would overlay a standard (111) projection upon Fig. 9-10, with $[1\bar{1}0]$ vertical upward, and read off the stress for [511]. As shown in Fig. 9-10, the Schmid factor for this simple case would be 0.42, and Eq. (9-6) indeed shows that this is the correct answer. For the same tensile axis, to determine the resolved shear stress on $[10\bar{1}](1\bar{1}1)$, one can read the result directly from a $(1\bar{1}1)$ standard projection, or one can transform the coordinates of this second system to the (111) projection used in the first instance and read the result there for the transformed tensile axis $[5\bar{1}1]$.

Equations (9-5) and (9-6) and the stereographic method of Fig. 9-10 also apply to compression tests, where the sign of σ_{11} is merely reversed, and to simple bending. For simple bending,[73] $\sigma_{11} = xE/R$, where x is the distance normal to the neutral stress surface, E is Young's modulus, and R is the radius of curvature of the neutral surface.

We emphasize that this method can be applied directly to any of the possible crystal structures. The only requirement is a standard projection along either the slip-plane normal or the slip direction. The graphical method is particularly useful for low-symmetry crystal structures, for which numerical computations of the resolved shear stresses are tedious. It is also quite useful for slip systems other than the primary slip system, for one can see at a glance which system will have the second highest resolved shear stress and can also

[70] E. Schmid, *Z. Elektrochem.*, **37**: 447 (1931); E. Schmid and W. Boas, "Plasticity of Crystals," Hughes, London, 1950.

[71] C. S. Hartley and J. P. Hirth, *Trans. Met. Soc. AIME*, **233**: 1415 (1965).

[72] Since the glide direction is normal to the glide-plane pole, the same results can also be obtained from a projection normal to $\langle h'k'l'\rangle$ with $\{hkl\}$ at the north pole.

[73] S. Timoshenko and J. N. Goodier, "Theory of Elasticity," McGraw-Hill, New York, 1951, p. 250.

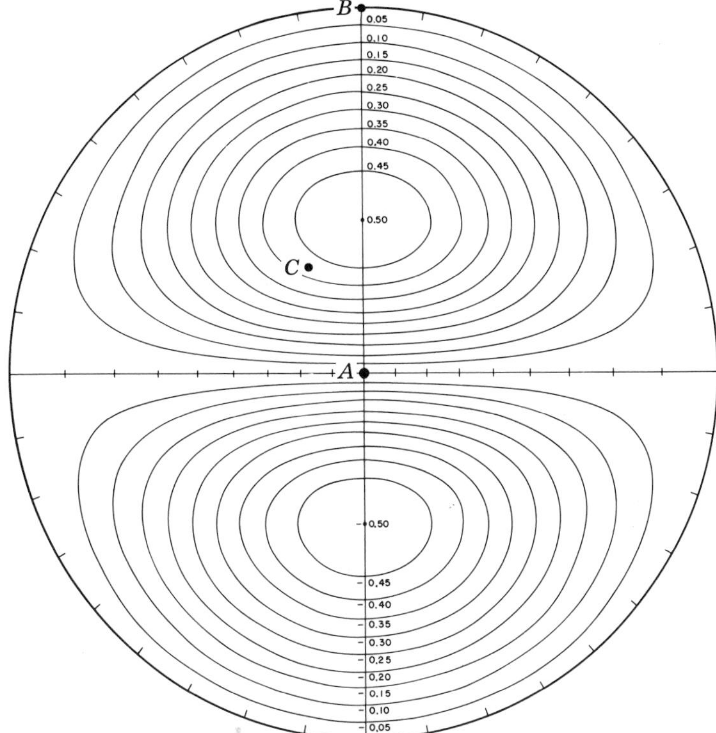

FIGURE 9-10. Plot of constant m contours (isostress contours) in a stereographic projection normal to the glide-plane pole. In general, the point A is the general glide-plane pole (hkl) and the point B is the general glide direction $[h'k'l']$. The Schmid factor on $(hkl)[h'k'l']$ is read off at the projection of the tensile axis. As a specific example, consider an fcc structure with A equal to (111) and B to $[1\bar{1}0]$. Then for the tensile axis C equal to [511], the Schmid factor is 0.42.

see graphically how the shear stresses on the various slip systems change as the tensile axis rotates.

Exercise 9-1. With the tensile axis parallel to [511] for an fcc single crystal, find the resolved shear stresses on all 12 possible glide systems.[74, 71]

Torsion and Simple Shear

Consider the case of torsion in a cylinder, as illustrated in Fig. 9-11. The choice of the x_1 axis in the cylinder is arbitrary, so it is chosen perpendicular to the specimen axis x_3 and to the glide-plane normal x_3'. The glide direction is x_1'. This arrangement corresponds to that of Fig. 2-3, with $\phi = 0$. Here θ is the

[74] J. Diehl, M. Krause, W. Offenhäuser, and W. Staubwasser, *Z. Met.*, **45**: 489 (1954); J. Diehl, *Z. Met.*, **47**: 331 (1956).

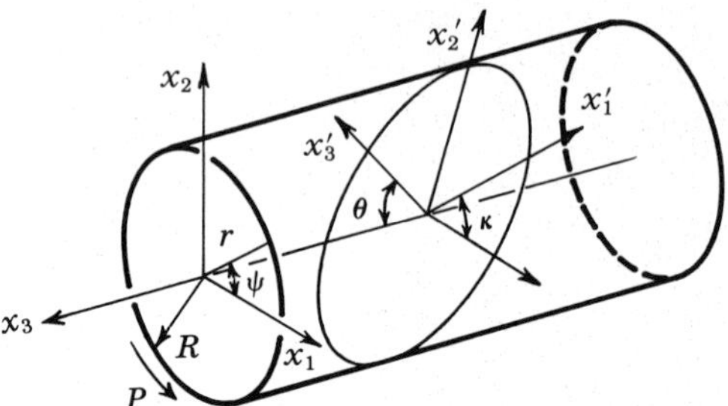

FIGURE 9-11. Coordinates for a single-crystal cylinder under torsion.

angle between x_3 and x_3' and κ that between x_1 and x_1'. The transformation matrix is thus given by Eq. (2-22). With a torque PR applied to the cylinder, the stress tensor is[75]

$$\sigma_{ij} = \begin{vmatrix} 0 & 0 & \sigma_{13} \\ 0 & 0 & \sigma_{23} \\ \sigma_{13} & \sigma_{23} & 0 \end{vmatrix} \tag{9-7}$$

where at a point (r, ψ) (Fig. 9-11)

$$\begin{aligned} \sigma_{13} &= -\frac{2Pr\sin\psi}{\pi R^3} = -\sigma_0 \sin\psi \\ \sigma_{23} &= \frac{2Pr\cos\psi}{\pi R^3} = \sigma_0 \cos\psi \end{aligned} \tag{9-8}$$

Equations (2-22) and (2-36) then give for the resolved shear stress

$$\begin{aligned} \sigma_{13}' &= \sigma_0(-\cos\kappa\cos\theta\sin\psi + \sin\kappa\cos2\theta\cos\psi) \\ &= \sigma_0(m_1\sin\psi + m_2\cos\psi) = \sigma_0 m(\psi) \end{aligned} \tag{9-9}$$

Thus in this case the resolved shear stress varies around the periphery of the cylinder. The factors m_1 and m_2 are plotted in Figs. 9-12 and 9-13. Again, choosing a standard projection normal to the slip plane and with the slip direction at the north pole, the resolved-shear-stress factors m_1 and m_2 can be read off the stereographic projection at the projection of the cylinder axis.

The maximum absolute value of $m(\psi)$ occurs at

$$\psi = \tan^{-1}[-\cot\kappa(\cos\theta/\cos2\theta)]$$

[75] S. Timoshenko and J. N. Goodier, "Theory of Elasticity," McGraw-Hill, New York, 1951, p. 249.

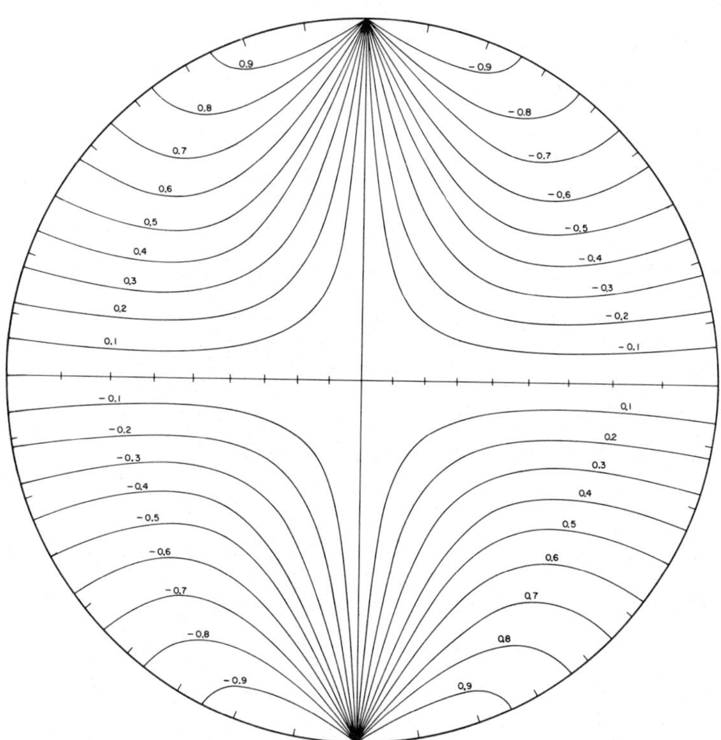

FIGURE 9-12. Stereographic plot of the resolved-shear-stress factor m_1, Eqs. (9-9) and (9-12). The plot is normal to the slip-plane pole, with the slip direction at the north pole.

where $m(\psi)$ has an extremum value of

$$m(\psi)=\pm(\sin^2\kappa\cos^2 2\theta+\cos^2\kappa\cos^2\theta)^{1/2} \tag{9-10}$$

which is fixed by the specimen orientation. This function has an absolute maximum and minimum of ±1 when $\theta=0$ (slip plane normal to the specimen axis). When $\theta=0$, $m(\psi)$ is independent of the choice of position of the x_1 axis; hence x_1 can be chosen perpendicular to x_1', so that $\kappa=\pi/2$ and $m(\psi)=\cos\psi$ in this special case.

In simple shear, the x_3 axis is normal to the direction of application of the shear force, as shown in Fig. 9-14. The x_1 axis is again chosen perpendicular to both x_3 and the x_3' glide-plane normal. Thus the coordinate scheme is the same as in the torsion case. In the shear case the stress tensor is again given by Eq. (9-7), but the stress components are

$$\sigma_{13}=\sigma\cos\alpha \qquad \sigma_{23}=\sigma\sin\alpha \tag{9-11}$$

where σ is the maximum shear stress (Fig. 9-14). The application of Eqs. (2-22)

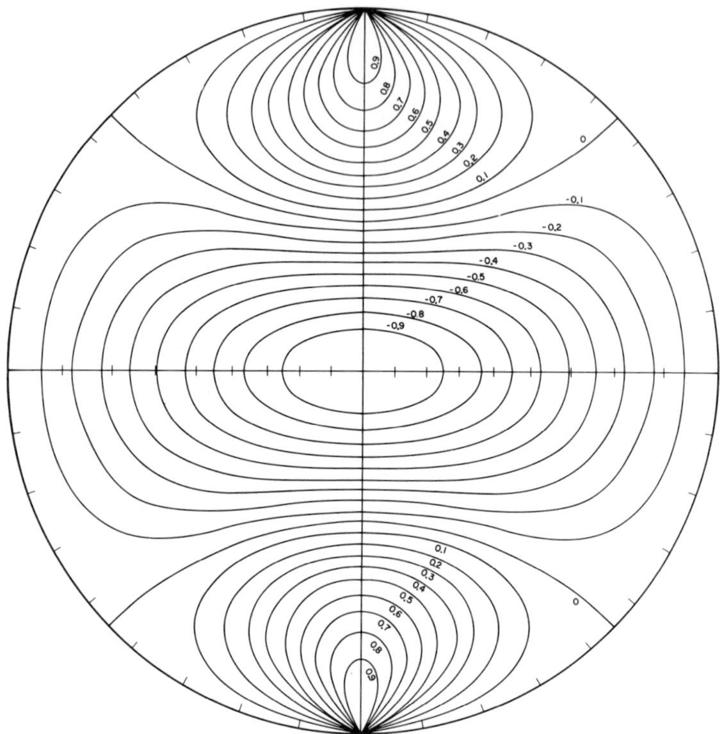

FIGURE 9-13. Stereographic plot of the resolved-shear-stress factor m_2, Eqs. (9-9) and (9-12). The plot is normal to the slip-plane pole, with the slip direction at the north pole.

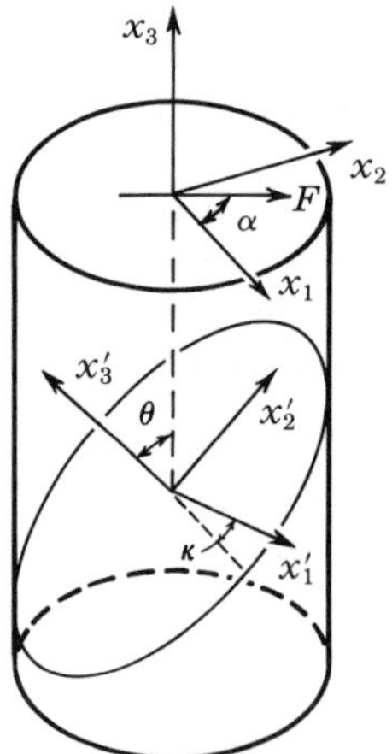

FIGURE 9-14. Coordinates for a single crystal being deformed in simple shear. The direction of the shear force F in the x_1x_2 plane is shown.

and (2-36) yields for this case

$$\sigma'_{13} = \sigma(-m_1 \cos\alpha + m_2 \sin\alpha) = \sigma m(\alpha) \tag{9-12}$$

Once again the resolved shear stresses can be determined from the plots in Figs. 9-12 and 9-13.

Thus, for both torsion and shear, which give somewhat more complex stress distributions than tension, the resolved shear stresses can be determined graphically for any crystal; only a standard projection normal to the glide plane or parallel to the glide direction is required.

Analogous stereographic plots have been presented for resolved normal stresses, of interest in dislocation climb problems.[76]

9-5. INDEPENDENT SLIP SYSTEMS

Slip Systems for General Deformation

In 1928 von Mises[77] demonstrated that five independent slip systems are required for a crystal to undergo general plastic deformation by slip. The development of von Mises' theory is of interest because of its application in studies of deformation near grain boundaries and of the accompanying stress concentrations there, and because of its application in predicting the ductility of polycrystalline specimens of a given crystal structure. When a polycrystal deforms, a grain within it is constrained to deform generally. If five independent slip systems are not available, pore formation, grain-boundary sliding, twinning, phase transformation, or fracture of the crystal will occur.

In the following discussion, von Mises' results are applied to several important crystal systems. A geometric treatment, essentially following the method of Groves and Kelly,[78] is adopted. Kocks[79] has developed a matrix method by which the same results can be found, and which relates directly to the yield surface for the crystal in question. However, in treating slip systems, we feel that the geometric method is easier to envision.

In the limit of infinitesimal homogeneous plastic strains, the state of plastic deformation can be described by a plastic-strain tensor, equivalent to Eq. (2-3),

$$\epsilon_{ij} = \frac{1}{2}\left(\frac{\partial u_i}{\partial x_j} + \frac{\partial u_j}{\partial x_i}\right) \tag{9-13}$$

[76] A. H. Clauer and J. P. Hirth, *Trans. Met. Soc., AIME*, **245**: 1075 (1969).

[77] R. von Mises, *Z. ang. Math. Mech.*, **8**: 161 (1928); see also G. I. Taylor, *J. Inst. Met.*, **62**: 307 (1938).

[78] G. W. Groves and A. Kelly, *Phil. Mag.*, **8**: 877 (1963).

[79] U. F. Kocks, *Phil. Mag.*, **10**: 187 (1964).

where the u_i are the plastic displacements. For larger strains second-order terms become important, and the mathematical description of the state of strain becomes cumbersome. Hence we restrict discussion to the limiting case of small homogeneous strains. However, a large strain can be thought of as being produced by successive increments of small strain, each of which can be described by linear theory. A finite state of plastic strain can be attained by summing such elements of incremental strain in many different possible sequences.

As shown in Chap. 3, the glide of dislocations that belong to a specific slip system, in the limit of infinitesimal strains, produces shear strains only in the glide direction on the glide plane. Let us represent the glide plane by a unit normal **n** parallel to the coordinate x_2', and the glide direction by a unit vector $\boldsymbol{\beta}$ parallel to x_1'. The plastic-strain tensor produced by a simple shear $\tan\alpha \simeq \alpha$ on the glide system $\boldsymbol{\beta}\mathbf{n}$ is

$$\epsilon_{ij}' = \frac{1}{2}\begin{vmatrix} 0 & \alpha & 0 \\ \alpha & 0 & 0 \\ 0 & 0 & 0 \end{vmatrix} \tag{9-14}$$

If the coordinate axes are transformed to a new set x_1, x_2, and x_3, the transformation matrix is

$$T_{ij} = \begin{vmatrix} \beta_1 & n_1 & p_1 \\ \beta_2 & n_2 & p_2 \\ \beta_3 & n_3 & p_3 \end{vmatrix} \tag{9-15}$$

where β_i and n_i are the components of $\boldsymbol{\beta}$ and **n**, respectively, in the new coordinates, and $\mathbf{p} = \boldsymbol{\beta} \times \mathbf{n}$. In the new coordinate system, ϵ_{ij}' transforms, according to Eq. (2-34), to the strain tensor

$$\epsilon_{ij} = \begin{vmatrix} \epsilon_{11} & \epsilon_{12} & \epsilon_{13} \\ \epsilon_{12} & \epsilon_{22} & \epsilon_{23} \\ \epsilon_{13} & \epsilon_{23} & \epsilon_{33} \end{vmatrix} = \frac{\alpha}{2}\begin{vmatrix} 2\beta_1 n_1 & \beta_1 n_2 + \beta_2 n_1 & \beta_1 n_3 + \beta_3 n_1 \\ \beta_1 n_2 + \beta_2 n_1 & 2\beta_2 n_2 & \beta_2 n_3 + \beta_3 n_2 \\ \beta_1 n_3 + \beta_3 n_1 & \beta_2 n_3 + \beta_3 n_2 & 2\beta_3 n_3 \end{vmatrix} \tag{9-16}$$

Since ϵ_{ij} is symmetrical in **n** and $\boldsymbol{\beta}$, the strain produced by unit slip in the direction $\boldsymbol{\beta}$ on the plane **n** is *identical* to the strain produced by unit slip in the direction **n** on the plane $\boldsymbol{\beta}$.

A general state of infinitesimal plastic deformation involves slip on k systems with strains[80] $\epsilon_{ij}^{(k)}$. The sum over the k systems gives the total strain

[80] The letter k is used as a superscript instead of a subscript to avoid confusion with the Einstein operator ϵ_{ijk}.

with the six strain components $\epsilon_{ij}^{(T)}$

$$\epsilon_{ij}^{(T)} = \sum_k \epsilon_{ij}^{(k)} \tag{9-17}$$

Only five of the strain components $\epsilon_{ij}^{(T)}$ are independent. Obviously, pure glide processes cannot change the crystal density to first order. In terms of macroscopic slip, this condition is referred to as the condition of constancy of volume and reads

$$\epsilon_{11}^{(T)} + \epsilon_{22}^{(T)} + \epsilon_{33}^{(T)} = 0 \tag{9-18}$$

Therefore there are only five independent total strain components to be determined in Eq. (9-17), and five independent slip systems producing strains α_k, with $k = 1$, 2, 3, 4, and 5, suffice to determine these. In other words, Eq. (9-17) represents five independent equations which can be solved simultaneously for the five unknowns α_k. In the transformed coordinates, the five unknowns α_k become the five unknowns $\epsilon_{ij}^{(k)}$. If the five slip systems are not independent, i.e., if any one can be expressed as a linear combination of the others, then Eq. (9-17) has no solution for an arbitrary given $\epsilon_{ij}^{(T)}$ satisfying Eq. (9-18).

Thus, as stated by von Mises, five independent slip systems are required for general slip, and no more than five independent slip systems can exist in a given crystal. We emphasize that this result is based on the approximation of linear, homogeneous plastic strain, and applies strictly only to an infinitesimal volume element. However, in the limit of volumes approaching atomic dimensions, the plastic strain is not homogeneous because of the localization of plastic strain to the core regions of dislocations, and on a somewhat larger scale, it is not homogeneous because of the motion of a large number of dislocations on a given slip plane forming a slip band. Hence, for the von Mises–Taylor model to be applicable, one must consider volume elements with dimensions large compared to slip-band spacings, so that the plastic shear can be considered to be homogeneous, but small enough that linear plastic strain is adequate. Thus a given grain in a polycrystal could deform on more than five slip systems even though only five systems operated in each little volume element within the grain. In general, one does not expect plastic strain to be homogeneously distributed.

Let us now determine the number of independent slip systems in several crystal systems to test whether they can undergo a general deformation by slip.

FCC Metals

The glide system in fcc crystals are conveniently represented by a tetrahedron,[81] as shown in Fig. 9-15. Resolved shear stresses are readily plotted on the

[81]Following N. Thompson, *Proc. Phys. Soc.*, **66B**: 481 (1953).

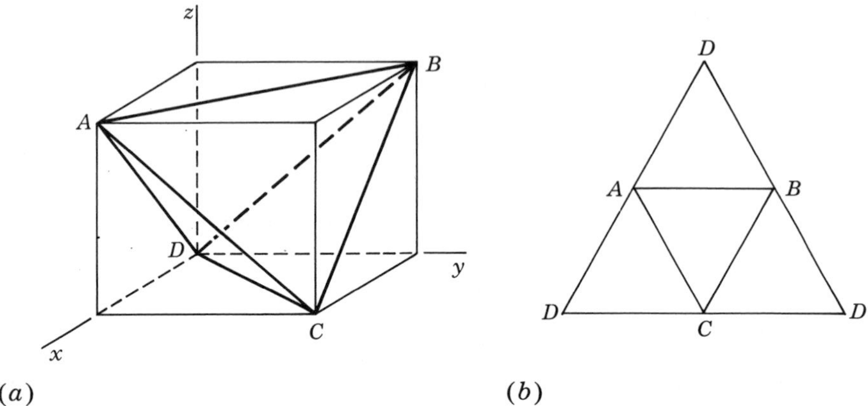

FIGURE 9-15. (*a*) A one-eighth unit cell in the fcc lattice with an enclosed tetrahedron, the faces of which are the possible glide planes and the edges of which are the possible glide directions. (*b*) A Thompson tetrahedron opened up at corner *D*.

opened-up tetrahedron, partial dislocations can be represented on it (as discussed in Chap. 10), and dislocation interactions are most easily depicted on it. Thus a geometric model related to the tetrahedron is employed here to represent the slip systems.

As can be seen in Fig. 9-15, there are four independent glide planes, the faces of the tetrahedron, each containing three slip directions, the tetrahedron edges, yielding a total of 12 physically distinct slip systems. The number of ways of choosing five members of a set of 12 is

$$_5C_{12}=\frac{12!}{7!5!}=792 \tag{9-19}$$

Of these, the combinations that do not contain five *independent* members must be eliminated. There are two types of combinations that involve interdependent slip systems: a set of slip directions that sum to zero in a common glide plane and sets of slip systems that correspond to pure rotations of the crystal.

As an example of the first combination, consider the three slip systems $[1\bar{1}0](111)$, $[\bar{1}01](111)$, and $[01\bar{1}](111)$. These are referred to coordinate axes coincident with the unit cell edges, so that $\beta_i=\pm\sqrt{2}/2$ and $n_i=\sqrt{3}/3$. Let the shear on one of these systems be α, and suppose that there are equal shears α on the other two systems. Then, according to Eq. (9-17), one obtains strains which sum to zero,

$$\sum_{k=1}^{3}\epsilon_{ij}^{(k)}=\frac{\alpha}{2\sqrt{6}}\begin{vmatrix}2 & 0 & 1\\ 0 & -2 & -1\\ 1 & -1 & 0\end{vmatrix}+\frac{\alpha}{2\sqrt{6}}\begin{vmatrix}-2 & -1 & 0\\ -1 & 0 & 1\\ 0 & 1 & 2\end{vmatrix}+\frac{\alpha}{2\sqrt{6}}\begin{vmatrix}0 & 1 & -1\\ 1 & 2 & 0\\ -1 & 0 & -2\end{vmatrix}$$
$$=0 \tag{9-20}$$

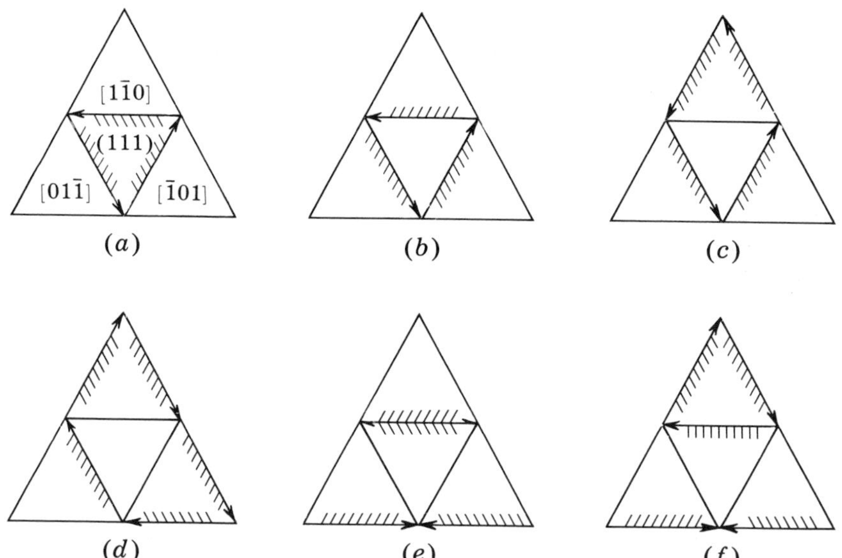

FIGURE 9-16. Graphical representation of interdependent sets of slip systems in fcc crystals. Hatched directions are the slip directions, and hatch marks lie on the related slip planes.

Thus the shear α on any one system can be expressed in terms of shears on the other two systems, and the slip systems are therefore interdependent. This type of reaction is represented graphically in Fig. 9-16. There are four planes on which such a set of three slip systems can be arranged, and for each set there are ${}_2C_9 = 36$ ways of choosing two other slip systems from the remaining nine. Thus 144 possible slip sets are interdependent because they involve combination (*a*) in Fig. 9-16.

Combination (*b*) in Fig. 9-16 represents a rigid rotation about $\langle 111 \rangle$ and, like (*a*), yields zero net plastic strain. Combination (*b*) has 144 permutations, but 12 of these correspond to combination (*a*), so that a net of 132 more sets are found to be interdependent.

The other interdependent combinations which yield zero net plastic strain are linear combinations of (*a*) and (*b*). Combination (*c*) in Fig. 9-16 contains four slip systems which are interdependent, because the array is a superposition of (*a*) and (*b*). (*c*) can occur in 12 ways, and the remaining slip system in the set of five can be chosen in seven of the eight remaining systems without forming a combination which contains (*a*) or (*b*). Thus combination (*c*) eliminates 84 sets as being interdependent. Combination (*d*) is a superposition of (*a*) and (*c*) and eliminates 12 more sets. Combination (*e*), representing a rigid rotation about $\langle 100 \rangle$, eliminates 24 sets, and its superposition with (*a*), combination (*f*), eliminates 12 more sets. Hence, of the 792 possible combinations, there are 384 different ways of choosing five independent slip systems in fcc structures. Any of the combinations illustrated in Fig. 9-16 are excluded.

With the large number of possible choices of slip systems, it is reasonable that polycrystals of fcc metals are highly ductile, as is generally observed.

BCC Metals

As shown above, slip on the plane **n** in the direction $\boldsymbol{\beta}$ produces a shear equivalent to slip on the plane $\boldsymbol{\beta}$ in the direction **n**. Thus, for $\langle 111\rangle\{110\}$ slip in bcc, there are 384 different ways of choosing five independent slip sets, equivalent to the $\langle 110\rangle\{111\}$ slip-set choices in fcc metals. If slip occurs on $\langle 111\rangle\{211\}$, there are four $\langle 111\rangle$ slip directions, each of which is a zone axis for three $\{211\}$ planes, so that again there are 12 physically distinct systems and 792 total possible choices of five. The only excludable combinations are the 144 corresponding to slip of a given slip direction on all three possible planes, so that 648 possibilities remain. If both $\{110\}$ slip and $\{211\}$ slip occur, there are 21,252 possible sets of five, some of which must be excluded, as above.

Most bcc crystals are known to be brittle at low temperatures. However, because of the large number of available systems, a lack of available systems alone cannot limit ductility in bcc metals.

HCP Metals

In order to use Eqs. (9-14) to (9-17) in their present form, the β_i and n_i components must be expressed in terms of orthogonal coordinates of equal measure. A possible set is illustrated in Fig. 9-17 (these coordinates are those used by Frank and Nicholas[82] in the discussion of dislocation reactions in hcp crystals). As an example, the Burgers vector with Miller-Bravais indices $\frac{1}{3}[\bar{1}2\bar{1}0]$ has components $a/2$, $-(\sqrt{3}\,a/2)$, and 0 in the orthogonal coordinate system, while the unit cell vector normal to the basal plane has components 0, 0, and Ka, where $K=c/a$ for the particular metal in question.

For $\langle 11\bar{2}0\rangle\{0001\}$ slip in hcp metals, there are three physically distinct slip systems, only two of which are independent. To illustrate this point, let us consider the three directions $[\bar{2}110]$, $[11\bar{2}0]$, and $[1\bar{2}10]$ in the (0001) plane. After transformation of these vectors to the orthogonal coordinates, Eq. (9-17) yields for the total strain

$$\sum_{1}^{3}\epsilon_{ij}^{(k)}=\frac{\alpha}{4}\begin{vmatrix}0 & 0 & -1\\ 0 & 0 & \sqrt{3}\\ -1 & \sqrt{3} & 0\end{vmatrix}+\frac{\alpha}{4}\begin{vmatrix}0 & 0 & 2\\ 0 & 0 & 0\\ 2 & 0 & 0\end{vmatrix}+\frac{\alpha}{4}\begin{vmatrix}0 & 0 & -1\\ 0 & 0 & -\sqrt{3}\\ -1 & -\sqrt{3} & 0\end{vmatrix}=0 \tag{9-21}$$

Thus basal glide in hcp crystals is insufficient to allow deformation in a general way by slip.

[82] F. C. Frank and J. F. Nicholas, *Phil. Mag.*, **44**: 1213 (1953).

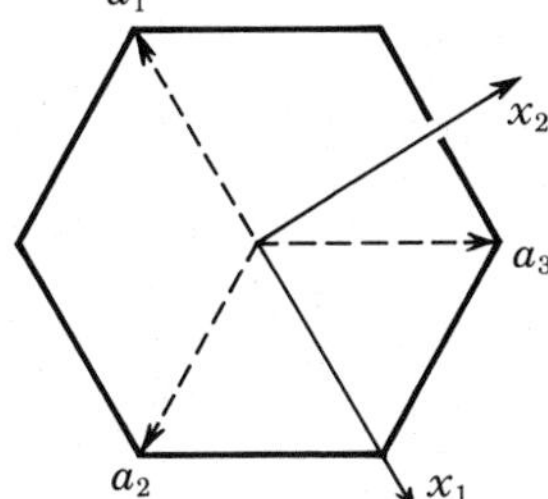

FIGURE 9-17. An orthogonal coordinate system for the hcp lattice.

For $\langle 11\bar{2}0\rangle\{10\bar{1}0\}$ slip there are again three slip systems, two of which are independent. When slip of $\langle 11\bar{2}0\rangle$ is possible on both $\{0001\}$ and $\{10\bar{1}0\}$, there are four independent slip systems, which can be chosen in nine different ways. If slip is possible on any plane for which $\langle 11\bar{2}0\rangle$ is a zone axis, there are still only four independent slip systems, which can be chosen in an infinite number of ways. With this infinite number of systems, it is not possible to produce a normal component of strain in the [0001] direction. In general, screw dislocations with a given slip direction (Burgers vector) cross slipping onto any number of planes can produce at most two independent slip systems. Thus the five independent slip systems that produce a general deformation must include at least three independent slip directions.

Exercise 9-2. By a consideration of the displacements produced by $\frac{1}{3}\langle 11\bar{2}0\rangle$ dislocations, rationalize the result that $\langle 11\bar{2}0\rangle$ slip cannot produce a strain in the [0001] direction.

There are six systems of the type $\langle 11\bar{2}3\rangle\{11\bar{2}2\}$, of which five are independent and can be chosen in six ways. A general deformation by slip is possible with this type of system, either alone or in combination with the others cited above.

The inability of $\langle 11\bar{2}0\rangle$ slip to provide a general glide deformation is doubtless associated with the prominence of twinning in polycrystalline hcp metals. There are a number of twinning systems in hcp structures, including $[\bar{1}011](10\bar{1}2)$, $[\bar{1}012](10\bar{1}1)$, $[\bar{1}\bar{1}23](11\bar{2}2)$, and others.[83, 84] As discussed in Chap. 23, twinning is associated with the motion of partial dislocations on the twinning plane. Thus twinning can contribute to a general deformation and reduce the requisite number of slip systems for such deformation. However, the analogy between slip and twinning is not complete, because twinning often can occur only with one sign of the displacement. An analysis of how many twinning systems of a given type are independent or of how many independent twinning systems are required for a general deformation has yet to be completed. However, twinning clearly can relax the requirement for slip systems.

[83] D. G. Westlake, in R. E. Reed-Hill, J. P. Hirth, and H. C. Rogers (eds.), "Deformation Twinning," Gordon and Breach, New York, 1964, p. 29.

[84] H. S. Rosenbaum, ibid., p. 43.

Since perfect dislocations with Burgers vectors not in the basal plane could have a high Peierls stress, it is feasible that twinning dislocations either can move before such perfect dislocations begin to glide or can move under only small stress concentrations at the stresses required to move the perfect dislocations.

Also, the inability of $\langle 11\bar{2}0\rangle$ slip to provide a general glide deformation is most likely associated both with the tendency toward brittle fracture of polycrystalline hcp metals, by the same reasoning as above, and with the observation that voids are sometimes formed in the plastic deformation of polycrystals.

Exercise 9-3. Use the coordinate system of Fig. 9-17 to prove that there are only two independent slip systems of the type $\langle 11\bar{2}0\rangle\{10\bar{1}0\}$.

Crystals with NaCl Structure

As a final example, consider the NaCl lattice. The main slip system for most crystals of this structure is of the type $\langle 110\rangle\{110\}$, illustrated in Fig. 9-18. There are six physically distinct slip systems: (1) $[101](\bar{1}01)$, $[01\bar{1}](011)$, and $[\bar{1}\bar{1}0](\bar{1}10)$ and (2) $[\bar{1}01](101)$, $[011](01\bar{1})$, and $[\bar{1}10](\bar{1}\bar{1}0)$. However, sets (1) and (2) each produce the same strain, because the two sets simply interchange slip plane and slip direction; therefore the two sets are interdependent. Applying Eq. (9-17) to set (1), one finds that the resultant strains sum to zero:

$$\sum_{1}^{3}\epsilon_{ij}^{(k)}=\frac{\alpha}{2}\begin{vmatrix}-2&0&0\\0&0&0\\0&0&2\end{vmatrix}+\frac{\alpha}{2}\begin{vmatrix}0&0&0\\0&2&0\\0&0&-2\end{vmatrix}+\frac{\alpha}{2}\begin{vmatrix}2&0&0\\0&-2&0\\0&0&0\end{vmatrix}=0 \qquad (9\text{-}22)$$

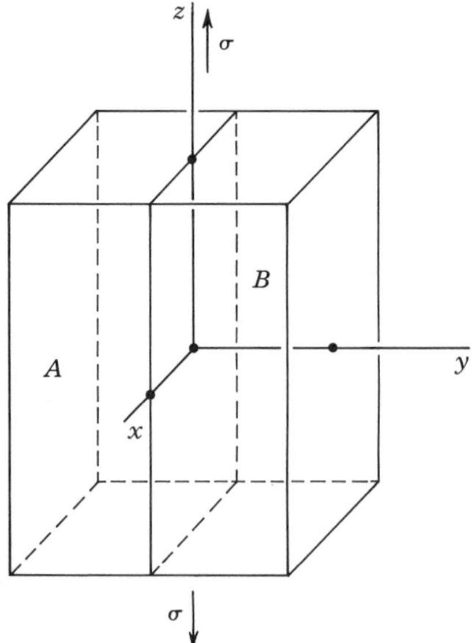

FIGURE 9-18. A bicrystal with the grain boundary normal to the y axis, and separating crystals A and B.

The simultaneous activity of the three slip systems produces only a rigid rotation about $[\bar{1}11]$. Thus only two of the six slip systems are independent, so that polycrystals of this structure cannot undergo a general deformation by $\langle 110\rangle\{110\}$ slip. In particular, because the off-diagonal terms in the strain tensors are all zero when referred to the unit cell axes, any deformation which would change the angle between the crystal axes cannot occur. For example, a crystal cannot be sheared on $\{001\}$, twisted about $\langle 001\rangle$, or extended along axes other than $\langle 001\rangle$. The two independent slip systems of the type $\langle 110\rangle\{110\}$ can be chosen in 12 different ways.

As noted in Table 9-5, $\langle 110\rangle\{001\}$ slip also occurs in crystals with the rock-salt structure. It is left as an exercise to show that there are six physically distinct slip systems of the type $\langle 110\rangle\{001\}$, from which three independent slip systems can be chosen in 16 different ways. Together, $\langle 110\rangle\{001\}$ and $\langle 110\rangle\{110\}$ slip provide five independent slip systems, which can be chosen in 192 different ways.

The well-known tendency[85–87] for brittle fracture in polycrystals of the rock-salt type, which tend to slip only on $\langle 110\rangle\{110\}$, is probably associated with the absence of five independent slip systems in such crystals. Also, the greater ductility of polycrystals such as silver chloride and sodium chloride,[87] which tend to glide on other systems in addition to $\langle 110\rangle\{110\}$, as noted in Table 9-5, is consistent with the greater number of independent slip systems available in such crystals. Similarly, the increased ductility of lithium fluoride at elevated temperatures,[87] where both $\{100\}$ and $\{110\}$ slip are observed, is consistent with the greater number of independent slip systems available under such conditions. In the latter case, grain-boundary sliding conceivably could play some role in relaxing the slip-system requirement.

9-6. SLIP SYSTEMS IN CRYSTALS WITH FREE SURFACES

Relaxation of Constraint

In crystals with free surfaces, the conditions discussed above are relaxed, and fewer than five independent slip systems are required in general for a specified deformation. The limiting case corresponds, for example, to that of a single crystal deformed by tension. Then only the normal strain along the tensile axis, ϵ_{zz}, is specified, and a single slip system suffices to produce the strain. Of course, other slip systems can appear if the plastic strain is inhomogeneous. One must bear in mind the assumption of linear plastic strain. It is well known[88] that single crystals sometimes shear to failure by slip entirely on one system.

[85] T. L. Johnston, in "Mechanical Behavior of Crystalline Solids," NBS monograph 59, 1963, p. 63.

[86] W. D. Kingery and R. L. Coble, ibid., p. 103.

[87] T. L. Johnston, R. J. Stokes, and C. H. Li, in "Strengthening Mechanisms in Solids," American Society of Metals, Cleveland, Ohio, 1960, p. 341.

[88] E. Schmid and W. Boas, "Plasticity of Crystals," Hughes, London, 1950.

The next step in complexity is represented by the deformation of a bicrystal (Fig. 9-18); this has been discussed by Chalmers and others.[89–92] With the planar boundary normal to the y axis, the compatibility conditions that must be satisfied at the boundary between crystals A and B are

$$\epsilon_{xx}^{A}=\epsilon_{xx}^{B} \qquad \epsilon_{zz}^{A}=\epsilon_{zz}^{B} \qquad \epsilon_{xz}^{A}=\epsilon_{xz}^{B} \tag{9-23}$$

These are required if the boundary is not to slide or open up during deformation. For specially symmetric bicrystals[89,90] one slip system in each crystal suffices to fulfill Eq. (9-23) and the external constraints. These are an infinitesimal fraction of the possible orientations, however. In general,[91] if the external constraints require that a single slip system operate in A, three more systems must then operate in B to fulfill Eq. (9-23). Notice that the existence of three independent slip systems in B does not guarantee that Eq. (9-23) is fulfilled. For example, if x were parallel to [0001] for an hcp B crystal, then $\langle 11\bar{2}0\rangle$ slip on any number of slip planes could not produce a strain component ϵ_{xx} in B. If the external constraints lead to the operation of two systems in A, two more are required in B.

In agreement with the above ideas, there have been a number of observations of metal bicrystal deformation wherein double slip occurs near the grain boundary.[93] The effects are particularly marked in highly anisotropic crystals,[92] where strains which would be compatible according to Eq. (9-23) in isotropic crystals become quite incompatible when anisotropy is taken into account. *Elastic* incompatibility, leading to image stresses near the boundary, is an important factor in the operation of slip systems near boundaries.[92] Figure 9-19 is an example from Hook's work. The bicrystal was stressed in compression until the resolved shear stress was slightly above the yield stress for slip on (011) and $(0\bar{1}1)$ in the $[\bar{4}19]$ crystal. Even though the resolved shear stress was only about one-half the yield stress in the [001] crystal, slip occurred on $(0\bar{1}1)$ and $(\bar{1}01)$ near the boundary to relieve the plastic and elastic stress concentration there.

In bicrystals of ionic crystals also, double slip is observed at grain boundaries.[85,87] Furthermore, in studies of MgO bicrystals,[94,95] cracks are observed near grain boundaries where only one slip band appears in each grain, or where two slip bands appear in one grain and none or one in the other. This latter finding presumably illustrates the absence of "microscopic" compatibility in the vicinity of a slip band, where the shear is inhomogeneous.

[89] J. D. Livingston and B. Chalmers, *Acta Met.*, **5**: 322 (1957); *Acta Met.*, **6**: 216 (1958).

[90] J. J. Hauser and B. Chalmers, *Acta Met.*, **9**: 802 (1961).

[91] U. F. Kocks, *Phil. Mag.*, **10**: 187 (1964).

[92] R. E. Hook and J. P. Hirth, *Trans. Jap. Inst. Met.*, (suppl.) **9**: 778 (1968).

[93] For a brief review of such experiments see J. J. Hauser and B. Chalmers, loc. cit.

[94] A. R. C. Westwood, *Phil. Mag.*, **6**: 195 (1961).

[95] T. L. Johnston, R. J. Stokes, and C. H. Li, *Phil. Mag.*, **7**: 23 (1962).

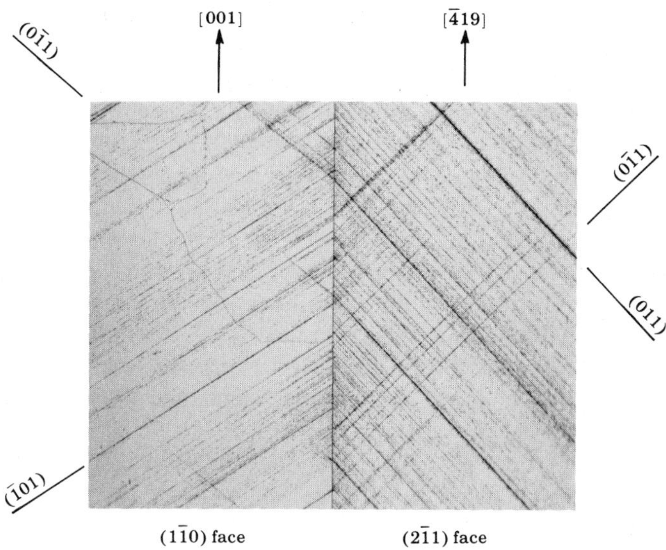

FIGURE 9-19. An Fe-3% Si bicrystal deformed 0.1 percent and then etch-pitted to reveal dislocations, 150× (R. E. Hook, private communication).

If local conditions of resolved shear stress and pinning of slip systems are such that four independent systems cannot locally operate in the two grains, then the stress concentration caused by the dislocation pileup cannot be relieved, and fracture ensues.

Studies[96] on coarse-grained polycrystals of aluminum, with two to twelve grains per cross section of the specimen, represent the next step toward general polycrystalline behavior.[97] In these studies, where most of the grains were adjacent to a free surface, one to three slip systems were observed in each grain. The occurrence of general deformation requiring five independent slip systems becomes likely only in fine-grained polycrystals.

Unusual Surface Slip

In thin crystals or where surface nucleation of slip is important, observations for both fcc[98] and bcc crystals[99, 100] reveal predominant slip on the most favored type of glide plane, {111} or {110} respectively, but not on the plane of maximum Schmid factor. Many of these observations are consistent with the image effects evidenced in Eqs. (5-93) and (8-53), which can favor unusual slip

[96] R. L. Fleischer and W. F. Hosford, Jr., *Trans. Met. Soc. AIME*, **221**: 244 (1961).

[97] R. W. Armstrong, *J. Mech. Phys. Solids*, **9**: 196 (1961).

[98] O. Lohne, *Phys. Stat. Solidi.*, **25a**: 209 (1974).

[99] C. J. Bolton and G. Taylor, *Phil. Mag.*, **26**: 1359 (1972).

[100] H. Matsui and H. Kimura, *Scripta Met.*, **9**: 971 (1975).

either by enhancing deformation on the unusual slip plane or impeding it on the plane with maximum Schmid factor. The former occurs for bcc metals[100] at low temperatures, where the Peierls stress is very large for screw dislocations, whose motion thus controls deformation. For dislocations that would be mixed if they lay normal to the surface the interactions of Eq. (8-53) would favor kink nucleation and hence favor glide. In fact, consistent with this idea, the slip system observed[101] is usually the one with **b** most nearly parallel to the surface (Vesely's rule[102]). Similarly, if they were screw and oblique to the surface, but in a glide plane nearly normal to the surface, image forces would dominate and again favor kink nucleation pulling the dislocation closer to the equilibrium angle prescribed by Eq. (5-93) and again favoring glide. The contrasting effect is observed in fcc Al[98] where the above forces produce more jogs on dislocations of the maximum Schmid factor slip system than on secondary systems and thus hinder slip on the former system.

PROBLEMS

9-1. In what range of orientations should a dislocation with $\mathbf{b}=[100]$ be stable in an fcc crystal? Include the variation of energy with screw-edge character in the analysis.

9-2. Explicitly estimate the dependence of the Peierls stress on the c/a ratio for the hcp slip systems listed in Table 9-3. What would be the effect of a superposed hydrostatic pressure on the Peierls stress?

9-3. With a uniaxial compressive stress acting in the $[11\bar{2}3]$ direction in an hcp single crystal, compute the resolved shear stresses on the slip systems of the type $\langle 11\bar{2}0\rangle\{10\bar{1}0\}$.

9-4. In the presence of a uniform torque applied about the $[11\underline{2}]$ axis, compute the resolved shear stresses on the $[1\bar{1}0](111)$ and $[011](1\bar{1}1)$ slip systems in an fcc crystal.

9-5. What is the number of independent slip systems in a simple cubic structure with the cube faces the allowed slip planes and the cube edges the allowed slip directions?

9-6. Consider three slip systems with slip vectors β_1, β_2, and β_3. Let β_3 be normal to β_1 and β_2, and let β_1 and β_2 have a common slip plane. Assume that the glide plane for β_3 is any plane containing β_3. Are the three slip systems independent?

9-7. With all conditions the same as in Prob. 9-6 except that β_3 is not normal to β_1 and β_2, are the slip systems independent?

[101] L. P. Kubin and J. L. Martin, in P. Haasen, V. Gerold and G. Kostorz (eds.), "Strength of Metals and Alloys," vol. 3, Pergamon, Oxford, 1980, p. 1639.

[102] D. Vesely, *Scripta Met.*, **6**: 753 (1972).

9-8. Is there a special choice of slip plane for β_3 in Prob. 9-6 for which a purely longitudinal strain in the β_3 direction would be impossible?

9-9. Discuss the effect of the activation of climb processes on the requirement for independent slip systems.

9-10. Determine the special orientations of a bicrystal, of the type shown in Fig. 9-17, for which only one slip system is required in each crystal. Would the presence of elastic anisotropy change the result?

BIBLIOGRAPHY

1. Barrett, C. S., "Structure of Metals," McGraw-Hill, New York, 1952.
2. Batchelor, G. K. (ed.), "Scientific Papers of G. I. Taylor," vol. I, "Mechanics of Solids," Cambridge University Press, New York, 1963.
3. Christian, J. W., and V. Vitek, *Prog. Phys.*, **33**: 307 (1970).
4. Kelly, A., and G. W. Groves, "Crystallography and Crystal Defects," Addison-Wesley, Reading, Mass., 1970.
5. Nabarro, F. R. N., Z. S. Basinski, and D. B. Holt, *Advan. Phys.*, **13**: 193 (1964).
6. Schmid, E., and W. Boas, "Plasticity of Crystals," Hughes, London, 1950.
7. Thomas, G., and J. Washburn (eds.), "Electron Microscopy and Strength of Crystals," Interscience, New York, 1963.
8. "Mechanical Behavior of Solids," NBS monograph 59, 1963.

10

Partial Dislocations in FCC Metals

10-1. INTRODUCTION

In particular crystal lattices for a restricted number of planes and Burgers vectors, the stacking fault equivalent to that in Fig. 9-2*c* has a low misfit energy compared to the misfit energy in the vicinity of the dislocation core. For these special cases the arguments of Chap. 9 do not apply, and the crystal in question possibly can contain *imperfect*, or *partial*, *dislocations* associated with the low-energy *stacking faults*.

Partial dislocations are important in twinning reactions, in phase transformations, and in the formation of dislocation barriers by intersecting dislocations. The extension of a perfect dislocation into partials bounding a stacking fault affects the climb and cross slip of dislocations. Stacking faults themselves are important barriers to dislocation motion.

This chapter deals with the possible stacking faults and partial dislocations in fcc crystals. The fcc lattice is selected for detailed study because stacking faults have definitely been observed, and because extended dislocation arrays are associated with specific mechanical properties in fcc structures. Some arrays, such as extended jogs and stacking-fault tetrahedra, are considered in detail. Whereas the primary emphasis is on intrinsic stacking-fault arrays, since these are the ones commonly observed, extrinsic faults are treated briefly. The chapter concludes with a discussion of the special considerations that are necessary in treating forces on partial dislocations.

10-2. STACKING FAULTS IN FCC CRYSTALS

For the close-packed fcc and hcp structures, the interatomic forces are such that it is a fair approximation to regard the atoms as hard spherical balls held together by attractive forces. These structures are generated by stacking close-packed layers on top of one another in the fashion illustrated in Fig. 10-1. Given a layer *A*, close packing can be extended by stacking the next layer so that its atoms occupy *B* or *C* sites. Here *A*, *B*, and *C* refer to the three

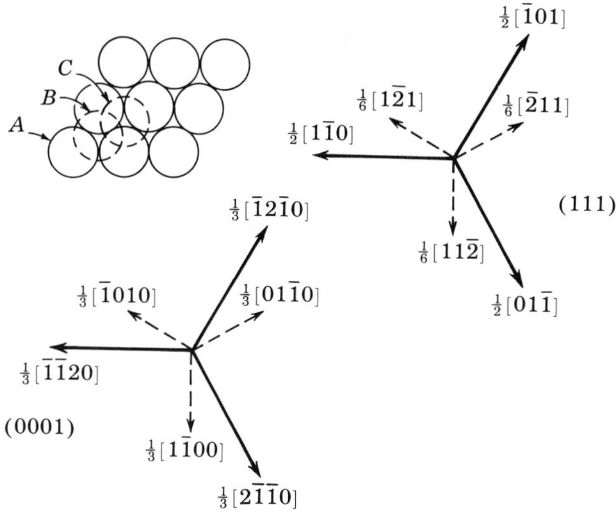

FIGURE 10-1. Projection normal to the (111) plane showing the three types of stacking positions *A*, *B*, and *C*. Fcc and hcp vector notations are also presented.

possible layer positions in a projection normal to the close-packed layers (Fig. 10-1). A close-packed structure is generated, provided that no two layers of the same letter index, such as *AA*, are stacked in juxtaposition to one another. The sequence[1] corresponding to an fcc crystal is $\cdots ABCABC \cdots$, or $\cdots CBACBA \cdots$; that for an hcp crystal is $\cdots ABABAB \cdots$, $\cdots BCBCBC \cdots$, or $\cdots CACACA \cdots$.

In fcc crystals the close-packed planes are $\{111\}$ planes, which are also the glide planes and coherent twin planes. The twin orientation corresponds to a 180° rotation in the $\{111\}$ plane or, equivalently, to a mirror plane reflection about $\{111\}$. Since the labels *A*, *B*, and *C* denote $\{111\}$ planes, the mirror symmetry, and hence the twin, can be readily represented in the notation. The stacking sequence for a twin is

$$ABCABCAB\overset{\dagger}{C}BACBAC \tag{10-1}$$

where the dagger denotes the twin plane and the dashed line the center of the fault.

The other $\{111\}$ faults are stacking faults, which are classified as *intrinsic* or *extrinsic*, as suggested by Frank.[2] In the intrinsic type the normal stacking

[1]An alternate representation is used sometimes, which is due to F. C. Frank [*Phil. Mag.*, **42:** 1014 (1951)]. He represents stacking in the successions *A* to *B*, *B* to *C*, or *C* to *A* by the symbol Δ, and their opposites by ∇. Thus normal fcc stacking is represented by ΔΔΔΔ or ∇∇∇, and an fcc twin by ∇∇∇ΔΔΔΔ.

[2]F. C. Frank, *Phil. Mag.*, **42:** 809 (1951).

sequence is maintained in the crystal on either side of the fault right up to the fault plane. In the extrinsic type planes are inserted at the center of the fault which are incorrectly stacked with respect to the layers on both sides of the fault.

The intrinsic stacking fault is formed when a layer of the atoms is removed from the normal sequence, leaving

$$\begin{matrix} & \dagger\,|\,\dagger & \\ ABCABC & | & BCABC \\ & | & \end{matrix} \qquad (10\text{-}2)$$

Notice that this fault can be regarded as being formed by two twinning operations (at the daggers) separated by one atomic layer. The extrinsic stacking fault is formed by the addition of a layer of atoms to the normal sequence, giving

$$\begin{matrix} & \dagger\,|\,\dagger & \\ ABCABC & B & ABCABC \\ & | & \end{matrix} \qquad (10\text{-}3)$$

This fault is equivalent to two twin planes separated by *two* atomic layers.

All the above faults preserve close packing, so that no nearest-neighbor bonds are distorted in the hard-ball model. The various interatomic-force laws proposed for close-packed lattices all predict that a large fraction of the atomic binding energy resides in nearest-neighbor bonds, so that one expects all the above faults to have low surface energies compared to surfaces with deformed or broken nearest-neighbor bonds, such as grain boundaries or free surfaces. Other faults, such as the one produced by the removal of two planes leaving

$$\begin{matrix} & | & \\ \cdots ABCABC & | & CABCA \\ & | & \end{matrix}$$

involve the violation of close packing on the fault plane and are expected to have relatively large surface energies.

One must realize that the low-energy faults can be produced by *shearing* operations on the $\{111\}$ planes. Suppose that the plane A in Fig. 10-1 is a plane in a perfect fcc crystal, and that the next plane above the A plane is a B plane. If the B plane and all planes above it are displaced by the vector $\frac{1}{6}[\bar{2}11]$, the B plane moves into a C position, and the planes above it undergo the transition $A \rightarrow B$, $B \rightarrow C$, and $C \rightarrow A$, relative to positions fixed on the original A plane. This shear displacement is represented by the arrows in the reaction

$$\begin{matrix} ABCA & BCABCABC \\ & \downarrow\downarrow\downarrow\downarrow\downarrow\downarrow\downarrow\downarrow \\ & CABCABCA \end{matrix} \qquad (10\text{-}4)$$

giving

```
  †¦†
ABCA¦CABCABC                (10-5)
    |
```

The resultant fault is an intrinsic stacking fault. If the plane *C* below *A* and all successively lower planes are displaced by $\frac{1}{6}[2\bar{1}\bar{1}]$,

```
     †¦†
CABCA¦CABCAB
↓ ↓ ↓ ↓  |                  (10-6)
BCAB     ¦
         |
```

an extrinsic stacking fault results:

```
  † | †
BCABACABCAB                 (10-7)
    |
```

On the other hand, if the displacement $\frac{1}{6}[\bar{2}11]$ is continued plane by plane above *A*,

```
ABCABCABCABC
       ↓ ↓ ↓ ↓ ↓
       CABCA
        ↓ ↓ ↓ ↓
        BCAB
         ↓ ↓ ↓              (10-8)
         ABC
          ↓ ↓
          CA
           ↓
           B
```

a coherent twin is formed,

```
        †
ABCABCACBACB                (10-9)
        |
        |
```

Finally, note that the high-energy fault

```
ABCABBCABCA
    ↓                       (10-10)
    C
```

can be transformed to an extrinsic fault by the shear of one layer with respect to the entire crystal as represented by the arrow in Eq. (10-10).

Theoretical Stacking-Fault Energy

The fault energy at 0°K can be represented by the number of pairs of separation N which are not in the proper stacking sequence, multiplied by the distortional energy ψ_N per pair. Formally, one should also consider the fault entropy and determine the specific surface free energy of the faults; however, the energy estimate itself is quite uncertain, and the more difficult entropy estimates are not undertaken here. The pair scheme is illustrated for the twin plane in Fig. 10-2. As indicated there, the twin has one pair of second-nearest-neighbor planes ($N=2$) in the wrong stacking sequence, two third neighbors, one fourth neighbor, etc. Proceeding in this manner, one finds that the energies γ_I, γ_E, and γ_T of the intrinsic, extrinsic, and twin faults, respectively, are given by

$$
\begin{aligned}
\gamma_I &= \sum_{N=1}^{\infty} [(3N-1)\psi_{3N-1} + 3N\psi_{3N}] = 2\psi_2 + 3\psi_3 + (0)\psi_4 + \cdots \\
\gamma_E &= \sum_{N=1}^{\infty} [2\psi_{3N-1} + (3N+1)\psi_{3N} + 3N\psi_{3N+1}] = 2\psi_2 + 4\psi_3 + 3\psi_4 + \cdots \\
\gamma_T &= \sum_{N=1}^{\infty} [2N\psi_{3N} + N(\psi_{3N-1} + \psi_{3N+1})] = \psi_2 + 2\psi_3 + \psi_4 + \cdots \qquad (10\text{-}11) \\
\gamma_H &= \sum_{N=1}^{\infty} [\psi_{6N-2} + \psi_{6N-3} + \psi_{6N-4}] = \psi_2 + \psi_3 + \psi_4 + \cdots
\end{aligned}
$$

Also listed is γ_H, the fault energy per close-packed plane of an hcp crystal, which is regarded as a completely faulted fcc crystal; that is, γ_H represents the fcc→hcp transformation energy per layer.

The energy of the fault in an atomic-bond model is composed of three kinds of terms: an energy term associated with the bonds across the fault plane

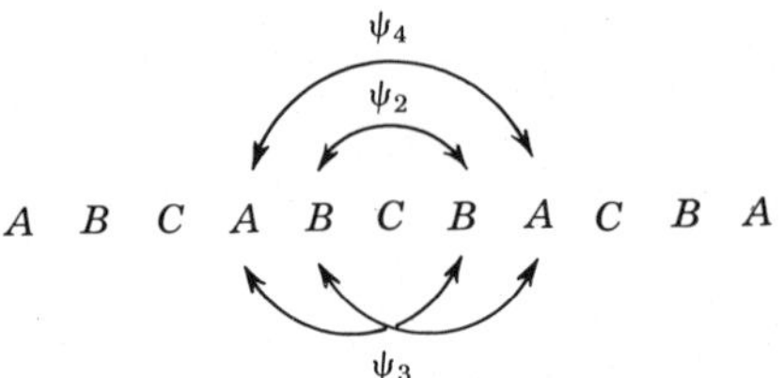

FIGURE 10-2. A representation of incorrectly stacked layers and ψ_N for a twin.

which are sheared by the fault, an energy term caused by dilatation normal to the close-packed layers, and a term arising from dilatation within a close-packed layer near the fault plane. The latter two are not so obvious as the first but are evidenced, for example, by the differences in layer spacing between layers and between atoms within a layer in the fcc and hcp structures of metals, such as cobalt, which exists in both forms.

Application of the various proposed semiempirical atomic force laws to the calculation of the above energy terms is not a satisfactory procedure because of the small magnitudes of the fault energies; the uncertainties in the force laws are likely to lead to energy uncertainties of the same order of magnitude as the fault energy. Numerous estimates of the interrelation among the various faults, however, have been based qualitatively on the results expected from the use of such force laws. Therefore a brief comparison is presented of the relative fault energies based on a model of hard spherical balls bound by *central forces*. We emphasize that all directional bonding is neglected. In this approximation the dilatation within and between close-packed layers vanishes, so that only the displaced central bonds contribute to the fault energy.

The energy ϕ_j of a jth-nearest-neighbor atom pair is associated with its atom-center separation r_j. Table 10-1 lists the perfect-lattice spacings r_j and the spacings as distorted by the shear displacements in the formation of the faults, $r_j \rightarrow r_j'$ and $r_j \rightarrow r_j''$. The eleven shortest bond spacings are listed. The sheared-bond vector separations r_j' are given by the original separations r_j plus the shear-displacement vector of the type $\frac{1}{6}\langle 112\rangle$. The central-force model predicts a decrease in bond energy with an increase in bond distance, so that the bond energies in Table 10-1 are listed in order of decreasing energy.

TABLE 10-1 Vectors Connecting Atom Centers $\mathbf{r}_j$, Their Magnitudes, and the Equivalent Bond Energy for the 11 Shortest Possible Bonds in FCC

Bond	$\mathbf{r}_j(\times 6)$	$\lvert\mathbf{r}_j\rvert(\times 6/a_0)$	Bond Energy
r_1	$\langle 033\rangle$	$\sqrt{18}$	ϕ_1
r_2	$\langle 006\rangle$	$\sqrt{36}$	ϕ_2
r_3'	$\langle 444\rangle$	$\sqrt{48}$	ϕ_3'
r_3	$\langle 336\rangle$	$\sqrt{54}$	ϕ_3
r_3''	$\langle 741\rangle$	$\sqrt{66}$	ϕ_3''
r_4	$\langle 660\rangle$	$\sqrt{72}$	ϕ_4
r_5	$\langle 039\rangle$	$\sqrt{90}$	ϕ_5
r_4'	$\langle 11(10)\rangle$	$\sqrt{102}$	ϕ_4'
r_6	$\langle 666\rangle$	$\sqrt{108}$	ϕ_6
r_6'	$\langle 774\rangle$	$\sqrt{114}$	ϕ_6'
r_5'	$\langle 24(10)\rangle$	$\sqrt{120}$	ϕ_5'
...	...	...	...

Including all bonds out to the eleventh nearest neighbors, one finds that the pair energies are given in terms of the bond energies by

$$\psi_2 = \phi_3' - 3\phi_3 + 6\phi_3'' - 3\phi_4 - 6\phi_5 + 6\phi_4' + 6\phi_5'$$

$$\psi_3 = -\phi_6 + 3\phi_6' \qquad (10\text{-}12)$$

$$\psi_N = 0 \qquad N > 3$$

Thus, according to Eq. (10-11),

$$\gamma_T = \phi_3' - 3\phi_3 + 6\phi_3'' - 3\phi_4 - 6\phi_5 + 6\phi_4' - 2\phi_6 + 6\phi_6' + 6\phi_5'$$

$$\gamma_H = \phi_3' - 3\phi_3 + 6\phi_3'' - 3\phi_4 - 6\phi_5 + 6\phi_4' - \phi_6 + 3\phi_6' + 6\phi_5'$$

$$\gamma_I = 2\phi_3' - 6\phi_3 + 12\phi_3'' - 6\phi_4 - 12\phi_5 + 12\phi_4' - 3\phi_6 + 9\phi_6' + 12\phi_5'$$

$$\gamma_E = 2\phi_3' - 6\phi_3 + 12\phi_3'' - 6\phi_4 - 12\phi_5 + 12\phi_4' - 4\phi_6 + 12\phi_6' + 12\phi_5' \qquad (10\text{-}13)$$

where the fault energies are expressed as the energy per atomic area in the fault.

In the hard-ball central-force model, both first- *and second*-nearest-neighbor bonds do not contribute to the fault energies. Also, if only the eight nearest-neighbor bonds from Table 10-1 are included, the twin-fault energy is found to equal the fcc→hcp energy of transformation per layer, while γ_I and γ_E are equal to each other and equal to twice γ_T or γ_H. Only the ninth-nearest-neighbor bonds and bonds of greater spacing lead to deviations from these equalities; in most simple central-force models such high-order bonds would have a negligible effect on the fault energy.

On an equivalent basis of reasoning, it is often assumed as a first approximation that the specific surface free energy of the intrinsic stacking fault is equal to either twice that of the coherent twin fault[3] or twice the free energy of transformation[4] per close-packed plane for the fcc→hcp case. As shown in Eq. (10-13), such an assumption self-consistently would predict that γ_I equals γ_E.

As discussed later, these rough estimates appear to be valid in some experimental results, but they must be regarded as very approximate. The uncertainties arise from the neglect of noncentral forces. The deviation of the c/a ratio for hcp crystals from the ideal ratio for close-packed spheres is an indication of the presence of noncentral forces. In particular, forces arising from the phase shift of conduction electrons at the fault surfaces, and associated electron scattering effects, are likely to contribute appreciably to the fault energies.

[3] W. T. Read, Jr., "Dislocations in Crystals," McGraw-Hill, New York, 1953, p. 94.

[4] R. D. Heidenreich and W. Shockley, in "Report of a Conference on Strength of Solids," Physical Society, London, 1948, p. 57.

Fundamental calculations of fault energies are progressing[5,6] and show promise of providing accurate estimates of fault energies. However, to date the calculated values tend to be low by a factor of about two relative to experimental measurements.[5,6] Calculations[6] using pseudopotentials[7] do indicate, in agreement with the hard-ball result, that $\gamma_I \sim \gamma_E \sim 2\gamma_T$.

Thus, despite all of its inherent uncertainties, the hard-ball model remains useful in rough estimates of fault energies. A list of experimentally determined fault energies, together with some values for surface energies and grain-boundary energies is presented in Appendix 2. The energies of coherent twin boundaries and intrinsic stacking faults are seen to be indeed small compared to surface or grain-boundary energies in fcc crystals, in agreement with expectation from the hard-ball model. The results show that the relation $\gamma_I \sim 2\gamma_T$ is followed approximately. In early studies, only intrinsic stacking faults were observed in fcc crystals, suggesting the possibility that the extrinsic fault has a higher energy. However, observations by Loretto[8] and subsequently by others as reviewed by Gallagher,[9] established the presence of extrinsic faults and suggest that $\gamma_I \sim \gamma_E$ in agreement with Eq. (10-13), the values for several alloys being $1.01 < \gamma_E/\gamma_I < 1.06$. Thus the absence of extrinsic faults in most extended arrays is associated with a formation constraint, not with large equilibrium energy.

10-3. PARTIAL DISLOCATIONS IN FCC CRYSTALS

Shockley Partials

Suppose that the close-packed layer shown in Fig. 10-3 corresponds to the bottom side of the glide plane of a dislocation, and that the atoms above the layer are originally in sites such as the one denoted by (a). The passage of the perfect dislocation $\mathbf{b} = \frac{1}{2}[\bar{1}01]$ from the bottom of the figure to the top results in the atom originally at (a) being displaced to (c). Motion of the atom along the straight path from (a) to (c) involves, in the hard-ball model, a larger dilatation normal to the slip plane, and hence a larger misfit energy than does motion along the path (a) to (b) to (c), so that the latter path should be favored. Referring to the previous section, one sees that in the intermediate position (b), the local atomic arrangement is precisely that of an intrinsic stacking fault and that the displacement $\frac{1}{6}[\bar{2}11]$ required to form configuration (b) is the shear displacement of Eq. (10-4). A crystal containing a partial dislocation is depicted in Fig. 10-4. The right side of the figure shows the

[5] C. C. Pei, *Phys. Rev.*, **18B:** 2583 (1978)

[6] J. P. Simon, *J. Phys. F*, **9:** 425 (1979)

[7] W. A. Harrison, "Pseudopotentials and the Theory of Metals," Benjamin, New York, 1966.

[8] M. H. Loretto, *Phil. Mag.*, **10:** 467 (1964); *Phil. Mag.*, **12:** 125 (1965).

[9] P. C. J. Gallagher, *Met. Trans.*, **1:** 2429 (1970).

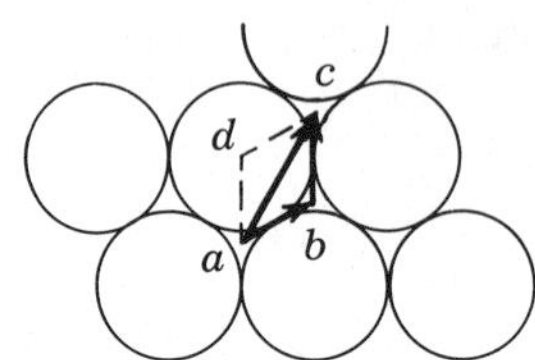

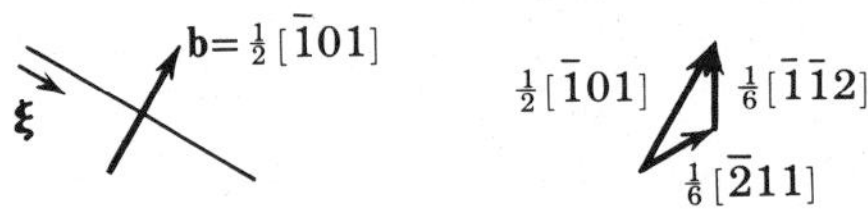

FIGURE 10-3. Atoms on the bottom side of a (111) glide plane. A perfect $\frac{1}{2}[\bar{1}01]$ dislocation and its component Shockley partials are also shown.

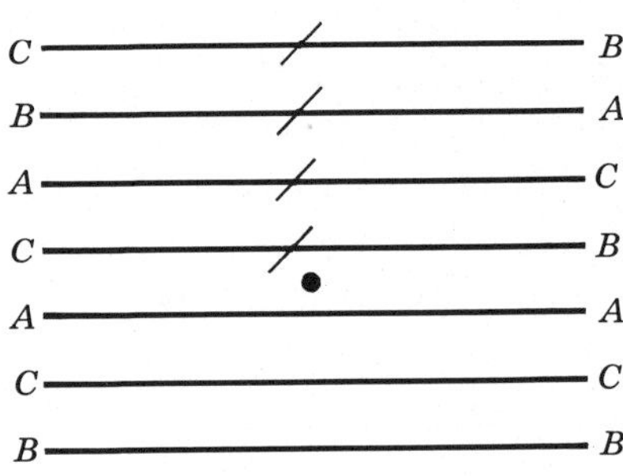

FIGURE 10-4. Representation of a crystal partially sheared by a Shockley partial. The dot represents the projection of the dislocation line, and the sloping lines indicate that the letter index changes upon passing from the unslipped to the slipped region.

unslipped lattice, and the left side shows the sheared lattice corresponding to Eq. (10-4). An actual picture of stacking faults bounded by partial dislocations is shown in Fig. 1-16. A picture of partials bounding a stacking fault in a three-layer bubble raft stacked in the fcc sequence is presented in Fig. 10-5.

Thus the intrinsic stacking fault and its associated shear can be produced by the motion of the *partial dislocation* $\frac{1}{6}[\bar{2}11]$; the fault is removed, restoring the

FIGURE 10-5. Partial dislocations bounding a stacking fault in a bubble raft (J. F. Nye, private communication).

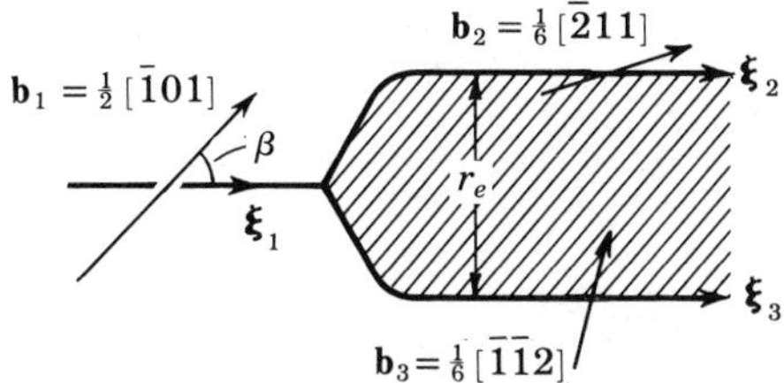

FIGURE 10-6. Dissociation of a perfect dislocation into Shockley partials.

perfect lattice arrangement behind the dislocation, by the subsequent glide of the partial $\frac{1}{6}[\bar{1}\bar{1}2]$. Partial dislocations of the type $\frac{1}{6}\langle 112\rangle$, which are glissile on $\{111\}$ planes, are called *Shockley partials.*[10] Figure 10-6 illustrates the *dissociation* of a perfect dislocation into an *extended dislocation* consisting of two Shockley partials and an enclosed stacking fault. Burgers circuits around the perfect and extended configurations are equivalent, so that $\mathbf{b}_1 = \mathbf{b}_2 + \mathbf{b}_3$. The partial-dislocation Burgers vectors can be determined individually by Burgers circuits, but special provision must be made for the fault plane which represents a "bad" region (Chap. 1) of infinitesimal thickness: *the Burgers circuit for a partial dislocation must start and end on the plane of the stacking fault.*

Application of the Frank criterion, Eq. (9-1), shows that the elastic strain energy in a crystal is reduced by the dissociation of a perfect dislocation into partials. Equivalently, Eq. (5-17) indicates that the partials in an extended dislocation repel one another by a force that varies as $1/r$, where r is the separation between partials. The formation of the fault between the partials produces an increase in energy $\gamma_I r$ per unit length; this energy increase leads to an attractive force per unit length between partials given by[11] $\partial(\gamma_I r)/\partial r = \gamma_I$. At the equilibrium separation r_e, the (attractive) force γ_I is equal and opposite to the (repulsive) elastic force [Eq. (5-17)], so that the equilibrium condition is

$$\gamma_I = \frac{\mu}{2\pi r_e}\left[(\mathbf{b}_2\cdot\xi_2)(\mathbf{b}_3\cdot\xi_3) + \frac{(\mathbf{b}_2\times\xi_2)\cdot(\mathbf{b}_3\times\xi_3)}{1-\nu}\right] \tag{10-14}$$

Exercise 10-1. The mixed perfect dislocation in Fig. 10-6 is inclined at an angle β to its Burgers vector. The partials $\mathbf{b}_2$ and $\mathbf{b}_3$ are inclined at $\beta \pm 30°$ to their respective lines. Apply Eq. (10-14) and show that in this case the equilibrium separation is given by[12]

$$r_e = \frac{\mu b_2^2}{8\pi\gamma_I}\frac{2-\nu}{1-\nu}\left(1 - \frac{2\nu\cos 2\beta}{2-\nu}\right) \tag{10-15}$$

[10] R. D. Heidenreich and W. Shockley, in "Report of a Conference on Strength of Solids," Physical Society, London, 1948, p. 57.

[11] This is a specific analogy to the general physical identity that a surface has a specific surface energy per unit area which is numerically equal to the surface tension, or force per unit length acting at the boundary of the surface.

[12] W. T. Read, Jr., "Dislocations in Crystals," McGraw-Hill, New York, 1953, p. 131.

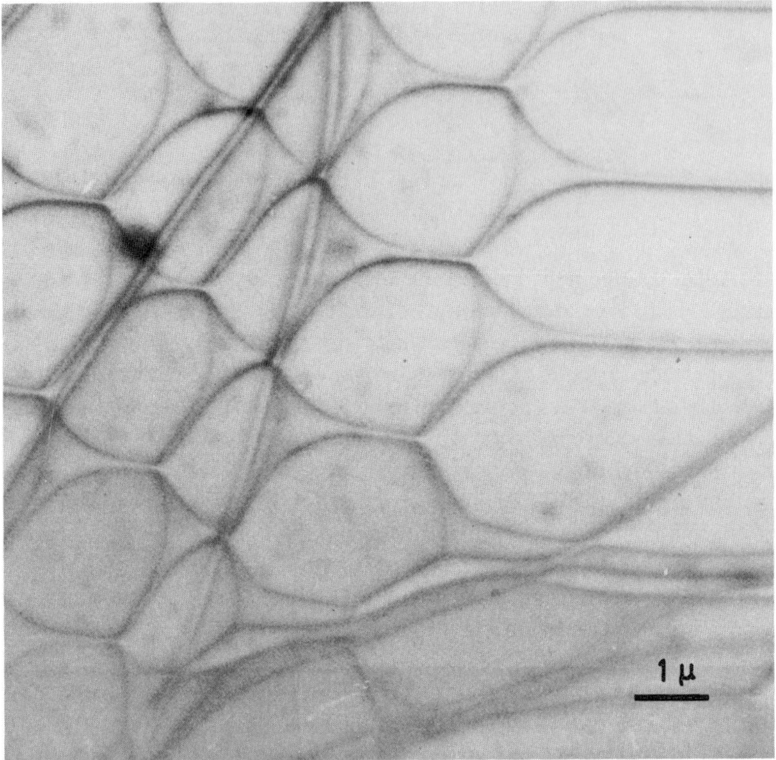

FIGURE 10-7. A dislocation dissociated into Shockley partials in graphite [as observed by P. Delavignette and S. Amelinckx, *J. Nucl. Mater.*, **5**:17 (1962)].

Thus r_e for an edge orientation of b_1 exceeds that for a screw orientation; for $\nu = \frac{1}{3}$, the ratio of the equilibrium separations is 7:3. The small values of r_e for metals hindered the direct observation of extension. However, extensions to $r_e \sim 3.0$ nm have been observed in metals (e.g., nickel[13]) with the advent of the weak-beam technique of electron microscopy.[14] In graphite, where the partial dislocations and stacking faults on the basal plane resemble those in fcc crystals, the partial separation is resolvable as indicated in Fig. 10-7. Delavignette and Amelinckx,[15] who were able to measure r_e as a function of β, found agreement with Eq. (10-15) and used the plot to determine the actual anisotropic average values of μ and ν for graphite.

Note that the partials must split in a specified way for an intrinsic fault to be formed. If the partial $\frac{1}{6}[\bar{1}\bar{1}2]$ should precede the other one in Fig. 10-3, the intermediate configuration would be (d), which has a large misfit energy.

[13]C. B. Carter and S. M. Holmes, *Phil. Mag.*, **35:** 1161 (1977).

[14]D. J. H. Cockayne, I. L. F. Ray, and M. J. Whelan, *Phil. Mag.*, **20:** 1265 (1969).

[15]P. Delavignette and S. Amelinckx, *J. Nucl. Mater.*, **5:** 17 (1962).

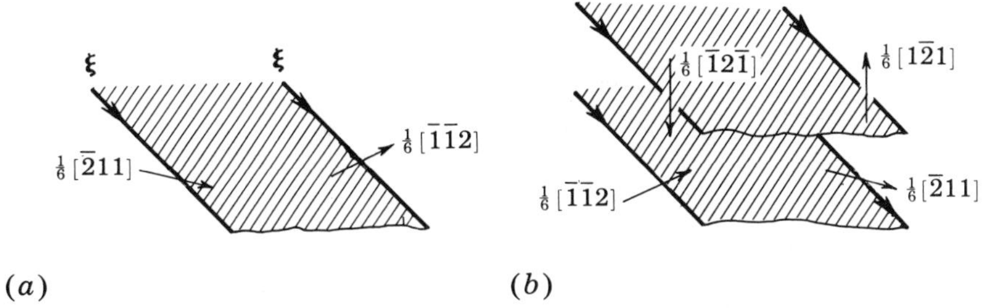

(*a*) (*b*)

FIGURE 10-8. (*a*) A high-energy stacking fault, which can be sheared to form (*b*) an extrinsic stacking fault. Arrows on lines in this and subsequent figures represent the sense vectors ξ.

Indeed, this configuration, shown in Fig. 10-8*a*, corresponds to the high-energy fault

$$\cdots ABCA|ABC\cdots$$

This high-energy fault can be transformed to the lower-energy extrinsic-fault arrangement of Fig. 10-8*b* by the nucleation of $\frac{1}{6}[1\bar{2}1]-\frac{1}{6}[\bar{1}2\bar{1}]$ partials on the high-energy-fault plane and on the plane immediately above it. The final arrangement of two pairs of Shockley partials bounding an extrinsic fault corresponds to the shear process of forming such a fault, as illustrated in Eqs. (10-4) to (10-7). The *net* Burgers vector of each of the pairs of partials bounding the extrinsic fault is the same as that of the corresponding single partial bounding the high-energy fault (Fig. 10-8).

Because of the high energy associated with the fault in Fig. 10-8*a*, and because the configuration of Fig. 10-8*b* has a somewhat higher fault energy and a larger core energy (two partials instead of one) than the intrinsic-fault dissociation of Fig. 10-6, the latter type of dissociation is expected to prevail in fcc crystals.

The Thompson Tetrahedron

In order to decide on the proper sequence of dissociation to form an intrinsic fault, it was necessary in the above treatment to consider detailed atomic arrangement, as in Fig. 10-3. Often this procedure is cumbersome, particularly for dislocation interactions. A convenient representation, which greatly simplifies this sequential-ordering operation and is useful in the treatment of other types of partials, has been introduced by Thompson.[16]

By joining the atoms in a one-eighth unit cell by straight lines, one forms the *Thompson tetrahedron*, shown in Fig. 10-9. The inner and outer faces of the tetrahedron represents the four possible {111} glide planes [eight if the

[16] N. Thompson, *Proc. Phys. Soc.*, **66B:** 481 (1953).

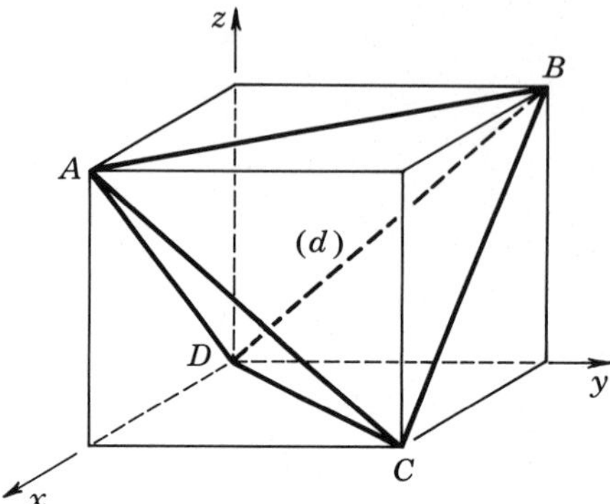

FIGURE 10-9. A one-eighth unit cell of edge length $\frac{1}{2}a_0$ of an fcc lattice, showing the Thompson tetrahedron *ABCD*.

plus-minus sense is distinguished, as for (111) and $(\bar{1}\bar{1}\bar{1})$] and the edges of the tetrahedron correspond to the six $\langle 110 \rangle$ glide directions of the fcc structure (12 with plus-minus sense). The atom at the origin is labeled *D* and the others are labeled *ABC* in the clockwise order indicated in Fig. 10-9.[17] The midpoints of the faces opposite *A*, *B*, *C*, and *D* are labeled α, β, γ, and δ, respectively, and the planes opposite *A*, *B*, *C*, and *D* are denoted by *a*, *b*, *c*, and *d*, respectively, on the outside surface, $\bar{a}$, $\bar{b}$, $\bar{c}$, and $\bar{d}$ on the inside surface. If the tetrahedron is opened up at *D*, it can be folded out into the planar arrangement displayed in Fig. 10-10. As indicated there, the $\{111\}$ planes are represented by letters—for example, (111) becomes (*d*); the perfect dislocations $\frac{1}{2}\langle 110 \rangle$ are represented by pairs of Roman letters—for example, $\frac{1}{2}[0\bar{1}\bar{1}]$ becomes **BD**; and Shockley partials are represented by Greek-Roman pairs—for example, $\frac{1}{6}[211]$ becomes **D$\boldsymbol{\beta}$**. Burgers vectors of the type $\langle 001 \rangle$ are represented, for example, by **DC**/**AB** for [001].

The various possible vectors can be described by the following convention: **PQ** *represents the vector connecting P to Q*; **PQ**/**RS** *represents a vector twice the length of the vector connecting the midpoint of PQ to the midpoint of RS. P, Q, R, and S can individually be either Greek or Roman letters.*

The vector algebra for the addition and subtraction of the Thompson vectors is defined by the set of relations:

$$
\begin{aligned}
&\mathbf{PQ} = -\mathbf{QP} \\
&\mathbf{PQ/RS} = -\mathbf{RS/PQ} \\
&\mathbf{PQ/RS} = \mathbf{QP/RS} = \mathbf{PQ/SR} = \mathbf{QP/SR} \qquad (10\text{-}16) \\
&\mathbf{PQ} + \mathbf{QR} = \mathbf{PR} \\
&\mathbf{PQ} + \mathbf{RS} = \mathbf{PR/QS}
\end{aligned}
$$

where again *P*, *Q*, *R*, and *S* can individually be Greek or Roman letters. In

[17]Others have ordered the atoms *ABC* in counterclockwise sequence; here Thompson's original notation is followed.

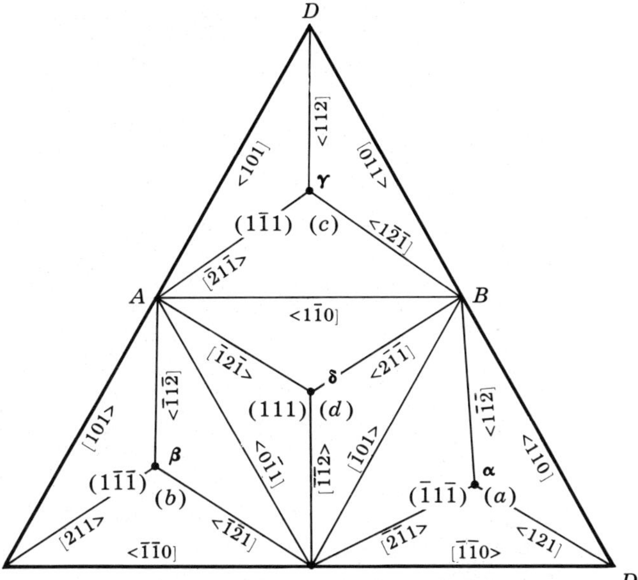

FIGURE 10-10. A Thompson tetrahedron opened up at corner *D*. Both the Thompson notation [(*a*) for glide plane, **AB** for Burgers vector of perfect dislocation, and **A**$\boldsymbol{\delta}$ for Burgers vector of partial dislocation] and one possible set of indices for the same planes and directions are presented. The notation $[1\bar{1}0\rangle$ is used, instead of the usual notation $[1\bar{1}0]$, to indicate the sense of the direction.

addition, for the particular dislocation of the type $\mathbf{CD}/\boldsymbol{\alpha\beta}$, the following relation holds:

$$\mathbf{CD}/\boldsymbol{\alpha\beta} = \boldsymbol{\alpha\beta}/\boldsymbol{\gamma\delta} = \boldsymbol{\gamma\delta}/\mathbf{AB}$$

The above relations are most easily verified by reference to a three-dimensional model of the fcc unit cell. The use of such a model is recommended for the solution of the following exercise.

Exercise 10-2. Consider Figs. 10-9 and 10-10 and verify the statement of Eq. (10-16) that (1) $\boldsymbol{\alpha\gamma} = -\boldsymbol{\gamma\alpha}$, (2) $\boldsymbol{\alpha\beta} + \mathbf{CD} = \boldsymbol{\alpha}\mathbf{C}/\boldsymbol{\beta}\mathbf{D}$, (3) $\boldsymbol{\alpha\delta} + \boldsymbol{\delta}\mathbf{B} = \boldsymbol{\alpha}\mathbf{B}$, and (4) $\boldsymbol{\alpha}\mathbf{B} + \boldsymbol{\delta}\mathbf{C} = \boldsymbol{\alpha\delta}/\mathbf{BC}$.

The proper sequential arrangement of partial dislocations to form an enclosed intrinsic stacking fault is given by the following axiom[18]:

Axiom 10-1: Viewing the perfect dislocation [say **AB**(*d*)] from outside the tetrahedron and along the positive sense of the line, the intrinsic stacking-fault

[18] The axiom as stated is valid only with the definition of **b** employed in this book; with the opposite definition of **b**, the arrangement of letters would be reversed.

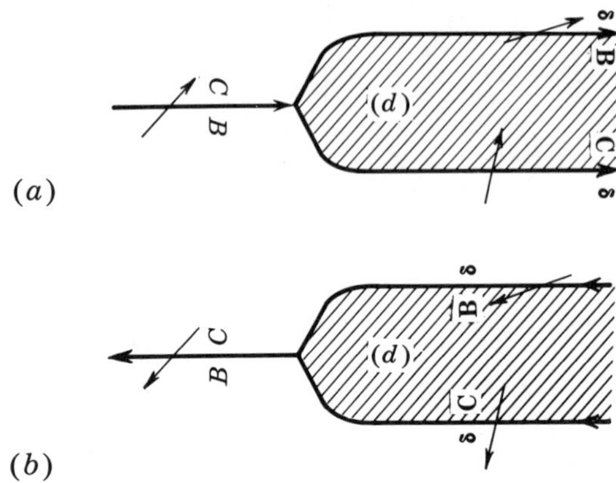

FIGURE 10-11. Two views of the dislocation **BC**(d) viewed from outside the tetrahedron.

arrangement is achieved by placing the Greek-Roman partial **δB** on the viewer's left and the Roman-Greek partial **Aδ** on the right; viewed in the positive sense from inside the tetrahedron, the intrinsic arrangement is achieved by placing the Roman-Greek partial **Aδ** on the left and **δB** on the right.

Thus the dislocations in Fig. 10-6, viewed on (d), i.e., outside the tetrahedron, become those listed in Fig. 10-11a when the sense vectors point to the right of the figure and those in Fig. 10-11b when the sense vectors point to the left. In both cases the Greek letters are to the outside of the fault, in accordance with Axiom 10-1.

The advantages of this scheme are not overwhelming for consideration of only the dissociation of a single dislocation, but it provides quite a reduction of labor in the consideration of more complex arrays such as nodes and extended jogs.

Stair-Rod Partial Dislocations

Stair-rod dislocations, introduced by Thompson,[19] are partial dislocations that are formed when a dissociated dislocation bends over from one glide plane to another or interacts with a dislocation on another glide plane. Consider the dislocation **AB**, which is glissile on (d) and (c). The dislocation line can bend from (d) to (c), forming an acute angle between the faults as in Figs. 10-12a and 10-12b, or forming an obtuse angle as in Fig. 10-12c. In the acute case, the dislocation segments in both glide planes are viewed from outside the tetrahedron, so the partials are arranged with Greek letters outside, as shown in the figure.

Applying Axiom 1-2 for the conservation of Burgers vectors one finds that the Burgers vectors are conserved at the fourfold node of Fig. 10-12a.

[19] N. Thompson, *Proc. Phys. Soc.*, **66B:** 481 (1953).

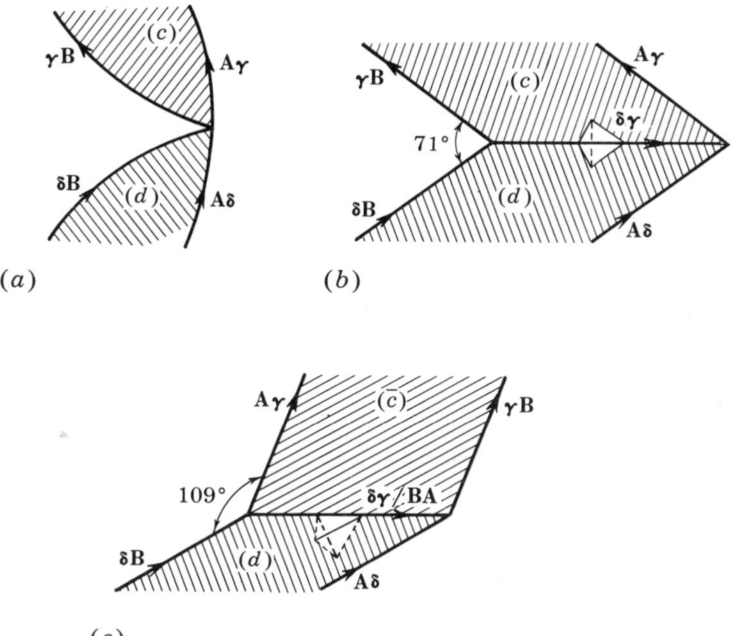

FIGURE 10-12. (*a*) A contracted acute bend, (*b*) an extended acute bend, and (*c*) an obtuse bend of **AB** from (*d*) to (*c*). In this and subsequent figures, the orientation of the Thompson tetrahedron is also shown in the drawing.

However, consideration of the interaction forces between $\boldsymbol{\delta}\mathbf{B}$ and $\boldsymbol{\gamma}\mathbf{B}$ indicates that these partials attract one another, particularly when they are nearly parallel. Similarly, $\mathbf{A}\boldsymbol{\gamma}$ and $\mathbf{A}\boldsymbol{\delta}$ attract one another, so that elastic interactions favor the formation of the configuration in Fig. 10-12*b*. The application of Axiom 1-2 to the latter array reveals that in order to conserve Burgers vectors at the nodes, a stair-rod partial dislocation $\boldsymbol{\delta}\boldsymbol{\gamma}$ must be present along the line of interaction of (d) and (c).

In the obtuse configuration, Fig. 10-12*c*, the plane ($\bar{c}$) is viewed from the point of view of an observer *inside* the tetrahedron, so that the opposite arrangement of the partials $\mathbf{A}\boldsymbol{\gamma}$ and $\boldsymbol{\gamma}\mathbf{B}$ is obtained. Proceeding as above, one finds that in the obtuse case the stair rod is $\boldsymbol{\delta}\boldsymbol{\gamma}/\mathbf{BA}$.

The other two stair-rod dislocation types of interest are depicted in Fig. 10-13. These stair rods form when dislocations on different glide planes interact. The stair rod $\boldsymbol{\delta}\mathbf{D}/\mathbf{C}\boldsymbol{\gamma}$ is formed by the combination of $\boldsymbol{\gamma}\mathbf{D}$ and $\boldsymbol{\delta}\mathbf{C}$, leaving an acute angle between their associated intrinsic faults. The stair rod $\boldsymbol{\delta}\boldsymbol{\alpha}/\mathbf{CD}$ is formed by the obtuse combination of $\boldsymbol{\delta}\mathbf{C}$ and $\mathbf{D}\boldsymbol{\alpha}$.

The likelihood of the formation of the stair-rod dislocations can be assessed by the application of condition (9-1) to the special case of reactant Shockley partials parallel to the product stair rod. As indicated in Table 10-2, the four stair-rod types considered above should be stable in the parallel case, based on

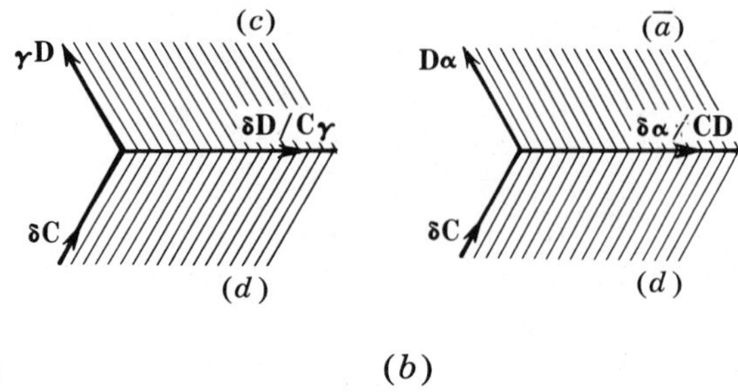

FIGURE 10-13. (*a*) The acute intersection of **δC** and **γD** to form **δD/Cγ**, and (*b*) the obtuse intersection of **δC** and **Dα** to form **δα/CD**.

TABLE 10-2 Stair-Rod Reactions and the Values of b^2 for the Reactant and Product Dislocations*;

Reaction	Shockley Reactants $\mathbf{b}_2+\mathbf{b}_3$	Stair-Rod Product $\mathbf{b}_1$	b_1^2 $(\times 36/a_0^2)$	$b_2^2+b_3^3$ $(\times 36/a_0^2)$
(1)	**δB+Bγ**	**δγ** $\frac{1}{6}\langle 110\rangle$	2	12
(2)	**δB+γA**	**δγ/BA** $\frac{1}{3}\langle 001\rangle$	4	12
(3)	**δC+Dγ**	**δD/Cγ** $\frac{1}{3}\langle 110\rangle$	8	12
(4)	**δB+γD**	**δγ/BD** $\frac{1}{6}\langle 013\rangle$	10	12
(5)	**δB+Dγ**	**δD/Bγ** $\frac{1}{6}\langle 123\rangle$	14	12
...	...	...	...	...

*The term a_0 is the lattice parameter. The Miller indices of the stair rods are also listed.

the b^2 criterion, while all other possible stair rods, such as the fifth entry in the table, should be unstable.[20] Considerations of the screw-edge character of the partials,[21] or of elastic anisotropy,[22,23] do not change the results of Table 10-2 markedly, so that only the first four stair rods are treated in the following discussion as possible stable stair rods.

[20] J. Friedel, *Phil. Mag.*, **46:** 1169 (1955).

[21] J. P. Hirth, *J. Appl. Phys.*, **32:** 700 (1961).

[22] L. J. Teutonico, *Phil. Mag.*, **10:** 401 (1964).

[23] T. Jøssang, C. S. Hartley, and J. P. Hirth, *J. Appl. Phys.*, **36:** 2400 (1965).

Exercise 10-3. Verify by the use of Eq. (5-17) that the inclusion of the screw-edge character in the treatment of the stair-rod reactions does not change the qualitative result of Table 10-2 that only the first four stair rods are stable.

Quantitative calculations have not been carried out for the cases illustrated in Figs. 10-12 and 10-13, where the reactant partials are not parallel to the product stair-rod partial. However, in view of the large energy reduction for configurations (b) and (c) in Fig. 10-12, it seems likely that some nonvanishing length of stair rod is present in each case. In the absence of calculations to settle this important question, the stair-rod configurations are assumed to be stable even in the nonparallel configurations of Figs. 10-12 and 10-13.

The motion in either the climb or glide direction of any of the stair-rod partials discussed above entails the formation of a high energy fault. Thus the stair rods are what Read has called *supersessile dislocations*. They resist both climb and glide forces, and as a consequence, they are important in providing barriers to dislocation motion.

Frank Partial Dislocations

The formation of a vacancy disc and its subsequent collapse to form a dislocation loop[24–27] provide an intrinsic-fault-formation mechanism that is equivalent to the removal of a close-packed layer. The vacancies are supposed to form a disc with its flat sides bounded by $\{111\}$ planes because of the low surface energy associated with such close-packed planes. As shown in Fig. 10-14, if the disc collapses by the displacement of the flat faces in a direction normal to the face, a loop of a *Frank partial dislocation*[28] bounding an intrinsic stacking fault is formed. Frank partials have Burgers vectors of the type $\frac{1}{3}\langle 111\rangle$, or $\mathbf{A}\boldsymbol{\alpha}$, $\boldsymbol{\beta}\mathbf{B}$, etc., in the Thompson notation.

As shown in Fig. 10-15, the nucleation and growth of a loop of Shockley partial in the faulted region leads to the annihilation of the fault. Also, the Shockley partial and the Frank partial combine to form a perfect dislocation. The perfect-dislocation loop would have been formed directly if the vacancy disc collapse had involved the normal displacement plus the offset or shear displacement of the flat faces of the disc.

Analogously, a precipitated interstitial disc corresponds to a Frank partial bounding an extrinsic stacking fault. The extrinsic fault can be removed and a

[24] F. C. Frank, in "Symposium on the Plastic Deformation of Crystalline Solids," U.S. Office of Naval Research, Washington, D.C., 1950, p. 89.

[25] F. Seitz, *Phys. Rev.*, **79:** 723, 890, 1002 (1950).

[26] F. C. Frank, in "Deformation and Flow of Solids," Springer, Berlin, 1965, p. 73.

[27] D. Kuhlmann-Wilsdorf, *Phil. Mag.*, **3:** 125 (1958).

[28] F. C. Frank, *Proc. Phys. Soc.*, **62A:** 202 (1949); F. C. Frank and J. F. Nicholas, *Phil. Mag.*, **44:** 1213 (1953).

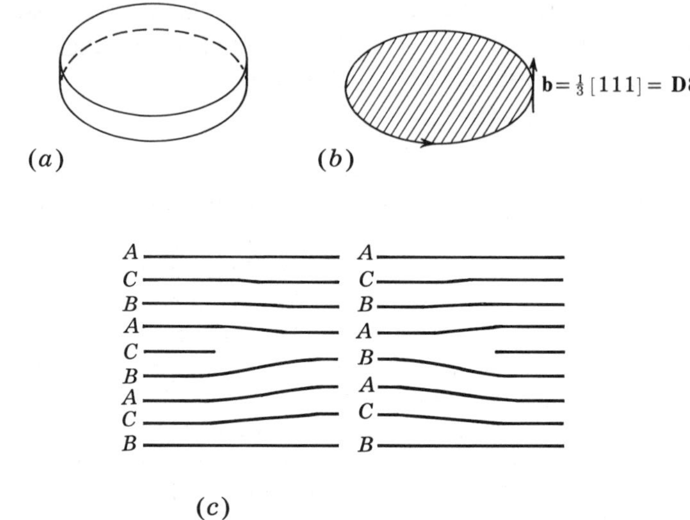

FIGURE 10-14. Collapse of (*a*) a vacancy disc to form (*b*) a loop of Frank sessile dislocation bounding an intrinsic stacking fault; (*c*) a cross section of the loop.

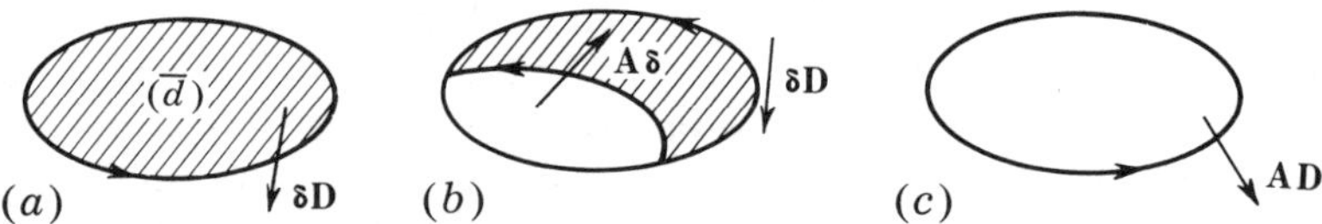

FIGURE 10-15. Nucleation and growth of a Shockley partial within a faulted Frank sessile loop.

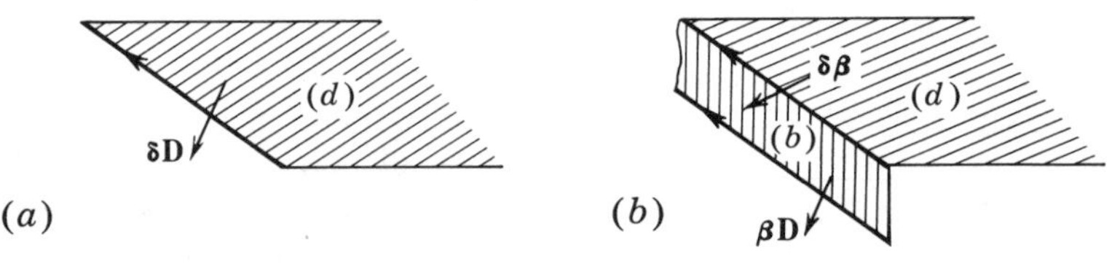

FIGURE 10-16. The dissociation $\boldsymbol{\delta D} \rightarrow \boldsymbol{\delta\beta} + \boldsymbol{\beta D}$.

perfect dislocation can be formed by the nucleation and growth of a pair of loops of Shockley partials.

The Frank partial is sessile, since its glide would produce a high-energy fault. However, as is evident in Fig. 10-14*c*, climb of a Frank partial is possible, because such climb simply extends or contracts the intrinsic (or extrinsic) stacking fault. Indeed, the process of formation of the Frank partial loop can be regarded as a climb process.

If the Frank partial-dislocation line lies along a $\langle 110 \rangle$ direction, it can dissociate with a decrease in elastic energy according to the reaction $\boldsymbol{\delta D} \rightarrow \boldsymbol{\delta\beta} + \boldsymbol{\beta D}$. Figure 10-16 illustrates this reaction for a dislocation lying along

AC. This type of dissociation is important in the formation of stacking fault tetrahedra, extended jogs, and faulted dipoles.

10-4. ARRAYS OF PARTIALS IN FCC CRYSTALS

Extended Superjogs

Superjogs were introduced in Sec. 8-9. The extension of such superjogs into faulted configurations has an important bearing on a number of deformation processes, including jog drag, cross slip, climb, and twinning, all discussed in later sections. Extended superjogs in the edge dislocation **BA**(d) are shown in Fig. 10-17. These arrays are simply combinations of two bends of the type depicted in Fig. 10-12. The stability of the superjogs in Fig. 10-17, of course, depends on the stability of the stair-rod dislocations. The line tension of the stair rods causes some constriction in the jogs as illustrated in Fig. 10-17. Should the stair rods prove to be unstable compared to the contracted node, contracted jogs such as the one shown in Fig. 10-18 would be the stable jog form. Recent studies by Carter and Hirsch[29] and Carter[30] indicate that while the degree of constriction is marked, there is some extension of jogs of all heights, particularly near screw orientation of the total dislocation. When the latter is near $\beta = 60°$ however, jogs appear completely contracted.[30] In this orientation one partial has edge character and, as shown by Eq. (6-84), weak resistance to bow-out; thus this factor plus the line tension of the stair rods contracts the jog toward the $\beta = 30°$ partial.[31] Calculations of elastic energies of jogs using the methods of Chap. 6 support the concept that jogs of all heights are extended and indicate that the maximum degree of constriction should occur for jogs of intermediate height, $\sim r_e/6$, where r_e is the separation of an unjogged dislocation [Eq. (10-15)].[30,32]

As the line of the dislocation containing the jog is rotated away from the edge orientation, a number of interesting possibilities arise, most of which were first discussed by Hirsch.[33] Figure 10-19*a* shows the projection normal to plane (d) of the obtuse edge superjog of Fig. 10-17. As the dislocation is rotated in a clockwise manner, the configuration of Fig. 10-19*b* forms. The length of the stair rods is decreased because of the line tension of **δA** and **Bδ**. When the rotation suffices for the achievement of the screw orientation, as in Fig. 10-19*c*, the superjog should be completely contracted by the line-tension forces.

[29] C. B. Carter and P. B. Hirsch, *Phil. Mag.*, **35:** 1509 (1977).

[30] C. B. Carter, *Phys. Stat. Solidi*, **54a:** 395 (1979).

[31] In the anistropic elastic case the edge partial can become unstable, corresponding to a negative line tension, and this trend is even more pronounced.

[32] T. Jøssang, *Can. J. Phys.*, **45:** 3357 (1967).

[33] P. B. Hirsch, *Phil. Mag.*, **7:** 67 (1962).

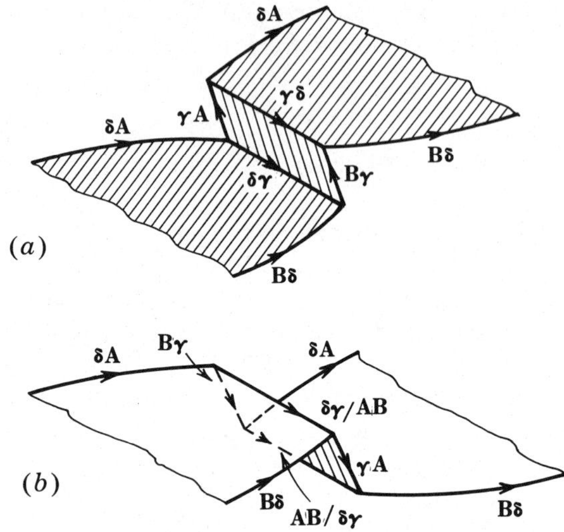

FIGURE 10-17. Superjogs: (*a*) acute and (*b*) obtuse.

Further rotation produces configuration (*d*), which becomes the acute jog of Fig. 10-17 for a rotation of π from the original orientation.

Hirsch suggested that configuration (*f*) would form for screw orientations, but to form (*f*) from (*c*) the intermediate configuration (*e*) must be attained so that **BA** can dissociate on (*b*). Work must be done to supply the extra line length in jog (*e*) in forming it from jog (*c*), so we do *not* expect jog (*f*) to form in the rotation operation. Rather, we expect that the contracted array (*c*) should be retained. However, as discussed in Chap. 12, jogs with the configuration (*e*) are formed in the process of dislocation intersection; in such a case jog (*f*) would then form from jog (*e*). Notice that one partial in the extended jog (*f*) is a Frank partial **B**$\boldsymbol{\beta}$. The inset shows a possible additional dissociation **B**$\boldsymbol{\beta} \rightarrow$ **B**$\gamma + \gamma\boldsymbol{\beta}$.

In the process of intersection, a dislocation **BA**(*d*) could acquire a jog whose line was along *CD*, as shown in Fig. 10-20. For the screw orientation of the dislocation, configuration (*a*) is stable. This jog contains the Frank partial

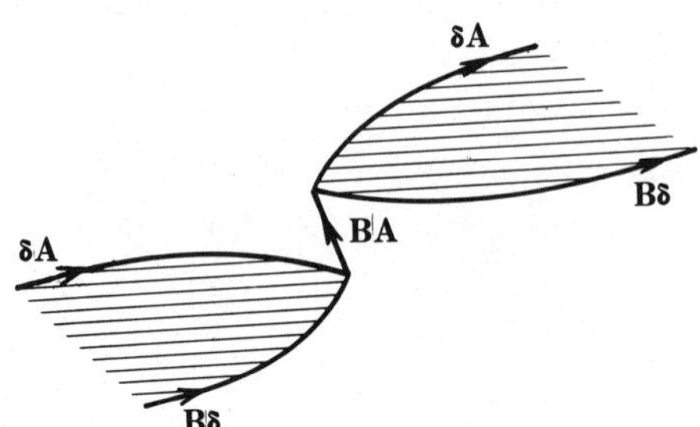

FIGURE 10-18. A contracted superjog.

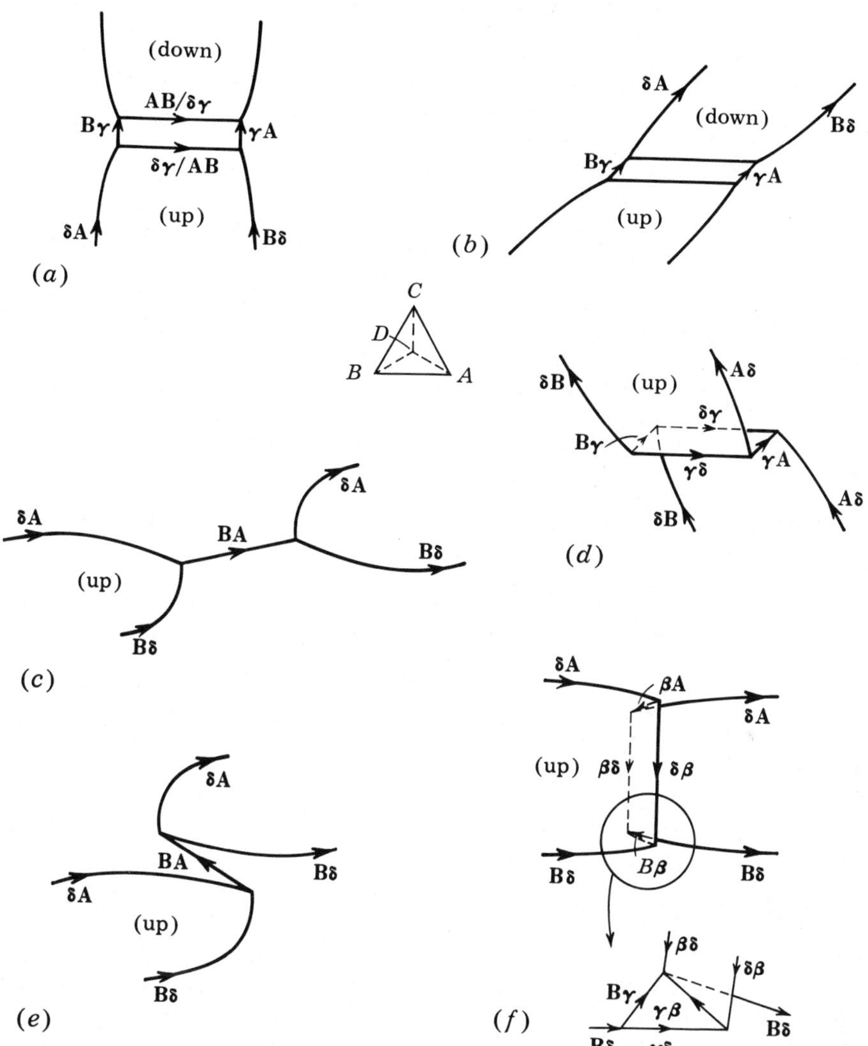

FIGURE 10-19. Projections normal to (*d*) of the obtuse superjog of Fig. 10-17*b*. The side labeled "up" is closer to the viewer than the "down" side. A projection of the Thompson tetrahedron is also shown.

αA, which can split further, as shown in the inset. As the dislocation is rotated clockwise, configuration (*b*) is formed in the edge orientation, and with further rotation, configuration (*c*), the mirror image of jog (*a*), is formed. If the line of jog (*a*) is rotated counterclockwise, the stair rods contract in length under line tension forces, as shown in Fig. 10-20*d*. Hirsch suggested that jog (*e*) would then form. Complete contraction of the jog is required in the formation process, but this would be facilitated by the edge orientation of **δA** in the jog (*d*) as discussed previously. Also, high-energy stair rods **AB**/**βδ** are present in

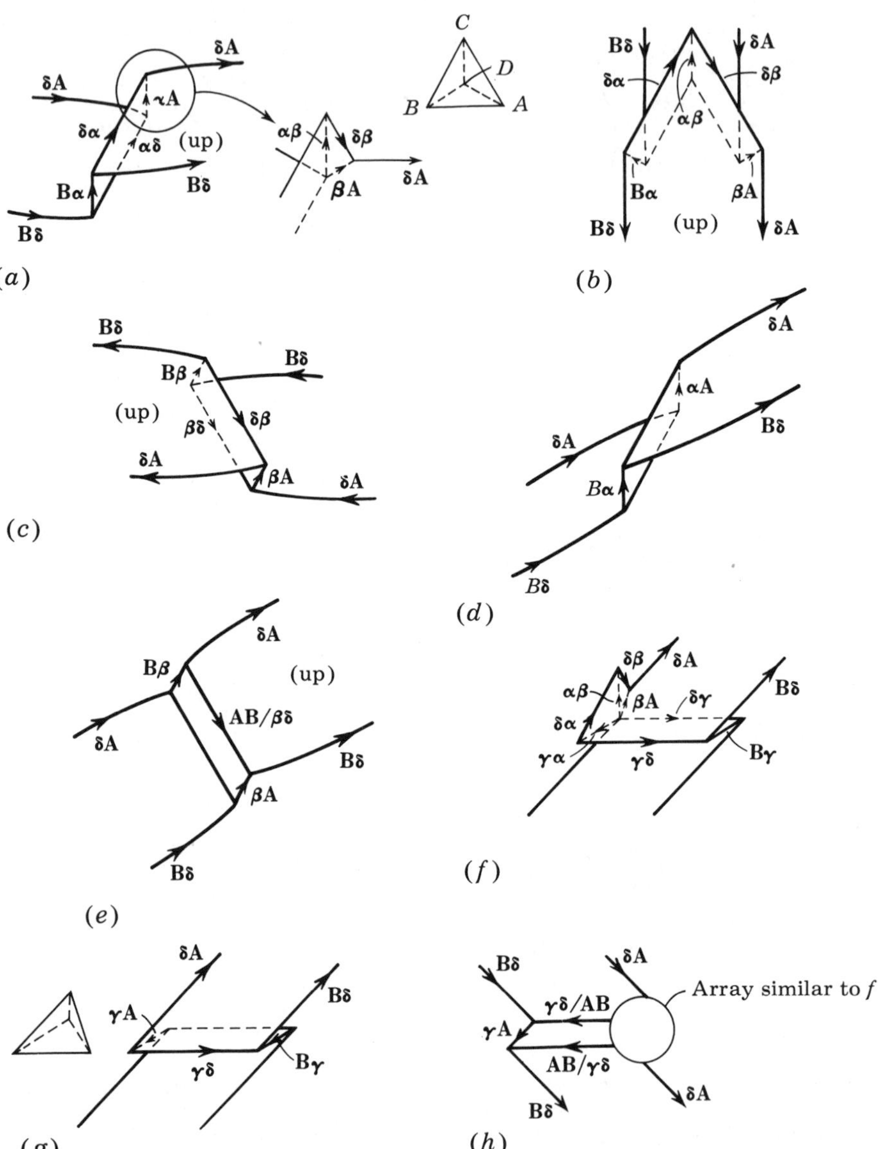

FIGURE 10-20. Projections normal to (d) of the acute superjog of Fig. 10-17a.

jog (e), so that it seems to us an unlikely configuration. It is more likely that the partial **B**$\boldsymbol{\alpha}$ dissociates instead on (c), **B**$\boldsymbol{\alpha} \rightarrow \boldsymbol{\gamma\alpha} + $**B**$\boldsymbol{\gamma}$, leading to configuration (f). The details at the left side of jog (f) are complicated, but with a net force on **BA**, configuration (g) is formed, leaving behind a stacking-fault tetrahedron. Finally, **B**$\boldsymbol{\alpha}$ in jog (b) can dissociate further to yield configuration (h), particularly when net forces act on **BA** because of an applied stress. The

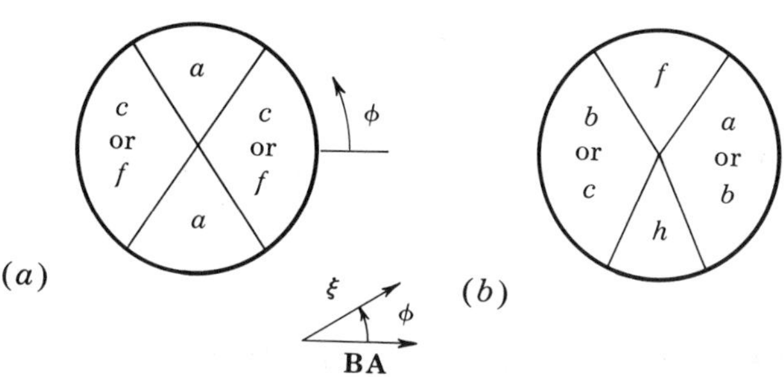

FIGURE 10-21. Regions of stability of the various configurations of (*a*) Fig. 10-19 and (*b*) Fig. 10-20 as a function of ϕ, the angle between **BA** and ξ.

regions of stability of the various configurations in Figs. 10-19 and 10-20 are presented schematically in Fig. 10-21 in the notation suggested by Hirsch.

These superjogs are important in the glide of jogged dislocations. In Fig. 10-19 the jogs (*a*) through (*e*) are glissile on (*c*). The partial $\boldsymbol{\beta}$**A** in jog (*f*) is glissile on (*b*), and jog (*c*) can dissociate into a partial **B**$\boldsymbol{\alpha}$ which is glissile on (*a*). The superjogs are sessile with respect to all other glide possibilities. Similar considerations apply to the superjogs in Fig. 10-20. All the jogs in screw dislocations are at least partly sessile, except jog (*c*) in Fig. 10-19; all the jogs in edge dislocations are glissile on one plane, except jog (*b*) in Fig 10-20.

Consider the stability of the pair of like-sign extended jogs shown in Fig. 10-22. Two unextended like-sign jogs repel, according to Eq. (8-93). For separations $l>h$ of the extended jogs, Eq. (8-93) still is valid (St. Venant's principle), so that there is a long-range elastic repulsion between like-sign extended jogs. However, for $l<h$, the attractive forces between $\boldsymbol{\gamma\delta}/\mathbf{AB}$ and $\mathbf{AB}/\boldsymbol{\delta\gamma}$ become important, and for $l\rightarrow 0$, the segments $\mathbf{B}\boldsymbol{\gamma}$ and $\mathbf{B}\boldsymbol{\gamma}$ also attract one another. Therefore there is a short-range elastic attraction between the extended jogs. Thus the accumulation of small jogs into a large superjog is unlikely in the absence of external forces, but once formed, a superjog is metastable with respect to dissociation.

Extended Unit Jogs or Jog Lines

In the case of extended unit jogs, the stair rods become a stair-rod dipole,[34] with a spacing about equal to a typical dislocation core diameter. In view of the misfit relaxations in the core region, it is questionable whether the extended-superjog models are applicable to unit jogs. However, the application of the superjog model is useful in suggesting that the "line of misfit" of the stair-rod dipole might have a sufficiently low energy for unit-jog extension to occur. It seems certain, though, that unit jogs, if extended are less extended

[34]A dislocation dipole consists of two parallel dislocations with equal and opposite Burgers vectors. Consequently, the elastic-strain energy of the pair is localized in the vicinity of the pair (St. Venant's principle) in analog to the localization of the electrostatic energy of an electric dipole.

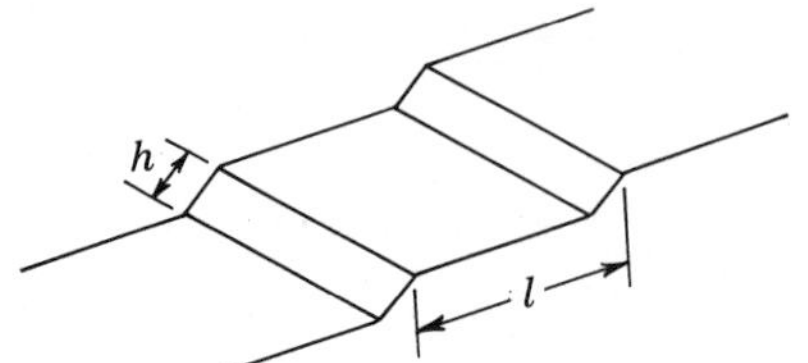

FIGURE 10-22. Two like-sign jogs in **BA**(d).

than superjogs, because the ratio of the repulsive forces associated with the Shockley partials bounding the jog to the attractive line-tension forces associated with the stair rods is less for unit jogs than for superjogs. Quantitative calculations which would clarify these points are difficult to perform because of the large uncertainties in the misfit energy (Chap. 8), so that we cannot meaningfully add to the above qualitative speculations.

A useful model of extended unit jogs in the hard-ball approximation of the fcc lattice has been proposed by Thompson,[35] who called the resulting configurations *jog lines*. Consider the projection in Fig. 10-23, as viewed along the line of a stair-rod dipole in a unit jog. The Burgers vectors are $\pm\frac{1}{6}[\bar{1}\bar{1}0]$, the atom spacing along the dipole line is $\frac{1}{2}[1\bar{1}0]$, and the normal distance between the "glide" planes of the stair rods is $\frac{1}{2}[001]$. Thus the vacant volume per atom site is

$$\left(\tfrac{1}{6}[\bar{1}\bar{1}0]\times\tfrac{1}{2}[1\bar{1}0]\right)\cdot\tfrac{1}{2}[001]=\tfrac{1}{12}a_0^{\,3}$$

The volume of an atom is $a_0^{\,3}/4$ so that the stair-rod dipole $\boldsymbol{\delta\gamma}$-$\boldsymbol{\gamma\delta}$ in the unit jog corresponds to a row of one-third vacancies. Similarly, the stair-rod dipole $\boldsymbol{\gamma\delta}$/**AB**-**AB**/$\boldsymbol{\gamma\delta}$ corresponds to a row of one-third interstitials, and $\boldsymbol{\delta}$**D**/**C**$\boldsymbol{\gamma}$-**C**$\boldsymbol{\gamma}$/$\boldsymbol{\delta}$**D** to a row of two-third interstitials; all of these dipoles contain only pure edge partials. The dipole $\boldsymbol{\delta\alpha}$/**AB**-**AB**/$\boldsymbol{\delta\alpha}$ corresponds to a row of two-third vacancies, because of the edge components of the partials, plus a screw dipole. According to Table 10-2, these include all the stable stair-rod possibilities. Some examples of these jog lines, following Hirsch, are shown in Fig. 10-24, but the reader can most easily envision them by constructing a marble model of his own.

All of the above discussion has dealt with jog lines lying along $\langle 110\rangle$ directions. In principle, the jog lines can lie in any direction normal to the $\{111\}$ pole of the dissociated dislocations, because the concept of an extension of the jog restricting it to a $\{111\}$ plane and the jog line to a $\langle 110\rangle$ direction breaks down for a unit jog. However, a consideration of the atomic arrangements, as in Fig. 10-24, indicates that the jog-line energy should be a minimum when it lies along a low-index direction. Jog lines lying along $\langle 112\rangle$ directions are another possible low-energy set, as discussed by Pfeffer et al.[36]

[35] N. Thompson, in "Defects in Crystalline Solids," Physical Society, London, 1955, p. 153.
[36] K. H. Pfeffer, P. Schiller, and A. Seeger, *Phys. Stat. Solidi*, **8:** 517 (1965).

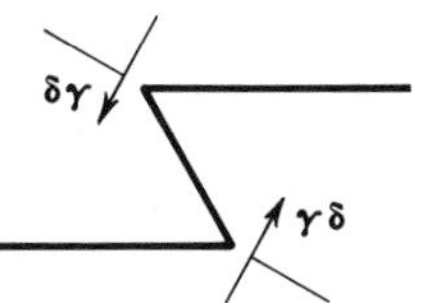

FIGURE 10-23. Projection along **AB** of the stair-rod dipole $\boldsymbol{\delta\gamma} - \boldsymbol{\gamma\delta}$.

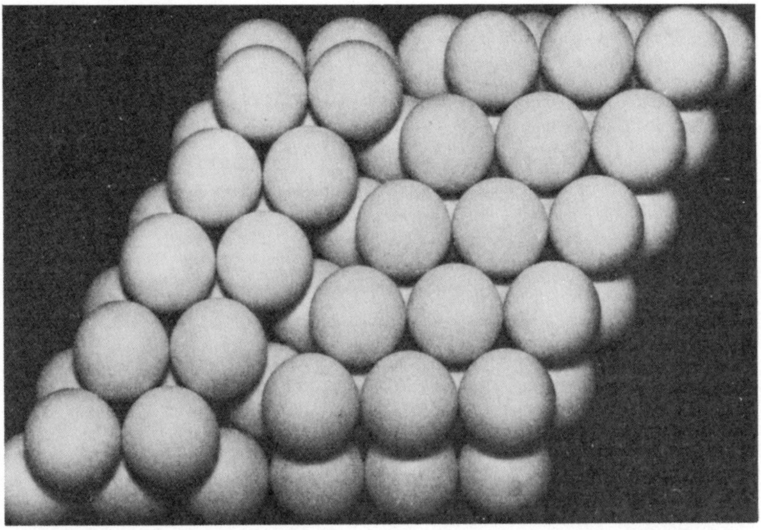

(*a*)

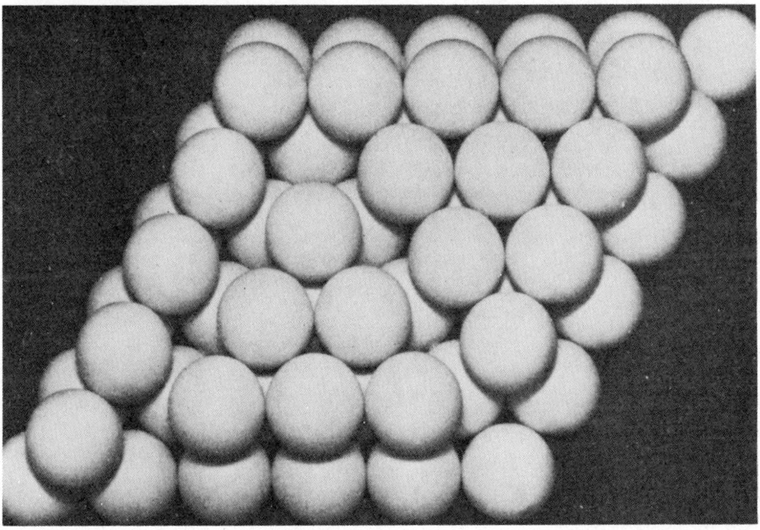

(*b*)

FIGURE 10-24. (*a*) View of the one-third-vacancy jog line along **BC**, shown as an extended jog in Fig. 10-20*a*. (*b*) View of Fig. 10-20*b* represented as two one-third-vacancy jog lines along **BC** and **CA**. [After P. B. Hirsch, *Phil. Mag.*, **7**: 67 (1962).]

Because of the uncertainties in the core structure, the energy of the jog lines is difficult to estimate. However, crude estimates can be made on the basis of the above models for the lines. If the jog lines are regarded as dislocation pairs, their energy should be roughly proportional to b^2. Table 10-2 indicates that the energies of $\boldsymbol{\delta\gamma}$, $\boldsymbol{\delta\gamma}/\mathbf{BA}$, $\boldsymbol{\delta}\mathbf{D}/\mathbf{C}\boldsymbol{\gamma}$, and $\boldsymbol{\delta\gamma}/\mathbf{BD}$ increase in the ratios 1:2:4:5. On the other hand, one can assert that the energies of the part vacancy or interstitial lines should be related to the respective formation energies of the point defects. Carter[37] has found evidence for unit jogs that could be interpreted as either lying along $\langle 112 \rangle$ or composed of pieces of $\langle 110 \rangle$ as in Fig. 10-20*b*. The formation energy of an interstitial is about three times that of a vacancy.[38,39] Thus the ratio of the above sequence in the latter approximation is 1:3:6:2. The screw dipole portion of $\boldsymbol{\delta\gamma}/\mathbf{BD}$ is not accounted for in this approximation, which might explain the anomalous position of $\boldsymbol{\delta\gamma}/\mathbf{BD}$ in the second ratio sequence compared to the first.

Stacking-Fault Tetrahedra

Stacking-fault tetrahedra are found in metals of low stacking-fault energy. Usually they are formed by vacancy condensation. Silcox and Hirsch[40] first observed such tetrahedra in quenched gold foils; they have also been found in quenched silver[41] and nickel-cobalt alloys.[42] Examples of tetrahedra in gold[43,44] are presented in Fig. 10-25. One mechanism of formation of tetrahedra for which there is supporting evidence is that of extension of a Frank partial-dislocation loop formed by vacancy condensation.

Consider a loop of intrinsic fault bounded by the Frank sessile $\delta\mathbf{D}$ as shown in Fig. 10-26*d*. Those portions of the loop which have line directions near $\langle 110 \rangle$ can dissociate by the reaction of Fig. 10-16, yielding the configuration of Fig. 10-26*a*. The decrease in energy accompanying the dissociation gives rise to forces that pull more of the original line of $\delta\mathbf{D}$ into $\langle 110 \rangle$; the attractive interaction forces between $\beta\mathbf{D}$, $\gamma\mathbf{D}$, and $\alpha\mathbf{D}$ also cause the length of the dissociated segments to increase. These forces, plus any caused by favorably oriented shear stresses, lead to the development of configurations (b) and (c) in sequence. The final result is a tetrahedron bounded completely by intrinsic stacking faults and stair-rod dislocations.

[37] C. B. Carter, *Phys. Stat. Solidi*, **54a:** 395 (1979).

[38] A. Seeger, D. Schumacher, W. Schilling, and J. Diehl, "Vacancies and Interstitials in Metals," North Holland, Amsterdam, 1970, p. 1.

[39] H. Wenzl, ibid., p. 363.

[40] J. Silcox and P. B. Hirsch, *Phil. Mag.*, **4:** 72 (1959).

[41] R. E. Smallman, K. H. Westmacott, and J. A. Coiley, *J. Inst. Met.*, **88:** 127 (1959).

[42] S. Mader and E. Simsch, *Proc. European Regional Conf. Electron Microscopy* (Delft), 1960, p. 379.

[43] J. Silcox and P. B. Hirsch, *Phil. Mag.*, **4:** 72 (1959).

[44] R. M. J. Cotterill, private communication.

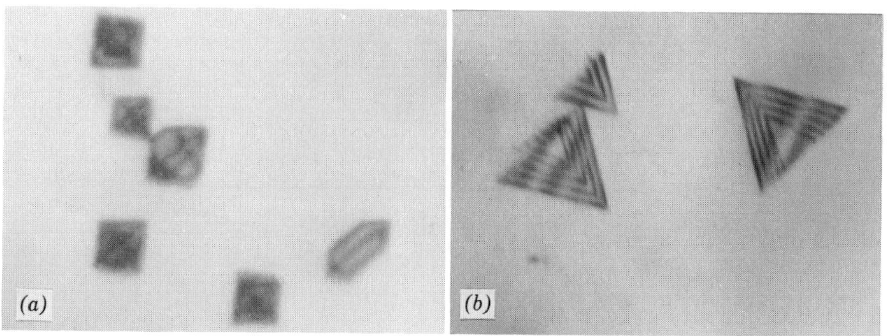

FIGURE 10-25. Stacking-fault tetrahedra in gold viewed by electron-transmission microscopy: (*a*) viewed along ⟨100⟩ [J. Silcox and P. B. Hirsch, *Phil. Mag.*, **4**:72 (1959)], (*b*) viewed along ⟨211⟩ [R. M. J. Cotterill, private communication]. 280,000×

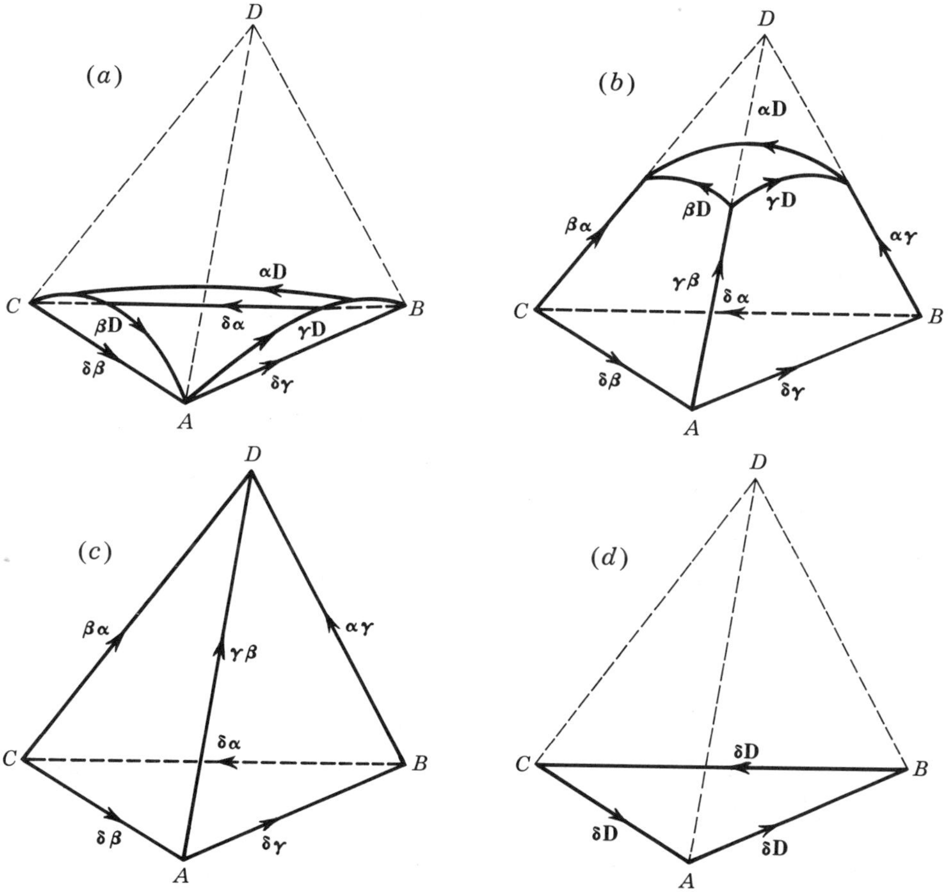

FIGURE 10-26. Various stages (*a*) to (*c*), in the formation of a stacking-fault tetrahedron from (*d*) a Frank sessile loop.

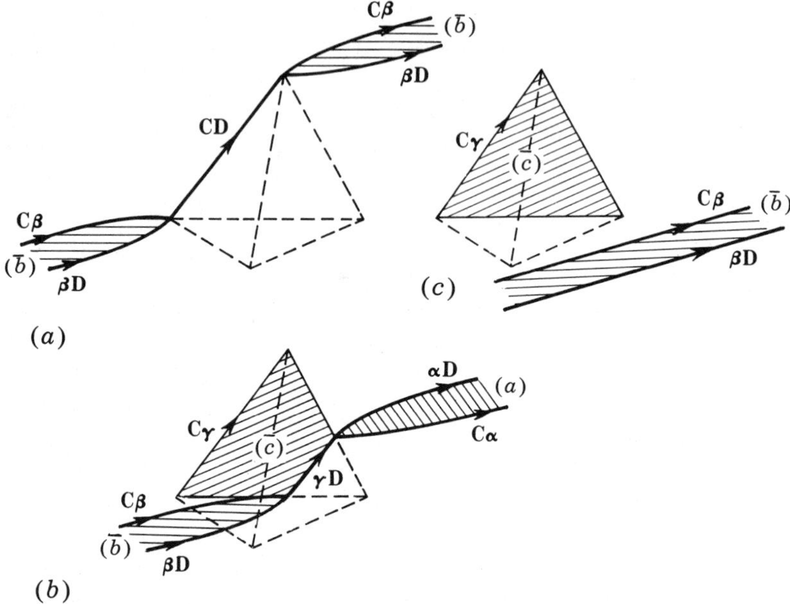

FIGURE 10-27. Formation of a Frank sessile triangle by cross slip at a jog. The triangle then transforms to a tetrahedron, as in Fig. 10-26.

Tetrahedra can also be formed by the glide of jogged screw dislocations. Loretto et al.[45] showed that tetrahedra were produced in deformed silver, gold, copper, copper alloy, gold alloy, and nickel-cobalt crystals. One mechanism of tetrahedron formation by glide is shown in Fig. 10-20*g*. A second possible glide mechanism is shown in Fig. 16-21. A third mechanism,[45] involving cross slip of a jogged screw dislocation, is presented in Fig. 10-27. Loretto et al.[45] find evidence for the last mechanism in that they observed Frank sessile loops only above a given size, 23.0 nm for gold, while tetrahedra were observed only up to 20.0 nm, suggesting that they formed from the Frank loops smaller than the minimum observed size. This finding supports the original idea of Silcox and Hirsch that the tetrahedra in quenched crystals can also grow by dissociation of a Frank sessile loop. The above result supports the idea that there is a formation constraint in creating extrinsic fault configurations. If the original sessile loop were to extend into extrinsic faults, obtuse bends between the fault planes would be formed, and the dissociation would produce a truncated pyramid with the original loop at its apex. The absence of such configurations indicates that their formation is energetically less favorable.

Finally, tetrahedra can nucleate from a relaxed trivacancy configuration and grow by the climb process of vacancy absorption at jog lines,[46,47] as shown in

[45] M. H. Loretto, L. M. Clarebrough, and R. L. Segall, *Phil. Mag.*, **11:** 459 (1965).

[46] M. De Jong and J. S. Koehler, *Phys. Rev.*, **129:** 49 (1963).

[47] H. Kimura, D. Kuhlmann-Wilsdorf, and R. Maddin, *Appl. Phys. Lett.*, **3:** 4 (1963).

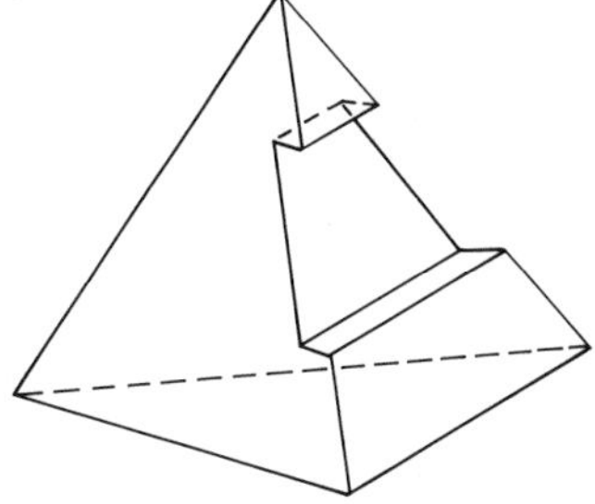

FIGURE 10-28. Jog lines forming steps on a tetrahedron face.

Fig. 10-28. Some evidence exists for such a mechanism,[48] but we defer discussion of it to Sec. 17-7.

The process of collapse of the tetrahedra is of interest, particularly with respect to stacking-fault-energy measurements. As the tetrahedra increase in size, they eventually become unstable with respect to collapse because of their increasing stacking-fault area. One possible mechanism of collapse is by the emission of vacancies at jog lines. Another collapse mechanism is the reverse glide process[49] of Fig. 10-26. The energies of the tetrahedron, of various possible collapsed arrays, and of the intermediate stage shown in Fig. 10-29 can be readily assessed by the methods of Chap. 6; provided the configurations actually involve straight line segments, the energies are exact. Even if the actual configuration is curved, as in Fig. 10-26, the straight-line approximation is a good one. Consideration of the variation of the energy of the truncated tetrahedron as a function of the reduced variable $x = L'/L$ enables one to determine the range of metastability of the tetrahedra.

Let us introduce the reduced variables

$$w = \frac{6\pi(1-\nu)W}{\mu b^3} \qquad \Gamma = \frac{6\sqrt{3}\,\pi(1-\nu)\gamma}{\mu b}$$

$$y = \frac{L}{b} \qquad b = \tfrac{1}{2}\langle 110\rangle \qquad x = \frac{L'}{L}$$

where, W = total energy of a given array, γ = stacking-fault energy, L = length of tetrahedron edge, and L' = length of the truncated edge (Fig. 10-27). The results are all normalized to arrays containing equal numbers of vacancies, and the line lengths of the various arrays are converted to equivalent lengths of tetrahedra edge. Thus L is a measure of the number of vacancies in a given configuration.[50] The formulas for the energies are then given by Eqs. (6-18), (6-32), and (6-40) to be

[48] R. L. Segall and L. M. Clarebrough, *Phil. Mag.*, **9:** 865 (1964).

[49] D. Kuhlmann-Wilsdorf, *Acta Met.*, **13:** 257 (1965).

[50] For example, a hexagon of edge length L'' contains an equal number of vacancies as a tetrahedron if $6(\sqrt{3}/4)L''^2 = (\sqrt{3}/4)L^2$. The lengths L'' that appear in the energy of the hexagon are converted to L by this expression.

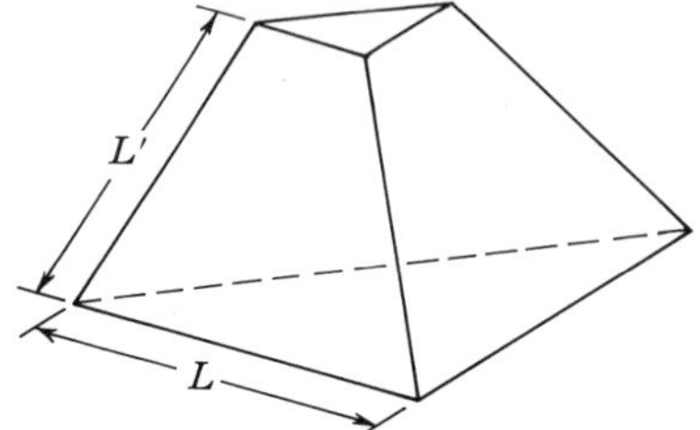

FIGURE 10-29. A truncated tetrahedron.

Frank sessile $\frac{1}{3}[111]$ triangle:

$$w = 3y(\ln \alpha y - 0.20) + \Gamma y^2/4$$

Frank sessile $\frac{1}{3}[111]$ hexagon:

$$w = 2.45y(\ln \alpha y + 0.16) + \Gamma y^2/4$$

Perfect $\frac{1}{2}[110]$ triangle:

$$w = 4.50y[\ln \alpha y - 1.41 + 2.88/(6-\nu)](1-\nu/6)$$

Perfect $\frac{1}{2}[110]$ hexagon:

$$w = 3.67y[\ln \alpha y - 1.06 + 4/(6-\nu)](1-\nu/6) \qquad (10\text{-}17)$$

Tetrahedron:

$$w = y(\ln \alpha y + 1.03 + 0.96\nu) + \Gamma y^2$$

Truncated tetrahedron:

$$w = 0.5y[2(2-x)\ln \alpha y + F(x) + \nu G(x)] + \Gamma y^2[1 - 0.75(1-x^2)]$$

The functions $F(x)$ and $G(x)$ are graphically presented in Fig. 10-30. Numerical results for the truncated tetrahedron were presented by Czjzek et al.[51] for the special case of gold. The above analytical results[52] are more generally applicable to any material. For the case of gold, the exact (isotropic elasticity) results[52] of Eq. (10-17) agree with the numerical results of Czjzek et al. within 25 percent for the energy but differ markedly with respect to the position of energy maximum with respect to x.

Energy results for the truncated tetrahedron for the case of gold are presented in Fig. 10-31. The pertinent elastic constants are the average

[51] G. Czjzek, A. Seeger, and S. Mader, *Phys. Stat. Solidi*, **2:** 558 (1962).

[52] T. Jøssang and J. P. Hirth, *Phil. Mag.*, **13:** 657 (1966). See also P. Humble, R. L. Segall and A. K. Head, *Phil. Mag.*, **15:** 281 (1967).

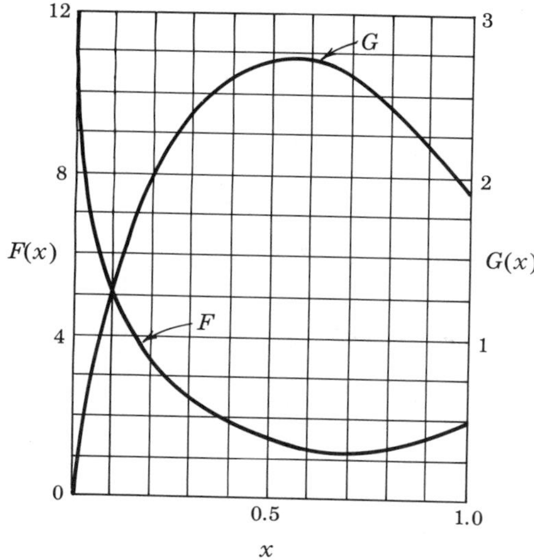

FIGURE 10-30. The functions $F(x)$ and $G(x)$ appearing in Eq. (10-17).

anisotropic modulus $\mu = 31.0$ GPa and Poisson's ratio $\nu = 0.412$, given by Eqs. (13-5) and (2-50), $\alpha = 4$, and $\gamma = 55$ mJ/m^2, determined as described below. As shown in Fig. 10-31, the tetrahedron is stable and should form without an activation barrier up to $L_c = 20.0$ nm. Above $L_c = 20.0$ nm, the tetrahedron is stable with respect to the Frank triangle but can form only if an activation barrier is surmounted. The tetrahedron becomes unstable when $L_E = 26.5$ nm but remains metastable with respect to collapse until $L_T = 580.0$ nm. Moreover, the activation-energy barriers Q are so large that the probability of thermal fluctuation over the barrier, $e^{-Q/kT}$, is essentially zero until the critical lengths cited above are reached.

The maximum resolved shear stresses τ_a required to force the Shockley partials over the energy barriers are listed in Fig. 10-31. For triangles with side lengths exceeding L_c by 5 percent ($L = 21.0$ nm) and by 15 percent ($L = 23.0$ nm), the required stresses are $(5 \times 10^{-4})\mu$ and $(2 \times 10^{-3})\mu$, respectively. The smaller of these exceeds the yield stress for gold single crystals by one order of magnitude. Thus, in the absence of marked cold work, these critical lengths should delineate the observed ranges of stability of tetrahedra.

In the experiments of Loretto et al. on gold,[53] the observed tetrahedra formed at room temperature by the extension of Frank triangles. The largest tetrahedron was $L = 20.0$ nm, and the smallest Frank triangle was $L = 23.0$ nm. Therefore, taking the critical length as 20.0 nm , Eq. (10-17) must be solved for a γ value that just gives an inflection point in Fig. 10-31. The result obtained from an iterative numerical analysis is $\gamma = 55$ mJ/m^2; γ values determined in

[53] M. H. Loretto, L. M. Clarebrough, and R. L. Segall, *Phil. Mag.*, **11:** 459 (1965).

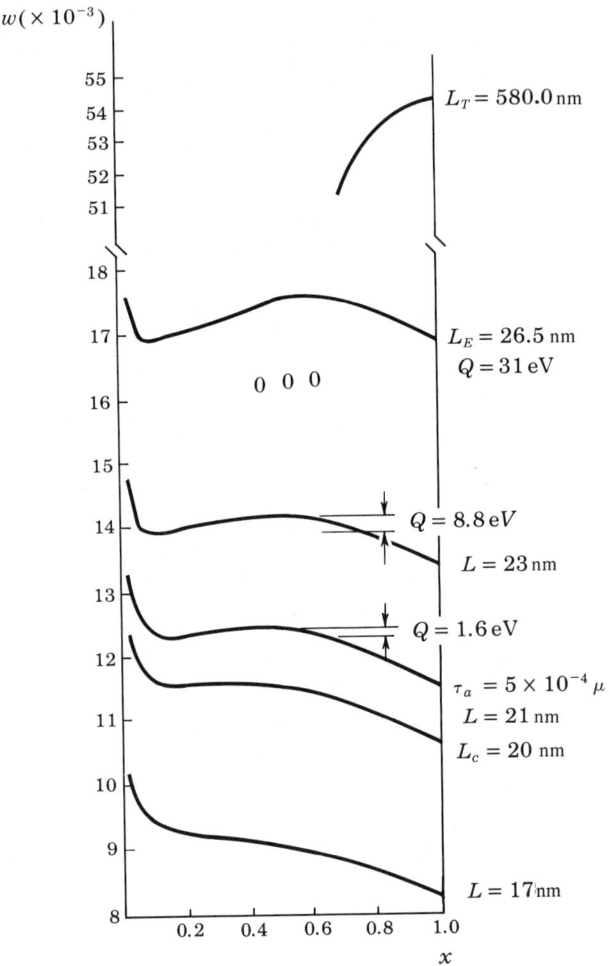

FIGURE 10-31. The reduced energy w versus x for various sizes of tetrahedra. For curves with maxima and minima, $Q = W_{max} - W_{min}$ is shown, together with the minimum resolved shear stress τ_a required to force the Shockley partials over the energy barrier. Note scale change at top.

this way are about 20 percent smaller than more recent values listed in Appendix 2 which are based on anisotropic elastic analysis.

Because the tetrahedra are metastable up to $L_T = 580.0$ nm once they are formed, one expects them to be able to grow to metastable sizes by vacancy absorption at jog lines.[54] A number of observations of such growth have been

[54]A. Seeger and D. Schumacher, in R. M. J. Cotterill et al. (eds.), "Lattice Defects in Quenched Metals," Academic Press, New York, 1965, p. 15.

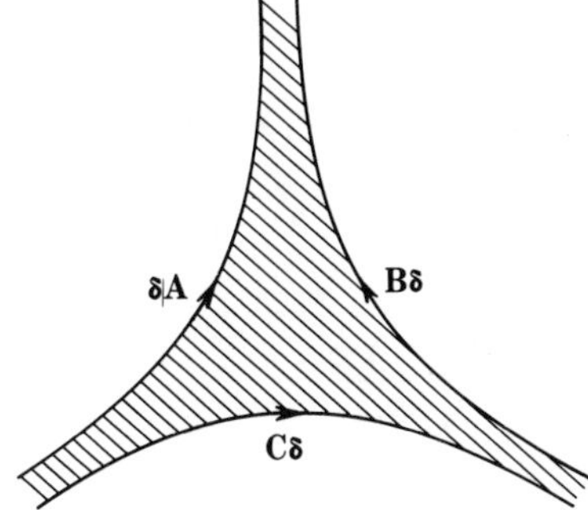

FIGURE 10-32. An extended dislocation node.

made.[55-58] Because the activation energy for glide collapse is prohibitively high for sizes approaching $L_T = 580.0$ nm, and observed maximum tetrahedra sizes[59] are much less than L_T, collapse by the inverse glide process of that in Fig. 10-26 appears to be ruled out. Application of nucleation theory[60] indicates that collapse by partial-dislocation nucleation in one of the faulted areas[61] is also unlikely. Thus the most likely collapse mechanisms are vacancy emission or interaction of a tetrahedron with a glissile dislocation.

Extended Nodes

Extended dislocation nodes also can be used to estimate stacking-fault energies.[62] Consider the threefold node connecting **AB**, **BC**, and **CA** on (d). If the dislocations are dissociated to form an intrinsic stacking fault, the extended configuration in Fig. 10-32 is developed. Provided that the node is isolated from free surfaces and other dislocations, the equilibrium arrangement is determined by a balance between the stacking-fault energy, which tends to constrict the node, and the self-energy and interaction energy of the partials, both of which tend to extend it. Using the method of Chap. 6 one approximates the node by the straight line segments shown in Fig. 10-33. The energy difference W between the extended straight-line array and the contracted

[55] M. De Jong and J. S. Koehler, *Phys. Rev.*, **129:** 49 (1963).

[56] R. L. Segall and L. M. Clarebrough, *Phil. Mag.*, **9:** 865 (1964).

[57] T. Mori and M. Meshii, *Acta Met.*, **12:** 104 (1964).

[58] W. Westdorp, H. Kimura, and R. Maddin, *Acta Met.*, **12:** 495 (1964).

[59] See R. M. J. Cotterill, M. Doyama, J. J. Jackson, and M. Meshii (eds.), "Lattice Defects in Quenched Metals," Academic Press, London, 1965.

[60] J. P. Hirth, in "Relation between Structure and Mechanical Properties of Metals," H.M.S.O., London, 1963, p. 217.

[61] M. Meshii and J. W. Kauffmann, *Phil. Mag.*, **5:** 939 (1960).

[62] M. J. Whelan, *Proc. Roy. Soc.*, **A249:** 114 (1959).

straight line configuration (dashed lines in Fig. 10-33) is[63]

$$\frac{W8\pi(1-\nu)}{3\mu b^2}=(1-\nu\cos^2\phi)\left[-\frac{d}{\sqrt{3}}\ln\frac{2^3u(w+u)(w+\sqrt{3}\lambda/2)^4}{\sqrt{3}\,(\sqrt{3}+1)^2d^6}\right.$$

$$+d\ln\frac{2^2(w+v)(w+\lambda+d/2)^2}{3\sqrt{3}\,d^2u}$$

$$\left.+\lambda\ln\frac{\sqrt{3}(w+v)\lambda^2}{e^2ud^2}\right]$$

$$\left.+\frac{\lambda}{\sqrt{3}}\ln\frac{e^6 2^3\sqrt{3}\,(\sqrt{3}-1)^{10}(w+u)d^6}{u(w+\sqrt{3}\lambda/2)^2\lambda^4}\right]$$

$$+\nu\cos 2\phi\left[\frac{\lambda}{\sqrt{3}}\ln\frac{\sqrt{3}\,(\sqrt{3}-1)^4(w+u)(w+\sqrt{3}\lambda/2)}{u\lambda}\right.$$

$$\left.-\frac{d}{\sqrt{3}}\ln\frac{2^3u(w+u)}{\sqrt{3}\,(\sqrt{3}+1)^2(w+\sqrt{3}\lambda/2)^2}\right]$$

$$+2(2-\sqrt{3})(1-\nu\cos^2\phi+\nu\cos 2\phi)\frac{\lambda}{\sqrt{3}}\ln\frac{\rho}{d}$$

$$+[2-\nu(1+2\cos 2\phi)]\frac{1}{4\sqrt{3}}\frac{\lambda^2}{d} \qquad (10\text{-}18)$$

where the following abbreviations have been used:

$$w=(\lambda^2+\lambda d+d^2)^{1/2}$$

$$u=\tfrac{1}{2}\sqrt{3}\,(\lambda+d)$$

$$v=\tfrac{1}{2}(\lambda-d)$$

[63]T. Jøssang, M. J. Stowell, J. P. Hirth, and J. Lothe, *Acta Met.*, **13:** 279 (1965); some minor typographical errors in the original paper are corrected here. Earlier approximate treatments of the node problem are reviewed in this article.

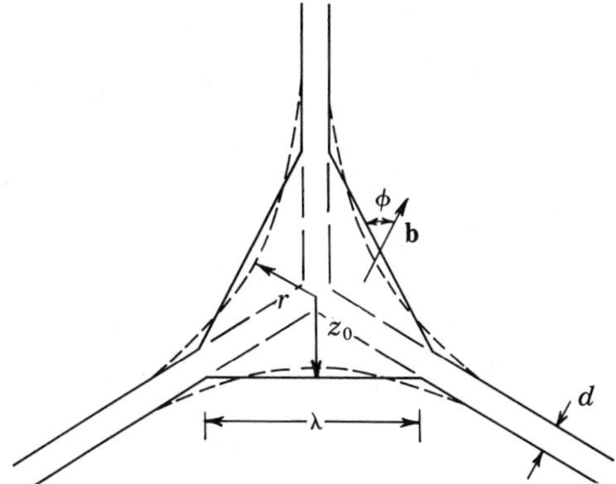

FIGURE 10-33. A straight-line approximation of an extended dislocation node.

Here d, the equilibrium spacing between parallel partials [Eq. (10-15)], is related to the core parameter $\rho = b/2\alpha$ by the expression

$$\frac{\rho}{d} = 4\pi \frac{\gamma_I}{\alpha\mu b} \frac{1-\nu}{2-\nu(1+2\cos 2\phi)} \tag{10-19}$$

and b is the modulus of Burgers vector of the Shockley partial. The extension of the node for a given value of γ_I is given in terms of the reduced parameter

$$\Sigma_0 = \sqrt{3}\,\frac{z_0}{d} = \frac{1}{2}\frac{\lambda}{d} + 1 \tag{10-20}$$

where z_0 is measured experimentally. z_0 is chosen so that the areas of stacking fault are the same for the exact configuration and the approximate straight-line configuration. In terms of the radius r of a circle inscribed in the exact configuration, $z_0 \sim 1.1r$.

At equilibrium, the variation of W with respect to z_0 (or λ) is zero. Equation (10-18) has been minimized numerically by Jøssang et al. Some of their results are presented in Table 10-3 for a screw node, defined by the condition $\phi = 0$ (Fig. 10-33). Considering the variation of node energy as a function of ϕ, one finds that screw nodes have the lowest energy; they are the type expected in experiment, but actual data indicate a prevalence of nodes rotated 30° from screw orientation.[64,65] Results for other node orientations are given in the original paper by Jøssang et al.[66]

In general, one knows ν, μ, and b for a given material, so that, selecting $\alpha \sim 1$ (Sec. 8-3), one knows Σ_0 as a function of γ_I from Table 10-3. The combination of Eqs. (10-19) and (10-20) gives another relation between γ_I and

[64] P. C. J. Gallagher, *J. Appl. Phys.*, **37:** 1710 (1966).

[65] T. Ericsson, *Acta Met.*, **14:** 853 (1966).

[66] T. Jøssang, M. J. Stowell, J. P. Hirth, and J. Lothe, *Acta Met.*, **13:** 279 (1965).

TABLE 10-3 Σ_0 Values for Equilibrium Screw Nodes as a Function of ν and of the Reduced Stacking-Fault-Energy Parameter $\gamma_I/\alpha\mu b$

	ν							
$\frac{\gamma_I}{\alpha\mu b}\times 10^4$	0.16	0.20	0.24	0.28	0.32	0.36	0.40	0.44
44.000	4.84	5.15	5.51	5.95	6.47	7.13	7.97	9.08
30.000	5.06	5.40	5.79	6.25	6.82	7.52	8.42	9.62
20.500	5.28	5.64	6.05	6.55	7.16	7.91	8.87	10.16
14.000	5.50	5.87	6.32	6.85	7.49	8.28	9.31	10.69
9.500	5.71	6.11	6.58	7.14	7.82	8.67	9.76	11.21
6.500	5.92	6.34	6.83	7.42	8.14	9.04	10.18	11.72
4.400	6.13	6.57	7.09	7.71	8.47	9.41	10.62	12.24
3.000	6.33	6.80	7.34	7.99	8.78	9.77	11.05	12.75
2.050	6.53	7.02	7.59	8.27	9.09	10.13	11.46	13.24
1.400	6.73	7.23	7.83	8.54	9.40	10.49	11.88	13.74
0.950	6.93	7.46	8.07	8.81	9.72	10.84	12.30	14.24
0.650	7.12	7.67	8.31	9.08	10.02	11.19	12.70	14.72
0.440	7.32	7.89	8.56	9.35	10.33	11.55	13.12	15.22
0.300	7.52	8.10	8.80	9.62	10.63	11.90	13.52	15.71
0.205	7.71	8.31	9.03	9.89	10.93	12.24	13.92	16.19
0.140	7.89	8.52	9.26	10.15	11.23	12.58	14.32	16.66
0.095	8.09	8.74	9.50	10.41	11.53	12.93	14.73	17.15
0.065	8.27	8.94	9.73	10.67	11.83	13.27	15.12	17.62

Σ_0 based on the measured value of z_0. Simultaneous solution of the two relations between Σ_0 and γ_I yields the value of γ_I that is consistent with the theoretical value of d and the measured distance z_0. In the less likely event that both z_0 and d can be measured experimentally, both unknowns γ_1 and α can be determined from the above relationships. This case provides one of the few means of experimentally determining the limits on the core parameter α.

Exercise 10-4. For the case of graphite,[67] a screw node yields the value $z_0 = 0.314\ \mu$m. The other pertinent parameters are $\nu = 0.24$, $\mu = 31.2$ GPa, and $b = 0.142$ nm. Use the above method to determine γ_I for the two cases $\alpha = 1$ and $\alpha = 2$. The results should be 0.61 and 0.64 mJ/m^2, respectively. Discuss the sensitivity of the results to the choice of the core parameter.

Two related theories have been developed in the isotropic elastic approximation. Brown and Tholen[68] used the method of Sec. 5-7 to make the net force vanish on 41 mesh points along one of the partials of Fig. 10-32 and presented equations that represented the results with numerical coefficients accurate to 10 percent. Siems[69] considered the variation of energy of a segment with

[67] R. Siems, P. Delavignette, and S. Amelinckx, *Z. Phys.*, **165:** 502 (1961).

[68] L. M. Brown and A. R. Tholen, *Disc. Faraday Soc.*, **38:** 35 (1964).

[69] R. Siems, *Disc. Faraday Soc.*, **38:** 42 (1964).

TABLE 10-4 Stacking-Fault Energies of Two FCC Alloys*

	$\gamma(\mathrm{mJ/m^2})$					
	Ag-12.5% In (Ref. 1)			Cu_2NiZn (Ref. 2)		
α						
0.5	4.6[3]	5.5[4]	5.5[5]	29[3]		
1	5.1[3]	6.0[4]	6.0[5]		31–36[4]	27–33[5]
2	5.5[3]	6.4[4]	6.5[5]	33[3]		
4	5.8[3]	—	—			

*Determined from node measurements[1,2] as a function of the core parameter α.

1. P. C. J. Gallagher, *J. Appl. Phys.*, **37:** 1710 (1966).
2. G. J. L. Van der Wegen, P. M. Bronsveld, and J. Th. M. De Hosson, *Met. Trans.*, **11A:** 1125 (1980).
3. T. Jøssang, M. J. Stowell, J. P. Hirth, and J. Lothe; *Acta Met.*, **13:** 279 (1965).
4. L. M. Brown, *Phil. Mag.*, **10:** 441 (1964).
5. R. Siems, *Disc. Faraday Soc.*, **38:** 42 (1964).

screw-edge character, approximated the interaction energy and minimized the energy in a variational method using the Euler-Lagrange equation. Of the three, the Brown-Tholen approach is, in principle, the most exact since it uses an exact energy-approximate configuration scheme with many segments approximating the shape, whereas Jøssang et al.[70] do the same with fewer segments and Siems uses an approximate energy. However, when approximated by the equation with coefficients, the relative accuracy of the Brown-Tholen method is less. The analytical results of Siems[69] and Jøssang et al.[70] have the simplest form.

Ruff[71] compared the theories for an hypothetical node shape and with a given value of γ_I, comparing values of the predicted size of an inscribed circle tangent to the three node partials. He found that for the same cutoff parameter the Brown-Tholen and Siems theories agreed fairly well but that the Jøssang results were lower by up to 20 percent. However, when the theories are compared for actual experimental nodes as in Table 10-4, the agreement is fairly good among all three. The larger of the double values[68,69] for Cu_2NiZn correspond to fits to the inscribed radius mentioned above, the smaller to fits to the minimum radius of curvature of the partials. For the Ag-12.5% In alloy, Gallagher[72] found $\gamma_I = 4.55\ \mathrm{mJ/m^2}$ from Eq. (10-21), a result independent of the core parameter α, in best agreement with the Jøssang result.

More recently, an anisotropic version of the node model using the methods described in Sec. 5-7 has been presented by Scattergood and Bacon.[73] An even

[70] T. Jøssang, M. J. Stowell, J. P. Hirth, and J. Lothe, *Acta Met.*, **13:** 279 (1965).

[71] A. W. Ruff, Jr., *Met. Trans.*, **1:** 2391 (1970).

[72] P. C. J. Gallagher, *J. Appl. Phys.*, **37:** 1710 (1966).

[73] R. O. Scattergood and D. J. Bacon, *Acta Met.*, **24:** 705 (1976).

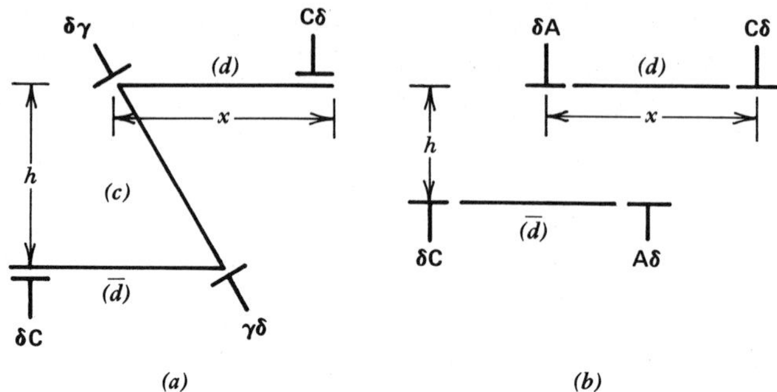

FIGURE 10-34. (*a*) Faulted *Z*-type dipole. (*b*) Extended but incompletely faulted dipole.

more elaborate model[74] including the orientation dependence of the core has been developed in the anisotropic elastic case. Saka[75] shows that these models both fit the node extension-orientation relationships well for Cu-13.4 percent Al, the latter[74] giving a somewhat better fit. A comparison of the resulting value of γ_I with that determined directly from the dissociation of straight dislocations gives an estimate of $\alpha = 0.5$ to 1.7. This type of prediction of the core parameter is perhaps the major advantage of the node theories now that direct dissociations can be observed in weakbeam electron microscopy. Together with the results of Table 10-4, the above result suggests $\alpha \sim 1$ for a Shockley partial dislocation.

Extended Dipoles

With the advent of weak-beam electron microscopy, two other dislocation arrays in addition to directly dissociated dislocations have become important in estimating γ_I. All these arrays comprise straight dislocations whose interactions are relatively easy to analyze in both the isotropic and anisotropic case. The configuration of Fig. 10-34*a* was first seen by Mader,[76] suggested to be long Frank loops by Hirsch,[77] and described by Seeger[78] as the extended array shown.

The *Z* dipole shown contains all edge partials and can form with either all intrinsic faults or all extrinsic faults. A corresponding *S* dipole with obtuse bends can form with mixed faults γ_I-γ_E-γ_I or γ_E-γ_I-γ_E. To date only intrinsic *Z*

[74] H. O. Kirchner and F. Prinz, cited in Ref. 75.

[75] H. Saka, *Phil. Mag.*, **42A:** 185 (1980).

[76] S. Mader, in G. Thomas and J. Washburn (eds.), "Electron Microscopy and Strength of Crystals," Interscience, New York, 1963, p. 183.

[77] P. B. Hirsch, in "The Relation Between the Structure and Mechanical Properties of Metals," HMSO, London, 1963, p. 48.

[78] A. Seeger, *Disc. Faraday Soc.*, **38:** 82 (1964)

dipoles have been observed,[79] consistent with the usual prevalence of intrinsic fault arrays, so only these are discussed further.

The interactions among the dislocations in the Z dipole are such that for small h, x increases with increasing h, passes through a maximum and then decreases with increasing h. The entire curve shifts with γ_I. Thus by either measuring x versus h, say, by direct lattice imaging or associating the maximum measured x with the maximum in the $x-h$ curve, one can determine γ_I from anisotropic elastic theory.[80] Values determined in this way are listed in Appendix 2.

Several models have been suggested for formation of the faulted dipoles. They can form by dissociations of the $\boldsymbol{\delta}\mathbf{A}$ and $\mathbf{A}\boldsymbol{\delta}$ partials as shown in Fig. 10-34b into stair rods and partials $\boldsymbol{\gamma}\mathbf{A}$ and $\mathbf{A}\boldsymbol{\gamma}$ on (c), which cross slip and annihilate.[81] They may also form during the process of dislocation intersection[82] (Chap. 22) by surface-enhanced cross slip as above, or by unzipping a closed-end dipole of the type shown in Fig. 10-34b by motion of an $A\gamma$ partial on glide plane (c).[83]

Analogously to the faulted dipole case, the extended but not completely faulted array shown in Fig. 10-34b can be measured to determine γ_I.[84] The ratio of x/h varies with h also for this type of array and can be analyzed by anisotropic elastic theory to give γ_I.[85] Unlike the faulted Z dipole, the character of the total dislocations can change continuously from screw to edge.

10-5. EXTRINSIC-FAULT CONFIGURATIONS

As discussed previously in this chapter, extended nodes in fcc crystals were found, until recently, to be extended to form intrinsic faults. Also, tetrahedra are found to be bounded by intrinsic faults. This experimental evidence indicated either that the extrinsic-fault energy greatly exceeded the intrinsic-fault energy, or that some mechanistic barrier existed that tended to prevent dissociation into extrinsic faults. The former explanation was usually assumed. However, extrinsic nodes have recently been observed in fcc alloys[86–88] and in pure silver.[88] An example of an extrinsic fault in an Au-4.8a/o Sn alloy from

[79] S. W. Chiang, C. B. Carter, and D. L. Kohlstedt, *Phil. Mag.*, **42A:** 103 (1980).

[80] J. W. Steeds, "Anisotropic Elastic Theory of Dislocations," Claredon, Oxford, 1973, p. 125.

[81] P. M. Hazzledine, H. P. Karnthaler, and E. Wintner, *Phil. Mag.*, **32:** 81 (1975).

[82] J. W. Steeds and P. M. Hazzledine, *Disc. Faraday Soc.*, **38:** 103 (1964).

[83] E. Wintner and H. P. Karnthaler, *Acta Met.*, **26:** 941 (1978).

[84] E. Wintner and H. P. Karnthaler, *Phil. Mag.*, **36:** 1317 (1977).

[85] A. J. Morton and C. T. Forwood, *Cryst. Lattice Defects*, **4:** 165 (1973).

[86] M. H. Loretto, *Phil. Mag.*, **10:** 467 (1964); *Phil. Mag.*, **12:** 125 (1965).

[87] P. R. Swann, *Acta Met.*, **14:** 76 (1966).

[88] P. C. J. Gallagher, *Phys. Stat. Solidi*, **16:** 95 (1966).

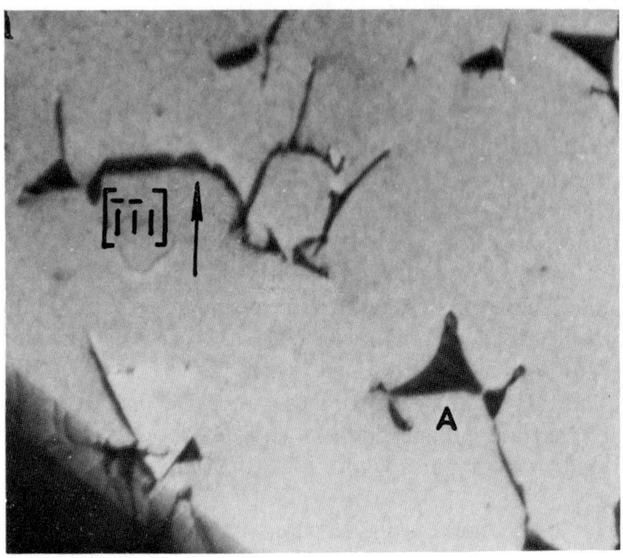

FIGURE 10-35. An intrinsic-extrinsic node pair in Au-4.8a/o Sn alloy, 60,000×. [M. H. Loretto, *Phil. Mag.*, **10**: 467 (1964).]

Loretto's work is shown in Fig. 10-35. Extrinsic faults have also been observed in obtuse dislocation bends[86] and in parallel dislocation interactions.[88]

The latter observations by Gallagher are of particular interest with respect to the extrinsic-fault energy. An example of his findings is presented in Fig. 10-36*a* and the assignment of Burgers vectors in Fig. 10-36*b*. A similar structure had been observed earlier in layer compounds,[89] (e.g., see Fig. 10-7). He established from contrast effects that the fault pair (also called *double ribbon*) was extrinsic-intrinsic (fringe-shift measurements) and that the three partials had the same Burgers vector. In the parallel configuration the repulsive forces between the partials can be computed exactly in either isotropic or anisotropic theory. In the isotropic case, Eq. (10-14) shows that the fault energies are given in terms of the partial vectors $\mathbf{b}=\mathbf{C}\boldsymbol{\delta}$ by

$$\begin{aligned}\gamma_I &= \frac{\mu b^2}{2\pi(1-\nu)}\left[(1-\nu)\cos^2\beta+\sin^2\beta\right]\left(\frac{1}{r_I}+\frac{1}{r_I+r_E}\right)\\ \gamma_E &= \frac{\mu b^2}{2\pi(1-\nu)}\left[(1-\nu)\cos^2\beta+\sin^2\beta\right]\left(\frac{1}{r_E}+\frac{1}{r_I+r_E}\right)\end{aligned} \tag{10-21}$$

For an Ag-8% In alloy, Gallagher found $\gamma_E/\gamma_I=1.09$, and for an Ag-12.5% In alloy he found $\gamma_E/\gamma_I=1.03$. For the latter alloy he found $\gamma_I=4.55$ mJ/m^2, from Eq. (10-21).

[89] P. Delavignette and S. Amelinckx, *J. Nucl. Mat.*, **5**: 17 (1962).

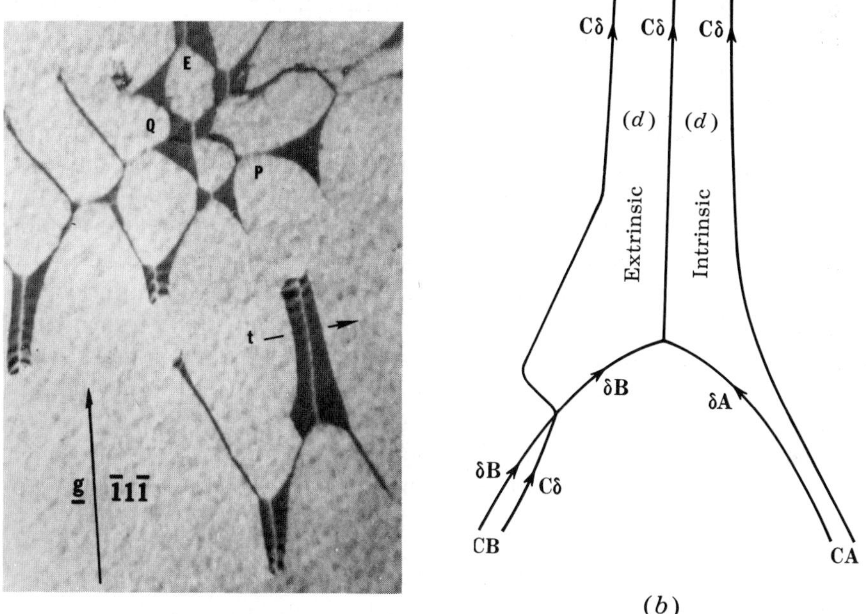

FIGURE 10-36. (*a*) Intrinsic-extrinsic nodes and parallel arrays in an Ag-12.5a/o In alloy, 80,000×. [P. C. J. Gallagher, *Phys. Stat. Solidi* (1966).] (*b*) The Burgers vector assignment for the parallel array at bottom of (*a*).

In any case, the above results suggest that $\gamma_I \cong \gamma_E$, in agreement with the approximation of Eq. (10-13). In turn, this conclusion leads to the implication that some other factor prevents the formation of extrinsic faults. One possible explanation is that the formation of the *pairs* of partials that bound an extrinsic stacking fault is kinetically more difficult than the formation of the single partials bounding an intrinsic fault. Consider the bending of such faults from one {111} plane to another. Figure 10-37 shows the stair rods associated with acute and obtuse bends from (*a*) to (*d*) of the dislocation **BC**. It can be

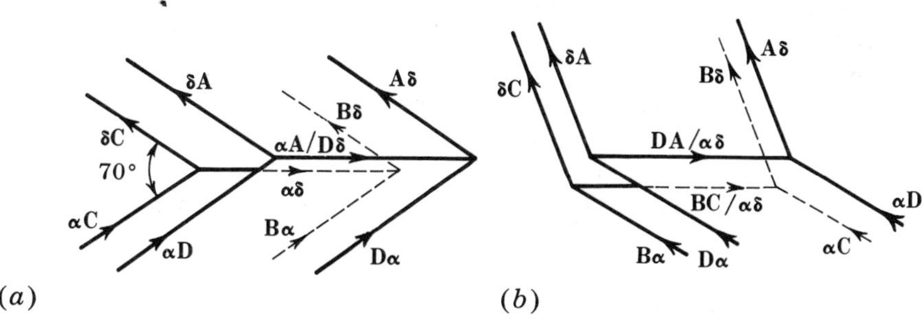

FIGURE 10-37. Bends of the extrinsically faulted dislocation **BC** from plane (*a*) to plane (*d*): *a* acute and *b* obtuse.

seen that such a bend involves two stair rods separated by about an interplanar spacing. The *net* stair-rod Burgers vectors for the acute and obtuse cases, respectively, are $\boldsymbol{\delta\alpha}$ and $\boldsymbol{\alpha\delta}/\mathbf{BC}$ (note that $\boldsymbol{\alpha}\mathbf{A}/\mathbf{D}\boldsymbol{\delta} = 2\boldsymbol{\delta\alpha}$ and that $\mathbf{DA}/\boldsymbol{\alpha\delta} = 2\boldsymbol{\delta\alpha}/\mathbf{BC}$).

Thus the elastic energies of the net stair rods for the extrinsic-fault case are about the same as for the intrinsic-fault case. However, the net stair rods have a higher core energy in the extrinsic-fault configuration of Fig. 10-37, so that they are less stable with respect to constriction than the stair rods of the intrinsic-fault extension of Fig. 10-12. Similarly, the double partials in an extrinsic node have a higher core energy than the partials for an intrinsic node. In general, then, we expect that core energies will favor the formation of intrinsic-node configurations, particularly for localized configurations where the core energy comprises an appreciable portion of the total energy. For less localized configurations, extrinsic faults are expected to appear.

Weertman[90] has suggested that jogs can dissociate into extrinsically faulted arrays. On the basis of the above reasoning, such extensions are expected only for superjogs.

Exercise 10-5. With the use of Fig. 10-37, construct the partial-dislocation configuration of jogs extended to form extrinsic faults. Consider both the case where all faults in the configuration are extrinsic, and the case where the main dislocation is dissociated intrinsically while only the jog is dissociated extrinsically. Discuss the likelihood of such extensions for unit jogs.

Exercise 10-6. Consider the elastic interaction of parallel dislocations **CB** and **CA** to form the array of Fig. 10-36*b*. Show that the dislocations repel one another at long range but attract at short range.[91]

10-6. FORCES ON PARTIAL DISLOCATIONS

Dynamic Effects

Most of the continuum theory of dislocations can be applied directly to partial dislocations. However, some differences exist in the concept of forces on partials. Consider a dislocation which is extended on the glide plane and moving with a velocity v_x, as shown in Fig. 10-38. The repulsive interaction force between the partials [Eq. (5-17)] is abbreviated to A/r; B, C, and E are the forces on $\mathbf{b}_1$, $\mathbf{b}_2$, and $\mathbf{b}_3$, respectively, because of the applied stress [Eq. (3-90)]; γ is the force exerted by the stacking fault, and D_1 and D_2 are the damping forces on the moving partials. Let us examine the net force on the moving partials to see if they dissociate or contract dynamically.

[90] J. Weertman, *Phil. Mag.*, **8:** 967 (1963).

[91] J. P. Hirth, *J. Appl. Phys.*, **32:** 700 (1961).

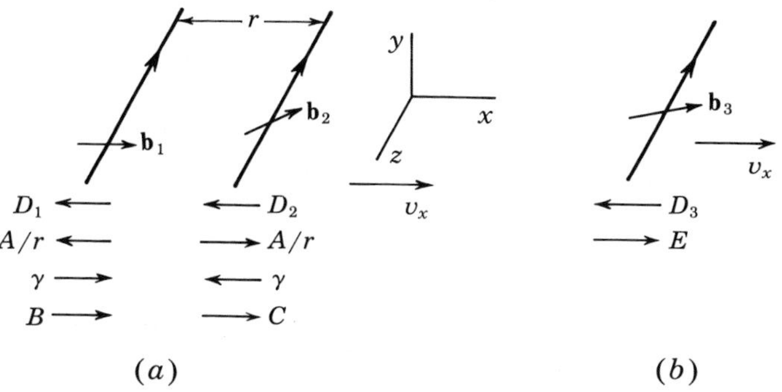

FIGURE 10-38. An extended dislocation moving with velocity v_x.

At steady state, where $r = r_e'$, the force balance on the partials yields the relations

$$\gamma + B = D_1 + \frac{A}{r_e'}$$
$$\gamma + D_2 = C + \frac{A}{r_e'} \tag{10-22}$$

To establish the sign of the dynamic effects, we neglect the difference between D_1 and D_2 caused by the screw-edge character of the partials, applying Eq. (7-74) to each partial. Then $D_1 = D_2$, so that Eqs. (10-22) can be solved simultaneously for the two unknowns D_1 and r_e', giving

$$D_1 = D_2 = \tfrac{1}{2}(B + C)$$
$$r_e' = \frac{A}{\gamma + \frac{1}{2}(B - C)} \tag{10-23}$$

One can verify by substitution into Eqs. (10-22) that Eqs. (10-23) represent a stable-equilibrium result. In the unextended configuration, it follows directly that $D_3 = E = B + C$. Thus the *net* total forces on the extended and unextended dislocation are the same.

Comparing Eqs. (10-23) to the static result that $r_e = A/\gamma$, one sees that the dissociation is less in the dynamic case if $B - C > 0$ and more if $B - C < 0$. Hence extended dislocations either extend or contract dynamically, depending on the magnitude of the local forces B and C. This effect should be distinguished from changes in the displacement fields of the dislocations at velocities approaching the speed of sound.

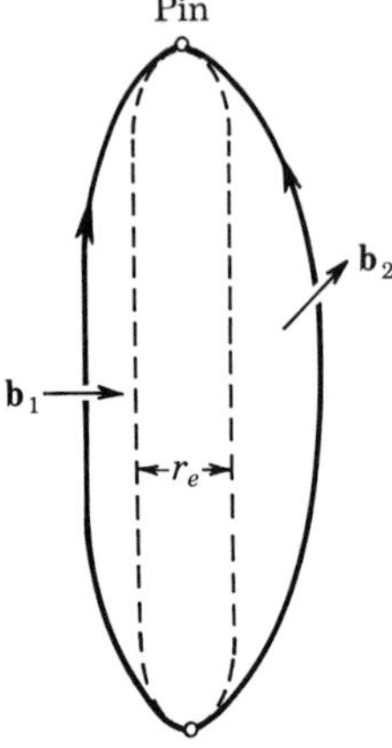

FIGURE 10-39. Bowed-out partial dislocations.

The Static Case

Suppose that a segment of the extended dislocation is pinned in the glide plane. Under an applied stress, the partials bow out until a constrained-equilibrium configuration like that of Fig. 10-39 is attained. Equations (10-22) apply to such a case, but line-tension forces replace D_1 and D_2. Even with the assumption of equal line tensions, the radii of curvature of the two partials will not be equal in general. A further complication is introduced if the variation of line tension with orientation [Eq. (6-82)] is considered. In analogy with the dynamic treatment, the bowed-out configuration has an equilibrium separation smaller than r_e if $B-C>0$ and larger than r_e if $B-C<0$. The example in Fig. 10-39 is an extreme example of such an effect for the case $C>0$ and $B<0$. Such effects are important in the process of tearing apart of partials by an applied stress. This process is important in the creation of sheets of stacking faults observed in heavily cold worked alloys that have low stacking-fault energies. It has been observed experimentally.[92]

10-7. THE PEIERLS BARRIER FOR A PARTIAL DISLOCATION

Qualitatively, the dissociation of a perfect dislocation into partials occurs because the decrease in elastic energy upon dissociation is sufficient to offset the increase in misfit energy associated with the stacking fault. Thus the *total* misfit energy is greater for the extended dislocation. Whether or not the Peierls energy arising from *variation* in misfit energy is larger for the extended dislocation is moot. A formal application of Eq. (8-38) would predict a lower Peierls energy for the extended configuration because the magnitudes of the

[92]See T. Malis, D. J. Lloyd, and K. Tangri, *Met. Trans.*, **6A:** 932 (1975).

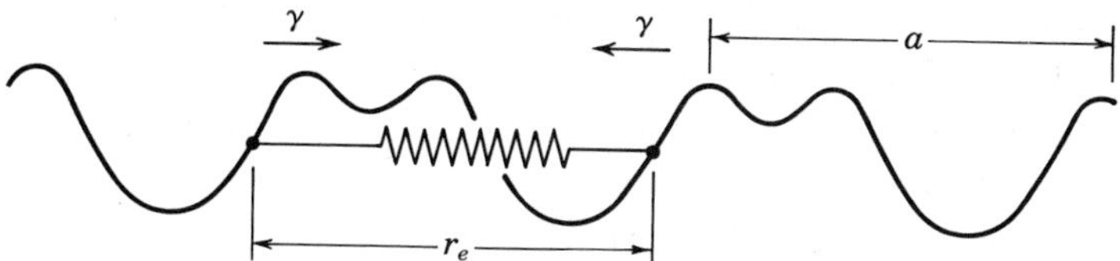

FIGURE 10-40. Representation of an extended dislocation by a ball-and-spring model on a potential substrate. The balls represent partial dislocations, the compressed spring represents the repulsive elastic interaction, and γ is the stacking-fault force. The subsidiary minimum is introduced into the variable misfit energy because of the stacking-fault orientation of the atoms.

Burgers vectors of the partials are smaller than that of the perfect dislocation. However, in addition to the reservations about Eq. (8-38) noted in Chap. 8, the presence of the stacking fault further complicates matters.

There is a subsidiary minimum, at the displacement corresponding to the fault, in the variation of local misfit energy with disregistry. The Peierls energy of both the extended and unextended dislocations would reflect this difference in misfit energy, so that their Peierls energy plots would resemble that of Fig. 8-9. However, in the absence of detailed atomic calculations, one is unable to deduce which would be affected more.

Some appreciation of the possible Peierls barriers for partials can be gained from the Frenkel-Kontorova type of model shown in Fig. 10-40. The partials are represented by balls, the elastic repulsion by a compressed spring, and the stacking-fault energy by the force γ. In the limit of high fault energy, the balls will move with essentially a rigid separation r. In this case the maximum effective Peierls energy would be $2W_m$ if $r_e = na$, $n = 1, 2, 3, \ldots$ but less than $2W_m$ if $r_e \neq na$. Similarly, the effective Peierls stress would be $2\sigma_m$ if $r_e = na$.

On the other hand, in the limit of low fault energy, the balls independently tend to lie in potential minima. In this case each ball moves independently, so that the effective Peierls energy and stress are reduced to W_m and σ_m. Thermal fluctuations suffice to maintain the partials at a separation $\sim r_e$. If the leading partial is torn away from the trailing partial (broken spring), the effective Peierls stress for the leading partial becomes $\sigma_m + \gamma$ and that for the trailing partial $\sigma_m - \gamma$.

One is tempted in the latter case to ascribe a lower Peierls stress to the partials than to the unextended perfect dislocation, but the opposite possibility cannot be excluded. Thus, in the absence of atomic calculations, one must resort to the phenomenological equations (8-40) and (8-41) to describe W_p and σ_p and regard these as parameters to be determined from experiment.

PROBLEMS

10-1. Determine the variation with orientation of the equilibrium separation of partials bounding an intrinsic stacking fault in copper.

10-2. Using the γ_I values from Appendix 2, find which fcc metals should have meaningful partial-dislocation extensions; that is, $r_e > 2b$.

10-3. In terms of nearest-neighbor bond energies φ_1, evaluate the stacking-fault energy associated with

- **a.** Climb of a Shockley partial
- **b.** Glide of the stair rod $\boldsymbol{\alpha\delta}$
- **c.** Glide of a Frank partial

10-4. Draw the various stages in the advance of a $\langle 110 \rangle$ jog line by absorption of a vacancy.

10-5. Discuss the possible reactions that would result in the formation of a tetrahedron from a triangular interstitial disc. What configuration would form if the Frank partial bounding the disc dissociated directly to form a Shockley partial bounding an intrinsic fault?

10-6. Label the dislocations in Fig. 10-28, using the Thompson notation.

10-7. Show that simultaneous glide collapse from two corners of an intrinsically faulted tetrahedron results in a perfect $\frac{1}{2}\langle 110 \rangle$ dislocation loop.[93]

10-8. Consider that the dislocation **AC**(d) is dissociated into intrinsic stacking fault over some length, after which it is dissociated into extrinsic stacking fault on the same $\{111\}$ plane. What partial crosses the dislocation where the change occurs?

10-9. Consider an extended dislocation node that is parallel to a free surface. What is the sign of the change in the apparent value of γ_I caused by image forces?

10-10. Consider the node between the dislocations **AC**, **BA**, and **CB** in the (d) plane. The dislocations are ribbons of intrinsic stacking fault, and the branches occur in such an order that the node would be contracted unless extrinsic stacking fault forms at the node. Show how the node could become extended by the formation of an extrinsic fault.

10-11. Consider the intrinsically dissociated dislocation **AC**(d). Describe the possible superjogs dissociated into extrinsic fault on the (c) plane.

10-12. Consider an intrinsically dissociated screw cross slipping onto a conjugate plane without contracting. Describe the intermediate dissociated configuration. Would cross slip be easier if the screw dissociated extrinsically on the conjugate plane?

10-13. Could twinning occur by the formation of successive layers of extrinsic fault? If so, what shear would accompany twinning?

10-14. Compare the magnitudes of the force caused by the Peierls stress, from Eq. (8-39), and that associated with the stacking fault for an isolated Shockley partial in copper.

10-15. Discuss the stability of the partial $\mathbf{D}\boldsymbol{\gamma}/\mathbf{AC}$ that can form by the reaction of **DC** and $\boldsymbol{\gamma}$**A**.

[93] D. Kuhlmann-Wilsdorf, *Acta Met.*, **13:** 257 (1965).

BIBLIOGRAPHY

1. Amelinckx, S., *Solid State Phys. Suppl.* 6 (1964).
2. Amelinckx, S., in F. R. N. Nabarro (ed.), "Dislocations in Solids," vol. 2, North-Holland, Amsterdam, 1979, p. 67.
3. Christian, J. W., and V. Vitek, *Rep. Prog. Phys.*, **33:** 307 (1970).
4. Hirsch, P. B., *Phil. Mag.*, **7:** 67 (1962).
5. Jøssang, T., M. J. Stowell, J. P. Hirth, and J. Lothe, *Acta Met.*, **13:** 279 (1965).
6. Kuhlmann-Wilsdorf, D., *Acta Met.*, **13:** 257 (1965).
7. Thompson, N., *Proc. Phys. Soc.*, **66B:** 481 (1953).
8. Proceedings of the Symposium on Measurement of Stacking Fault Energy, *Met. Trans.*, **1:** 2367–2486 (1970).

11

Partial Dislocations in Other Structures

11-1. INTRODUCTION

Many of the extended-dislocation configurations and fault properties in other crystal systems are generally analogous to those in fcc crystals, so a survey of partials in other lattices is not undertaken here. However, in some types of crystals the partials differ markedly from those in fcc structures, leading to different properties; we briefly consider examples of these types in this chapter. Crystals of hcp metals serve to introduce the concept of a *zonal dislocation*, bcc crystals indicate some of the possible high-energy types of stacking faults associated with twinning, and diamond crystals illustrate the existence of nonequivalent sets of dislocations in some crystal lattices that have a basis. The chapter concludes with a discussion of dislocations in ordered crystals, where special types of partial dislocations exist.

11-2. THE HCP LATTICE

Stacking Faults

In the hcp lattice the close-packed planes are (0002) planes. These basal planes are also the most frequently observed glide planes in hcp crystals, but unlike fcc crystals, they do not correspond to a twin plane. Refer again to the hard-ball model of Fig. 10-1. There are two kinds of intrinsic fault, I_1 and I_2, in hcp structures and one extrinsic fault E; these are of the low-energy type that do not disturb nearest-neighbor packing. The modes of formation of these faults are indicated below. The intrinsic fault I_1 can be formed by removing the B plane above the A plane (Fig. 10-1),

$$ABABABA \mid ABAB \tag{11-1}$$

and then shearing the remaining planes above the A plane by the displacement $\frac{1}{3}[\bar{1}100]$,

$$\begin{array}{l} ABABABA|ABAB \\ \qquad\qquad\quad |\downarrow\downarrow\downarrow\downarrow \\ \qquad\qquad\quad |CACA \end{array} \tag{11-2}$$

yielding

$$ABABABA|CACACA \tag{11-3}$$

The fault I_2 can be formed directly by shear

$$\begin{array}{r} ABABABABABA \\ \downarrow\downarrow\downarrow\ \downarrow\downarrow \\ CACAC \end{array} \tag{11-4}$$

yielding

$$ABABAB|CACACA \tag{11-5}$$

The extrinsic fault E is formed by inserting a C plane in an AB sequence,

$$ABABACBABABA \tag{11-6}$$

Notice that I_1 and E *cannot* be formed or removed by single shearing operations; they can be formed only by a sequence of shears, as is true for the twin in the fcc case. I_1 and E can be formed from one another by a single shear.

In terms of the ψ_i notation [Eq. (10-11)], the fault energies are

$$\begin{aligned} \gamma_{I_1} &= \sum_{N=1}^{\infty} N\psi_{2N} = \psi_2 + 2\psi_4 + \cdots \\ \gamma_{I_2} &= \sum_{N=1}^{\infty} (2N\psi_{2N} + N\psi_{2N+1}) = 2\psi_2 + \psi_3 + 4\psi_4 + \cdots \\ \gamma_E &= \sum_{N=1}^{\infty} [(2N+1)\psi_{2N} + 2N\psi_{2N+1}] = 3\psi_2 + 2\psi_3 + 5\psi_4 + \cdots \end{aligned} \tag{11-7}$$

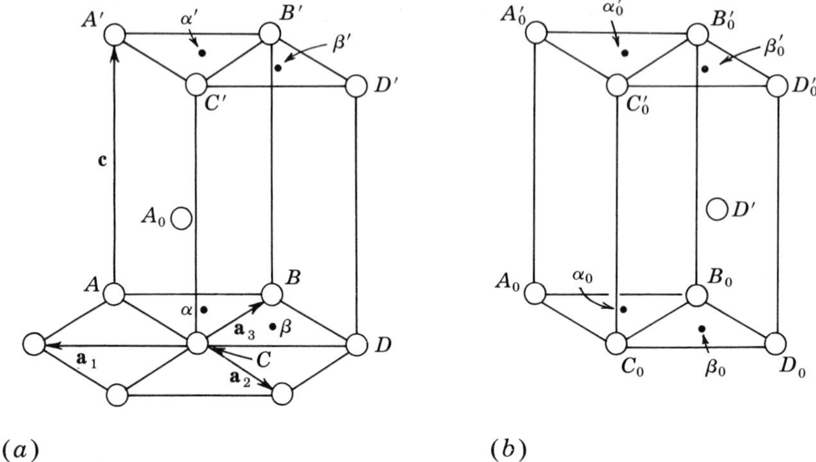

FIGURE 11-1. (*a*) A primitive hexagonal unit cell with the vector notation for hcp. (*b*) The cell indexed on A_0 in (*a*).

In the central-force approximation, the ψ_2 and ψ_3 values are the negatives of those given by Eq. (10-12), leading to the prediction $\gamma_{I_1} < \gamma_{I_2} < \gamma_E$. The same reservations about the central-force model apply here as for the fcc case. Nonetheless, the large differences in the numbers of "broken bonds" for these three faults lead us to expect that the fault energies should have the relative order given above. Because I_1 can be formed from E by a single shear (i.e., by nucleation of a Shockley partial), only the intrinsic faults are considered in the following discussion.

Vector Notation for the HCP Lattice

Several schemes have been suggested[1-3] for the representation of the hcp lattice in a notation resembling the Thompson notation in the fcc case. We shall adopt a variation of the model proposed by Damiano. This model involves the indexing of the simple hexagonal unit cell as shown in Fig. 11-1. There is no significance in the ordering of the Greek letters in the figure other than to yield Greek-Roman pairs for the partial dislocations. Examples of the dislocation Burgers vectors are listed in Table 11-1, along with the conventional Miller-Bravais notation. Glide planes are also listed, although there is little or no advantage in using the contracted notation for planes

[1] V. V. Damiano, *Trans. Met. Soc. AIME*, **227**: 788 (1963).

[2] A. Berghezan, A. Fourdeux, and S. Amelinckx, *Acta Met.*, **9**: 464 (1961).

[3] P. B. Price, in G. Thomas and J. Washburn (eds.), "Electron Microscopy and Strength of Crystals," Interscience, New York, 1963, p. 41.

TABLE 11-1 Examples of the Notation Depicted in Fig 11-1

Burgers Vector or Glide Plane	Contracted Notation	Standard Miller-Bravais Notation
(1) Perfect-dislocation Burgers vector	**AC**	$\frac{1}{3}[\bar{1}2\bar{1}0]$
	AA′	$[0001]$
	CB′	$\frac{1}{3}[\bar{1}\bar{1}23]$
(2) Glissile partials (Shockley partial)	$\mathbf{A}\boldsymbol{\alpha}$	$\frac{1}{3}[\bar{1}100]$
	$\mathbf{A}\boldsymbol{\beta}$	$\frac{2}{3}[\bar{1}100]$
(3) Sessile partials (Frank partial)	$\boldsymbol{\alpha}\mathbf{A}_0$	$\frac{1}{2}[0001]$
	$\mathbf{A}\boldsymbol{\beta}'$	$\frac{1}{3}[\bar{2}203]$
	$\mathbf{BA}_0$	$\frac{1}{6}[20\bar{2}3]$
	$\mathbf{C}\boldsymbol{\alpha}'$	$\frac{1}{3}[0\bar{1}13]$
(4) Partials on $\{10\bar{1}0\}$	$\boldsymbol{\epsilon}\mathbf{C}$	$\frac{1}{18}[42\bar{6}3]$
(5) Glide planes	(ABC)	(0001)
	$(ABB'A')$	$(0\bar{1}10)$
	$(AB'C')$	$(1\bar{1}01)$
	(CA_0B')	$(11\bar{2}2)$

Glissile Shockley Partials

A comparison of Figs. 10-1 and 11-1 shows that the partials $\mathbf{A}\boldsymbol{\alpha}$, $\mathbf{A}'\boldsymbol{\alpha}'$, $\mathbf{C}\boldsymbol{\beta}$, etc., are the equivalents of the Shockley partials in fcc structures. These partials lie in the basal plane; their glide on the basal plane produces the shears of Eqs. (11-2) and (11-4). In the basal plane, perfect dislocations of the type **AB** can dissociate into two Shockley partials bounding an intrinsic fault I_2. The separation of the partials is given by Eq. (10-15), with the substitution of γ_{I_2} (hcp) for γ_I (fcc).

The proper dissociation of the partials can be determined analogously to the fcc case. The notation is more complicated, though, because the cells in Figs. 11-1*a* and 11-1*b* are not equivalent; the "body" atom A_0 is in the left triangular prism ABC in Fig. 11-1*a*, while the "body" atom D' is in the right triangular prism $B_0C_0D_0$ in Fig. 11-1*b*. The axiom for dissociation is as follows:

Axiom 11-1: In order to dissociate a perfect dislocation to obtain Shockley partials bounding an intrinsic fault I_2, arrange the partials such that the Greek index letters are on the outside of the fault if the partials involve the index α, α', β_0, or β_0'; arrange them with the Greek letters inside if the partials involve the index α_0, α_0', β, or β'. This arrangement must be made from the point of view of an observer outside the cell of Fig. 11-1 and looking in the direction of $\boldsymbol{\xi}$.

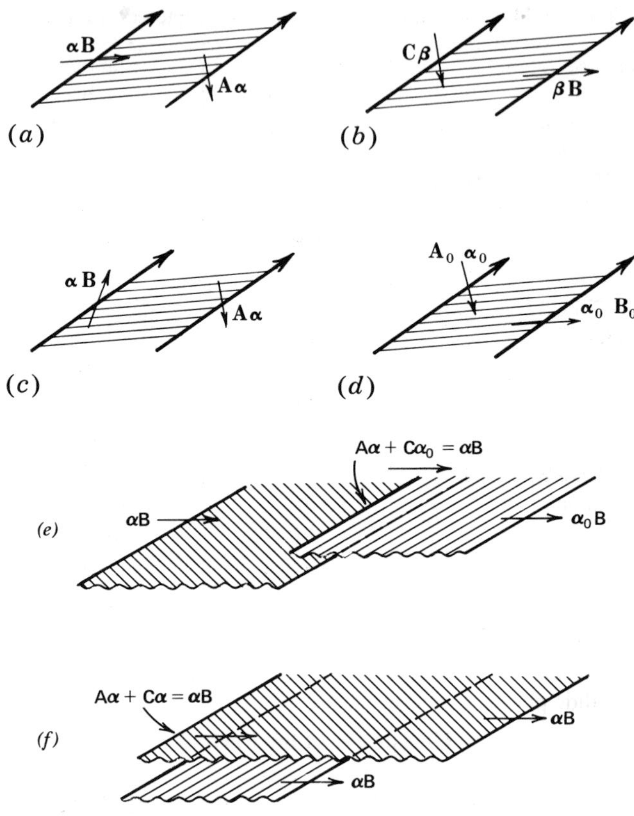

FIGURE 11-2. Dissociated dislocations in hcp.

Some examples of the application of this rule are shown in Fig. 11-2. The opposite order of dissociation would produce the high-misfit-energy fault *ABABA* ¦ *ACAC*. By a further shear of the lattice, achieved by the formation of an additional pair of Shockley partials, this high-energy fault can be transformed to a fault of the type I_2, with a *pair* of partials bounding each side of the fault. However, the direct formation of the I_2 fault according to Axiom 11-1 is much more likely. Note that an extended dislocation jogging from one basal plane to an adjacent one (or in general jogging over an odd number of planes) has the sequence of partials reversed, as in Figs. 11-2*a* and 11-2*d*, while jogging over an even number of planes does not produce a change in partial sequence.

The partial sequence change leads to two types of double-ribbon dislocations,[4] one of the same form as those in fcc crystals. As shown in Fig. 11-2*e*, two basal dislocations one plane apart can react to form a set of dislocations

[4]S. Amelinckx, in F. R. N. Nabarro (ed.), "Dislocations in Solids," vol. 2, North-Holland, Amsterdam, 1979, p. 67.

whose long-range elastic fields are all those of the same Burgers vector: the analog of the fcc reaction of Fig. 10-36. Both faults are of type I_1 in this case. If two dislocations two planes apart interact, again a double-ribbon fault is formed as in Fig. 11-2*f*. In this case the fault on the left side is of the I_2 type. Larger separation faulted dipoles can also form. Together, the configurations permit determination of γ_{I_1} and γ_{I_2}.

Frank Sessile Partials

The precipitation and collapse of a vacancy disc and the precipitation of an interstitial disc produce the configurations in Fig. 11-3 if only displacements normal to the basal plane accompany precipitation. The dislocation loops formed by such processes are bounded by Frank partials of the type $\alpha\mathbf{A}_0$. The fault formed by vacancy-disc collapse is a high-energy fault, while that formed by interstitial precipitation is the relatively high energy extrinsic fault. Thus a loop of Shockley partial dislocation is likely to nucleate in the faulted region. The growth of such a loop to the periphery of the fault produces the I_1 type of fault shown in Fig. 11-4. Equivalently, the loops in Fig. 11-4 could be formed directly if a shear offset were to accompany vacancy-disc collapse or interstitial precipitation. At the periphery of the above loops, the Shockley and Frank partials combine to form partials of the type $\mathbf{BA}_0$ by the reaction

$$\mathbf{B}\alpha + \alpha\mathbf{A}_0 \rightarrow \mathbf{BA}_0$$

Thus one expects that partials of the type $\alpha\mathbf{A}_0$ should be rare in the hcp case

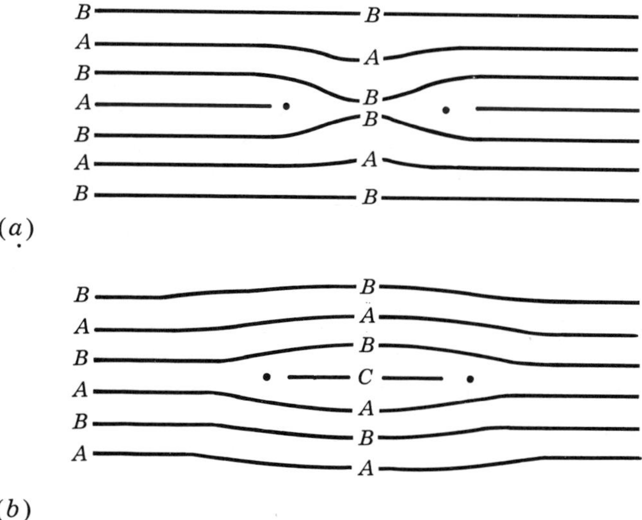

FIGURE 11-3. Cross sections of (*a*) a collapsed vacancy disc and (*b*) a precipitated interstitial loop.

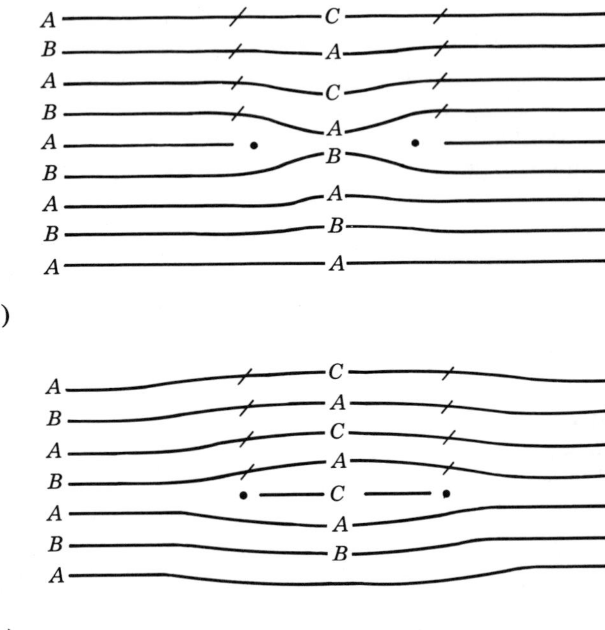

(b)

FIGURE 11-4. I_1-Type faults.

because of the high energy of the associated faults, but the partials of the type $\mathbf{BA}_0$ should be observed. Both partial types $\alpha\mathbf{A}_0$ and $\mathbf{BA}_0$ are sessile, but both can climb, as is evident from Figs. 11-3 and 11-4.

Other Partials

The remaining partials listed in Table 11-1 are related to dislocation dissociations or associations as discussed by Frank and Nicholas.[5] The partial $\mathbf{A}\boldsymbol{\beta}$ lies in the basal plan and bounds an I_2 fault. It is of marginal stability with respect to the dissociation reactions

$$\mathbf{A}\boldsymbol{\beta} \rightarrow \mathbf{AC} + \mathbf{C}\boldsymbol{\beta} \qquad \mathbf{A}\boldsymbol{\beta} \rightarrow \mathbf{AB} + \mathbf{B}\boldsymbol{\beta}$$

Dissociations of the type $\mathbf{A}\boldsymbol{\beta} \rightarrow 2\mathbf{A}\boldsymbol{\alpha}$ are unlikely, because a high-energy fault would be associated with the product partials. $\mathbf{A}\boldsymbol{\beta}$ is glissile on the basal plane.

$\mathbf{C}\boldsymbol{\alpha}'$ has marginal stability with respect to the dissociation

$$\mathbf{C}\boldsymbol{\alpha}' \rightarrow \mathbf{CC}' + \mathbf{C}'\boldsymbol{\alpha}'$$

$\mathbf{C}\boldsymbol{\alpha}'$ is sessile, while the products $\mathbf{CC}'$ and $\mathbf{C}'\boldsymbol{\alpha}'$ are glissile on $(ACC'A')$ and $(A'B'C')$, respectively.

[5] F. C. Frank and J. F. Nicholas, *Phil. Mag.*, **44**: 1213 (1953).

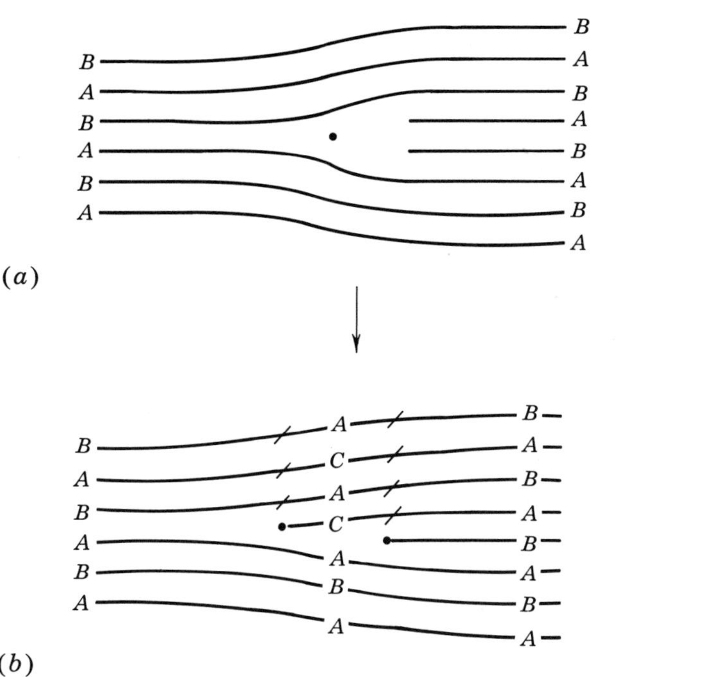

FIGURE 11-5. Dissociation of (*a*) $\mathbf{AA'}$ into (*b*) $\mathbf{AA_0} + \mathbf{A_0A'}$.

The perfect dislocation $\mathbf{AA'}$ is unstable with respect to the dissociation, as shown in Fig. 11-5:

$$\mathbf{AA'} \rightarrow \mathbf{AA_0} + \mathbf{A_0A'}$$

The dissociation

$$\mathbf{AB'} \rightarrow \mathbf{AA_0} + \mathbf{A_0B'}$$

produces the same configuration. These reactions require climb. They are expected to form as relaxed configurations by short-range vacancy or interstitial diffusion. These dissociations would, of course, produce sessile configurations from the glissile dislocations $\mathbf{AA'}$ and $\mathbf{AB'}$. Hence they should be important in restricting glide on nonbasal systems. Such dissociations can be thought of as contributing to an apparent high static Peierls barrier to the motion of $\mathbf{AA'}$ or $\mathbf{AB'}$. When prismatic loops of this type form by collapse of vacancy discs, sometimes with multiple loops, they form concentric dislocation rings with alternating stacking-fault contrast.[6]

[6]E. G. Tapetado, R. E. Smallman, and M. H. Loretto, *Cryst. Lattice Defects*, **5**: 199 (1974).

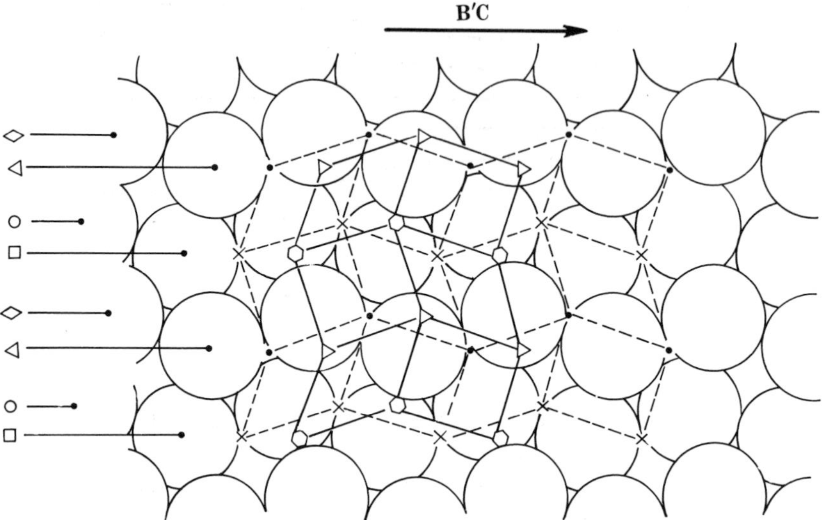

FIGURE 11-6. Atom positions in the normal hcp lattice in a projection normal to $\{11\bar{2}2\}$. The large complete circles are atoms in the plane of the paper. The incomplete lighter circles represent the atom layer immediately below the paper. × ● ⬡ and ▽ represent atoms in successively higher levels with respect to the plane of the paper. The symbols at the top of the diagram indicate the atom layers as seen in Fig. 11-9.

Zonal Dislocations

Low-energy stacking faults occur only in the basal plane in hcp crystals. Yet, as discussed in Chap. 9, glide is often observed on the system $\langle\bar{1}\bar{1}23\rangle(11\bar{2}2)$ and on other nonbasal planes in a manner inexplicable by the Peierls criterion. Also, $(10\bar{1}2)$, $(11\bar{2}2)$, and other nonbasal planes are observed as twin planes in hcp crystals. These results suggest that some dislocation extension occurs on these planes, and hence that stacking faults also occur. Invoking a hard-ball model of the hcp lattice, Thompson and Millard[7] deduced the atomic configuration near a "double step" in a $(10\bar{1}2)$ twin boundary. Kronberg[8] and Westlake[9] associated this step with a *zonal* twinning dislocation. Rosenbaum[10] has shown that the concept of zonal dislocations can account for fault formation, dislocation extension, and nonbasal slip in hcp crystals.

Consider slip on the system $[11\bar{2}\bar{3}](11\bar{2}2)$ or $\mathbf{B'C}(CA_0B')$. A view of the atom configuration on the (CA_0B') plane is shown in Fig. 11-6. Figure 11-7 indicates the next succession of atom movements during the advance of the

[7]N. Thompson and D. J. Millard, *Phil. Mag.*, **43**: 422 (1952).

[8]M. L. Kronberg, *Acta. Met.*, **9**: 970 (1961); *J. Nucl. Mater.*, **1**: 85 (1959).

[9]D. G. Westlake, *Acta. Met.*, **9**: 327 (1961).

[10]H. S. Rosenbaum, in R. E. Reed-Hill et al. (eds.), "Deformation Twinning," Gordon and Breach, New York, 1964, p. 43.

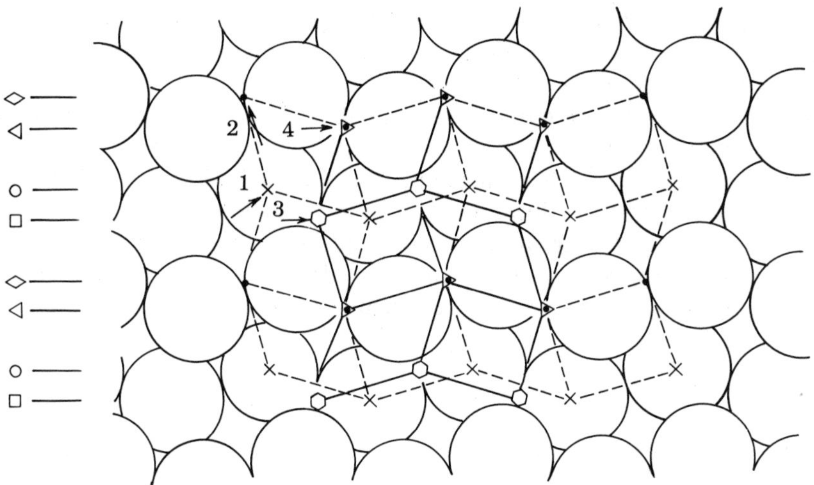

FIGURE 11-7. Displacements of the atom layers of Fig. 11-6 accompanying the partial advance of the dislocation **B′C**. The numbers 1, 2, 3, and 4 refer to the atom movements in the first, second, third, and fourth atom layers above the plane of the paper.

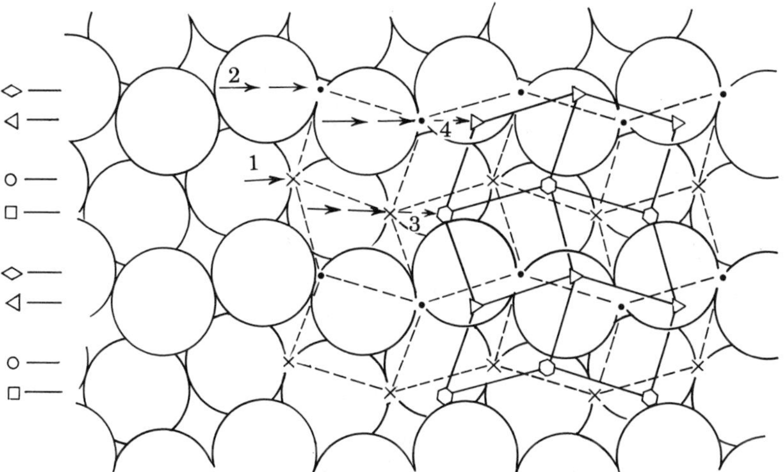

FIGURE 11-8. Further displacements of the atom layers of Fig. 11-7, which complete the advance of **B′C**.

dislocation **B′C** by the lowest-energy path in the hard-ball model. The atoms in the × positions in Fig. 11-7 are all displaced by $\sim\frac{1}{4}$**B′C**. In addition, they are displaced laterally in the **AD** direction, *and alternate atoms are displaced in opposite directions*. This zigzag displacement is called *shuffling*;[11] note that such

[11]The shuffle could be accomplished by the motion of a single dislocation followed by a dislocation dipole one plane high, but whether the motion is a cooperative shear process or a diffusive type reordering is not known, so we simply refer to the process as *shuffling*.

motion *cannot* be produced by the glide of single dislocations, but involves a local reordering of the atomic positions. The ○ atoms in the second layer are also shuffled, but the successively higher atom layers are all displaced by $\sim\frac{1}{4}\mathbf{B'C}$. Thus, except in the first two layers above the slip plane, the displacements are equivalent to those produced by a partial dislocation with a Burgers vector $\sim\frac{1}{4}\mathbf{B'C}$. The line defect which produces the displacements, including the shuffling, is the zonal dislocation. Unlike a normal dislocation, the zonal dislocation is spread over three close-packed layers and produces different displacements in each layer. The remainder of the atom movements required to complete the total dislocation displacement **B′C** are shown in Fig. 11-8. As

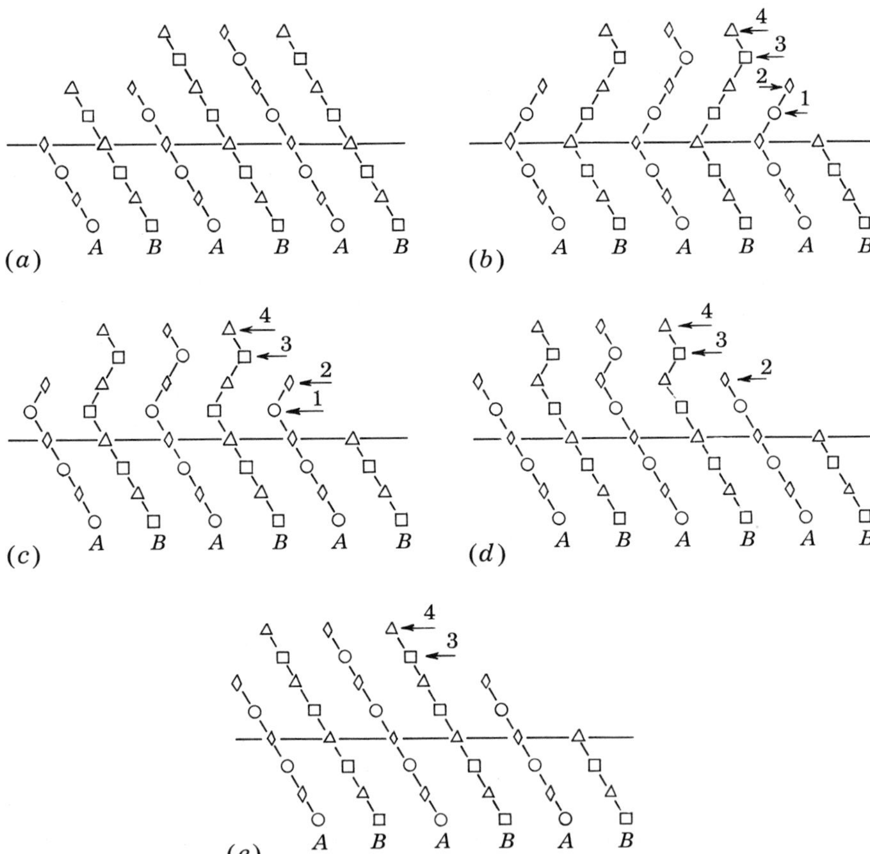

FIGURE 11-9. Side view of Figs. 11-6 to 11-8. The plane of the paper is $(\bar{1}100)$; $(11\bar{2}2)$ is perpendicular to the paper and appears as a horizontal trace. (0002) is also perpendicular to the paper and is indexed *A*, *B*, *A*, *B*, etc. ◇△○ and □ are atom positions at progressively lower levels with respect to the plane of the paper; (*a*) shows the normal structure, and (*b*) to (e) show the atom movements that complete the displacement. The arrows are the projections along $[11\bar{2}3]$ of the displacements in Figs. 11-6 to 11-8.

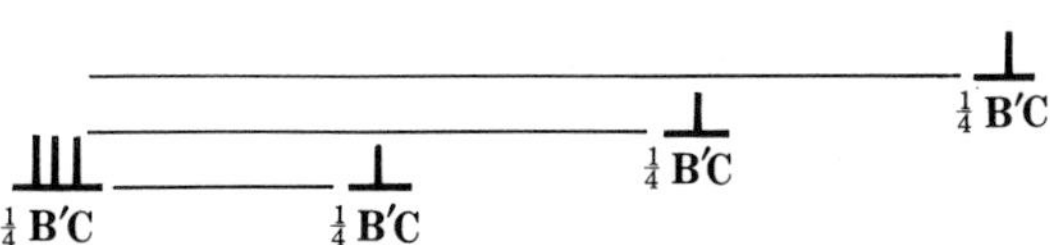

FIGURE 11-10. The dislocation **B′C** extended into a zonal dislocation ⊥⊥⊥ and three partials ⊥. The view is the same as that of Fig. 11-9.

seen there, these displacements are all parallel to **B′C** and can be produced by the motion of three partials with Burgers vectors $(1 \pm f)\frac{1}{4}$**B′C**, where f is a small fraction. The atomic displacements viewed parallel to the glide plane are depicted in Fig. 11-9. Notice that the total shear per atom plane (Fig. 11-9*e*) is *not* uniform.

Figure 11-10 illustrates the extension of the total dislocation **B′C** into a zonal dislocation and three partials. Consider again the faulted structure shown in Fig. 11-7. The displaced × atoms have moved from a position where they had five nearest neighbors below the slip plane to a position with only four neighbors. Thus the stacking fault associated with the zonal dislocation involves the distortion of one nearest-neighbor bond; therefore such faults should have higher energies than the basal-plane faults discussed earlier.

The glide of **AC** on $(ACC'A')$ or $[\bar{1}\bar{1}20](10\bar{1}0)$ is illustrated in Fig. 11-11. As shown there, **AC** can dissociate into two glissile partials on $(ACC'A)$, bounding a stacking fault, which again involves the distortion of one nearest-neighbor bond. For use in the next section, these partials are denoted as **Aϵ** and **ϵC**. The dissociated configuration is not stable in a computer simulation using Lennard-Jones potentials.[12] However, atomic scale relaxations, such as those in bcc dislocations, which lead to some extension of this type are possible.

Extended Jogs

Many of the extended-dislocation configurations in hcp crystals can be deduced from their equivalents in fcc crystals, so that only one example of such arrays is given here. Consider a jog in a Shockley partial dislocation lying in the basal plane. It is generally supposed that such jogs are constricted, because only the basal plane contains the intrinsic stacking fault I_2. However, if the fault of Fig. 11-11 exists in an extended arrangement, then the extended jog shown in Fig. 11-12 can also exist in the hcp case. The extended jog contains stair-rod dislocations **αϵ** analogous to the ones in fcc-crystals. The consequences of jog extension are the same in both cases.

[12] D. J. Bacon and J. W. Martin, in M. F. Ashby, R. Bullough, C. S. Hartley and J. P. Hirth (eds.), "Dislocation Modeling of Physical Processes," Pergamon, Oxford, 1981, p. 269.

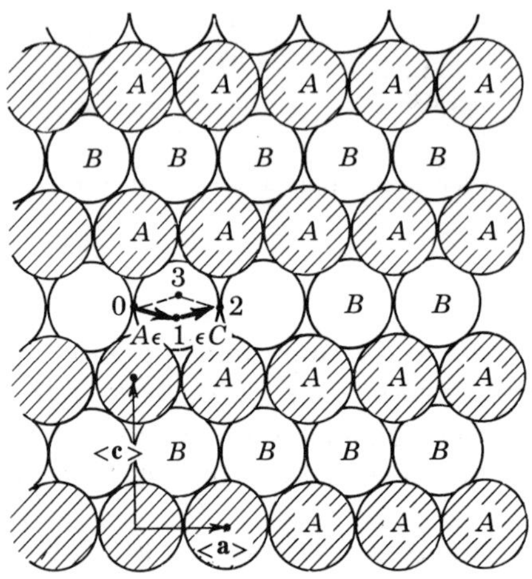

FIGURE 11-11. A $\{10\bar{1}0\}$ surface. Balls are labeled A or B to indicate their basal-plane stacking sequence. A glissile fault can be made by uniformly shearing the crystal so that a ball in the next layer out from the page and at 0 moves to 1 or to 3. The partial **Aϵ** and **ϵC** are also shown.

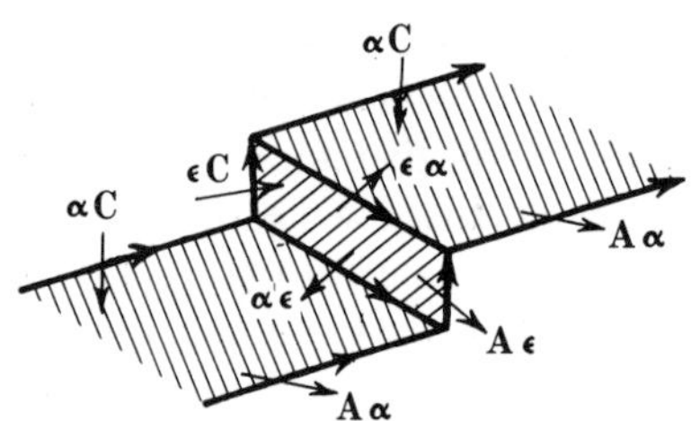

FIGURE 11-12. An extended jog.

11-3. THE BCC LATTICE

Stacking Faults

As reviewed by Vitek,[13] faults on {112}, and {310} planes in bcc crystals have all been suggested. The {112} plane is the coherent twin plane in bcc crystals and atomic calculations indicate that twin and multilayer stacking faults are stable on these planes.[13, 14] Extended faults have not been observed in electron microscopy, and atomic calculations[13] suggest that no stable faults exist other than on {112}. Thus, although there remain large uncertainties in the atomic potentials used for bcc calculations, all the above results are consistent with no faults other than {112} so we restrict discussion to that case.

As shown in Fig. 11-13, the bcc structure is generated by the stacking of {112} planes in the sequence

$$ABCDEFABCDEFAB$$

[13] V. Vitek, *Cryst. Lattice Defects*, **5**: 1 (1974).

[14] P. D. Bristowe and A. G. Crocker, *Phil. Mag.*, **33**: 357 (1976).

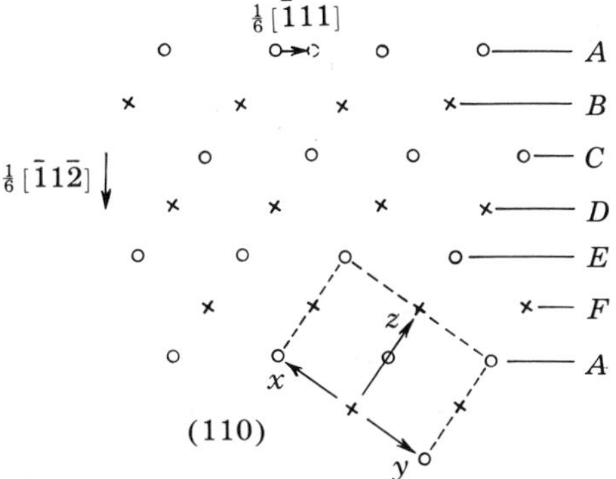

FIGURE 11-13. Projection normal to (110). ○ represents atoms in the plane of the paper and × those in the next plane below. A projection of the unit cell is also shown. The vector $\frac{1}{6}[\bar{1}11]$ at the top moves the atom from an A site to a C site.

where the letters refer to the six different positions of the atom layers projected normal to $\{112\}$. Analogous to the fcc and hcp cases, sequential shears by an amount $\frac{1}{6}[\bar{1}11]$ of layers in Fig. 11-13 or insertion, or removal of two adjacent layers, in principle could create intrinsic faults I_1 and I_2, an extrinsic fault E, and a twin fault T given by

$$I_1:\quad FEDCBAFE\overset{\dagger\;\dagger}{|}FEDCBA \qquad I_2:\quad FEDCBAFE|BAFEDCBA$$

$$E:\quad CDEFABE|FCDEFABC \qquad T:\quad FEDCBAF\overset{\dagger}{\underset{|}{E}}FABCDE \tag{11-8}$$

On the basis of the number of atomic bonds broken or distorted in the faulting operation, all other faults, such as the one produced by adding or removing a single layer from the normal lattice, are expected to have much higher fault energies and hence are of little importance. Indeed, atomic calculations[15] suggest that the simple faults I_1, and I_2, and E have such large energies that dislocation dissociation to create them should not occur. The only fault other than T that is predicted to be stable, in the sense of having energy sufficiently low to give dislocation dissociation or growth-accident faults, is the fault I_n composed of n successive $\{112\}$ layers in twin orientation with $n > 3$. For example,

$$I_3 = F\,E\,D\,C\,B\,A\,F\,\overset{\dagger}{\underset{|}{E}}\,F\,A\,\overset{\dagger}{\underset{|}{B}}\,A\,F\,E\,D\,C\,B\,A \tag{11-9}$$

[15] V. Vitek, *Cryst. Lattice Defects*, **5**: 1 (1974).

This fault can be formed from I_1 [Eq. (11-8)] by added shears of $\frac{1}{6}[\bar{1}11]$ between the layers FE and ED to the right of the I_1 fault or by sequential shears $\frac{1}{3}[1\bar{1}\bar{1}]$ and $\frac{1}{6}[\bar{1}11]$ between the same layers. With only first- and second-neighbor bonds contributing to the fault energy, the energy of an I_n fault is twice the energy of a twin fault, so the edges of the I_n fault can be treated simply as twin boundaries.

Further modifications at the twin interface are also possible. As shown in Fig. 11-13, the shear $\frac{1}{6}[\bar{1}11]$ compresses the nearest-neighbor bond PQ. This effect is minimized if instead the shear at the twin interface is $\frac{1}{12}[\bar{1}11]$ with subsequent shears being $\frac{1}{6}[\bar{1}11]$ in an I_n fault.[16] The unmodified I_n and modified I_n' configurations are comparable in energy for various empirical potentials and thus both should exist on the basis of present understanding.[16]

Central force fault energy recursion formulas analogous to Eqs. (10-12) and (10-13) can be developed for the bcc case.[17] However, there is little justification for a central force description for bcc crystals, so the result is not repeated here. One result of the development that is of interest in atomic calculations using pair potentials is that for shear-type displacements only, the faults of Eq. (11-8) have relative energies $2\gamma_T = \gamma_{I_1} = \gamma_{I_2} = 0.5\gamma_E$.

Partial Dislocations

The shear operations discussed above can be produced by glide of partial dislocations. Perfect dislocations $\frac{1}{2}[111]$ can dissociate into partials of the type $\frac{1}{3}[111]$, $\frac{1}{6}[111]$ and $\frac{1}{12}[111]$. A simple extension to form an I_1 fault is shown in Fig. 11-14*a*. The twinning displacements that have been observed for bcc iron are consistently described by a mechanism involving, for the particular geometry of Fig. 11-14*a*, motion of $\frac{1}{6}[111]$ to the right. The opposite motion of a leading $\frac{1}{6}[111]$ partial that would produce an I_2 fault, have never been observed. Thus many twinning models assume the configuration shown in Fig. 11-14*a*, implying that fault I_1 has low energy and that fault I_2 has a much higher energy. As previously mentioned, atomic calculations do not support the concept of a low-energy I_1 fault.

An alternative dissociation is shown in Fig. 11-14*b*, where a $\frac{1}{2}[111]$ dislocation dissociates into three $\frac{1}{6}[111]$ partials one plane apart, forming an I_3 fault, and a zonal dislocation with no net Burgers vector. The modified fault configuration can be formed by the dissociation shown in Fig. 11-14*c* with two $\frac{1}{12}[111]$ partials and one $\frac{1}{6}[111]$ partial bounding one side of a modified I_3' fault and a zonal partial dislocation $\frac{1}{6}[111]$ bounding the other side. Zonal dislocations of this type and $\frac{1}{12}[111]$ partials have been observed in atomic calculations.[18]

[16] P. D. Bristowe and A. G. Crocker, *Phil. Mag.*, **31**: 503 (1975). This paper clarifies an earlier suggestion by V. Vitek of such partial shears.

[17] J. P. Hirth and J. Lothe, "Theory of Dislocations," 1st ed., McGraw-Hill, New York, 1968, p. 348.

[18] P. D. Bristowe and A. G. Crocker, *Phil. Mag.*, **31**: 503 (1975).

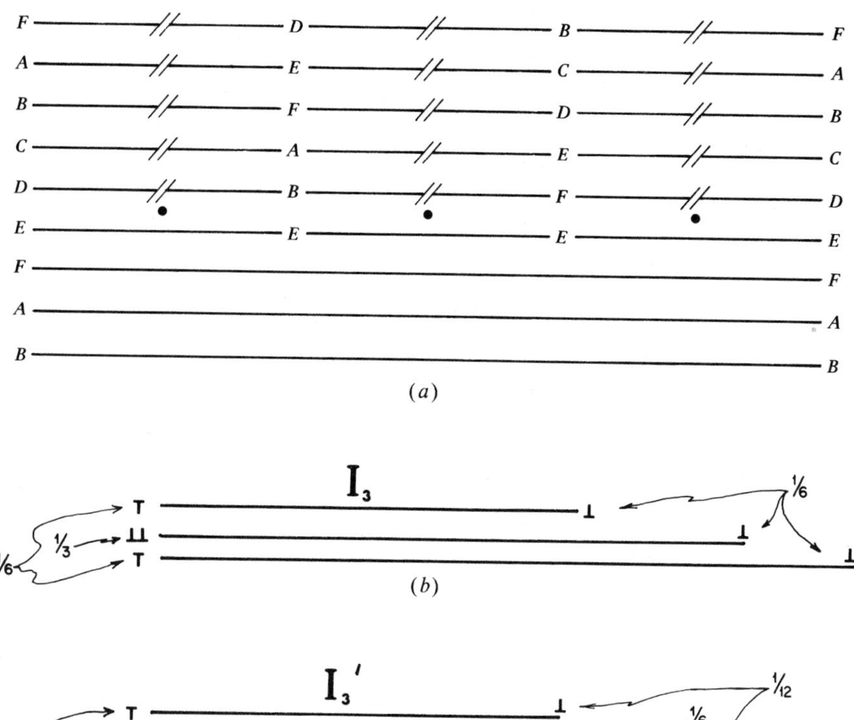

FIGURE 11-14. Sequences of partials and faults on the $(11\bar{2})$ glide plane. In (*b*) and (*c*) the partials $\frac{1}{3}[111]$, $\frac{1}{6}[111]$, and $\frac{1}{12}[111]$ are respectively designated simply $\frac{1}{3}$, $\frac{1}{6}$, and $\frac{1}{12}$, for simplicity.

Because of the three-fold symmetry of the $\langle 111 \rangle$ axis, screw dislocation can, in principle, dissociate into $\frac{1}{6}\langle 111 \rangle$ partials bounding two or three $\{112\}$ faults inclined to one another by an angle of $\pi/3$. Hirsch[19] reviews a number of such possibilities along with extensions on $\{112\}$ planes. However, as discussed next, none of these configurations appears to be stable, so they are not described further.

Fractional Screw Dislocations

Hirsch[20] proposed that the large Peierls stress for bcc screw dislocations was related to some degree of extension of screw dislocations in a nonplanar

[19] P. B. Hirsch, *Trans. Jap. Inst. Met.* (suppl.), **9**: xxx (1968).

[20] P. B. Hirsch, Fifth International Conference on Crystallography, Cambridge, U.K., 1960, p. 139.

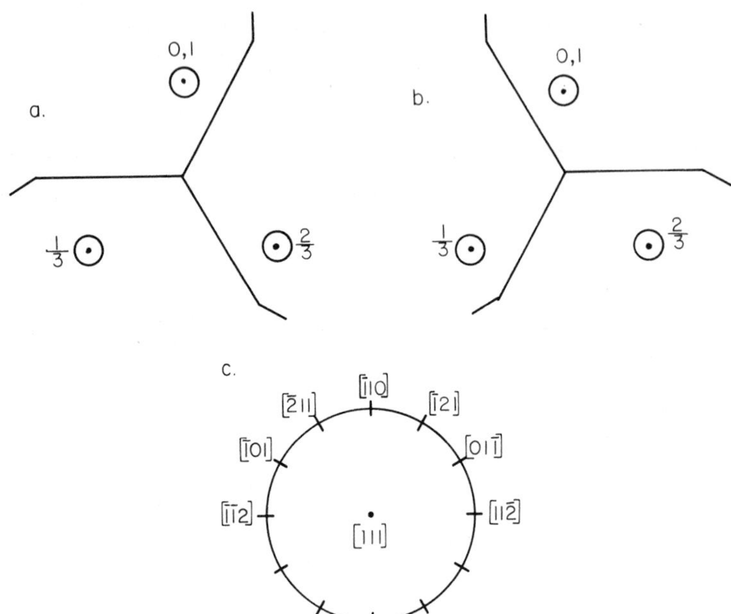

FIGURE 11-15. Equivalent right-handed screw dislocations fractionally extended on $\{110\}$ and $\{112\}$ planes with (a) + polarity and (b) − polarity. The orientation is given in (c) and $\boldsymbol{\xi}$ points out of the page. Atomic rows with their relative projections along $\boldsymbol{\xi}$ as a fraction of **b** are also depicted.

manner associated with the threefold $\langle 111\rangle$ symmetry. Atomic calculations,[21,22] showed that dissociations into stable faulted configurations did not occur, but that core-scale extensions did. As reviewed by Seeger and Wüthrich,[23] this finding has been verified in subsequent calculations by using potentials appropriate for both transition metals and alkali metals. The consequences for screw dislocation motion are those suggested by Hirsch.

Atomic calculations reveal that screw dislocations split into one of two equivalent configurations as shown in Fig. 11-15.[23,24] The stress-free equilibrium position is such that the dislocation is centered between three rows of atoms. One configuration can be produced from the other by a rigid body rotation by π about the $[\bar{1}10]$ axis so that the configurations have identical strain energies. Indeed, with a linear elastic displacement field, the two configurations would be indistinguishable. However, the nonlinear core field is such that there are added net core displacements parallel to the dislocation

[21] V. Vitek, R. C. Perrin, and D. K. Bowen, *Phil. Mag.*, **21**: 1049 (1970).

[22] Z. S. Basinski, M. S. Duesbery, and R. Taylor, *Phil. Mag.*, **21**: 1201 (1970).

[23] A. Seeger and C. Wüthrich, *Nuovo Cim.*, **33B**: 38 (1976).

[24] M. Puls, in "Dislocation Modeling of Physical Systems," Pergamon, Oxford, 1981, p. 249.

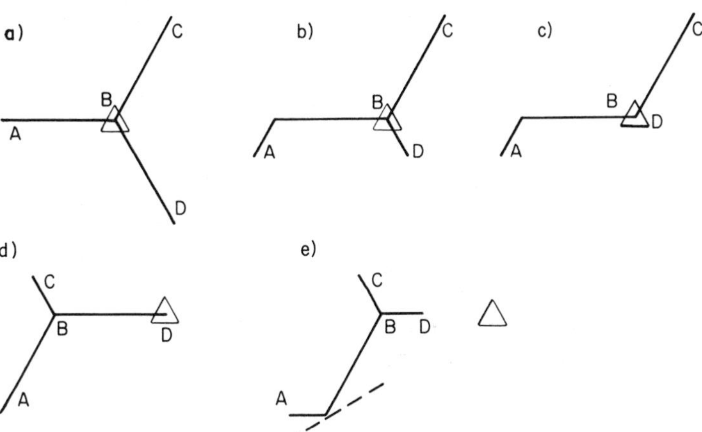

FIGURE 11-16. Sequence of motion of the right-handed screw dislocations shown in Fig. 11-15*a* with increasing stress: *A* to *D* represent fractional dislocations, △ represents the original dislocation position, and the orientation is that as in Fig. 11-15*c*.

line, along [111] for Fig. 11-15*a* and along $[\bar{1}\bar{1}\bar{1}]$ for Fig. 11-15*b*.[23] These displacements destroy the two-fold rotational symmetry about $[\bar{1}10]$ and make the configurations distinguishable, although they still have equal energies since one can be transformed into the other by the rigid body rotation. Seeger and Wüthrich[23] have characterized the added displacements as adding a polarity **p** to the dislocations. We define the polarity as $+(-)$ if the added displacements are in the direction of positive (negative) $\boldsymbol{\xi}$. We note that, like the Burgers vector, the polarity changes sign if the sense of $\boldsymbol{\xi}$ is reversed.

The displacements are such that the dislocation can be considered to be split into a dislocation $q[111]$ at the origin and three dislocations $(\frac{1}{6}-\frac{1}{3}q)$ [111] removed from the origin over core dimensions on $\{110\}$ planes. Because of the variable nature of q, these are called *fractional dislocations*.[25] As also shown in Fig. 11-15, the fractional dislocations have nascent extensions on $\{112\}$ planes, which are also observed to break the $\{110\}$ rotational symmetry.

The consequences for motion of a straight screw dislocation over the Peierls barrier are shown in Fig. 11-16.[25] The value of σ_p is large because the extended fractional dislocations must contract before motion can occur. A sequence of configurations with increased shear stress tending to move the dislocation to the left is presented in Figs. 11-16*a* to 11-16*e*. In Fig. 11-16*d* the dislocation has translated from one Peierls valley to the next and has changed to an equivalent configuration. With an increase of stress, the dislocation translates on the average on the dashed $\{\bar{2}11)$ plane by a mechanism of alternate small glide increments on $\{\bar{1}01)$ and $(\bar{1}10)$.

[25] V. Vitek, *Cryst. Lattice Defects*, **5**: 1 (1974).

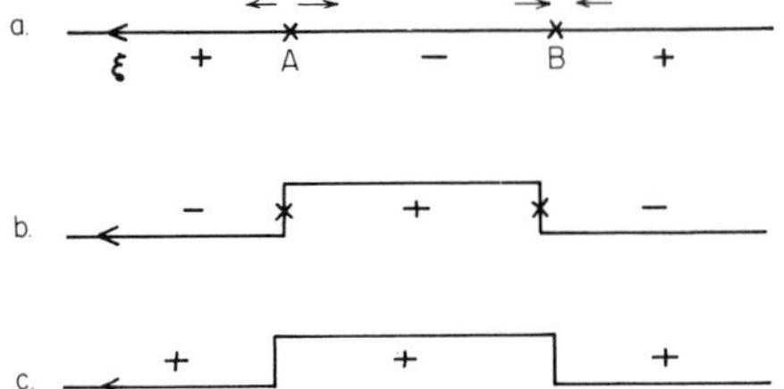

FIGURE 11-17. (*a*) Double flip formed by flips in polarity at *A* and *B*. Arrows designate added elastic displacements associated with the polarity, designated by + and −. (*b*) Double kink reversing polarity. (*c*) Double kink preserving polarity.

In the actual situation, glide over the Peierls barrier is likely to proceed instead by double kink nucleation. The situation for double kinks was elucidated by Seeger and Wüthrich.[23] An additional defect, shown in Fig. 11-17*a*, must be considered. It is a site such as *A* and *B*, where the polarity changes sign whereas the dislocation remains straight and in a Peierls valley. The defects have strain energy because the reversal of the extra displacements leaves a center of dilatation at *A* or compression at *B*. Wüthrich[26] analyzed the energies of these defects for an α-iron potential and found them to have energies (0.6 to 0.7 eV) comparable to the energies of various types of kinks (0.6 to 1.2 eV). The polarity changing defects were called *constrictions*,[26] but we feel that this term is better reserved for configurations where stacking faults are removed, and whether that is the case here is not known. Hence we call the defects *flips*, indicating the plus-minus (+ −) flip of the polarity.

Double kinks can occur with either polarity reversal, as in Fig. 11-17*b*, or with polarity preservation, as in Fig. 11-17*c*; similar defects occur with the signs of all polarities reversed. Screw dislocation motion can occur by (1) continued formation of polarity preserving double kinks, *cccc* ---, (2) sequential kink and flip formation, *bababa*----, or (3) alternating polarity reversing double-kink formation, *b* and its counterpart. The calculations by Wüthrich[26] suggest that model (3) has the lowest double-kink formation energy (0.6 eV), but as he notes, there is some reservation about this result because of the nature of the potential.

The configuration of the screw dislocation in Fig. 11-15 shows that motion in the $[0\bar{1}1]$ is easier; i.e., σ_p is lower, than in the $[01\bar{1}]$ direction. Even without dissociations a hard-ball model of Fig. 11-13 would indicate a lower σ_p for motion of layer *A* over *B* in the $[\bar{1}11]$ direction than for the opposite motion (this is true for either edge or screw dislocations). Both cases are consistent with the widely observed slip asymmetry in bcc metals at low temperatures, with slip in the so-called easy twinning direction favored.

Vector Notation

A scheme[27] for indexing the bcc lattice provides the basis for a rule for dissociation into intrinsic faults. The rule is not as simple as in the case of the

[26] C. Wüthrich, *Phil. Mag.*, **35**: 337 (1977).

[27] J. P. Hirth, *Acta. Met.*, **14**: 1394 (1966).

fcc lattice, but it is much simpler than the alternative of directly considering atom arrangements in deciding upon the proper dissociation for fractional dislocations or twinning partials (Chap. 23).

Because the reciprocal of the bcc lattice is fcc, one can use the fcc Thompson tetrahedron to index bcc slip systems, remembering that directions in the tetrahedron correspond to planes in the bcc lattice, while tetrahedron planes correspond to $\langle 111 \rangle$ bcc slip directions. Thus, with the tetrahedron oriented as in Fig. 10-9, the plane (d) corresponds to the Burgers vector $\frac{1}{2}[111]$, $(\underline{d})$ corresponds to $\frac{1}{2}[\bar{1}\bar{1}\bar{1}]$, and the direction $\mathbf{B}\boldsymbol{\delta}$ represents the glide plane $(2\bar{1}\bar{1})$.

The rule for formation of the fault I_1 by dissociation of a $\frac{1}{2}\langle 111 \rangle$ dislocation is given in the above notation as follows:

Axiom 11-2: Choose $\mathbf{b}$ to point out of the tetrahedron. Then the intrinsic fault I_1 is produced by moving a right-handed screw in the direction

$$\mathbf{s} = \mathbf{RG} \times \mathbf{b} = \mathbf{b} \times \mathbf{GR}$$

a left-handed screw in the $-\mathbf{s}$ direction. Here $\mathbf{RG}$ indicates the glide-plane vector arranged in the Roman-Greek sequence, such as $\mathbf{B}\boldsymbol{\delta}$. The fault I_2 is formed by motion in the opposite direction.

11-4. THE DIAMOND CUBIC LATTICE

Stacking Faults

The diamond cubic lattice, shown in Fig. 11-18, corresponds to two interpenetrating fcc lattices, one of which is displaced by $(\frac{1}{4}, \frac{1}{4}, \frac{1}{4})$ with respect to the other. Atoms in the two lattices do not have identical surroundings, so that the structure can be described as an fcc structure with a basis of two atoms per lattice point. The primitive-unit-cell vectors remain those associated with a single fcc lattice, $\frac{1}{2}\langle 110 \rangle$. Thus the perfect-dislocation Burgers vectors for the diamond cubic lattice are the same as those in the fcc lattice.

The layer structure of diamond contains $\{111\}$ planes in the sequence *AaBbCcAaBbCc*, as shown in Fig. 11-19. In the projection normal to $\{111\}$, the layers *A* and *a* are seen to project to the same type of position, as do *B* and *b* and *C* and *c*. Twins and stacking faults involve the insertion or removal of *pairs* of layers of the same index, *Aa*, etc. These low-energy faults involve no change in the four nearest-neighbor covalent bonds in the lattice; all other faults disturb the nearest-neighbor bonding and are expected to be high-energy faults.

Because the layers must be added or removed in pairs, one can drop the double-index notation and describe the packing by the sequence $\cdots ABCABC \cdots$, where each letter refers to a pair of layers. A comparison of

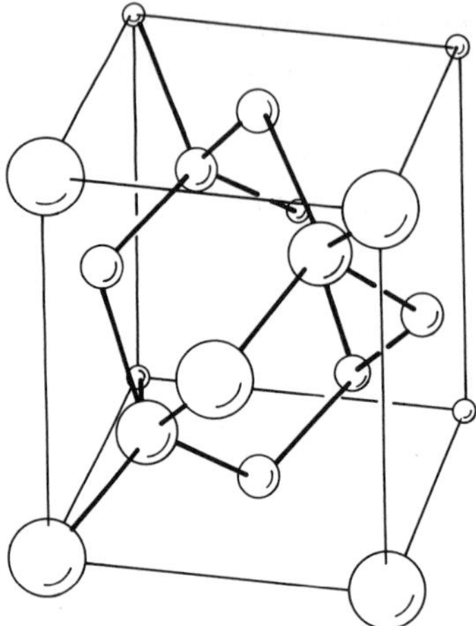

FIGURE 11-18. A diamond cubic unit cell.

this sequence with that for fcc structures reveals that the stacking faults in the diamond lattice are the same as in the fcc lattice; i.e., the intrinsic fault is $\cdots$ ABC ¦ BCAB $\cdots$, etc. Intrinsic (I) and extrinsic (E) stacking faults are illustrated in Fig. 11-20 in the same representation as that of Fig. 11-19.

In the covalently bonded diamond structure most of the binding energy resides in nearest-neighbor bonds. Also, because only double layers are out of registry in the faulted arrangement, only bonds between fourth-nearest-neighbor planes are distorted in the stacking fault. Therefore, one expects the stacking-fault energy to be lower, relative to the surface energy and grain-boundary energy, in the diamond lattice than in fcc or hcp lattices. The relative fault

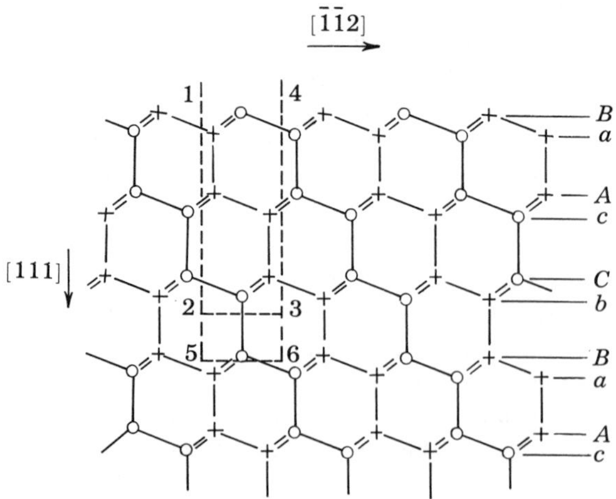

FIGURE 11-19. A diamond cubic lattice projected normal to $(1\bar{1}0)$. O represents atoms in the plane of the paper and + represents atoms in the plane below. (111) is perpendicular to the plane of the paper and appears as a horizontal trace.

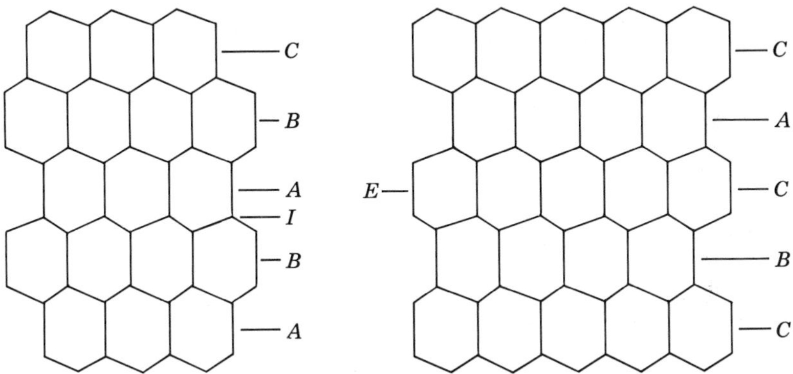

FIGURE 11-20. (*a*) Intrinsic and (*b*) extrinsic stacking faults viewed in a projection normal to $(1\bar{1}0)$.

energy should decrease as the bonding becomes more covalent, so that the ratio of fault energy to surface energy should decrease in the order gray tin, germanium, silicon, diamond. Although there remains some disagreement in current measurements of fault energies, the most recent measurements give $\gamma_I = 60$ mJ/m^2 for germanium[28] and 69 mJ/m^2 for silicon,[29] consistent with both of the above expectations. Also, the results indicate that γ_E is about the same or somewhat less than γ_I, with values of $\gamma_E = 60$ mJ/m^2 for silicon[29] and $\gamma_E = 30$ mJ/m^2 for germanium[28] or, alternatively, of $\gamma_E \sim \gamma_I$ in germanium.[29]

Perfect Dislocations

The glide plane in the diamond cubic case is $\{111\}$ and the perfect dislocations have Burgers vectors $\frac{1}{2}\langle 110\rangle$. Presumably because of a large Peierls barrier with deep troughs along $\langle 110\rangle$ directions, glide dislocations lie primarily along $\langle 110\rangle$ directions when the dislocation density is low. At higher dislocation densities, curved dislocation lines are observed, doubtless because of the presence of dislocation interaction forces. Glide dislocations lying along $\langle 110\rangle$ directions are either pure screw or 60° dislocations, so-called because **b** is inclined at an angle of 60° to $\boldsymbol{\xi}$. Because of their prominence in the early stages of glide, these two dislocation types have been analyzed geometrically in detail.

The model adopted to represent dislocations in the diamond cubic lattice was proposed by Hornstra.[30] This model is essentially that of balls connected by nearest-neighbor bonds (the covalent bonds). The misfit energy is supposed to be quite localized in view of the high energies entailed in breaking covalent bonds and in distorting their bond angles. Hence the depicted dislocations are

[28] A. Gomez, D. J. H. Cockayne, P. B. Hirsch, and V. Vitek, *Phil. Mag.*, **31**: 105 (1975).
[29] H. Foll and C. B. Carter, *Phil. Mag.*, **40**: 497 (1979).
[30] J. Hornstra, *J. Phys. Chem. Solids*, **5**: 129 (1958).

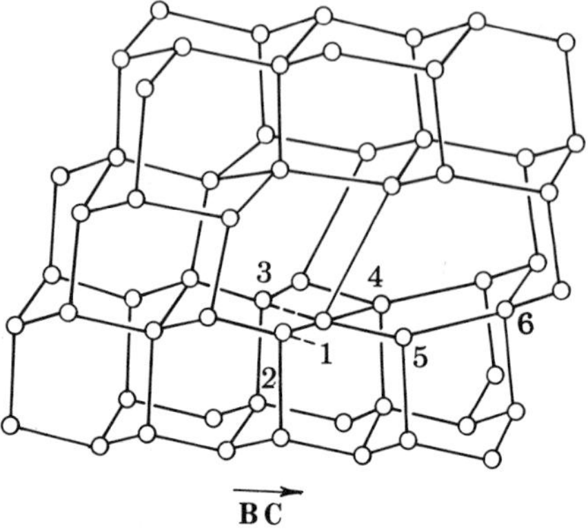

FIGURE 11-21. The 60° dislocation **BC**.

quite narrow. Again, one must maintain some reservation about the accuracy of these models in depicting the core structure because of the uncertainty in the distribution of misfit and elastic energy in the core. However, the models are useful in suggesting possible modes of dislocation dissociation.

Because of the double-layer atomic arrangement, shown in Fig. 11-19, there are two inherently different sets of dislocations in the diamond cubic lattice, with one member of each set having a Burgers vector identical to one dislocation in the other set. Study of Fig. 11-19 helps clarify this difference. A 60° dislocation of the *glide set* is formed by cutting out the material bounded by the surface 1-5-6-4 and then displacing the sides of the cut so that they join. The distinctive feature of dislocations of this type, as is evident from the position of the cut surface, is that their extra plane of atoms terminates between different letter index layers, for example, between a and B in the example of Fig. 11-19. Notice that all the cut bonds on the surfaces 1-5 and 4-6 are re-formed after the displacement, so that the perfect-lattice bonding is restored. On surface 5-6, however, one bond per atom site along the dislocation is left free; such extra bonds are called *dangling bonds*. A possible three-dimensional projection of the resultant dislocation and its row of dangling bonds is shown in Fig. 11-21. Bond resonance is expected between the 1-2 positions and the 1-3 positions, possibly even with the 1-4 and 1-5 positions. Such bonds resonance could lead to a decrease in energy from the configuration of rigid dangling bonds.

On the other hand, 60° dislocations of the *shuffle set* (so defined for reasons discussed later) are formed by cutting out material bounded by the surface 1-2-3-4 in Fig. 11-19. In this case the extra plane of atoms ends between layers of the *same* letter index, for example, b and B in Fig. 11-19. Again, after the cut surfaces are joined, all bonds are re-formed except one dangling bond on surface 2-3. A possible projected view of this dislocation is shown in Fig. 11-22. Bond resonance is again expected between the 1-2 and 1-3 positions. Hornstra[30] has discussed this type of dislocation and its dissociation in detail but apparently has neglected the glide-set type of Fig. 11-21.

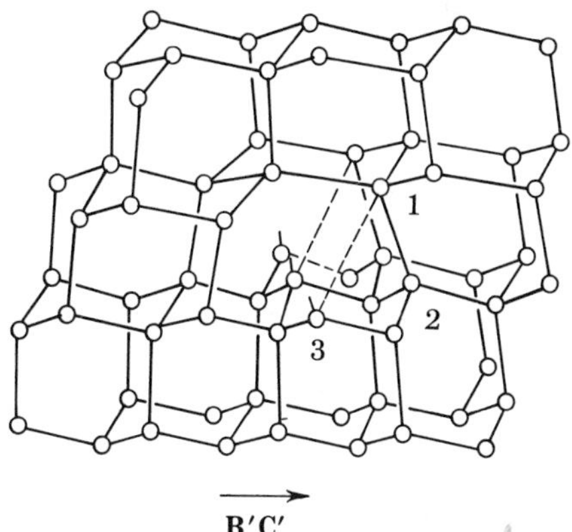

FIGURE 11-22. The 60° dislocation **B′C′**.

The 60° dislocations were used in the above examples because of their ease of depiction. However, it is apparent from Fig. 11-19 that the same distinction between glide plane (between layers a and B for the glide set and between b and B for the shuffle set) holds for screw or edge dislocations or for intermediate orientations. Also, as one can see by shifting the cut surface along the glide plane in Fig. 11-19 or by shifting the bonds in Figs. 11-21 and 11-22, perfect dislocations in either set are glissile. Of course, differences are expected between the Peierls energies in the two cases.

Vector Notation

One might be tempted to use the fcc Thompson tetrahedron directly to represent the Burgers vectors, in view of the analog between stacking faults. However, in order to distinguish between the two sets of dislocations with identical Burgers vectors, the notation of Fig. 11-23 is adopted. The double tetrahedron represents the two possible associations of the fcc tetrahedron with the two interpenetrating fcc lattices in the diamond cubic structure. The

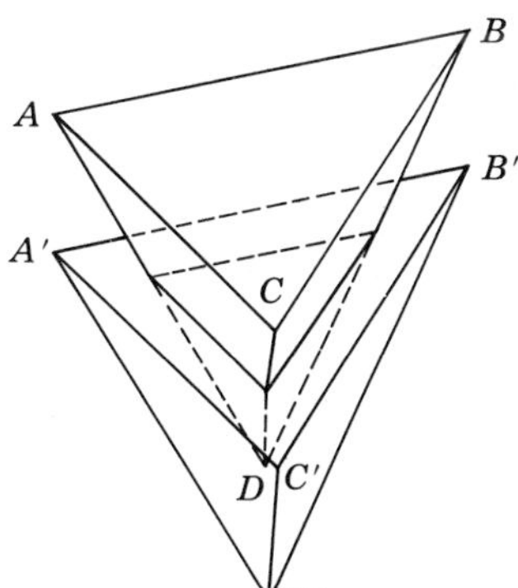

FIGURE 11-23. Interpenetrating Thompson tetrahedra with the vector notation for diamond cubic structures.

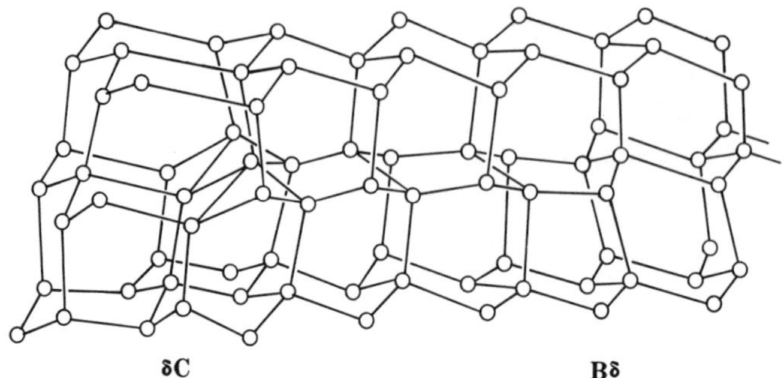

FIGURE 11-24. The dissociation of **BC** into **δC** and **Bδ**.

convention is adopted that dislocations of the shuffle set of Fig. 11-22 have Burgers vectors on the tetrahedron $A'B'C'D'$, while those of the glide set of Fig. 11-21 lie on $ABCD$. The perfect and partial dislocations of each set are related to the vector notation $[hkl]$ in a manner similar to that illustrated in Fig. 10-10 for the fcc case.

The Glide Set of Dislocations

Dislocations on $ABCD$ are termed the *glide set* because, in addition to the perfect dislocation, the Shockley partial dislocations **δA**, **Bδ**, etc., of this set are glissile; they can undergo glide as it is defined for a continuum dislocation. Figure 11-24 is a projected view of the dislocation **BC** dissociated into **δC** and **Bδ**. As indicated in Fig. 11-25, because the dislocation glides between layers b and C, the dissociation into **δC** and **Bδ** causes the formation of an intrinsic stacking fault between the partials. Finally, a view of the same dissociation projected normal to the glide plane is shown in Fig. 11-26.

The total dislocation is in screw orientation for **δC** between sites 6 and 7 and **Bδ** between 1 and 2; it is in the 60° orientation over the remainder of the portrayed length. A kink in **Bδ** is shown at 3. The misfit energy of the dislocation is represented by the distorted and dangling bonds in the figure. The actual configuration involves the elastic relaxation of the arrangement in Fig. 11-26 to reduce the misfit energy; possibly the relaxation involves bond rearrangements to eliminate dangling bonds, the so-called *reconstruction*.[31] The dissociations, interactions, stair rods, etc., associated with the dislocations of the glide set $ABCD$ are identical to those discussed for the fcc case.

[31] The possibility that ease of reconstruction energetically favors the glide set is discussed further by P. B. Hirsch in J. Narayan and T. Y. Tan (eds.), "Defects in Semiconductors," North-Holland, Amsterdam, 1981, p. 257. He also reviews atomic and electronic calculations for the different partial dislocation cores.

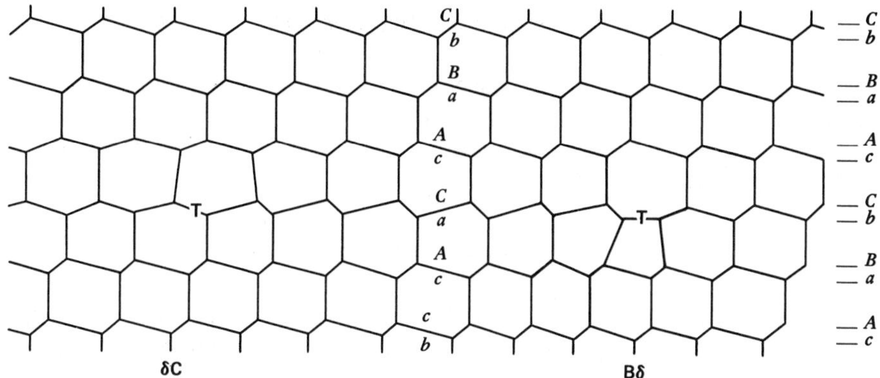

FIGURE 11-25. Projection of Fig. 11-24 perpendicular to (1$\bar{1}$0).

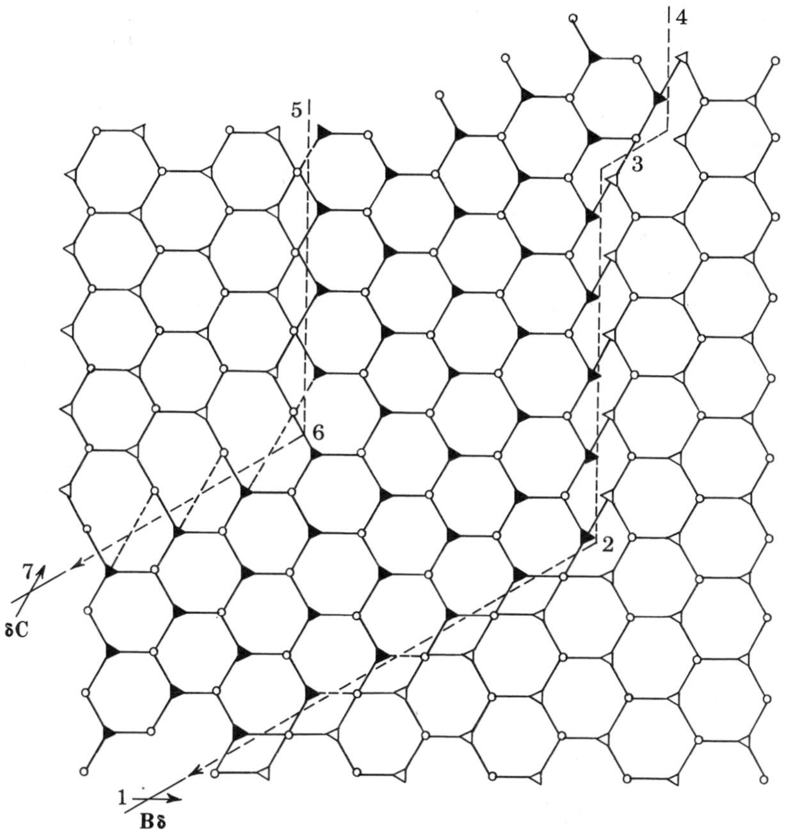

FIGURE 11-26. Projection of Fig. 11-24 normal to the glide plane (111) or (d): ○ represents atoms below the glide plane, △ represents atoms above the glide plane in the perfect lattice, and ▲ represents atoms above the glide plane in the faulted region. The dashed lines between 6 and 7 represent dangling bonds and the possibility of bond resonance.

The Shuffle Set of Dislocations

Dislocations in the set $A'B'C'D'$ have glide planes lying between layers of the same index, such as b and B in Fig. 11-19. Therefore these dislocations *cannot* dissociate directly into partials, because such dissociation would produce a high-energy fault of the type

$$CcAaB|aBbCc$$

The only possible dissociation of $\mathbf{B'C'}$ is of the type indicated in Fig. 11-27. This type involves the nucleation of a pair of partials $\boldsymbol{\delta}\mathbf{B}$ and $\mathbf{B}\boldsymbol{\delta}$ on a plane displaced from (d') by $\frac{1}{2}\boldsymbol{\delta}\mathbf{D}$. The resulting dislocation has one glissile partial $\mathbf{B}\boldsymbol{\delta}$ from the set $ABCD$ bounding one side of an intrinsic stacking fault. The partial bounding the other side of the fault is formally $\boldsymbol{\delta}\mathbf{B}+\mathbf{B'C'}$, or equivalently $\boldsymbol{\delta}\mathbf{C}+(\mathbf{CB}+\mathbf{B'C'})$. The bracketed dislocation pair is a dislocation dipole equivalent to a row of interstitials. A comparison of Figs. 11-24 and 11-27 indeed reveals that $\boldsymbol{\delta}\mathbf{C}$ can be transformed to $\boldsymbol{\delta}\mathbf{C}+(\mathbf{CB}+\mathbf{B'C'})$ by the addition of a row of interstitials. Alternatively, if the pair $\boldsymbol{\delta}\mathbf{B}$ and $\mathbf{B}\boldsymbol{\delta}$ were nucleated on a plane displaced from (d') by $\frac{1}{2}\mathbf{D}\boldsymbol{\delta}$, the result would be a partial $\mathbf{B}\boldsymbol{\delta}$ and a partial $\boldsymbol{\delta}\mathbf{C}+(\mathbf{CB}+\mathbf{B'C'})$, where the bracket dipole now would be equivalent to a row of vacancies (Fig. 11-28). For brevity, we write $\boldsymbol{\delta}\mathbf{C}^i$ for the partial in Fig. 11-27 and $\boldsymbol{\delta}\mathbf{C}^v$ for that in Fig. 11-28.

The partial dislocations $\boldsymbol{\delta}\mathbf{C}^v$ and $\boldsymbol{\delta}\mathbf{C}^i$ *cannot* glide in the sense defined for the continuum dislocation. As can be seen in Fig. 11-27, motion of the partial $\boldsymbol{\delta}\mathbf{C}^i$ to the right involves the shear of the bonds between 1 and 3 to form bonds between 1 and 2, but *in addition*, it requires the rearrangement of rows 3 and 4 by the process depicted in Fig. 11-29. This arrangement can be regarded as a short-range shear process (note that the shear is opposite to that producing net dislocation motion), or as motion of the row of the interstitials by a core-diffusion process. In accordance with the nomenclature developed for hcp zonal dislocations, this type of motion is called *shuffling*. Hence the set of dislocations on $A'B'C'D'$ is the *shuffle* set. The various partial dislocations, such as $\boldsymbol{\delta}\mathbf{C}^i$ and $\boldsymbol{\delta}\mathbf{C}^v$, in the shuffle set and their interactions have been

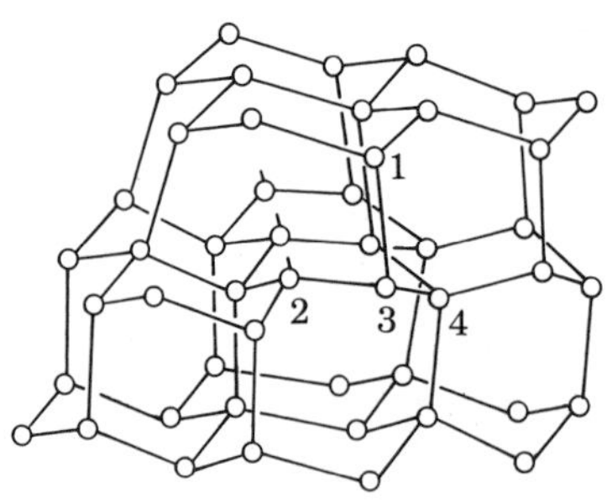

FIGURE 11-27. The partial $\boldsymbol{\delta}\mathbf{C}^i=\boldsymbol{\delta}\mathbf{C}+(\mathbf{CB}+\mathbf{B'C'})$.

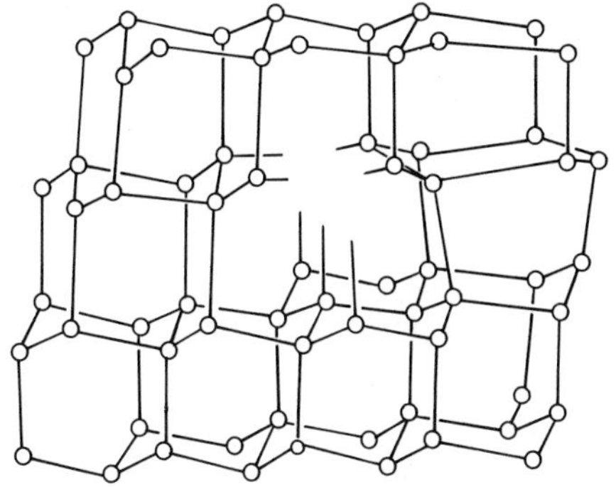

FIGURE 11-28. The partial $\boldsymbol{\delta}\mathbf{C}^v = \boldsymbol{\delta}\mathbf{C} + (\mathbf{CB} + \mathbf{B'C'})$.

FIGURE 11-29. Shuffle rearrangement accompanying the glide of $\boldsymbol{\delta}\mathbf{C}^i$.

discussed extensively by Hornstra,[30] to whose article the reader is referred for detail. The rules for dissociation presented for fcc crystals apply, but one must remember that one of the resulting partials is of the Shockley type $\boldsymbol{\delta}\mathbf{B}$, while the other is of the complicated type $\boldsymbol{\delta}\mathbf{C}^i$ or $\boldsymbol{\delta}\mathbf{C}^v$.

As is evident from the above discussion, the transformation of the shuffle-set dislocations to the glide set and the reverse process both can occur by climb. For example, **BC** in Fig. 11-21 transforms to **B′C′** in Fig. 11-22 upon acquiring a row of either interstitials or vacancies. Addition of a row of interstitials above rows 4 and 5 in Fig. 11-21 creates the configuration of Fig. 11-22 directly. Removal of rows 4 and 5 and bonding between rows 1 and 6 with accompanying relaxation creates the configuration of Fig. 11-22 but one double-plane lower in the drawing.[32] Referring back to Sec. 8-4, one sees that the adsorption of *part* of a row of vacancies to **BC** produces the modified core arrangement that Haasen[33] suggested should increase the Peierls energy.

The Peierls energies should be different for the two sets of dislocations, so that partial transformation of a dislocation from the low Peierls energy set to

[32] J. Blanc [*Phil. Mag.*, **32**: 1023 (1975)] has suggested that the removal of a row of atoms instead creates a different configuration, that of Fig. 11-22 with rows 3 and 4 removed and three sets of dangling bonds. He justifies the difference between this "vacancy" configuration and the interstitial configuration of Fig. 11-22 by the different free energies of formation of bulk interstitials and vacancies. Since both types of defect can be formed at any of the jogs in Fig. 11-30, this argument is invalid. However, there are uncertainties in core structure, as also discussed by Blanc, so configurations such as those in Figs. 11-21 and 11-22 and discussions in terms of tetrahedral "bonds" require validation.

[33] P. Haasen, *Acta. Met.*, **5**: 598 (1957).

the other set is expected to raise the average Peierls energy of the resultant configuration. Such effects should be important in dislocation pinning, work hardening, and yield-point phenomena.

Recent electron microscopic studies have shown unequivocally that dislocations in silicon and germanium are extended and glide in the extended configuration.[34, 35] Wessel and Alexander,[35] using the concepts illustrated in Fig. 10-38, showed that the partial dislocations had different mobilities as a function of their orientation and whether they were leading or trailing. Although the details remain to be elucidated, these results would be consistent with different Peierls energies for the partials, whether an intrinsic effect or one associated with point defect interactions as mentioned above. In view of the uncertainties in core energy and configurations, one can only speculate about the relative Peierls energies of the two sets of dislocations. On the basis of the more complicated shuffling motion of the $\boldsymbol{\delta C}^i$ type of partials, we suppose that the Peierls energy of the *ABCD* set will be lower when the dislocations are dissociated, but we leave the question open for the case of the undissociated dislocations. Again, quantitative answers to these questions must await detailed atomic calculations.

Jogs

Dislocation interactions within the glide set or shuffle set cause extended jogs to form. These jogs are composed of partials which can be described formally in the same notation as in the fcc case. In addition, a new type of jog forms when a dislocation moves from one set to the other. A jog in $\mathbf{B'C'}$ is shown in Fig. 11-30*a*; a jog in **BC** is shown in Fig. 11-30*b*. The small jog formed when the dislocation moves from $A'B'C'D'$ to *ABCD* is shown in Fig. 11-30*c*. The latter configuration also can be formed by climb of either of the other jogs or by the rearrangement of the jog of fig. 11-30*a* by a core-diffusion process.

Nature of Slip System

Although the glide and shuffle sets can be defined, there is still no unequivocal evidence for which set predominates. As discussed by Speake et al.,[36] many authors assume that glide occurs on the shuffle set, mostly on the basis of the wider separation between shearing atoms across the glide plane. The interpretation of the differential mobility of partials in silicon[35] as relating to point defect interactions would tend to favor the shuffle set. Contrariwise, the interpretation[37] of Fig. 1-12 suggests that the dislocation is dissociated in the

[34]A. M. Gomez and P. B. Hirsch, *Phil. Mag.*, **36**: 169 (1977).

[35]K. Wessel and H. Alexander, *Phil. Mag.*, **36**: 169 (1977).

[36]C. C. Speake, P. J. Smith, T. R. Lomer, and R. W. Whitworth, *Phil. Mag.*, **38**: 603 (1978).

[37]A. Olsen and J. C. H. Spence, *Phil. Mag.*, **43**: 945 (1981). Figure 1-12 views a 60° dislocation dissociated into an edge partial and a 30° partial in a 4-nm-thick Si wafer. A comparison of dynamical electron images with Fig. 1-12 indicates that the 30° partial belongs to the glide set.

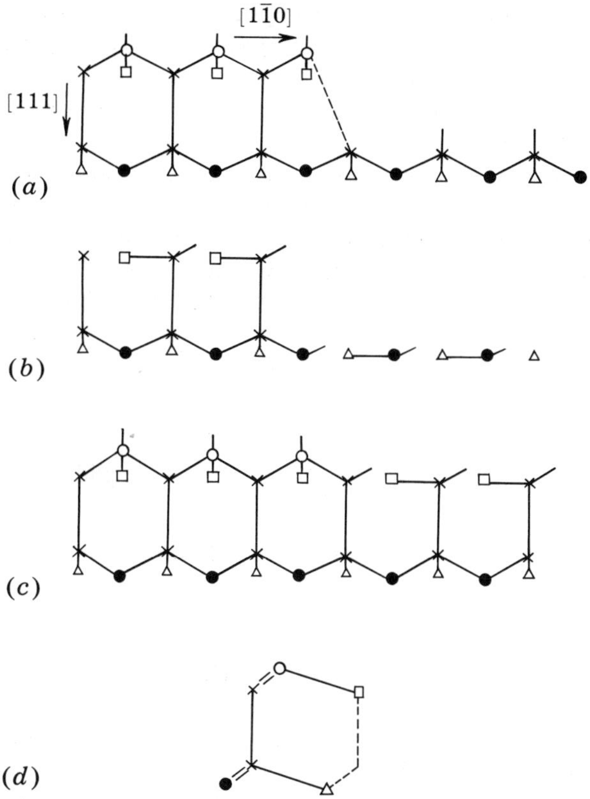

FIGURE 11-30. (*a*) A jog in **B′C′**, (*b*) a jog in **BC**, and (*c*) a jog between *A′B′C′D′* and *ABCD*. The plane of the paper is (11$\bar{2}$). In (*d*) the atom positions are keyed to the projection of Fig. 11-19.

glide set. Also, observations of constrictions in a faulted dipole in germanium indicate the presence of jogs of the type shown in Fig. 11-30*c*, the interpretation suggesting extension in the glide set, constriction in the shuffle set, and, of course, the existence of both sets.[38]

11-5. LAYER STRUCTURES

Stacking faults and coherent twin planes in other crystal structures can be analyzed in accordance with the precepts applied in the preceding sections. In general, stacking faults are expected in crystals which exhibit twinning behavior. Also, stacking faults are associated with crystals, such as carborundum, in which polytypism is observed, and crystals, such as mica and graphite, with

[38]S. W. Chiang, C. B. Carter, and D. L. Kohlstedt, *Scripta Met.*, **14**: 803 (1980).

marked layerlike anisotropy in their physical properties. Amelinckx[39, 40] has reviewed observations of extended dislocations and their associated faults in layer structures; his reviews are referred to for detail.

Crystals in these structures tend to have a mixed covalent-ionic nature of bonding with the consequence that dislocations can acquire electrical charge as discussed in Chap. 12. Studies of electrical currents associated with dislocation motion in cubic zinc sulfide indicate that slip occurs on the glide set.[41] These sphalerite crystals have the structure of Fig. 11-19 with zinc on $a, b, c \cdots$ and sulfur on $A, B, C \cdots$ and thus in principle contain the glide and shuffle sets as do diamond cubic crystals. Ice is a layer structure with a similar stacking and the possibility of both glide and shuffle dislocation sets. The possibility of motion on the glide set removes a quandary in the theoretical interpretation of flow in terms of dislocations.[42]

Exercise 11-1. Discuss the types of possible stacking faults and partial dislocations in fluorite, which has the layer-structure sequence

$$aAabBbcCcaAabBb \cdots$$

11-6. SUPERDISLOCATIONS IN ORDERED STRUCTURES

Ordered structures are introduced at this point because dislocations in ordered structures resemble partial dislocations in simple one-component crystals in their properties. Consider the two-dimensional ordered structure shown in Fig. 11-31*a* as an analog of the actual three-dimensional structure of Fig. 11-31*b*. In the perfectly ordered state, each A atom in this square lattice would be surrounded by four nearest-neighbor B atoms and vice versa, forming a *superlattice*.

Fault surfaces across which the crystal structure and crystal orientation remain unchanged but across which atoms have the wrong nearest-neighbor atoms are called *antiphase boundaries*. These boundaries separate *superlattice domains*; they can form when two domains that are out of phase impinge upon one another during the ordering process. In addition, antiphase boundaries are associated with dislocations as indicated in Fig. 11-31. For that simple structure, dislocations of the type [10], which are *perfect* dislocations in the disordered lattice, lie along the edges of fault surfaces. A *superdislocation* of the type [20], Fig. 11-31, has no fault associated with it; its motion leaves a

[39]S. Amelinckx, in F. R. N. Nabarro (ed.), "Dislocations in Solids," vol. 2, North-Holland, Amsterdam, 1979, p. 67.

[40]S. Amelinckx, *Solid State Phys. Suppl.* 6 (1964).

[41]V. F. Petrenko and R. W. Whitworth, *Phil. Mag.*, **41A**: 681 (1980).

[42]R. W. Whitworth, *Phil. Mag.*, **41A**: 521 (1980).

```
A  B  A  B  A  B  A  B  A  B  A  B
B  A  B  A  B  A  A  B  B  A  B  A
A  B  A  B  A  A  B  A  B  B  A  B
B  A  B  A  B  B  A  B  A  A  B  A
A  B  A  B  A  B  B  A  A  B  A  B
B  A  B  A  B  A  B  A  B  A  B  A
A  B  A  B  A  B  A  B  A  B  A  B
B  A  B  ⊥  B  A  B  A  ⊥  A  B  A
A  B  A     B  A  B  A     B  A  B
B  A     B  A  B  A  B  A  B  A
A  B     A  B  B  A  B  A  B  B
A  B     B  A  A  B  A  B  A  A
B  A     B  B  B  A  B  A  B  B
A  B     A  A  A  B  A  B  A  A
B  A     B  B  B  A  B  A  B  B
A  B     A  A  B  A  B  A  B  A
B  A     B  B  A  B  ⊥  B  A  B
A  B     A  B  A  B     A  B  A
B  A     B  A  B  A     B  A  B
A  B     A  B  A  B     A  B  A
```

(*a*)

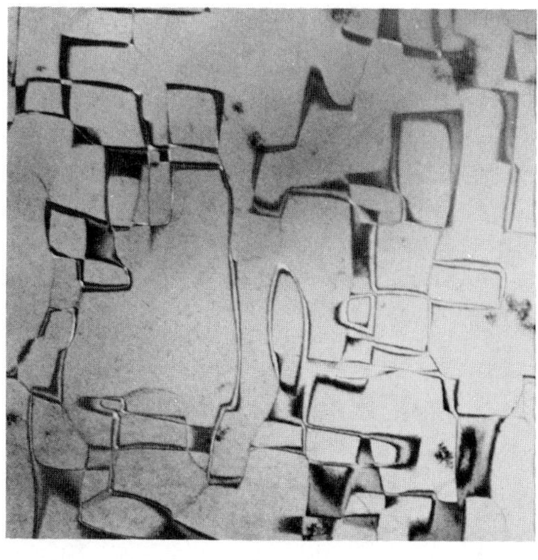

(*b*)

FIGURE 11-31. (*a*) A two-dimensional ordered structure with two superlattice domains completely bounded by a curved and by a straight-sided antiphase boundary. Two antiphase boundaries bounded by dislocations are also shown. (*b*) Actual antiphase boundaries in Cu_3Au, [001] foil orientation, 70,000× [M. J. Marcinkowski and L. Zwell, *Acta Met.*, **11**:373 (1963).]

perfectly ordered lattice in its wake. As shown in Fig. 11-31, a superdislocation can dissociate into two $\langle 10 \rangle$ dislocations bounding an antiphase boundary.

The disordered lattice before ordering is a simple square lattice with lattice parameter a_0. After ordering, each atom in the small square does not have identical surroundings, so that formally the ordered structure becomes the larger square lattice with lattice parameter $2a_0$, and with a pair of atoms A and B associated with each lattice point. Thus the superdislocation [20] is a perfect dislocation in the ordered lattice, the superdislocation [11] is the smallest perfect dislocation in the ordered lattice, and the dislocation [10] is a partial dislocation, all in accordance with our previous definitions. However, these dislocations are of most interest in cases where the alloy can exist in both the ordered and disordered states, so that it is convenient to use the terms "superdislocation" and "dislocation" for the [20] and [10] types to avoid confusion with perfect and partial dislocations in the disordered lattice.

An important distinction between the properties of dissociated superdislocations and those of the previously discussed dissociated perfect dislocations arises as a result of the difference in orientation dependence between stacking faults and antiphase boundaries. The stacking-fault energy is associated with wrong *numbers* and wrong orientations of near-neighbor bonds. As a consequence, only a few specific orientations have sufficiently low energies to appear as important faults in metals. On the contrary, antiphase boundaries have the correct numbers and orientations of nearest neighbors, but the bonds have the wrong chemical nature. As a result, the orientation dependence of the antiphase-boundary energy is small in some instances. Thus antiphase boundaries can be either smoothly curved, when the orientation dependence is small, or planar, when it is large.

The properties of extended superdislocations are analogous to those of extended perfect dislocations in that the fault surface exerts a force on the dislocations; extended jogs, extended barriers, and extended nodes can exist; cross slip and climb are affected by dissociation of superdislocations, etc. The configurations in the two specific ordered structures considered in the following discussions suffice to indicate the general properties of dislocations in ordered structures. Additional superlattices and their defect structures are discussed in the review articles by Marcinkowski and by Amelinckx.[39, 43]

11-7. THE β-BRASS SUPERLATTICE

Structure

The $B2$ type,[43] or β-brass, superlattice is depicted in Fig. 11-32. In the disordered state the crystal structure is bcc, but after ordering, the symmetry is

[43] M. J. Marcinkowski, in G. Thomas and J. Washburn (eds.), "Electron Microscopy and Strength of Crystals," Interscience, New York, 1963, p. 333.

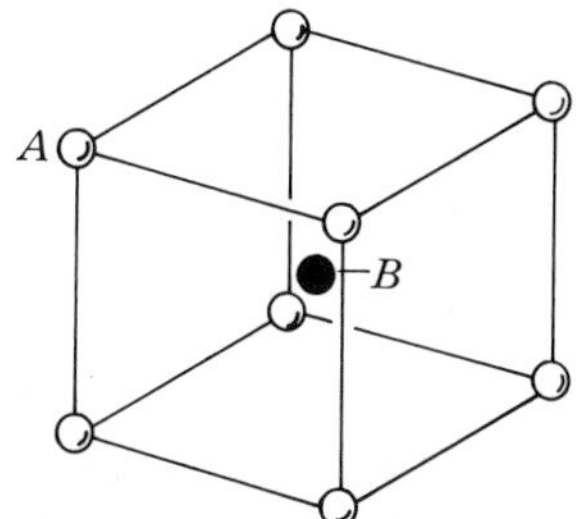

FIGURE 11-32. The β-brass superlattice unit cell. ○ and ● are A and B atoms, respectively.

lowered to simple cubic, with A and B atoms, respectively, occupying two interpenetrating simple cubic lattices displaced from one another by $(\frac{1}{2},\frac{1}{2},\frac{1}{2})$. Thus two types of domains exist in a crystal; type I, where A atoms occupy the $(0,0,0)$ sites, and type II, where A atoms occupy the $(\frac{1}{2},\frac{1}{2},\frac{1}{2})$ sites. There can be only one type of antiphase boundary between these two types of domains. The fault surface can completely enclose a region within the crystal, or it can terminate at a dislocation, a grain boundary, or a free surface. Antiphase-boundary networks cannot form because the mutual impingement of two fault surfaces leads to their annihilation; a three-boundary junction is impossible.

Across an antiphase boundary, like atoms are out of registry by vectors of the type $\frac{1}{2}\langle 111\rangle$. These vectors are perfect-lattice vectors in the disordered lattice and are the Burgers vectors of the dislocations bounding antiphase boundaries, as in Fig. 11-31. We adopt the conventional notation $[111]$-(hkl) to represent an antiphase boundary on an (hkl) plane separating domains out of phase by $\frac{1}{2}[111]$.

Shear-Antiphase-Boundary Energy

Shear-type antiphase boundaries produced by dislocation glide are of primary importance. The energy of such boundaries can be determined straightforwardly. Let us assume that the fault energy is determined by incorrect nearest-neighbor bonds only. The method could be extended to higher-order bonds, but terms involving elastic relaxations and point-defect interactions are of equal importance and should be included in a more refined description. These considerations would take us far afield from our main theme, so only the simpler case is treated.

In a shear process of antiphase-boundary formation, there are equal numbers of $A-A$ and $B-B$ bonds formed, and a corresponding number of $A-B$ bonds broken across the shear plane. All $A-B$ bonds cut by the shear plane contribute to the fault energy. This is so because all $\frac{1}{2}\langle 111\rangle$ antiphase vectors move an A atom from a type I site to a type II site, as shown in Fig. 11-32. Per $A-B$ bond broken, the energy of the crystal increases by

$$\phi=\tfrac{1}{2}(W_{AA}+W_{BB})-W_{AB} \tag{11-10}$$

where W_{AA}, W_{BB}, and W_{AB} are the bond energies of $A-A$, $B-B$, and $A-B$

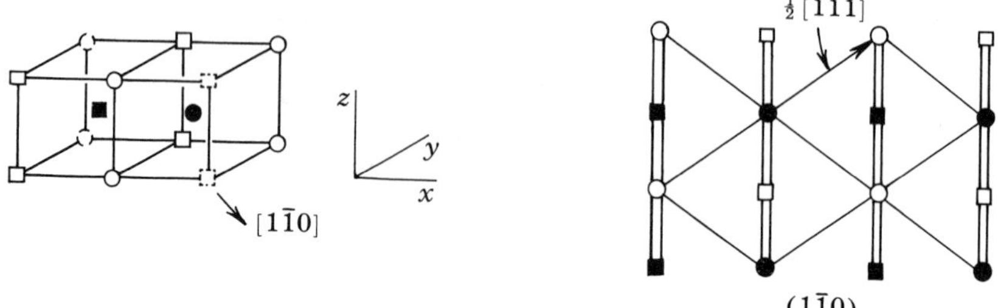

FIGURE 11-33. A projection normal to (1$\bar{1}$0). ○ and ● represent A and B atoms, respectively, in the plane of the paper, and □ and ■ represent A and B in the plane below. A projection of the unit cell is also shown.

bonds, respectively. Consider the (1$\bar{1}$0) projection in Fig. 11-33. Suppose that the plane of round atoms is sheared by $\frac{1}{2}[111]$ with respect to the underlying plane of square atoms. Initially each A atom has two $A—B$ bonds across the sheared plane, as does each B atom. After the shear, each A atom has two $A—A$ bonds, and each B atom two $B—B$ bonds. Thus the net increase in energy for a pair of atoms A and B in the round atom plane is 4ϕ. The area of (1$\bar{1}$0) plane per pair is $\sqrt{2}\,a_0^2$, so that the energy per unit area of [111]-(1$\bar{1}$0) boundary is

$$\gamma=\frac{2\sqrt{2}\,\phi}{a_0^2} \tag{11-11}$$

Now consider the energy of a general shear boundary. A shear on an (hkl) plane containing the shear vector $\frac{1}{2}[111]$ breaks every $A—B$ bond lying along $[1\bar{1}\bar{1}]$, $[\bar{1}11]$, or $[11\bar{1}]$ that is cut by the (hkl) plane. The density of cut bonds of the type of bond vector $\mathbf{l}_i$ is given by the expression

$$\frac{n_l(\mathbf{l}_i\cdot\mathbf{n})}{|\mathbf{l}_i|}=N_0(\mathbf{l}_i\cdot\mathbf{n})$$

where n_l = number of bonds of type $\mathbf{l}_i$ per unit area normal to $\mathbf{l}_i$, $N_0=2/a_0^3$, the atom density,

$$\mathbf{l}_i=\frac{a}{2}\langle 111\rangle$$

$$\mathbf{n}=\text{unit vector normal to } (hkl)$$

Thus the density of cut bonds of the $\frac{1}{2}\langle 111\rangle$ type is[44]

$$N_0(\mathbf{l}_i\cdot\mathbf{n})=\frac{\langle 111\rangle\cdot(hkl)}{a_0^2|(hkl)|} \tag{11-12}$$

[44] To verify this expression, note that the density of $\frac{1}{2}[\bar{1}\bar{1}\bar{1}]$ bonds normal to $(1\bar{1}\bar{1})$ is $\sqrt{3}/a_0^2$. The density on (hkl) is less because the area per bond is now $(\sqrt{3}/a_0^2)\cos\theta$, where θ is the angle between $\langle 111\rangle$ and $\langle hkl\rangle$. Substituting $\cos\theta=\langle 111\rangle\cdot\langle hkl\rangle/|\langle 111\rangle||\langle hkl\rangle|$, one obtains Eq. (11-12) directly.

Adding the bond densities of $[1\bar{1}\bar{1}]$, $[1\bar{1}1]$, and $[11\bar{1}]$ given by Eq. (11-12), one obtains the total bond density

$$D = \frac{3h - k - l}{a_0^2(h^2 + k^2 + l^2)^{1/2}} \tag{11-13}$$

In order that $\frac{1}{2}[111]$ lie on (hkl) so that the shear formation of the antiphase boundary on (hkl) is possible, it is required that $[111]\cdot(hkl)=0$ or $h=-(k+l)$. Therefore Eq. (11-13) becomes

$$D = \frac{4h}{a_0^2(h^2 + k^2 + l^2)^{1/2}} \tag{11-14}$$

Furthermore, if the other symmetric arrangements of the unit cell are selected, a similar expression results, with k or l replacing h. Thus one expression suffices for all orientations if $h>0$ and $h^2 \geqslant k^2 \geqslant l^2$ is specified. The energy of a [111]-(hkl) shear antiphase boundary is thus

$$\gamma = D\phi = \frac{4h\phi}{a_0^2(h^2 + k^2 + l^2)^{1/2}} \qquad h^2 \geqslant k^2 \geqslant l^2 \tag{11-15}$$

This result has been obtained in a different manner by Flinn.[45] He plotted γ stereographically for all orientations in the β-brass lattice. However, Eq. (11-15) is valid for γ or Eq. (11-10) for ϕ *only* if the antiphase boundary is a shear type of boundary. Thus his stereographic result is valid only for boundaries formed by dislocation glide.

The General Antiphase Boundary

A general type of fault is produced by the climb of a $\frac{1}{2}[111]$ edge dislocation with line direction $[1\bar{1}0]$, as shown in Fig. 11-34. Equation (11-12) still applies because the displacement vector remains $\frac{1}{2}\langle 111\rangle$. In this case *twice as many A* atoms are removed in climb as are B atoms. There is, therefore, an excess of B—B bonds (and of B atoms) along the resulting antiphase boundary. Applying Eq. (11-12) to this case, one finds that the energy[46] of the [111]-(111) boundary of Fig. 11-34 is

$$\gamma = \frac{\sqrt{3}}{3a_0^2}(5W_{BB} + W_{AA} - 6W_{AB}) \tag{11-16}$$

[45] P. A. Flinn, *Trans. Met. Soc. AIME*, **218**: 145 (1960).

[46] Relative to the state where each atom, A or B, has an energy $4W_{AB}$. The A and B atoms removed from the inserted plane are imagined to be deposited at a place where an energy $4W_{AB}$ is gained for each atom deposited, whether A or B.

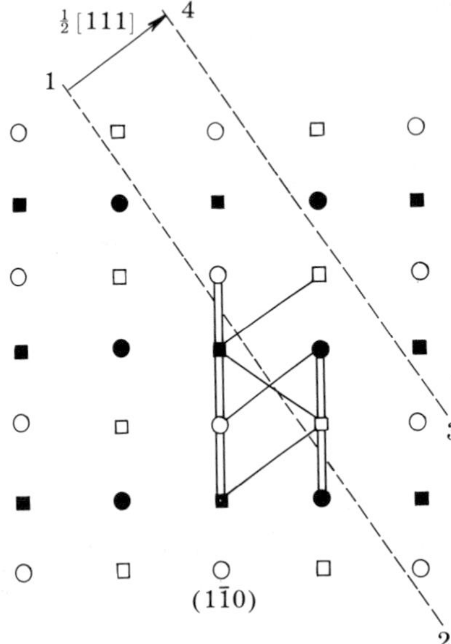

FIGURE 11-34. A projection normal to (1$\bar{1}$0). Climb of the edge dislocation $\frac{1}{2}$ [111] results in the removal of the material bounded by the surfaces 1-2 and 3-4. Atom symbols are the same as in Fig. 11-33.

If the fault in Fig. 11-34 were translated by $\frac{1}{6}$[111], A atoms would be in excess in the fault, and the energy would differ, with W_{AA} and W_{BB} transposed in Eq. (11-16). A comparison of Eqs. (11-10), (11-15), and (11-16) reveals, as stated previously, that neither Eq. (11-10) nor (11-15) applies unless the fault is a shear type of boundary.

The density of broken $A-B$ bonds for a general [111]-(hkl) antiphase boundary is given by Eq. (11-12) as

$$D = \frac{\pm(h-k+l)\pm(-h+k+l)\pm(h+k+l)\pm(h+k-l)}{a_0^2(h^2+k^2+l^2)^{1/2}} \quad (11\text{-}17)$$

where the plus or minus sign must be chosen *for each term* such that the term has its maximum possible positive value. In the special case that $W_{AA}=W_{BB}$, the general antiphase-boundary energy is given by D[Eq. (11-17)] times ϕ. One can verify readily that Eq. (11-17) reduces to Eq. (11-14) for the special case of a shear type of boundary.

Equations (11-16) and (11-17), involving only nearest-neighbor bond energies, were developed simply to indicate the procedure of determining a general fault energy. The rigorous determination of the energy of a general antiphase boundary produced by climb, by vacancy condensation, or by domain impingement is quite complicated for the following reasons. Climb and vacancy condensation involve long-range diffusion, while domain growth involves short-range diffusion. Hence in all cases a diffusional relaxation of the antiphase boundary to a lower free-energy state is possible. The free energy of a given boundary is lowered and the anisotropy of boundary energy with orientation is changed if *A* or *B* adsorb at the boundary. Such adsorption is simply a special case of Gibbs solute adsorption. For nonstoichiometric compositions, boundaries which have excesses of the more prevalent atom have low free energies relative to other orientations. In special cases, such boundaries can form spontaneously within a domain, with a zero increase in free energy.[47] This occurs because the decrease in bulk free energy of the domain, which occurs because its composition more nearly approaches the ordered stoichiometry, offsets the surface energy of the domain boundary. Finally, it should be noted that the configurational entropy of the boundaries becomes an important contribution to the boundary free energy at temperatures where diffusional rearrangement is possible.

In summary, one can reasonably approximate the energy of shear-type boundaries by Eq. (11-15) if the boundaries are formed by glide at low temperatures, where diffusional rearrangement is minimized. For all other cases, the boundary energy is more uncertain because of lack of knowledge about the degree of diffusional relaxation to an equilibrium state.

Dislocations in the β-Brass Superlattice

Dislocations of the type $\frac{1}{2}\langle 111\rangle$, perfect in the disordered lattice, are the common fault-producing dislocations in the ordered superlattice, while $\langle 100\rangle$, $\langle 110\rangle$, and $\langle 111\rangle$ are all possible superdislocations.

Equation (11-15) indicates that an edge dislocation $\frac{1}{2}[111]$ gliding on $(1\bar{1}0)$ produces an antiphase boundary with an energy $2.53(\phi/a_0^2)$, while that produced by an edge dislocation $\frac{1}{2}[111]$ gliding on $(2\bar{1}\bar{1})$ has an energy $3.27(\phi/a_0^2)$. The use of either Eq. (11-15) or Flinn's stereographic plot[45] reveals that a $\frac{1}{2}[111]$ screw dislocation produces a fault with energy $2.83(\phi/a_0^2)$ if it glides on $(1\bar{1}0)$, $(10\bar{1})$, or $(0\bar{1}1)$ and one with energy $3.27(\phi/a_0^2)$ if it glides on $(2\bar{1}\bar{1})$, with a continuous spectrum of energies between these extremes for intermediate orientations.

Since there is only a 15 percent range of difference in energy for these shear-type antiphase boundaries, screw dislocations should cross slip easily in the β-brass type of superlattice. In agreement with this expectation, Marcinkowski and Brown[48] have observed wavy slip lines on an ordered Fe_3Al alloy possessing the β-brass superlattice structure.[49] They associate this wavy slip with cross-slipping screws.

Because $\{110\}$ antiphase boundaries have the lowest energy, according to Eq. (11-15), $\langle 111\rangle$ superdislocations should dissociate on $\{110\}$.

[47] R. Kikuchi and J. W. Cahn, *J. Phys. Chem. Solids*, **20**: 94 (1961); *J. Phys. Chem. Solids*, **23**: 137 (1962).

[48] M. J. Marcinkowski and N. Brown, *Acta. Met.*, **9**: 764 (1961).

[49] With Fe atoms on *A* sites and a random distribution of 50 *a/o* Fe atoms and 50 *a/o* A*l* atoms on *B* sites.

Exercise 11-2. Apply Eq. (5-17) to the case of a mixed [111] superdislocation dissociating into two $\frac{1}{2}[111]$ dislocations on $(1\bar{1}0)$ in a β-brass superlattice. Show that the equilibrium width r_e of the antiphase boundary is given by

$$r_e = \frac{\mu b^2}{2\pi\gamma}\left(\cos^2\alpha + \frac{\sin^2\alpha}{1-\nu}\right) \tag{11-18}$$

where $\mathbf{b}$ is the vector $\frac{1}{2}[111]$ and α is the angle between the Burgers vector and $\boldsymbol{\xi}$ for the superdislocation. Marcinkowski[50] cites values of $\gamma[111]$-$(1\bar{1}0)$ equal to 83 mJ/m^2 for β-brass and $\sim$0 mJ/m^2 for Fe_3Al. Thus, according to Eq. (11-18), one expects minimal splitting of superdislocations in β-brass, but wide dissociation in Fe_3Al; this is what is observed experimentally.

11-8. THE Cu_3Au SUPERLATTICE

Superlattice Structure

The $L1_2$ type, or Cu_3Au type, of superlattice is shown in Fig. 11-35. The disordered fcc structure is lowered in symmetry, upon ordering, to a simple cubic structure with four atoms per primitive cell point. A B atom is at one site in the ordered structures, and A atoms occupy the other three sites. Thus four types of domains occur in a crystal; type I, with B at (0,0,0); type II, with B at $(0,\frac{1}{2},\frac{1}{2})$; type III, with B at $(\frac{1}{2},0,\frac{1}{2})$; and type IV, with B at $(\frac{1}{2},\frac{1}{2},0)$. These are sufficient to permit the formation of three-dimensional networks of antiphase boundaries in a soap-bubble type of cell structure. Also, in addition to the possibility of mutual annihilation, two impinging antiphase boundaries can combine to form a third. The remaining features of the boundaries are the same as for the β-brass case. The antiphase vectors correspond to perfect-dislocation Burgers vector in the disordered crystal. They are of the type $\frac{1}{2}\langle 110\rangle$.

Shear-Antiphase-Boundary Energy

Consider a shear-type antiphase boundary [110]-(hkl), produced by the glide of a $\frac{1}{2}[110]$ dislocation. The four antiphase vectors, $\frac{1}{2}[110]$, $\frac{1}{2}[1\bar{1}0]$, $\frac{1}{2}[\bar{1}\bar{1}0]$, and $\frac{1}{2}[\bar{1}10]$, are all equivalent in that they all move a B atom from a type I site to a type IV site, as shown in Fig. 11-35. Antiphase vectors such as $\frac{1}{2}[011]$ and $\frac{1}{2}[101]$ would transfer a B atom to a type II or type III site, respectively, so that they are *not* equivalent to $\frac{1}{2}[110]$. The [110] and $[\bar{1}\bar{1}0]$ bonds both lie in the glide plane and are not cut by a [110] shear. The vectors $[1\bar{1}0]$ and $[\bar{1}10]$ are equal and opposite, representing the same bonds, so that cut bonds of only one of the latter pair contribute to the fault energy. Cut bonds of the other $\langle 110\rangle$

[50] M. J. Marcinkowski, in G. Thomas and J. Washburn (eds.), "Electron Microscopy and Strength of Crystals," Interscience, New York, 1963, p. 333.

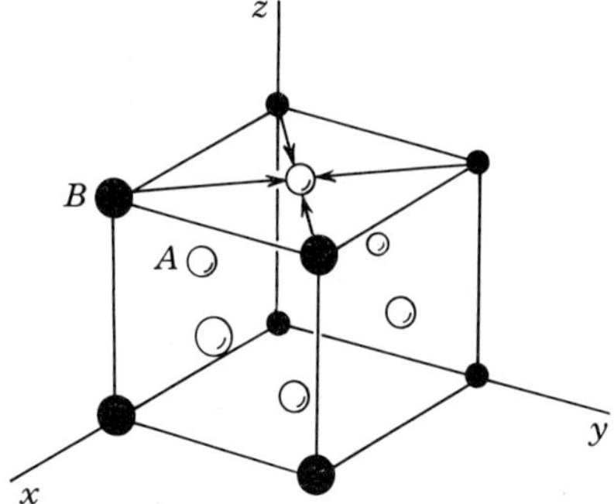

FIGURE 11-35. The Cu_3Au superlattice unit cell. ○ and ● are A and B atoms, respectively. Vectors show equivalent ways of moving a B atom from a type I site to a type IV site.

directions, such as $\frac{1}{2}[011]$ and $\frac{1}{2}[101]$, do not produce any *net* change in the numbers of $A—B$ and $A—A$ bonds and hence do not contribute to the fault energy. The density of cut $[1\bar{1}0]$ bonds is given by Eq. (11-12) to be

$$D=\frac{2[1\bar{1}0]\cdot(hkl)}{a_0^2(h^2+k^2+l^2)^{1/2}}=\frac{2(h-k)}{a_0^2(h^2+k^2+l^2)^{1/2}}=\frac{4h}{a_0^2(h^2+k^2+l^2)^{1/2}} \tag{11-19}$$

In the last step the relation $[110]\cdot(hkl)=0$ was used; this relation follows in order that [110] lie on the glide plane (hkl). By symmetry, the other symmetric orientations of the unit cell would give a similar expression for D, with k replacing h. Only one-half the total number of cut $[1\bar{1}0]$ bonds are $A—B$ bonds; the other half are $A—A$ bonds. After shear the cut $A—B$ bonds form equal numbers of $A—A$ and $B—B$ bonds, while the cut $A—A$ bonds form new $A—A$ bonds. Thus each cut $A—B$ bond contributes ϕ to the fault energy, giving[51]

$$\gamma=\tfrac{1}{2}D\phi=\frac{2\phi h}{a_0^2(h^2+k^2+l^2)^{1/2}} \qquad h\geqslant k \tag{11-20}$$

In analogy to the β-brass case, Eq. (11-20) is applicable *only* to shear-type faults; for the Cu_3Au superlattice, Eq. (11-19) gives the correct density of cut bonds for a general type of boundary, but there is an excess of $A—A$ or $B—B$ bonds, so that Eq. (11-10) is not valid unless the boundary is of shear type.

In addition to the above possibilities, there is a special type of antiphase boundary of the type $[211]$-$(1\bar{1}\bar{1})$ associated with what would be an intrinsic stacking fault in the *disordered* fcc lattice. Such an antiphase boundary is produced by the glide of the partial dislocation $\frac{1}{6}[211]$ in the ordered superlattice. As can be verified by the construction of a $\{111\}$ projection of the

[51]P. A. Flinn, *Trans. Met. Soc. AIME*, **218**: 145 (1960).

ordered alloy, the chemical contribution to the [211]-($1\bar{1}\bar{1}$) antiphase-boundary energy is $2\sqrt{3}\,\phi/3a_0^2$, which is *identical* to the energy of the boundary [110]-($1\bar{1}\bar{1}$). The total energy of the [211]-($1\bar{1}\bar{1}$) boundary involves the chemical contribution from nearest-neighbor bonds and higher-order bond contributions caused by the stacking fault. The latter contributions include both chemical effects and changes of the sort given in Eq. (10-13). The resulting energy is unlikely to be the sum of the intrinsic-fault energy and the first-order chemical term given above, but such an approximation is often made. The exact estimate of the [211]-($1\bar{1}\bar{1}$) energy again must involve elaborate atomic calculations.

Dislocations in the Cu_3Au Superlattice

According to Eq. (11-20), a $\frac{1}{2}[110]$ dislocation produces a fault with zero energy if it glides on (001). Of course, one expects the fault to have some energy attributable to second-nearest-neighbor bonds and other higher-order factors. A $\frac{1}{2}[110]$ dislocation produces a fault with energy $2\sqrt{3}\,\phi/3a_0^2$ if it glides on ($1\bar{1}\bar{1}$); if it glides on ($1\bar{1}0$), it produces the maximum-energy shear fault with energy $\sqrt{2}\,\phi/a_0^2$. Thus the antiphase-boundary-energy term clearly favors the extension of superdislocations onto {100} planes, and consequently favors glide on these planes.

In experiments, superdislocations are found to predominantly extend and glide on {111} planes in Cu_3Au[52] and Ni_3Mn,[53] although {100} glide is also reported for Cu_3Au.[54, 55] There are several possible explanations for {111} glide instead of {100}. First, if $\frac{1}{2}\langle 110\rangle$ dislocations are extended on {111} in the disordered alloy, the ordered lattice might inherit the {111} extension. The requirement for constriction of the $\frac{1}{6}\langle 211\rangle$ partials against their elastic repulsive forces could then provide an energy barrier to cross slip of screws onto {100}. This could stabilize {111} screws even if {100} screws had a lower energy. For edge dislocations, additional work against line-tension forces would be required to rotate the dislocation line into screw orientation before cross slip onto {100} could occur. Second, the superdislocation might have a lower total energy on {111} than on {100} because of the lower elastic energy in the former case caused by splitting into partials. Third, the Peierls barrier might be higher for {100} glide, so that {111} glide is preferred. There are other possibilities, but these serve to illustrate the complications that can arise when the dissociation of superdislocations into partials is possible.

Figure 11-36 shows a [110] superdislocation dissociated into partials on ($1\bar{1}\bar{1}$). One can compute r_1 and r from Eq. (5-17) if both fault energies are known. Assuming that the intrinsic-stacking-fault energy and the anti-

[52] M. J. Marcinkowski, N. Brown, and R. M. Fisher, *Acta. Met.*, **9**: 129 (1961).

[53] M. J. Marcinkowski and D. S. Miller, *Phil. Mag.*, **6**: 871 (1961).

[54] B. H. Kear, *Acta. Met.*, **14**: 659 (1966).

[55] D. E. Mikkola and J. B. Cohen, *Acta. Met.*, **14**: 105 (1966).

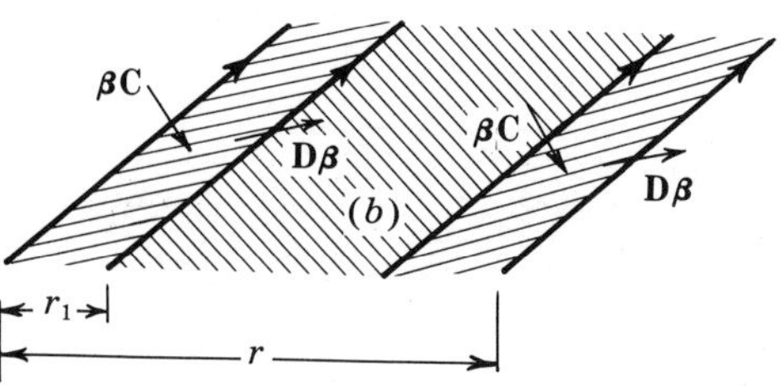

FIGURE 11-36. The superdislocation [110], or 2**DC**, dissociated into partials on $(1\bar{1}\bar{1})$, or (b).

TABLE 11-3 Extensions of the fault in Fig. 11.36.

Alloy	$\gamma_{[110]\text{-}(1\bar{1}1)}$ and $\gamma_{[211]\text{-}(1\bar{1}\bar{1})}$ mJ/m²	γ_I, mJ/m²	Screw r_1, nm	Screw r, nm	Edge r_1, nm	Edge r, nm
Cu_3Au	92	40	0.8	6.5	1.7	10.2
Ni_3Mn	114	20	1.2	7.8	2.7	12.0

phase boundary energy are additive in the region of the intrinsic fault, and using Marcinkowski's[50] values for the fault energies, one obtains the results in Table 11-3. Because of the elastic forces, $r-2r_1$ should always exceed r_1. Marcinkowski[50] has found experimental values of $r=13$ nm for Cu_3Au and $r=11$ to 17 nm for Ni_3Mn. In view of the approximations of isotropic elasticity, nearest-neighbor chemical bonds, etc., the agreement with the theoretical predictions of Table 11-3 is good.

Dislocations in other superlattices exhibit properties in general analogous to those of one of the two types discussed above. For discussion of other superlattices, the reader is referred to the review article by Marcinkowski.[50]

PROBLEMS

11-1. Why is not the basal plane in hcp metals a twin plane, like the close-packed plane in fcc metals? What is the formal definition of twinning?

11-2. Show that $(10\bar{1}2)$ is a possible twin plane in hcp structures.

11-3. Consider the dislocation **AC** in the basal plane of an hcp metal. If the dislocation contains a jog an uneven number of plane spacings high, can the jog extend in a fashion similar to the jog in Fig. 11-12? What will the partials be?

11-4. Show that the dislocations **AC** and $\mathbf{B_0C_0}$, lying on adjacent basal planes, can combine to form a configuration resembling Fig. 10-36*b*. What are the resultant partials and the types of stacking faults formed?

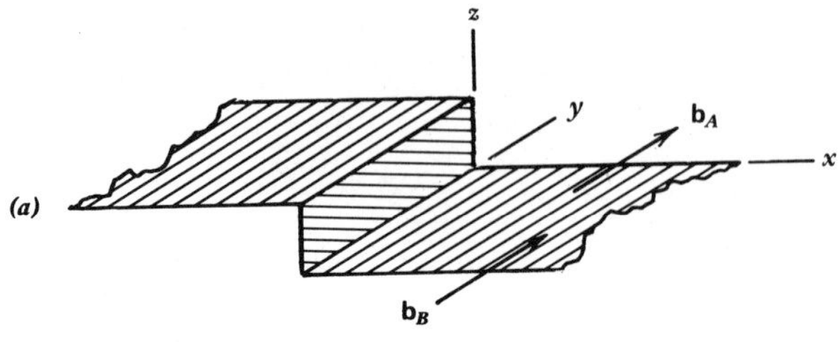

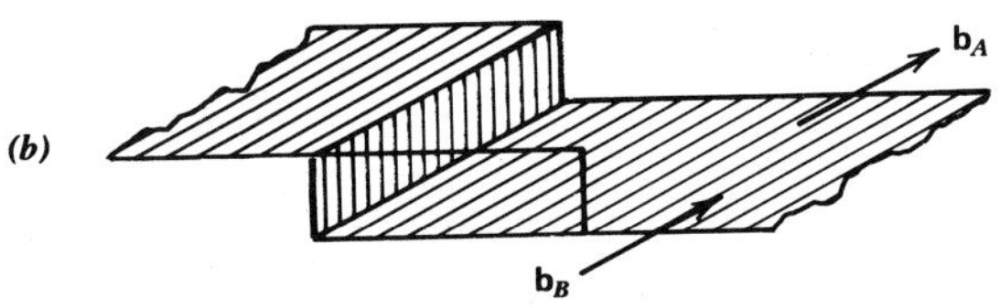

FIGURE 11-37. Perspective view of superdislocation composed of two jogged dislocations with collinear Burgers vectors. Jogs are offset in (*b*).

11-5. Draw models for the core of a pure screw dislocation $\mathbf{b}=\frac{1}{2}\langle 111\rangle$ in a bcc structure, with the core dissociated on the $(1\bar{1}0)$ plane. Show that the core structure is different above and below the slip plane.

11-6. Devise a rule for the proper stacking sequence in the construction of the extrinsic fault, Eq. (11-8). Why could not the inserted layer *EF* be put between *C* and *D*, say, producing the sequence *ABCEFDE*?

11-7. Apply Eq. (10-14) to the bcc dissociations of Fig. 11-14. Develop explicit expressions for the equilibrium partial spacing r_e and determine the proper dissociation to form the fault I_3.

11-8. Which jogs in Fig. 11-30 can be formed directly by an intersection process?

11-9. Could a junction of four antiphase boundaries exist in β-brass? Could it dissociate; i.e., would any two of the four domains be in phase? Are such junctions expected after deformation? After annealing, which allows atom regrouping?

11-10. Discuss the likelihood of dissociations to form $\frac{1}{6}\langle 111\rangle$ partials on $\{112\}$ in ordered β-brass.

11-11. The superdislocation containing dislocations with collinear Burgers vectors in Fig. 11-37*a* contains a jog and a stepped antiphase boundary (APB). The jogs in each component dislocation lie in the same (yz) glide plane, so if the dislocation moves in the y direction, dislocation *B* annihilates the APB created by dislocation *A*. The superdislocation in Fig. 11-37*b* has the jogs in the component dislocations misaligned in the x direction. Show that motion of this superdislocation creates a rectangular tube of APB.[56]

[56] A. E. Vidoz and L. M. Brown, *Phil. Mag.*, **7**: 1167 (1972).

BIBLIOGRAPHY

1. Amelinckx, S., *Solid State Phys. Suppl.* 6 (1964).
2. Amelinckx, S., in F. R. N. Nabarro (ed.), "Dislocations in Solids," North-Holland, Amsterdam, 1979, p. 67.
3. Reed-Hill, R. E., J. P. Hirth, and H. C. Rogers (eds.), "Deformation Twinning," Gordon and Breach, New York, 1964.
4. Seeger, A., and C. Wüthrich, *Nuovo Cim.*, **33B**: 38 (1976).
5. Thomas, G., and J. Washburn (eds.), "Electron Microscopy and Strength of Crystals," Interscience, New York, 1963.
6. Vitek, V., *Crystal. Lattice Defects*, **5**: 1 (1974).
7. Westbrook, J. H., "Intermetallic Compounds," Wiley, New York, 1967.

12

Dislocations in Ionic Crystals

12-1. INTRODUCTION

The presence of different types of ions of opposite charge engenders quite complicated electrical effects for dislocations in ionic crystals. The ionic bonding in the dislocation core is important in modifying the Peierls energy and in suppressing stacking-fault formation, as has been stressed in the preceding chapters. In addition, the dislocation core structure is associated with several specific electrical phenomena. These include the Gyulai-Hartly effect,[1] in which the electrical conductivity is observed to increase after compressive deformation of a crystal, and the Stepanov effect,[2] in which deformation is found to produce a transient electrostatic field. Also, the Joffé effect,[3] the decrease in brittleness accompanying the dissolution of the surface of rock salt, might be related to the charge on dislocations. Other effects[4] include an increase in electrical conductivity that is sustained only during deformation[5]; motion of dislocations produced by the application of an external electrostatic field[6]; enhancement of dislocation velocity in the presence of a field[7]; and, after the Gyulai-Hartly effect has decayed away, a loss in ionic conductivity following deformation.[8]

The Rehbinder effect,[4] the decrease in hardness and strength with the adsorption of surface active molecules, and the opposite effect, which has been extensively investigated by Westwood,[4] similarly might relate to dislocation

[1]Z. Gyulai and D. Hartly, *Z. Phys.*, **51:** 378 (1928).

[2]A. V. Stepanov, *Phys. Z. Sowj.*, **4:** 609 (1933).

[3]A. Joffé, M. W. Kirpitschewa, and M. A. Lewitsky, *Z. Phys.*, **22:** 286 (1924).

[4]For reviews see A. R. C. Westwood, in D. C. Drucker and J. J. Gilman (eds.), "Fracture of Solids," Interscience, New York, 1963, p. 553, and R. M. Latanision, in R. M. Latanision and J. T. Fourie (eds.), "Surface Effects in Crystal Plasticity," Noordhoff, Leyden, 1977, p. 3.

[5]P. Camagni and A. Manara, *J. Phys. Chem. Solids*, **26**: 449 (1965).

[6]R. L.Sproull, *Phil Mag.*, **5**: 815 (1960).

[7]V. P. Sergeev and L. B. Zuev, *Sov. Phys. Solid State*, **22**: 1028 (1980).

[8]R. W. Whitworth, *Adv. Phys.*, **24**: 203 (1980).

charge in ionic crystals (this type of effect is also observed in metal crystals). Finally, there are multifarious reactions between dislocations and charged point defects, one example being the photolytic reaction in silver bromide.[9] These effects, peculiar to ionic crystals, can be understood only through a detailed geometrical analysis of dislocation cores. In this chapter we consider dislocation core structure, jogs, and kinks in ionic crystals. Because of the special effects associated with charges on kinks and jogs, this chapter is also a convenient one in which to treat the topic of jogs and kinks formed by dislocation intersection. As an example, dislocations in NaCl-type crystals composed of singly charged ions are treated. The modifications for polyvalent ions are alluded to briefly. The method of analysis for ionic crystals other than the NaCl type would be similar, but for brevity they are not considered here.

The NaCl type of crystal is a simple model example of the electrical effects encountered in ionic crystals, excluding piezoelectricity, which is present only for crystals without a center of symmetry. Also, the NaCl crystals have been most widely studied experimentally with regard to dislocation properties. Of particular note are the beautiful etch-pit studies on lithium fluoride by Gilman and Johnston,[10] and the studies[11] of Mitchell, Amelinckx, and Dekeyser, in which they decorated dislocations so that they could be seen optically in transparent ionic crystals.

12-2. DISLOCATION STRUCTURE

The NaCl-Crystal Lattice

The NaCl type of lattice is actually an fcc crystal lattice with a pair of Na^+ and Cl^- ions as a basis. However, the lattice is customarily represented by a unit cube one-eighth the size of the fcc unit cell, as shown in Fig. 12-1. The corners of the little cube are occupied alternately by Na^+ and Cl^- ions. Free-surface edges and corners of crystals have effective charges whose sign depends on the type of ions that occupy corner and edge sites. Let us compute the magnitudes of these charges to establish the basis for the determination of the charge on a dislocation.

For simplicity, all interionic forces are assumed to be symmetrical for the two types of ions. This assumption is valid by definition for the pairwise interaction between ions of different sign, which is the dominant interaction. However, an Na^+-Na^+ overlap repulsion different from that for Cl^--Cl^- would destroy complete symmetry. Such asymmetries are neglected here. In addition, polarization effects are neglected. Charged defects polarize the surrounding dielectric medium and are partially shielded; the treatment of such

[9] J. M. Hedges and J. W. Mitchell, *Phil. Mag.*, **44:** 223, 357 (1953).

[10] Reviewed by J. J. Gilman and W. G. Johnston, *Solid State Phys.*, **13:** 147 (1962).

[11] Reviewed by S. Amelinckx, *Solid State Phys. Suppl.* 6 (1964).

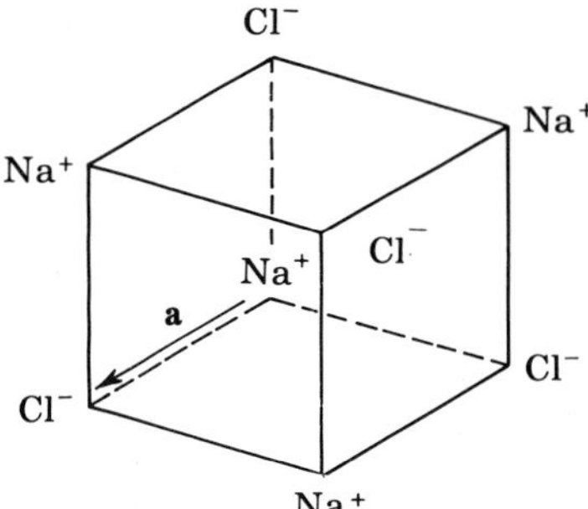

FIGURE 12-1. A simple cubic unit cell for NaCl.

polarization, a simple problem in electrostatics, should be included in a refined theory. Allowing for asymmetries and polarization, some of the following results for effective charges would change, but certain sums of charges would remain unchanged. For example, elementary jogs in an edge dislocation are found to have charges of $\pm e/2$. With asymmetries and polarization present, this is no longer true, but oppositely charged jogs still add up to zero net effective charge.

For the first effective-charge calculation, consider a long row of alternating positive and negative charges, as indicated in Fig. 12-2. At B, far removed from the ends, the field is small. In fact, if the chain were infinitely long, the field near B would be negligibly small except within a distance $\sim a$ from the chain. At the ends, on the other hand, the fields have a longer range. Consider the field at point D, with the end A *infinitely* far removed to the left. Let the field at D correspond to some effective charge q. Next, extend the chain by adding a positive ion; apart from a negligible translation, the field must change sign:

$$q+e=-q$$

or

$$q=\frac{-e}{2}$$

Therefore the end of a chain has the effective charge $e/2$ or $-e/2$, depending on whether the end ion has a positive or negative charge.

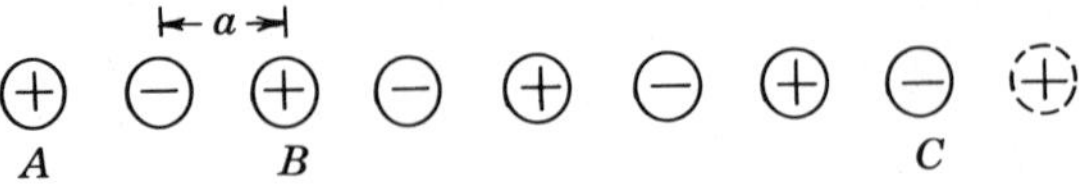

FIGURE 12-2. A semi-infinite row of alternating charges terminating at C.

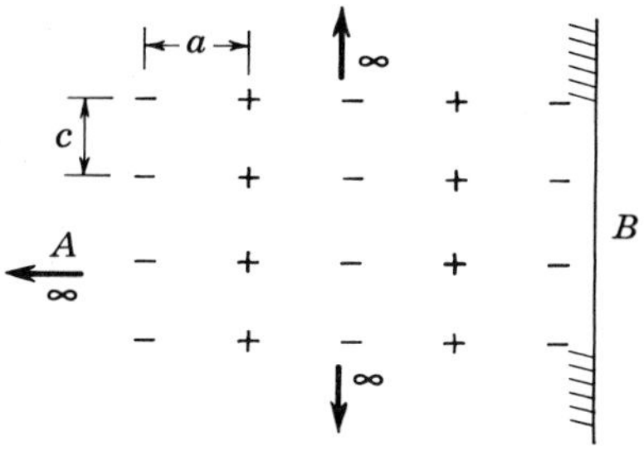

FIGURE 12-3. A semi-infinite plane of charges terminating at side B.

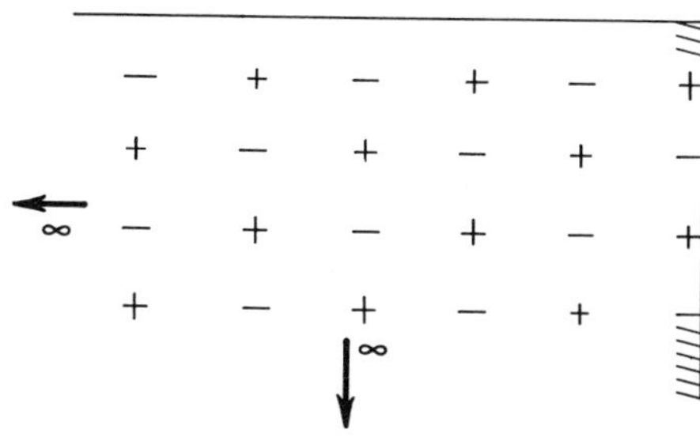

FIGURE 12-4. A partly infinite plane of charges terminating at a corner.

Next consider the planar array of Fig. 12-3. The edge on the right side of the semi-infinite plane produces the same field as a single column of charges $-e/2$ or a charged line with a linear charge density:

$$\lambda = -e/2c \tag{12-1}$$

For the planar array of Fig. 12-4, the edge on the right corresponds to alternating charges $\pm e/2$. With the addition of a row to the top of the array, the charge associated with the upper right corner changes to $q-(e/2)=-q$, so that the *corner* has a net charge $q=e/4$.

Similarly, a three-dimensional corner would have a net charge $\pm e/8$. These numbers, $\frac{1}{2}$, $\frac{1}{4}$, and $\frac{1}{8}$, arise because a line, a square, and a cube have 2, 4, and 8 corners (ends), respectively, to share a total charge unbalance of 0 or $\pm e$. For example, when the corners are alternately charged in a square, as in Fig. 12-5*a*, there is no net charge present, while with all corners of the same charge, as in Fig. 12-5*b*, an extra unit charge of the same sign is present.

For many purposes, effective charges are conveniently estimated by counting "broken bonds." The charge of each ion is considered to be distributed between six bonds, corresponding to the six nearest neighbors in the perfect crystal[12] (see Fig. 12-6). As an example, let us add up the charge for the cube in

[12] The broken-bond model was proposed by W. Kossel [*Naturw.*, **18:** 901 (1930)] and I. N. Stranski [*Z. Phys. Chem.*, **136:** 259 (1928)] and applied by them in crystal-growth problems. A similar model is used in the application of Pauling's rules to ionic crystals (see L. Pauling, "The Nature of the Chemical Bond," Cornell University Press, Ithaca, N. Y., 1948).

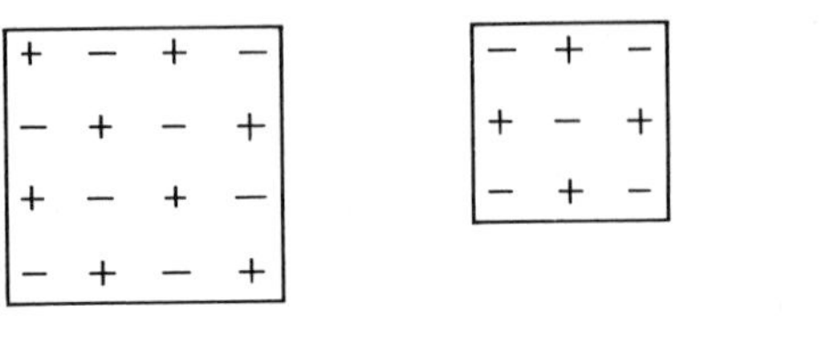

FIGURE 12-5. Two square arrays of charges.

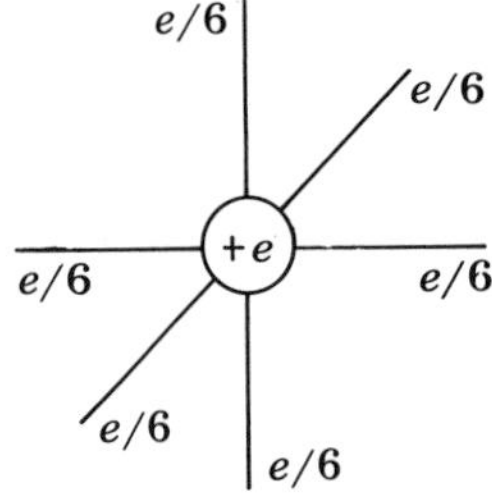

FIGURE 12-6. Nearest-neighbor bond charges in the broken-bond model.

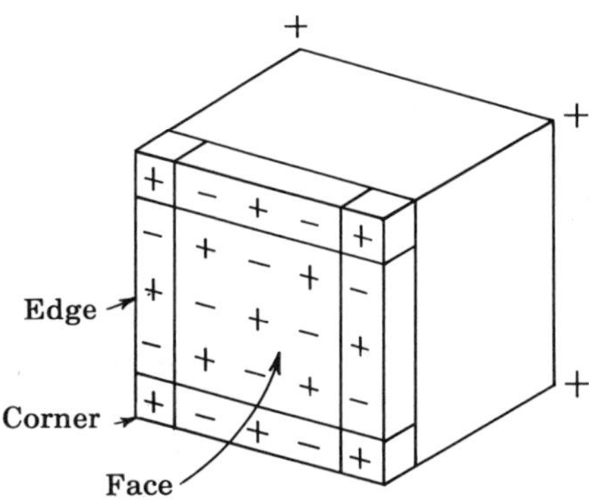

FIGURE 12-7. A charged cube.

Fig. 12-7, using the broken-bond model. There are eight corner ions with three broken bonds each,[13] which gives a total charge $8\times3\times(e/6)=4e$. There are twelve edges, with a net of two broken bonds each, which yields $12\times(-2)\times(e/6)=-4e$. Finally, the six faces, each with one net broken bond, give $6\times(e/6)=e$. These sum to a total excess charge of e, as they must.

Exercise 12-1. Show that the corners of a reentrant hole bounded by cube faces also have net effective charges $\pm e/8$.

[13]This does not mean that the effective charge of the corner is $3\times(e/6)=e/2$. The effective charge of the corner contains contributions from the corner, from the edges and from the faces. Adding these contributions, one gets for the effective charge of the corner $3\times(e/6)+3\times\frac{1}{2}[-2\times(e/6)]+3\times\frac{1}{4}(e/6)=e/8$, in agreement with the earlier result.

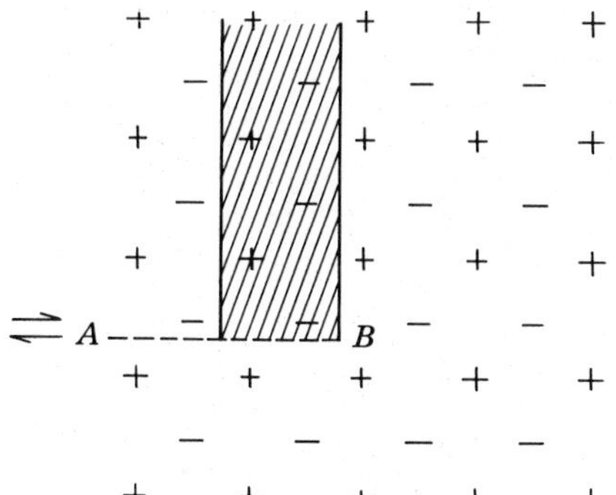

FIGURE 12-8. A (010) surface in the plane of the paper.

Charges on Dislocations

As discussed in Chap. 9, the predominant slip system in NaCl-type crystals is $\langle 110 \rangle \{1\bar{1}0\}$ (referred to the simple cubic unit cell of Fig. 12-1). Only dislocations of this slip system are considered here, but the method can easily be extended to other cases. Consider the pure edge dislocation $\mathbf{b}=[101]$ and $\boldsymbol{\xi}=[010]$. An initially perfect crystal is shown in Fig. 12-8. The glide plane is $(10\bar{1})$, and the view is of a surface layer of a (010) surface. In the next layer below the surface, all charge signs would change. Thus (+) designates the positive end of a row, perpendicular to the paper, of alternating positive and negative charges. An edge dislocation is formed by glide over the plane AB; the hatched slab is pushed in as an extra pair of planes, and a surface step is formed on the surface at the left side of the crystal. One relaxed symmetrical configuration II of the dislocation is shown in Fig. 12-9*a*[14]. If the dislocation is moved a distance $b/2$ to the right, the other symmetrical configuration I (Fig. 12-9*b*) is produced. The latter configuration could be retransformed to the former by removing row E in Fig. 12-9*b* and then peeling off one surface layer normal to the dislocation line to change all signs. Atomic calculations[15] indicate that, for KCl, NaCl and MgO, the configuration of Fig. 12-9*a* is one of minimum energy whereas that of Fig. 12-9*b* is one of maximum energy, corresponding to the top of the Peierls barrier.

Figure 12–9 shows that the dislocation is not symmetrical with respect to positive and negative charge. Some *net* charge is introduced on the surface at the point of emergence of the dislocation. If one surface layer of ions is removed from the (010) surface, the sign of all ions changes, and the net charge

[14] There is little agreement on notation, with either configuration being called I or II or A or B. We follow the original designation of H. B. Huntington, J. E. Dickey, and R. Thomson [*Phys. Rev.*, **100**: 1117 (1955)] of I and II in Fig. 12-9, also used by R. W. Whitworth [*Adv. Phys.*, **24**: 203 (1980)]. The "larger" anions in configuration I then have the simple arrangement of Fig. 1-4 with an inserted plane of anions terminating at the dislocation.

[15] Reviewed by M. P. Puls, in "Dislocation Modeling of Physical Systems," Pergamon, Oxford, 1981, p. 249.

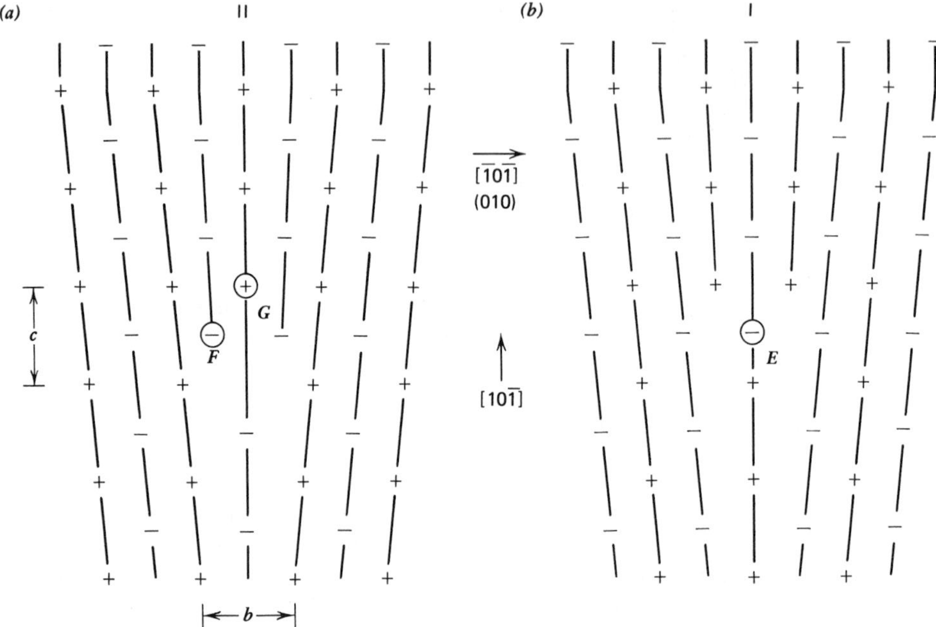

FIGURE 12-9. The (010) surface of Fig. 12-8 after slip has occurred on ($10\bar{1}$).

of the point of emergence also changes. In Fig. 12-9*a* all planes can be considered to neutralize one another except those terminating at F and G. These planes, which are of the type shown in Fig. 12-3, do not quite neutralize one another because F extends further down than G by a distance $c/2$. Thus, from Eq. (12-1), the point of emergence of the dislocation has a net effective charge

$$q=\lambda\frac{c}{2}=\frac{-e}{4} \tag{12-2}$$

Since the glide displacements of the pure edge dislocation are normal to the dislocation line, the glide of the edge dislocation *cannot produce any charge transport parallel to the dislocation and away from the* (010) *surface.* Thus the effective charge of the point of emergence of the [101] edge dislocation on the (010) surface *does not change* during glide, so that it is also $-e/4$ in the configuration of Fig. 12-9*b*. The removal of a (010) surface layer, or the climb of the dislocation by a distance $c/2$ would change the sign of the charge at the point of emergence. Hence the charge of the point of emergence of a pure edge dislocation normal to and with its Burgers vector lying in a $\{100\}$ surface is $\pm e/4$.

Exercise 12-2. The cube face containing the edge dislocation of Fig. 12-9 could have the corner charges shown in Fig. 12-10. The surface layer contains a

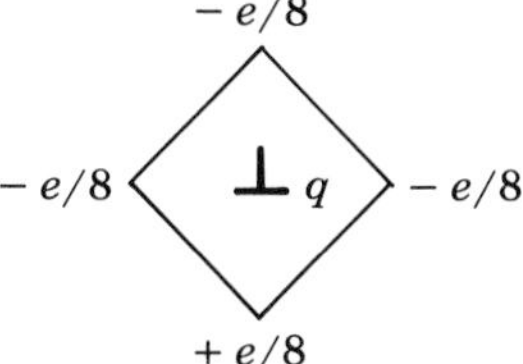

FIGURE 12-10. Corner charges on a cube bounded by {100} faces. An edge dislocation is shown emerging on (010).

net charge $-e$. Derive from this information the charge of the point of emergence.

Solution. Denote the charge of the point of emergence by q. If one surface layer is peeled off, all charges change sign. Charge balance requires that

$$\left[q+\left(\frac{e}{8}-\frac{3e}{8}\right)\right]-\left[-q+\left(-\frac{e}{8}+\frac{3e}{8}\right)\right]=-e$$

where $-e$ is the net charge removed. Thus $q=-e/4$.

Exercise 12-3. Show that the point of emergence of a $[\bar{1}10]$ edge dislocation on a (110) surface is also $\pm e/4$. Here $\boldsymbol{\xi}=(1/\sqrt{2})[110]$.

12-3. JOGS, KINKS AND VACANCIES IN EDGE DISLOCATIONS

Jogs

The extra double plane of Fig. 12-8 is redrawn as Fig. 12-11. With the removal of the lowest row, which starts with a negative ion at the surface, the dislocation moves up by one glide plane, and the charge changes to $e/4$. Suppose that only a part of the row is removed, starting at the surface, as shown in Fig. 12-12; then a jog is formed. In the example of Fig. 12-12 a net charge $-e$ is removed. At the same time, the point of emergence changes sign from $-e/4$ to $e/4$, and the jog of charge q is formed. By charge balance,

$$-\frac{e}{4}-\left(\frac{e}{4}+q\right)=-e$$

$$q=\frac{e}{2} \tag{12-3}$$

Thus elementary jogs in pure edge dislocations have charges $\pm e/2$. This result was first obtained by Seitz,[16] who applied the above type of reasoning to the process of creation of *pairs* of jogs in the interior.

[16] F. Seitz, *Phys. Rev.*, **80:** 239 (1950); *Adv. Phys.*, **1:** 43 (1952).

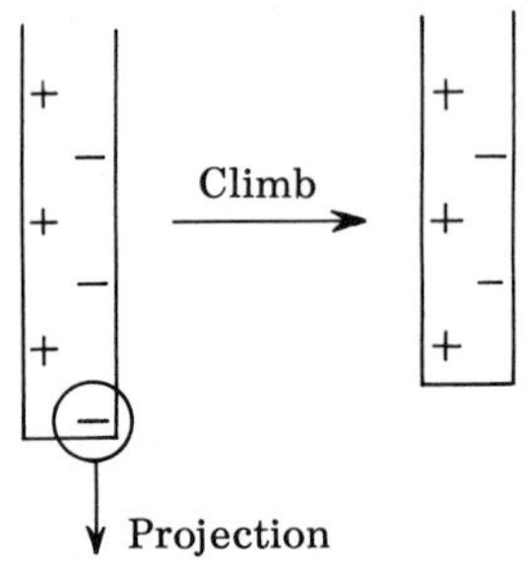

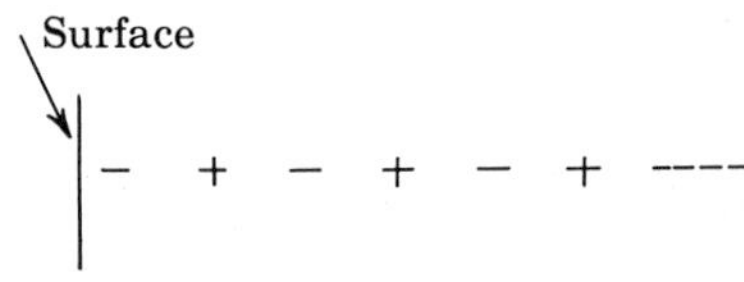

FIGURE 12-11. View parallel to the extra double plane in Fig. 12-8.

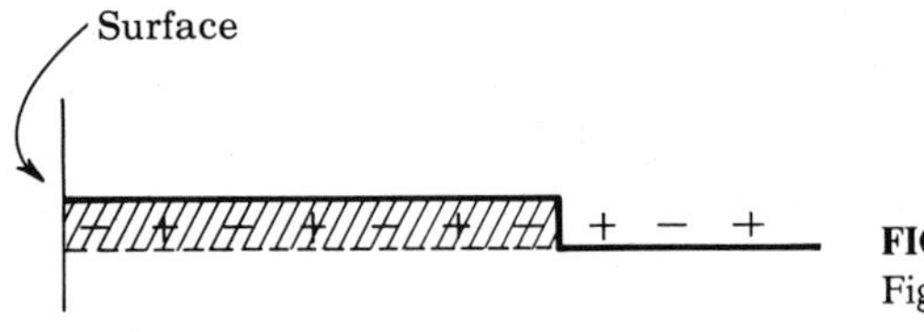

FIGURE 12-12. View perpendicular to Fig. 12-11, showing a jog.

Notice that jogs *cannot be neutralized* by point defects, ion vacancies, or interstitials, because their charge is half integral. Jogs of *twice* the elementary height, or more generally, jogs with a height an even number times the elementary height, are neutral. In Chap. 10 single jogs were shown to *repel* elastically, so that their combination to form a superjog either is energetically unfavorable or at least requires an activation energy in metals. This effect can be offset in ionic crystals because of the electrostatic attraction between opposite-sign jogs.

For example, consider NaCl with the average anisotropic moduli $\mu = 14.8$ GPa and $\nu = 0.25$, and $a = b/2$. Equation (8-93) then gives for the *elastic* repulsive force between jogs of like *geometric* sign

$$F_{\text{el}} = \frac{5.01 \times 10^{-30} Nm^2}{r^2} \tag{12-4}$$

The *electrostatic* force between jogs of opposite charge sign, with the dielectric constant $K = 6.12$, is

$$F_{\text{elec}} = \frac{e^2}{4Kr^2} = \frac{9.38 \times 10^{-30} Nm^2}{r^2} \tag{12-5}$$

Thus the electrostatic force is dominant, so that the jogs should combine by short-range core diffusion if such diffusion is possible. Furthermore, even the

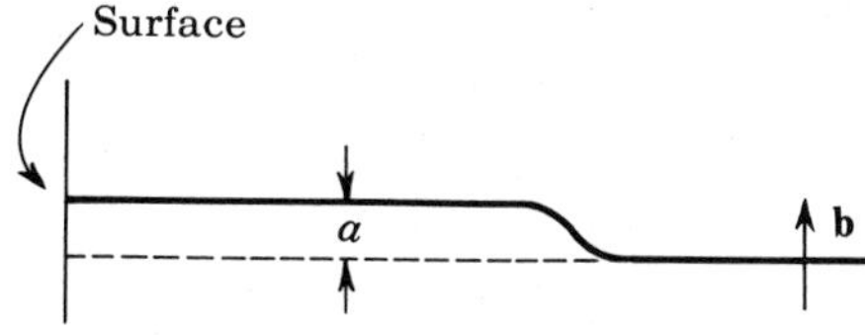

FIGURE 12-13. View perpendicular to the slip plane of an edge dislocation, showing a kink.

neutral superjogs contain an electric dipole, so that such superjogs exert a short-range electrostatic force on single jogs. Therefore we conclude that, unlike the case of metals, *superjogs can be stable in ionic crystals*; neutral double-step-height jogs should predominate over single-charged jogs on a relaxed dislocation line if $F_{elec} > F_{el}$. Braekhus and Lothe[17] examined the force balance for jogs and kinks for a number of ionic crystals by using Eq. (8-51) for the anisotropic elastic case. For NaCl, KCl, AgCl, and AgBr, they found that F_{elec} is larger so that neutral superjogs and superkinks should predominate on screw and edge dislocations. For LiF, F_{el} was found to dominate.

Since the point of emergence of an edge dislocation retains the sign of its charge as it moves, glissile jogs in edge dislocations also retain their charge sign as they move. As discussed in detail later, an edge dislocation can have a net charge at equilibrium because of an excess of one sign of charged jog.[18] A moving edge dislocation can acquire a charge by absorbing or desorbing charged point defects as it glides. In either case, edge dislocations clearly can transport electric charge by means of glissile jog motion.

Kinks

Consider a straight edge dislocation emerging perpendicular to a surface. If a segment *glides* by some distance a, shown in Fig. 12-13, a kink of step height a is created. Because no net charge is removed, and since the charge of the point of emergence remains unchanged, a charge balance indicates that the kink is neutral. *Kinks in pure edge dislocations are uncharged.*

Exercise 12-4. Show that the corners of the surface step produced by the creation of an edge dislocation in an originally perfect cube bounded by $\{100\}$ faces have charges $\pm e/4$. These charges just balance the charge of the created dislocation, as is required by charge neutrality.

Core Vacancies

At room temperature, where the time for thermal equilibration of defect concentrations is large, experiments suggest that moving dislocations acquire

[17] J. Braekhus and J. Lothe, *Phys. Stat. Solidi,* **51b:** 149 (1972).

[18] Net charges on edge dislocations appear to have been explicitly suggested first by F. Bassani and R. Thomson, *Phys. Rev.,* **102:** 1264 (1956). See also J. D. Eshelby, C. W. A. Newey, P. L. Pratt, and A. B. Lidiard, *Phil. Mag.,* **3:** 75 (1958).

charge by sweeping up cation vacancies. Studies on nominally pure and Mn^{++} doped NaCl indicate negative charges of e per 250 core sites and e per 2 core sites, respectively.[19] If these are present in the core, the density in the latter case is far too great to be accommodated at jogs. Such vacancies should persist, bound to the core and nominally equally spaced, because of electrostatic repulsion.

There are problems with such charges on a core if they remain uncompensated. A calculation[20] resembling that giving Eq. (12-5) shows that a length L of unit charges separated by $2b$ has a *negative* electrostatic contribution to the line tension $\mathcal{S}\cong -(e^2/4Kb^2)\ln(L/2b)$. If this outweighs the elastic line tension, $\mathcal{S}\cong[\mu b^2(1-2\nu)/4\pi(1+\nu)]\ln(L/b)$, the total line tension becomes negative and the dislocation becomes unstable with respect to spontaneous bow-out. Substituting typical values for μ, ν, and K, we find that one unit charge every other core site is beyond or close to the stability limit. Certainly, such charges at least reduce the line tension significantly. Thus perhaps a more realistic model for such large charges is a charge more spread out over an extended, near-core region. Of course, with nonlinear polarization included in the core region, the effective charge could be spread without spreading the vacancies. Also, part of the charge could be associated with aligned electrostatic dipoles, corresponding to cation vacancy-divalent cation impurity pairs, the alignment being induced by the dislocation motion. However, a significant reduction in line tension must in any case be expected. The critical stress for Frank-Read source operation (Chap. 20) or breakaway from pinning points (Chap. 18) for such dislocations is thereby reduced, making just these dislocations able to sweep up more charge from the lattice.

[19]A. Huddart and R. W. Whitworth, *Phil. Mag.*, **27:** 107 (1973).

[20]Consider n charges comprising $(n-1)$ pairs of spacing $2b$. These near-neighbor bonds contribute an energy $(n-1)e^2/2Kb$. There are $(n-2)$ second-neighbor bonds contributing $(n-2)e^2/4Kb$, $(n-3)$ third-neighbor bonds giving $(n-3)e^2/6Kb$, and so forth. Summing these contributions up to the one $(n-1)$th neighbor bond, the total energy is

$$W=\frac{e^2}{2Kb}\left\{\left[1+\frac{1}{2}+\frac{1}{3}+\cdots+\frac{1}{n-1}\right]n-n\right\}$$

$$=\frac{e^2n}{2Kb}\{\ln(n-1)+0.577-1\}\simeq\frac{e^2n}{2Kb}\ln n$$

$$=\frac{e^2n^2}{KL}\ln n$$

where the length of the segment $L=n2b$ is used in the last step and 0.577 is Euler's number, used in the approximation of the harmonic series Σn^{-1}. For fixed n and variable L, the electrostatic contribution to the line tension is

$$\mathcal{S}=\frac{\partial W}{\partial L}=\frac{-e^2n^2}{KL^2}\ln n=\frac{-e^2}{4Kb^2}\ln\frac{L}{2b}$$

[21]R. W. Whitworth, *Adv. Phys.*, **24:** 203 (1975).

Alternatively, the large core vacancy concentration could be stabilized kinetically. Whitworth,[21] at a time when Figs. 12-9*a* and 12-9*b* were both thought to be Peierls minima, suggested that vacancies could reduce the Peierls barrier. Then segments containing vacancies could move and multiply, acquiring more vacancies and leading to the observed result. Such a model could also apply with Fig. 12-9*b* as the Peierls maximum, provided that polarization reduced the large electrostatic energy. It is reasonable that the removal of cations from row *E* in Fig. 12-9*b* reduces the ion-core repulsions in that configuration, making the energies of the modified configurations of Figs. 12-9*a* and 12-9*b* more equal and thus reducing the Peierls barrier.

12-4. GEOMETRICAL CONSIDERATIONS

Jogs and Kinks in Screw Dislocations

Consider a pure screw dislocation with Burgers vector $\langle 101\rangle$ and $\boldsymbol{\xi}$ parallel to **b**. Along this direction the ions in a particular row all have the *same* sign. Viewed along $\langle 101\rangle$, the crystal appears as a set of alternating positively and negatively charged rods. Since the displacements associated with the screw are parallel to these rows, *the motion of a screw is accompanied by the motion of charge* back and forth along the dislocation line. When the screw moves by kink motion, the charge transport occurs by the motion of *charged kinks.*

In addition to the above distinction from edge-dislocation behavior, there is a surface step that terminates at the point of emergence of a screw at the surface; the effective charge of the point of emergence depends upon the orientation of the step. Because of these distinctions, the analysis of the screw is somewhat different from that of the edge. Consider the view normal to the (110) surface in Fig. 12-14. The signs (+) and (−) denote, respectively, positively and negatively charged rows perpendicular to the surface. Let a right-handed screw emerge at *B*, with **b** and $\boldsymbol{\xi}$ normal to the paper. If the screw glides counterclockwise about the closed path *BEDCB,* it "screws out" a single positive rod.[22] The glide steps *BE* and *DC* occur on the normal $(\bar{1}10)$ glide plane, while steps *ED* and *CB* occur on the cross-slip plane (001). A charge *e* is moved out at each turn, so that for a half turn,

$$q_g + q_c = \frac{e}{2} \tag{12-6}$$

where q_g is the charge transport in the normal glide step *BE* and q_c is the charge transport in the cross-slip step *ED*. Separate determinations of q_c and q_g would require detailed calculations involving the actual dislocation strains. As a rough approximation, one can put $q_c = q_g = e/4$.

[22] This operation is the same in principle as that evoked in the theory of whisker growth proposed by F. C. Frank [*Phil. Mag.*, **44:** 854 (1953)] and in the theory for extrusions in fatigue suggested by N. F. Mott [*Acta Met.*, **6:** 195 (1958)].

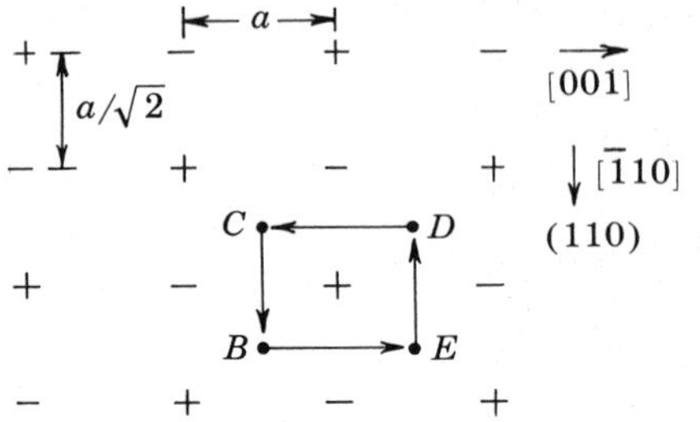

FIGURE 12-14. Projection of the lattice along [110].

The above concepts can readily be extended to determine the charges of kinks and jogs. If the right-handed screw segment *BP* in Fig. 12-15 glides to the position *DQ* (corresponding to the cycle *BED* in Fig. 12-14), while the segment *PO* is kept locked, a charge $e/2$ is moved to the surface. Since no charge transport takes place along *PO*, this must give rise to a net charge deficit $e/2$ in the region of the jog *PQ*. With the addition of a positive ion to the jog, its charge would change to $e/2$, and its configuration would change to that of the inset of Fig. 12-15. Local relaxations in the vicinity of the jog cannot change the effective charge of the jog. Only the absorption of charged point defects or the motion of the entire segments *DQ* or *PO* can change the net charge at the jog.

Proceeding in the same fashion, one can list the charges associated with a number of possible jog or kink configurations illustrated in Fig. 12-16. Charges $\pm e$ can be added to the charged jogs and kinks by point-defect reactions, so that no distinction is made with respect to sign. For each case the two screw segments occupy "equivalent" positions corresponding to possible Peierls potential valleys. The jogs and kinks which connect the screw segments then lie across Peierls energy hills. The charges associated with the various possibilities are listed in Table 12-1. As shown there, the charges can vary from zero to $e/2$.

The jogs and kinks exist in three categories, whose probability of existence depends on the actual minimum Peierls energy positions for the screw. The set in Fig. 12-16*a* can be Peierls valleys only for screws gliding on the primary glide plane $\{110\}$ and the set in Fig. 12-16*b* only for screws gliding on $\{100\}$.

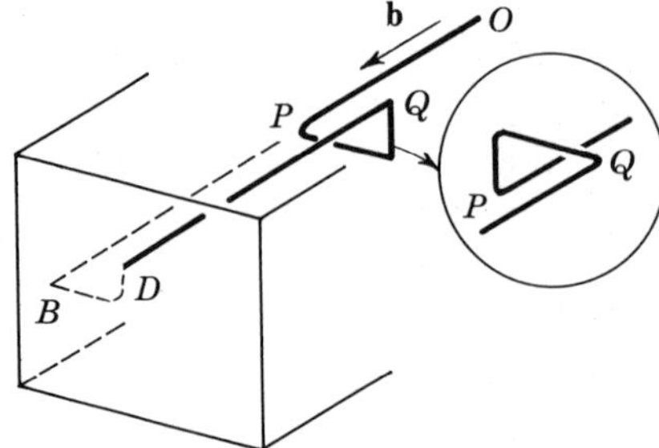

FIGURE 12-15. Projected view of a screw dislocation emerging on a (110) surface. The inset shows the jog configuration after a cation is absorbed.

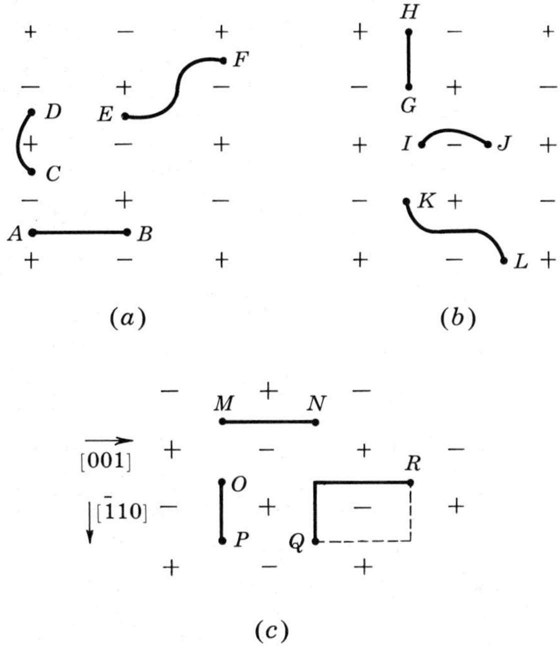

(a)

(b)

(c)

FIGURE 12-16. Projection of a lattice along [110], showing various possible jog and kink arrays.

The set in Fig. 12-16*c* could represent minimum Peierls energy positions for either type of glide. *OP*, for example, can be considered either as a kink for {100} glide or as a jog with respect to {110} glide. If set (*a*) represented the actual stable screw positions, kinks would be neutral but jogs charged. If set (*c*) represented the stable configurations, both kinks and jogs would be charged. Screws with *AB* [set (*a*)] and *MN* [set (*c*)] kinks are illustrated schematically in Figs. 12-17 and 12-18, respectively. Evidently, the screws in set (*a*) have oppositely charged ions very close together in the core. Although electostatically favorable, the short-range ion-ion repulsive forces make this an unlikely configuration for minimum energy. Thus set (*c*) is the most likely stable set of screw positions and *OP, MN*, and *QR* the most likely array of kinks and jogs. However, since atomic calculations will be necessary to settle this question conclusively, we have presented both possibilities. Similar arguments would apply to sets (*b*) and (*c*) as possibilities for {100} glide.

TABLE 12-1 Charges for Various Jog-Kink Segments

Charge	0	$\pm q_c$	$\pm q_g$	$\pm\frac{1}{2}e$
Segment	*AB*,*GH*	*OP*	*MN*	*CD*,*EF*,*IJ*,*KL*,*QR*

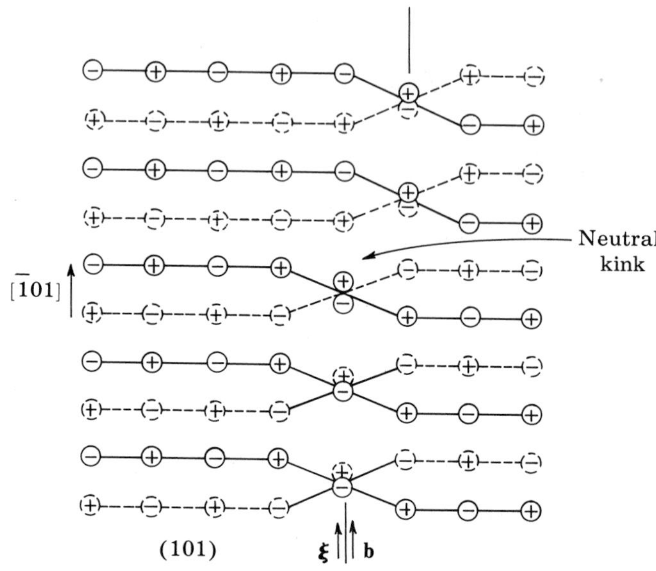

FIGURE 12-17. View normal to a screw lying on a (101) glide plane with a kink configuration AB of Fig. 12-16. Complete circles represent ions above the glide plane and dashed circles represent those below.

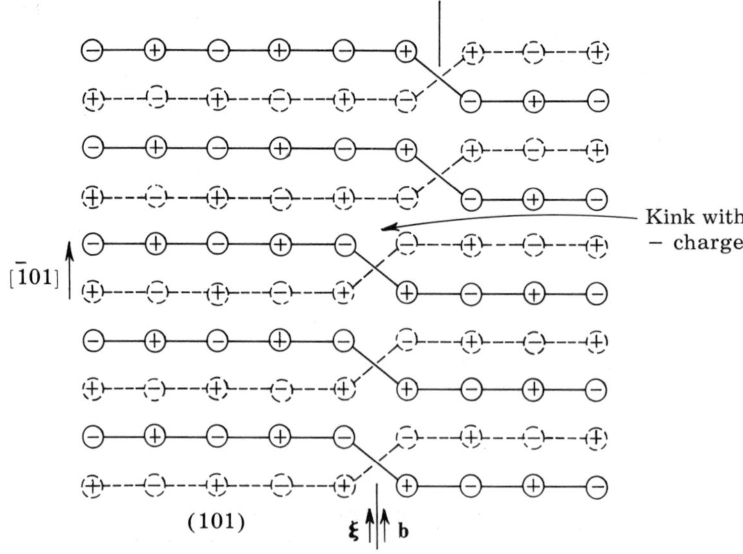

FIGURE 12-18. View normal to a screw lying on a (101) glide plane with a kink in configuration MN of Fig. 12-16. Complete circles represent ions above the glide plane and dashed circles represent those below.

The Moving Jogged Screw

Consider the motion of a screw containing a jog of the type *OP* in Fig. 12-16. Such a left-handed screw, with its associated ledges on the (101) surface normal to $\boldsymbol{\xi}$ is depicted in Fig. 12-19. The charge of the points of emergence is $\frac{1}{2}q_g$, which is left as an exercise to prove. With a positive ion at the end of the inserted row terminating at the jog, the effective charge of the jog is q_c. The jog can move to the right by emitting positive- and negative-ion interstitials alternately. The first elementary step for the jog in Fig. 12-19 involves the formation of a positive-ion interstitial, since a positive ion must be removed from the inserted row. For each elementary move, all effective charges change sign.

After the first elementary step to the right, the total effective charge is

$$-\tfrac{1}{2}q_g-\tfrac{1}{2}q_g-q_c+e=-(q_g+q_c)+e=\tfrac{1}{2}e \tag{12-7}$$

Here Eq. (12-6) is used, and the e occurring in the first line is the effective charge of the created positive ion interstitial. Prior to jog motion, the corresponding sum was

$$\tfrac{1}{2}q_g+\tfrac{1}{2}q_g+q_c=q_g+q_c=\tfrac{1}{2}e \tag{12-8}$$

Thus total effective charge is conserved during the process, but only if charge transport along the screw dislocation is included. The charge transport is manifested by the change in sign of the charge at the points of emergence of the screw. *The signs of the charge both on the points of emergence and on the jogs alternate as a screw dislocation moves by jog drag.*

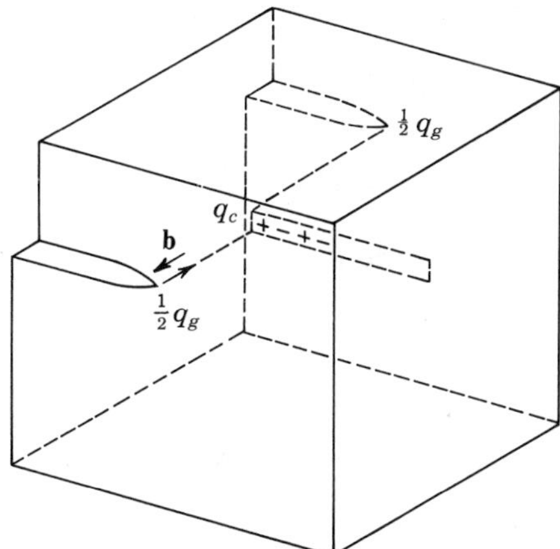

FIGURE 12-19. View of a screw with $\mathbf{b}=[101]$ emerging on (101) and containing a jog.

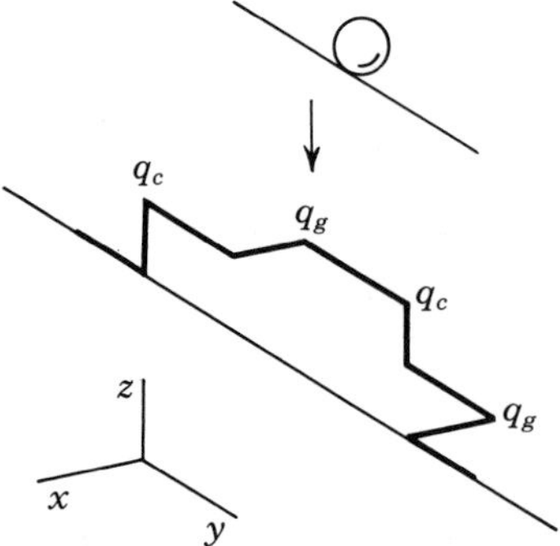

FIGURE 12-20. Decomposition of a vacancy adsorbed at a screw dislocation into four charged jogs and kinks.

Charged Screw Dislocations

As mentioned above, the fact that elementary jogs in edge dislocations are charged and should interact with charged point defects has been recognized for some time.[23] In addition, the possibility of a net charge on an edge dislocation is generally accepted.[24, 25] According to the prior development, screw dislocations should also interact strongly with point defects at charged jogs and kinks. Also, *screw dislocations can acquire a net charge;* this possibility seems to have been generally overlooked. Notice, as shown in Fig. 12-20, that a charged vacancy can decompose into four charged jogs and kinks along a screw. In dissociation of this type, the Coulomb energy is greatly reduced.[26] Thus, in a supersaturation of vacancies, a screw should acquire a net charge in the same way as an edge, via the formation of a preponderance of kinks and jogs of one sign. Furthermore, the formation energies of opposite-charge-sign vacancies differ, as do the energies of opposite-sign jogs and kinks, so that the net formation or annihilation energy of the opposite-sign vacancies at a screw should differ. Thus, even at equilibrium, a screw could acquire a net charge in the same way that Eshelby et al.[25] suggest for an edge.

A screw with a net charge that is moving by jog drag can also transport charge. The reasoning can be understood again with reference to Fig. 12-20. Suppose that a screw with a net charge of $25e/2$ moves through a neutral region of the lattice. For the first unit advance of the line, 25 excess negative-ion vacancies with a charge $25e$ form, and the charge on the dislocation becomes $-25e/2$, and so forth for successive steps. After a large number of advances, the alternate $25e$ and $-25e$ net charges of the point defects *cancel*, and the slipped region behind the dislocation retains a neutral long-range charge. If the number of unit steps advanced is even, the dislocation again has a net charge

[23] F. Seitz, *Adv. Phys.*, **1:** 43 (1952).

[24] F. Bassani and R. Thomson, *Phys. Rev.*, **102:** 1264 (1956).

[25] J. D. Eshelby, C. W. A. Newey, P. L. Pratt, and A. B. Lidiard, *Phil. Mag.*, **3:** 75 (1958).

[26] R. Thomson, in "Fundamental Aspects of Dislocation Theory," NBS Special Publication 317, 1970, p. 563, provides further justification for such decomposition of a vacancy.

$25e/2$; if it is odd, there is an excess row of negative-ion vacancies with charge $25e$, which together with the dislocation charge of $-25e/2$ still yields a net charge of $25e/2$ in the vicinity of the dislocation. Thus the excess charge of a screw dislocation is carried along with the screw if it moves by jog drag.

Conservative jog motion and kink motion along the dislocation line cannot remove the net charge on the dislocation. However, such motion does provide a mechanism for the exchange of charge between the crystal interior and the external surfaces.

12-5. CURVED, MIXED DISLOCATIONS

In the preceding development, the kinked edge dislocation was treated in a manner very different from the kinked screw dislocation. The method of presentation was chosed to indicate clearly that charge transport occurs *along* screw dislocations but not along edges. Actually, of course, edge and screw segments are created in the same operation, so that the effects for both types of dislocations must fit together in a consistent description of a curved dislocation. Suppose that slip is produced over a square portion of the slip plane, as in Fig. 12-21, so that one pure edge segment and one pure screw segment result. The effective charges are shown in the figure. Suppose for convenience that $q_g \sim e/4$. The dislocation could relax into the more general curved configuration of Fig. 12-22 by kink formation. If the charge scheme is self-consistent, it must be possible to run kinks in screws together to form longer edge segments connected by neutral kinks. Indeed, as indicated in Fig. 12-23, the neutral kinks in the edge segments are formally dipoles of two effective charges $\pm e/8$.

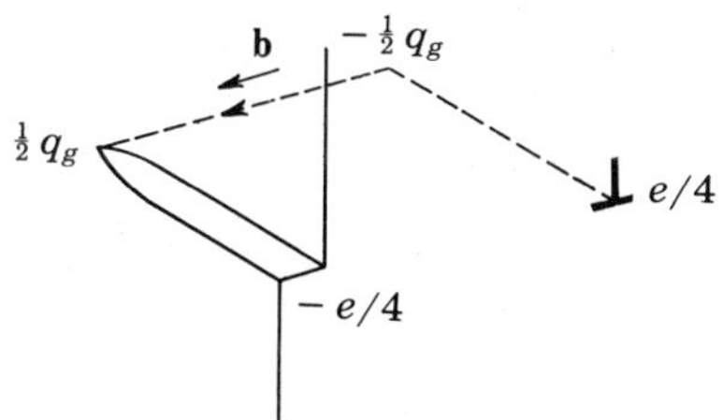

FIGURE 12-21. A crystal slipped over a square portion of the glide plane.

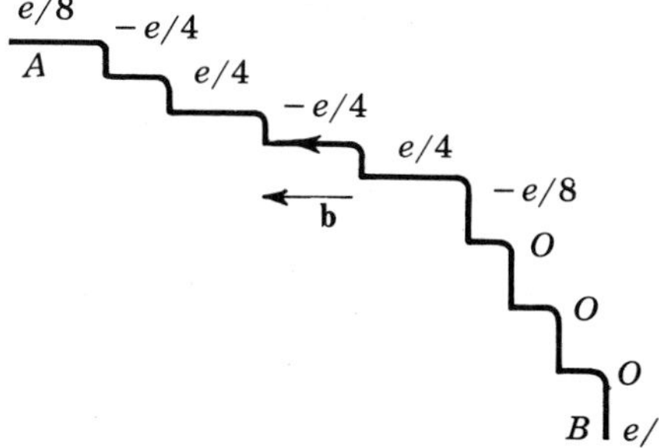

FIGURE 12-22. Relaxed configuration of Fig. 12-21 with jogs and kinks.

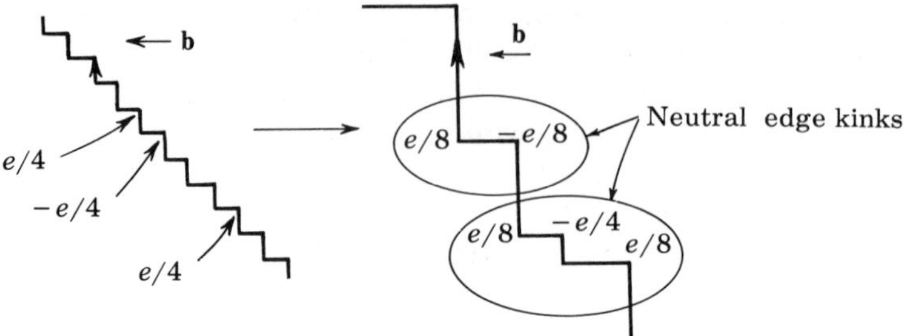

FIGURE 12-23. A kinked screw gliding to form a kinked edge segment.

These arise because each screw kink with charge $e/4$ has a charge $e/8$ at each end of the kink. One of the end $e/8$ charges of each screw kink is annihilated when an $e/4$ screw kink combines with a $-e/4$ screw kink. More complicated neutral kinks can also form in edge segments, an example being the $(e/8)-(e/4)+(e/8)$ neutral kink also shown in Fig. 12-23.

12-6. INTERSECTION JOGS AND KINKS

General Geometric Relations

A moving dislocation causes lattice displacements equal to its Burgers vector. A formal rule for the sign of the displacement is given in Axiom 1-3. Suppose that the moving dislocation cuts through a second, pinned dislocation. Clearly, the pinned dislocation acts as a marker fixed to specific lattice points. Thus, after being cut by the moving dislocation, the fixed dislocation acquires a displacement in its line equal to the overall lattice displacement, and hence equal to the Burgers vector of the moving dislocation. In general, neither dislocation need be pinned; upon intersection, each dislocation gains a displacement equal in magnitude to the Burgers vector of the other dislocation. If the displacement lies in the glide plane of its dislocation, it is an *intersection kink*, otherwise it is an *intersection jog.*

Usually, one can determine the sign of the kinks and jogs by inspection, but in some instances a formal procedure for determining the sign is valuable, as proposed by Hornstra.[27] Consider the two right-handed screws in Fig. 12-24a. Let dislocation (1) move toward (2) in a direction given by the vector $\mathbf{r}$, which is normal to $\boldsymbol{\xi}_1$ and lies in the glide plane of (1). After intersection, the dislocations have the configuration of Fig. 12-24b. The jog in dislocation (1) has a magnitude and direction $\pm\mathbf{b}_2$. With respect to the *positive* sense $\boldsymbol{\xi}_1$, the plus-or-minus sign of the displacement is given by [27]

$$\pm 1 = \frac{\boldsymbol{\xi}_1 \cdot (\mathbf{r} \times \boldsymbol{\xi}_2)}{|\boldsymbol{\xi}_1 \cdot (\mathbf{r} \times \boldsymbol{\xi}_2)|} \tag{12-9}$$

As is evident, the jogs in Fig. 12-24b agree with Eq. (12-9).

[27] J. Hornstra, *Acta Met.*, **10:** 987 (1962).

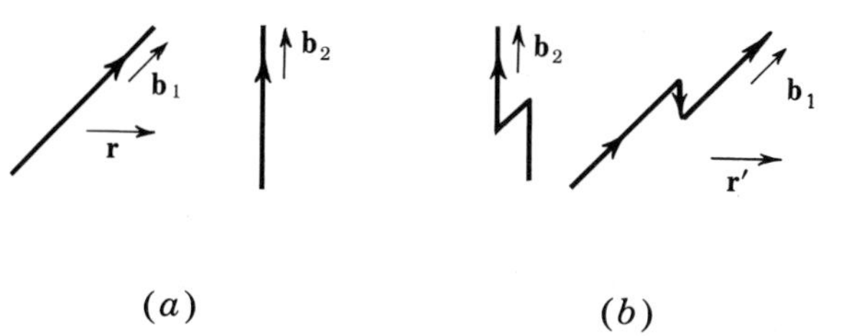

FIGURE 12-24. Two intersecting screws: (*a*) before intersection, (*b*) after intersection.

Often one wishes to know whether vacancies or interstitials are formed when the jogs move along with the screw dislocation. Let **r**′ (Fig. 12-24*b*) represent the distance and direction of motion of dislocation (1) *with respect to the crystal lattice* after intersection (note that the intersection in Fig. 12-24 could occur if both dislocations were moving to the left, with dislocation (2) moving more rapidly; in such a case **r**′ would point to the *left*). Such motion produces a number of point defects, given by [27]

$$N = \frac{[\boldsymbol{\xi}_1 \cdot (\mathbf{r} \times \boldsymbol{\xi}_2)][\mathbf{b}_1 \cdot (\mathbf{r} \times \mathbf{b}_2)]}{\Omega |\boldsymbol{\xi}_1 \cdot (\mathbf{r} \times \boldsymbol{\xi}_2)|} \tag{12-10}$$

where Ω is the atomic volume for metals and the volume of the pertinent ion or ion pair for ionic crystals. Interstitials are formed if the sign of Eq. (12-10) is positive, and vacancies are formed if it is negative. Usually **r**′ has the same direction as **r** for intersecting dislocations. In such a case, the intersection of either a pair of right-handed screws or a pair of left-handed screws produces interstitial-forming jogs in each screw; the intersection of a right-handed screw and a left-handed screw produces vacancy-forming jogs in each dislocation.

Application to Ionic Crystals

Since all intersection kinks and jogs are equal in length and direction to perfect-dislocation Burgers vectors, all possible cases can be illustrated in a projection along a screw-dislocation-line direction. Figure 12-25*a* shows such a projection of all possible perfect-dislocation Burgers vectors. Eight of the 12 possible **b** form charged jogs $\pm e/2$ of the type *QR* in Fig. 12-16. These jogs can decompose to charged jogs *OP* and charged kinks *MN*. Two others form *neutral* jogs twice the length of *OP*. The final two are parallel to the screw axis and would not create displacements. Figure 12-25*b* presents a similar projection along the axis of an edge dislocation. Inspection of this figure indicates that of the 12 possibilities, eight form charged jogs, two form neutral double jogs, and two form neutral kinks. Thus, for both the edge and the screw dislocation, eight of the 12 possible intersections of dislocations from other slip systems with a given dislocation produce charged jogs.[28]

[28] F. C. Frank, *Nuovo Cim. Suppl.* **7**: 386 (1958).

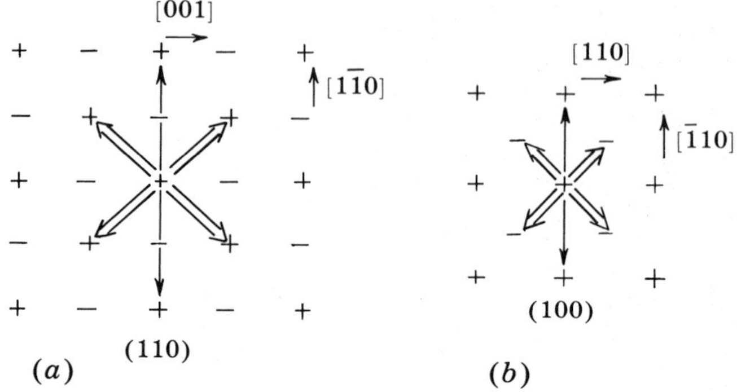

FIGURE 12-25. Possible perfect-dislocation Burgers vectors shown in projections along the axes of (*a*) a screw dislocation, $\xi = (\sqrt{2}/2)[101]$, and (*b*) an edge dislocation, $\xi = [010]$.

Let us consider some specific examples. Suppose that the maximum resolved shear stress acts on the glide system $[110](\bar{1}10)$. Then the system $[\bar{1}10](110)$ has an equal resolved shear stress on it, because, as shown in Fig. 12-26, the slip directions and plane normals in these two systems are mutually perpendicular, so that $\sigma_{xy} = \sigma_{yx}$. Thus intersections between these two systems should occur most frequently during deformation. The jogs formed in both sets of dislocations are the *neutral* jogs discussed above. Furthermore, rather than dissociating into oppositely charged jogs, these jogs are likely to remain neutral because of the elecrostatic attractive forces between the former [Eq. (12-5)]. Therefore, if vacancies, for example, are produced at jogs during deformation, most of them should appear as neutral associated Shockley pairs. If these pairs are to contribute to the Gyulai-Hartly effect, as suggested by Seitz,[29] the dissociation energy of the pair must be small enough that appreciable dissociation can occur at the temperature of observation. For NaCl one can use simple association theory to estimate an energy of dissociation to be about 0.4 eV. The binding energy of a neutral interstitial pair is more difficult to estimate, but again one would expect a negligible contribution to the current. Thus the degree of dissociation of neutral pairs at room temperature should be negligible. Hence vacancies formed at neutral jogs should *not* contribute to the electrical conductivity, but could lead to an enhanced diffusivity following deformation as discussed by Slifkin.[30]

Other intersections of the primary glide system with secondary slip systems do produce charged jogs. For example, if $[110](\bar{1}10)$ screw dislocations intersect with $[101](\bar{1}01)$ screws, jogs of the type *QR* with charges $\pm e/2$ form in both dislocations. If a $[110](\bar{1}10)$ edge intersects a $[101](\bar{1}01)$ screw, $\pm e/2$ intersection jogs are formed in both dislocations. In each of these cases the charges on the jogs, plus charges on any point defects formed, add up to give electrical

[29]F. Seitz, *Rev. Mod. Phys.*, **26:** 7 (1954).

[30]L. Slifkin, *Semicond. Insul.*, **3:** 393 (1978).

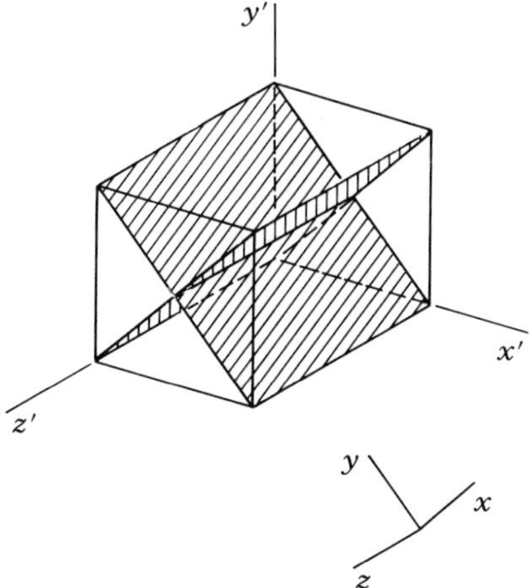

FIGURE 12-26. View parallel to [001], showing the equally stressed slip systems [110] $(\bar{1}10)$ and $[\bar{1}10](110)$.

charge neutrality. By Seitz's mechanism, only the small fraction of vacancy-forming jogs formed by such intersections with secondary slip systems can produce charged jogs and contribute to the Gyulai-Hartly effect.[31] Experimental results[31,32] do show that the true Gyulai-Hartly effect occurs only at large plastic strains and when multiple slip is occurring, consistent with the Seitz mechanism. As explained in Chap. 22, intersection processes generally lead to more interstitial-forming jogs than vacancy-forming jogs. However, interstitial mobilities are such [30] that they would anneal out essentially immediately, leaving only vacancies to produce the persistent change in conductivity of the Gyulai-Hartly effect. Also, atomic calculations suggest that interstitial formation energies are larger than vacancy formation energies by about 1 eV, so that interstitial-forming jogs would tend to remain immobile while the other jogs were moving and producing vacancies.

Exercise 12-5. Apply Eqs. (12-9) and (12-10) to the jogs formed on intersection of the $[110](\bar{1}10)$ and $[101](\bar{1}01)$ dislocations discussed above. Determine whether the jogs are interstitial forming or vacancy forming. Verify that charge neutrality is maintained.

The concept of charge neutrality is important with respect to charged dislocations. Suppose that a group of edge dislocations of the slip system $[110](\bar{1}10)$ intersects a group of screws from the slip system $[101](\bar{1}01)$. Then the charge sign of all the intersection jogs formed on the edge dislocations is the same. This charge is compensated for by the charged jogs on the screws, which, again, are all of the same sign. This provides another demonstration that screw dislocations can acquire a net charge.

[31] A. Taylor and P. L. Pratt, *Phil. Mag.*, **3:** 105 (1958).

[32] R. W. Whitworth, *Adv. Phys.*, **24:** 203 (1975).

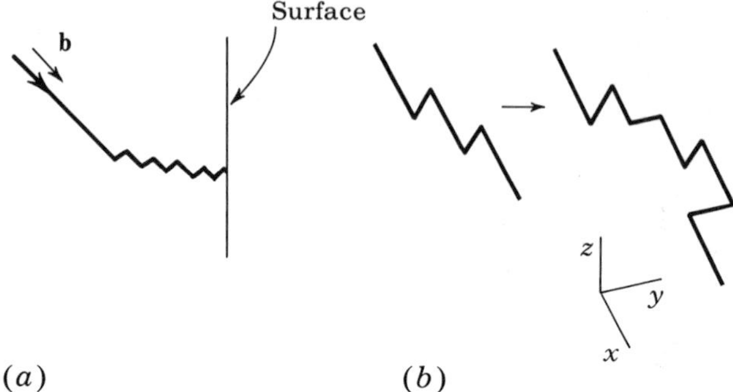

FIGURE 12-27. A kinked screw intersecting a {100} surface.

12-7. INFLUENCE OF CHARGE ON DEFORMATION

Although both edges and screw dislocations can transport charge, their relative importance differs. Jogs in edges are glissile; the jogs in screws are sessile, pin the screws, and restrict glide. Thus at low temperatures most of the charge transport observed in the deformation-associated effects enumerated in Sec. 12-1 should be associated with the glide of edge dislocations. At higher temperatures, where thermally activated jog drag occurs, screw dislocations also should contribute appreciably to charge transport. The effects of temperature on dislocation charge are considered further in Sec. 14-7.

While there once was considerable controversy about the charge on dislocations, the situation now appears to be resolved. As discussed by Whitworth,[32] a number of experiments below and at room temperatures show that the charge on dislocations in NaCl and KCl is negative when the crystals are pure or doped with divalent cations. Moreover, the evidence favors the charge being associated with the charged jogs and core cation vacancies discussed previously. The charged jogs would be formed by intersection and the vacancies would be swept up during deformation.

The experiments demonstrating the negative charge are those in which moving dislocations carry a charge, producing an electrostatic field after the deformation. The reciprocal effect also exists, the presence of an external electrostatic field enhancing the mobility of the dislocations.[33]

Near the surface, special image effects can occur. Screws cannot emerge normally to {100}, the principal cleavage plane. Image forces pull the dislocations out of screw orientation toward a perpendicular orientation. The resulting configuration, shown in Fig. 12-27*a*, has a number of kinks of alternate charge lying in the cross-slip plane. With these kinks already present, only two more elementary kinks or jogs are required to accommodate a charge e at the dislocation by point-defect absorption (Fig. 12-27*b*). Thus both edge and screw dislocations can easily acquire charge at the surface; they should do so to

[33] R. M. Turner and R. W. Whitworth, *Phil. Mag.*, **18:** 531 (1968).

equilibrate with the space-charge layer present at surfaces. This could be important in the Joffé effect, the removal of brittleness by the dissolution of a surface layer.

12-8. POLYVALENT IONIC CRYSTALS

Crystals containing polyvalent ions have properties closely resembling those of the rock-salt crystal. One important distinction is that for divalent ionic crystals, such as MgO and FeO, the elementary jogs have unit charges $\pm e$, twice those in NaCl. Thus, in these crystals, elementary jogs can be neutralized by interacting with singly charged point defects such as free electrons or electron holes. Hence one expects less of a tendency for charging of dislocations in divalent crystals. Trivalent crystals again have half-integral charges on the elementary jogs and kinks, while tetravalent ions are associated with doubly integral charges on the jogs. Obviously, the situation becomes complicated in mixed crystals composed of ions with various valences. In noncentrosymmetric crystals, piezoelectric interactions between elastic stresses and electrostatic fields appear. Weak effects of this type can occur in NaCl type crystals because of nonlinear elastic distortions in dislocation cores.

Crystals with mixed ionic-covalent nature are even more complicated, but many of the ideas discussed for the ionic crystals apply. Many such crystals such as ZnS, exhibit layer structures, and can have low-energy stacking faults as considered in Chap. 11. However, their ionic character also means that charge can be transported by perfect and partial dislocations in such crystals. A clear discussion of such effects is given by Petrenko and Whitworth.[34]

PROBLEMS

12-1. Suppose that a square prismatic loop of Burgers vector $\mathbf{b}=[\bar{1}01]$ and with edges along [010] and [101] has formed by vacancy condensation. What is the magnitude of the effective charge of the corners?

12-2. Suppose that the jog in the screw in Fig. 12-19 has an effective charge $q_c - e$, and that after moving one step forward it has the charge $-q_c + e$. How is the charge balanced with such motion of the dislocation?

12-3. If screws with Burgers vector $\mathbf{b}=[100]$ existed, would kinks and jogs in such screws be charged?

12-4. Consider an edge dislocation $\mathbf{b}=[101]$ moving through a monatomic surface ledge lying along [001] on (010). What would be the charge of the kink in the ledge? Derive the charge balance for the process.

12-5. Should the accumulation of jogs of one sign of charge for a screw dislocation gliding on (110) influence its tendency to cross slip onto (001)? Why?

[34] V. F. Petrenko and R. W. Whitworth, *Phil. Mag.*, **41A:** 681 (1980).

12-6. In an NaCl single crystal bounded by {100} faces and compressed along [001], suppose that slip occurs by the motion of only pure edge and pure screw dislocations.

- **a.** What type of intersection jog should be predominantly formed?
- **b.** Which set of dislocations is more mobile—the screws or the edges?
- **c.** Which set of dislocations, the screws or the edges, can produce transient electrical currents during glide?

12-7. Consider a CsCl or β-brass type of ionic crystal, with slip occurring on the systems $\langle 100 \rangle \{110\}$. Determine whether or not the following defects are charged: screw jogs, screw kinks, edge jogs, edge kinks.

12-8. Should superjogs tend to form in MgO? In PbS? Discuss the role of polarizability in superjog formation.

12-9. What is the charge on the (001) surface at the point of emergence of a screw dislocation along [101]? The surface ledge lies along [010].

BIBLIOGRAPHY

1. Amelinckx, S., *Solid State Phys. Suppl.* 6 (1964); in F.R.N. Nabarro (ed.), "Dislocations in Solids", North-Holland, Amsterdam, 1979, pp. 337-398.
2. Brantley W. A. and C. L. Bauer, *Phys. Stat. Solidi,* **18**:465 (1966).
3. Gilman, J. J., and W. G. Johnston, *Solid State Phys.,* **13**: 147 (1962).
4. Latanision, R. M., and J. T. Fourie, "Surface Effects in Crystal Plasticity," Noordhoff, Leyden, 1977.
5. Mitchell, J. W., in "Dislocations and Mechanical Properties of Crystals," Wiley, New York, 1957, p. 69.
6. Pratt, P. L., *Inst. Met. Monogr. and Repts. Ser.,* **23**: 99 (1957).
7. Sprackling, M. T., "The Plastic Deformation of Simple Ionic Crystals," Academic, New York, 1976.
8. Van Bueren, H. G., "Imperfections in Crystals," North-Holland, Amsterdam, 1961, pp. 512-582.
9. Whitworth, R. W., *Adv. Phys.,* **24**: 203 (1975).

13

Dislocations in Anisotropic Elastic Media

13-1. INTRODUCTION

The concept of an elastically isotropic crystal is an idealization; all real crystals are anisotropic. Yet in the preceding chapters, as in most of the dislocation literature, the theory is presented mainly in the isotropic elasticity approximation. Isotropic theory has led to useful results, but for some cases it is an inadequate approximation. There are two good reasons for these approximations. First, the mathematics, and in particular numerical evaluation, becomes more complicated when anisotropy is considered. Second, in many cases the errors involved in the isotropic approximation are about 20 to 30 percent, and these are submerged by other approximations in dislocation theory or by errors in experimental observation. However, there are increasing numbers of observations of sufficient accuracy to warrant comparison with more precise anisotropic calculations. Also, some effects, such as the instability of some straight dislocations with respect to break up into a zigzag shape, require anisotropic theory even for their qualitative explanation. Hence a discussion of the methods used in the anisotropic theory is important.

Anisotropic elastic constants are known for many crystals.[1] One wishes to know the best approximation for isotropic constants in terms of the anisotropic constants for use in the simpler isotropic theory. This problem is treated in the first part of the chapter. This material is also useful in providing an appreciation of the magnitude of the errors involved in the isotropic approximation.

The complexities of the anisotropic treatment are reduced considerably for certain configurations of straight dislocation lying along directions of high symmetry; indeed, the more exact theory is then quite simple. Hence it is convenient to use the more exact anisotropic theory for such dislocations. The original sextic theory of straight dislocations in anisotropic media is presented in the second part of the chapter, and the method is applied to some important

[1]A compendium of elastic constants is presented in Appendix 1.

but simple dislocation reactions. Most anisotropic calculations in the literature have used this form of the theory.

In recent years further developments have been made in two areas of anisotropic theory. First, a number of general results have been obtained in terms of straight dislocation parameters. Examples include the line tension expression (Sec. 6-5), the kink-kink interaction (Exercise 8-6), the image force at a free surface (Sec. 5-9), and, more generally, the treatment in Sec. 5-7 of the Brown formula and related work for the fields of curved dislocations. These developments are not repeated here. Second, a more complete theory for the sextic formalism that also leads to the alternative of so-called integral formalism has been developed to supersede the original sextic theory. The new methods particularly facilitate computer calculations that are necessary when a dislocation does not lie in a direction of high symmetry. These methods are treated in a separate section. Finally, the stability of straight dislocations is discussed.

Throughout this chapter we retain the assumptions of linear elasticity and of the independence of stresses on rotations. Some of the consequences of these assumptions are discussed by Huntington[2]; they are not important for our development.

With the recent developments, formal anisotropic theory has become almost as simple as isotropic theory. Simple line integrals exist for stresses and strains. However, so far no simple double line integral for the interaction energy between two arbitrary loops, corresponding to Blin's Eq. (4-40) for the isotropic case, has been developed. There had been no method for avoiding the calculation by surface integration of the work done by the stresses of one loop over the cut surface of the other loop as it is formed, but recent new methods are mentioned in Sec. 13-9.

13-2. AVERAGE ELASTIC CONSTANTS

In most problems involving dislocations of differing orientations and Burgers vectors, the complexity is such that isotropic average elastic constants must be used. The most appropriate values are usually those averaged over all possible orientations of the coordinate system relative to the crystal axes. In special cases one can do better, but these cases are most conveniently discussed in conjunction with the theory for straight dislocations in anisotropic media.

The question still remains of whether to average over the elastic constants c_{ijkl} or over the elastic compliances s_{ijkl}. The former is appropriate for a polycrystal in which the grains have the same state of strain, as in Fig. 13-1*a*; the latter for the case when they have the same stress, as in Fig. 13-1*b*. For most cases involving *local* strains around dislocations, uniform strain nearly obtains, and the Voigt averages over c_{ijkl} are most appropriate. In cases

[2] H. B. Huntington, *Solid State Phys.*, **7**: 213 (1958).

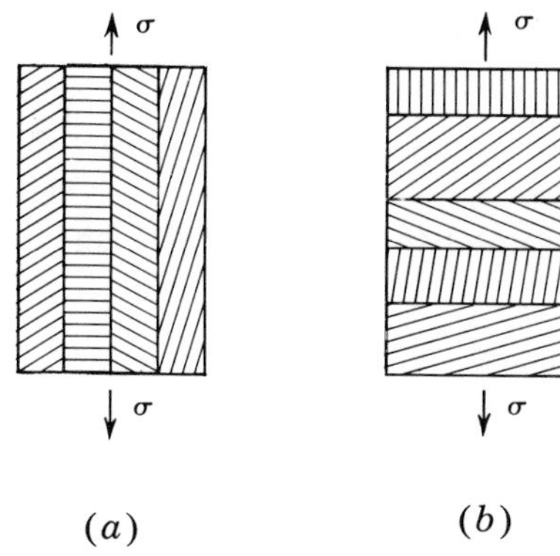

FIGURE 13-1. Polycrystals under simple tension. Grain boundaries lie (*a*) parallel to the tensile axis (uniform strain) or (*b*) perpendicular to the tensile axis (uniform stress).

involving long-range internal-stress fields, the Reuss averages over s_{ijkl} are applicable.[3] The results of these two methods provide upper and lower bounds for μ and B.[4] They agree to the *first order* in the anisotropy, but differ to second order. Both averages are developed in the following discussion.

The Voigt Average

The average elastic constants are found easily with the use of certain combinations of them which are invariant to rotation of the coordinate system.[5] As shown in Sec. 2-3, the 9×9 matrix for the elastic constants c_{ijkl} undergoes an orthogonal unitary transformation when the reference coordinate system is rotated. The *trace* of a matrix is invariant to orthogonal unitary transformations,[6] so that

$$I_1 = c_{ijij} = \text{const} \tag{13-1}$$

A second invariant follows from the invariance of strain-energy density. For the special case of homogeneous expansion, $\epsilon_{11} = \epsilon_{22} = \epsilon_{33} = \frac{1}{3}\epsilon_{ii}$, and $\epsilon_{ij} = 0$ for $i \neq j$, Eq. (2-14) gives

$$I_2 = c_{iijj} = \text{const} \tag{13-2}$$

Expanding Eq. (2-46) into a 9×9 matrix, one finds that the two invariants for an isotropic solid are

$$I_1 = 3(\lambda + 2\mu) + 6\mu \tag{13-3}$$

[3] W. Voigt, "Lehrbuch der Kristalphysik," Teubner, Leipzig, 1928, p. 716; A. Reuss, *Z. ang. Math. Mech.*, **9**: 55 (1929).

[4] R. Hill, *Proc. Phys. Soc.*, **65A**: 349 (1952).

[5] G. Leibfried, *Z. Phys.*, **135**: 23 (1953).

[6] H. Margenau and G. M. Murphy, "The Mathematics of Physics and Chemistry," Van Nostrand, New York, 1943, p. 304.

and

$$I_2 = 3(\lambda + 2\mu) + 6\lambda \tag{13-4}$$

Equating (13-1) to (13-3) and (13-2) to (13-4) and then solving for μ and λ gives the average values

$$\begin{aligned} \lambda + 2\mu &= \tfrac{1}{15}(2c_{ijij} + c_{iijj}) \\ \mu &= \tfrac{1}{30}(3c_{ijij} - c_{iijj}) \\ \lambda &= \tfrac{1}{15}(2c_{iijj} - c_{ijij}) \end{aligned} \tag{13-5}$$

The Reuss Average over Compliances

Thus far we have considered only the elastic constants c_{ijkl}, and not the inverse coefficients s_{ijkl}, the compliance constants. The relations between both sets is important in the treatment of the s_{ijkl}. As emphasized in Chap. 2, the c_{ijkl} really form the elements of a 9×9 matrix. In transformations this complete 9×9 matrix should be used instead of the shorthand 6×6 formulation. By simply inverting the matrix $\{c\}$, one would expect to find the matrix $\{s\}$, whose elements relate strain to stress.

$$\epsilon_{ij} = s_{ijkl}\sigma_{kl} \tag{13-6}$$

However, the 9×9 matrix $\{c\}$ cannot be inverted in the usual sense of matrix calculations, because generally the determinant of $\{c\}$ is zero. This is so because rows 7, 8, and 9 are repetitions of rows 4, 5, and 6, respectively, as are the columns, so that neither rows nor columns are linearly independent. The 9×9 matrices for both $\{s\}$ and $\{c\}$ are of this special structure.

The problem of inversion can be approached as follows. First consider the $6 \times 6\{c\}$ matrix [Eq. (2-19)],

$$\{c_{mn}\}_{6 \times 6} \tag{13-7}$$

In general, this matrix has a determinant different from zero, and can be inverted to

$$\{s_{mn}\}_{6 \times 6} \tag{13-8}$$

Multiplication of Eq. (13-8) into Eq. (2-19), with the use of the relation $\{c_{mn}\}_{6\times 6}\{s_{mn}\}_{6\times 6} = \{1\}$, reveals that the elements of $\{s_{mn}\}$ belong to the relation

$$\begin{bmatrix} \epsilon_{11} \\ \epsilon_{22} \\ \epsilon_{33} \\ 2\epsilon_{23} \\ 2\epsilon_{31} \\ 2\epsilon_{12} \end{bmatrix} = \begin{bmatrix} s'_{11} & s'_{12} & s'_{13} & s'_{14} & s'_{15} & s'_{16} \\ s'_{12} & s'_{22} & s'_{23} & s'_{24} & s'_{25} & s'_{26} \\ s'_{13} & s'_{23} & s'_{33} & s'_{34} & s'_{35} & s'_{36} \\ s'_{14} & s'_{24} & s'_{34} & s'_{44} & s'_{45} & s'_{46} \\ s'_{15} & s'_{25} & s'_{35} & s'_{45} & s'_{55} & s'_{56} \\ s'_{16} & s'_{26} & s'_{36} & s'_{46} & s'_{56} & s'_{66} \end{bmatrix} \begin{bmatrix} \sigma_{11} \\ \sigma_{22} \\ \sigma_{33} \\ \sigma_{23} \\ \sigma_{31} \\ \sigma_{12} \end{bmatrix} \tag{13-9}$$

The elements are symmetrical about the diagonal, as are the c_{mn}. As is evident, Eq. (13-9) can be rewritten in the form

$$\begin{bmatrix} \epsilon_{11} \\ \epsilon_{22} \\ \epsilon_{33} \\ \epsilon_{23} \\ \epsilon_{31} \\ \epsilon_{12} \end{bmatrix} = \begin{bmatrix} s_{11} & s_{12} & s_{13} & s_{14} & s_{15} & s_{16} \\ s_{12} & s_{22} & s_{23} & s_{24} & s_{25} & s_{26} \\ s_{13} & s_{23} & s_{33} & s_{34} & s_{35} & s_{36} \\ s_{14} & s_{24} & s_{34} & s_{44} & s_{45} & s_{46} \\ s_{15} & s_{25} & s_{35} & s_{45} & s_{55} & s_{56} \\ s_{16} & s_{26} & s_{36} & s_{46} & s_{56} & s_{66} \end{bmatrix} \begin{bmatrix} \sigma_{11} \\ \sigma_{22} \\ \sigma_{33} \\ 2\sigma_{23} \\ 2\sigma_{31} \\ 2\sigma_{12} \end{bmatrix} \tag{13-10}$$

which is also symmetrical about the diagonal. Here[7]

$$\begin{aligned} s_{mn} &= s'_{mn} \text{ if both } m \text{ and } n \text{ are } 1, 2, \textit{or } 3 \\ s_{mn} &= \tfrac{1}{2}s'_{mn} \text{ if either (but not both) } m \text{ or } n \text{ is } 4, 5, \text{ or } 6 \\ s_{mn} &= \tfrac{1}{4}s'_{mn} \text{ if both } m \text{ and } n \text{ are } 4, 5, \text{ or } 6 \end{aligned} \tag{13-11}$$

Notice that the s_{mn} are the coefficients that reduce directly to s_{ijkl} by the scheme of Eq. (2-16). Finally, the matrix in (13-10) is expanded into a 9×9 form, as in the case of the $\{c\}$ matrices:

$$\begin{bmatrix} \epsilon_{11} \\ \epsilon_{22} \\ \epsilon_{33} \\ \epsilon_{23} \\ \epsilon_{31} \\ \epsilon_{12} \\ \epsilon_{32} \\ \epsilon_{13} \\ \epsilon_{21} \end{bmatrix} = \begin{bmatrix} s_{11} & s_{12} & s_{13} & s_{14} & s_{15} & s_{16} & s_{14} & s_{15} & s_{16} \\ s_{12} & s_{22} & s_{23} & s_{24} & s_{25} & s_{26} & s_{24} & s_{25} & s_{26} \\ s_{13} & s_{23} & s_{33} & s_{34} & s_{35} & s_{36} & s_{34} & s_{35} & s_{36} \\ s_{14} & s_{24} & s_{34} & s_{44} & s_{45} & s_{46} & s_{44} & s_{45} & s_{46} \\ s_{15} & s_{25} & s_{35} & s_{45} & s_{55} & s_{56} & s_{45} & s_{55} & s_{56} \\ s_{16} & s_{26} & s_{36} & s_{46} & s_{56} & s_{66} & s_{46} & s_{56} & s_{66} \\ s_{14} & s_{24} & s_{34} & s_{44} & s_{45} & s_{46} & s_{44} & s_{45} & s_{46} \\ s_{15} & s_{25} & s_{35} & s_{45} & s_{55} & s_{56} & s_{45} & s_{55} & s_{56} \\ s_{16} & s_{26} & s_{36} & s_{46} & s_{56} & s_{66} & s_{46} & s_{56} & s_{66} \end{bmatrix} \begin{bmatrix} \sigma_{11} \\ \sigma_{22} \\ \sigma_{33} \\ \sigma_{23} \\ \sigma_{31} \\ \sigma_{12} \\ \sigma_{32} \\ \sigma_{13} \\ \sigma_{21} \end{bmatrix} \tag{13-12}$$

[7]There is no general agreement on which s_{mn} should be primed. Our notation is the opposite of that in J. F. Nye, "Physical Properties of Crystals," Oxford University Press, Fair Lawn, N.J., 1957, and H. B. Huntington, *Solid State Phys.*, **7**: 213 (1958), but the same as that in W. A. Wooster, "Crystal Physics," Cambridge University Press, New York, 1938. Thus one should be careful to identify the proper s_{mn} in using values from the literature.

Equation (13-12) is the form that should be used in transformations and in the definition of invariants.

In analogy to Eq. (13-1), the trace is invariant and in analogy to Eq. (13-2), a second invariant is found by considering the invariance of strain energy under a hydrostatic stress $\sigma_{11}=\sigma_{22}=\sigma_{33}=\frac{1}{3}\sigma_{ii}$ and $\sigma_{ij}=0$ for $i\neq j$, giving

$$I_1 = s_{ijij} = \text{const} \tag{13-13}$$

$$I_2 = s_{iijj} = \text{const} \tag{13-14}$$

Combining these expressions with Eqs. (2-51) and (13-12), one finds the averages over the compliances:

$$\begin{aligned} \frac{1}{E_R} &= \tfrac{1}{15}\left(2s_{ijij}+s_{iijj}\right) \\ \frac{\nu_R}{E_R} &= \tfrac{1}{15}\left(s_{ijij}-2s_{iijj}\right) \\ \frac{1}{\mu_R} &= \tfrac{1}{15}\left(6s_{ijij}-2s_{iijj}\right) \end{aligned} \tag{13-15}$$

Here the subscript R indicates the Reuss averages. With no subscript, Voigt averages [Eq. (13-5)] are understood.

Comparison of Average Elastic Constants

The product of an $\{s\}$ matrix and the corresponding $\{c\}$ matrix is of interest in comparing the above averages. The result cannot be an ordinary 9×9 unit matrix, for that would imply that $\{c\}$ or $\{s\}$ had simple inverses. Yet Eqs. (2-5) and (13-6) indicate that this product must resemble a unit matrix in its properties. The result of the multiplication must be

$$\{s\}\{c\}=\begin{bmatrix} 1 & 0 & 0 & 0 & 0 & 0 & 0 & 0 & 0 \\ 0 & 1 & 0 & 0 & 0 & 0 & 0 & 0 & 0 \\ 0 & 0 & 1 & 0 & 0 & 0 & 0 & 0 & 0 \\ 0 & 0 & 0 & \frac{1}{2} & 0 & 0 & \frac{1}{2} & 0 & 0 \\ 0 & 0 & 0 & 0 & \frac{1}{2} & 0 & 0 & \frac{1}{2} & 0 \\ 0 & 0 & 0 & 0 & 0 & \frac{1}{2} & 0 & 0 & \frac{1}{2} \\ 0 & 0 & 0 & \frac{1}{2} & 0 & 0 & \frac{1}{2} & 0 & 0 \\ 0 & 0 & 0 & 0 & \frac{1}{2} & 0 & 0 & \frac{1}{2} & 0 \\ 0 & 0 & 0 & 0 & 0 & \frac{1}{2} & 0 & 0 & \frac{1}{2} \end{bmatrix}\equiv\{1\} \tag{13-16}$$

This expression can be considered as a definition of a special unit matrix, limited to the 9×9 representation of elastic coefficients. Whenever it is multiplied into a complete matrix with the symmetries peculiar to elasticity, *it has the same effect as a proper unit matrix*. For example, with the above definition for $\{1\}$,

$$\{\epsilon_{ij}\}=\{1\}\{\epsilon_{ij}\} \tag{13-17}$$

Here the $\{\epsilon_{ij}\}$ are column matrices. Equation (13-17) is valid because

$$\epsilon_{23}=\tfrac{1}{2}\epsilon_{23}+\tfrac{1}{2}\epsilon_{32},\ldots \tag{13-18}$$

Also, multiplying Eq. (2-5) by $\{s\}$ gives Eq. (13-6), as it must:

$$\{s\}\sigma_{ij}=\{s\}\{c\}\epsilon_{kl}=\{1\}\epsilon_{kl}=\epsilon_{kl} \tag{13-19}$$

Thus with the above formal definition for $\{1\}$, and inverses, one can use all the usual rules of matrix algebra. In this scheme $\{c\}^{-1}\equiv\{s\}$, by definition,[8] because $\{s\}\{c\}=1$.

Now consider the difference between the Voigt and Reuss averages. Denoting the isotropic average of $\{c\}$ as $\{\bar{c}\}$, one can write

$$\{c\}=\{\bar{c}\}+\{\delta c\}=\left(\{1\}+\{\delta c\}\{\bar{c}\}^{-1}\right)\{\bar{c}\} \tag{13-20}$$

where the nonzero elements in the matrix $\{\delta c\}$ are a measure of the anisotropy. The inverse of Eq. (13-20) is

$$\{c\}^{-1}=\{\bar{c}\}^{-1}\left(\{1\}+\{\delta c\}\{\bar{c}\}^{-1}\right)^{-1} \tag{13-21}$$

Expansion in terms of the matrix $\{\delta c\}$ yields

$$\left(\{1\}+\{\delta c\}\{\bar{c}\}^{-1}\right)^{-1}=\{1\}-\{\delta c\}\{\bar{c}\}^{-1}+\{\delta c\}\{\bar{c}\}^{-1}\{\delta c\}\{\bar{c}\}^{-1} \tag{13-22}$$

so that

$$\{c\}^{-1}=\{\bar{c}\}^{-1}-\{\bar{c}\}^{-1}\{\delta c\}\{\bar{c}\}^{-1}+\{\bar{c}\}^{-1}\{\delta c\}\{\bar{c}\}^{-1}\{\delta c\}\{\bar{c}\}^{-1} \tag{13-23}$$

[8] D. S. Lieberman and S. Zirinsky [*Acta Cryst.*, **9**: 431 (1956)] present an alternative development that enables one to obtain inverses and transform $\{c\}$ or $\{s\}$. They contract Eq. (13-12), for example, to a reduced 6×6 matrix by the procedure

$$s'_{ij}=s_{ij}+s_{i+3,j+3} \qquad j=4,5,6$$
$$s'_{ij}=s_{ij} \qquad j=1,2,3$$

The remainder of their procedure is analogous to that presented here.

Averaging Eq. (13-23) over all orientation and noting that $\{\bar{c}\}^{-1}$ is invariant to rotation and that $\{\delta\bar{c}\}$ equals zero by definition, one finds

$$\overline{\{c\}^{-1}}=\{\bar{c}\}^{-1}+\overline{\{\bar{c}\}^{-1}\{\delta c\}\{\bar{c}\}^{-1}\{\delta c\}\{\bar{c}\}^{-1}} \tag{13-24}$$

The last term is of second order in the anisotropy, so to the *first order* in the anisotropy,

$$\overline{\{c\}^{-1}}=\{\bar{c}\}^{-1} \quad \text{or} \quad \{\bar{s}\}=\{\bar{c}\}^{-1} \tag{13-25}$$

That is, the Voigt and Reuss averages *agree* to the first order in the anisotropy.

Cubic Crystals

The expansion of Eq. (2-20) into a 9×9 representation reveals that for cubic crystals

$$c_{ijij}=3c_{11}+6c_{44} \quad \text{and} \quad c_{iijj}=3c_{11}+6c_{12}$$

Thus, by Eq. (13-5),

$$\begin{aligned} \lambda+2\mu &= c_{11}+\tfrac{2}{5}H \\ \mu &= c_{44}-\tfrac{1}{5}H \\ \lambda &= c_{12}-\tfrac{1}{5}H \end{aligned} \tag{13-26}$$

where H is the anisotropy factor

$$H=2c_{44}+c_{12}-c_{11} \tag{13-27}$$

Obviously, the condition for isotropy is $H=0$, in agreement with Eq. (2-44).

The equivalent averages from Eq. (13-15) are

$$\frac{1}{E_R}=s'_{11}-\tfrac{2}{5}J \qquad \frac{1}{\mu_R}=s'_{44}+\tfrac{4}{5}J \qquad \frac{\nu_R}{E_R}=-s'_{12}-\tfrac{1}{5}J \tag{13-28}$$

where

$$J=\left(s'_{11}-s'_{12}-\tfrac{1}{2}s'_{44}\right) \tag{13-29}$$

Values of elastic constants, elastic compliances, average elastic coefficients, H, and J are listed for selected cubic materials in Appendix 1. As shown there, the deviation between the Voigt and Reuss averages varies up to 100 percent, with a mean deviation of about 20 percent. Thus the difference is significant, and the appropriate average must be used in each case. In dealing with local interactions of dislocations, the Voigt averages are preferred.

Exercise 13-1. By inverting the 6×6 $\{c\}$ matrix, show that

$$s'_{11}=\frac{c_{11}+c_{12}}{c_{11}^2+c_{11}c_{12}-2c_{12}^2}$$

$$s'_{12}=\frac{-c_{12}}{c_{11}^2+c_{11}c_{12}-2c_{12}^2} \qquad (13\text{-}30)$$

$$s'_{44}=\frac{1}{c_{44}}$$

Exercise 13-2. Verify Eq. (13-16) for cubic crystals by direct matrix multiplication, using the result of Eq. (13-30).

Exercise 13-3. Compare $\bar{s}'_{44}=1/\mu_R$ and $\bar{c}_{44}=\mu$, Eqs. (13-26) and (13-28). Show that

$$\bar{s}'_{44}=\frac{1}{\bar{c}_{44}}+\frac{3H^2}{25c_{44}^2(c_{11}-c_{12})}\left(1-\frac{H}{5c_{44}}\right)^{-1}$$

This result demonstrates that the average results differ to the second order in the anisotropy, H.

Transformation of Orthogonal Axes

Straight dislocations in anisotropic media are most easily analyzed if one of the reference axes is oriented parallel to the dislocation line. In cubic crystals dislocations often lie in directions other than the cube axes, to which the cited elastic constants, Eq. (2-20), are referred. Also, in other crystal systems one often wishes to transform the elastic-constant matrix, referred to orthogonal axes of equal measure, to another orientation. As an example of such transformations, we present the important case of transforming to a system where two axes are in the (111) plane, with one in a $[\bar{1}01]$ direction. Other cases are left as exercises. A tabulation of transformed cubic elastic constants in a number of high-symmetry coordinate systems is presented by Waterman.[9]

[9] P. C. Waterman, *Phys. Rev.*, **113**: 1240 (1959).

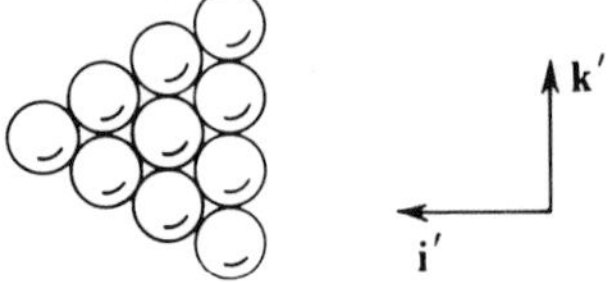

FIGURE 13.2. Projection normal to the (111) plane in an fcc crystal, showing axes $\mathbf{i}'=(1/\sqrt{6})[12\bar{1}]$, and $\mathbf{k}'=(1/\sqrt{2})[\bar{1}01]$. $\mathbf{j}'=(1/\sqrt{3})[111]$ points out of the page.

The constants c_{ijkl} refer to axes parallel to the cube axes and the coefficients c'_{ijkl} to a system with the $\mathbf{k}'$ axis along $[\bar{1}01]$ and the $\mathbf{i}'$ axis also in the (111) plane, as indicated in Fig. 13-2. In an fcc crystal the $\mathbf{k}'$ axis would be along a close-packed direction. Thus, as is evident in Fig. 13-2,

$$\mathbf{k}'=\frac{1}{\sqrt{2}}(-\mathbf{i}+\mathbf{k}) \qquad \mathbf{j}'=\frac{1}{\sqrt{3}}(\mathbf{i}+\mathbf{j}+\mathbf{k})$$

$$\mathbf{i}'=\mathbf{j}'\times\mathbf{k}'=\frac{1}{\sqrt{6}}(\mathbf{i}-2\mathbf{j}+\mathbf{k}) \tag{13-31}$$

Thus the transformation matrix $\{T_{ij}\}$ [Eq. (2-23)] is

$$\{T_{ij}\}=\frac{1}{\sqrt{6}}\begin{bmatrix} 1 & -2 & 1 \\ \sqrt{2} & \sqrt{2} & \sqrt{2} \\ -\sqrt{3} & 0 & \sqrt{3} \end{bmatrix} \tag{13-32}$$

Defining matrix elements for the 9×9 transformation matrix by

$$Q_{mnkl}=T_{km}T_{ln} \tag{13-33}$$

one can rewrite the transformation equation (2-43) in the form

$$c'_{ijkl}=Q_{ghij}c_{ghmn}Q_{mnkl} \tag{13-34}$$

With the introduction of the transpose $\tilde{Q}_{ijgh}$ of Q_{ghij}, Eq. (13-34) takes the form

$$c'_{ijkl}=\tilde{Q}_{ijgh}c_{ghmn}Q_{mnkl} \tag{13-35}$$

where the order of indices now conforms for standard matrix multiplication.

A direct, but tedious, calculation from Eqs. (13-32) and (13-33) yields

$$6\{Q\}=$$

$$\begin{bmatrix} 1 & 2 & 3 & -\sqrt{6} & -\sqrt{3} & \sqrt{2} & -\sqrt{6} & -\sqrt{3} & \sqrt{2} \\ 4 & 2 & 0 & 0 & 0 & -2\sqrt{2} & 0 & 0 & -2\sqrt{2} \\ 1 & 2 & 3 & \sqrt{6} & \sqrt{3} & \sqrt{2} & \sqrt{6} & \sqrt{3} & \sqrt{2} \\ -2 & 2 & 0 & \sqrt{6} & 0 & -2\sqrt{2} & 0 & -2\sqrt{3} & \sqrt{2} \\ 1 & 2 & -3 & -\sqrt{6} & \sqrt{3} & \sqrt{2} & \sqrt{6} & -\sqrt{3} & \sqrt{2} \\ -2 & 2 & 0 & 0 & 2\sqrt{3} & \sqrt{2} & -\sqrt{6} & 0 & -2\sqrt{2} \\ -2 & 2 & 0 & 0 & -2\sqrt{3} & \sqrt{2} & \sqrt{6} & 0 & -2\sqrt{2} \\ 1 & 2 & -3 & \sqrt{6} & -\sqrt{3} & \sqrt{2} & -\sqrt{6} & \sqrt{3} & \sqrt{2} \\ -2 & 2 & 0 & -\sqrt{6} & 0 & -2\sqrt{2} & 0 & 2\sqrt{3} & \sqrt{2} \end{bmatrix} \tag{13-36}$$

The expanded 9×9 representation of the matrix in Eq. (2-20) is

$$\{c\}=\begin{bmatrix} c_{11} & c_{12} & c_{12} & 0 & 0 & 0 & 0 & 0 & 0 \\ c_{12} & c_{11} & c_{12} & 0 & 0 & 0 & 0 & 0 & 0 \\ c_{12} & c_{12} & c_{11} & 0 & 0 & 0 & 0 & 0 & 0 \\ 0 & 0 & 0 & c_{44} & 0 & 0 & c_{44} & 0 & 0 \\ 0 & 0 & 0 & 0 & c_{44} & 0 & 0 & c_{44} & 0 \\ 0 & 0 & 0 & 0 & 0 & c_{44} & 0 & 0 & c_{44} \\ 0 & 0 & 0 & c_{44} & 0 & 0 & c_{44} & 0 & 0 \\ 0 & 0 & 0 & 0 & c_{44} & 0 & 0 & c_{44} & 0 \\ 0 & 0 & 0 & 0 & 0 & c_{44} & 0 & 0 & c_{44} \end{bmatrix} \tag{13-37}$$

The new elastic constants are now obtained by multiplying out the right-hand side of the expression

$$\{c'\}=\{\tilde{Q}\}\{c\}\{Q\} \tag{13-38}$$

where $\{\tilde{Q}\}$ is the transpose of (13-36).

Symmetry arguments alone show that c' must be of the form[10]

$$\{c'\}=\begin{bmatrix} c'_{11} & c'_{12} & c'_{13} & 0 & 0 & c'_{16} & 0 & 0 & c'_{16} \\ c'_{12} & c'_{22} & c'_{12} & 0 & 0 & 0 & 0 & 0 & 0 \\ c'_{13} & c'_{12} & c'_{11} & 0 & 0 & -c'_{16} & 0 & 0 & -c'_{16} \\ 0 & 0 & 0 & c'_{44} & -c'_{16} & 0 & c'_{44} & -c'_{16} & 0 \\ 0 & 0 & 0 & -c'_{16} & c'_{55} & 0 & -c'_{16} & c'_{55} & 0 \\ c'_{16} & 0 & -c'_{16} & 0 & 0 & c'_{44} & 0 & 0 & c'_{44} \\ 0 & 0 & 0 & c'_{44} & -c'_{16} & 0 & c'_{44} & -c'_{16} & 0 \\ 0 & 0 & 0 & -c'_{16} & c'_{55} & 0 & -c'_{16} & c'_{55} & 0 \\ c'_{16} & 0 & -c'_{16} & 0 & 0 & c'_{44} & 0 & 0 & c'_{44} \end{bmatrix} \tag{13-39}$$

where

$$2c'_{55}=c'_{11}-c'_{13} \tag{13-40}$$

This is the general form for the case where the y axis is a threefold symmetry axis, and a reflection plane is perpendicular to the z axis. Indeed, the actual multiplication of the matrices in Eq. (13-38) yields the above form and gives for the nonvanishing elements,

$$\begin{aligned} c'_{11}&=c_{11}+\tfrac{1}{2}H & c'_{44}&=c_{44}-\tfrac{1}{3}H \\ c'_{12}&=c_{12}-\tfrac{1}{3}H & c'_{55}&=c_{44}-\tfrac{1}{6}H \\ c'_{13}&=c_{12}-\tfrac{1}{6}H & c'_{16}&=\frac{\sqrt{2}}{6}H \\ &c'_{22}=c_{11}+\tfrac{2}{3}H \end{aligned} \tag{13-41}$$

with H given by Eq. (13-27). It is helpful to perform this multiplication as a check on an understanding of transformation. In calculations of this kind, one should predict the general form of the new elastic-coefficient matrix from the symmetry of the crystal relative to the new reference axes. This procedure provides a check for the matrix form resulting from multiplication.

[10]See J. F. Nye, "Physical Properties of Crystals," Oxford University Press, Fair Lawn, N.J., 1957, pp. 116 and 137, for a discussion of the symmetry method. We emphasize, however, that when the symmetry of elastic coefficients is considered, one can always *add* inversion symmetry to the real symmetry of the crystal, because the relation between stress and strain generally is invariant to inversion. Thus a twofold axis has the same consequence in the symmetry of the c_{ijkl} as a reflection plane. Often, understanding of this fact is tacitly assumed in the literature.

Exercise 13-4. Given a set of axes **i**, **j**, and **k** along the cube axes of a cubic crystal, show that for a rotation of 45° about the **i** axis, the elastic constants are given by the matrix

$$\{c'\}=\begin{bmatrix} c'_{11} & c'_{12} & c'_{12} & 0 & 0 & 0 \\ c'_{12} & c'_{22} & c'_{23} & 0 & 0 & 0 \\ c'_{12} & c'_{23} & c'_{22} & 0 & 0 & 0 \\ 0 & 0 & 0 & c'_{44} & 0 & 0 \\ 0 & 0 & 0 & 0 & c'_{55} & 0 \\ 0 & 0 & 0 & 0 & 0 & c'_{55} \end{bmatrix} \tag{13-42}$$

where

$$\begin{aligned} c'_{11} &= c_{11} & c'_{22} &= c_{11}+\tfrac{1}{2}H \\ c'_{12} &= c_{12} & c'_{23} &= c_{12}-\tfrac{1}{2}H \\ c'_{55} &= c_{44} & c'_{44} &= c_{44}-\tfrac{1}{2}H \end{aligned} \tag{13-43}$$

Exercise 13-5. Show that for cubic crystals, and with unprimed coefficients referring to coordinates fixed on the cube axes, Eq. (13-34) can be simplified to

$$c'_{ijkl}=c_{ijkl}-H\left(\sum_n T_{in}T_{jn}T_{kn}T_{ln}-\delta_{ij}\delta_{kl}\delta_{ik}\right)$$

Calculate Eqs. (13-41) and (13-43) from this formula. *Hint*: For cubic crystals, the relation analogous to Eq. (2-47) is

$$c_{ijkl}=c_{44}(\delta_{ik}\delta_{jl}+\delta_{il}\delta_{jk})+c_{12}\delta_{ij}\delta_{kl}-H\delta_{ij}\delta_{kl}\delta_{ik}$$

Hexagonal Crystals

The elastic-constant matrices for other crystal systems can be simplified by symmetry arguments similar to those presented for cubic crystals in Chap. 2. For the details of such work the reader is referred to Nye[10] and Wooster.[11] For hexagonal crystals, with the third axis perpendicular to the basal plane, the result is

$$\{c\}=\begin{bmatrix} c_{11} & c_{12} & c_{13} & 0 & 0 & 0 \\ c_{12} & c_{11} & c_{13} & 0 & 0 & 0 \\ c_{13} & c_{13} & c_{33} & 0 & 0 & 0 \\ 0 & 0 & 0 & c_{44} & 0 & 0 \\ 0 & 0 & 0 & 0 & c_{44} & 0 \\ 0 & 0 & 0 & 0 & 0 & c_{66} \end{bmatrix} \tag{13-44}$$

[11] W. A. Wooster, "Crystal Physics," Cambridge University Press, New York, 1938.

where

$$2c_{66} = c_{11} - c_{12} \tag{13-45}$$

Equation (13-45) [*cf.* Eq. (2-44)] indicates that hexagonal crystals are *isotropic in the basal plane*; $\{c\}$ is invariant to a rotation about the **k** axis.

Exercise 13-6. Rotate the above axes by an arbitrary angle α about the **k** axis. Prove that $\{c\}$ is invariant.

For hexagonal crystals

$$\begin{aligned} c_{ijij} &= 2c_{11} + c_{33} + 4c_{44} + 2c_{66} \\ &= 3c_{11} - c_{12} + c_{33} + 4c_{44} \\ c_{iijj} &= 2c_{11} + c_{33} + 2c_{12} + 4c_{13} \end{aligned}$$

Thus Eq. (13-5) gives the result

$$\begin{aligned} \lambda + 2\mu &= \tfrac{1}{15}(8c_{11} + 4c_{13} + 3c_{33} + 8c_{44}) \\ \mu &= \tfrac{1}{30}(7c_{11} - 5c_{12} + 2c_{33} + 12c_{44} - 4c_{13}) \\ \lambda &= \tfrac{1}{15}(c_{11} + c_{33} + 5c_{12} + 8c_{13} - 4c_{44}) \end{aligned} \tag{13-46}$$

Recall that this approximate result represents the Voigt average over all orientations; we shall show later that for the special case of a dislocation in the basal plane and with its Burgers vector in that plane, different average coefficients give *exact* results for the stress in the slip plane for *any* orientation of the dislocation in the slip plane.

The above examples of the cubic and hexagonal systems suffice to illustrate the methods for transforming elastic-coefficient matrices and for obtaining average coefficients. Other crystal systems can be treated by analogy but are not developed here.

13-3. STRAIGHT DISLOCATIONS IN ANISOTROPIC MEDIA

The classical sextic anisotropic elasticity theory of straight dislocations was developed by Eshelby et al.[12] and was extended by Foreman[13] to include anisotropy in energy calculations. This theory has been elaborated by Stroh,[14]

[12] J. D. Eshelby, W. T. Read, and W. Shockley, *Acta Met.*, **1**: 251 (1953).

[13] A. J. E. Foreman, *Acta Met.*, **3**: 322 (1955).

[14] A. N. Stroh, *Phil. Mag.*, **3**: 625 (1958).

Spence,[15] Chou,[16] and others. Somewhat different but equivalent methods were advanced by Seeger and Schoeck.[17] Here we shall follow the method of Eshelby et al. in outline and expand upon it somewhat.

The development is rather lengthy; therefore, the essential steps in the derivation are summarized briefly at the end of the section. The sextic theory is adequate for dislocations in high symmetry directions, for which analytical solutions are possible.

Displacements

Let us turn first to the problem of the elastic displacements about a straight dislocation. As a brief recapitulation of the results of Chap. 2, the basic equations are

$$\sigma_{ij} = c_{ijkl}\epsilon_{kl} \tag{13-47}$$

where

$$\epsilon_{kl} = \frac{1}{2}\left(\frac{\partial u_k}{\partial x_l} + \frac{\partial u_l}{\partial x_k}\right) \tag{13-48}$$

The σ_{ij} must fulfill the equations of equilibrium

$$\frac{\partial \sigma_{ij}}{\partial x_j} = 0 \qquad i = 1,2,3 \tag{13-49}$$

As in the work of Eshelby et al.,[12] the coordinate axes are oriented with the x_3 axis (z axis) parallel with the dislocation line, and the c_{ijkl} are referred to this system. The physical boundary conditions on the dislocation indicate that if images effects are excluded, the displacements, strains, and stresses are all independent of x_3. This allows use of the following convention to simplify the elasticity problem: The pair of indices α and β are understood to be one of the numbers 1 and 2, and the indices i, j, etc., are understood to be 1, 2, or 3, as usual. The Greek letters suffice in expressions involving $\partial/\partial x_i$, since $\partial/\partial x_3 \equiv 0$ for all cases considered. Thus, for our purposes, Eq. (13-49) can be written as

$$\frac{\partial \sigma_{i\alpha}}{\partial x_\alpha} = 0 \qquad i = 1,2,3 \tag{13-50}$$

Also, Eqs. (13-47) and (13-48) combine to yield

$$\sigma_{i\alpha} = c_{i\alpha k\beta}\frac{\partial u_k}{\partial x_\beta} \tag{13-51}$$

when the general symmetries in c_{ijkl} are invoked.

[15] G. B. Spence, *J. Appl. Phys.*, **33**: 729 (1962).

[16] Y. T. Chou, *J. Appl. Phys.*, **33**: 2747 (1962).

[17] A. Seeger and G. Schoeck, *Acta Met.*, **1**: 519 (1953).

The solution for the displacements u_k then follows directly. Insertion of Eq. (13-51) into (13-50) gives the result

$$c_{i\alpha k\beta}\frac{\partial^2 u_k}{\partial x_\alpha \partial x_\beta}=0 \qquad i=1,2,3 \tag{13-52}$$

Equation (13-52) represents three simultaneous equations $i=1$, 2, and 3 for the three functions u_k. The partial-differential equations in Eq. (13-52) are standard forms,[18] with solutions of the type

$$u_k = A_k f(\eta) \tag{13-53}$$

where

$$\eta = x_1 + px_2 \tag{13-54}$$

and where p and A_k are constants. Substitution into Eq. (13-52) yields

$$\left[c_{i1k1}+(c_{i1k2}+c_{i2k1})p+c_{i2k2}p^2\right]A_k\frac{\partial^2 f}{\partial \eta^2}=0 \tag{13-55}$$

or, after cancellation of the common factor $\partial^2 f/\partial\eta^2$,

$$a_{ik}A_k=0 \tag{13-56}$$

where

$$a_{ik}=c_{i1k1}+(c_{i1k2}+c_{i2k1})p+c_{i2k2}p^2 \tag{13-57}$$

The linear equations (13-56) have a nonzero solution for the A_k only when the determinant of the matrix $\{a_{ik}\}$ is zero,

$$|\{a_{ik}\}|=0 \tag{13-58}$$

Equation (13-58) is then a sixth-order equation in p with roots

$$p_n \qquad n=1,2,3,4,5,6$$

For each root p_n there is a set $A_k(n)$ that satisfies Eq. (13-56). The $A_k(n)$

[18] The analogy with the well-known fact that the wave equation

$$\frac{\partial^2\phi}{\partial x^2}=\frac{1}{c^2}\frac{\partial^2\phi}{\partial t^2}$$

has a general solution of the form

$$\phi=f(x+ct)+g(x-ct)$$

is obvious.

appear in ratios given by subdeterminants. Arbitrarily[19] setting $A_3(n)=1$, one finds the result

$$A_1(n)=\begin{vmatrix} a_{12}(n) & a_{13}(n) \\ a_{22}(n) & a_{23}(n) \end{vmatrix} \div \begin{vmatrix} a_{11}(n) & a_{12}(n) \\ a_{21}(n) & a_{22}(n) \end{vmatrix}$$

$$A_2(n)=-\begin{vmatrix} a_{11}(n) & a_{13}(n) \\ a_{21}(n) & a_{23}(n) \end{vmatrix} \div \begin{vmatrix} a_{11}(n) & a_{12}(n) \\ a_{21}(n) & a_{22}(n) \end{vmatrix} \tag{13-59}$$

$$A_3(n)=1$$

Eshelby et al. have shown[20] that the roots p_n of Eq. (13-58) are never real; since the coefficients in the polynomial are real, the roots must occur in *pairs* of complex conjugates. Let p_4 be the complex conjugate of p_1, etc.,

$$p_4=p_1^* \qquad p_5=p_2^* \qquad p_6=p_3^* \tag{13-60}$$

Then, also,

$$\begin{aligned} A_k(4)&=A_k(1)^* & \eta_4&=\eta_1^* \\ A_k(5)&=A_k(2)^* & \eta_5&=\eta_2^* \\ A_k(6)&=A_k(3)^* & \eta_6&=\eta_3^* \end{aligned} \tag{13-61}$$

and for an analytic function $f(\eta)$,

$$A_k(1)f(\eta_1)=[A_k(4)f(\eta_4)]^*,\ldots \tag{13-62}$$

Since the u_k [Eq. (13-53)] must be real, they must be composed of pairs of complex conjugates such as

$$\tfrac{1}{2}[A_k(1)f(\eta_1)+A_k(4)f(\eta_4)]=\mathrm{Re}[A_k(1)f(\eta_1)] \tag{13-63}$$

where Re means "the real part of." Thus one need only consider the three roots p_1, p_2, and p_3 and the corresponding sets $A_k(1)$, $A_k(2)$, and $A_k(3)$. Equation (13-53) becomes

$$u_k=\mathrm{Re}\left[\sum_{n=1}^{3} A_k(n)f_n(\eta_n)\right] \tag{13-64}$$

The f_n are three arbitrary analytic functions.

[19] The stresses and displacements depend only on the *relative* values of the A_k. In some cases A_3 could be zero and hence could not be normalized to 1. In such cases one must arbitrarily put $A_1=1$ or $A_2=1$, whichever is convenient. The general procedure is unchanged.

[20] The proof follows directly from the fact that the strain energy must be positive.

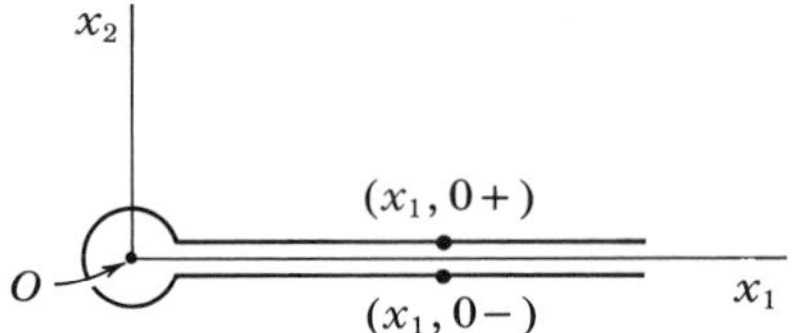

FIGURE 13-3. A cut encircling the dislocation lying along the x_3 axis. ξ points into the page.

The displacements u_k are multiple-valued in a region containing the dislocation. Figure 13-3 shows a cut encircling the dislocation core. The boundary conditions require u_k to be analytic and single-valued outside the cut and to have a discontinuity across the cut

$$\Delta u_k = \begin{cases} u_k(x_1,0^+) - u_k(x_1,0^-) = b_k & x_1 > 0 \\ 0 & x_1 < 0 \end{cases} \tag{13-65}$$

where b_k is the kth component of the Burgers vector **b**.

Consider now the form of $f(\eta)$. The stresses involve the functions $df(\eta)/d\eta$; since the stresses are single-valued and continuous except at the origin, the functions $df(\eta)/d\eta$ are also so. Thus the most general function for $f(\eta)$ is of the form[21]

$$f(\eta) = -\frac{D}{2\pi i}\ln\eta + \sum_{n=-\infty}^{\infty} a_n \eta^n \tag{13-66}$$

If the imaginary part of p is positive, then a counterclockwise path around the dislocation corresponds to a counterclockwise path in the η plane (Fig. 13-4). For a negative imaginary part of p, a counterclockwise path around the dislocation would be a clockwise path in the η plane. Thus across the cut

$$\Delta \ln\eta = \mp 2\pi i \tag{13-67}$$

where the minus sign is used when the imaginary part of p is positive, and vice versa. Thus, from Eq. (13-66),

$$\Delta f = \pm D \tag{13-68}$$

since all of the power-series terms are continuous across the cut. The power-series terms are not characteristic of the straight dislocation in an infinite

[21] Because $df(\eta)/d\eta$ is single-valued and continuous except at the origin, it can be expressed in a Laurent series (I.S. Sokolnikoff and R. M. Redheffer, "Mathematics of Physics and Modern Engineering," McGraw-Hill, New York, 1966, p. 560),

$$\frac{df}{d\eta} = \sum_{n=-\infty}^{\infty} a'_n \eta^n$$

Integration of this relation gives Eq. (13-66). The logarithmic term arises from integration of the term $a'_{-1}\eta^{-1}$.

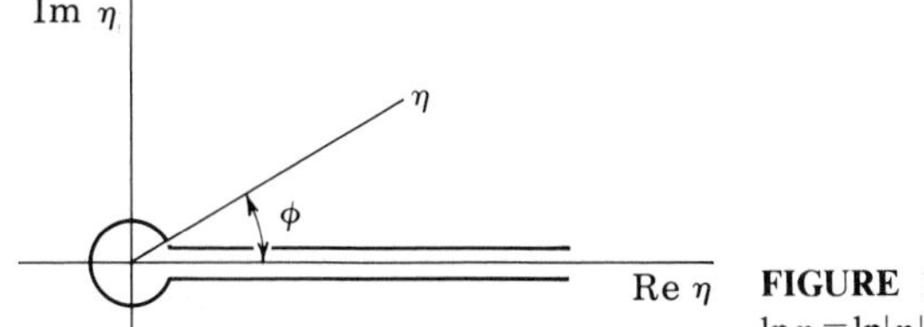

FIGURE 13-4. A cut in the complex plane, $\ln\eta=\ln|\eta|+i\phi$.

medium, but pertain only to image effects, $n\geqslant 1$, or to core anharmonicity effects, $n\leqslant -1$. The handling of these terms is analogous to the developments is Chap. 3 for the isotropic case. For brevity, these terms are not treated here; only the logarithmic term is retained in the subsequent discussion. Thus with

$$f_n(\eta_n)=-\frac{D(n)}{2\pi i}\ln\eta_n \tag{13-69}$$

Eqs. (13-64), (13-65), and (13-68) combine to the following expression for $D(n)$:

$$\operatorname{Re}\left[\sum_{n=1}^{3}\pm A_k(n)D(n)\right]=b_k \qquad k=1,2,3 \tag{13-70}$$

The plus sign is used when the imaginary part of p is positive and the minus sign when it is negative. Equation (13-70) is a set of three equations containing six unknowns, the real and imaginary parts of the $D(n)$ for $n=1$, 2, and 3. The three other equations needed to determine the $D(n)$ are provided by the requirement that there be no *net* force on the dislocation core.

The net force per unit length acting on the surface of a rod parallel to the dislocation and containing it (Fig. 13-5) is

$$\frac{\mathbf{F}}{L}=\oint\boldsymbol{\sigma}\cdot(d\mathbf{l}\times\mathbf{e}_3) \tag{13-71}$$

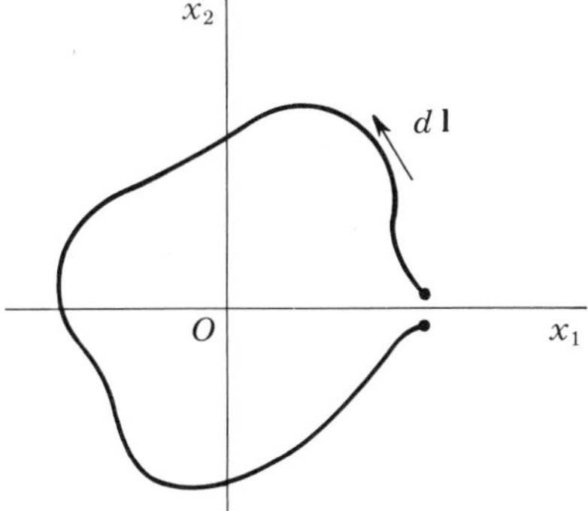

FIGURE 13-5. Path of integration along the surface, parallel to the dislocation, of a rod containing the dislocation along the x_3 axis.

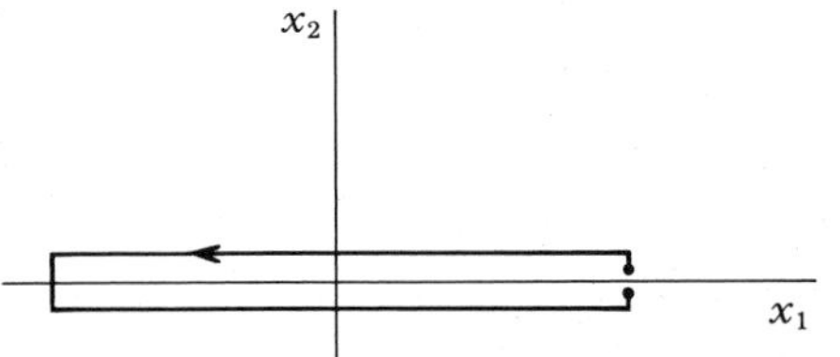

FIGURE 13-6. A deformed path of integration of Fig. 13-5.

where $\boldsymbol{\sigma}$ is the stress tensor and $d\mathbf{l}=\mathbf{e}_1\,dx_1+\mathbf{e}_2\,dx_2$. Thus, setting $F=0$, the condition for vanishing net force is

$$\oint(\sigma_{i1}\,dx_2-\sigma_{i2}\,dx_1)=0 \qquad i=1,2,3 \tag{13-72}$$

Insertion of Eq. (13-51) into this expression yields

$$\oint\left(c_{i1k1}\frac{\partial u_k}{\partial x_1}+c_{i1k2}\frac{\partial u_k}{\partial x_2}\right)dx_2-\oint\left(c_{i2k1}\frac{\partial u_k}{\partial x_1}+c_{i2k2}\frac{\partial u_k}{\partial x_2}\right)dx_1=0 \tag{13-73}$$

Since the equilibrium conditions are already satisfied in the material outside the core, one can deform the path of integration. A convenient path is a rectangle infinitely narrow in the limit, containing the dislocation (see Fig. 13-6). The integral over dx_2 then vanishes. After insertion of Eq. (13-64), Eq. (13-73) assumes the form[22]

$$\mathrm{Re}\left[\oint(c_{i2k1}+c_{i2k2}\,p_n)A_k(n)\frac{\partial f_n}{\partial x_1}dx_1\right]=0 \tag{13-74}$$

For the above path of integration, the integrals can be directly evaluated,

$$\oint\frac{\partial f_n}{\partial x_1}dx_1=-\Delta f_n=\mp D(n) \tag{13-75}$$

The result of Eq. (13-74) is then

$$\mathrm{Re}\left[\sum_{n=1}^{3}\pm B_{i2k}(n)A_k(n)D(n)\right]=0 \qquad i=1,2,3 \tag{13-76}$$

[22]Use is made of the relation

$$\frac{\partial f_n}{\partial x_2}=p_n\frac{\partial f_n}{\partial x_1}$$

where

$$B_{ijk}(n) = c_{ijk1} + c_{ijk2}\,p_n \tag{13-77}$$

The six equations (13-70) and (13-76) completely determine $D(n)$. All parameters needed in the expression for u_k, which becomes

$$u_k = \mathrm{Re}\left[\frac{-1}{2\pi i}\sum_{n=1}^{3} A_k(n)D(n)\ln\eta_n\right] \tag{13-78}$$

are now determined in principle.

Stresses and Energies

The stresses are directly obtained from Eqs. (13-47), (13-48), and (13-78) as

$$\sigma_{ij} = \mathrm{Re}\left[\frac{-1}{2\pi i}\sum_{n=1}^{3} B_{ijk}A_k(n)D(n)\eta_n^{-1}\right] \tag{13-79}$$

The energies can be determined from the work done in forming the dislocation by cutting and displacing the plane $x_2 = 0$. In this plane $\eta_n = x_1$, so that

$$\sigma_{ij} = \frac{-1}{2\pi x_1}\mathrm{Im}\left[\sum_{n=1}^{3} B_{ijk}A_k(n)D(n)\right] \tag{13-80}$$

where Im indicates "imaginary part of." The work done on the plane of the cut between $x_1 = r_0$ and $x_1 = R$ per unit length of dislocation as it is formed is

$$\frac{W}{L} = -\frac{1}{2}\int_{r_0}^{R}\sigma_{i2}b_i\,dx_1 \tag{13-81}$$

or, with the insertion of Eq. (13-80),

$$\frac{W}{L} = \frac{b_i}{4\pi}\ln\frac{R}{r_0}\mathrm{Im}\left[\sum_{n=1}^{3} B_{i2k}A_k(n)D(n)\right] \tag{13-82}$$

It is convenient to define a factor K,

$$Kb^2 = b_i\,\mathrm{Im}\left[\sum_{n=1}^{3} B_{i2k}A_k(n)D(n)\right] \tag{13-83}$$

so that Eq. (13-82) simply becomes

$$\frac{W}{L}=\frac{Kb^2}{4\pi}\ln\frac{R}{r_0} \tag{13-84}$$

This expression is from Foreman.[23] A comparison with Eqs. (3-13) and (3-52) reveals the utility of the definition of K; since the logarithmic term is quite insensitive to the exact values of R and r_0, and since the core energy usually is only a small fraction of the energy, the main effect of anisotropy on the dislocation energy is found in the factor K. K replaces the isotropic parameters μ for a screw dislocation, $\mu/(1-\nu)$ for an edge dislocation, etc. K is called the energy *coefficient*. In terms of K, the prelogarithmic energy *factor E* introduced in Chap. 5 is $E=Kb^2/4\pi$.

This completes the formal discussion. Some applications are considered next, but first we shall summarize the procedure developed in this section.

Summary of Method

The dislocation is along the x_3 axis. The sense ξ points in the negative x_3 direction.

1. Solve the equation

$$|\{a_{ik}\}|=0 \tag{13-85}$$

where

$$a_{ik}=c_{i1k1}+(c_{i1k2}+c_{i2k1})p+c_{i2k2}p^2 \tag{13-86}$$

Pick out three roots p_1, p_2, and p_3, one from each pair of complex conjugates.

2. Solve the equations

$$a_{ik}(n)A_k(n)=0 \tag{13-87}$$

for $A_k(n)$ for each of the three values of n. Arbitrarily put $A_3=1$ when $A_3\neq 0$. When $A_3=0$, normalize either A_1 or A_2 to 1.

3. With the $A_k(n)$ thus determined, solve the equations

$$\text{Re}\left[\sum_{n=1}^{3}\pm A_k(n)D(n)\right]=b_k \tag{13-88}$$

[23] A. J. E. Foreman, *Acta Met.*, **3**: 322 (1955).

and

$$\text{Re}\left[\sum_{n=1}^{3} \pm B_{i2k}(n)A_k(n)D(n)\right]=0 \tag{13-89}$$

for the six unknowns, the real and imaginary parts of the $D(n)$. Here

$$B_{ijk}(n)=c_{ijk1}+c_{ijk2}\,p_n \tag{13-90}$$

In Eqs. (13-88) and (13-89), the plus and minus signs are used when the imaginary part of p_n is positive or negative, respectively.

4. With the $A_k(n)$ and the $D(n)$ now determined, the displacements are completely determined from the formula

$$u_k=\text{Re}\left[-\frac{1}{2\pi i}\sum_{n=1}^{3} A_k(n)D(n)\ln\eta_n\right] \tag{13-91}$$

where

$$\eta_n=x_1+p_n x_2$$

5. The dislocation stresses are

$$\sigma_{ij}=\text{Re}\left[-\frac{1}{2\pi i}\sum_{n=1}^{3} B_{ijk}(n)A_k(n)D(n)\eta_n{}^{-1}\right] \tag{13-92}$$

6. The energy coefficient K in

$$\frac{W}{L}=\frac{Kb^2}{4\pi}\ln\frac{R}{r_0}$$

is

$$Kb^2=b_i\text{Im}\left[\sum_{n=1}^{3} B_{i2k}(n)A_k(n)D(n)\right] \tag{13-93}$$

13-4. SIMPLE SOLUTIONS TO ANISOTROPIC EQUATIONS

The general problem of the straight dislocation involves the solution of a sixth-order polynomial in p, which cannot be done in an elementary way.[24]

[24] This is true despite the general symmetry of the elastic constants in Eq. (2-6); see A. K. Head, *J. Elast.* **9**: 9 (1979).

Only polynomials of the fourth order or lower can be solved generally by algebraic methods. Computer programs for the numerical solution of the general sixth-order case have been worked out by Teutonico.[25] Here we shall restrict our discussion to the analytical solutions which occur for special orientations of the dislocation. For these orientations the problem simplifies enormously.

Separation into Screw and Edge Components

For an orientation such that the screw component gives rise only to displacements u_3 and the edge component only to displacements u_1 and u_2, we say that the solution consists of a pure screw and a pure edge part. Instead of using the full machinery of sixth-order polynomials and 3×3 matrices, one can solve the problem by decomposition into a screw part involving a second-order polynomial in p and 1×1 matrices, and an edge part involving a fourth-order polynomial and 2×2 matrices.

Consider the x_3 axis as parallel to the dislocation line. The c_{ijkl} are the elastic constants referred to this dislocation-oriented coordinate system $x_1x_2x_3$. The basic requirement for a simple solution of the pure edge-pure screw type is that the x_1x_2 plane be a reflection plane, or equivalently for elasticity, that the x_3 axis be an axis of evenfold symmetry; this equivalence of an axis of evenfold symmetry and a reflection plane follows from the inversion invariance of the relations between stress and strain. For such a case symmetry alone[26] indicates that for a pure edge dislocation, $u_3=0$ and $\sigma_{\alpha 3}=0$ when $\alpha=1$ and 2. Similarly for a pure screw dislocation, $u_1=u_2=0$ and $\sigma_{\alpha\beta}=0$ for $\alpha,\beta=1$ and 2. The 3×3 matrix is necessary in the general case because three displacements, u_1, u_2, and u_3, must be considered simultaneously. Thus by these symmetry considerations, one must expect that the problem can be reduced to a 2×2 matrix for the edge and a 1×1 matrix for the screw, as described above. Furthermore, the energy is expected to decompose into a pure screw part and a pure edge part; no cross term is present because the stresses associated with, say, the edge component do no work in a slip process in which the screw component is formed. As shown below, the results of the formal mathematical development agree with these expectations.

With x_1x_2 a symmetry plane, the following elastic constants equal zero,

$$
\begin{aligned}
c_{\alpha\beta\gamma 3}&=0 \qquad \alpha,\beta,\gamma=1,2\\
c_{\alpha 333}&=0 \qquad \alpha=1,2
\end{aligned}
\tag{13-94}
$$

[25] L. J. Teutonico, *Phys. Rev.*, **124**: 1039 (1961); *Phys. Rev.*, **127**: 413 (1962); *Acta Met.*, **11**: 1283 (1963). Teutonico includes in his work the case of moving dislocations.

[26] As an example, consider Fig. 13-7. Suppose that a shear stress σ_{23} exists at A, as shown there. Because x_1x_2 is a reflection plane, and because the dislocation itself has a twofold symmetry axis parallel to x_2, a 180° rotation about y must not change the situation, so that a shear stress $-\sigma_{23}$ must then exist at A'. Since points A and A' have identical surroundings on an infinite dislocation line, clearly, $\sigma_{23}=0$ at both points.

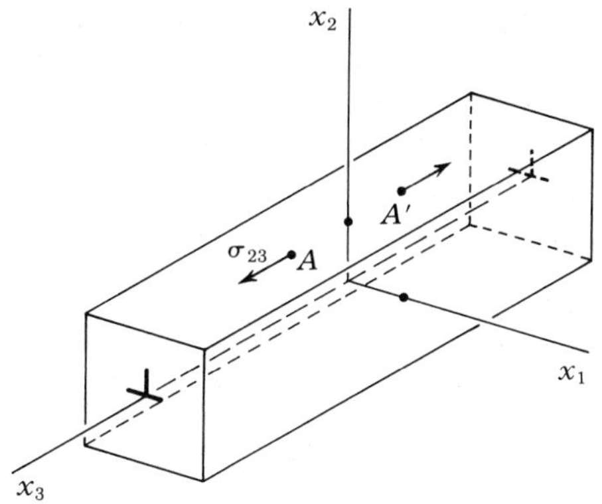

FIGURE 13-7. A crystal with an edge dislocation parallel to x_3.

Indeed, with Eq. (13-94), Eq. (13-87) reduces to two sets of equations, one set for A_1 and A_2 and one for A_3:

$$A_1(c'_{11}+2c'_{16}p_n+c'_{66}p_n^2)+A_2[c'_{16}+(c'_{12}+c'_{66})p_n+c'_{26}p_n^2]=0$$

$$A_1[c'_{16}+(c'_{12}+c'_{66})p_n+c'_{26}p_n^2]+A_2(c'_{66}+2c'_{26}p_n+c'_{22}p_n^2)=0 \quad (13\text{-}95)$$

with $p_n = p_2, p_3, p_5$, and p_6, and

$$A_3(c'_{55}+2c'_{45}p_n+c'_{44}p_n^2)=0 \quad (13\text{-}96)$$

with $p_n = p_1$ and p_4.

When Eq. (13-94) is fulfilled, the array of elastic constants is

$$\{c'\}=\begin{bmatrix} c'_{11} & c'_{12} & c'_{13} & 0 & 0 & c'_{16} \\ c'_{12} & c'_{22} & c'_{23} & 0 & 0 & c'_{26} \\ c'_{13} & c'_{23} & c'_{33} & 0 & 0 & c'_{36} \\ 0 & 0 & 0 & c'_{44} & c'_{45} & 0 \\ 0 & 0 & 0 & c'_{45} & c'_{55} & 0 \\ c'_{16} & c'_{26} & c'_{36} & 0 & 0 & c'_{66} \end{bmatrix} \quad (13\text{-}97)$$

This is the most general array that is compatible with a reflection plane normal to the x_3 axis. Foreman[27] points out that the most general matrix allowing separation into a pure edge and pure screw part is of the form of Eq. (13-97) with the constants c'_{34} and c'_{35} added.

Physical reasoning shows that these added constants do not affect the splitting into screw and edge parts. The constant c'_{34} can only produce stresses $\sigma_{yz}=c'_{34}\epsilon_{zz}$ and $\sigma_{zz}=2c'_{34}\epsilon_{yz}$. ϵ_{zz} and ϵ_{yz} are both zero for the pure edge (plane strain) so that there are no cross terms present. ϵ_{zz} is also zero for the screw.

[27]A. J. E. Foreman, *Acta Met.*, **3**: 322 (1955).

Shears ϵ_{yz} are present for the screw, giving rise to stresses σ_{zz}. But these stresses do not enter the force balance in the two-dimensional problem in x and y and do not change the displacements. Thus the displacements are unaffected by c'_{34}. Only in calculating the stresses must the extra term σ_{zz} be included. Similar reasoning holds for c'_{35}. Also, for the pure edge, neither displacements nor stresses are changed by the extra coefficients. One can readily check that c'_{34} and c'_{35} do not enter Eqs. (13-95) and (13-96).

Physical arguments identical with those given above show that the displacements are determined entirely by the reduced matrix

$$\{c'\}=\begin{bmatrix} c'_{11} & c'_{12} & 0 & 0 & 0 & c'_{16} \\ c'_{12} & c'_{22} & 0 & 0 & 0 & c'_{26} \\ 0 & 0 & 0 & 0 & 0 & 0 \\ 0 & 0 & 0 & c'_{44} & c'_{45} & 0 \\ 0 & 0 & 0 & c'_{45} & c'_{55} & 0 \\ c'_{16} & c'_{26} & 0 & 0 & 0 & c'_{66} \end{bmatrix} \tag{13-98}$$

Of course, this expression applies only for a dislocation lying along the x_3 axis.

We shall now write down explicitly the expressions for the displacements and stresses for dislocations lying along the x_3, or z, axis, and with the elastic constants given by Eq. (13-97), with c'_{34} and c'_{35} added for the screw dislocation and with c'_{16} and $c'_{26}=0$ for the edge dislocation. The corresponding energy factors are also given.

Pure Edge Dislocation

Let $\mathbf{b}=(b_x,0,0)$ and let $\boldsymbol{\xi}$ point in the negative z direction. For the edge dislocation, $A_3(2)=A_3(3)=0$ and $D(1)=0$. These equalities are compatible with Eq. (13-88), since

$$\sum_{n=1}^{3} \pm A_3(n)D(n)=0$$

Also, they are compatible with Eq. (13-89), since

$$\sum_{n=1}^{3} \pm B_{32k}(n)A_k(n)D(n)=0$$

because for $k=3$, $A_3(n)D(n)=0$, as above, while for $k=1$ and $k=2$, $B_{321}=B_{322}=0$, by Eq. (13-94). Thus only four relations, Eqs. (13-88) and (13-89), are left to determine the four unknowns, the real and imaginary parts of $D(2)$ and $D(3)$, as should be the case.

In the case of the edge dislocation, Eq. (13-95), one must solve a general quartic equation in p. This can be done algebraically, but the results are too lengthy to warrant inclusion here. The problem is much simpler if

$$c'_{16}=c'_{26}=0 \tag{13-99}$$

giving a quadratic equation in p^2. Equation (13-99) is true if either (1) one axis is an axis of sixfold symmetry, or (2) each axis is an axis of evenfold symmetry (when two axes have evenfold symmetry, the third does also). Conditions other than Eq. (13-99) might yield a second order equation in p^2, but these would not arise from symmetry. A number of important cases correspond to the simplifying condition (13-99), so the results are given explicitly.

With Eqs. (13-95) and (13-99), the determinantal equation (13-85) for p is

$$|\{a_{ij}\}|=\begin{vmatrix} c'_{11}+c'_{66}p^2 & (c'_{12}+c'_{66})p & 0 \\ (c'_{12}+c'_{66})p & c'_{66}+c'_{22}p^2 & 0 \\ 0 & 0 & c'_{55}+2c'_{45}p+c'_{44}p^2 \end{vmatrix}=0 \tag{13-100}$$

which reduces to the product of two subdeterminants,

$$\left(c'_{55}+2c'_{45}p+c'_{44}p^2\right)\left[\left(c'_{11}+c'_{66}p^2\right)\left(c'_{66}+c'_{22}p^2\right)-\left(c'_{12}+c'_{66}\right)^2p^2\right]=0 \tag{13-101}$$

the latter of which relates to the edge and reduces to

$$c'_{22}c'_{66}p^4+\left(c'_{11}c'_{22}-2c'_{12}c'_{66}-c'^{2}_{12}\right)p^2+c'_{11}c'_{66}=0 \tag{13-102}$$

The roots are of the form

$$p_2=\lambda e^{i\phi} \qquad p_5=\lambda e^{-i\phi}$$

$$p_3=-\lambda e^{i\phi} \qquad p_6=-\lambda e^{-i\phi} \tag{12-103}$$

A term-by-term comparison of the equation

$$\left(p-\lambda e^{i\phi}\right)\left(p+\lambda e^{i\phi}\right)\left(p-\lambda e^{-i\phi}\right)\left(p+\lambda e^{-i\phi}\right)=0 \tag{13-104}$$

with Eq. (13-102) shows that λ and ϕ are given by

$$\lambda=\left(\frac{c'_{11}}{c'_{22}}\right)^{1/4} \tag{13-105}$$

and

$$\phi=\tfrac{1}{2}\cos^{-1}\frac{c'^{2}_{12}+2c'_{12}c'_{66}-\bar{c}'^{2}_{11}}{2\bar{c}'_{11}c'_{66}} \tag{13-106}$$

where

$$\bar{c}'_{11} = (c'_{11}c'_{22})^{1/2} \tag{13-107}$$

In the isotropic case $\lambda = 1$ and $\phi = \pi/2$. Equation (13-106) gives real angles ϕ when

$$c'^2_{12} + 2c'_{12}c'_{66} - \bar{c}'^2_{11} > -2\bar{c}'_{11}c'_{66}$$

or, since $c'_{12} + \bar{c}'_{11}$ must be greater than zero for crystal stability, when

$$2c'_{66} + c'_{12} - \bar{c}'_{11} > 0 \tag{13-108}$$

We first treat the case that ϕ and λ are real. We can then select p_2 and p_3 as the fundamentally different roots, and $p_5 = p_2^*$ and $p_6 = p_3^*$ as the complex conjugates. Since $A_3(2) = A_3(3) = 0$, we arbitrarily put $A_1 = 1$ and obtain from Eq. (13-87) the result

$$A_1(2) \equiv 1 \qquad A_2(2) = -\lambda \frac{c'_{66}e^{i\phi} + \bar{c}'_{11}e^{-i\phi}}{c'_{12} + c'_{66}} \equiv A$$

$$A_1(3) \equiv 1 \qquad A_2(3) = -A_2(2) = -A \tag{13-109}$$

The conditional equations (13-88) and (13-89) now become, respectively,

$$\mathrm{Re}[D(2) - D(3)] = b_x$$

$$\mathrm{Re}\{A[D(2) + D(3)]\} = 0 \tag{13-110}$$

and

$$\mathrm{Re}\{(A + \lambda e^{i\phi})[D(2) + D(3)]\} = 0$$

$$\mathrm{Re}\{(c'_{12} + c'_{22}A\lambda e^{i\phi})[D(2) - D(3)]\} = 0 \tag{13-111}$$

The solution

$$D(2) = -D(3) = D \tag{13-112}$$

automatically satisfies two of the above equations, leaving the following conditions on D:

$$\mathrm{Re}\, D = \tfrac{1}{2}b_x$$

$$\mathrm{Re}[D(c'^2_{12} + c'_{12}c'_{66} - \bar{c}'^2_{11} - \bar{c}'_{11}c'_{66}e^{2i\phi})] = 0 \tag{13-113}$$

The solution of Eq. (13-113) is

$$D=\frac{-ib_x}{2\bar{c}'_{11}c'_{66}\sin 2\phi}\left(c'^2_{12}+c'_{12}c'_{66}-\bar{c}'^2_{11}-\bar{c}'_{11}c'_{66}e^{-2i\phi}\right)$$

or, with the use of Eq. (13-106), the more convenient form

$$D=\frac{ib_x}{2\bar{c}'_{11}\sin 2\phi}\left(c'_{12}-\bar{c}'_{11}e^{2i\phi}\right) \tag{13-114}$$

All the quantities needed for the solutions in Eqs. (13-91) to (13-93) are now given explicitly in Eqs. (13-90), (13-103), (13-109), and (13-114). The remainder of the calculation simply involves straightforward but tedious algebra, which is omitted here for brevity. The results for a Burgers vector component b_y can be obtained from the above results by rotation of the axes by $\pi/2$. Combining the results, one finds the general solution for an edge dislocation $\mathbf{b}=(b_x, b_y, 0)$.

The results are as follows. Two other convenient abbreviations are

$$q^2=x^2+2xy\lambda\cos\phi+y^2\lambda^2$$

and

$$t^2=x^2-2xy\lambda\cos\phi+y^2\lambda^2 \tag{13-115}$$

In terms of these quantities and the constants in Eqs. (13-105) to (13-107), the displacements given by Eq. (13-91) are

$$\begin{aligned} u_x=&-\frac{b_x}{4\pi}\left(\tan^{-1}\frac{2xy\lambda\sin\phi}{x^2-\lambda^2y^2}+\frac{\bar{c}'^2_{11}-c'^2_{12}}{2\bar{c}'_{11}c'_{66}\sin 2\phi}\ln\frac{q}{t}\right)\\ &-\frac{b_y}{4\pi\lambda\bar{c}'_{11}\sin 2\phi}\left[(\bar{c}'_{11}-c'_{12})\cos\phi\ln qt\right.\\ &\left.-(\bar{c}'_{11}+c'_{12})\sin\phi\tan^{-1}\frac{x^2\sin 2\phi}{\lambda^2y^2-x^2\cos 2\phi}\right] \end{aligned} \tag{13-116}$$

$$\begin{aligned} u_y=&\frac{\lambda b_x}{4\pi\bar{c}'_{11}\sin 2\phi}\left[(\bar{c}'_{11}-c'_{12})\cos\phi\ln qt\right.\\ &\left.-(\bar{c}'_{11}+c'_{12})\sin\phi\tan^{-1}\frac{y^2\lambda^2\sin 2\phi}{x^2-\lambda^2y^2\cos 2\phi}\right]\\ &-\frac{b_y}{4\pi}\left(\tan^{-1}\frac{2xy\lambda\sin\phi}{x^2-\lambda^2y^2}-\frac{\bar{c}'^2_{11}-c'^2_{12}}{2\bar{c}'_{11}c'_{66}\sin 2\phi}\ln\frac{q}{t}\right) \end{aligned}$$

Note that the arctan functions change by 4π on one revolution about the origin in xy space. As a check, the formulas for the isotropic case can be obtained by limiting procedures. Care is required in some cases; for example,

$$\frac{1}{\sin 2\phi}\ln\frac{x^2+2xy\lambda\cos\phi+\lambda^2y^2}{x^2-2xy\lambda\cos\phi+\lambda^2y^2}\to\frac{2xy}{x^2+y^2}\qquad \text{as } \phi\to\frac{\pi}{2}\quad\text{and}\quad\lambda\to 1$$

For the stresses, Eq. (13-92) yields

$$\begin{aligned}\sigma_{ij}=&-\frac{b_x\lambda(c'_{12}-\bar{c}'_{11})}{4\pi q^2t^2\bar{c}'_{11}c'_{66}\sin\phi}\Big\{c'_{ij11}\big[(\bar{c}'_{11}+c'_{12}+c'_{66})x^2y+\lambda^2c'_{66}y^3\big]\\&\quad-c'_{ij12}(c'_{12}+\bar{c}'_{11})(x^3-\lambda^2xy^2)\\&\quad-\frac{c'_{ij22}}{c'_{22}}\big[(c'^2_{12}+\bar{c}'_{11}c'_{12}+2c'_{12}c'_{66}+\bar{c}'_{11}c'_{66})x^2y-\bar{c}'_{11}c'_{66}\lambda^2y^3\big]\Big\}\\&+\frac{b_y\lambda(c'_{12}-\bar{c}'_{11})}{4\pi q^2t^2\bar{c}'_{11}c'_{66}\sin\phi}\Big\{c'_{ij22}\big[(\bar{c}'_{11}+c'_{12}+c'_{66})\lambda^2xy^2+c'_{66}x^3\big]\\&\quad-c'_{ij12}(c'_{12}+\bar{c}'_{11})(\lambda^2y^3-x^2y)\\&\quad-\frac{c'_{ij11}}{c'_{11}}\big[(c'^2_{12}+\bar{c}'_{11}c'_{12}+2c'_{12}c'_{66}+\bar{c}'_{11}c'_{66})\lambda^2xy^2-\bar{c}'_{11}c'_{66}x^3\big]\Big\}\end{aligned}\tag{13-117}$$

Other forms for the equations for σ_{ij} are possible. When only the stresses σ_{xx}, σ_{xy}, and σ_{yy} are required, as is often the case, the following expressions are quite convenient:

$$\begin{aligned}\sigma_{xx}=&\frac{Mb_x}{2\pi\rho^4c'_{22}}\left\{\big[(\bar{c}'_{11}-c'_{12})(\bar{c}'_{11}+c'_{12}+2c'_{66})-\bar{c}'_{11}c'_{66}\big]x^2y+\frac{\bar{c}'^2_{11}c'_{66}}{c'_{22}}y^3\right\}\\&+\frac{Mb_yc'_{66}}{2\pi\rho^4}\left(\frac{\bar{c}'_{11}}{c'_{22}}xy^2-x^3\right)\\\sigma_{yy}=&\frac{Mb_xc'_{66}}{2\pi\rho^4}\left(-x^2y+\frac{\bar{c}'_{11}}{c'_{22}}y^3\right)\\&-\frac{Mb_y}{2\pi\rho^4\bar{c}'_{11}}\{\big[(\bar{c}'_{11}-c'_{12})(\bar{c}'_{11}+c'_{12}+2c'_{66})-\bar{c}'_{11}c'_{66}\big]xy^2+c'_{22}c'_{66}x^3\}\\\sigma_{xy}=&\frac{Mb_xc'_{66}}{2\pi\rho^4}\left(-x^3+\frac{\bar{c}'_{11}}{c'_{22}}xy^2\right)+\frac{Mb_yc'_{66}}{2\pi\rho^4}\left(-x^2y+\frac{\bar{c}'_{11}}{c'_{22}}y^3\right)\end{aligned}\tag{13-118}$$

Here

$$M=(\bar{c}'_{11}+c'_{12})\left[\frac{\bar{c}'_{11}-c'_{12}}{c'_{22}c'_{66}(\bar{c}'_{11}+c'_{12}+2c'_{66})}\right]^{1/2}$$

and (13-119)

$$\rho^4=\left(x^2+\frac{\bar{c}'_{11}}{c'_{22}}y^2\right)^2+\frac{(\bar{c}'_{11}+c'_{12})(\bar{c}'_{11}-c'_{12}-2c'_{66})}{c'_{22}c'_{66}}x^2y^2$$

The energy coefficients K_{e_x} and K_{e_y}, associated with the respective b_x and b_y components, are given by Eq. (13-93) as

$$K_{e_x}=(\bar{c}'_{11}+c'_{12})\left[\frac{c'_{66}(\bar{c}'_{11}-c'_{12})}{(\bar{c}'_{11}+c'_{12}+2c'_{66})c'_{22}}\right]^{1/2}$$

$$K_{e_y}=(\bar{c}'_{11}+c'_{12})\left[\frac{c'_{66}(\bar{c}'_{11}-c'_{12})}{(\bar{c}'_{11}+c'_{12}+2c'_{66})c'_{11}}\right]^{1/2} \tag{13-120}$$

There is no cross term in the energy between the two edge components because $c'_{16}=c'_{26}=0$; that is, in Eq. (13-93) there are no terms containing the product b_xb_y.

When, instead of Eq. (13-108)

$$2c'_{66}+c'_{12}-\bar{c}'_{11}<0 \tag{13-121}$$

as it is for some bcc metals,[28] Eq. (13-106) gives complex angles and ϕ is of the form

$$\phi=\frac{\pi}{2}-i\delta \tag{13-122}$$

where δ is real and determined by the expression

$$\cosh 2\delta=\frac{\bar{c}'^2_{11}-c'^2_{12}-2c'_{12}c'_{66}}{2\bar{c}'_{11}c'_{66}} \tag{13-123}$$

In this case $p_2=p_3^*$, $p_5=p_6^*$, and p_2 and p_5 must be chosen as the fundamentally distinct roots,[29] making the intermediate calculations somewhat different from Eq. (13-110). However, Eqs. (13-116) and (13-117) can still be used

[28] K. Malén, *Scripta Met.*, **2**: 223 (1962).

[29] Alternatively, p_2 and p_6 could have been used as the fundamentally distinct roots in both cases. However, two sets of formulas for the cases of real and complex ϕ would still be required.

directly to yield correct, real results if the following substitutions are made.[28]

$$\cos 2\phi = -\cosh 2\delta, \qquad \sin 2\phi = i\sinh 2\delta$$

$$\cos\phi = i\sinh\delta, \qquad \sin\phi = \cosh\delta$$

$$q^2 = x^2 + y^2\lambda^2 + i2xy\lambda\sinh\delta$$

$$t^2 = x^2 + y^2\lambda^2 - i2xy\lambda\sinh\delta \tag{13-124}$$

$$\ln\left(\frac{q}{t}\right) = i\tan^{-1}\frac{2xy\lambda\sinh\delta}{x^2+\lambda^2y^2}$$

$$\tan^{-1} ia = \frac{1}{2i}\ln\frac{1-a}{1+a}$$

where a is real. Equations (13-118) and (13-120) apply directly.

Pure Screw Dislocation

$\mathbf{b}=(0,0,b_z)$ and ξ points in the negative z direction. Here, $A_1(1)=A_2(1)=0$, and one can choose $A_3(1)=1$. By arguments analogous to those for the edge dislocation, one can put $D(2)=D(3)=0$. The matrix for $\{c'\}$ is given by Eq. (13-98) with no restrictions on c'_{16} or c'_{26}. The second-order polynomial in Eq. (13-101), which applies to the screw even with $c'_{16}, c'_{26} \neq 0$, has one root,

$$p_1 = \frac{1}{c'_{44}}\left[-c'_{45} + i\left(c'_{44}c'_{55} - c'^2_{45}\right)^{1/2}\right] \tag{13-125}$$

With the imaginary part positive, Eqs. (13-88) and (13-89) yield the conditions to determine $D(1)$,

$$\mathrm{Re}[D(1)] = b_z$$

$$\mathrm{Re}[(c'_{45} + c'_{44}p_1)D(1)] = \mathrm{Re}\left[i\left(c'_{44}c'_{55} - c'^2_{45}\right)^{1/2} D\right] = 0 \tag{13-126}$$

Since the factor before $D(1)$ in the second equality is purely imaginary, D must be real, so that the solution to Eq. (13-126) is

$$D(1) = b_z \tag{13-127}$$

Inserting the above factors into Eq. (13-91), one finds that the displacement is

$$u_z = -\frac{b_z}{2\pi}\tan^{-1}\frac{\left(c'_{44}c'_{55} - c'^2_{45}\right)^{1/2} y}{c'_{44}x - c'_{45}y} \tag{13-128}$$

The stresses are

$$\sigma_{xz} = -\frac{b_z}{2\pi}\left(c'_{44}c'_{55} - c'^2_{45}\right)^{1/2} \frac{c'_{45}x - c'_{55}y}{c'_{44}x^2 - 2c'_{45}xy + c'_{55}y^2}$$

$$\sigma_{yz} = -\frac{b_z}{2\pi}\left(c'_{44}c'_{55} - c'^2_{45}\right)^{1/2} \frac{c'_{44}x - c'_{45}y}{c'_{44}x^2 - 2c'_{45}xy + c'_{55}y^2} \qquad (13\text{-}129)$$

$$\sigma_{zz} = -\frac{b_z}{2\pi}\left(c'_{44}c'_{55} - c'^2_{45}\right)^{1/2} \frac{c'_{34}x - c'_{35}y}{c'_{44}x^2 - 2c'_{45}xy + c'_{55}y^2}$$

The energy coefficient is

$$K_s = \left(c'_{44}c'_{55} - c'^2_{45}\right)^{1/2} \qquad (13\text{-}130)$$

Third-Order Solutions

Explicit solutions for the p_n are possible for somewhat lower symmetries.[30] With either the x or y axis a twofold axis, or equivalently, with either in a reflection plane containing the z axis, the equation for p is third order in p^2 and can be solved algebraically. With the roots p_n determined, one is still left with the problem of solving Eqs. (13-87) for $A_k(n)$ and, worse, Eqs. (13-88) and (13-89) for the six components in $D(n)$. In the case when edge and screw components do not separate these solutions are very cumbersome if one mechanically follows the recipe in Sec. 13-3. Instead of proceeding in such a manner, it is generally more convenient to directly use the advanced theory of Sec. 13-7. Hence we do not consider the third-order solutions in detail.

13-5. APPLICATIONS TO HEXAGONAL CRYSTALS

Dislocations in the Basal Plane

In hexagonal crystals the most prominent dislocations are those in the basal plane, as discussed in Chap. 9. Let xz be the basal plane, and let the dislocation coincide with the z axis, as shown in Fig. 13-8. With this orientation of the coordinate system, Eq. (13-35) yields the elastic-constant matrix

$$\{c'\} = \begin{bmatrix} c'_{11} & c'_{12} & c'_{13} & 0 & 0 & 0 \\ c'_{12} & c'_{22} & c'_{12} & 0 & 0 & 0 \\ c'_{13} & c'_{12} & c'_{11} & 0 & 0 & 0 \\ 0 & 0 & 0 & c'_{44} & 0 & 0 \\ 0 & 0 & 0 & 0 & c'_{55} & 0 \\ 0 & 0 & 0 & 0 & 0 & c'_{44} \end{bmatrix} \qquad (13\text{-}131)$$

[30]A. K. Head, *Phys. Stat. Solidi*, **6**: 461 (1964).

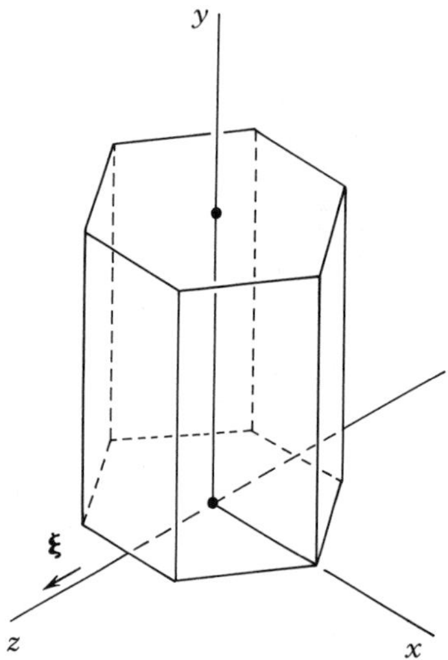

FIGURE 13-8. Coordinates for a dislocation lying in the basal plane.

where $2c'_{55}=c'_{11}-c'_{13}$. Relative to the standard elastic constants c_{ij}, which are referred to a coordinate system with z perpendicular to the basal plane [Eq. (13-44)], the constants are

$$c'_{11}=c_{11} \qquad c'_{44}=c_{44} \qquad c'_{22}=c_{33} \qquad c'_{12}=c_{13} \qquad c'_{13}=c_{12} \qquad c'_{55}=c_{66}$$

Evidently, Eq. (13-131) has the form of Eq. (13-98), with $c'_{16}=c'_{26}=0$. Thus the results for this case are given by Eqs. (13-116) to (13-130), with the additional simplifications that $c'_{66}=c'_{44}$, and that $c'_{34}=c'_{35}=c'_{45}=0$.

For an edge dislocation with its Burgers vector in the basal plane, the shear stress in this slip plane is given by Eq. (13-118) as

$$\sigma_{xy}(x,0)=\frac{-K_{e_x}b_x}{2\pi x} \tag{13-132}$$

where K_{e_x} is the energy coefficient, Eq. (13-120),

$$K_{e_x}=(\bar{c}'_{11}+c'_{12})\left[\frac{c'_{44}(\bar{c}'_{11}-c'_{12})}{c'_{22}(\bar{c}'_{11}+c'_{12}+2c'_{44})}\right]^{1/2} \tag{13-133}$$

These results are self-consistent, since $\sigma_{xy}(x,0)$ is the only stress component which contributes to the work when the dislocation is formed by deforming a cut in the slip plane.

For a screw dislocation in the basal plane, Eq. (13-129) gives the stresses. The shear stress in the basal plane is

$$\sigma_{yz}(x,0)=\frac{-K_s b_z}{2\pi x} \tag{13-134}$$

with the energy factor

$$K_s = (c'_{44} c'_{55})^{1/2} \tag{13-135}$$

Exercise 13-7. Determine the stress components and energy factor for the case of a dislocation lying in the basal plane but with its Burgers vector normal to the basal plane. Show that the energy of a dislocation in the basal plane can be decomposed into additive screw and edge parts,

$$\frac{W}{L} = \frac{1}{4\pi}\left(K_s b_s^2 + K_{e_x} b_x^2 + K_{e_y} b_y^2\right) \ln \frac{R}{r_0} \tag{13-136}$$

where

$$K_{e_y} = (\bar{c}'_{11} + c'_{12}) \left[\frac{c'_{44}(\bar{c}'_{11} - c'_{12})}{c'_{11}(\bar{c}'_{11} + c'_{12} + 2c'_{44})} \right]^{1/2} \tag{13-137}$$

Isotropy in the Basal Plane

The results of Exercise 13-6 prove that hexagonal crystals are isotropic in the basal plane. This isotropicity allows one to take anisotropy into account for *curved dislocations* for some special cases in hexagonal crystals. Let us revert to the standard notation, with the z axis normal to the basal plane and the elastic constants as in Eq. (13-44). Comparison with results expressed in terms of Eq. (13-131) is simple.

Consider a small circular loop in the basal plane and with Burgers vector in the basal plane. Let the Burgers vector be parallel with the **i** axis, $\mathbf{b} = (b_x, 0, 0)$, as illustrated in Fig. 13-9. Consider the shear stresses σ_{zx} and σ_{zy} in the plane a distance R from the infinitesimal loop. By symmetry, $\sigma_{zz} = 0$ in the plane of the

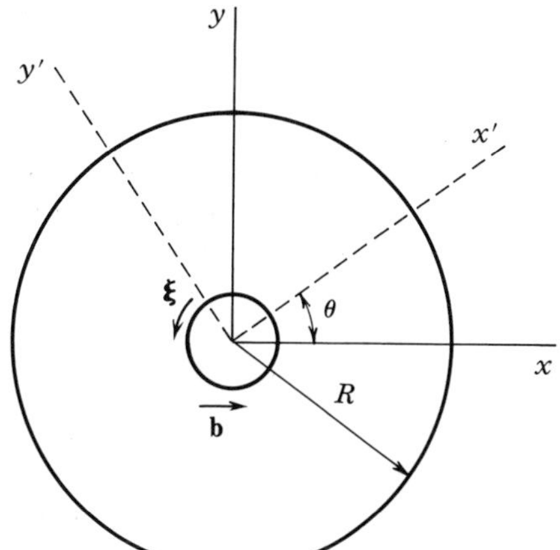

FIGURE 13-9. An infinitesimal loop of dislocation in the basal plane xy.

loop. By symmetry, $\sigma_{zy}=0$ when $\theta=0, \pi/2, \pi$, or $3\pi/2$. $\sigma_{zx}(R,\pi)=-\sigma_{zx}(R,0)$, and for $\theta=\pm\pi/2$, σ_{zx} will be some fraction α of its value for $\theta=0$ for the same distance,

$$\sigma_{zx}\left(R,\frac{\pi}{2}\right)=\alpha\sigma_{zx}(R,0) \tag{13-138}$$

The stresses at some arbitrary angle θ can be found as follows. Referred to a coordinate system $x'y'$ rotated an angle θ with respect to the system xy, the Burgers vector has the components

$$b'_x=b_x\cos\theta \qquad b'_y=-b_x\sin\theta \tag{13-139}$$

According to the superposition principle, the loop can be considered as the superposition of two loops, one with Burgers vector b'_x and one with Burgers vector b'_y. Thus, taking the isotropy in the basal plane into account, one finds from Eq. (13-138) that

$$\sigma'_{zx}(R,\theta)=\sigma_{zx}(R,0)\cos\theta$$

and (13-140)

$$\sigma'_{zy}(R,\theta)=-\alpha\sigma_{zx}(R,0)\sin\theta$$

Further, since

$$\sigma_{zx}=\sigma'_{zx}\cos\theta-\sigma'_{zy}\sin\theta$$

and (13-141)

$$\sigma_{zy}=\sigma'_{zx}\sin\theta+\sigma'_{zy}\cos\theta$$

one finally obtains

$$\sigma_{zx}(R,\theta)=\sigma_{zx}(R,0)(\cos^2\theta+\alpha\sin^2\theta)$$

and (13-142)

$$\sigma_{zy}(R,\theta)=\sigma_{zx}(R,0)(1-\alpha)\sin\theta\cos\theta$$

which is precisely the same dependence on angle θ as in isotropic theory, with a value for Poisson's ratio agreeing with α. Also, the dependence on R is the same as in isotropic theory [see Eq. (5-60) and the discussion following it],

$$\sigma\sim\frac{1}{R^3} \tag{13-143}$$

Since any curved dislocation in the basal plane and with the Burgers vector in the plane can be considered as made up of infinitesimal loops, each giving rise to increments of stress of the form Eqs. (13-142) and (13-143), it follows that σ_{zx} and σ_{zy} in the plane could be calculated from isotropic theory with the appropriate values for the elastic constants. In particular, agreement must obtain for straight dislocations, so by comparison with Eqs. (13-132) to (13-135), the appropriate elastic constants for isotropic theory are

$$\mu = K_s = (c_{44}c_{66})^{1/2}$$

$$\frac{\mu}{1-\nu} = K_e = (\bar{c}_{11} + c_{13})\left[\frac{c_{44}(\bar{c}_{11} - c_{13})}{c_{33}(\bar{c}_{11} + c_{13} + 2c_{44})}\right]^{1/2} \quad (13\text{-}144)$$

where

$$\bar{c}_{11} = (c_{11}c_{33})^{1/2} \quad (13\text{-}145)$$

Furthermore, since work is done only against the stresses σ_{zx} and σ_{zy} as the dislocation is formed by the operation of a cut in the basal plane, the energy expression of isotropic theory is also correct with this choice for μ and ν.

Similarly, different constants μ and ν can be defined such that the energy and distribution of stress σ_{zz} in the plane of the dislocation are given correctly by isotropic theory for a dislocation in the basal plane with Burgers vector normal to the plane. One then obtains the condition

$$\frac{\mu}{1-\nu} = (\bar{c}_{11} + c_{13})\left[\frac{c_{44}(\bar{c}_{11} - c_{13})}{c_{11}(\bar{c}_{11} + c_{13} + 2c_{44})}\right]^{1/2} \quad (13\text{-}146)$$

With this choice of $\mu/(1-\nu)$, the energy expression of isotropic theory also is correct. The above condition does not determine μ and ν independently. One more condition could be added, say, that the distribution of pressure $p = -\frac{1}{3}(\sigma_{xx} + \sigma_{yy} + \sigma_{zz})$ as given by isotropic theory holds true. When one is interested only in σ_{zz}, it is sufficient to choose an arbitrary value of ν and determine what μ must then be used for Eq. (13-146) to be fulfilled.

The energy of a curved dislocation in the basal plane can be split into two separate parts. One part is obtained by considering just the component of Burgers vector normal to the plane. There is no cross term in the energy between the two components of Burgers vector.

Exercise 13-8. Calculate the energy coefficient for a straight, prismatic edge dislocation with $\boldsymbol{\xi} = [0001]$ and $\mathbf{b} = \frac{1}{3}[11\bar{2}0]$.

Answer:

$$K_e(\text{prism}) = \frac{c_{11}{}^2 - c_{12}{}^2}{2c_{11}} \quad (13\text{-}147)$$

where the c_{ij} are those of Eq. (13-44).

Nonbasal Dislocations

As seen in Table 9-3, nonbasal slip is also of importance in hexagonal crystals. Because of the elastic isotropy of the basal plane, an axis perpendicular to the dislocation and in the basal plane is effectively a twofold axis insofar as the elasticity problem is concerned. Thus the characteristic polynomial for the roots p_n reduces to a third-order polynomial in p^2 for which explicit solutions are possible. As reviewed by Indenbom et al.,[31] the problem has been solved[32] and extended[33] to give convenient expressions for the energy factors as a function of dislocation orientation in the most important prismatic and pyramidal planes.

13-6. APPLICATIONS TO CUBIC CRYSTALS

FCC Crystals

Instead of discussing cubic crystals generally, we successively consider fcc, bcc, and NaCl crystals, because the most important types of dislocations vary with the crystal, as discussed in Chap. 9. In fcc crystals, the important dislocations are those in the close-packed {111} planes. The conditions for simple separation into screw and edge components are fulfilled for dislocations with $\boldsymbol{\xi} \| \langle 110 \rangle$. This case includes the pure screw dislocation and the 60° mixed dislocation among the glide dislocations and the sessile edge dislocation, such as the one with $\boldsymbol{\xi} \| \mathbf{AB}$ and $\mathbf{b} = \mathbf{CD}$ in the Thompson notation.

For the screw dislocation, the pertinent coordinates are those of Fig. 13-2, with the elastic constants given by Eq. (13-39). The energy coefficient is determined from Eq. (13-130) to be

$$K_s = \left(c'_{44} c'_{55} - c'^2_{16}\right)^{1/2} \tag{13-148}$$

Similarly, the stresses follow directly from Eq. (13-129).

For the dislocations with edge components, the applicable coordinates are those of Fig. 13-10 with c_{ij} given by Eq. (13-42). This coordinate system is easier to work with because the inconvenient c'_{16} terms, present in Eq. (13-39), disappear. In this case the stresses follow directly from Eq. (13-118), and, by Eq. (13-120), the energy coefficients are

$$K_{e_x} = (\bar{c}'_{11} + c'_{12}) \left[\frac{c'_{55}(\bar{c}'_{11} - c'_{12})}{c'_{22}(\bar{c}'_{11} + c'_{12} + 2c'_{55})} \right]^{1/2}$$

$$K_{e_y} = (\bar{c}'_{11} + c'_{12}) \left[\frac{c'_{55}(\bar{c}'_{11} - c'_{12})}{c'_{11}(\bar{c}'_{11} + c'_{12} + 2c'_{55})} \right]^{1/2} \tag{13-149}$$

[31] V. L. Indenbom, V. I. Alshits, and V. M. Chernov, in Yu. A. Ossipyan (ed.), "Defects in Crystals and their Computer Modeling," Nauka, Leningrad, 1980, p. 23.

[32] L. J. Teutonico, *Mat. Sci. Eng.*, **6**: 2747 (1970).

[33] M. M. Savin, V. M. Chernov, and A. M. Strokova, *Phys. Stat. Solidi*, **38a**: 747 (1976).

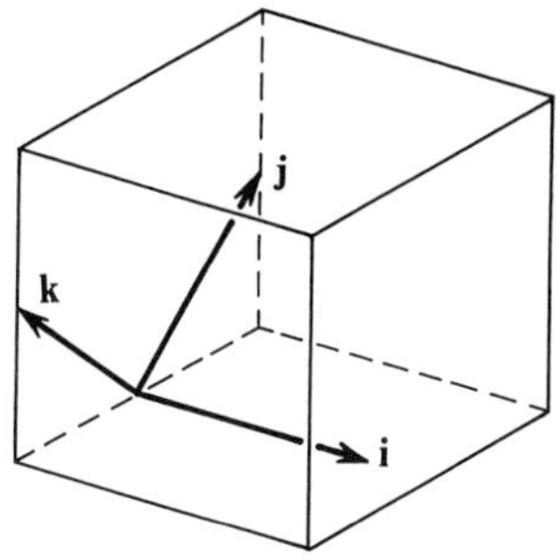

FIGURE 13-10. The coordinates $\mathbf{i}=[010]$, $\mathbf{j}=(1/\sqrt{2})[\bar{1}01]$, $\mathbf{k}=(1/\sqrt{2})[101]$, relative to the cubic unit cell.

Since $K_{e_x} \neq K_{e_y}$, because $c'_{11} \neq c'_{22}$, the cross term in the energy between the two edge components is zero *only* for the coordinates of Exercise 13-4. If the edge component of a dislocation lying along $[1\bar{1}0]$ were decomposed into one component in (111) and one component perpendicular to (111), there would be a cross term in the edge energy.

For the edge dislocation and the 30° mixed dislocation, the line direction is $\langle 112 \rangle$. In this case the edge and screw components do not separate. The solution is of the third-order type, with the detailed results so complicated that their development is not worthwhile here. The $\langle 112 \rangle$ dislocation has been treated by Teutonico,[34] who used numerical methods, and by Duncan and Kuhlmann-Wilsdorf,[35] who obtained an analytical solution.

The extension of glide dislocations into Shockley partials bounding an intrinsic stacking fault is an important phenomenon in fcc metals. The calculation of the width of extension as a function of orientation is an important problem which illustrates the difference between isotropic and anisotropic approaches. Seeger and Schoeck[36] first considered anisotropic effects in the splitting of the pure edge dislocation, while Teutonico[37] later numerically calculated the splitting as a function of character for a number of metals. Table 13-1, based on Teutonico's results, presents the splitting of the various dislocations relative to the splitting of the pure screw. The dissociation of the screw as a function of stacking-fault energy is presented in Exercise 13-9.

In the approximation of isotropic elasticity, the ratio of edge splitting to screw splitting is given by Eq. (10-15) as $(2+\nu)/(2-3\nu)$. For silver with the appropriate average value $\nu=0.35$ (Appendix 1), this ratio is 2.5, to be compared with the value 3.8 in Table 13-1. Similarly, for gold the isotropic value is 3.2, and the anisotropic value 5.1. The above results demonstrate that anisotropic effects are indeed appreciable. These results, together with other

[34] L. J. Teutonico, *Phys. Rev.*, **127**: 413 (1962); *Acta Met.*, **11**: 391 (1963).

[35] T. R. Duncan and D. Kuhlmann-Wilsdorf, *Bull. Am. Phys. Soc.*, **11**: 46 (1966).

[36] A. Seeger and G. Schoeck, *Acta Met.*, **1**: 519 (1953).

[37] L. J. Teutonico, *Acta Met.*, **11**: 1283 (1963).

TABLE 13-1 Widths of Dissociated Dislocations for Various FCC Metals Relative to the Width of a Dissociated Screw Dislocation*

	Metal							
Character	Al	Pd	Ni	Ag	Au	Cu	Th	Pb
Screw	1	1	1	1	1	1	1	1
30°	1.39	1.75	1.45	1.79	2.16	1.77	1.58	2.37
60°	2.17	3.12	2.23	3.11	4.20	2.96	2.48	4.53
Edge	2.29	3.75	2.59	3.75	5.09	3.49	2.86	5.48

*Results are based on the elastic constants listed in Appendix 1.

results[38–41] for energies of dislocation arrays in fcc crystals, such as the linear barriers discussed in Chap. 22, indicate that isotropic results generally agree with anisotropic results within about 50 percent, although in one instance[39] the difference is as large as a factor of 4.

Exercise 13-9. Derive an expression for the distance between the partials of a screw dissociated on a $\{111\}$ plane as a function of the stacking-fault energy γ.

Solution. The c'_{ij} are those of Eq. (13-42). In terms of these elastic constants, the energy coefficient for a pure screw is

$$K_s=(c'_{44}c'_{55})^{1/2}$$

Thus on the slip plane the shear stress produced by the screw is

$$\sigma_s=\frac{b_s}{2\pi r}(c'_{44}c'_{55})^{1/2}=\frac{K_s b_s}{2\pi r}$$

An edge dislocation with Burgers vector b_e in the slip plane has components

$$b_x=\sqrt{2/3}\,b_e \quad \text{and} \quad b_y=-\frac{1}{\sqrt{3}}b_e$$

Thus the coefficient for the total edge component is given by Eq. (13-120) as

$$K_e=\tfrac{2}{3}K_{e_x}+\tfrac{1}{3}K_{e_y}$$

$$=\frac{1}{3}\left(2+\frac{c'_{22}}{\bar{c}'_{11}}\right)(\bar{c}'_{11}+c'_{12})\left[\frac{c'_{55}(\bar{c}'_{11}-c'_{12})}{c'_{22}(\bar{c}'_{11}+c'_{12}+2c'_{55})}\right]^{1/2}$$

[38] A. J. E. Foreman, *Acta Met.*, **3**: 322 (1955).

[39] T. Jøssang, J. P. Hirth, and C. S. Hartley, *J. Appl. Phys.*, **36**: 2400 (1965).

[40] L. J. Teutonico, *Phil. Mag.*, **10**: 401 (1964).

[41] H. P. Karnthaler and E. Wintner, *Acta Met.*, **23**: 1501 (1975).

where $\bar{c}'_{11}=(c'_{11}c'_{22})^{1/2}$. Hence the edge shear stress in the slip plane is

$$\sigma_e=\frac{K_e b_e}{2\pi r}$$

Let the total Burgers vector of the dissociating screw be $\mathbf{b}=\frac{1}{2}[1\bar{1}0]$. It dissociates into two Shockely partials whose edge components

$$b_e=\frac{1}{2\sqrt{3}}b$$

attract each other and whose screw components

$$b_s=\tfrac{1}{2}b$$

repel each other. The equation of force balance with the stacking fault is thus

$$\frac{b^2}{2\pi r}\left(\frac{K_s}{4}-\frac{K_e}{12}\right)=\gamma$$

Hence the width r is

$$r=\frac{b^2}{8\pi\gamma}\left(K_s-\tfrac{1}{3}K_e\right) \tag{13-150}$$

BCC Crystals

The following discussion is confined to glide dislocations. As summarized in Chap. 9, the important glide dislocations have Burgers vectors $\frac{1}{2}\langle 111\rangle$ and glide on $\{110\}$ and $\{112\}$ planes. For dislocations in $\{110\}$ planes, the $35°16'$ mixed dislocation lies along a $\langle 110\rangle$ direction, so that the stresses and energy factors can be immediately obtained from the preceding analysis; the pure screw lies along $\langle 111\rangle$ and the edge along $\langle 112\rangle$, so that in neither case are the screw and edge components separable. For dislocations in a $\{112\}$ plane the pure edge dislocation lies along $\langle 110\rangle$ and can be treated easily, while the screw again lies along $\langle 111\rangle$, which is the line of intersection between $\{110\}$ and $\{112\}$. As discussed earlier, the treatment of the $\langle 112\rangle$ dislocation is too complicated to undertake here. However the $\langle 111\rangle$ dislocations are in the third-order category and can be discussed analytically. Head[42] solved the anisotropic elasticity problem for dislocations lying along $\langle 111\rangle$ directions using the theory of Stroh.[43] The energy factors and stresses given by Head are presented below. In addition, we add the explicit expression for a stress component not given by Head.

[42]A. K. Head, *Phys. Stat. Solidi*, **5**: 51 (1964); *Phys. Stat. Solidi*, **6**: 461 (1964).

[43]A. N. Stroh, *Phil. Mag.*, **3**: 625 (1958).

The pertinent coordinate system has **i** along $[\bar{1}2\bar{1}]$, **j** along $[\bar{1}01]$, and **k**, the dislocation-line direction, along [111]. This coordinate system is obtained from that of Fig. 13-2 by 90° rotation about the **i** axis. A comparison with Eq. (13-39) reveals that the matrix of elastic constants for this case is

$$\{c'\}=\begin{bmatrix} c'_{11} & c'_{12} & c'_{13} & 0 & c'_{15} & 0 \\ c'_{12} & c'_{11} & c'_{13} & 0 & -c'_{15} & 0 \\ c'_{13} & c'_{13} & c'_{33} & 0 & 0 & 0 \\ 0 & 0 & 0 & c'_{44} & 0 & -c'_{15} \\ c'_{15} & -c'_{15} & 0 & 0 & c'_{44} & 0 \\ 0 & 0 & 0 & -c'_{15} & 0 & c'_{66} \end{bmatrix} \qquad (13\text{-}151)$$

where, in terms of the standard cubic coefficients,

$$\begin{aligned} c'_{11}&=c_{11}+\tfrac{1}{2}H & c'_{44}&=c_{44}-\tfrac{1}{3}H \\ c'_{12}&=c_{12}-\tfrac{1}{6}H & c'_{66}&=c_{44}-\tfrac{1}{6}H \\ c'_{13}&=c_{12}-\tfrac{1}{3}H & c'_{15}&=-\frac{\sqrt{2}}{6}H \\ c'_{33}&=c_{11}+\tfrac{2}{3}H \end{aligned} \qquad (13\text{-}152)$$

and where $2c'_{66}=c'_{11}-c'_{12}$. Head lists the corresponding compliances s'_{ij} in terms of the standard compliances s_{ij}.

Head's theory is given in terms of the reduced compliances S_{ij} obtained by inverting the 5×5 matrix resulting from deleting the third row and third column in the matrix (13-151). The 5×5 matrix for S_{ij} is

$$\{S\}=\begin{bmatrix} S_{11} & S_{12} & 0 & S_{15} & 0 \\ S_{12} & S_{11} & 0 & -S_{15} & 0 \\ 0 & 0 & S_{44} & 0 & -2S_{15} \\ S_{15} & -S_{15} & 0 & S_{44} & 0 \\ 0 & 0 & -2S_{15} & 0 & S_{66} \end{bmatrix} \qquad (13\text{-}153)$$

The elements are given in terms of Eq. (13-152) as

$$S_{11}=\frac{c'_{11}c'_{44}-c'^2_{15}}{2(c'_{11}+c'_{12})(c'_{44}c'_{66}-c'^2_{15})} \qquad S_{12}=-\frac{c'_{44}c'_{12}+c'^2_{15}}{2(c'_{11}+c'_{12})(c'_{44}c'_{66}-c'^2_{15})}$$

$$S_{44}=\frac{c'_{66}}{c'_{44}c'_{66}-c'^2_{15}} \qquad S_{66}=\frac{c'_{44}}{c'_{44}c'_{66}-c'^2_{15}} \qquad S_{15}=-\frac{c'_{15}}{2(c'_{44}c'_{66}-c'^2_{15})} \qquad (13\text{-}154)$$

where $\frac{1}{2}S_{66} = S_{11} - S_{12}$. In terms of the compliances s'_{ij}, the reduced compliances are

$$S_{ij} = s'_{ij} - s'_{i3}s'_{j3}/s'_{33}$$

The s'_{ij} are the compliances of Eq. (13-11) referred to the same coordinates as in Eq. (13-151).

The equation for p is of the third-order form and can be expressed in terms of the factor M,

$$M = \left(\frac{S_{11}S_{44}}{S_{11}S_{44} - S_{15}{}^2} \right)^{1/2} \tag{13-155}$$

as can be verified from Eq. (13-85). For purposes of brevity, we omit the further derivation of the solution. The coefficients, derived by Head, are

$$K_s = \frac{M}{S_{44}}$$

$$K_e = K_{e_x} = K_{e_y} = \frac{M}{2S_{11}} \tag{13-156}$$

From these energy expressions, utilizing the fact that there are no cross terms, Head found the stress components for the [111] general dislocation,

$$\sigma_{r\theta} = K_e \frac{b_x \cos\theta + b_y \sin\theta}{2\pi r} \qquad \sigma_{\theta z} = K_s \frac{b_z}{2\pi r}$$

$$\sigma_{\theta\theta} = K_e \frac{-b_x \sin\theta + b_y \cos\theta}{2\pi r} \tag{13-157}$$

referred to a cylindrical coordinate system, with θ measured from the x axis. The sense ξ points in the positive z direction. Equations (13-157) are derived from (13-156) by the device discussed previously of equating the energies to the work done on cut surfaces in forming the dislocation.

Because of its possible importance in cross-slip processes, we have derived the additional stress component for a [111] screw dislocation,

$$\sigma_{rz} = -\frac{K_s(M^2 - 1)b_z}{\cot 3\theta + M^2 \tan 3\theta} \frac{1}{2\pi r} \tag{13-158}$$

$\sigma_{rz} = 0$ when $\theta = 0$ or $\theta = \pi/2$ and for the equivalent planes that follow from the threefold symmetry, i.e., on all {112} and {110} planes containing the screw dislocation.

Consider two parallel screws, A and B in Fig. 13-11, of the same sign and in the same {110} plane. Now move dislocation B the small distance from P to Q.

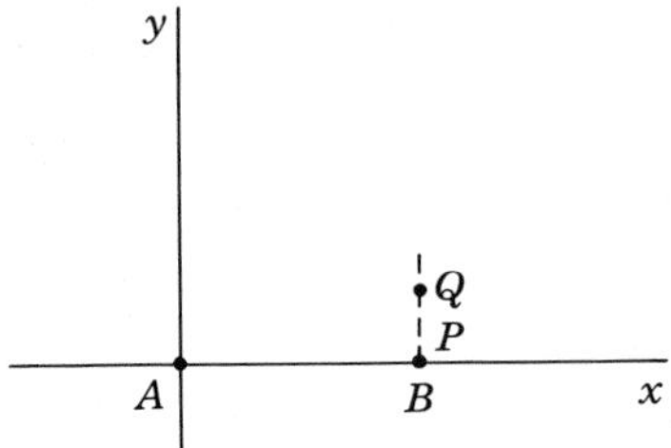

FIGURE 13-11. Two screw dislocations A and B parallel to the z axis.

At Q the stress σ_{rz} from dislocation A is finite, and its sign is such that it tends to displace B further. The radial force caused by $\sigma_{\theta z}$ adds to the instability. Extending this reasoning, one concludes that a pileup of screws on a $\{110\}$ plane is *elastically* unstable to cross slip. The pileup could be stabilized only by such effects as dissociation on $\{110\}$ planes.

For a pileup of screws along the y axis, i.e., on a $\{112\}$ plane, the situation differs somewhat. Here the stress component σ_{rz} tends to stabilize the pileup. A detailed treatment shows that if

$$M^2 > 1.5 \tag{13-159}$$

the pileup on $\{112\}$ is elastically stable against cross slip. When condition (13-159) is not fulfilled, pileups on $\{112\}$ planes still will be less unstable with respect to cross slip than those on $\{110\}$ planes.

An evaluation of M^2 for various bcc metals shows that M^2 is close to unity except for Na and K. For example, $M^2 = 1.1$ for both Fe and Nb, so that for these metals the σ_{rz} terms are not very important. For Na, $M^2 = 1.7 > 1.5$, while for K, $M^2 = 1.55 > 1.5$. The K case is uncertain; slightly different elastic constants could give $M^2 < 1.5$. On the other hand, Na is a clear case where the $\langle 111 \rangle \{112\}$ slip system should yield long straight slip lines at low temperatures, indicating the absence of cross slip.

NaCl Structures

In this case the Burgers vectors are $\langle 110 \rangle$ and the most prominent slip planes $\{110\}$. Let us consider only glide dislocations. The pure screw lies in a $\langle 110 \rangle$ direction, so that the energy coefficient is

$$K_s = (c'_{44} c'_{55})^{1/2} \tag{13-160}$$

with the elastic constants given by Eq. (13-42). The pure edge dislocations lies in a $\langle 100 \rangle$ direction and has the energy coefficient

$$K_e = K_{e_x} = K_{e_y} = (c_{11} + c_{12}) \left[\frac{c_{44}(c_{11} - c_{12})}{c_{11}(c_{11} + c_{12} + 2c_{44})} \right]^{1/2} \tag{13-161}$$

in terms of the standard constants [Eq. (13-37)]. The mixed dislocation lying along $\langle 111 \rangle$ is a case for which Head's analysis would apply.

13-7. ADVANCED STRAIGHT DISLOCATION FORMALISM

When a straight dislocation is not lying in a high symmetry direction, computer calculations are necessary to describe the elastic field. Recent progress has led to better organization of the elasticity problem for such calculations as well as providing an improved basis for theoretical analysis. The recent advances have been reviewed comprehensively.[44, 45] The cornerstone in the more advanced theory is the work by Stroh.[46]

The Stroh Theory

In order to follow the customary literature notation for both cases, we use a notation different from that in Sec. 13-3. The indices α and β replace the index n rather than referring to coordinate axes. Also, although some quantities are similar in the two theories, for example, A_α and $A(n)$, they are normalized differently. Thus the quantities cannot be directly carried from one theoretical form to the other.

Let **a** and **b** be two real vectors. Then (ab) is by definition an abbreviated form for the matrix with elements

$$(ab)_{jk} = a_i c_{ijkl} b_l \tag{13-162}$$

Because of the general symmetries in c_{ijkl} given by Eqs. (2-6) and (2-13)

$$(ab) = (ba)^T \tag{13-163}$$

where T means the transpose. Relative to the sense vector ξ, two other orthogonal, unit vectors **m** and **n** are defined by $\boldsymbol{\xi} = \mathbf{m} \times \mathbf{n}$ as indicated in Fig. 13-12.

Guided by Eqs. (13-53) and (13-69), we seek partial solutions to Eq. (13-52) of the type[47]

$$\mathbf{u} = \frac{D\mathbf{A}}{2\pi i} \ln \eta \tag{13-164}$$

[44] D. J. Bacon, D. M. Barnett, and R. O. Scattergood, *Prog. Mat. Sci.*, **23**: 51 (1978).

[45] V. L. Indenbom, V. I. Alshits and V. M. Chernov, in Yu. A. Ossipyan (ed.), "Defects in Crystals and Their Computer Modeling," Nauka, Leningrad, 1980, p. 23.

[46] A. N. Stroh, *J. Math. Phys.*, **41**: 77 (1962).

[47] Note the change of sign in Eq. (13-164) relative to Eq. (13-69), corresponding to the change in sign of $\boldsymbol{\xi}$ between Figs. 13-3 and 13-12 when we let **m** correspond to x_1 and **n** to x_2.

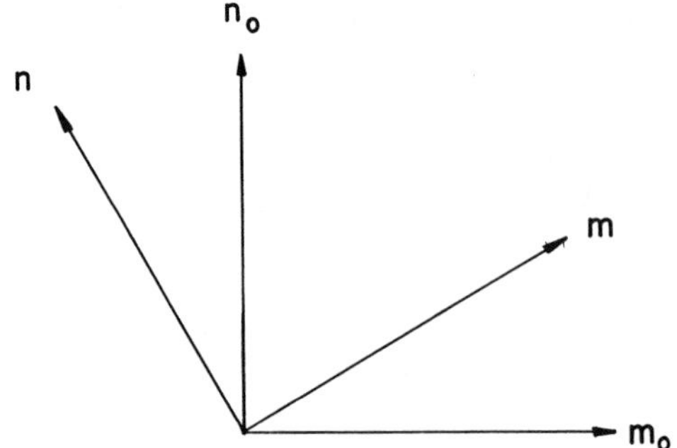

FIGURE 13-12. Orthogonal coordinate systems **m**, **n**, ξ and $\mathbf{m}_0$, $\mathbf{n}_0$, ξ; ξ points out of the page.

where, corresponding to Eq. (13-53)

$$\eta = \mathbf{m}\cdot\mathbf{x} + p\mathbf{n}\cdot\mathbf{x} \tag{13-165}$$

Substitution into Eq. (13-52) and multiplication by η^2 yields

$$\{(mm)+[(mn+nm)]\,p+(nn)p^2\}\mathbf{A}=0 \tag{13-166}$$

for the determination of p and **A**.

If, in particular, $\mathbf{m}=\mathbf{e}_1$, $\mathbf{n}=\mathbf{e}_2$, and $\xi=\mathbf{e}_3$, Eqs. (13-54) and (13-56) are recovered. However, the above formulation does not imply that the c_{ijkl} are referred to a dislocation-oriented system, that is, **m**, **n**, ξ, and **x** can each have three nonzero components.

So far, the derivation is a slightly generalized rewriting of the original sextic theory. The departure comes with the definition of a second vector **L**, related to **A** by

$$\mathbf{L} = -[(nm)+p(nn)]\mathbf{A} \tag{13-167}$$

Also **A** and **L** are combined to form the six-dimensional vector ζ with components (A_1, A_2, A_3, L_1, L_2, L_3). With these definitions, Eq. (13-166) can be written as the six-dimensional eigenequation

$$\mathbf{N}\cdot\zeta = p\zeta \tag{13-168}$$

Here **N** is the 6×6 matrix comprised of four 3×3 blocks, $-(nn)^{-1}(nm)$ in the upper left, $-(nn)^{-1}$ in the upper right, $-[(mn)(nn)^{-1}(nm)-(mm)]$ in the lower left, and $(mn)(nn)^{-1}$ in the lower right portions. The equivalence with Eq. (13-166) can be verified directly by substitution of Eq. (13-167) into (13-168) and use of the property $(nn)^{-1}(nn)=1$.

In the six-dimensional space, some useful 6×6 matrices involving the unit matrix **I**, are

$$\mathbf{U}=\begin{Bmatrix} I & 0 \\ 0 & I \end{Bmatrix},\mathbf{V}=\begin{Bmatrix} 0 & I \\ I & 0 \end{Bmatrix},\mathbf{I}=\begin{Bmatrix} 1 & 0 & 0 \\ 0 & 1 & 0 \\ 0 & 0 & 1 \end{Bmatrix} \tag{13-169}$$

The eigenvalues, p_α, $\alpha=1,2,\ldots,6$ are solutions of the determinental relation

$$\|\mathbf{N}-p_\alpha\mathbf{U}\|=0 \tag{13-170}$$

or, the equivalent from Eq. (13-166)

$$\|(mm)+[(mn)+(nm)]\,p_\alpha+(nn)p_\alpha^2\|=0 \tag{13-171}$$

The roots occur as pairs of complex conjugates. As before, let p_1, p_2, p_3 be the roots with positive imaginary part, and let p_4, p_5, and p_6 be their respective complex conjugates.

N is not a symmetric matrix. However, it has the symmetry

$$\mathbf{N}^T\cdot\mathbf{V}=\mathbf{V}\cdot\mathbf{N} \qquad \text{or} \qquad \mathbf{N}^T=\mathbf{V}\cdot\mathbf{N}\cdot\mathbf{V} \tag{13-172}$$

Let p_α and p_β be two different eigenvalues and ζ_α and ζ_β be the associated eigenvectors:

$$\mathbf{N}\cdot\zeta_\alpha=p_\alpha\zeta_\alpha, \qquad \mathbf{N}\cdot\zeta_\beta=p_\beta\zeta_\beta \tag{13-173}$$

Operating on the first form of this relation with $\zeta_\beta\cdot\mathbf{V}$ gives[48]

$$\zeta_\beta\cdot\mathbf{V}\cdot\mathbf{N}\cdot\zeta_\alpha=p_\alpha\zeta_\beta\cdot\mathbf{V}\cdot\zeta_\alpha \tag{13-174}$$

But, Eq. (13-172) shows that

$$\zeta_\beta\cdot\{\mathbf{V}\cdot\mathbf{N}\}\cdot\zeta_\alpha=\zeta_\alpha\cdot\{\mathbf{N}^T\cdot\mathbf{V}\}\cdot\zeta_\beta=\zeta_\alpha\cdot\{\mathbf{V}\cdot\mathbf{N}\}\cdot\zeta_\beta=p_\beta\zeta_\alpha\cdot\mathbf{V}\cdot\zeta_\beta \tag{13-175}$$

with Eq. (13-174) used in the last step. A comparison of the above two relations reveals that

$$[p_\beta-p_\alpha][\zeta_\beta\cdot\mathbf{V}\cdot\zeta_\alpha]=0$$

[48]When multiplied by a matrix on the left-hand side, ζ is understood as a column vector, when multiplied on the right-hand side it is understood as a row vector.

In other words, the orthogonality relation

$$\zeta_\beta \cdot \mathbf{V} \cdot \zeta_\alpha = 0 \qquad \text{when} \qquad p_\alpha \neq p_\beta \tag{13-176}$$

exists. Written out in detail, Eq. (13-176) states that

$$\mathbf{A}_\alpha \cdot \mathbf{L}_\beta + \mathbf{A}_\beta \cdot \mathbf{L}_\alpha = \delta_{\alpha\beta} \tag{13-177}$$

when the magnitudes of $\mathbf{A}_\alpha$ and $\mathbf{L}_\alpha$ are adjusted such that

$$2\mathbf{A}_\alpha \cdot \mathbf{L}_\alpha = 1, \qquad \text{no summation over } \alpha \tag{13-178}$$

The orthogonality relation (13-177) is the pivotal result of the Stroh theory. It greatly facilitates the determination of $D(\alpha)$, Eq. (13-88) and Eq. (13-89).

With a solution of the form of Eq. (13-91), the superposition of partial solutions of the type in Eq. (13-164) gives

$$\mathbf{u} = \frac{1}{2\pi i} \sum_{\alpha=1}^{6} D(\alpha) \mathbf{A}_\alpha \ln \eta_\alpha \tag{13-179}$$

The boundary conditions that the Burgers circuit has a discontinuity **b** and that there be no external force at the core become[49]

$$\sum_{\alpha=1}^{6} \pm D(\alpha) \mathbf{A}_\alpha = \mathbf{b} \tag{13-180}$$

$$\sum_{\alpha=1}^{6} \pm D(\alpha) \mathbf{L}_\alpha = 0 \tag{13-181}$$

These two equations repeat Eqs. (13-88) and (13-89).

Multiplying Eq. (13-180) by $\mathbf{L}_\beta$ and (13-181) by $\mathbf{A}_\beta$, summing, and using the orthogonality relation (13-177), we readily determine that

$$D(\alpha) = \pm \mathbf{L}_\alpha \cdot \mathbf{b} \tag{13-182}$$

Thus the Stroh theory gives an explicit form for $D(\alpha)$ and the complete solution (13-179) is determined. This is precisely the point in which the classical theory is inconvenient: $D(n)$ is determined implicitly only by Eqs. (13-88) and (13-89), and in the simple high-symmetry solutions of Sec. 13-4, D is deduced by inspection. An alternative method for explicit solutions, employing Fourier analysis has been developed by Willis.[50]

[49] The case of an external force **f** per unit length along the core can be readily handled, $-\mathbf{f}$ then replacing 0 on the right side of Eq. (13-181).

[50] J. R. Willis, *Phil. Mag.*, **21**: 931 (1970).

With $D(\alpha)$ known, the solution for the displacements [Eq. (13-179)] becomes

$$\mathbf{u}=\frac{1}{2\pi i}\sum_{\alpha=1}^{6}\pm\mathbf{A}_\alpha[\mathbf{L}_\alpha\cdot\mathbf{b}]\ln\eta_\alpha \tag{13-183}$$

The distortions $\partial u_k/\partial x_l$ follow directly

$$\frac{\partial u_k}{\partial x_l}=\frac{1}{2\pi i}\sum_{\alpha=1}^{6}\pm[m_l+p_\alpha n_l]A_{\alpha_k}[\mathbf{L}_\alpha\cdot\mathbf{b}]\frac{1}{\eta_\alpha} \tag{13-184}$$

as do the strains from Eq. (2-3) and the stresses from Eq. (2-7),

$$\sigma_{ij}=\frac{1}{2\pi i}\sum_{\alpha=1}^{6}\pm c_{ijkl}[m_l+p_\alpha n_l]A_{\alpha_k}[\mathbf{L}_\alpha\cdot\mathbf{b}]\frac{1}{\eta_\alpha} \tag{13-185}$$

From Eqs. (13-185) and (13-167), the resolved stress in the plane $\mathbf{n}\cdot\mathbf{x}=0$ (the plane containing $\boldsymbol{\xi}$ and $\mathbf{m}$) produces an in-plane force $\boldsymbol{\sigma}\cdot\mathbf{n}$, given by

$$\boldsymbol{\sigma}\cdot\mathbf{n}=-\frac{1}{2\pi i}\sum_{\alpha=1}^{6}\pm\mathbf{L}_\alpha[\mathbf{L}_\alpha\cdot\mathbf{b}]\frac{1}{\mathbf{m}\cdot\mathbf{x}} \tag{13-186}$$

When the dislocation is formed by the creation of a displacement discontinuity $\mathbf{b}$ in this plane, work is done against the force of Eq. (13-186) in an amount $\mathbf{b}\cdot\boldsymbol{\sigma}\cdot\mathbf{n}/2$, so it follows by analogy to the developments in Chap. 3 that the energy per unit length of the dislocation is

$$\frac{W}{L}=-\frac{1}{4\pi i}\left(\sum_{\alpha=1}^{6}\pm[\mathbf{b}\cdot\mathbf{L}_\alpha][\mathbf{L}_\alpha\cdot\mathbf{b}]\right)\ln\frac{R}{r_0} \tag{13-187}$$

Thus the prelogarithmic energy factor E is

$$E=-\frac{1}{4\pi i}\sum_{\alpha=1}^{6}\pm[\mathbf{b}\cdot\mathbf{L}_\alpha][\mathbf{L}_\alpha\cdot\mathbf{b}] \tag{13-188}$$

and, in dyadic notation, the energy coefficient tensor $\mathbf{K}$ is

$$\mathbf{K}=i\sum_{\alpha=1}^{6}\pm\mathbf{L}_\alpha\otimes\mathbf{L}_\alpha \tag{13-189}$$

These relations completely solve the dislocation problem in the sextic formalism. Before concluding, we derive some completeness relations that are useful in the integral theory that follows. When all the p_α are different, the eigenvectors $\boldsymbol{\zeta}_\alpha$ of Eq. (13-168) can be asserted to form a complete set. Then any complex vector ζ can be expressed by a linear combination of the ζ_α,

$$\zeta = \sum_1^6 a_\alpha \zeta_\alpha \tag{13-190}$$

Let us define a dyadic tensor $\mathbf{U}$

$$\mathbf{U} = \sum_{\alpha=1}^{6} \zeta_\alpha \otimes [\mathbf{V} \cdot \zeta_\alpha] \tag{13-191}$$

Operating on ζ with $\mathbf{U}$, and using Eqs. (13-176) and (13-177), we find that

$$\mathbf{U} \cdot \zeta = \zeta \tag{13-192}$$

Therefore, $\mathbf{U}$ must be the unit tensor as already defined in Eq. (13-169). With this identification of $\mathbf{U}$, the written-out form of Eq. (13-191) gives the completeness relations

$$\sum_{\alpha=1}^{6} \mathbf{A}_\alpha \otimes \mathbf{A}_\alpha = \sum_{\alpha=1}^{6} \mathbf{L}_\alpha \otimes \mathbf{L}_\alpha = 0 \tag{13-193}$$

$$\sum_{\alpha=1}^{6} \mathbf{A}_\alpha \otimes \mathbf{L}_\alpha = \sum_{\alpha=1}^{6} \mathbf{L}_\alpha \otimes \mathbf{A}_\alpha = \mathbf{I} \tag{13-194}$$

The relation (13-193) can be decomposed to

$$\sum_1^3 \mathbf{L}_\alpha \otimes \mathbf{L}_\alpha + \sum_1^3 \{\mathbf{L}_\alpha \otimes \mathbf{L}_\alpha\}^* = 0$$

which shows that the two parts are purely imaginary. Thus Eq. (13-189) simplifies to

$$\mathbf{K} = 2i \sum_{\alpha=1}^{3} \mathbf{L}_\alpha \otimes \mathbf{L}_\alpha \tag{13-195}$$

By a development like that in Eqs. (13-190) to (13-192), we also understand that an alternative form for the matrix operator $\mathbf{N}$ is

$$\mathbf{N} = \sum_{\alpha=1}^{6} p_\alpha \zeta_\alpha \otimes [\mathbf{V} \cdot \zeta_\alpha] \tag{13-196}$$

In accord with the orthogonality Eq. (13-176), the operator satisfies Eq. (13-173) for all eigenvectors ζ_α. Thus a comparison of Eqs. (13-196) and

(13-168) yields the sum rules

$$\sum_{\alpha=1}^{6} p_\alpha \mathbf{A}_\alpha \otimes \mathbf{L}_\alpha = -(nn)^{-1}(nm) \tag{13-197}$$

$$\sum_{\alpha=1}^{6} p_\alpha \mathbf{A}_\alpha \otimes \mathbf{A}_\alpha = -(nn)^{-1} \tag{13-198}$$

$$\sum_{\alpha=1}^{6} p_\alpha \mathbf{L}_\alpha \otimes \mathbf{L}_\alpha = -\left[(mn)(nn)^{-1}(nm)-(mm)\right] \tag{13-199}$$

$$\sum_{\alpha=1}^{6} p_\alpha \mathbf{L}_\alpha \otimes \mathbf{A}_\alpha = -(mn)(nn)^{-1} \tag{13-200}$$

The Integral Method

The Stroh theory relies on the solution of the sextic eigenvalue problem. One can reformulate the theory so that the sextic problem is replaced by one of evaluating a set of definite integrals. This was first recognized by Barnett and Swanger,[51] who constructed an integral theory for the dislocation energy factors. Later,[52] the invariance relations of the Stroh theory were found to be a convenient starting point for a more general integral theory which is followed here.

The Stroh theory began with the coordinate set $\mathbf{m}$, $\mathbf{n}$, $\boldsymbol{\xi}$ in Fig. 13-12. However, only $\boldsymbol{\xi}$ is fixed for a dislocation, and a set $\mathbf{m}$, $\mathbf{n}$ rotated by ϕ about $\boldsymbol{\xi}$ relative to a set $\mathbf{m}_0$, $\mathbf{n}_0$ is an equally valid choice as also indicated in Fig. 13-12. The energy of the dislocation must be independent of ϕ; thus $\Sigma_{\alpha=1}^{3}\mathbf{L}_\alpha \otimes \mathbf{L}_\alpha$ in Eq. (13-195) must be an invariant. The issue is, then, to determine whether the invariants are the individual $\mathbf{L}_\alpha$ values, or only certain sums such as $\Sigma_{\alpha=1}^{3}\mathbf{L}_\alpha \otimes \mathbf{L}_\alpha$.

The operator $\mathbf{N}$ [Eq. (13-168)] is a function of ϕ. The issue now is to determine how the eigenvectors and eigenvalues in the following equation vary with ϕ:

$$\mathbf{N}(\phi)\cdot \boldsymbol{\zeta}_\alpha(\phi) = p_\alpha(\phi)\boldsymbol{\zeta}_\alpha(\phi) \tag{13-201}$$

Simple vector theory shows that the vectors $\mathbf{m}$ and $\mathbf{n}$ vary in the manner $\partial \mathbf{m}/\partial\phi = \mathbf{n}$, $\partial \mathbf{n}/\partial\phi = -\mathbf{m}$. Thus the matrices such as (mn), (nn), and $(nn)^{-1}$

[51] D. M. Barnett and L. A. Swanger, *Phys. Stat. Solidi*, **48b**: 419 (1971).

[52] D. M. Barnett and J. Lothe, *Phys. Norvegica*, **7**: 13 (1973).

vary as

$$\frac{\partial}{\partial\phi}(mn)=\left(\frac{\partial m}{\partial\phi}n\right)+\left(m\frac{\partial n}{\partial\phi}\right)=(nn)-(mm)$$

$$\frac{\partial}{\partial\phi}(nn)^{-1}=-(nn)^{-1}\frac{\partial}{\partial\phi}(nn)(nn)^{-1}=-(nn)^{-1}[(mn)+(nm)](nn)^{-1}$$

$$(13\text{-}202)$$

and so forth. With the use of such expressions, $\partial\mathbf{N}/\partial\phi$ can be determined directly from Eq. (13-168), resulting in a very lengthly matrix. However, as can be verified by substitution of Eq. (13-168), the matrix reduces simply to

$$\frac{\partial\mathbf{N}}{\partial\phi}=-\{\mathbf{U}+\mathbf{N}\cdot\mathbf{N}\} \tag{13-203}$$

This result leads to a simple analysis of the invariance problem. To first order in ϕ, Eq. (13-203) gives

$$\mathbf{N}(\phi)=\mathbf{N}(0)-\{\mathbf{U}+\mathbf{N}(0)\cdot\mathbf{N}(0)\}\phi \tag{13-204}$$

Now, $\zeta_\alpha(0)$ is the eigenvector of $\mathbf{N}(0)$. It is thus also an eigenvector for $\mathbf{N}(0)\cdot\mathbf{N}(0)$, and since it is an eigenvector for $\mathbf{U}$ with a eigenvalue of unity, it is an eigenvector of $\mathbf{N}(\phi)$ [Eq. (13-204)], with an eigenvalue $p_\alpha(0)-[1+p_\alpha^2(0)]\phi$, or

$$\mathbf{N}(\phi)\cdot\zeta_\alpha(0)=\{p_\alpha(0)-[1+p_\alpha^2(0)]\phi\}\zeta_\alpha(0) \tag{13-205}$$

For this relation to be compatible with Eq. (13-201), it is necessary that

$$\frac{\partial\zeta_\alpha}{\partial\phi}=0, \qquad \frac{\partial p_\alpha}{\partial\phi}=-(1+p_\alpha^2) \tag{13-206}$$

These relations are valid for all ϕ because the choice of origin for ϕ is arbitrary. Thus ζ_α is invariant, depending only on ξ and not on ϕ, or, in terms of its components,

Axiom 13-1: Both $\mathbf{A}_\alpha$ and $\mathbf{L}_\alpha$ are invariant.

This is an important feature of the Stroh theory itself, proving that $\mathbf{K}$ in Eq. (13-195) depends only on ξ as it should.

Integration of Eq. (13-206) gives

$$p_\alpha(\phi)=\tan(\psi_\alpha-\phi) \tag{13-207}$$

where ψ_α is a complex integration constant. The positive complex part of ψ_α

corresponds to that in p_α. The average value of p_α is

$$\langle p_\alpha\rangle = \frac{1}{2\pi}\int_0^{2\pi} p_\alpha(\phi)\,d\phi = \frac{1}{2\pi}\ln\cos(\psi_\alpha-\phi)\Big|_0^{2\pi} \tag{13-208}$$

With positive complex part in ψ_α, $\cos(\psi_\alpha-\phi)$ encircles the origin in complex space once anticlockwise, giving an increment $2\pi i$ to the logarithmic term. Thus

$$\langle p_\alpha\rangle = \pm i \tag{13-209}$$

where the $\pm$ sign corresponds to the sign of the imaginary part of p_α.

The remainder of the integral theory now follows directly. The average value of **N** is

$$\langle \mathbf{N}\rangle = \frac{1}{2\pi}\int_0^{2\pi} \mathbf{N}(\phi)\,d\phi = \begin{Bmatrix} S & Q \\ B & S^T \end{Bmatrix} \tag{13-210}$$

where **S**, **Q**, and **B** are 3×3 matrices that follow from the definition of **N** in Eq. (13-168). They are given by

$$\mathbf{S} = -\frac{1}{2\pi}\int_0^{2\pi} (nn)^{-1}(nm)\,d\phi \tag{13-211}$$

$$\mathbf{Q} = -\frac{1}{2\pi}\int_0^{2\pi} (nn)^{-1}\,d\phi \tag{13-212}$$

$$\mathbf{B} = -\frac{1}{2\pi}\int_0^{2\pi} \left[(mn)(nn)^{-1}(nm)-(mm)\right]d\phi \tag{13-213}$$

$$\mathbf{S}^T = -\frac{1}{2\pi}\int_0^{2\pi} (mn)(nn)^{-1}\,d\phi \tag{13-214}$$

They are definite integrals over ϕ from 0 to 2π and thus depend only on ξ. All physical quantities of interest can be expressed in terms of these matrixes. The sextic Stroh problem is replaced by that of computing these integrals.

Consider first the energy coefficient **K**. Substitution of Eq. (13-199) into (13-213) together with the result of (13-209) shows that

$$\mathbf{B} = \sum_{\alpha=1}^{6} \pm i\mathbf{L}_\alpha \otimes \mathbf{L}_\alpha \tag{13-215}$$

But this is precisely the result for **K** [Eq. (13-189)], so **B**=**K** and Eq. (13-213) is the integral expression for **K**.

Consider next the distortions $\partial u_j/\partial x_k$ at the point (r,ϕ) in a cylindrical coordinate system with the dislocation at the origin. A convenient choice of coordinates is with **m** pointing toward the field point so that $\mathbf{m}\cdot\mathbf{x}=r$, $\mathbf{n}\cdot\mathbf{x}=0$. Equation (13-184) then becomes

$$\frac{\partial u_j}{\partial x_k}=\frac{1}{2\pi ir}\sum_{\alpha=1}^{6}\pm[m_k+p_\alpha n_k]A_{\alpha_j}L_{\alpha_l}b_l \tag{13-216}$$

From Eqs. (13-197), (13-209), and (13-211), we obtain

$$\sum_{\alpha=1}^{6}\pm i\mathbf{A}_\alpha\otimes\mathbf{L}_\alpha=\mathbf{S} \tag{13-217}$$

Inserting Eq. (13-167) into Eq. (13-215), we find

$$\mathbf{B}=-(nm)\sum_{\alpha=1}^{6}\pm i\mathbf{A}_\alpha\otimes\mathbf{L}_\alpha-(nn)\sum_{\alpha=1}^{6}\pm ip_\alpha\mathbf{A}_\alpha\otimes\mathbf{L}_\alpha \tag{13-218}$$

or, using Eq. (13-197) and (13-211),

$$\sum_{\alpha=1}^{6}\pm ip_\alpha\mathbf{A}_\alpha\otimes\mathbf{L}_\alpha=-(nn)^{-1}(nm)\mathbf{S}-(nn)^{-1}\mathbf{B} \tag{13-219}$$

Finally, substituting Eqs. (13-217) and (13-219) into (13-216), we obtain

$$\frac{\partial u_j}{\partial x_k}(r,\phi)=-\frac{b_l}{2\pi r}\left\{m_kS_{jl}-n_k\left[(nn)^{-1}(nm)\right]_{jo}S_{ol}-n_k(nn)^{-1}_{jo}B_{ol}\right\} \tag{13-220}$$

This is a convenient expression. **S** and **B** are independent of ϕ. All of the ϕ dependence in $\partial u_j/\partial x_k$ resides in the terms m_k, n_k, $(nn)^{-1}$ and (nm) and can be expressed explicitly. The stresses follow directly from $\partial u_j/\partial x_k$ by Eq. (2-7).

We mention without proof that the point Green's tensor function **G** that replaces the isotropic u_{ij} of Eq. (2-70) is given, with $\boldsymbol{\xi}$ chosen parallel to $\boldsymbol{r}$, by[53]

$$\mathbf{G}=-\frac{\mathbf{Q}}{4\pi r} \tag{13-221}$$

More formulas, including integral expressions for displacements and angular derivatives of energy factors for both dislocations and line forces, are available.[54, 55] Results have also been obtained in the integral formalism for

[53] K. Malén, *Phys. Stat. Solidi*, **44b**: 661 (1971).

[54] R. J. Asaro, J. P. Hirth, D. M. Barnett, and J. Lothe, *ibid*., **60b**: 261 (1973).

[55] D. J. Bacon, D. M. Barnett, and R. O. Scattergood, *Prog. Mat. Sci*., **23**: 51 (1978).

half-spaces and bicrystals.[56] For dislocations in the vicinity of cracks, however, results are available only in the sextic formalism.[57]

The matrices **S**, **Q**, and **B** are interrelated. From Eq. (13-201), the invariance of $\boldsymbol{\zeta}_\alpha$, and Eq. (13-209), we find

$$\langle \mathbf{N} \rangle \cdot \boldsymbol{\zeta}_\alpha = \pm i \boldsymbol{\zeta}_\alpha \tag{13-222}$$

Operating on this result with $\langle \mathbf{N} \rangle$, we obtain

$$\langle \mathbf{N} \rangle \cdot \langle \mathbf{N} \rangle \cdot \boldsymbol{\zeta}_\alpha = - \boldsymbol{\zeta}_\alpha \tag{13-223}$$

Since this is true for any eigenvector $\boldsymbol{\zeta}_\alpha$ out of the complete set, it must be true that

$$\langle \mathbf{N} \rangle \cdot \langle \mathbf{N} \rangle + \mathbf{U} = 0$$

or, in detail,

$$\mathbf{S} \cdot \mathbf{S} + \mathbf{Q} \cdot \mathbf{B} + \mathbf{I} = 0, \qquad \mathbf{S} \cdot \mathbf{Q} + \mathbf{Q} \cdot \mathbf{S}^T = 0$$

$$\mathbf{B} \cdot \mathbf{S} + \mathbf{S}^T \cdot \mathbf{B} = 0, \qquad \mathbf{B} \cdot \mathbf{Q} + \mathbf{S}^T \cdot \mathbf{S}^T + \mathbf{I} = 0 \tag{13-224}$$

Applications

Whether the sextic Stroh theory or the integral approach is more convenient depends on the specific problem. The two formalisms are strongly interrelated and really constitute one unified theory with formulas that can easily be switched from one approach to the other. In the sextic representation the invariance of the eigenvectors is the key factor, whereas in the integral method it is the independence of the arbitrary choice of reference vectors.

The new methods for straight dislocations combined with the Brown-Indenbom-Orlov approach (Sec. 5-7) make many new applications possible. Bacon et al.[55] have used the integral expressions for extensive computer calculations of the tractions in the slip plane as functions of dislocation orientation for the most important glide systems in a number of fcc, bcc, and hcp crystals. They have fitted the results to analytical Fourier expressions that can be used in the Brown theory[58] for planar curved dislocation problems. The applications are numerous: extended node configurations, the critical stress to operate a Frank-Read source, and loop and intersection problems, for example. A number of such applications are discussed by Bacon et al.[55]

[56] D. M. Barnett and J. Lothe, *J. Phys. F*, **4**: 1618 (1974).

[57] C. Atkinson, *Int. J. Fract. Mech.*, **2**: 567 (1966).

[58] L. M. Brown, *Phil. Mag.*, **15**: 363 (1967).

13-8. DISLOCATION STABILITY

For a straight dislocation in an isotropic medium, the line tension given by Eq. (6-81) was shown to control stability of the dislocation with respect to fluctuation to form a zigzag shape. The expression is somewhat of an approximation in that it neglects interactions with other dislocations, leading to some uncertainty in the logarithmic factor. Expressed in terms of the energy factor E, Eq. (6-81) reads

$$\mathfrak{S}=\left(E+\frac{\partial^2 E}{\partial \beta^2}\right)\ln\frac{L}{e\rho}=\mathfrak{E}\ln\frac{L}{e\rho} \qquad (13\text{-}225)$$

where L is the wavelength of the deviation (see Fig. 6-10). For isotropic solids, $\mathfrak{S}$ is positive and straight segments are stable. For anisotropic solids, $\mathfrak{S}$ can be small or even negative so that zigzag shapes can form at equilibrium. Head[59] first observed such glide instabilities and explained them as an elastic anisotropy effect giving negative $\mathfrak{S}$.

With the reglect of the small uncertainties in the logarithmic term, a consideration of the sign of the reduced factor $\mathfrak{E}$ suffices to determine instability. In the discussion of Fig. 6-10, we already used the analog with instability of surfaces with respect to faceting. The analogy is general, and several powerful theories for surfaces apply directly to dislocations. A plot[60] of $E(\beta)$ is analogous to a plot of surface energy versus orientation, and the Gibbs-Wulff construction applies.[61] This provides a simple geometric test for instability as proved by Herring.[61] An even simpler plot of reciprocal surface energy[62] has an analog $E^{-1}(\beta)$.

A portion of an $E^{-1}(\beta)$ plot is presented in Fig. 13-13. Consider unit length of dislocation of orientation β defined by a unit vector **e** in the glide plane. For the situation depicted, we show that it is unstable with respect to formation of length L_A of orientation β_A defined by unit vector $\mathbf{e}_A$ and L_B of orientation β_B with unit vector $\mathbf{e}_B$. Frank[63] showed that E_z^{-1} for such a zigzag configuration is given by the line OC, the intersection of a line along **e** with the line connecting A and B, the intersections with the E^{-1} surface for orientations β_A and β_B. If OD lies within a line OC connecting to the tangent to the E^{-1} surface as in

[59] A. K. Head, *Phys. Stat. Solidi*, **19b**: 185 (1967).

[60] The choice of reference angle in the plane is arbitrary, but it is convenient to choose it as the angle β relative to **b**.

[61] C. Herring, *Phys. Rev.*, **82**: 87 (1951).

[62] In the present context Herring noted that $E=E_A L_A+E_B L_B$ whereas $\mathbf{e}=L_A\mathbf{e}_A+L_B\mathbf{e}_B$. Defining reciprocal vectors by $\boldsymbol{\tau}_A\cdot\mathbf{e}_A=\boldsymbol{\tau}_B\cdot\mathbf{e}_B=1$, $\boldsymbol{\tau}_A\cdot\mathbf{e}_B=\boldsymbol{\tau}_B\cdot\mathbf{e}_A=0$, we can define a vector **N** by $\mathbf{N}=E_A\boldsymbol{\tau}_A+E_B\boldsymbol{\tau}_B$. As seen in Fig. 13-13, **N** is normal to line AB because $\mathbf{OA}=E_A^{-1}\mathbf{e}_A$ and $\mathbf{OB}=E_B^{-1}\mathbf{e}_B$ both have the same component along **N**. That is $\mathbf{OA}\cdot\mathbf{N}=\mathbf{OB}\cdot\mathbf{N}=1$. Also, **OC** then has the same component along **N**, $1=\mathbf{OC}\cdot\mathbf{N}=|\mathbf{OC}|\mathbf{e}\cdot\mathbf{N}=|\mathbf{OC}|E=1$ or $|\mathbf{OC}|=E^{-1}$ and $\mathbf{OC}=E^{-1}\mathbf{e}$.

[63] F. C. Frank, in "Metal Surfaces," American Society of Metals, Metals Park, Ohio, 1963, p. 1.

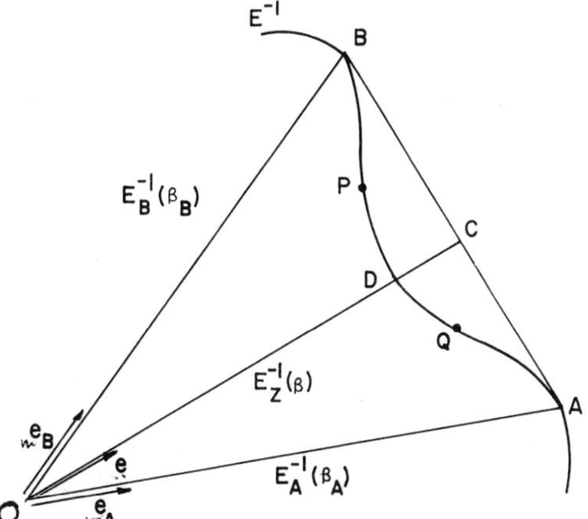

FIGURE 13-13. Portion of a plot of $E^{-1}(\beta)$ as a function of angle β in the glide plane.

Fig. 13-13, the orientation β is thermodynamically unstable with respect to zigzag lengths of orientation β_A and β_B.

The factor $\mathfrak{S}$ is negative, and hence $\mathscr{S}$ is negative, only between the points of inflection P and Q. Thus between P and Q the dislocation is absolutely unstable, whereas the discussion of Fig. 6-10 would suggest metastability and a barrier to faceting between B and P and Q and A. Whether the dislocations can still decompose spontaneously by a core fluctuation for these regions,[64] whether they can decompose from nodes or corners, or whether they require thermal fluctuation remains an open question.

Exercise 13-10. Prove that the points of inflection P and Q in Fig. 13-13 are points where $\mathfrak{S} = \mathscr{S} = 0$.

Head's original work on bcc β brass showed instability regions in both $\{110\}$ and $\{112\}$ planes for $\frac{1}{2}\langle 111\rangle$ dislocations.[64] Since then such instabilities have been studied widely,[65, 66] with instabilities of the type Head found common in bcc metals,[65] absent,[65] with the borderline exception of Pb, for fcc glide dislocations $\frac{1}{2}\langle 110\rangle\{111\}$, present[67] for fcc Lomer dislocations $\langle 100\rangle\{100\}$, present[65] for fcc Shockley partials $\frac{1}{6}\langle 112\rangle\{111\}$, and present for $\frac{1}{3}\langle 1\bar{1}20\rangle$ hcp dislocations in prismatic and pyramidal planes.[66] Steeds[65] and

[64] A. K. Head, in "Fundamental Aspects of Dislocation Theory," Special Publication 317, 1970, p. 5.

[65] J. W. Steeds, "Introduction to Anisotropic Elastic Theory of Dislocations," Oxford University Press, 1973.

[66] M. M. Savin, V. M. Chernov, and A. M. Strokova, *Phys. Stat. Solidi*, **35a**: 747 (1976).

[67] S. M. Holmes, P. M. Hazzledine and H. P. Karnthaler, *Phil. Mag.*, **39**: 277 (1979).

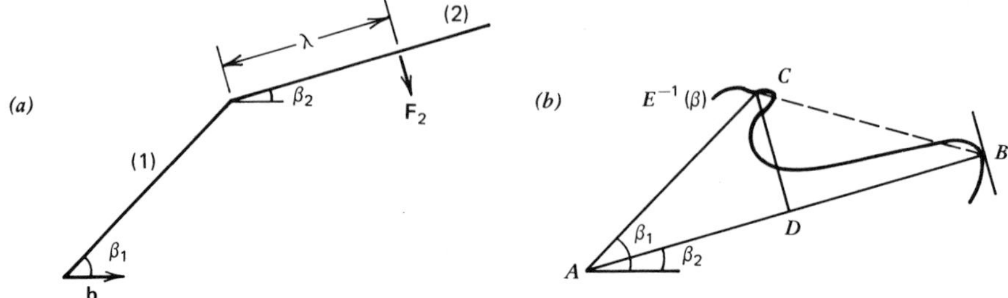

FIGURE 13-14. (*a*) Dislocation bend. (*b*) Corresponding plot of $E^{-1}(\beta)$.

Indenbom et al.[68] present tables for $E(\beta)$ for wide ranges of elastic constants, in cubic materials, which can be used to test for stability.

Instabilities with respect to climb are also possible, of course. They have received little attention. A three-dimensional E^{-1} plot, analogous to the three-dimensional surface energy plot, would then be needed to test for stability.

The preceding discussion proceeded in terms of the line-tension approximation. Earlier chapters presented some exact laws, developed from the theories developed in Sec. 5-7. A comparison of an exact result with the line-tension results is of interest. A relevant case is the bend of Fig. 13-14*a*. Equation (5-85) applied to this situation gives the result

$$dF_2 = \left[\frac{E(\beta_1)}{\sin(\beta_1-\beta_2)} - \frac{E(\beta_2)}{\tan(\beta_1-\beta_2)} + \frac{\partial E(\beta_2)}{\partial\beta_2}\right]\frac{d\lambda}{\lambda} = \frac{M_2}{\lambda}d\lambda \tag{13-226}$$

This is also the result of Exercise 5-9 with the θ coordinate rotated. The quantity M_2 has been called the *moment of force per unit length*.[69] It can be determined from an $E^{-1}(\beta)$ plot as shown in Fig. 13-14*b*.[69]

At point B a tangent is constructed to the $E^{-1}(\beta)$ curve. A line parallel to this tangent is drawn through C, intersecting AB at D. The moment M_2 is given by

$$M_2 = \frac{DB}{2A_{ABC}} \tag{13-227}$$

where A_{ABC} is the area of the triangle ABC.

Exercise 13-11. Prove that the geometric construction of Eq. (13-227) gives the result in Eq. (13-226).

[68] V. L. Indenbom, V. I. Alshits, and V. M. Chernov in Yu. A. Ossipyan (ed.), in "Defects in Crystals and Their Computer Modeling," Nauka, Leningrad, 1980.

[69] V. L. Indenbom and S. S. Orlov, *Kristallografiya*, **12**: 971 (1967).

As is apparent from the construction in Fig. 13-14*b*, for branch (2) to be force-free with $M_2=0$, for a given orientation of branch (1), CB must be tangent to $E^{-1}(\beta)$ at B. Then $DB=0$. However, the moment M_1 on branch (1) is not zero also unless CB is tangent to $E^{-1}(\beta)$ at C as well. Thus, for local equilibrium, CB must be a double tangent, and the equilibrium orientation of the branches is precisely that given by Fig. 13-13 based on the line-energy approximation. The line-energy approximation is remarkably good. We now understand that the zigzags are very sharp, with any local rounding at the corners caused by core energy terms.

The analysis of the stability of bends can be extended to other configurations, such as nodes. The result is that a triple node can be in equilibrium with straight branches only for specific orientations of the three branches. When long-range constraints such as network interactions are such that this configuration cannot be realized, the node experiences a net torque. The torque tends to rotate the node toward the characteristic equilibrium configuration near the node, with the consequence that the node arms are no longer straight.[70]

13-9. DISLOCATION SEGMENTS

Expressions for the distortion, stress and interaction forces produced by a dislocation loop in the anisotropic elastic case are presented in Sec. 5-7 in the form of explicit line integrals containing distortion or energy factors. An alternative line integral formalism is to directly modify Eq. (4-6). This relation has been converted to a form closely resembling the integral formalism of Sec. 13-7, but leaving the integrand in a Fourier integral form, by Leibfried[71] and further developed by Idenbom and Orlov[72]. The result can be written as

$$\mathbf{u}(\mathbf{r})=-\frac{\Omega\mathbf{b}}{4\pi}-\frac{1}{8\pi^2}\oint\left[\int_0^{2\pi}(nn)^{-1}(nq)\,d\phi\right]\frac{\mathbf{b}\,dl}{R} \qquad (13\text{-}228)$$

where $\mathbf{q}=\mathbf{n}\times\boldsymbol{\xi}$, $\mathbf{n}$ is perpendicular to $\mathbf{t}$, $\mathbf{t}=\mathbf{R}/R$, and $\mathbf{m}=\mathbf{n}\times\mathbf{t}$, i.e. $\mathbf{t}$ replaces $\boldsymbol{\xi}$ in the coordinate set of Fig. 13-12. The expression for the interaction energy of two segments, analogous to Eq. (4-40), has been derived by Lothe[73].

PROBLEMS

13-1. Compute the compliance matrix s'_{ij} for a cubic crystal with the coordinates of Fig. 13-2. Verify that $\{s'\}$ is the inverse of $\{c'\}$ for this case.

[70] V. L. Indenbom and G. N. Dubnova [*Sov. Phys. Solid State*, **9**: 117 (1967)] first discussed such effects in the isotropic elastic approximation.

[71] G. Leibfried, *Z. Phys.*, **135**: 23 (1953)

[72] V. L. Indenbom and S. S. Orlov, "Proceedings of the Kharkov Conference on Dislocation Dynamics" Acad. Sci. USSR, Moscow, 1968, p. 406.

[73] J. Lothe, submitted to *Phil. Mag.*, 1981.

13-2. Determine the elastic-constant matrix appropriate for determining the energy of a screw dislocation of the system $\langle 11\bar{2}3\rangle\{11\bar{2}2\}$ in an hcp crystal. What class of solution for p_n applies in this case?

13-3. Present a polar plot of σ_{xx} and σ_{xy} for an edge dislocation $\mathbf{b}=\frac{1}{2}[101]$ and $\boldsymbol{\xi}=[010]$ in NaCl. Compare the result with the isotropic results of Fig. 3-12.

13-4. The following hcp elastic constants[74] are given in units of 10 GPa.

Metal	c_{11}	c_{33}	c_{12}	c_{13}	c_{44}
Cd	12.1	5.13	4.81	4.42	1.85
Co	30.7	35.85	16.5	10.3	7.53
Mg	5.97	6.17	2.62	2.17	1.64
Zn	16.1	6.10	3.42	5.01	3.83

Compute the energy coefficients K_s and K_e for $\langle 11\bar{2}0\rangle(0001)$ dislocations for these metals.

13-5 **a.** Show that for a 60° mixed dislocation in fcc crystals the formula equivalent to Eq. (13-150) for the splitting into partials is

$$r=\frac{K_e b^2}{12\pi\gamma}$$

b. Use this result to verify the results of Table 13-1 for the 60° dislocation.

c. Compare the relative splittings given in Table 13-1 to those given by Eq. (10-15) for Al, Cu, Au, and Pb, using the Voigt averages for μ and ν.

d. Compare the above results with those obtained with the Reuss averages for μ and ν.

e. Compare the above results with those obtained with the crude approximation $\mu=c_{44}$ and $\nu=\frac{1}{3}$.

13-6. Derive the formula for the equilibrium dissociation of a $\frac{1}{2}\langle 111\rangle$ screw dislocation into partials $\frac{1}{3}\langle 111\rangle$ and $\frac{1}{6}\langle 111\rangle$ bounding an intrinsic stacking fault in a bcc crystal.

13-7. Show that for a dislocation lying along the threefold axis of a trigonal crystal, the solution for p_n does *not* reduce to a third-order equation in p^2. Why not?

13-8. For a screw dislocation lying along a $\langle 110\rangle$ direction in a cubic crystal, discuss whether a simple image construction yields the correct force acting on the screw because of the presence of a free surface parallel to it. Does the same reasoning apply for a $\langle 111\rangle$ screw dislocation?

[74] H. B. Huntington, *Solid State Phys.*, **7**: 213 (1958).

13-9. For Cu, using a force balance, assess the likelihood of the dissociation of a $\frac{1}{3}[111]$ partial dislocation lying along $[1\bar{1}0]$ into the two partials $\frac{1}{6}[112]$ and $\frac{1}{6}[110]$.

BIBLIOGRAPHY

1. Bacon, D. J., D. M. Barnett, and R. O. Scattergood, *Prog. Mat. Sci.*, **23**: 51 (1978).
2. Eshelby, J. D., W. T. Read, and W. Shockley, *Acta. Met.*, **1**: 251 (1953).
3. Hearmon, R. F. S., *Rev. Mod. Phys.*, **18**: 409 (1946); *Advan. Phys.*, **5**: 323 (1956).
4. Huntington, H. B., *Solid State Phys.*, **7**: 213 (1958).
5. Indenbom, V. L., V. I. Alshits, and V. M. Chernov, in Yu. A. Ossipyan (ed.), "Defects in Crystals and Their Computer Modeling," Nauka, Leningrad, 1980, p. 23.
6. Leibfried, G., *Z. Phys.*, **135**: 23 (1953).
7. Nye, J. F., "Physical Properties of Crystals," Oxford University Press, Fair Lawn, N.J., 1957.
8. Steeds, J. W., "Introduction to Anisotropic Elastic Theory of Dislocation," Oxford University Press, 1973.
9. Stroh, A. N., *Phil. Mag.*, **3**: 625 (1958).

PART 3

DISLOCATION–POINT-DEFECT INTERACTIONS AT FINITE TEMPERATURES

14

Equilibrium Defect Concentrations

14-1. INTRODUCTION

In Parts 1 and 2 we considered the elastic and geometric properties of dislocations at 0°K. In Part 3, comprising the next five chapters, we consider crystals at finite temperatures, treating equilibrium and nonequilibrium concentrations of point defects and their influence on dislocation dynamics and dislocation configurations. This chapter deals with equilibrium concentrations of point defects. First, kinks and jogs, which can be thought of as point defects on a dislocation line, are discussed. Vacancies and interstitials in the crystal lattice are considered, together with their interaction with dislocations; the concept of core vacancies and core interstitials is introduced. Impurity interactions with dislocations are then developed. Finally, as an example of more complicated cases, point defects in ionic crystals are treated.

For all the point-defect equilibria considered here there are various levels of approximation to an exact treatment, analogous to the successive approximations of linear isotropic elasticity, linear anisotropic elasticity, and nonlinear elasticity in deriving the elastic properties of dislocations. The development of a complete, rigorous statistical treatment of dislocation–point-defect equilibria would require a book in itself. Furthermore, some theoretical refinements are in a state of either controversy or continuing refinement. Thus, rather than attempting a general treatment, we develop a sequence of simple cases of equilibrium and avoid the most complex statistical-mechanical methods. More complicated treatments would follow the same physical reasoning as that presented here.

14-2. THERMAL KINKS

Free Energy of Formation of Kinks

Let us consider a crystal in which dislocations encounter Peierls barriers of the type discussed in Chap. 8. At 0°K a dislocation lying parallel to a Peierls

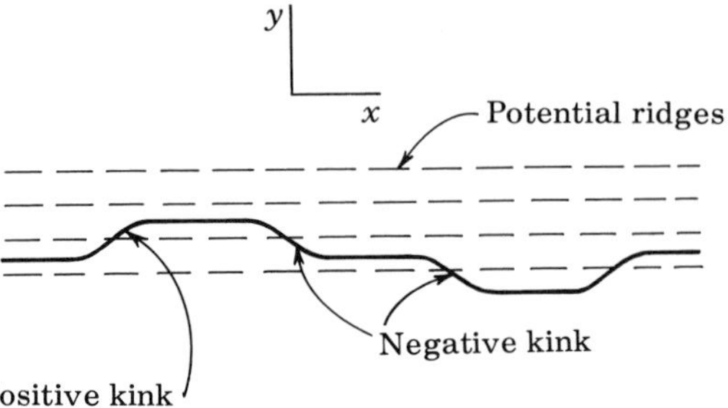

FIGURE 14-1. Double kinks in a dislocation line.

trough is at equilibrium if it is straight and lies in the bottom of the Peierls trough everywhere. However, at finite temperatures, the minimum free-energy-equilibrium configuration includes a number of randomly positioned double kinks which increase the entropy of the system. Such kinks, shown in Fig. 14-1, would not be present at 0°K; hence the term *thermal kinks*. The concentration of thermal kinks is determined by their free energy of formation, which is derived as follows.

The Helmholtz free energy $2F_k$ of a kink pair is composed generally of an energy part $2U_k$ and an entropy part $2S_k$,

$$2F_k = 2U_k - 2TS_k \tag{14-1}$$

Other contributions to $2U_k$ are negligible compared to the elastic and core misfit energy $2W_k$, given by Eq. (8-46). This energy includes kink formation energies, given by Eq. (8-47) or (8-50), and interaction energy, Eq. (8-48) or (8-50). For closely spaced kinks, the interaction energy is an important fraction of $2W_k$. However, a comparison of the above equations reveals that the interaction energy becomes negligible compared to the formation energy when the separation between kinks is large compared to the kink width w. This condition is fulfilled in most cases of interest, and even when kinks are closely spaced, the interaction energy can be neglected to a good approximation; in the limit that the separation between kinks equals w, the kink model breaks down, and a continuous-string model must be invoked for the dislocation. Thus, when a kink model is appropriate,

$$2U_k \cong 2W_k \cong \text{const} \tag{14-2}$$

with W_k given by Eq. (8-47) or (8-50).

The kink entropy depends on whether the kink is localized or not. First consider the case where each kink is localized at a minimum energy position, spaced at intervals a along the dislocation line, where a is the shortest crystal-symmetry repeat distance along the line (Fig. 14-2). The entropy is then connected with vibrational modes. In the high-temperature limit, the entropy

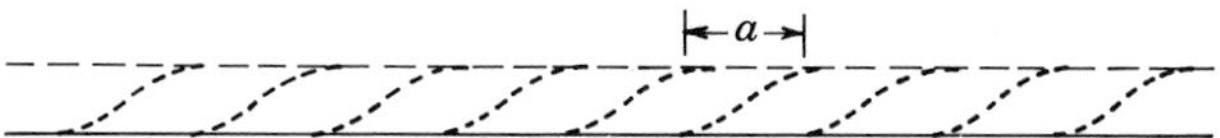

FIGURE 14-2. Equilibrium kink positions separated by intervals a along the dislocation line. Only one geometric-sign kink is shown for clarity; an opposite-sign kink could equally well occupy any site.

per oscillator of circular frequency ω is[1]

$$S = -k \ln\left[1 - \exp\left(-\frac{\hbar\omega}{kT}\right)\right] \cong k \ln \frac{kT}{\hbar\omega} \qquad kT \gg \hbar\omega$$

In the formation of the kink, a vibrational mode of frequency ω_k, the frequency of vibration of a kink about its equilibrium position, comes into existence. Since the total number of vibrational modes in the crystal is constant, one crystal or dislocation mode must disappear as the kink mode appears. We suppose that one oscillator of frequency ω_p is removed, where ω_p is the dislocation vibrational frequency in the Peierls valley. Thus the total entropy change S_k is given by

$$S_k \cong k\left(\ln \frac{kT}{\hbar\omega_k} - \ln \frac{kT}{\hbar\omega_p}\right) = k \ln \frac{\omega_p}{\omega_k} \tag{14-3}$$

This is a high-temperature expression. In a temperature range $\hbar\omega_p \gg kT \gg \hbar\omega_k$, only the kink vibrations would be excited classically, while the Peierls barrier vibration would have negligible entropy, so that

$$S_k \cong k \ln \frac{kT}{\hbar\omega_k} \tag{14-4}$$

The details about changes in vibrational modes when kinks are created are not understood well enough to warrant further elaboration. With reasonable values for the vibrational frequencies, TS_k is quite unimportant compared to U_k, so that for this case

$$2F_k \cong 2W_k \tag{14-5}$$

to a fair approximation.

In many cases, particularly for close-packed metals, the activational energy for lateral translation of a kink is expected to be so small that the kink can translate freely along the dislocation under thermal excitation. Indeed, experimental evidence from both microcreep studies[2,3] and internal-friction work[4,5]

[1] The circular frequency is related to the normal vibrational frequency ν by $\omega = 2\pi\nu$. Also $\hbar = h/2\pi$, where h is Planck's constant.

[2] J. W. Glen, *Phil. Mag.*, **1**: 400 (1956).

[3] R. F. Tinder, *J. Metals*, **16**: 94 (1964); R. F. Tinder and J. Washburn, *Acta Met.* **12**: 129 (1964).

[4] G. Fantozzi, C. Esnouf, W. Benoit, and I. G. Ritchie, *Prog. Mat. Sci.* (in press).

[5] R. H. Chambers, *Physical Acoustics*, **3a**: 123 (1966).

indicates that dislocation segments can move at very low stresses down to very low temperatures. Also, even at 0°K, the zero-point motion might suffice to prevent the kinks from being localized.[6] The simplest model for the entropy involves the removal of a vibrational mode and its replacement by a translational mode, giving

$$S_k = k\ln\left\{\frac{a}{h}(2\pi m^*kT)^{1/2}\left[1-\exp\left(-\frac{\hbar\omega_p}{kT}\right)\right]\right\} \tag{14-6}$$

Here m^*, the effective mass of the kink, is given, by analogy with Eq. (7-21), as $m^* = W_k/C_t^2$. The factor a is introduced to normalize the kink concentration relative to a standard state of one kink per site, $1/a$ (Fig. 14-2), so that Eqs. (14-3) and (14-6) are self-consistent. More refined models,[7,8] which include contributions from the total vibrational spectrum of the dislocation line, yield results within about $4k$ of Eq. (14-6) at moderate and high temperatures. This difference is probably not more than the uncertainty in a calculation for any one of the models. At low temperatures the entropy factors can be quite different, but there the value of S_k matters less because of the decrease in T, which multiplies S_k in the free-energy expression. Once again, Eq. (14-5) provides a fair approximation for $2F_k$ in this case.

Equilibrium Kink Concentration

The equilibrium concentration of kinks is related directly to the free energy of formation $2F_k$. As discussed in Chap. 8, one cannot specify how the free energy $2F_k$ should be split between the two individual kinks without specifying how the kinks can be formed individually at the surface. The free energy of a double-kink pair, however, always equals $2F_k$. Only this total free energy $2F_k$ enters the following discussion, since the kinks can appear or disappear only in *pairs* under thermal fluctuations in the crystal interior.

Consider a straight dislocation segment of length L lying in a Peierls valley and containing $N_k^+ = c_k^+ L$ positive kinks (see Fig. 14-1). The segment L contains $N = L/a$ kink sites for the positive kinks. The N_k^+ kinks can be distributed among the N sites in

$$P^+ = \frac{N!}{(N-N_k^+)!N_k^+!} \tag{14-7}$$

[6] J. Lothe, *J. Appl. Phys.*, **33**: 2116 (1962).

[7] J. Lothe and J. P. Hirth, *Phys. Rev.*, **115**: 543 (1959); A. Seeger and P. Schiller, *Acta Met.*, **10**: 348 (1962). The results for S_k (expressed in terms of partition functions) differ by a factor of $k\ln 3$ in the two papers. However, this is not a serious discrepancy, and is the result of the more approximate model used in the first calculation.

[8] Other developments of the partition function for a kink differ in detail from the results of reference 7, but do not differ in principle. For example, see V. Celli, M. Kabler, T. Ninomiya, and R. Thomson, *Phys. Rev.*, **131**: 58 (1963); G. Alefeld, R. H. Chambers, and T. E. Firle, *Phys. Rev.*, **140**: A1771 (1965).

ways. With the use of Stirling's approximation, $\ln x! = x(\ln x - 1)$, Eq. (14-7) becomes

$$\ln P^{+} = -Lc_k^{+}(\ln c_k^{+}a - 1) \qquad c_k^{+}a \ll 1 \tag{14-8}$$

The corresponding configurational entropy term is $k \ln P^{+}$ Thus the total configurational entropy contribution per unit length to the free energy from both positive and negative kinks is

$$-TS_c = kT\left[c_k^{+}(\ln c_k^{+}a - 1) + c_k^{-}(\ln c_k^{-}a - 1)\right] \tag{14-9}$$

At internal equilibrium the total free energy per unit length F must be a minimum with respect to a change in the number of kink pairs

$$\delta F = 2F_k\,\delta c - T\delta S_c = 0 \tag{14-10}$$

where $\delta c = \delta c_k^{+} = \delta c_k^{-}$. The combination of Eqs. (14-9) and (14-10) yields

$$2F_k + kT\ln c_k^{+}a + kT\ln c_k^{-}a = 0 \tag{14-11}$$

In an exponential form, this expression becomes

$$c_k^{+}c_k^{-} = \frac{1}{a^2}\exp\left(\frac{-2F_k}{kT}\right) \tag{14-12}$$

Equation (14-12) represents a mass-action law for positive and negative kinks. In a kinetic derivation, one would equate the kink-kink annihilation rate, proportional to $c_k^{+}c_k^{-}$, to the kink-pair creation rate, proportional to $\exp(-2F_k/kT)$. When entropy of formation effects are ignored, leading to the approximation of Eq. (14-5), Eq. (14-12) reduces to the simple form first proposed by Weertman.[9]

Skew Dislocation

If the average direction of the dislocation line makes an angle θ with the direction of the Peierls valley (Fig. 14-3), then

$$c_k^{+} - c_k^{-} = \frac{\theta}{d} \tag{14-13}$$

where d is the separation between neighboring valleys. Equations (14-12) and (14-13) combine to

$$c_k^{-}\left(\frac{\theta}{d} + c_k^{-}\right) = \frac{1}{a^2}\exp\left(\frac{-2F_k}{kT}\right) \tag{14-14}$$

[9] J. Weertman, *J. Appl. Phys.*, **28**: 1185 (1957).

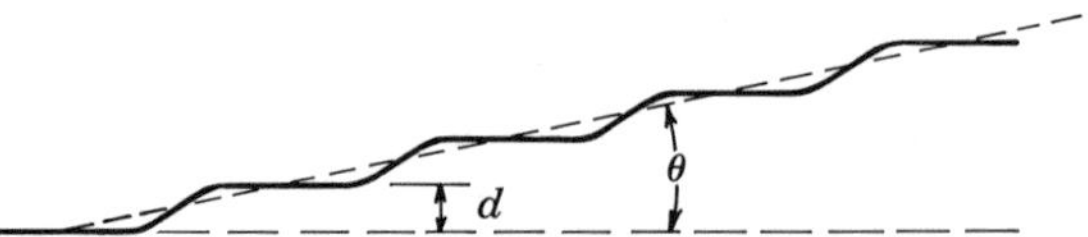

FIGURE 14-3. Kinks in a dislocation line inclined at an angle θ to the direction of the Peierls valleys.

For inclinations so large that

$$\frac{\theta}{d} \gg \frac{1}{a} \exp\left(\frac{-F_k}{kT}\right)$$

the kink densities become

$$c_k^+ = \frac{\theta}{d} \qquad c_k^- = \frac{1}{c_k^+ a^2} \exp\left(\frac{-2F_k}{kT}\right) \tag{14-15}$$

The negative kinks are "quenched" out by the much higher concentration of positive kinks. The entropy is reduced greatly, and the kinks present are mainly those which have to be present by geometric necessity.

The String Model

For high kink densities, kink-kink interactions cannot be neglected. For an average angle with the Peierls valley

$$\theta \gtrsim \frac{d}{w} \tag{14-16}$$

where w is the kink width, the kinks overlap severely. Instead of a discrete kink model, the dislocation resembles a taut string with small wiggles reflecting the presence of the Peierls valleys (Fig. 14-4). In this case the dislocation is most conveniently considered as a continuous string with line tension $\mathcal{S}$ given by Eq. (6-83) or (6-84), depending on the dislocation character. The small wiggles influence the cutoff parameter ρ, but an estimate of this effect cannot be made with present theory. Equation (6-13) still can be used as a reasonable approximation.

Most estimates[10] give kink widths of the order $w \sim 5d$ to $10d$ for slip dislocations in close-packed metals. Thus for angles $\theta \gtrsim 0.1$ the kinks already lose their individuality. Although several preferred orientations with noticeable Peierls valleys must exist, the taut-string model should be appropriate over most of a generally curved dislocation. This expectation is in agreement with

[10] J. Lothe and J. P. Hirth, *Phys. Rev.*, **115**: 543 (1959); A. Seeger and P. Schiller, *Acta Met.*, **10**: 348 (1962); *Phys. Acoust.*, **3a**: 361 (1966).

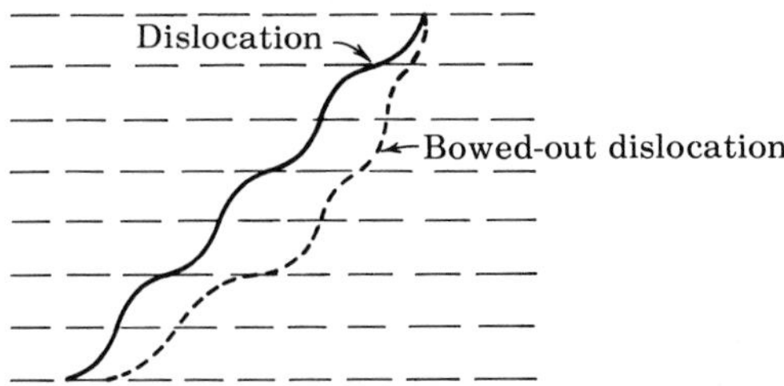

FIGURE 14-4. A dislocation inclined at a large angle to the Peierls valleys.

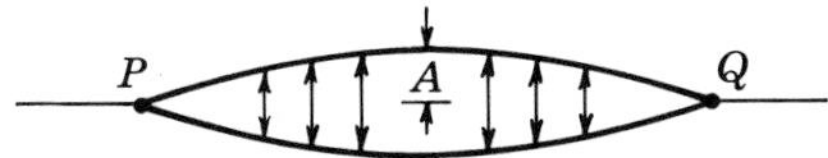

FIGURE 14-5. String model of a dislocation vibrating with amplitude A between two pinning points P and Q.

electron-microscopic observations on the common metals. The dislocations are continuously curved for dislocation densities $\lesssim 10^7$ cm^{-2}, such as usually observed in thin films; dislocation interactions and image forces completely obscure any effects caused by the Peierls barrier. For very low dislocation densities, 10^2 to 10^3 cm^{-2}, where dislocation interactions are weak, dislocations in aluminum appear as straight lines along low-index crystallographic directions,[11] suggesting Peierls barrier influence. Similarly, in silicon, which has a relatively large Peierls energy, dislocations are observed to lie in low-index directions for somewhat larger dislocation densities.[12]

The concept of an equilibrium concentration of kinks is meaningless for the dislocation in Fig. 14-4. Instead, the dislocation is considered to be a continuous string in thermal vibration. Consider the dislocation segment pinned at P and Q, as in Fig. 14-5. Let us treat the lowest mode of thermal excitation, that of a half-wave of amplitude A. In the line-tension model, the energy of the excitation is

$$W = \mathcal{S}(L' - L) \tag{14-17}$$

where L' is the length of the segment in the bowed-out configuration. With a sinusoidal form,

$$L' - L = \frac{\pi^2 A^2}{4L} \tag{14-18}$$

for small bow-outs. Thus

$$W = \frac{\pi^2 \mathcal{S} A^2}{4L} \tag{14-19}$$

[11] B. Nøst, *Phil. Mag.*, **11**: 183 (1965); B. K. Basu and C. Elbaum, *Phil. Mag.*, **9**: 533 (1964); A. Authier, C. B. Rogers, and A. R. Lang, *Phil. Mag.*, **12**: 547 (1965).

[12] W. C. Dash, *J. Appl. Phys.*, **27**: 1193 (1956).

Equating this expression to $\frac{1}{2}kT$, one finds that the root-mean-square amplitude of thermal vibration is

$$\left(\overline{A^2}\right)^{1/2} = \frac{1}{\pi}\left(\frac{2LkT}{\mathcal{S}}\right)^{1/2} \tag{14-20}$$

With the typical values $\mathcal{S} \sim (\mu b^2/4\pi)\ln(L/b)$, $\mu b^3 \sim 5$ eV, and $kT \sim 0.025$ eV (room temperature), Eq. (14-20) yields

$$\begin{aligned} \left(\overline{A^2}\right)^{1/2} &\sim 1.4b \quad \text{for } L \sim 10^3 b \\ \left(\overline{A^2}\right)^{1/2} &\sim 4.2b \quad \text{for } L \sim 10^4 b \end{aligned} \tag{14-21}$$

The above development does not involve any assumptions about the mobility of the dislocation. When the mobility is high, the dislocation can vibrate quite freely at essentially its eigenfrequency, determined by $\mathcal{S}$, and by its effective mass [see Eq. (7-48)]. When severely damped, the dislocation undergoes a slow diffusional motion. In any case, Eq. (14-20) applies for the mean deviation caused by thermal fluctuations. Evidently, the same argument is applicable in the calculation of the mean thermal fluctuation of a dislocation in a climb process, involving much higher frictional drag than in glide processes.

High Kink Densities

Two extreme cases of noninteracting kinks and of overlapping kinks are treated above. For kink densities that are high, but not high enough to produce overlap, kink-kink interactions are important, particularly for skew dislocations. Let us consider a skew dislocation pinned at P and Q, as in Fig. 14-6, and suppose that the kinks are the freely translating type. At absolute zero, the kinks will be equally spaced to minimize elastic repulsion. At finite temperature, the system is set in thermal motion. Because of the kink-kink interaction, the thermal motion is analyzed best in terms of correlated motion. The kink-kink interaction arises from two sources. One of these is the elastic repulsion of the kinks, which leads to a line tension given by Eq. (8-87). The other involves an entropy contribution. At finite temperatures the line tension formally represents the change in free energy with change in length. A variation in the kink distribution accompanying bow-out changes the configurational entropy [(Eq. (14-9)], and this change contributes to the line tension.[13, 14] In most cases the elastic part of the line tension is dominant.

At higher temperatures, the thermal fluctuations are large enough that kink-kink interactions can be ignored, while at even higher temperatures, the

[13] G. Alefeld, *J. Appl. Phys.*, **36**: 2633, 2642 (1965).

[14] E. Bode, *Phil. Mag.*, **13**: 275, 493 (1966).

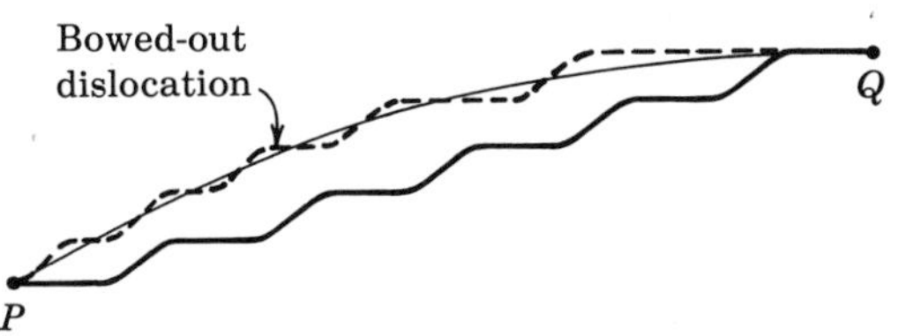

FIGURE 14-6. A bowed-out dislocation containing discrete kinks.

nucleation of kink pairs complicates the discussion. A rigorous and useful treatment of all these effects is difficult and is not undertaken here.

In summary, three cases of kink equilibria are distinguishable:

1. Dislocations nearly parallel to Peierls valleys, with the *thermal* concentration of kinks dominant, where the effect of temperature is mainly in the nucleation of kink pairs.
2. Skew dislocations containing essentially only *geometric* kinks, where kink nucleation is unimportant. The effect of temperature is that of excitation of correlated thermal motion of the kinks, with the mode of agitation depending on kT compared with the kink-kink interaction.
3. Dislocations so skew that they behave like smooth vibrating strings under thermal agitation.

Only a narrow range of dislocation orientations can correspond to cases (1) and (2). At the present state of theory, no clear criterion can be given to predict which case should apply; there must be a wide spectrum of intermediate cases.

The results of this section are of interest in creep phenomena[15] and internal friction.[16] Some of the applications are discussed in the next chapter.

14-3. THERMAL JOGS

Thermal agitation also causes a dislocation to deviate out of its original slip plane by the formation of thermal jog pairs. Much of the analysis in this case is directly analogous to the theory of thermal kinks. Consider an initial edge dislocation which is straight and lies in its glide plane under no net force. Let a double-jog pair be formed by removing (or adding) atoms from the dislocation and restoring (or removing) them at a ledge site on the crystal surface, so that

[15]See J. Weertman, *J. Appl. Phys.*, **28**: 185 (1957); J. Lothe and J. P. Hirth, *Phys. Rev.*, **115**: 543 (1959).

[16]See A. Seeger, H. Donth, and F. Pfaff, *Disc. Faraday Soc.*, **23**: 19 (1957); J. Lothe, *Phys. Rev.*, **117**: 704 (1960); Conf. on Internal Friction, *Acta Met.*, **10**: 267–500 (1962); G. Fantozzi, C. Esnouf, W. Benoit, and I. G. Ritchie, *Prog. Mat. Sci.* (in press).

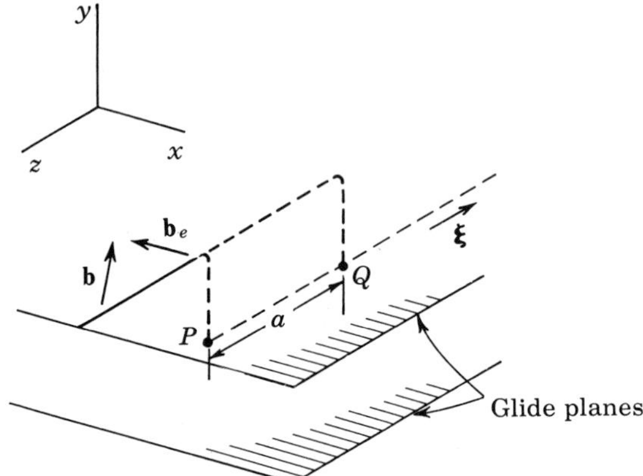

FIGURE 14-7. A jog in a mixed dislocation.

they regain the equilibrium chemical potential of the system (see Sec. 3-4). The atoms could equally well be put into a vapor in *equilibrium* with the crystal, or interstitially if the interstitial concentration has the equilibrium value. The net free-energy change in the above process is the free energy of formation of the jog pair $2F_j$. The formation energy of the jogs [Eq. (8-90)] is much larger than that of a kink pair, while the interaction energies are about the same in either case. The entropy of formation is given by an expression such as Eq. (14-3). Thus, to an even better approximation than in the case of kinks, one can set

$$2F_j = 2W_j \tag{14-22}$$

with W_j constant and equal to the jog formation energy [Eq. (8-90)].

Consider the jog configuration in a mixed dislocation, as shown in Fig. 14-7. Here d is the spacing between slip planes. With the removal of one atom at the jog, it moves from P to Q, a distance

$$a = \frac{v_a}{b_e d} \tag{14-23}$$

where v_a is the atomic volume. Hence possible jog positions are a distance a apart. For crystals such as ionic crystals, the jogs at P and Q are not identical. Complications such as these are neglected for the present but are treated at the end of the chapter.

The remainder of the development proceeds exactly like that for kinks. The concentration of thermal jogs is given by

$$c_j^+ c_j^- = \frac{1}{a^2} \exp\left(-\frac{2F_j}{kT}\right) \tag{14-24}$$

In the special case of a dislocation lying on one slip plane except for thermal jogs, Eq. (14-24) reduces to

$$c_j^+ = c_j^- = \frac{1}{a}\exp\left(-\frac{F_j}{kT}\right) \tag{14-25}$$

The long-range jog-jog interaction is similar to that for kinks (see Sec. 8-9), so its discussion is the same as that of the preceding section. Indeed, the theories of kinks and jogs are identical with respect to equilibrium properties. The chief difference between jogs and kinks is a difference in mobility; kinks can glide easily and are highly mobile, while jogs can move only by climb, which requires diffusion. However, the mobility has no effect on the *equilibrium* behavior of jogs or kinks.

Since the formation energy of jogs is much larger than that for kinks, the concentration of jogs is less than that of kinks at equivalent temperatures. Also, the jogs are believed to be quite abrupt, with much smaller widths than kinks (Chap. 8). Thus the concept of individual jogs interacting via long-range stresses has more general validity than the equivalent idea applied to kinks.

The theory of this section is very important in considerations of climb processes.[17]

14-4. INTRINSIC POINT DEFECTS IN EQUILIBRIUM WITH DISLOCATIONS

In addition to dislocations, crystals contain other defects that interact with the dislocations. This interaction is one of the most important topics in the discussion of real crystals at finite temperatures. Even in infinitely pure crystals, vacancies and interstitials are present at finite temperatures as intrinsic point defects that lower the free energy of the crystal. Divacancies, trivacancies, etc., also exist as intrinsic defects, but, except at temperatures near the melting point, they are generally present in very small concentrations compared to single vacancies. For this reason, and for simplicity, the following discussion is restricted to single point defects.

The intrinsic point defects are defined by the operations illustrated in Fig. 14-8. The removal of an atom A from the crystal shown in Fig. 14-8 creates a *vacancy* at A, while an extra atom squeezed into an interstitial position, such as C, is called as *interstitial.* Unlike the case of kinks and jogs, where only the free energies of formation of pairs of defects could be well defined, the free energy of the interstitial-vacancy pair can be divided into free energies of formation for the individual defects. Let us first treat the case of interstitial equilibrium.

[17]See J. Lothe, *J. Appl. Phys.*, **31**: 1086 (1960); R. M. Thomson and R. W. Balluffi, *J. Appl. Phys.*, **33**: 803, 817 (1962); R. W. Balluffi and A. V. Granato, in F. R. N. Nabarro (ed.), "Dislocations in Solids," vol. 4, North-Holland, Amsterdam, 1979, p. 1.

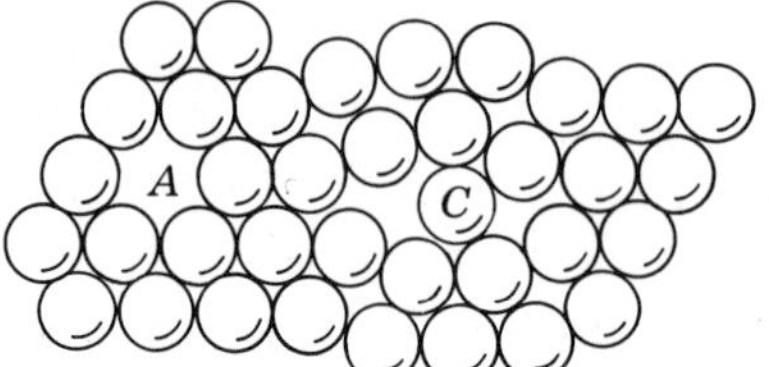

FIGURE 14-8. A vacancy at A and an interstitial at C in a two-dimensional section of a close-packed lattice.

Interstitial Equilibria

Figure 14-8 shows obviously that the interstitial is a strong center of repulsion, producing compressive stresses in the lattice. Indeed, in the model that the interstitial is a ball inserted into a hole too small for it in an isotropic continuum, the results of Eq. (2-79) indicate that the insertion of an interstitial into a stress-free crystal produces an external volume expansion

$$\delta V = V_i = v_a \frac{1+(4\mu/3B)}{1+(4\mu/3B')} \tag{14-26}$$

where B' is the "bulk modulus" of the interstitial atom. The *local* volume change in the vicinity of the interstitial is

$$v_i = v_a \frac{1}{1+(4\mu/3B')} \tag{14-27}$$

while the remainder of the volume change is an expansion of the entire lattice under image forces

$$v_e = v_a \frac{4\mu/3B}{1+(4\mu/3B')} \tag{14-28}$$

Consider now the formation process illustrated in Fig. 14-9, where an atom is removed from a surface ledge on a stress-free reference crystal A and placed into the crystal in question B.[18] If the crystal B is initially stress free, the change in free energy in this process, composed of the potential energy of the broken bonds in the process and the distortion energy surrounding the interstitial, is the Gibbs free energy of formation[19] of the interstitial G_i.

[18] W. Kossel [*Nach. Ges. Wiss.* (*Göttingen*), 135 (1927)] first pointed out that a condensing atom gains its entire free energy of condensation when it is added to a surface ledge at a half-crystal position or surface kink. This concept is very useful in studies of surface structure, surface diffusion, and evaporation, as reviewed in Chaps. 1 to 6 in N. A. Gjostein and W. D. Robertson (eds.), "Metal Surfaces," American Society of Metals, Cleveland, Ohio, 1963. The reader is referred to this review for further discussion of surface ledges.

[19] In treating constant-pressure processes, the Gibbs free energy is more convenient to use than the Helmholtz free energy F. They are related by $G = F + PV$.

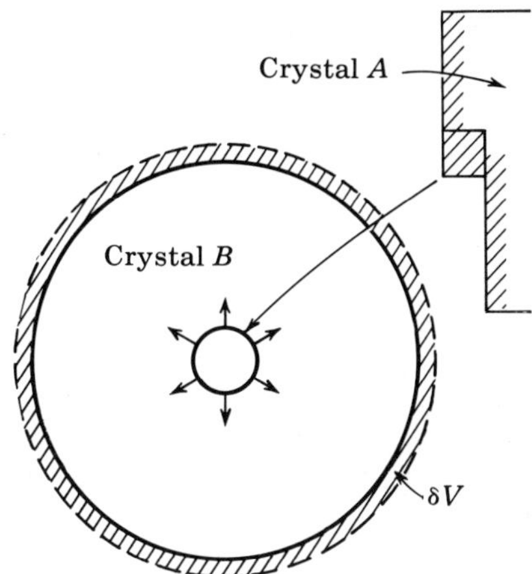

FIGURE 14-9. Formation of an interstitial by the removal of an atom from the surface of reference crystal A and its insertion into crystal B. In general, both crystals A and B can be acted upon by external stresses.

Including an entropy-of-mixing term in analogy with previous developments, the equilibrium concentration of interstitials is given in terms of n_i, the concentration of interstitial sites as

$$c_{i_0}{}^0 = n_i \exp\left(-\frac{G_i}{kT}\right) \tag{14-29}$$

Let us now generalize to the case where B in Fig. 14-9 represents a region of crystal under an external stress σ_{ij} with a hydrostatic component $P = -\frac{1}{3}\sigma_{ii}$ and under an internal stress with a hydrostatic component p. Then, in the insertion of the interstitial into the crystal, there is an additional work term $(P+p)v_i^*$. This term arises partly from the extra work $(P+p)v_i$ done in the expansion of the hole to accommodate the interstitial and partly from the extra work of compressing the interstitial to the pressure $P+p$. Thus v_i^* generally differs from v_i. Detailed calculations show that the two contributions in v_i^* simply add to give the *external volume change* V_i (see Probs. 14-13 and 14-14),

$$v_i^* = V_i$$

The quantity V_i is a thermodynamic formation volume in the context of being the coefficient of the pressure in the reversible work process of inserting the interstitial. If the removal of atom A from the reference surface, perhaps now on the same crystal as B where a pressure P exists, entails extra work W_e, the total free energy of formation becomes[20]

$$G = G_i + (P+p)V_i + W_e \tag{14-30}$$

[20] Work PV_i is done against external pressure. However, there is a negative interaction energy $-Pv_e$ between external pressure and the stress field of the interstitial (Sec. 2-7), leaving a net term $P(V_i - v_e) = Pv_i$. This net term is indeed one of the contributions to the reversible work done

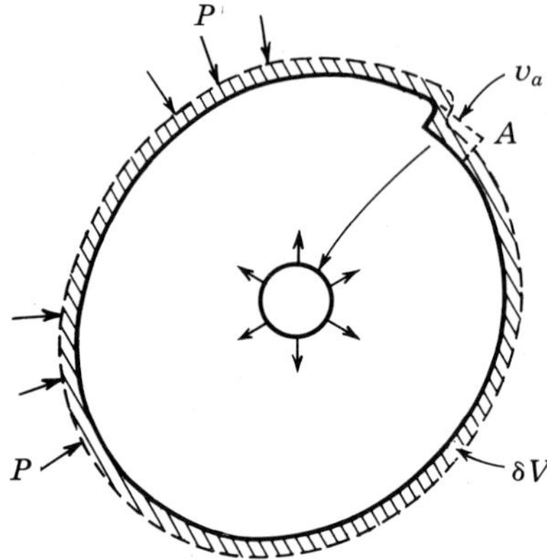

FIGURE 14-10. Creation of an interstitial in a crystal under hydrostatic pressure P by the removal of a surface atom at A and its insertion into the crystal.

The equilibrium concentration of interstitials then becomes

$$c_i = c_i^0 \exp\left(-\frac{W_e}{kT}\right) \tag{14-31}$$

where

$$c_i^0 = n_i \exp\left[-\frac{G_i + (P+p)V_i}{kT}\right] \tag{14-32}$$

c_i^0 is defined as the reference-state, or standard-state, concentration of interstitials. This definition, including some of the P-V work in the reference free energy, is most convenient in the description of *climb forces* in the following chapters. The definition also has the advantage that the chemical potential $\overline{G} = kT\ln(c_i/c_i^0)$ is constant throughout an internally equilibrated crystal with an interstitial supersaturation relative to $c_{i_0}{}^0$, Eq. (14-29), but with no sources or sinks present, and in which the internal pressure is variable.[21]

As an application of the above formulas, consider the case of a crystal under a uniform external hydrostatic pressure P, and let the interstitials be formed at the crystal surface (Fig. 14-10). The removal of a surface atom yields external work

$$W_e = -Pv_a \tag{14-33}$$

locally in inserting the interstitial. Thus, although a term PV_i is present in the free energy of the interstitial formed under external pressure P, while no such term is present for internal pressure p, the free-energy change caused by either type of pressure can be determined from the reversible work done locally in inserting the interstitial, in agreement with the general principles of thermodynamics.

[21] J. Lothe and J. P. Hirth, *J. Appl. Phys.*, **38**: 845 (1967).

so that the interstitial concentration is given by Eq. (14-31) as

$$c_i = c_i^0 \exp\left(\frac{Pv_a}{kT}\right) \tag{14-34}$$

Inserting $B = B'$ in Eq. (14-26), one finds $V_i = v_a$. Thus, in a rough approximation, Eq. (14-34) reduces to

$$c_i = n_i \exp\left(-\frac{G_i + pv_a}{kT}\right) \tag{14-35}$$

The interstitial concentration is *independent* of external pressure with the approximation that $V_i = v_a$.

Vacancy Equilibria

Figure 14-8 indicates that the vacancy is surrounded by less distortion than the interstitial. In fact, in the continuum theory for vacancies, the vacancy would be produced by simply cutting out a sphere, and the surrounding material would be *absolutely stress free*. In actual crystals, the vacancy generally tends to contract the lattice about it and hence is a moderately strong center of hydrostatic tension[22]; image stresses of the type discussed in Sec. 2-7 would tend to accentuate this tension effect. In the continuum model the vacancy is represented by inserting a ball into a hole that is larger than the ball, stretching the two surfaces to coincide, and gluing the surfaces together. The approximations involved in this model are more questionable than for the interstitial, but the model serves to clarify the relation between the external volume contraction V_v and the amount of the local contraction v_v. Analogous to the case of interstitials, V_v is composed of the local volume contraction v_v and an image contraction of the entire crystal v_c. The actual local vacant volume is

$$\Delta v_v = v_a - v_v$$

Consider that a vacancy is formed by the reverse of the process of Fig. 14-9; an atom is removed from crystal B and placed at a ledge site on the reference crystal A. The development proceeds analogously to the case of the interstitial. In the most general case of a crystal under external stresses with a hydrostatic component P and internal stresses with a hydrostatic component p, the total free energy of formation is

$$G = G_v - (P + p)v_v^* + W_e \tag{14-36}$$

where G_v is the free energy of formation under zero pressure and W_e is the

[22]A. Seeger [*J. Phys. F*, **3**: 248 (1973)] lists experimental values for metals indicating a relaxation V_v / v_a in the range 0.25 to 0.08.

extra external work done in placing the atom on the surface of crystal A. The equilibrium concentration of vacancies is

$$c_v = c_v{}^0 \exp\left(-\frac{W_e}{kT}\right) \tag{14-37}$$

where

$$c_v{}^0 = n_v \exp\left[-\frac{G_v - (P+p)v_v^*}{kT}\right] \tag{14-38}$$

$c_v{}^0$ is the reference-state concentration of vacancies, and n_v is the concentration of possible vacancy sites. Again, this reference state is chosen for convenience in treatments of climb forces and to give a constant chemical potential

$$\bar{G} = kT \ln\left(\frac{c_v}{c_v{}^0}\right)$$

for vacancies equilibrated in a crystal in which the internal pressure p is variable.

Detailed calculations show that v_v^* differs from v_v and is equal to the external volume contraction

$$v_v^* = V_v$$

analogous to the result for interstitials. The formation volume differs, however, as we demonstrate for vacancies equilibrated at an external surface under hydrostatic pressure P. Forming a vacancy by removing an atom from the interior and placing it on the surface leads to external work in the final step

$$W_e = P v_a$$

Thus the vacancy concentration in equilibrium with the surface is

$$c_v = c_v{}^0 \exp\left(-\frac{P v_a}{kT}\right) \tag{14-39}$$

With $v_v^* = V_v$, the insertion of Eq. (14-38) into (14-39) yields the result that

$$c_v = n_v \exp\left[-\frac{G_v + P(v_a - V_v) - pV_v}{kT}\right] \tag{14-40}$$

The coefficient of internal pressure p is the relaxation volume V_v, whereas that of the external pressure P is the vacancy volume as measured externally. The thermodynamic formation volume is thus $\Delta V_v = v_a - V_v$.

Instead of being formed separately, a vacancy-interstitial pair can be formed by removing an atom from a normal site and placing it in an interstitial position removed from the initial site sufficiently that the interaction energy between the two is negligible. Mass action requires that the product $c_v c_i$ be constant. Since the external work terms W_e are zero in this formation process, the local equilibrium condition is, from Eqs. (14-32) and (14-38),

$$c_v c_i = c_v{}^0 c_i{}^0 = n_v n_i \exp\left[-\frac{G_i + G_v + (P+p)(V_i - V_v)}{kT}\right] \tag{14-41}$$

Point-Defect–Dislocation Equilibria

In general, point defects can equilibrate at dislocations because the latter can act as sources and sinks for point defects. Near a dislocation, then, the point defects are in quasiequilibrium with the dislocation, while closer to the surface they are in quasiequilibrium with the surface. If the dislocations and the surface did not produce *identically* the same chemical potential of vacancies or interstitials throughout the volume of the crystal, diffusive flows of point defects between dislocations and to the surface would be produced. The dislocations would *climb*. Such nonequilibrium situations are studied in Chap. 15. In equilibrium, when dislocation motion has ceased, the point defects are in equilibrium with the surface, the dislocations, and other possible sources and sinks *individually*.

Consider the change in point-defect concentration produced by the internal stress field of the dislocation. In the approximation of the above model that the vacancy and interstitial are centers of *isotropic* contraction or expansion, respectively, the point defects interact only with the hydrostatic component p of the internal stress. For a straight dislocation the hydrostatic stress is produced by the edge component b_e only. Equation (3-44) gives the result

$$p = -\tfrac{1}{3}(\sigma_{rr} + \sigma_{\theta\theta} + \sigma_{zz}) = \frac{\mu b_e}{3\pi}\,\frac{1+\nu}{1-\nu}\,\frac{\sin\theta}{r} \tag{14-42}$$

Inserting Eq. (14-42) into (14-32) and (14-38), one finds that there is an increased concentration of interstitials but a decreased concentration of vacancies in the region of tension; the opposite holds for the region of compression. This result is illustrated in Fig. 14-11.

The above treatment is in the approximation of rigidly straight dislocations. However, the stress field about a point defect described as a center of expansion is one of pure shear [Eq. (2-81)]. These shear stresses induce wiggles on the dislocations that are actually flexible. The wiggles modify the situation so that a center of dilatation interacts with an initially straight screw dislocation also.[23] The induced wiggles have some nonscrew character with an

[23] R. J. Arsenault and R. de Wit, *Scripta Met.* **15**: 615 (1981).

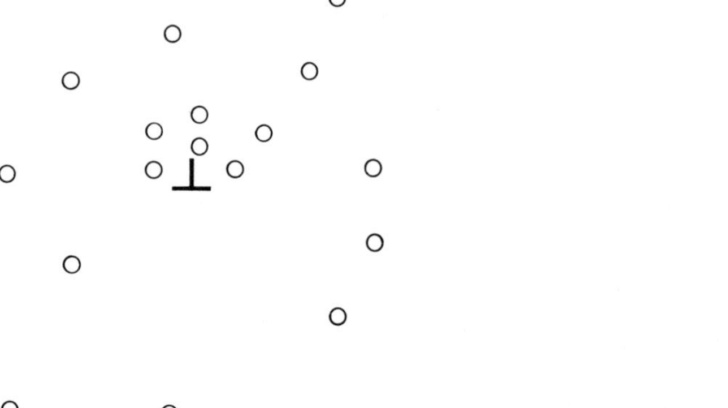

FIGURE 14-11. Distribution of vacancies about an edge dislocation.

associated p field and a $p\,\delta v$ interaction with the point defects. The interaction energy is of the order $\sim 0.1\mu(\delta v)^2/r^3$ and is thus relatively weak and short-ranged.[24] This type of effect also leads to interaction among point defects: the wiggles produced by one defect give a hydrostatic pressure that causes interaction with another defect. The determination of atmospheres formed in such a manner about screw dislocations is thus a difficult many-body problem. It appears that the flexible line can interact effectively only with isolated point defects near the core, which can be considered roughly as interacting independently with the flexible screw. We do not elaborate further on this case. One should keep in mind that size-effect interactions between screw dislocations and point defects may be of some significance.

Higher-Order Approximations

Near the dislocation line, where the internal stresses are high, second-order terms in p^2 are also important. When a vacancy is created in a region of internal stress, the strain energy that was stored in that atom site, wv_v, where w is the strain-energy density [Eq. (2-14)], is released. Here w is proportional to p^2. Actually, since the relaxation occurs in a somewhat larger volume than v_v, the total relaxation energy is perhaps about $2wv_a$. Equation (14-38) is modified by this term to the form

$$c_v{}^0 = n_v \exp\left[-\frac{G_v-(P+p)v_v^*-2wv_a}{kT}\right] \tag{14-43}$$

This second-order effect tends to cause vacancies to be attracted to all regions

[24]For the model of Fig. 6-8 with $\tan\phi \cong \phi = 2h/L$, the bow-out energy is $2h^2\mathcal{S}/L - \sigma bLh/2$, with the first term associated with the increased line length, the second the work done by the local shear stress in sweeping the area $Lh/2$. Minimizing this energy with respect to h and using the approximation $\mathcal{S} \cong \mu b^2/2$, we find the relaxation energy on bow-out to be $-\sigma^2L^3/16\mu$. With the stress produced by the point defect $\sigma \sim \mu\,\delta v/\pi r^3$ and $r \sim L \sim$ the distance of the point defect from the dislocation, we obtain the above result.

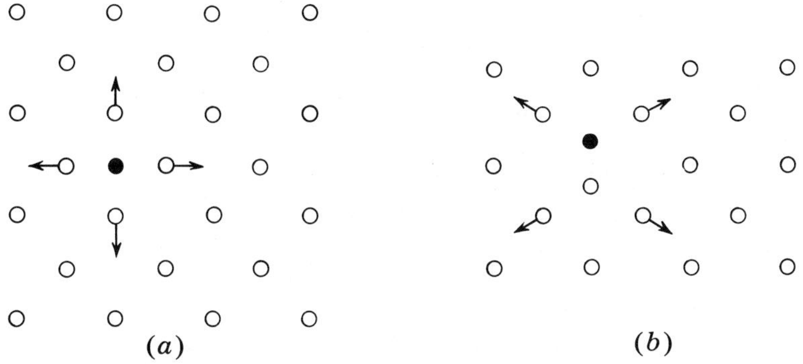

FIGURE 14-12. The {100} plane in an fcc crystal showing an interstitial atom (*a*) in a normal octahedral position and (*b*) in a dumbbell configuration aligned along ⟨100⟩.

of internal stress, regardless of sign or whether it is shear or dilatational, and causes an appreciable interaction with both screw and edge dislocations. For an order-of-magnitude estimate, with $2wv_a \sim \mu b^5/4\pi^2r^2$, $\mu b^3 \sim 5$ eV, and $kT \sim 0.025$ eV (room temperature), then $2wv_a/kT \sim 1$ at a distance $r \sim 2.5b$ from the dislocation. Except at low temperatures, the second-order term is important only very close to dislocations.

Similar second-order effects are present for the interstitial. According to Bullough and Willis,[25] an interstitial produces a small surrounding region of low shear rigidity, in which internally and externally induced shear stresses are relaxed. Although the direct first-order interaction is much stronger, the second-order interaction is important in creep under irradiation.

Another type of correction involves asymmetries in the volume change produced by a point defect. In the isotropic continuum approximation, the interstitial is assumed to produce isotropic dilatation only. In actual crystals because of both local nonlinear elastic effects and elastic anisotropy, the real, stable interstitial configuration can have tetragonal symmetry, even in the fcc lattice.[26] A schematic illustration of such a dumbbell-shaped defect is presented in Fig. 14-12. A given interstitial in a "normal" configuration of cubic symmetry can react into the dumbbell configuration with three possible mutually perpendicular orientations. An interstitial in such a tetragonal distortion would interact with shear stresses as well as normal stresses. In a stress field such as that around a dislocation, the three orientations of the defect would have different energies of interaction with the stress field. At equilibrium the defects would be distributed over the three orientations in accordance with a Boltzmann factor containing the interaction energy. Divacancies and other clusters of defects, the properties of which are known in detail,[27] can also have nonisotropic distortions associated with them.[28]

[25] R. Bullough and J. R. Willis, *Phil. Mag.*, **31**: 855 (1975).

[26] W. Schilling, *J. Nucl. Mat.*, **69/70**: 465 (1978).

[27] R. W. Siegel, J. Nucl. Mater., **69/70**: 117 (1978); R. W. Balluffi, ibid., p. 240.

[28] A. Seeger, Cryst. Latt. Defects, **4**: 221 (1973).

Finally, very near the core and in the core, all theories involving elasticity break down, and in most cases atomistic theories are too crude to yield accurate estimates of "interstitial" and "vacancy" energies in the core. There is ample evidence of enhanced self-diffusion along dislocations,[29-31] with an activation energy about half that for bulk diffusion. In terms of core structure, one would say that the concentration and/or mobility of "vacancies" or "interstitials" in the core must be much higher than in the bulk.[32, 33] However, there are difficulties in rationalizing experiments by such models, because they would require extremely high defect concentrations in the core region.[31, 32] Alternatively, the core might be regarded as being hollow,[31] or as consisting of core point defects which act as incipient jogs. For example, in a pure screw dislocation a core "vacancy" can be resolved, at least formally, into the incipient spiral containing four kinks, as shown in Fig. 12-20.

In the common metals the vacancies have much lower formation energies than the interstitials—for noble metals typically 1 eV versus 5 eV. The thermal concentration of interstitials is thus negligible compared to that of vacancies at all temperatures below the melting point. Therefore, in *equilibrium* problems it suffices to consider vacancies alone. For all metals the vacancy concentration just below the melting point is typically of the order $c_v \sim 10^{-3}$ to $10^{-4} v_a^{-1}$. Interstitials can be introduced in significant concentrations under nonequilibrium conditions of plastic flow or irradiation. In the treatment of such effects in a quasiequilibrium or steady-state diffusion approximation, the preceding formulae for interstitials have important applications. In irradiation creep under external load, the second-order interaction between interstitials and stress becomes important.[34] This interaction provides coupling between external stress and internal dislocation stress and thereby gives a mechanism for preferential interstitial absorption by dislocations with certain orientations relative to the external stress field.

14-5. SOLUTE-ATOM EQUILIBRIA

Isotropic Size Effect

Solute-atom–dislocation interactions are, of course, important in alloys. Even in pure crystals, impurity solutes can preferentially adsorb at dislocations and appreciably affect the dislocation properties. Solute atoms can be present

[29] See D. Lazarus, *Solid State Phys.*, **10**: 71 (1960) for a review of early work in this area.

[30] R. N. Tucker and P. Gibbs, *J. Appl. Phys.*, **29**: 1375 (1958).

[31] R. N. Tucker, A. Laskar, and R. Thomson, *J. Appl. Phys.*, **34**: 445 (1963).

[32] J. Lothe, *J. Appl. Phys.*, **31**: 1077 (1960).

[33] G. R. Love, *Acta Met.*, **12**: 731 (1964).

[34] R. Bullough, D. W. Wells, J. R. Willis, and M. H. Wood, in "Dislocation Modeling of Physical Systems," Pergamon, Oxford, 1981 p. 116.

either substitutionally, replacing a solvent atom on a lattice site, or interstitially in an interstice in the solvent matrix. In either case, the impurity is a center of expansion or contraction. To a first approximation, let us again describe the solute atom in the isotropic continuum model of Sec. 2-7, where it is regarded as a sphere placed in a spherical hole of a different size.[35] This approximation appears to be good for substitutional solutes, particularly in close-packed structures. One would expect uncertainties caused by the uncertain "compressibility" of the solute atom, chemical-valency effects, and local relaxations. However, for appreciable solubility, the size mismatch is less than 15 percent, according to the Hume-Rothery rule,[36] so that size differences are small. Indeed, experiments indicate that volume changes are proportional to the atomic volume of the solute,[37] and that the variation of lattice parameter with solute concentration agrees well with the continuum model.[38]

In the continuum approximation of Eq. (2-91), then, the external volume change produced by a solute atom is

$$\delta V \cong (v_s - v_a)\frac{1+(4\mu/3B)}{1+(4\mu/3B')} \cong v_s - v_a \tag{14-44}$$

where v_s and v_a are the atomic volumes of solute and solvent atoms, respectively. The strength of the center of expansion is

$$\delta v = \frac{v_s - v_a}{1+4\mu/3B'}$$

If the solute is inserted in a field of pressure p, reversible work $p\,\delta v$ must be done. However, with the refinement of including the energy associated with the increased pressure p in the inserted solute atom, one finds that the total reversible work of inserting the solute atoms is $p\,\delta V$. If, in addition, $B \sim B'$, the energy of interaction becomes $p(v_s - v_a)$. In the presence of the internal stress field of a dislocation, the solute concentration is, then, in complete analogy to Eq. (14-32),

$$c = c_0 \exp\left[-\frac{p(v_s - v_a)}{kT}\right] \tag{14-45}$$

where c_0 is the solute concentration in regions of zero internal pressure.

In the continuum model the solute atoms interact negligibly except through their image fields; only the image fields give rise to hydrostatic pressure. To the

[35] J. D. Eshelby, *Phil. Trans. Roy. Soc.*, **A244**: 87 (1951); *Solid State Phys.*, **3**: 79 (1956).

[36] W. Hume-Rothery and G. V. Raynor, "The Structure of Metals and Alloys," Institute of Metals, London, 1956, p. 100.

[37] J. Friedel, *Phil. Mag.*, **46**: 514, 1169 (1955).

[38] J. D. Eshelby, *Solid State Phys.*, **3**: 79 (1956).

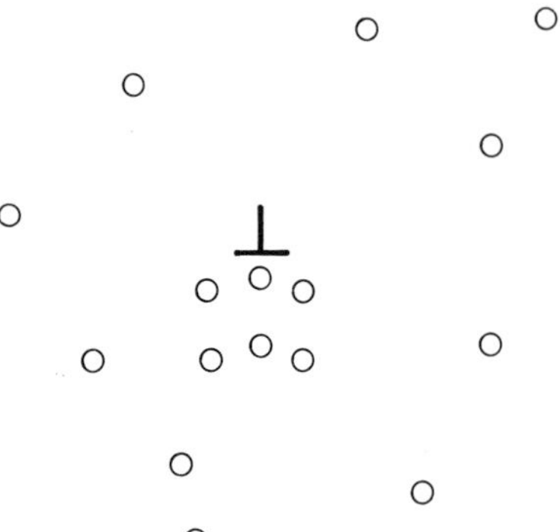

FIGURE 14-13. Distribution of solute atoms in the stress field of an edge dislocation for the case $v_s > v_a$.

same approximation, the presence of a solute concentration does not screen the hydrostatic stress field of the dislocation; the hydrostatic component of the dislocation stress field is the same with or without solute atoms present. Therefore, for the solute distribution around a dislocation with an edge component, Eqs. (14-42) and (14-45) combine to yield

$$c = c_0 \exp\left(- \frac{\beta \sin\theta}{rkT} \right) \tag{14-46}$$

where

$$\beta = \frac{\mu b_e}{3\pi} \frac{1+\nu}{1-\nu} (v_s - v_a) \tag{14-47}$$

Near the core of an edge dislocation, $r \sim b$, and with $\nu \sim 0.3$, $\theta = -\pi/2$, $v_s = 1.1 v_a$, $v_a \sim b^3$, and $\mu b^3 \sim 5$ eV, the interaction energy is ~ 0.1 eV. With $v_s > v_a$, there is a depletion of impurities on the compression side of the edge dislocation and accumulation in the dilational region (Fig. 14-13).

Often one is interested in the total capacity of a dislocation for solute atoms, in the sense of solute atoms per unit length of dislocation line *above* that which would exist for a uniform concentration c_0. Consider again the edge dislocation in Fig. 14-13, and let us calculate the net extra number of solute atoms between the radial distances r_0 and R, that is, for the moment the core region is excluded. If r_0 is big enough that $\beta < r_0 kT$, Eq. (14-46) can be expanded in a power series,

$$c - c_0 = -\frac{c_0 \beta \sin\theta}{rkT} + \frac{c_0 \beta^2 \sin^2\theta}{2r^2k^2T^2} - \frac{c_0 \beta^3 \sin^3\theta}{3!r^3k^3T^3} + \cdots \tag{14-48}$$

The net number of extra solute atoms per unit length is

$$\frac{N}{L} = \int_{r_0}^{R} r\,dr \int_0^{2\pi} (c - c_0)\, d\theta \tag{14-49}$$

The odd terms on the right-hand side of Eq. (14-48) integrate to zero, but the even terms gives a contribution

$$\frac{N}{L}=\frac{\pi\beta^2 c_0}{k^2T^2}\left\{\frac{1}{2}\ln\frac{R}{r_0}+\sum_{n=1}^{\infty}\frac{1}{n[2^{n+1}(n+1)!]^2}\left[\left(\frac{\beta}{r_0kT}\right)^{2n}\left[1-\left(\frac{r_0}{R}\right)^{2n}\right]\right]\right\} \tag{14-50}$$

Numerical evaluations of the sum[39] show that the leading logarithmic term gives a good approximation of N/L except for values of β so large that the near core region is saturated so that a Fermi-Dirac expression must also be used instead of Eq. (14-46). Thus we restrict further discussion to the leading term only. The result for N/L diverges as $R\to\infty$, so that a straight dislocation in an infinite medium has an *infinite* capacity for impurity atoms.[40] In real crystals, R is roughly the spacing between opposite-sign dislocations, according to St. Venant's principle. With $R\sim10^4b$ and $r_0\sim2b$, then $\ln(R/r_0)\sim8.5$. With the same values for the atomic parameters as above, and with $kT\sim0.05$ eV, $N/L\cong50b^2c_0$. In a 2 percent alloy, this would amount to about one solute atom per atomic plane intersected by the dislocation. In pure crystals, the effect would be negligible. For reasonable dislocation densities, it is marginal whether or not an effect on solubility would be detectable. In the above example, even for the extremely high dislocation density of 10^{12} cm^{-2}, which would correspond to $R\sim10^2b$, the total increase in solute content in an alloy with a matrix composition of 2 percent would be ~0.1 percent.

Exercise 14-1. For the cases of Zn in Cu and Cu in Al, plot the concentration-distance curves for a 3 percent alloy according to Eqs. (14-46) and (14-48) for $\theta=\pi/2$ and $\theta=-\pi/2$. Determine N/L for the former case, with $R=10^4b$. Discuss the choice of a cutoff distance r_0 in view of the results.

Exercise 14-2. Consider a dissociated dislocation of width d in an fcc crystal. Let the dislocation lie along the z axis, with the x axis in the plane of the fault and the origin midway between the partials. Superpose the stress fields of the two partials and show that the hydrostatic stress component is

$$p=\frac{\mu y}{3\pi}\frac{1+\nu}{1-\nu}\left[\frac{b_{e_1}}{(x-d/2)^2+y^2}+\frac{b_{e_2}}{(x+d/2)^2+y^2}\right]$$

where b_{e_1} and b_{e_2} are the edge components of the partials. In particular, for a

[39] J. P. Hirth and B. Carnahan, *Acta Met.*, **26**: 1795 (1978). As discussed in this work and by W. R. Tyson [*Corrosion*, **36**: 441 (1980)], other series evaluations of Eq. (14-49) are in error.

[40] R. Thomson, *Acta Met.*, **6**: 23 (1958).

dissociated screw dislocation with total Burgers vector **b**,

$$p=\frac{\sqrt{3}\,\mu yb}{18\pi}\frac{1+\nu}{1-\nu}\frac{2xd}{\left[(x-d/2)^2+y^2\right]\left[(x+d/2)^2+y^2\right]}$$

Plot the concentration-distance curves with the same parameters as in Exercise 14-1 and compare the N/L values. The results of this exercise show that dissociated dislocations of *all* characters interact hydrostatically with solutes in fcc metals.

Concentrated Solutions

The preceding theory is applicable for dilute solutions. Some modifications are necessary for concentrated solutions. Let us consider finite concentrations of *noninteracting* solute atoms. The approximation of no interaction is not as bad as one might think. As discussed in Sec. 2-7, a simple size mismatch causes *no* hydrostatic pressure in the surrounding medium in the absence of image forces; in this approximation, a second solute atom would not interact with the first one. When image terms are included, there are interaction terms proportional to $(\delta V)^2$ arising from the interaction of the image strain fields.[41] However, this interaction does not manifest itself as a simple two-body interaction of impurity atoms; the first simple considerations above are correct in that there is no large interaction term that prevents the impurities from being randomly distributed on a local scale. There are short-range repulsions, which are accounted for roughly in the following treatment by requiring that no two solute atoms can occupy the same site.

Consider a crystal with N atom sites and cV solute atoms, substitutional or interstitial. Let $n=N/V$ be the number of possible sites per unit volume. When no two solute atoms are allowed to occupy the same site, the impurities can be arranged in

$$P=\frac{(nV)!}{(nV-cV)!(cV)!} \tag{14-51}$$

ways. Employing Stirling's formula, one obtains

$$\ln P=-V\left[c\ln\frac{c}{n}+(n-c)\ln\left(1-\frac{c}{n}\right)\right] \tag{14-52}$$

In Eq. (14-8) the term corresponding to the final term in Eq. (14-52) was expanded to give the -1; the retention of the logarithmic term here accounts for the statistical effects of finite concentrations. The entropy per unit volume

[41] J. D. Eshelby, *Solid State Phys.*, **3**: 79 (1956).

becomes

$$S = -k\left[c\ln\frac{c}{n} + (n-c)\ln\left(1-\frac{c}{n}\right)\right] \tag{14-53}$$

This entropy term produces a contribution to the chemical potential $\bar{G}$, the change in free energy with the addition of a solute atom, given by

$$\bar{G} = -T\frac{\partial S}{\partial c} = kT\ln\frac{c}{n-c} \tag{14-54}$$

Let us now rederive an expression for the distribution of solute atoms in a field of internal pressure p in a more exact form than Eq. (14-45). With c_0 the concentration in the region where $p=0$, the equilibrium requirement that the chemical potential be the same everywhere leads to the result

$$p(v_s - v_a) + kT\ln\frac{c}{n-c} = kT\ln\frac{c_0}{n-c_0} \tag{14-55}$$

or, in place of Eq. (14-45),

$$\frac{c(n-c_0)}{c_0(n-c)} = \exp\left[-\frac{p(v_s-v_a)}{kT}\right] \tag{14-56}$$

which can be rearranged to the form

$$\frac{c}{n} = \frac{1}{1+\exp\{[p(v_s-v_a)-\bar{G}_0]/kT\}} \tag{14-57}$$

where

$$\bar{G}_0 = kT\ln\frac{c_0}{n-c_0} \tag{14-58}$$

Equation (14-57) has the form of the Fermi-Dirac distribution, with $\bar{G}_0$ as the Fermi energy. The requirement that no two solute atoms occupy the same site is analogous to the Pauli principle of quantum mechanics.[42]

Exercise 14-3. Expand Eq. (14-57) to give

$$\frac{c}{n} = \sum_{j=1}^{\infty}(-1)^{j+1}\left(\frac{c_0}{n-c_0}\right)^j \exp\left[-\frac{jp(v_s-v_a)}{kT}\right] \tag{14-59}$$

Verify from the formula that the logarithmic divergence of Eq. (14-50) is not a spurious effect peculiar to the approximation of infinite dilution.

[42] D. N. Beshers, *Acta Met.*, **6**: 521 (1958).

Core Segregation

Obviously, in addition to solute interactions with the elastic strain fields of dislocations, solute states also are possible in the dislocation core $r < r_0$, where elastic calculations break down. Such core states could attract impurities more strongly than could any site in the elastic region. Unfortunately, atomic calculations for possible core states are very difficult to perform. Here we postulate that solute sites exist at intervals $a \sim b$ along the dislocation core. Solute atoms in these sites are assumed to be noninteracting[43] and to have a free energy, exclusive of entropy of mixing, given by

$$F = F_0 - F_B \tag{14-60}$$

where F_0 is the free energy of a solute atom at a bulk site far removed from the dislocation, and F_B represents the free energy of *binding*. To the same kind of approximation as Eq. (14-22), F_B usually is considered to be a simple energy term

$$F_B \sim U_B \tag{14-61}$$

Such core binding states are expected to be present for *both* edge and screw dislocations.

Although more than one type of core state is possible, for simplicity, only one type is assumed here. Because of anisotropies on an atomic scale, the most favorable core states are realized only when the dislocation line has particular crystallographic directions. Furthermore, with a given direction of the dislocation line, these states are possible only for certain dislocation positions regularly spaced, similar to Peierls valleys,[44] as shown in Fig. 14-14. Thus the impurity interaction tends to align the dislocation in the same manner as does the Peierls barrier, but the preferred dislocation positions or directions need not be the same for the two effects. The consequences of the solute interaction with regard to kink structure, etc., are analogous to those discussed in Chap. 8.

The derivation of the concentration of impurities at the core proceeds analogously to the treatment in the preceding section. By analogy to Eq (14-57), the core concentration is

$$c_d = \frac{1}{a}\,\frac{1}{1+\exp\left[-\left(F_B+\bar{G}_0\right)/kT\right]} \tag{14-62}$$

with $\bar{G}_0$ relating to c_0, the concentration in bulk solution, by Eq. (14-58). For

[43] H. Reiss [*J. Chem. Phys.*, **40**: 1783 (1964)] has presented a model that includes nearest-neighbor interactions for solutes within the core. The mathematics for such a case becomes quite lengthy. Because of this, and because of the uncertainties in our understanding of core states, we present the simpler model here.

[44] J. Lothe, *Acta Met.*, **10**: 663 (1962).

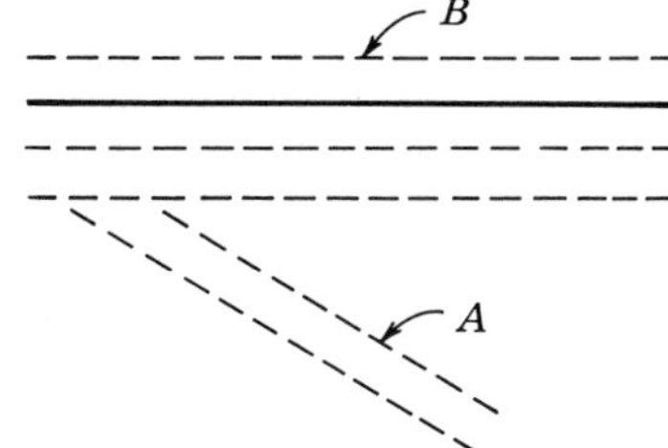

FIGURE 14-14. A dislocation lying parallel to the direction of positions of maximum impurity accommodation, set B. These directions are not in general the same as the normal Peierls valleys for a pure crystal, set A.

the case $F_B \ll -\overline{G}_0$ and $c_0 \ll n$ (dilute bulk concentration), Eq. (14-62) reduces to[45]

$$c_d \cong \frac{c_0}{an}\exp\left(\frac{F_B}{kT}\right) \tag{14-63}$$

The core has only a dilute concentration of impurities in this case. In the other limit, when $F_B \gg -\overline{G}_0$, then $c_d \sim 1/a$; the core is *saturated*. To a better approximation, the latter case yields

$$1 - ac_d = \exp\left(-\frac{F_B + \overline{G}_0}{kT}\right) \tag{14-64}$$

which gives the probability of *holes*, i.e., unoccupied core sites, along the dislocation. Since a hole in an otherwise saturated string and a solute atom in solution can be created in the same operation, a mass-action law should relate holes and solute atoms in solution. Indeed, substituting Eq. (14-58) into Eq. (14-64), one obtains

$$(1 - ac_d)\frac{c_0}{n} = \exp\left(-\frac{F_B}{kT}\right) \tag{14-65}$$

Usually, c_d is dilute for a sufficiently high temperature, while at lower temperatures it is saturated, except in extremely pure crystals. For a given c_0 and F_B, one can define a *transition temperature* T_0 that divides the regions of dilute and saturated behavior as the temperature at which half the core sites are occupied. From Eq. (14-62), $ac_d = \frac{1}{2}$ when $\overline{G}_0 = -F_B \cong -U_B$, or

$$T_0 = -\frac{U_B}{k\ln[c_0/(n - c_0)]} \tag{14-66}$$

At some temperature below T_0, the next-lowest energy-impurity levels outside the core would start to saturate, etc. In the region of core saturation, several of these levels are saturated in general. Thus a realistic model in the saturation

[45]A. H. Cottrell and B. A. Bilby, *Proc. Phys. Soc.*, **A62**: 49 (1949).

range would have to include a large number of parameters, so that a convincing assignment of specific values to parameters such as U_B from experimental observations is usually very difficult. The entire development is useless, of course, when actual precipitation of a new phase takes place at the dislocation core.

Other Elastic Interaction Effects

In addition to the size effect, solutes interact with internal stresses in the hardball–hole model because they are small spheres with different elastic constants from that of the matrix. Within the approximations of linear elasticity, the strained volume of the solute atom is the same as the volume of the solvent atom it replaces, so that one can simply consider the different amount of energy stored in a sphere of size v_a with different elastic constants. With the strain-energy density given by Eq. (2-14), the difference in energy caused by the difference in elastic constant is[46]

$$\Delta W = \tfrac{1}{2}\epsilon_{ij}\epsilon_{kl}v_a\left(c'_{ijkl} - c_{ijkl}\right) \tag{14-67}$$

where the primed constants refer to the solute atom and the strains are those produced by the internal stress field with which the solute interacts. For example, the interaction with the hydrostatic pressure field p of an edge dislocation is

$$\Delta W = \tfrac{1}{2}p^2 v_a \frac{B' - B}{BB'} \tag{14-68}$$

Notice that Eq. (14-67) includes interaction with shear-stress fields, so that this effect causes interaction with both screw and edge dislocations.[47–49] In either case, the solute concentration is changed by

$$c = c_0 \exp\left(\frac{\Delta W}{kT}\right) \tag{14-69}$$

The approximation that the solute atom has bulk elastic properties is questionable, so that Eq. (14-67) should not be taken too literally.

For many solutes, crystalline anisotropy leads to nonisotropic distortions around solute atoms. The classical example of such an effect is the tetragonal distortion about a carbon atom dissolved interstitially in iron (Fig. 14-15).

[46] Strictly valid only when $|c'_{ijkl} - c_{ijkl}| = |\Delta c| \ll c_{ijkl} = c$. However, the relation is still a reasonable rough estimate when $\Delta c = -c$ (vacancy) or $\Delta c = c$ (solute twice as stiff as solvent). For $\Delta c > c$, second-order terms become important.

[47] J. D. Eshelby, *Phil. Trans. Roy. Soc.*, **A244**: 87 (1951).

[48] J. Friedel, "Les Dislocation" Gauthier-Villars, Paris, 1956.

[49] R. L. Fleischer, *Acta Met.*, **9**: 996 (1961).

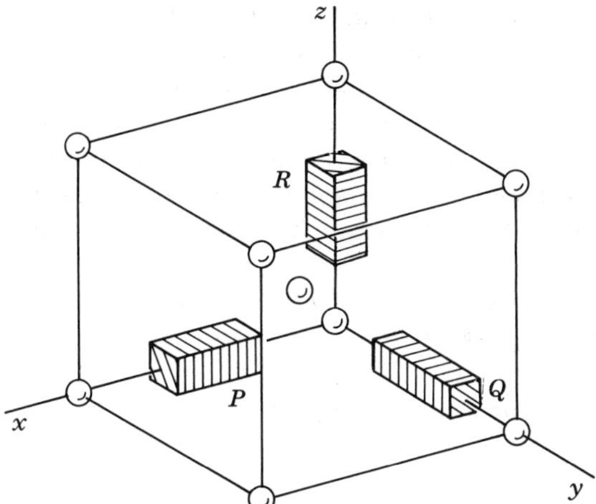

FIGURE 14-15. Possible interstitial sites P, Q, and R for carbon dissolved in bcc iron. The carbon atom is idealized in the continuum model as a tetragonal defect with sides $(\mathbf{a}, \mathbf{a}, \mathbf{c})$.

Such defects interact to the first order in δV with both shear stresses and normal stresses, with the same consequences as those discussed above for the dumbbell interstitial defect. Such asymmetric defects can be described, in principle, in terms of Green's functions. Green's functions for a point force in the anisotropic case are available[50] and have been applied for point defect dislocation interactions.[51] The problem with the results is that the strength of the point forces are generally unknown. Thus we restrict our discussion to the simpler isotropic elastic case, using Eq. (2-70). The principal manifestation of the tetragonality is in a preference for one of the three possible orientations of the tetragonal distortion. For example, in Fig. 14-15, in a stress field of simple tension along the x axis, interstitials in configuration P would have a lower energy than those in configurations Q or R. In thermal equilibrium, more carbon atoms would be in P positions than in Q or R positions. Such rearrangements are important in the interaction of dislocations, particularly in bcc metals, with interstitial carbon, nitrogen, or hydrogen.[52, 53] Interstitials in $(\frac{1}{2}, \frac{1}{4}, 0)$ tetrahedral positions in bcc metals also produce tetragonal distortion,[54] but not so pronouncedly as interstitials in the $(\frac{1}{2}, 0, 0)$ octahedral positions, such as P, Q, and R in Fig. 14-15. Remarkably, although the partial molar volume of interstitial hydrogen is nearly as large as that of carbon and

[50] D. M. Barnett and J. Lothe, *Phys. Norv.*, **8**: 13 (1975); R. J. Asaro, J. P. Hirth, D. M. Barnett, and J. Lothe, *Phys. Stat. Solidi*, **60b**: 261 (1973).

[51] R. A. Masamura and G. Sines, *J. Appl. Phys.*, **41**: 3930 (1970).

[52] J. Snoeck, *Physica*, **6**: 591 (1939); *Physica*, **8**: 711 (1941); *Physica*, **9**: 862 (1942).

[53] A. W. Cochardt, G. Schoeck, and H. Wiedersich, *Acta Met.*, **3**: 533 (1955).

[54] D. N. Beshers, *J. Appl. Phys.*, **36**: 290 (1965).

nitrogen, experiments[55] indicate that its strain field is isotropic in many bcc metals and can be treated by the ball-in-hole model.

Unlike the result of the continuum theory, when tetragonal distortions are present, solute-solute interactions occur. Also, screening of the strain field of a dislocation by a solute atmosphere is possible even in the linear elastic case. This effect, which should be important in the indirect influence of solute atoms on dislocation-dislocation interaction, has received little attention.

Exercise 14-4. Compute the interaction energy of an interstitial atom in configuration P in Fig. 14-15 with a screw dislocation.

Outline of solution. Represent the tetragonal distortion by three double forces of the type shown in Fig. 2-5, only two of which are equal. The displacement field of a single double force is

$$u_x = A\left[(1-4\nu)\frac{x}{r^3} + \frac{3x^3}{r^5}\right] \qquad u_y = -A\left(\frac{y}{r^3} - \frac{3x^2 y}{r^5}\right) \qquad u_z = -A\left(\frac{z}{r^3} - \frac{3x^2 z}{r^5}\right)$$

The strain tensor produced by these displacements reduces, to first order, to the three components ϵ_{xx}, ϵ_{yy}, and ϵ_{zz}. Comparison with lattice parameter measurements[56,57] suggests that at the surface of the unit cell containing the carbon atom $\epsilon_{xx} \sim 0.38$, $\epsilon_{yy} = \epsilon_{zz} \sim -0.026$. Let the axes of Fig. 14-15 specifically be $x \| [100]$, $y \| [010]$, and $z \| [001]$. The stress tensor of a screw dislocation lying along $[111]$ is given in terms of the coordinates $x'y'z'$ referred to $\mathbf{i}' = (1/\sqrt{2})\,[\bar{1}10]$, $\mathbf{j}' = (1/\sqrt{6})\,[\bar{1}\bar{1}2]$, and $\mathbf{k}' = (1/\sqrt{3})\,[111]$ by Eq. (3-3) as

$$\sigma_{x'z'} = -\frac{\mu b_s \sin\theta}{2\pi r} \qquad \sigma_{y'z'} = \frac{\mu b_s \cos\theta}{2\pi r}$$

where θ is measured from the x' axis, $r = (x'^2 + y'^2)^{1/2}$, and $\mathbf{b}_s$ is the Burgers vector of the screw dislocation. Transformed to the xyz coordinates, the stress tensor becomes

$$\sigma_{xx} = -\frac{\sqrt{6}}{3}\sigma_{x'z'} - \frac{\sqrt{2}}{3}\sigma_{y'z'} \qquad \sigma_{yy} = \frac{\sqrt{6}}{3}\sigma_{x'z'} - \frac{\sqrt{2}}{3}\sigma_{y'z'} \qquad \sigma_{zz} = \frac{2\sqrt{2}}{3}\sigma_{y'z'}$$

The first-order work term in creating the tetragonal defect in the presence of the stress field is

$$W \sim v_s(\sigma_{xx}\epsilon_{xx} + \sigma_{yy}\epsilon_{yy} + \sigma_{zz}\epsilon_{zz}) = 0.406\, v_s \frac{\mu b_s}{6\pi r}\left(\sqrt{6}\sin\theta - \sqrt{2}\cos\theta\right)$$

Note that this energy would be zero if $\epsilon_{xx} = \epsilon_{yy} = \epsilon_{zz}$. Comparing this result with that of Eq. (14-46), one sees that the interaction of the tetragonal defect with a screw dislocation is roughly of the same order of magnitude as the hydrostatic interaction with an edge dislocation.

[55] Reviewed by J. P. Hirth, *Met. Trans.*, **11A**: 861 (1980).

[56] G. Kurdjumov and E. Kaminsky, *Z. Phys.*, **53**: 696 (1929).

[57] K. Jack, *Proc. Roy. Soc.*, **208A**: 200 (1951).

Suzuki Segregation

Suzuki[58] pointed out that dissociated dislocations could interact with impurities or solute atoms by means of adsorption on the stacking-fault ribbon. In general, the energy of a solute atom at the stacking fault is different from its energy in the regular lattice, so that at equilibrium the concentration of solute in the fault differs from that throughout the bulk phase. Quantitatively, such adsorption can be treated in the same manner as adsorption at grain boundaries.[59-61] For an equilibrated stacking fault, the results are expressed in terms of the *Gibbs adsorption isotherm*,

$$\left(-\frac{\partial\gamma}{\partial\bar{G}_2}\right)_T = \Gamma_2 - \frac{x_2}{x_1}\Gamma_1 \qquad (14\text{-}70)$$

where x_2 and x_1 are the atom fractions of solute and solvent, respectively, in the bulk phase, $\bar{G}_2$ is the chemical potential of solute, γ is the stacking-fault energy, and Γ_2 and Γ_1 are the surface excess concentrations of solute and solvent, respectively, at the fault. Usually, one is interested in dilute solutions where $\Gamma_2 > (x_2/x_1)\Gamma_1$ and the adsorption isotherm reduces to

$$\left(-\frac{\partial\gamma}{\partial\bar{G}_2}\right)_T = \Gamma_2 \qquad (14\text{-}71)$$

Equation (14-71) shows that when solute accumulates at the stacking fault, the equilibrium stacking-fault energy decreases with increasing solute content (adsorption), while if solute is repelled from the fault, the stacking-fault energy decreases with decreasing solute content (desorption). In either case, adsorption or desorption, the resulting equilibrium stacking-fault energy is lower than that for a fault with the same composition as the bulk.

For example, a substitutional solute that makes the fcc solvent crystal less stable against a transition to the hcp structure would be expected to accumulate in the stacking fault, which, to first order, is a thin strip of hcp material (Chap. 10), and lower its energy. The effect is important in alloys, but it is not a very important impurity effect in pure crystals. The Suzuki effect is obviously present for both screw and edge dislocations; it depends only on the presence of the stacking fault.

The effect manifests itself as a change in the width d of an extended dislocation with adsorption. To first order, solute adsorption does not change the partial-dislocation strain fields, but affects only γ, lowering it and thus

[58] H. Suzuki, *Sci. Repts. Tohoku Univ.*, **A4**: 455 (1952); *J. Phys. Soc.* (*Japan*), **17**: 322 (1962).

[59] D. McLean, "Grain Boundaries in Metals," Oxford University Press, Fair Lawn, N.J., 1957, p. 143.

[60] T. Ericsson, *Acta Met.*, **14**: 1073 (1966).

[61] J. P. Hirth, *Met. Trans.*, **1**: 2367 (1970).

leading to an increase in width d [Eq. (10-14)]. To a higher order of approximation, adsorption in the elastic strain fields changes the strain fields of the partials, and this effect also influences d. In such a case the exact determination of d is a difficult problem; one cannot even predict with confidence whether adsorption will increase or decrease d. These latter effects are particularly important for solutes that produce tetragonal distortions.

14-6. FLUCTUATIONS AND PINNING IN SEGREGATED REGIONS

Many impurity phenomena are associated with impurities adsorbed to the dislocation core rather than with the extended atmospheres treated previously, and in such situations fluctuations are particularly important. Fluctuations provide weakly pinned segments in otherwise firmly pinned dislocations and must be considered in yield-point phenomena. Knowledge of the distribution of lengths of free segments between pinning points in nearly clean dislocations is important in internal friction experiments. These two limiting situations of nearly saturated and almost clean dislocations are discussed in turn.

Nearly Saturated Cores

Consider first the simple case of a saturated dislocation core. At a given time there is a distribution of clean segments of various lengths that constitute weak points where an applied stress can tear the dislocation free. This problem was treated first by Louat in connection with carbon pinning of dislocations in mild steel.[62, 63] Equation (14-64) indicates that the probability for a hole at a given site is

$$P_h = \exp\left(-\frac{F_B + \bar{G}}{kT}\right) \tag{14-72}$$

where $\bar{G}$ is the bulk chemical potential of solute. The probability that a *particular* segment of length l be clean is simply the probability that l/a consecutive sites be empty,

$$P_l = P_h^{l/a} = \exp\left[-\frac{l(F_B + \bar{G})}{akT}\right] \tag{14-73}$$

Since a segment of length l can be chosen in L/a different ways[64] in a long

[62] N. Louat, *Proc. Phys. Soc.*, **B69**: 459 (1956); *Proc. Phys. Soc.*, **71**: 444 (1958).

[63] J. Lothe, *Acta Met.*, **10**: 663 (1962).

[64] This probability term will differ somewhat for short lengths L. The probability can be determined by the theory of runs in such a case (M. Fisz, "Probability Theory and Mathematical Statistics," Wiley, New York, 1963, p. 418). The result of the more complex analysis leads only to a small change in the preexponential of Eq. (14-74).

segment of length L, the number of clean segments of length l per unit length of dislocation is

$$c_l = \frac{1}{a}\exp\left[-\frac{l(F_B+\bar{G})}{akT}\right] \tag{14-74}$$

The thermodynamic meaning of the various terms in Eq. (14-74) is apparent. The term $(l/a)F_B$ is the free energy of desorption of the segment l, that is, the free energy required to clean the segment by putting the impurities into solution, *not* counting the gain in entropy of mixing as they are put into solution. The free-energy decrease with the gain in entropy of mixing in bulk solution appears as the term $(l/a)\bar{G}$, since $\bar{G}$ is the chemical potential, with the energy U defined to be zero for an impurity in solution [Eq. (14-58)]. The entropy-of-mixing term for the core is zero, since the core is saturated.

Extending this reasoning, one sees that Eq. (14-74) is a special case of the more general formula

$$c_l \cong \frac{1}{a}\exp\left[-\frac{l(F_{\text{des}}+c_d\bar{G})}{kT}\right] \tag{14-75}$$

where the $\cong$ sign appears because of some uncertainty in the preexponential. F_{des} is the free energy of desorption per unit length, not counting the entropy of mixing in bulk solution, and c_d is the average number of impurities per unit length of dislocation. Equation (14-75) is valid quite generally, since it does not involve the assumptions of a very specific model.

Exercise 14-5. Show that Eq. (14-75) is valid for the case of a dilute core concentration of impurities, and that for this case

$$F_{\text{des}} = \frac{1}{a}\{ac_dF_B - kT[ac_d\ln ac_d + (1-ac_d)\ln(1-ac_d)]\} \tag{14-76}$$

This relation includes an entropy-of-mixing term for the core. The results of this exercise are important for such problems as the initial yield strength and the amplitude-dependent internal friction for crystals containing substitutional impurities.

Throughout this section thermal equilibrium is assumed. In solids at ambient and lower temperatures, *mobilities* are often so low that thermal equilibrium is not attained.

Almost Clean Dislocations

Consider a long straight dislocation string containing impurity pinning points with a *mean* separation L. The segments will vibrate by thermal excitation as discussed in Sec. 14-2 [Eq. (14-20)]. Alefeld[65] and Bauer[66] pointed out that the vibrational entropy of the segments provides a driving force for pinning point clustering and thus influences the statistics of segment length distribution.

Consider for simplicity a straight segment ABC, where A, B, and C are sequential pinning points. Consider A and C to be fixed. Calculations now show that the sum of the free energies of vibration of the segments AB and BC is a *maximum* with B midway between A and C; that is, the free energy is lowered with B close to A or C rather than in the middle.

However, a crucial assumption for this argument is that B is *on* the straight line AC. But according to Eq. (14-21), with the length $AC \sim 10^4 b$, and with B in the middle region, about eight different atomic positions for B are with about equal probability in the direction of bow-out. Near A and C this transverse space for B is not available. Put differently, relaxing the condition of strictly linear configurations ABC, allowing slightly bent configurations that do not increase the energy by more than $\sim kT$, we have an effect that increases the occupational probability for B in the middle region.

It would appear that the two effects almost balance. Until final clarification, the impurities are best considered as randomly distributed.

The effect does not imply transverse mobility of B; it could be a frozen-in effect. Frozen-in *linear* configurations ABC have their B pin closer to A or C, in agreement with the Alefeld-Bauer theory, but in nonlinear configurations ABC, B would preferentially be in the middle region. On the average, averaging over all configurations ABC, linear and nonlinear, B is probably anywhere with about the same probability.

14-7. THE CHARGED EDGE DISLOCATION

As an example of more complicated interactions between point defects and dislocations, we conclude this chapter with a discussion of a jogged edge dislocation in an NaCl-type crystal interacting with positive- and negative-ion vacancies at sufficiently high temperatures that local equilibrium concentrations of bulk ionic defects are maintained. The vacancies are assumed to form or annihilate only at jogs, with a negative-ion vacancy attaching itself to a negative jog and changing it to a positive jog with charge $e/2$, etc. Divalent impurities are supposed only to change the relative bulk concentrations of positive- and negative-ion vacancies and not to interact directly with the dislocation. This model is only an approximation of the real situation, but it has interesting features that are expected to be qualitatively the same as for an exact model. Although the treatment deals with an edge dislocation, screws should interact with vacancies in essentially the same way.

[65] G. Alefeld, *Phil. Mag.*, **11**: 809 (1965).

[66] C. L. Bauer, *Phil. Mag.*, **11**: 827 (1965).

Eshelby et al.[67] first realized that positive- and negative-ion vacancies do not interact equally strongly with a dislocation. Hence the dislocation acquires a net charge balanced out by a surrounding, compensating Debye-Hückel atmosphere of opposite charge within some radius λ_D, the Debye-Hückel radius. As a consequence, an electric potential builds up around the dislocation. At equilibrium the electric potential repels those ion vacancies of the sign which has the higher attraction to the dislocation line in the absence of the atmosphere and attracts the opposite-sign vacancies, so that a balance between core attraction and the electric forces around the dislocation exists. With divalent impurities present, there is a temperature where the dislocation is uncharged. That temperature, which is a function of impurity concentration, is called the *isoelectric point*. As the temperature changes through the isoelectric point, the charge on the dislocation changes sign.

The Charged Tube

As an illustration of the physical significance of the various terms in the dislocation case, the problem of a charged tube in an NaCl crystal is an instructive example. Let us determine the charge distribution around a hollow tube of radius R, at which surface ion-vacancy concentrations can equilibrate. This is the two-dimensional analog to Lehovec's treatment[68] of a planar surface. Suppose that the free energy of formation of a cation vacancy F_c is less than that of an anion vacancy F_a in the absence of an electric field. The free energy of formation F_c is that of the cation vacancy together with its associated dislocation defect, a positive jog formed from a negative jog. Thus F_c differs from the value F_c' for the free-surface case where the equivalent surface defect is a positive surface kink,[69] and the dipole strength and isoelectric temperatures of the surface differ from those for the dislocation. With a lower value for F_c than for F_a, the anion free energy, a positive charge builds up at the surface of the hollow tube and it is compensated by a negative Debye-Hückel charge. Let us determine the space charge in the Debye-Hückel potential field ϕ, measured relative to the potential at $r=\infty$; that is, $\phi=0$ when $r=\infty$. Adjacent to the positively charged surface, this potential field has some value ϕ_s. The concentrations of cation and anion vacancies, c_c^- and c_a^+, respectively, are given by

$$c_c^- = n\exp\left[-\frac{F_c - e(\phi-\phi_s)}{kT}\right] \tag{14-77}$$

$$c_a^+ = n\exp\left[-\frac{F_a + e(\phi-\phi_s)}{kT}\right] \tag{14-78}$$

[67] J. D. Eshelby, C. W. A. Newey, P. L. Pratt, and A. B. Lidiard, *Phil. Mag.*, **3**: 75 (1958).

[68] K. Lehovec, *J. Chem. Phys.*, **21**: 1123 (1953).

[69] D. W. Short, R. A. Rapp, and J. P. Hirth, *J. Chem. Phys.*, **57**: 1381 (1972).

where n is the number of available sites for either cations or anions. The plus and minus superscripts indicate the effective charges of the vacancies. At an infinite distance from the tube, $c_c^- = c_a^+$ and $\phi = 0$, so that

$$\phi_s = \frac{F_a - F_c}{2e} \tag{14-79}$$

The substitution of Eq. (14-79) into (14-77) and (14-78) yields

$$c_c^- = n\alpha \exp\left(\frac{e\phi}{kT}\right) \tag{14-80}$$

$$c_a^+ = n\alpha \exp\left(\frac{-e\phi}{kT}\right) \tag{14-81}$$

where

$$\alpha = \exp\left(-\frac{F_a + F_c}{2kT}\right) \tag{14-82}$$

The charge per unit volume

$$q = e(c_a^+ - c_c^-) = -2en\alpha \sinh\frac{e\phi}{kT} \tag{14-83}$$

The potential is related to the charge density by Poisson's equation

$$\nabla^2\phi = \frac{-4\pi q}{\epsilon} \tag{14-84}$$

where ϵ is the dielectric constant. In this case Eq. (14-84) can be solved analytically only for the limiting case where $\sinh(e\phi/kT) \sim e\phi/kT$.

Since we are interested only in a physical description of a typical case, we proceed to describe this limiting case analytically. Equation (14-84) assumes the form

$$\nabla^2\phi = \left(\frac{\partial^2}{\partial r^2} + \frac{1}{r}\frac{\partial}{\partial r}\right)\phi = \frac{\phi}{\lambda^2} \tag{14-85}$$

where

$$\lambda^2 = \frac{\epsilon kT}{8ne^2\pi\alpha} \tag{14-86}$$

The solution to Eq. (14-85), with the boundary conditions that $\phi = \phi_s$ at $r = R$

and $\phi = 0$ at $r = \infty$, is[70]

$$\phi = \phi_s \frac{K_0(r/\lambda)}{K_0(R/\lambda)} \tag{14-87}$$

where K_0 is the modified Bessel function of zero order of the second kind.

The total positive charge per unit length on the surface is equal to the integral negative charge distributed throughout the volume from $r = R$ to $r = \infty$,

$$Q = -\int_R^\infty q 2\pi r \, dr \tag{14-88}$$

Substituting Eqs. (14-83) and (14-87) into (14-88), one obtains

$$Q = \frac{\epsilon R \phi_s}{2\lambda} \frac{K_1(R/\lambda)}{K_0(R/\lambda)} \tag{14-89}$$

where K_1 is the modified Bessel function of first order of the second kind. The compensating negative charge is effectively limited to a volume within the Debye-Hückel distance λ from the tube. The potential ϕ within this region is effectively the same as that which would be produced by distributing the negative charge $-Q$ in an infinitesimally thick cylindrical shell of radius $R + \lambda$.

Asymptotically, when $R/\lambda \ll 1$, Eq. (14-89) becomes

$$Q = -\frac{\epsilon \phi_s}{2 \ln(R/\lambda)} \tag{14-90}$$

The total charge diverges logarithmically as $R \to 0$. Thus some cutoff procedure defining an appropriate R must be devised in applying this model to dislocations. This cutoff problem has not yet been solved in a satisfactory way.

The presence of a divalent cation impurity would affect the concentration of cation vacancies in the crystal interior. The analysis in such a case is more complicated, but proceeds similarly to the above treatment.[71]

The Charged Edge Dislocation

Consider now the charge and potential associated with an edge dislocation in an NaCl crystal.

[70] G. N. Watson, "A Treatise on the Theory of Bessel Functions," Cambridge University Press, New York, 1952.

[71] See J. D. Eshelby, C. W. A. Newey, P. L. Pratt, and A. B. Lidiard, *Phil Mag.*, **3**: 75 (1958).

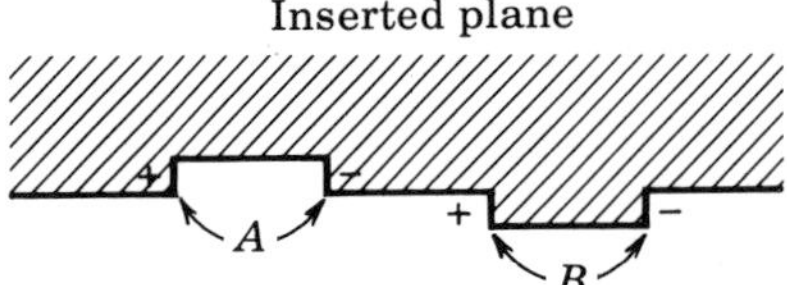

FIGURE 14-16. The four different types of jogs in an edge dislocation in an ionic crystal. The two positively charged jogs (+) are not identical; the one in set A is geometrically positive, and that in set B is geometrically negative.

The dislocation contains four kinds of jogs with charges $\pm e/2$ (Chap. 12). In a more exact treatment one would distinguish between isolated jogs and jogs paired by the electrostatic interaction of Eq. (12-5), but we neglect this complication here. Let c_+^+ be the concentration of geometrically positive jogs positively charged, c_+^-, the concentration of geometrically positive jogs negatively charged, etc., as shown in Fig. 14-16. Opposite jogs within set A can mutually annihilate, as can opposite jogs within set B. Thus there are two equations of mass action, which, with the neglect of small entropy terms, as in Eq. 14-5, can be expressed as

$$c_+^+ c_-^- = \frac{1}{a^2}\exp\left(-\frac{2W_j}{kT}\right)$$

$$c_-^+ c_+^- = \frac{1}{a^2}\exp\left(-\frac{2W_j}{kT}\right) \tag{14-91}$$

where a is the atomic spacing along the dislocation line. The energy of formation $2W_j$ for the pair in set A is assumed to be the same as that for the pair in set B. When the dislocation is neutral, $c_+^+ = c_+^-$ and $c_-^- = c_-^+$, so that the two mass-action equations are equivalent.

Let the formation energy of a pair of positive- and negative-ion vacancies be $2W_v$. The mass-action equation for these vacancies is

$$c_a^+ c_c^- = \frac{1}{v_a^2}\exp\left(-\frac{2W_v}{kT}\right) \tag{14-92}$$

where v_a is the volume per ion pair. With a concentration c^{++} of divalent impurities present, charge neutrality requires that in bulk solution

$$c^{++} + c_a^+ - c_c^- = 0 \tag{14-93}$$

Theory[72] and experiment[73] indicate that the energy of formation of a single cation vacancy is less than that of a single anion vacancy. Thus if a cation vacancy moves to a positively charged jog, the charge on the jog becomes negative and there is an energy change

$$\delta W = -W_v + \Delta \tag{14-94}$$

[72] N. F. Mott and M. J. Littleton, *Trans. Faraday Soc.*, **34**: 485 (1938).

[73] L. Slifkin, *Semicond. Insul.*, **3**: 393 (1978).

where Δ is some positive incremental energy change. The energy change is assumed to be the same whether the jog is geometrically positive or negative. Similarly, when an anion vacancy is absorbed at a negatively charged jog, the jog becomes positively charged and the energy change is

$$\delta W = -W_v - \Delta \tag{14-95}$$

so that the two energy changes add up to $-2W_v$, as they must. The term Δ includes dipole relaxations around the point defects, image interactions, and differences in elastic energies and cannot be estimated very easily; in terms of the preceding section, $F_c \sim W_v - \Delta$ and $F_a \sim W_v + \Delta$.

The charged jogs are surrounded by an electric field, which is screened off within some Debye-Hückel radius λ_D. Provided that the spacing between jogs is small compared to λ_D, the Debye-Hückel atmosphere is cylindrically distributed around the dislocation, and λ_D is effectively the *same* as the radius λ given by Eq. (14-86). When ion vacancies are removed from a region in the solution outside λ_D and placed in the field ϕ of the Debye-Hückel atmosphere, work $\pm e\phi$ must be done, analogous to that in the case of the tube.

All quantities required to describe equilibrium between the dislocation and vacancies in solution now are defined. Balancing free-energy changes for the processes of moving vacancies to the jogs, one obtains

$$kT \ln \frac{ac_+^+}{ac_+^- v_a c_a^+} - W_v - \Delta + e\phi = 0 \tag{14-96}$$

$$kT \ln \frac{ac_-^+}{ac_-^- v_a c_a^+} - W_v - \Delta + e\phi = 0 \tag{14-97}$$

$$kT \ln \frac{ac_+^-}{ac_+^+ v_a c_c^-} - W_v + \Delta - e\phi = 0 \tag{14-98}$$

and

$$kT \ln \frac{ac_-^-}{ac_-^+ v_a c_c^-} - W_v + \Delta - e\phi = 0 \tag{14-99}$$

The subtraction of Eq. (14-98) from (14-96), together with the use of Eq. (14-92), yields

$$\frac{c_+^-}{c_+^+} = v_a c_c^- \exp \frac{W_v + e\phi - \Delta}{kT} \tag{14-100}$$

and similarly,

$$\frac{c_-^-}{c_-^+} = v_a c_c^- \exp \frac{W_v + e\phi - \Delta}{kT} \tag{14-101}$$

These two equations, together with Eqs. (14-91), do determine all jog concentrations, and hence the charge on the dislocation. In principle the derivation proceeds in the same manner as that for the charged tube. However, although the results are the same as those for the tube in the region $r \cong \lambda$, they differ markedly near the dislocation core, because the field is localized around points (the jogs) rather than around a line, so that it is difficult to assign a meaningful boundary condition to the core. We do not treat this difficult problem, but note that the results qualitatively resemble those for the hollow tube.

Even without the detailed solution, one can derive the condition for neutrality and hence determine the isoelectric point. The combination of Eqs. (14-91), (14-100), and (14-101) gives

$$c_+^- c_-^- = \frac{v_a c_c^-}{a^2} \exp\left(\frac{W_v + e\phi - \Delta - 2W_j}{kT}\right)$$

$$c_+^+ c_-^+ = \frac{1}{a^2 v_a c_c^-} \exp\left(\frac{-W_v - e\phi + \Delta - 2W_j}{kT}\right) \qquad (14\text{-}102)$$

When the dislocation is neutral $\phi = 0$, and for a given geometric type of jog, equally many are positively and negatively charged, $c_+^- = c_+^+$ and $c_-^- = c_-^+$. Equation (14-102) then yields the condition for neutrality

$$v_a c_c^- = \exp\left(-\frac{W_v - \Delta}{kT}\right) \qquad (14\text{-}103)$$

Except at high temperatures, the concentration c_c^- is impurity controlled. In this *extrinsic* range, Eqs. (14-92) and (14-93) give

$$c_c^- \cong c^{++} \qquad (14\text{-}104)$$

so that the isoelectric point T_0 is given by Eq. (14-103) as

$$T_0 = -\frac{W_v - \Delta}{k \ln v_a c^{++}} \qquad (14\text{-}105)$$

Even for very pure crystals, the isoelectric point is a few hundred degrees above room temperature for the common alkali halides. In the intrinsic range $T > T_0$ the dislocation is positively charged because of the assumed larger formation energy for anion vacancies than for cation vacancies. In the extrinsic range $T < T_0$ the dislocation is negatively charged, because essentially only cation vacancies are present to interact with the dislocation. In the low extrinsic range, if the divalent impurity precipitates out of solution, $c^{++} \rightarrow 0$, and the ion vacancies approach their thermal-equilibrium concentrations again, so that the dislocation can again become positively charged in this quasi-intrinsic range.[74] Of course, at low temperatures, thermal equilibrium might not obtain and the charge would then be determined kinetically as in the vacancy sweep-up model mentioned in Chap. 12.

The above treatment involves no assumption about the direction of the dislocation line. The isoelectric point is the same for dislocations that are on the average straight or for those that are initially skew and contain geometric jogs. At low temperatures, where diffusion is slow, compensating Debye-Hückel atmospheres around charged dislocations would tend to lock the dislocations because of the electrostatic and elastic interaction. Once this electrostatic interaction was thought to control the flow stress,[74] but Pratt[75] showed that the interaction was a factor of 10 or more too low. Flow is now thought to be controlled by jog drag, by interaction with divalent cation-cation vacancy pairs and larger complexes. A complete treatment including elastic and electrostatic effects, the latter including dipole and multipole interactions, is lacking. Consequently, isoelectric points have been associated with *minima* in yield strengths as a function of temperature for doped ionic crystals.[75, 76, 77]

As already mentioned, experiments are controversial regarding the sign of the charge on dislocations in ionic crystals. The preceding theory is based on a simplified model, with the intent of emphasizing the physical parameters involved; a number of complications might be required to rationalize experimental findings. For example, diffusion and rearrangement of divalent impurities and their association with dislocations are likely to be important. In addition, *dynamic* factors, which are not treated in the preceding *equilibrium* theory, can affect the dislocation charge. Under deformation, dislocations can be charged by various mechanisms; sweeping up of cation vacancies at low temperatures as discussed in Chap. 12, for example. An example at higher temperatures would be the acquisition of a positive charge by a dislocation, even in the extrinsic range, in the case where negatively charged jogs emit a cation vacancy to become positively charged more often than positively charged jogs emit an anion vacancy. Interactions with surfaces are complicated because the strength of the Debye-Hückel atmospheres differ for dislocations and surfaces. At low temperatures dislocations may act as conduits for ions to form the compensating atmospheres for surfaces; deformation has been found to produce surface charge in alkali halides.[78]

[74] L. M. Brown, *Phys. Stat. Solidi*, **1**: 585 (1961).

[75] P. L. Pratt, *Proc. Br. Ceram. Soc.*, **1**: 177 (1964).

[76] J. D. Eshelby, C. W. A. Newey, P. L. Pratt, and A. B. Lidiard, *Phil. Mag.*, **3**: 75 (1958).

[77] J. S. Koehler, D. Langreth, and B. von Turkovich, *Phys. Rev.*, **128**: 573 (1962).

[78] G. Turchanyi, I. Foldvari, and I. Tarjan, *J. Phys. Suppl 12*, **C7**: 604 (1976).

In the developments of this section only interactions of charged dislocations with vacancies and divalent impurities are considered. Conduction electrons also interact with charged dislocations, with the expected formation of optical centers analogous to the well-known color centers,[74] which involve associations between electrons and anion vacancies. An example of the influence of electronic interactions is the effect of photoexcitation on the flow stress of ZnS.[79]

PROBLEMS

14-1. **a.** For $\omega = 10^{11}$ and 10^{12} $\sec^{-1}$, compute the temperature at which the entropy of the vibrational mode approaches its high-temperature limit.
b. If $\omega_p = 10^{12}$ $\sec^{-1}$ and $\omega_k = 10^{11}$ $\sec^{-1}$, compute the kink entropy S_k. Compare the entropy contribution to F_k to the energy contribution at room temperature if $2W_k = 0.3$ eV. At what temperature are the two contributions equal?
c. With the effective mass of the kink $m^* = W_k / C_t^2$ and the typical value $C_t \sim 10^5$ cm/sec, compare the result of Eq. (14-6) with that of (**b**) at room temperature.

14-2. Use Eqs. (8-38) and (8-75) to compute the approximate kink widths in copper, silver, and gold. Show in a polar plot of orientations in the slip plane the range of orientations where a discrete kink model should apply for these metals.

14-3. For the case of copper, compute the concentrations of thermal kinks and thermal jogs in an otherwise straight dislocation line at 100, 300, and 900°K.

14-4. **a.** Utilize the results of Chap. 8 and include the kink-kink interaction energy in the free energy of formation of a kink pair. For copper at 200°K, plot the equilibrium concentration of double kinks of a given separation h as a function of h. Over what distance does the interaction energy appreciably affect the double-kink concentration?
b. Repeat the steps in (**a**) for jog pairs.

14-5. Consider two solute atoms near the center of a large spherical crystal and separated by a distance h. Let $v_s > v_a$. Use the results of Sec. 2-7 to show that there is *no* hydrostatic interaction between the two atoms in the absence of image stresses. Compute the hydrostatic interaction when image terms are present.

14-6. Suppose the isotropic continuum model for the external volume contraction associated with a vacancy $V_v = v_v + v_c$. Identify which volume terms of the *continuum model* would be measured in the following experiments, and

[79] V. F. Petrenko and R. W. Whitworth, *Phil. Mag.*, **41A**: 681 (1980).

whether these terms would all be in agreement.

a. Dilatometric determination of the lattice contraction accompanying the annealing out of vacancies at room temperature under zero external stress.[80]

b. Pressure dependence of the quenched-in resistivity following quenching from elevated temperatures under various pressures.[81]

c. High-temperature dilatometric measurement with the effect of thermal expansion of the bulk lattice subtracted out.[82]

14-7. Consider a cylindrical crystal under uniform simple tension σ_{xx} parallel to the cylinder axis. Suppose *local* vacancy equilibrium at the crystal surfaces, and compute the local equilibrium concentration at the surface where the stress σ_{xx} is applied and at the lateral surface under zero external stress.

14-8. For the case of gold, compare the vacancy concentrations as a function of distance from a pure edge dislocation as given by Eqs. (14-38) and (14-43), respectively.

14-9. Compare the magnitudes of the energy terms in Eqs. (14-45) and (14-68) for the cases of Al dissolved in Cu, Zn in Cu, Ag in Au, and Cu in Ni.

14-10. Determine from Eq. (14-74) the integral concentration of free lengths between l' and some value l'' which represents the average length of dislocation segments in a crystal. For the case of bcc iron with one carbon atom per 10^4 possible bulk interstitial sites and the reasonable value $F_B \sim 0.5$ eV, compute the concentration of all lengths between $l' = 10^{-6}$ cm and $l'' = 10^{-3}$ cm at 100, 300, and 700°K.

14-11. Determine the divalent impurity concentration that would give $T_0 =$ 300°C in NaCl. Take $v_a = 4.4 \times 10^{-23}$ cm^3 and $W_v = 1.01$ eV,[83] and assume $\Delta = 0.20$ eV.

14-12. Equation (14-86) applies to the Debye-Hückel radius in the intrinsic range. Consider an extrinsic range with a divalent impurity concentration c^{++} and suppose all cation vacancies to be quenched out. If both divalent impurities and anion vacancies can rearrange themselves to provide an atmosphere, what would be the Debye-Hückel radius?

14-13. Consider that an incompressible atom, bigger by δv than a host atom, is substituted for the latter.

a. Calculate the external expansion δV.

b. Show that when an external pressure P is present, an extra amount of work $P\delta V$ is required for the insertion of the atom. *Hint:* With the crystal initially compressed, a volume $\delta v + Pv_a/B$ must be opened up to accommodate the inserted atom.

[80] F. J. Fraikor and J. P. Hirth, *J. Appl. Phys.*, **38**: 2312 (1967).

[81] R. P. Huebener and C. G. Homan, *Phys. Rev.*, **129**: 1162 (1963).

[82] R. O. Simmons and R. W. Balluffi, *Phys. Rev.*, **125**: 862 (1962).

[83] H. W. Etzel and R. J. Maurer, *J. Chem. Phys.*, **18**: 1003 (1950).

14-14. Consider the same situation as in Prob. 14-13, but with a compressible substitutional atom, $B' \neq B$. Show that again an extra amount of work $P\,\delta V$ is required to insert the atom when external pressure P acts.

BIBLIOGRAPHY

1. Eshelby, J. D., *Solid State Phys.*, **3**: 79 (1956).
2. Eshelby, J. D., C. W. A. Newey, P. L. Pratt, and A. B. Lidiard, *Phil. Mag.*, **3**: 75 (1958).
3. Fleischer, R. L., in D. Peckner (ed.), "Strength of Metals," Reinhold, New York, 1965, p. 93.
4. Pratt, P. L., *Inst. Met. Monog. and Rept. Ser.*, **23**: 99 (1957).
5. Rosenfield, A. R., A. L. Bement, and G. T. Hahn (eds.), "Dislocation Dynamics," McGraw-Hill, New York, 1968.
6. Seitz, F., *Rev. Mod. Phys.*, **23**: 328 (1951); *ibid.*, **26**: 7 (1954).

15

Diffusive Glide and Climb Processes

15-1. INTRODUCTION

At finite temperatures dislocations contain equilibrium concentrations of kinks and jogs, as discussed in the previous chapter. Under small applied stresses these equilibrium concentrations are not changed appreciably, but the kinks and jogs undergo a diffusive drift under the applied stresses, producing glide and climb, respectively. In the presence of larger stresses, kinks and jogs are swept to the ends of dislocation segments and pile up there at pinning points. Two types of deformation processes are then possible. First, if the relaxation time for dragging along the pinning point is short compared to that for nucleating a double kink or double jog, the nucleation and diffusive lateral propagation of the kinks and jogs control dislocation motion. This diffusion process, together with the drift process, comprises the subject matter of this chapter. If the relative magnitudes of the relaxation times are reversed, dragging of the pinning points is rate controlling; this process is discussed in Chap. 16.

The distinction between diffusion and drift is apparent in many types of transport processes. For example, a kink on a dislocation or an impurity atom in solution undergoes *random* movement under the action of thermal agitation; they diffuse. The diffusion coefficient or diffusivity is given by

$$D = \beta a^2 \omega \tag{15-1}$$

where a is the jump distance, ω is the jump frequency to a given site, and β is a numerical factor of the order of unity which depends on correlation effects and on the number of possible jump sites. After a time t, a defect initially at $x=0$ is probably in the interval $-\bar{x}<x<\bar{x}$, where $\bar{x}$ is the root-mean-square diffusion distance given by the Einstein relation[1]

$$\bar{x} = \langle x^2 \rangle^{1/2} = \sqrt{2Dt} \tag{15-2}$$

Any position within the interval is about equally probable.

[1]A. Einstein, "Investigations on the Theory of Brownian Movement," Methuen, London, 1926, p. 9. For two- or three-dimensional random walk, $\bar{x}$ becomes $2\sqrt{Dt}$ or $\sqrt{6Dt}$, respectively.

When acted upon by a constant small force F, the motion of the diffusing defect is biased. The jumps to one side are aided by the force, while those to the other side are retarded, so that a net *drift* velocity in the direction of the force results. The drift velocity v_D is related to the mobility D/kT by the Einstein mobility relation[1]

$$v_D = \frac{D}{kT} F \tag{15-3}$$

F need not be a mechanical force; it can be a general thermodynamic force. In a concentration gradient in an ideal solution, for example, F is defined as the negative derivative of the chemical potential,

$$F = -\frac{\partial}{\partial x} kT \ln \frac{c}{c_0} = -\frac{kT}{c}\frac{\partial c}{\partial x} \tag{15-4}$$

which, together with Eq. (15-3), yields the current density

$$J = cv_D = -D\frac{\partial c}{\partial x} \tag{15-5}$$

Equation (15-5) is known as Fick's first law. For a more thorough discussion of diffusion and drift the reader is referred to Shewmon.[2]

15-2. GLIDE OVER THE PEIERLS BARRIER

Kink Mobility

When kinks have equilibrium positions of lowest free energy at intervals of length a along the dislocation, *thermal activation* is required to move a kink from one position to a neighboring one. If the activation-energy barrier to kink motion is W_m, the jump frequency is

$$\omega \cong \nu \exp\left(-\frac{W_m}{kT}\right) \tag{15-6}$$

Here ν is the attempt frequency for a jump to a particular site, and the exponential gives the fraction of the attempts that are successful. ν is typically of the order of the Debye frequency ν_D,

$$\nu \sim \nu_D \sim 10^{12} \text{ to } 10^{13} \text{ sec}^{-1}$$

More generally, one should put F_m, the free energy of activation, into Eq. (15-6), but for such a refinement to be meaningful, one also would have to

[2] P. G. Shewmon, "Diffusion in Solids," McGraw-Hill, New York, 1963.

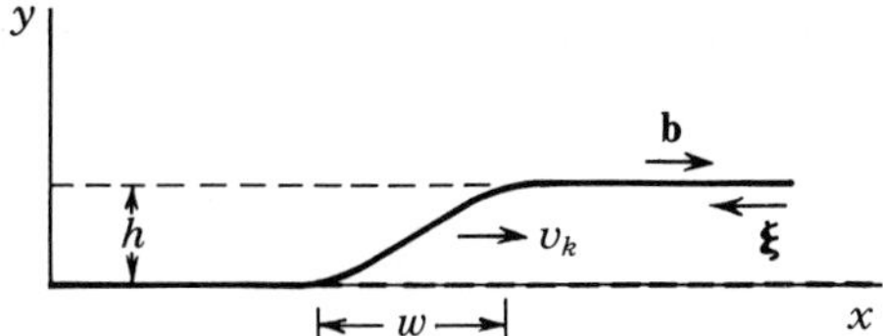

FIGURE 15-1. A kink in the glide plane xy moving with a velocity v_k under an applied stress $\sigma = \sigma_{xz}$.

determine rather uncertain entropy terms and define more precisely what constitute preexponential as opposed to exponential terms.[3] Equation (15-6) is adequate for our purposes. Equations (15-1) and (15-6) combine to give the kink diffusivity

$$D_k \cong \nu_D a^2 \exp\left(-\frac{W_m}{kT}\right) \tag{15-7}$$

An external stress produces a force F on a kink. Consider for definiteness a kink in a pure screw dislocation, and let σ be the shear stress component in the glide plane, acting in the direction of **b**. The lateral force component on the kink (Fig. 15-1) is then

$$F = \sigma b h \tag{15-8}$$

so that from Eq. (15-3) the lateral velocity v_k of the kink is

$$v_k = D_k \frac{\sigma b h}{kT} \tag{15-9}$$

Equation (15-9) is valid in general for small stresses, regardless of the nature of kink diffusion. Equation (15-7), however, is based on the concept of discrete kink positions with an energy of activation $W_m \gg kT$. In cubic metals there is evidence that dislocations remain mobile down to temperatures approaching absolute zero.[4,5] A reasonable interpretation of these results is that the mobility is associated with kink motion, suggesting that $kT \gtrsim W_m$ down to temperatures where quantum effects become important. When the activation barrier to motion is negligible, the mobility of the kink is determined by damping terms associated with phonon scattering (Sec. 7-7). As a rough approximation, consider a kink of width w moving with velocity v_k as a segment of length w moving normal to itself with a velocity $v_k h/w$ and for

[3] For discussions of the analogous problem in atomic diffusion see C. Wert and C. Zener, *Phys. Rev.*, **76**: 1169 (1949); G. Vineyard, *J. Phys. Chem. Solids*, **3**: 121 (1957); S. Glasstone, K. J. Laidler, and H. Eyring, "Theory of Rate Processes," McGraw-Hill, New York, 1941.

[4] R. H. Chambers, *Physical Acoustics*, **3A**: 123 (1966); D. H. Niblett, ibid., p. 77.

[5] G. Fantozzi, C. Esnouf, W. Benoit, and I. G. Ritchie, *Prog. Mat. Sci.* (in press).

which Eq. (7-74) applies. The effective force per unit length on the segment is $F/h = \sigma b$, so

$$v_k = \frac{\sigma b w}{Bh}$$

A comparison with Eq. (15-9) reveals that

$$D_k = \frac{wkT}{Bh^2} \tag{15-10}$$

At $T \sim \theta_D$, the Debye temperature, with Eq. (7-77) for B, $h \sim b$, $k\theta_D \sim 0.005\mu b^3$, $C_t \sim \nu_D b$, Eq. (15-10) yields for a kink of width $w \sim 5b$ the result

$$D_k \sim \nu_D b^2, \qquad T \sim \theta_D \tag{15-11}$$

With $a \sim b$, this is also the limiting form of Eq. (15-7) when $W_m \ll kT$, but the correspondence is accidental.

The above is a rough estimate. Eshelby[6] and Lothe[7] have considered kink mobility controlled by the flutter mechanism, but according to Seeger and coworkers[8, 9] the strain field scattering is more important. This scattering is included in a rough way in the above estimate as discussed in Sec. 7-7. However, at low temperatures ($T < 0.1\theta_D$) the interaction of the kink with the vibrations (heavy phonons) in the Peierls valley of the dislocation line itself are dominant. This effect is neglected in Eq. (15-10). The heavy-phonon contribution remains appreciable down to the Peierls temperature $\theta_p = h\nu_p/k$ and for lower temperatures it freezes out. Here ν_p is the lowest frequency of vibration in the Peierls valley. Typically, $\theta_p \sim 10^{-2}\theta_D$ in metals.

A finite W_m should exist in all cases, so that one might expect that, at a sufficiently low temperature $T < W_m/k$, Eq. (15-7) and the model of thermal activation would again apply. However, one cannot be sure that the kinks will ultimately crystallize into well-defined positions as $T \to 0°K$. Since the zero-point motion is appreciable in solids, it could conceivably obliterate a small energy barrier. An analogy to such a possibility exists in the fact that He^2 stays liquid as $T \to 0°K$ because the zero-point energy is large in comparison to the energy of crystallization. Whether there is an energy barrier W_m, how big it is, and below what temperature it is noticeable can be decided only by experiment. On the basis of measurements of internal friction and modulus defects, W_m appears to have a negligible effect above 4°K for fcc metals,[10] above 4°K

[6] J. D. Eshelby, *Proc. Roy. Soc.*, **A266**: 222 (1962).

[7] J. Lothe, *J. Appl. Phys.*, **33**: 2116 (1962).

[8] A. Seeger and P. Schiller, *Acta Met.*, **10**: 348 (1962).

[9] A. Seeger and H. Engelke, in A. R. Rosenfield, G. T. Hahn, A. L. Bement, and R. I. Jaffee (eds.) "Dislocation Dynamics," McGraw-Hill, New York, 1968, p. 623.

[10] G. Fantozzi, C. Esnouf, W. Benoit, and I. G. Ritchie, *Prog. Mat. Sci.* (in press).

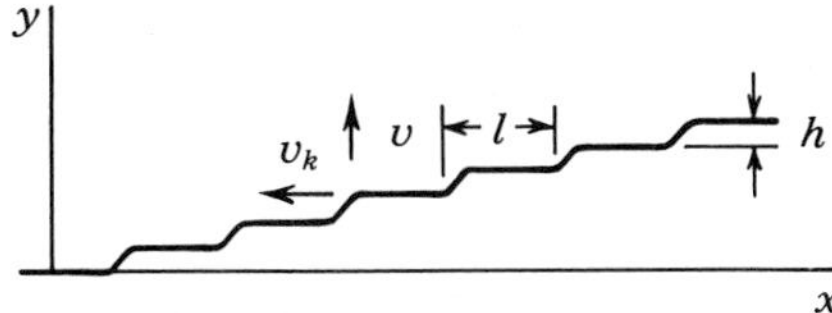

FIGURE 15-2. A skew dislocation containing kinks spaced at intervals l.

for mixed dislocations, but above ~25°K for screws in iron,[11] as examples of cubic metals. For strongly bonded ionic and covalent materials, the kink barrier should be more important. Once it was thought to control dislocation motion in silicon, but the situation is unresolved because of complications associated with dislocation extension.[12]

Single-Kink Drift

Consider the skew dislocation in Fig. 15-2, which has long unkinked segments in screw orientation. If the equally spaced kinks move to the left with a velocity v_k, the dislocation as a whole moves normal to itself with a velocity

$$v = \frac{h}{l} v_k \tag{15-12}$$

In the presence of a shear stress σ, which gives the kinks the drift velocity of Eq. (15-9), the dislocation velocity is

$$v = D_k \frac{\sigma b h^2}{lkT} \tag{15-13}$$

For a dislocation so skew that the kinks overlap, a more realistic model for the dislocation is a continuous string, as discussed in Chap. 14. In such a case, the motion of the dislocation as a whole is determined by phonon scattering, with the velocity given by Eq. (7-78) as

$$v \cong \frac{50 \sigma C_t}{\mu g(T/\theta)} \tag{15-14}$$

where C_t = transverse sound velocity and the factor $g(T/\theta)$ is as given in Fig. 7-16.

Equation (15-13) was developed with the assumption that new kinks are constantly supplied at the right side as the kinks move left. Consider now a skew dislocation pinned at A and B, as in Fig. 15-3. Under stress the kinks present initially move laterally, pile up against the pinning point A, and leave a long straight segment in the Peierls valley. This straight segment can then

[11] A. Seeger and C. Wüthrich, *Nuovo Cim.*, **33B**: 38 (1976).

[12] P. Haasen, *J. Phys.*, *Colloq. C6*, **40**: 111 (1979).

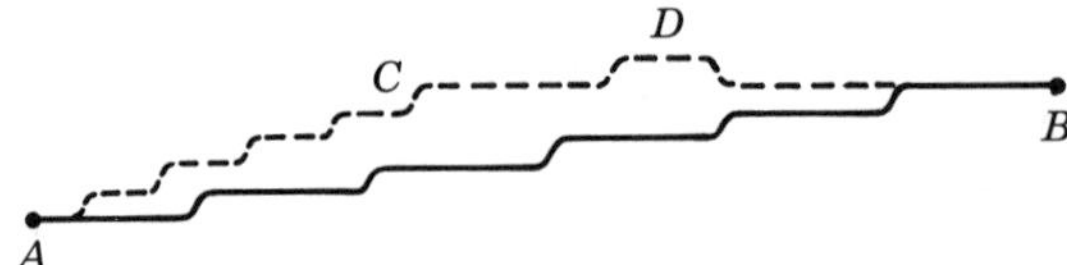

FIGURE 15-3. Skew dislocation segment AB bowed out to the configuration ACB, with a nucleated double kink at D.

advance further, bowing out the dislocation line, only by the *nucleation* of kink pairs and subsequent sidewise motion of the kinks. Such a configuration is expected quite generally for bow-outs in the glide plane. Even with the segment so skew that it should be considered as a continuous string, as in Fig. 15-4, the portion of the line near B approaches the Peierls valley orientation on bow-out. For increasing bow-outs, longer and longer straight segments, BC_1, BC_2, BC_3, etc., develop. Motion beyond the Peierls valley passing through B again requires kink-pair nucleation.

In either of the above cases, two possible glide processes can result from continued bow-out. If the relaxation time for dragging of the pinning points is long compared to the relaxation time required to completely bow out the line to a configuration such as A in Fig. 16-7, the kink-motion processes will only affect the initial stages of glide, and subsequent glide will be controlled by pinning-point drag, as discussed in Chap. 16. If the magnitudes of these relaxation times are reversed, the kink-nucleation process will control glide. Let us now consider this latter case of motion of dislocation segments which are straight along Peierls valleys except for thermal kinks.

Kink Glide at Low Stresses

In the limit of small external stresses, the concentration of thermal kinks in a straight segment in a Peierls valley remains nearly the same as the equilibrium concentration, given by Eq. (14-12) as

$$c_k = \frac{2}{a} \exp\left(-\frac{F_k}{kT}\right) \tag{15-15}$$

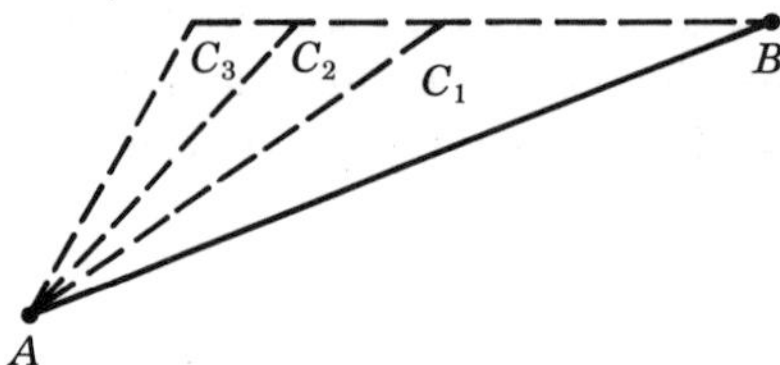

FIGURE 15-4. Highly skew dislocation segment AB bowed out successively to AC_1B, AC_2B, and AC_3B. The segments BC_1, BC_2, and BC_3 lie in Peierls valleys.

for this case where $c_k^+ + c_k^- = c_k$. All these kinks have a small drift velocity [Eq. (15-9)], with the positive and negative kinks drifting in opposite directions. The net velocity of the dislocation normal to itself is then [*cf*. Eq. (15-12)]

$$v = hc_k v_k \tag{15-16}$$

Inserting for c_k and for v_k from Eq. (15-9), one obtains

$$v = \frac{2\sigma bh^2}{akT} D_k \exp\left(-\frac{F_k}{kT}\right) \tag{15-17}$$

The condition that indicates how small the stress must be for the above near-equilibrium treatment to hold depends on the mean lifetime of a kink. When no stress is present, the mean time between the creation of a kink and its annihilation with an opposite-sign kink is the mean diffusion lifetime

$$\tau_{\text{diff}} \sim \frac{1}{c_k{}^2 D_k} \tag{15-18}$$

This expression is simply the Einstein result, Eq. (15-2), for the time required to diffuse a distance equal to the mean interkink spacing $1/c_k$. With a drift velocity v_k arising from an applied stress, the kink would travel a distance $1/c_k$ in a time

$$\tau_{\text{drift}} = \frac{1}{c_k v_k} \tag{15-19}$$

When $\tau_{\text{drift}} \gg \tau_{\text{diff}}$, the motion of individual kinks is dominated by the random-diffusion process, so that the balance between nucleation and annihilation cannot differ much from the balance with no stress present. Thus the condition for the near-equilibrium result of Eq. (15-17) to be valid is that $\tau_{\text{drift}} \gg \tau_{\text{diff}}$, which reduces to

$$\sigma \ll \frac{kT}{bh} c_k = \frac{2kT}{bah} \exp\left(-\frac{F_k}{kT}\right) \tag{15-20}$$

Double-Kink Nucleation at High Stresses

The small-stress condition (15-20) ensures that the somewhat intuitively developed Eq. (15-17) is valid. Let us now consider the double-kink nucleation process under higher external stresses. We shall see that Eq. (15-17) can be approximately valid up to quite high stresses, such as those typically encountered in gross plastic deformation. In the presence of a large stress, only double kinks formed in the direction favored by the applied stress, such as AB in Fig. 15-5, need be considered. Kink pairs of the type CD are forced together by the applied stress and their concentration is suppressed, so that they can be neglected.

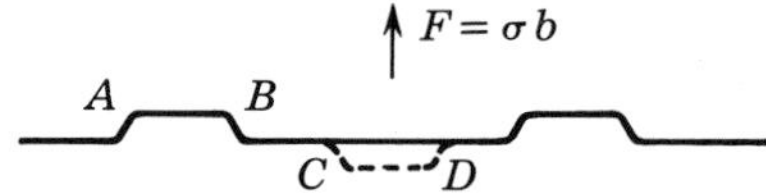

FIGURE 15-5. Double-kink pairs AB and CD.

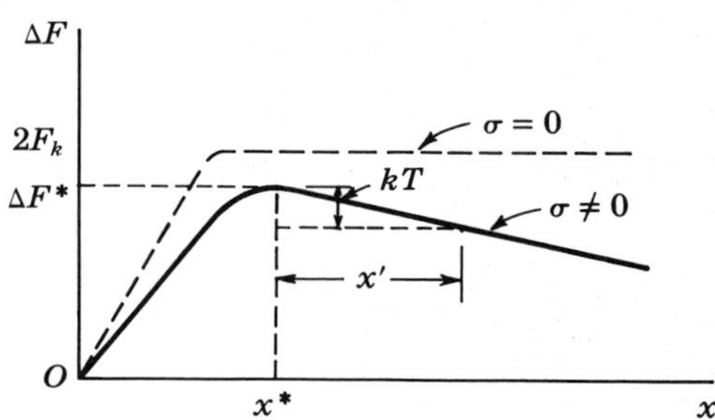

FIGURE 15-6. Free energy of formation of a kink pair as a function of kink separation x for the case where the applied stress is zero and the case where it is finite.

With a stress σ present, the free energy ΔF of a kink pair as a function of kink separation x has the general form shown in Fig. 15-6. At some critical separation x^*, the kink-kink attraction balances the external force σbh tending to tear the kinks apart. At x^* the free energy of the pair has its maximum value ΔF^*. The kinks in a pair with separation greater than x^* tend to separate further with a decrease in free energy, while those with separation less than x^* tend to mutually annihilate. This type of free-energy variation and growth-annihilation behavior is precisely that treated in classical steady-state nucleation theory.[13–15] Thus the results of nucleation theory can be applied directly to the double-kink problem.

First consider the effective width of the activated state or saddle point. Let us define x' as the distance beyond x^* where the free energy of the kink pair is $\Delta F^* - kT$ (Fig. 15-6). Kink pairs of separation greater than $x^* + x'$ have a negligible probability of self-annihilation, since that would require a thermal activation larger than kT to move the pair back over the saddle point. Kink pairs separated beyond x^* but not beyond $x^* + x'$ cannot be considered to

[13]See L. Farkas, *Z. phys. Chem.*, **125**: 236 (1927); J. B. Zeldovich, *Acta Physicochem. URSS*, **18**: 1 (1943); J. Frenkel, "Kinetic Theory of Liquids," Oxford University Press, Fair Lawn, N.J., 1946. These are all *dynamic* theories which treat the steady-state diffusion flux over the free-energy barrier.

[14]Often nucleation theory is formally expressed in terms of the *equilibrium* theory of M. Volmer ("Kinetik der Phasenbildung," Steinkopff, Dresden, 1939), and the kinetic depletion effects of the dynamic theory are expressed as correction factors; see D. Turnbull, *Solid State Phys.*, **3**: 225 (1956); J. P. Hirth and G. M. Pound, "Condensation and Evaporation," Pergamon, New York, 1963. Here we follow the developments of the dynamic theories directly.

[15]For a review of the dynamic theory, its terminology, and its relation to the equilibrium theory, see J. Feder, K. C. Russell, J. Lothe, and G. M. Pound, *Adv. Phys.*, **15**: 111 (1966).

have passed the activation barrier successfully; a fraction of them diffuse back over the barrier against the applied stress, annihilate, and hence never contribute to the net motion of the dislocation. Thus the *net* nucleation rate of new kink pairs is approximately the number of kink pairs of separation greater than x^*+x' appearing per unit length of dislocation per unit time. These kink pairs expand under the applied stress until the kinks annihilate with kinks from *other* pairs, or until they are stopped at pinning points, depending on the specific boundary conditions.

The net nucleation flux is formulated conveniently as a general diffusion problem. The diffusing species are the kink pairs, which diffuse in the space of double-kink length x. Because a double kink can change its length x by the independent diffusion of either component kink, the diffusivity for the pair is

$$D_p = 2D_k \tag{15-21}$$

The concentration for this diffusion problem is defined so that $c_p(x)\,dx$ is the number of double kinks of length between x and $x+dx$ per unit length of dislocation. Notice that the concentration c_p thus has the dimension $(\text{length})^{-2}$. With neither kink-kink interaction nor applied stress present, the random concentration c_p is determined as follows. Given a positive kink, the probability that a negative kink is in the interval x to $x+dx$, with x measured relative to the positive kink, is $c_k^-\,dx$. When a negative kink is found in dx, a pair of kinks with length between x and $x+dx$ exists. There are altogether c_k^+ positive kinks per unit length that could be considered as the left-hand kink in the pair. Thus the probability per unit length of a kink pair existing with length between x and $x+dx$ is determined from Eq. (14-12) to be

$$c_p^{\,0}\,dx = c_k^+ c_k^-\,dx = \frac{1}{a^2}\exp\left(-\frac{2F_k}{kT}\right)dx \tag{15-22}$$

With kink-kink interaction and external stress present, Eq. (15-22) is generalized to

$$c_p^{\,0}(x) = \frac{1}{a^2}\exp\left[-\frac{F(x)}{kT}\right] \tag{15-23}$$

Here $c_p^{\,0}(x)$ is the constrained-equilibrium concentration of size x that would exist if a current of kinks were prevented from flowing.

The diffusion model giving the *net* nucleation rate is depicted in Fig. 15-7. The form of the steady-state concentration relative to the constrained-equilibrium concentration is that given by nucleation theory.[16, 17] In the region $x \ll x^*$, the balance between incipient nucleation and annihilation is nearly the same as with no stress present, so that $c_p \sim c_p^{\,0}$. Double kinks reaching the

[16]L. Farkas, *Z. phys. Chem.*, **125**: 236 (1927).

[17]J. B. Zeldovich, *Acta Physicochem. URSS*, **18**: 1 (1943).

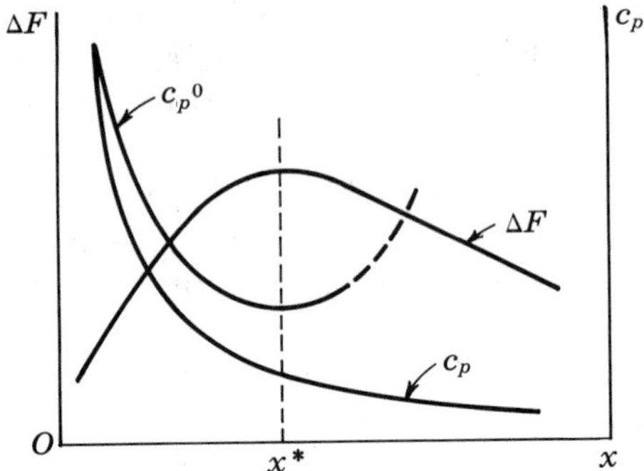

FIGURE 15-7. Constrained equilibrium concentration $c_p^{\,0}$ and actual concentration c_p of kink pairs as a function of kink separation x. The ΔF-x curve from Fig. 15-6 is also shown.

length $x^* + x'$ effectively disappear from the system, so that $c_p \sim 0$ at $x = x^* + x'$. At the saddle point, the Zeldovich[16, 17] treatment yields[18] $c_p(x^*) \sim \frac{1}{2} c_p^{\,0}(x^*)$. Thus in the region x^* to $x^* + x'$, there is a concentration gradient

$$\frac{\partial c_p}{\partial x} \cong \frac{c_p(x^* + x') - c_p(x^*)}{x'} = -\frac{c_p^{\,0}(x^*)}{2x'} \tag{15-24}$$

Thus the diffusion flux, which is equal to the net double-kink nucleation rate, as noted above, is

$$J = \frac{D_p c_p^{\,0}(x^*)}{2x'} \tag{15-25}$$

To a first approximation, the potential drop of kT over the interval x' is produced by the applied stress

$$\sigma b h x' = kT \tag{15-26}$$

With x' so defined, and with the use of Eqs. (15-21) and (15-23), Eq. (15-25) becomes

$$J = \frac{\sigma b h}{a^2 kT} D_k \exp\left(-\frac{F^*}{kT}\right) \tag{15-27}$$

With the additional approximation, to be justified shortly, that

$$F^* \cong 2F_k \tag{15-28}$$

[18] It is often stated in the literature that the Zeldovich factor is a measure of the depletion, relative to the constrained-equilibrium concentration, in the activated state when a current is flowing. Actually, the concentration depletion is only a factor of $\frac{1}{2}$, while the main contribution to the Zeldovich factor involves the *gradient* of concentration at the saddle point. The present treatment preserves all factors in the complete Zeldovich treatment; see J. Feder, K. C. Russell, J. Lothe, and G. M. Pound, *Adv. Phys.*, **15**: 111 (1966).

Eq. (15-27) reduces to[19]

$$J=\frac{\sigma bh}{a^2kT}D_k\exp\left(-\frac{2F_k}{kT}\right) \tag{15-29}$$

Modifications of the Double-Kink Theory

The derivation of Eq. (15-29) involved several approximations that merit discussion. The complete form[20] of $F(x)$ includes the kink-kink interaction energy [Eq. (8-48)] and the work done by the applied stress,

$$F(x)=2F_k-\frac{\mu b^2h^2}{8\pi x}-\sigma bhx \tag{15-30}$$

when the kink pair is separated by a distance x that is greater than the kink width w. The kink-kink interaction is assumed to have a typical value intermediate between the pure screw and pure edge values (Sec. 8-5). The condition for the determination of x^* is then

$$\left[\frac{\partial F(x)}{\partial x}\right]_{x=x^*}=\frac{\mu b^2h^2}{8\pi x^{*2}}-\sigma bh=0 \tag{15-31}$$

giving

$$x^*=\left(\frac{\mu bh}{8\pi\sigma}\right)^{1/2} \tag{15-32}$$

and

$$F^*=2F_k-\left(\frac{\mu\sigma b^3h^3}{2\pi}\right)^{1/2} \tag{15-33}$$

For the case of $\sigma=10^{-4}\ \mu$, $h=b$, and $\mu b^3=5$ eV, the last term in Eq. (15-33) amounts to 0.02 eV, or of the order of kT at $T\sim200°$K. With a typical Peierls stress of the order $\sigma_p\sim10^{-3}\ \mu$, $2F_k$ is expected to be of the order ~0.2 eV (Chap. 8). Thus in this example the Seeger-Schiller correction of including the kink-kink interaction energy modifies the Lothe-Hirth result [Eq. (15-29)] by 10 percent, and for lower applied stresses the correction would be less. With $\sigma\sim10^{-4}\ \mu$, $x^*=20b$, so that the kinks are well separated compared to w (Chap. 8) in the critical configuration. For stresses greater than $10^{-4}\ \mu$, the kink model would be meaningful only if σ_p were correspondingly larger, and these changes would leave the relative magnitudes of the last two terms in Eq.

[19]J. Lothe and J. P. Hirth, *Phys. Rev.*, **115**: 543 (1959).

[20]A. Seeger and P. Schiller, *Acta Met.*, **10**: 348 (1962).

(15-30) unchanged. Thus, provided that $x^* \gtrsim w$, the kink-kink interaction correction is of the order of the uncertainties already present in Eq. (15-29), so that this simpler equation can consistently be used.[21]

Consider now the estimation of x'. In the standard steady-state nucleation theory, instead of the approximation of Eq. (15-26), x' is determined by the curvature of the potential at the saddle point,

$$-\tfrac{1}{2}x'^2\left[\frac{\partial^2 F(x)}{\partial x^2}\right]_{x=x^*} = kT \tag{15-34}$$

From Eq. (15-30),

$$\left[\frac{\partial^2 F(x)}{\partial x^2}\right]_{x=x^*} = -\frac{\mu b^2 h^2}{4\pi x^{*3}} \tag{15-35}$$

so that

$$x' = \frac{kT}{\sigma b h}\left(\frac{\sigma \mu b^3 h^3}{8\pi k^2 T^2}\right)^{1/4} \tag{15-36}$$

The factor in parentheses in Eq. (15-36) represents the correction to Eq. (15-26). However, Eq. (15-34) is applicable only if the curvature $\partial^2 F/\partial x^2$ is quite uniform in the region about x^*. When $\sigma \sim 10^{-4}$ μ and $b = h$, the kink-kink interaction at the saddle point amounts to $\sim$0.01 eV, which is appreciably less than $kT \sim 0.025$ eV at room temperature. Thus, in this range of parameters, the change in curvature at the saddle point is marked because the main part of the potential drop kT is associated with the last term in Eq. (15-30). In this case, Eq. (15-26) is a better estimate for x' than is Eq. (15-36). For higher ratios of σ/T, in the range where the factor in parentheses in Eq. (15-36) is greater than unity, Eq. (15-36) is the better estimate, yielding a *different* type of stress dependence in J.

As mentioned previously, another condition for the validity of Eq. (15-29) is that $x^* > w$, so that the critical configuration resembles *AB* in Fig. 15-8 rather than, say, configuration *CD*. From Eqs. (8-38), (8-75), and (8-77), a value for w intermediate between the screw and edge value is

$$w \sim \left(\frac{\mu b h}{4\sigma_p}\right)^{1/2} \tag{15-37}$$

Using this expression and Eq. (15-32), one sees that the condition $x^* > w$ reduces to

$$\sigma_p \lesssim 6\sigma \tag{15-38}$$

[21] However, retention of the term containing σ can be important in interpreting internal friction phenomena such as the broadening of the Bordoni internal friction peak.

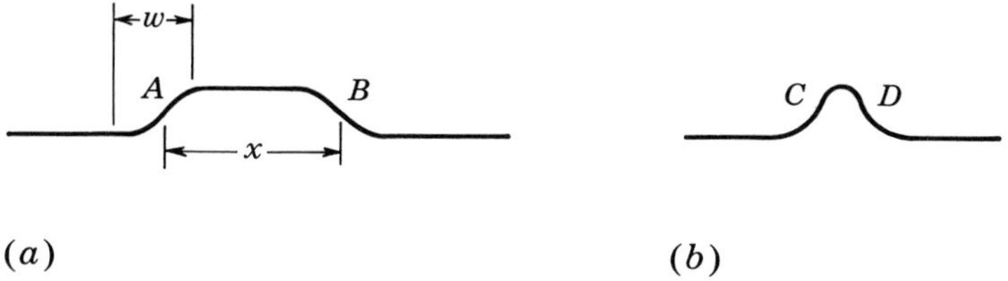

FIGURE 15-8. Two possible critical-sized double-kink configurations AB and CD. In case (a) $x > w$; in case (b) $w > x$.

The numerical factor in Eq. (15-38) is rather uncertain because of the uncertainties discussed in Sec. 8-4, but the result shows that the kink-drift model of Eq. (15-29) should be valid except for high stresses within about one order of magnitude of σ_p. When condition (15-38) does not hold, $F(x)$ can no longer be separated into individual kink energies $2F_k$ and interaction energy, as in Eq. (15-30). Instead, one must minimize the energy of configurations such as CD in Fig. 15-8 as a function of bow-out y, including the elastic energy, and the potential energy of the Peierls barrier. The energies for such a case have been estimated by Celli et al.[22] and by Dorn and Rajnak.[23] Since the size of the critical configurations is of the order of a few core radii, these energy calculations are only rough approximations; the estimation of the elastic energies of such small configurations is difficult and atomic calculations are needed.

For materials with very large Peierls energies, such as germanium, the Peierls energy dominates in determining the critical configuration, which is expected[22] to resemble CD in Fig. 15-8. Again, the above treatments[22, 23] would apply in such a case. For such materials, W_m is expected to exceed kT up to temperatures approaching the melting point, so that D_k is given by Eq. (15-7). Celli et al[22] point out for this case that presence of the activation barriers for kink diffusion produces bumps in the potential that a diffusing kink experiences (Fig. 15-9). When x' exceeds a, the spacing between bumps, the flux over the barrier is further modified, and an additional stress dependence appears in the expression for J.

In summary, different approximations, yielding types of stress dependence different from Eq. (15-29), are appropriate for different ranges of parameters. The systematic treatment of all possibilities would be quite space consuming and confusing. In the following discussion we use only Eq. (15-29), well aware of the approximations involved. Other forms of J can be derived on the basis of the preceding discussion when needed. The literature contains examples of the application of specific expressions for J outside the ranges of parameters where the conditions for their application are valid, so that caution must be observed in applying the various expressions.

[22]V. Celli, M. Kabler, T. Ninomiya, and R. Thomson, *Phys. Rev.* **131**: 58 (1963).

[23]J. E. Dorn and S. Rajnak, *Trans. Met. Soc. AIME*, **230**: 1052 (1964).

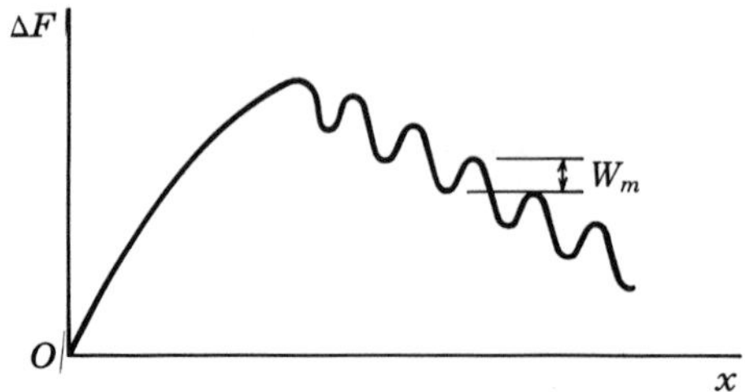

FIGURE 15-9. Free-energy-separation plot for a double-kink pair in the case when the activation energy for motion exceeds kT.

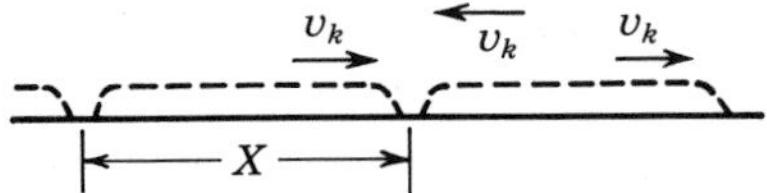

FIGURE 15-10. Double kinks annihilating at a separation X.

Dislocation Velocity

Consider an infinite dislocation line moving uniformly under stress by kink-pair nucleation and subsequent sidewise motion of the kinks. Let X be the average distance along the dislocation swept out by one kink pair before it annihilates with the kinks from other pairs (Fig. 15-10). The velocity of the dislocation normal to itself is

$$v = hXJ \tag{15-39}$$

The distance X is given by

$$X = 2v_k\tau \tag{15-40}$$

where τ is the mean lifetime of a kink pair.[24] By the requirement that steady state be maintained, τ must also be the average time required for a successful nucleation event in the newly exposed line segment within a growing kink pair,

$$\tau = \frac{1}{J(X/2)} \tag{15-41}$$

where $\frac{1}{2}X$ is the average segment available for nucleation during the lifetime of

[24]Alternatively, one can think of a double kink being *annihilated* when a new double kink nucleates within a growing kink pair; this model also yields Eqs. (15-39) to (15-41); J. Lothe and J. P. Hirth, *Phys. Rev.*, **115**: 543 (1959).

the kink pair.[25] Combining Eqs. (15-40) and (15-41), one finds

$$X = 2\left(\frac{v_k}{J}\right)^{1/2} = 2a\exp\left(\frac{F_k}{kT}\right) \tag{15-42}$$

The insertion of this relation into Eq. (15-39) gives

$$v = 2h(Jv_k)^{1/2} = \frac{2\sigma bh^2}{akT} D_k \exp\left(-\frac{F_k}{kT}\right) \tag{15-43}$$

This result is *identical* to Eq. (15-17). The condition for Eq. (15-43) to be valid is that $X \gg x'$, so that the diffusion sink for kinks is outside the nucleation saddle point. This condition is just the opposite of the condition (15-20), which ensures that Eq. (15-17) is valid. Thus Eq. (15-43) is valid in the entire range of linear stress dependence.

If the dislocation segment has a finite length L, such that the kink pairs expand only to the length L, the normal velocity of the segment is given by the modification of Eq. (15-43) to the form

$$v = 2h(Jv_k)^{1/2} \frac{L}{L+X} \tag{15-44}$$

When $L \ll X$, that is, when the segment is shorter than the average distance between thermal kinks, Eq. (15-44) reduces to

$$v = hLJ = \frac{\sigma bh^2 L}{a^2 kT} D_k \exp\left(-\frac{2F_k}{kT}\right) \tag{15-45}$$

Notice that the exponent in Eq. (15-45) is *twice as large* as that in Eq. (15-43).

Exercise 15-1. The dislocation velocity is related to the strain rate by the expression

$$\dot{\epsilon} \cong \rho b v$$

where ρ is the total length of active dislocation per unit volume. Take, as an example, the physical constants for Al, and assume $\rho = 10^6$ cm/cm^3, $2F_k = 0.15$ eV, $\sigma = 10^{-5}\,\mu$, $T = 300$°K, and $\dot{\epsilon} = 10^{-4}$ sec^{-1}. Determine whether Eq. (15-43) or (15-45) more consistently applies for these deformation parameters.[26]

[25] A more detailed model for a random distribution of segments replaces $X/2$ by X in Eq. (15-41) and 2 by $\sqrt{2}$ in Eq. (15-42), U. Bertocci, *Surface Sci.*, **16**: 286 (1969). The factor $\sqrt{2}$ was verified in Monte-Carlo calculations by Bertocci, which we have duplicated. The result is useful also for crystal growth by a ledge model and for several soliton problems including conductance of one dimensional molecules.

[26] J. L. Lytton, L. A. Shepard, and J. E. Dorn, *Trans. AIME*, **212**: 220 (1958).

15-3. APPLICATIONS TO INTERNAL FRICTION

The application of the double-kink glide theory to creep and constant-strain-rate deformation is straightforward, as illustrated in Exercise 15-1. In other applications the relation between the double-kink theory and deformation-rate equations is less obvious and deserves some comment. An example is the phenomenon of internal friction,[27] where double-kink nucleation is postulated to produce a *relaxation* type of internal-friction process at low temperatures and in the kilocycle-megacycle frequency range. The resulting phenomena are classified as Bordoni relaxations[28] although a Niblett-Wilks peak[29] is also sometimes distinguished.

Internal friction also provides a useful distinction between the double-kink type of behavior and the continuous-string type of dislocation behavior. The Granato-Lücke[30] vibrating-string model has been quite successful in explaining amplitude-independent internal friction in the megacycle region. Because of the apparent wide applicability of the Granato-Lücke model, this type of process of forced vibrations is also presented.

The Granato-Lücke Theory

Consider a crystal containing a dislocation network, with segments in skew orientation relative to the Peierls valleys. These segments behave as continuous strings, with mobilities roughly given by the Leibfried formula, Eq. (15-14). Let the average segment length between pinning points be L. The appropriate line tension for bow-out between the end pinning points is

$$\mathcal{S} \simeq \frac{\mu b^2(2-\nu)}{8\pi(1-\nu)} \ln \frac{L}{\rho e^2} \tag{15-46}$$

which is the average between the screw and edge values, Eqs. (6-83) and (6-84). The average effective mass per unit length of the dislocation is

$$m = \frac{\mathcal{S}}{C_t^2} \tag{15-47}$$

by analogy with Eq. (7-21). The dislocation mobility is M; that is, a resolved shear stress σ is required to move the dislocation uniformly at a velocity v,

$$v = M\sigma b = \sigma b/B \tag{15-48}$$

[27]For a general discussion of internal friction, see C. Zener, "Elasticity and Anelasticity of Metals," University of Chicago Press, Chicago, 1948; A. S. Nowick and B. S. Berry, "Anelastic Behavior in Crystalline Solids," Academic, New York, 1972.

[28]P. G. Bordoni, *J. Acoust. Soc. Am.*, **26**: 495 (1954).

[29]D. H. Niblett and J. Wilks, *Phil. Mag.*, **1**: 415 (1956).

[30]A. Granato and K. Lücke, *J. Appl. Phys.*, **27**: 583 (1956).

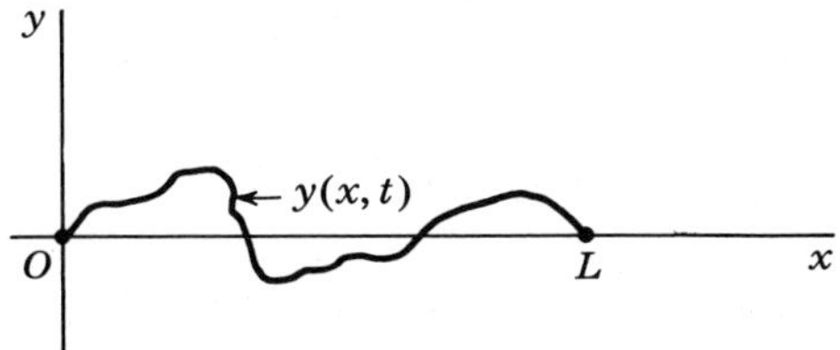

FIGURE 15-11. Configuration of a vibrating dislocation segment of length L.

In terms of the above parameters, the equation of dynamic equilibrium for the vibrating dislocation string is[31]

$$m\frac{\partial^2 y}{\partial t^2}+\frac{1}{M}\frac{\partial y}{\partial t}-\mathcal{S}\frac{\partial^2 y}{\partial x^2}=\sigma b \tag{15-49}$$

where $y(x,t)$ is the deflection at x at time t (Fig. 15-11). Equation (15-49) is conveniently rewritten as

$$\frac{\partial^2 y}{\partial t^2}+\frac{1}{mM}\frac{\partial y}{\partial t}-\frac{\mathcal{S}}{m}\frac{\partial^2 y}{\partial x^2}=\frac{\sigma b}{m} \tag{15-50}$$

Since there are nodes at $x=0$ and $x=L$, the normal modes of the pinned dislocation are given by $\sin(n\pi x/L)$. Only the odd modes can be excited by a uniform applied stress $\sigma=\sigma_0 e^{i\omega t}$, because only these modes yield a net strain. Expanding the applied stress in terms of the dislocation modes, one finds

$$\sigma_0=\sum_n \sigma_n \sin\frac{n\pi x}{L} \qquad n=1,3,5,\ldots \tag{15-51}$$

where, by Fourier analysis,

$$\sigma_n=\frac{4\sigma_0}{\pi n} \tag{15-52}$$

Consider now the forced vibration caused by the component

$$\sigma_n \sin\frac{n\pi x}{L}e^{i\omega t}$$

which must satisfy the relation

$$\frac{\partial^2 y_n}{\partial t^2}+\frac{1}{mM}\frac{\partial y_n}{\partial t}-\frac{\mathcal{S}}{m}\frac{\partial^2 y_n}{\partial x^2}=\frac{b}{m}\sigma_n \sin\frac{n\pi x}{L}e^{i\omega t} \tag{15-53}$$

As can be verified directly by substitution, the solution to Eq. (15-53) is

$$y_n=a_n \sin\frac{n\pi x}{L}e^{i\omega t} \tag{15-54}$$

[31]For a simple development of the vibrating-string problem, see K. R. Symon, "Mechanics," Addison-Wesley, Cambridge, Mass., 1953, p. 47.

where

$$a_n=\frac{\sigma_n b}{m\left(\Omega_n^2-\omega^2+i\omega/Mm\right)}$$

$$=\frac{\sigma_n b\left(\Omega_n^2-\omega^2-i\omega/Mm\right)}{m\left[\left(\Omega_n^2-\omega^2\right)^2+\left(\omega/Mm\right)^2\right]}$$

$$=\frac{\sigma_n b\left(\cos\phi_n-i\sin\phi_n\right)}{m\left[\left(\Omega_n^2-\omega^2\right)^2+\left(\omega/Mm\right)^2\right]^{1/2}}$$

$$=\frac{\sigma_n b}{m\left[\left(\Omega_n^2-\omega^2\right)^2+\left(\omega/Mm\right)^2\right]^{1/2}}\exp(-i\phi_n) \tag{15-55}$$

Here

$$\Omega_n^2=\frac{\mathcal{S}n^2\pi^2}{mL^2}=\frac{C_t^2n^2\pi^2}{L^2} \tag{15-56}$$

and ϕ_n is the phase lag defined by

$$\tan\phi_n=\frac{\omega}{Mm\left(\Omega_n^2-\omega^2\right)} \tag{15-57}$$

Differentiating Eq. (15-54) and making the substitution $i=\exp(i\pi/2)$, one finds

$$\frac{\partial y_n}{\partial t}=\frac{\sigma_n b\omega}{m\left[\left(\Omega_n^2-\omega^2\right)^2+\left(\omega/Mm\right)^2\right]^{1/2}}\sin\frac{n\pi x}{L}\exp\left[i\left(\omega t-\phi_n+\frac{\pi}{2}\right)\right] \tag{15-58}$$

The work done per unit time on the dislocation by the applied stress is

$$\frac{\partial W_L}{\partial t}=\mathrm{Re}(\sigma)Lb\,\mathrm{Re}\left(\frac{\partial y_n}{\partial t}\right)$$

$$=\sigma_n\sin\frac{n\pi x}{L}\cos(\omega t)Lb\,\mathrm{Re}\left(\frac{\partial y_n}{\partial t}\right)$$

$$=\frac{\sigma_n^2Lb^2\omega}{m\left[\left(\Omega_n^2-\omega^2\right)^2+\left(\omega/Mm\right)^2\right]^{1/2}}\sin^2\frac{n\pi x}{L}\cos(\omega t)\cos\left(\omega t-\phi_n+\frac{\pi}{2}\right) \tag{15-59}$$

where Re indicates "real part of." The average value of the $\sin^2$ factor is $\frac{1}{2}$, while the average value of the cosine factors is $\frac{1}{2}\sin\phi_n$, so that the average value of the power input is

$$\frac{\partial W_L}{\partial t} = \frac{\sigma_n^2 L b^2 \omega^2}{4m^2 M\left[(\Omega_n^2 - \omega^2)^2 + (\omega/Mm)^2\right]} \tag{15-60}$$

Summed over all n nodes, the average energy loss from the segment L per cycle of period $2\pi/\omega$ is

$$\Delta W_L = \frac{8\alpha^2 \sigma_a^2 L b^2 \omega}{\pi m^2 M} \sum_{n(\text{odd})} \frac{1}{n^2\left[(\Omega_n^2 - \omega^2)^2 + (\omega/Mm)^2\right]} \tag{15-61}$$

where α^2 is an orientation factor, so that $\langle \sigma_0^2 \rangle = \alpha^2 \sigma_a^2$, and σ_a is the amplitude of the applied stress. With a number of segments N, the energy loss is $\Delta W = N \Delta W_L$. If the applied stress is σ_a, the total energy in the crystal of volume V is

$$W = \frac{V\sigma_a^2}{2\mu} \tag{15-62}$$

so that the internal friction, or relative loss per cycle, is

$$Q^{-1} = \frac{\Delta W}{W} = \frac{16\alpha^2 \mu \rho b^2 \omega}{\pi m^2 M} \sum_{n(\text{odd})} \frac{1}{n^2\left[(\Omega_n^2 - \omega^2)^2 + (\omega/mM)^2\right]} \tag{15-63}$$

where $\rho = NL/V$ is the total active length of dislocation per unit volume. When $\omega \ll \Omega_1$, the approximate result is

$$Q^{-1} = \frac{16\alpha^2 \mu \rho b^2 \omega}{\pi m^2 M \Omega_1^4} \tag{15-64}$$

The modulus defect (or the change in sound velocity) that follows from the same model can be calculated in a similar, straightforward way. Consistency between measured values of Q^{-1} and the modulus defect is one of the tests for applicability of the theory.

With $L \sim 10^4$ to $10^5 b$, the first maximum Ω_1 typically occurs in the range 100 to 1000 megahertz. Thus in most experimental situations the first term in Eq. (15-63) suffices to give a good approximation and the "low-frequency" limit in Eq. (15-64) is adequate. In the resonance region, one must consider a distribution of lengths L, rather than an average length; this has been done by Granato and Lücke[32] and Koehler.[33]

[32] K. Lücke and A. Granato, in J. C. Fisher et al. (eds.), "Dislocations and Mechanical Properties of Crystals," Wiley, New York, 1957, p. 425.

[33] J. S. Koehler, in "Imperfections in Nearly Perfect Crystals," Wiley, New York, 1952, p. 197.

A number of experiments have been rationalized in terms of the above theory.[34] The analysis of experiments in terms of the theory often yield frequency-dependent values of M and L, in contradiction with the notion of M as an intrinsic coefficient. An effective M depending on additional weak pinning points that are dragged along in the forced vibrations of the segments has been suggested.[35] The subtraction of a frequency- and temperature-dependent relaxation contribution of this type led to a consistent interpretation[36] of experimental results for copper for a wide range of frequency and temperature, with an intrinsic drag coefficient $B(T)$ agreeing with Eq. (15-14).

Exercise 15-2. Compute the relaxation time τ_1 corresponding to the fundamental mode Ω_1 of the dislocation line. That is, with the dislocation initially bowed out to its maximum amplitude A, compute the time for it to have relaxed to an amplitude A/e in the absence of stress. Show that in the case of strong overdamping

$$\tau_1 = \frac{1}{mM\Omega_1{}^2}$$

so that the internal friction associated with Ω_1 can be written

$$Q^{-1} = \frac{16\alpha^2\mu\rho b^2}{\pi m\Omega_1{}^2}\,\frac{\omega\tau_1}{\left[1-(\omega/\Omega_1)^2\right]^2+(\omega\tau_1)^2}$$

The Bordoni Peak

Following the original suggestion of Mason,[37] several versions of a Peierls barrier model for the Bordoni peak have been proposed.[38–44] The basic requirement for a relaxation phenomenon is the existence of different but

[34] D. Lenz and K. Lücke, "Fifth International Conference on Internal Friction and Ultrasonic Attenuation in Crystalline Solids," vols. 1 and 2, Springer-Verlag, Berlin, 1975.

[35] H. M. Simpson and A. Sosin, *Phys. Rev.*, **B5**: 1382 (1972).

[36] N. P. Kobelev, Ya. M. Soifer, and V. I. Alshits, *Sov. Phys. Solid State*, **21**: 680 (1979).

[37] W. P. Mason, *J. Acoust. Soc. Am.*, **27**: 643 (1955).

[38] A. Seeger, *Phil. Mag.*, **1**: 651 (1956).

[39] H. Donth, *Z. Phys.*, **149**: 111 (1957).

[40] T. Jøssang, K. Skylstad, and J. Lothe, in "Relation between the Structure and Mechanical Properties of Metals," H.M. Stationery Office, London, 1963, p. 528. These authors showed that the Donth theory gave an incorrect expression for the activation energy.

[41] J. Lothe, *Phys. Rev.*, **117**: 704 (1960).

[42] A. Seeger and P. Schiller, *Acta Met.*, **10**: 348 (1962).

[43] G. Alefeld, in R. Hasiguti (ed.), "Lattice Defects and Their Interactions," Physical Society of Japan, Tokyo, 1967.

[44] G. Fantozzi, C. Esnouf, W. Benoit, and I. G. Ritchie, *Prog. Mat. Sci.*, in press.

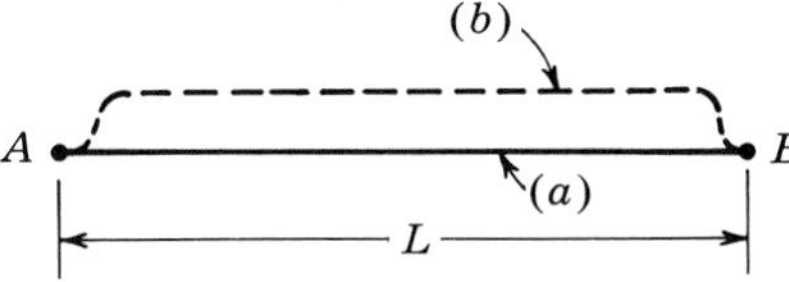

FIGURE 15-12. Two configurations of a dislocation segment AB.

equally probable dislocation configurations, between which the dislocation can jump under thermal activation. Consider a segment of length L, pinned at points A and B, and lying in a Peierls valley (Fig. 15-12). In the presence of an internal stress σ_i of just such a magnitude that

$$\sigma_i bhL = 2W_k \tag{15-65}$$

configurations (a) and (b) have the same energy and are equally probable, apart from minor entropy terms considered in the more refined theory due to Alefeld.[43] If these were the only two possible configurations for the segment, the dislocation would jump between them, by thermal activation over the Peierls barrier in going from (a) to (b), and by annihilation against the internal stress σ_i in going from (b) to (a). By the principle of detailed balance, the rates of the two processes are the same. Paré[45] first stated Eq. (15-65) clearly, so it is usually called the *Paré condition*.

Let us then compute the jump frequency for the simple model. Consider a segment in configuration (a) of Fig. 15-12. In a time $1/\nu_d$ the segment on the average will jump into configuration (b). The jump frequency is determined from the relation

$$\nu_d = LJ$$

where J is the double-kink nucleation rate per unit length. With J given by Eq. (15-29), the jump frequency becomes

$$\nu_d = \frac{\sigma_i Lbh}{a^2 kT} D_k \exp\left(-\frac{2F_k}{kT}\right) \tag{15-66}$$

Inserting Eq. (15-65), one obtains, finally,

$$\nu_d = \frac{2W_k}{a^2 kT} D_k \exp\left(-\frac{2F_k}{kT}\right) \tag{15-67}$$

independent of internal stress in the present approximation.

With an incremental applied stress superposed on the internal stress, configuration (b) is more probable than (a) by a small bias. In an assembly of segments, the concentration of segments in configuration (b) is then higher

[45] V. K. Paré, doctoral dissertation, Cornell University, Ithaca, N.Y., 1955; *J. Appl. Phys.*, **32**: 332 (1961).

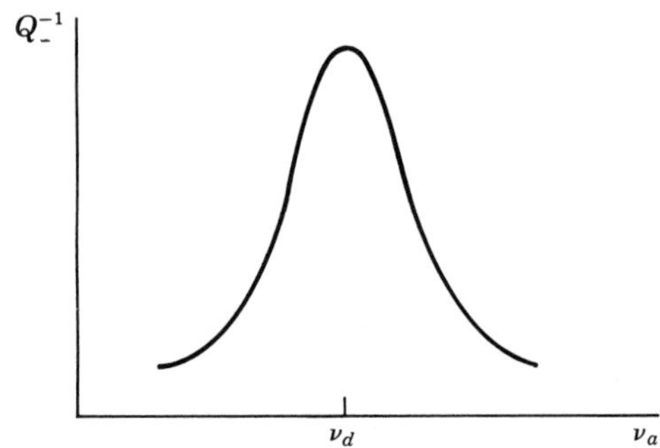

FIGURE 15-13. Internal friction Q^{-1} as a function of ν_a.

than the number in (a). In the presence of an oscillating applied stress, the dislocation assembly can rearrange the relative concentrations in phase with the applied stress for frequencies much lower than the jump frequency ν_d, while for frequencies much higher than ν_d the dislocations do not have time to respond to the applied stress. As is well known,[46] the internal friction is smaller in either extreme and is a maximum when the frequency of the applied stress ν_a equals the dislocation jump frequency

$$\nu_a = \nu_d \tag{15-68}$$

The quantitative relation of the internal friction to the relaxation time $1/\nu_d$, illustrated in Fig. 15-13, can be found in any standard treatise on internal friction, such as Zener's. A plot of the internal friction as a function of ν_a is presented in Fig. 15-13.

Consistent with the above theory, very perfect crystals with long, straight dislocations pinned by irradiation show only a weak Bordoni relaxation: internal stresses that could fulfill the Paré condition are not present.[47] However, in the absence of internal stresses, the Paré condition can be satisfied by a suitable external static bias stress. Esnouf et al.[47] observed that the Bordoni peak appeared when a bias stress was applied whereas when the static bias stress was removed, the crystal returned to its original state. These experiments convincingly substantiate the Paré concept.[48]

The Bordoni peak is well developed in crystals deformed to a state of appreciable internal stresses. One would think that the above model is too simple to describe such a case, since only for very few segments would the internal stresses be just such that the Paré condition (15-65) would be satisfied. Then one would expect only a weak Bordoni peak, contrary to experimental evidence. However, consider Fig. 15-14 as a more realistic picture of the longer

[46] C. Zener, "Elasticity and Anelasticity of Metals," University of Chicago Press, Chicago, 1948.

[47] C. Esnouf, G. Fantozzi, and P. F. Gobin, *Phys. Stat. Solidi*, **32a**: 441 (1975).

[48] G. Fantozzi, C. Esnouf, W. Benoit, and I. G. Ritchie, *Prog. Mat. Sci.* (in press).

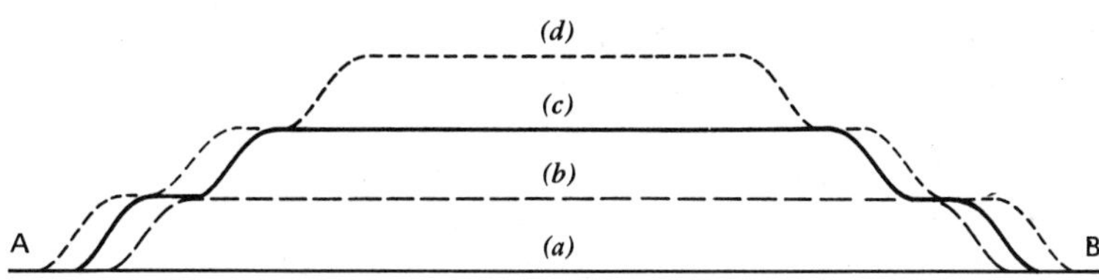

FIGURE 15-14. Various configurations of a dislocation segment AB.

bow-outs expected in crystals with large internal stresses. Let (c) be the equilibrium configuration under internal stress σ_i. In a first approximation, the neighboring configurations (b) and (d) have nearly the same energy as the lowest energy configuration (c). This is so because the smoothed $W-x$ plot, where x is a continuous reaction coordinate for the sequence $a, b, c, d, \ldots$, has a minimum at (c) where $\partial W/\partial x = 0$. But this, in turn, gives precise fulfillment of the Paré condition in a general sense: the existence of different but equally probable dislocation configurations. That is, for a given internal stress, a segment takes up just such a kinked configuration that the neighboring configurations have nearly the same energy, with adjacent configurations being separated by free energy barriers $\sim 2F_k$.

The above description is a first-order discussion of the model of Fig. 15-14. Fantozzi et al.[48] have analyzed the model more quantitatively and conclude that the Paré condition is not exactly satisfied with an arbitrary, sufficiently large, internal stress, but that it is satisfied periodically as a function of σ_i. With this refinement, together with a stress-dependent activation energy $2F_k^*$ instead of the rough approximation $2F_k$, they satisfactorily explain both the appearance of Bordoni peaks in fcc metals and the fact that they are two to four times broader than a simple, standard relaxation peak.

The Bordoni peak in fcc metals was resolved into two peaks by Niblett and Wilks.[49] The peak at the lower frequency is associated with edge dislocations, the other with screw dislocations.

The double-kink model for the Bordoni peak predicts a correlation between the Bordoni peak relaxation and suitable microcreep experiments, for example those described by Exercise 15-1. Experiments have been performed which demonstrate such a correlation.[47, 50]

Skew Dislocation

As mentioned several times previously, either the simple kink description of Eq. (15-29) or the continuous-string model are adequate to rationalize most deformation behavior, including microcreep and the limiting internal-friction

[49] D. H. Niblett and J. Wilks, *Phil. Mag.*, **1**: 415 (1956).

[50] G. Alefeld, J. Filloux, and H. Harper, in A. R. Rosenfield, G. T. Hahn, A. L. Bement, Jr., and R. I. Jaffee (eds.), "Dislocation Dynamics," McGraw-Hill, New York, 1968, p. 191.

cases discussed above. For the intermediate case of a kinked skew line, as in Fig. 15-3, kink-kink interaction and kink entropy are important in the description of *internal friction*; we emphasize that in the case of *creep*, double-kink nucleation rapidly predominates for a skew line, as in Fig. 15-4, so that these refinements are unnecessary. Both kink-kink interaction and kink entropy have been included in detail in the internal-friction theory of Alefeld.[51] With both factors, the theory is quite complicated. However, the kink-kink interaction is the predominant term, so that a good approximation of the theory is obtained by neglecting kink entropy. With the neglect of the entropy term, the intermediate case and the result of Alefeld can be derived readily from our treatment of the Granato-Lücke theory.

For small bow-outs of the skew line, Eq. (15-49) still applies, but $\mathcal{S}$ and m differ from the continuous-string case. The appropriate line tension is given by Eq. (8-87), Eq. (8-88), or an average of these, depending on the dislocation character. The effective mass m of the dislocation line is derived from the model of Fig. 15-2. Considered as a continuous line, the kinetic energy per unit length is $\frac{1}{2}mv^2$. Considered as a kinked line, the kinetic energy per unit length is the product of the kinetic energy of a single kink $\frac{1}{2}m_k v_k^2$ and the number of kinks per unit length $1/l$. Equating these two expressions and using Eq. (15-12), one finds

$$m = \frac{m_k}{l}\left(\frac{v_k}{v}\right)^2 = \frac{m_k l}{h^2} \tag{15-69}$$

With these definitions of $\mathcal{S}$ and m, the results of the Granato-Lücke theory, Eqs. (15-63) and (15-64), apply directly to the skew-dislocation case and reduce to the Alefeld result. The natural frequencies of the dislocation are given by Eq. (15-56) in its exact form.

Since the kink spacing l appears logarithmically in $\mathcal{S}$ but linearly in m, the mass effect predominates in phenomena which change with l. As the kink spacing decreases, i.e., as the skewness increases, the resonant frequency Ω_1 increases and the internal friction decreases; the relaxation time τ_1 also decreases, but only logarithmically.

For large bow-outs of the skew line, such as in Fig. 15-4, the previous theory is inapplicable even when double-kink nucleation is excluded; the internal friction becomes amplitude dependent. The configuration of the bowed-out line contains a kink pileup which is much stiffer than the unbowed line because of kink-kink interaction. Without treating the problem in detail, one can make some qualitative predictions. As the amplitude increases, the average stiffness increases, and the internal friction should decrease. Also, since a bias stress produces a configuration such as Fig. 15-4 in the *relaxed* condition, the internal friction should decrease with increasing bias stress.[52]

[51] G. Alefeld, *J. Appl. Phys.*, **36**: 2633, 2642 (1965).

[52] N. F. Mott, *Proc. Phys. Soc.*, **B64**: 729 (1951).

15-4. DIFFUSION-CONTROLLED CLIMB

In principle, dislocation climb by jog nucleation and growth is similar to dislocation glide by kink nucleation and growth. Important differences exist, however, in that climb occurs by point-defect emission or annihilation, so that *osmotic forces* act in addition to elastic forces on climbing dislocations. In order to clearly elucidate the role of these osmotic forces, we first consider the case of climb under pure diffusion control; local equilibrium is supposed to exist between dislocations and point defects in their vicinity. Dislocation climb under diffusion control has been treated by a number of writers.[52–56] Balluffi and Seidman[56, 57] compared a variety of experiments involving climb with theoretical expectation and conclude that in a number of cases, particularly when the dislocations have relatively small dissociations, climb occurs under diffusion control. Thus the diffusion-control case is a realistic example of a dislocation climb process.

For simplicity, let us first consider climb of the edge dislocation and then generalize the results to a mixed dislocation. Also, suppose that only vacancies transport matter to and from the dislocation, because this is the defect that predominates in the diffusion of mass. The results evidently can be generalized to include the interstitials.

Climb Force on a Straight Dislocation

Consider the edge dislocation shown in Fig. 15-15, with a stress σ_{xx} acting over its core. The results of Sec. 3-4 show that if matter is removed from the surface of a *stress-free* crystal and deposited along the dislocation so that it moves upward by one atomic distance h, the total energy change per unit length is

$$\frac{\delta W}{L} = -\sigma_{xx} bh \tag{15-70}$$

The corresponding *elastic* force per unit length acting on the dislocation is

$$\frac{F_{y_{\text{el}}}}{L} = \sigma_{xx} b \tag{15-71}$$

Now suppose that $\sigma_{xx} = 0$ at the dislocation core. The dislocation can climb by a distance h if atoms are removed from lattice sites near the dislocation, thus creating vacancies, and deposited along the dislocation line. There is no

[53] J. Weertman, *J. Appl. Phys.*, **26**: 1213 (1955).

[54] J. Lothe, *J. Appl. Phys.*, **31**: 1077 (1960).

[55] R. M. Thomson and R. W. Balluffi, *J. Appl. Phys.*, **33**: 803, 817 (1962).

[56] D. N. Seidman and R. W. Balluffi, *Phys. Stat. Solidi*, **17**: 531 (1966).

[57] R. W. Balluffi, *Phys. Stat. Solidi*, **31**: 443 (1969).

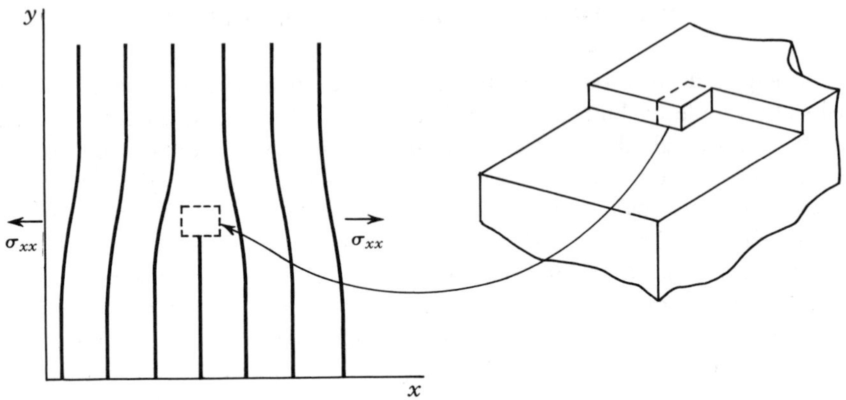

FIGURE 15-15. Schematic representation of dislocation climb by the removal of an atom from the surface of a stress-free crystal and its deposition on an edge dislocation. Here $\mathbf{b}=(b_x,0,0)$ and $\xi=(0,0,-1)$.

change in elastic energy in this case because $\sigma_{xx}=0$. However, there is a free-energy change associated with the formation of the vacancies; per unit length the free-energy change is

$$\frac{\delta G}{L}=\frac{\bar{G}bh}{v_a} \tag{15-72}$$

where $\bar{G}$ is the chemical potential of the vacancies

$$\bar{G}=kT\ln\frac{c}{c^0} \tag{15-73}$$

referred to the standard-state concentration c^0 given by Eq. (14-38). Corresponding to the above free-energy change, there is an *osmotic* force per unit length acting on the dislocation,

$$\frac{F_{y_{os}}}{L}=-\frac{\bar{G}b}{v_a}=-\frac{kTb}{v_a}\ln\frac{c}{c^0} \tag{15-74}$$

In the presence of both a stress σ_{xx} at the core and a vacancy concentration different from c^0, the total force on the dislocation is

$$\frac{F_y}{L}=\frac{F_{y_{el}}}{L}+\frac{F_{y_{os}}}{L}=\sigma_{xx}b-\frac{kTb}{v_a}\ln\frac{c}{c^0} \tag{15-75}$$

At equilibrium $F_y=0$, so that Eq. (15-75) yields the local equilibrium concentration of vacancies near the dislocation as

$$c=c^0\exp\left(\frac{\sigma_{xx}v_a}{kT}\right) \tag{15-76}$$

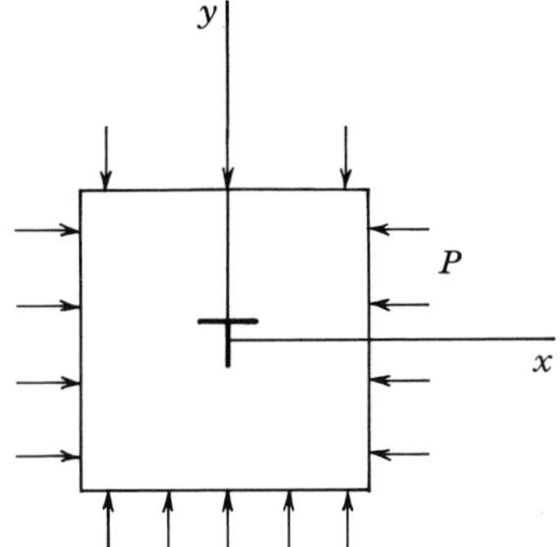

FIGURE 15-16. An edge dislocation in a crystal under hydrostatic pressure P.

with a corresponding vacancy chemical potential given by Eq. (15-73). Note that c^0 also depends on σ_{xx}, according to Eq. (14-38). If the *chemical potential* of the vacancies is different from that far away from the dislocation, a diffusive flow of vacancies down the gradient will occur, and the dislocation will climb by vacancy emission or annihilation in order to maintain the local equilibrium concentration [Eq. (15-76)].[58]

Let us consider some examples. Consider first a block under hydrostatic pressure containing one edge dislocation, as depicted in Fig. 15-16. Image stresses are neglected. According to Eq. (15-76), the local chemical potential of vacancies in equilibrium with the dislocation is

$$\overline{G} = \sigma_{xx} v_a = -Pv_a$$

The concentration at the external surface is given by Eq. (14-37); since $W_e = Pv_a$ in the present case, the concentration at the external surface is

$$c = c^0 \exp\left(-\frac{Pv_a}{kT}\right)$$

corresponding to a chemical potential $\overline{G} = -Pv_a$, the same as at the dislocation core. Thus the dislocation in Fig. 15-16 will not climb.

Consider now the situation shown in Fig. 15-17. The chemical potential of vacancies near the dislocation is again $\sigma_{xx} v_a$. At surface A the chemical potential is also $\sigma_{xx} v_a$. However, at surface B, $W_e = 0$, and the chemical potential of the vacancies is zero. Thus diffusive vacancy flow occurs both to

[58] We emphasize that the vacancy chemical potential must differ for vacancy diffusion to occur. Notice that the standard-state vacancy concentration c^0 [Eq. (14-38)] *varies* with total pressure $p+P$. Thus, in a field of varying internal pressure, concentration gradients exist when the chemical potential is constant throughout the system. No diffusive flow takes place in the presence of the concentration gradients in such a case; only chemical-potential gradients produce a diffusion current.

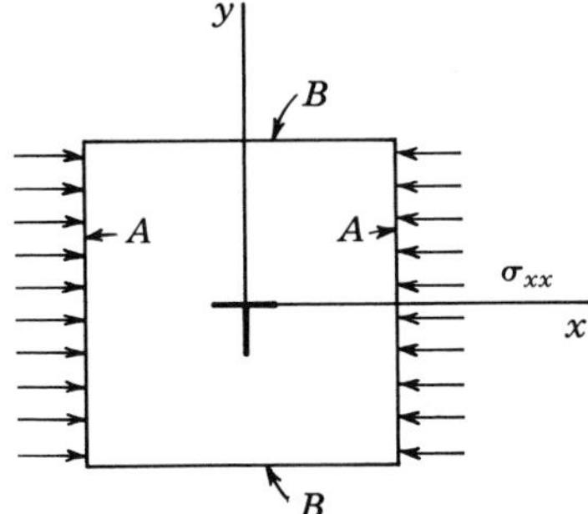

FIGURE 15-17. An edge dislocation in a crystal under uniaxial compression σ_{xx}.

the dislocation and from surface B to surface A. This example is a somewhat idealized version of so-called Nabarro-Herring creep.[59]

The generalization of Eq. (15-75) for a mixed dislocation is straightforward. The total climb force per unit length in the direction $\mathbf{b}\times\boldsymbol{\xi}$ is

$$\frac{F}{L}=\frac{F_{\rm el}}{L}+\frac{F_{\rm os}}{L} \tag{15-77}$$

where $F_{\rm el}$ is given by Eq. (3-90) as

$$\frac{F_{\rm el}}{L}=\frac{(\mathbf{b}\cdot\boldsymbol{\sigma}\times\boldsymbol{\xi})\cdot(\mathbf{b}\times\boldsymbol{\xi})}{|\mathbf{b}\times\boldsymbol{\xi}|} \tag{15-78}$$

The osmotic force is given by

$$\frac{F_{\rm os}}{L}=-\frac{kTb_e}{v_a}\ln\frac{c}{c^0} \tag{15-79}$$

where b_e is the edge component defined by

$$b_e=\frac{(\mathbf{b}\times\boldsymbol{\xi})\cdot(\mathbf{b}\times\boldsymbol{\xi})}{|\mathbf{b}\times\boldsymbol{\xi}|} \tag{15-80}$$

Notice that $F_{\rm el}$ in general includes a contribution from the force on the screw component of the dislocation. Thus, in general, *both normal stresses and shear stresses can produce climb of a mixed dislocation.* Equation (15-77) can be thought of as a generalization of the Peach-Koehler equation [Eq. (3-90)].[60]

[59] F. R. N. Nabarro, in "Report on the Conference on the Strength of Solids," Physical Society, London, 1947, p. 75; C. Herring, *J. Appl. Phys.*, **21**: 437 (1950).

[60] J. Weertman [*Phil. Mag.*, **11**: 1217 (1965)] has discussed the generalization of the Peach-Koehler equation for the climb case. His result and that given here are identical, although they appear to differ because of a difference in the definition of the vacancy standard state in the two treatments. Instead of Eq. (14-38), he defines

$$c^0=n_v\exp\left(-\frac{G_v-pv_v^*+Pv_v}{kT}\right)$$

Exercise 15-3. Consider a 45° mixed dislocation with ξ parallel to the z axis and $\mathbf{b}$ in the xz plane. Compare the magnitudes of the climb force associated with the screw and edge components of the dislocation if the applied stress is such that $\sigma_{xx}=\sigma_{xz}$. What is the local equilibrium vacancy concentration at the dislocation core if only $\sigma_{xz}\neq 0$?

Exercise 15-4. Show that in the presence of an interstitial supersaturation c_i/c_i^0, the osmotic force in the $\mathbf{b}\times\xi$ direction is

$$\frac{F_{\mathrm{os}}}{L}=\frac{kTb_e}{v_a}\ln\frac{c_i}{c_i^0}$$

In the presence of both vacancies and interstitials, equilibrated according to Eq. (14-41), should the above force be added to Eq. (15-79) to give the total osmotic force, or does either expression give the osmotic force independently? *Hint*: An atom removed from the inserted plane at a dislocation core can *either* annihilate a vacancy *or* create an interstitial, but cannot do both.

Diffusion-Controlled Climb Rate

As a simple example of a climb process, let us treat the case of a mixed dislocation near the center of a cylinder and parallel to its axis, as shown in Fig. 15-18. Suppose that an externally applied stress $\sigma=\sigma_{xz}$ is present, and that image stresses are negligible. Only tangential surface forces are required to maintain the stress σ_{xz}, so that the hydrostatic component of the external stress is zero and the concentration of vacancies in local equilibrium with the surface is c^0. The shear stress exerts a force per unit length σb_s on the dislocation, so that near the core the local equilibrium vacancy concentration is given by Eq. (15-79) as

$$c=c^0\exp\left(\frac{\sigma b_s v_a}{b_e kT}\right) \tag{15-81}$$

That is, he factors out the work term Pv_a. Thus his expression for c is, instead of Eq. (14-37),

$$c=c^0\exp\left(-\frac{W_e-Pv_a}{k}\right)$$

With the derivation proceeding as above, his method gives Eq. (15-77), Eq. (15-79) with his definition of c^0 and an expression similar to Eq. (15-78) for F_{el}, but with $\boldsymbol{\sigma}-(P\mathbf{I})$ replacing $\boldsymbol{\sigma}$. Here $\mathbf{I}$ is the unit tensor, so that $-P\mathbf{I}$ is the hydrostatic component of $\boldsymbol{\sigma}$. We prefer our method because it clearly indicates the parallel effects of external and internal pressure on vacancy chemical potential. [See J. Lothe and J. P. Hirth, *J. Appl. Phys.*, **38**: 845 (1967)]. R. W. Balluffi and A. V. Granato, in F. R. N. Nabarro (ed.), "Dislocations in Solids," vol. 4, North-Holland, Amsterdam, 1979, p. 1, present another equivalent method. Instead of separately defining a standard state chemical potential and concentration c^0, they directly incorporate c^0 in the definition of c, for example, by combining the above two expressions.

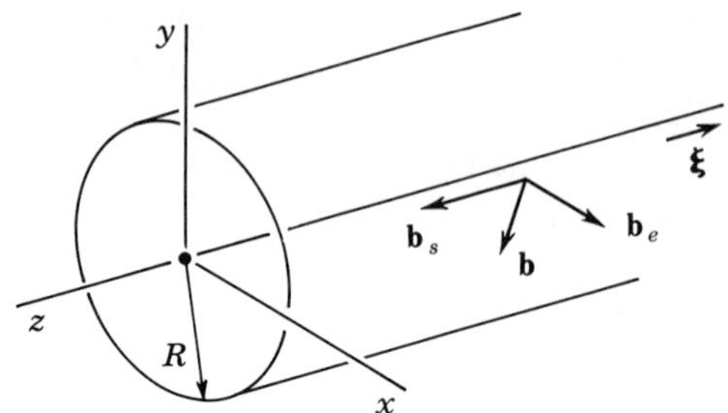

FIGURE 15-18. Mixed dislocation lying along the axis of a cylinder and under an external stress $\sigma = \sigma_{xz}$.

Thus a vacancy chemical-potential gradient exists between the dislocation and the surface.

The diffusion flux **J** of vacancies is given by the product of the vacancy concentration c and the drift velocity [Eq. (15-3)], with the force equal to the negative gradient of the chemical potential, yielding

$$\mathbf{J} = -\frac{D_v c}{kT}\nabla\overline{G} = -D_v c\nabla\ln\left(\frac{c}{c^0}\right) \tag{15-82}$$

where D_v is the vacancy diffusivity. The steady-state condition is that the divergence of **J** vanishes, yielding

$$\nabla\cdot\mathbf{J} = -\nabla\cdot\left(D_v c\nabla\ln\frac{c}{c^0}\right) = 0 \tag{15-83}$$

The solution to Eq. (15-83) is mathematically very cumbersome because of the variation of D_v and c^0 with internal pressure and hence with position. The variation of c^0 according to Eq. (14-43) has been included in a numerical solution of the problem[61] that shows that the results of the approximate analytical treatment in this section underestimates the climb velocity by a factor of 1.3 for typical cases. The variation of D_v has not been considered. Analogous calculations with somewhat different boundary conditions have been performed for metals under radiation damage.[62] Because of these complications, unlike the equilibrium treatment of the previous section, where the retention of the internal pressure dependence of c^0 is important, we make the simplifying assumptions here that c^0 and D_v are constant. These assumptions are actually quite good except in the region near the dislocation core. Equation

[61] M. A. Burke and W. D. Nix, *Phil. Mag.*, **37**: 479 (1978).

[62] See F. A. Nichols, in "Dislocation Modeling of Physical Systems," Pergamon, Oxford, 1981, p. 147 and H. Rauh and D. Simon, *Phys. Stat. Solidi*, **46a**: 499 (1978), who obtain an analytic solution in the form of an infinite series.

(15-83) then reduces to Laplace's equation

$$\nabla^2 c = \frac{1}{r}\frac{\partial}{\partial r} r \frac{\partial c}{\partial r} = 0 \tag{15-84}$$

where the derivatives with respect to θ and z are zero by symmetry. With the assumptions that $\sigma b_s v_a \ll b_e kT$, so that the first term in a power-series expansion of the exponential in Eq. (15-81) is adequate, and that the concentration of Eq. (15-81) is maintained at a radius $r = b$, the boundary conditions are

$$\begin{aligned} c - c^0 &= 0 && \text{at } r = R \\ c - c^0 &= c' = \frac{\sigma b_s v_a}{b_e kT} c^0 && \text{at } r = b \end{aligned} \tag{15-85}$$

The solution to Eq. (15-84) that satisfies these boundary conditions is

$$c - c^0 = c' \frac{\ln(R/r)}{\ln(R/b)} \tag{15-86}$$

The net vacancy current per unit length away from the dislocation is

$$\begin{aligned} I &= -2\pi r D_v \frac{\partial}{\partial r}(c - c^0) \\ &= \frac{2\pi D_v c'}{\ln(R/b)} = \frac{2\pi D_v \sigma b_s v_a c^0}{b_e kT \ln(R/b)} \end{aligned} \tag{15-87}$$

The velocity of climb of the dislocation is

$$v_y = \frac{I v_a}{b_e} \tag{15-88}$$

Substituting Eq. (15-87) into (15-88) and invoking the definition of the atomic self-diffusion coefficient

$$D_s = v_a c^0 D_v \tag{15-89}$$

one obtains

$$v_y = \frac{2\pi D_s v_a \sigma b_s}{b_e^2 kT \ln(R/b)} \tag{15-90}$$

The logarithmic term is essentially the same as that which appears in energy expressions; it depends on an external cutoff parameter R as well as an internal cutoff parameter b.

In the generalization of the above result, R is the distance from the dislocation to the region where $c \sim c^0$. In an annealing experiment or a creep experiment, with some dislocations emitting vacancies and some absorbing them, a reasonable value for R is the spacing between dislocations, typically $R \sim 10^4 b$. Also, the force is given in general by Eq. (15-78), including the contribution from σ_{xx}. Thus the velocity in the $\mathbf{b} \times \boldsymbol{\xi}$ direction is

$$v = \frac{2\pi D_s v_a (F_{\text{el}}/L)}{b_e^2 kT \ln(R/b)} \cong \frac{D_s v_a (F_{\text{el}}/L)}{b_e^2 kT} \tag{15-91}$$

Since v is rather insensitive to the precise value of R, Eq. (15-91) is a reasonable estimate for a wide variety of typical experimental conditions. The use of the approximate form of Eq. (15-91) for a typical creep mobility is equally as meaningful as the use of Eq. (6-95) for a typical line tension. Some of the uncertainties involved in Eq. (15-91) are made evident in the next section.

Vacancy Source-and-Sink Interaction

Consider two opposite-sign edge dislocations in the middle region of a cylinder with their separation l much smaller than R (Fig. 15-19). Ignore image forces. The two dislocations attract one another with a force

$$\frac{F}{L} = \frac{\mu b^2}{2\pi(1-\nu)l} \tag{15-92}$$

and tend to climb together and annihilate. Both dislocations climb by vacancy emission and both tend to establish a vacancy concentration near their cores

$$c' = c - c^0 = c^0 \frac{F v_a}{LbkT} \tag{15-93}$$

Both dislocations are centers of vacancy distributions such as Eq. (15-86),

$$c - c^0 = A \ln \frac{R}{r} \tag{15-94}$$

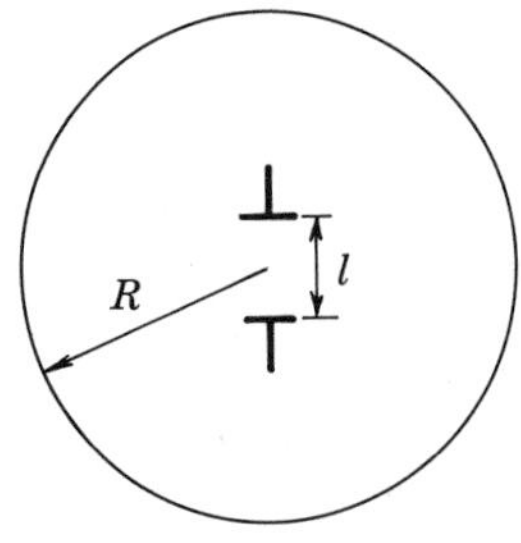

FIGURE 15-19. Two edge dislocations near the center of a cylinder.

where A is a parameter to be determined. By superposition, the two terms of the form of Eq. (15-94) give the vacancy distribution with both dislocations present. With the requirement that the concentration be that of Eq. (15-93) at the distance $r=b$ from either dislocation, one obtains the relation for A,

$$A\left(\ln\frac{R}{b}+\ln\frac{R}{l}\right)=2A\ln\frac{R}{(bl)^{1/2}}=\frac{c^0 F v_a}{LbkT} \tag{15-95}$$

Proceeding in the manner of the preceding section, one finds that the creep velocity of either dislocation is

$$v=\frac{\pi D_s v_a F}{b^2 LkT\ln\left[R/(bl)^{1/2}\right]} \tag{15-96}$$

or *half* the velocity that would be predicted by Eq. (15-91), except for a different logarithmic term. With $R\sim 10^4 b$ and $l\sim 10^2 b$, the ratio of $\ln(R/b)$ to $\ln[R/(bl)^{1/2}]$ is 1.3, so that the factor $\frac{1}{2}$ is the more important difference.

In general, one must consider interaction between climbing dislocations.[63] Vacancies emitted by one dislocation influence the vacancy concentration in the vicinity of another one. Climbing dislocations cannot always be treated independently, even in an approximate way. The factor $\frac{1}{2}$ that appeared in the above example was not a very large factor. However, a group of N dislocations, all emitting or all absorbing vacancies, would entail a factor $1/N$. Such factors could be important, for example, in the climb relaxation of a dislocation pileup.

Climb Forces on Curved Dislocations

In the preceding examples only the climb of *straight* dislocations acted upon by stresses is treated. *Curved* dislocations can climb under the action of line-tension forces. Consider the important example of the contraction of a sessile loop by climb (Fig. 15-20). The energy W of the loop and the line tension $\mathcal{S}$ are given, respectively, by Eqs. (6-52) and (6-79). If the loop contracts by an amount δR, the number of vacancies emitted is

$$\delta N=\frac{2\pi Rb}{v_a}\delta R \tag{15-97}$$

The elastic energy released per vacancy emitted is

$$w=\frac{\partial W}{\partial N}=\frac{\partial W}{\partial R}\frac{\partial R}{\partial N}=2\pi\mathcal{S}\frac{\partial R}{\partial N}=\frac{\mathcal{S}v_a}{Rb} \tag{15-98}$$

[63]An example where this interaction has led to some controversy is discussed by P. M. Marquis and R. E. Smallman, *Phil. Mag.*, **34**: 903 (1976).

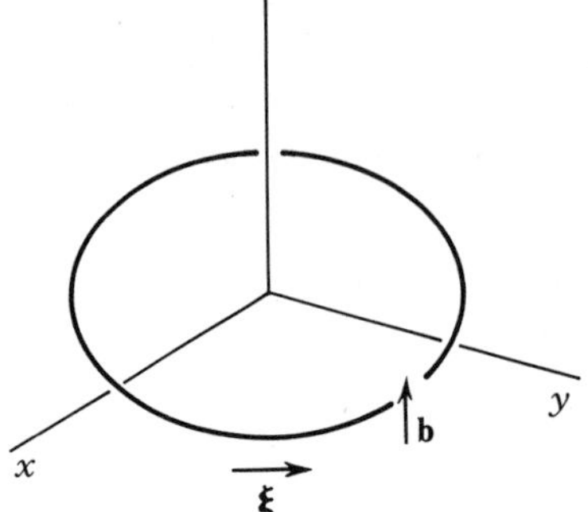

FIGURE 15-20. A circular dislocation loop formed by the collapse of a vacancy disc.

Consider now the boundary conditions on the diffusion problem. For the loop to be in local equilibrium with the vacancies, w must equal the local value of $\bar{G}$, or

$$w = kT \ln \frac{c}{c^0}$$

which reduces to

$$c' = c - c^0 = c^0\left[\exp\left(\frac{w}{kT}\right) - 1\right] \cong \frac{c^0 \mathbb{S} v_a}{RbkT} \tag{15-99}$$

when $w \ll kT$. The superconcentration c' is assumed to be maintained within a distance $r = b$ of the dislocation line. Far from the dislocation, $c \to c^0$ as $r \to \infty$.

Although the loop is a finite source, its effect is that of a point source in the region $r \gg R$. The steady-state solution to the form of Laplace's equation, $\nabla^2 c = 0$, appropriate for a point source is

$$c - c^0 = \frac{B}{r} \tag{15-100}$$

where B is a parameter to be determined. The corresponding net current of vacancies away from the loop is

$$I = -4\pi r^2 D_v \frac{\partial}{\partial r}\frac{B}{r} = 4\pi D_v B \tag{15-101}$$

In actuality, the "point" source of Eq. (15-100) is composed of a number of point sources of equal strength β, uniformly distributed along the circumference of the loop at intervals of the order of b. Thus

$$B = \frac{2\pi r}{b}\beta \tag{15-102}$$

β is determined by the condition that within a distance b of the loop itself, the

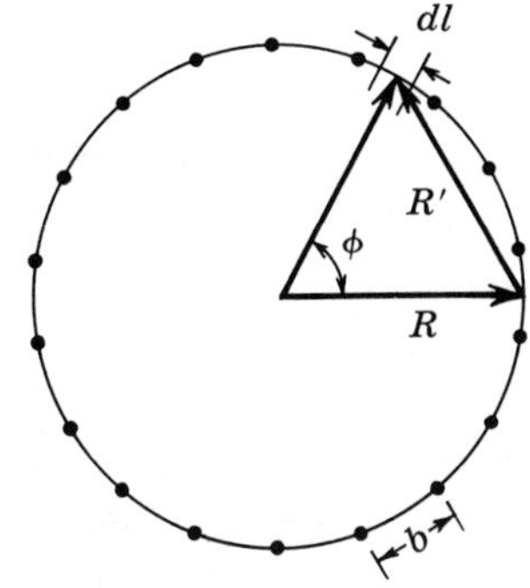

FIGURE 15-21. The prismatic dislocation loop of Fig. 15-20 idealized as an array of point sources of strength β at intervals b around the loop.

elementary contributions

$$c - c^0 = \frac{\beta}{r} \tag{15-103}$$

add up to give the concentration c'. Specifically, c' must be the concentration midway between the point sources β, or, with the parameters of Fig. 15-21,

$$c' = \beta \int_{b/2}^{2\pi R - b/2} \frac{dl}{bR'} = \frac{\beta}{2b} \int_{b/2R}^{2\pi - b/2R} \frac{d\phi}{\sin(\phi/2)}$$

$$\cong \frac{2\beta}{b} \ln \frac{8R}{b} \tag{15-104}$$

provided that $2R \gg b$. Combining Eqs. (15-89) and (15-99) through (15-104), one finds

$$I = \frac{4\pi^2 D_s \mathbb{S}}{bkT \ln(8R/b)} \tag{15-105}$$

The line tension $\mathbb{S}$ is given by Eq. (6-79); a convenient reduction for the present case is

$$\mathbb{S} = \mathbb{S}_0 \ln \frac{4R}{\rho} \tag{15-106}$$

that is, $\mathbb{S}_0$ is the line tension when the logarithmic term is unity and is an *intrinsic* parameter of the dislocation. When $R \gg b$, the logarithmic term in $\mathbb{S}$ practically cancels the logarithmic term in Eq. (15-105), which reduces to

$$I = \frac{4\pi^2 D_s \mathbb{S}_0}{bkT} \tag{15-107}$$

where all cutoff parameters have vanished. The climb velocity normal to the circumference is given by

$$v = \frac{2\pi D_s v_a \mathbb{S}_0}{Rb^2 kT} \tag{15-108}$$

Shrinkage rates closely corresponding to Eq. (15-108) have been observed for prismatic dislocation loops in aluminum,[64] indicating that the climb is under pure diffusion control in this case.

The model leading to Eq. (15-104) is only one of several possibilities, all of which yield essentially the same logarithmic behavior with a cutoff parameter of the order of b. For example, the dislocation line could be considered to be a torus-shaped source of vacancies, and Eq. (15-104) would again result, except that the cutoff parameter b would correspond to the distance from the center of the dislocation to the torus surface. Other shapes of the loop are possible, and in many cases a finite boundary condition would replace the condition $c \to c^0$ at $r \to \infty$. We do not consider these more complex cases but note that the results would resemble those of Eq. (15-105), differing by numerical factors. These diffusion problems have exact analogs in the treatment of heat flow and in electrostatic problems involving the capacitance of a conducting loop. Many quantitative results for complex loop shapes can be found in the literature on electrostatics.[65]

Another application of this type of approach arises in the problem of the climb of an arbitrarily bowed-out dislocation toward its center of curvature. The analysis is analogous to that above but is more complicated because the local vacancy chemical potential varies along the dislocation line, and because both local sources and local sinks exist along the line. Nonetheless, one expects that the logarithmic factor in the appropriate line tension (Sec. 6-5) should cancel the diffusive logarithmic factor of the type of Eq. (15-104), so that a result analogous to Eq. (15-108) should apply to the local climb velocity.

Climb in a Moving Reference System

All the solutions in Sec. 15-4 involve the assumption of quasi-steady state; i.e., static solutions are applied for any instantaneous position of the climbing dislocation, so that the motion of the dislocation is neglected in the diffusion solution. The condition that must be fulfilled for the static solution to be valid is that the root-mean-square diffusion distance, $\bar{x} = 2\sqrt{D_v t}$, where $t = R/v$ is

[64] J. Silcox and M. J. Whelan, *Phil. Mag.*, **5**: 1 (1960); P. S. Dobson, P. J. Goodhew, and R. E. Smallman, *Phil. Mag.*, **16**: 9 (1967). P. M. Marquis and R. E. Smallman [*Phil Mag.*, **34**: 903 (1976)] considered criticisms of their earlier analysis and support the interpretations of purely diffusion controlled climb.

[65] D. N. Seidman and R. W. Balluffi [*Phil. Mag.*, **13**: 649 (1966)] specifically obtain a result exactly equivalent to Eq. (15-105) by an electrostatic analog (their torus case). O. K. Kirchner [*Phil. Mag.*, **31**: 87 (1975)] has treated the case of two concentric loops in a thin foil by the electrostatic analog.

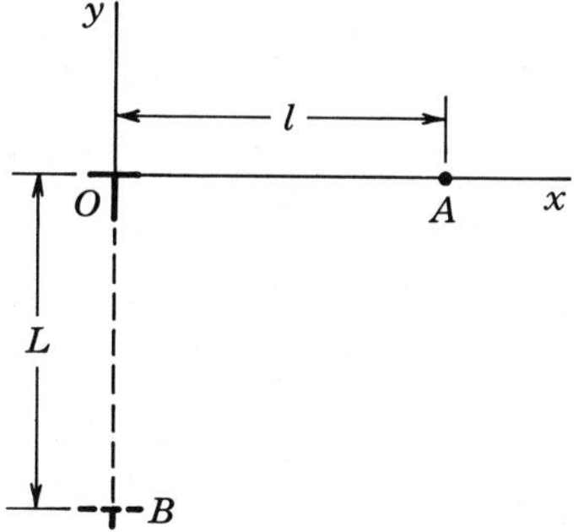

FIGURE 15-22. An edge dislocation at position O in a crystal. A vacancy, emitted when the dislocation was at point B, is at point A.

the time of travel between source and sink, exceeds the spacing R between source and sink, or equivalently,

$$R < \frac{D_v}{v} \tag{15-109}$$

for the present case of dislocation climb. Detailed derivations of Eq. (15-109) have been presented for the specific case of dislocation climb[66, 67] and for general diffusion problems.[68] We shall not present a formal derivation of Eq. (15-109) here. However, the following derivation of the limiting climb velocity of a dislocation in an infinite medium does indicate the physical basis for an expression of the above type.

Consider the edge dislocation in Fig. 15-22, which is climbing by the emission of vacancies. The quasi-steady-state solution, Eq. (15-90), contains a term $\ln(R/b)$, where R is the source-sink spacing and hence cannot be applied to the dislocation in an infinite medium. Suppose that a vacancy at point A in the figure had been emitted by the dislocation. Then, according to Eq. (15-2), it was probably emitted a time

$$t \geqslant \frac{l^2}{2D_v} \tag{15-110}$$

earlier in order to have diffused a distance l laterally. Thus the vacancy was probably emitted when the dislocation was at point B, a distance L removed from its present position, where

$$L = vt \tag{15-111}$$

But then the vacancy must have diffused the distance L in the y direction in a

[66] J. Lothe, *J. Appl. Phys.*, **31**: 1077 (1960).

[67] R. W. Balluffi and D. N. Seidman, *J. Appl. Phys.*, **36**: 2708 (1965).

[68] M. J. Turunen and V. K. Lindroos, *Phil. Mag.*, **27**: 81 (1973).

time L/v, which is probable only if

$$\frac{L}{v} \geqslant \frac{L^2}{2D_v} \tag{15-112}$$

Combining Eqs. (15-110) to (15-112), one finds that the vacancies emitted by the dislocation are confined to a range in the x direction defined by the condition

$$l \leqslant l_c = \frac{2D_v}{v} \tag{15-113}$$

Vacancies farther away from the dislocation would probably not have originated at the dislocation. It is reasonable to conclude that within a radius l_c the vacancy distribution is close to that for a static source with the boundary condition $c = c^0$ at $r = l_c$; indeed, along the y axis, $c - c^0 \sim 0$ when $r > l_c$. The climb velocity would be roughly that of Eq. (15-91) with $l_c = 2D_v/v$ replacing R,[69]

$$v \cong \frac{2\pi D_s v_a (F/L)}{b^2 kT \ln(2D_v/vb)} \tag{15-114}$$

This transcendental relation for the intrinsic climb mobility cannot be developed in a power-series expansion in terms of v.

This development leads to the conclusion that the static approximation is valid if the actual source-sink spacing R is less than l_c; indeed, this is precisely the result of the more detailed treatments of the diffusion problem. Let us find the smallest values of l_c that might be encountered in experiments. Equations (15-91) and (15-113) combine approximately to give

$$l_c \sim \frac{1}{b^2 c^0} \frac{kT}{\sigma b^3} \tag{15-115}$$

Near the melting point typical values are $kT \sim 2 \times 10^{-2} \mu b^3$ and $c^0 \sim 10^{-4}(1/b^3)$. For the very high stress $\sigma \sim 2 \times 10^{-3}\ \mu$, Eq. (15-115) then yields $l_c \sim 10^5 b$, well above reasonable values of R for most practical cases. With decreasing T, l_c would increase rapidly because of the temperature dependence of c^0. Thus dislocations act as static sources in steady-state climb except in very special circumstances.

Furthermore, the climb velocity during the initial transient upon the sudden application of an external force should not exceed the steady-state velocity by more than a factor of ~ 10. The maximum possible climb velocity is determined by the difference between the gross rate of evaporation of vacancies

[69]A formal solution in a moving frame of reference by M. J. Turunen and V. K. Lindroos [*Phil. Mag.*, **27**: 81 (1973)] yields a similar relation with a factor 2.25 replacing the factor 2 in the argument of the logarithm so the approximate treatment is accurate.

from a dislocation core and the rate of impingement on the core from an equilibrium population c^0. Equation (15-91) is close to the result obtained from such a model. The relaxation time to reach steady state is $\tau \cong R^2/D_v$. However, because the climb velocity depends only logarithmically on R, the climb rate approaches the steady-state rate closely for much shorter times.

15-5. CLIMB OF JOGGED DISLOCATIONS

Climb motion by the nucleation of jog pairs and the propagation of single jogs along the dislocation line is in principle directly analogous to glide by kink nucleation and growth. However, an analysis of the degree of vacancy equilibration near jogs is necessary in order to determine the conditions under which dislocations equilibrate with vacancies along their entire length, giving pure diffusion control, as discussed in Sec. 15-4. A vacancy moved to the edge of a straight edge dislocation (or removed from it) creates a *core vacancy* (core interstitial), which is also an incipient double jog (Fig. 15-23). A vacancy moved to a preexisting jog simply translates the jog (Fig. 15-23); the configuration is translated, but no change in core energy, core structure, or jog configuration occurs. Jogs in dislocations have the same role in atom transport as ledge kink sites on a free surface. In this section we first examine the problem of motion of a single jog and then turn to the general problem of groups of jogs on a dislocation line.

Single-Jog Motion

Consider a single jog on the otherwise straight edge dislocation in Fig. 15-24. If a climb force F/L acts on the dislocation, the force on the jog is

$$F_{\text{jog}} = \frac{Fh}{L} \tag{15-116}$$

The elastic work done in the emission of a vacancy by jog motion is

$$W = \frac{Fha}{L} = \frac{F}{L}\frac{v_a}{b} \tag{15-117}$$

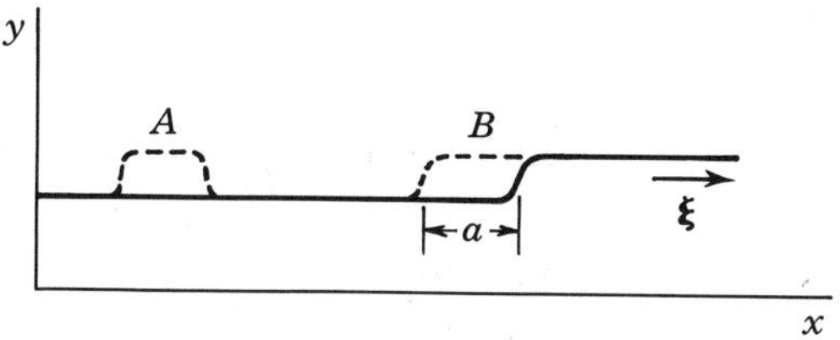

FIGURE 15-23. A vacancy moved to position A on an edge dislocation, forming an incipient double jog, or to position B, causing a jog to translate. Here $\mathbf{b}=(0,0,-b_z)$.

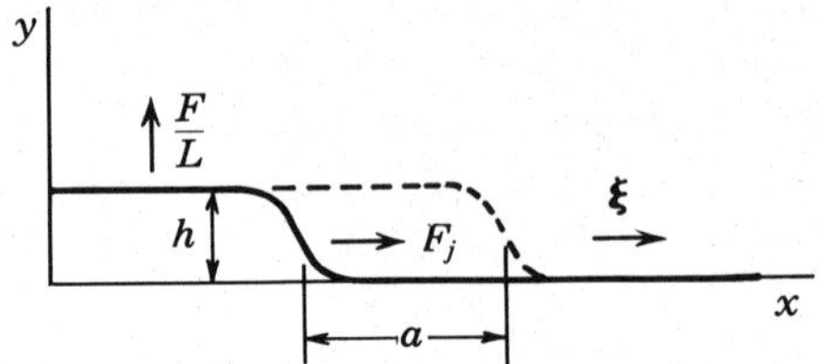

FIGURE 15-24. A single jog in an edge dislocation. The force per unit length on the dislocation is F/L. Here $\mathbf{b}=(0,0,b_z)$.

Thus the vacancy superconcentration in local equilibrium with a jog is

$$c' = c - c^0 = c^0\left[\exp\left(\frac{Fv_a}{LbkT}\right) - 1\right]$$

$$\cong c^0 \frac{Fv_a}{LbkT} \tag{15-118}$$

This relation agrees with Eq. (15-85) of the preceding section as it should; the vacancy concentration in local equilibrium with a fixed dislocation is independent of the dislocation structure.

Suppose that the jog is a point source where quasi-equilibrium is maintained within a surface of radius $\sim b$ around the jog. Preferential vacancy diffusion along the dislocation core is thus ignored for the moment. By arguments similar to those of Sec. 15-4, Lothe[70] has shown that the translation of the jog can be neglected and a quasi-steady-state solution can be applied to the vacancy-diffusion problem. The static diffusion field about the jog is then [*cf.* Eq. (15-100)]

$$c - c^0 = c' \frac{b}{r} \tag{15-119}$$

giving rise to a vacancy current

$$I = -4\pi r^2 D_v \frac{\partial c}{\partial r} = 4\pi D_v c' b \tag{15-120}$$

and a jog drift velocity

$$v_j = Ia = 4\pi abc^0 D_v \frac{Fv_a}{LbkT}$$

$$= \frac{4\pi D_s aF}{LkT} \tag{15-121}$$

[70] J. Lothe, *J. Appl. Phys.*, **31**: 1077 (1960).

This expression was first suggested by Mott.[71] By the Einstein relation [Eq. (15-3)], the corresponding diffusivity for random motion of a jog along a dislocation line is

$$D_j = \frac{4\pi a}{h} D_s \tag{15-122}$$

With $a \sim b \sim h$, Eq. (15-122) can be derived approximately by the following kinetic argument. With $4\pi \sim 12$ possible vacancy sites adjacent to the jog, and with a fraction $c^0v_a = \exp(-W_v/kT)$ of these vacant, the frequency of jumps which move the jog is

$$\omega = \left[4\pi \exp\left(-\frac{W_v}{kT}\right)\right] \nu \exp\left(-\frac{W_v'}{kT}\right)$$

where ν is the atomic vibrational frequency and W_v' is the activation energy for vacancy motion. Thus the jog diffusivity is

$$D_j \cong b^2\omega = 4\pi\nu b^2 \exp\left(-\frac{W_v + W_v'}{kT}\right)$$

$$\cong 4\pi D_s$$

where $W_v + W_v' = W_s$ is the activation energy for self-diffusion. Equation (15-122) is more general, however, in that the local values of W_v and W_v' need not correspond to the bulk values; it is sufficient to assume quasi-equilibrium with the bulk at some small radius r.

When the dislocation core is a line of easy diffusion, as it is expected to be in general,[72–74] Eq. (15-121) is an underestimate of the jog mobility. Because of the high core diffusivity, the jog is assumed to maintain quasi-equilibrium with the surrounding material over a length $\bar{z}$, where $\bar{z}$ is the root-mean-square distance of diffusion of a core vacancy along the dislocation prior to its evaporation into the lattice. Let the energy of a core vacancy be $W_v - \Delta W_v$, where ΔW_v is the binding energy of a vacancy to the dislocation, and let the energy for migration of a core vacancy along the dislocation be $W_v' - \Delta W_v'$. The core vacancy diffusivity D_c is given by

$$D_c = a^2 \nu \exp\left(-\frac{W_v' - \Delta W_v'}{kT}\right) \tag{15-123}$$

Consider all sites outside the core as bulk sites, so that the energy that must be

[71] N. F. Mott, *Proc. Phys. Soc.*, **B64**: 729 (1951).

[72] D. Turnbull and R. E. Hoffman, *Acta Met.*, **2**: 419 (1954).

[73] G. R. Love, *Acta Met.*, **12**: 731 (1964).

[74] R. W. Balluffi, *Phys. Stat. Solidi*, **42**: 11 (1970).

supplied to evaporate a vacancy from the core is $\Delta W_v + W_v'$; that is, the activation barrier to motion of a vacancy *into* the core is W_v', the same as in the bulk lattice. The average lifetime of a core vacancy before it evaporates is thus

$$\tau = \frac{1}{\nu}\exp\left(\frac{\Delta W_v + W_v'}{kT}\right)$$

The mean free path along the dislocation line, the z axis, prior to evaporation is then

$$\bar{z} = (2D_c\tau)^{1/2} = \sqrt{2}\,a\exp\left(\frac{\Delta W_v + \Delta W_v'}{2kT}\right) \tag{15-124}$$

Defining ΔW_s to be the difference between activation energy for self-diffusion in bulk W_s and that in the core W_c, one obtains

$$\bar{z} = \sqrt{2}\,a\exp\left(\frac{\Delta W_s}{2kT}\right) \tag{15-125}$$

Equation (15-125) is more general than the above derivation indicates; it can be derived in a less specific model by quasi-equilibrium considerations.[75]

Asymptotically, when $r \gg \bar{z}$, the jog maintains local equilibrium with the bulk lattice like a point source, but the length $\sim 2\bar{z}$ over which quasi-equilibrium exists along the dislocation line determines the strength of the source. The length $2\bar{z}$ should be considered as a line of interacting elementary point sources. Proceeding as in the case of Eq. (15-104), one finds that a jog produces a vacancy distribution characteristic of a "point" source of strength[75]

$$c - c^0 = \frac{c'\bar{z}}{r\ln(\bar{z}/b)} \tag{15-126}$$

which exceeds that of Eq. (15-119) by a factor

$$\frac{\bar{z}}{b\ln(\bar{z}/b)} \cong \frac{1}{\ln(\bar{z}/b)}\exp\left(\frac{\Delta W_s}{2kT}\right)$$

Thus, with pipe diffusion along the core taken into account, Eqs. (15-121) and (15-122) become

$$v_j = \frac{4\pi D_s aF}{LkT\ln(\bar{z}/b)}\exp\left(\frac{\Delta W_s}{2kT}\right) \tag{15-127}$$

and

$$D_j = \frac{4\pi a D_s}{h\ln(\bar{z}/b)}\exp\left(\frac{\Delta W_s}{2kT}\right) \tag{15-128}$$

[75] J. Lothe, *J. Appl. Phys.*, **31**: 1077 (1960).

Available data[74] for undissociated dislocations suggested that $W_c \sim 0.4W_s$, or $\Delta W_s/2 \sim W_s/4$, so that the enhancement of jog mobility caused by pipe diffusion is expected to be appreciable. For dissociated dislocations, $W_c \sim 0.6$ to $0.8W_s$, and the enhancement is less dramatic.[74, 76]

Climb of Skew Dislocations

Let us first consider skew dislocations that contain a number of geometric jogs, or dislocations that for other reasons already are so jogged that jog nucleation is unimportant. If the spacing between jogs λ is less than $\bar{z}$,

$$\lambda < \bar{z} = \sqrt{2}\, a \exp\left(\frac{\Delta W_s}{2kT}\right) \qquad (15\text{-}129)$$

then quasi-equilibrium exists along the entire dislocation line. *The results of Sec.* 15-4 *would then apply directly to the jogged dislocation.*

Condition (15-129) may apply quite generally. With $\Delta W_s \sim W_s/2 \sim W_v$, and $\exp(W_v/kT) \sim 10^4$ near the melting point, one finds $\lambda \sim 10^2 a$ near the melting point and $\lambda \sim 10^4 a$ in a temperature region about one-half the melting point. In view of the uncertainties in the model, this may be an overestimate of λ. Nevertheless, the above result suggests that quite low jog densities suffice to fulfill condition (15-129).

When $\lambda > \bar{z}$, one might expect the creep velocity to be proportional to both the jog concentration $1/\lambda$ and v_j. However, because of overlap between the diffusion fields, the result is not so simple. The diffusion field of an individual jog is, by modification of Eq. (15-126),

$$c - c^0 = \frac{Ac'\bar{z}}{r\ln(\bar{z}/b)} \qquad (15\text{-}130)$$

where the parameter A is determined by the degree of overlap. The individual diffusion fields do not extend farther than $\sim R$, the source-sink distance. Thus jogs within a distance $\sim R$ of a given jog contribute to the vacancy concentration in the region of the jogs. The equilibrium condition at a given jog is then

$$Ac'\left[1 + \frac{2\bar{z}}{\ln(\bar{z}/b)} \sum_{n=1}^{R/\lambda} \frac{1}{n\lambda}\right] = c'$$

The sum is approximately given by

$$\sum_{n=1}^{R/\lambda} \frac{1}{n\lambda} \cong \frac{1}{\lambda} \ln \frac{R}{\lambda}$$

[76] T. P. Darby and R. W. Balluffi, *Phil. Mag.*, **36**: 53 (1977).

so that the condition becomes

$$A\left[1+\frac{2\bar{z}}{\lambda\ln(\bar{z}/b)}\ln\frac{R}{\lambda}\right]=1 \tag{15-131}$$

When

$$\bar{z}>\lambda\frac{\ln(\bar{z}/b)}{2\ln(R/\lambda)} \tag{15-132}$$

then Eq. (15-131) becomes

$$A\sim\frac{\lambda\ln(\bar{z}/b)}{2\bar{z}\ln(R/\lambda)} \tag{15-133}$$

Thus, with the use of Eqs. (15-130) and (15-133), one finds that Eq. (15-127) is modified to the form

$$v_j=\frac{2\pi D_s a\lambda F}{LbkT\ln(R/\lambda)} \tag{15-134}$$

The total dislocation velocity is

$$v=\frac{h}{\lambda}v_j=\frac{2\pi D_s ahF}{LbkT\ln(R/\lambda)}=\frac{2\pi D_s v_a F}{Lb^2kT\ln(R/\lambda)} \tag{15-135}$$

This relation is identical to Eq. (15-91), except for the different cutoff parameter in the logarithm, λ instead of b. Saturation-type behavior has occurred; $\bar{z}$ and λ have disappeared from Eq. (15-135), except for the relatively insignificant cutoff term in the logarithm.

The condition (15-132) replaces (15-129) as a saturation condition. For large R, saturation behavior occurs for larger values of the jog spacing λ. Only when Eq. (15-132) is not fulfilled is the dislocation velocity proportional to both the jog concentration and v_j. In such a case $A=1$, so that v_j is given by Eq. (15-127), and the dislocation velocity is

$$v=\frac{4\pi D_s v_a F}{\lambda LbkT\ln(\bar{z}/b)}\exp\left(\frac{\Delta W_s}{2kT}\right) \tag{15-136}$$

The treatment of this section is not exact because of the approximate treatment of the overlapping diffusion fields, Eqs. (15-126) and (15-131). Equation (15-135) should reduce exactly to Eq. (15-91) when $\lambda<\bar{z}$, while it actually differs by a small numerical factor $\ln(R/b)/\ln(R/\lambda)$ because of the different cutoff parameter in the logarithm. An uncertainty of this order in the velocity is not critical; it is of the order of the uncertainty in the model anyway. Thus the above treatment usefully predicts the *physical* behavior of a jogged dislocation in climb. An exact treatment would be very difficult, because it must include the moving frames of reference in both the dislocation pipe and in the bulk. Even in the static case, the problem of pipe diffusion in a single core or a number of parallel cores (a grain boundary) accompanied by flow into the bulk lattice is quite complex.[77–79] Nevertheless, we expect that the above results do not differ[80] from the exact results by more than a numerical factor of 1 to 5.

The activation energies for climb are all roughly W_s, the exponential in D_s, except for the case of Eq. (15-136). For the latter equation to apply, however, $\bar{z}$ must be small, approaching b, which requires that ΔW_s be small and approach zero in the limit $\bar{z} \sim b$. Thus, even for Eq. (15-136), the climb activation energy should be close to the activation energy for self-diffusion. These activation energies do not correspond exactly because of the small temperature dependence of the factor v_a/bkT in the various formulas for the climb velocity.

Climb by Local Core Diffusion

In some circumstances climb can be controlled by core diffusion. In the above examples of long, straight dislocations climbing under conditions requiring net mass transport to or from dislocations, the activation energies corresponded to those for D_s. However, for the rearrangement of curved dislocations moving

[77] R. T. Whipple, *Phil. Mag.*, **45**: 1225 (1954).

[78] T. Suzuoka, *Trans. Jap. Inst. Met.*, **2**: 25 (1961).

[79] D. Gupta, D. R. Campbell, and P. S. Ho, in J. M. Poate, K. N. Tu, and J. W. Mayer (eds.), "Thin Films—Interdiffusion and Reaction," Wiley, New York, 1978, p. 161.

[80] To illustrate the insensitivity of the result of Eq. (15-132) to the exact model, consider the result of R. W. Balluffi and R. M. Thomson [*J. Appl. Phys.*, **33**: 817 (1962)]. They assume vacancy diffusion from jogs along dislocation pipes and flow into the bulk, with a cylindrically symmetrical sink at R. However, they assume only radial flow in the bulk, so that they underestimate the bulk-diffusion overlap discussed above. With local equilibrium of core vacancies at jogs, their result for the average core-vacancy concentration $\bar{n}$ divided by the local equilibrium concentration at jogs $n(0)$ is, in our notation,

$$\frac{\bar{n}-n(\infty)}{n(0)-n(\infty)}=\frac{\bar{z}B}{\lambda}\tanh\frac{\lambda}{\bar{z}B}$$

where $n(\infty)$ is the core concentration in equilibrium with the bulk at a distance $z=\infty$ from a jog and B is a numerical factor

$$B=\left(1+2\ln\frac{R}{r_0}\right)^{1/2}$$

Considering the power-series expansion of $\tanh x$, one sees that their condition equivalent to Eq. (15-132) is that saturation occurs when $\bar{z}>\lambda/B$, which qualitatively depends on R in the same way as (15-132) and differs from it by only a small numerical factor. Thus, even with their widely different assumed model, their results are nearly the same as ours. R. W. Balluffi [*Phys. Stat. Solidi*, **31**: 443 (1969)] elaborates on the agreement between the two approaches.

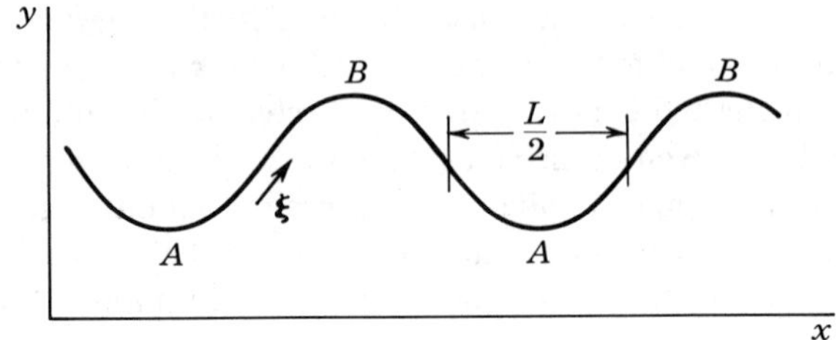

FIGURE 15-25. A bowed-out edge dislocation with $\mathbf{b}=(0,0,b_z)$.

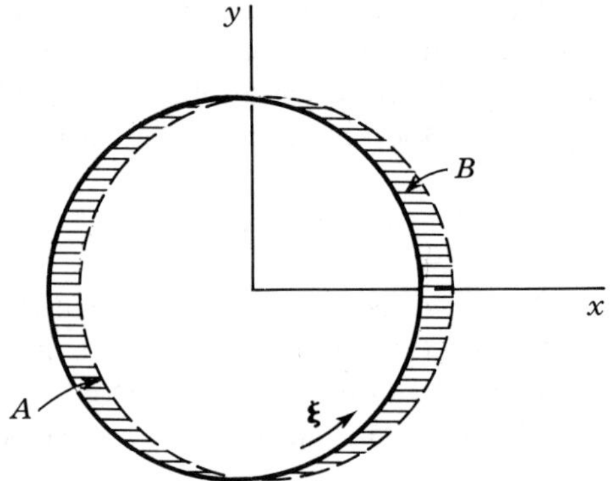

FIGURE 15-26. Translation of a prismatic dislocation loop by removal of matter at B, core diffusion, and deposition at A. Here $\mathbf{b}=(0,0,b_z)$.

under line-tension forces, such as that in Fig. 15-25, core diffusion dominates. If the dislocation can straighten out by removing matter from regions A and depositing it at regions B, no net transport away from the dislocation is required. If $\bar{z}>L/2$, where L is the length of the bowed-out segments, then core diffusion is rate controlling, and the activation energy for climb is $\sim W_c$. When $L/2>\bar{z}$, bulk diffusion is rate controlling and Eq. (15-108) should apply.

Another example is the translation of a prismatic loop (Fig. 15-26). At such a low temperature that bulk diffusion is negligible, the loop cannot shrink. However, the loop can translate in the plane of the loop by core diffusion with an activation energy $\sim W_c \ll W_s$. Such behavior has been observed in dilute quenched alloys of Mg in Al[81] and in quenched Cd.[82]

In general, core diffusion can be quite important in dislocation kinetics. Easy core diffusion makes *local* forces acting at a point on a dislocation

[81] A. Eikum and G. Thomas, *J. Phys. Soc.* (*Japan*) *Suppl.* 3, **18**: 98 (1963).

[82] F. Kroupa and P. B. Price, *Phil. Mag.*, **6**: 243 (1961).

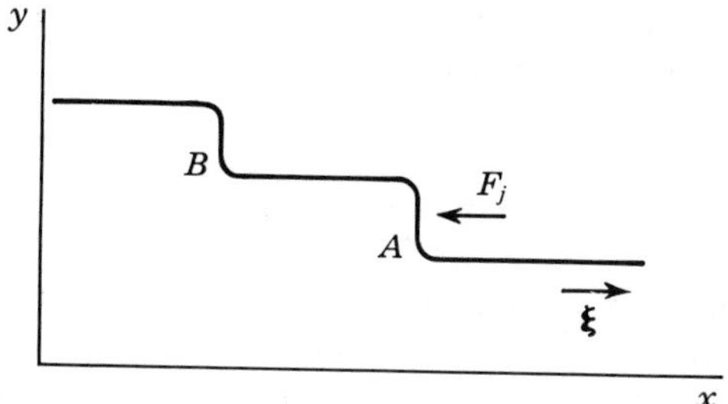

FIGURE 15-27. Two jogs at A and B on a dislocation line. Here $\mathbf{b}=(0,0,b_z)$.

effective at *other* points within a distance $\bar{z}$. As a simple illustration, consider the two jogs A and B in Fig. 15-27, with a local force on A tending to move it to the left. Jog A would then produce a supersaturation of vacancies, which in turn would produce an osmotic force tending to move B to the right if B is closer to A than a distance $\bar{z}$. With no external force on B, it would approach A until the jog-jog repulsion force, Eq. (8-93), balances the osmotic force. According to Eq. (8-93) the repulsive force corresponds to that produced by a shear stress $\sigma \sim (2\times 10^{-3})\mu$ when the jogs are $5b$ apart, and increases inversely with separation for shorter distances. A supersaturation $c/c^0 \cong 5$ would be required to balance the repulsive force for a separation $5b$, when $kT \sim 0.05$ eV. Thus, even in the presence of large osmotic forces, superjogs are not expected to be stable. For materials with low stacking-fault energy, of course, superjogs are stabilized by dissociation, as discussed in Chap. 10.

An interesting example of the importance of pipe diffusion was revealed by Junqua and Grilhe.[83] Analogously to studies of the instability of growing planar interfaces, they considered the breakup of a dislocation loop into a dendritic shape, which has been observed in some alloys. The periodic instability in the loop shape is opposed by line-tension forces but favored by the point effect of diffusion when the loop is growing by climb in a vacancy supersaturation. For a typical pure metal case, they calculate critical radii for breakup of $R \sim 2.5b$ when pipe diffusion is not included in the analysis and $R \sim 5\times 10^4 b$ when it is. The absence of dendritic breakup for pure metals is thus evidence of the important influence of pipe diffusion.

Double-Jog Nucleation

On straight dislocations lying in a single glide plane only *thermal* jogs are present. The climb motion of such straight segments occurs by the nucleation and lateral drift of jog pairs. Since the analogy to the kink case is obvious, no extensive derivations are presented. The approximations are also the same as in the kink case.

[83] N. Junqua and J. Grilhe, *Phil. Mag.*, **42A**: 621 (1980).

The equilibrium concentration of thermal jogs is given by Eq. (14-25), or in terms of the equilibrium spacing λ_e,

$$\lambda_e = a\exp\left(\frac{F_j}{kT}\right) \tag{15-137}$$

Saturation-type climb [Eq. (15-135)] obtains, provided that condition (15-132) is fulfilled, with $\lambda = \lambda_e$. With reasonable values of the parameter R, the condition for saturation-type climb reduces essentially to $F_j < \Delta W_s/2$. If $F_j > \Delta W_s/2$, Eq. (15-132) would not be fulfilled, except possibly near the melting point, where the preexponential factors are more important. Typical values of F_j [Eq. (8-91)] are of the order of 1 to 3 eV, while typically $\Delta W_s/2 \sim W_s/4 \sim 0.2$ to 0.5 eV. Thus saturation-type behavior is not expected in general. Some reservation about this conclusion must be maintained because of the uncertainties in F_j.

When condition (15-132) is not fulfilled, the climb velocity of an infinite dislocation is given by Eq. (15-136), with $\lambda = \lambda_e$, or

$$v = \frac{4\pi D_s v_a F}{LabkT\ln(\bar{z}/b)} \exp\left(\frac{\Delta W_s - 2F_j}{2kT}\right) \tag{15-138}$$

This relation is the analog of Eqs. (15-17) and (15-43) for the kink case. For a finite segment of length L, Eq. (15-138) still applies if $L \gg \lambda_e$. However, if $L < \lambda_e$, then in analogy with Eq. (15-45) for the kink case, the velocity is lower than that of Eq. (15-138) by a factor L/λ_e, or

$$v = \frac{4\pi L D_s v_a}{a^2 bkT\ln(\bar{z}/b)} \frac{F}{L} \exp\left(\frac{\Delta W_s - 4F_j}{2kT}\right) \tag{15-139}$$

For creep described by either Eq. (15-138) or Eq. (15-139), the activation energy should greatly exceed that for self-diffusion.

The Quasi-equilibrium Assumption

The preceding theories are all quasi-equilibrium theories in that local vacancy equilibrium is assumed either at jogs or along the entire dislocation line. The assumption of quasi-equilibrium is not always valid. Let us consider the conditions for quasi-equilibrium, choosing the equilibration of vacancies at jogs as an example. For simplicity, core diffusion is neglected.

The creation of a vacancy at a jog and its motion into the bulk lattice entails a series of energy jumps such as those in Fig. 15-28. W_{12} is the energy difference between positions 1 and 2, etc.; W_{15} is equal to W_v, the formation energy of a vacancy in the bulk. Beyond site 5, the vacancy is considered to be in the perfect bulk lattice. In an assembly of jogs at equilibrium, a fraction $\exp(-W_{13}/kT)$ have vacancies next to them in sites 3. Imagine for the

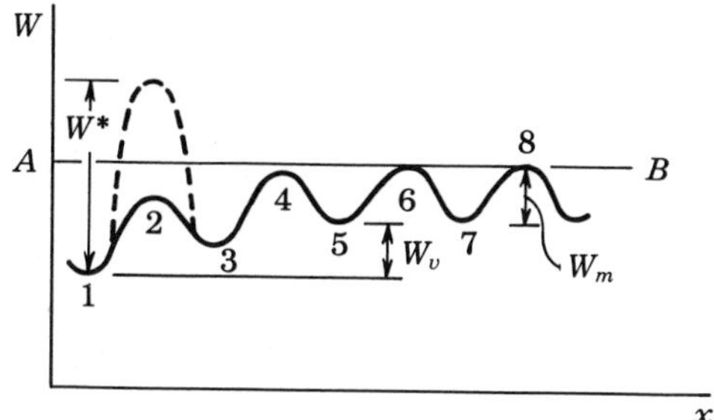

FIGURE 15-28. Energy-displacement curve for a vacancy at a jog site (1) and in successive positions removed from the jog into the bulk lattice. The dashed line represents an alternative curve.

moment that jumps from 3 to 5 do not occur. If the relative distribution between sites 1 and 3 is displaced from equilibrium, the relaxation time to restore equilibrium is given by

$$\frac{1}{\tau} = \nu \exp\left(-\frac{W_{12}}{kT}\right) + \nu \exp\left(-\frac{W_{32}}{kT}\right)$$

$$\cong \nu \exp\left(-\frac{W_{32}}{kT}\right)$$

If $W_{32} < W_{34}$, then the time τ is shorter than the time for a jump from 3 to 5 to occur; i.e., vacancies in sites 3 are in quasi-equilibrium with the jog. Similarly, vacancies in sites 5 equilibrate with sites 3. The sites beyond site 3 are all bulk sites, so the usual equation for macroscopic diffusion would apply in this range, with the boundary condition of quasi-equilibrium with the jog at site 5. Possibly one would have to go out to site 7 to find a bulk-type site. As long as the energy barriers in the direction 1 to 2, 3 to 4, etc., are appreciably greater than those in the reverse direction for all sites out to the first "bulk" site, and provided that no energy barrier is higher than the line AB, quasi-equilibrium with the jog approximately obtains at the bulk site.

The general criterion for quasi-equilibrium of vacancies with a nearby source is that an energy path to annihilation at the source exists that does not entail configurations with higher energies than the energy of the activated state for motion in bulk. The extension of this condition to jog–core-vacancy equilibrium, double-kink equilibrium, etc., is obvious.

For the case of point sources, the kinetics depend on the specific spherical radius at which quasi-equilibrium is postulated. Assuming quasi-equilibrium over a surface $r \sim 2b$ rather than b, in order to exclude more of the highly strained material from the bulk-diffusion treatment, one finds twice as large a diffusion current [in Eq. (15-120), for example]. However, uncertainties of this magnitude in the preexponential in the climb velocity are not serious. For *line* sources the inner radius appears only in a logarithm, so that the creep velocity is insensitive to its precise value.

If the creation of a vacancy at a jog involved an energy barrier, such as the dashed one in Fig. 15-28, then motion over that barrier would be rate controlling. In such a case, the method of quasi-equilibrium would be inapplicable. Such barriers are likely only for metals with low stacking-fault energies, and hence with widely extended dislocations.

15-6. CLIMB OF EXTENDED DISLOCATIONS

Jog Mobility

As discussed in Sec. 10-4, experimental evidence[84] indicates that most jogs of all heights can be extended in fcc metals with the possible exception of dislocations with $\beta = 60°$, which are contracted at the resolution of electron microscopy. For extended jogs, the energy of vacancy formation in materials of low stacking-fault energy could involve an energy barrier such as W^*, the dashed line in Fig. 15-28, in which case quasi-equilibrium would not obtain. When the jogs are contracted, as in Fig. 10-18, quasi-equilibrium should obtain and the preceding formulas should hold. If the jogs are extended, as in Fig. 10-17, one suggested possibility[85] is that the extended jog must contract to the form of Fig. 10-18 before vacancy formation or annihilation can occur. This contraction would require appreciable elastic energy and would lead to a large W^*, composed of W_s plus the constriction energy. The requirement for contraction is unlikely, however[86]; extended jogs, or jog lines, can climb directly, by the process illustrated in Fig. 15-29 for an acute extended jog in an edge dislocation in fcc. The absorption of a vacancy, equivalent to the small loop **AB**, produces the configuration in Fig. 15-29*b*. The partials $\mathbf{B}\gamma$ and $\mathbf{B}\delta$ are glissile and can glide to the left, reproducing the configuration of Fig. 15-29*a*, with the net result that one vacancy has been absorbed. Note that the vacancy could equally well have been absorbed initially at the partial $\delta\mathbf{B}$. Two such configurations, one formed at each end, could not meet and annhilate in the middle of the jog line; they would have to diffuse out to one or the other end and repeat a single formation process there.

With a unit jog, the identification of the sequence $\boldsymbol{B\gamma}, \boldsymbol{\delta B}, \boldsymbol{\gamma B}, \boldsymbol{B\delta}$ as partial dislocations is problematical, and it is better to regard it as a core-level defect. Argon and Moffatt[87] constructed a marble model of a jog line and showed that the configuration is equivalent to that of a partial vacancy. Whether one

[84] C. B. Carter, *Phys. Stat. Solidi*, **54a**: 395 (1979).

[85] A. N. Stroh, *Proc. Phys. Soc.*, **B67**: 427 (1954); A. Seeger, *Phil. Mag.*, **45**: 771 (1954).

[86] N. Thompson, "Report of a Conference on Defects in Solids," Physical Society, London, 1955, p. 153; H. Kimura, D. Kuhlmann-Wilsdorf, and R. Maddin, *Appl. Phys. Letters*, **3**: 4 (1963).

[87] A. S. Argon and W. C. Moffatt, *Acta Met.*, **29**: 293 (1981).

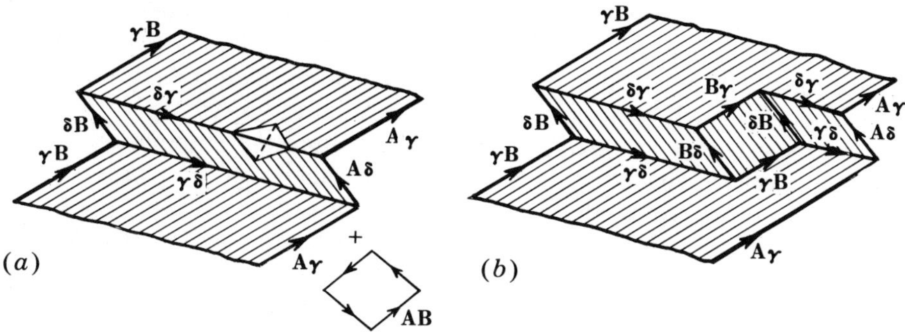

FIGURE 15-29. An extended dislocation jog: (*a*) before vacancy absorption, (*b*) after vacancy absorption. Arrows indicate the sense of ξ.

envisions the motion of the array as glide of partials or diffusion of a partial vacancy, under low driving forces the motion is diffusive in type and could provide a weak constraint to climb in the sense that a partial vacancy would have to diffuse out of the way for continued vacancy absorption. Balluffi and Granato[88] considered a very similar model of a jog line and suggested that a vacancy could be directly absorbed onto a jog line and split into two partial vacancies, a model that would make the constraint to climb even less.

Exercise 15-5. Show that the obtuse extended jog of Fig. 10-19*b* can climb by vacancy absorption and partial-dislocation glide in a manner exactly analogous to that shown in Fig. 15-29.

The above results indicate that extended dislocation jogs may able to maintain near quasi-equilibrium vacancy concentrations by local climb in the vacancy-absorption case. In the less likely event that an activation barrier is involved, it would appear in the initial stage of combining the vacancy with the partial **Aδ** in Fig. 15-29*a*. The jog mobility would then be the advance distance a times the frequency of surmounting the barrier W^*, or

$$v_j \cong a\nu \exp\left(-\frac{W^*}{kT}\right) \tag{15-140}$$

and the corresponding dislocation climb velocity would be

$$v \cong \frac{ha\nu}{\lambda} \exp\left(-\frac{W^*}{kT}\right) \tag{15-141}$$

For vacancy emission, the equivalent to Fig. 15-29*b* would be a partial interstitial. The hard-ball model or the analog with a lattice interstitial suggests

[88] R. W. Balluffi and A. V. Granato, in F. R. N. Nabarro (ed.), "Dislocations in Solids," vol. 4, North-Holland, Amsterdam, 1979, p. 102.

the possibility of a large barrier to formation (atomic calculations are needed to confirm this hypothesis). Then climb could require an alternative process[87] of diffusion of a partial vacancy from one side to another before vacancy emission occurs. In this case W^* would include the energy W_f to form the partial vacancy in an interior of the jog line from the terminus. Also, the partial vacancy would have to occupy a site adjacent to the emission site, which in a diffusive model would lower the preexponential factors in Eqs. (15-140) and (15-141) by a factor (a/r_e) since there are (r_e/a) random occupation sites for the vacancy. If W_f is zero, the probability of two vacancies being at one end of the jog line is $2\,(b/r_e)^2$ and the preexponential would be lowered by this amount as in the result due to Argon and Moffatt.[89] The implication of the above discussion is that there should be a bias factor at extended dislocations, with vacancy emission or interstitial absorption being more difficult than vacancy absorption under conditions far removed from equilibrium,[89] an effect that could be important in both climb and radiation damage.

Jog Nucleation

The nucleation of double jogs is also complicated by dislocation extension. A number of interesting possibilities have been suggested. One possibility is that contraction is required for jog nucleation.[90] This would require a large activation energy, but as discussed above, the requirement for contraction is unlikely. A more probable situation is that an extended configuration nucleates directly on the extended dislocation; several possible mechanisms of such nucleation have been proposed by Thomson and Balluffi[91] and by others.[92–94] Available experimental evidence does not suggest any means of distinguishing between the various possibilities, and energy calculations are difficult to perform because the critical nucleus is of the order of core dimensions. Thus we present only one model,[91] and discuss it qualitatively.

A possible nucleation sequence is shown in Fig. 15-30. A row of vacancies has formed along the partial **B**γ in Fig. 15-30*a*. These react to form the dislocation loop **BA**, which dissociates by glide, as shown in Fig. 15-30*c*. Further glide produces a complete double jog with an acute extended jog on the left and an obtuse extended jog on the right. The critical configuration, i.e., the maximum-energy configuration, is difficult to define. However, the critical energy of formation W_{crit} is a function of elastic energy, applied stress,

[89] Near equilibrium a process and its inverse are equally probable, by the principle of microscopic reversibility, so the above processes are equally likely in that limit.

[90] A. N. Stroh, *Proc. Phys. Soc.*, **B67**: 427 (1954).

[91] R. M. Thomson and R. W. Balluffi, *J. Appl. Phys.*, **33**: 803 (1962).

[92] B. Escaig, *Acta Met.*, **11**: 595 (1963).

[93] H. Kimura, R. Maddin, and D. Kuhlmann-Wilsdorf, *Acta Met.*, **7**: 145, 154 (1959).

[94] J, Grilhe, M. Boisson, K. Seshan, and R. J. Gaboriaud, *Phil. Mag.*, **36**: 923 (1977).

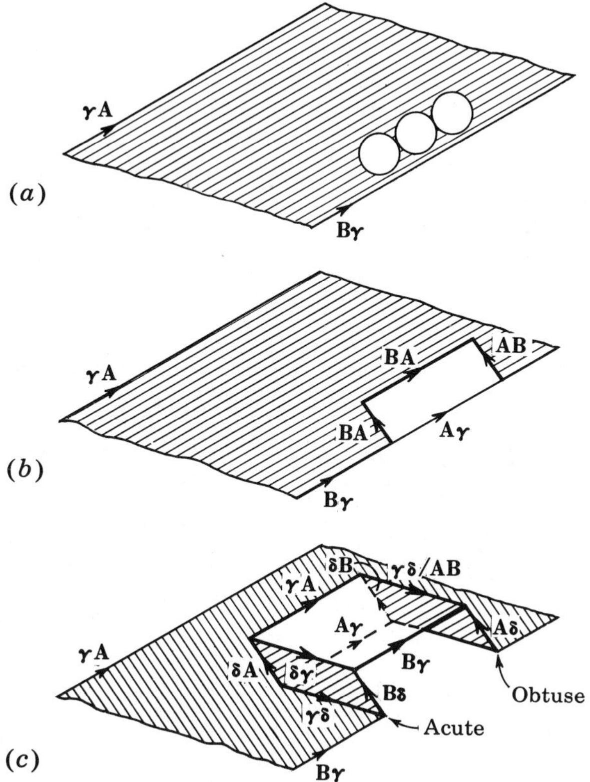

FIGURE 15-30. Three stages in the nucleation of a double jog by vacancy adsorption: (*a*) vacancies absorbed on the partial **B**γ, (*b*) collapse to form a prismatic loop, and (*c*) glide extension of the loop. Arrows indicate the sense of $\boldsymbol{\xi}$. In (*c*) the fault bounded by γ**A** and **B**γ is not hatched for clarity.

temperature, and vacancy supersaturation. Grilhe et al.[94] determined W_{crit} by methods of the type presented in Chap. 6. They found that, provided that the loop size is the order of the width r_e of the dislocation, the critical step is that of forming the loop in Fig. 15-30*b*. Nucleation of a closed loop resembling this one is discussed in Sec. 17-3. The way in which the various parameters enter W_{crit} is made evident there. With W_{crit} known, the nucleation rate of double jogs per unit length of dislocation line per unit time is given by

$$J \cong \frac{\nu}{b} \exp\left(-\frac{W_{\text{crit}}}{kT}\right) \tag{15-142}$$

This nucleation rate gives rise to a dislocation climb velocity of an infinite straight dislocation [*cf.* Eq. (15-43)] of

$$v = 2h\left(Jv_j\right)^{1/2} \tag{15-143}$$

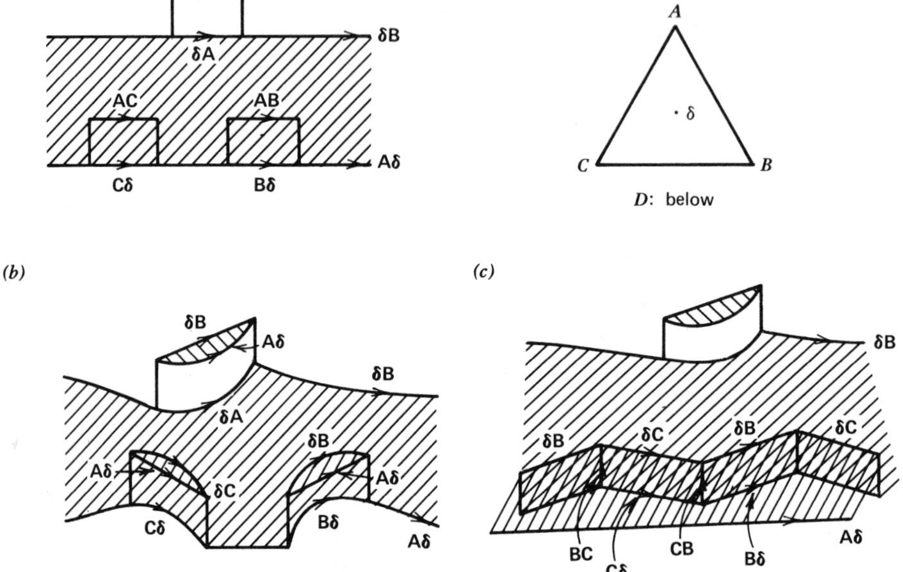

FIGURE 15-31. Extended climb configuration for a $\beta = 60°$ dislocation **AB**(d) [D. Cherns, P. B. Hirsch and H. Saka, Proc. Roy. Soc., 371A:213 (1980)].

with v_j given by Eq. (15-127), (15-134), or (15-140), whichever is appropriate. W_{crit} is roughly expected to be of the form (see Sec. 17-3)

$$W_{\text{crit}} \propto \frac{(W_{\text{el}}/L)^2}{kT\ln(c/c^0)}$$

where W_{el} is the elastic energy of the critical configuration. Thus nucleation-controlled climb is characterized by a strong dependence of climb rate on temperature and vacancy supersaturation. Quite a different mechanism operates under large supersaturations for widely dissociated dislocations, and perhaps for narrower dissociations as well.[95] When the partials are widely separated, the elastic interaction with a single partial dominates and the nucleation of a foreign dislocation **AC** becomes as likely as that of the self dislocation **AB**, as shown in Fig. 15-31a. This leads to more complex jog configurations, as illustrated in Fig. 15-31b and 15-31c when climb forces on the jog associated with osmotic forces are included. The jogs can grow as discussed for Fig. 15-24.[95] Related arrays involving the foreign dislocation **AC** can occur for screw and edge dislocations as well.[95] Transmission electron

[95] D. Cherns, P. B. Hirsch, and H. Saka, *Proc. Roy. Soc.*, **371A**: 213 (1980).

micrographs illustrating these various possibilities for the case of interstitial absorption are presented by Cherns et al.[95]

Balluffi and Seidman[96] have reviewed a number of experiments on dislocation climb, with the following conclusions. With large vacancy supersaturations produced by severe quenches, $\bar{G}=0.7$ to 5 eV, dislocations act as quasi-equilibrium line sources or sinks even in metals of low stacking-fault energy. With moderate supersaturations produced by down-quenching or up-quenching, dislocations in low-stacking-fault-energy gold were still nearly perfect sinks. Under low supersaturations the efficiency of dislocations as perfect line sources was as low as 0.1 percent in the case of climb in Cu–Ni alloys[97] and in the case of dislocations acting as vacancy sources for helium-bubble growth.[98] In sum, these results indicate that double-jog nucleation is relatively easy even on widely extended dislocations.

PROBLEMS

15-1. Verify by direct substitution that

$$c=(4\pi Dt)^{-1/2}\exp\left(-\frac{x^2}{4Dt}\right)$$

the solution for one-dimensional diffusion with all particles concentrated at $x=0$ at $t=0$, satisfies the diffusion equation

$$\frac{\partial c}{\partial t}=D\frac{\partial^2 c}{\partial x^2}$$

where c is concentration.

15-2. Discuss the time dependence of the solution in Prob. 15-1. Show that

$$\langle x^2\rangle=2Dt$$

15-3. Assume for a metal the parameters $D_v \cong b^2\nu\exp(-W_v'/kT)$, $\nu=10^{13}$ sec^{-1}, $b=2\times10^{-8}$ cm, $W_v'=0.2\mu b^3$, and the melting point $T_m=(2\times 10^{-2})\mu b^3/k$. How far will a vacancy diffuse in 1 sec at a temperature just below the melting point?

15-4. Suppose that D_k is given by Eq. (15-11). According to the Einstein mobility relation, how fast will the kink move relative to C_t at room temperature when acted upon by a force $F\cong10^{-4}\mu b^2$, corresponding to a stress $\sigma\sim10^{-4}\mu$? Assume $\mu b^3=5$ eV.

[96] D. N. Seidman and R. W. Balluffi, *Phys. Stat. Solidi*, **17**: 531 (1966); R. W. Balluffi, *Phys. Stat. Solidi*, **31**: 443 (1969).

[97] R. S. Barnes and D. J. Mazey, *Acta Met.*, **6**: 1 (1958); H. Fara and R. W. Balluffi, *J. Appl. Phys.*, **30**: 325 (1959).

[98] R. S. Barnes, *Phil. Mag.*, **5**: 635 (1960).

15-5. With one kink per 30 kink sites in a screw dislocation, what would be the velocity of motion of the dislocation under the conditions of Prob. 15-4? Assume $a = h = b = 2 \times 10^{-8}$ cm.

15-6. Consider a straight dislocation segment of length L lying in a Peierls valley. Suppose that it can move to a double-kink configuration with kink separation $\sim L$. For $kT \sim \mu b^3/400 \sim 0.012$ eV, how large a resolved shear stress is required to change the energy of the kinked configuration by kT relative to the unkinked configuration? How large is this stress for $L = 10^3 b$? For $L = 10^2 b$? Assume $h = b$.

15-7. Discuss the significance of the inception of amplitude dependence in a Bordoni peak experiment in terms of the results of Prob. 15-6.

15-8. **a.** Consider a crystal that is subjected suddenly to a hydrostatic compressive stress P under isothermal conditions. Before the vacancy concentration changes, is there a super- or undersaturation c/c^0 of vacancies according to the definition of c^0 in Eq. (14-38)?

b. Before a new equilibrium concentration is established, will an edge dislocation in the crystal move transiently?

c. Derive the expression for the initial total force on the dislocation.

15-9. Consider a region of crystal where the pressure is $P + p$, and where the vacancy concentration is that in equilibrium with local internal sources and sinks in that region. Is there a super- or undersaturation of vacancies in the region?

15-10. Discuss the mechanism by which vacancy diffusion tends to diminish internal stresses when internal sources and sinks are present.

15-11. Derive the expression for the time for a circular prismatic dislocation loop, formed by collapse of a vacancy disc, to shrink to one-half its original radius R.

15-12. Express Eqs. (15-135) and (15-136) in a general vector notation for $\mathbf{v}$ in terms of a force produced by the stress tensor $\boldsymbol{\sigma}$.

15-13. **a.** Derive the growth velocity analogous to Eq. (15-108) for a square prismatic dislocation loop formed by vacancy disc collapse. Assume quasi-equilibrium with jogs everywhere on the dislocation line.

b. Note that the square shape of the loop suggests that geometric jogs are absent in growth because they grow out to the corners of the square and vanish. Discuss the mechanism of growth of the loop. What condition must obtain for the quasi-equilibrium assumption of (a) to be valid?

c. When such a loop shrinks, its shape tends to be circular, in contrast to its square (or more generally polygonal) growth shape. Discuss the reason for this difference in shape. There is a direct analog to this shape difference in the topography of crystal growth and dissolution.[99]

[99]See F. C. Frank and M. B. Ives, *J. Appl. Phys.*, **31**: 1996 (1960).

15-14. Discuss the situations analogous to that in Fig. 15-29*b* for the cases where a vacancy is absorbed in the middle of the jog line and where a vacancy is absorbed at each end of the jog line at the same time.

BIBLIOGRAPHY

1. Balluffi, R. W., and A. V. Granato, in F. R. N. Nabarro (ed.), "Dislocations in Solids," vol. 4, North-Holland, Amsterdam, 1979, p. 1.
2. Peterson, N. L., and R. W. Siegel, "Properties of Atomic Defects in Metals," North-Holland, Amsterdam, 1978.
3. Robinson, M. T. and F. W. Young (eds.), "Fundamental Aspects of Radiation Damage in Metals," Nat. Tech. Inf. Service, Springfield, Virginia, 1975.
4. *Vacancies '76*, the Metals Society, London, 1976.

16

Glide of Jogged Dislocations

16-1. INTRODUCTION

The glide of a jogged dislocation is a quite specific case of dislocation motion that actually consists of mixed glide and climb. The treatment of this topic has a more general application, however. Such glide involves the concept of pinning points restraining bowed-out dislocations, and the idea of point forces acting at pinning points. These concepts and the equations describing dislocation motion for such cases are generally applicable for *any* type of local pinning of the dislocation. Also, the somewhat nebulous concept of an "activation area," which has caused some confusion in experimental analysis, is introduced and discussed in this chapter.

The initial portion of the chapter deals with undissociated dislocations pinned by vacancy-forming jogs, by interstitial-forming jogs, and by both types simultaneously. Geometric and kinetic modifications associated with extended dislocations are treated in the final portion of the chapter.

16-2. VACANCY-PRODUCING JOGS

Diffusion-Controlled Glide

A jog in a mixed dislocation can move conservatively in the manner shown in Fig. 16-1, so that the entire dislocation is glissile; such jogs would not *pin* the dislocation. A jog in a pure screw dislocation, however, cannot move forward except by the production of point defects. The lateral conservative motion of a screw jog has no component in the direction of motion of the screw as a whole. Let us consider the motion of a jogged screw dislocation when *climb* is the only possible motion of the jogs. Core diffusion and the possible effect of conservative sidewise motion are neglected for the moment, as are possible effects of a Peierls barrier on glissile motion. Kink nucleation and motion are expected to be rapid compared to jog climb in most cases anyway.

Consider the screw in Fig. 16-2 with jogs of one sign at a spacing l and height h. The jogs would produce vacancies if the dislocation moved as

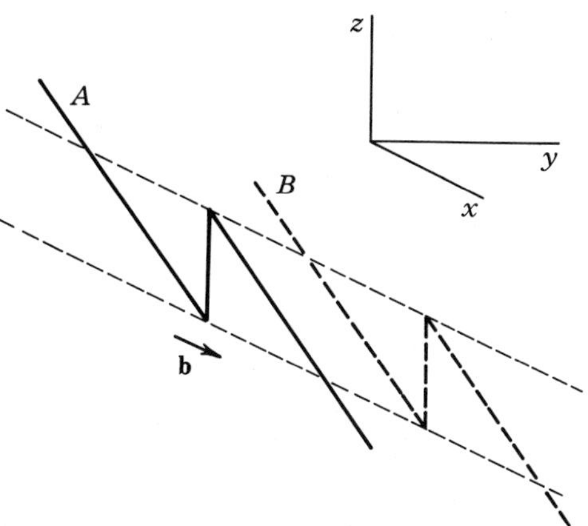

FIGURE 16-1. Conservative glide of a jog in a mixed dislocation as the latter glides from A to B. $\mathbf{b}=(b_x,0,0)$, the dislocation glide plane is xy, and the jog glide plane is xz.

depicted and interstitials for the opposite motion. In the presence of the resolved shear stress σ, the force on the dislocation is $F/L=\sigma b$ per unit length. For stresses below a critical value, the dislocation bows out between the jogs but is pinned at the jogs. By means of thermal activation, vacancies are produced at the jogs and the dislocation glides forward, as shown in Fig. 16-3. If the entire bowed-out dislocation configuration moves forward uniformly by the emission of one vacancy from each jog, the distance of advance is

$$\delta x = \frac{v_a}{bh} = a \tag{16-1}$$

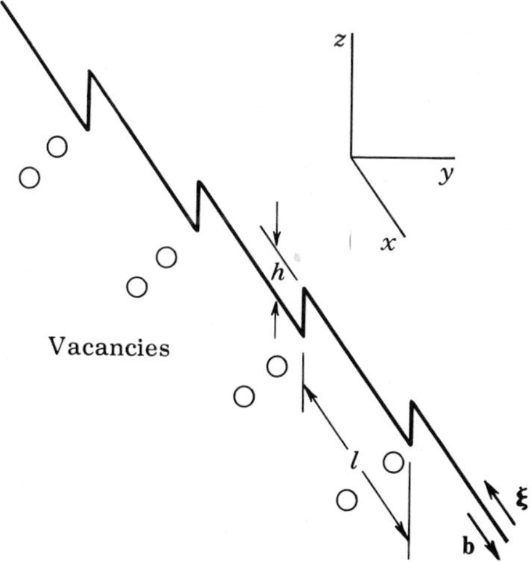

FIGURE 16-2. A jogged left-handed screw dislocation which produces vacancies when it moves to the right.

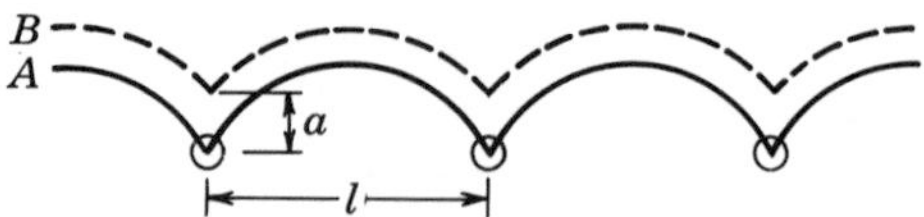

FIGURE 16-3. Advance of a pinned screw dislocation from A to B by a distance $\delta x = a$, with the formation of a vacancy at each jog. View is normal to the glide plane.

The free-energy change per length l of dislocation segment is

$$\delta F = kT\ln\frac{c}{c^0} - \sigma bla \tag{16-2}$$

At local equilibrium $\delta F = 0$, so that

$$\frac{c}{c^0} = \exp\left(\frac{\sigma bal}{kT}\right) \tag{16-3}$$

When the vacancy superconcentration c/c^0 is in quasi-equilibrium with the jogs over spheres of radius $\sim b$, the jog drift velocity is given by Eqs. (15-118) and (15-121). Since the jog drift velocity equals the dislocation velocity in this case, the latter is given by

$$v = \frac{4\pi D_s}{h}\left[\exp\left(\frac{\sigma bal}{kT}\right) - 1\right] \tag{16-4}$$

For small stresses the velocity becomes

$$v \cong \frac{4\pi D_s \sigma bal}{hkT} \tag{16-5}$$

while in the high-stress limit it is

$$v \cong \frac{4\pi D_s}{h}\exp\left(\frac{\sigma bal}{kT}\right) \tag{16-6}$$

For $\mu b^3 \sim 5$ eV and $\sigma \cong 10^{-3}$ μ at room temperature the exponential is $\sim\exp(0.2l/b)$. Since usually $l \lesssim 10^2 b$, Eq. (16-6) is most often the applicable form for v.

We emphasize that Eq. (16-4) is derived by quite general quasi-equilibrium arguments and does not entail assumptions about the activation process. Although the curved dislocations pull on the pinning points with line-tension forces, no approximation involving the line tension is necessary. Also, the term *bal* in the exponential of Eq. (16-4) is related to the activation area of slip, $A^* = al$, in moving from one equilibrium position to the next.

Often in the literature, linear laws for small forces such as Eq. (16-5) have been generalized to hyperbolic sine laws to replace Eq. (16-4). Such a procedure is valid only when the bias caused by an applied stress acts at both the source and sink for point defects. An example of such a case is the climbing apart of two like-sign edge dislocations by vacancy emission and annihilation, where the repulsive force is active at both source and sink. If climb occurs by jog motion, Eqs. (15-118) and (15-121) would yield for such a case

$$v \propto \sinh \frac{\sigma bah}{kT}$$

Thermally Activated Glide

In the high-stress limit, jog motion can be so rapid that steady-state diffusion fluxes are not established. In such a case, motion is controlled by thermal activation. Consider the case illustrated in Fig. 16-4. As the dislocation advances from configuration A to C, forming a vacancy at the jog, it passes through some activated state B. The probability of this event is proportional to $\exp[-(W^* - \sigma bla^*)/kT]$. However, provided that the energy of position B is less than that of position E, *configuration C does not correspond to a final net advance of the dislocation line*. When the energy barrier from C to B is less than that from C to E, the vacancy and jog tend to recombine, restoring the original configuration A. Only when the vacancy jumps into the lattice, forming configuration D, does net advance of the dislocation line occur.[1]

The dislocation velocity in such a case is determined as follows. The number of jogs per unit length in the intermediate state C is given by[2]

$$n_j = \frac{1}{l} \exp\left(- \frac{W_v - \sigma bla}{kT} \right)$$

Note that this is *independent* of the energy and configuration of state B in Fig. 16-4. The number of successful jumps to state D is given by n_j times the vacancy-diffusion jump frequency,

$$n_j \nu \exp\left(- \frac{W_v'}{kT} \right)$$

Finally, the dislocation velocity is given by the above term times the average area of advance per event, al, giving

$$\begin{aligned} v &= a\nu \exp\left(- \frac{W_v + W_v'}{kT} \right) \exp\left(\frac{\sigma bla}{kT} \right) \\ &\cong \frac{D_s}{a} \exp\left(\frac{\sigma bla}{kT} \right) \end{aligned} \tag{16-7}$$

[1] This process was suggested first by N. F. Mott, *Phil. Mag.*, **43**: 1151 (1952).

[2] To a better approximation, one could include the stress dependence of W_v itself. To first order, however, the free energy of formation of the vacancy is the same as that in the absence of an applied stress.

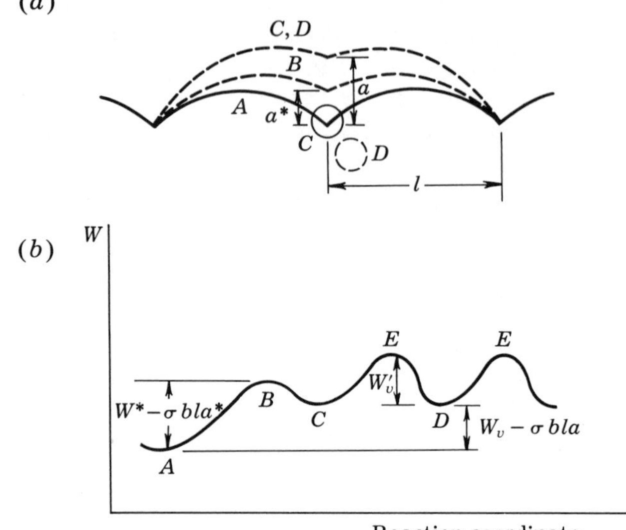

FIGURE 16-4. (*a*) Stages in the advance of a screw at a jog pinning point: initial state, *A*; activated state, *B*; vacancy formation and jog advance, *C*; jump of vacancy into bulk, *D*. (*b*) Corresponding energy-displacement curve.

This expression has the same temperature dependence, i.e., the same activation energy, and is nearly the same numerically as the quasi-equilibrium expression, Eq. (16-6). The work done by the applied stress now involves the advance of a segment of length $2l$ by an average distance $\frac{1}{2}a$, however.

Suppose now that the energies of the various configurations are those of Fig. 16-5 instead of Fig. 16-4. The frequency of passage over the activation barrier *B* is rate controlling in this case, giving a dislocation velocity

$$v \cong a\nu_d \exp\left(-\frac{W^* - \sigma bla^*}{kT}\right) \tag{16-8}$$

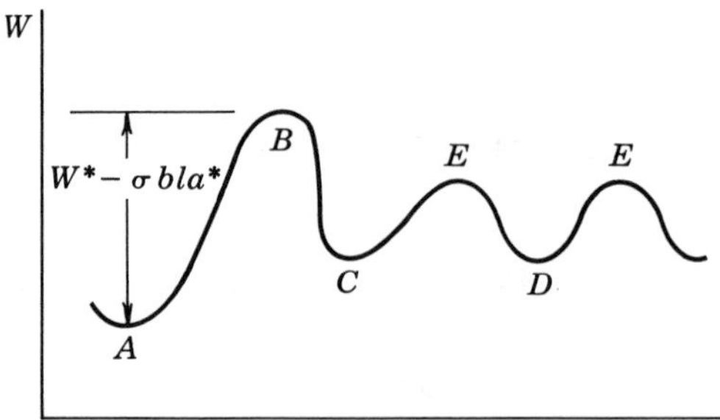

FIGURE 16-5. Alternative energy-displacement curve to Fig. 16-4*b*.

where ν_d is the vibrational frequency of the bowed dislocation loops, and a^* is the distance that the jog moves in reaching the top of the barrier (Fig. 16-4*a*). Since the term bla^* has the dimensions of volume, it is sometimes defined as an *activation volume* V^*; it is not a physical volume, and this term is better reserved for the coefficient of the hydrostatic stress in influencing activated events. A better terminology is the use of *activation area* A^* for the quantity la^*. W^* in general depends on both the reaction distance x and on σ, since the *local* energy and the *local* configuration in the vicinity of the jog depend on the exact line tension near the jog and on local atomic relaxations near the jog. Different W-x curves result for different values of the product σl. For a given constant value of σ, a^* is determined by the relation

$$\left[\frac{\partial W(\sigma,x)}{\partial x}\right]_{x=a^*}=\sigma bl \tag{16-9}$$

and W^* is the value of W at $x=a^*$; one must, of course, choose the position satisfying Eq. (16-9) where $\partial^2W/\partial x^2<0$. Thus the value of a^* (or A^*) is itself stress dependent.

Often, mechanisms of dislocation motion are predicted on the basis of the "activation area" determined from measurements of $\partial \ln\dot{\epsilon}/\partial\sigma$. Here the strain rate is $\dot{\epsilon}=\rho bv$, where the dislocation density ρ is assumed to be stress independent, so that only the stress dependence of v is supposedly measured. In the case of Eq. (16-6) or (16-7), the activation area would correspond to the quantity la, and would relate to the actual dislocation geometry quantities l and a. However, in the case of Eq. (16-8) the apparent activation area determined by the above method would be

$$\frac{kT}{b}\frac{\partial \ln v}{\partial\sigma}=la^*+\sigma l\frac{\partial a^*}{\partial\sigma}-\frac{\partial W^*}{b\,\partial\sigma} \tag{16-10}$$

This activation area is not generally related to geometric dislocation quantities.[3] Hence extreme caution must be exercised in the use of activation-area measurements to establish dislocation-motion mechanisms.

In all the above cases the jog spacing l is assumed to be uniform. Actually, a spectrum of spacings l_j is expected to be present. The above formulas should still hold for individual jogs, except for a small correction for the average distance of advance per vacancy formed. Thus the above velocity formulas can be applied to such a case. The spectrum of jog spacings, either initially inherited from an equilibrated, unstressed dislocation or produced by dislocation intersections, must be treated statistically. Also under a stress bias, small lengths l tend to shrink under unbalanced line tension forces, and vice-versa for large l values. As lengths disappear, superjogs are formed, or, for opposite

[3] J. P. Hirth and W. D. Nix [*Phys. Stat. Solidi*, **35**: 177 (1969)] show that the last two terms in Eq. (16-10) do sum to zero for simple forms of $W(x)$, but not in general.

sign jogs, jog annihilation takes place. The aggregate problem is complicated and is not discussed here.

Point Forces on Dislocations

In the preceding sections the quantity bla^* is related to a segment of length $2l$ advancing by a distance $\frac{1}{2}a^*$. In general, the energy σbla^* is released even when the activated configuration is *localized* to the vicinity of the jog. This concept, which is discussed subsequently, leads to the definition of *point forces* on dislocations at pinning points.

Consider the screw dislocation in Fig. 16-6, bowed out between the pinning points by stress. The equilibrium configuration is that which gives the lowest total energy under the constraint that the pinning points are fixed. The total energy is a minimum with respect to small variations $\delta\eta(y)$ in configuration, $\delta W/\delta\eta = 0$. Thus the dashed line in Fig. 16-6 has the same energy as the equilibrium configuration, *to first order* in $\delta\eta(y)$. Obviously, the dislocation-line energy is less for the dashed position; on the other hand, the stress has done work on a smaller slipped area, implying a higher potential energy of the external mechanisms maintaining stress. To first order, the two effects cancel.

Imagine the two possible activated configurations in Fig. 16-7, which have the same activation energy W^*. By the above principle, the total energy of states B and C is the same to first order. Thus the energy released in moving configuration A to B is the *same* as in moving from A to C, or σbla^*. Hence, provided the forward motion of the jog involves a smooth, gradual variation in line configuration, the activation energy is independent of the degree of localization of the activated configuration to first order.

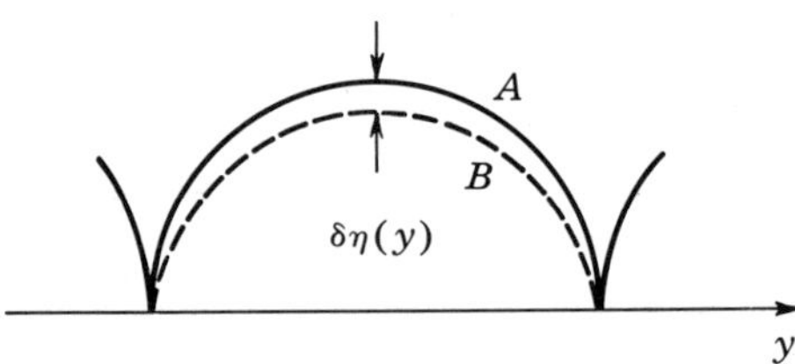

FIGURE 16-6. View of the slip plane of a bowed-out screw, showing a configuration B that varies from the equilibrium configuration A.

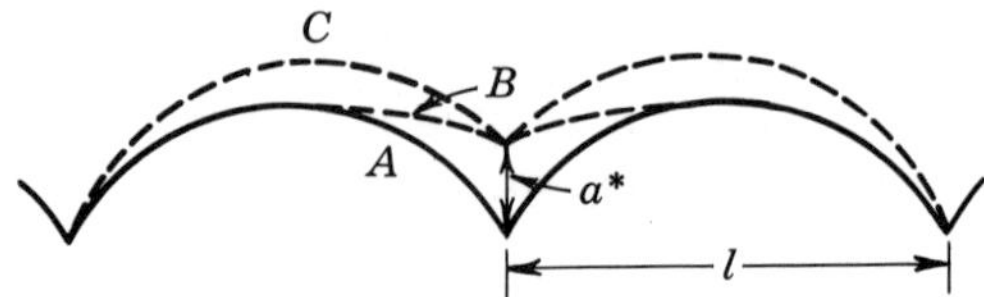

FIGURE 16-7. Quasi-equilibrium bow-out of a pinned screw from an initial state A to an activated state C, and a local bow-out B.

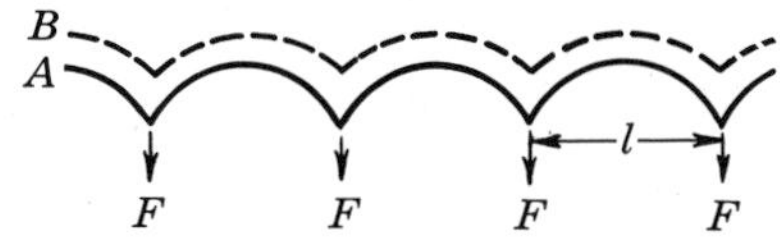

FIGURE 16-8. Bow-out of a uniformly jogged screw against point forces F.

The same result obtains if the pinning points are considered to act on the dislocation with point forces

$$F = \sigma b l \tag{16-11}$$

These forces maintain equilibrium with the distributed force $\sigma b l$, as shown in Fig. 16-8. When a pinning point is displaced by δx, the energy of the configuration changes by

$$\delta W = -F\delta x \tag{16-12}$$

In general, the component F_η acting at a pinning point on a dislocation is defined by

$$F_\eta = -\frac{\delta W}{\delta \eta} \tag{16-13}$$

where $\delta\eta$ is the general displacement indicated in Fig. 16-9. By the above arguments, δW can be calculated to first order for *any* virtual displacement small in $\delta\eta$, and the same point force will result. Some possible configurations are shown in Fig. 16-9. Thus if one knows the initial equilibrium configuration, one can calculate the point force F_η to first order without having to know the exact displaced configuration.

In general, quite complicated energy calculations must be performed to determine the initial configuration. Also, when abrupt changes in dislocation configuration occur on bow-out, the first-order approximation of Eq. (16-13) can yield large inaccuracies unless the precise displaced configuration is known. On the other hand, when no change in configuration, but only

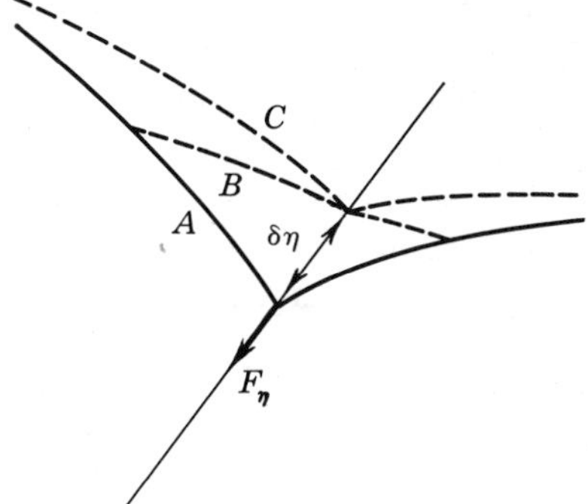

FIGURE 16-9. Possible virtual displacements B and C of the initial configuration A, consistent with a displacement $\delta\eta$ on the pinning point.

translation of the configuration, occurs, the result is exact. An example is the uniformly pinned dislocation of Fig. 16-8; Eq. (16-11) is exact.

Breakaway from Pinning Points

For the vacancy processes illustrated in Figs. 16-4 and 16-5, breakaway of the jogs can occur above a critical stress. When

$$W^* - \sigma bla^* < 0 \qquad (16\text{-}14)$$

jog-vacancy recombination does not occur, and vacancy jumps are not required for net jog motion; the jog moves *without thermal activation*, leaving a string of vacancies behind. A string of vacancies also results when

$$W_v - \sigma bla < 0 \qquad (16\text{-}15)$$

In this case a thermal-activation barrier can still be present if condition (16-14) is not fulfilled, but the details are difficult to analyze very accurately. Actually, instead of W_v, one should put the formation energy of a vacancy *in a vacancy string* into Eq. (16-15); the equation is only a rough estimate. Approximately, then, the critical stress above which a jogged, vacancy-producing screw can move without thermal activation is

$$\sigma_{\text{crit}} \cong \frac{W^*}{bla^*} \sim \frac{W_v}{bla} \qquad (16\text{-}16)$$

Such athermal vacancy production could be important in contributing to the onset of the Portevin-LeChatelier effect, discussed in Chap. 18, and in causing transitions in behavior in low and intermediate temperature creep. The concept has received little attention.

16-3. INTERSTITIAL-PRODUCING JOGS

Kinetic behavior differs for dislocations containing only jogs that produce interstitials when dragged in the direction favored by the applied stress. Interstitials usually have much larger formation energies than vacancies, typically a factor of 5 larger for the common metals, so that vacancies are the dominant defect in self-diffusion. Thus, except at very high stresses, the jogs move by absorbing vacancies rather than by creating interstitials.

Under an applied stress, the vacancy undersaturation at a jog is

$$\frac{c}{c^0} = \exp\left(-\frac{\sigma bal}{kT}\right) \qquad (16\text{-}17)$$

Proceeding as in the development of Eq. (16-4), one finds that the dislocation velocity in this case is

$$v=\frac{4\pi D_s}{h}\left[1-\exp\left(-\frac{\sigma bal}{kT}\right)\right] \tag{16-18}$$

At low stresses, the velocity is again given by Eq. (16-5) in direct analogy to the vacancy-production case. At higher stress levels, however, the analogy with the previous case no longer holds. While increasing stress can produce very large vacancy supersaturations, $c/c^0 \gg 1$, at vacancy-producing jogs, leading to Eq. (16-6), an interstitial-producing jog cannot move faster by vacancy absorption than when it is a *perfect* sink for vacancies. Perfect sink behavior is approached asymptotically at high stresses, in which case Eq. (16-18) reduces to

$$v=\frac{4\pi D_s}{h} \tag{16-19}$$

Only when the stress reaches such a value that *interstitial* production becomes important does the velocity again increase with stress; the dislocation velocity would be given by Eq. (16-8) in such a case, with the appropriate values of W^* and a^* for the interstitial. For stresses higher than

$$\sigma_{\text{crit}} \sim \frac{W_i}{bla} \tag{16-20}$$

the jogs can move without thermal activation. In most cases the segments l would operate as Frank-Read sources (Chap. 20) at stresses lower than σ_{crit}, so that spontaneous interstitial production would not occur.

A dislocation-velocity-stress curve for the entire range of stresses is depicted in Fig. 16-10 for the case of interstitial-producing jogs. A particularly important aspect of the plateau region of Fig. 16-10 is that the apparent

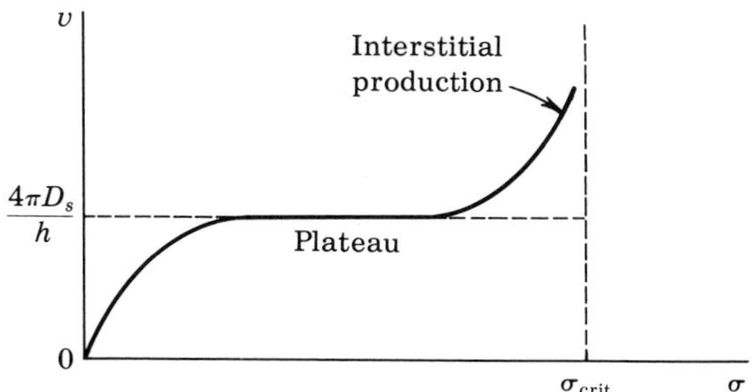

FIGURE 16-10. Velocity-stress curve for a screw pinned by interstitial-forming jogs. Interstitial production commences when the velocity exceeds $4\pi D_s/h$.

activation area is

$$\frac{kT}{b}\frac{\partial \ln v}{\partial \sigma} \cong la \exp\left(-\frac{\sigma bla}{kT}\right) \tag{16-21}$$

Thus the apparent activation area is extremely small in the plateau region.

16-4. SCREWS WITH JOGS OF BOTH SIGNS

Saturation effects could have been treated in Secs. 16-2 and 16-3, because the jogs were either all vacancy producing or interstitial producing. However, each case was idealized and presented to develop the limiting rate equations. The case important in practice is that of a screw with about equal numbers of jogs of the two kinds. In this case saturation effects do not arise. Let us consider the case illustrated in Fig. 16-11 of a screw with jogs of alternating sign separated by a mean spacing l. A steady-state jog concentration can be achieved by a balance between jog formation by nucleation or dislocation intersection and jog-jog annihilation.

For stresses above a critical stress

$$\sigma_{\text{crit}} \sim \frac{W_v + W_i}{2lba} \tag{16-22}$$

the dislocation can move without the aid of thermal fluctuations. For smaller stresses one expects the dislocation mobility to be controlled by the slower-moving interstitial jogs. The vacancy-producing jogs would also move, leaving vacancies behind, but would exert a smaller dragging force on the dislocation. However, several complications associated with conservative lateral jog motion and jog-jog annihilation can be envisioned. Of these, the possible effects of conservative motion are treated first.

An alternative to the requirement that a vacancy that is created at a jog jump into the bulk to avoid recombination is that the jog move laterally by glide to prevent recombination. Because such lateral glide is expected to require little thermal activation, the creep velocity is obtained by simply dropping the factor $\exp(-W_v'/kT)$ from Eq. (16-7), yielding

$$v \cong a\nu \exp\left(-\frac{W_v - \sigma bla}{kT}\right) \tag{16-23}$$

More generally, Eq. (16-8) is expected to obtain, because the local equilibrium

FIGURE 16-11. Screw dislocation with alternating-sign jogs spaced at a distance l.

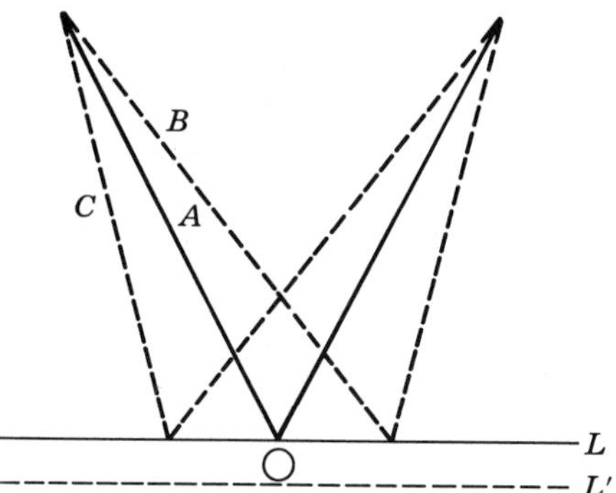

FIGURE 16-12. A bowed-out screw that has advanced at a jog, creating a vacancy, configuration A. Two possible lateral glide configurations, B and C, are shown.

argument leading to Eq. (16-7) might not be applicable. A similar expression is expected for interstitial-producing jogs in the range of interstitial production. Formulas such as Eq. (16-8) have been used in many theories for the flow stress[4] instead of formulas of the type of Eq. (16-7).

However, there are reservations about the model that yields Eq. (16-8). If the force on the jog has no lateral component, i.e., if the bowed-out loops pull at the jog in a symmetrical fashion, the jog could glide off to *either* side from, say, a vacancy, as shown in Fig. 16-12. Having done so, it would not continue to glide constantly farther away, but would diffuse freely back and forth along the line and would tend to recombine with the vacancy if the latter had not diffused into the bulk. Thus, in this example the lateral motion of the jog does not facilitate jog climb, and the original equation [Eq. (16-7)] should apply. For lateral jog motion to aid in jog climb, the jog must be in *unstable* elastic equilibrium with respect to lateral glide, so that once sidewise motion had begun, it would continue. But such sidewise motion could equally easily occur along the line L' (Fig. 16-12) *before* the jog created a vacancy. Thus jog climb is facilitated by lateral glide only before the system is in a relatively stable configuration with respect to glide. Put differently, facilitated climb occurs only as long as the dislocation can move easily by glide processes such as the one depicted in Fig. 16-13. There one bowed-out segment increases at the expense of a shorter one, leading to a net increase in slipped area and a net advance of the dislocation. We conclude that if jog climb is rate controlling, it is not facilitated by lateral jog motion, so that Eq. (16-7) should describe dislocation motion. Jog climb cannot be rate controlling before the possibilities for dislocation advance by the easier process of lateral glide have been exhausted.

One might think that jogs could disappear by sidewise motion so easily that they would not appreciably impede the advance of the screw. This would

[4]A. Seeger, *Phil. Mag.*, **46**: 1194 (1955).

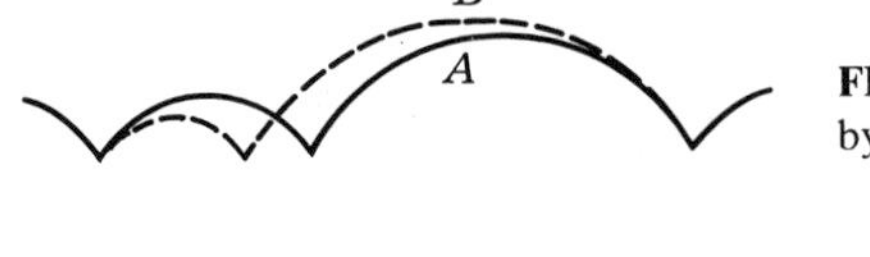

FIGURE 16-13. Bow-out of a screw from A to B by lateral jog glide.

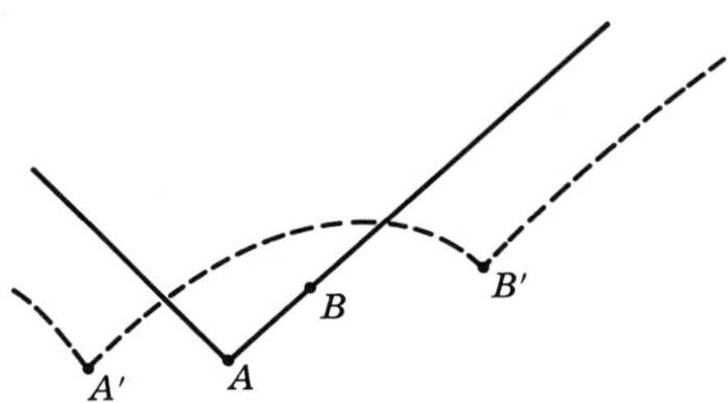

FIGURE 16-14. A screw dislocation about to intersect two screws perpendicular to it at A' and B' (dashed line), and the resulting pinned configuration after lateral glide of the vacancy-forming jog A and the interstitial-forming jog B.

certainly be so if the jogs moved all the way to the ends of a moving segment, so that it could proceed to bow out as a very long Frank-Read source. In most cases, however, jog-jog annihilation within the moving segment is required in order that the screw not be gradually immobilized. Such complete jog-jog annihilation is *not* possible in general without jog climb. The jogs that are created by dislocation intersection are generally not aligned, so that they can annihilate by glide only. A typical example is presented in Fig. 16-14.

In general, groups of jogs are expected to form, with the one farthest behind acting as the pinning point and the others exerting negligible point forces on the dislocation. In the example of Fig. 16-15, jog A can annihilate with B only by climb. Jog A climbs forward by vacancy emission, while jog B glides in conservatively. In position (3) they annihilate, and the dislocation segment can then freely leap forward to a new quasi-equilibrium position. The trailing jog would equally often be an interstitial-producing jog. The climb of both types of jogs is important in jog-jog annihilation processes.

Easy core diffusion between opposite-sign jogs enhances the dislocation mobility but does not contribute to jog-jog annihilation. If vacancies are

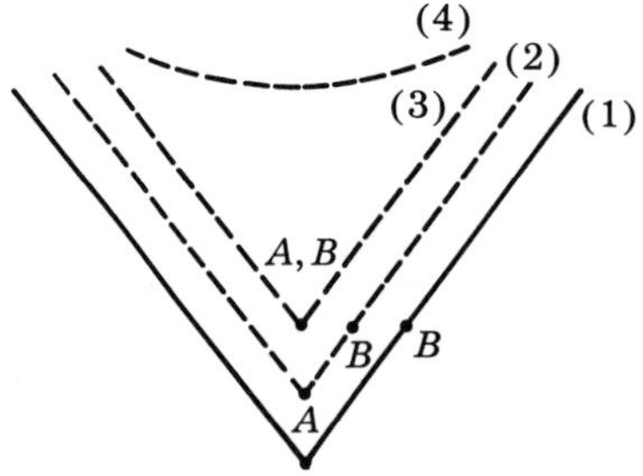

FIGURE 16-15. Advance of jog A by vacancy emission, accompanied by conservative glide of the interstitial-forming jog B. Annihilation of the jogs occurs in position (3).

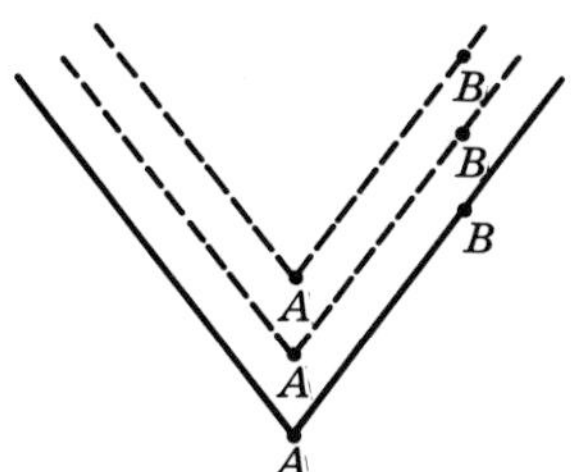

FIGURE 16-16. Advance of a pinning point by core diffusion. Vacancies are formed at A and annihilated at B.

emitted at jog A, core diffuse, and annihilated at jog B, the entire configuration advances uniformly, as shown in Fig. 16-16. The velocity would be given by Eq. (16-7) in such a case, with the appropriate core values substituted for the vacancy-formation and migration energies.

With the groups of jogs present as in Fig. 16-15, the trailing jogs must provide the entire reaction to the external force favoring dislocation motion. If the mean jog separation is l, and if in the mean there are n jogs in each group, the mean point force on the trailing jog is

$$F = n\sigma bla \tag{16-24}$$

Thus the grouping of jogs yields a higher overall dislocation velocity and a lower critical stress for breakaway than when the jogs are uniformly spaced.

If superjogs form, they are likely to become the trailing jogs because more vacancies must be transferred for them to advance by an atomic distance a; hence they are harder to move.

16-5. EFFECTS OF JOG EXTENSION

This section deals with the effects of jog extension on the glide of a jogged dislocation, with the fcc lattice as an example. The unit jogs are treated throughout as if they had the superjog type of configuration discussed in Chap. 10. One must remember that unit jogs might better be treated as misfit lines having the same properties with respect to point-defect production as the limiting form of the superjogs. The present treatment at least suggests possibilities that should be considered in detailed atomic calculations of unit-jog behavior.

If the extended jogs are all contracted, as in Fig. 10-18, the prior discussion for unextended dislocations applies *in toto*, and this case is not discussed further. If the jogs themselves are extended, important differences arise in the glide behavior of the jogged dislocations, as discussed below.

Glissile Extended Jogs

Consider again the extended-jog configurations in Figs. 10-19 and 10-20. Provided that the jog dissociates only into Shockley partials, it is glissile and its

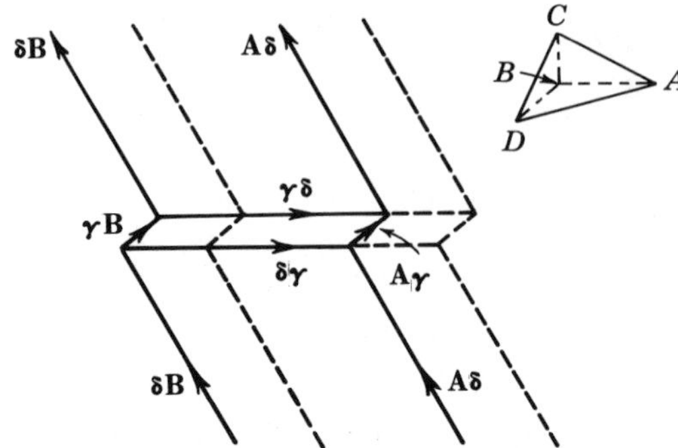

FIGURE 16-17. Conservative glide of an acute extended jog. The orientation of the Thompson tetrahedron is shown. Arrowheads indicate the sense of ξ.

motion corresponds to the conservative glide motion of an unextended jog, as depicted in Fig. 16-1. Jogs (a), (b), (c), and (d) in Fig. 10-19 and jog (g) in Fig. 10-20 are all of this type. As an example, Fig. 16-17 illustrates the conservative glide of a jog in the dislocation **BA**(d). Edge-dislocation jogs extended to include a sessile stair-rod partial, as for jog (f) in Fig. 10-20, can become glissile by the process illustrated in (f), (g), and (h) in Fig. 10-20, leaving behind a stacking-fault tetrahedron.

Sessile Edge Jogs

The motion of the extended edge jog, Fig. 10-20(b), can produce the sessile arrangement shown in Fig. 16-18. Motion to the right leaves the sessile Frank

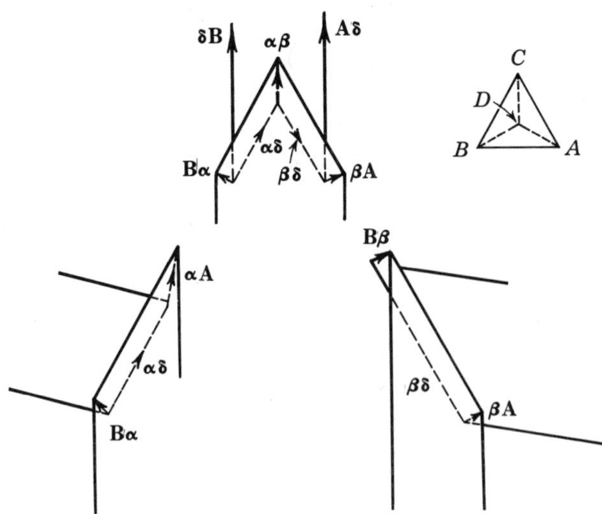

FIGURE 16-18. An extended jog in an edge dislocation in an equilibrium position and after glide to the right or left.

partial $\mathbf{B}\boldsymbol{\beta}$ in the trailing position, while motion to the left leaves the sessile $\boldsymbol{\alpha}\mathbf{A}$ in the trailing position. Study of the geometry of the bowed-out arrangement reveals that *in either case*, motion of the sessile partial in the direction of bow-out can proceed only by vacancy production. Thus sessile edge jogs form vacancy-producing jogs in the trailing position, independent of the direction of motion of the dislocation. This result is important in theories of the flow stress, as discussed in Chap. 22.

Sessile Screw Jogs

A jogged screw can contain both acute and obtuse jogs. Alternatively, all the extended jogs might be acute, as shown in Fig. 16-19(1). The downward motion of the jog on the left of this configuration is equivalent to the upward motion of the one on the right, and vice versa, so that a consideration of the glide of one of the two in both directions suffices. Motion of the dislocation to configuration (2) leaves the sessile partial $\boldsymbol{\alpha}\mathbf{A}$ in the trailing position. $\boldsymbol{\alpha}\mathbf{A}$ produces vacancies if it moves in the bow-out direction.

Hirsch and Mott[5] have suggested that configuration (3) results for bow-out in the opposite direction because of the higher point force restraining the now interstitial-forming partial $\boldsymbol{\alpha}\mathbf{A}$. Furthermore, they have developed a work-hardening theory based on configuration (3), arguing that interstitial-forming jogs constrict and glide conservatively until they annihilate at a vacancy-forming jog. In their model, vacancy formation at extended vacancy-forming jogs is rate controlling in the motion of a jogged screw. We propose, on the contrary, that the jog (3) always forms an extended configuration wherein the interstitial jog partial is trailing. As the dislocation bows out, both the line-tension forces from $\boldsymbol{\delta}\mathbf{A}$ and the interaction force between $\mathbf{B}\boldsymbol{\alpha}$ and $\boldsymbol{\alpha}\mathbf{A}$ favor the dissociation of configuration (4) in Fig. 16-19. As $\mathbf{B}\boldsymbol{\alpha}$ glides toward $\boldsymbol{\alpha\beta}/\mathbf{AC}$, configuration (5) can form; here the trailing jog partial $\mathbf{B}\boldsymbol{\beta}/\mathbf{AC}$ is interstitial producing. The stair-rod partials $\boldsymbol{\delta\beta}/\mathbf{AC}$ are of the second-lowest energy type in Table 10-2, so that the mode of extension of (5) is likely under bow-out forces. There is a weak repulsion between $\mathbf{B}\boldsymbol{\alpha}$ and $\boldsymbol{\alpha\beta}/\mathbf{AC}$, so that their combination to form $\mathbf{B}\boldsymbol{\beta}/\mathbf{AC}$ may or may not occur under stress. $\mathbf{B}\boldsymbol{\beta}/\mathbf{AC}$ is of the type $\frac{1}{6}\langle 411\rangle$ and is unstable in the absence of an applied stress. However, even if $\mathbf{B}\boldsymbol{\alpha}$ and $\boldsymbol{\alpha\beta}/\mathbf{AC}$ did not combine, $\boldsymbol{\alpha\beta}/\mathbf{AC}$ could climb directly, as shown in configuration (6). The lateral glide of $\mathbf{C}\boldsymbol{\alpha}$, $\boldsymbol{\delta}\mathbf{C}$, $\boldsymbol{\alpha}\mathbf{C}$, and $\mathbf{C}\boldsymbol{\delta}$ in this configuration would restore configuration (4). Thus the net process would still entail the climb of the extended configuration by a and the annihilation of a vacancy (or formation of an interstitial). In either case, the trailing partial in the extended configuration would be interstitial producing and sessile, and would pin the dislocation, contrary to the model of Hirsch and Mott.

Small extensions of the Frank sessiles of the type of configuration (f) in Fig. 10-19 are ignored in Fig. 16-19. If such configurations exist, they can still climb by the process of Fig. 16-19(6).

[5]N. F. Mott, *Trans. AIME*, **218**: 962 (1960); P. B. Hirsch, *Phil. Mag.*, **7**: 67 (1962).

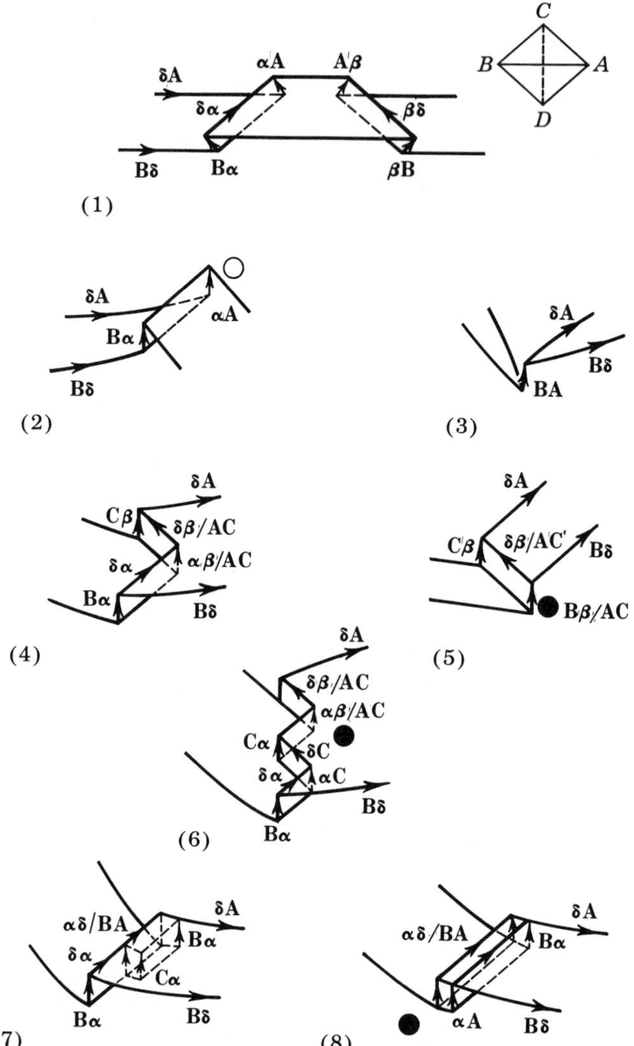

FIGURE 16-19. Various configurations of extended acute jogs in the screw dislocation **BA**(d). Open circles represent vacancy formation, and closed circles represent interstitial formation.

All the configurations discussed above involve intrinsic stacking faults. As an alternative for bow-out in the interstitial-forming direction, Weertman[6] has suggested that the sessile **αA** of configuration (1) in Fig. 16-19 can dissociate to form dissociated jogs with extrinsic stacking faults as shown in configurations (7) and (8), in which the interstitial-forming partial is again trailing. Double lines bounding the jogs are inserted to indicate that *pairs* of partials are

[6] J. Weertman, *Phil. Mag.*, **8**: 967 (1963).

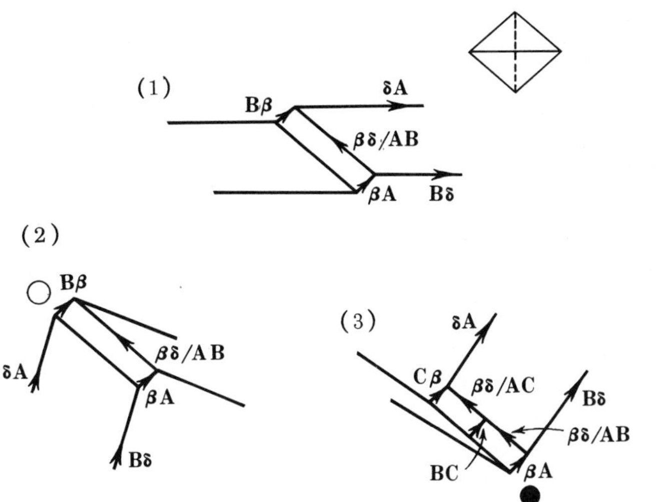

FIGURE 16-20. Configurations of obtuse jogs in the screw dislocation **BA**(d). Notation same as Fig. 16-19.

involved, but only the *net* Burgers vectors of the pairs are listed for brevity. The higher fault energy of the extrinsic fault and the high core energy of the pairs of stair-rod partials are factors favoring the appearance of intrinsic fault configurations. On the other hand, the elastic energy of the stair rods is often less for the extrinsically faulted jogs (the extrinsic version of Fig. 16-20 has net stair rods of the type **B$\boldsymbol{\delta}$**). Thus, particularly under bow-out forces, both intrinsic and extrinsic jogs are possible.

Obtuse extended jogs, such as the one in Fig. 16-20(1), can also exist on a jogged screw. Motion of the dislocation downward results in the vacancy-forming partial **B$\boldsymbol{\beta}$** pinning the trailing partial, as in configuration (2). Motion in the opposite direction produces the intermediate configuration (3), which upon continued motion again forms configuration (5) of Fig. 16-19, with the interstitial-forming partial **B$\boldsymbol{\beta}$/AC** trailing and pinning the dislocation.

To summarize briefly these geometric effects, extended jogs are either glissile or sessile, with the sessile jog partial in the trailing position. The various configurations are *not* equilibrium arrangements, but form only under the action of bow-out forces.

Tetrahedron Formation

Another interesting possibility for the motion of a sessile jog partial involves the formation of stacking-fault tetrahedra. Consider configuration (1) in Fig. 16-21, and suppose that the screw **BA**(d) is in the *primary slip system*. The resolved shear stresses on the partials can be determined readily from Fig. 9-10. Both the reduction of energy of the partial **$\boldsymbol{\alpha}$A** on splitting *and* the

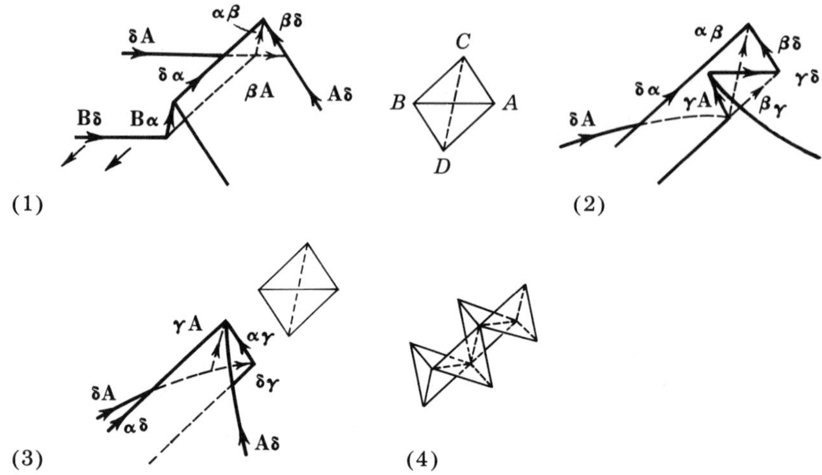

FIGURE 16-21. Formation of stacking-fault tetrahedra by jog glide in the screw dislocation **BA**(d). In (4), the view is normal to plane (d).

applied stress favor the formation of configuration (1). When the line of **βA** becomes parallel to **AD**, the resolved shear stress on **γA** and **δA** favors the formation of configuration (2). Continued glide of **γA** forms one complete stacking-fault tetrahedron, with its apex pointing toward D. The further development of configuration (3) in the same manner produces an adjacent tetrahedron, with its apex pointing away from D. Thus the applied stress and interaction forces can lead to the motion of the sessile jog partial to form a trail of tetrahedra instead of a row of vacancies.

Such a mechanism possibly occurs for unit jogs, although it is less likely for such a case. A single tetrahedron formed by a unit jog has a vacant volume equivalent to one-half a vacancy, so that two unit tetrahedra correspond to a single vacancy. The relaxation of a single vacancy into two unit tetrahedra is shown in Fig. 16-22. If such a relaxed configuration is possible for a single vacancy, then a unit tetrahedron trail of the type of Fig. 16-21 should be a possible consequence of unit jog motion. There is no direct evidence for such a relaxed vacancy configuration.[7]

Seidman and Balluffi[8] have found tetrahedra along ⟨110⟩ directions in gold that was quenched, deformed, and annealed. The observed tetrahedra were of the order of 30 nm in size and were formed during annealing. They showed that the tetrahedra trails were associated with trails left behind by gliding dislocations. A plausible explanation of these results is that tetrahedra trails of the type of Fig. 16-21 formed at superjogs or unit jogs dragged along during

[7]N. L. Peterson and R. W. Siegel, "Properties of Atomic Defects in Metals," North-Holland, Amsterdam, 1978.

[8]D. N. Seidman and R. W. Balluffi, *Phil. Mag.*, **10**: 1067 (1964).

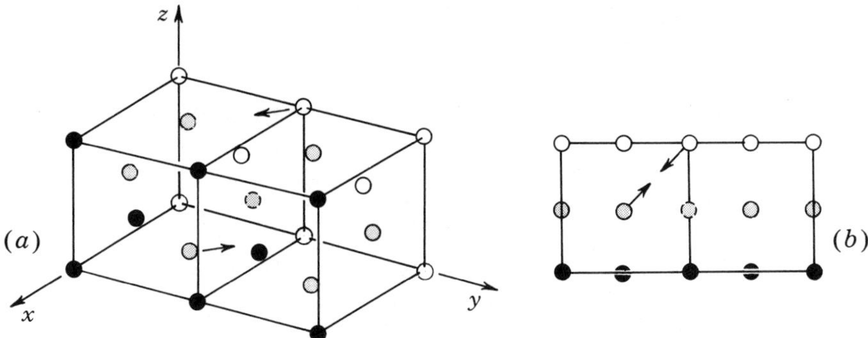

FIGURE 16-22. Formation of two unit tetrahedra at a vacant site in an fcc crystal: (*a*) perspective view, (*b*) view along [001]. Arrows represent atomic displacements. Gradation in shading of balls represents a variation in their *x* coordinate.

deformation, and subsequently grew to the observed size by a local agglomeration process involving vacancy absorption along jog lines on the tetrahedra. Support for this hypothesis is provided by the work of Kubin et al.[9] They computed from Eq. (4-40) that the elastic interaction between tetrahedra favored alignment along $\langle 100 \rangle$ and observed such alignment in irradiated gold. Thus the $\langle 110 \rangle$ alignment is connected with the dislocation motion.

Motion of Dislocations with Extended Jogs

In the absence of core diffusion, there are several possible effects of jog extension on dislocation motion. The resolved shear stress $\sigma_{(1)}$ required to pull one partial $\mathbf{b}_1$ away from another pinned one $\mathbf{b}_2$ in a dissociated pair is given by the condition

$$\sigma_{(1)} b_1 \geqslant \gamma - F_{\text{int}} \cong \gamma \tag{16-25}$$

where F_{int} is the interaction force between partials and γ is the stacking-fault energy. The approximation is fairly good, because F_{int} decreases inversely with the separation of the partials [Eq. (10-14)]. The force on the trailing partial is $\sigma_{(2)} b_2 + \gamma$. $\sigma_{(2)}$ is unequal to $\sigma_{(1)}$ in general because $\mathbf{b}_1$ is nonparallel to $\mathbf{b}_2$. The condition that the trailing partial remain pinned is

$$\sigma_{(2)} b_2 + \gamma < \sigma_{\text{crit}} \tag{16-26}$$

where σ_{crit} is given by Eq. (16-16) or (16-20), depending on whether the sessile jog partial is vacancy forming or interstitial forming. The partial Burgers vector must be substituted for b in either case, of course.

[9] L. P. Kubin, A. Rocher, M. O. Ruault, and B. Jouffrey, *Phil. Mag.*, **33**: 293 (1976).

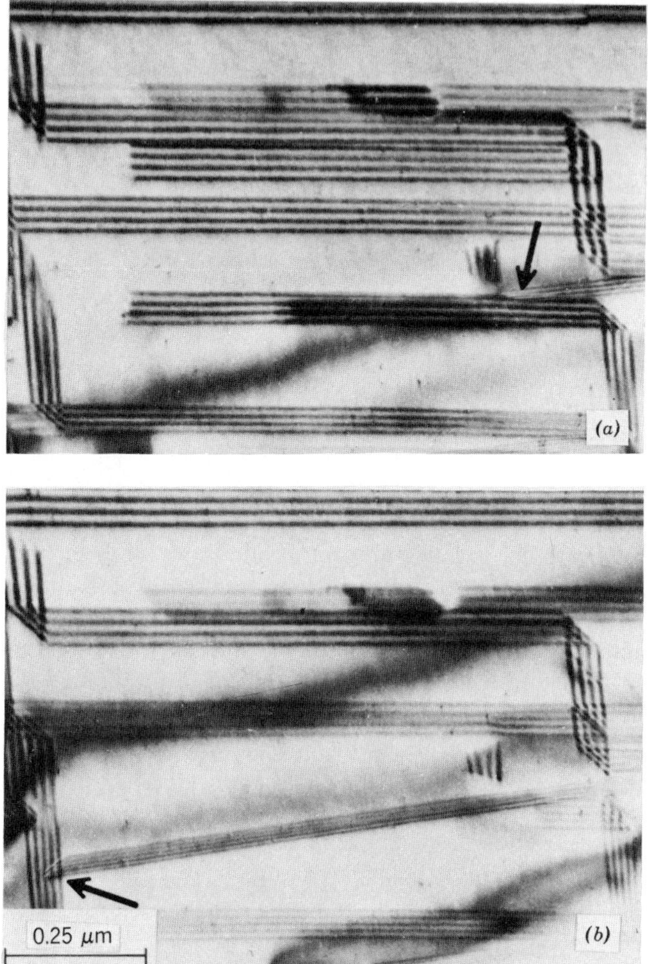

FIGURE 16-23. Formation of stacking faults in a Cu-10% Fe alloy by partial-dislocation glide. The partial at the arrow in (*a*) has translated, annihilating an intervening fault, to a new position in (*b*), where it is restrained by another fault. (P. R. Swann, doctoral dissertation, Cambridge University, Cambridge, 1959.)

When conditions (16-25) and (16-26) are both fulfilled, the leading partial tears away from the trailing partial, and the motion of the former is rate controlling. Alternatively, the stress can force the two partials in opposite directions as in Fig. 10-39. In either case, profuse production of stacking faults during deformation would result. Such processes probably account for the observations in electron-transmission microscopy of large regions of stacking

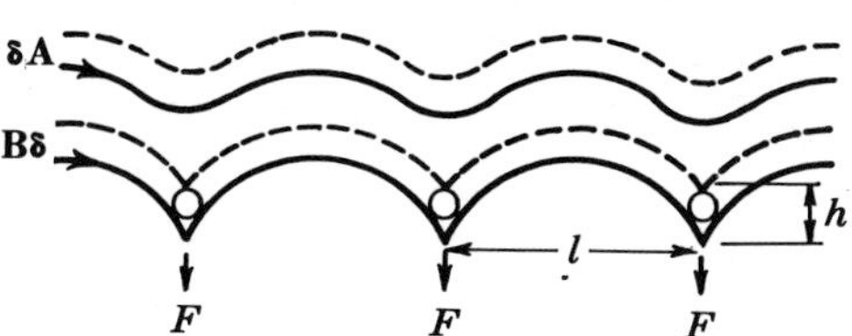

FIGURE 16-24. An extended screw dislocation **BA**(d), restrained by point forces at jogs on the partial **B**δ. The dislocation advances to the dashed configuration by vacancy emission.

faults on {111} planes, bounded by partial dislocations. An example of such a configuration from the work of Swann[10] is shown in Fig. 16-23. X-ray measurements of stacking-fault probability in heavily cold-worked powders are often used to estimate stacking-fault energies.[11] The faults most likely form by the above processes. Since $\sigma_{(1)}$ is likely to vary through a specimen and to depend on local stress concentrations, condition (16-25) is not an accurate fault-nucleation condition to use in determining γ. Also, the kinetics of partial motion subsequent to tearaway will affect the total fault area found. Thus this method is not expected to be accurate in the determination of absolute fault energies, although it is useful in assessing *relative* fault energies.

If condition (16-26) is fulfilled but Eq. (16-25) is not, then both partials are constrained by the pinning points on the trailing partial. As discussed in Sec. 10-6, the total force on both partials in such a case is identical to the force σb which would act on the total dislocation **b** if it were undissociated. Thus, if the extended dislocation advances by a distance a because of vacancy formation, say, at the jogs (Fig. 16-24), the energy change is given by Eq. (16-2). Thus, when the extended partials move together, the dislocation velocity is given by equations identical to those developed for the motion of undissociated dislocations.

PROBLEMS

16-1. **a.** For the case of silver at 800°C, and with $l = 100b$, compute the value of σ for which Eq. (16-5) approximates Eq. (16-4) within 5 percent. Within 1 percent.

b. With $D_s = a_0^2 \nu \exp(-W_s/kT)$, $\nu = 10^{13}\,\text{sec}^{-1}$, and $W_s = 1.9$ eV, compute v for $\sigma = 10^{-3}\,\mu$ at 800°C. What creep strain rate does this v correspond to if the active dislocation length is $\rho = 10^8$ cm/cc?

[10] P.R. Swann, doctoral dissertation, Cambridge University, Cambridge, 1959.

[11] B. E. Warren and E. P. Warekois, *Acta Met.*, **3**: 473 (1955); see also the review in L. F. Vassamillet and T. B. Massalski, *J. Appl. Phys.*, **35**: 2629 (1964).

16-2. Assume an energy-displacement curve for vacancy formation at a jog of the form of Fig. 16-5. With

$$W(\sigma,x)=1.5(W_v+W_v')\left(1-\frac{10\sigma}{\mu}\right)\sin^2\left(\frac{\pi x}{a}\right)$$

compute W^* and a^*. Compare the magnitudes of the various terms in Eq. (16-10) for this case.

16-3. **a.** How are the point forces at jogs in screw dislocations affected if the jogs are superjogs?
b. Derive the equation analogous to Eq. (16-7) for the superjog case.
c. Describe a mechanism whereby superjogs can advance by single-vacancy emission. Draw the appropriate jog configurations.

16-4. For the case of silver with $W_v \sim 1.0$ eV and $W_i \sim 4$ eV, compute the critical stresses for motion of vacancy-forming and interstitial-forming jogs without thermal activation. Compare these stresses with the stress required to move a dislocation at a velocity of 10^{-6} cm/sec [Eq. (16-18)]. Use $l=10^2 b$ and $T=800°C$.

16-5. In the case where an extended screw dislocation in an fcc crystal intersects a *random* array of dislocations, what are the probabilities of forming acute and interstitial jogs? Describe a mechanism by which an acute extended jog can become an obtuse extended jog. Which type of jog is expected to predominate in an *equilibrium* array?

16-6. Suppose that an acute extended jog in a screw dislocation is a superjog. Show mechanistically how such a jog can climb by single-vacancy absorption.

17

Dislocation Motion in Vacancy Supersaturations

17-1. INTRODUCTION

In most cases of macroscopic deformation, such as creep at elevated temperature, vacancy concentrations are maintained in quasi-equilibrium, so that the developments of Chaps. 14 to 16 adequately describe these processes. Dislocation motion in large vacancy supersaturations is a somewhat specialized problem in which quasi-equilibrium kinetics do not suffice to describe the deformation. This problem is of interest from the fundamental viewpoint that the nucleation and incremental growth of dislocation configurations such as prismatic loops and stacking-fault tetrahedra can be studied under carefully controlled experimental conditions by quenching in vacancies from elevated temperatures. Large vacancy supersaturations can also be maintained in sintering experiments and in the interface region of diffusion couples.

Large vacancy supersaturations are seldom maintained for a sufficient period for a quasi–steady-state treatment of dislocation motion to be valid. In a quenching experiment, for example, the initial superconcentration of vacancies decays roughly exponentially as the vacancies are absorbed at dislocations and other sinks. The details of the decay kinetics depend on the dislocation density, the dislocation distribution, the grain size, etc. The local dislocation climb rates change continuously throughout such an experiment. In the initial period of supersaturation, all dislocations absorb vacancies, and new dislocations are often nucleated, whereas later in the decay only a fraction of the dislocations continue to absorb vacancies, and some dislocations anneal out.

In this situation, the rates of climb cannot be treated in a general way. Such a variety of conditions is present in experiments that the rates are best discussed in conjunction with a particular experiment. Thus we do not present a general discussion of climb rates for this case; model discussions of vacancy decay after quenching are presented, for example, by Kimura et al.[1] and by

[1] H. Kimura, R. Maddin, and D. Kuhlmann-Wilsdorf, *Acta Met.*, 7: 145, 154 (1959).

Cotterill[2], and complications due to clustering are discussed by Balluffi and Ho.[3]

A general discussion of the dislocation structure produced in vacancy supersaturations is useful, however. The types of structure formed under supersaturation forces can be discussed without detailed consideration of overall rates of the processes. Some discussion of the *relative* rates of specific processes is important in classifying the structure types. In this chapter we consider, in sequence, the forces produced by vacancy supersaturations, the nucleation of vacancy aggregates, and dislocation configurations produced by vacancy superconcentration. These configurations include Bardeen-Herring sources, dislocation spirals, and stacking-fault tetrahedra.

17-2. CLIMB FORCES

After a quench, very strong osmotic forces are produced by vacancy supersaturation, much stronger than the forces usually attainable under external stresses. Suppose that a crystal is quenched from a temperature near the melting point T_m, where the equilibrium vacancy concentration is

$$c=\frac{1}{v_a}\exp\left(\frac{-W_v}{kT_m}\right) \tag{17-1}$$

to a temperature $T_0=\frac{1}{2}T_m$, where the equilibrium vacancy concentration is

$$c^0=\frac{1}{v_a}\exp\left(\frac{-W_v}{kT_0}\right) \tag{17-2}$$

If all vacancies are retained during the quench, the supersaturation would be

$$\frac{c}{c_0}=\exp\left(\frac{W_v}{2kT_0}\right) \tag{17-3}$$

corresponding to a chemical potential

$$\overline{G}=kT\ln\frac{c}{c^0}=\tfrac{1}{2}W_v \tag{17-4}$$

In other words, the free energy decreases by $\frac{1}{2}W_v$, with the disappearance of each vacancy immediately after quenching. With the typical values $W_v \sim 0.2\mu b^3$

[2] R. M. J. Cotterill, in R. M. J. Cotterill, M. Doyama, J. J. Jackson, and M. Meshii (eds.), "Lattice Defects in Quenched Metals," Academic Press, New York, 1965, p. 97.

[3] R. W. Balluffi and P. S. Ho, in "Diffusion", American Society Metals, Metals Park, Ohio, 1973, p. 83.

and $v_a \sim b^3$, the corresponding force on an edge dislocation is given by Eq. (15-74) as

$$\frac{F}{L} \sim \frac{\mu b}{10} \tag{17-5}$$

Thus the osmotic force generated upon quenching is equivalent to that produced by a compressive stress $\sigma \sim \mu/10$, which greatly exceeds attainable external stresses except for whiskers.

There is some evidence[4, 5] in quenching experiments that vacancies agglomerate to divacancies and larger complexes before annihilation at dislocations and other sinks becomes the predominant vacancy interaction. Such agglomeration would lower the vacancy chemical potential somewhat from the value given by Eq. (17-4). Nevertheless, the forces in the initial period after quenching should be of the order of those in Eq. (17-5). These forces are sufficiently large to overcome all the reaction barriers to dislocation motion that would be important under near-equilibrium vacancy concentrations. For example, an initially straight edge dislocation should be equally effective as a sink for vacancies as a dislocation with a high density of geometric jogs. The initial climb force suffices to nucleate jog pairs easily, independently of the degree of dissociation of the dislocation. Immobilization of dislocations because of reaction-barrier constraints occurs only after appreciable decay of the vacancy supersaturation.

17-3. NUCLEATION OF VACANCY AGGREGATES

The osmotic forces produced upon quenching are sufficient in some cases to *nucleate* new vacancy aggregates. Mechanisms for the formation of prismatic dislocation loops and of stacking-fault tetrahedra by vacancy condensation are discussed in Chap. 10. There are numerous observations of such dislocation configurations in quenched metals.[6] Specific examples are shown of nucleated tetrahedra in Fig. 10-25 and of nucleated prismatic loops in Fig. 17-1.

The interpretation of the experimental data in terms of nucleation theory is questionable at present; simple homogeneous nucleation theory is inadequate to explain the results. In the following discussion we shall briefly review nucleation theory and discuss some of the problems in its application to experiments.

[4] J. E. Bauerle and J. S. Koehler, *Phys. Rev.*, **107**: 1493 (1957).

[5] See the reviews in the Conference Proceedings listed in the bibliography for this chapter, in particular, the discussion by R. W. Balluffi [*J. Nucl. Mat.*, **69/70**: 240 (1978)] is directly relevant to Sec. 17-3.

[6] A recent review of such observations is presented in *ibid*.

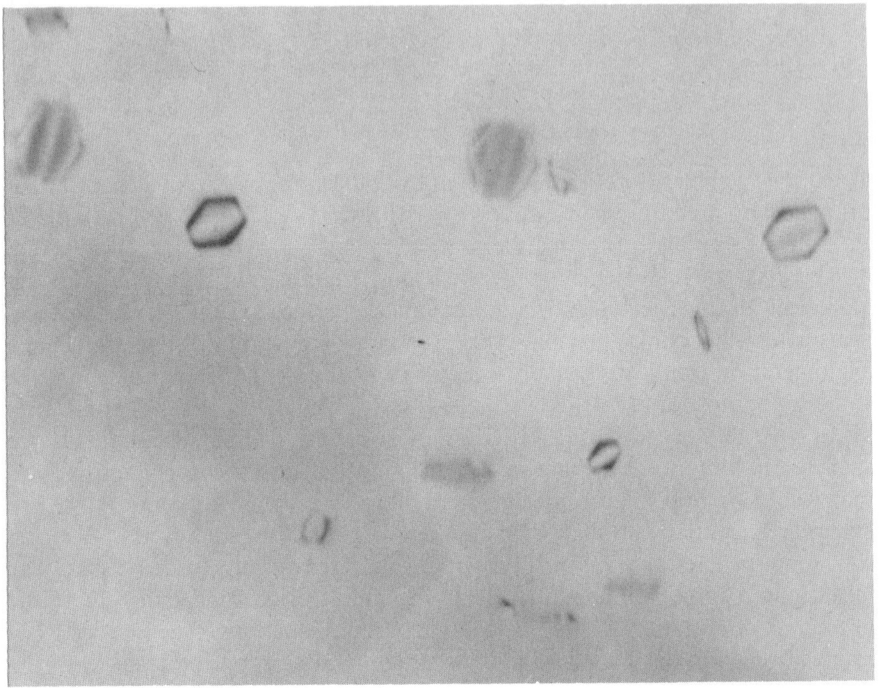

FIGURE 17-1. Prismatic partial-dislocation loops nucleated by vacancy condensation in zone-refined aluminum, 100,000×. (T. L. Davis, private communication.)

Nucleation Theory

The details of nucleation theory are presented elsewhere.[7] Here we treat only the specific example of prismatic loop nucleation in the high-stacking-fault-energy fcc metal aluminum. A typical plot of free energy of formation versus size for vacancy aggregation is shown in Fig. 17-2. The change in free energy with increase in size is initially positive because the change in elastic energy and surface energy for small clusters is larger than the decrease in free energy [Eq. (17-4)] associated with vacancy annihilation. Eventually the term of Eq. (17-4) becomes predominant and the free energy of formation has a critical maximum value ΔG^* at a size i^*. Embryos grow to the critical size by a series of bimolecular reactions. The rate $i^*-1 \rightarrow i^*$ is slower than the other steps, so that a steady-state treatment of this flux approximates the nucleation rate. Also, because the earlier steps $i^*-2 \rightleftharpoons i^*-1$ are near equilibrium, the type of vacancy aggregate that should predominate in the nucleation process is

[7]T. L. Davis and J. P. Hirth, *J. Appl. Phys.*, **37**: 2112 (1966); earlier work is reviewed in this article. *See also* J. A. McComb and M. Meshii, *J. Appl. Phys.*, **38**: 2388 (1967) and T. L. Davis and J. P. Hirth, *J. Appl. Phys.*, **38**: 2390 (1967).

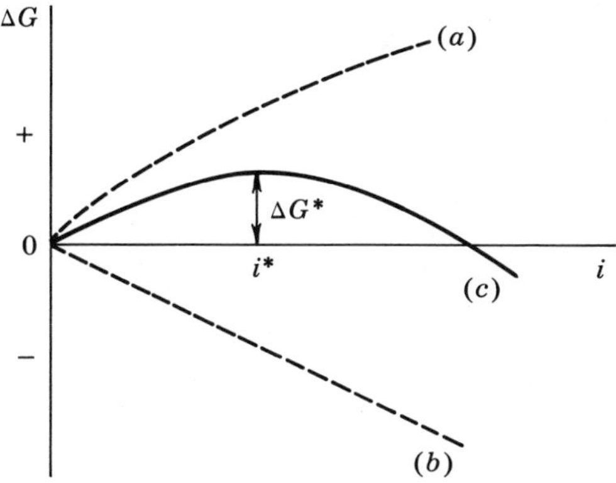

FIGURE 17-2. Free energy of formation of a vacancy cluster versus size i: (a) positive contribution associated with elastic and surface terms, (b) negative contribution associated with vacancy annihilation, and (c) total free energy.

that with the lowest value of ΔG^*, and hence the lowest value of i^*. Considering the configurational energies given in Eq. (10-17) as well as that for a spherical void, one concludes that perfect $\frac{1}{2}\langle 110 \rangle$ prismatic dislocation loops should nucleate in aluminum.

The steady-state nucleation rate[8] is given by

$$J = Z\omega n_c \tag{17-6}$$

where Z is the nonequilibrium factor representing the ratio of the actual flux through size i^* to the flux from an *equilibrium* population of size i^*. As discussed in connection with Eqs. (15-24) and (15-34), $Z = \frac{1}{2}[(\partial^2 G/\partial x^2)/2kT]^{1/2}$; in the present application $Z \sim 0.1$; ω is the frequency with which vacancies diffuse to join a critical-sized cluster, given by the product of the number of vacancies next to a nucleus, $\sim 10\pi r^* c v_a / b$, and the vacancy jump frequency,

$$\omega = \frac{10\pi r^* c v_a}{b} \nu \exp\left(-\frac{W_v'}{kT}\right) \tag{17-7}$$

Here the radius of the critical-sized loop is $r^* = (i^* v_a / \pi d)^{1/2}$, and d is the spacing between $\{111\}$ planes. Finally, n_c is the equilibrium concentration of critical-sized loops, given by

$$n_c = \frac{1}{v_a} \exp\left(-\frac{\Delta G^*}{kT}\right) \tag{17-8}$$

[8]Formally, a time-lag factor in the achievement of steady state should be included. However, the factor is near unity for cases of practical interest [T. L. Davis and J. P. Hirth, *J. Appl. Phys.*, **37**: 2112 (1966)].

The free energy of formation of a loop of size r is

$$\Delta G = 2\pi r \frac{W}{L} - \frac{\pi r^2 d}{v_a} \bar{G}$$

The crucial approximation in simple nucleation theory is that the critical sized nucleus can be described in terms of macroscopic quantities; linear elastic energies for dislocations, equilibrium surface energies for voids and stacking faults. For the loop, then, W/L is the appropriate elastic energy per unit length, given by Eqs. (6-51) and (6-52). Thus the maximum value is

$$\Delta G^* = \frac{\pi v_a}{\bar{G} d}\left[\left(\frac{W}{L}\right)^2 - r^{*2}\left(\frac{\partial(W/L)}{\partial r}\right)^2\right] \tag{17-9}$$

with

$$r^* = \frac{v_a}{\bar{G} d}\left(\frac{W}{L} + r^* \frac{\partial(W/L)}{\partial r}\right) \tag{17-10}$$

All quantities necessary to predict J by Eq. (17-6) are now explicitly defined.

Comparison with Experiment

For aluminum quenched from 873°K, the maximum nucleation rate predicted by Eq. (17-6) is at about 300°K. With the elastic constants given in Appendix 1, and with[9] $W_v = 0.76$ eV and $W_v' = 0.62$ eV, one finds $i^* = 60$, $r^* = 0.4$ nm and $\Delta G^* = 3.25$ eV, giving a completely negligible predicted homogeneous nucleation rate.

In the presence of an applied stress σ_{ij}, the free energy of formation of a loop is reduced by an amount $\sigma_{ij} n_j b_i \pi r^2$, ΔG^* is lowered, and the nucleation rate is increased. Here **n** is a unit vector normal to the plane of the loop. The value of J required to give loop densities such as that in Fig. 17-1 is about 10^{11} nuclei per cm^3-sec. In order to attain theoretical values of J equal to 10^{11} per cm^3-sec, the stress required is $\sim 0.1\ \mu$. Thus the experimentally observed nucleation rates can be explained by simple homogeneous nucleation theory only if stresses of the order of the theoretical strength are present.

For gold, Eq. (10-17) indicates that the most likely nucleus is a stacking-fault tetrahedron. However, an analysis similar to the above[10] indicates that homogeneous nucleation in this case again occurs only at sufficient rates to explain experimental results when stresses of the order of the theoretical strength are present. Other metals are expected to give similar results.

[9] T. Federighi, in R. M. J. Cotterill, M. Doyama, J. J. Jackson, and M. Meshii (eds.), "Lattice Defects in Quenched Metals," Academic Press, New York, 1965, p. 217.

[10] T. L. Davis and J. P. Hirth, *J. Appl. Phys.*, **37**: 2112 (1966).

There are four possible reasons for this discrepancy:

1. All observed nucleation of vacancy aggregates is heterogeneous instead of homogeneous. There is strong evidence for this, with some indications that ubiquitous single impurity atoms can catalyze nucleation[11, 12]; they certainly have binding energies for vacancies and vacancy clusters.[13] Direct observations of heterogeneities when vacancy clusters dissolve,[14] and repeated loop nucleation at the same site[15] also support this effect.

2. The elastic energy of the critical-sized nuclei is overestimated because core relaxations are important in such small dislocation configurations; hence ΔG^* is overestimated. This is almost certainly the case at large supersaturations where a dislocation loop, for example, would be of near core dimensions. The problem is analogous to that for other types of nucleation. For critical sized configurations containing more than a number of vacancies in the range 20 to 50, macroscopic descriptions such as those used in Eq. (17-9) become reasonable approximations for the critical free energy. For such configurations containing one to perhaps three or four vacancies, statistical mechanical descriptions are superior since binding energies are known. For intermediate sizes, corresponding to many experimental cases, neither approach is satisfactory; atomic calculations can be performed, but the appropriate potentials are not well developed as discussed in Chap. 8.

3. Clusters are present in thermodynamic equilibrium at elevated temperature (unstable with respect to growth of course) and are retained during quenching so that clusters of size i^* are present after quenching, presenting *athermal* nucleation sites.[16]

4. Some clustering takes place during quenching, and these clusters then act as athermal nuclei at the final temperature achieved after the quench.

All these factors may have some role in nucleation of vacancy aggregates.[17]

Heterogeneous nucleation is well established for impurity aggregates, precipitates, and inclusions. The heterogeneous nucleation of dislocation loops at an inclusion is shown in Fig. 1-10. The nucleation of dislocations at precipitate-matrix interfaces has been observed by a number of authors.[18]

In summary, the nucleation of vacancy aggregates under osmotic forces produced by quenching has been observed. The mechanism for nucleation is

[11] R. W. Balluffi and D. N. Seidman, in J. W. Corbett and L. C. Ianiello (eds.), "Radiation-Induced Voids in Metals," National Technical Information Service, Springfield, Virginia, 1972, p. 563.

[12] B. L. Eyre, *J. Phys. F*, **3**: 422 (1973).

[13] P. S. Ho and R. Benedek, *J. Nucl. Mat.*, **69/70**: 730 (1978).

[14] I. A. Johnston, P. S. Dobson, and R. E. Smallman, *Cryst. Lattice Defects*, **2**: 127 (1971).

[15] J. W. Edington and R. A. Smallman, *Phil. Mag.*, **11**: 1109 (1965).

[16] J. C. Fisher, J. H. Hollomon, and D. Turnbull, *J. Appl. Phys.*, **19**: 775 (1948); numerical calculation shows that athermal nucleation is a marginal possibility in the present case.

[17] A further discussion of problems in nucleation calculations is presented in the first 13 papers in M. J. Makin (ed.), "The Nature of Small Defect Clusters," H. M. Stat. Off., London, 1966.

[18] See A. Kelly and R. B. Nicholson, *Prog. Materials Sci.*, **10**: 149 (1963).

moot, but there is evidence that the nucleation is heterogeneous; perhaps it is always so. Because of the possibility of heterogeneous nucleation, energy calculations such as those in Eq. (10-17) cannot be used to predict which type of vacancy aggregate should nucleate in a given case. The energy calculations are useful in predicting cluster types only if transitions to the lowest energy form occur during cluster growth.

Exercise 17-1. Consider cluster nucleation in gold, with $\gamma_I = 55\ mJ/m^2$, $W_v = 0.98$ eV, and $W_v' = 0.82$ eV. Plot the elastic energy as a function of size i for the vacancy aggregates (a) stacking-fault tetrahedron, (b) perfect $\frac{1}{2}\langle 110\rangle$ hexagonal loop, and (c) faulted $\frac{1}{3}\langle 111\rangle$ hexagonal loop. For a quench from 1273 to 300°K, compute i^*, ΔG^*, and J for the stacking-fault tetrahedron, which is the aggregate with the lowest free energy of formation.

17-4. CLIMB OF STRAIGHT DISLOCATIONS

Super- or undersaturations of vacancies or interstitials cause dislocation climb. As in Sec. 15-4, consider the dislocation lying along the axis of the cylinder in Fig. 17-3, and suppose that a vacancy concentration $c \neq c^0$ is maintained at the cylinder surface but that there are no external stresses present. If an initial superconcentration $c > c^0$ is imposed throughout the cylinder, the initial climb force is given by Eq. (15-79),

$$\frac{F}{L} = \frac{kTb_e}{v_a} \ln \frac{c}{c^0} \tag{17-11}$$

Under this force the initial transient flux of vacancies to the dislocation is the maximum possible flux, given by the product of the number of vacancies adjacent to unit length of core, $\sim 8cv_a/a$, and the vacancy jump frequency, $\omega = \nu \exp(-W_v'/kT)$; the initial vacancy-emission flux can be neglected when $c \gg c^0$. The initial climb rate is given by the above product times v_a/b_e,

$$\begin{aligned} v &\cong \frac{8c^0 v_a{}^2}{ab_e} \nu \exp\left(-\frac{W_v'}{kT}\right)\frac{c}{c^0} \\ &= \frac{8D_s}{b_e}\frac{c}{c^0} \end{aligned} \tag{17-12}$$

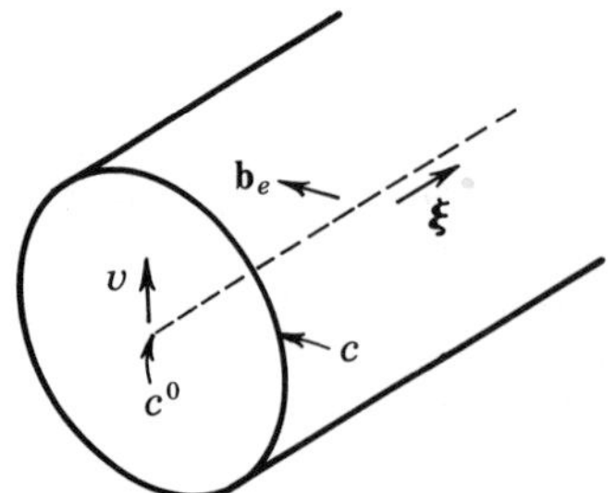

FIGURE 17-3. A dislocation with edge component $\mathbf{b}_e$ lying along the axis of a cylinder.

Eventually, a local equilibrium concentration c^0 is approached at the core when the emission and annihilation rates nearly cancel. Then the flux of vacancies from the surface to the core reaches a steady-state value [*cf.* Eq. (15-90)]

$$v = \frac{2\pi D_s}{b_e \ln(R/b)}\left(\frac{c}{c^0} - 1\right) \tag{17-13}$$

Note that when the osmotic force is substituted into Eq. (15-91), with F_{os} replacing F_{el}, the correct result of Eq. (17-13) is obtained provided that the osmotic forces first be linearized,[19] $\ln(c/c^0) \cong (c/c^0) - 1$.

Similarly, when a large osmotic force is produced by the imposition of an initial underconcentration $c \ll c^0$ on the dislocation, the initial emission rate is the maximum possible rate, giving an initial climb velocity

$$v \cong -\frac{8D_s}{b_e} \tag{17-14}$$

The dislocation cannot climb faster than the rate of evaporation of vacancies from the core permits, independent of the degree of undersaturation. After an initial transient period, v relaxes to the steady-state value

$$v = -\frac{2\pi D_s}{b_e \ln(R/b)}\left(1 - \frac{c}{c^0}\right) \tag{17-15}$$

In general, all the formulas for the various cases of dislocation climb and jog climb treated in Chaps. 15 and 16 are valid for climb under osmotic forces, with the osmotic force substituted for the force caused by an external stress, provided that the osmotic forces can be linearized, $\ln(c/c^0) \cong (c/c^0) - 1$. Thus there is an important distinction between climb under the two types of forces. The entire magnitude of the force produced by an external stress is manifested in the climb velocity of a dislocation; however, under pure osmotic forces, only the linear component of the force affects the climb rate.

17-5. BARDEEN-HERRING SOURCES

Except for the initial period of high vacancy supersaturation immediately following a quench, nucleation of new dislocation loops is negligible. However,

[19] This provides the correct result and is required because the low-stress formulas of Chaps. 15 and 16, such as Eq. (15-91), contain the approximation $c - c^0 \cong (F/L)v_a/bkT$. Exact solutions for $c - c^0$ contain the force factor in an exponential term as in Eqs. (16-4) and (16-18). With the latter type formulas, the full osmotic forces, without linearization, produce exact results such as Eqs. (17-13) and (17-15) as exemplified in the reduction of Eq. (15-82) to the first form in Eq. (15-87). The approximations of constant D_s and ideal solution of vacancies of course apply in all cases.

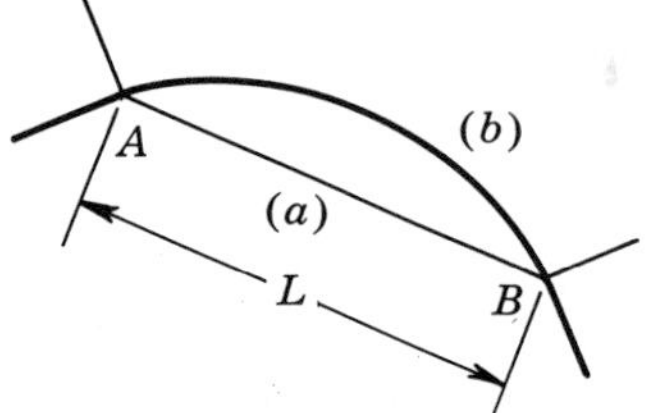

FIGURE 17-4. A pure-edge-dislocation segment pinned at nodes A and B, bowing out by climb to a metastable equilibrium position (b).

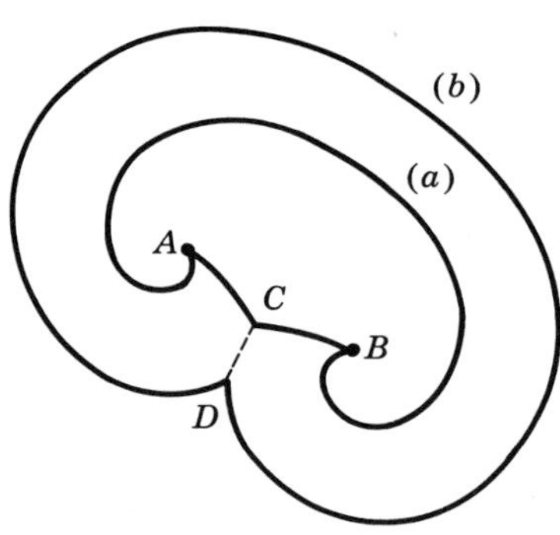

FIGURE 17-5. Two successive positions (a) and (b) of activated climb of the segment AB in Fig. 17-4. In (b) annihilation has occurred over the length CD.

new loops can be introduced at relatively low supersaturations by the operation of *Bardeen-Herring sources*.[20] Consider the segment AB of edge dislocation shown in Fig. 17-4. The ends are pinned, corresponding, for example, to dislocation network nodes immobilized by impurity precipitation. Under an osmotic driving force, the segment bows out by climb. As bow-out proceeds, the radius of curvature decreases and the line-tension forces that tend to restore the loop to a straight configuration increase (see Sec. 6-5). For osmotic forces less than a critical value, the bow-out reaches a metastable equilibrium position, such as (b) in Fig. 17-4, at which the line-tension force exactly balances the osmotic force. For osmotic forces greater than the critical value, the maximum line-tension forces associated with the minimum radius of curvature $r^* = L/2$ of the bowed-out loop are insufficient to achieve metastable equilibrium; the loop grows continuously. Provided that the expanding loop does not intersect dislocations with a net screw component normal to the plane of the loop, the unfolding loop recombines over a section of its length, as illustrated in Fig. 17-5. Thus a closed expanding loop and a restored Bardeen-Herring source segment capable of forming a second loop result.

For the expanding loop to grow through the critical size $r^* = L/2$, the energy change produced by the osmotic force [Eq. (15-74)] with an infinitesimal displacement must exceed that produced by the line-tension force

[20] J. Bardeen and C. Herring, in "Imperfections in Nearly Perfect Crystals," Wiley, New York, 1952, p. 261. These climb sources had an antecedent in the Frank-Read sources for glide loops, discussed in Chap. 20, and are configurationally similar to them.

TABLE 17-1 Minimum Values of Supersaturation c/c^0 and of Vacancy Chemical Potential $\bar{G}$ Required to Satisfy Condition (17-16) as a function of L and T

L/b	T, °K	$\bar{G}$, eV	c/c^0
10^2	300	6.4×10^{-2}	12.0
10^2	600	6.4×10^{-2}	3.4
10^3	300	9.2×10^{-3}	1.4
10^3	600	9.2×10^{-3}	1.2
10^4	300	1.2×10^{-3}	1.05
10^4	600	1.2×10^{-3}	1.02

[Eq. (6-77)],

$$\frac{F_{os}}{L}\pi r\,\delta r \geqslant \mathbb{S}\pi\,\delta r$$

or

$$\frac{kTb}{v_a}\ln\frac{c}{c^0} \geqslant \frac{\mu b^2}{2\pi L(1-\nu)}\ln\frac{L\alpha}{1.8b} \tag{17-16}$$

One might think that thermal fluctuations could activate metastable bow-out loops and enable them to grow for osmotic forces less than those of condition (17-16). However, Nabarro[21] and Shemenski[22] have shown that the contribution of thermal fluctuation forces is completely negligible except for segment lengths $L \lesssim 20b$. In all practical cases, the average loop lengths greatly exceed $20b$, so that thermal forces can be neglected altogether.

With the typical values $\mu b^3 = 5$ eV, $\alpha = 4$, and $\nu = 1/3$, the critical supersaturations required to satisfy condition (17-16) are those listed in Table 17-1 for various values of L and T. In all cases the necessary supersaturations are small compared to those produced initially after rapid quenching, which are of the order $c/c^0 \cong 10^3$ to 10^4. Thus the Bardeen-Herring sources are expected to operate practically throughout the entire period of annealing out of vacancies resulting from a quench.

Let us consider the total amount of climb produced in a quenching experiment if all of the vacancies are absorbed at dislocations. Assume that the dislocation density is ρ and that the average edge component is $\frac{1}{2}b$. If the dislocations in a volume V climb an average distance l, the number n of vacancies absorbed is

$$n = \frac{V\rho l b}{2v_a} \tag{17-17}$$

[21] F. R. N. Nabarro, *Advan. Phys.*, **1**: 269 (1952).

[22] R. M. Shemenski, *Trans. Quart. ASM*, **58**: 360 (1965).

Equating n to the excess number of quenched-in vacancies to be annealed out, cV, one finds

$$l = \frac{2cv_a}{\rho b} \tag{17-18}$$

For a quench from just below the melting point, $cv_a \cong 10^{-4}$. With $\rho = 10^8$ cm^{-2} and $b = 3 \times 10^{-8}$ cm, the average climb distance after such a quench would be $l \cong 6 \times 10^{-5}$ cm$= (2 \times 10^3)b$. This value of l is of the order of the segment spacing L in a typical network of density $\rho \cong 10^8$ cm^{-2}. Thus, if roughly all the segments in such a typical network operate as Bardeen-Herring sources, the sources would on the average generate one or just a few loops in a quench from the highest possible temperatures. Many of the sources would simply bow out somewhat without generating a new loop.

Spiral Source

When only one end of a climbing edge-dislocation segment is pinned, or for one end of a double source like that in Fig. 17-5, when L is large and $L/2 > r^*$, a *spiral source* such as that in Fig. 17-6 is formed. After some transient period, the spiral achieves a steady-state shape that rotates rigidly with a constant angular velocity ω. Far from the spiral center, the curvature of the line approaches zero, and the climb velocity normal to the line is given by Eq. (17-13) or (17-15). Near the center, the curvature increases and the back force associated with the line tension reduces the net climb force and hence the climb velocity normal to the line. At the center, the radius of curvature ρ_0 equals r^* [Eq. (17-10)] and the velocity is zero. The exact solution for the steady-state spiral shape is very complicated, involving boundary conditions

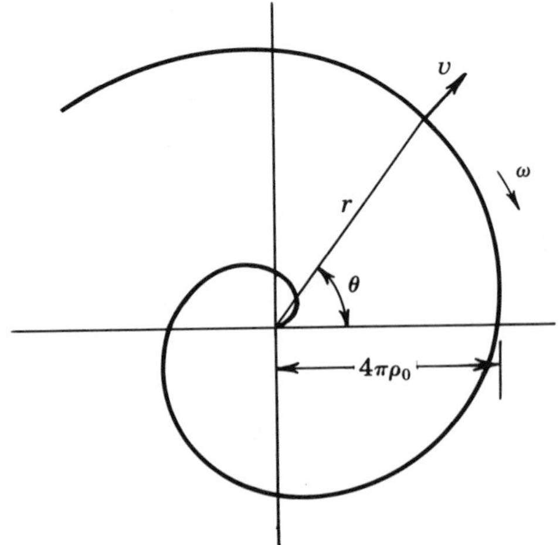

FIGURE 17-6. Spiral configuration of a climbing edge dislocation.

along the line which are complex functions of r and θ, involving elastic interactions between the turns of the spiral, and involving saturation effects; the exact solution has not yet been derived. However, an archimedean spiral is a steady-state shape which qualitatively has the above properties, and which, if one ignores interaction and saturation effects, satisfies the boundary-condition curvatures along the line within ~20 percent. Since errors of this order are involved in the isotropic elasticity approximation and in internal stress distributions, etc., we present this approximate solution rather than attempting an exact one.

The equation for the spiral is

$$r = 2r^*\theta = 2\rho_0\theta \tag{17-19}$$

The climb velocity normal to the line at any point is

$$v = 2\rho_0\omega\frac{\theta}{(1+\theta^2)^{1/2}} = 2\rho_0\omega\frac{r}{(4\rho_0^2 + r^2)^{1/2}} \tag{17-20}$$

The radius of curvature at any point is

$$\rho = 2\rho_0\frac{(1+\theta^2)^{3/2}}{2+\theta^2} \tag{17-21}$$

Exercise 17-2. Verify that the above solution satisfies the boundary condition at $r=0$ and $r=\infty$. Add the osmotic force and the line-tension force for a radius of curvature ρ. Compute the climb velocity from Eq. (15-91) under this net force and show that the result agrees with Eq. (17-20) within 20 percent for all $r > 2\rho_0$.

Bardeen-Herring sources of both the double-ended and the spiral types have been considered, for pedagogical convenience, to form under osmotic forces. Clearly, these sources can also activate under climb forces produced by external stresses or under mixed osmotic and externally produced forces. Double-ended sources have been observed experimentally in both ionic crystals[23] and in metal alloys.[24]

17-6. DISLOCATION HELICES

The climb bow-out of a dislocation segment with screw or mixed character by vacancy absorption or emission produces a dislocation helix. There have been

[23] W. Bontinck, *Phil. Mag.*, **2**: 561 (1957).

[24] K. H. Westmacott, R. S. Barnes, and R. E. Smallman, *Phil. Mag.*, **7**: 1585 (1962); G. Edelin and V. Levy, *Phil. Mag.*, **27**: 487 (1973).

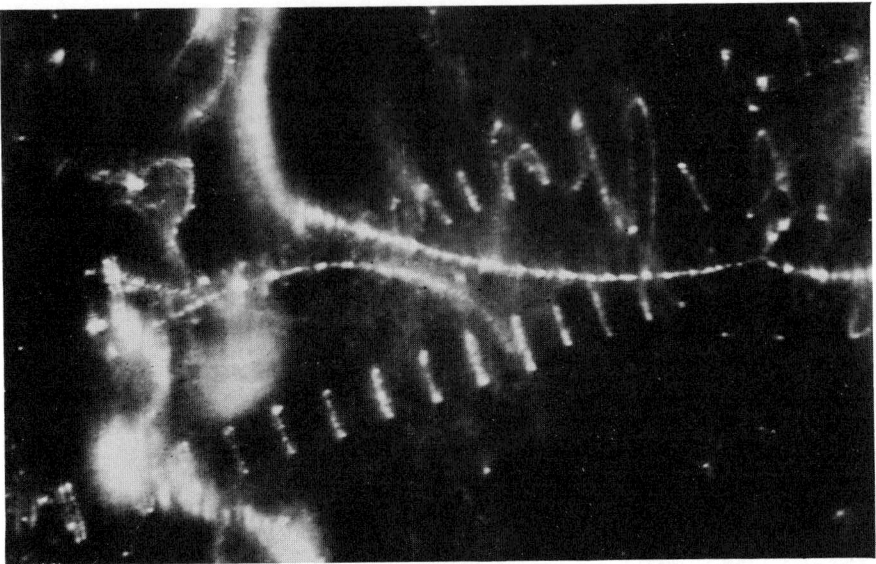

FIGURE 17-7. Helical dislocations decorated in CaF_2. [W. Bontinck and S. Amelinckx, *Phil. Mag.*, **2**:94 (1957)].

numerous experimental observations of such helices; one example is shown in Fig. 17-7. The theory of dislocation helices also has been discussed extensively.[25–28]

The association of helix formation with nonconservative climb motion is demonstrated easily. Consider a straight left-handed screw dislocation touching a prismatic dislocation loop with the same Burgers vector, as shown in Fig. 17-8*a*. The loop corresponds to one formed by vacancy absorption. The configuration can react by *glide* to form one turn of a left-handed helix, as shown in Fig. 17-8*b*. The process of Fig. 17-8 is one of many possible reversible cycles for the formation of such a helix from a straight screw dislocation; any one of these processes requires the absorption of the number of vacancies equal to those contained in the prismatic loop in Fig. 17-8. A general rule is:

Axiom 17-1: The reaction of a left-handed screw into a left-handed helix or of a right-handed screw into a right-handed helix involves vacancy absorption (or matter rejection); the reaction of a left-handed screw into a right-handed helix or of a right-handed screw into a left-handed helix involves vacancy emission.

[25] J. Weertman, *Phys. Rev.*, **107**: 1259 (1957); *Trans. AIME*, **227**: 1439 (1963).

[26] R. de Wit, *Phys. Rev.*, **116**: 592 (1959); *Trans. AIME*, **227**: 1443 (1963).

[27] S. Amelinckx, W. Bontinck, W. Dekeyser, and F. Seitz, *Phil. Mag.*, **2**: 355 (1957).

[28] J. W. Mitchell, *J. Appl. Phys.*, **33**: 406 (1962).

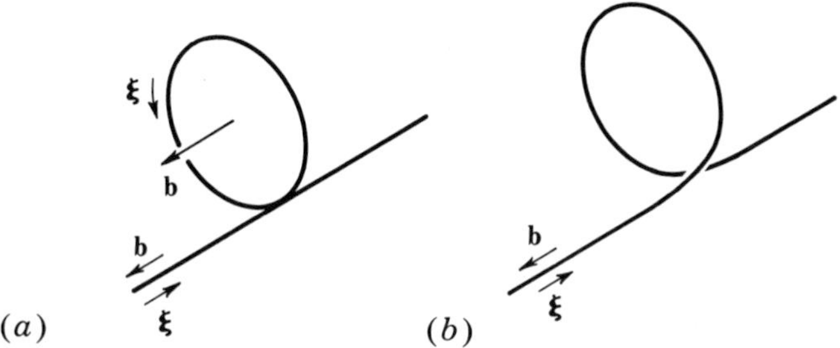

FIGURE 17-8. (*a*) A prismatic, circular dislocation loop reacting with a straight left-handed screw dislocation to form (*b*) one turn of a helix.

An example of the application of this rule is provided in the study of whiskers. Webb[29] observed dislocation helices in palladium whiskers by x-ray methods. He also determined the Eshelby twist, Eq. (3-8), by x-ray diffraction and deduced that the right-handed helices formed from right-handed screws along the whisker axes. Thus he was able to deduce that the helices formed by vacancy condensation, and hence that the whisker growth involved the trapping of vacancies at the growth interface.

By the method illustrated in Fig. 4-3, the total volume of matter transported to or from a helix *per turn* is

$$V = A'b \tag{17-22}$$

where A' is the projected area of the turn of the helix on a screen perpendicular to **b**. Motion that does not change the projection is conservative; the dislocation can rearrange itself by glide on the cylindrical surface upon which the helix is wound.

Mechanisms of Helix Formation

There are a number of possible mechanisms for the initiation of helix formation on a straight dislocation. The kinetics of formation probably influence the *pitch* of the resulting helices. Consider the case of a pure screw dislocation containing a few jogs of both signs, as depicted in Fig. 17-9. In a supersaturation of vacancies, the jogs absorb vacancies and move perpendicular to the screw in a direction determined by the sign of the jog. The jog motion produces the zigzag configuration of Fig. 17-9*b*, where all segments have developed some edge component. The signs of the edge components differ, so that they move in different directions normal to the paper as they climb by vacancy absorption. The net result of the process is the formation of the left-handed helix shown in Fig. 17-9*c*.

[29] W. W. Webb, *J. Appl. Phys.*, **36**: 214 (1965).

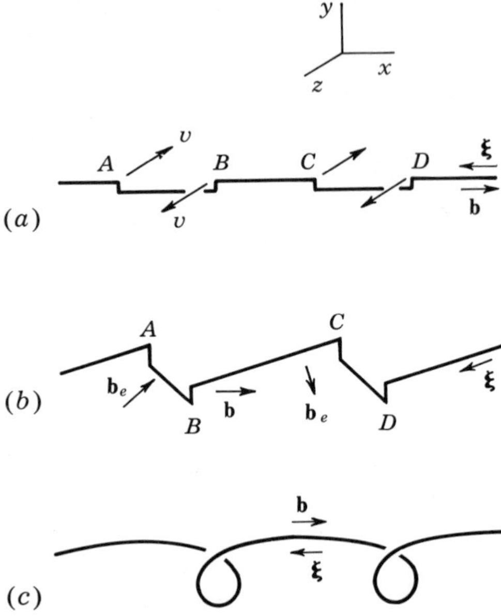

FIGURE 17-9. Climb of a jogged screw from configuration (a) to (b), and subsequent formation (c) of a left-handed spiral.

As suggested by Frank,[30] a thermal fluctuation or internal stresses can cause a small deviation from screw orientation which develops edge components and hence which can initiate helix formation (Fig. 17-10). Frank also postulated that an adsorbed vacancy could dissociate on a screw into the jog-kink configuration of Fig. 12-20. Obviously, the latter array is an incipient spiral itself, and can grow to form a larger spiral.

Now consider the mixed dislocation segment AB, pinned at its ends as drawn in Fig. 17-11. Because of the presence of the edge component, the dislocation can climb directly. The climb motion appears quite simple in a projection normal to **b**, as shown in Figs. 17-11b and 17-11c. Spirals of many turns develop at A and B. The *inner* parts of the spiral projection achieve the steady-state form given by Eqs. (17-19) or (17-21). In this case, a local-equilibrium helix configuration can be achieved, in which the forces arising from curvatures and coil-coil interactions balance the osmotic force.

The force components in the x direction, or glide direction, in Fig. 17-11, produced by both repulsive interactions between turns [Eq. (5-25)] and by line-tension forces, tend to force the spirals at each end into roughly conical helices. If the pitch of the helix were small compared to the diameter, the interaction forces between turns would become attractive and oppose the line-tension force, but this case is not observed experimentally and hence is not

[30] F. C. Frank, *Nuovo Cim. Suppl.* **7**: 386 (1958).

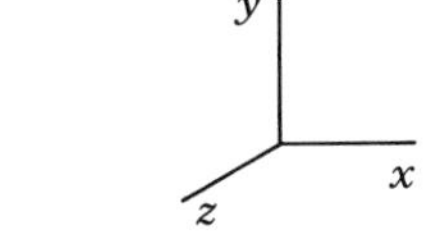

FIGURE 17-10. Fluctuation in the xz plane on a screw dislocation (solid line), and subsequent climb (dashed line).

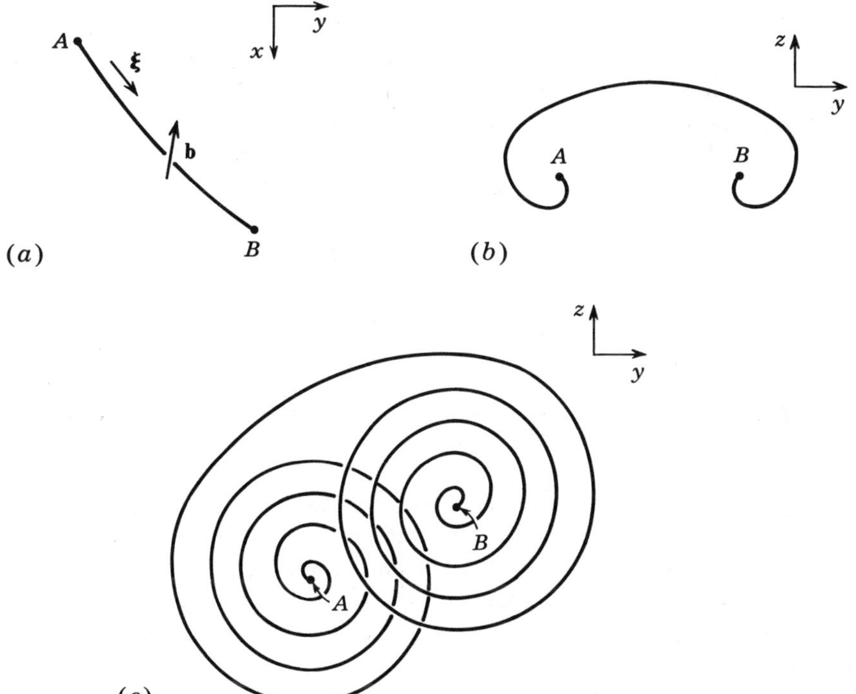

FIGURE 17-11. (a) A mixed straight dislocation AB; (b) and (c), projections along **b** of AB as it climbs.

of interest. In a three-dimensional projection, the helix is expected to have the form shown in Fig. 17-12. Such helicoids (for example, that in Fig. 17-7) were first observed by Bontinck and Amelinckx[31]; Amelinckx et al.[32] gave essentially the above explanation for their formation. Later observations of helices

[31]W. Bontinck and S. Amelinckx, *Phil. Mag.*, **2**: 94 (1957).

[32]S. Amelinckx, W. Bontinck, W. Dekeyser, and F. Seitz, *Phil. Mag.*, **2**: 355 (1957).

FIGURE 17-12. Perspective view of the helicoid of Fig. 17-11*c*.

in metal-alloy crystals are reviewed by Amelinckx.[33] In all these examples, the observed helicoidal dislocations are likely to be pinned by point defects. Helicoids such as the one in Fig. 17-12 are *not* equilibrium configurations. Equilibrium forms are discussed in the next section.

Weertman[34] has pointed out that helical dislocations can form in the absence of osmotic forces. Suppose that the dislocations in Figs. 17-9 and 17-10 correspond initially to pure edge dislocations, $\mathbf{b}=(0, b_y, 0)$, which are bowed out in their glide planes. The zigzag pieces then develop screw character, alternately left-handed and right-handed. Under a shear stress σ_{xy}, forces acting on the screw components push the segments in opposite directions along the z axis. The edge components move along with the screw components by short-range core diffusion; no net transport of matter to the dislocation is required. Thus a helix can form and achieve a constrained equilibrium shape in which line-tension forces balance the stresses σ_{xy} acting over the dislocation cores.

For a mixed dislocation in the general case, the total force balance on the helix involves the entire stress tensor acting over the core via Eq. (3-90), the osmotic force, and the line-tension force. In the following section, for simplicity, mainly osmotic forces and line-tension forces are considered. Weertman[34] and de Wit[35] have considered some of the effects of external stresses on the properties of helical dislocations.

Uniform Cylindrical Helices

Let us now consider the equilibrium shape of a helical dislocation. As shown subsequently, helices can be in a constrained equilibrium with respect to a change in pitch, but *not* with respect to a change in radius, except in very unusual distributions of internal stress. The general solution for the energy of a

[33] S. Amelinckx, *Solid State Phys. Suppl.* 6 (1964).

[34] J. Weertman, *Trans. Met. Soc. AIME*, **227**: 1439 (1963).

[35] R. de Wit, *Trans. Met. Soc. AIME*, **227**: 1443 (1963).

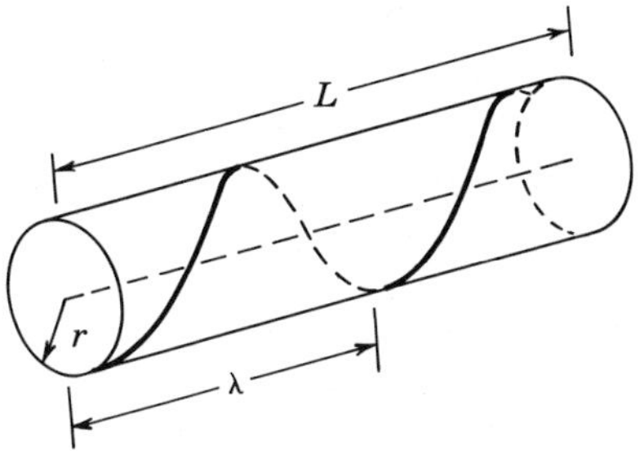

FIGURE 17-13. Cylindrical-helical dislocation of pitch λ and radius r.

uniform helix is outlined by de Wit.[36] However, he showed that the solution is intractable except in the asymptotic limits of loose winding and tight winding, the former of which was discussed earlier in a simple line-tension model by Weertman.[37]

Consider a helix formed from a straight screw with the geometry shown in Fig. 17-13. Here λ is a measure of the pitch of the helix, $n=\lambda^{-1}$ is the number of turns per unit length, and r is the radius of the cylinder on which the helix is wound. Let us first treat the loose-winding case $\lambda \gg r$. The appropriate line tension is estimated by setting $\rho = b/8$ and $L=\lambda/2$ in Eq. (6-83), yielding

$$\mathcal{S} \cong \frac{\mu b^2}{4\pi}\frac{1+\nu}{1-\nu}\ln\frac{\lambda}{3b} \tag{17-23}$$

The increase in length when the screw coils up into a helix is

$$\Delta L = L\left\{\frac{1}{\lambda}\left[(2\pi r)^2+\lambda^2\right]^{1/2}-1\right\} \tag{17-24}$$

Thus in the line-tension approximation the increase in energy per unit length is

$$\frac{\Delta W}{L} = \mathcal{S}\left\{\frac{1}{\lambda}\left[(2\pi r)^2+\lambda^2\right]^{1/2}-1\right\} \tag{17-25}$$

The number of vacancies absorbed per unit length in creating the helix is

$$\frac{N}{L} = \frac{\pi r^2 b}{\lambda v_a} \tag{17-26}$$

The equilibrium condition for the helix is thus

$$\frac{\partial(\Delta W/L)}{\partial r} = \frac{\partial(\bar{G}N/L)}{\partial r} \tag{17-27}$$

[36] R. de Wit, *Phys. Rev.*, **116**: 592 (1959).

[37] J. Weertman, *Phys. Rev.*, **107**: 1259 (1957).

where $\bar{G}$ is the vacancy chemical potential. Equation (17-27) reduces to the expression for the value of $\bar{G}$ that would maintain equilibrium with the helix,

$$kT\ln\frac{c}{c^0}=\frac{2\pi v_a \mathcal{S}}{b\left[(2\pi r)^2+\lambda^2\right]^{1/2}} \tag{17-28}$$

For $r\to 0$, the critical pitch length that satisfies Eq. (17-28) is

$$\lambda^*=\frac{2\pi v_a \mathcal{S}}{bkT\ln(c/c^0)} \tag{17-29}$$

The equilibrium radius r^* for any λ is given by Eqs. (17-28) and (17-29) as

$$r^*=\frac{1}{2\pi}\left(\lambda^{*2}-\lambda^2\right)^{1/2} \tag{17-30}$$

If $\lambda>\lambda^*$, the free energy of the helix decreases monotonically with increasing r, there is no equilibrium radius, and the helix will grow *spontaneously* from a straight screw. If $\lambda<\lambda^*$, a nucleation barrier exists, and thermal energy must be supplied to achieve the unstable equilibrium radius r^*, above which spontaneous growth would again occur. Thus fluctuations such as the one in Fig. 17-10 can nucleate helix formation directly only if their wavelength exceeds λ^*. Because the glide surface of a helix formed from a screw is the cylinder on which it is wound, nascent helical coils of length $\lambda<\lambda^*$ can adjust to the supercritical size $\lambda>\lambda^*$ by a *glide* fluctuation. Thus the kinetics of helix formation depend on the frequency with which fluctuations of magnitude $\lambda>\lambda^*$ occur. Once a helix starts to form, independent of its mode of nucleation, it is *unstable* with respect to radial growth.

Consider now the effect of external stresses σ_{ij} acting over the dislocation core. Choose the axis of the helix along z, as in Fig. 17-13, with $\mathbf{b}=(b_x,0,b_z)$. We are interested mainly in pure helical bow-out, so we impose the condition that $\mathbf{F}$ vanish for the initially straight dislocation, i.e., when $\boldsymbol{\xi}=(0,0,\xi_z)$. For the helix, $\boldsymbol{\xi}=(0,\xi_\theta,\xi_z)$. Equation (3-90) then reduces to

$$\begin{aligned}\frac{\mathbf{F}}{L}&=(\sigma_{xz}b_x+\sigma_{zz}b_z)(-\xi_y\mathbf{i}+\xi_x\mathbf{j})\\&=-(\sigma_{xz}b_x+\sigma_{zz}b_z)\xi_\theta\mathbf{e}_r\\&=\frac{F_r}{L}\xi_\theta\mathbf{e}_r\end{aligned} \tag{17-31}$$

where $\mathbf{e}_r$ is a unit vector in the radial direction and $\xi_\theta=[1+(\lambda/2\pi r)^2]^{-1/2}$. The force has a component only in the r direction, F_r. Thus the work done per unit length in forming a helix of radius r is $\pi r^2F_r/L\lambda$. With this term included,

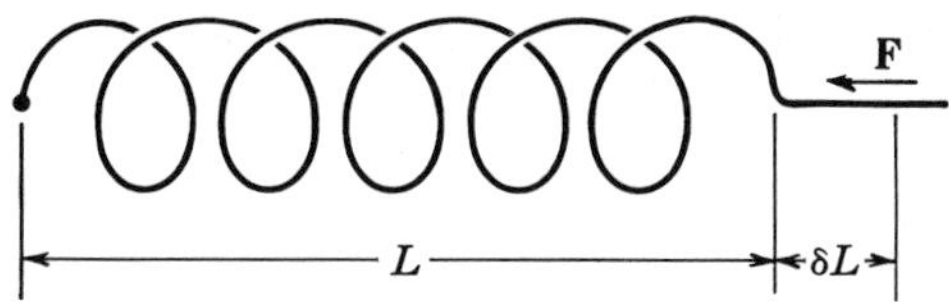

FIGURE 17-14. Compressive force **F** constraining a helical dislocation of length L.

Eq. (17-28) is modified to

$$\frac{F_r v_a}{L b_z} + kT \ln \frac{c}{c^0} = \frac{2\pi v_a \mathcal{S}}{b_z\left[(2\pi r)^2 + \lambda^2\right]^{1/2}} \tag{17-32}$$

Thus the consequences with respect to helix formation, the existence of r^*, λ^*, etc., are the same as in the earlier case; the only difference is that the term $F_r v_a / L b_z$ appears on equal footing with $\overline{G}$. For an initially pure screw dislocation, the effective stress component in F_r is the normal stress σ_{zz}, as anticipated from the treatment of the osmotic force, while for an initially pure edge, the shear stress σ_{xz} is the effective stress component. Both stresses exert forces on a helix formed from a mixed dislocation. Mechanistically, a helix formed from a mixed dislocation differs from the screw case in that the *glide plane* is inclined to the axis of the helix, so that a change in pitch requires dislocation climb.

In a cold-worked metal, one expects in general that osmotic forces are present because of vacancy formation at jogs, and that internal stresses σ_{ij} are present. Thus, as first suggested by Weertman,[38] *all* straight dislocation segments of length $L > \lambda^*$ should be unstable with respect to helix formation in a cold-worked metal. This phenomenon is important in the process of work hardening in general, and specifically in the formation of dislocation tangles.

Fluctuations in Helix Geometry

Equation (17-32) is the equilibrium condition for a *uniform* cylindrical helix. Consider now the stability of the helix with respect to local fluctuations in geometry. First let us treat the stability of the helix formed from a screw against fluctuations in pitch occurring by glide. The repulsive forces within a coil determine the stability. These forces can be estimated from the process illustrated in Fig. 17-14.

If the helix length is changed by an amount δL under the condition that the number of turns $N = L/\lambda$ remain constant, the increase in energy of the helix is

$$\delta W = \frac{\partial \Delta W}{\partial L}\delta L = \frac{1}{N}\frac{\partial \Delta W}{\partial \lambda}\delta L \tag{17-33}$$

[38] J. Weertman, *Trans. Met. Soc. AIME*, **227**: 1439 (1963).

with ΔW given by Eq. (17-25). The compressive force F is then defined by the expression (including the differential of Eq. (17-23))

$$F = -\frac{\delta W}{\delta L} = -\frac{1}{N}\frac{\partial \Delta W}{\partial \lambda}$$

$$= \mathbb{S}\left\{1 - \frac{\lambda}{\left[(2\pi r)^2 + \lambda^2\right]^{1/2}}\right\}\left\{1 - \frac{\left[(2\pi r)^2 + \lambda^2\right]^{1/2}}{\lambda \ln(\lambda/3b)}\right\}$$

$$\cong \mathbb{S}\frac{2\pi^2 r^2}{\lambda^2} \qquad (17\text{-}34)$$

the approximate form holding when $\lambda \gg 2\pi r$. This expression shows the expected behavior of decreasing "stiffness" of helix with looser winding, $F \to 0$ as $\lambda \to \infty$. The compressive force, or "spring tension," in the helix is of the order of a line-tension force $\mathbb{S}$ when λ is of the order of the cylinder diameter, but decreases rapidly with increasing λ.

Because $\partial F/\partial \lambda$ is negative, the uniform coil is *stable* with respect to fluctuations in pitch. If a few coils in an otherwise uniform helix fluctuate, so that the coil spacings on either side are $\lambda + \delta\lambda$ and $\lambda - \delta\lambda$, the compressive force on the latter side is greater and the original uniform helix is restored, in agreement with intuitive expectation in the spring analog.

Now consider the thermodynamic equilibrium between two helices of different radii r_1 and r_2, which have equilibrated with respect to glide (Fig. 17-15). Such glide equilibrium requires that $F_1 = F_2$, so that with the neglect of small differences in the logarithmic factor, Eq. (17-34) yields the condition

$$\frac{\lambda_1}{r_1} = \frac{\lambda_2}{r_2} \qquad (17\text{-}35)$$

Since Eq. (17-28) can be written in the form

$$\bar{G} = \frac{2\pi v_a \mathbb{S}}{b\left[(2\pi)^2 + (\lambda/r)^2\right]^{1/2}}\frac{1}{r} \qquad (17\text{-}36)$$

it follows from Eqs. (17-23) and (17-36) in the same approximation that the

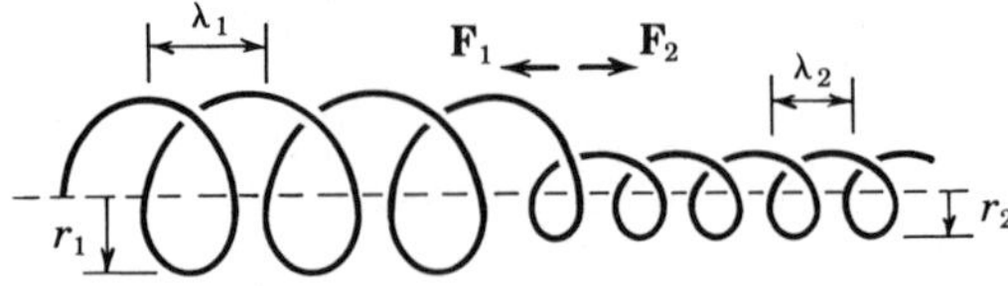

FIGURE 17-15. Two helices of different radius in glide equilibrium.

chemical potentials of vacancies in equilibrium with the two helices are in the ratio

$$\frac{\bar{G}_2}{\bar{G}_1} = \frac{r_1}{r_2} \tag{17-37}$$

Thus vacancies would flow from the coil of smaller r to make helix 1 grow and helix 2 shrink. Hence a uniform coil is *unstable* with respect to fluctuations in r. The equilibrium form of a helix is expected to be a degenerate helix of one turn.

The Tight-Winding Case

In the limit $\lambda \ll r$, the elastic interaction between loops is dominant over line-tension forces in the coil. Equation (5-26) for the interaction between coaxial loops applies in this case; it indicates that the repulsive force between two turns, $F \sim \mu b^2 r/(1-\nu)\lambda$, is already larger than typical line tensions $\mathcal{S} \sim \mu b^2$. The interaction energy per unit length in a stack of loops, corresponding to the helix when $\lambda \ll r$, was found by de Wit[39] by summing Eq. (5-25),

$$\frac{W_{\text{int}}}{L} = \frac{\pi\mu b^2}{1-\nu}\frac{r^2}{\lambda^2} \qquad r \ll \lambda \tag{17-38}$$

This term is the dominant term in $\Delta W/L$ for a closely wound helix. Inserting Eq. (17-38) into (17-33), one finds

$$F = \frac{\pi\mu b^2}{1-\nu}\frac{r^2}{\lambda^2} \tag{17-39}$$

Thus the form of the F-λ curve is the same as that for the loose-winding case, Eq. (17-34).

The interaction forces also add to the instability of the coil with respect to a fluctuation in r in the tight-winding case. Indeed, these forces and the compressive forces of Eq. (17-39) are so large that tightly wound coils are not expected to exist. The present treatment does serve to indicate the expected influence of the interaction forces in the loose-winding case when $\lambda \to r$. Estimates of $\Delta W/L$ for the region intermediate between tight winding and loose winding would be very valuable.

Other Effects

Helices observed experimentally are often remarkably uniform in radius and pitch. The observed pitches vary from about the order of magnitude of the

[39] R. de Wit, *Phys. Rev.*, **116**, 592 (1959).

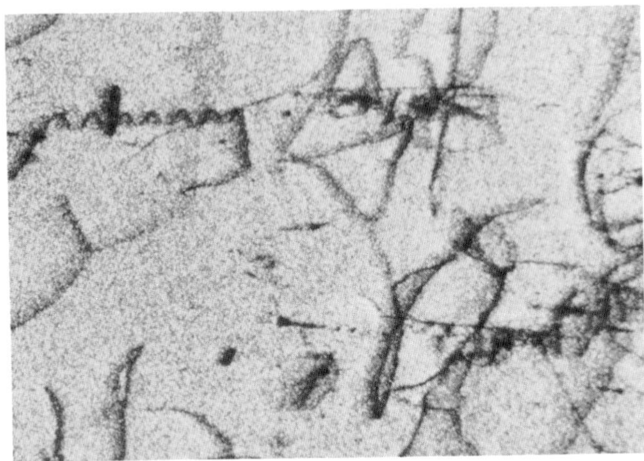

FIGURE 17-16. Helical dislocation in zone-refined aluminum foil, 1 mm thick; cooled from 510°C at 4°C/hr. (220) x-ray reflection at 50×. (B. Nøst, private communication).

helix diameter to larger values; tightly wound helices are not seen. The above treatment indicates clearly that uniform helices are in unstable equilibrium with respect to fluctuation, whether by volume or core diffusion, while slightly nonuniform coils are not in equilibrium. Thus the uniform loops must represent a frozen-in stage in kinetic development. Loops in ionic crystals and metal alloys, such as the one in Fig. 17-7, could be pinned by point defects. However, helices are also observed, although rarely, in high-purity metals, as illustrated in Fig. 17-16 for aluminum.

With uniformity of radius as an observed fact, compressive forces within the coil tend to establish a uniform pitch, by glide for the initially pure screw, by glide and climb for the initially mixed dislocation. The compressive forces at the ends of the helical segment are of the order of line-tension forces and can be balanced by local interactions at pinning points or nodes. Also, a return loop along the axis (Fig. 17-17) can provide a force $\sim \mathbb{S}$ to balance the compressive force $F \sim \mathbb{S}$ of the helix. Thus the observation of helices with $\lambda \sim 2r$, corresponding to $F \sim \mathbb{S}$, is consistent with the expected strengths of end constraints.

Finally, we suggest one possible mechanism that might limit the magnitude of the ratio λ / r. A pure screw is expected to have a lower core energy than a generally skew dislocation by an amount ΔW_{core}. The straight segment in Fig.

FIGURE 17-17. A dislocation helix with a portion of the dislocation line returning along the axis of the helix.

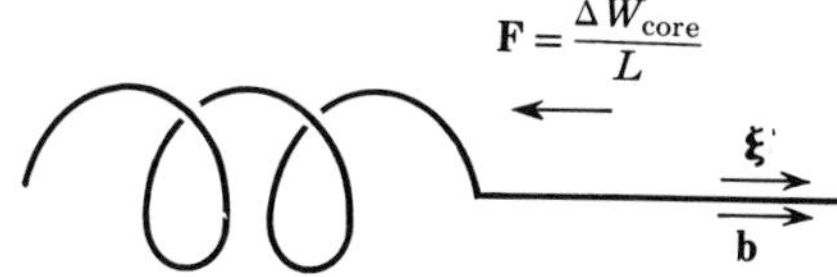

FIGURE 17-18. Core forces constraining a helical dislocation adjoining a straight screw segment.

17-18 would then exert a force

$$F = \frac{\Delta W_{\text{core}}}{L} \tag{17-40}$$

on the helix. This force could stabilize a loosely wound helix in the configuration shown in Fig. 17-18. With $\Delta W_{\text{core}}/L \sim 0.05\mathbb{S}$, the limiting pitch in Fig. 17-18 would be $\lambda/r \sim 6\pi$, according to Eq. (17-34).

17-7. STACKING-FAULT TETRAHEDRA

The discussion of tetrahedra in Sec. 10-4 centered on their formation and annihilation by glide processes. The results of Sec. 17-3 indicate that tetrahedra can also form by nucleation and growth. All the problems associated with nucleation apply to the case of tetrahedra. Plausible atomistic mechanisms have been proposed for nucleation via small clusters of vacancies,[40, 41] but as discussed in Sec. 17-3, the *rate* of nucleation is not explainable by simple homogeneous nucleation theory, and there are strong indications that nucleation is heterogeneous.[42]

Once formed, tetrahedra can grow by vacancy absorption at jog lines, by the process illustrated in Fig. 15-29 for an acute jog line. This process is precisely the process of climb of extended jogs discussed in Chap. 15. Evidence exists that such climb occurs readily under low driving forces. The climb of extended dislocations occurs with nearly perfect sink efficiency under moderate and low vacancy supersaturations.[43] In addition, tetrahedra overage; that is, large tetrahedra grow at the expense of small ones by vacancy diffusion.[44] Since the results of Eq. (10-17) indicate that the driving force for such overaging is small, the results again indicate that growth at jog lines occurs easily. For gold, the small deviation from perfect sink behavior has been found to be associated

[40] M. deJong and J. S. Koehler, *Phys. Rev.*, **129**: 40 (1963).

[41] R. M. J. Cotterill and M. Doyama, in R. M. J. Cotterill et al (eds.), "Lattice Defects in Quenched Metals," Academic Press, New York, 1965, p. 653.

[42] B. L. Eyre, *J. Phys. F*, **3**: 422 (1973).

[43] D. N. Seidman and R. W. Balluffi, *Phys. Stat. Solidi*, **17**: 531 (1966).

[44] R. M. J. Cotterill, *J. Phys. Soc. (Japan) Suppl.* 3, **18**: 48 (1963)

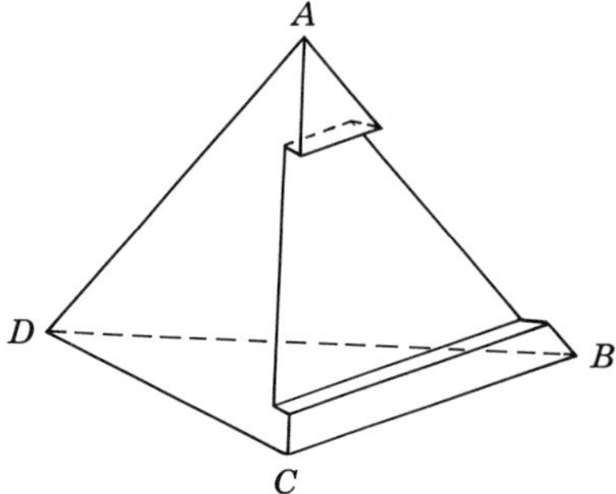

FIGURE 17-19. A stacking-fault tetrahedron with an acute jog line nucleated at corner A and an obtuse jog line nucleated at edge BC.

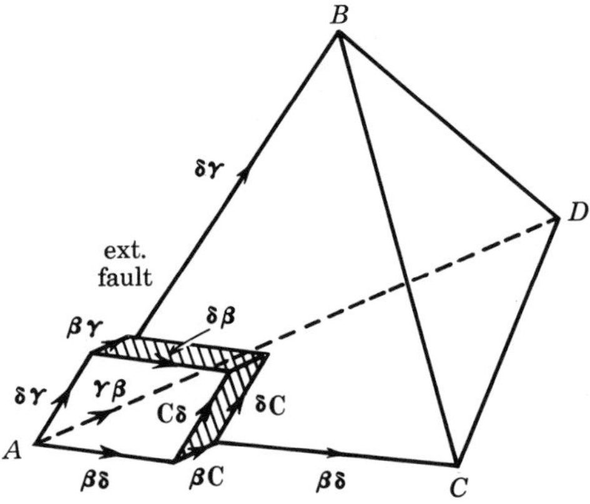

FIGURE 17-20. Nucleation of a partial obtuse jog line at corner A. The small shaded area bounded by $\mathbf{C\delta}$ and $\boldsymbol{\delta}\mathbf{C}$ contains no stacking fault.

with an activation energy of 0.07 eV.[45] There are indications[46] of difficulty in the annealing out of large tetrahedra, but the mechanism in this case might be *glide* collapse.

The nucleation of jog lines on tetrahedra is of theoretical interest. Kuhlmann-Wilsdorf[47] has noted that acute jog lines nucleate at tetrahedron corners during vacancy absorption and at edges during vacancy emission, as shown in Fig. 17-19. Obtuse jogs nucleate at corners during vacancy emission and at edges during vacancy absorption. If the critical nucleus configuration involves a jog line completely traversing the tetrahedron face, corner nucleation is favored because the line length is much smaller. Because the obtuse jog lines have larger elastic energies than the acute ones, jog-line nucleation should then be easier during vacancy absorption than during emission,[47] and this could account for the above-mentioned difficulty in the annealing out of tetrahedra. Contrariwise, if the critical nucleus involves only part of a jog line,

[45] K. C. Jain and R. W. Siegel, *Phil. Mag.*, **25**: 105 (1972).

[46] M. Meshii and J. W. Kauffman, *Phil. Mag.*, **5**: 687, 939 (1960).

[47] D. Kuhlmann-Wilsdorf, *Acta Met.*, **13**: 257 (1965).

then nucleation can always occur at a corner, as shown in Fig. 17-20 for an obtuse jog. In this case nucleation should occur under similar driving forces in either vacancy absorption or emission. Because the entire configuration is of core dimensions, we do not attempt to assess the latter possibility quantitatively.

PROBLEMS

17-1. Consider an edge dislocation along the axis of a right circular cylinder of radius $R=10^4 b$ and composed of silver. If the cylinder is rapidly up-quenched from room temperature to near the melting point, compute the maximum possible climb rate and the initial quasi–steady-state climb rate of the dislocation. Assume $W_v=1.0$ eV and $W_v'=0.9$ eV.

17-2. For an aluminum crystal quenched from 650 to 300°C, compute the nucleation rate of circular prismatic $\frac{1}{2}\langle 110\rangle$ loops in the presence of normal stresses $\sigma=10^{-6}$, 10^{-4}, 10^{-2}, and $10^{-1}E$.

17-3. With $\gamma_I=200$ mJ/m^2, compute the nucleation rate of circular, Frank partial, prismatic $\frac{1}{3}\langle 111\rangle$ loops after quenching from 650 to 300°C. Compare the results to those of Prob. 17-2. Discuss the experimental observation of $\frac{1}{3}\langle 111\rangle$ faulted loops in high-purity quenched aluminum.

17-4. Suppose that one arm of an operative Bardeen-Herring source swings around several screw dislocations so that it passes over the opposite arm by three interplanar distances. Consider the likelihood of annihilation in such a case, and the resultant dislocation configuration.

17-5. What are the forces acting on a helical dislocation formed from a screw and lying normal to a free surface? What is the result of the glide of the helix into the surface? Assess the likelihood of such glide quantitatively.

17-6. Suppose that a straight mixed dislocation line exists in an unstressed crystal. Show that when an external hydrostatic pressure is applied to the crystal, the dislocation will initially tend to transform into a helical shape, but will eventually straighten out again. Discuss the forces acting on the dislocation in each stage.

17-7. Suppose that a mixed dislocation transforms into a helical dislocation in the absence of net mass transport to or from the dislocation. Show that the helix formation requires both local core diffusion and net climb of the dislocation, i.e., the axis of the helix is displaced from the original position of the straight dislocation. Compute the distance of climb as a function of λ, r, and b_e.

17-8. Construct a sequence of diagrams showing the stages of the following:

a. Formation of an acute jog line at a tetrahedron corner by vacancy emission

b. Annihiliation of an interstitial at an obtuse jog line on a tetrahedron

17-9. Verify, using the method of Chap. 6, that the interaction energy between a triangular Frank partial and a hexagonal Frank partial, within which the triangle is concentrically situated, is positive, so that the nucleation of such triangles is unfavorable, indicating the possibility of heterogeneous nucleation.

BIBLIOGRAPHY

1. Arsenault, R. J. (ed.), *Nucl. Met.*, **18**: 1 (1973).
2. Balluffi, R. W., and A. V. Granato, in F.R.N. Nabarro (ed.), "Dislocations in Solids," vol. 4, North Holland, Amsterdam, 1979, p. 1.
3. Bullough, R., (ed.), *J. Phys. F*, **3**: 233 (1973).
4. Cotterill, R. M. J., M. Doyama, J. J. Jackson, and M. Meshii, "Lattice Defects in Quenched Metals," Academic, New York, 1965.
5. Makin, M. J., "The Nature of Small Defect Clusters," H. M. Stat. Off., London, 1966.
6. Peterson, N. L., and R. W. Siegel, "Properties of Atomic Defects in Metals," North-Holland, Amsterdam, 1978.
7. Robinson, M. T., and F. W. Young, Jr., "Fundamental Aspects of Radiation Damage in Metals," National Technical Information Service, Springfield, Virginia, 1975.
8. Seeger, A., D. Schumacher, W. Schilling, and J. Diehl, "Vacancies and Interstitials in Metals," North-Holland, Amsterdam, 1970.

18

Effects of Solute Atoms on Dislocation Motion

18-1. INTRODUCTION

Equilibrium distributions of solute atoms or impurities about dislocations are treated in Chap. 14. Here we consider the effects of solute on dislocation motion. This topic is very broad and includes a large number of possible types of behavior, so that for brevity we restrict this treatment primarily to those cases which introduce new concepts or new types of behavior. Some other very important cases are alluded to only briefly, either because they are so complex that their exact treatment is difficult or because they can be handled easily by the theory presented in previous chapters.

As an example of the latter type of process, consider a dislocation pinned by a dilute concentration of solute atoms in the core, as shown in Fig. 18-1. Suppose that the relaxation time for the solute atom to diffuse along with the dislocation is less than that for breakaway of the dislocation from the solute. The motion of the dislocation then entails motion of the solute pinning point along with the dislocation, with net advance corresponding to motion from configuration A to C in Fig. 18-1. The activated position corresponds to position B in the figure. This case is similar to that of jog pinning points (Fig. 16-5) and is described by an expression such as Eq. (16-8). The jump frequency in the direction favored by the bow-out is

$$\nu_d \exp\left(-\frac{W^* - \sigma b l a^*}{kT}\right) \tag{18-1}$$

while the jump frequency in the reverse direction is

$$\nu_d \exp\left(-\frac{W^* + \sigma b l a^*}{kT}\right) \tag{18-2}$$

where now l is the spacing of solute atoms along the core. Proceeding as in the

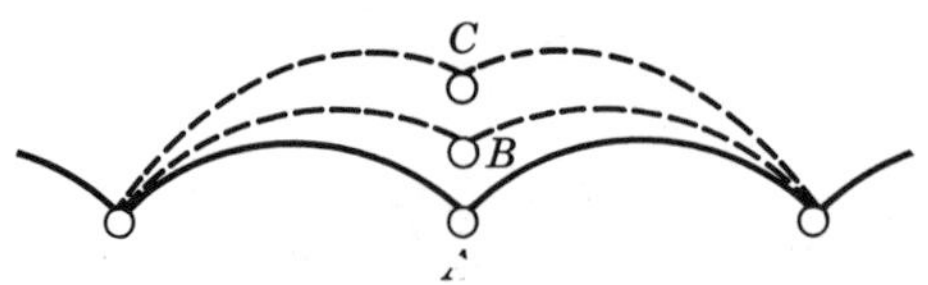

FIGURE 18-1. A bowed-out dislocation pinned by solute atoms in its core.

derivation of Eq. (16-8), one finds that the dislocation velocity is

$$v \cong 2a\nu_d \exp\left(-\frac{W^*}{kT}\right)\sinh\left(\frac{\sigma bla^*}{kT}\right) \tag{18-3}$$

which for small stresses, $\sigma \ll kT/bla^*$, reduces to

$$v \cong 2a\nu_d \frac{\sigma bla^*}{kT}\exp\left(-\frac{W^*}{kT}\right) \tag{18-4}$$

For the case of Fig. 18-1, W^* is expected to be of the order of the activational free energy for core diffusion of the solute, but as discussed in Sec. 16-2, the exact value of W^* depends on stress and the reaction coordinate x, and hence on the details of the local activated configuration.

Other cases cannot be treated by analogy to previously developed theories, but they involve new derivations; these comprise most of the present chapter. We first consider the motion of dislocations which drag along Cottrell or Snoek solute atmospheres, and then we treat breakaway phenomena. We conclude with a brief discussion of some other phenomena, including the Mott-Nabarro solid-solution-hardening theory, Suzuki hardening, some effects of ordering, and atmospheres in ionic crystals.

Physically, the types of behavior involved in solute drag are easy to rationalize. In the presence of a dislocation potential of effective width r_1, *diffusion* of the solute atoms controls dislocation mobility, provided that the relaxation time r_1/v for the dislocation to move through a distance r_1 is greater than the relaxation time r_1^2/D for diffusion over such a distance. Otherwise *drift* of the solute in the potential field is rate controlling. The drag force on a dislocation is a maximum at about the transition velocity between diffusion and drift control. The detailed analysis of such processes is mathematically lengthy, however, because the asymptotic behavior must be carefully considered to prove that divergent behavior does not occur.

18-2. DIFFUSION ASSOCIATED WITH A MOVING POTENTIAL WELL

To set the stage for the analytic treatment of the drag of a solute atmosphere by a moving dislocation, we first treat the general problem of diffusion accompanying the motion of a potential well (or hill). The solution to the two-dimensional problem involves infinite series of Bessel and Mathieu functions and is quite complex.[1] The complexity arises from the $r^{-1}\sin\theta$ dependence of the interaction parameter, [Eq. (14-47)], driving the drift of the solute atoms. Therefore we treat only the simpler one-dimensional problem. The resulting solution is applied approximately to the two-dimensional dislocation problem in Sec. 18-3, giving valuable insight into the asymptotic behavior of dislocation motion.

Diffusion Solutions

Consider one-dimensional diffusion in a system where c is the number of solute particles per unit length. Let a potential well $W(x)$ be present, which can be translated along the line.[2] In the limit $x \to \infty$, where $W(x)=0$, $c=c_0$. When the well is at rest,

$$c = c_0 \exp\left[-\frac{W(x)}{kT}\right] \tag{18-5}$$

The net excess number of particles attracted by the well is

$$N = \int_{-\infty}^{\infty} (c - c_0)\, dx = c_0 \int_{-\infty}^{\infty} \left\{ \exp\left[-\frac{W(x)}{kT}\right] - 1 \right\} dx \tag{18-6}$$

With certain types of potential, the result of Eq. (18-6) *diverges* logarithmically. Specifically, consider the potential

$$W(x) = -\frac{A}{(x^2 + B^2)^{1/2}} \tag{18-7}$$

which resembles a dislocation type of potential in the limit. Asymptotically, in the limit $|x| \gg B$, Eq. (18-7) becomes

$$W(x) \cong -\frac{A}{|x|} \tag{18-8}$$

The asymptotic value of c [Eq. (18-5)] is then

$$c = c_0\left(1 + \frac{A}{kT}\frac{1}{|x|}\right) \tag{18-9}$$

[1] R. Fuentes-Samaniego, Ph.D. thesis, Stanford University, Stanford, Calif., 1979. Some inconsistencies in a similar treatment by A. H. Cottrell and M. A. Jaswon [*Proc. Roy. Soc.*, **A199:** 104 (1949)] are corrected in this work.

[2] The one-dimensional theory is directly applicable to the mobility of *interfaces*, an important problem in physical metallurgy; see J. W. Cahn, *Acta Met.*, **10:** 789 (1962); M. Hillert and B. Sundman, *Acta Met.*, **25:** 11 (1977).

With this value for c, Eq. (18-6) for N diverges *logarithmically*; there is no limit to the magnitude of the solute atmosphere attracted by the potential well.

Consider now the steady-state diffusion solution in a coordinate system moving with the well at a constant velocity v. For such a case the solution for N turns out to be convergent. The conservation equation, Fick's second law, in the moving frame of reference is

$$D\frac{\partial^2 c}{\partial x^2}+\frac{D}{kT}\frac{\partial}{\partial x}c\frac{\partial W}{\partial x}+v\frac{\partial c}{\partial x}=0 \tag{18-10}$$

The first term is well known, the second term is the divergence of the drift current cv_D [Eq. (15-3)], and the last term is the accumulation term $-\partial c/\partial t$ transformed to the moving coordinate system. The integration of Eq. (18-10) yields the result

$$D\frac{\partial c}{\partial x}+\frac{Dc}{kT}\frac{\partial W}{\partial x}+v(c-c_0)=0 \tag{18-11}$$

After division by c, the homogeneous differential equation corresponding to this expression is

$$\frac{D}{c}\frac{\partial c}{\partial x}+\frac{D}{kT}\frac{\partial W}{\partial x}+v=0 \tag{18-12}$$

which can be directly integrated to give

$$c=\exp\left(-\frac{W}{kT}-\frac{vx}{D}\right) \tag{18-13}$$

A particular solution to Eq. (18-11) is of the type

$$c=C\exp\left(-\frac{W}{kT}-\frac{vx}{D}\right) \tag{18-14}$$

Inserting this expression into Eq. (18-11), one finds

$$\frac{dC}{dx}=\frac{vc_0}{D}\exp\left(\frac{W}{kT}+\frac{vx}{D}\right) \quad \text{or} \quad C=\frac{vc_0}{D}\int_{-\infty}^{x}\exp\left(\frac{W}{kT}+\frac{vx}{D}\right)dx \tag{18-15}$$

Thus the complete solution is

$$c=\frac{vc_0}{D}\exp\left(-\frac{W}{kT}-\frac{vx}{D}\right)\int_{-\infty}^{x}\exp\left(\frac{W}{kT}+\frac{vx}{D}\right)dx \tag{18-16}$$

This result does satisfy the boundary condition[3] that $c=c_0$ at $x=\pm\infty$ and is thus the desired solution.

[3] In either limit, $W\to 0$ and the integral is elementary.

As an example, consider the potential of Eq. (18-8), with v small but finite. Consider first the case $x<0$. If the quantity exp (W/kT) is expanded to the second order, the integral in Eq. (18-16) becomes

$$\int_{-\infty}^{x} \exp\left(\frac{W}{kT}+\frac{vx}{D}\right) dx = \frac{D}{v}\exp\left(\frac{vx}{D}\right)+\frac{A}{kT}\int_{-\infty}^{x}\frac{1}{x}\exp\left(\frac{vx}{D}\right) dx$$
$$+\frac{A^2}{2k^2T^2}\int_{-\infty}^{x}\frac{1}{x^2}\exp\left(\frac{vx}{D}\right) dx \qquad (18\text{-}17)$$

The second term on the right-hand side of Eq. (18-17) is the exponential integral $Ei(vx/D)$. This integral can be developed into a series expansion by partial integration. The series is semiconvergent and accurately represents the integral for values of the argument vx/D greater than unity, i.e., when[4]

$$|x| \gg D/v \qquad (18\text{-}18)$$

The last term in Eq. (18-17) can be developed similarly. The result of the expansion, to the second order in $1/x$, is

$$\int_{-\infty}^{x} \exp\left(\frac{W}{kT}+\frac{vx}{D}\right) dx \cong \exp\left(\frac{vx}{D}\right)\left(\frac{D}{v}+\frac{AD}{vxkT}+\frac{AD^2}{v^2x^2kT}+\frac{A^2D}{2vx^2k^2T^2}\right)$$
$$(18\text{-}19)$$

Inserting this result into Eq. (18-16) and also expanding the term $\exp(-W/kT)$ to the second order in $1/x$, one obtains

$$c=c_0\left(1+\frac{DA}{vx^2kT}\right) \qquad (18\text{-}20)$$

A similar development for the case $x>0$ (see Exercise 18-1) gives the result

$$c=c_0\left(1-\frac{DA}{vx^2kT}\right) \qquad (18\text{-}21)$$

Thus

$$c=c_0+0\left(\frac{1}{x^2}\right) \qquad (18\text{-}22)$$

[4]See P. M. Morse and H. Feshbach, "Methods of Theoretical Physics," McGraw-Hill, New York, 1953, p. 434, for a discussion of the asymptotic behavior of this series, which is semiconvergent. As an example, the first step in the expansion yields the partial-integration result

$$\int_{-\infty}^{x}\frac{1}{x}\exp\left(\frac{vx}{D}\right) dx = \frac{D}{vx}\exp\left(\frac{vx}{D}\right)+\frac{D}{v}\int_{-\infty}^{x}\frac{1}{x^2}\exp\left(\frac{vx}{D}\right) dx$$

provided that $DA/vx^2kT \ll 1$ or

$$|x| > \left(\frac{DA}{vkT}\right)^{1/2} \tag{18-23}$$

Condition (18-23) defines the region in which solute redistribution by *drift* in the potential gradient can be neglected.

Thus the solute concentration is $c \sim c_0$ in the asymptotic region where both conditions (18-18) and (18-23) are satisfied; the question of which of these is more constraining depends on the specific conditions of a given situation. Usually, for dislocation applications Eq. (18-18) is the critical condition; that is, $D/v > (DA/vkT)^{1/2}$. Comparing the result of Eq. (18-22) with the asymptotic behavior of the static solution, Eq. (18-9), one sees that the logarithmic divergence in N *disappears* in the case of a moving potential well.

Now let us turn to the concentration in the region $|x| < D/v$. With v sufficiently small, then $\exp(W/kT) \sim 1$ over most of the interval contributing to the integral in Eq. (18-16). Explicitly, the integral is given by

$$\int_{-\infty}^{x} \exp\left(\frac{W}{kT}\right)\exp\left(\frac{vx}{D}\right)dx \cong \int_{-\infty}^{x} \exp\left(\frac{vx}{D}\right)dx = \frac{D}{v}\exp\left(\frac{vx}{D}\right) \tag{18-24}$$

With the result of Eq. (18-24), the concentration given by Eq. (18-16) is simply

$$c \cong c_0 \exp\left(-\frac{W}{kT}\right) \tag{18-25}$$

identical to the static solution. Thus, provided that the velocity is small enough that the interval D/v is large compared to the range of the potential, the concentration in most of the region in the moving reference frame is given by Eqs. (18-22) and (18-25). Near the origin, where the potential is such that $W \gg kT$, the condition on how small the velocity must be for Eq. (18-25) to describe the concentration may be more restrictive, but for sufficiently small v, Eq. (18-25) holds everywhere; this region near the origin is discussed in the following sections. The excepted cases must be treated by numerical integration.

Exercise 18-1. Derive Eq. (18-21). *Hint*: For the case $x > 0$, the major contribution to the integral in Eq. (18-16) involves the range of x near the upper limit x. Thus $\exp(W/kT)$ can be developed in a series expansion in terms of the variable x' about the upper limit x,

$$\exp\left(\frac{W}{kT}\right) = \exp\left(\frac{-A}{x'kT}\right) = \exp\left(\frac{-A}{xkT}\right) + \frac{A}{x^2kT}\exp\left(\frac{-A}{xkT}\right)(x'-x) + \cdots$$

and the resultant integral can be evaluated term by term. The fact that Eq. (18-21) is antisymmetrical to Eq. (18-20) is a coincidence; to higher order, the concentration distributions in the ranges $x > 0$ and $x < 0$ are unsymmetrical.

Exercise 18-2. Develop the concentration dependence from Eq. (18-16) for the case of the slowly convergent potential

$$W = -\frac{\alpha}{(x^2+\beta^2)^{1/4}} \tag{18-26}$$

Show that $c \sim c_0$ for $|x| > D/v$, and that for small v and $x<0$,

$$c = c_0\left\{1 + \frac{\alpha}{kT\sqrt{-x}} - \frac{2\alpha}{kT}\left(\frac{v}{D}\right)^{1/2}\left[\frac{1}{2}\sqrt{\pi} - \left(-\frac{vx}{D}\right)^{1/2} + \cdots\right]\right\} \tag{18-27}$$

which reduces to the static solution when $-x < D/\pi v$.

Square-Well Potential

In the above considerations of asymptotic behavior, the treatment of the concentration in the region in the range of the potential is omitted. As an illustration of the physical effects in this range, let us consider the moving, square, repulsive potential well.[5] This can be thought of as a limiting case of a sloping potential.

At rest, the concentration profile is that of Fig. 18-2. When the well is moving at velocity v, the integral result of Eq. (18-13) indicates that concentrations are present with a functional dependence $\exp(-vx/D)$. Equation (18-16) further indicates that to the left of the potential well, $c = c_0$. This result is readily rationalized physically by a consideration of the steady state, since $\exp(vx/D)$ increases catastrophically for $x \to -\infty$. With $vl/D \ll 1$, the gradient over the potential well is constant, see. Fig. 18-3, where the parameters c_1, c_2, and l are defined. The condition of continuity of mass at the potential well boundaries then yields the result for the gradient

$$D\frac{\partial c}{\partial x} = (c_0 - c_1)v \tag{18-28}$$

The lower concentration at B is then $c_1 + \Delta c$, where

$$\Delta c = \frac{(c_0 - c_1)vl}{D} \tag{18-29}$$

The requirement for quasi-equilibrium of the concentrations on either side of the potential drops leads to the condition

$$\frac{c_2}{c_1 + \Delta c} = \frac{c_0}{c_1} \tag{18-30}$$

[5] This example is theoretically analogous to problems of steady-state crystal growth in a binary system; see W. A. Tiller, in "Liquid Metals and Solidification," American Society of Metals, Cleveland, Ohio, 1958, p. 276.

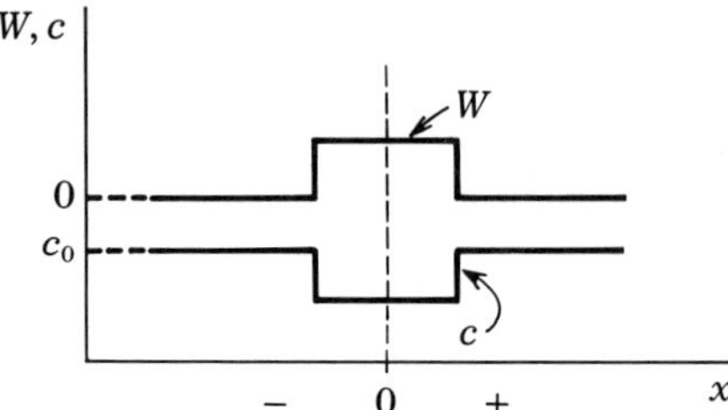

FIGURE 18-2. Solute concentration c versus distance in the case of a square-well potential.

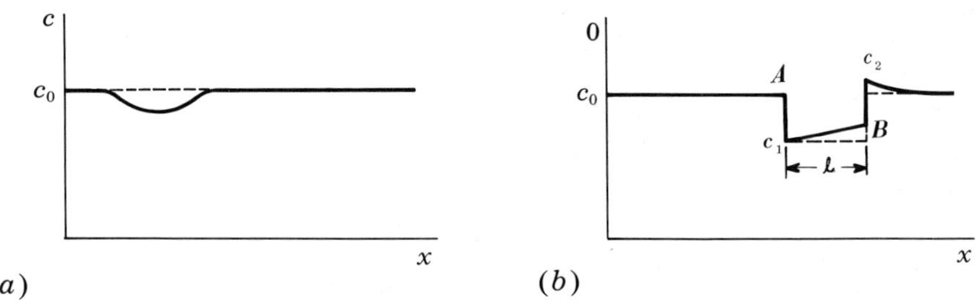

FIGURE 18-3. Square-well-concentration profiles: (*a*) at the origin after the potential well has moved away, (*b*) at the moving potential well in steady state.

so that

$$c_2 - c_0 = \frac{c_0}{c_1}\Delta c \tag{18-31}$$

Finally, this superconcentration decays as $\exp(-vx/D)$ to the right of the well, giving an extra number of impurities to the right of the well,

$$\frac{c_0}{c_1}\Delta c \int_0^\infty \exp\left(-\frac{vx}{D}\right) dx = \frac{c_0 D \Delta c}{c_1 v} = \frac{c_0 l}{c_1}(c_0 - c_1) \tag{18-32}$$

Because $\sim(c_0 - c_1)l$ impurities are excluded by the potential itself, the net excess number of impurities in the moving frame is

$$N = \frac{(c_0 - c_1)^2 l}{c_1} \tag{18-33}$$

for small v, that is, when

$$v \ll \frac{2D(c_0 - c_1)}{c_1 l} \tag{18-34}$$

For large v compared with condition (18-34) the excess impurities in the well would have to be considered. Also for large v, the gradient in the well would no longer be linear.

The corresponding concentration profile in the moving reference frame is shown in Fig. 18-3*b*. One might ask how this steady-state excess N arises. As the well starts from rest and slowly accelerates to velocity v, a deficit of impurities is left behind at the origin, as shown in Fig. 18-3*a*. This "tail" does not interact with the potential once it is left behind, and thus does not enter into the above solution. The asymmetry in the concentration in the vicinity of the potential well is evident in Fig. 18-3.

Forces in Moving Potential Fields

According to the Einstein mobility relation, Eq. (15-3), the force required to move the N excess atoms associated with a potential well at velocity v is

$$F=\frac{kTv}{D}N \tag{18-35}$$

which is also the result obtained from Eq. (18-11) by direct integration of the expression

$$F=-\int_{-\infty}^{\infty} c\frac{\partial W}{\partial x}dx \tag{18-36}$$

As discussed in the previous sections, one cannot use the static result for N in Eq. (18-35), for logarithmic divergences would appear. Also, the use of the static result for the case of a repulsive potential, where $N<0$, would lead to the absurd result that F is negative for a positive v. *One must use the result for N determined in the moving frame of reference in force considerations*. Thus, for the case of the moving square potential well, N is given by Eq. (18-33), so that

$$F=\frac{kTvl}{D}\frac{(c_0-c_1)^2}{c_1} \tag{18-37}$$

In general, the complex asymptotic behavior of c must be considered in the moving frame of reference, as illustrated in the treatment of Eq. (18-16). However, much of the inconvenience of such calculations is removed by the development of a *dissipation theorem*. This development also removes the ambiguities associated with repulsive potentials discussed above.

The dissipation theorem is derived as follows. Multiply Eq. (18-36) through by v and insert $c-c_0$ in place of c; this does not change the equality, because

$$\int_{-\infty}^{\infty} c_0\frac{\partial W}{\partial x}dx=c_0\int_{-\infty}^{\infty}\frac{\partial W}{\partial x}dx=c_oW\Big|_{-\infty}^{\infty}=0$$

The result is

$$Fv=-\int_{-\infty}^{\infty} v(c-c_0)\frac{\partial W}{\partial x}dx \tag{18-38}$$

Equation (18-11) can be written simply as

$$J = v(c - c_0) \tag{18-39}$$

where J is the steady-state atom flux, including both diffusion and drift terms. Substituting for $\partial W/\partial x$ in Eq. (18-38) from Eq. (18-11), one obtains

$$Fv = \frac{kT}{D}\int_{-\infty}^{\infty} \frac{v^2(c - c_0)^2}{c}\,dx + kT\int_{-\infty}^{\infty} \frac{v(c - c_0)}{c}\frac{\partial c}{\partial x}\,dx \tag{18-40}$$

The last term in Eq. (18-40) vanishes, because

$$\int_{-\infty}^{\infty} \left(v\frac{\partial c}{\partial x} - vc_0 \frac{\partial \ln c}{\partial x} \right) dx = (vc - vc_0 \ln c)\Big|_{-\infty}^{\infty} = 0$$

Thus Eqs. (18-39) and (18-40) combine to yield

$$Fv = \frac{kT}{D}\int_{-\infty}^{\infty} \frac{J^2}{c}\,dx \tag{18-41}$$

On the left side of Eq. (18-41) is the work done on the system per unit time, while the right side represents the rate of dissipation of free energy associated with the diffusion currents[6]; Eq. (18-41) is the desired dissipation expression.

Equation (18-41) is most convenient; in this expression one can assume for dislocation problems that, to first order, the concentration in the moving frame of reference is the *same as at rest*, $c(v) = c(0)$, giving

$$J \cong v[c(0) - c_0] \tag{18-42}$$

and

$$F \cong \frac{kTv}{D}\int_{-\infty}^{\infty} \frac{[c(0) - c_0]^2}{c(0)}\,dx \tag{18-43}$$

[6]The diffusion flux in an element δx is related to the chemical potential gradient $\partial\bar{G}/\partial x$ by the Einstein relation

$$J = -\frac{Dc}{kT}\frac{\partial \bar{G}}{\partial x}$$

The rate of dissipation of free energy in the element is

$$-J\frac{\partial \bar{G}}{\partial x}\delta x = \frac{J^2 kT}{Dc}\delta x$$

after substituting from the first expression. Equation (18-41) then follows upon integration.

Small currents proportional to v^2 which flow over regions D/v *do not contribute* to the second order in v in Eq. (18-43). The details of the diffusion problem in the moving coordinate system need not be treated; Eq. (18-43) contains only equilibrium concentrations.

In the example of the square-well potential, Eq. (18-43) directly gives the result of Eq. (18-37). The complex contribution of the concentration excess accumulated ahead of the moving well is automatically and simply, accounted for by substituting the rest concentration $c(0)=c_1$ into Eq. (18-43). This example graphically illustrates the usefulness of the dissipation theorem.

For the potential of Eq. (18-7), which gives divergent behavior for N in the static case, Eq. (18-43) gives the result

$$F=\frac{kTvc_0}{D}\int_{-\infty}^{\infty} 4\sinh^2\frac{A}{2kT(x^2+B^2)^{1/2}}dx \tag{18-44}$$

which is *convergent*. Equation (18-44) in general must be integrated numerically. However, in the limit $A \ll 2kT(x^2+B^2)^{1/2}$, an expansion of the sinh function yields the analytical solution

$$F=\frac{vc_0}{DkT}\int_{-\infty}^{\infty}\frac{A^2}{x^2+B^2}dx=\frac{\pi vc_0A^2}{DBkT} \tag{18-45}$$

For the more slowly convergent potential of Eq. (18-26), the result of Eq. (18-43) would diverge, so that this case must be handled asymptotically, as in Exercise 18-2, to show that convergence occurs in the moving system. In any case, the result of Eq. (18-41) clearly extends the range of slowly convergent potentials that can be handled in the static approximation.

Exercise 18-3. Extend Eq. (18-41) to the three-dimensional case and prove that

$$\mathbf{F}\cdot\mathbf{v}=\frac{kT}{D}\int_v\frac{\mathbf{J}\cdot\mathbf{J}}{c}dV \tag{18-46}$$

where V is the volume.

Exercise 18-4. In estimating the force required to maintain motion, the currents in the regions $c\sim c_0$, $|x|>D/v$, are usually neglected. Show that these currents contribute negligibly to F (to the order $1/x^2$) for the result of the potential of Eq. (18-45).

Outline of solution. In the regions $c\sim c_0$, where there is no gradient, the currents must be purely drift currents

$$J=v(c-c_0)\cong-\frac{D}{kT}c_0\frac{\partial W}{\partial x}=\pm\frac{Dc_0A}{kTx^2}$$

This result can be inserted into Eq. (18-41) and integrated from D/v to ∞ and $-D/v$ to $-\infty$. Note that since $J=v(c-c_0)$, c is given asymptotically by

$$c=c_0\left(1\pm\frac{DA}{vkTx^2}\right)$$

in agreement with Eq. (18-22), as must be the case for self-consistency.

18-3. DRAG OF A COTTRELL ATMOSPHERE BY A DISLOCATION

Interaction Forces

The results of Sec. 14-5 indicate that an edge dislocation has a so-called Cottrell atmosphere of solute atoms described by the relation

$$c=c_0\exp\left(-\frac{\beta\sin\theta}{rkT}\right) \tag{18-47}$$

where c_0 is the concentration at $r=\infty$ and β is given by Eq. (14-47). At equilibrium, the region on the compression side of the dislocation is diluted in solute for $\beta>0$, while the tension side is enriched in solute, as shown in Fig. 18-4. If the solute atmosphere is frozen in while the dislocation is displaced through a distance x by glide (Fig. 18-5), the total interaction energy $W(x)$ between the dislocation and the impurity atoms is raised. In the displaced configuration, more impurities are in the region A' of high-impurity energy, and fewer in the region B' of low-impurity energy than in the case of the equilibrium distribution. An external shear stress

$$\sigma b=\frac{F}{L}=\frac{\partial}{\partial x}\frac{W(x)}{L} \tag{18-48}$$

is required to keep the dislocation displaced by x (Fig. 18-5). The maximum force F_{max}/L occurs at r_{max}, of the order of several b.

For this case, the net force exerted on the dislocation by the impurity atoms is equal and opposite to that exerted on the impurity atoms by the dislocation.

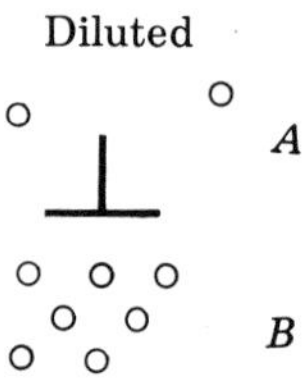

FIGURE 18-4. Adsorbed solute atmosphere about an edge dislocation in the case that the size of the solute atom exceeds that of the solvent.

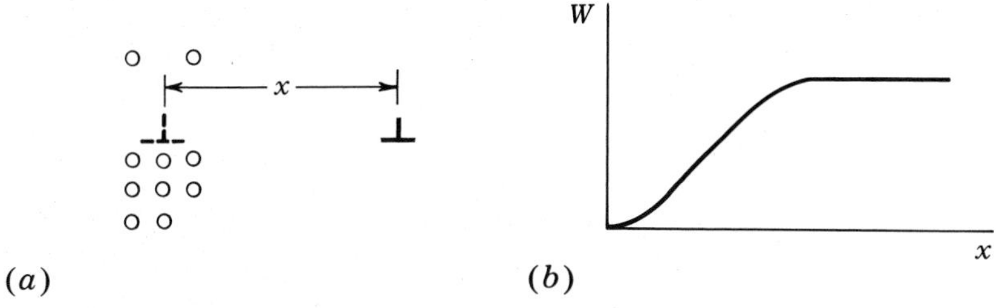

FIGURE 18-5. (*a*) A dislocation displaced from its solute atmosphere, and (*b*) the corresponding energy-displacement curve.

To illustrate this point, consider an edge dislocation and one impurity atom at positions $\mathbf{r}_d$ and $\mathbf{r}_i$, respectively. The interaction energy depends on the relative coordinate $\mathbf{r}_i - \mathbf{r}_d$,

$$W_{\text{int}} = W(\mathbf{r}_i - \mathbf{r}_d) \tag{18-49}$$

The x component of the force on the impurity is by definition

$$F_{i_x} = -\frac{\partial}{\partial x_i} W_{\text{int}}$$

while the x component of the force on the dislocation is

$$F_{d_x} = \frac{\partial W_{\text{int}}}{\partial x_d}$$

Since only the relative coordinate enters Eq. (18-49), the force components are equal and opposite,

$$F_{i_x} = -F_{d_x} \tag{18-50}$$

and similarly for the other components. W_{int} is derived in Sec. 14-5 by considering the work done against the hydrostatic component of the dislocation stresses in making space for the impurity. The interaction energy could equally well be derived from the work done in moving the dislocation in its glide plane against the shear stresses caused by the solute atom.

For external forces less than F_{max}/L, the dislocation cannot leave the solute atmosphere without thermal activation. For smaller stresses the possibility of thermal activation can be ignored. The dislocation can then move only by dragging the solute atmosphere in a motion controlled by diffusion of the solute atoms. Because of the equivalency of interaction forces, Eq. (18-50), the force on the dislocation can be determined by the method of Sec. 18-2.

Such solute drag, important in creep at temperatures where solute diffusion is not too slow, is treated below. Atoms in the core region, which are not describable by Eq. (18-47), are excluded throughout the discussion. This is a reasonable approximation, because solute atoms in the core region should be able to move easily with the dislocation by core diffusion when the motion is controlled by *bulk* diffusion of the solute.

The Diffusion Solution

The diffusion problem associated with a moving dislocation is two dimensional. The solution to the two-dimensional form of Eqs. (18-11) and (18-12) is extremely complicated for the case of the edge dislocation. Cottrell and Jaswon[7] have solved the equations partly analytically and partly numerically. However, they cut off their calculations arbitrarily when the concentration approached c_0. As a consequence, they missed a logarithmically divergent term in the force on a dislocation, and therefore predicted a constant mobility at low velocities, which is so only under special circumstances. Instead of repeating the complex two-dimensional analysis here, we present an approximate treatment, based on the one-dimensional analysis of Sec. 18-2, which indicates the form of the solution for the asymptotic regions of low and high velocity. In the intermediate-velocity range, we rely on the results of Fuentes-Samaniego.[8] In cartesian coordinates the potential of Eq. (18-47) assumes the form

$$W=\frac{\beta y}{x^2+y^2} \tag{18-51}$$

Assume temporarily that solute atoms can diffuse only in the x direction as the edge dislocation moves. The problem is then one dimensional, of the type treated in Sec. 18-2.

Consider diffusion in the strip between y and $y+\delta y$ (Fig 18-6). Equation (18-43) in this case is of the form of Eq. (18-44). For a sufficiently large y that

$$\exp\left(-\frac{W}{kT}\right)=1-\frac{W}{kT} \tag{18-52}$$

the force expression is

$$\frac{\delta F}{L}=\frac{vc_o\beta^2\delta y}{DkT}\int_{-\infty}^{\infty}\frac{y^2dx}{(x^2+y^2)}=\frac{\pi vc_o\beta^2\delta y}{2DkTy} \tag{18-53}$$

for the dragging force produced by the impurities in the strip. Next, integrating over y one finds that the total dragging force is

$$\frac{F}{L}=\frac{\pi vc_o\beta^2}{2DkT}\ln\frac{y_2}{y_1} \tag{18-54}$$

[7]A. H. Cottrell and M. A. Jaswon, *Proc. Roy. Soc.*, **A199:** 104 (1949).

[8]R. Fuentes-Samaniego, Ph.D. thesis, Stanford University, Stanford, Calif., 1979.

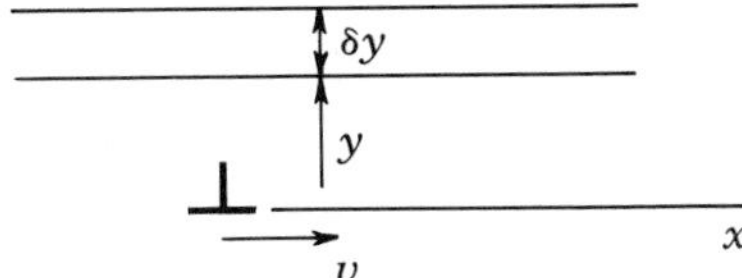

FIGURE 18-6. Dislocation in moving coordinate system.

This expression *does not converge* for $y_2 \to \infty$. However, the analogy with the one-dimensional case of Sec. 18-2 suggests that for $y \gg D/v$ the concentration becomes uniform, $c \sim c_0$, so that $y_2 \doteq D/v$ is a reasonable upper limit. Adding an identical contribution from the region $-y_1 > y > -y_2$, one therefore obtains the result

$$\frac{F}{L} = \frac{\pi v c_0 \beta^2}{DkT} \ln \frac{D}{v y_1} \tag{18-55}$$

Equation (18-55) is not linear in v as $v \to 0$ because of the logarithmic singularity in v. *The mobility of a single edge dislocation in an infinite medium containing diffusing solute atoms does not approach a constant value in the limit of small velocities*, contrary to the result of Cottrell and Jaswon.[9]

One might suspect that this anomalous behavior would disappear when the restriction of one-dimensional diffusion, introduced for mathematical convenience, is removed. However, the following reasoning indicates that Eq. (18-55) is probably correct. As the dislocation in Fig. 18-7 moves, solute atoms must be transported from the regions A, A', A'', and A''' to the regions B, B', B'', and B'''. Qualitatively, one expects currents J_x to predominate in quadrants II and IV and currents J_y in quadrants I and III, with the magnitude of the currents as a function of distance r from the dislocation the same for both components of J. Since only the *magnitude* of J enters the general expression for F/L [Eq. (18-46)], the estimate in Eq. (18-55) is reasonable.

To further demonstrate the above conclusion, suppose that diffusion in only the y direction occurs. The equation of continuity then is

$$\frac{\partial J_y}{\partial y} = -v \frac{\partial (c - c_0)}{\partial x} = \frac{v c_0 \beta}{kT} \frac{\partial}{\partial x} \frac{y}{x^2 + y^2}$$

$$= \frac{v c_0 \beta}{kT} \frac{\partial}{\partial y} \frac{x}{x^2 + y^2}$$

so that

$$J_y = \frac{v c_0 \beta x}{kT(x^2 + y^2)} \tag{18-56}$$

[9]A. H. Cottrell and M. A. Jaswon, *Proc. Roy. Soc.*, **A199:** 104 (1949); the above result is confirmed in the exact solution by R. Fuentes-Samaniego, Ph.D. thesis, Stanford University, Stanford, Calif., 1979.

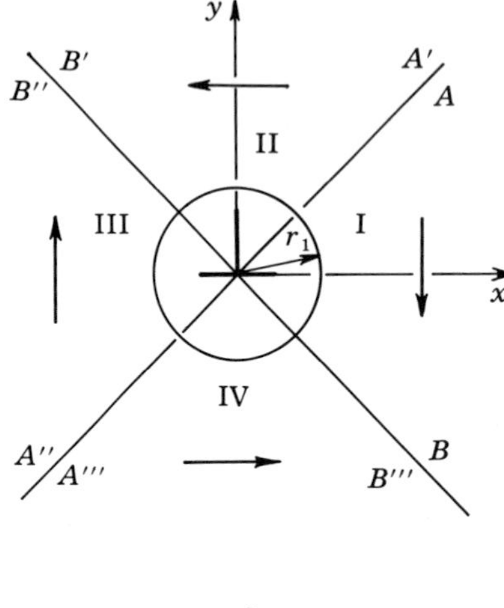

FIGURE 18-7. Large arrows indicate the directions of solute transport as the dislocation moves.

Inserting this relation into Eq. (18-46) and integrating from $y = -\infty$ to ∞ and then from $x = y_1$ to $x = D/v$, one obtains identically the same result as Eq. (18-55). Because the same logarithmic singularity occurs whether the approximation of diffusion in only the x direction or in only the y direction is imposed, the singular behavior is not expected to disappear in the case of isotropic diffusion.

In view of the above arguments, the two-dimensional diffusion case can be reasonably approximated by the integration of Eq. (18-46) over the hatched area in Fig. 18-8, with only currents J_x in quadrants II and IV and only currents J_y in quadrants I and III. Continuity at the quadrant boundaries requires that J_x and J_y vanish there. Thus

$$J_x = -\frac{vc_0\beta}{kT}\left(\frac{y}{x^2+y^2} - \frac{1}{2y}\right) \qquad |x|<|y|$$

$$J_y = \frac{vc_0\beta}{kT}\left(\frac{x}{x^2+y^2} - \frac{1}{2x}\right) \qquad |x|>|y| \tag{18-57}$$

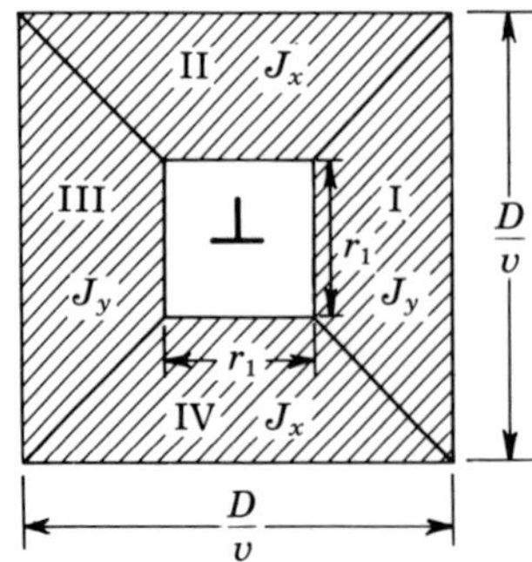

FIGURE 18-8. Regions of integration for Eq. (18-57).

The contribution of quadrant II is

$$\frac{F_{\mathrm{II}}}{L}=\frac{vc_0\beta^2}{DkT}\int_{r_1}^{D/v}dy\int_{-y}^{y}\left(\frac{y}{x^2+y^2}-\frac{1}{2y}\right)^2dx$$

$$=\frac{vc_0\beta^2}{DkT}\,\frac{4-\pi}{4}\ln\frac{D}{vr_1}$$

Multiplying by 4, because all quadrants contribute equally, one finds for the total drag force[10]

$$\frac{F}{L}\cong\frac{vc_0\beta^2}{DkT}\ln\frac{D}{vr_1} \tag{18-58}$$

For simplicity, Eq. (18-58) is also considered to be the estimate of the force per unit length caused by solute atoms outside a *cylinder* of radius r_1. The cutoff distance r_1 is chosen to be the value of r beyond which the expansion (18-52) is valid,

$$r_1=\frac{\beta}{kT} \tag{18-59}$$

Throughout this section we have assumed that $c\sim c_0$ in the denominator of Eq. (18-46) and that Eq. (18-52) is valid; both assumptions break down for $r<r_1$.

For velocities so small that D/v is larger than the specimen dimension R or the distance R between dislocations in the crystal, D/v is not a meaningful outer cutoff and must be replaced by the appropriate distance R,

$$\frac{F}{L}\cong\frac{vc_0\beta^2}{DkT}\ln\frac{R}{r_1} \tag{18-60}$$

In this case the mobility is constant for small velocities, as suggested by Cottrell and Jaswon.[11] If r_1 in Eq. (18-59) is smaller than the core cutoff radius r_0, r_1 must be replaced by r_0 in Eq. (18-58) or Eq. (18-60). The effect of solute atoms outside the core is then completely included in the drag force.

Forces in the Range of the Dislocation Potential

In the case that $r_1>r_0$, solute atoms between the cylinders defined by r_0 and r_1 must be considered. According to the definition (18-59), this is the region where the Boltzmann factor $\exp(-W/kT)$ is much different from unity,

[10]In the exact treatment (see reference 9, above), a factor $\pi/2$ appears in the prefactor in Eq. (18-58).

[11]A. H. Cottrell and M. A. Jaswon, *Proc. Roy. Soc.*, **A199:** 104 (1949).

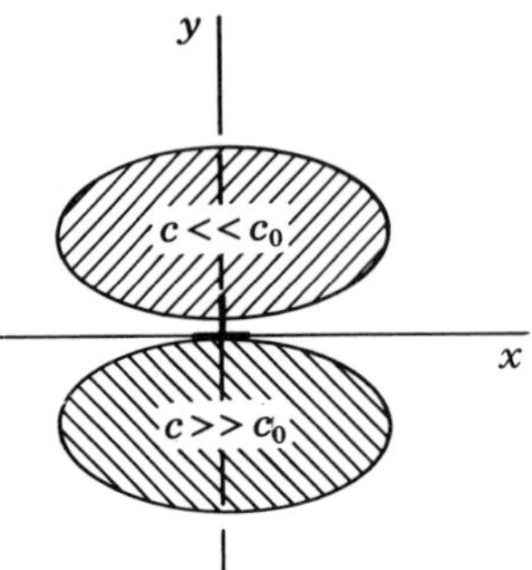

FIGURE 18-9. Regions of solute aggregation and depletion.

except for a narrow region adjoining the x axis (Fig. 18-9). In the region of depletion,

$$c = c_0 \exp\left(-\frac{W}{kT}\right) \ll c_0$$

so that

$$\frac{[c(0)-c_0]^2}{c(0)} \cong \frac{c_0^{\,2}}{c_0 \exp(-W/kT)} = c_0 \exp\left(\frac{W}{kT}\right)$$

The region of depletion thus contributes to the force in the amount

$$\frac{F}{L} \cong \frac{kTv}{D} \int\int c_0 \exp\left(\frac{W}{kT}\right) dx\, dy \tag{18-61}$$

analogous to Eq. (18-53). In the region of solute enhancement,

$$c = c_0 \exp\left(\frac{W}{kT}\right) \gg c_0$$

so that

$$\frac{[c(0)-c_0]^2}{c(0)} \cong \frac{c_0^{\,2} \exp(2W/kT)}{c_0 \exp(W/kT)} = c_0 \exp\left(\frac{W}{kT}\right)$$

and Eq. (18-61) again results. The regions of solute depletion and enhancement contribute *equally* to the drag force.

In analogy with the development of the preceding section for the total drag force, Eq. (18-61) is evaluated for the region of depletion, and the result is multiplied by 2, to include the effect of the region of enhancement, and divided by 3, to allow for the increase in mobility if J_y currents were included. Equation (18-61) can be solved more easily in the coordinates of Fig. 18-10. In

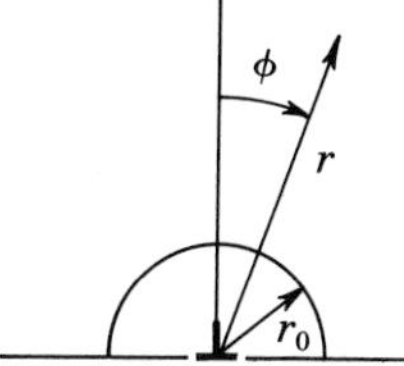

FIGURE 18-10. Coordinates for the integral in Eq. (18-63).

these coordinates the dislocation potentials is

$$W = \frac{\beta \cos \phi}{r} \tag{18-62}$$

and the total drag force is two-thirds of the force from the region of depletion, or

$$\frac{F}{L} = \frac{2kTvc_0}{3D} \int_{r_0}^{r_1} r\,dr \int_{-\pi/2}^{\pi/2} \exp\left(\frac{\beta \cos \phi}{rkT}\right) d\phi \tag{18-63}$$

Because only the region $\phi \cong 0$ contributes strongly, the cosine factor can be expanded to $\cos \phi = 1 - \frac{1}{2}\phi^2$ and the limits extended to $\pm\infty$,

$$\frac{F}{L} = \frac{2kTvc_0}{3D} \int_{r_0}^{r_1} r \exp\left(\frac{\beta}{rkT}\right) dr \int_{-\infty}^{\infty} \exp\left(-\frac{\beta\phi^2}{2rkT}\right) d\phi$$

$$= \frac{2kTvc_0}{3D} \left(\frac{2\pi kT}{\beta}\right)^{1/2} \int_{r_0}^{r_1} r^{3/2} \exp\left(\frac{\beta}{rkT}\right) dr$$

Introducing the change of variable $z = \beta / rkT$, one finally obtains

$$\frac{F}{L} = \frac{vc_0\beta^2}{DkT} I(z_0) \tag{18-64}$$

where

$$I(z_0) = \int_1^{z_0} \frac{2\sqrt{2\pi}}{3} z^{-7/2} e^z \, dz \tag{18-65}$$

and

$$z_0 = \frac{\beta}{kTr_0} = \frac{W_{max}}{kT} \tag{18-66}$$

W_{max} is the maximum binding energy to the dislocation on the side of solute enhancement. The integrand in Eq. (18-65) has a minimum at $z = 3.5$ and rises rapidly above $z = 7$, as shown in Fig. 18-11. As a rough estimate, the integrand

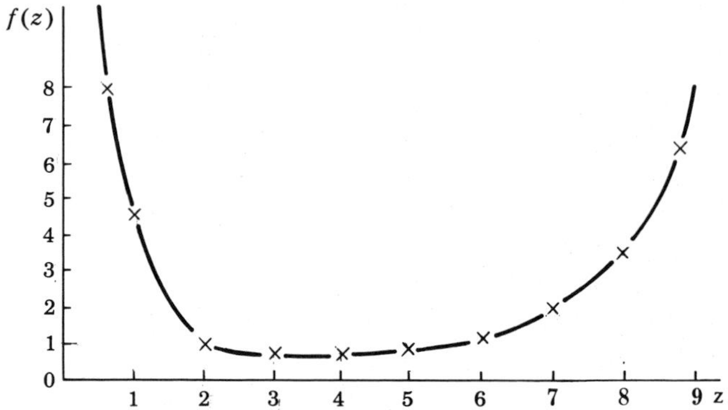

FIGURE 18-11. Integrand of Eq. (18-65), $f(z)$, versus z.

is assumed to have the average value ~ 1.5 throughout the interval $1 < z < 7$, so that

$$I(z_0) \cong 1.5(z_0 - 1) \quad 1 < z_0 < 7 \tag{18-67}$$

For $z_0 \gg 7$, the asymptotic formula

$$I(z_0) = \frac{2\sqrt{2\pi}}{3} \frac{z_0^{-7/2} e^{z_0}}{1 - (7/2z_0)} \cong \frac{2\sqrt{2\pi}}{3} z_0^{-7/2} e^{z_0} \tag{18-68}$$

applies.

Because the potential [Eq. (18-62)] is highly singular at the core, there is a large uncertainty associated with the core cutoff in z_0 [Eq. (18-66)]. Since the asymptotic formula (18-68) receives its main contribution from the region adjacent to the core, the currents may be quite different from the estimate of Eq. (18-64). Another obvious source of uncertainty for large z_0 involves the variation of D in the core region. In summary, the contribution of Eq. (18-64) is more uncertain than that of Eqs. (18-58) or (18-60), particularly for large z_0.

A comparison of the magnitudes of the two contributions to F/L is useful at this juncture. Creep motion controlled by solute diffusion occurs at appreciable rates typically at the temperature where $kT \sim 0.05$ eV. With a maximum binding energy $W_{\max} \sim 0.1$ eV, typical of a substitutional solute, $z_0 = 2$. The solute atoms in the inner region $r < r_1$ then cause a drag force

$$\frac{F}{L} = \frac{1.5 v c_0 \beta^2}{DkT} \tag{18-69}$$

The solute in the outer region, with the reasonable value $R \sim 10^4 r_0$, causes a drag force

$$\frac{F}{L} = \frac{8.5 v c_0 \beta^2}{DkT} \tag{18-70}$$

which is larger but of the same order of magnitude.

Let us now determine typical values of D/v for such creep data. With the above parameters, and with the additional typical values $F/L = 10^{-5}\mu b$, $c_0 = 0.1b^{-3}$, and $\beta = (3\times 10^{-2})\mu b^4$, Eqs. (18-69) and (18-70) yield the result

$$\frac{D}{v} \cong \frac{10c_0\beta^2}{kT(F/L)} = (6\times 10^3)b$$

which is about the same as the initially assumed values of R. Thus for this case the use of Eq. (18-60) is appropriate. On the other hand, if $c_0 = 0.01b^{-3}$, $D/v = (6\times 10^2)b$ and Eq. (18-58) should then be used for self-consistency. These results show that Eq. (18-58) is important and indicate that solute atoms quite far away from the core contribute to F/L.

All in all, the above results indicate that the uncertain contribution of the solute near the core is generally negligible compared to the outer contribution. The fact that the "core" D in Eq. (18-69) can be orders of magnitude larger than the D in Eq. (18-70) further supports this conclusion. Thus Eq. (18-58) or (18-60) should approximately describe creep under solute drag forces, and the temperature dependence of the mobility should be about that of the solute diffusivity D. Expressed in terms of the creep velocity v, these equations are

$$v = \frac{(F/L)DkT}{c_0\beta^2\ln(y_2/r_1)} \tag{18-71}$$

with y_2 equal to either D/v or R. The expression derived by Cottrell and Jaswon[11] is similar to Eq. (18-71), differing in that a numerical factor ~ 4 replaces $\ln(y_2/r_1)$. The agreement with the exact treatment is even closer as discussed in connection with Eq. (18-58). There is some experimental evidence that an expression of the form of Eq. (18-71) describes alloy creep.[12]

High-Velocity Creep

At high velocities severe deformation of the solute cloud takes place, and the preceding development is not valid. The region around the dislocation where $c \neq c_0$ decreases in extent with increasing v. Beyond some critical velocity v_c, the decrease is so rapid that the drag stress itself decreases with increasing v. Figure 18-12 illustrates the velocity dependence of the concentration and the stress.

In the velocity range $v > v_c$, the extension of the atmosphere is drift controlled rather than diffusion controlled. To illustrate this, reconsider the one-dimensional concentration expression, Eq. (18-16). In the range of the potential, for sufficiently large v for a given upper limit x, the factor $\exp(vx/D)$ increases so rapidly with x that the factor $\exp(W/kT)$ can be set equal to its value at the upper limit and removed from the integrand, canceling the factor

[12] J. Weertman, *Trans. Met. Soc. AIME*, **218:** 207 (1960).

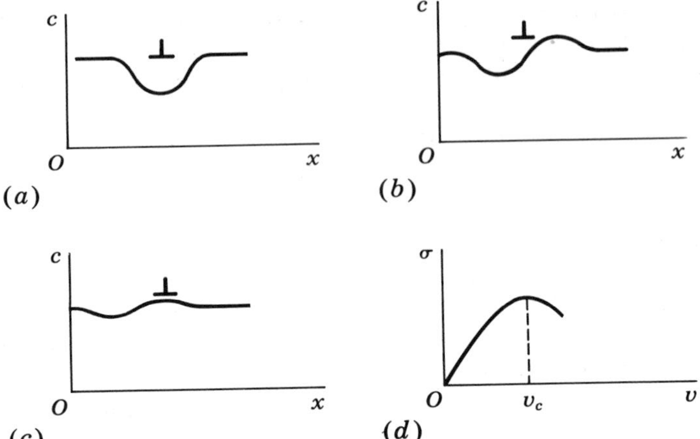

FIGURE 18-12. Concentration distance curves for (*a*) $v \ll v_c$, (*b*) $v \sim v_c$, and (*c*) $v \sim 2v_c$. Dislocation position is also depicted. (*d*) The corresponding σ-v plot.

$\exp(-W/kT)$. The integral is thus

$$c \cong \frac{vc_0}{D} \exp\left(-\frac{vx}{D}\right) \int_{-\infty}^{x} \exp\left(\frac{vx}{D}\right) dx = c_0 \tag{18-72}$$

within the range of the potential, $W > kT$. For the potential $W = -A/x$, the rate of change of the integrand is

$$\frac{\partial}{\partial x} \exp\left(-\frac{A}{xkT}\right) \exp\left(\frac{vx}{D}\right) = \left(\frac{A}{x^2kT} + \frac{v}{D}\right) \exp\left(-\frac{A}{xkT}\right) \exp\left(\frac{vx}{D}\right)$$

Thus the condition for which $c \cong c_0$ is

$$\frac{A}{x^2kT} < \frac{v}{D} \tag{18-73}$$

Indeed, for high enough velocities, $c \sim c_0$ and the motion is drift controlled.

For the dislocation potential, β/r replaces A/x, and the inner radius of the drift region, corresponding to Eq. (18-73), is

$$r_2 = \left(\frac{D\beta}{vkT}\right)^{1/2} \tag{18-74}$$

Drift control occurs when r_2 is less than the radius of the static cloud $r_1 = \beta/kT$ [Eq. (18-59)]. Thus the velocity for which transition from diffusion to drift control occurs, i.e., where $r_1 = r_2$, is

$$v_t = \frac{DkT}{\beta} \tag{18-75}$$

In the drift region, $c=c_0$ and currents

$$J_r = -\frac{c_0 D\beta \sin\theta}{kTr^2}$$

$$J_\theta = \frac{c_0 D\beta \cos\theta}{kTr^2} \tag{18-76}$$

exist in the strain field of the dislocation. Hence

$$\mathbf{J}\cdot\mathbf{J} = \frac{c_0{}^2 D^2 \beta^2}{k^2T^2r^4} \tag{18-77}$$

Such currents are present in the region $r > r_2$ for high velocities. In such a case the dissipation theorem, Eq. (18-46), yields the contribution to F/L from the outer region,

$$\frac{F}{L} = \frac{kT}{Dv}\int_{r_2}^{\infty} \frac{\mathbf{J}\cdot\mathbf{J}}{c_0} 2\pi r\, dr$$

$$= \frac{\pi c_0 D\beta^2}{vkTr_2{}^2} \tag{18-78}$$

Inserting for r_2 from Eq. (18-74), one finds

$$\frac{F}{L} \cong \pi\beta c_0 \tag{18-79}$$

Thus in the high-velocity region, the drag force has a significant contribution that is independent of velocity. For velocities so high that

$$v > \frac{D\beta}{kTb^2} = v_c' \tag{18-80}$$

Eq (18-74) would predict $r_2 < b$, which is meaningless. In this region, there is *no* cloud, and b must be substituted for r_2 in Eq. (18-78), giving the force[13]

$$\frac{F}{L} \cong \frac{\pi c_0 D\beta^2}{b^2 vkT} \tag{18-81}$$

which decreases with increasing velocity. In the region $v_c < v < v_c'$, diffusion in the region $r < r_2$ gives a contribution to the drag force, approximately given by

[13] This expression agrees with the limiting result of the exact treatment of R. Fuentes-Samaniego, Ph.D. thesis, Stanford University, Stanford, Calif., 1979, and with numerical solutions by S. Takeuchi and A. S. Argon, *Phil. Mag.*, **40:** 65 (1979).

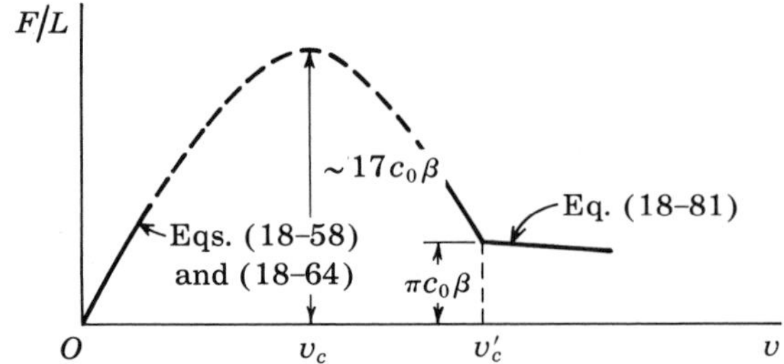

FIGURE 18-13. Force-velocity curve for the case of Cottrell drag.

Eq. (18-64), with r_2 replacing r_1 as the integration limit, as

$$\frac{F}{L}=\frac{3vc_0\beta^3}{Dk^2T^2}\frac{r_2-r_0}{r_2r_0}$$

This contribution also decreases with increasing velocity as $r_2 \to r_0$. In the limit of Eq. (18-80) the contribution vanishes, so that Eq. (18-81) gives the *total* drag force.

Thus the above analysis provides reasonable theoretical estimates of the force-velocity relationship in the limits of low and high dislocation velocities, as shown in Fig. 18-13. In the intermediate region the analytical treatment breaks down, for the reasons discussed in Sec. 18-2; one can only conclude from the limiting behavior that a critical velocity v_c must exist. For the intermediate region we must refer to the more exact calculations. Cottrell and Jaswon[14] estimate that for a dilute atmosphere, such as is typical for substitutional solutes at creep temperatures, the critical velocity is about $4v_t$ [Eq. (18-75)],

$$v_c \cong \frac{4DkT}{\beta} \tag{18-82}$$

More extensive analytical[8] and numerical[15, 16] solutions give a factor ranging from ~0.7 to ~10 replacing the factor 4. At the velocity of $4v_t$ the calculated maximum force[14] is

$$\frac{F_{\max}}{L} \cong 17c_0\beta \tag{18-83}$$

or about 4 times the drift force [Eq. (18-79)]. The other treatments[8, 16] give

[14]A. H. Cottrell and M. A. Jaswon, *Proc. Roy. Soc.*, **A199:** 104 (1949); A. H. Cottrell, "Dislocations and Plastic Flow in Crystals," Oxford University Press, Fair Lawn, N.J., 1953, p. 137.

[15]H. Yoshinaga and S. Morozumi, *Phil. Mag.*, **23:** 1367 (1971).

[16]S. Takeuchi and A. S. Argon, *Phil. Mag.*, **40:** 65 (1979).

results within about a factor of 5 of this for $\beta/bkT \sim 1$ to 3. Therefore, the total force-velocity curve is expected to be about like the one in Fig. 18-13, with the major uncertainty the magnitude of F_{max}.

Exercise 18-5. Show that for a slowly moving dislocation with diffusion-controlled extension of the cloud, the dissipation in the outer region $c \sim c_0$ can be ignored in a rough calculation.

Outline of solution. Insert $r_2 = D/v$ in Eq. (18-78) and compare the result to Eq. (18-58) for typical cases. For the parameters leading to Eq. (18-70) the drift term is about one-fourth of the total, which is appreciable, but not important in a rough estimate.

Limitations and Extensions of the Model

Throughout this section the assumptions have been stressed. The one-dimensional approximate model is presented to avoid the enormous complexities of the exact treatment; the approximate model does exhibit the *method* of approach applicable in the more complex case. There are implicit in the model, however, additional assumptions that are noteworthy. An obvious one is the assumption of isotropic elasticity. The effects of flexibility, which also provides a size-effect interaction with screw dislocations (Sec. 14-4), are neglected. The possibility that a screw dislocation might drag nearby impurities along with it cannot be dismissed. Second-order terms in the interaction potential are neglected. One effect of including either anisotropic elasticity or second-order terms in the development is that the solute atmosphere would partially screen the strain field of the dislocation. As a consequence, the concentration in all cases would tend to converge more rapidly with an increase in distance from the dislocation. Whether or not the divergent behavior would be removed by including such terms can be determined only by calculation for each specific case. Another effect of anisotropy is to introduce the possibility of *local* rearrangements of solute atoms into preferred orientations, as discussed in Chap. 14. For large values of β, Fermi-Dirac statistics should be used for the near-core region. A final assumption of note is that of a concentration-independent diffusivity D, which in general is valid only for relatively low solute concentrations.

With regard to other applications of the method, the method of Sec. 18-2 is directly applicable to the drag of solute atoms by moving grain boundaries and by interphase boundaries. Less evidently, the results of Sec. 18-3 can be used to describe thermoelastic damping of dislocation motion, with phonon diffusion replacing solute diffusion as the physical means of free-energy dissipation. A more subtle point is that the free energy per unit length of dislocation is lowered by the amount of the net interaction energy with the solute atoms. The line tension resisting length change is lowered accordingly if the atmosphere moves with the dislocation. Thus in the low-stress, high-temperature limit, dislocations in solid solutions should bow out more easily with consequent implications for anelastic behavior and creep.

Atmosphere drag computations have been extended to the case of climb.[16] In the low-velocity limit the result for climb or mixed glide-climb agrees with Eq. (18-58). For higher velocities, the results differ from those for glide but are of the same form and magnitude. The attendant climb viscosity is of the order of or greater than that of vacancy emission or absorption (Chap. 15) and should be included in the explanation of creep of solid solutions.

18-4. DRAG OF A SNOEK ATMOSPHERE BY A DISLOCATION

As discussed in Sec. 14-5, interstitial solutes in bcc order into preferential sites in the stress field of a dislocation in a so-called Snoek atmosphere. Similar effects occur for solute atom pairs in fcc crystals. The interstitial distribution about a moving dislocation can be understood with reference to the moving square-well potential of Fig. 18-14. In this model, long-range diffusion of interstitials is excluded, and only local rearrangement is considered; the overlap with the diffusing Cottrell atmosphere is introduced later. In contrast to the model of Fig. 18-3, no concentration enhancement arises ahead of the moving well. Atoms order in the region of the well and disorder to a random distribution behind the well with a relaxation time $\tau_r = \omega^{-1}$, where ω is the mean diffusion jump frequency. For very slow velocities, the equilibrium distribution is nearly maintained within the well, while for very high velocities, $c \sim c_0$ throughout, both cases corresponding to a relatively low drag force. The drag force is expected to be a maximum when the relaxation time τ_r is equal to the time r_1/v_c for the well to translate a distance r_1 equal to the width of the well. This condition occurs at a critical velocity

$$v_c = \frac{r_1}{\tau_r} = r_1 \omega \tag{18-84}$$

Schoeck and Seeger[17] have considered this problem, and derive Eq. (18-84) as the critical condition. We present instead a treatment by analogy to the preceding section, obtaining a result of the same form as theirs, but differing somewhat.

The potential field for interstitials about a screw dislocation is derived in Exercise 14-4 and elsewhere.[17, 18] The potential field as a function of θ for the three interstitial positions is shown in Fig. 18-15. For a mixed or edge dislocation, the potential would be more complex. Rather than treat this complex case, we treat only the screw case approximately to illustrate the method. The potential of Fig. 18-15*a* is reasonably approximated by the

[17] G. Schoeck and A. Seeger, *Acta Met.*, **7**: 469 (1959).

[18] A. W. Cochardt, G. Schoeck, and H. Wiedersich, *Acta Met.*, **3**: 533 (1955).

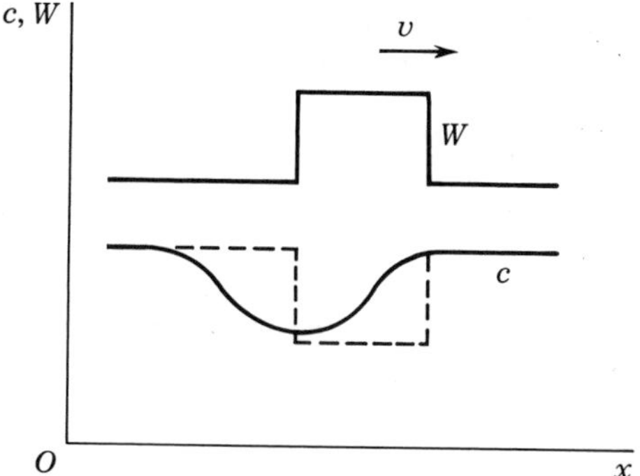

FIGURE 18-14. Concentration profile for a Snoek atmosphere about a moving square potential.

three-level potential of Fig. 18-15*b*, with different positions favored in each sector of $2\pi/3$. Thus the potential is

$$W_1 = -W_2 = W = -\frac{\beta}{r} = -\frac{\beta}{(x^2+y^2)^{1/2}} \tag{18-85}$$

For the specific case of carbon in iron (Exercise 14-4),

$$\beta = 0.05\mu b v_s \tag{18-86}$$

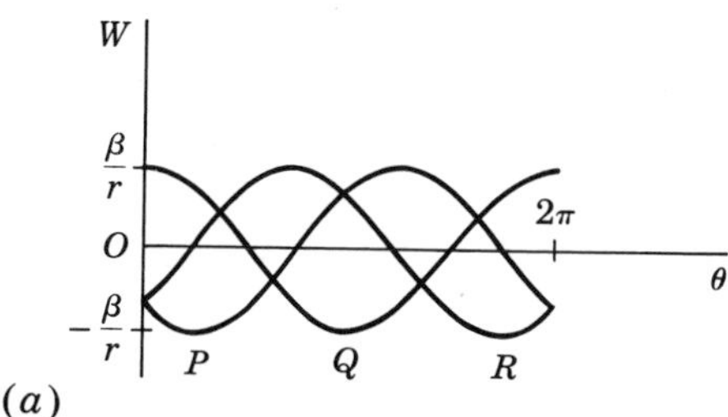

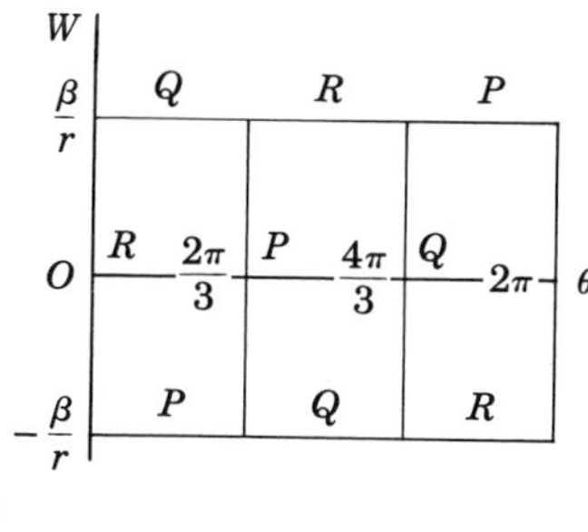

FIGURE 18-15. (*a*) Interaction energy for interstitials in positions *P*, *Q*, and *R* of Fig. 14-15 as a function of the angle θ about a screw dislocation (Exercise 14-4). (*b*) Three-level approximation of (*a*).

The equilibrium concentration in the ith level is related to the mean concentration c_∞ at $r=\infty$ by the expression

$$c_{i_e}=\frac{c_\infty \exp(-W_i/kT)}{\sum_i \exp(-W_i/kT)}=\frac{c_\infty \exp(-W_i/kT)}{1+\exp(W/kT)+\exp(-W/kT)} \tag{18-87}$$

As with the Cottrell atmosphere, let us first consider the region outside the range of the potential, $r>r_1$, where

$$r_1=\frac{\beta}{kT} \tag{18-88}$$

In this range the potential can be expanded and linearized, so that the concentrations are

$$\begin{aligned} c_{0_e} &= c_0=\tfrac{1}{3}c_\infty \\ c_{1_e} &= c_{0_e}\left(1-\frac{W}{kT}\right) \\ c_{2_e} &= c_{0_e}\left(1+\frac{W}{kT}\right) \end{aligned} \tag{18-89}$$

Hence we can consider the drag force produced by atoms in level 1 and multiply the answer by 2 for the total since levels 1 and 2 contribute equally.

The conservation of mass in the moving coordinate system yields the conservation equation

$$v\frac{\partial c}{\partial x}=\omega(c-c_e)=-\dot{c} \tag{18-90}$$

The term $v(\partial c/\partial x)$ is the depletion per unit time in the moving coordinate system, while $\dot{c}$ is the diffusive relaxation jump flux into level 1.

Equation (18-90) resembles Eq. (18-12) and can be solved by analogous methods, giving

$$c=\frac{c_0\omega}{v}\exp\left(\frac{\omega x}{v}\right)\int_x^\infty \exp\left(-\frac{\omega x}{v}\right)\exp\left(-\frac{W}{kT}\right)dx \tag{18-91}$$

which indeed reproduces the distribution of Fig. 18-14.

The force expression analogous to Eq. (18-38) is

$$\frac{F}{L}=-\int_A (c-c_e)\frac{\partial W}{\partial x}\,dA \tag{18-92}$$

where A is the area normal to the dislocation line. In the region where the

Boltzmann factor in Eq. (18-87) can be linearized, Eq. (18-92) reduces to a general dissipation equation analogous to Eq. (18-41),

$$\frac{F}{L}=\frac{kT}{\omega v}\int_A \frac{\dot{c}^2}{c}\,dA \tag{18-93}$$

The right side of Eq. (18-93) is related to the rate of dissipation of free energy.

Let us now determine the drag force at low velocity produced by the interstitials in the range outside the potential, $r>r_1$. Here [*cf.* Eq. (18-73)] Eq. (18-91) indicates that $c=c_e$ for x negative and v sufficiently small that

$$|x|>r_2=\frac{v}{\omega} \tag{18-94}$$

Hence, from Eq. (18-90),

$$\dot{c}=-v\frac{\partial c}{\partial x}\cong -v\frac{\partial c_e}{\partial x}=\frac{vc_e}{kT}\frac{\partial W}{\partial x}$$

so that Eq. (18-93) becomes

$$\frac{F}{L}=\frac{v}{\omega kT}\int_A c_e\left(\frac{\partial W}{\partial x}\right)^2 dA \tag{18-95}$$

From Eq. (18-85),

$$\frac{\partial W}{\partial x}=\frac{\beta}{r^2}\cos\theta$$

so that the drag force is

$$\frac{F}{L}=\frac{vc_0\beta^2}{\omega kT}\int_{r_1}^{\infty}\frac{dr}{r^3}\int_0^{2\pi}\cos^2\theta\,d\theta=\frac{\pi vc_0\beta^2}{2\omega kTr_1{}^2} \tag{18-96}$$

Multiplying by 2 to give the total drag force and substituting from Eq. (18-88) for r_1, one obtains

$$\frac{F}{L}=\frac{\pi vc_\infty kT}{3\omega}\qquad r>\frac{\beta}{kT}>\frac{v}{\omega} \tag{18-97}$$

Inside the range of the potential, Eq. (18-87) reduces to

$$c_{1_e}\cong c_\infty[1-\exp(W/kT)]$$

$$c_{0_e}\cong c_{2_e}\cong 0 \tag{18-98}$$

In this region

$$\dot{c} = -v\frac{\partial c}{\partial x} = \frac{vc_\infty}{kT}\exp\left(\frac{W}{kT}\right)\frac{\partial W}{\partial x}$$

Thus, from Eqs. (18-90) and (18-92), one obtains a drag force

$$\frac{F}{L} = \frac{vc_\infty \beta^2}{\omega kT}\int_b^{r_1}\frac{1}{r^3}\exp\left(-\frac{r_1}{r}\right)dr\int_0^{2\pi}\cos^2\theta\, d\theta$$

$$= \frac{\pi vc_\infty \beta^2}{\omega kT}\int_b^{r_1}\frac{1}{r^3}\exp\left(-\frac{r_1}{r}\right)dr$$

This integral can be approximately evaluated as follows:

$$\int_b^{r_1}\frac{1}{r^3}\exp\left(-\frac{r_1}{r}\right)dr \cong \frac{1}{r_1}\int_b^{r_1}\frac{1}{r^2}\exp\left(-\frac{r_1}{r}\right)dr = \frac{1}{r_1^2}\exp\left(-\frac{r_1}{r}\right)\Bigg]_b^{r_1} \cong \frac{1}{3r_1^2}$$

The drag force is thus

$$\frac{F}{L} = \frac{\pi vc_\infty kT}{3\omega} \qquad \frac{\beta}{kT} > r \tag{18-99}$$

which is equal to the result of Eq. (18-97). The sum of the two contributions gives for the total force[19]

$$\frac{F}{L} = \frac{2\pi vc_\infty kT}{3\omega} \tag{18-100}$$

Now let us consider the high-velocity limit. In the region where $b < v/\omega < \beta/kT$, the integration is complex and must be performed numerically. In the limit $v/\omega = r_2 \gg \beta/kT$, on the other hand, the drift contribution to the force can be readily calculated. Drift control occurs in the region $r < v/\omega$. This is *opposite* to the case of the Cottrell atmosphere, where drift occurred for $r > D/v$; this opposite behavior arises because the diffusion relaxation time $1/\omega$ is constant in the former case, while it is r dependent, r^2/D, in the latter case. In the drift region, Eq. (18-91) indicates that $c \cong c_0 = \frac{1}{3}c_\infty$ for both levels

[19]The asymptotic expansion of the result of G. Schoeck and A. Seeger [*Acta Met.*, 7: 469 (1959)] for low v gives

$$\frac{F}{L} = \frac{2\pi vc_\infty kT}{3\omega}\ln\frac{RkT}{\beta}$$

where R is an outer cutoff [*cf.* Eq. (18-60)]. The difference appears to arise from their use of the model of a static cloud, rigidly displaced relative to the moving dislocation, a model which gives a logarithmically divergent expression.

1 and 2, so that the atom flux is given by the drift term [Eq. (18-90)]

$$\dot{c} = \frac{\omega c_0 W}{kT}$$

The total drift contribution to the drag force from the two levels is then given by Eq. (18-93) as

$$\frac{F}{L} = \frac{2c_0\omega}{vkT}\int_b^{v/\omega}\left(\frac{\beta}{r}\right)^2 r\,dr\int_0^{2\pi} d\theta$$

$$= \frac{4\pi c_\infty \omega\beta^2}{3vkT}\ln\frac{v}{\omega b} \tag{18-101}$$

In the region $r > v/\omega$, the concentration is $c = c_e$. The diffusion contribution to the drag force from this region is given by the same integral as Eq. (18-96), with v/ω replacing r_1 as an integration limit. The result is

$$\frac{F}{L} = \frac{\pi c_\infty \beta^2\omega}{3vkT} \tag{18-102}$$

which generally is negligible compared to the contribution of Eq. (18-101). For velocities so large that v/ω exceeds the appropriate cutoff distance R, the latter must be inserted as an upper limit, giving

$$\frac{F}{L} = \frac{4\pi c_\infty \omega\beta^2}{3vkT}\ln\frac{R}{b} \tag{18-103}$$

Thus, as in the case of the Cottrell atmosphere, we have reasonable estimates of the asymptotic force-velocity curves for the Snoek atmosphere, as shown in Fig. 18-16. A lower cutoff velocity is shown in Fig. 18-16 to indicate that effects caused by the Cottrell interstitial atmosphere become important in the low-velocity range. The ratio of the critical velocity of the Snoek atmosphere to that of the Cottrell atmosphere is given by Eqs. (18-82) and (18-84)

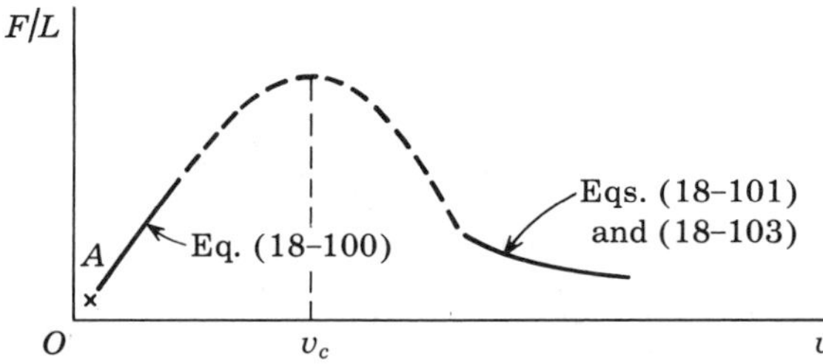

FIGURE 18-16. Force-velocity curve for Snoek drag. The lower cutoff at *A* indicates the onset of Cottrell drag.

to be

$$\frac{v_c(S)}{v_c(C)} = \frac{\beta(S)\beta(C)}{4b^2k^2T^2} \tag{18-104}$$

18-5. DISLOCATION CORE EFFECTS

Large Core Drag

In Sec. 18-3 and 18-4 we assumed that core impurities could diffuse with the dislocation and hence could be neglected in the temperature range where bulk diffusion jumps were important. Under other conditions the core impurities can control the dislocation velocity. Let us adopt the model of Sec. 14-5, involving favored line directions for impurity adsorption, with a resultant Peierls barrier type of behavior. To simplify the calculations, the core impurities are assumed to be noninteracting; there is some experimental support for such a model in iron.[20] As a specific example, let us consider carbon in iron at 200°C, with a heat of solution of[21] $H_s \cong W_s \cong 0.8$ eV and a binding energy between carbon atoms and the core of $W_B \cong 1.0$ eV.[22]

In a typical case the interstitial carbon in bulk iron would be in equilibrium with cementite, so that Eq. (14-58) reduces to

$$\frac{c_\infty}{n} = \exp\left(-\frac{W_s}{kT}\right) \tag{18-105}$$

The concentration of holes, or vacant impurity sites, in the core is then given by Eq. (14-64) as[23]

$$c_h = \frac{1}{b}\exp\left(-\frac{W_B - W_s}{kT}\right) \tag{18-106}$$

Consider now a kink in the dislocation line, as shown in Fig. 18-17. Because the kink itself does not lie in the direction favorable for impurity adsorption, it is clean. The impurities exert an osmotic force tending to decrease the kink

[20] J. Chipman, *Met. Trans.*, **3**: 55 (1972).

[21] H. A. Wriedt and L. S. Darken, *Trans. Met. Soc. AIME*, **233**: 111, 122 (1965).

[22] Much smaller values (~0.3 to 0.5 eV) have been deduced from internal friction data. However, as discussed by A. Seeger [*Phys. Stat. Solidi*, **55a**: 457 (1979)] and J. P. Hirth [*Met. Trans.*, **11A**: 861 (1980)], the lower values are instead likely to be connected with double-kink nucleation. Because static and dynamic strain aging of iron indicate that W_B must exceed W_s and by analogy with the experimental finding that $W_B > W_s$ for hydrogen in iron as discussed by Hirth, we assign 1.0 eV as a rough probable value.

[23] If carbon were preequilibrated at elevated temperatures and a concentration c_∞ were quenched in, W_s would vanish for Eq. (18-106) and the preexponential factor would be simply c_∞/b.

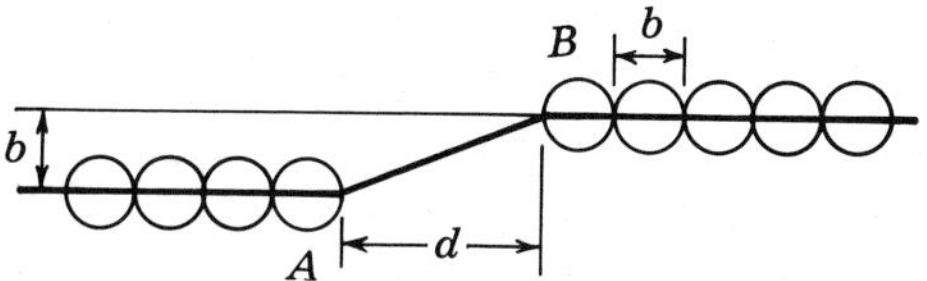

FIGURE 18-17. A kink in a dislocation line pinned by core solute atoms.

width d, while the line tension tends to increase d. Equation (6-84) gives

$$\mathcal{S} \cong \mu b^2/5$$

for the line tension appropriate for the increase in dislocation length $b^2/2d$ associated with the kink. The free energy required to clean the length d of impurities is $F = d(W_B - W_s)/b$, according to Eq. (18-106). Thus the free energy of the kink is

$$F = \frac{\mu b^4}{10d} + \frac{d}{b}(W_B - W_s) \tag{18-107}$$

The equilibrium condition $\partial F/\partial d = 0$ yields for the equilibrium kink width and energy the values

$$d_k = b\left[\frac{\mu b^3}{10(W_B - W_s)}\right]^{1/2} \tag{18-108}$$

and

$$F_k = \left[\frac{2\mu b^3(W_B - W_s)}{5}\right]^{1/2}$$

At thermal equilibrium, the kink concentration is

$$c_k = \frac{1}{b}\exp\left(-\frac{F_k}{kT}\right) \tag{18-109}$$

The kink can diffuse in a number of ways, but a likely mechanism involves annihilation of a hole at A and creation of a hole at B. The jump frequency for such an event is

$$\omega = c_h b \nu \exp\left(-\frac{W_m}{kT}\right) \tag{18-110}$$

where ν is the Debye frequency and W_m the activation energy for the jump of a hole in the core. Thus the kink diffusivity is

$$D_k = b^2\omega = c_h b^3 \nu \exp\left(-\frac{W_m}{kT}\right) \tag{18-111}$$

The derivation of the dislocation velocity now proceeds exactly as for the case of double-kink nucleation over the Peierls barrier, discussed in Sec. 15-2. The velocity is given by Eq. (15-45), with the above expressions substituted for D_k and F_k. The derivation of Eq. (15-45) for the present application was presented by Lothe.[24] Louat[25] originally suggested a similar mechanism, which differs only in the double-kink nucleation step. He supposed that the core must be cleaned of impurity over a length sufficient that the segment can bow out into the next potential valley, where it is again pinned. As discussed by Lothe, the present mechanism should predominate, except perhaps in the high-stress limit; Louat's considerations are more important in the breakaway region, to be discussed next.

Let us now reflect on the relative conditions where Snoek or Cottrell drag and core drag might be important. Because Snoek and Cottrell drag have the same type of force-velocity relation, a treatment of one of the two indicates the general behavior, so for brevity only the former is considered. The ratio of the core drag velocity v_D to the Snoek drag velocity v_S is given by Eqs. (15-45) and (18-100) as

$$\frac{v_D}{v_S} \cong \frac{L}{b} \exp\left[\frac{W_D-(W_B+W_m+2F_k)}{kT}\right] \tag{18-112}$$

where W_D is the activation energy for bulk diffusion of the solute. For carbon in iron $W_B=1.0$ eV, $\mu b^3=8$ eV, $F_k=0.8$ eV, $W_D=0.9$ eV, and we guess $W_m \cong 0.1$ eV and $L \cong 10^4 b$. Thus

$$\frac{v_D}{v_S} \cong 10^4 \exp\left(-\frac{Q}{kT}\right) \tag{18-113}$$

with $Q=1.9$ eV. At low temperatures $v_D < v_S$, so that the force-velocity relation appears as in Fig. 18-18*a*. Drag of the core impurities is rate controlling at all stress levels; the critical force for breakaway is in general large compared to F_{max} for the Snoek atmosphere (see Sec. 18-6). At higher temperatures $T > Q/9k$, $v_D > v_S$, and the curve is that of Fig. 18-18*b*. In this case, drag of the Snoek atmosphere is rate controlling at low velocities. However, near $F = F_{max}$ dislocations can escape the Snoek atmosphere and accelerate to a higher velocity, where motion is controlled by core drag.

Small Core Drag

The above treatment is based on the model of Fig. 18-17, which leads to the largest possible values of F_k. An alternative kink model is shown in Fig. 18-19. Here sites A correspond to the highest binding energies for solute W_B, and sites

[24] J. Lothe, *Acta Met.*, **10:** 663 (1962).

[25] N. Louat, *Proc. Phys. Soc.*, **B69:** 459 (1956); *Proc. Phys. Soc.*, **71:** 444 (1958).

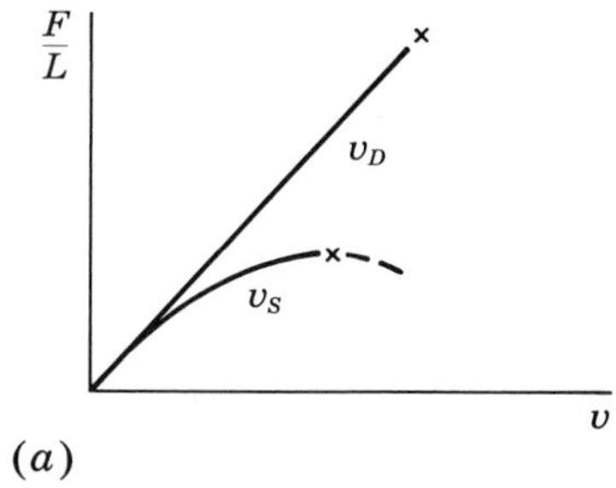

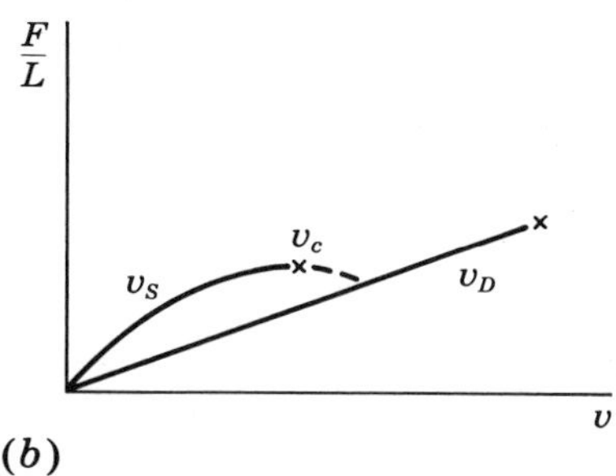

FIGURE 18-18. Force-velocity curves for core-solute drag: (*a*) low temperatures, (*b*) high temperatures.

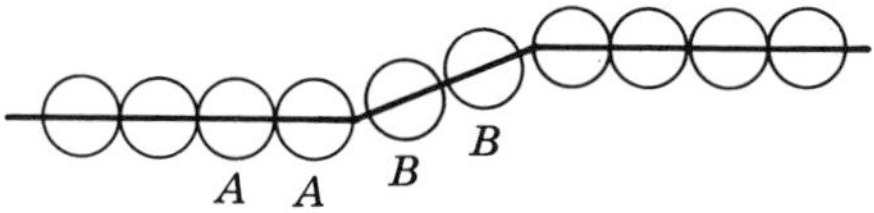

FIGURE 18-19. A kink in a dislocation line pinned by core solute atoms. Sites *A* have a higher binding energy than sites *B*.

B correspond to binding sites with a lower binding energy W_B'. The kink energy would then approach the smaller value characteristic of the energy of a kink over the Peierls barrier, $F_k \cong 0.1$ eV. Also, the kink diffusivity would have an activation energy about equal to that for core diffusion, $\sim W_D/2$. For this case,

$$\frac{v_D}{v_S} \cong 10^4 \exp\left(\frac{\frac{1}{2}W_D - 2F_k - W_s}{kT}\right) \tag{18-114}$$

Here one can have $v_D > v_S$ down to quite low temperatures, perhaps even for all temperatures if $W_D > 4F_k + 2W_s$. In such a case, the curve of Fig. 18-18*b* would hold for all temperatures, but the difference between v_D and v_S would decrease with increasing temperature.

The analysis by Hirth[26] of the cold-work internal friction peak for hydrogen in iron is consistent with the model shown in Fig. 18-19. With the relaxation time τ_1 for bow-out as in Exercise 15-2 and Eq. (15-45) defining *B* in Eq. (15-48) for the double-kink process, the activation energy becomes that of *B*, i.e., that in Eq. (15-45). A fit of this activation energy to that of τ_1 gives $F_k = 0.15$ eV for hydrogen in iron, intermediate between the values for pure iron of $F_k = 0.06$ eV for edge dislocations and $F_k = 0.6$ eV for screw dislocations.[27] A similar analysis for the data for nitrogen and carbon in iron compiled by Seeger[28] would give $F_k \cong 0.6$ eV for the cold-work peak in these cases.

Evidently, in general one expects both alloy systems exhibiting high core-drag behavior and those exhibiting low core drag. Even for the case of carbon in

[26] J. P. Hirth, *Met. Trans.*, **11A:** 861 (1980).

[27] V. Hivert, P. Groh, W. Frank, I. Ritchie, and P. Moser, *Phys. Stat. Solidi*, **46a:** 89 (1978).

[28] A. Seeger, *Phys. Stat. Solidi*, **55a:** 457 (1979).

iron, the core-energy parameters are so uncertain that neither case can be absolutely excluded. In either case, however, the drag models suffice to rationalize the overall deformation behavior, as discussed presently.

Breakaway from Core Atmospheres

At sufficiently low temperatures, solute or impurity atoms become so immobile in both the bulk crystal and the dislocation core that the solute drag velocities given by any of the preceding equations approach zero. In such a low-temperature region, dislocations can become mobile only if they tear away from their atmospheres. Let us first consider breakaway from solute segregated to the core. At low temperatures, Eq. (14-62) shows that the equilibrated core is *saturated*, $c_d \sim 1/a$. For a saturated core, the stress required to tear a dislocation away from the atmosphere is roughly given by the relation

$$\sigma b = \frac{F_{\text{max}}}{L} \cong \frac{c_d W_B}{b} \cong \frac{W_B}{ab} \tag{18-115}$$

since the energy W_B must be supplied over a potential range of b. The requisite stress generally is very large. Typically, for interstitial carbon or nitrogen in iron, $W_B \sim \mu b^3/10$ and $\sigma \sim \mu/10$, while for substitutional solutes $\sigma \sim \mu/100$. In either case, the stress far exceeds normal yield strengths after strain-aging,[29] so that such breakaway can occur only in regions of large stress concentration.

Fisher[30] assessed the probability of local breakaway under thermal activation. The pertinent configuration is shown in Fig 18-20. In order to free the dislocation over a length l, an energy

$$W_1 = \frac{l}{a} W_B \tag{18-116}$$

must be provided. The balance between the line-tension force $2\mathcal{S}\sin(\theta/2)$ and that caused by the applied stress $\sigma b r \theta$ gives a radius of curvature $r \cong \mathcal{S}/\sigma b$ for the bowed line and $\theta \cong l/r$. With $\mathcal{S} \cong \mu b^2/10$, according to Eq. (6-84), and an increase in line length $r\theta^3/24$, the increase in line energy upon bow-out is

$$W_2 = \frac{\sigma^2 b^2 l^3}{24\mathcal{S}} \tag{18-117}$$

The applied stress does work during bow-out in the amount

$$W_3 = \frac{\sigma^2 b^2 l^3}{12\mathcal{S}} \tag{18-118}$$

[29]For other cases such as yielding in heat-treated steels or quench-aged steels, interactions with precipitate particles or generation of dislocations at grain boundaries are important, so the breakaway process discussed here applies to static or dynamic strain aging only.

[30]J. C. Fisher, *Trans. ASM*, **47:** 451 (1955).

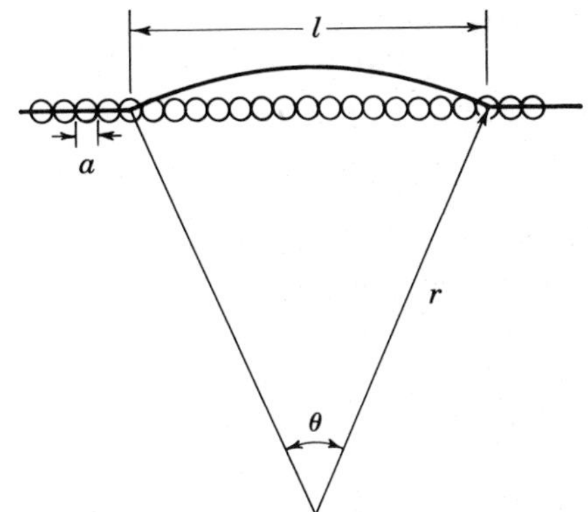

FIGURE 18-20. Thermally excited configuration for Fisher breakaway. A segment l of radius of curvature r has bowed away from its impurity atmosphere.

so that the total energy change on bow-out is

$$W = W_2 - W_3 = -\frac{\sigma^2 b^2 l^3}{24\mathcal{S}} \tag{18-119}$$

When the force $-\partial W/\partial l$ exceeds the force $\partial W_1/\partial l$, the dislocation can continue to unzip from the core atmosphere without thermal activation; this occurs for lengths greater than a critical length

$$l_c = \left(\frac{8W_B\mathcal{S}}{\sigma^2 b^2 a}\right)^{1/2} \tag{18-120}$$

Thus, in the critical configuration, the energy that must be supplied by thermal activation is

$$W_c = \frac{l_c W_B}{a} - \frac{\sigma^2 b^2 l_c^3}{24\mathcal{S}} = \frac{2l_c}{3a} W_B \tag{18-121}$$

In order that the strain produced by breakaway be measurable, $\dot{N}$, the number of breakaway events per cubic centimeter per second, must be about unity, or

$$\dot{N} = \frac{\rho\nu}{b}\exp\left(-\frac{W_c}{kT}\right) = 1\,\text{sec}^{-1}\,\text{cm}^{-3}$$

where ρ is the dislocation density and ν is the Debye frequency. With $\rho \sim 10^8$ cm^{-2}; this requires

$$W_c = \frac{2l_c^*}{3a} W_B \leqslant 65kT \tag{18-122}$$

At room temperature for carbon in iron, with $W_B = 1.0$ eV$=\mu b^3/6$, the fulfillment of this condition requires that $l_c^* \sim 3a$, or $\sigma^* = \mu/12$. For the other example of $W_B = \mu b^3/50$, $l_c^* = 25a$ and $\sigma^* = \mu/125$. Thus the critical stress for breakaway from a saturated core is not appreciably reduced by thermal fluctuations. General breakaway can be achieved only in regions of large stress concentration.

The above development considers breakaway at an *arbitrary* point on a pinned dislocation line. Breakaway can occur more easily at *statistical weak spots* in the segregated solute atmosphere, as suggested by Louat[31] and elaborated upon by Lothe.[32] The probability of a *hole* existing at a given core site is given by Eq. (14-64) as $c_h a$. The probability of l_c/a successive sites being vacant, thus creating a free length l_c, is

$$(c_h a)^{l_c/a} = \left(\frac{n}{c_0}\right)^{l_c/a} \exp\left(-\frac{l_c W_B}{akT}\right) \tag{18-123}$$

or, in the usual case that c_0 is given by Eq. (18-105),

$$(c_h a)^{l_c/a} = \exp\left[-\frac{l_c}{akT}(W_B - W_s)\right] \tag{18-124}$$

The frequency of growth from the size $l_c - a$ to l_c equals the frequency of decay from l_c to $l_c - a$, given by

$$\omega = \nu \exp\left(-\frac{W_m}{kT}\right) \tag{18-125}$$

The breakaway rate $\dot{N}$ is then the product of Eqs. (18-124) and (18-125) multiplied by the possible sites for such an event per unit volume, ρ/b, or

$$\dot{N} = \frac{\rho\nu}{b} \exp\left(-\frac{W_c'}{kT}\right) \tag{18-126}$$

Here [*cf*. Eq. (18-122)], for an appreciable strain rate, the activation energy must satisfy the condition

$$W_c' = \frac{l_c^*}{a}(W_B - W_s) + W_m \leqslant 65kT \tag{18-127}$$

with $W_B = 1.0$ eV, $W_s = 0.8$ eV, and $W_m = 0.1$ eV as before for carbon in iron; at room temperature, condition (18-127) requires that $l_c^* \sim 10a$ and $\sigma^* = \mu/50$. Thus the critical stress is less than that given by Eq. (18-122) but is still large compared to observed flow stresses after strain aging. The only case in which breakaway in iron is likely in the absence of stress concentrations is the situation where the core atmosphere has not relaxed to its equilibrium value.

[31] N. Louat, *Proc. Phys. Soc.*, **B69:** 459 (1956).

[32] J. Lothe, *Acta Met.*, **10:** 663 (1962).

This could occur if a more dilute atmosphere is quenched in from elevated temperature, if fresh dislocations are created without sufficient time for subsequent diffusive formation of an equilibrium atmosphere, or if lower energy sites for solute exist at sites such as nodes, which would then tend to drain the core of solute. This expectation agrees with experimental observation in mild steel[33, 34]; aged dislocations in general remain pinned at room temperature, and fresh dislocations must be created (presumably at stress concentrations) to enable deformation to proceed. In the case of substitutional solutes, Eq. (18-127) predicts breakaway at nominal yield strengths in the typical case. In any case, Eq. (18-127) describes the most likely of the possible breakaway mechanisms.

Exercise 18-6. Compute from Eq. (18-127) the value of $W_B - W_s$ required to give $\sigma^* = 10^{-3}\ \mu$; $\sigma^* = 10^{-4}\ \mu$. Compute σ^* at room temperature for carbon in iron with the parameters given above in the case that the hole concentration is quenched in from 720°C; that is, the hole concentration is that characteristic of the equilibrium value at 720°C.

Breakaway from a Cottrell Atmosphere

Although a Cottrell (or Snoek) atmosphere extends over distances large compared to b, most of the contribution to the force resisting breakaway is associated with solute atoms adjacent to the core.[35] Hence we treat such breakaway as essentially a core phenomenon. Consider an edge dislocation displaced a distance x from its atmosphere as shown in Fig. 18-5, in the case of a weak interaction in the sense that $\beta < bkT$. The increase in energy of the system as a function of x is

$$\begin{aligned}\frac{\Delta W}{L} &= \int_A (c-c_0)[W(r',\theta') - W(r,\theta)]\,dA \\ &= \int_A (c-c_0)\left(\frac{\beta\sin\theta'}{r'} - \frac{\beta\sin\theta}{r}\right)dA \\ &= \beta c_0 \int_b^\infty r\,dr \int_0^{2\pi} \frac{\beta\sin\theta}{rkT}\left(\frac{r\sin\theta}{r^2+x^2-2rx\cos\theta} - \frac{\sin\theta}{r}\right)d\theta \end{aligned} \tag{18-128}$$

where c and W are given by Eq. (18-47) and use is made of the fact that

[33] A. S. Keh, in "Relation between Structural and Mechanical Properties of Metals," H. M. Stationery Office, London, 1963, p. 436.

[34] J. D. Baird, *Iron and Steel*, **36:** 450 (1963); *Met. Rev.*, **16:** 1 (1971).

[35] H. Suzuki, in J. C. Fisher et al. (eds.), "Dislocations and Mechanical Properties of Crystals," Wiley, New York, 1957, p. 361.

$r\sin\theta = r'\sin\theta'$. The solution to Eq. (18-128) is

$$\frac{\Delta W}{L} = \frac{\pi c_0 \beta^2}{4kT}\left[\frac{\pi}{2} - \tan^{-1}(z^4+1)^{1/2} + (z^4+1)^{1/2} - z^2\right] \qquad (18\text{-}129)$$

where $z = b/x$. At equilibrium, the external force required to maintain the displacement x is

$$\frac{F}{L} = \frac{\partial(\Delta W/L)}{\partial x} = \frac{\pi c_0 \beta^2}{4bkT}\left[\frac{2z^5}{(z^4+1)^{1/2}}\left(\frac{1}{z^4+2} - 1\right) + 2z^3\right] \qquad (18\text{-}130)$$

This force has a maximum value

$$\frac{F_{\max}}{L} \cong \frac{c_0\beta^2}{bkT} \qquad (18\text{-}131)$$

at $x = 0.65b$. At $x = 2b$, the force has dropped to one-seventh of this value, so that it is sharply peaked near the core, as suggested at the beginning. This result cannot be deduced without such a detailed analysis.

The above result must mean that the force peak can be derived essentially from the interaction with solute atoms adjacent to the core. Consider the row nearest to the core, at a distance $\sim b$, on the side of agglomeration. The extra number per unit length is $c_0 b^2(\beta/bkT)$, each with a binding energy β/b. This binding energy is released over a distance $\sim b$, so the required force is

$$\frac{F}{L} \sim c_0 b^2 \frac{\beta}{bkT}\frac{\beta}{b}\frac{1}{b} = \frac{c_0\beta^2}{bkT}$$

in agreement with Eq. (18-131).

When $\beta > bkT$, ΔW is difficult to evaluate directly. However, the above results indicate that the binding is primarily with the solute row nearest the core, because the concentration $c - c_0$ decreases more rapidly with r than for the weak interaction. The nearest row is usually nearly saturated for the case $\beta > bkT$, so that Eq. (18-115) should apply to this case, with β/b replacing W_B.

18-6. APPLICATION TO DEFORMATION BEHAVIOR

Several macroscopic plastic-flow phenomena are associated with solute drag. One example is the observation of blue brittleness in mild steel,[36] wherein a decrease in ductility and the onset of serrated stress-strain behavior in tensile

[36] F. R. N. Nabarro, "Report of the Conference on Strength of Solids," Physical Society, London, 1948, p. 38.

tests occur at about 300°C, where carbon diffusion is appreciable. Flow-velocity relations produced by either Cottrell or Snoek atmospheres can be used to rationalize blue brittleness, so that discussion of one case illustrates both possibilities. Let us assume that the Snoek effect is controlling and that its drag is related to core drag in the fashion depicted in Fig. 18-18*b*.

In a simple approximate model, sufficient for our present purposes the macroscopic strain rate is related to the dislocation velocity by the expression

$$\dot{\epsilon} = m\rho b v \tag{18-132}$$

where ρ is the density of moving dislocations and m is a geometric strain resolution factor ~ 0.5. Thus in a constant-strain-rate test such as a tensile test, the active dislocation density is inversely proportional to v. Stress-velocity relationships and corresponding dislocation-density-strain curves are shown for three temperatures in Fig. 18-21. At $T \ll T_c$, the atmosphere is immobile, and flow can occur only by breakaway. Hence flow in this region is characterized by an upper yield point, and then by Lüders flow at a lower stress determined by a mechanism other than solute drag. A possibility would be phonon drag, which would yield a much higher mobility, as shown in Fig. 18-21*a*. Thus, once breakaway occurs, v is large, and one expects ρ to be *relatively* small at a given strain.

As $T \to T_c$, drag of the Snoek atmosphere becomes rate controlling, the mobility is reduced, and ρ is relatively large at a given strain, as shown in Fig. 18-21*b*. At $T \sim T_c$, thermal activation reduces F_{max}, so that flow is occurring

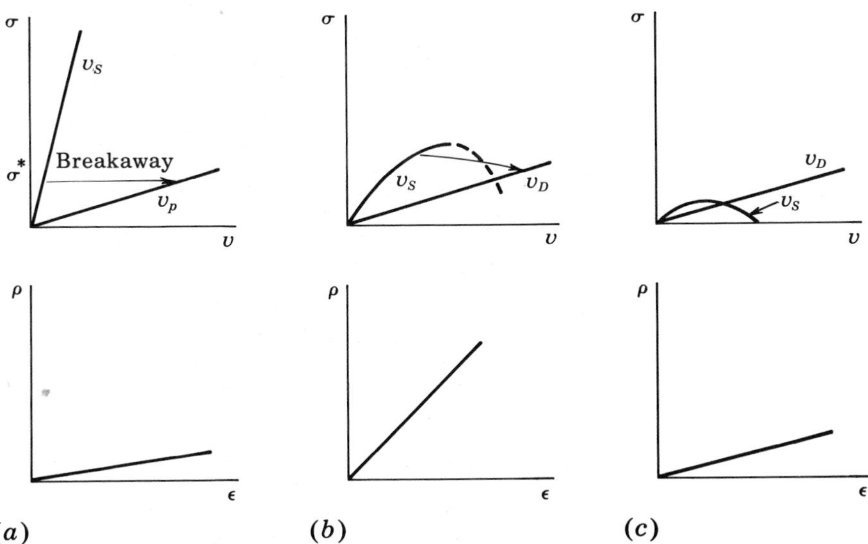

FIGURE 18-21. Stress-velocity curves and corresponding dislocation-density-strain curves (*a*) at $T \ll T_c$, (*b*) at $T \sim T_c$, and (*c*) at $T \gg T_c$. The velocity v_p indicates phonon drag control.

with $v_s \sim v_c$, and thermal activation can excite dislocations from the region where $d\sigma/dv_s$ is positive to the region where it is negative. In the latter region, flow is unstable and dislocations accelerate to a higher velocity v_D and lower required force, controlled by core drag. Such breakaway can, of course, be manifested as a Lüders band propagation. When these fast dislocations are restrained by internal stress concentrations or other obstacles, the Snoek atmosphere again forms, and the dislocation reverts to the slow velocity v_s. Hence, in this region $T \sim T_c$, velocity and stress transitions occur, explaining the serrated stress-strain behavior. Since flow for temperatures approaching T_c corresponds to the largest $d\rho/d\epsilon$, as shown in Fig. 18-21, this region should also exhibit the lowest ductility. At a given strain, the higher value of ρ implies greater internal stresses and hence a greater tendency to fracture. Thus blue brittleness and serrated stress-strain behavior should occur at about the same temperature range $T \sim T_c$.

For $T \gg T_c$, core atmosphere drag controls flow, but v_D is relatively high, so that again ρ is relatively small. Thus the ductility is again higher than in the blue-brittle range, where $d\rho/d\epsilon$ is a maximum. Keh[37] has observed that $d\rho/d\epsilon$ is a maximum in the blue-brittle range, supporting the above postulation.

A number of theories[38, 39] have been proposed for serrated flow that are similar to the above qualitative discussion but that are given in more quantitative detail. The onset of serrations after a critical strain, the Portevin-Le Chatelier effect,[40] can occur if the yield point or Lüders band phenomena are suppressed, and the effect is then influenced by the decrease of ρ caused by work hardening. In fcc crystals the Portevin-Le Chatelier effect always occurs only after a critical strain. One rationalization[41] of this effect is that the vacancies produced by cold work enhance the diffusivity of substitutional solute atoms sufficiently to produce a dynamic Cottrell atmosphere, and hence drag, following the behavior shown in Fig. 18-13. Another possibility is that such vacancies associate with solute atoms, forming a pair with tetragonal distortion, and leading to Snoek type drag. Finally,[38] the effect has been associated with pipe diffusion from a forest dislocation to a temporarily arrested mobile dislocation waiting to intersect it.

The variation of T_c in iron-carbon alloys is also pertinent to the present discussion. The critical temperature for serrated flow, T_c in Fig. 18-21, is reduced if the strain rate is lowered.[42, 43] This result is consistent with Eq.

[37]A. S. Keh, in "Relation between Structural and Mechanical Properties of Metals," H. M. Stationery Office, London, 1963, p. 436; Y. Nakada and A. S. Keh, *Acta Met.*, **16:** 903 (1968).

[38]Reviewed by R. A. Mulford and U. F. Kocks, *Acta Met.*, **27:** 1125 (1979); P. Wycliffe, U. F. Kocks, and J. D. Embury, *Scripta Met.*, **14:** 1349 (1980).

[39]P. G. McCormick, *Scripta Met.*, **12:** 197 (1978).

[40]A. Portevin and F. Lechatelier, *C. R. Acad. Sci.*, **176:** 507 (1923).

[41]A. H. Cottrell, in "Relation of Properties to Microstructure," American Society of Metals, Cleveland, Ohio, 1954, p. 131; *Phil. Mag.*, **44:** 829 (1953).

[42]J. D. Baird, *Met. Rev.*, **16:** 1 (1971).

[43]P. G. McCormick, *Scripta Met.*, **12:** 197 (1978).

(18-127) because a lower $\dot{\epsilon}$ leads to a lower $\dot{N}$ required to maintain flow, and hence to a numerical factor lower than 65 in Eq. (18-127) and a lower T_c. T_c is also lowered if the alloy is quenched after holding at 700°C prior to testing. This effect is associated with the increase in c_h discussed in Exercise 18-6. Moreover, the critical strain for serrated yielding and for $d\sigma/d\dot{\epsilon}$ have similar dependences on strain rate, stress, and temperature consistent with the qualitative discussion of Fig. 18-21.[43]

18-7. OTHER SOLUTE EFFECTS

There are a number of other possible impurity effects that are quite important in plastic deformation. For many of these the theory is so complex that we do not undertake its discussion here. Many such possibilities are reviewed by Friedel,[44] Suzuki,[45] and Haasen.[46] Here we briefly consider only a few prominent examples.

Low-Temperature Hardening

At temperatures where drag effects occur, solute can diffusively redistribute about a dislocation, so that the dislocation relaxes to a straight configuration under line-tension forces. At temperatures so low that diffusion is negligible, a dislocation becomes wiggly, adjusting its configuration to conform to the internal stresses of the immobile solute atoms. The essential features of the theory of dislocation motion in such a case were propounded by Mott and Nabarro,[47] with more recent extensions.[48, 49] An exact theory for such hardening is extremely difficult, so we only outline the concepts of the theory and enumerate the problems in the theoretical development.

In the presence of a dilute solute concentration, a dislocation under stress bows out, as shown in Fig. 18-22*b*. The activation barrier for the breakaway of the dislocation is that of Fig. 16-5, and the kinetics resemble those discussed in Chap. 16. Point forces corresponding to large bow-out, $\sim\mu b^2$, cannot be sustained by solute atoms, and breakaway occurs for small bow-outs, such as the motion from A to B in Fig. 18-22*b*; in terms of Eqs. (16-11) and (16-16),

[44] J. Friedel, "Dislocations," Pergamon, New York, 1964, pp. 351–447.

[45] H. Suzuki, in P. Haasen, V. Gerold and G. Kostorz (eds.), "Strength of Metals and Alloys," Vol. 3, Pergamon, Oxford, 1980, p. 1595.

[46] P. Haasen, in F. R. N. Nabarro (ed.), "Dislocations in Solids," vol. 4, North-Holland, Amsterdam, 1979.

[47] N. F. Mott and F. R. N. Nabarro, *Proc. Phys. Soc.*, **52:** 86 (1940); "Report of the Conference on Strength of Solids," Physical Society, London, 1948, p. 1.

[48] R. Labusch, *Phys. Stat. Solidi*, **41:** 659 (1970).

[49] F. R. N. Nabarro, *Phil. Mag.*, **35:** 613 (1977).

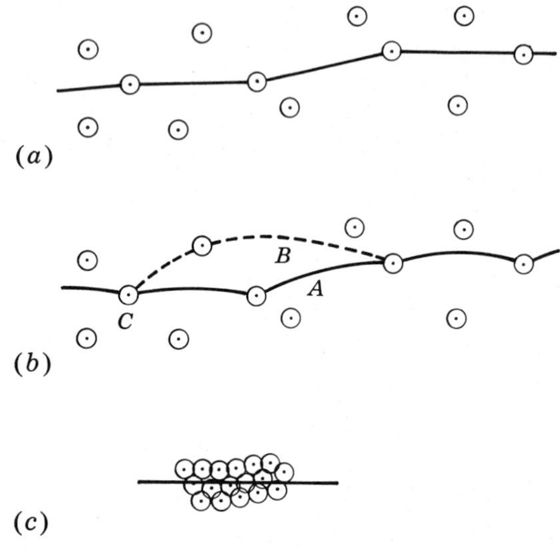

FIGURE 18-22. Interaction of a dislocation with solute atoms, with atoms represented by dots and their strain fields by circles: (*a*) dilute solution, (*b*) dilute solution under stress, (*c*) concentrated solution.

the maximum point forces are

$$F^* = \frac{W^*}{a^*} \ll \mu b^2 \tag{18-133}$$

The critical stress for breakaway is

$$\sigma_{\text{crit}} = \frac{F^*}{bl} = \frac{W^*}{a^* bl} \tag{18-134}$$

Transforming Eq. (2-81) to cartesian coordinates, one finds that the shear stress on the dislocation glide plane produced by the solute atom is

$$\sigma_{xy} = \frac{\mu \delta v}{\pi r^3} \sin\theta \cos\theta \tag{18-135}$$

which falls off rapidly with increasing r. Since the dislocation essentially interacts only with solute atoms in the glide plane because of the short range of the interaction, the spacing between pins is $l \sim (c_0 b)^{-1/2}$, where c_0 is the solute concentration. Thus

$$\sigma_{\text{crit}} = \frac{W^*}{a^* b}(bc_0)^{1/2} \tag{18-136}$$

This is the Fleischer-Orowan equation for solid solution hardening.[50] It can

[50] R. L. Fleischer, in D. Peckner (ed.), "The Strengthening of Metals," Reinhold, New York, 1964, p. 93.

alternatively be derived as the Friedel equation by relating the area swept in the above $A-B$ process to the mean free path between solute atoms and introducing stress as it influences the area via bow-outs of the participating dislocation segments.[51]

The computation of the energy W as a function of displacement along the reaction coordinate, and hence of W^* and a^*, is a difficult problem and has not been satisfactorily solved. As discussed previously, the maximum interaction occurs when the separation between the dislocation and the solute atom is of the order of core dimensions, where elastic calculations break down and interaction energies are not accurately known. However, W^*/a^* should be of the order of magnitude of W_B/b, which justifies condition (18-133), since typical ratios of W_B/b are one or two orders of magnitude smaller than the line-tension force μb^2. Because breakaway occurs for small bow-outs, W^*/a^* is expected to be roughly independent of l, and hence of c_0. Thus, for the dilute solution, Eq. (18-136) predicts a $c_0^{1/2}$ dependence of σ_{crit}.

Thermal activation can contribute to breakaway according to an expression like Eq. (16-8), the only difference being that the distance of advance per breakaway, $\sim l/2$, replaces a. The temperature dependence of flow is discussed in detail in Chap. 22. Here we simply note that with typical values for the various parameters, the breakaway process should be rate controlling at temperatures less than those where solute drag is important. When the intrinsic mobility is high, inertial effects such as those suggested in connection with the softening effect of the normal-superconducting transition (see Sec. 7-7) can also be significant in the process of overcoming obstacles.

Even if the above approximations are valid, the model is crude. In an actual case, there is a statistical distribution of lengths l. Also, the interaction is more general than Eq. (18-135) and involves tetragonal distortions and second-order effects, as discussed in Chap. 14. Breakaway should occur first at weak points, where l is large. Such breakaway can then produce kinklike configurations such as (C) in Fig. 18-22b. The line-tension forces are greater at these "kinks," so that they are expected to be preferential breakaway sites, leading to net motion by kink-type behavior. The calculation of an exact flow equation in such a case involves statistical distributions of solute core energies and complicated line-tension integrations and is most formidable. Numerical calculations have been performed for random point obstacles and varying strengths W^* in the constant line tension case.[52–54] The results give flow stresses of the form of the Friedel version of Eq. (18-136) but modified by a constant factor varying by up

[51] J. Friedel, "Dislocations," Pergamon, Oxford, 1964, p. 224. This consideration of bow-outs actually modifies Eq. (18-136) and brings in an added factor $(F^*/\mu b^2)^{1/2}$. The important point is that the stress is proportional to $c_0^{1/2}$ for dilute solutions. Assumptions of constant line tension and of small bow-out in the area swept remain in the treatment. A detailed discussion is given by U. F. Kocks, A. S. Argon, and M. F. Ashby, *Prog. Mat. Sci.*, **19:** 55 (1975).

[52] U. F. Kocks, *Can. J. Phys.*, **45:** 737 (1967).

[53] A. S. Argon, *Phil. Mag.*, **25:** 1053 (1972).

[54] K. Hanson, S. Altintas, and J. W. Morris, Jr., *Nucl. Met.*, **20:** 917 (1976).

to 30 percent. Modifications including anisotropic elasticity and dislocation interactions in the line tension have been developed,[55] but the utilization for the solute case is impeded by the problem in defining W^* and a^*. Finally as previously mentioned, inertial effects as a function of intrinsic mobility should be included. Calculations by Schwarz and Labusch[56] indicate the presence of significant inertial effects.

As the concentration of solute increases, l decreases and σ_{crit} increases. However, it does not increase without bound. At high concentrations various models give linear concentration dependence. At such concentrations, the strain fields of the solute atoms overlap, as shown in Fig. 18-22*c*. In such a case the long-range elastic stress acting on the dislocation is the algebraic mean of the solute stress fields and hence is *zero*. The dislocation is again nominally straight under such conditions, while the interaction is primarily a core interaction. As in the case of short-range order, discussed next, the energy per unit length required to advance the dislocation a distance δx involves the number of chemical bonds cut by the core,

$$\delta W = W_{ch} c_0 d\, \delta x$$

Here W_{ch} is the net energy associated with solute-solvent bonds cut and formed in the core region, and d is the interplanar spacing normal to the slip plane. Thus the required stress for such motion is

$$\sigma_{\text{crit}} = \frac{1}{b}\frac{\delta W}{\delta x} = \frac{W_{ch} c_0 d}{b} \tag{18-137}$$

The critical stress is expected to depend linearly on c_0 for low-temperature flow in a concentrated solid solution. Suzuki[57] has developed a model for bcc metals where solute hardening occurs through its influence on kink mobility D_k in Eq. (15-17), which also gives a dependence on $c_0(1-c_0)$, linear for small to moderate c_0.

In the intermediate region between dilute and concentrated behavior, Mott[58] has proposed a model involving breakaway from many pinning points, small bow-outs, and a statistical distribution of random forces on lengths l. This model has been elaborated by Labusch[59] and Nabarro,[60] who find a $c_0^{2/3}$ dependence of the flow stress. In the limit $c \to 0$ this law goes over to the limiting $c_0^{1/2}$ form.[56]

[55] R. O. Scattergood and D. J. Bacon, *Phil. Mag.*, **31:** 179 (1975).

[56] R. B. Schwarz and R. Labusch, *J. Appl. Phys.*, **49:** 5174 (1978).

[57] H. Suzuki, in F. R. N. Nabarro (ed.), "Dislocations in Solids," vol. IV, North-Holland, Amsterdam, 1980, p. 191.

[58] N. F. Mott, in "Imperfections in Nearly Perfect Crystals," Wiley, New York, 1952, p. 173.

[59] R. Labusch, *Phys. Stat. Solidi*, **41:** 659 (1970).

[60] F. R. N. Nabarro, *Phil. Mag.*, **35:** 613 (1977).

In summary, the solid solution models involve some approximations and are quite complicated except in somewhat idealized cases. They do indicate that low-temperature solute hardening should vary as $c_0{}^{1/2}$ in very dilute solutions and as $c_0{}^{2/3}$ or c_0 in more concentrated solutions. All these types of dependence have been observed experimentally.

The above model is directly applicable to precipitation hardening and inclusion or dispersion hardening. These defects, of course, can withstand larger point forces, of the order of or greater than μb^2, so that large bow-outs can occur. This case is discussed in Chap. 20.

Ordered Crystals

In concentrated solid solutions of metal B in A, there is generally a tendency for clustering or short-range order, depending on the quantity

$$\phi=\tfrac{1}{2}(U_{AA}+U_{BB})-U_{AB} \tag{18-138}$$

where U_{AA} is the energy of an $A-A$ bond, etc. If ϕ is positive, the ordering parameter α is negative and short-range order occurs; i.e., the fraction $(1-\alpha)X_B$ of nearest neighbors of an A atom that are B atoms exceeds the fraction X_B of B atoms in the alloy; the same is true for A neighbors of a B atom. If ϕ is negative, then α is positive, and clustering occurs, since A atoms tend to group together, as do B atoms. In either case, the motion of a dislocation involves breaking of the favored type of bond and hence a tendency toward randomization of the solution. The stress required for such motion is[61]

$$\sigma_c=\frac{C|\alpha\phi|}{b^3} \tag{18-139}$$

where C is a numerical factor near unity depending on the atomic-bond distribution in the glide plane and depending weakly on α. As slip proceeds and the solution becomes random, $\alpha\to 0$ and $\sigma\to 0$; as many favorable bonds are formed as are cut in the glide process. Thus one expects either a yield-point phenomenon, wherein the stress to produce flow drops with increasing strain, or a decrease in the rate of hardening with increasing strain in such solid solutions. The temperature dependence of σ_c is directly related[62] to that of the order parameter α

$$\alpha\cong X_A X_B\left[\exp\left(-\frac{2\phi}{kT}\right)-1\right] \tag{18-140}$$

[61] In general, there is also a contribution caused by the size mismatch of the A and B atoms; see P. S. Rudman, *Acta Met.*, **13:** 387 (1965). This would appear in a parallel way to ϕ.

[62] T. Muto and Y. Takagi, *Solid State Phys.*, **1:** 193 (1955).

There are several interesting possibilities associated with the presence of long-range order. At low temperatures, one possible mechanism of deformation involves the motion of a single dislocation, with the accompanying generation of an antiphase boundary with energy γ (Sec. 11-6). The stress required for such motion is simply

$$\sigma = \frac{\gamma}{b} \tag{18-141}$$

For high-boundary-energy alloys, this stress is prohibitively high in the absence of stress concentrations, so that only superdislocations are glissile. The expected larger Peierls barrier for superdislocations and the cutting of antiphase boundaries provide two possible mechanisms for control of superdislocation motion. In the latter case, for example, the advance δx of unit length of superdislocation through an antiphase boundary network of size l increases the boundary by an amount $\delta xb/l$ (Fig. 18-23a) and requires a stress[63]

$$\sigma = \frac{1}{b}\frac{\delta W}{\delta x} = \frac{1}{b}\frac{\delta(xb\gamma/l)}{\delta x} = \frac{\gamma}{l} \tag{18-142}$$

Such processes increase the fault area on the glide planes, as shown in Fig. 18-23b. These new boundaries make slip on secondary systems difficult, accounting for rapid hardening of polycrystalline ordered alloys. Also, as slip continues, configurations such as Fig. 18-23c form, leading to hardening of the primary system. When the network size l is less than the equilibrium-fault spacing of the superdislocation, configurations like that in Fig. 18-23c are present initially, and the flow-stress expression is more complicated than Eq. (18-142) predicts.[64]

Heating to elevated temperatures produces several effects that are peculiar to ordered alloys. Unlike normal stacking faults, the antiphase-boundary energy does not increase markedly if the boundary is displaced out of the glide plane [Eq. (11-17)]. Thus configurations such as Fig. 18-24a can form by local climb and glide, effectively pinning the superdislocation with respect to low-temperature glide. Indeed, if the dislocations were aligned vertically in a tilt array, with the connecting APB on {100}, the configuration would correspond to the Kear-Wilsdorf lock, which is important in controlling the flow stress in Ni_3Al.[65] Second, the boundary can fluctuate to the configuration of Fig. 18-24b, again pinning the dislocation. Finally, adsorption of A or B to the boundary can lower its energy,[66] particularly for nonstoichiometric alloys,

[63] A. H. Cottrell, in "Relation of Properties to Microstructure," American Society of Metals, Cleveland, Ohio, 1954, p. 131.

[64] W. D. Biggs and T. Broom, *Phil. Mag.*, **45:** 246 (1954).

[65] E. Kuramoto and D. P. Pope, *Acta Met.*, **26:** 207 (1978).

[66] R. Kikuchi and J. W. Cahn, *J. Phys. Chem. Solids*, **20:** 94 (1961)

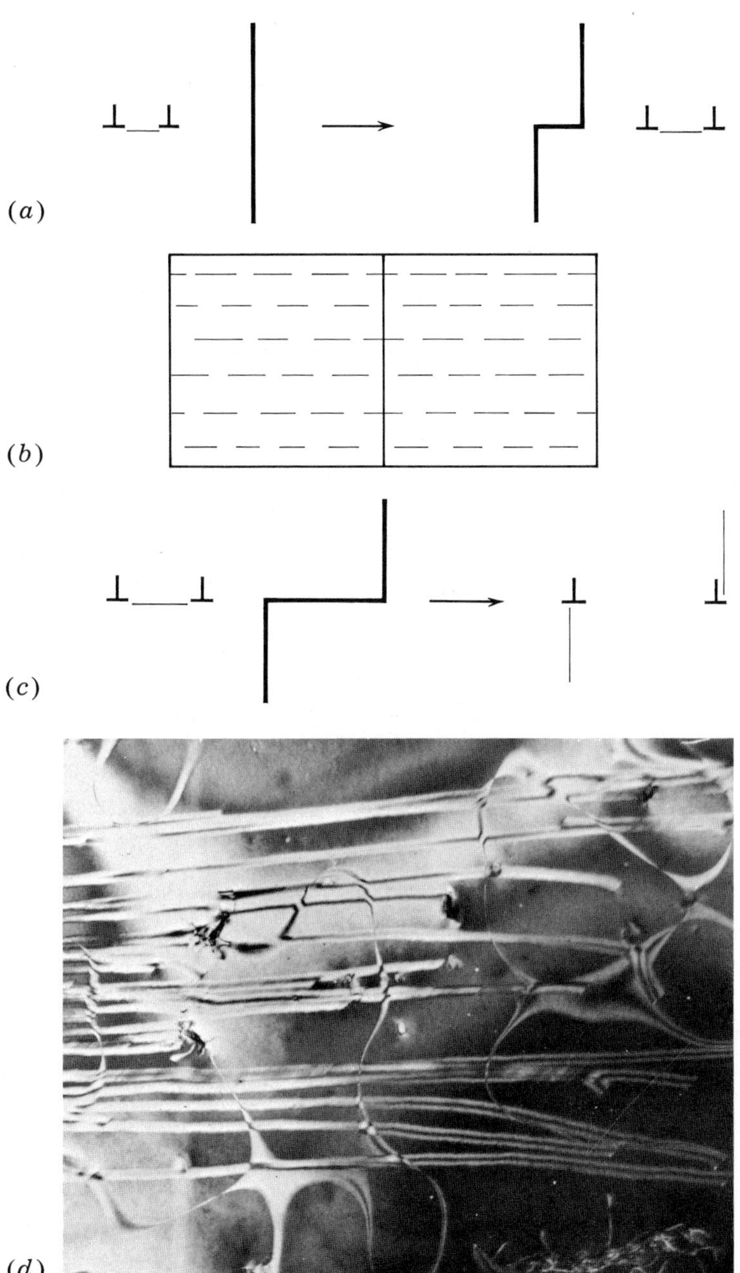

FIGURE 18-23. (*a*) The process of antiphase boundary formation by glide. (*b*) Original antiphase boundaries (solid lines) and new boundaries formed by superdislocation glide (dashed lines). (*c*) Locking of the superdislocation at the boundary. (*d*) Examples of (*b*) and (*c*) in Fe_3Al, 24,000×. [M. J. Marcinkowski and N. Brown, Acta Met., **9**:764 (1961).]

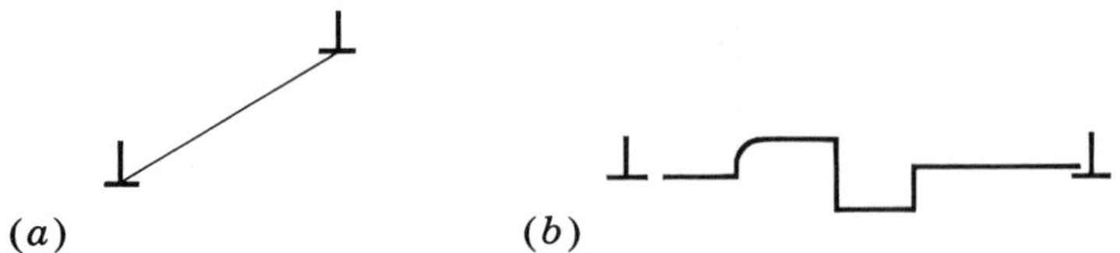

FIGURE 18-24. (*a*) A superdislocation extended so that the antiphase boundary does not lie in the glide plane. (*b*) Fluctuations of a boundary out of the glide plane.

providing locking resembling the Suzuki locking discussed next. All these mechanisms can produce yield-point phenomena and aging phenomena.

For ordered alloys which disorder at a critical temperature T_c, an increase in hardness or flow stress occurs near but below T_c, followed by a drop above T_c. The mechanisms cited contribute dynamically to such effects. Also, however, normal Snoek and Cottrell atmosphere drag and Suzuki drag can contribute. Such effects are clearly possible, because disordering requires short-range diffusion, and the onset of such diffusion is associated with these drag effects. Near T_c the antiphase boundaries become diffuse and spread over several lattice planes.[67] This effect could also produce hardening similar to Fig. 18-24*b*. Finally, a number of mechanisms can be envisioned that involve diffusive motion of a boundary along with a superdislocation—the glide motion of the array of Fig. 18-24*a*, for example. Such processes produce diffusive drag forces on the dislocations.

Other aspects of dislocation motion in superlattices have been reviewed extensively.[68] In general, many of the properties are analogous to those associated with extended dislocations in pure metals.

Suzuki Locking

As pointed out in Sec. 14-5, segregation to (or desegregation from) a stacking fault lowers its energy. This process can lead to dislocation pinning, as first suggested by Suzuki.[69] Consider first the variational process illustrated in Fig. 18-25*a*, where the width r of the faulted region is changed infinitesimally and reversibly and the excess solute in the unfaulted region is reversibly distributed in the bulk phase. The surface excess free energy of the fault per unit length

[67] *Ibid.*; R. Kikuchi and J. W. Cahn, *J. Phys. Chem. Solids*, **23:** 137 (1962).

[68] P. A. Flinn, in D. Peckner (ed.), "Strengthening of Metals," Reinhold, New York, 1964, p. 219; J. H. Westbrook, in "Mechanical Properties of Intermetallic Compounds," Wiley, New York, 1960, p. 1; M. J. Marcinkowski, in G. Thomas and J. Washburn (eds.), "Electron Microscopy and Strength of Solids," Interscience, New York, 1963, p. 333; S. Amelinckx, in F. R. N. Nabarro (ed.), "Dislocations in Solids," vol. 2, North-Holland, Amsterdam, 1979, p. 67.

[69] H. Suzuki, *Sci. Repts. Tohoku Univ.*, **A4:** 455 (1952).

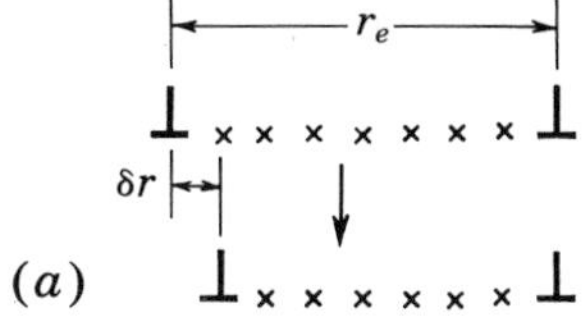

FIGURE 18-25. Various configurations of an extended dislocation with a Suzuki atmosphere. Segregated atoms are represented by crosses.

changes by[70]

$$dG = \gamma_1\, dr + \sum_i \overline{G}_i\, dn_i = \gamma_1\, dr \tag{18-143}$$

where γ_1 is the stacking-fault surface energy and $\overline{G}_i$ and n_i are, respectively, the chemical potential and number of atoms of the ith species. The term containing the chemical potential vanishes, because the ith chemical potential is constant throughout the system at equilibrium and the number of ith atoms removed from the fault exactly equals the number that appear in the bulk. Thus the equilibrium fault energy γ_1 of the segregated fault can be inserted directly to determine the equilibrium separation r_e from

$$F_r = \frac{C}{r_e} = \gamma_1 \tag{18-144}$$

where F_r is the repulsive force between partials and C is defined as in Eq. (10-15).

Now consider the variation under an applied stress, as shown in Fig. 18-25*b*. The change in free energy, with **b** the Burgers vector of the partial, is

$$\delta G = -F_r\,\delta r + \gamma_2\,\delta r - \sigma b\,\delta r = (-\gamma_1 + \gamma_2 - \sigma b)\,\delta r$$

Here γ_2 is the energy of the new stacking fault formed in the unsegregated

[70] W. W. Mullins, in N. A. Gjostein and W. D. Robertson (eds.), "Metal Surfaces," American Society of Metals, Cleveland, Ohio, 1963, p. 17.

region. For the fluctuation to be stable, the applied stress must be

$$\sigma b = \gamma_2 - \gamma_1 \tag{18-145}$$

If the back partial remains pinned, the front partial will continue to extend as the stress is increased, F_r will decrease, and eventually the front partial will break away and advance independently of the trailing partial when

$$\sigma b = \gamma_2 \tag{18-146}$$

Consider now the displacement of the trailing partial, Fig. 18-25*c*. The change in free energy is

$$\delta G = F_r \delta r - \gamma_1 \delta r + \gamma_{ch} \delta r - \sigma b\, \delta r = (\gamma_{ch} - \sigma b)\, \delta r$$

The term $-\gamma_1 \delta r$ arises from a reversible displacement such as that in Fig. 18-25*a*. In this case, however, the excess solute in the region freed from stacking fault is not reversibly distributed in the bulk, but appears as a segregated region, with mole fractions X_A^s and X_B^s of components A and B, respectively, instead of the bulk composition X_A and X_B. Thus there is an additional contribution to the free energy defined by

$$\gamma_{ch}\, \delta r = \frac{2\delta r d}{v_a}\left[\int_{X_B}^{X_B^s} \overline{G}_B\, dX_B + \int_{X_A}^{X_A^s} \overline{G}_A\, dX_A\right]$$

$$= \frac{2\delta r d}{v_a}\int_{X_B}^{X_B^s} (\overline{G}_B - \overline{G}_A)\, dX_B \tag{18-147}$$

where $2d$ is the thickness of the region of enhanced solute and v_a is the average atomic volume.[71] For this fluctuation to be stable, then, the stress must be

$$\sigma b = \gamma_{ch} \tag{18-148}$$

If γ_{ch} exceeds $\gamma_2 - \gamma_1$, the trailing partial remains pinned while the front one advances, F_r decreases, and the stress required to move the trailing partial approaches the limiting value

$$\sigma b = \gamma_{ch} - \gamma_1 \tag{18-149}$$

Comparing the above expressions, one sees that two situations are possible. If $\gamma_{ch} - \gamma_1$ exceeds γ_2, the front partial will tear away from the trailing partial at a stress $\sigma = \gamma_2 / b$, while the trailing partial remains pinned. If $\gamma_{ch} - \gamma_1$ is less

[71]Equation (18-147) implies a relatively sharp line of demarcation between the segregated and unsegregated regions. For diffuse segregates the free-energy integral would be more complex but would be similar to the above expressions.

than γ_2, a stress will be reached where both partials advance together, as in Fig. 18-25*d*. When the trailing partial escapes from the atmosphere, the dislocation will accelerate and can then move under a lower stress, corresponding to a yield-point phenomenon. The critical stress for simultaneous motion of the two partials with a total Burgers vector **b** is

$$\sigma b = \gamma_2 - \gamma_1 + \gamma_{ch} \tag{18-150}$$

because an infinitesimal advance will create fault γ_2 and create equal area of segregate γ_{ch} while annihilating the equal area of fault γ_1.

Alternatively, the motion of the trailing partial can be regarded as simply annihilating stacking fault *in the segregated region*. Let us define γ_1' as the stacking-fault energy in an unsegregated alloy with the same bulk composition $X_A{}^s$, $X_B{}^s$ as the segregated region in the actual case. Then evidently $\gamma_1' = \gamma_1 - \gamma_{ch}$, and Eq. (18-150) can be written

$$\sigma b = \gamma_2 - \gamma_1' \tag{18-151}$$

Note that $\gamma_1' = \gamma_1 - \gamma_{ch}$ can be *negative*, so that breakaway of the leading partial is physically possible. Imagine a case of limited solubility of B in A, with the activity coefficient of B large and with the B segregating almost exclusively to the fault, creating $A-B$ bonds across the fault plane. Motion of the trailing partial would then cut favorable $A-B$ bonds and create unfavorable $A-A$ and $B-B$ bonds, resulting in a negative γ_1', as discussed above. Negative γ_1' values are thus expected for cases of limited solubility of the solute. Pinning of the trailing partial in such a case can lead to the breakaway of the leading partial and accompanying profuse stacking-fault formation. This process can contribute to the extensive fault formation observed in heavily cold-worked alloys. On the other hand, for solute with large solubility, the case of Fig. 18-25*d* should obtain and no fault formation should occur, but there should be yield-point phenomena. Various possibilities for relations of the above stresses to yield phenomena are discussed by Hirth.[72]

Since Suzuki segregation requires at least short-range diffusion, Cottrell and Snoek locking of the partials is, of course, also possible. Suzuki segregation can lead to drag-type behavior, where the segregate diffuses along with a gliding dislocation, but this effect is difficult to separate from Cottrell or Snoek drag.

18-8. IONIC CRYSTALS

Many of the effects discussed in this chapter can occur in ionic crystals with the added complication of electrostatic interactions as discussed in Chaps. 12 and 14. As an example of the ionic case, we consider NaCl doped with divalent

[72] J. P. Hirth, *Met. Trans.*, **1:** 2367 (1970).

cation impurity M^{2+}. The M^{2+} in bulk is combined mainly with cation vacancies to form electrostatic dipoles that also have tetragonal elastic fields analogous to those discussed in Exercise 14-4. Grown-in dislocations tend to be immobile, probably pinned by M^{2+} and its aggregates,[73] which control microyield. Free dislocations have negative charges and, if then equilibrated, compensating atmospheres.[74] An example of an atmosphere would be a cloud of M^{2+} impurities formed from dipoles by stripping away cation vacancies. Yield in nominally pure crystals and those with small amounts of M^{2+} in the parts per million range was once thought to be controlled by breakaway of the charged dislocation from its electrostatic atmosphere, but this effect was shown to be one to two orders of magnitude too small. The possibility remains of breakaway being controlled by the above electrostatic interaction plus the Cottrell interaction with the cation vacancies plus a Snoek interaction with the dipoles. This combined mechanism possibility has not been examined quantitatively.

There is strong evidence[73-75] that in crystals with larger amounts of M^{2+} (>100 ppm), the macroyield stress is associated with moving dislocations interacting with dipoles. The interaction should be both elastic and electrostatic. Haasen[76] has calculated the electrostatic interaction for edge dislocations in NaCl as 0.2, 3.4, and 5.7 eV on $\{110\}$, $\{100\}$, and $\{111\}$ planes, respectively, and advances this as another reason for the prevalence of $\{110\}$ slip. With aging at room temperature or above, aggregates of dipoles form, producing the so-called Suzuki phase,[77] coherent with the matrix, which gives precipitation hardening. The dipole hardening follows a law of the form of Eq. (18-136).[75]

The dipoles produce maximum hardening when they are aligned with their axis inclined to glide plane, presumably related to cutting or reorientation by the dislocation, and minimum hardening when the axis is parallel to **b**. After alignment of dipoles in LiF by annealing in the presence of a field and quenching, the flow stress is found to be 1.85, 2.45, and 2.35 MPa, respectively, when the field vector **E** is parallel to **b**, inclined to **b**, and when the field is zero.[78] Moreover, single slip persists to much larger strains when $\mathbf{E} \| \mathbf{b}$, thus leading to much larger ductility. Similar effects are found when LiF is irradiated optically to produce Schottky pairs under an elastic load.[79] The stress orients the cation-anion vacancy dipoles that also have both elastic and

[73] M. T. Sprackling, "The Plastic Deformation of Simple Ionic Crystals," Academic, New York, 1976.

[74] R. W. Whitworth, *Adv. Phys.*, **24:** 203 (1975).

[75] J. Soullard, research in progress, National University of Mexico, Mexico City, December, 1980.

[76] P. Haasen, *J. Phys. Suppl.*, **34:** 295 (1974).

[77] K. Suzuki, *J. Phys. Soc. Jap.*, **16:** 67 (1961).

[78] M. Paperno and M. V. Galustashvili, *Sov. Phys. Solid State*, **18:** 151 (1976).

[79] E. L. Andronikashvili, I. M. Paperno, M. V. Galustashvili, E. M. Barkhudarov, and M. I. Taktakishvili, *Sov. Phys. Solid State*, **21:** 1575 (1979).

electrostatic tetragonal symmetry. The irradiation produces hardening in all cases, but relative to the unstressed case, alignment of dipole axes parallel and inclined to **b** produces additional softening and hardening, respectively.

As in the simpler metal case, drag of M^{2+} and vacancy atmospheres is possible at higher temperatures. There is evidence for such effects, but the details of the process are not yet understood.[74, 80]

PROBLEMS

18-1. Show that Eq. (18-35) follows from Eqs. (18-11) and (18-36).

18-2. Compute the creep velocity in mild steel for $\sigma \sim 10^{-4}\,\mu$ at $T=600°\mathrm{K}$ as predicted by Eq. (18-100). Assume that carbon equilibrates with cementite.

18-3. Compute the creep velocity in mild steel for $\sigma \sim 10^{-4}\,\mu$ at $T=600°\mathrm{K}$ as predicted by Eq. (18-64). Assume that carbon equilibrates with cementite.

18-4. Consider that $W_B \gg kT$, but that c_0 is so low that the core is not saturated. Generalize the theory of Sec. 18-5 to describe such conditions. In particular, derive the equations equivalent to Eqs. (18-115), (18-120), and (18-121).

18-5. Discuss the possibility of using the results of Prob. 18-4 to construct a theory for amplitude-dependent internal friction in very dilute substitutional alloys.

18-6. The stress field of a dissociated pure screw dislocation interacts with impurities. Will the problem of logarithmic divergencies, as discussed in Sec. 18-3, arise in this case? Derive an approximate formula for the Cottrell drag in this case.

18-7. Consider Cottrell drag for an edge dislocation in an AB alloy that is not dilute. Discuss what diffusion coefficient one should use in this case.

18-8. Discuss the qualitative effects of a binding energy *between* core impurities upon the ease of breakaway of a dislocation from the core solute.

18-9. According to Eq. (14-43), vacancies can form a Cottrell atmosphere about a dislocation. Compute the drag force produced by such an atmosphere on a screw dislocation and compare the results to that of Eq. (16-4). Is the neglect of the second-order Cottrell atmosphere effect justified in this case?

18-10. Discuss the various factors that would contribute to the binding of a dislocation to a tube of second phase precipitated on the dislocation. Calculate the binding energy as a function of displacement for the case of the hollow dislocation core (Prob. 3-9).

18-11. Consider the possibility of vacancies producing Suzuki type locking of an extended dislocation in a pure metal.

[80] R. A. Menezes and W. D. Nix, *Mat. Sci. Engin.*, **16:** 57,67 (1974).

18-12. Roughly what order of magnitude of binding energy between solute and dislocation would be required for the spontaneous generation of dislocations in a supersaturated solid solution?

18-13. Compute the drag force for the glide motion of the configuration in Fig. 18-24*a*.

BIBLIOGRAPHY

1. Baird, J. D., *Met. Rev.*, **16:** 1 (1971).
2. Haasen, P., V. Gerold, and G. Kostorz, "Strength of Metals and Alloys," Pergamon, Oxford, 1980.
3. Kocks, U. F., A. S. Argon, and M. F. Ashby, *Prog. Mat. Sci.*, **19:** 1 (1975).
4. Rosenfield, A. R., A. L. Bement, and G. T. Hahn (eds.), "Dislocation Dynamics," McGraw-Hill, New York, 1968.
5. Thomas, G., and J. Washburn (eds.), "Electron Microscopy and Strength of Crystals," Interscience, New York, 1963.

PART 4

GROUPS OF DISLOCATIONS

19

Grain Boundaries

19-1. INTRODUCTION

In the final chapters aggregates of dislocations are treated. The topic is introduced in this chapter with a discussion of grain boundaries, and particularly their dislocation structure.

Sources of glide dislocations are dealt with in Chap. 20. Dislocation pileups, a natural consequence of dislocation-source operation, provide the subject matter of Chap. 21. The book then concludes with a brief discussion of general dislocation interactions, Chap. 22, and of twinning, which is the result of source operation and specific pileups of partial dislocations, Chap. 23.

A *grain boundary* is the interface where two single crystals of different orientation join in such a manner that the material is continuous across the boundary. Polycrystalline materials are composed of a number of small single crystal grains bonded together by grain boundaries. In general, a grain boundary can be curved (Fig. 19-1), but in thermal equilibrium it is planar in order to minimize the boundary area and hence the boundary energy. In complete analogy with the usual definition of surface tension, the grain-boundary tension γ is defined as the reversible work of formation of unit area of the boundary, other variables being held constant,

$$\gamma = \frac{\partial G}{\partial A} \tag{19-1}$$

where G is the free energy. Sometimes γ is termed the interfacial free energy; however, γ, in general, is not equal to the free energy per unit area of surface G/A; a clear exposition of the various definitions of general surface tensions is presented by Mullins.[1] If γ depends only on the change in orientation across the boundary, and not on the orientation of the boundary plane, the simple boundary-tension concept suffices in equilibrium considerations.

In general, γ also depends on the boundary-plane orientation, giving rise to *twist* terms that tend to rotate the boundary into an orientation of minimum

[1] W. W. Mullins, in N. A. Gjostein and W. D. Robertson (eds.), "Metal Surfaces," American Society of Metals, Cleveland, Ohio, 1963, p. 17.

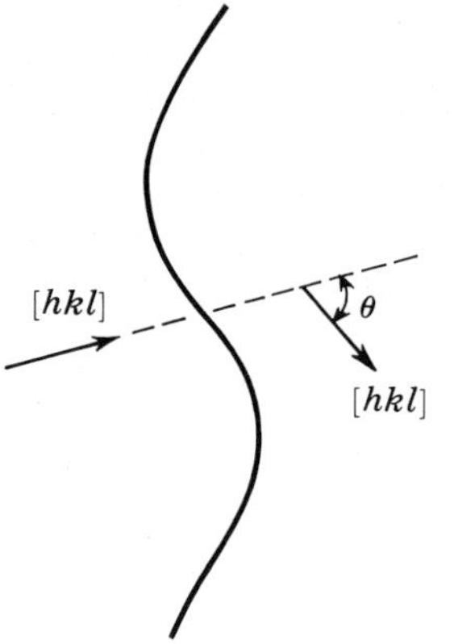

FIGURE 19-1. An arbitrarily curved boundary for an angle of tilt θ.

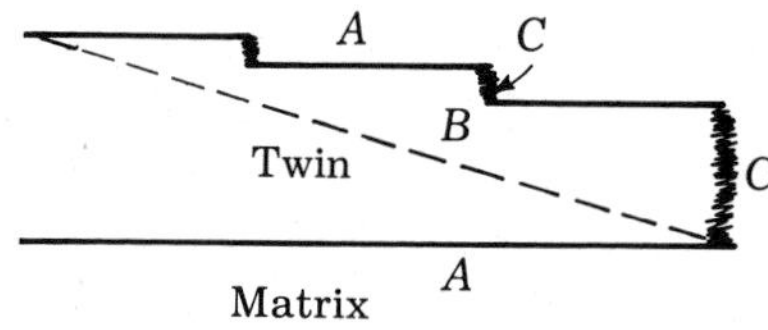

FIGURE 19-2. Coherent-twin-boundary (horizontal) and incoherent-twin-boundary segments (vertical).

surface tension. An example is the twin boundary shown in Fig. 19-2 (see Chap. 23). Because of the low energy of the coherent twin planes A compared to incoherent planes B or C, the faceted structure AC forms with lower total surface energy than the minimum-area surface B. For boundaries with large orientation differences between grains, these complicating twist terms are usually quite small.

Grain boundaries joining stress-free crystallites obviously cannot end within a specimen; they must either branch into other grain boundaries or close on themselves, as shown in Fig. 19-3. If configuration ABC represents the equilibrium position of boundaries pinned at points A, B, and C far removed from O, the equilibrium conditions can be determined by determining the change in total energy with the virtual displacements δx and δy; for example, $\delta W/\delta x = 0$. Performing such a virtual work balance, one finds the equilibrium condition[2]

$$\sum_i \left[\gamma_i \mathbf{t}_i + (\mathbf{t}_i \times \mathbf{s}) \frac{\partial \gamma_i}{\partial \theta_i} \right] = 0 \tag{19-2}$$

As shown in Fig. 19-3, the $\mathbf{t}_i$ are unit vectors in the boundary planes normal to

[2]Equation (19-2) was first derived by C. Herring (in "Physics of Power Metallurgy," McGraw-Hill, New York, 1951, p. 143). Another form of the expression is useful for grain-boundary-energy estimates from dihedral angle measurements, and as shown later, for dislocation node measurements. Let us define $\epsilon_i = (\partial \gamma_i / \partial \theta_i)/\gamma_i$. The variation δy in Fig. 19-3 then yields the relation

$$\gamma_1 = -\gamma_2(\cos\phi_3 - \epsilon_2 \sin\phi_3) - \gamma_3(\cos\phi_2 + \epsilon_3 \sin\phi_2)$$

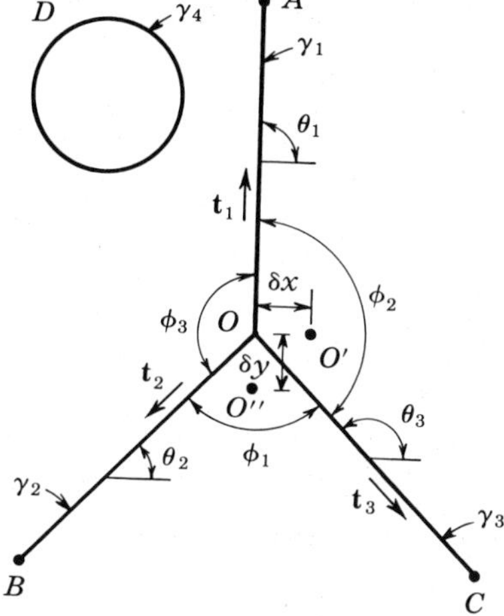

FIGURE 19-3. Junction of three grain boundaries, *A*, *B*, and *C*, and a closed boundary *D*. The vector **s** points into the page.

the vector **s** that lies along the line of intersection of the three boundaries. The term containing $\partial\gamma_i/\partial\theta_i$ is the twist term arising from the variation of γ with orientation θ. When twist terms are absent, the variation δx in Fig. 19-3 leads directly to the result $\gamma_2 \sin\phi_3 = \gamma_3 \sin\phi_2$, which holds cyclically for all boundary pairs, leading to the well-known result in terms of the dihedral angles ϕ_i,

$$\frac{\gamma_1}{\sin\phi_1} = \frac{\gamma_2}{\sin\phi_2} = \frac{\gamma_3}{\sin\phi_3} \tag{19-3}$$

Equation (19-3) also follows directly from Eq. (19-2).

Exercise 19-1. Perform the variations in Fig. 19-3 and derive Eq. (19-2).

A closed boundary such as *D* in Fig. 19-3 is not thermodynamically stable; under thermal activation *D* would shrink until the enclosed grain vanished. Thus at internal equilbrium the grain boundaries would form a cell structure

---

while the variation δx yields

$$\frac{\partial\gamma_1}{\partial\theta_1} = \gamma_1\epsilon_1 = -\gamma_2(\sin\phi_3 + \epsilon_2\cos\phi_3) + \gamma_3(\sin\phi_2 - \epsilon_3\cos\phi_2)$$

Substituting the first of these expressions into the second to eliminate γ_1, and applying the result cyclically to pairs of energies, one finds

$$\frac{\gamma_1}{(1+\epsilon_2\epsilon_3)\sin\phi_1 + (\epsilon_3 - \epsilon_2)\cos\phi_1} = \frac{\gamma_2}{(1+\epsilon_1\epsilon_3)\sin\phi_2 + (\epsilon_1 - \epsilon_3)\cos\phi_2}$$

$$= \frac{\gamma_3}{(1+\epsilon_1\epsilon_2)\sin\phi_3 + (\epsilon_2 - \epsilon_1)\cos\phi_3}$$

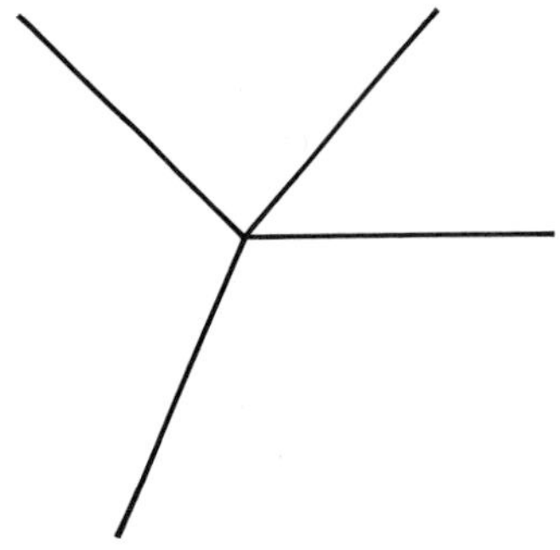

FIGURE 19-4. Junction lines meeting at a node. Three grains meet along each junction line.

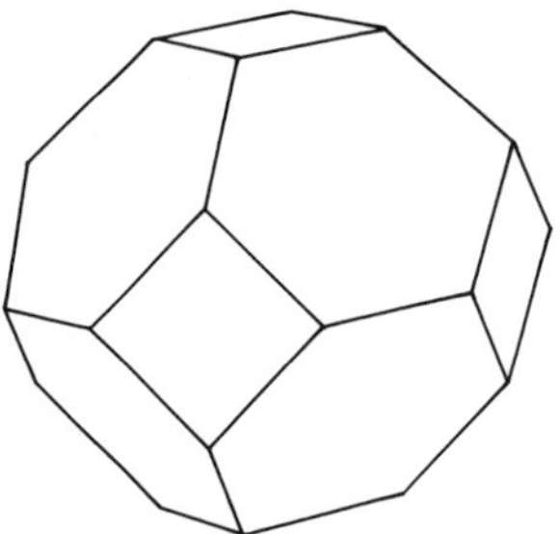

FIGURE 19-5. A unit cell in a stable foam.

resembling a foam. The cells must fill all of the volume, boundary tensions must balance at all junction lines, and the junction-line tensions must balance at nodes, where four junction lines meet. At nodes, four grains and six grain boundaries meet (Fig. 19-4). When all grain-boundary tensions are nearly the same, and if the junction-line tensions are the same, a possible space-filling foam is a regular lattice structure whose unit cells are slightly distorted truncated octahedra,[3] consisting of eight hexagonal faces and six square faces (Fig. 19-5). However, experiments[4] show that neither the cells in a real foam nor the grains in annealed polycrystals have the ideal form of the truncated octahedron, or equivalently, the ideal hexagonal cells in a two-dimensional cross section. Five-sided grains are often found; Cahn and Padawer[5] pointed out, interestingly, that such five-sided grains represent "dislocations" in the foam lattice, as shown in Fig. 19-6. The defects need not extend very far normal to the page, and in that sense they could be thought of as dislocation dipoles, or, in the limit, as inserted "rods" of cells. Hence, the foam dislocations are retained in annealing, just as normal crystal dislocations are. For further discussion of foam models see the review by McLean.[6]

[3] Or tetrakaidecahedra, well known as the first Brillouin zone in fcc metals. This foam structure was propounded by Lord Kelvin, *Phil. Mag.*, **24**: 503 (1887).

[4] D. McLean, "Grain Boundaries in Metals,' Oxford University Press, Fair Lawn, N.J., 1957.

[5] J. W. Cahn and G. E. Padawer, *Acta Met.*, **13**: 1091 (1965).

[6] D. McLean, "Grain Boundaries in Metals," Oxford University Press, Fair Lawn, N.J., 1957.

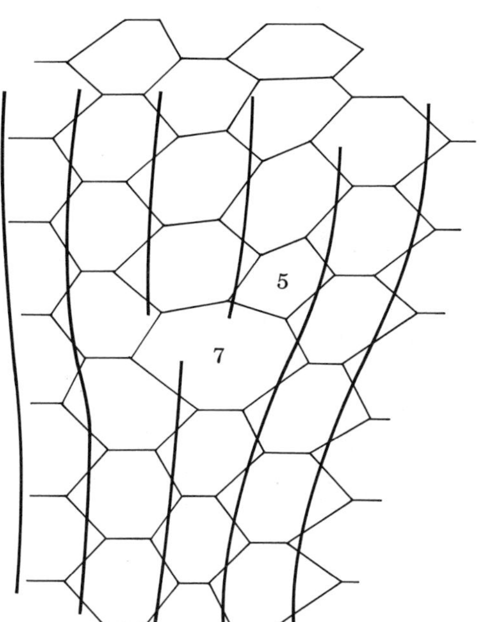

FIGURE 19-6. "Dislocation" in a foam lattice.

Individual grains are usually not perfect single crystals but are composed of *subgrains*, slightly rotated in relation to each other and joined by *subboundaries* (Fig. 19-7). As discussed below, such subboundaries are planar arrangements of individual dislocations. The energy of a crystallite containing dispersed dislocations is lowered when the dislocations arrange themselves into boundaries, so that upon annealing, subboundaries form to accommodate those dislocations which are not annihilated by other dislocations or at free surfaces. A particularly striking example is that of polygonization. A single crystal bent about one axis typically has the dislocation distribution of Fig. 19-8*a*. Upon annealing, the dislocations rearrange into *tilt* boundaries (Fig. 19-8*b*) separating stress-free blocks tilted about the bending axis relative to one another. Etch-pit studies and decoration studies have directly corroborated the model of Fig. 19-8 as reviewed by Amelinckx.[7]

The dislocation density in a grain boundary increases with increasing misorientation, until, for misorientations greater than 20 to 25°, the concept of individual dislocations in the boundary becomes meaningless. Formally, in such a case, the boundary can still be regarded as the superposition of dislocation boundaries. Our treatment here is restricted mainly to the theory of low-angle grain boundaries for which dislocation models are applicable. Overwhelming experimental evidence has accumulated to prove the validity of the dislocation models. The density of etch pits revealing the emerging edge

[7]S. Amelinckx, *Solid State Phys.*, Suppl., **6**: 1 (1974).

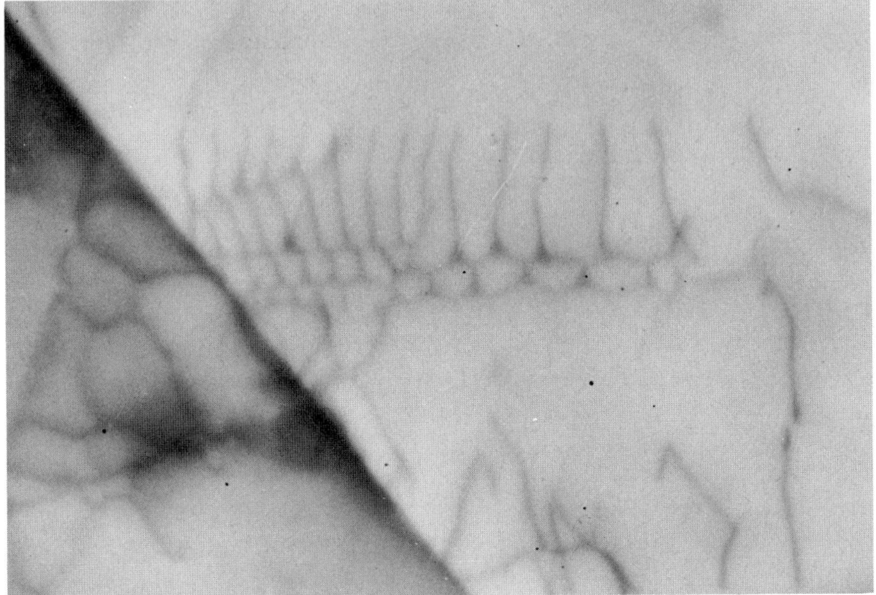

FIGURE 19-7. Subboundary junction with a grain boundary in austenitic stainless steel, 65,000× (P. B. Hirsch, private communication).

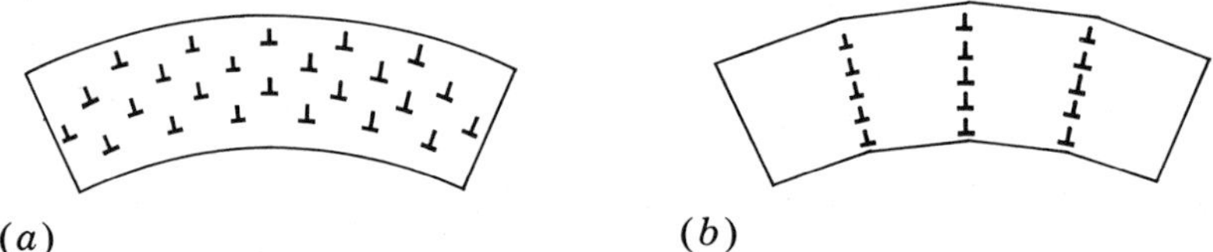

FIGURE 19-8. Rearrangement of (*a*) glide dislocations into (*b*) tilt boundaries.

dislocations in small-angle grain boundaries checks with the misorientations as measured by x-ray diffraction.[8] Also, electron-microscopy studies[9] and observations[10] on decorated dislocations in transparent crystals directly show the dislocation structure of the boundaries as exemplified in Fig. 1-15.

For a boundary of a given orientation, producing a given misorientation between grains, we discuss such important problems as boundary structure, boundary energy, and modes of boundary motion in terms of dislocations. Conversely, the properties of grain boundaries are shown to yield some information about individual dislocation properties, such as core parameters and core diffusion, which are difficult to obtain in other ways. Next, we

[8]F. L. Vogel, W. G. Pfann, H. E. Corey, and E. E. Thomas, *Phys. Rev.*, **90**: 489 (1953).

[9]P. B. Hirsch, R. W. Horne, and M. J. Whelan, *Phil. Mag.*, **1**: 677 (1956).

[10]J. M. Hedges and J. W. Mitchell, *Phil. Mag.*, **44**: 223 (1953).

consider the types of boundaries that can form from given sets of dislocations introduced by deformation, a problem important in the understanding of cell formation during creep and low-temperature deformation. Among high-angle boundaries, special coincidence lattice boundaries exist,[11] across which a small fraction of the atoms on either side of the boundary are in perfect crystallographic order, forming a sort of superlattice. Such boundaries have low boundary tensions relative to general high-angle boundaries. They also contain special grain boundary dislocations as discussed in Sec. 19-6.

The following discussion is based in part on the review article by Amelinckx and Dekeyser,[12] to which we refer for more details, particularly for discussions of observations. A broader treatment of the general properties of grain boundaries is presented by McLean.[13]

19-2. DISLOCATION MODELS OF GRAIN BOUNDARIES

Simple Boundaries

If grain B can be brought into the same orientation as grain A by a rotation $-\boldsymbol{\omega}$, then grain B is rotated by the vector $\boldsymbol{\omega}$, which defines the rotation axis and the amount of the rotation. If $\boldsymbol{\omega}$ is contained in the boundary, the grain boundary is a *tilt boundary* (Fig. 19-9), while if $\boldsymbol{\omega}$ is perpendicular to the boundary, it is a pure *twist boundary* (Fig. 19-10). In the general case a boundary is of mixed character, containing both tilt and twist components.

The simplest possible boundary is a tilt boundary composed of only one set of edge dislocations. Consider a simple cubic crystal with the boundary in a symmetrical position with respect to the (100) planes of the two crystals, and with the rotation axis parallel to [001]. One might expect that the (100) planes of such crystals would meet in pairs, as shown in Fig. 19-11*a*. The boundary could then be considered as an array of parallel edge dislocations, with the Burgers vector 2**b** of an inserted double plane (Fig. 19-11*b*). However, the elastic energy is lowered by displacing the boundary so that (100) planes from grains A and B end in the boundary at alternating intervals (Fig. 19-12*a*), yielding a tilt boundary composed of elementary edge dislocations with Burgers vector **b** (Fig. 19-12*b*). With an angle of tilt θ, the separation D between the edge dislocations is fixed. All (100) planes intersecting the surface within the region ABC must end as incomplete planes at the boundary (Fig. 19-13). The number of incomplete planes is

$$n_p = \frac{2h}{b}\sin\frac{\theta}{2} \tag{19-4}$$

[11] M. L. Kronberg and F. H. Wilson, *Trans. AIME*, **185**: 501 (1949).

[12] S. Amelinckx and W. Dekeyser, *Solid State Phys.*, **8**: 325 (1959).

[13] D. McLean, "Grain Boundaries in Metals," Oxford University Press, Fair Lawn, N.J., 1957.

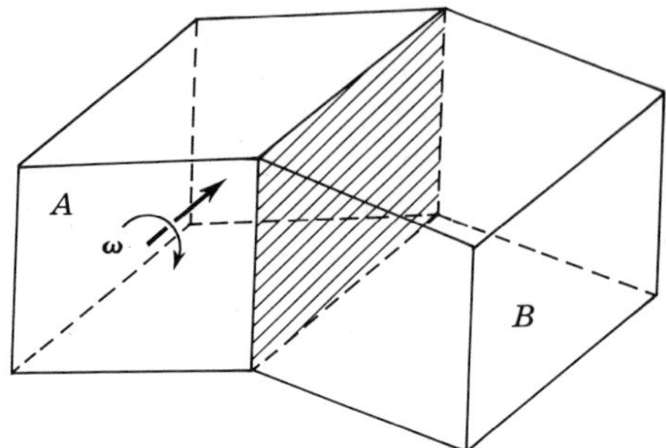

FIGURE 19-9. A tilt boundary defined by the rotation vector ω in the plane of the boundary.

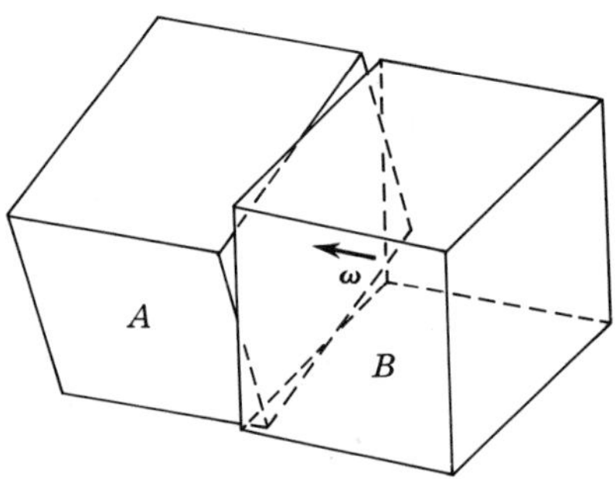

FIGURE 19-10. A twist boundary, where ω is normal to the boundary plane.

(a) (b)

FIGURE 19-11. An unstable tilt boundary.

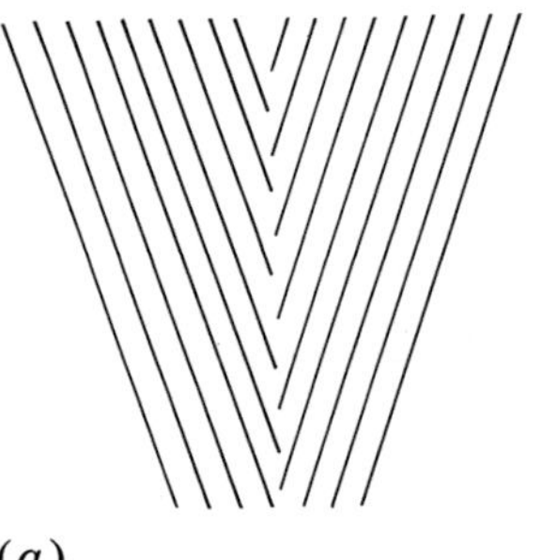

(a) (b)

FIGURE 19-12. A stable tilt boundary.

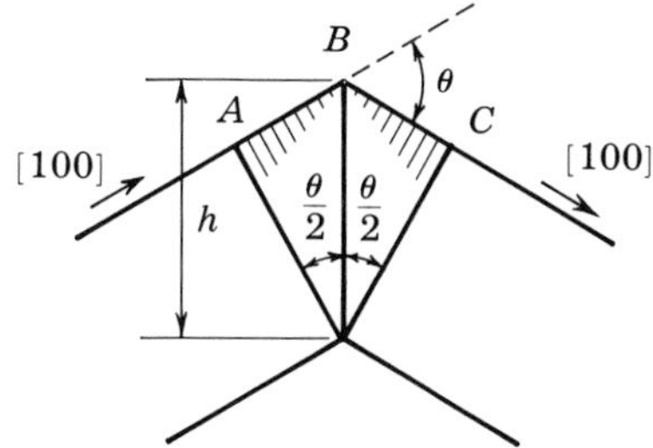

FIGURE 19-13. A (100) tilt boundary.

Hence, the mean separation between dislocations in the boundary is

$$D = \frac{h}{n_p} = \frac{b}{2\sin(\theta/2)} \tag{19-5}$$

or, in the limit of small angles θ,

$$D \cong \frac{b}{\theta} \tag{19-6}$$

With an angle ψ between the plane of the boundary and the symmetric orientation of Fig. 19-13, *two* sets of dislocations are required to construct even a simple tilt boundary. Let the subscripts 1 and 2 refer to dislocations of Burgers vector [100] and [010], respectively. As is evident in Fig. 19-14*a*, both (100) and (010) planes must terminate in the boundary, so that dislocations of both sets $\mathbf{b}_1$ and $\mathbf{b}_2$ are present. Because a number $(h/b)\sin(\psi+\theta/2)$ of planes of type (100) intersect AB, while $(h/b)\sin(\psi-\theta/2)$ planes intersect EF, the number of (100) planes ending at the boundary is

$$\begin{aligned} n_{p_1} &= \frac{h}{b}\left[\sin\left(\psi+\frac{\theta}{2}\right)-\sin\left(\psi-\frac{\theta}{2}\right)\right] \\ &\cong \frac{h\theta}{b}\cos\psi \end{aligned} \tag{19-7}$$

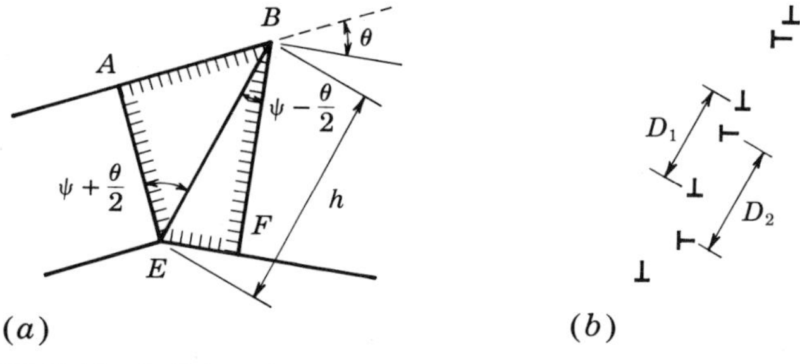

FIGURE 19-14. A tilt boundary inclined at an angle ψ to the symmetric position.

The corresponding separation between dislocations is

$$D_1 = \frac{b}{\theta \cos \psi} \tag{19-8}$$

Similarly, for the dislocations associated with (010) planes,

$$D_2 = \frac{b}{\theta \sin \psi} \tag{19-9}$$

The dislocation model for the resultant boundary is depicted in Fig. 19-14*b*. Note that the two sets of dislocations do not truly lie in a plane because of local interactions that tend to make the dislocations locally lie along a line inclined at 45° to their Burgers vectors.

The simple twist boundary always requires at least two sets of dislocations for its representation. Consider first a wall of one set of parallel screw dislocations, introduced in an orthogonal-section plane of a bar of square cross section (Fig. 19-15). The two parts of the crystal must meet at this boundary, with cross sections *deformed elastically in shear*. Therefore this arrangement cannot represent a grain boundary joining stress-free crystals. However, with another set of parallel screw dislocations in the boundary, orthogonal to the first set and with the same mean separation for the same magnitude of Burgers vector, the cross sections meeting at the boundary are simply rotated relative to each other, without change in shape. Such a crossed grid of screw dislocations indeed forms a pure *twist* boundary (Fig. 19-16). In general, when the twist boundary is composed of nonorthogonal dislocations, they are not pure screw dislocations.

Exercise 19-2. Show that the mean separation of screw dislocations in either set in Fig. 19-16 is given by Eqs. (19-5) or (19-6).

Net Dislocation Density in an Arbitrary Small-Angle Boundary

Consider a wall containing several sets of dislocations, with the dislocations in each set parallel and evenly spaced and of the same Burgers vector (Fig. 19-17). Choose a vector **V** in the boundary, and then specify that vector in terms of the

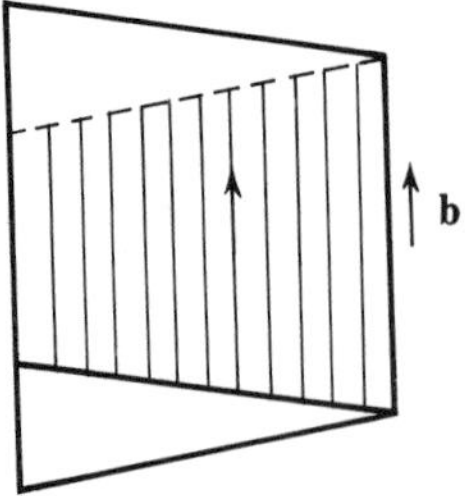

FIGURE 19-15. A single set of parallel screw dislocations in a planar array.

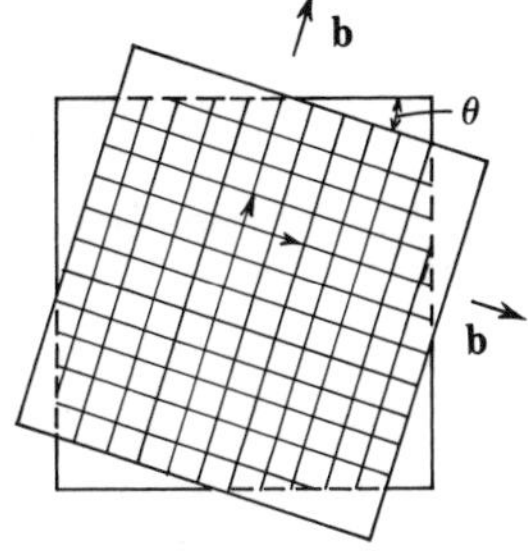

FIGURE 19-16. A crossed grid of screw dislocations.

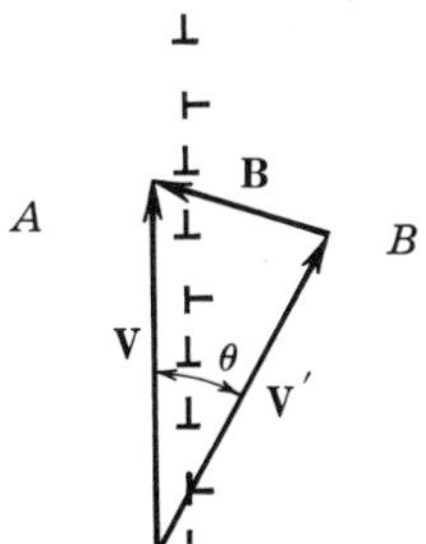

FIGURE 19-17. Closure failure **B** for a small-angle boundary.

crystallographic coordinates of grain A. Next draw from the same origin O the vector $\mathbf{V}'$ that has the same crystallographic indices with respect to grain B. If there were no dislocations in the wall, $\mathbf{V}'$ and $\mathbf{V}$ would be identical, and the two grains would be two parts of one perfect crystal. Thus the closure failure caused by dislocations in the wall is

$$\mathbf{B} = \mathbf{V} - \mathbf{V}' \tag{19-10}$$

Equation (19-10) is simply a vector description of the operations discussed in connection with Fig. 19-14. In terms of the i sets of dislocations, $\mathbf{B}$ is the resultant Burgers vector of all the dislocations cut by $\mathbf{V}$,

$$\mathbf{B}(\mathbf{V}) = \sum_i c_i(\mathbf{V})\mathbf{b}_i \tag{19-11}$$

where c_i is the number of dislocations of Burgers vector $\mathbf{b}_i$ cut by $\mathbf{V}$. A dislocation cut by $\mathbf{V}$ is counted as a positive contribution to $\mathbf{B}$ if $\mathbf{V} \times \mathbf{n}$ has a positive component along $\boldsymbol{\xi}$, where $\mathbf{n}$ is the unit normal to the boundary pointing from grain A to grain B. Equation (19-11) is generally valid for a wall of dislocations, whether the wall is a true grain boundary between stress-free crystals or whether it is associated with long-range stresses, as is the case for the boundary in Fig. 19-17.

If the dislocation wall is a true grain boundary, crystal B is merely rotated with respect to crystal A and is not deformed. For small angles, it is convenient to introduce the expression

$$\boldsymbol{\omega} = \theta \mathbf{a} \tag{19-12}$$

where $\mathbf{a}$ is a unit vector along the axis of rotation. In the small-angle approximation, to the same order that $2\sin(\theta/2) \simeq \theta$,

$$\mathbf{B} = \mathbf{V} \times \boldsymbol{\omega} = \theta(\mathbf{V} \times \mathbf{a}) \tag{19-13}$$

Those arrangements of boundary dislocations which satisfy both Eqs. (19-11) and (19-13) for all possible $\mathbf{V}$ in the boundary represent grain boundaries of rotation $\boldsymbol{\omega}$. Equations (19-11) and (19-13) constitute Frank's formula[14] for small-angle grain boundaries. The dislocations defined by $\mathbf{B}$ constitute a true grain boundary array with no long-range strain field. We define these as *intrinsic* grain boundary dislocations. Added dislocations that retain their long-range stress fields are defined as *extrinsic* grain boundary dislocations.

Arbitrary Large-Angle Boundary

Large-angle grain boundaries can be specified by $\mathbf{B}$, as in the previous case, but one must be more specific about the definition of $\mathbf{V}$ and of $\mathbf{b}_i$. The Burgers vectors $\mathbf{b}_i$ are now referred to a perfect reference lattice by the **FS**/*RH* convention of Fig. 1-20. The closure failure $\mathbf{B}$ is also referred to the reference lattice by an **FS**/*RH* Burgers circuit, but with the circuit being taken in a right-handed manner about $\boldsymbol{\omega}$. An example of the procedure is presented in Fig. 19-18. Relative to the reference crystal, grain B is produced by a rotation $\theta_1\mathbf{a}$, while grain A is produced by a rotation $-\theta_2\mathbf{a}$. $\mathbf{V}$ is a vector in the plane of the boundary that produces the vector $\mathbf{V}'$ when rotated by $\theta_1\mathbf{a}$, or the vector $\mathbf{V}''$ when rotated by $-\theta_2\mathbf{a}$. Here $\theta_1 + \theta_2 = \theta$. The Burgers circuit is drawn in Fig. 19-18*b*, yielding the closure failure $\mathbf{B}$. Inspection of Fig. 19-18*b* shows that $\mathbf{B}$ is again given by the result of Eq. (19-11). Indeed, in the case of Fig. 19-18*b*, $\mathbf{B} = 7\mathbf{b}_2 + \mathbf{b}_1$, and referring to Fig. 19-18*a*, one sees that seven planes of type (100) and one of type (010) end along $\mathbf{V}$ in the boundary.

Once $\mathbf{V}'$ and $\mathbf{V}''$ are selected the above procedure gives a well-defined $\mathbf{B}$ referred to the reference crystal. There is a degree of arbitrariness in $\mathbf{B}$ in that a different $\mathbf{V}'$ or $\mathbf{V}''$ could have been selected, for example, $\mathbf{V}'$ at $\theta_1 + \pi/2$ in Fig. 19-18. Usually an obvious choice is the set that minimizes θ_1 and the other equivalent angles for a general boundary. For a given grain boundary the reference crystal can always be oriented such that the grain boundary is produced by rotations $\theta_1 = -\theta_2 = \frac{1}{2}\theta$ (Fig. 19-18*c*). In this situation $\mathbf{V}$ is orthogonal to $\mathbf{B}$, which in turn is related to θ by the expression

$$\mathbf{B} = 2\sin\frac{\theta}{2}(\mathbf{V} \times \mathbf{a}) \tag{19-14}$$

[14] F. C. Frank, "Report of the Symposium on the Plastic Deformation of Crystalline Solids," Carnegie Institute of Technology, Pittsburgh, 1950, p. 150.

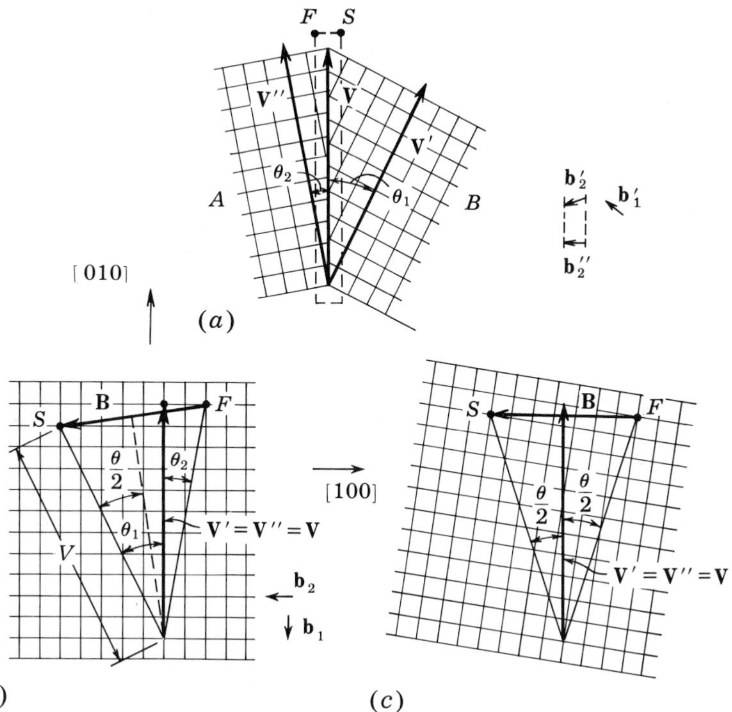

FIGURE 19-18. Closure failure **B** for a large-angle boundary: (*a*) actual boundary, (*b*) Burgers circuit in reference crystal, (*c*) reference crystal tilted relative to (*b*), so that **V** is orthogonal to **B**; **B** is the same in both cases, but **V** is not, when referred to the reference crystal. $\boldsymbol{\xi}$ points into the plane of the page.

The method of Fig. 19-18*c* is thus the most convenient one for relating the large-angle development to that for the small-angle boundary.

With the above definitions, Eqs. (19-11) and (19-14) represent Frank's formula[14] for large-angle grain boundaries. A dislocation boundary described by these equations is a grain boundary of rotation ω, separating crystals free from long-range stress fields.

We emphasize that the Burgers vectors must be referred to the reference crystal to avoid ambiguity.[14, 15] For example, the dislocation content in Fig. 19-18*b* could, in principle, be referred to the Burgers vectors in the rotated crystals, $\mathbf{b}'$ or to $\mathbf{b}''$, the components of $\mathbf{b}'$ normal to the boundary plane.[16] Only the **b** referred to the reference crystal, however, can be consistently used in the following grain-boundary formulas.

[15] For a discussion of the relations among **b**, $\mathbf{b}'$, and $\mathbf{b}''$ see R. Bullough, *Phil. Mag.*, **12**: 1139 (1965).

[16] A. W. Sleeswyk, *Acta Met.*, **11**: 1192 (1963).

A somewhat different general treatment of grain boundaries in terms of *surface dislocations* has been developed by Bilby, Bullough, and Smith.[17–19] Their result is equivalent to our treatment for the case of grain boundaries, but they express the dislocation content **B** of the boundary in terms of a general displacement tensor; their result is important in the description of martensitic transformation boundaries and other interfaces. In terms of the notation used here, their result is

$$\mathbf{B}=\boldsymbol{\beta}\cdot(\mathbf{V}\times\mathbf{n}) \quad \text{or} \quad B_i=\beta_{ij}\epsilon_{jkl}V_k n_l \tag{19-15}$$

The tensor β_{ij} represents the ith component of the total Burgers vector of boundary dislocations cut by a line in the boundary perpendicular to the unit vector **j** and of length $\cos\alpha_j$, where α_j is the inclination of **j** to the boundary. In general, β_{ij} relates to the general deformation matrix e_{il} across the boundary,

$$\beta_{ij}=\epsilon_{jkl}n_k e_{il}$$

For grain boundaries of small rotation $\boldsymbol{\omega}$,

$$e_{il}=\epsilon_{mil}\omega_m$$

and β_{ij} is given by

$$\beta_{ij}=\mathbf{i}\cdot\boldsymbol{\beta}_j \tag{19-16}$$

where

$$\boldsymbol{\beta}_j=(\mathbf{j}\times\mathbf{n})\times\boldsymbol{\omega} \tag{19-17}$$

Dislocation Spacing in the Boundary

The Frank equations for general grain boundaries can be reduced to convenient parametric expressions for the dislocation density by the introduction of a reciprocal vector notation. If the spacing between boundary dislocations $\mathbf{b}_i$ is D_i, then N_i is defined by

$$N_i=\frac{1}{2D_i\sin(\theta/2)} \tag{19-18}$$

and the vector $\mathbf{N}_i$ by

$$\mathbf{N}_i=N_i(\mathbf{n}\times\boldsymbol{\xi}_i) \tag{19-19}$$

[17] B. A. Bilby, in "Report of the Conference on Defects in Crystalline Solids," Physical Society, London, 1955, p. 124.

[18] B. A. Bilby, R. Bullough, and E. Smith, *Proc. Roy. Soc.*, **A231**: 263 (1955).

[19] R. Bullough, "Dislocations," A.E.R.E., Harwell, 1964.

where $\boldsymbol{\xi}_i$ is the sense vector for dislocations of type i. The vector $\mathbf{N}_i$ is a reciprocal vector lying in the boundary and perpendicular to the dislocation lines of type i. Given the values of $\mathbf{N}_i$ and $\mathbf{n}$, the directions of all dislocations are determined from

$$\boldsymbol{\xi}_i = \frac{\mathbf{N}_i}{N_i} \times \mathbf{n} \tag{19-20}$$

and the separations between the dislocations in each set are given by Eq. (19-18). The complete description of a general grain boundary is thus reduced to the determination of the possible $\mathbf{N}_i$ values for a given $\mathbf{n}$ and $\mathbf{a}$.

In terms of $\mathbf{N}_i$,

$$c_i(\mathbf{V}) = (\mathbf{N}_i \cdot \mathbf{V}) 2 \sin\frac{\theta}{2} \tag{19-21}$$

as can be verified by constructing a parallel set of dislocations cut by an arbitrary vector $\mathbf{V}$ in the plane of the set. Hence, after cancellation of the common factor $2\sin(\theta/2)$, Eqs. (19-11) and (19-14) combine to the result

$$\mathbf{V} \times \mathbf{a} = \sum_i \mathbf{b}_i (\mathbf{N}_i \cdot \mathbf{V}) \tag{19-22}$$

for all vectors $\mathbf{V}$ in the boundary.

If Eq. (19-22) is satisfied for *two* independent directions of $\mathbf{V}$ in the boundary, it is satisfied for all $\mathbf{V}$ in the boundary. For each $\mathbf{V}$, the vector equation has three components. Thus, with a given $\mathbf{a}$ and $\mathbf{b}_i$, Eq. (19-22) constitutes six equations for the determination of the values of $\mathbf{N}_i$. For the case of three sets of dislocations with independent Burgers vectors, the three $\mathbf{N}_i$ values represent six unknowns, because each $\mathbf{N}_i$ has two components in the boundary. Therefore, with three independent $\mathbf{b}_i$, a unique set of three $\mathbf{N}_i$ can be found for the general boundary.[20] With more than three sets of $\mathbf{b}_i$, the solution is indeterminate; i.e., there are many possible dislocation constructions for the boundary. With less than three independent sets of $\mathbf{b}_i$, only special boundaries are possible, as discussed in the next section.

Given $\mathbf{a}$, $\mathbf{n}$, and three independent $\mathbf{b}_i$, the solution for $\mathbf{N}_i$ is as follows. Let us define the reciprocal Burgers vectors $\mathbf{b}_i^*$ as

$$\begin{aligned} \mathbf{b}_1^* &= \frac{\mathbf{b}_2 \times \mathbf{b}_3}{\mathbf{b}_1 \cdot (\mathbf{b}_2 \times \mathbf{b}_3)} \\ \mathbf{b}_2^* &= \frac{\mathbf{b}_3 \times \mathbf{b}_1}{\mathbf{b}_2 \cdot (\mathbf{b}_3 \times \mathbf{b}_1)} \\ \mathbf{b}_3^* &= \frac{\mathbf{b}_1 \times \mathbf{b}_2}{\mathbf{b}_3 \cdot (\mathbf{b}_1 \times \mathbf{b}_2)} \end{aligned} \tag{19-23}$$

[20] That three independent $\mathbf{b}_i$ are required in general can be rationalized by imagining a mixed boundary composed of a simple tilt boundary (one set) superposed upon a simple twist boundary (two more sets).

Equation (19-23) indicates that

$$\mathbf{b}_i^* \cdot \mathbf{b}_j = \delta_{ij} = \begin{cases} 0 & i \neq j \\ 1 & i = j \end{cases} \tag{19-24}$$

Hence the multiplication of Eq. (19-22) by $\mathbf{b}_i^*$ yields the result

$$\mathbf{N}_i \cdot \mathbf{V} = \mathbf{b}_i^* \cdot (\mathbf{V} \times \mathbf{a}) = (\mathbf{a} \times \mathbf{b}_i^*) \cdot \mathbf{V} \tag{19-25}$$

Since $\mathbf{V}$ is any vector in the boundary, $\mathbf{N}_i$ is the component of $\mathbf{a} \times \mathbf{b}_i^*$ in the boundary,

$$\mathbf{N}_i = \mathbf{a} \times \mathbf{b}_i^* - \mathbf{n}[\mathbf{n} \cdot (\mathbf{a} \times \mathbf{b}_i^*)] \tag{19-26}$$

Equation (19-26), the requisite expression for defining a general boundary, was derived originally by Read and Shockley.[21]

The indeterminacy with more than three sets of dislocations can now be discussed more directly. Because the Burgers vector of a fourth set $\mathbf{b}_4$ can be expressed as a linear combination of $\mathbf{b}_1$, $\mathbf{b}_2$, and $\mathbf{b}_3$, we can treat without loss of generality the simpler case of splitting of dislocations of Burgers vector $\mathbf{b}_1$ into two subsets, $\mathbf{N}_1'$ and $\mathbf{N}_1''$. If $\mathbf{N}_1'$ and $\mathbf{N}_1''$ are chosen such that

$$\mathbf{N}_1' + \mathbf{N}_1'' = \mathbf{N}_1 \tag{19-27}$$

with $\mathbf{N}_1$ given by Eq. (19-26), Eq. (19-22) still is satisfied for all $\mathbf{V}$. Therefore the arrangements of Figs. 19-19*a* and 19-19*b* are equivalent with regard to the dislocation contribution to rotation and the cancellation of long-range stresses. When several dislocation models for a boundary are possible, the correct model is the one with lowest energy. To the first approximation, the various sets of dislocations contribute to the boundary energy in proportion to the density of the dislocations and the square of the Burgers vector. In this approximation, all dislocations of the same $\mathbf{b}$ remain in one set as in Fig. 19-19*b*, rather than splitting into two sets. Splitting into two sets is likely only when "faceting" of a single dislocation is energetically favorable, i.e., when a straight line is not the minimum-energy form.[22] *Local* reactions of crossing straight dislocations to form short segments of dislocations with different $\mathbf{b}$ in mesh networks are sometimes energetically favorable; such rearrangements are discussed in Sec. 19-4. In summary, then, any general grain boundary can be generated by moving three independent sets of dislocations into the boundary; other component dislocations in an actual boundary are produced by local reactions within the boundary.

[21] W. T. Read and W. Shockley, in "Imperfections in Nearly Perfect Crystals," Wiley, New York, 1952, p. 352.

[22] Such instability would occur when the line tension [Eq. (13-225)] becomes negative. This cannot occur in isotropic crystals, but can occur in anisotropic crystals.

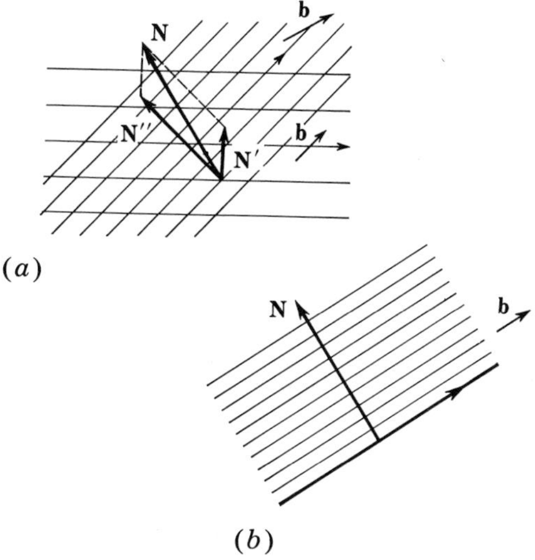

FIGURE 19-19. (*a*) A two-dislocation net equivalent to (*b*) a one-dislocation net.

Exercise 19-3. Show that when two subsets of dislocations are formed according to Eq. (19-27), the total density of dislocations per unit area is always greater than that of the original set $\mathbf{N}_1$.

19-3. BOUNDARIES POSSIBLE UNDER RESTRICTED CONDITIONS

Because three independent sets of dislocations, i.e., sets with three noncoplanar Burgers vectors, are required for a general boundary, only special boundaries can form when dislocations of only one or two sets are introduced into a crystal. Some boundary formation occurs during deformation at low temperatures, or during recovery, under conditions where no climb takes place, so that boundaries found under the restriction of glide only being allowed are of importance.[23, 24] The subsequent relaxation of such boundaries under recovery conditions such that climb occurs is also of interest.[25] Restricted boundaries of both types are treated systematically in this section. The results of these considerations are important in the understanding of recovery, polygonization, cell formation during deformation, recrystallization *in situ*, and work hardening.

[23] C. J. Ball and P. B. Hirsch, *Phil. Mag.*, **46**: 1343 (1955).

[24] C. J. Ball, *Phil. Mag.*, **2**: 977 (1957).

[25] S. Amelinckx, *Physica*, **23**: 663 (1957).

One Set of Dislocations

Suppose that only one kind of dislocation of Burgers vector $\mathbf{b}$ is present in a boundary, as is the case after deformation involving only one slip system. According to Eq. (19-22), $\mathbf{V}\times\mathbf{a}$ must then be parallel to $\mathbf{b}$ for any $\mathbf{V}$,

$$\mathbf{V}\times\mathbf{a}\|\mathbf{b} \tag{19-28}$$

so that $\mathbf{b}$ must be perpendicular to both $\mathbf{V}$ and $\mathbf{a}$. Thus, the plane of the boundary is normal to $\mathbf{b}$ and the rotation axis $\mathbf{a}$ is contained in the boundary; the boundary is a pure *tilt* boundary. Since the closure failure $\mathbf{B}\propto\mathbf{V}\times\mathbf{a}$ is zero when $\mathbf{V}$ is parallel to the line directions $\boldsymbol{\xi}$, $\mathbf{a}$ must also be parallel to $\boldsymbol{\xi}$,

$$\mathbf{a}\|\boldsymbol{\xi} \tag{19-29}$$

With climb allowed, $\mathbf{a}$ can have any direction in the boundary. If the dislocation motion is restricted to glide only, the dislocations in the boundary must be parallel to $\mathbf{p}\times\mathbf{b}$, where $\mathbf{p}$ is the unit normal to the glide plane. Therefore, by condition (19-29), $\mathbf{a}$ must also be parallel to this direction,

$$\boldsymbol{\xi}\|\mathbf{a}\|\mathbf{p}\times\mathbf{b} \tag{19-30}$$

yielding the unique tilt boundary of the type in Fig. 19-12. When the glide plane itself is not unique, i.e., when cross slip or pencil glide occur, $\mathbf{a}$ once again can have any direction in the boundary, as in the climb case.

Two Sets of Dislocations

The introduction of two sets of dislocations in deformation is a common case. In NaCl-type structures, for example, the two slip systems $[101](\bar{1}01)$ and $[\bar{1}01](101)$ are always equally stressed, so that such a pair of slip systems is often activated in deformation. Consider, then, two sets of dislocations with Burgers vectors $\mathbf{b}_1$ and $\mathbf{b}_2$. Equation (19-22) becomes

$$\mathbf{V}\times\mathbf{a}=\mathbf{b}_1(\mathbf{N}_1\cdot\mathbf{V})+\mathbf{b}_2(\mathbf{N}_2\cdot\mathbf{V}) \tag{19-31}$$

Multiplying Eq. (19-31) by $\mathbf{b}_1\times\mathbf{b}_2$, one finds

$$\mathbf{V}\cdot[\mathbf{a}\times(\mathbf{b}_1\times\mathbf{b}_2)]=0 \tag{19-32}$$

One of the two types of boundaries satisfying condition (19-32) is that where $\mathbf{V}$ is perpendicular to $\mathbf{a}\times(\mathbf{b}_1\times\mathbf{b}_2)$. Since $\mathbf{V}$ is perpendicular to the boundary normal $\mathbf{n}$ for *any* $\mathbf{V}$, according to Eq. (19-32),

$$\mathbf{n}\|\mathbf{a}\times(\mathbf{b}_1\times\mathbf{b}_2) \tag{19-33}$$

Thus, both $\mathbf{a}$ and $\mathbf{b}_1 \times \mathbf{b}_2$ lie in the plane of the boundary; it is a pure tilt boundary. The vector $\mathbf{a}$ can have any orientation in the boundary plane. If one chooses the arbitrary boundary vector $\mathbf{V}$ to be equal to $\mathbf{a}$, Eq. (19-31) reduces to

$$\mathbf{b}_1(\mathbf{N}_1 \cdot \mathbf{a}) + \mathbf{b}_2(\mathbf{N}_2 \cdot \mathbf{a}) = 0 \tag{19-34}$$

Since $\mathbf{b}_1$ and $\mathbf{b}_2$ are not collinear, Eq. (19-34) requires that

$$\mathbf{N}_1 \cdot \mathbf{a} = \mathbf{N}_2 \cdot \mathbf{a} = 0$$

Hence condition (19-29) holds for this case, the dislocations of both sets are all parallel to $\mathbf{a}$, and the tilt boundary is of the type shown in Fig. 19-14. The magnitudes of $\mathbf{N}_1$ and $\mathbf{N}_2$ are found by inserting $\mathbf{V} = \mathbf{N}_1 / N_1$ into Eq. (19-31), taking the vector product of the result with $\mathbf{b}_2$, and then finding the scalar product of the latter result and $\mathbf{a}$, which yields

$$N_1 = \frac{\mathbf{b}_2 \cdot \{[(\mathbf{N}_1/N_1) \times \mathbf{a}] \times \mathbf{a}\}}{\mathbf{a} \cdot (\mathbf{b}_2 \times \mathbf{b}_1)} \qquad N_2 = \frac{\mathbf{b}_1 \cdot \{[(\mathbf{N}_2/N_2) \times \mathbf{a}] \times \mathbf{a}\}}{\mathbf{a} \cdot (\mathbf{b}_1 \times \mathbf{b}_2)} \tag{19-35}$$

If only glide can contribute to boundary formation, and if the dislocations have different glide planes with normals $\mathbf{p}_1$ and $\mathbf{p}_2$, the dislocations in the boundary must be parallel to $\mathbf{p}_1 \times \mathbf{p}_2$, so that condition (19-29) yields the result

$$\mathbf{a} \| \boldsymbol{\xi} \| \mathbf{p}_1 \times \mathbf{p}_2 \tag{19-36}$$

When $\mathbf{p}_1 \times \mathbf{p}_2$ is not collinear with $\mathbf{b}_1 \times \mathbf{b}_2$, only one boundary plane is possible,

$$\mathbf{n} \| (\mathbf{p}_1 \times \mathbf{p}_2) \times (\mathbf{b}_1 \times \mathbf{b}_2) \tag{19-37}$$

If $\mathbf{p}_1 \times \mathbf{p}_2$ and $\mathbf{b}_1 \times \mathbf{b}_2$ are collinear, all planes parallel to $\mathbf{b}_1 \times \mathbf{b}_2$ are possible boundary planes. The final special case of the glide-restricted type occurs when both sets of dislocations have the same glide plane $\mathbf{p}$. Then

$$\mathbf{p} \| \mathbf{b}_1 \times \mathbf{b}_2 \tag{19-38}$$

so that the possible boundary planes are normal to the glide plane, since the boundary plane must contain $\mathbf{b}_1 \times \mathbf{b}_2$. Both the dislocation lines and $\mathbf{a}$ are parallel to the line of intersection between the boundary plane and glide plane.

The other class of boundary that satisfies condition (19-32) is that where

$$\mathbf{a} \| \mathbf{b}_1 \times \mathbf{b}_2 \qquad \text{or} \qquad \mathbf{a} = \frac{\mathbf{b}_1 \times \mathbf{b}_2}{|\mathbf{b}_1 \times \mathbf{b}_2|} \tag{19-39}$$

All possible boundaries of this type have a common rotation axis for a given $\mathbf{b}_1$

and $\mathbf{b}_2$. The vector product of $\mathbf{V}$ and $\mathbf{a}$, as given by Eq. (19-39), is

$$\mathbf{V}\times\mathbf{a}=\frac{\mathbf{b}_1(\mathbf{b}_2\cdot\mathbf{V})-\mathbf{b}_2(\mathbf{b}_1\cdot\mathbf{V})}{|\mathbf{b}_1\times\mathbf{b}_2|} \tag{19-40}$$

Comparing this result with Eq. (19-31) for all $\mathbf{V}$ in the boundary, one finds that the reciprocal vectors $\mathbf{N}_1$ and $\mathbf{N}_2$ are simply the components in the boundary plane of $\mathbf{b}_2/|\mathbf{b}_1\times\mathbf{b}_2|$ and $-\mathbf{b}_1/|\mathbf{b}_1\times\mathbf{b}_2|$, respectively,

$$\mathbf{N}_1=\frac{\mathbf{b}_2-\mathbf{n}(\mathbf{n}\cdot\mathbf{b}_2)}{|\mathbf{b}_1\times\mathbf{b}_2|} \qquad \mathbf{N}_2=\frac{-\mathbf{b}_1+\mathbf{n}(\mathbf{n}\cdot\mathbf{b}_1)}{|\mathbf{b}_1\times\mathbf{b}_2|} \tag{19-41}$$

Thus the lines $\boldsymbol{\xi}_1$ and $\boldsymbol{\xi}_2$ are nonparallel and form a crossed grid in the boundary.

When, in particular,

$$\mathbf{n}=\frac{\mathbf{b}_1\times\mathbf{b}_2}{|\mathbf{b}_1\times\mathbf{b}_2|} \tag{19-42}$$

the boundary is a twist boundary, so that Eq. (19-41) reduces to

$$\mathbf{N}_1=\frac{\mathbf{b}_2}{|\mathbf{b}_1\times\mathbf{b}_2|} \qquad \mathbf{N}_2=\frac{-\mathbf{b}_1}{|\mathbf{b}_1\times\mathbf{b}_2|} \tag{19-43}$$

Thus the vectors $\mathbf{N}_i$ for arbitrary $\mathbf{n}$ are the projections onto that boundary of the $\mathbf{N}_i$ for the pure twist boundary. Conversely, as shown in Exercise 19-4, the dislocation grid for a given $\mathbf{n}$ and θ projects onto the plane of the pure twist boundary as the correct twist-boundary grid.[26]

If only glide boundaries are allowed, the dislocations must be parallel to $\mathbf{p}\times\mathbf{n}$, so that

$$(\mathbf{p}_1\times\mathbf{n})\cdot\mathbf{N}_1=(\mathbf{p}_2\times\mathbf{n})\cdot\mathbf{N}_2=0 \tag{19-44}$$

which, together with Eq. (19-41), gives

$$(\mathbf{p}_1\times\mathbf{n})\cdot\mathbf{b}_2=(\mathbf{b}_2\times\mathbf{p}_1)\cdot\mathbf{n}=0$$

$$(\mathbf{p}_2\times\mathbf{n})\cdot\mathbf{b}_1=(\mathbf{b}_1\times\mathbf{p}_2)\cdot\mathbf{n}=0 \tag{19-45}$$

Thus only one boundary plane is possible,

$$\mathbf{n}\,\|\,(\mathbf{b}_1\times\mathbf{p}_2)\times(\mathbf{b}_2\times\mathbf{p}_1) \tag{19-46}$$

[26] F. C. Frank, in "Report of the Conference on Defects in Crystalline Solids," Physical Society, London, 1955, p. 159.

However, if

$$\mathbf{b}_1 \times \mathbf{p}_2 \| \mathbf{b}_2 \times \mathbf{p}_1 \tag{19-47}$$

further analysis is necessary. Taking scalar products of condition (19-47) with $\mathbf{b}_1$, and $\mathbf{b}_2$, one finds that

$$(\mathbf{b}_1 \times \mathbf{p}_2) \cdot \mathbf{b}_2 = (\mathbf{b}_2 \times \mathbf{p}_1) \cdot \mathbf{b}_1 = 0$$

or (19-48)

$$\mathbf{p}_1 \cdot (\mathbf{b}_1 \times \mathbf{b}_2) = \mathbf{p}_2 \cdot (\mathbf{b}_1 \times \mathbf{b}_2) = 0$$

Equation (19-48) implies that $\mathbf{p}_1$ and $\mathbf{p}_2$ are linearly dependent on $\mathbf{b}_1$ and $\mathbf{b}_2$, so that

$$\mathbf{b}_1 \times \mathbf{p}_2 \| \mathbf{b}_2 \times \mathbf{p}_1 \| \mathbf{b}_1 \times \mathbf{b}_2 \tag{19-49}$$

Thus, according to Eq. (19-45), $\mathbf{n}$ is perpendicular to $\mathbf{b}_1 \times \mathbf{b}_2$ and the boundary is a tilt boundary, as already discussed in connection with Eq. (19-33). Also, if

$$\mathbf{p}_1 \| \mathbf{b}_2 \tag{19-50}$$

the derivation leading to Eq. (19-46) is not valid. When $\mathbf{p}_1 \| \mathbf{b}_2$, the glide planes are orthogonal. In such a case Eq. (19-45) leaves as the only condition on $\mathbf{n}$ that

$$\mathbf{n} \cdot (\mathbf{b}_1 \times \mathbf{p}_2) = 0$$

that is, the boundary is parallel to $\mathbf{b}_1 \times \mathbf{p}_2$. If also

$$\mathbf{p}_2 \| \mathbf{b}_1 \tag{19-51}$$

any boundary-plane orientation is allowed.

The final case consistent with Eq. (19-39) is that where both dislocations have the same glide plane $\mathbf{p}$. In such a case, only pure twist boundaries in the glide plane can form.

Exercise 19-4. Prove that the pure-twist-boundary dislocation grid is the projection of grids for arbitrary $\mathbf{n}$ for boundaries of the class described by Eq. (19-39).

Outline of proof. With given values of $\mathbf{N}_1$, $\mathbf{N}_2$, and $\mathbf{n}$, let us calculate the vectors $\mathbf{L}_1$ and $\mathbf{L}_2$ for the dislocation grid shown in Fig. 19-20. The unit vector in the direction of $\mathbf{L}_1$ is $(\mathbf{N}_2/N_2)\times\mathbf{n}$, while the length of $\mathbf{L}_1$ is $D_1/\sin\phi$, where

$$\sin\phi = \left(\frac{\mathbf{N}_2}{N_2}\times\mathbf{n}\right)\cdot\frac{\mathbf{N}_1}{N_1}$$

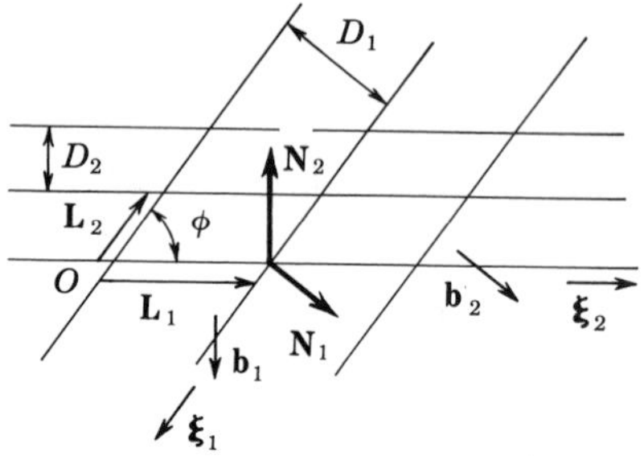

FIGURE 19-20. A lozenge-shaped dislocation boundary.

Combining these definitions with Eq. (19-18), one finds

$$2\mathbf{L}_1 \sin\frac{\theta}{2} = \frac{\mathbf{N}_2 \times \mathbf{n}}{\mathbf{N}_1 \cdot (\mathbf{N}_2 \times \mathbf{n})}$$

and similarly,

$$2\mathbf{L}_2 \sin\frac{\theta}{2} = \frac{\mathbf{N}_1 \times \mathbf{n}}{\mathbf{N}_2 \cdot (\mathbf{N}_1 \times \mathbf{n})}$$

The set of vectors $2\mathbf{L}_1\sin(\theta/2)$, $2\mathbf{L}_2\sin(\theta/2)$, and $\mathbf{n}$ are *reciprocal* to the set $\mathbf{N}_1$, $\mathbf{N}_2$, and $\mathbf{n}$.

Now consider the component $\mathbf{L}_1'$ produced by the projection of $\mathbf{L}_1$ onto the pure twist plane defined by the normal

$$\mathbf{n}' = \frac{\mathbf{b}_1 \times \mathbf{b}_2}{|\mathbf{b}_1 \times \mathbf{b}_2|}$$

Clearly,

$$\mathbf{L}_1' = \mathbf{L}_1 - (\mathbf{n}' \cdot \mathbf{L}_1)\mathbf{n}'$$

Inserting Eq. (19-41) into the above definition of $2\mathbf{L}_1\sin(\theta/2)$, one finds

$$2\mathbf{L}_1 \sin\frac{\theta}{2} = \frac{\mathbf{n} \times \mathbf{b}_1}{\mathbf{n} \cdot \mathbf{n}'} \qquad (19\text{-}52)$$

Thus

$$2\mathbf{L}_1'(\mathbf{n} \cdot \mathbf{n}')\sin\frac{\theta}{2} = (\mathbf{n} \times \mathbf{b}_1) - [(\mathbf{n} \times \mathbf{b}_1) \cdot \mathbf{n}']\mathbf{n}'$$

The last term is the component of $\mathbf{n} \times \mathbf{b}_1$ perpendicular to the pure twist boundary. Because $\mathbf{b}_1$ is contained in the pure twist boundary, the above is also the contribution to $\mathbf{n} \times \mathbf{b}_1$ produced by the component of $\mathbf{n}$ in the pure twist projection plane. Hence the difference in the above expression is the contribution produced by the component of $\mathbf{n}$

normal to the projection plane, $(\mathbf{n}\cdot\mathbf{n}')\mathbf{n}'\times\mathbf{b}_1$, so that the expression reduces to

$$2\mathbf{L}_1'\sin\frac{\theta}{2}=\mathbf{n}'\times\mathbf{b}_1$$

and similarly,

$$2\mathbf{L}_2'\sin\frac{\theta}{2}=\mathbf{n}'\times\mathbf{b}_2 \tag{19-53}$$

The projected lengths are independent of the original $\mathbf{n}$ and do project onto the pure twist plane as the proper network for the pure twist boundary. Since boundaries of the class described by Eq. (19-33) have both sets of dislocations parallel with one another, *all* boundaries formed by two sets of dislocations in a crossing network are of the type considered in this exercise.

Three Sets of Dislocations

In the general case that climb is allowed, any boundary orientation is possible with three sets of independent boundary dislocations. When glide is restricted to one plane, only two of the three sets are independent, so that the previous treatment holds. The first new case involves the restriction of glide to only two intersecting glide planes. In a slip plane containing dislocations with two independent Burgers vectors, a "superdislocation" consisting of a group of parallel dislocations can have a total Burgers vector of any direction in the slip plane. In many practical cases, there are two independent Burgers vectors in the glide plane, as for {111} slip in fcc, {110} slip in bcc, and (0001) slip in hcp crystals. For such crystals, any boundary can be analyzed in terms of two sets of dislocations with arbitrary Burgers vector in the slip plane for either set, because of the superdislocation concept.

Consider first the boundaries containing crossing dislocation grids. Such boundaries must satisfy Eq. (19-39), where $\mathbf{b}_1$ and $\mathbf{b}_2$ are, respectively, the Burgers vectors of the dislocations in glide planes $\mathbf{p}_1$ and $\mathbf{p}_2$. Given $\mathbf{a}$, the directions of the Burgers vectors are

$$\mathbf{b}_1\parallel\mathbf{p}_1\times\mathbf{a}\qquad \mathbf{b}_2\parallel\mathbf{p}_2\times\mathbf{a} \tag{19-54}$$

Inserting this result into Eq. (19-46), one obtains the result

$$\mathbf{n}\parallel[(\mathbf{p}_1\times\mathbf{a})\times\mathbf{p}_2]\times[(\mathbf{p}_2\times\mathbf{a})\times\mathbf{p}_1] \tag{19-55}$$

Therefore, for a given $\mathbf{a}$, only the one boundary plane defined by Eq. (19-55) is possible.

Conversely, given the boundary plane, only one $\mathbf{a}$ is possible, as now discussed. Because $\mathbf{L}_1\parallel\boldsymbol{\xi}_2$ and $\mathbf{L}_2\parallel\boldsymbol{\xi}_1$ (Fig. 19-20), the directions of the grid vectors are

$$\mathbf{L}_1\parallel\mathbf{p}_2\times\mathbf{n}\qquad \mathbf{L}_2\parallel\mathbf{p}_1\times\mathbf{n}, \tag{19-56}$$

With the use of the fact that $\mathbf{b}_1 \cdot \mathbf{p}_1 = \mathbf{b}_2 \cdot \mathbf{p}_2 = 0$, Eq. (19-52) yields the result

$$\mathbf{L}_1 \times \mathbf{p}_1 \| (\mathbf{n} \cdot \mathbf{p}_1)\mathbf{b}_1 \| \mathbf{b}_1 \qquad \mathbf{L}_2 \times \mathbf{p}_2 \| (\mathbf{n} \cdot \mathbf{p}_2)\mathbf{b}_2 \| \mathbf{b}_2 \tag{19-57}$$

Thus

$$\mathbf{a} \| \mathbf{b}_1 \times \mathbf{b}_2 \| (\mathbf{L}_1 \times \mathbf{p}_1) \times (\mathbf{L}_2 \times \mathbf{p}_2)$$

or, with the insertion of condition (19-56),

$$\mathbf{a} \| [(\mathbf{p}_2 \times \mathbf{n}) \times \mathbf{p}_1] \times [(\mathbf{p}_1 \times \mathbf{n}) \times \mathbf{p}_2] \tag{19-58}$$

As a final possibility for two glide planes, tilt boundaries can occur with all dislocations parallel. Both the dislocation lines $\boldsymbol{\xi}$ and the rotation vector **a** must be parallel to the line of intersection of the glide planes, as shown in Eq. (19-36).

The final possible special grain boundary is that formed by three independent sets of dislocations on three different glide planes. In this case Eq. (19-22) becomes

$$\mathbf{V} \times \mathbf{a} = \mathbf{b}_1(\mathbf{N}_1 \cdot \mathbf{V}) + \mathbf{b}_2(\mathbf{N}_2 \cdot \mathbf{V}) + \mathbf{b}_3(\mathbf{N}_2 \cdot \mathbf{V}) \tag{19-59}$$

Multiplying this expression by $\mathbf{b}_2 \times \mathbf{b}_3$ and selecting $\mathbf{V} = \boldsymbol{\xi}_1$, one obtains

$$\boldsymbol{\xi}_1 \cdot [\mathbf{a} \times (\mathbf{b}_2 \times \mathbf{b}_3)] = \boldsymbol{\xi}_1 \cdot \mathbf{q}_1 = 0 \tag{19-60}$$

where $\mathbf{q}_1$ is an abbreviation for the bracketed factor. Since also $\boldsymbol{\xi}_1 \cdot \mathbf{n} = 0$,

$$\boldsymbol{\xi}_1 \| \mathbf{q}_1 \times \mathbf{n} \tag{19-61}$$

Multiplying this condition by $\mathbf{p}_1$ and noting that $\mathbf{p}_1 \cdot \boldsymbol{\xi}_1 = 0$, one finds

$$\mathbf{p}_1 \cdot (\mathbf{q}_1 \times \mathbf{n}) = 0 \qquad \text{or} \qquad \mathbf{n} \cdot (\mathbf{p}_1 \times \mathbf{q}_1) = 0 \tag{19-62}$$

Similarly,

$$\mathbf{n} \cdot (\mathbf{p}_2 \times \mathbf{q}_2) = \mathbf{n} \cdot (\mathbf{p}_3 \times \mathbf{q}_3) = 0 \tag{19-63}$$

The boundary normal is thus determined by

$$\mathbf{n} = (\mathbf{p}_1 \times \mathbf{q}_1) \times (\mathbf{p}_2 \times \mathbf{q}_2) \tag{19-64}$$

All the vectors $\mathbf{p}_i \times \mathbf{q}_i$ are normal to **n**, and are thus coplanar, yielding an expression which restricts the choice of **a**,

$$[(\mathbf{p}_1 \times \mathbf{q}_1) \times (\mathbf{p}_2 \times \mathbf{q}_2)] \cdot (\mathbf{p}_3 \times \mathbf{q}_3) = 0 \tag{19-65}$$

The special cases in which the above derivation is not valid, such as when $\mathbf{q}_1 \| \mathbf{n}$, so that Eq. (19-61) is indeterminate, reduce to simpler cases treated earlier (a tilt boundary, for example). Thus all possible grain boundaries are now described.

Exercise 19-5. Verify that the vectors **a** that satisfy Eq. (19-65) generate a cone in crystallographic space.[27]

19-4. LOCAL INTERACTIONS TO FORM STABLE BOUNDARIES

General Considerations

The preceding theory treats straight parallel or crossing dislocations. However, these models often do not predict the true dislocation configuration for a given boundary, because the crossing dislocations interact and react to form new configurations of lower energy.[28] The possible reactions depend upon the self- and interaction energies of the reacting dislocations, and thus depend upon both the crystal structure and the character of the boundary. We consider boundary rearrangements in a general way and then treat the fcc case as a specific example, particularly to illustrate effects associated with extended dislocations. For details on grain-boundary configurations in other structures the reader is referred to Amelinckx and Dekeyser.[29]

First, let us consider what rearrangements are possible without changing the boundary character or producing long-range stresses. As explained in connection with Fig. 19-19, some such rearrangement is possible, entailing the splitting of sets of parallel dislocations into subsets of different directions. In addition, local changes, repeated regularly throughout the network, which do not change the *average* direction of the original dislocations, neither change the boundary character nor create long-range stresses. Such rearrangements do not change the number c_i of dislocations $\mathbf{b}_i$ cut by a vector $\mathbf{V}$, so that Eq. (19-11) is valid and provides a formal proof of the above remark. For a physically more direct explanation, consider identical closed dislocation loops distributed regularly in a boundary. As a simple example, suppose that they are circular shear loops (Fig. 19-21); actually, the area of the loops, not their form, is important for long-range stress effects, according to St. Venant's principle. The two crystals separated by such a wall are sheared relative to one another, but *no orientation difference* between the two crystals is produced. Similarly, a wall of regularly spaced, identical, prismatic loops merely displaces the two crystals

[27] C. J. Ball, *Phil. Mag.*, **2**: 977 (1957).

[28] F. C. Frank, in "Report of the Conference on Defects in Crystalline Solids," Physical Society, London, 1954, p. 159.

[29] S. Amelinckx and W. Dekeyser, *Solid State Phys.*, **8**: 325 (1959).

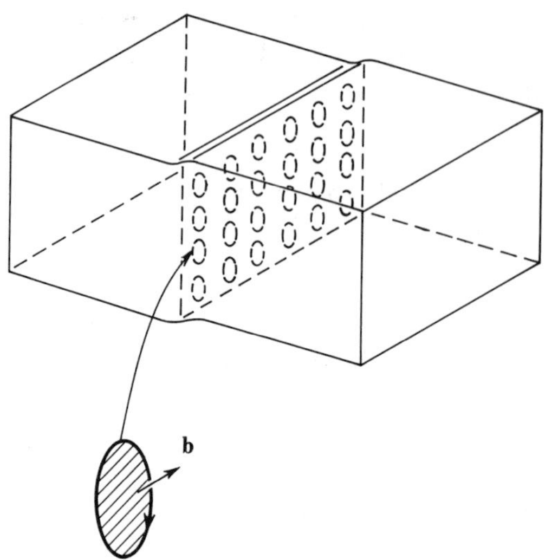

FIGURE 19-21. Offset with no long-range stress field produced by a set of shear-dislocation loops in a plane.

normal to the wall. The boundary resulting from a regular rearrangement of a network of straight dislocations can be regarded as a super-position of the original network and a wall, such as that in Fig. 19-21. Thus the rearrangement produces neither orientation changes nor long-range stresses.

The type of rearrangement that occurs in a given structure is associated with the types of dislocation *nodes* that form. Consider two crossing dislocations of Burgers vector $\mathbf{b}_1$ and $\mathbf{b}_2$, as shown in Fig. 19-22. In general, whether or not reaction can occur to form segments AB or CD must be decided by applying the methods of Chap. 6 to the determination of the energy of the entire configuration as a function of the lengths CD or AB. Such a procedure is quite complicated, even for symmetric cases. Hence, while we recommend the general method for specific cases of experimental interpretation, we adopt more expedient methods here for brevity. When interaction energies are neglected, the energy per unit length of a dislocation is analogous to the surface tension of a boundary, so that equilibrium at a node can be determined by analogy to Eq. (19-2),[30]

$$\sum_i \left[\left(\frac{W_s}{L} \right)_i \boldsymbol{\xi}_i + \frac{\boldsymbol{\xi}_i \times (\boldsymbol{\xi}_i \times \mathbf{b}_i)\, \partial (W_s/L)_i}{|\boldsymbol{\xi}_i \times \mathbf{b}_i|\, \partial \beta_i} \right] = 0 \qquad (19\text{-}66)$$

[30]Note that the force-balance expression differs from the line tension $\mathcal{S}$ appropriate for bow-out [Eqs. (6-80) and (6-81)]. This occurs because in bow-out the torque term $\partial(W_s/L_i)/\partial\beta_i$ acts in the direction of bow-out for part of the loop and opposes the bow-out of the other portion, so that only the torque derivative $\partial^2(W_s/L)_i/\partial\beta_i^2$ appears in $\mathcal{S}$. A directly analogous situation exists in the case of surface energies; see W. W. Mullins, in N. A. Gjostein and W. D. Robertson (eds.), "Metal Surfaces," American Society of Metals, Cleveland, Ohio, 1963, p. 17; C. Herring, in W. E. Kingston (ed.), "The Physics of Powder Metallurgy," McGraw-Hill, New York, 1951, p. 143; Y. T. Chou and J. D. Eshelby, *J. Mech. Phys. Solids*, **10**: 27 (1962).

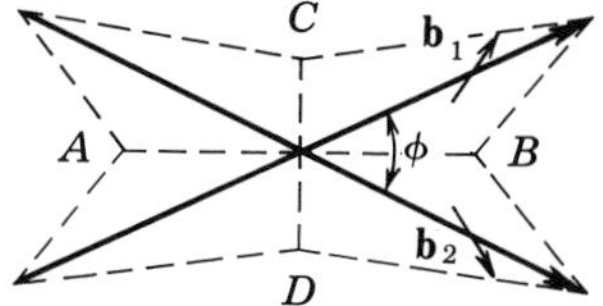

FIGURE 19-22. Crossing dislocations and possible reacted configurations AB and CD.

with $\boldsymbol{\xi}_i$ defined to point away from the node and (W_s/L) and β_i defined by Eq. (6-18). Unlike the grain-boundary case, the torque vectors are not necessarily in the plane of the boundary. Substituting from Eq. (6-18) and setting the logarithmic terms equal, one obtains the reduced form

$$\sum_i b_i^2\left[(1-\nu\cos^2\beta_i)\boldsymbol{\xi}_i+2\nu\cos\beta_i\sin\beta_i\frac{\boldsymbol{\xi}_i\times(\boldsymbol{\xi}_i\times\mathbf{b}_i)}{|\boldsymbol{\xi}_i\times\mathbf{b}_i|}\right]=0 \qquad (19\text{-}67)$$

When the plane of the boundary is the glide plane for all dislocations in the boundary, the second vector in Eq. (19-67) is simply $\boldsymbol{\xi}_i\times\mathbf{n}$. In the cruder approximation that the variation of energy with screw-edge character is neglected, Eq. (19-67) reduces to

$$\sum b_i^2\boldsymbol{\xi}_i=0 \qquad (19\text{-}68)$$

When reaction is very favorable, as for the fcc reaction $\mathbf{AB}+\mathbf{BC}\rightarrow\mathbf{AC}$, Eq. (19-68) suffices to determine whether or not reaction will occur. However, in marginal cases such as the fcc reaction $\mathbf{AC}+\mathbf{DB}\rightarrow\mathbf{AD}/\mathbf{BC}$, the more accurate Eq. (19-67) must be used.

When the formation of a segment such as AB or CD in Fig. 19-22 is energetically favorable, the application of Eq. (19-67) establishes the equilibrium angle in the plane of the boundary ϕ_e (Fig. 19-23). Thus, if $\phi<\phi_e$ in Fig. 19-22, dislocation $\mathbf{b}_3$ will form, while if $\phi>\phi_e$, no reaction will occur. In general, both possibilities AB and CD must be tested for stability at a fourfold node. With regard to the neglect of interaction forces in Eq. (19-67), we note that when the configuration AB of Fig. 19-22 is stable, repulsive interaction forces exist between branches $\mathbf{b}_1$ and $-\mathbf{b}_2$ and between branches $-\mathbf{b}_1$ and $\mathbf{b}_2$.

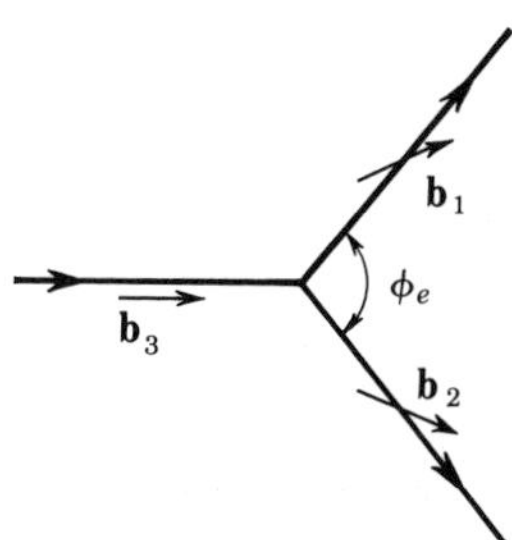

FIGURE 19-23. Equilibrium angle ϕ_e for a node lying in a plane.

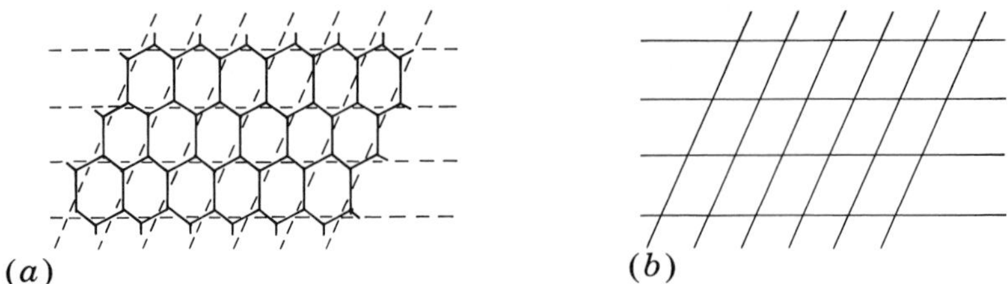

FIGURE 19-24. (*a*) Hexagonal and (*b*) lozenge-shaped dislocation nets.

These forces make the reacted configuration more stable. Also, interaction forces can lead to "puckering" of the boundary, in which alternate nodes are displaced in opposite directions normal to the average boundary plane; the torque terms in Eq. (19-67) can produce puckering as well if a component of the torque vector is normal to the boundary plane.

Reactions between crossing dislocations produce a dislocation net with hexagonal meshes (Fig. 19-24*a*). When reaction does not occur, the network has a lozenge-shaped mesh (Fig. 19-24*b*). An actual example of a hexagonal mesh is shown in Fig. 19-7, and Fig. 1-15 shows a boundary that is partially lozenge-shaped. In general, either type of mesh can be puckered, as discussed above.

Interestingly, dislocations in pure twist boundaries should always form hexagonal networks unless their lines are within a few degrees of being orthogonal. Consider, for example, the bcc twist boundary shown in Fig. 19-25. In order to satisfy both Eqs. (19-19) and (19-43), the $\boldsymbol{\xi}_i$ and $\mathbf{b}_i$ must be those depicted. Indeed, the dislocations in the acute angle are those which react to form the dislocation $\mathbf{b}_3=[001]$, which is stable according to Eq. (19-67), while those in the obtuse angle would combine to form the unstable dislocation $\mathbf{b}=[\bar{1}10]$. The same reasoning holds for twist boundaries in any structure. When the crossed straight dislocations are not in the pure twist plane, the segments favorable for reaction can be in either the acute or obtuse angle.

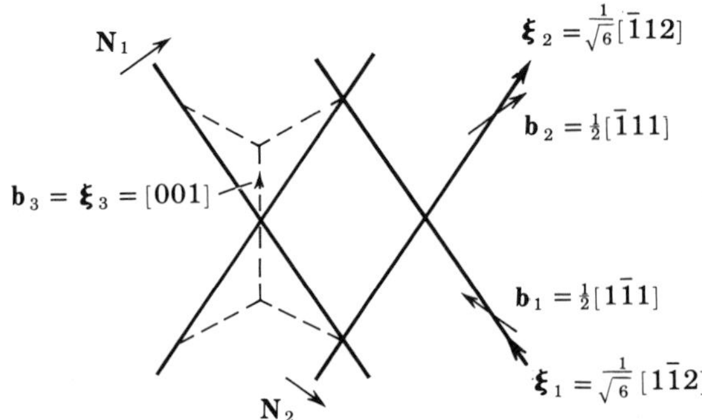

FIGURE 19-25. Twist boundary on a (110) plane in a bcc crystal, showing a reaction in a lozenge-shaped net to form the dislocations $\mathbf{b}_3=[001]$.

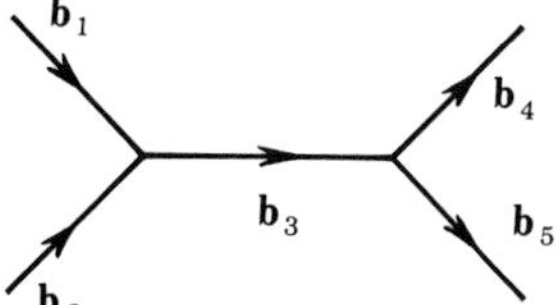

FIGURE 19-26. A general-dislocation node pair.

More general nodes, such as that shown in Fig. 19-26, are possible in dislocation networks. However, they cannot form directly from crossing straight dislocations, and are expected only as links between different sections of regular networks.

Exercise 19-6. Consider the twist boundary formed from [100] and [010] dislocations in a simple orthorhombic crystal. What is the critical value of the lattice-parameter ratio $\mathbf{a}_1 : \mathbf{a}_2$ for which a lozenge-shaped network first breaks up into a hexagonal network according to Eq. (19-67)?

Grain Boundaries in FCC Crystals

The Burgers vectors for fcc crystals are, as usual, referred to the Thompson tetrahedron of Fig. 10-10. In this application to grain boundaries, however, a more detailed notation suggested by Frank[31] is useful. The two component letters of the Burgers vector are placed on either side of the dislocation, so that they read in the correct order to give **b** from the viewpoint of the observer looking along positive ξ. Thus the dislocation in Fig. 19-27 is **AB** when viewed from the bottom of the page and **BA** when viewed from the top.

With this notation, two types of nodes with different dissociation properties can be defined. For convenience, we consider mainly nodes viewed as if the observer were outside the Thompson tetrahedron.[32] The K nodes are those for which all Burgers vectors are defined by placing one letter in each sector (Fig. 19-28a). A K node can be classified further as a K_S node (Fig. 19-28b) or a K_U node (Fig. 19-28c). K_S nodes can be rotated, without changing the sequence of branches, into a symmetrical configuration of pure screw character: hence the subscript S for symmetrical. K_U, or unsymmetrical nodes, cannot be rotated into such a configuration.

The P nodes are those shown in Fig. 19-29, where pairs of letters, separated by the extension of the line opposite, appear in each sector. Subcategories P_S

[31] F. C. Frank, in "Report of the Conference on Defects in Crystalline Solids," Physical Society, London, 1954, p. 159.

[32] All the following rules and developments hold for nodes viewed as if the observer were inside the tetrahedron; the only difference is that for such orientations, K nodes have pairs of letters in one sector, while P nodes have only one letter in each sector.

FIGURE 19-27. Projection of dislocation segment **AB** onto the plane of the page.

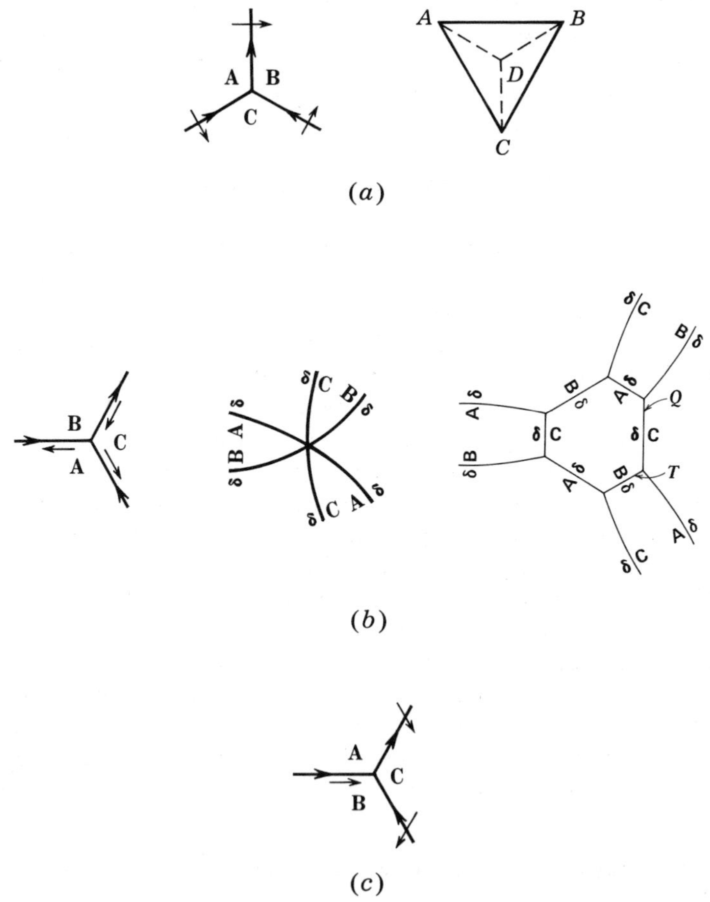

FIGURE 19-28. (*a*) A *K* node and the corresponding orientation of the Thompson tetrahedron. (*b*) A K_S node with unextended branches and with branches extended to form an intrinsic stacking fault, or further extended to form a central extrinsic fault hexagon. (*c*) A K_U node.

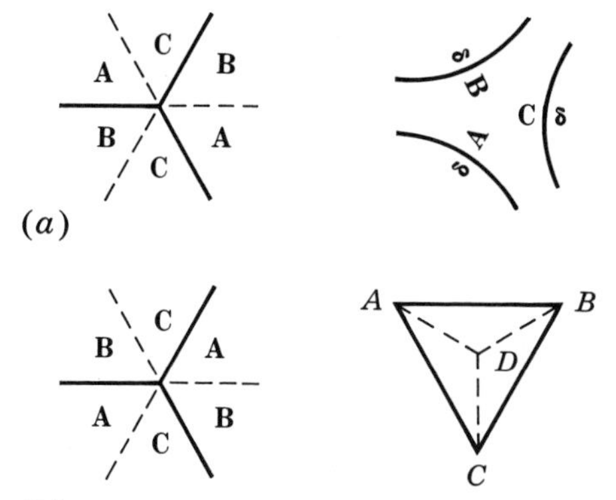

FIGURE 19-29. (*a*) A P_S node, unextended and extended to form an intrinsic stacking fault. (*b*) A P_U node. The orientation of the Thompson tetrahedron is also shown.

and P_U also exist for P nodes, as shown in Figs. 19-29*a* and 19-29*b*. The various nodes can be distinguished by the following rule:

Axiom 19-1: When the boundary is viewed from outside the Thompson tetrahedron, K_S or P_U nodes are those with the letters *ABC* in a clockwise sequence. K_U or P_S nodes are those with *ABC* in a counterclockwise sequence; the situation is reversed when the boundary is viewed as if from inside the tetrahedron.

Since the extension of dislocations is also related to the tetrahedron, Axiom 19-1 leads directly to the conclusion that for nodes lying in the glide plane *K* nodes are contracted, while *P* nodes are extended if only intrinsic stacking fault is permitted (Figs. 19-28 and 19-29). *K* nodes can extend only if extrinsic stacking fault is formed. An extended K_S mode is shown in Fig. 19-28. The central hexagon encloses extrinsic stacking fault. The partials labeled *T* are cross-tie partials with single cores, and those labeled *Q* are double-core partials of the type of Fig. 10-8*b*.[33] A triangular central region is also possible if the *Q* partials shrink to zero length. Many networks have been observed with contracted *K* nodes, and it was once thought that this must always be so. Suggested reasons[34] were the larger core energy of the *Q* partials or the presence of jogs at the node[34] (the γ_E and γ_I fault energies are nearly equal and hence are not a significant factor; see Chap. 10). However, extended K_S and K_U nodes also have been observed,[33, 35–37] albeit with greater degrees of constriction than extended *P* nodes. Completely contracted *K* and *P* nodes

[33] C. B. Carter, *Phil. Mag.*, **41A**: 619 (1980).

[34] A. Cullis, *J. Microsc.*, **98**: 191 (1973).

[35] M. H. Loretto, *Phil. Mag.*, **10**: 467 (1964); *Phil. Mag.*, **12**: 125 (1965).

[36] P. R. Swann, *Acta Met.*, **14**: 76 (1966).

[37] P. C. J. Gallagher, *J. Appl. Phys.*, **37**: 1710 (1966).

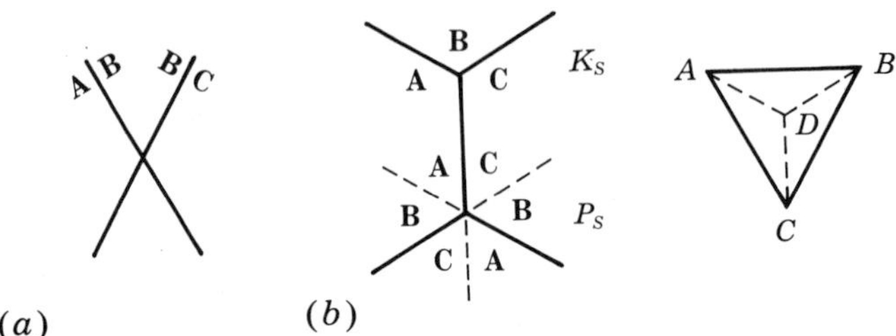

FIGURE 19-30. Reaction of **AB** and **BC** to form a $K_S P_S$ node pair.

were observed together with extended nodes in Carter's work and were associated with superjogs at the contractions. Hence, the degree of constriction can be associated with Q partial core energy and perhaps with unit jogs, whereas complete contraction of K nodes can be associated with superjogs or with the difficulty in nucleating the T and Q partials to form the extrinsic fault. In the remainder of the discussion, we consider only completely contracted K nodes.

Exercise 19-7. Construct the completely extended form of the K_U node in Fig. 19-28.

Let us now consider node arrangement in networks. The dislocations **AB** and **BC** react to form either a pair of one K_S and one P_S node or a pair of one K_U and one P_U node, as shown in Figs. 19-30 and 19-31. With the above definitions for the nodes, the pairs $P_U K_S$, $P_S K_U$, $K_U K_U$, and $P_S P_S$ are physically impossible; the letters cannot be arranged to yield such pairs. The pairs $K_U K_S$ and $P_U P_S$ are possible, as demonstrated for the former in Fig. 19-32*a*. However, such pairs are energetically unstable with respect to the reaction shown in Fig. 19-32*b* unless they are so widely separated that line-tension forces provide an activation barrier and the $K_U K_S$ pair is metastable. Thus meshes of dislocations composed of the elementary perfect dislocations of the type **AB** are composed either of alternating $K_U P_U$ pairs or $K_S P_S$ pairs. The pairs $K_U K_S$ and $P_U P_S$ should occur only rarely as bridges between meshes of the above types.

A few particular boundaries are now considered. A twist boundary formed by the action of the two slip systems **AB**(d) and **BC**(d) will react by the general mechanism demonstrated in Fig. 19-25 to produce the network shown in Fig. 19-33. Here alternate K_S and P_S nodes are constricted or extended, and the perfect dislocations are in pure screw orientation. Such configurations have been observed in electron-transmission microscopy[38] (see Figs. 10-7 and 19-7). In some cases the configuration of Fig. 19-33 has been observed,[39, 40] but the

[38] P. B. Hirsch, P. G. Partridge, and R. L. Segall, *Phil. Mag.*, **4**: 721 (1959).

[39] P. C. J. Gallagher, *J. Appl. Phys.*, **37**: 1710 (1966).

[40] T. Ericsson, *Acta Met.*, **14**: 853 (1966).

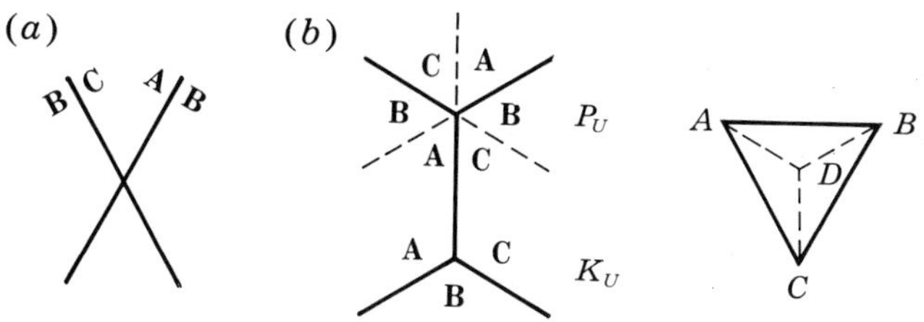

FIGURE 19-31. Reaction of **AB** and **BC** to form a $K_U P_U$ node pair.

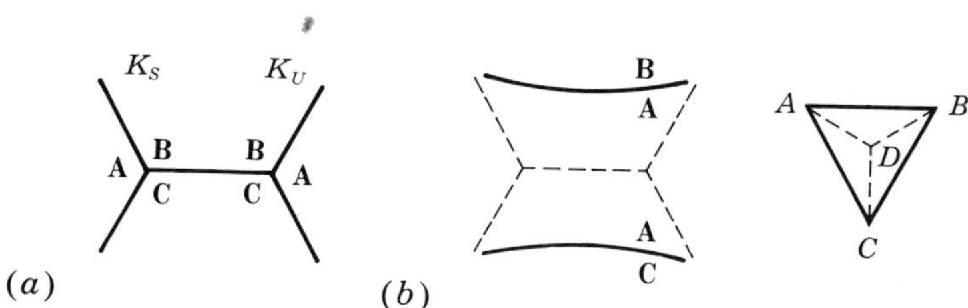

FIGURE 19-32. Reaction of a $K_S K_U$ node pair to annihilate the nodes.

dislocations are rotated 30° from pure screw orientation; this is difficult to rationalize in terms of grain-boundary theory, because the regular hexagonal boundary is stress free only when the dislocations are pure screw. The pure screw orientation remains the minimum energy configuration in the typical anisotropic elastic case.[41] However, the change in energy of the Shockley partials becomes small[41] in the range $0° < \beta < 30°$. Thus, if the formation mechanism leads to initial values of $\beta > 30°$, rotation to values of $\beta < 30°$ might be blocked by weak effects such as friction forces or variations of core energy with orientation. Alternatively, the formation of nodes from screw dislocations on parallel glide planes separated from one another normal to the glide plane would lead to superjogs with attendant contractions.[42, 43] Hexagonal networks are observed to have warped planes, implying the presence of such jogs at nodes.[43] Thus the 30° orientation could be stabilized by pinning by superjogs. Alternatively, boundaries rotated 30° from screw orientation are expected from the local interaction of a screw grid formed from the glide systems **AC**(b) and **BC**(a), with one set of screws right-handed and one left-handed; such a boundary, however, would consist of unsymmetric nodes and would have a long-range stress field.

Exercise 19-8. Demonstrate that only symmetrical nodes are produced in twist boundaries in fcc crystals. Unsymmetrical node networks have long-range stress fields associated with them.

[41] R. O. Scattergood and D. J. Bacon, *Acta Met.*, **24**: 705 (1976).

[42] C. B. Carter, *Phil. Mag.*, **41A**: 619 (1980).

[43] D. Grinner and G. Packeiser, *Phil. Mag.*, **42A**: 645, 661 (1980).

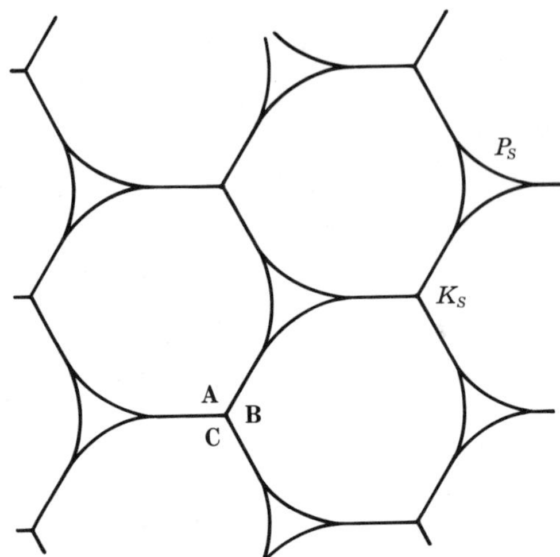

FIGURE 19-33. A pure-screw twist boundary in a {111} plane.

A stable square-grid twist boundary can form from the screw dislocations **AB** and **CD** in the {100} plane containing both vectors. This boundary is stable because of the torque terms in Eq. (19-67). Such a boundary in fcc crystals involves the interaction of screws from different glide planes (c) and (a), however, and hence is less likely than in NaCl-type crystals, where {100} glide occurs. Also, dislocations of the type **AB**/**CD** are possible in fcc meshes, but not for pure twist boundaries. Interactions of single dislocations with regular boundary meshes produce irregularities in the meshes that can include **AB**/**CD** dislocations. We do not discuss the many possibilities of this type, but refer the reader to the work of Frank[44] and of Amelinckx,[45] who observed such irregular boundaries.

The degree of extension of dislocations in small-angle boundaries depends on the boundary structure. Consider pure tilt boundaries in the fcc case, such as the one-dislocation boundary. The edge dislocations remain dissociated in their respective glide planes, as shown in Fig. 19-34a. For small angles of rotation and large separations D_i, the partials maintain the equilibrium spacing r_e characteristic of an isolated dislocation. For tilt angles so high that D is less than r_e, the dislocations contract to partial separations of the same order as D_i (Fig. 19-34b).[46] This contraction can be understood by an application of St. Venant's principle, as discussed in the next section, or from the formal theory presented there. For very small D, the dislocations can interact by local climb

[44] F. C. Frank, in "Report of the Conference on Defects in Crystalline Solids," Physical Society, London, 1955, p. 159.

[45] S. Amelinckx, *Acta Met.*, **6**: 34 (1958); in "Dislocations and Mechanical Properties of Crystals," Wiley, New York, 1957, p. 3.

[46] W. T. Read, "Dislocations in Crystals," McGraw-Hill, New York, 1953, p. 171.

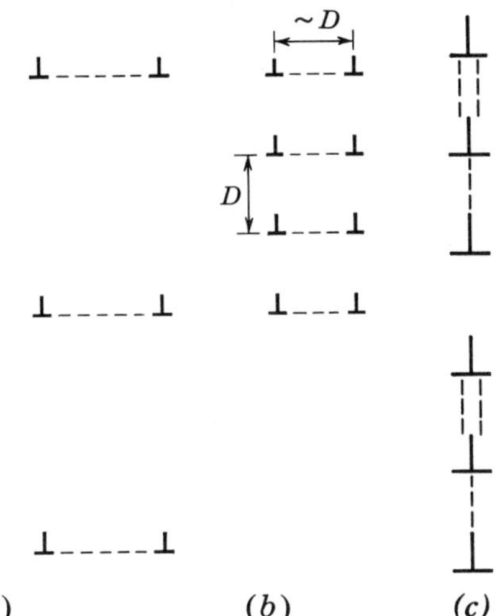

FIGURE 19-34. A pure tilt boundary in an fcc crystal dissociated into Shockley partials (a) and (b) or Frank partials (c). In (c) the single and double dashed lines represent intrinsic and extrinsic stacking faults, respectively. In case (a), the partials have the separation $r \cong r_e \ll D$, nearly equal to that for an isolated dislocation. In case (b), $r \cong D < r_e$.

rearrangement and dissociation to form the Frank partial array in Fig. 19-34c. This array, as well as some other arrays involving two sets of dislocations, has been observed in Si.[47] In the observed case, a (110) tilt boundary faceted into segments of (211) and (111) boundaries. A dislocation reaction that would accomplish this is $3\times\frac{1}{2}[110]\rightarrow\frac{1}{2}[11\bar{2}]+3\times\frac{1}{3}[111]$. The boundary could also form by reactions of the type $\frac{1}{2}[110]+\frac{1}{2}[011]+\frac{1}{2}[101]\rightarrow 3\times\frac{1}{3}[111]$. In each case, local climb would be required.

Because of the analogy between atom packing in the fcc glide plane and that in diamond cubic and hexagonal-layer structures, as discussed in Chaps. 10 and 11, most of the above discussion also applies to boundaries in diamond, NaCl-type crystals, hcp crystals, and other hexagonal-layer structures. In particular, the discussion of dislocation extension applies directly.

19-5. STRESSES NEAR GRAIN BOUNDARIES

By definition, grain boundaries separate stress-free crystallites. Indeed, the Frank criterion, Eq. (19-14), does give only those boundaries that do not produce long-range stress fields. However, near the boundary the stress fields of the individual dislocations are appreciable. In this section, we compute the stress field for a simple tilt boundary and demonstrate explicitly that no

[47] A. Bourret and J. Desseaux, *Phil. Mag.*, **39A**: 405, 419 (1979).

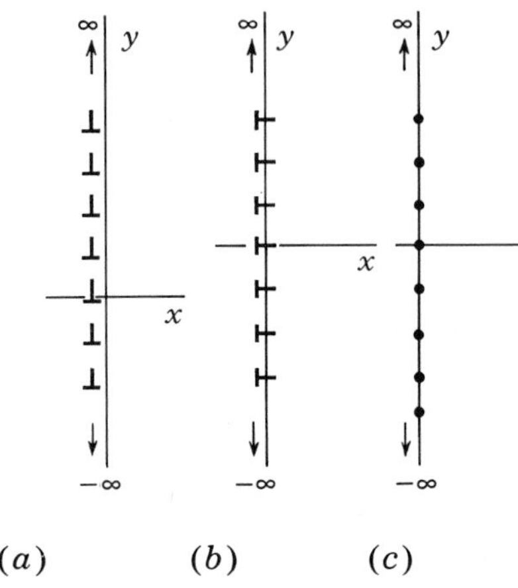

FIGURE 19-35. (*a*) A pure tilt boundary. (*b*) An edge-dislocation array with **b** parallel to the y axis. (*c*) A right-handed screw-dislocation array with **b** and ξ parallel to the z axis.

long-range stresses arise. The results of this analysis are typical for more general grain boundaries.

Consider the stress σ_{xy} for the boundary drawn in Fig. 19-35*a*. Summing the contributions of the individual dislocations as given by Eq. (3-43), one finds

$$\sigma_{xy} = \frac{\mu b}{2\pi(1-\nu)} \sum_{n=-\infty}^{\infty} \frac{x\left[x^2-(y-nD)^2\right]}{\left[x^2+(y-nD)^2\right]^2} \tag{19-69}$$

The basic formula used in the sum is

$$\sum_{-\infty}^{\infty} \frac{1}{n+a} = \pi \cot \pi a \tag{19-70}$$

which is proved in texts on analysis.[48] Adding the expressions given by Eq. (19-70) for $a=p+iq$ and $a=p-iq$, one obtains

$$\sum_{-\infty}^{\infty} \frac{n+p}{q^2+(n+p)^2} = \frac{\pi}{2}\left[\cot \pi(p+iq)+\cot \pi(p-iq)\right]$$

$$= \frac{\pi \sin 2\pi p}{\cosh 2\pi q - \cos 2\pi p} \tag{19-71}$$

Subtraction of the same expressions gives

$$\sum_{-\infty}^{\infty} \frac{1}{q^2+(n+p)^2} = \frac{\pi}{q}\,\frac{\sinh 2\pi q}{\cosh 2\pi q - \cos 2\pi p} \tag{19-72}$$

[48]See, for instance, P. M. Morse and H. Feshbach, "Methods of Theoretical Physics," vol. I, McGraw-Hill, New York, 1953, p. 383.

which is the sum used in solving Eq. (8-37). Differentiating Eqs. (19-71) and (19-72) with respect to p, one obtains, respectively,

$$\sum_{-\infty}^{\infty} \frac{q^2-(n+p)^2}{\left[q^2+(n+p)^2\right]^2} = 2\pi^2 \frac{\cosh 2\pi q \cos 2\pi p - 1}{(\cosh 2\pi q - \cos 2\pi p)^2} \tag{19-73}$$

and

$$\sum_{-\infty}^{\infty} \frac{n+p}{\left[q^2+(n+p)^2\right]^2} = \frac{\pi^2}{q} \frac{\sinh 2\pi q \sin 2\pi p}{(\cosh 2\pi q - \cos 2\pi p)^2} \tag{19-74}$$

Equation (19-73), used in summing Eq. (19-69), was first derived by Burgers[49]; the sums were developed explicitly by Cottrell.[50]

Performing the sum in Eq. (19-69) by the above method, one finds

$$\sigma_{xy} = \sigma_0 2\pi X(\cosh 2\pi X \cos 2\pi Y - 1) \tag{19-75}$$

where

$$\sigma_0 = \frac{\mu b}{2D(1-\nu)(\cosh 2\pi X - \cos 2\pi Y)^2} \tag{19-76}$$

while $X = x/D$ and $Y = y/D$ are reduced variables. The other stress components are similarly found from Eqs. (3-43) and (19-71) to (19-74) to be

$$\begin{aligned} \sigma_{xx} &= -\sigma_0 \sin 2\pi Y(\cosh 2\pi X - \cos 2\pi Y + 2\pi X \sinh 2\pi X) \\ \sigma_{yy} &= -\sigma_0 \sin 2\pi Y(\cosh 2\pi X - \cos 2\pi Y - 2\pi X \sinh 2\pi X) \end{aligned} \tag{19-77}$$

For $x \gg D/2\pi$, Eq. (19-75) is approximated by the form

$$\sigma_{xy} \cong \frac{2\pi\mu b x}{(1-\nu)D^2} \exp\left(-\frac{2\pi x}{D}\right) \cos \frac{2\pi y}{D} \tag{19-78}$$

In this region, the stresses decrease exponentially to zero with increasing x; there are no long-range stresses. In fact, for $x = 2D$, σ_{xy} is 10^{-2} times the stress at equivalent x for a single dislocation. This result provides an explicit demonstration of St. Venant's principle and also justifies the structure shown in Fig. 19-34*b*. Near the boundary, $|x| \ll D/2\pi$, the stresses σ_{xy} are of the order

[49] J. M. Burgers, *Proc. Kon. Ned. Akad. Wetenschap.*, **42**: 293 (1939).

[50] A. H. Cottrell, "Dislocations and Plastic Flow in Crystals," Oxford University Press, Fair Lawn, N.J., 1953, p. 94.

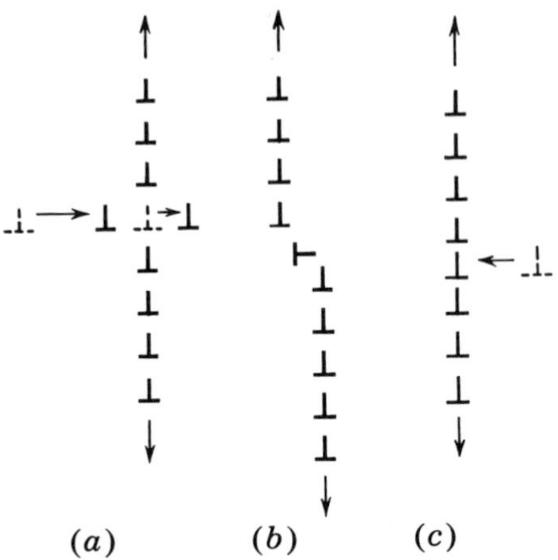

FIGURE 19-36. Some local interactions of glide dislocations with a tilt boundary.

of the stresses from the first one to three nearest dislocations, depending on the value of y. The other stress components behave similarly to σ_{xy}.

The above results can be applied directly to determine the interaction of single dislocations with tilt boundaries. It is left as an exercise to show that the addition of a glide dislocation will produce the local equilibrium configurations of Fig. 19-36. Configurations such as these have been directly observed, for example, by Mitchell et al.[51] in silver chloride crystals.

Grain boundaries of other types, such as that in Fig. 19-25, have no long-range stress fields. However, their stress fields are most easily determined by the superposition of boundaries which do have long-range stresses—two sets of edge dislocations and two sets of screws, corresponding to the edge and screw components of the mixed dislocations in the specific example of Fig. 19-25. By the same procedure as above, we obtain for the edge-dislocation array of Fig. 19-35*b*

$$\begin{aligned}\sigma_{xy} &= \sigma_0 \sin 2\pi Y(\cosh 2\pi X - \cos 2\pi Y - 2\pi X \sinh 2\pi X)\\ \sigma_{xx} &= -\sigma_0 2\pi X(\cosh 2\pi X \cos 2\pi Y - 1)\\ \sigma_{yy} &= -\sigma_0[2\sinh 2\pi X(\cosh 2\pi X - \cos 2\pi Y) - 2\pi X(\cosh 2\pi X \cos 2\pi Y - 1)]\end{aligned} \tag{19-79}$$

and for the screw-dislocation array of Fig. 19-35*c*

$$\begin{aligned}\sigma_{xz} &= -\frac{\mu b}{2D}\frac{\sin 2\pi Y}{\cosh 2\pi X - \cos 2\pi Y}\\ \sigma_{yz} &= \frac{\mu b}{2D}\frac{\sinh 2\pi X}{\cosh 2\pi X - \cos 2\pi Y}\end{aligned} \tag{19-80}$$

[51] D. A. Jones and J. W. Mitchell, *Phil. Mag.*, **2**: 1047 (1957); A. S. Parasnis and J. W. Mitchell, *Phil. Mag.*, **4**: 171 (1959).

Any boundary composed of sets of straight dislocations can be analyzed in terms of Eqs. (19-75) to (19-80).[52] The results of Eqs. (19-79) and (19-80) are of interest because they show explicitly that the dislocation arrays that they describe have long-range stress fields; σ_{yy} in Eq. (19-79) converges to $\mu b/D(1-\nu)$ for $x \to \infty$, and σ_{yz} in Eq. (19-80) converges to $\mu b/2D$ in the same limit. For anisotropic elastic materials, the behavior is similar.[53] The results of this analysis also can be applied to boundaries containing dissociated dislocations, since, as shown in Fig. 19-25*a*, extended boundaries can be considered as two walls composed of partials and separated by a distance r. Even though the net boundary has no long-range stress field, the individual partial walls do have long-range stresses.

Long-range stresses arise for infinite dislocation walls separating incompatible crystals, the incompatibility arising from different crystal structure with different elastic properties in the isotropic elastic case or from elastic anisotropy for the same crystal structure.[54] Even arrays such as those in Fig. 19-25 or 19-35*a* can then have long-range fields. The long-range stresses can be found in the anisotropic case in terms of a reduced elastic coefficient.[55] In a true grain boundary these stresses must be relaxed. This can be accomplished without incompatibility problems arising at the boundary by the superposition of additional elastic solutions.[56] Incomplete boundaries do set up long-range stresses; one example is a tilt boundary that ends within a grain. Such boundaries are not proper grain boundaries and are better thought of as dislocation pileups. The stress fields of pileups are developed in Chap. 21.

Exercise 19-9. Consider a pure tilt boundary composed of $\mathbf{b}=\frac{1}{2}[1\bar{1}0]$ dislocations lying along $\boldsymbol{\xi}=[11\bar{2}]$ in an fcc structure. Show that if the dislocations dissociate into partials bounding stacking fault γ_I, the equilibrium separation r is determined by the expression.[57, 58]

$$\gamma_I = \frac{\mu b^2}{24D(1-\nu)}\left[\frac{3\pi r}{D}\operatorname{csch}^2\left(\frac{\pi r}{D}\right) - (1-\nu)\coth\left(\frac{\pi r}{D}\right)\right] \tag{19-81}$$

Compare the dissociation predicted by this expression with that of Eq. (10-15).

[52] A compilation of stress tensors for mixed dislocation grids is presented by J. C. M. Li in G. Thomas and J. Washburn (eds.), "Electron Microscopy and Strength of Crystals," Interscience, New York, 1963, p. 713.

[53] C. Rey and G. Saada, *Phil. Mag.*, **33**: 825 (1976); G. Saada, ibid., p. 639.

[54] J. Dundurs and G. P. Sendeckyj, *J. Appl. Phys.*, **36**: 3353 (1965).

[55] S. S. Orlov and V. L. Indenbom, *Kristallografiya*, **14**: 780 (1969).

[56] J. P. Hirth, D. M. Barnett, and J. Lothe, *Phil. Mag.*, **40**: 39 (1979).

[57] A. Seeger, in "Handbuch der Physik," vol. 7, Springer, Berlin, 1955, p. 654.

[58] J. C. M. Li and B. Chalmers, *Acta Met.*, **11**: 243 (1963).

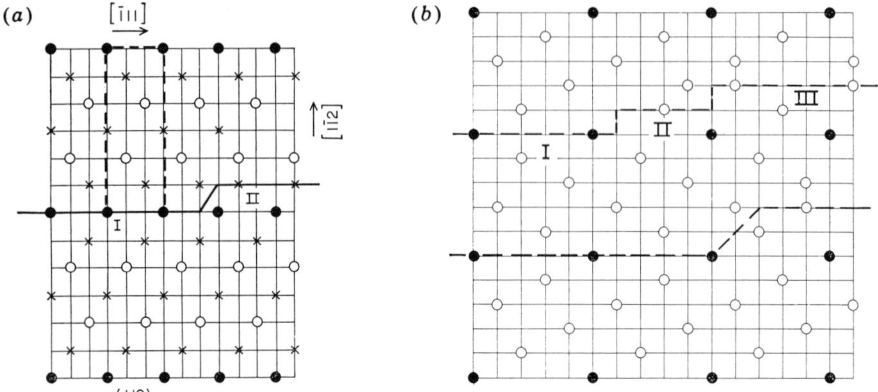

FIGURE 19-37. Coincidences lattices (solid circles) and DSC lattices (fine lines): (*a*) bcc lattice projected along [110] parallel to a stepped (112) twin plane, with a unit CSL cell shown by dashed lines; (*b*) single cubic lattice projected along [001] parallel to stepped (210) twin boundaries.

19-6. SECONDARY GRAIN BOUNDARY DISLOCATIONS

Schober and Bulluffi[59] observed grain boundary dislocations (GBDs) in twist boundaries in gold and found them to have Burgers vectors only a fraction of those of lattice dislocations. They are called *secondary GBDs* for this reason. The vectors are related to the geometric description of coincidence site lattices (CSL) developed by Bollmann as summarized in his book.[60] The formal description is beyond the scope of our discussion, but we present a physical overview of the geometry and its relation to GBDs. For further detail, as well as a summary of a number of applications in this burgeoning field, we refer to Bollmann and coworkers[60,61] and to two sets of reviews.[62,63]

Figure 19-37 presents two examples of coincidence lattices. An associated lattice is the DSC lattice (displacement shift complete), which is the coarsest lattice of highest symmetry in CSL orientation containing both constituent crystal lattices as sublattices. The DSC lattices are also shown in Fig. 19-37. Σ represents the number of lattice points in the CSL, so Σ^{-1} is the fraction of lattice points (atoms in our examples) in coincidence. Some properties of these lattices are[61]: (1) the DSC lattice and the CSL are reciprocal lattices for simple cubic crystals; and (2) the volumes of the CSL, the parent lattice and the DSC lattice vary in the ratios $\Sigma : 1 : \Sigma^{-1}$ for simple cubic crystals and $\Sigma : 1 : \frac{1}{2}\Sigma^{-1}$ for

[59] T. Schober and R. W. Balluffi, *Phil. Mag.*, **21**: 109 (1970); **24**: 165 (1971).

[60] W. Bollmann, "Crystal Defects and Crystalline Interfaces," Springer-Verlag, Berlin, 1970.

[61] H. Grimmer, W. Bollmann, and D. H. Warrington, *Acta Cryst.*, **30A**: 197 (1974).

[62] G. A. Chadwick and D. A. Smith, "Grain Boundary Structure and Properties," Academic, New York, 1976.

[63] "Grain Boundary Structure and Kinetics," American Society for Metals, Metals Park, Ohio, 1980.

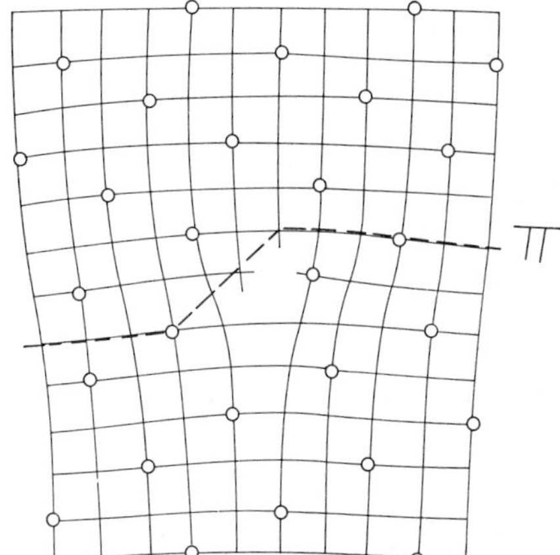

FIGURE 19-38. Projection of simple cubic lattice along $[00\bar{1}]$ parallel to stepped (210) twin boundary containing an extrinsic GBD, $\frac{2}{5}[1\bar{2}0]$.

bcc and fcc crystals, as can be verified from Fig. 19-37. The Burgers vectors of the GBDs are DSC lattice vectors.

The intrinsic[64] GBD content, **B**, of the boundaries in Fig. 19-37 can be determined as illustrated in Fig. 19-18. Here **B** can be expressed in terms of lattice dislocations or in terms of secondary GBDs as may be appropriate. The DSC description is easy to use with the model of Fig. 19-18.[63] Extrinsic dislocations (twinning dislocations for the examples in Fig. 19-37) can also form, for example by the intersection model in Fig. 23-17. An example is presented in Fig. 19-38. The Burgers vector of this extrinsic GBD is $\frac{2}{5}$ [120] and, as verified from Fig. 19-37*b*, this is indeed a DSC lattice vector.

For small deviations from CSL orientation of a twin, the boundary energy is minimized if dislocations are superposed on the twin boundary. For example, a small tilt of Fig. 19-37*b* could be accommodated by an intrinsic GBD tilt boundary superposed on the (210) twin boundary. An example of both intrinsic and extrinsic GBDs is presented in the transmission electron micrograph of stainless steel in Fig. 19-39.[65] The boundary plane between crystals *A* and *B* has the orientation $(232)_A$ and $(223)_B$. The crystals are slightly misoriented relative to the $\Sigma = 9$ CSL orientation of 39.94° rotation about [011]. The deviation from exact CSL orientation is accommodated by three sets of intrinsic GBDs with Burgers vectors referred to lattice *B* of $\mathbf{b}_{\rm I} = \frac{1}{18}[4\bar{1}1]_B$, $\mathbf{b}_{\rm II} = \frac{1}{9}[12\bar{2}]_B$, and $\mathbf{b}_{\rm III} = \frac{1}{18}[127]_B$. In Fig. 19-39 only the set with $\mathbf{b}_{\rm III}$ is in contrast. The extrinsic dislocations of the pileup impinging on the boundary have $\mathbf{b} = \frac{1}{2}[0\bar{1}1]_B$. At ambient temperature these dislocations interacted only elastically with the GBDs as shown. Subsequent heating to $\sim 0.75T_m$ leads to a

[64]We follow J. P. Hirth and R. W. Bulluffi, *Acta. Met.*, **21**: 929 (1973), in defining intrinsic dislocations as those required to give a **B** satisfying Frank's formula [Eq. (19-14)], and producing a true grain boundary without long-range stresses. Thus the perfect lattice dislocations in Fig. 19-35*a* are intrinsic. Any additional dislocation, perforce with a long-range strain field is called *extrinsic*.

[65]W. A. T. Clark and D. A. Smith, *Phil. Mag.*, **38**: 367 (1978).

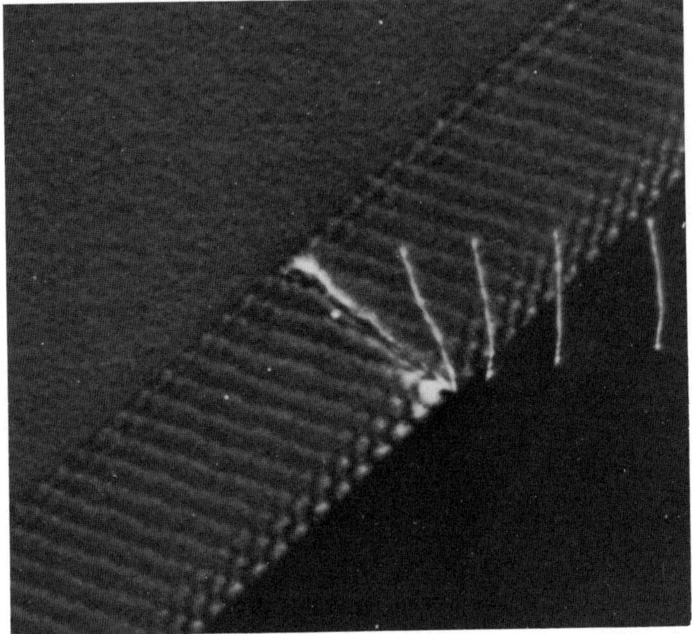

FIGURE 19-39. A grain boundary in type 316 stainless steel. Intrinsic grain boundary dislocations with $b_{\mathrm{III}} = \frac{1}{6}[112]_A = \frac{1}{18}[127]_B$ are in contrast in the boundary. A pileup with extrinsic dislocations with $\mathbf{b} = \frac{1}{2}[0\bar{1}1]_B$ is impinging on the boundary. The diffraction vector is $\mathbf{g} = [\bar{1}1\bar{3}]_A = [11\bar{3}]_B$. [(W. A. T. Clark and D. A. Smith, *Phil. Mag.*, **38:** 367 (1978).]

dissociation reaction of the form, referred to lattice B,

$$\tfrac{1}{2}[0\bar{1}1]_B \rightarrow \tfrac{1}{18}[4\bar{1}1]_B + 2\times\tfrac{1}{9}[\bar{1}\bar{2}2]_B \tag{19-82}$$

All the preceding formulas apply for intrinsic secondary GBDs just as for intrinsic lattice GBDs. Thus a general grain boundary could be described in terms of three independent secondary GDBs. The three DSC vectors, in Fig. 19-37, for example, are the logical first choice. Multiple DSC vectors, as in Fig. 19-38, or partial DSC vectors are possible, and the latter have been observed.[66] Indeed, perfect lattice dislocations are combinations of primitive DSC vectors, as suggested by the reaction of Eq. (19-82). The detailed choice depends on considerations of energy. As one example, steps on the twin interface in Fig. 19-37*b* create twin boundaries with different configurations and therefore different fault energies.[67] Thus GBDs of unit DSC vector may form to transform a high-energy fault to a lower-energy one.

All possible grain boundary orientations are within a few degrees of a CSL if one goes to high enough $\Sigma \sim 150$. Whether very high Σ CSLs are meaningful

[66] R. C. Pond and V. Vitek. *Proc. Roy. Soc.*, **357A**: 453 (1977).

[67] J. P. Hirth, *Acta Met.*, **22**: 1023 (1974).

again depends on energy considerations. In any case, the CSL cases greatly extend the range of boundaries for which the preceding formalism is appropriate: low-angle boundaries for perfect lattice dislocations and near-CSL boundaries for secondary GBDs. For high-angle boundaries of either class, the formal dislocation description can be applied, but the concept loses physical significance. Similarly, the intrinsic GBD content of the twin boundaries of Fig. 19-37 can be described by the formalism, but it has no physical significance.

Extrinsic lattice dislocations have been observed to dissociate into extrinsic secondary GBDs upon entering a grain boundary.[68] In some cases an extrinsic dislocation is absorbed into a boundary and becomes invisible in electron microscopy. This has led to a controversy about whether the secondary GBDs spread their cores or whether the strain contrast is insufficient for visibility.[69] The issue depends on core-scale phenomena that can be decided only by atomic calculations, and these have deficiencies as discussed in Sec. 8-3. The relevant configurations would appear to be on the borderline between orientations where the GBD description is physically meaningful or not, so the issue does not have great physical significance.

There are many ramifications of the DSC-GBD models with respect to mechanical properties. We mention only two, particularly relevant to the discussion in other chapters. First, extrinsic secondary GBDs have been observed to climb and act as sources (sinks) for vacancies[70, 71] with quite high efficiencies. Second, the dissociation of a lattice dislocation as it enters a grain boundary has obvious implications for the ease of passage of a dislocation through the boundary and hence for hardening.

The DSC-CSL model is not complete in the sense of predicting the configuration of Fig. 19-37*a*, for example. Local atomic interactions at the twin interface may lead to a minimum energy configuration in which one crystal is sheared or moved in the [112] direction relative to the other, or vacancies might be present periodically along the boundary. If the vacancies had the period of the CSL (corresponding to the climb removal of a DSC plane), the resulting orientation would be one that could also be produced by a shear parallel to the boundary; if the vacancy period were greater, the configuration could not be produced by any shear operation. Also, atomic interactions in the fault plane may produce dilatations normal to the boundary.[72] A consequence of the normal and shear displacements would be that atoms no longer reside on

[68] W. Bollmann, B. Michaut, and G. Sainfort, *Phys. Stat. Solidi*, **13a**: 637 (1972).

[69] A. P. Sutton and V. Vitek, *Scripta Met.*, **14**: 563 (1980); H. Gleiter, ibid.: 569.

[70] R. C. Pond and D. A. Smith, *Phil. Mag.*, **36**; 353 (1977).

[71] R. W. Balluffi, in "Grain Boundary Structure and Kinetics," American Society of Metals, Metals Park, Ohio, 1980, p. 297.

[72] G. Nouet and P. Delavignette [*Phys. Stat. Solidi*, **62a**: 187 (1980)] have demonstrated dilatational displacements of $\frac{1}{36}\langle 144 \rangle$ normal to a $\{321\}$ habit plane and $\frac{1}{36}\langle 243 \rangle$ normal to a $\{110\}$ habit plane for a [211] twin in vanadium.

lattice points of the DSC lattice, but the DSC-CSL description would still hold. These issues can be decided only by atomic calculations. A number of such calculations have been performed,[73] but they suffer from the problems discussed in Sec. 8-3. Of course, they are very useful in demonstrating classes of behavior not specific to a given material. One particular result is the structural unit model,[74] which is complementary to the DSC model. For high-order twin boundaries such as those in Fig. 19-37, after atomic relaxation, atom groupings are observed to repeat periodically along the boundary. They resemble Bernal holes in amorphous materials. The way in which they repeat is useful in predicting structures near steps such as those in Fig. 19-37*b*, and in predicting faceting in the boundaries.[75, 76]

Finally, we consider other nomenclature for the grain boundary defects. GBDs were once called *ledges*, but this is unsatisfactory. As shown in Figs. 19-37*b* and 19-38, steps or ledges in boundaries can occur with or without the presence of extrinsic GBDs. Similarly, intrinsic or extrinsic GBDs can occur with or without ledges. Thus ledges and GBDs should be regarded as separate entities describing the boundary structure.

Interphase interfaces can also be described by the CSL-DSC-GBD model. The difference in lattice parameter, of course, is a factor additional to rotation in determining lattice coincidence. The applications for precipitates and second phases are perhaps broader than those for grain boundaries.

19-7. GRAIN-BOUNDARY ENERGIES

Perfect Lattice Dislocation Boundaries

As an example of grain-boundary-energy calculations, we again consider the one-dislocation-set tilt boundary. A pair of such boundaries of opposite sign can be formed in an infinite crystal. The specific energy of formation of a well-separated pair is twice the energy per unit area of a single boundary. In accordance with Eq. (6-13), the energy of a pair of dislocations separated by a distance $r_0 = 2\rho = b/\alpha$ is set equal to zero, and the energy for greater separations is then computed from the interaction energy. The core energy is approximately included in the factor α, as discussed in Sec. 8-3.

For the simple tilt boundary, the stress σ_{xy} in a glide plane containing one of the boundary dislocations is given by Eq. (19-75), with $y = 0$,

$$\sigma_{xy} = \frac{\pi \mu b x}{2D^2(1-\nu)\sinh^2(\pi x/D)} \tag{19-83}$$

[73]See the reviews in "Grain Boundary Structure and Kinetics," American Society of Metals, Metals Park, Ohio, 1980.

[74]G. H. Bishop and B. Chalmers, *Scripta Met.*, **2**: 133 (1968).

[75]H. J. Frost, M. F. Ashby, and F. Spaepen, *Scripta Met.*, **14**: 1051 (1980).

[76]A. Brokman, P. D. Bristowe, and R. W. Balluffi, *Scripta Met.*, **15**: 201 (1981).

The attractive force per unit length on an opposite-sign dislocation in the slip plane is then $\sigma_{xy}b$. Hence the energy per unit length per dislocation in one boundary as computed by the above scheme is one-half the interaction energy, or

$$\frac{W}{L}=\frac{1}{2}\int_{b/\alpha}^{\infty}\sigma_{xy}b\,dx=\frac{\mu b^2}{4\pi(1-\nu)}\int_{\eta_0}^{\infty}\frac{\eta\,d\eta}{\sinh^2\eta} \tag{19-84}$$

where $\eta=\pi x/D$ and $\eta_0=\pi b/\alpha D$. The integral is a standard one and can be evaluated explicitly, yielding

$$\gamma=\frac{W}{LD}=\frac{\mu b^2}{4\pi(1-\nu)D}\left[\eta_0\coth\eta_0-\ln(2\sinh\eta_0)\right] \tag{19-85}$$

In the small-angle limit, when $\eta_0\ll 1$, the grain-boundary energy becomes

$$\gamma\cong\frac{\mu b^2}{4\pi(1-\nu)D}\ln\frac{e\alpha D}{2\pi b} \tag{19-86}$$

The physical interpretation of this formula is obvious. The stress field of an individual dislocation extends roughly throughout a cylinder of radius $\sim D/2$, according to St. Venant's principle. The energy per unit length stored in the cylinder is roughly $[\mu b^2/4\pi(1-\nu)]\ln(\alpha D/2b)$. Because there are $1/D$ dislocations per unit area of boundary, the form of Eq. (19-86) follows. Introducing the angle of tilt θ in the approximation of Eq. (19-6), the small-angle formula, Eq. (19-86) assumes the form[77]

$$\gamma=\gamma_0\theta\ln\frac{e\theta_m}{\theta} \tag{19-87}$$

where

$$\gamma_0=\frac{\mu b}{4\pi(1-\nu)}\qquad\text{and}\qquad\theta_m=\frac{\alpha}{2\pi} \tag{19-88}$$

Actually, these calculations are rigorously correct only for an idealized continuum-dislocation model, or, in real crystals, for the discrete angles corresponding to a mean dislocation spacing equal to an integral number of interplanar spacings. For other angles, some dislocations must be separated by a distance larger than the theoretical spacing and some by a lesser distance; only the *mean* separation agrees with the theoretical separation. Thus the actual energy should exceed that of Eq. (19-87), except at cusps where the spacing between dislocations is an integral number of lattice spacings, so that Eq. (19-87) is valid. At high temperatures, thermal vibrations and point defects

[77] W. T. Read and W. Shockley, *Phys. Rev.*, **78**: 275 (1950).

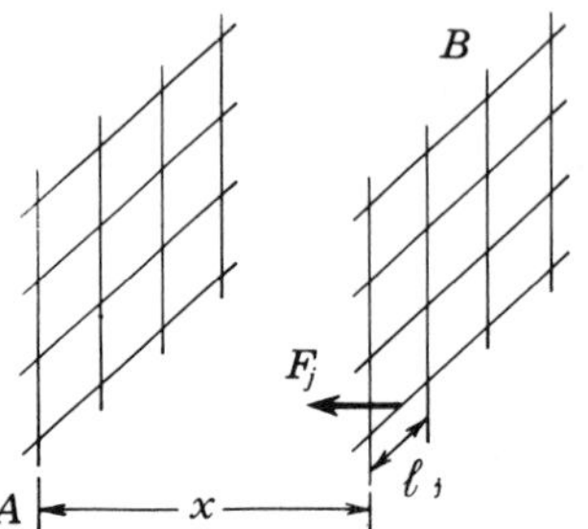

FIGURE 19-40. Perspective view of two boundaries separated by a distance x and lying normal to the x axis.

in both the boundary and the dislocation core should tend to smooth out the cusps. In any case, the effect is small in low-angle boundaries and would be noticeable only in boundaries of extreme perfection.

Although derived for a very simple boundary, Eq. (19-87) is valid for any grain boundary wherein $\mathbf{n}$, $\mathbf{N}_i$, and $\mathbf{b}_i$ are fixed. Both γ_0 and θ_m depend on the character of the boundary and the form of the dislocation arrangement. In general, θ_m is difficult to calculate, because it depends on the core parameters of the dislocation sets and on the core interaction energy. However, the constant γ_0, which is more important for low-angle boundaries, can be computed straightforwardly.[78]

Consider a general boundary network, either hexagonal or lozenge-shaped, with sides l_j on the elementary mesh polygon. As before, consider a pair of boundaries, of opposite-sign dislocations, which would annihilate upon coinciding. When they are separated by a distance x (Fig. 19-40), boundary A exerts a force F_j on a segment l_j in boundary B. As shown by Eqs. (19-75) *et seq.*, the force per unit length acting on l_j is such that, when multiplied by the length $l_j \propto D$, the distance l_j always appears with x as the reduced variable $X_j = x/l_j$. Thus the curve of F_j versus X_j is independent of the size of the mesh and depends only upon the character of the dislocations and their arrangement. For a specific boundary, F_j can be computed exactly from Eqs. (19-75) to (19-80). However, such detail is not necessary in the computation of γ_0, as shown next.

For convenience, we divide the force F_j into a portion that includes the singular behavior at the origin and another term which does not. According to Eqs. (19-75), (19-79), and (19-80), F_j varies typically as $X_j/\sinh^2 \pi X_j$ and in all cases converges to zero as $x \to \infty$. A form which has this general type of behavior is

$$F_j' = \frac{C_j}{X_j} \exp(-X_j) \tag{19-89}$$

where C_j is given by the interaction-force formula (5-17) as

$$C_j = \frac{\mu}{2\pi}\left(b_{s_j}{}^2 + \frac{1}{1-\nu} b_{e_j}{}^2\right) \tag{19-90}$$

[78] W. T. Read, "Dislocations in Crystals," McGraw-Hill, New York, 1953, p. 176.

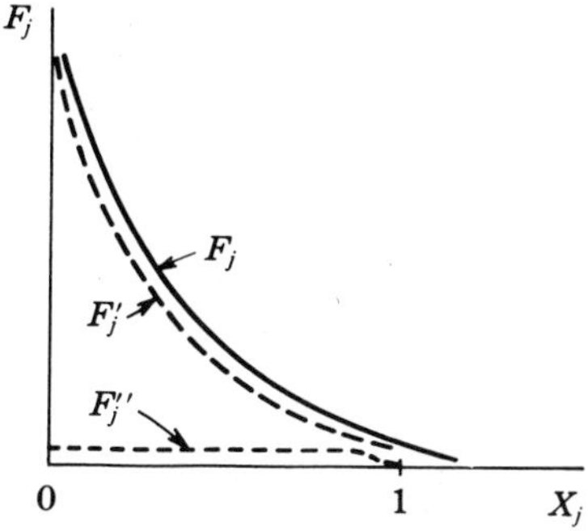

FIGURE 19-41. Plot of F_j, F_j', and F_j'' as a function of X_j.

in order to give the correct limiting behavior for $X_j \to 0$. The difference between the true total F_j and F_j' as defined by Eqs. (19-89) and (19-90) is defined to be F_j''. Because F_j depends only on X_j and converges to zero as $X_j \to \infty$, and because F_j' is defined to depend only on X_j and to converge to zero as $X_j \to \infty$, the difference F_j'' must also depend only on X_j and converge to zero as $X_j \to \infty$. Unlike F_j', F_j'' is nonsingular at the origin (Fig. 19-41). For small X_j, F_j must be antisymmetric about the origin and so, to fit the typical form of Eqs. (19-75) *et seq*., must have the form $C_j/X_j + C_j'X_j + \cdots$. For small X_j, F_j' expands to $C_j/X_j - C_j$. Thus as $X_j \to 0$, $F_j'' \to C_j$.

With these definitions of forces, the boundary energy can be determined by the same procedure as for the tilt boundary. The energy of formation of a segment l_j in one member of a boundary pair is

$$W_j = \tfrac{1}{2}\int_{b_j/\alpha}^{\infty} F_j\,dx = \tfrac{1}{2}\int_{b_j/\alpha}^{\infty} F_j'\,dx + \tfrac{1}{2}\int_{b_j/\alpha}^{\infty} F_j''\,dx \qquad (19\text{-}91)$$

To a good approximation in the small-angle limit $b_j/\alpha l_j \ll 1$, the first integral is

$$\int_{b_j/\alpha}^{\infty} F_j'\,dx = C_j l_j \ln\frac{\alpha l_j}{\Gamma b_j} \qquad (19\text{-}92)$$

where $\Gamma = 1.78$ is Euler's constant. Further, for small-angle boundaries, the second integral can be extended to $x = 0$,

$$\int_{b_j/\alpha}^{\infty} F_j''\,dx \simeq l_j \int_0^{\infty} F_j''\,dX_j = l_j \beta_j \qquad (19\text{-}93)$$

where β_j is independent of the mesh size, being constant for a given boundary character. Thus, finally, one obtains for the boundary energy

$$\gamma = \frac{1}{2A}\sum_j W_j = \frac{1}{4A}\sum_j l_j\left(C_j \ln\frac{\alpha l_j}{\Gamma b_j} + \beta_j\right) \qquad (19\text{-}94)$$

where A is the polygon area and the summation is over all sides in one polygon. The factor $\frac{1}{2}$ accounts for each segment being a side in two polygons. According to Eq. (19-52), the l_j are all proportional to $1/\theta$ in the small-angle limit, so that

$$l_j = \frac{\lambda_j}{\theta} \tag{19-95}$$

Thus the area is proportional to $(1/\theta)^2$,

$$A = \frac{A_0}{\theta^2} \tag{19-96}$$

With these definitions, Eq. (19-94) contracts to the form of Eq. (19-87). In this case

$$\gamma_0 = \frac{1}{4A_0} \sum_j \lambda_j C_j \tag{19-97}$$

and

$$\theta_m = \prod_j \left(\frac{\alpha \lambda_j}{e \Gamma b_j} \right)^{\phi_j} \exp\left(\frac{\beta_j \lambda_j}{4A_0 \gamma_0} \right) \qquad \phi_j = \frac{\lambda_j C_j}{4A_0 \gamma_0} \tag{19-98}$$

As asserted at the outset, all the uncertain terms that would require more detailed calculations for their explicit determination are in the factor θ_m. The factor γ_0 is determined explicitly in terms of the geometry of the dislocation arrangement and the energy factors for straight dislocations (in C_j). Indeed, anisotropy could be accounted for exactly by the use of the appropriate energy factors for straight dislocations in anisotropic media in C_j.

For straight crossing dislocations producing a lozenge-shaped mesh, the above result can be conveniently expressed in terms of the results of Exercise 19-4. The λ_j are the moduli of vectors defined in terms of $\mathbf{L}_j$ [Eq. (19-52)] as

$$\boldsymbol{\lambda}_1 = \mathbf{L}_2 \theta \qquad \boldsymbol{\lambda}_2 = \mathbf{L}_1 \theta$$

Thus the vectors $\boldsymbol{\lambda}_2$, $\boldsymbol{\lambda}_1$, and $\mathbf{n}$ are reciprocal to $\mathbf{N}_1$, $\mathbf{N}_2$, and $\mathbf{n}$, and $A_0 = \mathbf{n} \cdot (\boldsymbol{\lambda}_2 \times \boldsymbol{\lambda}_1)$. Explicitly,

$$\mathbf{N}_1 = \frac{\boldsymbol{\lambda}_1 \times \mathbf{n}}{A_0} \qquad \mathbf{N}_2 = \frac{\boldsymbol{\lambda}_2 \times \mathbf{n}}{A_0}$$

so that Eq. (19-97) can be expressed in terms of the N_j values[79] as

$$\gamma_0 = \tfrac{1}{2} \sum_j N_j C_j \tag{19-99}$$

[79] Ibid., p. 186.

As a specific example, $\gamma_0 = \mu b/2\pi$ for a pure twist boundary composed of orthogonal screw dislocations.

Early experimental work[80] on grain-boundary energies as a function of θ indicated that the Read-Shockley equation (19-87) explained data for twist and tilt boundaries, but that the fit with theory occurred up to angles of ~15 to 50°, far above those where the small-angle approximation involved in the equation should be valid. However, little of the data in these experiments were obtained in the small-angle region $\theta \lesssim 5°$. Later work,[81] in which the small-angle region was studied in some detail, indicated that the fit of high-angle energy data by the Read-Shockley equation was fortuitous, but that the equation did describe the low-angle data, with values of γ_0 and θ_m different from the high-angle values. A plot of Gjostein's results for $\langle 100 \rangle$ tilt boundaries in copper is shown in Fig. 19-42; the low-angle fit of Eq. (19-87) and the fortuitous high-angle fit are both shown. Also, Van der Merwe's result is shown to give a better fit at high angles than the Read-Shockley equation, as anticipated.

A comparison of the results for copper and for germanium[81] is of interest with respect to the core parameter α. For copper the fit with Eq. (19-86) was good up to ~5°, while for germanium the fit was good up to ~10 to 15°; the difference is in agreement with expectation for a smaller core extension in the covalently bonded germanium. Also, for copper θ_m was found to be ~10°, which gives $\alpha \sim 1$ by Eq. (19-88), while for germanium $\theta_m \sim 40°$ and $\alpha \sim 4$, in agreement with the prediction of Sec. 8-3. Again, this result is consistent with the expectation that α should be smaller for the "soft-core" metal than for the covalently bonded germanium.

Coincidence Lattice Orientations

The atom matching at boundaries such as those of Fig. 19-37 leads us to expect cusps (local sharp minima) in plots of surface energy versus boundary orientation for coincidence lattices with low-index boundary orientations. Indeed, atomic calculations support such a viewpoint for simple model potentials.[82] Experimentally, cusps have been observed directly[83] in surface energy measurements, and indirectly in experiments where single crystal balls are allowed to sinter and rotate to form a low-energy orientation on a single crystal substrate or when plate crystals come in to contact and rotate into low-energy orientations. As reviewed by Goodhew,[83] cusps have been identified, for

[80]Reviewed by D. McLean, "Grain Boundaries in Metals," Oxford University Press, Fair Lawn, N.J., 1957, and by S. Amelinckx and W. Dekeyser, *Solid State Phys.*, **8**: 325 (1959).

[81]N. A. Gjostein and F. N. Rhines, *Acta Met.*, **7**: 319 (1959); N. A. Gjostein, *Acta Met.*, **8**: 263 (1960). R. S. Wagner and B. Chalmers, *J. Appl. Phys.*, **31**: 581 (1960).

[82]V. Vitek, A. P. Sutton, D. A. Smith, and R. C. Pond, in "Grain Boundary Structure and Kinetics," American Society of Metals, Metals Park, Ohio, 1980, p. 115.

[83]See the review by P. J. Goodhew, ibid., p. 155.

example in fcc metals for $\Sigma = 5$, 9, 11, 13, 17, 25, 33, and 83 coincidence lattices. These give further credence to the concept of existence of secondary GBDs for orientations near the low-energy cusps.

19-8. OTHER GRAIN-BOUNDARY EFFECTS

Grain-Boundary Mobility

Hypothetically, a boundary can move normal to itself by pure slip through motion of the individual dislocation segments on their glide planes, provided none of the segments with edge components has glide planes coinciding with the boundary plane. However, such motion in general would produce a boundary configuration quite different from the equilibrium one; the boundary would move through high-energy configurations, so that motion over large distances would require high driving forces. For example, the tilt boundary of Fig. 19-43, containing two sets of dislocations with orthogonal Burgers vectors, would have to be forced through a stage of coalescence of pairs of dislocations in the course of motion. Furthermore, some stress distributions would produce separation of the boundary-dislocation sets, e.g., when the maximum resolved shear stress acts on the slip system of each set, a process that again would require large driving forces except for boundaries of very low angle,[84] $\theta \simeq 10^{-3}$. Thus low-angle grain boundaries are generally both immobile in glide and stable against dissociation.

The one-dislocation tilt boundary, on the other hand, is an example of a glissile boundary. The boundary advances normal to itself by glide on slip planes normal to the boundary plane. Such motion has been observed[85] for tilt boundaries normal to the basal plane in zinc, providing striking support for the dislocation model of grain boundaries. In principle, the crossed grid of pure screws and the hexagonal network containing only screw segments should be mobile in glide, but such motion has not been observed. The absence of such observations might be attributable to dissociation into partials in the boundary plane or to a high Peierls stress for glide normal to the boundary plane.

Usually, then, a low-angle boundary can move at commonly applied stress levels only at elevated temperatures, where climb can occur. As a grain boundary advances, the volume of one grain increases as much as the volume of the other decreases. Thus forward boundary motion requires no net transport of matter to or from the boundary but only exchange of matter between the dislocations in the boundary. Some segments in the boundary act as sources and some as sinks. The theory of the mobility of individual dislocations in climb is not directly applicable to boundary mobility problems because dislocation-dislocation interactions are as important as the jog self-energies in

[84] W. T. Read, "Dislocations in Crystal," McGraw-Hill, New York, 1953, p. 199.

[85] J. Washburn and E. R. Parker, *Trans. AIME*, **194**: 1076 (1952).

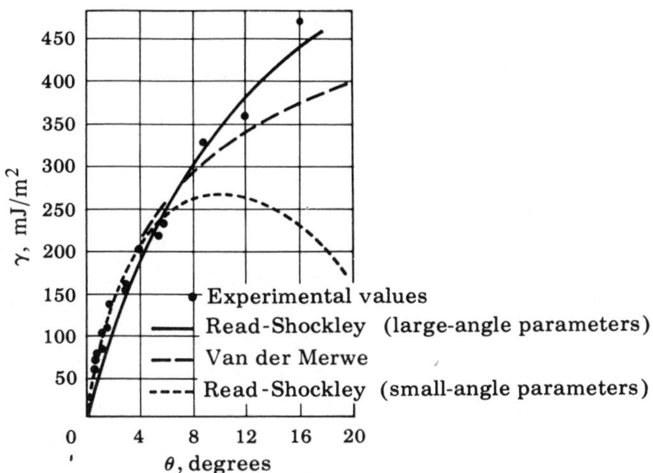

FIGURE 19-42. Grain-boundary energy of a (100) tilt boundary in copper as a function of angle of tilt. [N. A. Gjostein and F. N. Rhines, *Acta Met.*, **7**: 319 (1959).] Theoretical fits to the Read-Shockley equation (19-87) for the low-angle data and for the high-angle data are shown, as is the fit of the formula of Van der Merwe.

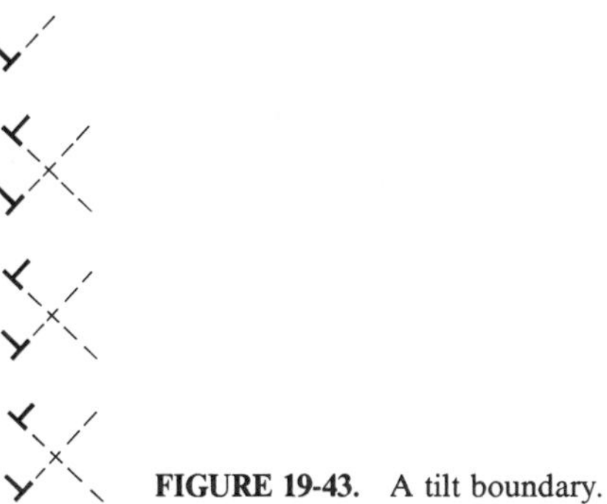

FIGURE 19-43. A tilt boundary.

determining the jog concentration in the boundary.[86] At sufficiently high temperatures, saturation occurs; the individual dislocations are effectively continuous line sinks and sources, as discussed in Chap. 15. The uncertainties in point-defect formation and migration energies in boundaries are about the same as those for point defects in the core; the reader is referred to Chap. 15 for a discussion of mobility equations to describe boundary motion.

Boundary Resistance to Dislocation Motion

In Sec. 9-7 we discussed the resistance to glide at a grain boundary because of plastic incompatibility and because of elastic incompatibility, which gives rise

[86] J. Lothe, *J. Appl. Phys.*, **31**: 1077 (1960).

to image effects in the anisotropic theory. Evidently, the short-range stresses discussed in Sec. 19-7 do work on a dislocation as it approaches the boundary, producing local forces on the dislocation. Perhaps more important, local interactions with the boundary dislocations can effectively pin a dislocation, preventing its glide through the boundary. Obviously, local interactions producing rearrangements such as that of Fig. 19-22 cause the configuration to relax toward metastable equilibrium, so that a force is required to unpin a reacted dislocation. Another local effect is the formation of intersection jogs and kinks in the boundary dislocations when a glide dislocation cuts through the boundary; the energy of formation of these intersection segments must be supplied in the cutting operation. For a high-angle boundary, the extension of this concept is the formation of a step in the boundary equal to $\mathbf{b} \cdot \mathbf{n}$ in magnitude, where $\mathbf{b}$ is the Burgers vector of the glide dislocation. Energy must be supplied to increase the grain-boundary area by the amount of the step. Dissociation of a dislocation into extrinsic secondary GBDs or a smeared continuous distribution in a boundary would also be a hardening effect. All these effects produce resistance to glide at boundaries, whether low angle or high angle. Grain boundaries are expected to be effective barriers to dislocation glide.

Anisotropic Elasticity Modifications of Boundary Theory

As a final topic, a few words on anisotropic elasticity are appropriate. All the geometric developments of Secs. 19-1 to 19-3 and 19-6 apply exactly for anisotropic crystals. Also, the stresses of grain boundaries in anisotropic crystals can be summed by the same methods as used in Sec. 19-5, while Sec. 19-7 applies with the anisotropic stress tensors and energy factors substituted. The other case where anisotropic modifications are important is the local-rearrangement problem considered in Sec. 19-4. The variation of W_s with orientation in anisotropic crystals can be sufficient for reactions which are stable according to isotropic theory, such as *CD* in Fig. 19-22, to be unstable in anisotropic crystals; possibly even the opposite type of reaction, such as *AB* in Fig. 19-22, can become stable in anisotropic theory. In any case, the equilibrium angle ϕ_e in Fig. 19-22 is expected to differ in the isotropic and anisotropic cases. An interesting example of the latter effect is presented by Asaro and Hirth.[87] For the boundary of Fig. 19-25 in the case of α-iron, they find a predicted value of $\phi_e = 92°$ compared to an experimental value of $\phi_e = 91°$ and an isotropic theory prediction of $\phi_e = 103°$.[87, 88]

[87] R. J. Asaro and J. P. Hirth, *J. Phys. F*, **3**: 1659 (1973).

[88] S. M. Ohr and D. N. Beshers, *Phil. Mag.*, **8**: 1343 (1963).

PROBLEMS

19-1. Derive $\mathbf{N}_1$ for a simple tilt boundary consisting of one set of edge dislocations.

19-2. Determine $\mathbf{N}_1$ and $\mathbf{N}_2$ for a simple twist boundary consisting of two sets of orthogonal screw dislocations.

19-3. Consider two sets of dislocations whose Burgers vectors are inclined to one another at a 45° angle. Determine $\mathbf{N}_1$ and $\mathbf{N}_2$ for a pure twist boundary in their common glide plane. What is the dislocation density in the two sets for an angle of rotation θ?

19-4. For a simple cubic crystal, a grain boundary is composed of the dislocations $\mathbf{b}_1=[100]$, $\mathbf{b}_2=[010]$, and $\mathbf{b}_3=[001]$. Determine $\mathbf{N}_1$, $\mathbf{N}_2$, and $\mathbf{N}_3$ when

$$\mathbf{a}=\frac{1}{\sqrt{14}}[123] \qquad \mathbf{n}=[100]$$

What are the directions of the dislocations? What are the dislocation densities when $\theta=5°$?

19-5. If only the slip systems $[101](10\bar{1})$ and $[10\bar{1}](101)$ are active in an NaCl-type structure, what boundaries can form by glide?

19-6. If only the slip systems $[101](10\bar{1})$ and $[110](1\bar{1}0)$ are active in an NaCl-type structure, what boundaries can form by glide? Is this a realistic case?

19-7. Generalize the example of Fig. 19-34 to fcc tilt boundaries formed by glide on intersecting slip planes.

19-8. Derive the stress distribution about the boundary of Prob. 19-3. Show explicitly that no long-range stresses exist.

19-9. Demonstrate that the general theory for grain-boundary energies (Sec. 19-7) agrees with the results of Prob. 19-8. Calculate the values of F_j and F_j'' for that example and discuss the limiting behavior as $X_j \to 0$.

19-10. What type of applied stress tensor could cause a pure twist boundary to glide

- **a.** In the most general case?
- **b.** When the sets of dislocations in the boundary are pure screw?

19-11. Consider the interaction of a single dislocation **DC**(a) when it intersects the pure twist boundary composed of the pure screws **AB**, **BC**, and **CA** on glide plane (d) in an fcc crystal.

BIBLIOGRAPHY

1. Amelinckx, S., *Solid State Phys. Suppl.* 6 (1964).
2. Amelinckx, S., and W. Dekeyser, *Solid State Phys.*, **8**: 325 (1959).

3. Chadwick, G. A., and D. A. Smith, "Grain Boundary Structure and Properties," Academic, New York, 1976.
4. Chaudhari, P., and J. W. Matthews, "Grain Boundaries and Interfaces," North-Holland, Amsterdam, 1972.
5. McLean, D., "Grain Boundaries in Metals," Oxford University Press, Fair Lawn, N.J., 1957.
6. Orlov, A. N., V. N. Perevesentsev, and V. V. Rybin, "Grain Boundaries in Metals," Metallurgiya, Moscow, 1980.
7. Read, W. T., "Dislocations in Crystals," McGraw-Hill, New York, 1953.
8. "Grain Boundary Structure and Kinetics," American Society of Metals, Metals Park, Ohio, 1980.

20
Dislocation Sources

20-1. INTRODUCTION

In well-annealed single crystals, the dislocation density is typically[1] $\rho = 10^6$ cm/cc but can approach zero in special cases.[2] During the early stages of deformation, up to strains of ~ 10 percent, the dislocation density increases to a typical value[1] $\rho = 10^{10}$ cm/cc. The problem of dislocation multiplication is thus important in the theory of crystal deformation. There are a number of possible mechanisms for dislocation multiplication at a source. Most of these mechanisms have been verified in specific experiments, but at present it is not possible to assess which mechanism is predominant in a given macroscopic slip process, such as easy glide in fcc crystals or Lüders band propagation in bcc crystals. Indeed, several mechanisms might operate simultaneously.

In this chapter we discuss the various multiplication mechanisms. The Frank-Read dislocation source is treated first. Then other mechanisms, similar to the Frank-Read one but different in detail, are considered. Finally, the less likely possibilities of nucleation of dislocations and of kinematic sources are discussed.

20-2. FRANK-READ SOURCES

Frank-Read sources[3] in glide closely resemble the Bardeen-Herring climb sources discussed in Sec. 17-5, and, in fact, inspired them. Consider the segment of dislocation AB in Fig. 20-1, whose ends are pinned, corresponding to nodes in a network, precipitates, or sites where the dislocation leaves the glide plane. Under an applied stress the segment bows out by glide. As bow-out proceeds, the radius of curvature of the line decreases and the line-tension forces tending to restore the line to a straight configuration increase. For stresses less than a critical value, a metastable equilibrium

[1] W. G. Johnston and J. J. Gilman, *J. Appl. Phys.*, **30**: 129 (1959).

[2] W. C. Dash, *J. Appl. Phys.*, **29**: 736 (1958); in R. H. Doremus, B. W. Roberts, and D. Turnbull (eds.), "Growth and Perfection of Crystals," Wiley, New York, 1958, p. 361.

[3] F. C. Frank and W. T. Read, in "Symposium on Plastic Deformation of Crystalline Solids," Carnegie Institute of Technology, Pittsburgh, 1950, p. 44.

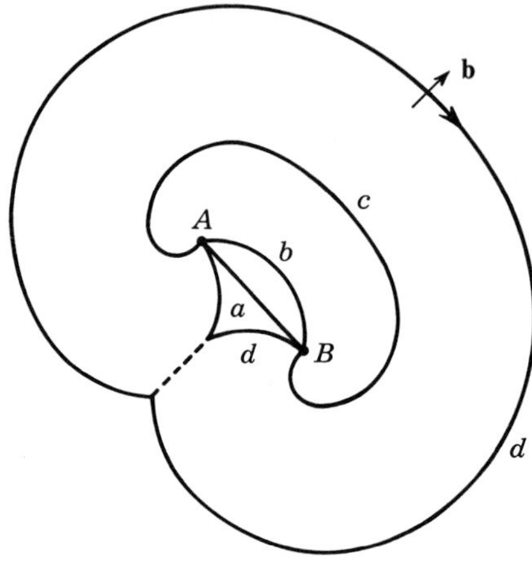

FIGURE 20-1. Dislocation segment (a) pinned at A and B, (b) bowing out in its glide plane, and (c) and (d), operating as a Frank-Read source.

configuration such as (b) in Fig. 20-1 is attained, in which the line-tension force balances that caused by the applied stress. For the large bow-out case, Eqs. (6-71) and (6-72) yield the equilibrium condition [*cf.* Eq. (17-16)]

$$b\sigma = \frac{\mathbb{S}}{r} = \frac{\mu b^2}{4\pi r(1-\nu)}\left\{\left[1-\frac{\nu}{2}(3-4\cos^2\beta)\right]\ln\frac{L}{\rho}-1+\frac{\nu}{2}\right\} \quad (20\text{-}1)$$

where r is the radius of curvature of the loop. The radius of curvature is a minimum when $r = L/2$. Hence the maximum stress for which local equilibrium is possible is given by Eq. (20-1), with $r = L/2$. For the typical case that $L = 10^3\rho$ and $\nu = 0.33$, the critical stress for a dislocation initially pure edge and pure screw, respectively, is $\sigma^* = 0.5\mu b/L$ and $\sigma^* = 1.5\mu b/L$. As shown for the Bardeen-Herring sources, thermal activation does not reduce σ^* by a significant amount, except for the unlikely cases where $L \lesssim 20b$.[4,5] When the net local resolved shear stress (the applied stress plus the internal stress) exceeds σ^*, the loop has no stable equilibrium configuration but passes through the successive positions shown in Fig. 20-1. Provided that the expanding loop neither jogs out of the original glide plane because of intersections with other dislocations nor is obstructed from rotating about the pinning point, it will annihilate over a portion of its length, creating a complete closed loop and restoring the original configuration. A sequence of loops then continues to form from the source until sufficient internal stresses are generated for the net resolved shear stress at the source to drop below σ^*. Dash's[6] classical picture of a Frank-Read source in silicon is presented in Fig. 20-2.

[4] F. R. N. Nabarro, *Advan. Phys.*, **1**: 269 (1952).

[5] R. M. Shemenski, *Trans. Quart. ASM*, **58**: 360 (1965).

[6] W. C. Dash, *J. Appl. Phys.*, **27**: 1193 (1956).

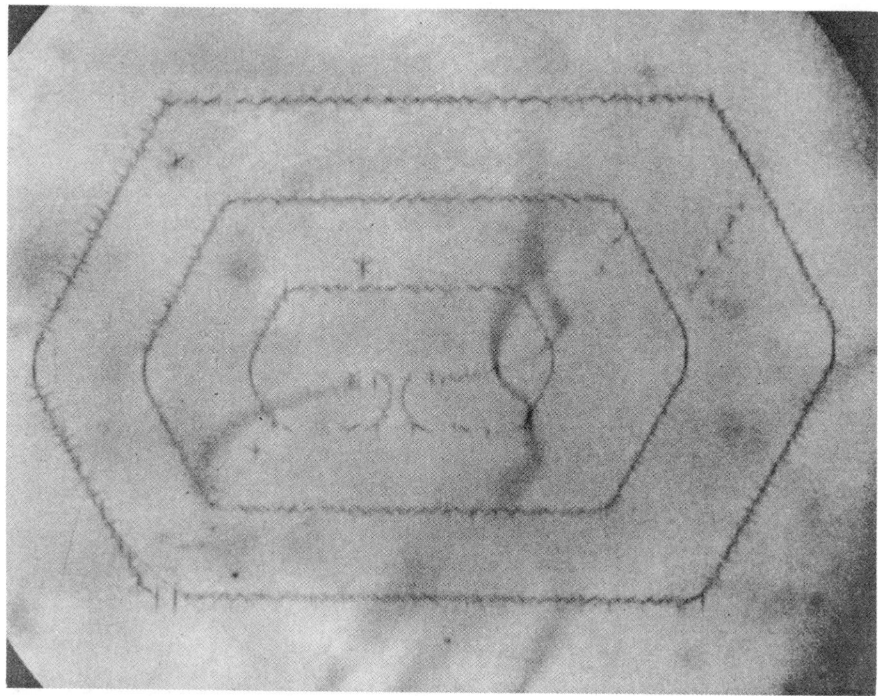

FIGURE 20-2. A Frank-Read source in silicon decorated with copper. [W. C. Dash, *J. Appl. Phys.*, **27**: 1193 (1956).]

In actuality, the bowed-out configuration of a loop is not an arc of a circle, because the local value of the line tension varies with orientation.[7] A possible true configuration is shown in Fig. 20-3*a*. Another effect arises when adjacent segments also bow out, as in Fig. 20-3*b*. In this case, the negative interaction energy of a given loop and its neighboring loops reduces its line tension and hence reduces σ^*. The first of the above effects is a second-order effect in the energy and is not very important in the calculation of the critical stress, while the latter effect is important only for a dislocation line lying in a single glide plane and pinned by regularly spaced obstacles. Thus, in most cases, the description of Eq. (20-1) suffices, and we do not consider these refinements further.

As in the example of climb, singly pinned sources, or the ends of double sources for large L, assume spiral configurations like that in Fig. 17-6 as glide proceeds. To the same approximation as for the Bardeen-Herring climb spiral, however, the Frank-Read glide spiral can be described as the archimedean spiral of Eqs. (17-19) to (17-21). The exact solution for the spiral shape is in principle quite complex because of interaction between spiral turns and because of variations in internal stress fields across the spiral. Also, unlike the

[7]G. de Wit and J. S. Koehler, *Phys. Rev.*, **116**: 1113 (1959).

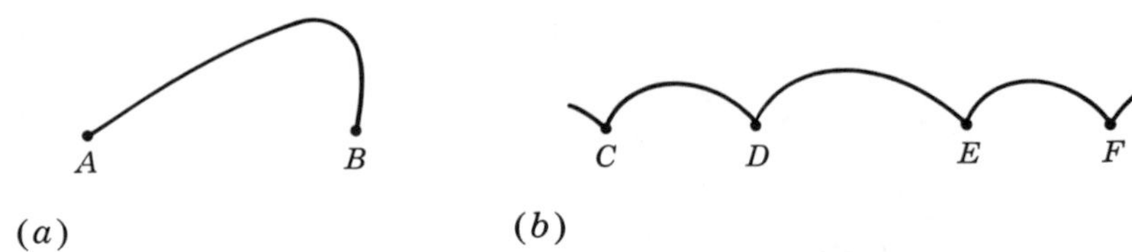

FIGURE 20-3. (*a*) Asymmetric bow-out. (*b*) Adjacent bowed-out loops.

Bardeen-Herring sources where dynamic effects are negligible because of the large climb viscosity, acceleration and damping effects are important in the operation of a Frank-Read source. The problem has been considered in several approximations,[8-10] but most recently by Steif and Clifton.[11] They found that for values of mobility B typical for fcc metals [Eq. (7-74)] damping at the source was dominant. The process was controlled by production of loops at the source where the net driving force is least because of back forces from line tension. While they used a constant loop shape approximation to simplify the problem, the results indicate that inertial effects[8] can be neglected as can interactions among loops until pileup back stresses become important at the end of the process. Thus the problem can be treated in the straight line segment approximation[9] without complications from loop-loop interactions.

The results[8] for nucleation time τ are presented in terms of a universal curve in reduced coordinates. For a typical reduced stress $\sigma L/\mu b \sim 4$, the reduced time is $\tau\sigma b/BL \sim 10$. With typical values $L = 1\ \mu\text{m}$, $\sigma = 10^{-3}\ \mu$ and B from Eq. (7-77) with $C_t = 10^3$ m/sec, we find $\tau \sim 0.2\ \mu$sec as a typical period to generate one loop.[12]

If the pinning point is a node, with the other nodal dislocations having screw components relative to the glide plane of the source dislocation, the dislocation will wind into a conical helix, as shown in Fig. 20-4. The pitch of the helix equals the screw component of the "pole" dislocation around which the source dislocation winds. Such pole mechanisms are important in twinning, as discussed in Chap. 23. Carter[13] reviews the observations of both types of Frank-Read source in electron microscopy. Their dynamic operation has been displayed cinematographically.[14]

[8] J. D. Campbell, J. A. Simmons, and J. E. Dorn, *J. Appl. Mech.*, **28**: 447 (1961).

[9] H. J. Frost and M. F. Ashby, *J. Appl. Phys.*, **42**: 5273 (1971).

[10] T. Yokobori, A. T. Yokobori, Jr., and A. Kamei, *Phil. Mag.*, **30**: 367 (1974).

[11] P. S. Steif and R. J. Clifton, *Mat. Sci. Eng.*, **41**: 251 (1979).

[12] Of course, when dislocation motion is thermally activated, as in double-kink formation in bcc metals at low temperatures, B is much smaller and the situation is analogous to that discussed for the Bardeen-Herring source.

[13] C. B. Carter, *Phil. Mag.*, **35**: 75 (1977).

[14] Reviewed by L. P. Kubin and J. L. Martin, in P. Haasen, V. Gerold, and G. Kostorz (eds.), "Strength of Metals and Alloys," vol. 3, Pergamon, Oxford, 1980, p. 1639; and by T. Imura, *Electron Microsc.*, **4**: 280 (1980).

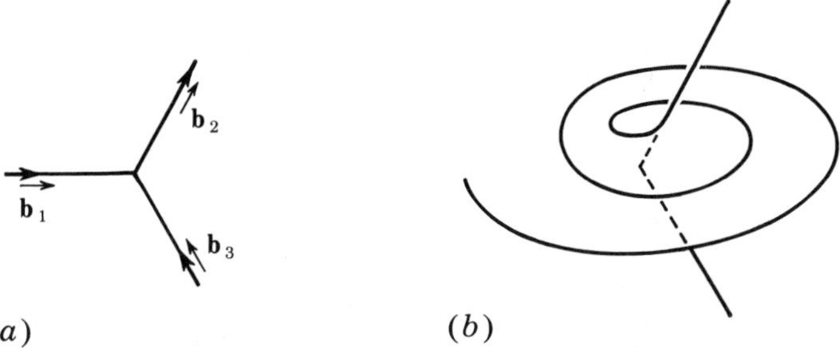

FIGURE 20-4. (a) A dislocation node; $\mathbf{b}_2$ and $\mathbf{b}_3$ have a component normal to the glide plane of $\mathbf{b}_1$. (b) Operation of $\mathbf{b}_1$ as a helical source; the helix advances by an amount equal to the normal component of $\mathbf{b}_2$ with each turn.

Exercise 20-1. Consider a dislocation segment pinned at one point at a distance L below a clean free surface and lying in a glide plane normal to the surface. Show that the critical stress required to operate the loop as a single-ended Frank-Read source is roughly one-half that given by Eq. (20-1).

For cases where glide planes are not nearly normal to the surface, the situation is more complicated. Image stresses still influence near-surface source operation,[15,16] but image forces can create jogs that preferentially harden certain slip systems. These effects are important for the so-called anomalous slip systems discussed in Sec. 9-6. Even for glide planes parallel to the surface, image effects can reduce σ^*.[17]

20-3. THE DOUBLE-CROSS-SLIP MECHANISM

Multiplication by double cross slip, suggested by Koehler,[18] is closely related to the Frank-Read mechanism. Consider the fcc screw dislocation **AB** gliding on the plane (d), corresponding to the maximum local resolved shear stress, as shown in Fig. 20-5a. If the dislocation encounters an obstacle, it tends to cross slip onto the lower resolved-shear-stress plane (c) (Fig. 20-5b). Once the cross-slipping segment is out of the range of the obstacle, it tends to cross slip back onto the original glide plane (Fig 20-5c). The doubly cross-slipped segment can now operate exactly as a Frank-Read source (Fig. 20-5d). The segments of the original dislocation that did not cross slip could in principle operate as a Frank-Read source bowing out in the opposite direction, but they are prevented from doing so by the impenetrable obstacle. Another possibility

[15] O. Lohne, *Phys. Stat. Solidi*, **25a**: 709 (1974).

[16] F. Minari, B. Pichaud, and L. Capella, *Phil. Mag.*, **31**: 275 (1975).

[17] R. M. Latanision and R. W. Staehle, *Acta Met.*, **17**: 1973 (1969).

[18] J. S. Koehler, *Phys. Rev.*, **86**: 52 (1952).

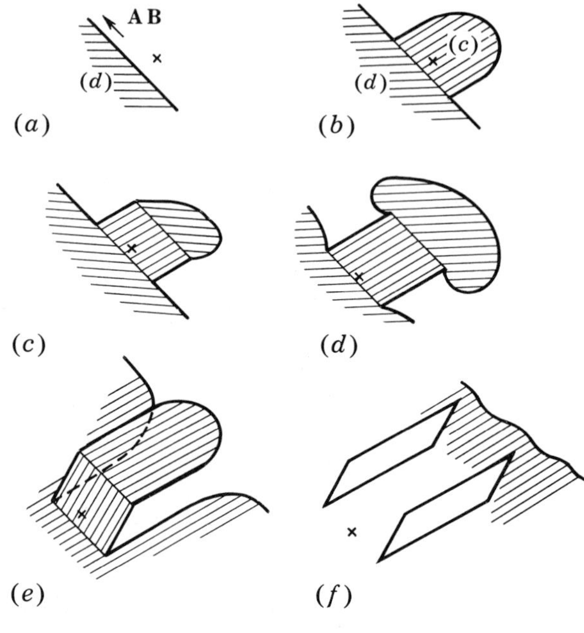

FIGURE 20-5. (*a*) Screw dislocation **AB** gliding on plane (*d*). (*b*) Cross slip at an obstacle. (*c*) Double cross slip. (*d*) Operation of cross-slipped segment as a Frank-Read source. (*e*) Dipole formation. (*f*) Closed-dipole-loop formation.

arises when the distance h of cross slip is small. The attractive interaction between the cross-slipped and non-cross-slipped segments can constrain the segments to remain parallel, as in Fig. 20-5*e*, forming a dislocation dipole. Cross slip back onto the original glide plane can produce a closed dipole loop, as depicted in Fig. 20-5*f*. Definite evidence for a double cross slip was obtained by Gilman and Johnston[19] in successive etching experiments on deformed lithium fluoride. Dipoles have been observed to form by the above mechanism, in electron-transmission microscopy of various materials.[20,21] The double-cross-slip mechanism appears to be the most likely multiplication mechanism in Lüders band formation during crystal deformation.

Single-ended variants resembling the double-cross-slip source mechanism and the dipole-formation model can also form. If intersection jogs of one sign accumulate into a superjog on a screw dislocation, the resulting configurations resemble those of one end of the models of Fig. 20-5. Kuhlmann-Wilsdorf et al.[22] have determined the critical stress for the two rotating segments to pass

[19] W. G. Johnston and J. J. Gilman, *J. Appl. Phys.*, **31**: 632 (1960).

[20] R. L. Segall, P. G. Partridge, and P. B. Hirsch, *Phil. Mag.*, **6**: 1493 (1961).

[21] J. R. Low and A. M. Turkalo, *Acta Met.*, **10**: 215 (1962).

[22] D. Kuhlmann-Wilsdorf, J. H. van der Merwe, and H. G. F. Wilsdorf, *Phil. Mag.*, **43**: 632 (1952).

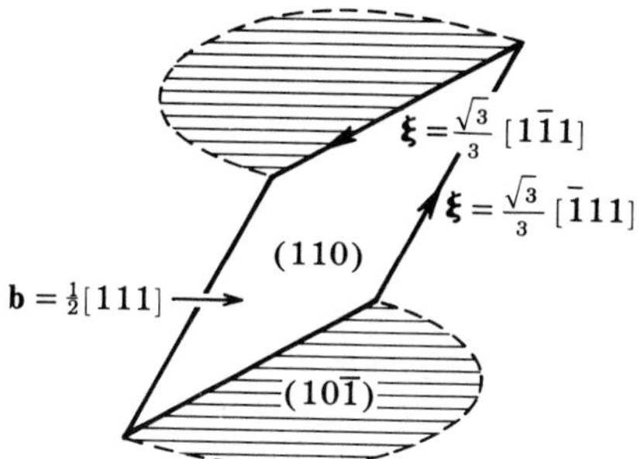

FIGURE 20-6. Prismatic dislocation loop $\mathbf{b} = \frac{1}{2}[111]$ formed from a vacancy disc condensed onto a (110) plane, and its bow-out under stress on $(10\bar{1})$.

one another as a function of their distance of separation h. The calculation is a quasi-equilibrium one; when the mobility is large, inertial effects will also favor passing.

20-4. VACANCY DISC NUCLEATION

The results of Sec. 17-3 indicate that prismatic dislocation loops are the result of vacancy aggregate nucleation in high-stacking-fault-energy materials. Nabarro[23] pointed out that such discs could operate as dislocation sources. Consider as an example the prismatic loop on a bcc (110) plane (Fig. 20-6). With a high resolved shear stress on the $(10\bar{1})$ plane but a small stress on the $(01\bar{1})$ plane, the loop can bow out onto the $(10\bar{1})$ plane and form Frank-Read sources at either end. If the $(01\bar{1})$ plane were also highly stressed, the loop would glide as a whole and not act as a source. Equivalent sources are possible in other crystal structures. Because this source mechanism requires that the crystal be highly supersaturated with vacancies to nucleate the loops prior to deformation, it is likely only in special situations.

20-5. NUCLEATION OF GLIDE LOOPS

Homogeneous nucleation theory was first extended to the treatment of nucleation of glissile dislocation loops by Frank.[24] The formalism of the nucleation theory[24] closely parallels that of Sec. 17-3 and is not developed in detail again here. The nucleation rate in number per unit volume per unit time is

$$J = Z\omega n_c \cong 0.1\omega n_c \tag{20-2}$$

[23]F. R. N. Nabarro, in "Report of the Conference on Strength of Solids," Physical Society, London, 1949, p. 75.

[24]F. C. Frank, in "Symposium on Plastic Deformation of Crystalline Solids," Carnegie Institute of Technology, Pittsburgh, 1950, p. 89; J. P. Hirth, in "Relation between Structure and Strength in Metals and Alloys," H. M. Stationery Office, London, 1963, p. 218.

Here the frequency factor ω is given by the probability that an atom on the periphery of the critical nucleus vibrates to join the nucleus,

$$\omega = \frac{8\pi r^* \nu}{b} \tag{20-3}$$

where ν is the Debye frequency. At low temperatures, where the segments of the loop are expected to lie in Peierls valleys, ω would involve double-kink nucleation over the Peierls barrier, would be given by Eq. (15-29), and would be much smaller than indicated by Eq. (20-3). The equilibrium concentration of critical nuclei is given by Eq. (17-8) as

$$n_c = n_0 \exp\left(\frac{-\Delta G^*}{kT}\right) \tag{20-4}$$

where n_0 is the concentration of atom sites and ΔG^* is the free energy of formation of a critical-sized loop. We emphasize that Eq. (20-4) represents the usual quasi-equilibrium development of nucleation theory; embryos of various subcritical sizes are nearly in local equilibrium with one another and are continually fluctuating in size by small increments. The critical nuclei do *not* form by one instantaneous fluctuation with no intermediate stages, as is sometimes erroneously assumed in discussing nucleation; such a large fluctuation would lead to a very large negative entropy of activation, which in actuality does not appear in ΔG^*. ΔG^* is the maximum of the free energy of formation of a loop of size r,

$$\Delta G = 2\pi r W - \pi r^2 b\sigma + \pi r^2 \gamma \tag{20-5}$$

where W is the pertinent elastic energy per unit length given by Eq. (6-51), and the last term applies only for partial-dislocation nucleation, γ being the stacking-fault energy.

TABLE 20-1 Critical Resolved Shear Stress σ^* for an Appreciable Dislocation-Nucleation Rate at Several Types of Sites in Aluminum and Copper*

	Al		Cu	
	T, °K		T, °K	
Reaction	300	900	300	900
Perfect dislocation				
Homogeneous nucleation	7.0	4.6	7.4	5.1
Nucleation at surface step	4.6	2.6	3.6	2.4
Partial dislocation				
Homogeneous nucleation	8.5	5.8	5.4	2.9
Nucleation at surface step	6.0	4.8	2.1	1.3

*Presented as the ratio $(\sigma^*/\mu)\times 10^2$.

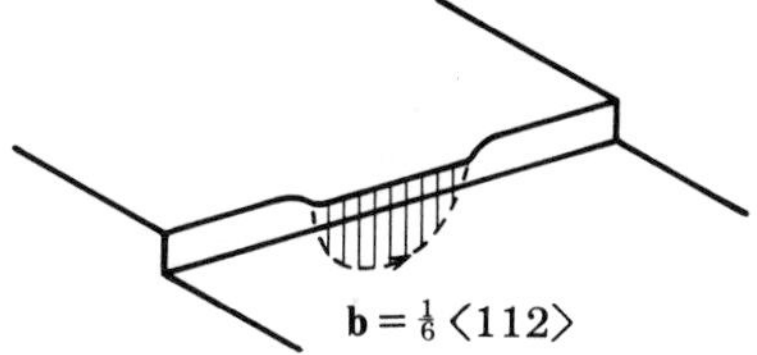

FIGURE 20-7. Formation of a half-loop of partial dislocation in an fcc crystal at a free surface by punching in a portion of a surface ledge.

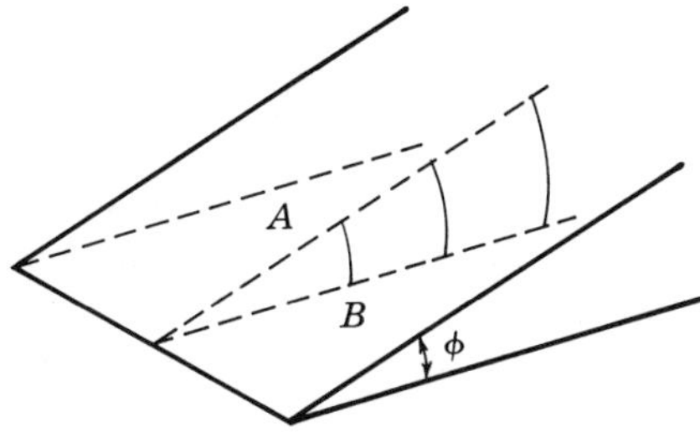

FIGURE 20-8. Nucleation of several dislocations on the same glide plane in a wedge-shaped crystal.

All data necessary to evaluate J by Eq. (20-2) are now explicitly defined. The actual determination of ΔG^* and r^* involves the solution of the transcendental equation (20-5). The equation has been solved graphically for a number of typical cases.[24] Results for some of these cases are summarized in Table 20-1. As shown there, nucleation at appreciable rates occurs only for stresses approaching the theoretical strength of perfect crystals, so that it contributes negligibly to initial yielding and can occur only in regions of marked stress concentration.

Nucleation is more favorable (1) at free surfaces, where image effects lower ΔG^*; (2) when surface ledges are annihilated by nucleation, lowering ΔG^* by an amount equal to the ledge surface energy; and (3) when partial dislocations are nucleated, provided the stacking-fault energy is low. The most favorable case is that illustrated in Fig. 20-7, when the glide plane is normal to the surface. The results for such nucleation also are presented in Table 20-1. The critical stresses are lowered, but are still much higher than typical yield stresses. The results do correspond closely to the yield strengths observed for metal whiskers, which are presumably nearly perfect. If the glide plane were inclined at a small angle to the surface instead of being normal to it, the image stresses would further reduce ΔG^*.[25] For angles of inclination less than about 10°, the critical stresses approach the values of typical yield stresses for single crystals. For surfaces with adsorbed impurity layers, the image stresses and surface energies are changed, in general in a manner which increases ΔG^*.

As a final example,[26] consider nucleation in the wedge-shaped crystal of Fig. 20-8. The critical radius of curvature of such a configuration is roughly equal to that for a closed loop, but ΔG is reduced from that for the closed loop

[25] J. Friedel, private communication.

[26] P. B. Hirsch, *J. Inst. Met.*, **87**: 406 (1959).

by a factor $\phi/2\pi$. Thus nucleation in wedge-shaped crystals can occur at arbitrarily small stresses in the limit $\phi \to 0$. Realistically, ϕ is limited to values greater than 5 or 10° by the stiffness of the thin wedge. Nucleation in wedge-shaped crystals has been directly observed in electron-transmission microscopy.[27]

20-6. GRAIN-BOUNDARY SOURCES

Grain boundaries can act as dislocation sources in several ways. Clearly, small angle nets can act directly as Frank-Read sources, a segment operating between node-pinning points. Another possibility is associated with incompatabilities at grain boundaries. Large stress concentrations can arise at grain boundaries because of plastic incompatibility (glide pileups) or elastic incompatibility, producing image stresses. Under these large stress concentrations, Frank-Read sources or nucleation sources are expected to operate preferentially at the boundary. An example of such an effect is shown in Fig. 9-18.

Li[28] suggested that edge grain-boundary dislocations could be emitted into a grain, thereby acting as a source. A variety of other possibilities for grain boundary sources have been suggested as reviewed by Hirth[29] and Malis and Tangri.[30] An example is the inverse of the reaction in Eq. (19-82). The latter authors[30] also cite a number of observations of grain boundary source operation.

20-7. KINEMATIC SOURCES

There are possible dislocation sources associated with high-velocity dislocations. Frank[31] considered the possibility that a moving dislocation could acquire sufficient kinetic energy to form an opposite-sign dislocation when it intersected the surface (Fig. 20-9). A detailed analysis of this problem would require knowledge of the rate of acceleration of the dislocation under image stresses as it approached the surface and of the damping coefficient in this high-velocity region, both of which quantities are most uncertain, as discussed in Chap. 7. We can, however, roughly estimate the required stresses as follows. At a stress level of $\sigma \sim 10^{-3}\mu$, the image stresses become more important than the applied stress at a distance $l \sim 100b$ below the surface. Within the surface region defined by l, image stresses accelerate the incident dislocation, but if

[27]M. J. Whelan, P. B. Hirsch, R. W. Horne, and W. Bollman, *Proc. Roy. Soc.*, **A240**: 524 (1957).

[28]J. C. M. Li, *Trans. AIME*, **227**: 239 (1963).

[29]J. P. Hirth, *Met. Trans.*, **3**: 3047 (1972).

[30]T. Malis and K. Tangri, *Acta Met.*, **27**: 25 (1979).

[31]F. C. Frank, in "Report of the Conference on Strength of Solids," Physical Society, London, 1948, p. 46.

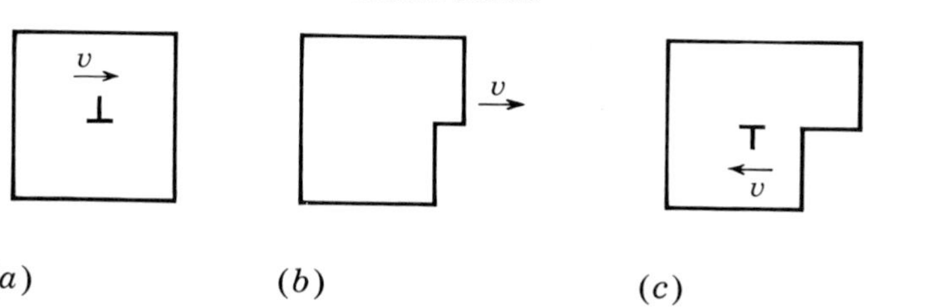

FIGURE 20-9. Three stages in the operation of the Frank kinematic source. The dislocation velocity is v.

damping is small in the surface region, the image stresses decelerate the reflected dislocation by the same amount. The deceleration process can be regarded as a transformation of the large kinetic energy immediately following reflection into elastic energy, increasing with separation from the surface. Thus in this approximation, for the reflected dislocation to reach a depth below the surface where the applied stress can move it farther in, the kinetic energy of the incident dislocation at $l = 100b$ must be sufficient to supply any damping loss plus the energy of a double step,

$$\frac{W}{L} = 2\gamma b \cong \frac{\mu b^2}{5} \tag{20-6}$$

Here γ is the step surface energy, and we have inserted the typical value $\gamma \sim \mu b/10$. Combining Eq. (20-6) and, for a rough estimate, the expression for the kinetic energy of a screw, Eq. (7-18), one finds that the expression for the critical required value of the velocity is

$$\frac{\mu b^2}{5} = \mu b^2 \left[\left(1 - \frac{v^2}{C_t^2} \right)^{-1/2} - 1 \right] \tag{20-7}$$

Equation (20-7) yields a critical value $v \sim 0.6C_t$. As discussed in connection with the damping formulas [Eqs. (7-74) and (7-77)], such velocities of edge dislocations can be attained only at stresses approaching $\mu/10$. For unjogged screw dislocations, no surface steps are formed during reflection, so that this process is more likely than for edge dislocations. However, as discussed in Chap. 16, screws are likely to be highly damped by jogs, so, all in all, the reflection process is unlikely. Thus the Frank kinematic mechanism appears likely only for shock-loaded materials, where stress pulses are very large and velocities approaching C_t are possible.

For velocities above the Rayleigh velocity for edge dislocations [Eq. (7-33)], another mechanism possibly could be applicable in shock-loaded crystals. Above the Rayleigh velocity, like-sign edges attract and opposite-sign edges repel. Thus, if a configuration such as that in Fig. 20-10 can form by core fluctuation, the opposite-sign dislocations would spontaneously move apart,

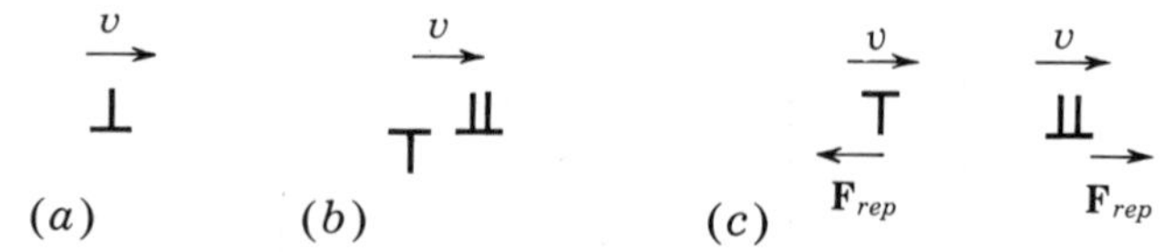

(a) (b) (c) $\mathbf{F}_{rep}$ $\mathbf{F}_{rep}$

FIGURE 20-10. (*a*) An edge dislocation moving at velocity v. (*b*) Pair nucleation by core fluctuation. (*c*) Opposite-sign dislocations move apart under repulsive forces.

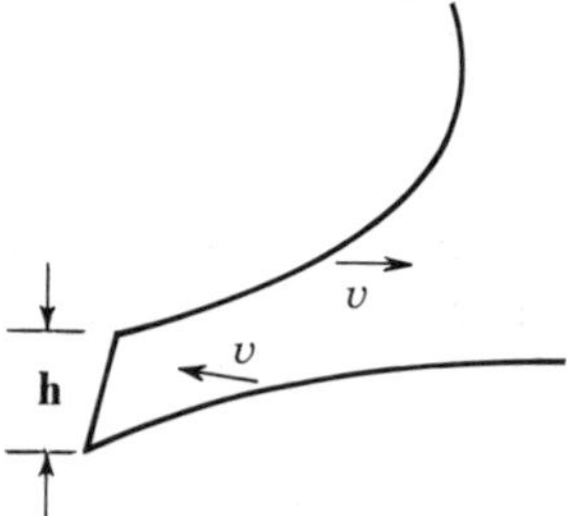

FIGURE 20-11. Arms of two Frank-Read sources moving past one another with velocity v.

resulting in dislocation multiplication. Such effects have been observed in a two-dimensional atomic simulation.[32]

A final kinematic problem is the determination of the probability that two arms of a double Frank-Read helix source have sufficient kinetic energy to overcome their interaction energy as they sweep by one another (Fig. 20-11). In the absence of damping, the acceleration under attractive forces as they moved together would exactly balance the deceleration as they moved apart, and the arms would always sweep by one another. With damping, the problem resembles that of Fig. 20-9 for the screw case. The kinetic energy must supply the change in interaction energy between $x=0$ and the value of x where the force caused by the applied stress equals the attractive interaction force. The attractive interaction energy between the branches decreases with increasing h. Thus the dislocations are more likely to pass one another for large h and for small damping forces, but will be constrained to a dipole for small h and large damping forces. The problem has been treated for the quasi-static case,[33] which is applicable when B in Eq. (7-74) is large. It has not been studied in detail for the small B case where damping and inertial effects are important. However, the critical value of h for transition from one type of behavior to the other is expected to be in the range 10 to $100b$.

PROBLEMS

20-1. Estimate the critical resolved shear stress for yielding of a crystal with a dislocation network of average segment length $l=10^4 b$.

[32] J. H. Weiner and M. Pear, *Phil. Mag.*, **31**: 679 (1975).

[33] For example, see D. Kuhlmann-Wilsdorf, J. H. van der Merwe, and H. G. F. Wilsdorf, *Phil. Mag.*, **43**: 632 (1952).

20-2. Consider a prismatic loop formed by vacancy condensation on a (111) plane in an fcc structure, reacted so that no stacking fault is present in the plane of the loop. Can this loop be a source for slip? If so, on which planes and in which slip directions?

20-3. Which slip systems can operate from a source consisting of a pure twist boundary of screw segments in a hexagonal mesh on a (111) plane in an fcc crystal?

20-4. For $\sigma < \sigma_{crit}$, compute the positions of the stable equilibrium configuration and the metastable equilibrium configuration of a bowed loop with the geometry of Fig. 20-1. Determine the activation energy for achieving the latter from the former, and show that the process is unlikely under thermal activation unless $\sigma \sim \sigma_{crit}$

20-5. Use the methods of Chap. 6 to compute the degree of bow-out as a function of stress for a loop constrained to have its component segments in $\langle 110 \rangle$ directions on a $\{111\}$ plane in germanium, as in Fig. 20-2. Compare the results with those of Eq. (20-1). Assume that the original unbowed segment is pure screw.

20-6. For pinned segments uniformly spaced along a screw dislocation, compute the contribution of interaction between adjacent loops to the bow-out energy for loops constrained to be composed of straight segments, as in Prob. 20-5.

20-7. Show that the applied stress tensor causes closed-dipole-loop formation by the mechanism of Fig. 20-5*f* to be less likely in an fcc crystal than in a simple cubic crystal. For which other crystal systems is the process likely?

20-8. Perfect low-index surfaces are called singular surfaces, while those near low-index orientation which contain surface ledges and surface kinks are called vicinal surfaces. On which type of surface is dislocation nucleation more likely? For which type of surface is edge dislocation reflection by Frank's mechanism more likely? Why?

21

Dislocation Pileups

21-1. INTRODUCTION

When a queue of dislocations is forced against some obstacle by the applied stress, the queue is called a *dislocation pileup*. An example is the pileup of edge dislocations pushed against a grain boundary which blocks the motion of the leading dislocation (Fig. 21-1). The leading dislocation is acted upon not only by the applied stress, but also by the interaction force with the other dislocations in the pileup. The pileup *concentrates* on the leading dislocation a large force proportional to the applied stress and to the number of dislocations in the pileup. When the pileup contains many dislocations, stresses of the order of the theoretical shear stress can develop at the spearhead of the pileup at moderate applied stresses, so that the pileup either can initiate yielding on the other side of the grain boundary or can nucleate a crack at the boundary.

Pileups are thus important in the initiation and propagation of deformation in polycrystalline materials. In single crystals as well, obstacles such as dislocation barriers (discussed in Chap. 22) can block slip. Pileups then form until the force on the leading dislocations is sufficient to make them break through the barrier, or, for screw-dislocation pileups, until the force suffices to cause cross slip of the leading dislocations. Hence pileup theory is an important topic in the deformation behavior of both single crystals and polycrystals.

A consideration of the interaction forces among discrete dislocations in a pileup reduces the equilibrium pileup problem to a set of nonliner algebraic equations for the dislocation positions.[1] Methods for approximate and asymptotic solution for these equations are reviewed by Chou and Li.[2] With the advent of fast computers, solutions of this type can be performed expediently by numerical methods.

An alternative method, useful in obtaining analytical results accurate except in the near vicinity of the pileup dislocations, is the use of a continuous

[1]J. D. Eshelby, F. C. Frank, and F. R. N. Nabarro, *Phil. Mag.*, **42**: 351 (1951).

[2]Y. T Chou and J. C. M. Li, in T. Mura (ed.), "Mathematical Theory of Dislocations," American Society of Mechanical Engineering, New York, 1969, p. 116.

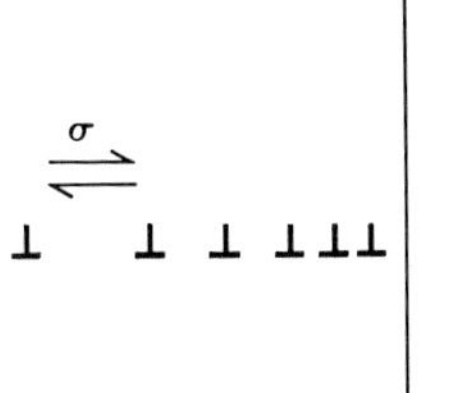

Boundary **FIGURE 21-1.** Edge dislocations piled up at a grain boundary.

distribution of infinitesimal dislocations on a glide plane, as used by Eshelby[3] in the Peierls-Nabarro description in Chap. 8. With the additional use of transform theory,[4] the continuous dislocation model readily describes the strain fields of pileups. Various methods for describing such a model are reviewed by Chou and Li.[2]

The methods of continuous dislocations are equivalent to continuum mechanical descriptions of shear cracks (so-called mode II and mode III cracks), many of which originated in the work of Muskelishvili.[5] An account of this type of methodology is given by Rice.[6] The shear-crack continuous-dislocation model of pileups has been extended widely; for example, to shear-type earthquake nuclei[7]. Other analogous methods also exist, such as conformal mapping in a hydrodynamic model.[8]

In this chapter we adopt the continuous-dislocation method and shear-crack analog. In the following sections we develop the dislocation distribution and strain fields of double-ended and single-ended pileups. The same methods then are extended to the treatment of discontinuous tilt boundaries, which can be thought of as climb pileups. Pileup theory is related to macroscopic deformation behavior through the Petch analysis of the grain-size dependence of strength properties. Finally, we consider tensile cracks, both brittle and ductile.

21-2. PILEUP AS A SUPERDISLOCATION

Consider a pileup of N edge dislocations in the presence of a resolved external shear stress σ (Fig. 21-2). Suppose the leading dislocation to be fixed, or equivalently, suppose that it is interacting with an obstacle whose internal stress gradient is so steep that the leading dislocation moves only a negligible

[3] J. D. Eshelby, *Phil. Mag.*, **40**: 903 (1949).

[4] G. Leibfried, *Z. Phys.* **130**: 214 (1951).

[5] N. I. Muskelishvili, "Some Basic Problems in the Mathematical Theory of Elasticity," Noordhoff, Groningen, 1953.

[6] J. R. Rice, in H. Liebowitz (ed.), "Fracture," vol. II, Academic, New York, 1968, p. 191.

[7] J. Weertman, *Bull. Seismol. Soc. Am.*, **54**: 1035 (1964); *Bull. Seismol. Soc. Am.*, **55**: 945 (1965).

[8] H. H. Johnson, *J. Appl. Phys.* **37**: 1763 (1966).

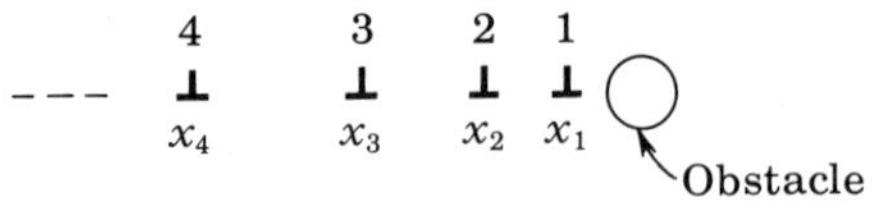

FIGURE 21-2. Indexing of edge-dislocation pileup.

distance with an increase in σ. Let the coordinate of dislocation (1) be x_1, the coordinate of dislocation 2, x_2, etc. The repulsive forces between the dislocations produce an interaction energy which depends only on their relative spacings,

$$W_{\text{int}} = W(x_2 - x_1, x_3 - x_1, \ldots, x_N - x_1) \tag{21-1}$$

At equilibrium, the net force on the individual dislocations 2 to N is zero,

$$\sigma b = \frac{\partial W}{\partial x_2} = \frac{\partial W}{\partial x_3} = \cdots = \frac{\partial W}{\partial x_N} \tag{21-2}$$

Since the only relative spacings matter in W, the force exerted by all the other dislocations upon dislocation (1) is

$$-\frac{\partial W}{\partial x_1} = \frac{\partial W}{\partial x_2} + \frac{\partial W}{\partial x_3} + \cdots + \frac{\partial W}{\partial x_N} = (N-1)\sigma b \tag{21-3}$$

The external stress exerts a force σb on dislocation (1); so, in all, a force

$$\frac{F}{L} = \sigma N b \tag{21-4}$$

is acting on the leading dislocation. At equilibrium this force is balanced by that of the internal stress field of the obstacle.

The force of Eq. (21-4) is the same force that would act on a superdislocation of strength Nb; in fact, the above result could be derived by an argument involving the virtual displacement of the entire pileup as if it were a superdislocation. A pileup of screw or mixed dislocations can be treated in the same way. In a homogeneous external stress field the force on the leading dislocation is always N times the force on the individual dislocations. As one would anticipate from St. Venant's principle, the stress field of a pileup converges to that of a superdislocation at distances greater than the pileup length. Thus the pileup can be usefully envisioned as a superdislocation.

21-3. DOUBLE PILEUP

Dislocation Distribution

Although Eq. (21-4) gives the force at the spearhead of a pileup containing N dislocations, in most physical problems the length of the pileup is specified.

FIGURE 21-3. Electron-transmission micrograph of a dislocation pileup at a grain boundary in a Cu–4.5% Al alloy. [P. R. Swann and J. Nutting, *J. Inst. Met.*, **50:** 133 (1961).]

The distribution of dislocations in a pileup of given length and stress is thus of importance. Eshelby et al.[9] solved the problem of the positions of the individual dislocations in a pileup under uniform external stress by an approximate analytical method. The problem was later solved by numerical methods as reviewed by Chou and Li.[10] The distribution of dislocations in several small pileups observed experimentally by etch-pitting techniques[11,12] and by electron-transmission microscopy[13] agreed closely with the theory of Eshelby et al.[9] An actual example of such a pileup is presented in the electron-transmission micrograph of Fig. 21-3. However, there are possible sources of error in the application of the discrete dislocation theory. Image stresses are difficult to determine for grain boundaries, and the image-stress distribution affects the pileup distribution. Also, the presence of internal stress gradients influences the dislocation arrangement. In many cases, particularly for pileups containing a large number of dislocations, the approximation of a *continuous* distribution in a pileup is about as good an approximation as the discrete model. A spread-out Peierls-Nabarro core structure and dissociation into

[9]J. D. Eshelby, F. C. Frank, and F. R. N. Nabarro, *Phil. Mag.*, **42**: 351 (1951).

[10]Y. T. Chou and J. C. M. Li, in T. Mura (ed.), "Mathematical Theory of Dislocations," American Society of Mechanical Engineering, New York, 1969, p. 116.

[11]F. W. Young and T. S. Noggle, *J. Appl. Phys.*, **31**: 604 (1960).

[12]W. G. Johnston, *Prog. Ceramic Sci.*, **2**: 1 (1962).

[13]M. J. Whelan, P. B. Hirsch, R. W. Horne, and W. Bollman, *Proc. Roy. Soc.*, **A240**: 524 (1957).

partial dislocations are other factors that help to make the continuum approximation a reasonable one. Because of these factors, and because the continuous pileup is easier to treat mathematically, we adopt the continuum model for our description of pileups. For pileups containing so few dislocations that the discreteness cannot be ignored, the discrete theory of Eshelby et al. must be consulted.[14]

G. Leibfried[15] showed that the continuous distribution of dislocations in a pileup could be handled readily by means of the Hilbert transform. Because of its usefulness, we briefly review the transformation theory.[16] The Hilbert transform is

$$H_x[f(y)] \equiv \frac{1}{\pi}\int_{-1}^{1}\frac{f(y)\,dy}{y-x} \tag{21-5}$$

where the integral on the right is defined by its principal value [see Eq. (8-7)]. The integral transforms the function $f(y)$ into a function of x by the transform. Now consider the functions

$$T_n(\cos\theta) \equiv \cos n\theta \qquad U_n(\cos\theta) \equiv \frac{\sin(n+1)\theta}{\sin\theta} \tag{21-6}$$

With $x = \cos\theta$, Eqs. (21-6) define two types of polynomials, $T_n(x)$ and $U_n(x)$, called Tschebyscheff polynomials. Several of the polynomials are

$$\begin{aligned} &T_{-1}=x \quad T_0=1 \quad T_1=x \quad T_2=2x^2-1 \\ &U_{-1}=0 \quad U_0=1 \quad U_1=2x \quad U_2=4x^2-1 \end{aligned} \tag{21-7}$$

These polynomials satisfy the orthogonality relations

$$\int_{-1}^{1}\frac{T_m T_n\,dx}{(1-x^2)^{1/2}} = \int_{-1}^{1}(1-x^2)^{1/2}U_m U_n\,dx = \begin{cases} 0 & m \neq n \\ \pi/2 & m = n \neq 0 \\ \pi \text{ and } \pi/2 & m = n = 0 \end{cases} \tag{21-8}$$

where m and n are integers. The polynomials are very important in connection

[14] See P. M. Hazzledine and P. B. Hirsch, *Phil. Mag.*, **15**: 121 (1967), for a graphical presentation of a discrete calculation. Within distances from the pileup of the order of the interdislocation spacing, the stress field deviates markedly from the continuum result and can even change sign.

[15] G. Leibfried, *Z. Phys.*, **130**: 214 (1951).

[16] The theory of the transform and the polynomials is discussed further in P. M. Morse and H. Feshbach, "Methods of Theoretical Physics," McGraw-Hill, New York, 1953, p. 544, and F. G. Tricomi, "Vorlesungen über Orthogonalreihen," Springer, Berlin, 1955, p. 188.

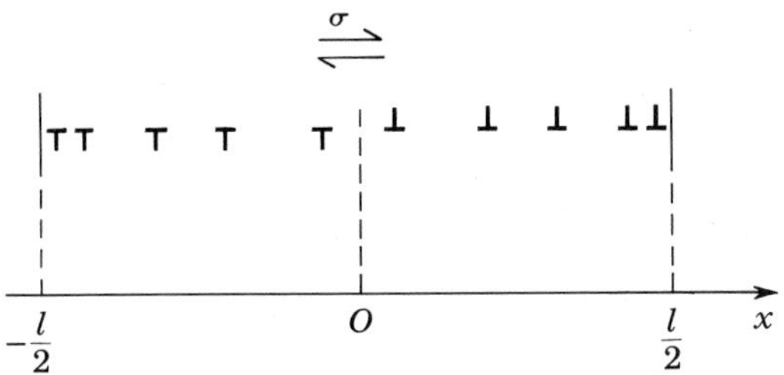

FIGURE 21-4. Double pileup of edge dislocations.

with the Hilbert transform (21-5) because they interconnect by it,

$$H_x\left[\frac{T_n(y)}{(1-y^2)^{1/2}}\right]=U_{n-1}(x)$$
$$H_x\left[(1-y^2)^{1/2}U_{n-1}(y)\right]=-T_n(x) \tag{21-9}$$

As an example of the transform method, let us consider a double pileup of length l (Fig. 21-4). The pileup forms by the operation of a Frank-Read source at the center of the interval, which has dislocation barriers at both ends. The dislocation density is described by the quantity

$$n(x)=\pm\frac{1}{b}\frac{db}{dx}$$

that is, $n(x)\,dx$ is the number of dislocations between x and $x+dx$. A negative $n(x)$ refers to dislocations with negative Burgers vectors.

For the individual dislocations in the pileup to be in equilibrium, the force produced by external stress and that caused by interaction with the other dislocations must balance,

$$\sigma b=\frac{\mu b^2}{2\pi(1-\nu)}\int_{-l/2}^{l/2}\frac{n(x')\,dx'}{x'-x}\qquad -\frac{l}{2}\leqslant x\leqslant\frac{l}{2} \tag{21-10}$$

The integral is defined by its principal value, for the same reasons as explained in the theory of the Peierls-Nabarro core structure in Sec. 8-2. Introducing the variable

$$\eta=\frac{2x}{l} \tag{21-11}$$

and defining $f(\eta)$ as

$$f(\eta)=n\left(\frac{l\eta}{2}\right) \tag{21-12}$$

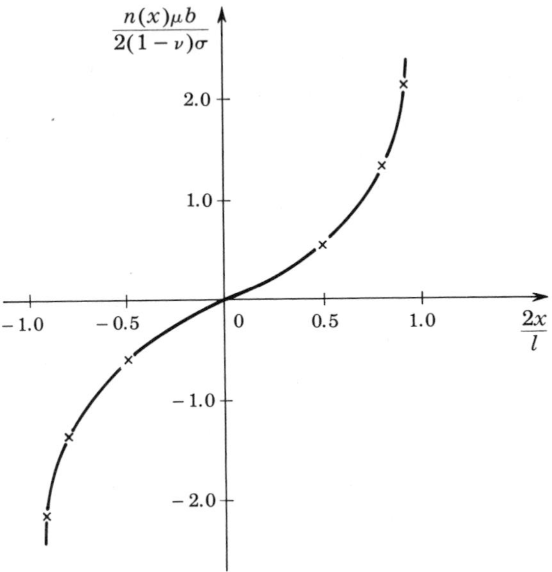

FIGURE 21-5. Plot of Eq. (21-15).

one can rewrite Eq. (21-10) in the form

$$\frac{2(1-\nu)\sigma}{\mu b}=\frac{1}{\pi}\int_{-1}^{1}\frac{f(\eta')\,d\eta'}{\eta'-\eta}\qquad -1\leqslant\eta\leqslant 1 \tag{21-13}$$

Equations (21-7) and (21-9) then reveal that $f(\eta)$ can be solved for directly[17] in terms of the Tschebyscheff polynomial $T_1(\eta)$

$$f(\eta)=\frac{2(1-\nu)\sigma}{\mu b}\frac{T_1(\eta)}{(1-\eta^2)^{1/2}}=\frac{2(1-\nu)\sigma}{\mu b}\frac{\eta}{(1-\eta^2)^{1/2}} \tag{21-14}$$

or, by Eq. (21-12),

$$n(x)=\frac{2(1-\nu)\sigma}{\mu b}\frac{x}{\left[(l/2)^2-x^2\right]^{1/2}} \tag{21-15}$$

Figure 21-5 is a plot of the dislocation density $n(x)$ given by Eq. (21-15).

The total number N of dislocations of either sign in the pileup is found by a standard integral. The result is

$$N=\int_0^{l/2}n(x)\,dx=\frac{(1-\nu)l\sigma}{\mu b} \tag{21-16}$$

[17]Explicitly, the integral on the right side of Eq. (21-13) is seen to be the Hilbert transform of $f(\eta')$, while the left side is a constant A times the polynomial $U_0(\eta)=1$. Since $U_0(\eta)$ is the Hilbert transform of $T_1(\eta')/(1-\eta'^2)^{1/2}$, it follows that $f(\eta')$ equals $AT_1(\eta')/(1-\eta'^2)^{1/2}$, which yields Eq. (21-14).

Force on the Leading Dislocation

The determination of the force on the leading dislocations at the pileup ends by integration of the interaction forces is complicated. However, the force is given expediently by an indirect approach, treating the double pileup as a shear crack. The total shear ϵ in forming the pileup is the integral of the area x swept out by unit length of infinitesimal dislocation, times the magnitude of its Burgers vector $bn(x)\,dx$,

$$\epsilon = \int_{-l/2}^{l/2} bxn(x)\,dx = \frac{\pi(1-\nu)l^2\sigma}{4\mu} \tag{21-17}$$

Note that b has disappeared from this result, which signifies that the continuum approximation for the pileup is equivalent to a shear crack.

The formation of the crack is envisioned as follows. Make a cut of length l and maintain shear stresses $-\sigma$ over the cut surfaces to balance the external shear stress σ. Next, gradually decrease the internal surface shear stresses $-\sigma$ to zero, so that the shear crack relaxes in a reversible manner to its equilibrium configuration. The average internal stress maintained in the process is $-\sigma/2$. Hence, during the process, the hypothetical device maintaining shear forces on the cut surfaces extracts an energy $\frac{1}{2}\sigma\epsilon$ from the system of the crystal and the external mechanisms maintaining constant external stress. Thus the total energy per unit depth released in the formation of the crack is

$$\frac{W}{L} = \tfrac{1}{2}\sigma\epsilon = \frac{\pi(1-\nu)l^2\sigma^2}{8\mu} \tag{21-18}$$

Consequently, the force per unit length acting on a dislocation at the tip of the crack is

$$\frac{F}{L} = \frac{\partial(W/L)}{\partial l} = \frac{\pi(1-\nu)l\sigma^2}{4\mu} \tag{21-19}$$

As a function of the number of dislocations N [Eq. (21-16)], the force is

$$\frac{F}{L} = \frac{\pi}{4}Nb\sigma \tag{21-20}$$

The force is somewhat less than that for a superdislocation Nb because the two pileups of opposite sign attract; i.e., the analog is actually that of two superdislocations of opposite sign.

21-4. UNSTRESSED SINGLE PILEUP

Consider the case of a number of dislocations of the same sign locked in an interval l in the absence of an external applied stress. Such a situation could

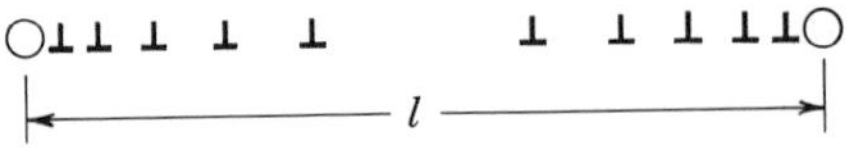

○⊥⊥ ⊥ ⊥ ⊥ ⊥ ⊥ ⊥ ⊥⊥○

$\vdash$——— l ———$\dashv$

FIGURE 21-6. Single pileup in the absence of stress.

arise, for example, in unloading when a lock forms in the rear of a pileup built up during loading of a specimen. The dislocations in such a pileup repel one another and accumulate at the two ends (Fig. 21-6). By analogy with Eq. (21-13), the dislocation distribution must satisfy the force balance

$$0=\frac{1}{\pi}\int_{-1}^{1}\frac{f(\eta')\,d\eta'}{\eta'-\eta} \tag{21-21}$$

and the solution must be nontrivial, $f(\eta)\neq 0$. The solution is given by Eqs. (21-7) and (21-9) to be

$$f(\eta)=\frac{T_0(\eta)}{(1-\eta^2)^{1/2}}=\frac{1}{(1-\eta^2)^{1/2}} \tag{21-22}$$

or, in terms of x,

$$n(x)=\frac{A}{\left[(l/2)^2-x^2\right]^{1/2}} \tag{21-23}$$

The constant A is determined from the boundary condition

$$N=\int_{-l/2}^{l/2}n(x)\,dx=\pi A \tag{21-24}$$

Therefore the complete distribution function becomes

$$n(x)=\frac{N}{\pi\left[(l/2)^2-x^2\right]^{1/2}} \tag{21-25}$$

which is plotted in Fig. 21-7.

Unlike the double pileup, the forces at the ends of the single pileup must be determined by integration; the simple shear-crack analogy is not applicable. The single pileup resembles a single superdislocation Nb with a spread-out core, so the energy of the configuration must depend on an outer cutoff R. The work introduced in the glide plane between $-R$ and $-l/2$ in forming the superdislocation (Fig. 21-6) is

$$\frac{W}{L}=\tfrac{1}{2}Nb\int_{-R}^{-l/2}\sigma\,dx=\frac{\mu Nb^2}{4\pi(1-\nu)}\int_{-R}^{-l/2}dx\int_{-l/2}^{l/2}\frac{n(x')\,dx'}{x'-x} \tag{21-26}$$

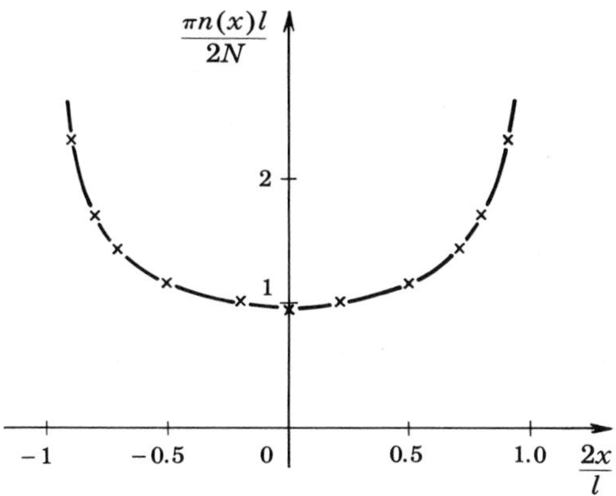

FIGURE 21-7. Plot of Eq. (21-25).

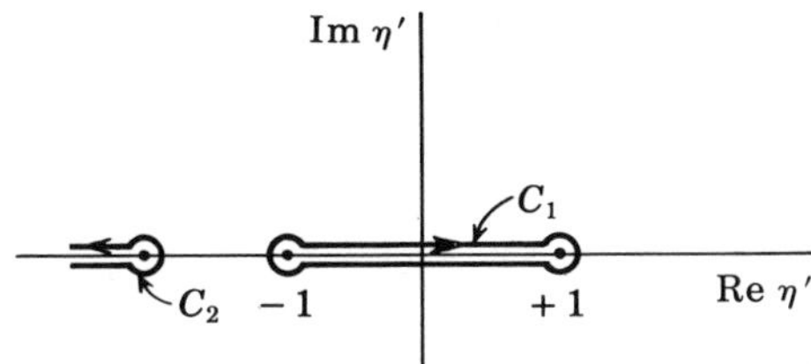

FIGURE 21-8. Line-integral paths in the complex plane.

or, with Eqs. (21-11) and (21-23),

$$\frac{W}{L} = \frac{\mu N^2 b^2}{4\pi^2(1-\nu)} \int_{-2R/l}^{-1} d\eta \int_{-1}^{1} \frac{d\eta'}{(1-\eta'^2)^{1/2}(\eta'-\eta)} \tag{21-27}$$

The integral can be accomplished in the complex plane (Fig. 21-8). Since the square root changes sign on one revolution around either of the poles 1 or -1, the integral in Eq. (21-27) can be written

$$\int_{-1}^{1} \frac{d\eta'}{(1-\eta'^2)^{1/2}(\eta'-\eta)} = \frac{1}{2}\oint_{C_1} \frac{d\eta'}{(1-\eta'^2)^{1/2}(\eta'-\eta)}$$

where the path of the line integral C_1 is around the cut from -1 to 1. This path is then deformed to the path C_2 enclosing the pole η. Because only the circular path around the singular point contributes, the solution is given by the residue at this pole,[18]

$$\int_{-1}^{1} \frac{d\eta'}{(1-\eta'^2)^{1/2}(\eta'-\eta)} = \pi i (1-\eta^2)^{-1/2} = \frac{\pi}{(\eta^2-1)^{1/2}} \qquad \eta < -1 \tag{21-28}$$

[18]Note that with η in the interval $-1<\eta<1$, the integral is zero in agreement with Eq. (21-21).

Thus

$$\frac{W}{L}=\frac{\mu N^2b^2}{4\pi(1-\nu)}\int_{-2R/l}^{-1}\frac{d\eta}{(\eta^2-1)^{1/2}}=\frac{\mu N^2b^2}{4\pi(1-\nu)}\ln\left\{\frac{2R}{l}+\left[\left(\frac{2R}{l}\right)^2-1\right]^{1/2}\right\}$$

$$\cong\frac{\mu N^2b^2}{4\pi(1-\nu)}\ln\frac{4R}{l} \tag{21-29}$$

in the limit $R\gg l$. The force tending to push the leading dislocation out of the pileup is

$$\frac{F}{L}=-\frac{\partial}{\partial l}\left(\frac{W}{L}\right)=\frac{\mu N^2b^2}{4\pi(1-\nu)l} \tag{21-30}$$

Therefore the barriers at the pileup ends must withstand at least forces of this magnitude. Note that the treatment of this section could be used to determine the variation of the *elastic* energy of a single dislocation with the degree of core splitting (Chap. 8).

21-5. STRESSED SINGLE PILEUP

Dislocation Distribution

The single pileup pushed against an obstacle by a shear stress σ is treated easily by superposition of the results of Secs. 21-3 and 21-4. Superposing Eqs. (21-15) and (21-23) and adjusting A so that the dislocation density is zero at $x=-l/2$, one finds that the dislocation distribution in the single pileup of length l is

$$n(x)=\frac{2(1-\nu)\sigma}{\mu b}\left[\frac{(l/2)+x}{(l/2)-x}\right]^{1/2} \tag{21-31}$$

This distribution is shown in Fig. 21-9. The total number of dislocations in the distribution of Fig. 21-9 is

$$N=\int_{-l/2}^{l/2}n(x)\,dx=\frac{\pi(1-\nu)l\sigma}{\mu b} \tag{21-32}$$

For a given stress σ and a given number N, the length of the pileup is

$$l=\frac{\mu Nb}{\pi(1-\nu)\sigma} \tag{21-33}$$

Some of the inadequacy of the continuum model appears in the above result. The total length of the pileup involves the last few, widely spaced

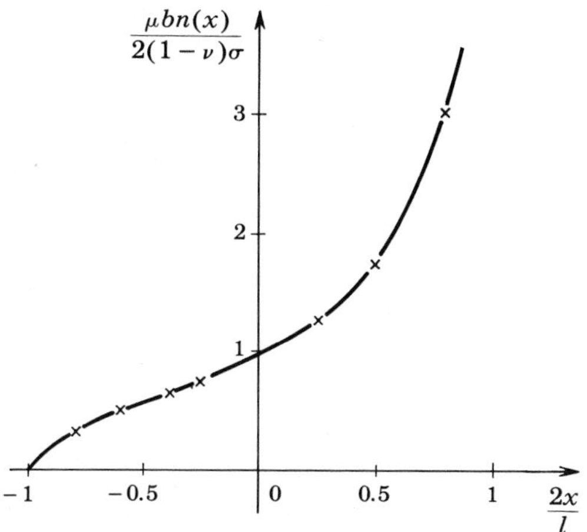

FIGURE 21-9. Plot of Eq. (21-31).

dislocations, which are ill-represented in the continuum model. On the other hand, the length of the pileup depends strongly upon whether it is defined as the distance from the spearhead to the last dislocation or the distance to the next to last dislocation. Nevertheless, the distribution function (21-31) with l given by Eq. (21-33) does predict accurately how most of the dislocations in the pileup are arranged. We adopt this value of l as the *definition* of the length of the single pileup. In the next section this model is shown to yield a reasonable result for the back stress at the rear of the pileup, with no serious singularity at $x=-l/2$.

Back Stress of the Pileup

A knowledge of the back stress produced by the pileup is useful in such problems as the exhaustion of a Frank-Read source. The back stress σ_B in the region $x<-l/2$ is

$$\sigma_B=-\frac{\mu b}{2\pi(1-\nu)}\int_{-l/2}^{l/2}\frac{n(x')\,dx'}{x'-x}$$

$$=-\frac{\sigma}{\pi}\int_{-1}^{1}\left(\frac{1+\eta'}{1-\eta'}\right)^{1/2}\frac{d\eta'}{\eta'-\eta} \tag{21-34}$$

In the complex plane of Fig. 21-8,

$$\sigma_B=-\frac{\sigma}{2\pi}\oint_{C_1}\left(\frac{1+\eta'}{1-\eta'}\right)^{1/2}\frac{d\eta'}{\eta'-\eta}$$

Deforming the path as in the solution of Eq. (21-27), one finds that the residue at the pole η gives a contribution

$$2\pi i\left(\frac{1+\eta}{1-\eta}\right)^{1/2}=2\pi\left(\frac{\eta+1}{\eta-1}\right)^{1/2} \qquad \eta<-1$$

In this case there is also a contribution from the outer return loop in path C_2, which, since $[(1+\eta')/(1-\eta')]^{1/2}=i$ on this path with $|\eta'|=\infty$, is simply

$$i\oint\frac{d\eta'}{\eta'}=i(-2\pi i)=2\pi$$

Thus, all in all, the result of the integration is

$$\sigma_B=-\sigma\left[1-\left(\frac{\eta+1}{\eta-1}\right)^{1/2}\right] \qquad \eta<-1$$

$$=-\sigma\left[1-\left(\frac{x+l/2}{x-l/2}\right)^{1/2}\right] \qquad x<-\frac{l}{2} \tag{21-35}$$

At $x=-l/2, \sigma_B=-\sigma$; by the definition of the problem, this is the correct value of the back stress in the interval $-l/2<x<l/2$ (Fig. 21-10). Asymptotically, for $x\ll -l/2$, σ_B reduces to

$$\sigma_B=\frac{l\sigma}{2x}=\frac{\mu Nb}{2\pi(1-\nu)x} \tag{21-36}$$

which is the back stress from a superdislocation Nb.

The Stress Field Ahead of the Pileup

The shear stress field $\sigma_{xy}(x,0)$ ahead of a pileup is of interest from the viewpoint of breaking down barriers on a glide plane within a single crystal. When the pileup is at a grain boundary, the entire stress tensor ahead of the pileup is required to assess the probability of slip on a nonparallel system in the next grain, or the probability of a crack nucleus forming in the next grain. The shear stress field $\sigma_{xy}(x,0)$ is found by a derivation exactly analogous to

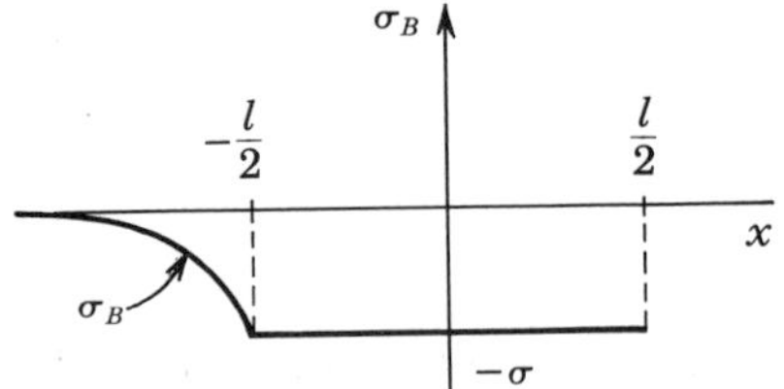

FIGURE 21-10. Plot of Eq. (21-35).

that for σ_B, yielding the result

$$\sigma_{xy}(x,0)=\sigma\left[\left(\frac{x+l/2}{x-l/2}\right)^{1/2}-1\right] \tag{21-37}$$

Again, asymptotically, the result reduces to that for a superdislocation

$$\sigma_{xy}(x,0)\cong\frac{\mu Nb}{2\pi(1-\nu)x} \qquad x\gg\frac{l}{2} \tag{21-38}$$

Near the spearhead of the pileup,[19] where $x-l/2\ll l/2$,

$$\sigma_{xy}(x,0)\cong\sigma\left(\frac{l}{x-l/2}\right)^{1/2} \tag{21-39}$$

The stresses can attain quite high values close to the spearhead. Of course, the results of the continuum model cannot be used validly at distances closer than the separation between the leading dislocations.

The general stress tensor ahead of the pileup is now developed. From Eqs. (3-43) and (21-31),

$$\begin{aligned}\sigma_{xx}&=-\frac{\sigma}{\pi}\int_{-l/2}^{l/2}\left(\frac{l/2+x'}{l/2-x'}\right)^{1/2}\frac{y\left[3(x-x')^2+y^2\right]}{\left[(x-x')^2+y^2\right]^2}dx'\\ &=-\frac{\sigma}{2\pi}\oint_{C_1}\left(\frac{1+\eta'}{1-\eta'}\right)^{1/2}\frac{\zeta\left[3(\eta'-\eta)^2+\zeta^2\right]}{\left[(\eta'-\eta)^2+\zeta^2\right]^2}d\eta'\end{aligned} \tag{21-40}$$

in the complex plane, where

$$\zeta=\frac{2y}{l}$$

Deforming the path of integration to C_2, as shown in Fig. 21-11, one finds contributions from the residues at $\eta'=\eta\pm i\zeta$. There is no contribution from the outer path.

Because the singular points are double poles, the calculation of the residues is somewhat complicated. Consider the singular point $\eta_+=\eta+i\zeta$. Apart from a factor $-\sigma\zeta/2\pi$, the Laurent series expansion of the integrand about the

[19] With a change of origin to the head of the pileup, the result is recognized to be equivalent to the continuum mechanics solution for a mode II crack, $\sigma_{xy}=K_{II}/(\pi x)^{1/2}$, with $K_{II}=\sigma(\pi l)^{1/2}$ the stress-intensity factor. See F. A. McClintock and A. S. Argon, "Mechanical Behavior of Materials," Addison-Wesley, Reading, Mass., 1966, Chap. 11.

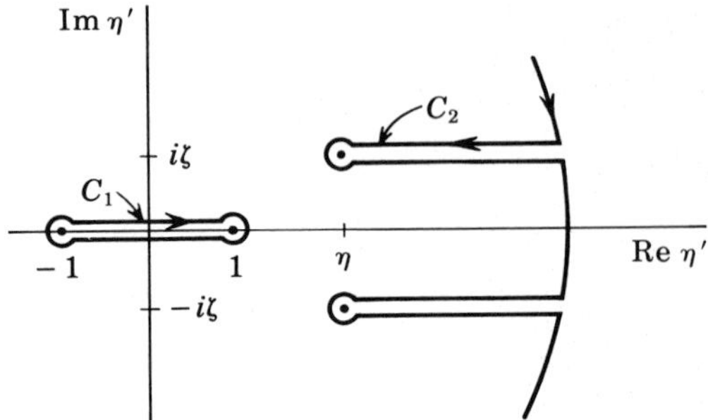

FIGURE 21-11. Line-integral paths for Eq. (21-40) in the complex plane.

point η_+ is

$$\frac{1}{2}\left(\frac{1+\eta_+}{1-\eta_+}\right)^{1/2}\frac{1}{(\eta'-\eta_+)^2}+\left[\frac{1}{2(1-\eta_+)^2}\left(\frac{1-\eta_+}{1+\eta_+}\right)^{1/2}+\frac{1}{i\zeta}\left(\frac{1+\eta_+}{1-\eta_+}\right)^{1/2}\right]\frac{1}{\eta'-\eta_+}+\cdots$$

The contribution to the integral from the residue is thus

$$2\pi i\left[\frac{1}{2(1-\eta_+)^2}\left(\frac{1-\eta_+}{1+\eta_+}\right)^{1/2}+\frac{1}{i\zeta}\left(\frac{1+\eta_+}{1-\eta_+}\right)^{1/2}\right]$$

$$=\frac{\pi i}{\zeta}\left(\frac{1-\eta_+}{1+\eta_+}\right)^{1/2}\frac{\zeta-2i(1-\eta_+{}^2)}{(1-\eta_+)^2} \tag{21-41}$$

The square root can be written

$$\left(\frac{1-\eta_+}{1+\eta_+}\right)^{1/2}=\frac{\left[(1-\eta^2-\zeta^2)^2+4\zeta^2\right]^{1/4}}{\left[(1+\eta)^2+\zeta^2\right]^{1/2}}\left(\cos\frac{\psi}{2}+i\sin\frac{\psi}{2}\right)$$

where

$$\psi=\tan^{-1}\frac{2\zeta}{\eta^2+\zeta^2-1}-\pi \qquad \eta^2+\zeta^2>1$$

The π is included to make the square-root expression real and positive on the upper side of the cut from -1 to 1 when defined by analytical continuation

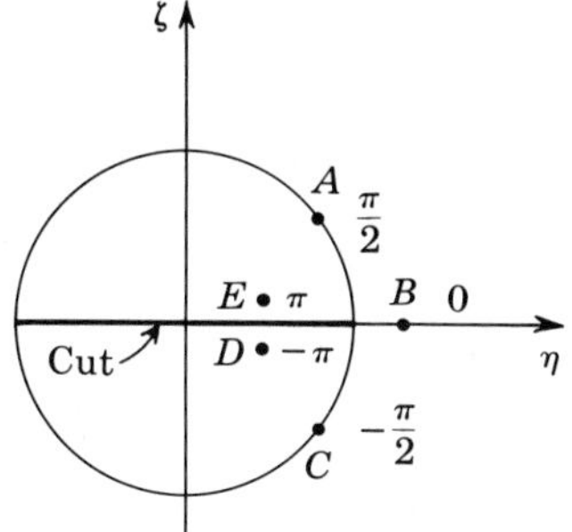

FIGURE 21-12. Variation of the function ϕ in Eq. (21-42). At points A, B, C, D, and E, ϕ assumes the values $\pi/2$, 0, $-\pi/2$, $-\pi$, and π, respectively.

along a path not crossing the cut. In terms of an angle ϕ,[20]

$$\phi = \tan^{-1}\frac{2\zeta}{\eta^2+\zeta^2-1} \qquad \eta^2+\zeta^2>1 \tag{21-42}$$

the result is

$$\left(\frac{1-\eta_+}{1+\eta_+}\right)^{1/2} = \frac{\left[(1-\eta^2-\zeta^2)^2+4\zeta^2\right]^{1/4}}{\left[(1+\eta)^2+\zeta^2\right]^{1/2}}\left(\sin\frac{\phi}{2}-i\cos\frac{\phi}{2}\right) \tag{21-43}$$

The real and imaginary parts of the contribution at η_+ can now be found in a straightforward but tedious manner. The contribution at η_- adds another real part identical with the first one and an imaginary part canceling the first one. Thus, in total,

$$\sigma_{xx} = -\sigma G\left\{\zeta\left[5(\eta-1)^2+3\zeta^2\right]\cos\frac{\phi}{2}\right.$$
$$\left.-2\left[(1+\eta)(\eta-1)^3+(2\eta-1)(\eta-1)\zeta^2+\zeta^4\right]\sin\frac{\phi}{2}\right\} \tag{21-44}$$

and similarly,

$$\sigma_{yy} = -\sigma G\left\{\zeta\left[\zeta^2-(\eta-1)^2\right]\cos\frac{\phi}{2}-2\zeta^2(\eta-1)\sin\frac{\phi}{2}\right\} \tag{21-45}$$

$$\sigma_{xy} = -\sigma-\sigma G\left\{-\zeta\left[3(\eta-1)^2+\zeta^2\right]\sin\frac{\phi}{2}\right.$$
$$\left.+\left[-(\eta+1)(\eta-1)^3-2\zeta^2(\eta-1)^2-\zeta^4\right]\cos\frac{\phi}{2}\right\} \tag{21-46}$$

[20]The variation of ϕ is shown in Fig. 21-12. The value of ϕ at the points A, B, and C is given directly by Eq. (21-42). The meaning of continuation in the upper half-plane is that $\phi=\pi$ on the upper side of the cut and $\phi=-\pi$ on the lower side of the cut.

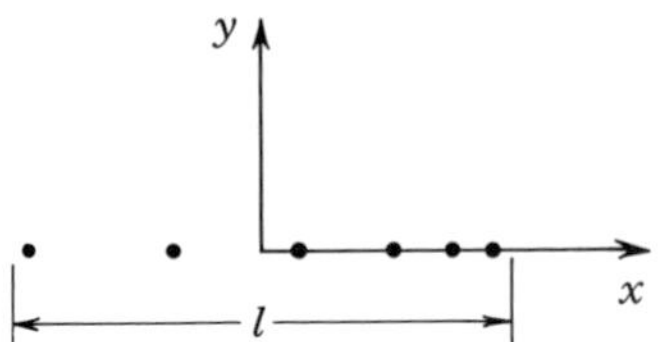

FIGURE 21-13. Pileup of screw dislocations whose lines are emerging normal to the plane of the page.

where

$$G = \frac{\left[(\eta^2+\zeta^2-1)^2+4\zeta^2\right]^{1/4}}{\left[(\eta+1)^2+\zeta^2\right]^{1/2}\left[(\eta-1)^2+\zeta^2\right]^2} \tag{21-47}$$

In the expression for σ_{xy}, the term $-\sigma$ is the contribution from the outer circular path in the integral.

Screw-Dislocation Pileup

The theory for pileup of screw dislocations is completely analogous to the above results; the formulas for the edge pileups can be applied to the screw case simply by replacing the factor $(1-\nu)$ by 1. The stress σ now denotes an *applied* shear stress σ_{yz} (Fig. 21-13). The general stress tensor of the screw pileup is calculated to be

$$\sigma_{xz} = -\frac{\sigma\left[(1+\eta)^2+\zeta^2\right]^{1/2}\sin\phi/2}{\left[(1-\eta^2-\zeta^2)^2+4\zeta^2\right]^{1/4}} \tag{21-48}$$

and

$$\sigma_{yz} = \frac{\sigma\left[(1+\eta)^2+\zeta^2\right]^{1/2}\cos\phi/2}{\left[(1-\eta^2-\zeta^2)^2+4\zeta^2\right]^{1/4}} - \sigma \tag{21-49}$$

For both screw pileups and edge pileups, elastic anisotropy can also be included in the analysis directly.[21] The various forms for $n(x)$; for example, in Eqs. (21-15) and (21-31) and in Fig. 21-9, still hold, but with an anisotropic elastic factor Z replacing μ. Z has a fairly simple form for pure edge or screw dislocations lying along high-symmetry directions but can be complex for other cases. These cases can still be treated expediently by the methods discussed in

[21] For example, see D. M. Barnett and R. J. Asaro, *J. Mech. Phys. Solids*, **20**: 353 (1972).

Sec. 13-7. Two dimensional fields such as the one in Eq. (21-44) become more complicated, with Z a function of ϕ and the anisotropic elastic constants.

Exercise 21-1. Verify that the general results, Eqs. (21-44) to (21-49), reduce to the superdislocation result at distances much greater than l away from the pileup. Show that the edge result for σ_{xy} reduces to Eq. (21-37) on the plane $y=0$.

Dynamical Arrays

Numerical[22] and analytical[23] solutions have been obtained for dynamical arrays of dislocations emitted from a source. The behavior is taken to follow Eq. (7-74) with the mobility $B^{-1} \propto (\sigma + \sigma_i)^n$ with σ and σ_i the resolved applied and back stresses, respectively. Physically, the dislocations experience accelerations as a consequence of a decrease in line-tension force as loops expand. Leading dislocations are preferentially accelerated as a consequence of the interaction forces with other dislocations in the array. As a consequence, the arrays resemble inverse pileups, with the smallest spacings near the source.[22,23] The results indicate that the average effective stress $(\sigma + \sigma_i)$ on and the average velocity of a dislocation in the array exceed those on single dislocations by a factor from 1 to 2. The pulse of strain accompanying the motion is reflected as a plastic instability in the strain distribution.[24]

21-6. THE DISCONTINUOUS TILT BOUNDARY

In a manner similar to the extension of the edge-dislocation-pileup results to the screw case, the above results can also be applied to tilt boundaries that terminate at a barrier or within a crystal; such boundaries can be thought of as climb pileups. Indeed, such a model is invoked in Sec. 21-7 to describe a tensile crack. However, in general, discontinuous tilt boundaries do not assume an equilibrium configuration. A more likely type of discontinuous tilt boundary is one in which the component edge dislocations are essentially regularly spaced. Such arrays can occur during polygonization or when a grain boundary pulls away from a low-angle tilt boundary. An example of the latter case is shown in Fig. 21-14.

Consider, for example, the long-range stress of the half-infinite tilt wall of Fig. 21-15a. In this case the dislocation density is $n(y)=1/D$, so that the

[22]A. R. Rosenfield and M. F. Kanninen, *Phil. Mag.*, **22**: 143 (1970); T. Yokobori, A. T. Yokobori, Jr., and A. Kamei, ibid., **30**: 367 (1974).

[23]A. K. Head, ibid., **26**: 65 (1972); X. Markenscoff, Proceedings of the Seventh Canadian Congress of Applied Mechanics, Sherbrooke, 1979.

[24]A. Korbel and A. Pawelek, in "Dislocation Modeling of Physical Systems," Pergamon, Oxford, 1981, p. 332.

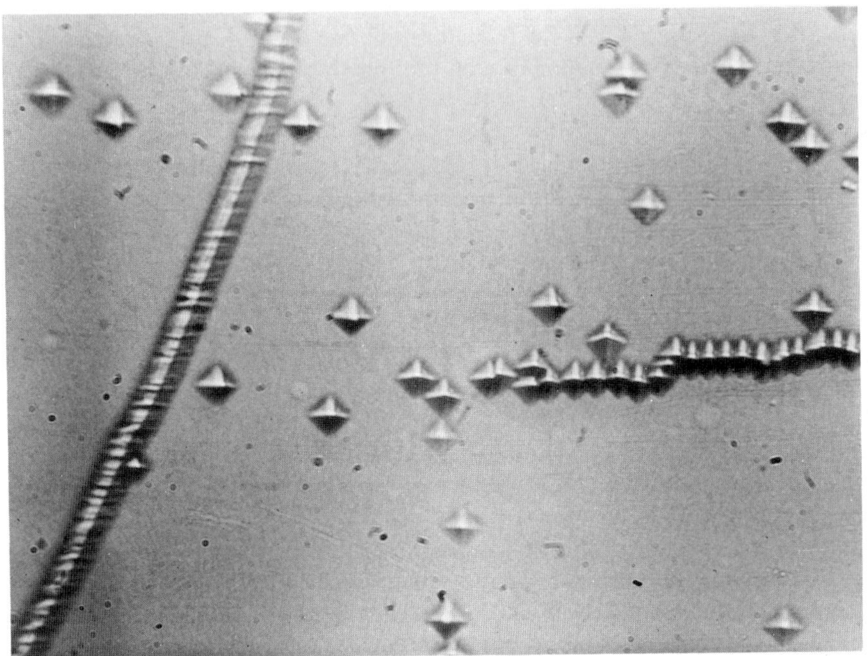

FIGURE 21-14. An etch-pitted {100} surface of a polygonized lithium fluoride crystal, showing dislocations being repelled from the end of a tilt boundary. One of the local-equilibrium configurations of Fig. 19-36 is also shown. Magnification 50×.

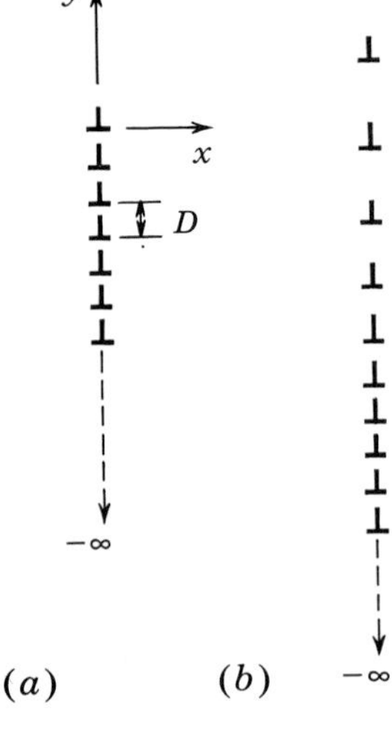

FIGURE 21-15. (a) A regularly spaced semi-infinite tilt wall of edge dislocations. (b) A partially relaxed semi-infinite tilt wall.

stress component σ_{xy} is

$$\sigma_{xy} = \frac{\mu b}{2\pi(1-\nu)D} \int_{-\infty}^{0} \frac{x\left[x^2-(y-y')^2\right]}{\left[x^2+(y-y')^2\right]^2} dy'$$

$$= -\frac{\mu bxy}{2\pi(1-\nu)D(x^2+y^2)} \tag{21-50}$$

This glide stress at the edge of an incomplete boundary is insensitive to how the boundary terminates at the other end. A similar calculation for the climb stress σ_{xx}, however, shows that it does not converge, but diverges logarithmically with the boundary length, as expected from the superdislocation analog.

For a *finite* tilt boundary of length l and containing dislocations uniformly separated by a distance D, the climb stress is

$$\sigma_{xx} = -\frac{\mu b}{2\pi(1-\nu)D} \int_{-l}^{0} \frac{3x^2(y-y')+(y-y')^3}{\left[x^2+(y-y')^2\right]^2} dy'$$

$$= -\frac{\mu b}{2\pi(1-\nu)D}\left[\frac{x^2}{x^2+y^2} - \frac{x^2}{x^2+(y+l)^2} + \frac{1}{2}\ln\frac{x^2+(y+l)^2}{x^2+y^2}\right] \tag{21-51}$$

Thus the climb force is very large at the tip of the tilt wall and tends to force the leading dislocations out of the wall. A typical nonequilibrium configuration after some climb is shown in Fig. 21-15*b*. Such configurations have been observed in the course of polygonization.[25] Figure 21-14 is another example in which the leading dislocations are being pushed out of the tilt wall.

21-7. THE TENSILE CRACK

Brittle Cracks

In the preceding discussion the analogy between the continuum theory of glide pileups and shear cracks is emphasized. In a related way, the tensile crack can be treated by the continuum theory for climb pileups. Consider a crack of length l in a field of normal stress σ (Fig. 21-16*a*). The crack is characterized by lack of cohesion across its surface, so the crack opens up when a tensile stress is present.

[25] C. G. Dunn and W. R. Hibbard, *Acta Met.*, **3**: 409 (1955).

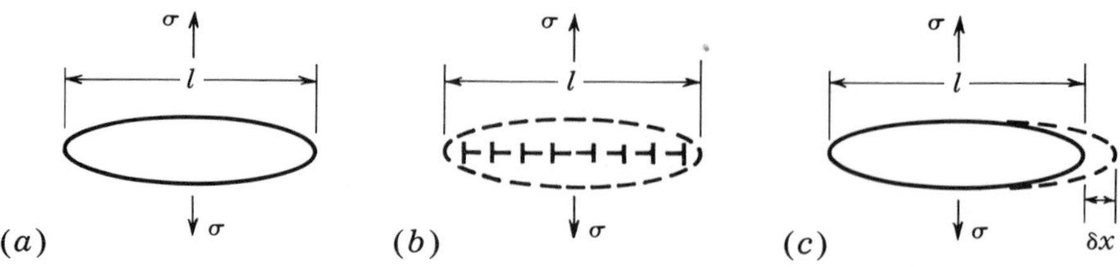

FIGURE 21-16. (*a*) Cross section of a cylindrical crack. (*b*) Dislocation analog of (*a*). (*c*) Infinitesimal extension of (*a*).

Equilibrium under stress would not be disturbed by filling the opened-up crack with matter. Formally, the matter can be considered to be supplied by a Bardeen-Herring source at the center of the crack (Fig. 21-16*b*). Hence, the filled-in material is comprised of a double climb pileup of edge dislocations. Equilibrium of all the dislocations with respect to climb under both their mutual interaction and the external normal stress σ is equivalent to equilibrium for the crack.

The analogy with the treatment for the shear crack is obvious. Because the normal stress exerts a climb force σb on each dislocation, and because the dislocations interact with forces $\mu b^2/2\pi x(1-\nu)$, Eqs. (21-15) and (21-19) apply exactly to the present case, with σ now meaning the normal stress,

$$n(x)=\frac{2(1-\nu)\sigma}{\mu b}\frac{x}{\left[(l/2)^2-x^2\right]^{1/2}} \tag{21-52}$$

and

$$\frac{F}{L}=\frac{\pi(1-\nu)l\sigma^2}{4\mu} \tag{21-53}$$

The quantity $(F/L)\,\delta x$ is the total change of elastic energy and potential energy of external mechanisms, per unit depth L, released when the crack length increases by δx (Fig. 21-16*c*).

In an actual brittle crack, growth of the crack by δx would create an amount of surface energy $2\gamma\,\delta x$ per unit depth. Thus, for the crack to grow spontaneously, the stress must be large enough that

$$\frac{F}{L}\geqslant 2\gamma$$

or, with the insertion of Eq. (21-53),

$$\sigma\geqslant\left[\frac{8\mu\gamma}{\pi(1-\nu)}\right]^{1/2} l^{-1/2} \tag{21-54}$$

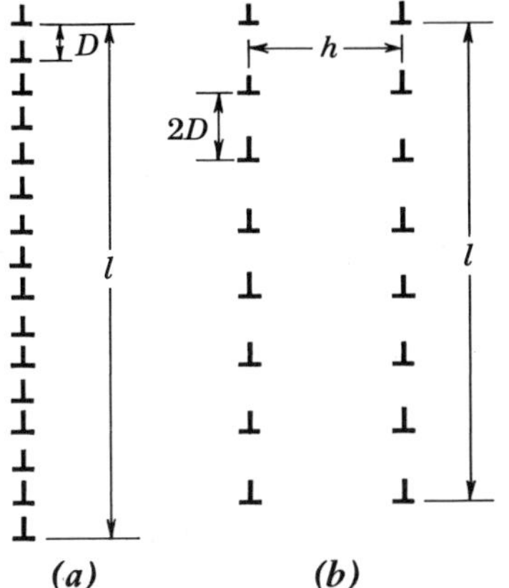

FIGURE 21-17. An incomplete tilt boundary. (*b*) Idealization of the blunting of boundary (*a*).

Equation (21-54) is called the *Griffith criterion*[26] and is very important in the discussion of the strength of brittle materials.

As is obvious from Fig. 21-16, the Griffith crack is a long cylindrical one, $L \gg l$. The results for an ellipsoidal crack,[27,28] or its analog of a pileup of circular dislocation loops,[28] however, do not differ significantly; the stress analogous to Eq. (21-54) is larger by a factor $(\pi/2)$.

Ductile Cracks

Local slip at the tip of a tensile crack relieves the stress concentration there and effectively blunts the tip of the crack. If such plastic flow is sufficient to lower the stress at the crack tip below that required to propagate the crack in a brittle manner, the crack becomes a ductile crack. For a ductile crack to grow as in Fig. 21-16*c*, the stress must be large enough to supply both the surface energy and the work required to expand the plastic zone[29]; the latter term is usually dominant. The stress required to propagate a ductile crack is therefore greater than that given by Eq. (21-54). As the crack grows, the plastic zone also grows, and a larger stress is required to propagate the crack. Hence the crack opens up gradually with increasing stress, in a ductile fashion. Considerations of ductile fracture are beyond the scope of this work. We do note that blunting has been treated in terms of either single or superdislocations emanating from a crack. The fields of dislocations in the vicinity of cracks are known in both the isotropic[30] and anisotropic[31] elastic approximations.

[26] A. A. Griffith, *Phil. Trans. Roy. Soc.*, **A221**: 163 (1920).

[27] R. A. Sack, *Proc. Phys. Soc.*, **58B**: 729 (1946).

[28] G. Leibfried, *Z. Phys.*, **130**: 214 (1951).

[29] E. Orowan, *Repts. Prog. Phys.*, **12**: 185 (1948).

[30] J. P. Hirth and R. H. Wagoner, *Int. J. Solids Struct.*, **12**: 117 (1976).

[31] C. Atkinson, *Int. J. Fract. Mech.*, **2**: 567 (1966).

Exercise 21-2. As an idealization of the blunting of a crack, suppose that a single climb pileup of the type described by Eq. (21-51) divides into two such pileups, each containing half as many dislocations (Fig. 21-17). Verify that the stress concentration is reduced by such blunting.

21-8. SOME APPLICATIONS TO MACROSCOPIC FLOW PHENOMENA

Blunting of Pileups

Just as for the tensile crack, glide pileups can also be blunted by plastic flow (Fig. 21-18). Cross slip or operation of new dislocation sources near the tip can contribute to such processes. Indeed, regions where pileups would be expected are found experimentally to consist of tangles of dislocations.[32, 33] The Burgers vectors of the dislocations in the tangles are largely the same. Here the superdislocation analogy is very useful. Despite the fact that the dislocations in a tangle do not at all resemble the ideal pileup, the long-range stress field at distances greater than the length of the tangled pileup is still given by the superdislocation limit of the pileup results. Only at the spearhead of the tangled pileup are the stresses reduced by blunting.

Because blunting depends largely on cross-slip processes at ordinary temperatures, and because the likelihood of cross slip is greatly influenced by dislocation dissociation, blunting is expected to depend on the stacking-fault energy. In materials with low stacking-fault energy, such as many of the fcc metals and alloys, blunting should not occur easily; an example of an ideal type of pileup in a low-stacking-fault-energy Cu–4.5% Al alloy is shown in Fig. 21-3. In metals of high stacking-fault energy, such as bcc crystals, blunting should occur more easily. These concepts are verified roughly by experiment. Grain-boundary pileups appear to be more easily blunted in copper than in lower-stacking-fault-energy α-brass, to be easily blunted in high-purity bcc metals (where interstitial effects are minimized), and to be quite stable against blunting in hcp metals, where cross slip is difficult.[34] Factors other than the blunting of pileups affect the cited experiments, so that the correlation with ease of cross slip is not unequivocal.

Fracture and Yielding Mechanisms at Pileups

There are a number of plausible mechanisms for the initiation of fracture or yielding at a pileup spearhead, but at present one usually cannot distinguish

[32] J. E. Bailey and P. B. Hirsch, *Phil. Mag.*, **5**: 485 (1960).

[33] A number of tangle observations are reviewed by P. B. Hirsch, in "Relation between Structure and Mechanical Properties of Metals," H.M. Stationery Office, London, 1963, p. 39.

[34] R. W. Armstrong, I. Codd, R. M. Douthwaite, and N. J. Petch, *Phil. Mag.*, **7**: 45 (1962).

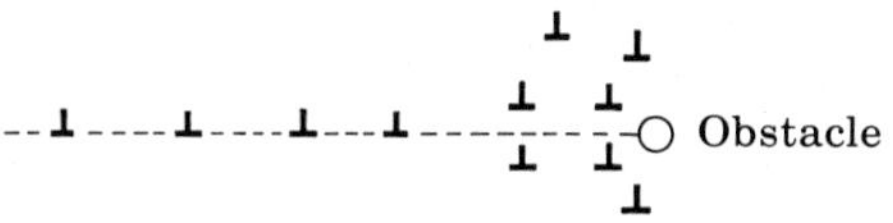

FIGURE 21-18. Blunting of a glide pileup.

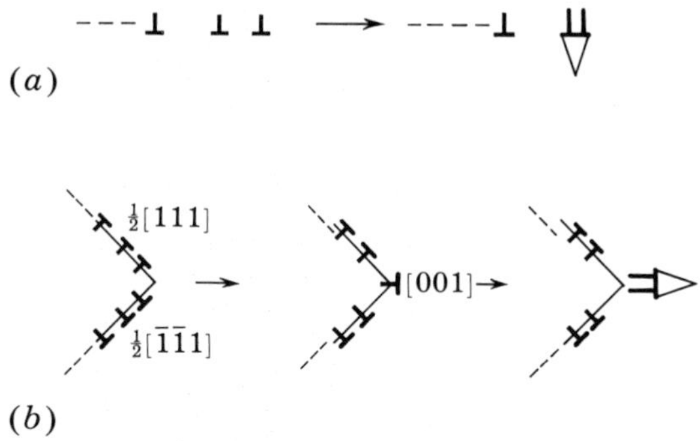

FIGURE 21-19. (*a*) Stroh's mechanism of coalescence of dislocations at the head of a pileup into a crack. (*b*) Cottrell's mechanism of reaction of $\frac{1}{2}\langle 111\rangle$ dislocations in a bcc crystal to form [001] dislocations, which then coalesce to form a crack.

among them experimentally. Two prominent fracture mechanisms are shown in Fig. 21-19. Stroh[35] suggested that an incipient crack forms when the two leading dislocations in a pileup are forced to within a distance b of one another (Fig. 21-19*a*). Cottrell[36] proposed, specifically for bcc metals, that a crack is formed when two intersecting $\langle 111\rangle$ glide systems interact to form $\langle 001\rangle$ dislocations (Fig. 21-19*b*). When the $\langle 001\rangle$ dislocations are sessile, the coalescence of the $\langle 111\rangle$ dislocations produces a crack nucleus. Direct evidence for both the Stroh mechanism[37] and the Cottrell mechanism[38] has been found experimentally.

The stresses required to force idealized pileups into configurations such as those in Fig. 21-19 can readily be calculated by the theory of this chapter. However, such calculations are uncertain because of the possibility of blunting, and because the configurations require knowledge of dislocation core interactions, which are known only in an approximate way.[39] Also, three-dimensional dislocation arrays and distributions of long-range internal stresses are complicating factors in crack nucleation. Such calculations, therefore, are not presented here. Additional dislocation mechanisms for fracture are reviewed by Low.[40]

[35]A. N. Stroh, *Advan. Phys.*, **6**: 418 (1957).

[36]A. H. Cottrell, *Trans. AIME*, **212**: 192 (1958).

[37]R. J. Stokes, T. L. Johnston, and C. H. Li, *Phil. Mag.*, **3**: 718 (1958).

[38]J. Washburn, A. E. Gorum, and E. R. Parker, *Trans. AIME*, **215**: 230 (1959).

[39]H. Kuan and J. P. Hirth, *Mat. Sci. Eng.*, **22**: 113 (1976).

[40]J. R. Low, *Prog. Mat. Sci.*, **12**: 1 (1963).

The nucleation of slip at the spearhead of a pileup is important in the theory of yielding of polycrystalline materials. Several mechanisms are possible for yielding, as is the case for fracture. Consider a pileup at a grain boundary. The pileup produces a stress concentration because of plastic incompatibility across the boundary. In addition, for anisotropic materials, elastic incompatibility stresses are present (Chap. 9). Slip can be initiated if the leading dislocation is forced through the boundary into the next grain; this process entails supplying the energy of the grain-boundary step (or of jogs in the boundary dislocations) and of the residual dislocation left in the boundary if the Burgers vectors are not collinear in the two grains. A second mechanism entails forcing a boundary dislocation to act as a glide source under the pileup stress concentration. Third, the stress concentration can activate a Frank-Read source near the boundary in the unslipped grain. The requisite stresses for these mechanisms can be calculated from the preceding theory, assuming ideal pileups. Again, detailed computations are not undertaken because of the uncertainties in the model arising from blunting.

The Hall-Petch Relation

Hall[41] and Petch[42] pointed out that the yield stress of polycrystalline α-iron is proportional to $d^{-1/2}$, where d is the grain diameter. This form of relation also describes quite generally the grain-size dependence of flow stress and the fracture stress in many materials.[43,44] Petch[42] showed how such a relation could be explained by dislocation pileup theory. There are, however, alternative rationalizations[45] of a $d^{-1/2}$ relation. With the realization that these alternatives exist, and with appreciation of the uncertainties arising from blunting, etc., as discussed above, we develop the Hall-Petch relation for an idealized case of yielding as an example.

Consider a grain of diameter d, with a Frank-Read source in the middle of the grain. Suppose that the stress required to operate the source has been achieved, but that the resolved shear stress in the adjacent grains is insufficient to cause slip. As discussed in connection with Eq. (21-54), the resultant pileups can be approximated accurately by the simplified model of Fig. 21-20. For such a pileup, the force at the spearhead is given by Eq. (21-19). Suppose that the pileup will induce slip in the adjoining grain when the force reaches a critical value F^*/L. The requisite external stress σ^* is then found by solving Eq. (21-19) for σ,

$$\sigma^* = Kd^{-1/2} \tag{21-55}$$

[41] E. O. Hall, *Proc. Phys. Soc.*, **B64**: 747 (1951).

[42] N. J. Petch, *J. Iron Steel Inst.*, **174**: 25 (1953).

[43] R. W. Armstrong, *Adv. Mat. Res.*, **4**: 101 (1970); *Can. Met. Quart.*, **13**: 187 (1974).

[44] A. W. Thompson, *Met. Trans.*, **8A**: 833 (1977).

[45] See J. P. Hirth, *Met. Trans.*, **3**: 3047 (1972); H. Yaguchi and H. Margolin, *Scripta Met.*, **15**: 449 (1981).

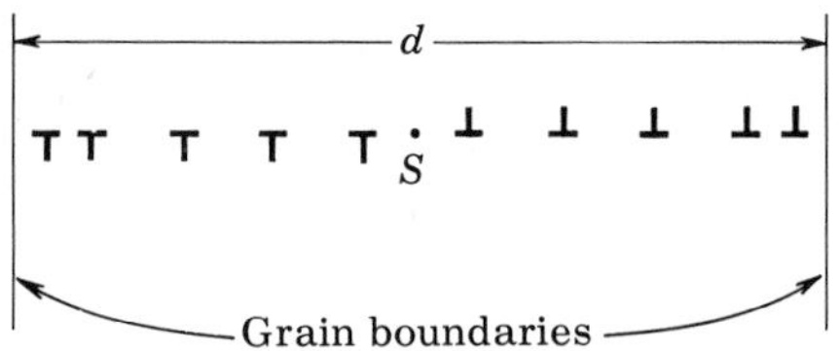

FIGURE 21-20. A double pileup formed from dislocations emanating from a Frank-Read source S and being blocked at grain boundaries.

where

$$K = m\left[\frac{4\mu}{\pi(1-\nu)}\frac{F^*}{L}\right]^{1/2} \tag{21-56}$$

and m is a factor resolving the external stress tensor into a resolved shear stress.

Experimental data for σ^* versus $d^{-1/2}$ show constant slopes but do not extrapolate to $\sigma^*=0$ as $d\to\infty$. Instead, the experimental data fit a relation

$$\sigma^* = \sigma_0 + Kd^{-1/2} \tag{21-57}$$

Such a wide variety of data fits Eq. (21-57) that it must have fundamental significance, but the factor σ_0 is difficult to rationalize. Two possible explanations are that σ_0 is a friction stress for motion of dislocations when glide is nucleated and that σ_0 represents an internal back stress. The first explanation has the drawback that pileups producing the stress concentration generally are envisioned to exist over a long time period in typical applications so that friction forces can be overcome by creep. Strong pinning points with local interactions with the dislocations or long-range back stresses seem a more likely cause of σ_0. However, even these do not provide a simple explanation of some effects such as the influence of interstitials on σ_0 in bcc metals. With the second rationalization, the effect of interstitials on σ_0 in bcc metals cannot be explained readily. Further work is required to resolve the quantity σ_0.

PROBLEMS

21-1. Consider a glide pileup of three edge dislocations against a barrier under an applied resolved shear stress σ. Calculate the positions of the three dislocations with the leading dislocation considered to be fixed.

21-2. Suppose that the frictional stress on the glide plane is σ_f. At 0°K, how wide will a single glide pileup of N edge dislocations be when no applied stress is present?

21-3. Use the preceding problem to estimate dislocation core widths. Assume that the critical shear stress for slip of a perfect crystal is $\mu/20$.

21-4. Let the shear stress required to operate a source of edge dislocations be σ^*, and suppose that there is a barrier at a distance L in front of the source but none in the rear of the source. When a resolved shear stress $\sigma > \sigma^*$ is present, how many dislocations will pile up behind the barrier? What is the resultant force on the barrier?

21-5. In a brittle solid with surface energy $\gamma = \mu b/10$, how large a microcrack must be present for the solid to fail under a tensile stress $\sigma = 10^{-3}\,\mu$? Assume $\nu = \frac{1}{3}$.

21-6. Consider a surface source of screw dislocations which operates at a negligibly small stress level. If a pileup forms at a distance L below the source as it operates, what is the force on the obstacle in the presence of an applied stress σ?

21-7. Treat the problem of a tensile crack whose surface lies normal to a free surface by analogy with Prob. 21-6. Derive the equivalent of the Griffith criterion for this case.

21-8. Determine the plane of maximum normal stress ahead of a pileup of edge dislocations.

21-9. Verify, using a continuum theory, the result of Chap. 19 that a pure tilt boundary composed of a single set of uniformly spaced edge dislocations has no long-range stress field. Would there be no long-range stresses if the dislocations were not uniformly spaced?

21-10. Show how the results of this chapter could be applied to the determination of the stress field around a coherent, plate-shaped precipitate.

BIBLIOGRAPHY

1. Averbach, B. L., D. K. Felbeck, G. T. Hahn, and D. A. Thomas (eds.), "Fracture," Wiley, New York, 1959.
2. Chou, Y. T., and J. C. M. Li, in T. Mura (ed.), "Mathematical Theory of Dislocations," American Society of Mechanical Engineering, New York, 1969, p. 116.
3. Drucker, D. C., and J. J. Gilman (eds.), "Fracture of Solids," Interscience, New York, 1963.
4. Hahn, G. T., and W. F. Flanagan, in "Dislocation Modeling of Physical Systems," Pergamon, Oxford, 1981, p. 1.
5. Rice, J. R., in H. Liebowitz (ed.), "Fracture," vol. II, Academic, New York, 1968, p. 191.
6. Seeger, A., in "Handbuch der Physik," vol. VII, part I, Springer, Berlin, 1955, p. 537.
7. Taplin, D. M. R., "Fracture 1977," vol. 2, University of Waterloo Press, Waterloo, Ontario, 1977.

22

Dislocation Interactions

22-1. INTRODUCTION

General interactions among aggregates of dislocations are complex and difficult to analyze. In Chaps. 5 and 6 interactions between straight dislocations were developed. However, as shown there, the mathematics becomes very cumbersome for curved dislocations, or even for the interaction between nonparallel straight dislocations. On the other hand, deformed crystals contain very large numbers of dislocations, 10^8 to 10^{12} cm/cc, which are arranged in complex, tangled, curved arrays. Thus there is an enormous gap between the theory developed for the interaction between a few simple geometry dislocations and a description of macroscopic deformation. For example, consider the Orowan expression for the plastic strain rate produced by gliding dislocations,

$$\dot{\epsilon} = m\rho b v \tag{22-1}$$

where ρ is the mobile dislocation density, m is an average Schmid strain resolution factor, and v is the average dislocation velocity. This relation is often used to describe plastic flow. A more accurate description for a local segment density contributing a dislocation density $\boldsymbol{\rho} = \rho\boldsymbol{\xi}$ is

$$\dot{\epsilon}_{jk} = \epsilon_{nop} m_{jkln} b_l \rho_o v_p(\eta_{ln})$$

$$\eta_{ln} = m_{lnqr}\left(\sigma_{qr} + \sigma_{qr}{}^i\right) \tag{22-2}$$

with m_{jkln} and m_{lnqr} as strain and stress resolution factors. The operator ϵ_{nop} is given by Eq. (4-11) and σ_{qr} and $\sigma_{qr}{}^i$ are the applied and internal stresses, respectively. The overall strain rate is then the average volume integral $V^{-1}\int \dot{\epsilon}_{jk}\, dV$ with $\sigma_{qr}{}^i$, ρ_0 and the local crystal orientation all position dependent quantities.

At present there is no hope of determining the overall $\sigma - \epsilon$ relation in such a way. Theories of work hardening[1] are perforce highly idealized if based on an

[1] F. R. N. Nabarro, Z. S. Basinski, and D. B. Holt, *Adv. Phys.*, **13**: 193 (1964); an excellent review of the subject of work hardening is presented by these authors.

explicit model or are phenomenological in nature and attempt to describe an average state of the crystal, such as in Eq. (22-1).

In the tensile deformation of single crystals, relatively simple coplanar arrays form during easy glide. As the stage II, rapid work-hardening regime is entered, however, multiple slip occurs, a form of dynamic recovery initiates, and dislocation cells are formed.[2] The cells become finer and are broken down by bands of shear instability, cross-slip becomes prominent, and the rate of work hardening decreases in stage III. Eventually the material behaves much like an ideal plastic material. Similarly, in fatigue, coplanar arrays form at low strain amplitudes and successively transform in a form of plastic instability into dislocation cells and eventually to subgrains.[3] In polycrystals the easy glide stage in simple tension and the coplanar array stage in reversed loading tend to be suppressed.

Only the coplanar array stages are amenable to a detailed dislocation description of the dynamical events at this time, although cell structures and the like can be analyzed after deformation ceases. Rather than attempt a general analysis of complex dislocation interactions, we present a few examples which typify the processes and illustrate the difficulties involved in the analysis of macroscopic deformation. In the following discussion we treat the intersection of nonparallel dislocations, the types of point defects formed in intersection processes, extended dislocation-barrier configurations, and the process of cross slip.

22-2. INTERSECTION OF PERFECT DISLOCATIONS

One of the important mechanisms restricting the motion of a dislocation in its glide plane and contributing to hardening is that of "forest intersection." This process consists of a glide dislocation cutting through a forest dislocation that does not lie in the same glide plane (Fig. 22-1). This important but difficult problem has yet to be analyzed in quantitative detail. The stages of the intersection process are (1) long-range interaction of elastic strain fields and (2) interaction on a core level, including the formation of kinks and/or jogs in the dislocations. Let us consider each of these stages in turn.

As indicated in Fig. 22-1, the local equilibrium configuration under the elastic interaction consists of curved dislocations. Some insight into the type of dislocation arrangement can be elicited from the theory of straight non-parallel dislocations.[4,5] The interaction force between a perpendicular pair of screw dislocations, one right-handed and one left-handed, is shown in Fig. 22-2. Under pure shear stresses, the local equilibrium dislocation configuration

[2] H. Mecking, in M. F. Ashby, R. Bullough, C. S. Hartley and J. P. Hirth (eds.), "Dislocation Modeling of Physical Systems," Pergamon, Oxford, 1981, p. 197.

[3] C. Laird, in A. W. Thompson (ed.), "Work Hardening in Tension and Fatigue," *Met. Soc. AIME*, Warrendale, PA., 1977, p. 150.

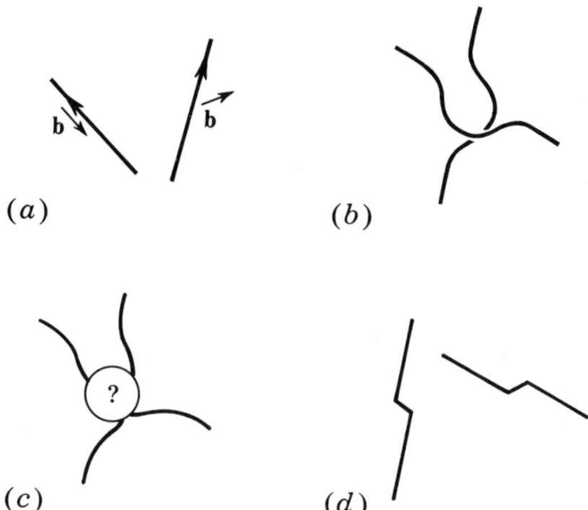

FIGURE 22-1. Stages of intersection: (*a*) a screw and an edge dislocation, (*b*) long-range elastic interaction, (*c*) unknown critical configuration, (*d*) result of intersection, showing a jog in the edge dislocation and a kink in the screw.

should resemble the force distribution of Fig. 22-2, since the dislocations should be pushed farthest apart along z at the point of greatest repulsion. In such a case, the numerical calculation of the equilibrium configuration from the differential form of Eq. (5-38) or (5-45) is feasible. However, in general, the local external-stress tensor (including long-range internal stresses from other dislocations) will contain normal stress components. In such a case, the motion of the dislocations out of pure screw orientation provides a nucleus for the formation of a dislocation helix,[6] as discussed in connection with Eq. (17-31). The presence of any excess point defects formed during deformation could also contribute to helix formation.[7] Such helices would be sessile with respect to glide on the original plane; energy would be required to restore the dislocation to the glide plane before glide could proceed. Hence, even for the relatively simple case of Fig. 22-2, the interaction analysis can become very complicated.

A more complicated interaction-force distribution is presented in Fig. 22-3. Obviously, the screw dislocation can again form into a helix. In addition, elastic interaction can lead to dipole formation, as shown in Fig. 22-4. Furthermore, the interaction stresses can evidently cause the screw to cross slip.[8] These interactions doubtless contribute to the formation of complex, tangled arrays during crystal deformation, but the exact calculation of the elastic interaction energy or of the local-equilibrium configuration is quite complex.

[4]C. S. Hartley and J. P. Hirth, *Acta Met.*, **13**: 79 (1965).

[5]R. Bullough and J. V. Sharp, *Phil. Mag.*, **11**: 605 (1965).

[6]J. Weertman, *Trans. AIME*, **227**: 1439 (1963).

[7]D. Kuhlmann-Wilsdorf, *Trans. AIME*, **224**: 1047 (1962).

[8]J. Washburn, *Appl. Phys. Lett.*, **7**: 183 (1965).

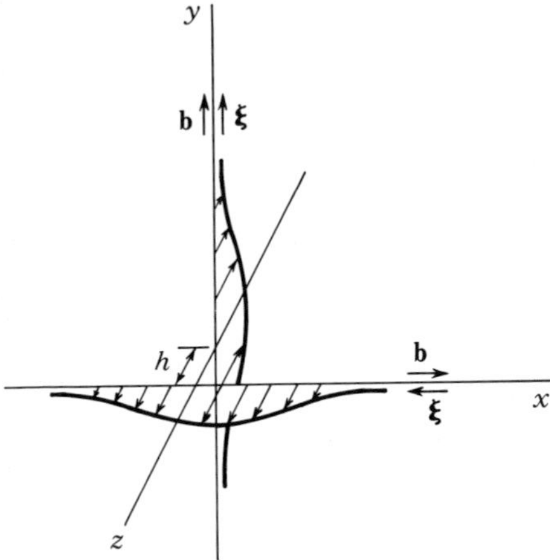

FIGURE 22-2. Local interaction forces between perpendicular screw dislocations, one right-handed and one left-handed, separated by a distance h. [After C. S. Hartley and J. P. Hirth, *Acta Met.*, **13**: 79 (1965).]

Even if one could calculate the elastic portion of the energy-displacement curve, the core interaction would still have to be considered. The net work caused by core interactions involves the formation energy of the jogs and/or kinks formed in the intersection process. However, the entire formation energy W_k or W_j does not enter into the activation energy for the process, since the critical configuration of maximum energy almost certainly involves only partial formation of the jogs and kinks. The energy calculations in the core region are complicated by all the uncertainties discussed in Chap. 8. Thus about the best one can do is to postulate a curve of the form of Fig. 22-5. This curve is analogous to that in Fig. 16-5. The activation energy and the activation area for the intersection process are given by Eqs. (16-8) to (16-10). As discussed in

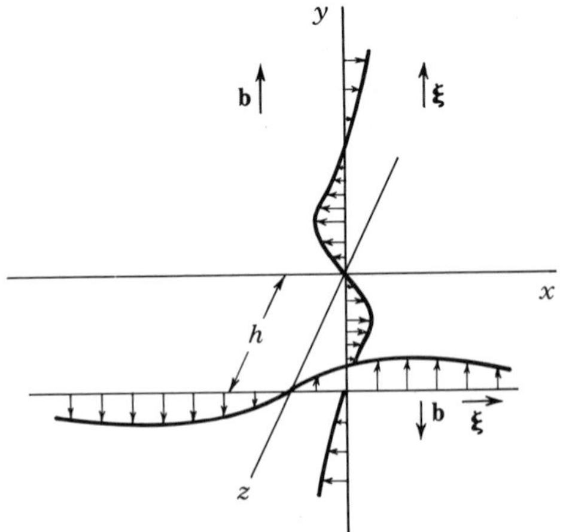

FIGURE 22-3. Local interaction forces between perpendicular screw and edge dislocations with common Burgers vector.

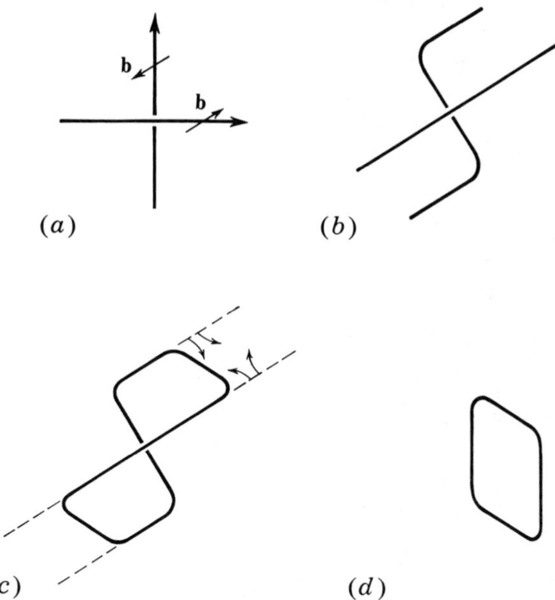

FIGURE 22-4. Dipole formation by local interaction of perpendicular mixed dislocations (*a*). (*b*) Dislocations twist into screw orientation. (*c*) Cross slip and some annihilation occur. (*d*) The closed loop rearranges into a planar dipole loop. [After A. S. Tetelman, *Acta Met.*, **10**:813 (1962).]

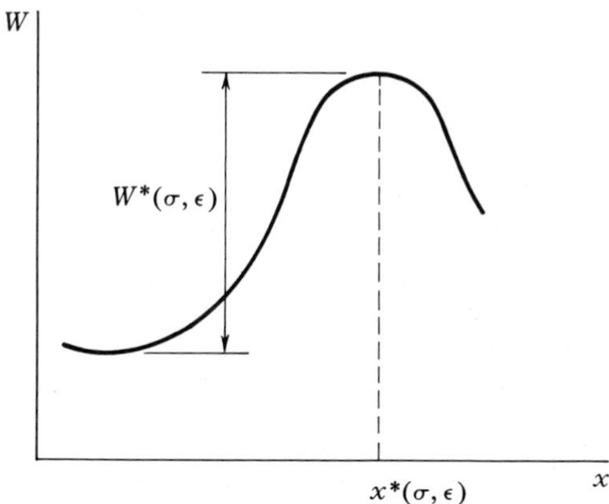

FIGURE 22-5. Energy-reaction coordinate plot for a dislocation intersection process.

Chap. 16, these quantities are certain to be stress dependent and, because the local net stress depends on the long-range back stress, to be dependent on strain and dislocation distribution.

A further complication occurs when the intersecting dislocations are attractive and react locally in the manner discussed in Sec. 19-4. Consider the bcc reaction in Fig. 22-6, for example. The dislocations locally relax to the

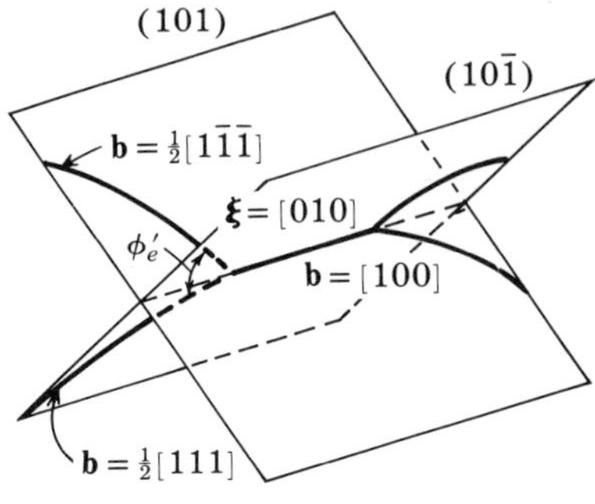

FIGURE 22-6. The bcc reaction $\frac{1}{2}[111] + \frac{1}{2}[1\bar{1}\bar{1}] \rightarrow [100]$ forming along the line of intersection of (101) and (10$\bar{1}$). The glide plane for the product dislocation is (001).

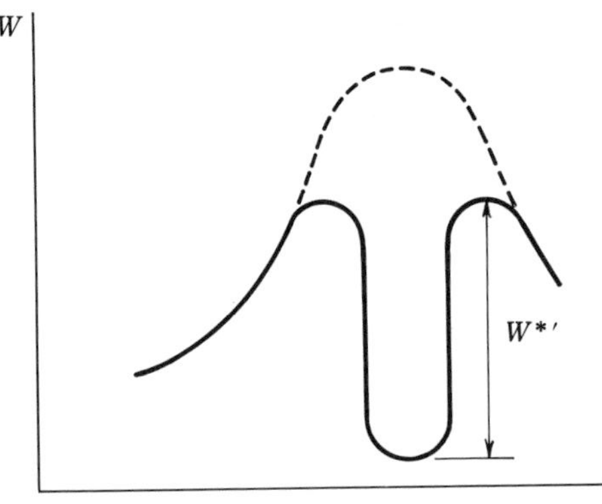

FIGURE 22-7. Modification of the energy-distance curve of Fig. 22-5 (dashed line) because of local relaxation.

equilibrium angle ϕ_e' (Fig. 19-23), determined by interaction forces and by the forces caused by the external stress tensor. This relaxation in the configuration changes the energy-distance curve from that of Fig. 22-5 to that of Fig. 22-7. In order to force the intersecting dislocations through one another, the applied stresses must "unzip" the relaxed configuration.[9, 10] Such an unzipping requires an energy $W^{*\prime}$ (Fig. 22-7), which in general depends on stress and strain, and which in general can be determined only by numerical analysis. Those slip systems which would form such relaxed arrays have been cataloged for both fcc crystals[10, 11] and for bcc crystals.[11, 12] The inverse of the unzipping process is the "knitting" process, which leads to cell and subgrain formation in low temperature and creep deformation. The knitting process has been extensively studied by Lindroos and Miekk-oja.[13]

[9]G. Saada, *Acta Met.*, **8**: 841 (1960).

[10]J. P. Hirth, *J. Appl. Phys.*, **32**: 700 (1961); D. Kuhlmann-Wilsdorf, *Acta Met.*, **14**: 439 (1966).

[11]B. Gale, in "Relation between Structure and Mechanical Properties of Metals," H.M. Stationery Office, London, 1963, p. 100.

[12]L. J. Teutonico, *Phys. Stat. Solidi*, **14**: 457 (1966); C. S. Hartley, *Phil. Mag.*, **14**: 7 (1966).

[13]V. K. Lindroos and H. M. Miekk-oja, *Phil. Mag.*, **20**: 329 (1969).

The above discussion makes it evident that no simple model can completely describe crystal deformation controlled by dislocation intersection processes. Indeed, in view of the complexities, it is amazing that experimentally determined activation energies and activation volumes are roughly reproducible in the temperature range where intersection processes are thought to be important.

22-3. INTERSECTION-JOG FORMATION

As shown in Chap. 16, the rate equations describing crystal flow differ for interstitial-producing jogs and vacancy-producing jogs when jog drag is rate controlling. The types of jogs produced at different stages of deformation are thus of interest. Cottrell[14] pointed out that in the usual deformation test (tension, compression, simple shear) most of the intersection jogs formed are interstitial producing; we show presently that this is so only in the early stages of deformation. For a given crystal under a specified state of stress, Cottrell's hypothesis can be verified by determining from Eq. (3-90) the direction of motion of dislocations on each possible glide system and then determining from Eq. (12-10) the type of jog produced on intersection of each pair of systems.

Consider the NaCl-type single crystal of Fig. 22-8, with a simple tensile stress along [001]. Loops that expand under such a stress are shown on (011) and $(0\bar{1}1)$ planes, respectively. The stress forces the right-handed screw segments B and E toward one another, leading to an interstitial-forming jog (Sec. 12-6). Similarly, intersections formed by left-handed screws such as A and F moving together create interstitial-forming jogs. Also, if the edges G or H cut through B, they create an interstitial-forming jog in it. A consideration of the other highly stressed glide systems on (101) and $(\bar{1}01)$ shows that screws from these systems that cut through A or B also produce interstitial-forming jogs. Thus, early in deformation, when most of the dislocation segments that intersect are moving relative to one another in the direction favored by the applied stress, i.e., when $\mathbf{r}'$ and $\mathbf{r}$ are equal in Eq. (12-10), interstitial-forming jogs are formed in the screw dislocations. However, later in the deformation process, when loops are pinned in place by interaction and by back stresses, right-handed screw segments such as E cut through left-handed segments such as A, which, if it were free, would have moved away under the applied stress and hence would have avoided the intersection. The latter type of intersection creates vacancy-forming jogs. Thus late in the deformation process about equal numbers of both types of jogs should form.

Jogs produced in undissociated edge dislocations are glissile. However, as shown in Fig. 16-18, jogs in edge dislocations can be sessile and hence

[14]A. H. Cottrell, in J. C. Fisher et al. (eds.), "Dislocations and Mechanical Properties of Crystals," Wiley, New York, 1957, p. 509.

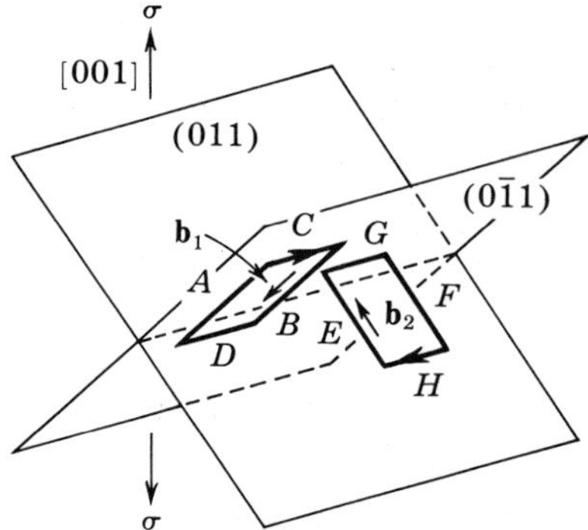

FIGURE 22-8. Expanding glide loops in an NaCl-type crystal under simple tension along [001].

point-defect forming. If only intrinsic-type jogs (or, equivalently, the jog lines corresponding to intrinsic-type superjogs) form, all the edge jogs will be vacancy producing. If extrinsic-type jogs form, they will be interstitial producing. As discussed in Sec. 10-5, the core energy of the extrinsic-type jogs is higher than that of the intrinsic type, so that, in any case, the latter is expected to predominate.

An estimate of the type of jog formed as a function of strain is summarized in Fig. 22-9. For undissociated dislocations, where only screw jogs form point defects, the ratio of interstitial to vacancy jogs starts out greater than unity for small strains (i.e., small dislocation densities) and decreases asymptotically to unity. For dissociated dislocations, the ratio depends in detail on the relative mobilities and jog densities on screws and edges. In particular, if the edges are much more mobile than the screws, as appears often to be the case experimentally, vacancy jogs can predominate.

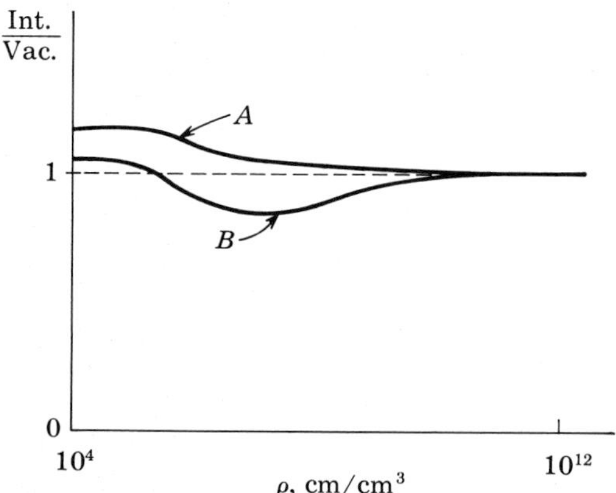

FIGURE 22-9. Ratio of interstitial-producing to vacancy-producing jogs formed by dislocation intersection as a function of dislocation density. Curve *A* is for undissociated dislocations; curve *B* is a possible curve for extended dislocations.

Exercise 22-1. Utilize Eqs. (3-90) and (12-10) to verify that the intersection jogs formed by the loops of Fig. 22-8 are interstitial forming.

22-4. EXTENDED-DISLOCATION BARRIERS AND INTERSECTION JOGS

Low-Fault-Energy Barriers

Local reactions such as that of Fig. 22-6 are of special interest in fcc crystals because the product dislocation is extended[15] on two planes and can provide a strong barrier to dislocation glide. Lomer[16] first noted that the reaction

$$\mathbf{BA}(d)+\mathbf{DB}(a)\rightarrow\mathbf{DA} \tag{22-3}$$

is favorable energetically in fcc crystals.[17] Cottrell[18] pointed out that the product dislocation could be extended,

$$\mathbf{BA}(d)+\mathbf{DB}(a)\rightarrow\boldsymbol{\delta}\mathbf{A}(d)+\mathbf{D}\boldsymbol{\alpha}(a)+\boldsymbol{\alpha\delta} \tag{22-4}$$

with a stair-rod dislocation (Table 10-2) at the line of intersection of the glide planes. The product of Eq. (22-4) is called the Lomer-Cottrell barrier. Other possible barriers of this type, involving the other stable stair rods of Table 10-2, have been considered by Friedel[19] and Whelan[20] in the approximation that the line tension is $\sim\mu b^2$ for the reacting partials. The equilibrium configurations and energies of the entire arrays, including screw-edge character via Eq. (5-16), were first obtained by Hirth.[21]

When the reactant and product dislocations are straight and parallel to the line of intersection of the glide planes, as in Fig. 22-10, the energies can be determined accurately from Eq. (5-16).[22] Figure 22-11 depicts the possible

[15]Throughout this section we consider only dislocations extended to form intrinsic stacking faults.

[16]W. M. Lomer, *Phil. Mag.*, **42**: 1327 (1951).

[17]With Thompson's notation, shorthand rules are easily established to give the possible reactions which form Lomer-Cottrell or the other types of barriers. For example, the reaction of glide systems described by cyclic permutations of the *letters CAB* form Lomer-Cottrell barriers, that is, **CA**(b)+**BC**(a), **BC**(a)+**AB**(c), or **AB**(c)+**CA**(b). The development of the rules of this game is left as an exercise.

[18]A. H. Cottrell, *Phil. Mag.*, **43**: 645 (1952).

[19]J. Friedel, *Phil. Mag.*, **46**: 1169 (1955).

[20]M. J. Whelan, *Proc. Roy. Soc.*, **A249**: 114 (1958).

[21]J. P. Hirth, *J. Appl. Phys.*, **32**: 700 (1961).

[22]Indeed, the line of intersection is $\langle 110\rangle$, which is one of the directions for which the anisotropic elastic theory of dislocations is analytically tractable (Chap. 13). The various fcc barrier dislocations have been analyzed in the anisotropic model by L. J. Teutonico [*Phil. Mag.*, **10**: 401 (1964)] and by T. Jøssang, C. S. Hartley, and J. P. Hirth [*J. Appl. Phys.*, **36**: 2400 (1965)].

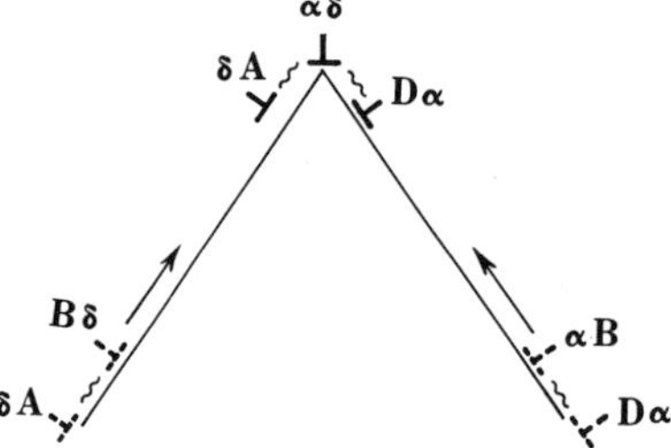

FIGURE 22-10. View along the straight-dislocation lines of **DB** and **BA** reacting to form a Lomer-Cottrell barrier.

barriers,[22] identified in terms of Eq. (22-5),

$$\begin{aligned}
&1.\ \mathbf{B\delta}(d)+\boldsymbol{\delta}\mathbf{A}(d)+\mathbf{D}\boldsymbol{\alpha}(a)+\boldsymbol{\alpha}\mathbf{B}(a)\to\boldsymbol{\delta}\mathbf{A}+\mathbf{D}\boldsymbol{\alpha}+\boldsymbol{\alpha\delta}\\
&2.\ \mathbf{B\delta}(d)+\boldsymbol{\delta}\mathbf{A}(d)+\mathbf{D}\boldsymbol{\alpha}(a)+\boldsymbol{\alpha}\mathbf{C}(a)\to\mathbf{B\delta}+\boldsymbol{\alpha}\mathbf{C}+\boldsymbol{\delta}\mathbf{D}/\mathbf{A}\boldsymbol{\alpha}\\
&3.\ \mathbf{B\delta}(d)+\boldsymbol{\delta}\mathbf{A}(d)+\mathbf{C}\boldsymbol{\alpha}(a)+\boldsymbol{\alpha}\mathbf{D}(a)\to\boldsymbol{\delta}\mathbf{A}+\boldsymbol{\alpha}\mathbf{D}+\mathbf{BC}/\boldsymbol{\alpha\delta}\\
&4.\ \mathbf{B\delta}(d)+\boldsymbol{\delta}\mathbf{A}(d)+\mathbf{D}\boldsymbol{\alpha}(a)+\boldsymbol{\alpha}\mathbf{B}(a)\to\boldsymbol{\alpha}\mathbf{B}+\mathbf{B\delta}+\boldsymbol{\delta}\mathbf{D}/\mathbf{A}\boldsymbol{\alpha}\\
&5.\ \mathbf{B\delta}(d)+\boldsymbol{\delta}\mathbf{A}(d)+\mathbf{D}\boldsymbol{\alpha}(a)+\boldsymbol{\alpha}\mathbf{C}(a)\to\boldsymbol{\delta}\mathbf{A}+\boldsymbol{\alpha}\mathbf{C}+\mathbf{BD}/\boldsymbol{\delta\alpha}\\
&6.\ \mathbf{B\delta}(d)+\boldsymbol{\delta}\mathbf{A}(d)+\mathbf{B}\boldsymbol{\gamma}(c)+\boldsymbol{\gamma}\mathbf{D}(c)\to\mathbf{B\delta}+\mathbf{B}\boldsymbol{\gamma}+\boldsymbol{\delta\gamma}/\mathbf{AD}
\end{aligned}\qquad(22\text{-}5)$$

Barriers (5) and (6) are asymmetric and are only marginally stable, involving high-energy stair rods $\frac{1}{6}\langle 130\rangle$; hence they are not discussed here. The relative equilibrium dissociations of barriers (1) to (4) are shown in Fig. 22-12; the original Lomer-Cottrell barrier (1) and its inverted form (4) are least dissociated among the symmetric barriers. The barriers all have dissociations large compared to b in most cases; typical examples of equilibrium widths are listed

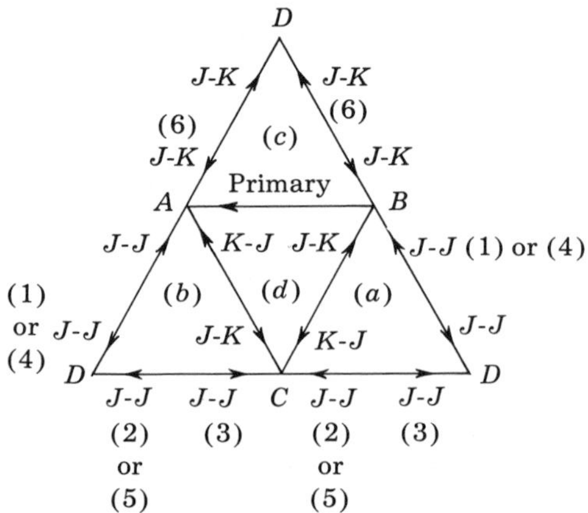

FIGURE 22-11. Possible barriers [Eq. (22-5)] formed by reaction with the primary glide system **BA**(d) displayed on the Thompson tetrahedron. The type of defect, jog J or kink K, formed on intersection with the primary system is also shown.

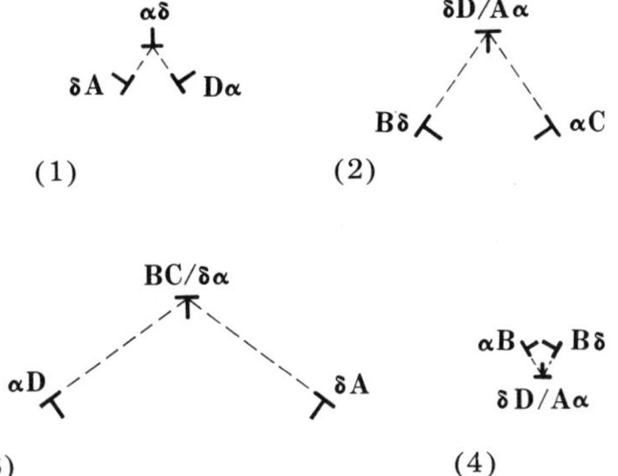

FIGURE 22-12. Relative extensions of barriers (1) to (4) for the example of copper. [T. Jøssang, C. S. Hartley, and J. P. Hirth, *J. Appl. Phys.*, **36**: 2400 (1965).]

TABLE 22-1 Ratios of Equilibrium Extension to the Burgers Vector of a Perfect Dislocation for Several FCC Metals*

	r/b				
Configuration	Ag	Au	Cu	Al	Ni
Barrier (1)	9.7	2.7	1.6	0.6	0.9
Barrier (2)	23.3	6.9	7.4	1.6	2.3
Barrier (3)	34.4	11.5	10.6	2.5	3.2
Screw	6.9	1.7	2.3	0.7	0.9

*Based on the values of stacking-fault energy listed in Appendix 2, and on the anisotropic elasticity solution of T. Jøssang, C. S. Hartley, and J. P. Hirth [*J. Appl. Phys.*, **36**: 2400 (1965)]. The ratio for the dissociation of a pure screw dislocation in an infinite medium is listed also for comparison.

in Table 22-1. There have been numerous observations of barrier (1), a lesser number of barrier (3), a few of barrier (4), and none of barrier (2).

There are a number of possible ways in which the barriers can block dislocation motion, each case involving a different activation energy. As first pointed out by Whelan,[20] and as verified extensively in experiment,[23] the barriers are unlikely to form over long lengths. Most dislocations intersect at fairly high angles, so that line-tension forces constrain them to configurations such as that of Fig. 22-13. Ways that such barriers can block dislocations are shown in Fig. 22-14. Calculations of energy-reaction-distance curves for any of these cases are very difficult, for the same reasons as those discussed in Sec. 22-2, and have yet to be accomplished. However, the effectiveness of the

[23] For example, see H. Strunk, *Phil. Mag.*, **21**: 857 (1970); V. K. Lindroos and H. M. Miekk-oja, ibid., **16**: 593 (1967); H. P. Karnthaler and E. Wintner, *Acta Met.*, **23**: 1501 (1975).

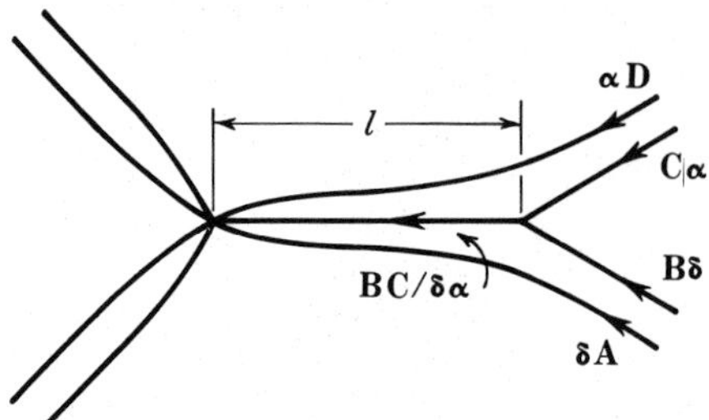

FIGURE 22-13. Configuration of barrier (3) formed over a length l.

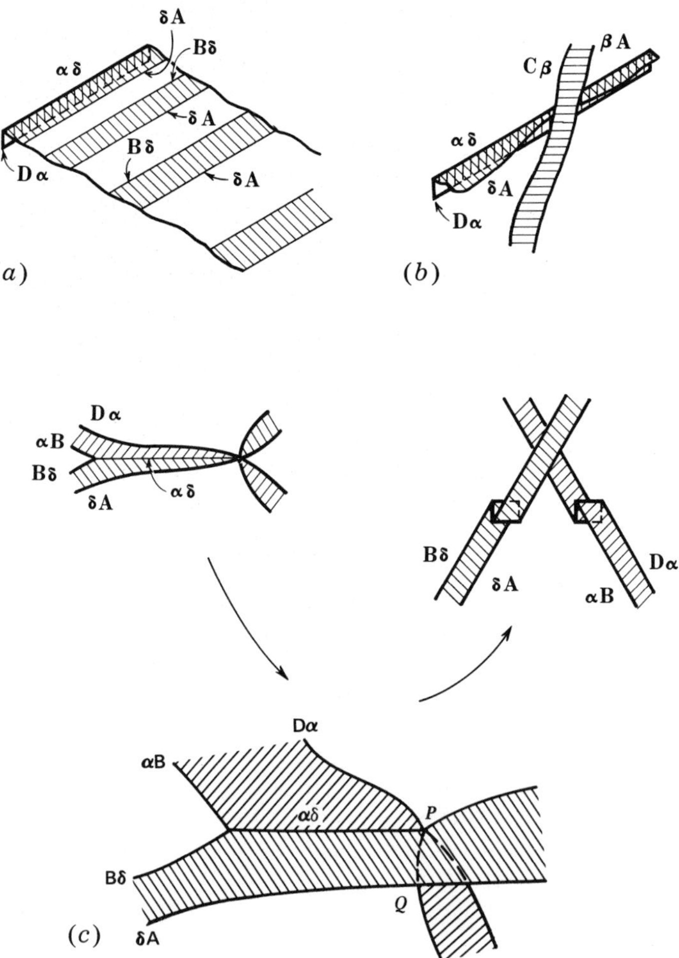

FIGURE 22-14. (a) Lomer-Cottrell barrier blocking a pileup of dislocations **BA**(d). (b) Intersection of barrier by the dislocation **CA**(b). (c) Unzipping of a barrier dislocation.

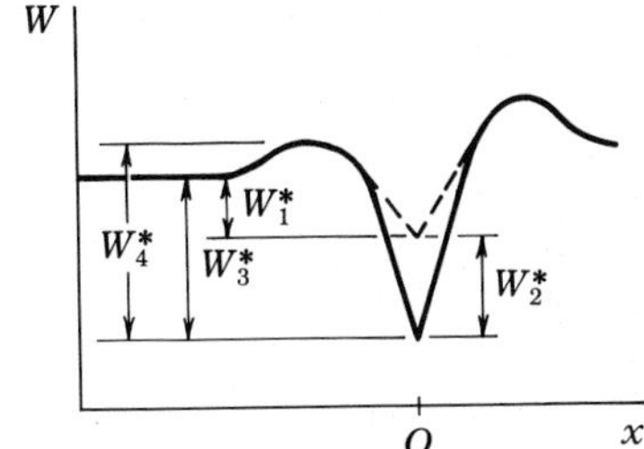

FIGURE 22-15. Energy-reaction coordinate curves for barrier formation and dissociation processes.

TABLE 22-2 Values of W_2^*, W_3^* and F^* for Barriers (1) to (4) in Copper.*

Barrier	$W_2^*, \mu b^2$	$W_3^*, \mu b^2$	$F^*, \mu b$
(1)	0.003	0.86	0.019
(2)	0.13	0.22	0.011
(3)	0.26	~0	0.014
(4)	~0	~0	0.011

*Values for W_3^* from J. P. Hirth, *J. Appl. Phys.*, **32**: 700 (1961); values for F^* from T. Jøssang, C. S. Hartley, and J. P. Hirth, *J. Appl. Phys.*, **36**: 2400 (1965).

various barriers should qualitatively relate to the energies of idealized straight dislocation interactions.

A schematic representation of the interaction energy for long straight reactants and barrier products is shown in Fig. 22-15. Energy W_1^* represents the net long-range attraction between reactant dislocations constrained to remain undissociated. Energy W_2^* represents the *local* relaxation energy to the equilibrium configuration of the barrier. Thus the redissociation of the barrier to its original unreacted configuration of two well-separated, undissociated dislocations requires an energy $W_3^* = W_1^* + W_2^*$, while the collapse of the barrier requires an energy W_2^*. The well-separated reactants are, of course, actually dissociated, but the barriers, except for (1), are widely dissociated, so that W_3^* is a reasonable indicator of the stability against redissociation. A parameter in addition to W_3^* which indicates the stability of the barriers against redissociation and which indicates to some extent the influence of the activation barrier W_4^* in Fig. 22-15 is the slope $(\partial W/\partial x)_{x=b} = F^*$. The values of these parameters are listed in Table 22-2.

Now consider the various processes of Fig. 22-14. The collapse of long straight barriers under pileup stresses (Fig. 21-14a) has been considered as a mechanism in work hardening by Seeger et al.[24] Once the barrier dislocation is collapsed to the perfect dislocation **DA** or **BD**/**AC**, it can glide out of the way on $\{100\}$ under the stress concentration of the pileup. The energy for such

[24]A. Seeger, J. Diehl, S. Mader, and H. Rebstock, *Phil. Mag.*, **2**: 323 (1957).

collapse is related to W_2^*; hence barriers (2) and (3) should be most resistant to such a process. Figure 21-14*b* shows a dislocation cutting through a barrier. Again, the energy is related to W_2^*, so that barriers (2) and (3) should be strongest. The energy for the unzipping process of Fig. 21-14*c* is related to W_3^*; thus barrier (1) is most resistant to unzipping.[25, 26] Finally, the energy to redissociate the barriers under stress reversal is related to W_3^*, so that barrier (1) is strongest. We emphasize that the above conclusions are qualitative in nature and are based on an analogy of a complex reaction with a hypothetical straight-dislocation configuration. Further work is needed to determine the complex activation energies for the processes of Fig. 22-14.

Latent hardening measurements on Al and Cu[27, 28] indicate that intersection of the slip systems that would produce the Lomer-Cottrell barrier (1) produce the greatest hardening, followed by systems that could form the parallel array of Fig. 10-36 and barrier (3). Also, the previously mentioned observations of barriers in the configuration in Fig. 22-13 show that the lengths l are much greater for (1) than those for (3). Thus, as emphasized by Kuhlmann-Wilsdorf,[29] the process of Fig. 22-14*c*, the extended version of the hardening reaction of Saada[30] of the type shown in Fig. 22-6, is dominant, and W_3^* governs the hardening behavior. Indeed, the similarity of hardening for Al and Cu, with high and intermediate stacking-fault energies, suggests that extension does not have a great effect on hardening. The attribution of hardening to the length of reaction (1) is reinforced by the observation[27] that latent hardening does not correlate with jog versus kink formation by the intersecting systems. For very low stacking-fault energy alloys, where W_2^* should be relatively more important, barriers (3) and possibly (2) may be more important for hardening. In such a case both length and extension of the barrier could influence hardening.

In summary, there are a number of ways that extended barriers can affect hardening (Fig. 22-15). In some cases the length of the barrier is related to strength, while in others the degree of extension of the barriers is related to strength. Neither length nor extension should, therefore, be used as a sole criterion for barrier strength. In all cases detailed calculations are required to determine exact energy-displacement relations.

Intersection Jogs

The intermediate stage of formation of jogs in extended dislocations which do not favor barrier formation involves uncertain degrees of constriction and is

[25] J. P. Hirth, *J. Appl. Phys.*, **32**: 700 (1961).

[26] D. Kuhlmann-Wilsdorf, *Acta Met.*, **14**: 439 (1966).

[27] P. J. Jackson and Z. S. Basinski, *Can. J. Phys.*, **45**: 707 (1967).

[28] P. Franciosi, M. Berveiller, and A. Zaoui, *Acta Met.*, **28**: 273 (1980).

[29] D. Kuhlmann-Wilsdorf, *Acta Met.*, **14**: 439 (1966).

[30] G. Saada, *Acta Met.*, **8**: 841 (1960).

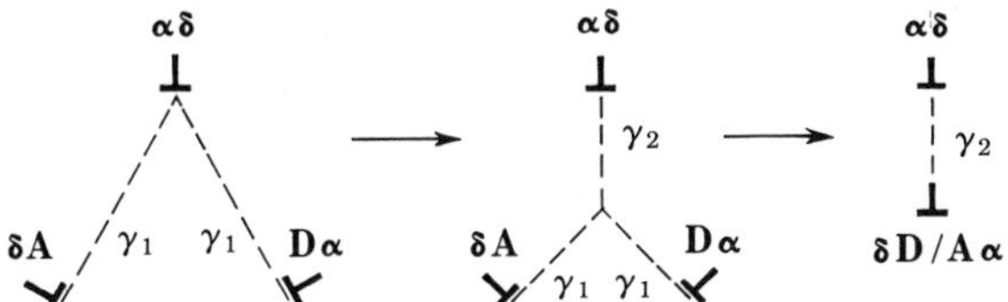

FIGURE 22-16. Formation of a climb barrier from the Lomer-Cottrell barrier (1). γ_1 represents a low-energy intrinsic stacking fault and γ_2, a high-energy fault.

equally uncertain as that in Fig. 22-1*c*. However, for dislocations that form barriers, the situation has been clarified by Carter and Hirsch.[31] They deduced from electron micrographs that the intermediate steps were first formation of barrier (1) and then formation of the partially unzipped configuration in Fig. 22-14*c*. There a jog line has already formed on **BA** between the points *P* and *Q*. The stair-rod $\boldsymbol{\alpha\delta}$ must then unzip, creating the final configuration in Fig. 22-14*c* with jog lines on both dislocations. They also discuss possible intermediate configurations for intersection kink formation.

High-fault-energy barriers

As shown in Table 22-1, barriers in relatively low-stacking-fault-energy fcc metals are unquestionably extended, $r \gg b$. Another class of barriers exists whose extension (or, in fact, existence) is marginal. One such possibility[32] is the extension in fcc crystals of the Lomer-Cottrell reactant of Eq. (22-3) into a *climb* barrier (Fig. 22-16). The climb forces in question are larger than the glide forces of the previous section and favor formation of the barrier. On the other hand, the fault energy is quite large, and at present indeterminate experimentally; it should be larger than the intrinsic-fault energy but less than that of a high-angle grain boundary. For the metal copper as an example, the climb fault energy γ must satisfy the condition $\gamma \gtrsim 240$ mJ/m^2 for the extension r to exceed $2b$. Comparing this requirement with the fault and grain-boundary energies of Appendix 2, one sees that the existence of the fault is possible, but only marginally so, and if so it occurs only over atomic dimensions resembling the fractional dislocation arrays described in Sec. 11-3. Other systems do have fault energies sufficiently low for climb barriers observable in electron microscopy to form. Examples include the climb dissociation $\frac{1}{3}\langle 1120\rangle \rightarrow \frac{1}{3}\langle 1010\rangle + \frac{1}{3}\langle 0110\rangle$ in $Al_2 0_3$[33] and the Kear-Wilsdorf barrier[34] composed of two $\frac{1}{2}\langle 110\rangle$ dislocations with an associated (100) antiphase boundary in the ordered $L1_2$ structure.

[31] C. B. Carter and P. B. Hirsch, *Phil. Mag.*, **35**: 1509 (1977).

[32] J. P. Hirth, *J. Appl. Phys.*, **33**: 2286 (1962).

[33] T. E. Mitchell, B. J. Pletka, D. S. Phillips, and A. H. Heuer, *Phil. Mag.*, **34**: 441 (1976).

[34] B. H. Kear and H. G. F. Wilsdorf, *Trans. Met. Soc. AIME*, **224**: 382 (1962).

Similarly, barriers resembling the fcc ones are possible in bcc crystals,[35–38] but the fault energies are so high in bcc metals that the existence of the barriers is of marginal probability. Because of the large number of possible glide planes in bcc crystals, the number of possible barriers is correspondingly greater. We consider only one example.

Teutonico[35] considered barriers extended on $\{112\}$ planes. About the most likely of his barriers is

$$\tfrac{1}{2}[111]_{(1\bar{2}1)}+\tfrac{1}{2}[\bar{1}\bar{1}1]_{(1\bar{2}\bar{1})}\rightarrow\tfrac{1}{6}[111]_{(1\bar{2}1)}+\tfrac{1}{6}[\bar{1}\bar{1}1]_{(1\bar{2}\bar{1})}+\tfrac{1}{3}[002] \qquad (22\text{-}6)$$

with the reactants and products lying along $\boldsymbol{\xi}=[210]$. Using Axiom 11-2, one finds that the barrier forms an acute configuration if both faults are of the same kind, either I_1 or I_2. For the extension of the barrier to exceed $2b$, the fault energies for several typical metals must be[35] less than 210, 672, and 20mJ/m^2, respectively, for niobium, tungsten, and lithium. The respective values for an edge dislocation to dissociate by $2b$ in these metals are 310, 1,080, and 47mJ/m^2; the edge is the orientation with the most widely separated partials on $\{211\}$. The existence of such small stacking-fault energies is problematical. Extensions have not been found in atomic simulations for pure bcc transition metals. They are possible both for alloys such as W-Re solutions, which do have relatively lower fault energies,[39] and for alkali metals where there are some theoretical indications of lower fault energies.

22-5. CROSS SLIP

As a final example of complex dislocation interactions, we consider the process of cross slip of extended dislocations. The onset of stage III in the work hardening of an fcc single crystal, in which the curvature $\partial^2\sigma/\partial\epsilon^2$ of a true-stress–true-strain curve becomes negative, is definitely associated with the onset of cross slip, as verified many times by experiment.[40] Specific idealized mechanisms for cross slip have been used as the basis for theories of work hardening. However, there are a number of possible mechanisms for cross slip, and, most likely, several or all of them can actually operate in the final stages of work hardening. Since the determination of the activation energies for these processes is as complex as in the previous examples of this chapter, the detailed analysis of stage III remains uncertain.

[35] L. J. Teutonico, *Phys. Stat. Solidi*, **10**: 535 (1965); **14**: 457 (1966).

[36] R. Priestner and W. C. Leslie, *Phil. Mag.*, **11**: 895 (1965).

[37] C. S. Hartley, *Phil. Mag.*, **14**: 7 (1966); *Acta Met.*, **14**: 1133 (1966).

[38] F. Kroupa and V. Vitek, *Czech. J. Phys.*, **B14**: 337 (1964); V. Vitek, *Phys. Stat. Solidi*, **15**: 557 (1966).

[39] S. Mahajan, *Phil. Mag.*, **26**: 161 (1972).

[40] A. Seeger, J. Diehl, S. Mader, and H. Rebstock, *Phil. Mag.*, **2**: 323 (1957).

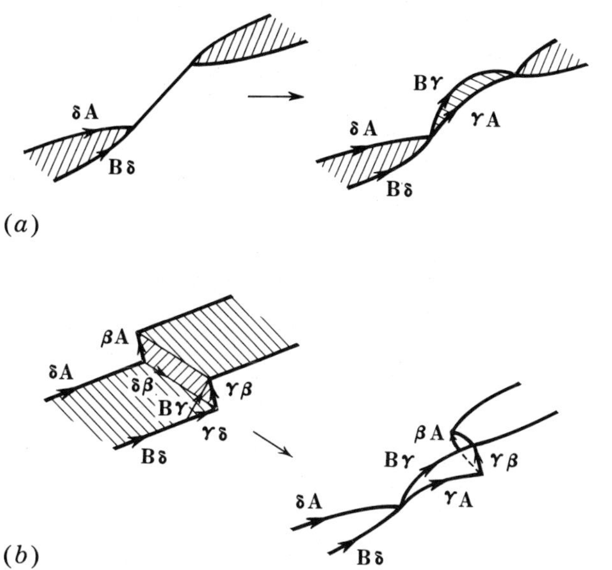

FIGURE 22-17. (*a*) Cross slip at a contracted superjog; the line direction ξ of the jog is along **DB**. (*b*) Cross slip at an extended superjog.

One mechanism for cross slip is already suggested by Fig. 22-3. The local interaction stresses approach $\sim\mu/10$ as nonparallel dislocations such as these near each other. Internal stresses of this magnitude suffice to nucleate cross slip onto another plane, as suggested by the force distribution of Fig. 22-3. If the cross-slip system has a large resolved shear stress acting on it, the nucleus will grow and produce appreciable cross slip. Similarly, the large internal stress fields near inclusions or other obstacles to glide can cause cross slip. Superjogs can also provide sources for cross slip. Stroh[41] noted that a contracted jog in a screw dislocation is a segment which could directly bow out on the cross-slip plane with a sufficient stress resolved there (Fig. 22-17*a*). Hirsch[42] showed that a similar mechanism applied for extended superjogs (Fig. 22-17*b*). With a sufficient stress on the cross-slip plane, segment **B**$\boldsymbol{\gamma}$ is stabilized on the cross-slip plane; subsequently, $\boldsymbol{\delta}$**A** reacts with $\boldsymbol{\gamma\delta}$ to form $\boldsymbol{\gamma}$**A** on the cross-slip plane. Another mechanism[43] simply involves a pileup on the primary glide plane, producing a large enough stress concentration to activate a Frank-Read source on the cross-slip plane.

Schoeck and Seeger[44] proposed a cross-slip model with the configuration of Fig. 22-18*b* assumed to be the critical configuration of maximum energy in the

[41] A. N. Stroh, *Proc. Phys. Soc.*, **B67**: 427 (1954).

[42] P. B. Hirsch, *Phil. Mag.*, **7**: 67 (1962).

[43] D. H. Avery and W. A. Backofen, *Trans. Met. Soc. AIME*, **227**: 835 (1963).

[44] G. Schoeck and A. Seeger, in "Report of the Conference on Defects in Solids," Physical Society, London, 1955, p. 340.

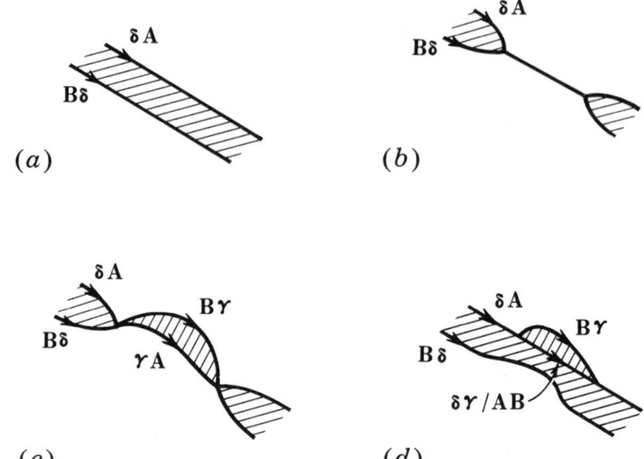

FIGURE 22-18. Stages (*a*) to (*c*) in cross slip of a straight screw dislocation. (*d*) Alternative intermediate configuration to that of (*b*).

cross-slip process. The basis for this assumption is the work of Donth[45] showing that the most likely activation path is not in general the minimum-energy saddle-point path. However, Lothe,[46,47] showed that dislocation activation processes follow the usual Einsteinian model in which the saddle-point configuration is the likely activation path. Thus, as first suggested by Friedel,[48] and developed by Escaig,[49] the actual maximum-energy configuration should resemble Fig. 22-18*c*. Indeed, under large internal stress concentrations and for widely dissociated dislocations, the configuration of Fig. 22-18*d* is a possible maximum-energy array. The energies of the configurations in Figs. 22-18*c* and 22-18*d* are difficult to compute for the reasons discussed in Sec. 22-2, and have yet to be determined. Experimental data[50] developed to test cross-slip models by biasing the stress on the cross-slip plane to form the configurations in Figs. 10-39 and 22-18*c* favor the Friedel-Escaig model. The model in Fig. 22-18*b* would be possible if it represented a saddle point to form an intermediate configuration prior to that of Fig. 22-18*c*, as might be possible for the fractional extensions for bcc metals, (Sec. 11-3), but it is not relevant for the fcc case.

[45] H. Donth, *Z. Phys.*, **149**: 111 (1957).

[46] J. Lothe, *Z. Phys.*, **157**: 457 (1960).

[47] T. Jøssang, K. Skylstad, and J. Lothe, in "Relation between Structural and Mechanical Properties of Metals," H.M. Stationery Office, London, 1963, p. 527.

[48] J. Friedel, in J. C. Fisher et al. (eds.), "Dislocations and Mechanical Properties of Crystals," Wiley, New York, 1957, p. 330.

[49] B. Escaig, *J. Phys.*, **29**: 225 (1968).

[50] B. Escaig and J. Bonneville, in P. Haasen, V. Gerold, and G. Kostorz (eds.), "Strength of Metals and Alloys," vol. 1, Pergamon, Oxford, 1980, p. 3.

In summary, there are multifarious dislocation reactions which can contribute to macroscopic deformation phenomena. Each of these reactions involves dislocation configurations whose energies are difficult to calculate. In many cases, several possible mechanisms, which cannot at present be decided among, can plausibly explain a given deformation phenomenon. Our purpose in presenting this largely qualitative discussion of these complex phenomena is to emphasize the difficulty in extending the analysis of simple dislocation reactions to the description of large numbers of dislocations and to suggest the energy calculations still needed to describe some of the proposed hardening mechanisms. Simple models can be tested by careful experiments, as for the latent hardening experiments discussed in connection with barrier formation. Also, dislocation configurations formed in easy glide in tension are becoming well defined by electron microscopy,[51] are relatively simple, and are amenable to analysis in terms of attractive-junction reactions of dislocations together with longer range dipolar and multipolar elastic interaction. Similarly, the coplanar structure in fatigue, in particular the "ladder" structure in persistent slip bands, is now well defined by electron microscopy,[52, 53] and is also amenable to analysis in terms of simple dislocation models.[54]

For other cases such as polycrystal deformation, the situation is too complicated to analyze by simple dislocation models, or, say, by Eq. (22-2). Thus, at present the average phenomenological descriptions of complex processes such as work hardening probably are most appropriate, but one must bear in mind that they are phenomenological in nature.

PROBLEMS

22-1. Compute the distribution of interaction force between two orthogonal edge dislocations, separated by a distance h, and each with its Burgers vector parallel to the line of the other dislocation.

22-2. Indicate how the two dislocations in Prob. 22-1 would relax (a) if only glide were allowed and (b) if both climb and glide could occur.

22-3. Classify those interactions in bcc crystals which yield dislocations of the $\langle 100 \rangle$ type in terms of the screw-edge character of the reacting dislocations and the line direction of the product dislocation. If the [001] dislocation is glissile on (110) and $(1\bar{1}0)$ but sessile on other planes, which of the reactions yield sessile [001] dislocations?

[51] H. Mughrabi, in R. M. Latanision and J. T. Fourie (eds.), "Surface Effects in Crystal Plasticity," Noordhoff, Leyden, 1977, p. 479.

[52] C. Laird, in A. W. Thompson (ed.), "Work Hardening in Tension and Fatigue," *Met. Soc. AIME*, Warrendale, PA., 1977, p. 150.

[53] H. Mughrabi, in P. Haasen, V. Gerold and G. Kostorz (eds.), "Strength of Metals and Alloys," Pergamon, Oxford, 1980, p. 1615.

[54] For example, see P. Neumann, *Acta. Met.*, **22**: 1155 (1974).

22-4. Consider the possibility that an intrinsically dissociated screw dislocation cross slips into an extrinsically dissociated screw through an intermediate state analogous to that of Fig. 22-18*d*. According to the Frank b^2 criterion for the energy of the partials, would this be an easier cross-slip process than that in Fig. 22-18*d* if γ_E were only slightly larger than γ_I? What reservations should be made concerning the application of the Frank criterion in this case?

22-5. Derive the extrinsically faulted barriers corresponding to barriers (1) to (4) in Fig. 22-12.

22-6. With a tensile stress along [001] in an NaCl-type crystal, verify that dislocation jogs formed by intersections between the slip systems $[\bar{1}01](101)$ and $[0\bar{1}1](011)$ are interstitial forming.

22-7. Using the criterion that the degree of operation of a slip system is proportional to the resolved shear stress on it, show that in fcc tensile tests, the formation of barriers (1) and (4) is relatively favored when the tensile axis is near [211], while that of barriers (2) and (3) is favored by orientations near [120]. In both cases the axes lie within the unit stereographic triangle [100], [110], [111].

22-8. Verify that barrier (5) in Eq. (22-5) should have an asymmetric configuration in its equilibrium form.

22-9. Which barriers in Table 22-1 can be considered to be meaningfully extended? What is the implication of your conclusion with respect to the size of pileup that a barrier can sustain if $\langle 110 \rangle \{001\}$ slip can occur easily in fcc?

22-10. Draw the product barrier of Eq. (22-6) extended to form an obtuse angle barrier. Identify the types of faults in the barrier.

BIBLIOGRAPHY

1. Basinski, Z. S., and F. Weinberg (eds.), "Conference on Deformation of Crystalline Solids," *Can. J. Phys.*, **45**: 453 (1967).
2. Haasen, P., V. Gerold, and G. Kostorz, "Strength of Metals and Alloys," vols. 1–3, Pergamon, Oxford, 1980.
3. Hirth, J. P., and J. Weertman (eds.), "Work Hardening," Gordon and Breach, New York, 1967.
4. Mecking, H., in M. F. Ashby, R. Bullough, C. S. Hartley and J. P. Hirth (eds.), "Dislocation Modeling of Physical Systems," Pergamon, Oxford, 1981, p. 197.
5. Nabarro, F. R. N., Z. S. Basinski, and D. B. Holt, *Advan. Phys.*, **13**: 193 (1964).
6. Thompson, A. W., "Work Hardening in Tension and Fatigue," *Met. Soc., AIME*, Warrendale, PA., 1977.
7. *Disc. Faraday Society*, **38** (1964).
8. "Relation between Structural and Mechanical Property of Metals," H. M. Stationery Office, London, 1963; in particular the articles by P. B. Hirsch and by A. Seeger.

23

Deformation Twinning

23-1. INTRODUCTION

Our final topic is that of deformation twinning. This subject is appropriately discussed in connection with groups of dislocations, because the tip of an advancing twin corresponds to a dislocation pileup and sources of the Frank-Read type are prominent in originating twinning dislocations.

Twinning is a particularly important deformation mechanism in crystals with only a limited number of slip systems. As discussed in Sec. 9-6, the general deformation of a polycrystal requires five independent shear systems. In crystals with only a few slip systems, twin systems can operate to provide the five systems. Thus deformation twinning is pronounced in hcp crystals. In deformation at low temperatures, twinning is also observed in higher-symmetry crystals—bcc and fcc, for example. The latter observations are attributed to a more rapid rise of the flow stress for slip with decreasing temperature than for the flow stress for twinning. A plausible explanation of this difference is that the Peierls stress for perfect dislocations increases more rapidly with decreasing temperature than that for the partial dislocations responsible for twinning. In addition to providing an alternate mechanism for shear, twins are important as barriers to the motion of slip dislocations, providing sites for pileups, and consequently for fracture nucleation and cross slip.

The coincidence lattices described in Sec. 19-6 include the commonly observed first-order twins as well as higher-order twins. The formalism presented here applies to the higher-order twins. The examples in this chapter are mainly restricted to the first-order twins. In this chapter we emphasize dislocation mechanisms for twinning. Macroscopic phenomena associated with twinning are treated elsewhere—for example, in the reviews listed in the bibliography.

The initial portion of the chapter consists of a brief review of the formal crystallography of twinning. This topic, treated extensively elsewhere,[1-3] is

[1] B. A. Bilby and A. G. Crocker, *Proc. Roy. Soc.*, **A288:** 240 (1965).

[2] R. W. Cahn, *Advan. Phys.*, **3:** 363 (1954).

[3] M. V. Klassen-Neklyudova, "Mechanical Twinning of Crystals," Consultants Bureau, New York, 1964.

outlined only in sufficient detail to introduce the concept of dislocation mechanisms of twinning, of primary interest in this book. Dislocation twinning mechanisms in bcc, hcp, and fcc crystals are then discussed in succession to illustrate various aspects of twinning theory. The chapter concludes with a discussion of the role of twinning in macroscopic deformation phenomena, including the role of twinning in low-temperature deformation and the effectiveness of twin boundaries as dislocation barriers.

23-2. CRYSTALLOGRAPHY OF TWINNING

In general, twins in crystals are of two classes, growth twins and deformation twins, which can be inherently different in nature. To illustrate this difference, let us first consider growth twins in α-quartz,[4] which has the trigonal symmetry (32). Such quartz crystals are enantiomorphic; i.e., they can exist in both right-handed and left-handed forms. In crystal growth of α-quartz, *reflection twins* sometimes form,[4] having mirror symmetry across the twin plane. The left-handed crystal comprising one half of the twin cannot be deformed or rotated to correspond with the other, right-handed crystal. Such reflection twins can form only during growth processes, such as crystal growth from solution or recrystallization.[5] In high-symmetry crystals, including most metals, reflection twins happen to correspond also to rotation or deformation twins, as discussed below.

Deformation twins formally correspond to rotation, or axial, twins. If one half of the crystal is rotated by π about an axis normal to the twin plane (type I twin) or about the shear direction in the twinning plane (type II twin), it will join with the other half to form an unfaulted single crystal. Equivalently, the twin can be formed by shear, as depicted in Fig. 23-1. The *elements* of this mode of twinning are shown in Fig. 23-2. Because the deformation is a pure shear, the plane parallel to the large twin interface remains undistorted. This plane is denoted K_1 and is called the *twin plane* or *composition plane*. The plane K_1 may or may not be the glide plane for the twinning dislocations, and must not be confused with the glide plane for the shear deformation. The plane K_2, which intersects K_1 in a line perpendicular to the shear direction η_1, and which makes equal angles with K_1 before and after the shear, is also undistorted. The shear plane is, by definition, that plane which is normal to K_1 and contains η_1. The direction η_2 lies along the line of intersection of the shear plane and K_2.

The requirement that the crystal lattice, or, less restrictively, a superlattice in the crystal, not be changed by the shear leads to the condition that either K_1 and η_2 both have rational, low-index elements (type I twin) or K_2 and η_1 both

[4]A. V. Shubnikov, cited by Klassen-Nekyludova, *ibid*., p. 87.

[5]Although such twinning occurs with macroscopic change in form, the twin can be made to grow by the application of stress. The driving force is connected with elastic anisotropy and is quadratic in the applied stress; see L. A. Thomas and W. A. Wooster, *Proc. Roy. Soc.*, **A208:** 43 (1951).

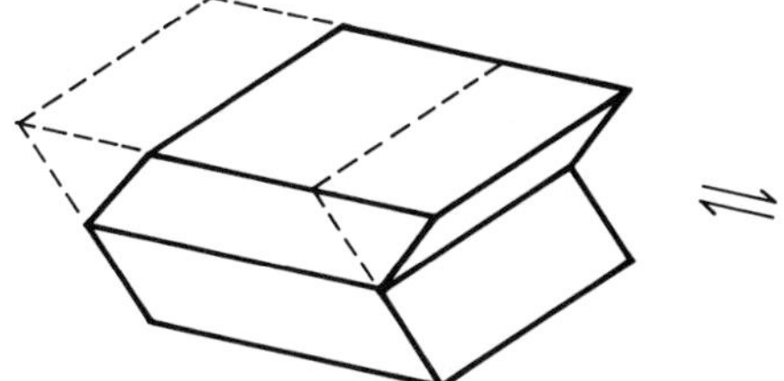

FIGURE 23-1. Formation of a twin (solid lines) from an undistorted single crystal (dashed lines) by shear.

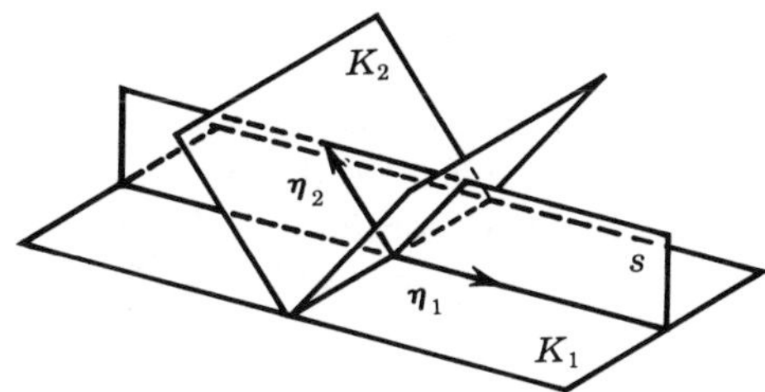

FIGURE 23-2. Twinning elements η_1 lying in K_1, η_2 lying in K_2, and the shear plane s, perpendicular to K_1 and K_2 and containing η_1 and η_2. The sheared position of K_2 after a shear η_1 on K_1 is also shown.

have rational, low-index elements (type II twin). Proofs of this condition are given by Cahn[6] and by Bilby and Crocker.[7] If all four indices are rational, the twin is called a *compound twin*. The twin orientation can then be considered as resulting from either a 180° rotation about the normal to K_1 or a 180° rotation about η_1. Most twins in the simple metals are of the compound type.

The elements K_1 and η_2 suffice to define a twin, but usually all four elements are cited in the order K_1, K_2, η_1, and η_2. Corresponding to this twin, there is a different possible twin defined by K_1', K_2', η_1', and η_2', where $K_1' = K_2$, $K_2' = K_1$, $\eta_1' = \eta_2$, and $\eta_2' = \eta_1$, which is said to be *reciprocal* to the original twin. For many twinning systems in the simple metals, K_1 and K_2 are equivalent planes and η_1 and η_2 are equivalent directions. A twin and its reciprocal are then equivalent, but differently oriented with respect to the parent crystal. In general, the reciprocal twin to a type I twin is a type II twin.

Let us now relate the above formalism to dislocation models of twinning. Initially, we consider in only a general way those dislocation models that can account for the macroscopic deformation accompanying twinning, i.e., the change in external shape and superlattice orientation. The problem of the correct positioning of *all* atoms in the twin by shuffling is returned to later. Obviously, the shear of Fig. 23-1 can be accomplished by the successive motion of twinning dislocations, each on a glide plane one interplanar spacing

[6] R. W. Cahn, *Advan. Phys.*, **3:** 363 (1954).

[7] B. A. Bilby and A. G. Crocker, *Proc. Roy. Soc.*, **A288:** 240 (1965).

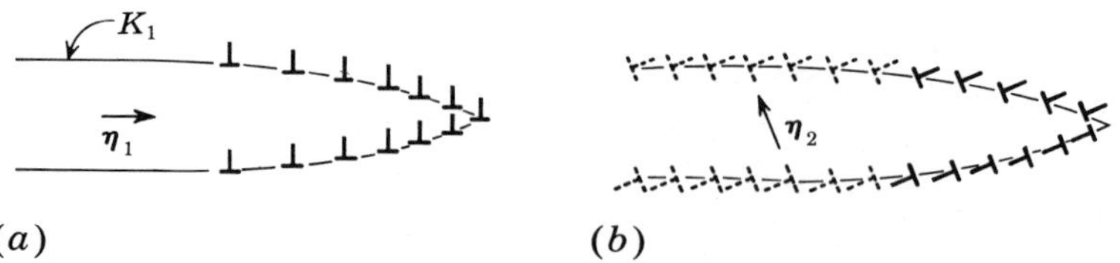

FIGURE 23-3. (*a*) Twin of type I formed by dislocations with **b** parallel to $\boldsymbol{\eta}_1$ gliding on K_1. (*b*) The same twin formed by dislocations with **b** parallel to $\boldsymbol{\eta}_2$ gliding on K_2. The dashed dislocations indicate that the coherent interface parallel to K_1 can be regarded as a special dislocation boundary.

removed from its predecessor.[8] In most metals the twins are compound. Thus they can be formed either by glide of dislocations on K_1 with the rational vector $\boldsymbol{\eta}_1$ parallel to the twinning-dislocation Burgers vector (Fig. 23-3*a*) or by glide on K_2 of dislocations with Burgers vectors parallel to $\boldsymbol{\eta}_2$ (Fig. 23-3*b*).[9] In the second mechanism the dislocation boundaries parallel to K_1 actually describe the coherent twin interface in a formal way. Thus, if a crystal twins by the latter mechanism, it not only shears in the direction $\boldsymbol{\eta}_2$, but it also rotates about an axis perpendicular to the shear plane by an amount determined by the angle of tilt of the tilt boundaries. Only the shear is depicted in Fig. 23-3*b*, the rotation associated with the dashed dislocations comprising the tilt boundaries rotates the vector $\boldsymbol{\eta}_2$ to the right as in Fig. 23-2. The overall effect is shear in the $\boldsymbol{\eta}_1$ direction, as for the first mechanism. The two twinning modes are alternative ways of forming the *same twin*. The real difference between the models lies in the description of the incoherent leading tip of the twin and its mode of motion. Whether the coherent interface is formally described as a dislocation boundary or not is an academic question; the coherent interface is physically the same interface in both models. The configuration of Fig. 23-3*a* is the one supported generally by experimental evidence, but some evidence for that of Fig. 23-3*b* exists.[10]

Type I twins can be described in general by the motion of two sets of glide dislocations on K_1. If the number of dislocations in the incoherent twin interface is N_1 and N_2, respectively, for sets with Burgers vectors $\mathbf{b}_1$ and $\mathbf{b}_2$ (Fig. 23-4*a*), the overall shear direction is given by

$$\boldsymbol{\eta}_1 \parallel N_1\mathbf{b}_1 + N_2\mathbf{b}_2 \tag{23-1}$$

In any case where K_1 is rational, two nonparallel partial-dislocation Burgers vectors $\mathbf{b}_1$ and $\mathbf{b}_2$, which can produce a given fault displacement, always exist (Fig. 23-4*b*). Thus $\boldsymbol{\eta}_1$ can be in a direction with very high-index, essentially

[8] Twinning was first described in terms of dislocations by J. Frenkel and T. Kontorova [*J. Phys. Moscow*, **1:** 137 (1939)].

[9] R. Bullough, *Proc. Roy. Soc.*, **A241:** 568 (1957).

[10] See the review by R. W. Cahn, in R. E. Reed-Hill, J. P. Hirth, and H. C. Rogers (eds.), "Deformation Twinning," Gordon and Breach, New York, 1963, p. 1.

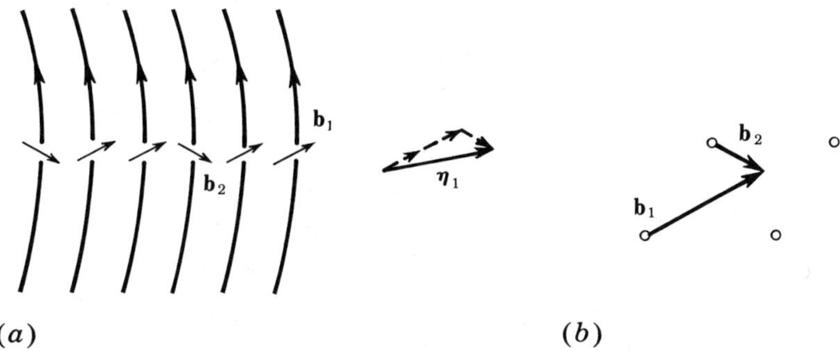

FIGURE 23-4. (*a*) View normal to the twinning plane, showing a twin shear $\boldsymbol{\eta}_1$ produced by two sets of dislocations $\mathbf{b}_1$ and $\mathbf{b}_2$. (*b*) Basal plane of a triclinic crystal, showing two possible twinning dislocation vectors.

irrational, elements. Alternatively, type I twins can form by motion of a single set of dislocations with Burgers vectors parallel to the low-index direction $\boldsymbol{\eta}_2$. In this case, the screw-twin-dislocation segments cross slip onto the high-index plane K_2, and the edge segments are highly jogged, on the average lying on K_2. The final configuration resembles Fig. 23-3*b*.

Simply applying the above reasoning to the reciprocal twin mode K_1', K_2', $\boldsymbol{\eta}_1'$, and $\boldsymbol{\eta}_2'$, one sees that type II twins can form by cross slipping dislocations $\mathbf{b} \| \boldsymbol{\eta}_1'$ on K_1' or by the operation of two slip systems on K_2'. In either case, the high-index twin boundary K_1' is expected to be a higher-energy boundary than the boundary K_1 in type I twins. This is one probable reason for the rare observation of type II twins.[11]

In some cases the motion of single twinning dislocations translates some but not all of the matrix atoms into the twin orientation. The remainder of the atoms then move into the correct twin orientation by shuffling.[12] The total motion, shear plus shuffling, is best understood in the dislocation context by the consideration of *zonal dislocations*,[13] introduced in connection with hcp partials in Sec. 11-2. The prediction of the crystallography of possible shuffles for a given twin has been accomplished by Bilby and Crocker.[14]

Based upon the above models, crystallographic theories for the prediction of possible twinning modes have been developed by Kiho,[15] Jaswon and Dove,[16] and in a quite general manner by Bilby and Crocker.[17] These models are based

[11] One example of type II twinning in uranium has been found by R. W. Cahn [*Acta Met.*, **1:** 49 (1953)].

[12] P. Niggli, *Z. Krist.*, **71:** 413 (1929).

[13] M. L. Kronberg, *Acta Met.*, **9:** 970 (1961); *J. Nucl. Mater.*, **1:** 85 (1959).

[14] B. A. Bilby and A. G. Crocker, *Proc. Roy. Soc.*, **A288:** 240 (1965).

[15] H. Kiho, *J. Phys. Soc.* (*Japan*), **9:** 739 (1954); **13:** 269 (1958).

[16] M. A. Jaswon and D. B. Dove, *Acta Cryst.*, **9:** 621 (1956); **10:** 14 (1957); **13:** 232 (1960).

[17] B. A. Bilby and A. G. Crocker, *Op. Cit.*

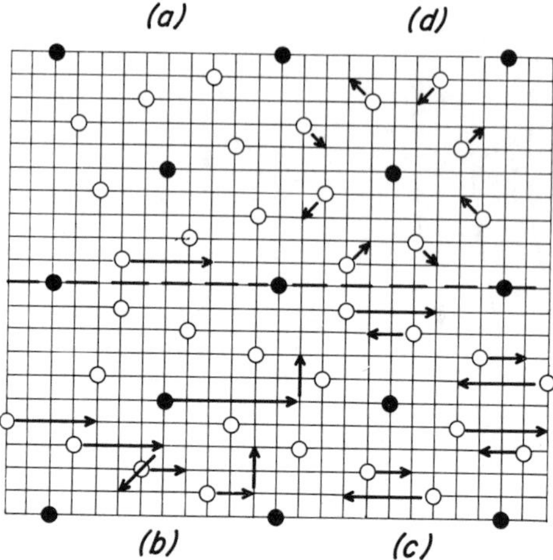

FIGURE 23-5. Projection of a simple cubic crystal along $[00\bar{1}]$, parallel to a (310) twin boundary. Grid is DSC lattice and coincidence lattice is illustrated by solid balls. Arrows represent four ways in which twin boundary could be translated normal to itself by ten DSC lattice planes: (*a*) pure glide, (*b*) glide plus shuffle, (*c*) pure shuffle parallel to boundary, and (*d*) pure shuffle.

on the hypotheses that (1) the twinning shear should be small (low-accommodation strain, low Peierls stress for a given dislocation), (2) the shuffle mechanism should be simple and the shuffle magnitudes small (minimization of distortion of atomic bonds), and (3) shuffles should be parallel to the twinning shear direction (minimization of bond breaking). The reader is referred to the formal theories for further discussion of the crystallography of twinning modes.

Wasilewski[18] has suggested that twinning could proceed by pure shuffle and observed twins in bcc crystals forming without a shear. Hirth[19] noted that couple stresses (nonlinear stresses associated with rotations; see Sec. 2-2) could provide a driving force for twinning by pure shuffle. The various possible combinations of shuffle and glide are illustrated in Fig. 23-5. Essentially pure shuffle models have also been developed for hcp twins.[20]

Dislocation configurations exist that are closely related to twins but are not formally twins. Examples of arrays resembling those of Fig. 23-3*b* have been found[21] in which the misorientation across the boundary varies along the boundary. These boundaries constitute *kink bands*, which are sheared regions that can be bounded by partial dislocations or perfect dislocations irregularly spaced, in an array similar to Fig. 23-3*b*.[22] While such arrays can form by

[18] R. J. Wasilewski, *Met. Trans.*, **1:** 1617 (1970).

[19] J. P. Hirth, *Acta Met.*, **22:** 1023 (1974).

[20] A. Doubertret and A. LeLann, *Phys. Stat. Solidi*, **60a:** 145 (1980).

[21] E. J. Freise and A. Kelly, *Proc. Roy. Soc.*, **A264:** 269 (1961).

[22] R. W. Cahn, in R. E. Reed-Hill, J. P. Hirth, and H. C. Rogers (eds.), "Deformation Twinning," Gordon and Breach, New York, 1963, p. 1.

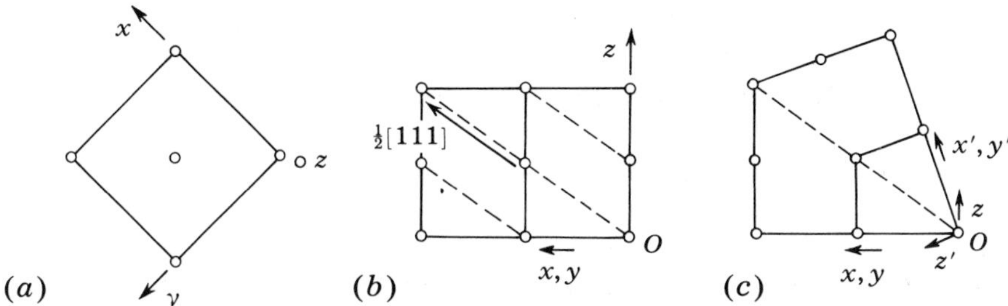

FIGURE 23-6. (*a*) Projection of bcc lattice on (001) and on $(\bar{1}10)$. The latter projection is viewed parallel to the twinning plane $(\bar{1}\bar{1}2)$. (*b*) $[111](\bar{1}\bar{1}2)$ twin formed by a rotation of π about [111].

mechanisms like those of twin formation, the arrays are not discussed further here.

The orientation relationships between parent and twin are of interest in developing dislocation interactions. Let us consider the bcc twin $\boldsymbol{\eta}_1=[111]$ and $K_1=(\bar{1}\bar{1}2)$ as an example. The twin can formally be produced by a rotation of π about the [111] twin direction (Fig. 23-6). The transformation matrix can be developed easily by decomposing the unit vectors in the parent crystal into components normal to (111) and lying in (111):

$$\mathbf{i}=[100]=\tfrac{1}{3}[111]+\tfrac{1}{3}[2\bar{1}\bar{1}] \qquad \mathbf{j}=[010]=\tfrac{1}{3}[111]+\tfrac{1}{3}[\bar{1}2\bar{1}]$$
$$\mathbf{k}=[001]=\tfrac{1}{3}[111]+\tfrac{1}{3}[\bar{1}\bar{1}2] \tag{23-2}$$

These vectors can be expressed in the twin indices by changing the sign of components lying in the (111) plane:

$$\mathbf{i}'=\tfrac{1}{3}[111]_t+\tfrac{1}{3}[\bar{2}11]_t \qquad \mathbf{j}'=\tfrac{1}{3}[111]_t+\tfrac{1}{3}[1\bar{2}1]_t$$
$$\mathbf{k}'=\tfrac{1}{3}[111]_t+\tfrac{1}{3}[11\bar{2}]_t \tag{23-3}$$

The transformation matrix from parent to twin notation is then simply given by vector multiplication

$$\mathbf{i}'=(\mathbf{i}'\cdot\mathbf{i})\mathbf{i}+(\mathbf{i}'\cdot\mathbf{j})\mathbf{j}+(\mathbf{i}'\cdot\mathbf{k})\mathbf{k}, \qquad \text{etc.,}$$

yielding the transformation matrix

$$\mathbf{T}=\tfrac{1}{3}\begin{pmatrix} -1 & 2 & 2 \\ 2 & -1 & 2 \\ 2 & 2 & -1 \end{pmatrix} \tag{23-4}$$

This matrix, of course, applies also to the transformation from twin to parent coordinates, since $\mathbf{T}\cdot\mathbf{T}=\mathbf{T}\cdot\mathbf{T}^{-1}=\mathbf{I}$, see Eq. (4-31).

23-3. POLE MECHANISMS FOR TWINNING

The Cottrell-Bilby Pole Mechanism in BCC Crystals

One prominent class of dislocation sources for twinning dislocations is that of pole mechanisms, which produce spiral dislocations, as illustrated in Fig. 20-4. Let us first consider bcc crystals. Evidently, if the spiral dislocation $\mathbf{b}=\frac{1}{6}\langle 111\rangle$ has a pitch equal to the interplanar spacing of $\{112\}$ planes and moves in a direction to produce fault I_1, a twin will be formed, according to Eq. (11-8). As discussed in Sec. 11-3, dissociations to form I_1 faults or the somewhat related I_3 faults are most important in deformation twinning although I_2 faults could have some role.

Cottrell and Bilby[23] proposed a pole mechanism that fulfills the above requirements for bcc crystals. Their model is illustrated in Fig. 23-7*a*. A $\frac{1}{2}[\bar{1}\bar{1}\bar{1}]$ dislocation lying on (112) dissociates by the reaction

$$\tfrac{1}{2}[\bar{1}\bar{1}\bar{1}]\rightarrow\tfrac{1}{3}[\bar{1}\bar{1}\bar{2}]+\tfrac{1}{6}[\bar{1}\bar{1}1] \tag{23-5}$$

which involves a zero change in elastic energy to first order, but that can occur under bow-out stresses. The partial $\frac{1}{6}[\bar{1}\bar{1}1]$ cross slips onto $(1\bar{2}\bar{1})$, as shown in Fig. 23-7*a*. The dissociation is such that I_1 faults are formed on both (112) and $(1\bar{2}\bar{1})$, according to Axiom 11-2. The partial can then wind around the $\frac{1}{3}[\bar{1}\bar{1}\bar{2}]$ pole dislocation. The pole dislocation has a right-handed screw component normal to $(1\bar{2}\bar{1})$ and equal to the interplanar spacing of $(1\bar{2}\bar{1})$ planes. Thus the partial forms a spiral such as that in Fig. 20-4, and generates a twin with $K_1=(1\bar{2}\bar{1})$, $\eta_1=[\bar{1}\bar{1}1]$. Note that the stacking fault on (112) involves a relative shear parallel to the Burgers vector of the rotating twinning dislocation, but in the sense opposite to that in Fig. 23-7*a*, so that the fault is an I_2 fault. Whether it acts as an impediment to twin growth depends on inertial and damping effects, Chap. 7. Sleeswyk[24] suggests that the fault is removed by the nucleation of a $\frac{1}{6}[\bar{1}\bar{1}1]$ partial, which also annhilates the twinning partial, however, thus stopping the twinning.

The Cottrell-Bilby model can also be considered as a ratchet mechanism, as shown in Fig. 23-7*b*. This variation is of interest in introducing twinning in fcc crystals and exemplifies the role of superjogs as twinning sources. Suppose that a superjog has formed on the $\frac{1}{2}[\bar{1}\bar{1}\bar{1}]$ dislocation because of the accumulation of intersection jogs formed by interaction with $\frac{1}{2}[11\bar{1}]$ dislocations. A layer of I_1 fault can then form on $(1\bar{2}\bar{1})$ by reaction (23-5), as shown in Fig. 23-7*b*. However, as illustrated in Fig. 23-8, this configuration cannot directly wind into a spiral dislocation. Figure 23-8*a* is a cross section of the undissociated superjog, while Fig. 23-8*b* shows the dissociation of Fig. 23-7*b* producing fault I_1. The portion of the partial $\frac{1}{6}[\bar{1}\bar{1}1]$ that winds around pole *B* in Fig. 23-7*b*

[23]A. H. Cottrell and B. A. Bilby, *Phil. Mag.*, **42:** 573 (1951).

[24]A. W. Sleeswyk, *Phil. Mag.*, **29:** 407 (1974).

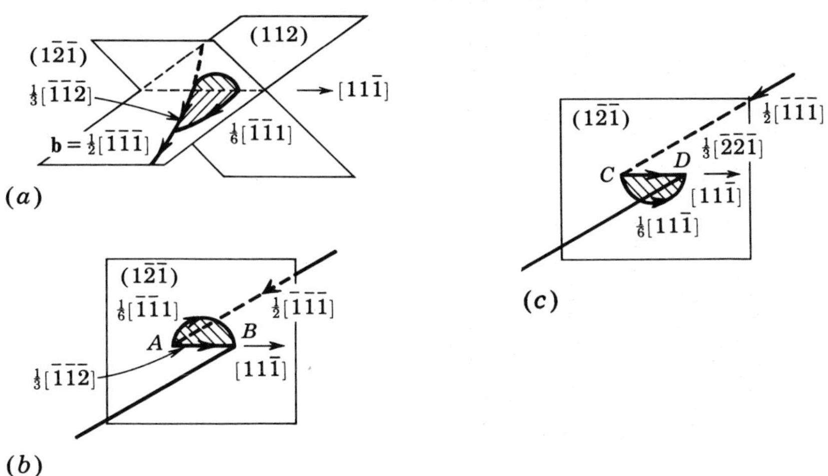

FIGURE 23-7. (*a*) The Cottrell-Bilby ratchet-pole mechanism in a bcc crystal. (*b*) Jog version of (*a*). (*c*) Alternate pole mechanism.

leaves along path 1 in Fig. 23-8*b* and returns along path 3; that which winds around pole *A* leaves along path 1 and returns along path 2. In either case, the returning partial is blocked when it recombines with the sessile partial $\frac{1}{3}[\bar{1}\bar{1}\bar{2}]$; further motion would take the partial $\frac{1}{6}[\bar{1}\bar{1}1]$ back along path 1 and would produce the fault I_2. In order to continue to form a twin, the recombined total $\frac{1}{2}[\bar{1}\bar{1}\bar{1}]$ dislocation must cross slip by one interplanar $(\bar{1}21)$ distance, as shown in Fig. 23-8*b*, so that it can again dissociate by reaction (23-5) and send a partial out along path 4. Thus a twin is formed by successive ratchet steps of a twin loop forming and the recombined perfect dislocation cross slipping. The equivalency of the mechanisms of Figs. 23-7*a* and 23-7*b* is obvious from the figure.

A pure pole mechanism for the same superjog is illustrated in Fig. 23-7*c*. The twin partial $\frac{1}{6}[11\bar{1}]$ is opposite in sign to that of Fig. 23-7*b*, but it moves in the opposite direction on $(1\bar{2}\bar{1})$, so that an I_1 fault is again formed, Eq. (11-8). The cross section of Fig. 23-7*c* is shown in Fig. 23-8*c*. The partial moving about pole *C* leaves along path 1 and returns along path 3; it is not blocked, but can simply move on along path 5, propagating a spiral source and continuing to form a twin. Similarly, the partial moving about pole *D* leaves along path 1, returns along path 2, leaves along path 4, and also propagates a twin. The dissociation reaction in this case is

$$\tfrac{1}{2}[\bar{1}\bar{1}\bar{1}] \rightarrow \tfrac{1}{3}[\bar{2}\bar{2}\bar{1}] + \tfrac{1}{6}[11\bar{1}] \tag{23-6}$$

with $\frac{1}{3}[\bar{2}\bar{2}\bar{1}]$ transforming to $[001]_t$ in the twin. This reaction is unfavorable energetically, however. The sessile partial is higher in energy than that of reaction (23-5). Thus, while propagation of reaction (23-6) is perhaps easier than that of (23-5), because no recombination and cross slip is required upon

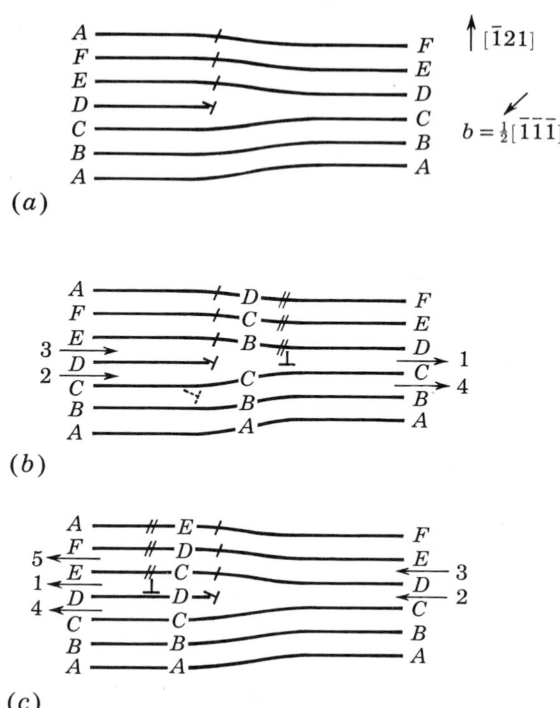

FIGURE 23-8. Cross sections of (a) the unreacted dislocation in Fig. 23-7b, (b) the configuration of Fig. 23-7b, and (c) the configuration of Fig. 23-7c. The vector $[11\bar{1}]$ points normally into the plane of the paper.

each revolution of the partial, the nucleation of the former is much less likely. Only in the presence of very large bow-out stresses favoring dissociation (23-6) is its operation likely.

The pole in Fig. 23-8c is a perfect one in the sense described by Sleeswyk.[24] He suggests an alternative that can also be described by Fig. 23-7c with the replacements $\frac{1}{2}[\bar{1}\bar{1}\bar{1}] \rightarrow [00\bar{1}]$, $\frac{1}{6}[11\bar{1}] \rightarrow 2\times\frac{1}{6}[11\bar{1}]$, and $\frac{1}{3}[\bar{2}\bar{2}\bar{1}] \rightarrow \frac{1}{3}[\bar{1}\bar{1}\bar{2}]$. The two glissile partials leave along paths 1 and 4 in Fig. 26-6c, two free partials return along paths 3 and 6 and can propagate the twin, and the other partials recombine and return along path 2, reacting by $\frac{1}{3}[\bar{1}\bar{1}\bar{2}] + \frac{1}{6}[11\bar{1}] \rightarrow \frac{1}{6}[\bar{1}\bar{1}\bar{5}]$, which is the perfect dislocation $\frac{1}{2}[111]_t$ in the twin. Thus this pole is also a perfect one. The dissociation to form the I_3 fault of Fig. 11-14 is equivalent to a simple dissociation as in Figs. 23-7 and 23-8 plus dipolar pairs. Thus the ratchet or pure pole models in these illustrations can be converted to versions initially involving the I_3 fault by adding the dipolar pairs in the initial dissociation. The consequences after the first revolution about the poles are similar to those for the Sleeswyk case. The minimum stress required to nucleate twinning by the above mechanisms is that required to supply the fault energy for the first turn of the source,

$$\sigma = \frac{\gamma}{b} \tag{23-7}$$

For iron with the fault energy $\gamma \sim 200$ mJ/m^2, this gives $\sigma = 0.02\mu$. For the mechanism of Figs. 23-7*a* or 23-7*b*, work must also be supplied to bow-out the original loop. Once the first loop has formed, no new fault is formed as the dislocation moves, and only the elastic energy need be supplied. For the mechanism of Fig. 23-7*c* to nucleate, the stress must overcome the attractive force between the partials $\frac{1}{3}[\bar{2}\bar{2}\bar{1}]$ and $\frac{1}{6}[11\bar{1}]$; at $r = b$ this requires a stress $\sigma = 0.8\mu$ in addition to the stress required to produce the fault. In this case, an interaction stress between the moving partial and the sessile $\frac{1}{3}[\bar{2}\bar{2}\bar{1}]$ partial, decreasing in magnitude with each turn, would have to be overcome on each successive turn. Actual applied stresses at which twinning occurs are typically[25, 26] about $10^{-3}\mu$, but stress concentrations appear to be required to nucleate the twins.[26] Thus all one can say at present is that the above critical stresses for operation of the pole mechanisms can plausibly be attained in experiment, so that such mechanisms are possible. There have been no direct observations of the operation of pole sources.

Crystallographically, the elements of twinning in bcc crystals have been found to be $K_1 = (112)$, $\boldsymbol{\eta}_1 = [11\bar{1}]$, $K_2 = (11\bar{2})$, $\boldsymbol{\eta}_2 = [111]$. This bcc twinning mode thus exemplifies the high symmetry of twinning in simple metals; the reciprocal twins are equivalent in this system. Paxton[27] has verified that the shear deformation accompanying twinning in α-iron corresponds exactly to that expected if a $\frac{1}{6}\langle 111 \rangle$ dislocation moves on each successive $\{112\}$ plane, as predicted by the twinning model.

Exercise 23-1. Compute the ideal twinning shear expected if a set of $\frac{1}{6}[111]$ dislocations produces a twin on $(11\bar{2})$.

Poles in HCP Crystals

Thompson and Millard[28] presented the first hcp pole mechanism for $\{10\bar{1}2\}$ twinning in hcp crystals. Since we have already described the zonal partial dislocation for $\{11\bar{2}2\}$ twinning in Chap. 11, however, we treat this case first. Figure 11-9(c) shows one layer of twin formed by motion of the zonal partial dislocation $\sim \frac{1}{4}\mathbf{B'C}$ and shows the shuffling that accompanies zonal dislocation motion. Clearly, a pole mechanism must require a Burgers vector component

[25] J. J. Cox, R. F. Mehl, and G. T. Horne, *Trans. ASM*, **49:** 118 (1957).

[26] R. W. Cahn, in R. E. Reed-Hill, J. P. Hirth, and H. C. Rogers (eds.), "Deformation Twinning," Gordon and Breach, New York, 1963, p. 1.

[27] H. W. Paxton, *Acta Met.*, **1:** 141 (1953).

[28] N. Thompson and D. J. Millard, *Phil. Mag.*, **43:** 422 (1952).

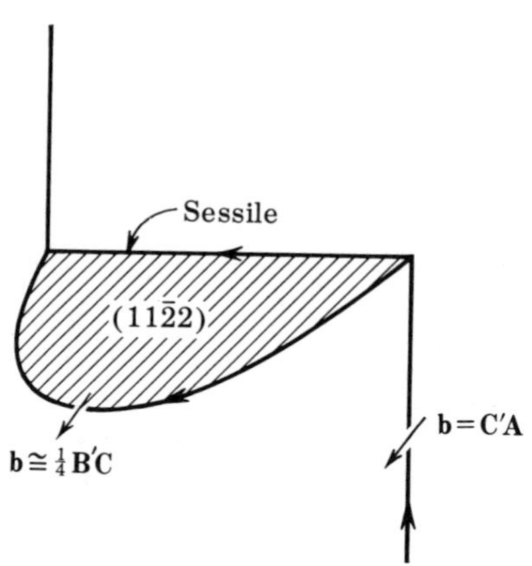

FIGURE 23-9. Pole mechanism for [11$\bar{2}\bar{3}$] (11$\bar{2}$2) twinning in an hcp crystal.

of the pole dislocation equal to three interplanar (11$\bar{2}$2) planes for the twin of Fig. 11-9(c) to be propagated. A pole that meets these requirements is depicted in Fig. 23-9. The perfect dislocation **C′A**, glissile on (1$\bar{2}$12), dissociates into a zonal partial dislocation $\sim\frac{1}{4}$**B′C** and a sessile dislocation on (11$\bar{2}$2). **C′A** has a left-handed screw component normal to (11$\bar{2}$2) and equal to three interplanar (11$\bar{2}$2) spacings in magnitude. Thus the requirements for a pole are satisfied. The complete reaction is

$$\mathbf{C'A} \rightarrow \sim\tfrac{1}{4}\mathbf{B'C} + \text{sessile} \tag{23-8}$$

or

$$\tfrac{1}{3}[1\bar{2}1\bar{3}] \rightarrow \sim\tfrac{1}{12}[11\bar{2}\bar{3}] + \sim\tfrac{1}{4}[1\bar{3}2\bar{3}]$$

The more prominent case of twinning in hcp crystals is the system $K_1 =$ (10$\bar{1}$2), $\boldsymbol{\eta}_1 = [\bar{1}011]$, $K_2 = (10\bar{1}\bar{2})$, and $\boldsymbol{\eta}_2 = [10\bar{1}1]$, in which the reciprocal twin is equivalent. The zonal dislocation in this case[29, 30] is two "rumpled" (10$\bar{1}$2) planes thick, as shown in Fig. 23-10. The pole mechanism is illustrated in Fig. 23-10b. Actually, this is a ratchet pole, because the glissile partial recombines with the sessile partial after one revolution, and the recombined perfect [000$\bar{1}$] dislocation must cross slip before twin propagation can occur, analogous to the case of Fig. 23-8b. The reaction is

$$[000\bar{1}] \rightarrow \alpha[10\bar{1}\bar{1}] + \text{sessile} \tag{23-9}$$

where α is a small fraction, roughly $\frac{1}{12}$. A continuing pole not requiring cross-slip ratcheting could be formed by the reaction

$$[0001] \rightarrow \alpha[10\bar{1}\bar{1}] + \text{sessile}' \tag{23-10}$$

[29] N. Thompson and D. J. Millard, *ibid.*

[30] D. G. Westlake, *Acta Met.*, **9:** 327 (1961).

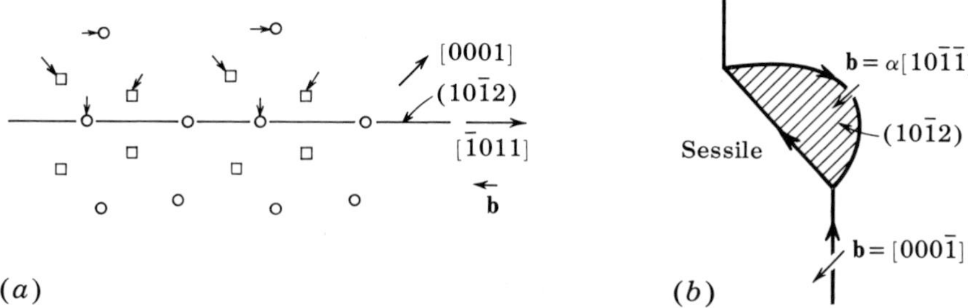

FIGURE 23-10. (*a*) Atom movements for [$\bar{1}$011](10$\bar{1}$2) twinning in an hcp crystal. (*b*) Ratchet-pole mechanism for the twin in (*a*).

but this reaction is elastically unfavorable because of the high-energy, sessile partial.

Pole sources have not been directly observed for hcp metals. However, the twinning shear does correspond to the magnitude expected if twinning dislocations of the type $\alpha[10\bar{1}\bar{1}]$ are operative.[31]

FCC Pole Mechanisms

Pole mechanisms for twinning in fcc crystals can be understood directly by analogy with the bcc and hcp cases. We present the fcc examples principally because there is unequivocally dislocation extension in fcc crystals, so that extended superjog sources are definitely stable in fcc crystals, whereas their stability is moot in the other crystal systems.

Examples of pole sources for twinning on the primary glide plane (*a*) are shown in Fig. 23-11. Suppose that a single crystal under tension is oriented such that **BC**(*a*) is the primary glide system; twins are observed experimentally on plane (*a*) for orientations of the tensile axis near [111].[32] The mechanism corresponding to the ratchet mechanism was proposed by Venables[33] and is shown in Figs. 23-11*c* and 23-11*d* for undissociated and dissociated superjogs, respectively, lying along **DB**. The pole mechanism not requiring ratcheting was proposed by Hirth[34] and is shown in Figs. 23-11*a* and 23-11*b*. The extended superjogs in Fig. 23-11 are those previously shown in Fig. 10-19 and in configuration (4) of Fig. 16-19. Venables' jog has the lower-energy stair-rod partials of the two, so that it should be extended more at equilibrium. On the other hand, the forces produced by the applied stress, for orientations near

[31] For a review see the article by H. S. Rosenbaum in R. E. Reed-Hill, J. P. Hirth, and H. C. Rogers (eds.), "Deformation Twinning," Gordon and Breach, New York, 1964, p. 43.

[32] See the review by J. A. Venables, ibid., p. 77.

[33] J. A. Venables, *Phil. Mag.*, **6**: 379 (1961).

[34] J. P. Hirth, in R. E. Reed-Hill, J. P. Hirth, and H. C. Rogers, (eds.), "Deformation Twinning," Gordon and Breach, New York, 1964, p. 112.

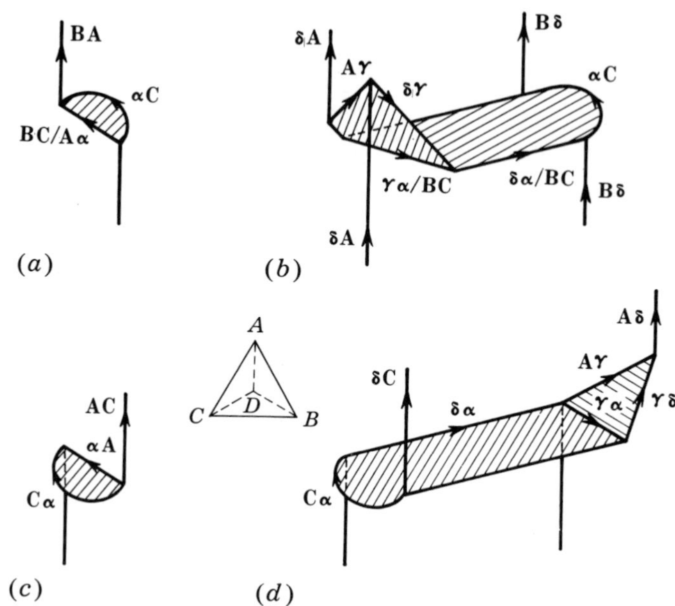

FIGURE 23-11. Pure pole mechanism for twinning in fcc crystals nucleated at (*a*) an undissociated superjog and (*b*) a dissociated superjog. (*c*) and (*d*) are equivalent examples for the ratchet-pole mechanism.

[111], aid the dissociation of the jog proposed by Hirth, but not that of Venables. Hence both are plausible nuclei for pole twin sources. Sleeswyk[35] shows that a perfect pole, in the sense of Fig. 23-11*a*, can also form from a threefold node in the method similar to that illustrated in Exercise 20-1. The difference is that the rotating arm dissociates into partials by the reaction shown in Fig. 10-39, with one partial rotating up one pole and the other down the other pole. He uses a different notation, but the pole left in the twin is identical to **BC/Aα** in Fig. 23-11*a*, $\frac{1}{6}[\bar{4}\bar{1}1]$, in Miller indices, which becomes $\frac{1}{2}[10\bar{1}]_t$ in the twin. Thus his result is a double-ended version of Fig. 23-11*a*.

If nucleated, the growth of the twins by the two mechanisms resembles that of the bcc ratchet-and-pole pair. In the ratchet case, the glissile and sessile partials must recombine and cross slip before a new layer of twin can propagate; in some stress distributions this step can be energetically unfavorable. In the pure pole case, the partials moving around the two pole branches must swing past one another and must overcome the attractive interaction energy between them; the problem resembles that discussed in Sec. 20-7. Which mechanism, the ratchet or the pure pole, is favored depends on the local stress distribution and on complicated calculations of bow-out energies.

[35] A. Sleeswyk, *Phil. Mag.*, **29:** 407 (1974).

[36] T. H. Blewitt, R. R. Coltman, and J. K. Redman, *J. Appl. Phys.*, **28:** 651 (1957).

Twinning in fcc crystals provides some of the strongest, though indirect, support for pole mechanisms. Blewitt et al.[36] have found the twin shear in copper to be consistent with motion of a single set of $\frac{1}{6}\langle 112\rangle$ partials on $\{111\}$ planes. This finding supports a pole mechanism rather than a nucleation mechanism of twinning, because in the latter case one would expect alternate nucleation of all three possible $\frac{1}{6}\langle 112\rangle$ partials within a given $\{111\}$ plane in order to minimize the elastic strain field of the incoherent twin interface.

Exercise 23-2. By considering the screw components of the pole dislocations relative to the twin plane, verify that Fig. 23-11*a* represents a pure pole, while Fig. 23-11*c* represents a ratchet. Draw the planar cross sections analogous to those in Fig. 23-8 and verify (*a*) that the faults in Fig. 23-11 are intrinsic and (*b*) that the ratchet pole of Fig. 23-11*c* cannot continue to operate without cross slip.

23-4. OTHER TWIN-SOURCE MECHANISMS

Nucleation

Homogeneous nucleation of the first loop of partial dislocation in twin formation would require the stresses listed in Table 20-1, $\sigma \cong 0.05$ to 0.1μ. As also shown in Table 20-1, nucleation at surfaces could occur at stresses down to $\sigma \cong 0.01\mu$. In either case, nucleation of the first loop appears likely only in the presence of stress concentrations. Once the first loop has nucleated, however, nucleation of succeeding loops propagates the coherent twin interface but does not create new area of twin interface (Fig. 23-12). Thus the term $\pi r^2\gamma$ drops out of Eq. (20-8) for these later loops, and the critical stress is lowered. For the example of copper at 300°K, the critical stress is lowered to $\sigma \sim 0.03\mu$.

Price[37] studied the formation of twins in dislocation-free platelets of zinc. He found by electron microscopy that twins nucleated at surfaces, particularly at sites where stress raisers such as etch pits were present. He found that nucleation occurred at a stress of $\sigma \sim 0.02\mu$, in reasonable agreement with the above predictions from nucleation theory. Hull[38] observed nucleation of twins near free surfaces and at grain boundaries in an Mo-35% Re alloy. In one case he actually observed the twinning dislocations emanating from a source near the surface.[39]

Thus definite evidence for nucleation sources of twinning exists. Whether nucleation supplants all other mechanisms in all cases remains conjectural and seems unlikely.

[37] P. B. Price, *Proc. Roy. Soc.*, **A260:** 251 (1961).

[38] D. Hull, in R. E. Reed-Hill, J. P. Hirth, and H. C. Rogers (eds.), "Deformation Twinning," Gordon and Breach, New York, 1964, p. 121.

[39] *Ibid.*, p. 135.

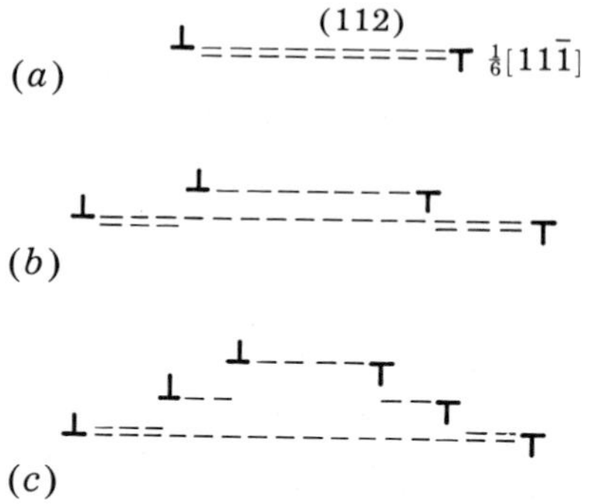

(112)

(a) $\frac{1}{6}[11\bar{1}]$

(b)

(c)

FIGURE 23-12. Cross sections of successive stages in twin formation by dislocation nucleation. Double-dashed and single-dashed lines represent stacking-fault and twin-fault boundaries, respectively. After the first nucleation event, no new fault area is added.

Partial-Dislocation Breakaway

There are several possible mechanisms for twin formation associated with motion of a single partial dislocation. Under stress concentrations, a single partial dislocation can break away from an extended configuration provided that the applied stress exceeds γ/b [Eq. (23-7)], where γ is the fault energy, and that the rest of the configuration remains pinned under such a stress. Configurations that are possible sources for such a partial are (1) the barrier reactions of Sec. 22-4;[40] (2) the Suzuki lock, discussed in Sec. 18-7; and (3) simply two partials pulled apart by the applied stress (Fig. 10-39). Once free, a single partial can wind about any forest dislocation with a Burgers vector component normal to the twin glide plane and equal in magnitude to the requisite number of lattice spacings, thus providing a pole mechanism.[41] Alternatively, the single partial can activate the reflection mechanism discussed below. In both Mo-Re alloys and Fe-Cr-Co alloys, Mahajan[42] has observed three-layer microtwins that he views as emanating from the array in Fig. 11-14. He suggests that coplanar microtwins of this sort combine, in some cases aided by cross slip, to form macrotwins. If incompletely combined or further reacted, they could also account for the emissary dislocations discussed in Sec. 23-5. He proposes that most twinning occurs by such a mechanism. The difficulty in this general application is in explaining observed twinning shears, which correlate well with the pole model.

A barrier reaction of interest is that proposed by Priestner and Leslie[43] for bcc crystals,

$$\frac{1}{2}[\bar{1}11]_{(1\bar{1}2)} + \frac{1}{2}[1\bar{1}1]_{(\bar{1}01)} \rightarrow [001] \rightarrow \frac{1}{6}[1\bar{1}5] + \frac{1}{6}[\bar{1}11] \rightarrow \frac{1}{3}[1\bar{1}2] + 2\times\frac{1}{6}[\bar{1}11]$$
$$\rightarrow \frac{1}{2}[1\bar{1}1] + 3\times\frac{1}{6}[\bar{1}11] \qquad (23\text{-}11)$$

[40] H. Suzuki and C. S. Barrett, *Acta Met.*, **6:** 156 (1958).

[41] A. Seeger, *Proc. Phys. Soc.*, (*Lond.*), **68:** 328 (1955).

[42] S. Mahajan, in M. F. Ashby, R. Bullough, C. S. Hartley and J. P. Hirth (eds.), "Dislocation Modeling of Physical Systems," Pergamon, Oxford, 1981.

[43] R. Priestner and W. C. Leslie, *Phil. Mag.*, **11:** 895 (1965).

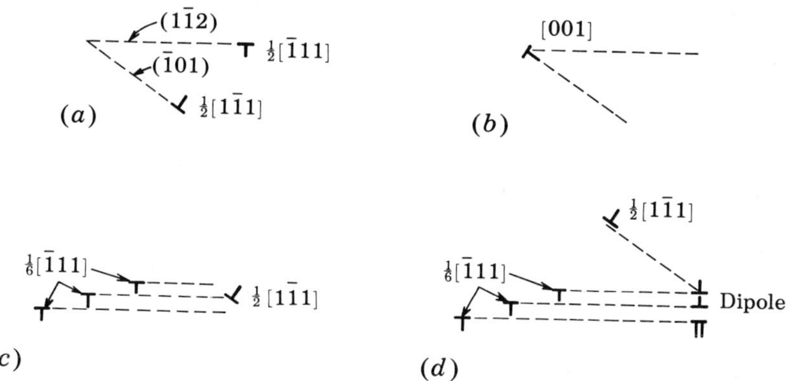

FIGURE 23-13. Successive stages in twin formation at a [001] barrier dislocation. The vector $[\bar{1}\bar{3}\bar{1}]$, the line of intersection of $(1\bar{1}2)$ and $(\bar{1}01)$, points into the plane of the paper.

The difference between the reactants and the products is that the $\frac{1}{2}[\bar{1}11]$ dislocation rests on a single glide plane, while the three-product $\frac{1}{6}[\bar{1}11]$ partials lie on successive $(1\bar{1}2)$ glide planes forming a three-layer twin and leaving a dipole behind, as shown in Fig. 23-13. Experimental evidence for twins originating at the intersections of slip bands supports mechanism (23-11). However, this is not the only mechanism that requires the intersection of slip bands and a stress concentration for its operation. A plausible alternative is the formation of a pole mechanism of the type shown in Fig. 23-7. In the latter case, intersection jogs produced by slip-band intersection could provide the nucleus for operation of the pole.

Cross Slip at an Obstacle

Cohen and Weertman[44] suggested that fcc twinning could originate at a pileup behind a Lomer-Cottrell barrier, as illustrated in Fig. 23-14. The partials $\boldsymbol{\alpha}\mathbf{B}$ in the pileup dissociate into $\boldsymbol{\alpha\delta}$ and $\boldsymbol{\delta}\mathbf{B}$ on the twinning plane to provide twinning dislocations. More generally, a mechanism of this kind could operate at any pileup behind an obstacle. Elastic interactions prevent twinning dislocations from being emitted on every twin plane. Thus this mechanism either produces a highly faulted twin or requires an additional mechanism, such as limited pole operation, to complete a more perfect twin.

The Reflection Mechanism

Once a single partial dislocation has been generated by any of the previously discussed mechanisms, it can propagate by reflection at free surfaces or grain boundaries.[45] The model resembles closely that of Fig. 20-11 and is subject to

[44] J. B. Cohen and J. Weertman, *Acta Met.*, **11:** 997, 1368 (1963).

[45] J. W. Christian, *Proc. Roy. Soc.*, **A206:** 51 (1951).

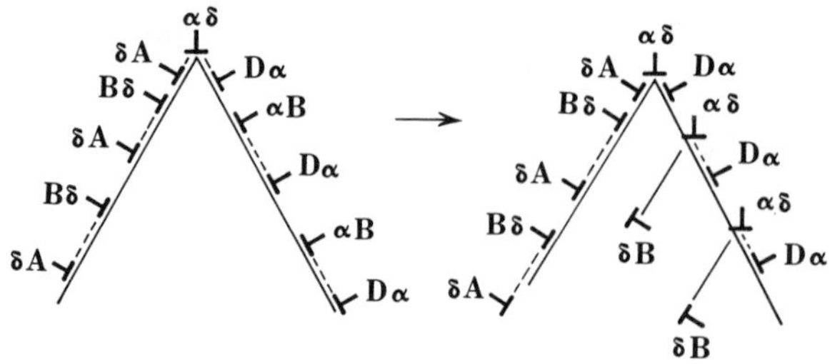

FIGURE 23-14. Twinning dislocations formed at a pileup behind a Lomer-Cottrell barrier in an fcc crystal.

the same uncertainties in description. One difference in this case is that the reflected dislocation must reflect on a different plane from the glide plane of the incident twinning dislocation in order to propagate the twin. The twin boundary energy is a factor which would tend to cause such an alternation in glide plane. In the case of the high-symmetry fcc crystal, twins can be propagated by reflecting the three possible $\frac{1}{6}\langle 112\rangle$ dislocations alternately as the twin propagates, so that no long-range stress fields are developed at the reflection boundary (grain boundary). The observation of other types of sources, as discussed above, suggests that the reflection mechanism is not of general applicability. The reflection mechanism, however, could be important in the very rapid twin formation that is observed in shock-loaded metals.[46] Finally, various boundary sources are possible. Direct emission of partial dislocations from grain boundary sources (Chap. 20), as observed for cobalt and its alloys,[47] could produce twins. Twin fault arrays were also observed in this work that would be consistent with the above reflection mechanism operating when partial dislocations impinged on an already existing fault. Partials have been observed to emanate from a coherent twin boundary (in nickel) as well.[48]

In summary, there are many possible mechanisms for twin nucleation and propagation. At present these cannot be clearly distinguished, except for the few cases where the dynamic occurrence of twinning was observed in electron-transmission microscopy. An outstanding problem in the understanding of twinning mechanisms involves the question of the speed of formation. Sonic measurements of stress pulses accompanying twin formation, together with twin-length measurements, suggest that the velocity of twin boundaries during twin formation approaches the speed of transverse sound waves;[49, 50] similar

[46] C. S. Smith, *Trans. Met. Soc. AIME*, **212:** 574 (1958).

[47] E. M. Kennedy and R. Speiser, in "Physical Chemistry in Metallurgy," U.S. Steel, Pittsburgh, Pennsylvania, 1976, p. 345.

[48] T. Malis, D. J. Lloyd, and K. Tangri, *Phil. Mag.*, **26:** 1081 (1972).

[49] C. N. Reid, G. T. Hahn, and A. Gilbert, in R. E. Reed-Hill, J. P. Hirth, and H. C. Rogers (eds.), "Deformation Twinning," Gordon and Breach, New York, 1964, p. 386.

[50] R. F. Bunshah, in *ibid.*, p. 390.

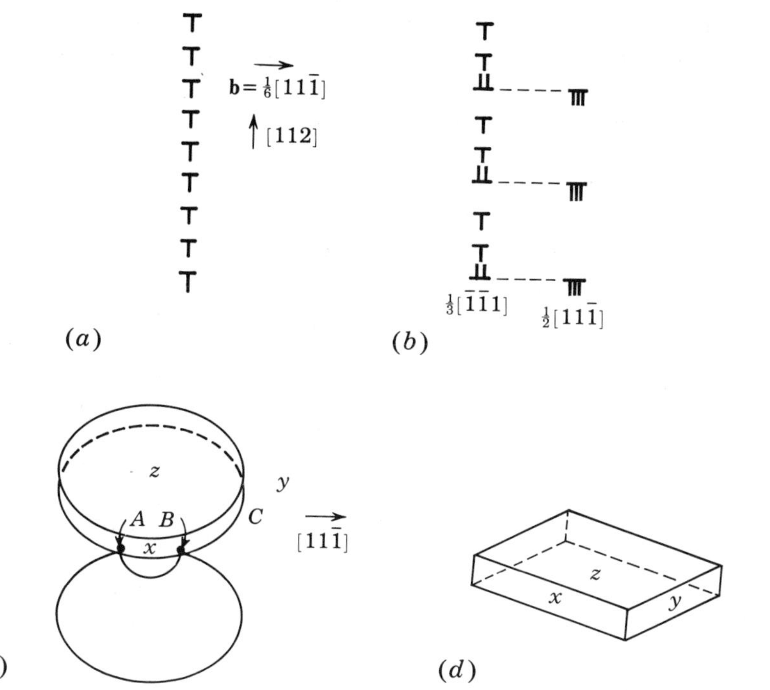

FIGURE 23-15. Reaction of (*a*) an edge-dislocation twin interface to form (*b*) an incoherent twin interface with no long-range stress field and a set of emissary dislocations. (*c*) Perspective view of emissary dislocations forming by cross slip. (*d*) Resultant twin following the emission of emissary dislocation.

velocities are expected in shock loading. Most of the dislocation mechanisms, particularly the pole mechanisms, appear inadequate to explain such rates of formation. On the other hand, the pole models correlate best with observed twinning shears in many cases. Further work is required to resolve this question.

23-5. TWIN-BOUNDARY PHENOMENA

Emissary Dislocations

Many of the macroscopic effects of twinning on mechanical properties are associated with twin-boundary phenomena. We conclude by discussing several examples of such phenomena. Sleeswyk[51] introduced the concept of *emissary* dislocations. These are perfect dislocations that are emitted from the incoherent tip of an advancing twin front to relieve the long-range stress field of the front. The emissary dislocations are thus an indicator of the stress field of the

[51]A. W. Sleeswyk, *Acta Met.*, **10:** 705 (1962).

twin boundary and provide strong support for the dislocation models of twinning. Emissary dislocations are observed primarily in bcc metals, but some evidence exists for them in hcp crystals.[52]

Consider the twin front of Fig. 23-15*a*, which shows twinning dislocations in the edge orientation. By the theory of Chap. 21, such a configuration has a long-range stress field. The twin boundary can lower its energy if it reacts to the form of Fig. 23-15*b*. Here every third partial has reacted to form an emissary dislocation and an opposite-sign edge partial

$$\tfrac{1}{6}[11\bar{1}]\rightarrow\tfrac{1}{3}[\bar{1}\bar{1}1]+\tfrac{1}{2}[11\bar{1}] \qquad (23\text{-}12)$$

The reaction produces a set of emissary slip dislocations and an incoherent twin boundary with a zero long-range stress field. Reaction (23-12) is, of course, unfavorable from an elastic-energy viewpoint and can occur only under the stress concentration of the pileup at the twin front. An alternative[53, 54] to the direct formation of emissary dislocations by reaction (23-12) is shown in Fig. 23-15*c*. Here twinning dislocations in the screw orientation cross slip to form a perfect $\frac{1}{2}[11\bar{1}]$ dislocation along AB. This perfect dislocation then bows out as shown in the figure, until it reacts at C with a $\frac{1}{6}[11\bar{1}]$ partial to form a $\frac{1}{3}[\bar{1}\bar{1}1]$ partial. The net result is the same as in Fig. 23-15*b*. If all the partials along the x face cross slip and produce emissary dislocations, the net result should be that of Fig. 23-15*d*. The z face (112) and the x face $(1\bar{2}\bar{1})$ will both be coherent, dislocation-free faces, while the y face will comprise sets of dipoles whose cross section is that of Fig. 23-15*b*; long-range stresses are absent.

The shear strain produced by the emissary dislocations is equivalent to that of the original twin interface from which they emerged. Thus the concept of emissary dislocations satisfactorily explains the blocky appearance of bcc deformation twins, as illustrated in Fig. 23-16. The shear is constant along the entire length of the band of width h, being propagated by emissary dislocations in the untwinned regions of the band; no long-range stresses are present. The presence of emissary dislocations has been verified, for example, by electron-transmission microscopy[55] and by observing the offset of subboundaries ahead of a twin front,[56] as also depicted in Fig. 23-16.

Accommodation Bands

Another demonstration of the long-range stress field of a twin front is provided by the presence of accommodation bands around twins.[57] The

[52]See the review by R. W. Cahn, in R. E. Reed-Hill, J. P. Hirth, and H. C. Rogers (eds.), "Deformation Twinning," Gordon and Breach, New York, 1964, p. 1.

[53]D. Hull, *Acta Met.*, **9:** 909 (1961).

[54]Discussion to the article by D. Hull, in Reed-Hill, J. P. Hirth, and H. C. Rogers (eds.), "Deformation Twinning," Gordon and Breach, New York, 1964, p. 155.

[55]E. Votava and A. W. Sleeswyk, *Acta Met.*, **10:** 965 (1962).

[56]A. W. Sleeswyk, *Acta Met.*, **10:** 705 (1962).

[57]Accommodation bands were proposed by D. C. Jillson, *Trans. AIME*, **188:** 1005 (1950).

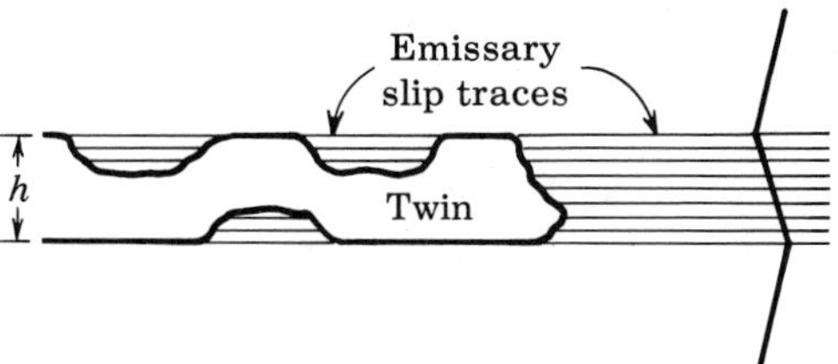

FIGURE 23-16. Cross section of a twinned region. Displaced line at right represents a tilt boundary sheared by the emissary dislocations that propagate the twinning shear ahead of the twin interface.

presence of the stress fields can be rationalized by either the shear-crack analogy or the superdislocation analogy of Chap. 21. The bands have the appearance illustrated in Fig. 23-17. They have been studied extensively in zinc.[58] As shown in Fig. 23-17, the boundaries of the accommodation bands correspond to kink-band boundaries.

There are several possible formation mechanisms for the accommodation bands. First, they can be formed from cross-slipping emissary dislocations. Second, the stress concentration at the tip of the twin can nucleate the slip dislocations which glide on a plane nonparallel with the original twin plane to form the band. Third, if the dislocation model of the twin boundary is that of Fig. 23-3*b*, the accommodation band can form from emissary dislocations with Burgers vectors parallel to η_2.

Whether or not accommodation bands form from emissary dislocations, it is certain that emissary dislocations are left behind when twins anneal out. Rosenbaum[59] has shown that when twins such as that of Fig. 23-17 anneal out, the shear displacement of the crystal is retained, being accounted for by the presence of the emissary dislocations.

Interaction of Glide Dislocations with Twin Boundaries

Twin boundaries can act as barriers to perfect glide dislocations. Also, the intersection of glide dislocations with twin boundaries can produce twinning dislocations and propagate the twin. The latter mechanism is more likely in high-symmetry crystals; it was first proposed as a twin phenomenon in bcc crystals.[60]

Applying the transformation matrix of Eq. (23-4) to the four possible glide-dislocation Burgers vectors in the parent crystal, one finds that they

[58] Representative pictures are presented by A. J. W. Moore, *Proc. Phys. Soc.*, **B65:** 956 (1952); *Acta Met.*, **3:** 163 (1955).

[59] H. S. Rosenbaum, *Acta Met.*, **9:** 742 (1961); see also F. F. Lavrentev and V. I. Startsev, *Fiz. Met. i Metalloved.*, **13:** 441 (1962).

[60] A. W. Sleeswyk and C. A. Verbraak, *Acta. Met.*, **9:** 917 (1961).

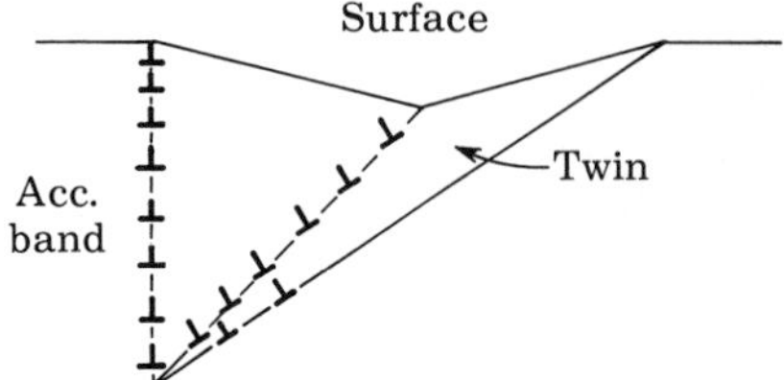

FIGURE 23-17. Cross section of a twin intersecting a free surface, and the accommodation band associated with the twin.

transform as follows:

$$\tfrac{1}{2}[111]\rightarrow\tfrac{1}{2}[111]_t \qquad \tfrac{1}{2}[1\bar{1}1]\rightarrow\tfrac{1}{6}[\bar{1}5\bar{1}]_t$$
$$\tfrac{1}{2}[11\bar{1}]\rightarrow\tfrac{1}{6}[\bar{1}\bar{1}5]_t \qquad \tfrac{1}{2}[\bar{1}11]\rightarrow\tfrac{1}{6}[5\bar{1}\bar{1}]_t \tag{23-13}$$

Now, as can be seen by inspection of Fig. 23-6*b*, the intersection of the $(\bar{1}\bar{1}2)$ twin plane by the dislocations of Eq. (23-13) produces steps in the boundary of 0, 2, 1, and 1 interplanar $(\bar{1}\bar{1}2)$ distances, respectively, for the dislocations $\frac{1}{2}[111]$, $\frac{1}{2}[11\bar{1}]$, $\frac{1}{2}[1\bar{1}1]$, and $\frac{1}{2}[\bar{1}11]$. As demonstrated in Fig. 11-14, these steps must correspond to $\frac{1}{6}\langle 111\rangle$ twinning dislocations or the complementary partials $\frac{1}{3}\langle 111\rangle$. Selecting the partials to give the proper fault to propagate the twin according to Axiom 11-2, one finds that the product dislocations in Eq. (23-13) can dissociate further to give a perfect dislocation in the twin and a residual twinning dislocation (or dislocations) as follows:[61]

$$\begin{aligned}\tfrac{1}{6}[\bar{1}\bar{1}5]_t&\rightarrow\tfrac{1}{2}[\bar{1}\bar{1}1]_t+\tfrac{1}{6}[111]_t+\tfrac{1}{6}[111]_t\\ \tfrac{1}{6}[\bar{1}5\bar{1}]_t&\rightarrow\tfrac{1}{2}[\bar{1}1\bar{1}]_t+\tfrac{1}{3}[111]_t\\ \tfrac{1}{6}[5\bar{1}\bar{1}]_t&\rightarrow\tfrac{1}{2}[1\bar{1}\bar{1}]_t+\tfrac{1}{3}[111]_t\end{aligned} \tag{23-14}$$

A complete sequence of reactions for the first case in Eq. (23-14) is presented in Fig. 23-18. In the presence of a shear stress σ which tends to expand the twin, the products of the reaction cause the twin to grow by four layers. The reactions producing $\frac{1}{3}[111]_t$ partials cause the twin to shrink if the $\frac{1}{3}[111]_t$ partials glide when acted upon by the same stress.

Thus the intersection of a twin by a glissile dislocation is a possible mechanism for twin growth (or shrinkage). However, we note that the dissociations of Eq. (23-14) are energetically unfavorable and can occur only under large applied stresses. Furthermore, the Peierls stress for motion of $\frac{1}{3}\langle 111\rangle$ partials appears to be quite large, as discussed in Chap. 11. As a final complication, under favorable stress distributions, particularly when the glide plane of the incident dislocation is inclined at only a small angle to the twin plane, the glissile dislocation that is formed in the twin may not glide on the

[61] This process is a special case of the general example discussed in Chap. 19, where passage of a dislocation through a grain boundary is shown to require that a residual dislocation remain in the boundary.

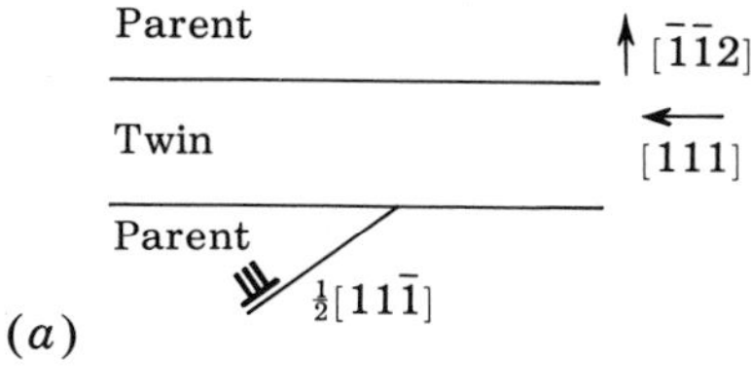

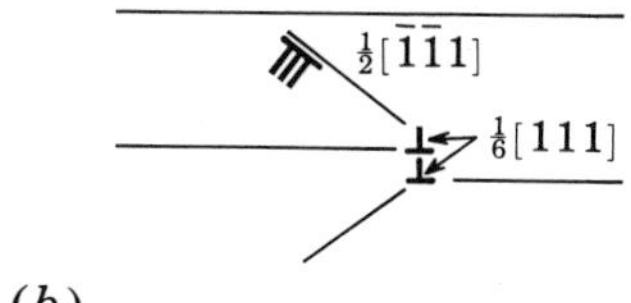

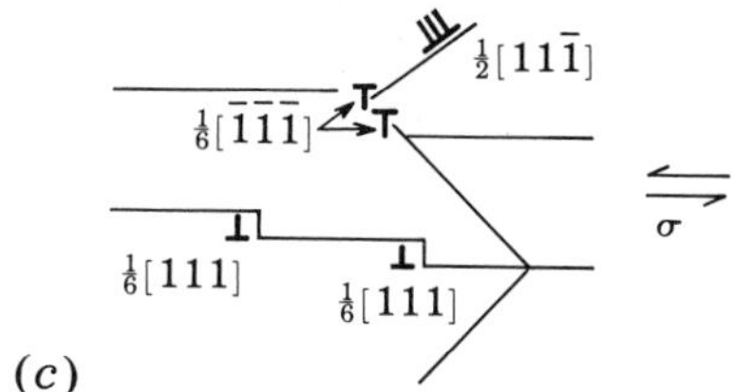

FIGURE 23-18. Sequence showing the interaction of a $\frac{1}{2}[11\bar{1}]$ dislocation with twin boundaries.

plane in mirror symmetry to the incident-dislocation glide plane. A possible example is the reaction

$$\tfrac{1}{6}[\bar{1}\bar{1}5]_t \to \tfrac{1}{2}[1\bar{1}1]_t + \tfrac{1}{3}[\bar{2}11]_t \tag{23-15}$$

illustrated in Fig. 23-19. In this case the partial $\frac{1}{3}[\bar{2}11]$ is not glissile along the interface, so reaction (23-15) would not contribute to lateral twin growth. In summary, there are reservations about the applicability of the mechanism; the reaction producing the $\frac{1}{6}\langle 111\rangle$ partials is the most likely of the possible reactions to occur. Strong evidence has been found for a mechanism of the general type of Eq. (23-14) in α-iron.[62]

Although the above discussion deals only with the bcc example, the mechanism applies to any crystal system. Whenever a glide dislocation passes through the coherent twin boundary and reacts to form a glissile, perfect dislocation on the mirror-symmetry plane in the twin, the residual dislocations at the steps formed in the boundary will be glissile twinning dislocations.

Whenever conditions are such that reactions such as Eq. (23-14) do not occur for the reasons discussed above, the twin boundary should act as a barrier, so that pileups of glide dislocations form there. Such pileups can then

[62] A. W. Sleeswyk, *Acta Met.*, **12:** 669 (1964).

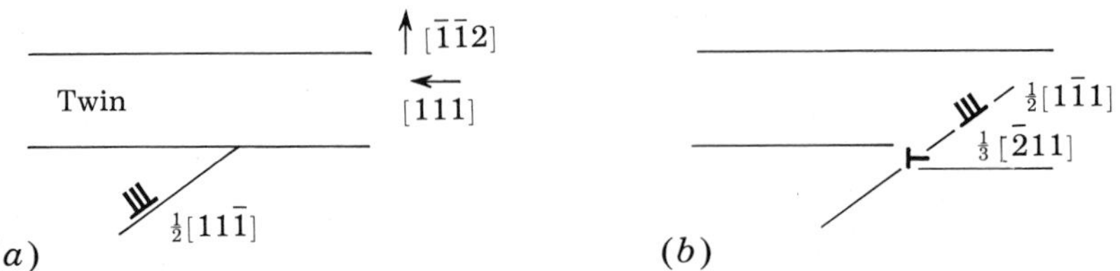

FIGURE 23-19. Sequence showing the interaction of a $\frac{1}{2}[11\bar{1}]$ dislocation with twin boundaries.

contribute to hardening or fracture by the mechanisms discussed in Chap. 21. A review of the role of twinning in fracture is presented by Armstrong.[63]

PROBLEMS

23-1. In principle, could $(1\bar{1}0)$ be a twin plane in an fcc crystal?

23-2. Consider a twin lamella of thickness h forming by $\frac{1}{6}[111]$ glide on $(11\bar{2})$, the K_1 plane. What is the superdislocation at the spearhead of the twin? Estimate the stress at a distance $\sim h$ ahead of the spearhead, assuming that no emissary glide has occurred.

23-3. Consider the same twin as in Prob. 23-2, but formed by Bullough's mechanism. Are the stresses at the spearhead different in this case?

23-4. Would compressive stresses normal to K_1 influence twinning if it proceeded as assumed in Prob. 23-2? As in Prob. 23-3?

23-5. Would shear stresses other than that resolved on the twinning plane and in the twinning direction be important for twin nucleation in the models of Figs. 23-7*a* and 23-7*b*?

23-6. Discuss possible inadequacies of the concept of a critical resolved shear stress for twinning. Probs. 23-4 and 23-5 are relevant to this question.

23-7. Consider Fig. 23-16. During annealing, should the twin shrink or grow to consume the entire slipped zone? Why?

23-8. Derive the possible reactions for dislocations of the $\frac{1}{2}\langle 110\rangle$ type gliding through coherent twin interfaces in fcc crystals, leaving partials in the boundary.

BIBLIOGRAPHY

1. Cahn, R. W., *Adv. Phys.*, **3:** 363 (1954).
2. Clark, R., and G. B. Craig, *Prog. Met. Phys.*, **3:** 115 (1953).
3. Hall, E. O., "Twinning," Butterworths, London, 1954.
4. Klassen-Neklyudova, M. V., "Mechanical Twinning of Crystals," Consultants Bureau, New York, 1964.
5. Mahajan, S., and D. F. Williams, *Int. Met. Rev.*, **18:** 43 (1973).
6. Reed-Hill, R. E., J. P. Hirth, and H. C. Rogers (eds.), "Deformation Twinning," Gordon and Breach, New York, 1964.

[63] R. W. Armstrong, in R. E. Reed-Hill, J. P. Hirth, and H. C. Rogers (eds.), "Deformation Twinning," Gordon and Breach, New York, 1964, p. 356.

APPENDIX 1

Elastic Constants

For convenience in performing the numerical calculations throughout the book, we list values of the anisotropic elastic constants and compliances for some common cubic crystals. In addition, values are listed for the Reuss average compliances $1/E_R$, $1/\mu_R$, and ν_R [Eq. (13-15)]; for the compliance anisotropy factor J [Eq. (13-29)]; for the Voigt average elastic constants μ and λ [Eq. (13-5)], together with ν calculated from Eq. (2-50); for the anisotropy factor H [Eq. (13-27)]; and for the anisotropy ratio A [Eq. (2-45)]. The anisotropy factors are given by

$$A = \frac{2c_{44}}{c_{11} - c_{12}} \qquad H = 2c_{44} + c_{12} - c_{11} \qquad J = s'_{11} - s'_{12} - \tfrac{1}{2}s'_{44}$$

The values for the various crystals are those listed by Huntington,[1] except for niobium,[2] vanadium,[2] tantalum,[2] chromium,[3] tungsten,[4] and iron.[5] All elastic constants are listed in units of 10^{10} Pa; all elastic compliances in units of 10^{-11} Pa^{-1}.

[1] H. B. Huntington, *Solid State Phys.*, **7:** 213 (1958).

[2] D. I. Bolef, *J. Appl. Phys.*, **32:** 100 (1961).

[3] D. I. Bolef and J. de Klerk, *Phys. Rev.*, **129:** 1063 (1963).

[4] D. I. Bolef and J. de Klerk, *J. Appl. Phys.*, **35:** 2311 (1962).

[5] W. P. Mason, "Piezoelectric Crystals and Their Applications to Ultrasonics," Van Nostrand, New York, 1950.

Crystal	s'_{11}	$-s'_{12}$	s'_{44}	J	$\frac{1}{\mu_R}$	$\frac{1}{E_R}$	$\frac{\nu_R}{E_R}$	ν_R	μ_R
Al	1.57	0.568	3.51	0.38	3.81	1.42	0.492	0.347	2.63
Ag	2.29	0.983	2.17	2.18	3.91	1.42	0.537	0.378	2.56
Au	2.33	1.065	2.38	2.20	4.14	1.45	0.625	0.431	2.42
Cr	0.300	0.043	0.992	−0.153	0.87	0.392	0.073	0.186	11.5
Cu	1.498	0.629	1.326	1.464	2.496	0.913	0.337	0.369	4.00
Fe	0.760	0.287	0.892	0.601	1.37	0.399	0.167	0.419	7.30
Ge	0.978	0.266	1.490	0.499	1.89	0.778	0.166	0.214	5.30
K	82.3	37.0	38.0	100.3	46.0	42.3	17.0	0.402	0.218
Mo	0.28	0.078	0.91	−0.102	0.83	0.32	0.098	0.306	12.1
Na	48.6	21.0	17.1	61.0	65.9	24.2	8.80	0.364	0.152
Nb	0.660	0.233	3.48	−0.85	2.80	1.17	0.393	0.336	3.57
Ni	0.734	0.274	0.802	0.607	1.29	0.491	0.153	0.312	7.78
Pb	9.28	4.24	6.94	10.05	14.98	5.26	2.23	0.424	0.667
Ta	0.686	0.258	1.212	0.338	1.48	0.483	0.190	0.394	6.76
Th	2.72	1.07	2.09	2.74	4.28	1.62	0.52	0.321	2.34
Si	0.767	0.214	1.256	0.353	1.54	0.63	0.143	0.227	6.49
V	0.683	0.234	2.348	−0.257	2.14	0.837	0.285	0.340	4.68
W	0.244	0.068	0.624	0.00	0.624	0.244	0.068	0.278	16.0
AgBr	3.13	1.17	13.9	−2.65	11.8	4.18	1.70	0.406	0.85
KCl	2.62	0.35	16.0	−5.03	12.0	4.62	1.36	0.294	0.833
LiF	1.135	0.31	1.59	0.65	2.11	0.875	0.18	0.206	4.74
MgO	0.408	0.095	0.676	0.165	0.808	0.342	0.062	0.181	12.4
NaCl	2.29	0.465	7.94	−1.21	6.97	2.78	0.707	0.245	1.43
PbS	0.864	0.164	4.03	−0.99	3.24	1.26	0.363	0.288	3.09
Diamond	0.095	0.0099	0.174	0.018	0.188	0.088	0.0063	0.072	53.3

Crystal	c_{11}	c_{12}	c_{44}	H	A	μ	λ	ν
Al	10.82	6.13	2.85	1.01	1.21	2.65	5.93	0.347
Ag	12.40	9.34	4.61	6.16	3.01	3.38	8.11	0.354
Au	18.6	15.7	4.20	5.5	2.9	3.10	14.6	0.412
Cr	35.0	5.78	10.1	−10.0	0.69	12.1	7.78	0.13
Cu	16.84	12.14	7.54	10.38	3.21	5.46	10.06	0.324
Fe	24.2	14.65	11.2	12.9	2.36	8.6	12.1	0.291
Ge	12.89	4.83	6.71	5.36	1.66	5.64	3.76	0.200
K	0.457	0.374	0.263	0.443	6.35	0.174	0.285	0.312
Mo	46.0	17.6	11.0	−6.4	0.775	12.3	18.9	0.305
Na	0.603	0.459	0.586	1.03	8.15	0.380	0.253	0.201
Nb	24.6	13.4	2.87	−5.46	0.51	3.96	14.5	0.392
Ni	24.65	14.73	12.47	15.02	2.52	9.47	11.7	0.276
Pb	4.66	3.92	1.44	2.14	3.90	1.01	3.48	0.387
Ta	26.7	16.1	8.25	5.9	1.56	7.07	14.9	0.339
Th	7.53	4.89	4.78	6.92	3.62	3.40	3.51	0.254
Si	16.57	6.39	7.96	5.74	1.56	6.81	5.24	0.218
V	22.8	11.9	4.26	−2.37	0.78	4.73	12.4	0.352
W	52.1	20.1	16.0	0.00	1.00	16.0	20.1	0.278
AgBr	5.63	3.3	0.720	−0.89	0.618	0.87	3.45	0.401
KCl	3.98	0.62	0.625	−2.11	0.372	1.045	1.04	0.250
LiF	11.12	4.20	6.28	5.64	1.82	5.15	3.07	0.187
MgO	28.6	8.7	14.8	9.7	1.49	12.9	6.8	0.173
NaCl	4.87	1.24	1.26	−1.11	0.694	1.48	1.46	0.248
PbS	12.7	2.98	2.48	−4.76	0.510	3.43	3.93	0.267
Diamond	107.6	12.5	57.6	20.1	1.21	53.6	8.50	0.068

APPENDIX 2

Surface Energies

Here we list values of the intrinsic stacking-fault energy γ_I for some fcc metals together with selected values of grain-boundary energies and crystal-vapor surface energies. There have been large uncertainties in the values of such interfacial energies. For example, the values of γ_I for copper quoted in the literature have ranged from 20 to 160 mJ/m^2 and those for nickel from 70 to 415 mJ/m^2. As of 1968[1] the values were converging on the basis of node and stacking-fault tetrahedron measurements and on estimates from twin-boundary energies. With the advent of weak-beam electron microscopy,[2] direct measurements of faulted dipoles, double ribbons, and extended single dislocations became possible, as discussed in Chap. 10, and these techniques gave lower values than the node-tetrahedron values by 20 to 30 percent, except for very low stacking-fault energy materials such as silver, where agreement was good. The higher values may be associated with pinning effects, as discussed below, or with the use of more accurate anisotropic elasticity for the straight dislocation arrays. In the same period measured twin boundary energies were lower,[3] however, so the earlier trend that $2\gamma_T \sim \gamma_I \sim \gamma_E$ persists with the newer data. Rather than compile an exhaustive list of values from the literature, we have selected values based on the most recent direct observation methods and list them in Table A-1.

There are large uncertainties in the temperature dependence of the various interfacial free energies, which are not discussed here. The cited values are generally elevated temperature values for the grain-boundary and crystal-vapor free energies, and thus include entropy contributions. The stacking-fault data are generally lower-temperature data and are expected to have less of an entropy contribution. For a discussion of entropy effects see the review by Inman and Tipler.[4] Cases where the variation with temperature is measured at

[1]See this book, first edition, 1968.

[2]D. J. H. Cockayne, I. L. F. Ray, and M. J. Whelan, *Phil. Mag.*, **20:** 1265 (1969).

[3]L. E. Murr, "Interfacial Phenomena in Metals and Alloys," Addison-Wesley, Reading, Mass., 1975, p. 138.

[4]M. C. Inman, and H. R. Tipler, *Met. Rev.*, **8:** 105 (1963).

TABLE A-1 Selected Values of Intrinsic Stacking-Fault Energy γ_I, Twin-Boundary Energy γ_T, Grain-Boundary Energy γ_G, and Crystal-Vapor Surface Energy γ for Various Materials (in mJ/m^2)

Metal	γ_I	γ_T	γ_G	γ
Ag	16[1]	8[2]	790[3]	1140[4]
Al	166[2]	75[2]	325[2]	980[2]
Au	32[5]	15[2]	364[3]	1485[4]
Cu	45[6]	24[2]	625[2]	1725[4]
Fe			780[3]	1950[3]
Ni	125[7]	43[2]	866[2]	2280[2]
Pd	180[8]			
Pt	322[2]	161[2]	1000[9]	3000[9]
Rh	~750[8]			
Th	115[8]			
Ir	300[2]			
W				2800[2]
Cd	175[2]			
Mg	125[2]			
Zn	140[2]		340[2]	

1. H. Saka, T. Iwata and T. Imura, *Phil. Mag.*, **37A:** 273 (1978).
2. L. E. Murr, "Interfacial Phenomena in Metals and Alloys," Addison-Wesley, Reading, Mass., 1975.
3. M. C. Inman and H. R. Tipler, *Met. Rev.*, **8:** 105 (1963).
4. D. McLean, "Grain Boundaries in Metals," Oxford University Press, Fair Lawn, N.J., 1957, p. 76.
5. M. L. Jenkins, *Phil. Mag.*, **42A:** 185 (1980).
6. C. B. Carter and I. L. F. Ray, *Phil. Mag.*, **35:** 189 (1977).
7. C. B. Carter and S. M. Holmes, *Phil. Mag.*, **35:** 1161 (1977).
8. I. L. Dillamore and R. E. Smallman, *Phil. Mag.*, **12:** 191 (1965).
9. M. McLean and H. Mykura, *Surface Sci.*, **5:** 466 (1966).

elevated temperature[3] usually involve determinations over a limited temperature range, so linear extrapolation is dangerous. Other factors can also produce uncertainties in γ values. These include image-force effects for unpinned dislocation arrays in thin films; effects of internal stresses; impurity effects, even in nominally high purity crystals, which can either give erroneously low values of γ because of adsorption to fault surfaces or high or low values because of pinning of dislocations on cooling from an equilibration temperature; and uncertainties in the analysis of dislocation images in electron microscopy. It appears that the total uncertainties are of the order of 10 to 15 percent for even the best recent data.

Author Index

Subject Index